SOIL MECHANICS
AND
FOUNDATION ENGINEERING

SOIL MECHANICS
AND
FOUNDATION ENGINEERING
[GEOTECHNICAL ENGINEERING]
[IN SI UNITS]

Dr. K.R. ARORA
B.E. (Civil); M.E. (Hons.); Ph. D. (IITD)
F.I.E.; M.I.G.S., FISDT; MIWRS
Formerly Professor and Head of Civil Engg. Department
Engineering College, KOTA (Raj.)

STANDARD PUBLISHERS DISTRIBUTORS
1705-B, NAI SARAK, POST BOX No.: 1066, DELHI-110006
Phones : 23262700, 23285798, Fax: 23243180
email: stpub@vsnl.com www.engineeringbooks.co.in www.standardpublishers.com

Published by :
A. K. Jain
For Standard Publishers Distributors
1705-B, Nai Sarak, Delhi-110006.

First Edition : 1987
Second Edition : 1989
Third Edition : 1992
Fourth Edition : 1997
Fifth Edition : 2000
Reprint : 2001
Reprint : 2002
Sixth Edition : 2003
Seventh Edition : 2008
Reprint : 2009
Reprint : 2010
Reprint : 2011
Reprint : 2014
Reprint : 2015
Reprint : 2016
Reprint : 2017
Reprint : 2018
Reprint : 2019
Reprint : 2021
Reprint : 2022
Reprint : 2025

Price: Rs. 650-00

ISBN: 81-8014-112-8
ISBN: 978-81-8014-112-6

Laser Typesetting by : Bhargave Laser Printers, Delhi.
Printed by : Lomus Offset Press, Delhi.

PREFACE TO THE SEVENTH EDITION

In this edition, the text has been revised and updated. A number of changes have been made in the text to make it more useful to readers. Chapter 33 on Ground Improvement Techniques and Chapter 34 on Foundation on Expansive and Collapsible Soils have been added to incorporate latest research and practice in the field of geotechnical engineering.

A large number of new objective-type question and Miscellaneous illustrative examples have been included for the students appearing for competitive examination.

The author thanks his wife Mrs. Rani Arora for assistance in proof reading. The author also thanks Sh. Bhagwan Sawroop Sharma, Draughtsman, Engineering College, Kota, for drawings.

Suggestions for further improvement of the text will be appreciated.

—Dr. K.R. ARORA

KOTA (Raj.)
July, 2007

PREFACE TO THE SIXTH EDITION

In this edition, the text has been revised and updated. A new chapter on 'Geotechnical Earthquake Engineering' has been included to introduce the readers to the recent developments. The importance of geotechnical aspects of earthquake engineering has considerably increased in recent years, especially after the Bhuj Earthquake of 2001. On the suggestions received from readers, this chapter has been included in this text book.

The author heartily thanks his wife Mrs Rani Arora and son-in-law Dr. B.P. Suneja, Lecturer (Selection scale) in Civil Engineering, Engineering College, Kota for the assistance provided. The author also thanks Sh. Bhagwan Sawroop Sharma, Droughtsman, Engineering College, Kota for excellent drawings.

The author gratefully acknowledges the courtesy of the Bureau of Indian Standards, 9 Bahadur Shah Zafar Marg, New Delhi-110002 to include two figures from IS: 1893 (Part 1)—2002.

Suggestions for further improvement of the text will be gratefully acknowledged.

—Dr. K.R. ARORA

KOTA (Raj.)
July, 2003

PREFACE TO THE FOURTH EDITION

The basic aim of the fourth edition of *Soil Mechanics and Foundation Engineering* is the same as that of the earlier three editions, namely, to present the fundamentals of the subject in a simplified manner. In this edition, a number of improvements and additions have been incorporated to make the text more useful.

A large number of multiple-choice questions and objective type questions (with answers) have been added at the end of each chapter. Chapter 30 gives the detailed procedure for conducting nineteen common laboratory experiments. Chapter 31 covers the basic principles of Rock Mechanics. Appendix A gives the glossary of common terms for ready reference. Selected references and a list of relevant publications of Bureau of Indian Standards are given at the end for further study.

It is gratifying that the book has been appreciated by students, teachers and practising engineers throughout the country. The book has established itself as a useful text in most of the engineering colleges and technical institutions of the country. The author is grateful to the teachers and students who have sent their comments, suggestions and letters of appreciation.

The author thanks his colleagues Dr. R. C. Mishra and Sh. N. P. Kaushik for their help in proof reading.

The author also thanks his wife Mrs. Rani Arora for her assistance in the revision of the book. The help received from Shri Bhagwan Sawroop Sharma, Draughtsman, in improving the diagrams is appreciated.

Efforts made by the publisher Sh. N. C. Jain and his sons Sh. Ajay Kumar Jain and Sh. Atul Kumar Jain for bringing out this edition in a short time and in a good form are appreciated.

In spite of every care taken to ensure accuracy, some errors might have crept in. The author will be grateful to the readers for bringing such errors, if any, to his notice. Suggestions for the improvements of the text will be gratefully acknowledged.

KOTA (Raj.)
February 26, 1997

—Dr. K.R. ARORA

PREFACE TO THE FIRST EDITION

Soil mechanics and Foundation engineering (geotechnical engineering) is a fast developing discipline of civil engineering. Considerable work has been done in the field in the last 6 decades. A student finds it difficult to have access to the latest literature in the field. The author has tried to collect the material from various sources and to present in the form of a text.

The text has been divided into two parts. The first part deals with the fundamentals of soil mechanics. The second part deals with earth retaining structures and foundation engineering. The subject matter has been presented in a logical and organised manner such that it may be taken up serially without any loss of continuity. The book covers the syllabi of undergraduate courses inn Soil Mechanics and Foundation Engineering prescribed by most Indian universities and institutes.

An attempt has been made to explain the fundamentals in a simple, lucid language. Basic concepts have been emphasised throughout. The author, who has about 25 years of teaching experience, has paid special attention to the difficulties experienced by students. A large number of illustrative examples have been given to show the application of the theory to field problems. Numerical problems, with answers, have been given for practice. Some objective type questions have also been given at the end of each chapter. The text is profusely illustrated with diagrams and charts. Latest IS codes have been followed, as far as possible. References are given at the end of each chapter. As complete switch over to SI units has not taken place in India, both MKS and SI units have been used.

The book will be useful for the undergraduate students. The students appearing for various competitive examinations and AMIE will also find the text useful. A large number of charts and tables have been included to make the text useful for practising engineers.

The author is grateful to Prof. Alam Singh of Jodhpur University who introduced the subject to him about 3 decades ago as a student at M.B.M. Engineering College, Jodhpur. The author is indebted to Prof. A. Varadarajan of IIT, Delhi, who helped him in understanding some of the intricate problems during his doctoral programme. The author thanks the faculty of Geotechnical Division of IIT, Delhi, for the help extended. The author also thanks his fellow research scholars, Dr. K.K, Gupta, Dr. B. Shankeriah, Dr. T.S. Rekhi, Dr. B.S. Satija, and Dr. R.N. Shahi for the fruitful discussions.

The author is grateful to Prof. A.V. Ramanujam, Principal, Engineering College, Kota for constant encouragement. The author thanks his colleagues at Engineering College, Kota, especially Sh. Amin Uddin, Draughtsman. The author also thanks his wife Mrs. Rani Arora who helped in proof reading and other works related with this text. The help received from his daughter Sangeeta Arora and son Sanjeev Arora is also acknowledged.

In spite of every care taken to ensure accuracy, some errors might have crept in. The author will be grateful to readers for bringing such errors to his notice. Suggestions for improvement of the text will be acknowledged with thanks.

KOTA (Raj.)
January 4, 1987

—K.R. ARORA

NOTATIONS

The notations have been explained wherever they appear. The following notations have been more commonly used.

A = Pore pressure parameter
= Activity of soils
A_v = Area of voids
A° = Angstrom
A_c = Air content
a_v = Coefficient of compressibility
B = Pore pressure parameter
C_c = Compression index
= Coefficient of curvature
C_u = Uniformity Coefficient
= Coefficient of elastic uniform compression
c = Unit cohesion
c' = Effective unit cohesion
c_u = Apparent cohesion
c_v = Coefficient of consolidation
D_{10} = Effective size
D_f = Foundation depth
D_r = Relative density
E = Modulus of elasticity
e = Void ratio
FS = Factor of safety
f = Friction
G = Specific gravity of particles
g = Acceleration due to gravity
h = Hydraulic head
I = Moment of inertia
I_p = Plasticity index
i = Hydraulic gradient
= Angle of surcharge
K = Coefficient of absolute permeability
K_0 = Coefficient of earth pressure at rest
K_a = Coefficient of active pressure
K_p = Coefficient of passive pressure
k = Coefficient of permeability
= Coefficient of subgrade reaction
k_s = Coefficient of subgrade reaction
k_p = Coefficient of percolation
N = Number of blows (SPT)
= Percent finer
= Normal component
n = Porosity
n_a = Percentage air voids
P = Force
P_a = Active pressure force
P_p = Passive pressure force
p = Pressure
p_a = Active pressure
p_p = Passive pressure
p_h = Horizontal pressure
Q = Force, Load
= Total quantity of water
Q_a = Allowable load
Q_u = Ultimate load
q = Surcharge
= Intensity of Load
= Discharge
q_c = Static cone resistance
q_n = Net footing pressure capacity
q_{ns} = Net safe bearing capacity
q_{np} = Net safe settlement pressure
q_{na} = Allowable bearing pressure
q_u = Ultimate bearing capacity
= Unconfined compressive strength
S = Degree of saturation
= Surface area
S_n = Stability no.
S_t = Sensitivity
s = Shear strength
= Settlement
T = Tangential component
= Temperature
T_s = Surface tension
t = Time
U = Degree of consolidation
= Total pore water pressure
u = Pore water pressure
$\bar{u}$ = Hydrostatic excess pore pressure
V = Volume, total volume, Velocity
V_d = Volume of dry soil
V_a = Volume of air
V_w = Volume of water
V_v = Volume of voids
V_s = Volume of solids
v = Velocity
v_c = Critical velocity
v_s = Seepage velocity
W = Weight, total weight
W_w = Weight of water
W_s = Weight of solids
W_q = Water table factor
W_γ = Water table factor
w = Water content
M = Mass, total mass
M_w = Mass of water
M_s = Mass of solids
w_l = Liquid limit
w_p = Plastic limit
w_s = Shrinkage limit
γ = Bulk unit weight
γ_d = Dry unit weight
γ_{sat} = Saturated unit weight
γ' = Submerged unit weight
γ_s = Unit weight of solids
γ_w = Unit weight of water
δ = Angle of wall friction
ε = Strain
η = Coefficient of viscosity
μ = Poisson's ratio
= Micron
= Coefficient of viscosity
ρ = Displacement
= Settlement
ρ_f = Final settlement
σ = total stress
$\bar{\sigma}$ = Effective stress
$\sigma_1, \sigma_2, \sigma_3$ = Principal stresses
$\bar{\sigma}_1, \bar{\sigma}_2, \bar{\sigma}_3$ = Effective principal stresses
σ_c = Preconsolidation pressure
σ_z, σ_v = Vertical stress
σ_x, σ_h = Horizontal stress
τ = Shear stress
τ_m = Mobilised shear strength
ϕ = Angle of shearing resistance
ϕ' = Effective angle of shearing resistance
ϕ_u = Apparent angle of shearing resistance
ϕ_m = mobilised angle of shearing resistance
ρ = Bulk density
ρ_d = Dry density
ρ' = Submerged density

CONVERSION FACTORS

(a) MKS to SI Units

From	*To*	*Multiply by*	*Equivalence*
kgf	N	9.81	1 kgf = 9.81 N
gmf	N	9.81×10^{-3}	1gmf = 0.00981N
tonne	kN	9.81	1 t = 9.81 kN
kgf/cm^2	kN/m^2	98.1	1 kgf/cm^2 = 98.1 kN/m^2
kgf/cm^2	N/mm^2	9.81×10^{-2}	1 kgf/cm^2 = 0.0981 N/mm^2
gmf/cm^2	N/m^2	98.1	1 gmf/cm^2 = 98.1 N/m^2
t/m^2	kN/m^2	9.81	1 t/m^2 = 9.81 kN/m^2
kgf/m^3	kN/m^3	9.81×10^{-3}	1 kgf/m^3 = 0.00981 kN/m^3
t/m^3	kN/m^3	9.81	1 t/m^3 = 9.81 kN/m^3
gmf/cm^3	kN/m^3	9.81	1 gmf/cm^3 = 9.81 kN/m^3
kgf/m	N/m	9.81	1 kgf/m = 9.81 N/m
kgf-m	N-m	9.81	1 kgf–m = 9.81 N–m
kgf-sec/m^2	N-s/m^2	9.81	1 kgf–sec/m^2 = 9.81 N–s/m^2

(b) SI to MKS Units

From	*To*	*Multiply by*	*Equivalence*
N	kgf	0.102	1 N = 0.102 kgf
N	gmf	102.0	1 N = 102 gmf
kN	tonne	0.102	1 kN = 0.102 t
kN/m^2	kgf/cm^2	0.102×10^{-1}	1 kN/m^2 = 0.0102 kgf/cm^2
N/mm^2	kgf/cm^2	10.2	1 N/mm^2 = 10.2 kgf/cm^2
N/m^2	gmf/cm^2	0.102×10^{-1}	1 N/m^2 = 0.0102 gmf/cm^2
kN/m^2	t/m^2	0.102	1 kN/m^2 = 0.102 t/m^2
kN/m^3	kgf/m^3	0.102×10^{3}	1 kN/m^3 = 102.0 kgf/m^3
kN/m^3	t/m^3	0.102	1 kN/m^3 = 0.102 t/m^3
kN/m^3	gmf/m^3	0.102	1 kN/m^3 = 0.102 gmf/cm^3
N/m	kgf/m	0.102	1 N/m = 0.102 kgf/m
N-m	kgf-m	0.102	1 N–m = 0.102 kgf–m
N-s/m^2	kgf-sec/m^2	0.102	1 N–s/m^2 = 0.102 kgf–sec/m^2

Note : 1 poise = 0.1 N-s/m^2 = 1.02×10^{-2} kgf-sec/m^2

1 bar = 100 kN/m^2

CONTENTS

PART II. EARTH RETAINING STRUCTURES AND FOUNDATION ENGINEERING

PART—I

FUNDAMENTALS OF SOIL MECHANICS

1

Introduction

1.1. DEFINITION OF SOIL

The word 'soil' is derived from the latin word *solium* which, according to Webster's dictionary, means the upper layer of the earth that may be dug or plowed; specifically, the loose surface material of the earth in which plants grow. The above definition of soil is used in the field of agronomy where the main concern is in the use of soil for raising crops. In geology, earth's crust is assumed to consist of unconsolidated sediments, called mantle or regolith, overlying rocks. The term 'soil' is used for the upper layer of mantle which can support plants. The material which is called soil by the agronomist or the geologist is known as top soil in geotechnical engineering or soil engineering. The top soil contains a large quantity of organic matter and is not suitable as a construction material or as a foundation for structures. The top soil is removed from the earth's surface before the construction of structures.

The term 'soil' in soil engineering is defined as an unconsolidated material, composed of solid particles, produced by the disintegration of rocks. The void space between the particles may contain air, water or both. The solid particles may contain organic matter. The soil particles can be separated by such mechanical means as agitation in water.

A natural aggregate of mineral particles bonded by strong and permanent cohesive forces is called 'rock'. It is an indurated material that requires drilling, wedging or blasting for its removal from the earth's surface. Since the terms weak and strong have different interpretations, the boundary between soil and rock is rather arbitrary. In case of a partially disintegrated rock, it is extremely difficult to locate the boundary between soil and rock.

Fig. 1.1 shows a cross-section through the earth's surface, indicating the nomenclature used in geology,

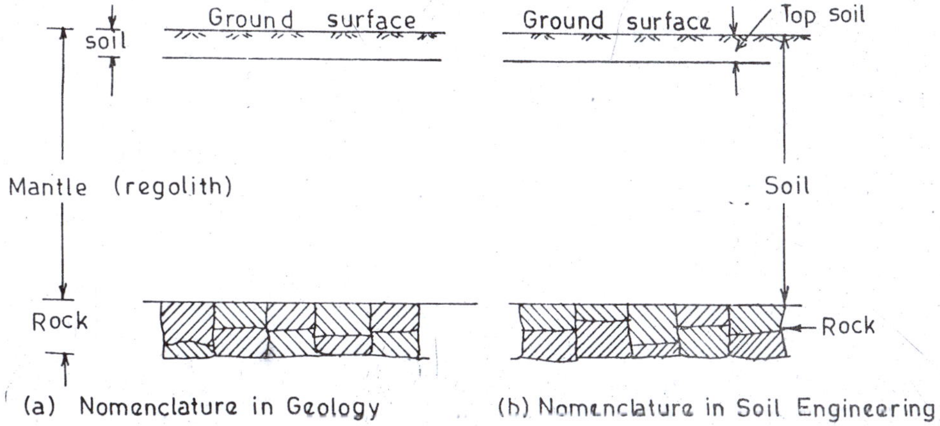

Fig. 1.1. Nomenclature.

and in Soil Engineering. It may be noted that the material which is called mantle (regolith) in geology is known as soil in Soil Engineering.

1.2. DEFINITION OF SOIL MECHANICS

The term 'soil mechanics' was coined by Dr. Karl Terzaghi in 1925 when his book *Erdbaumechanic* on the subject was published in German. According to Terzaghi, 'Soil mechanics is the application of the laws of mechanics and hydraulics to engineering problems dealing with sediments and other unconsolidated accumulations of solid particles produced by the mechanical and chemical disintegration of rock, regardless of whether or not they contain an admixture of organic constituents'. Soil mechanics is, therefore, a branch of mechanics which deals with the action of forces on soil and with the flow of water in soil.

The soil consists of discrete solid particles which are neither strongly bonded as in solids nor they are as free as particles of fluids. Consequently, the behaviour of soil is somewhat intermediate between that of a solid and a fluid. It is not, therefore, surprising that soil mechanics draws heavily from solid mechanics and fluid mechanics. As the soil is inherently a particulate system. soil mechanics is also called *particulate mechanics.*

Rock mechanics is the science dealing with the mechanics of rocks.

1.3. DEFINITION OF SOIL ENGINEERING AND GEOTECHNICAL ENGINEERING

Soil engineering in an applied science dealing with the applications of principles of soil mechanics to practical problems. It has a much wider scope than soil mechanics, as it deals with all engineering problems related with soils. It includes site investigations, design and construction of foundations, earth-retaining structures and earth structures.

Geotechnical engineering is a broader term which includes soil engineering, rock mechanics and geology. This term is used synonymously with soil engineering in this text.

1.4. SCOPE OF SOIL ENGINEERING

Soil engineering has vast application in the construction of various civil engineering works. Some of the important applications are as under :

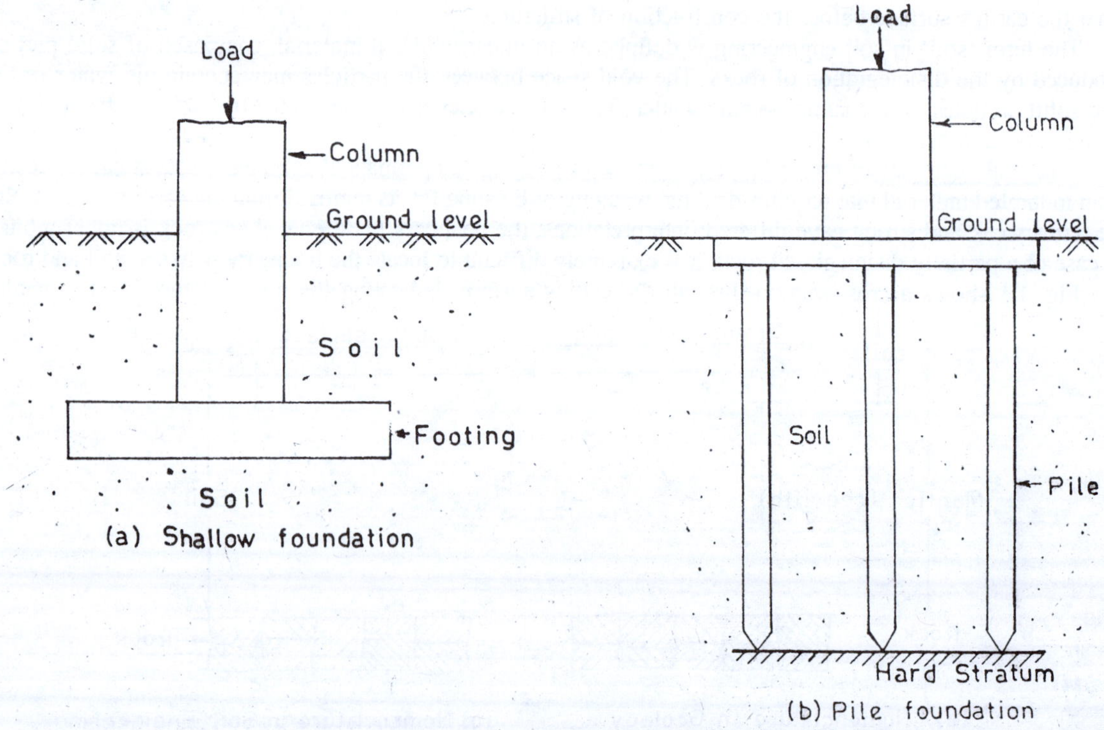

Fig. 1.2. Different types of foundations.

(1) Foundations—Every civil engineering structure, whether it is a building, a bridge, or a dam, is founded on or below the surface of the earth. Foundations are required to transmit the load of the structure to soil safely and efficiently.

A foundation is termed shallow foundation when it transmits the load to upper strata of earth. A foundation is called deep foundation when the load is transmitted to strata at considerable depth below the ground surface (Fig. 1.2). Pile foundation is a type of deep foundation. Foundation engineering is an important branch of soil engineering.

(2) Retaining Structures—When sufficient space is not available for a mass of soil to spread and form a safe slope, a structure is required to retain the soil. An earth retaining structure is also required to keep the

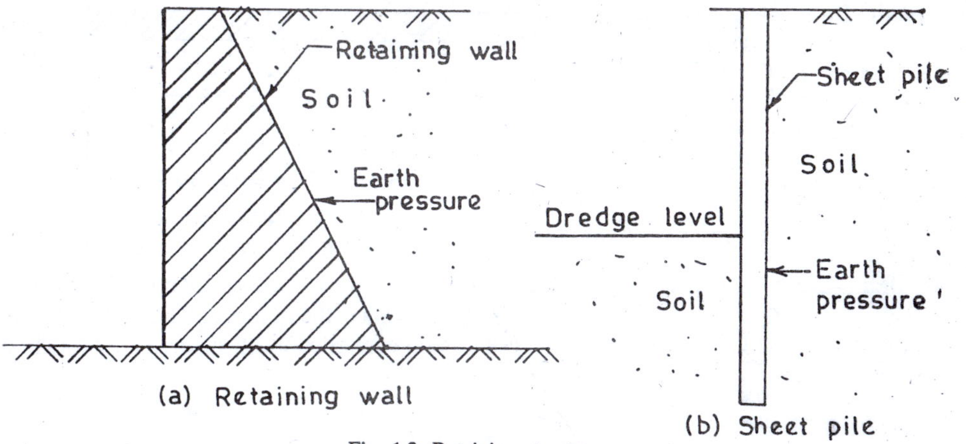

Fig. 1.3. Retaining structures.

soil at different levels on its either side. The retaining structure may be a rigid retaining wall or a sheet pile bulkhead which is relatively flexible (Fig. 1.3). Soil engineering gives the theories of earth pressure on retaining structures.

(3) Stability of Slopes—If soil surface is not horizontal, there is a component of weight of the soil which

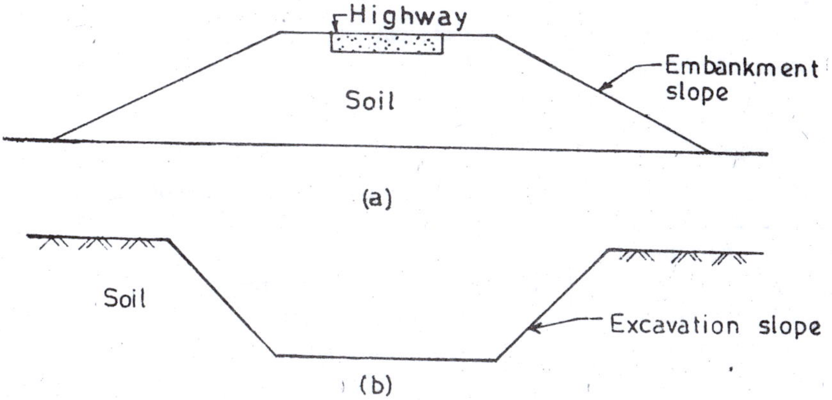

Fig. 1.4. Slopes in (*a*) filling and (*b*) cutting.

tends to move it downward and thus causes instability of slope. The slopes may be natural or man-made Fig. 1.4 shows slopes in filling and cutting. Soil engineering provides the methods for checking the stability of slopes.

(4) Underground Structures—The design and construction of underground structures, such as tunnels, shafts, and conduits, require evaluation of forces exerted by the soil on these structures. These forces are discussed in soil engineering. Fig. 1.5 shows a tunnel constructed below the ground surface and a conduit laid below the ground surface.

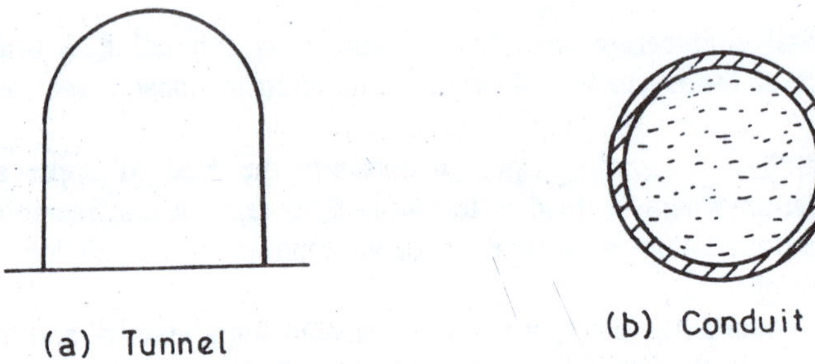

Fig. 1.5. Underground structures.

(5) Pavement Design—A pavement is a hard crust placed on soil (subgrade) for the purpose of providing a smooth and strong surface on which vehicles can move. The pavement consists of surfacing, such as a bitumen layer, base and subbase (Fig. 1.6). The behaviour of subgrade under various conditions of loading and environmental changes is studied in soil engineering.

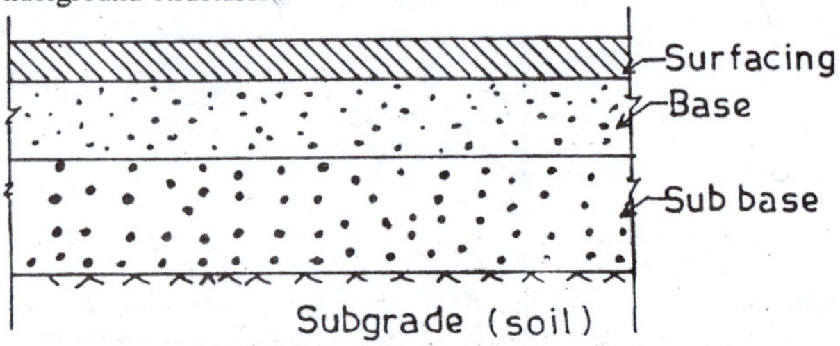

Fig. 1.6. Pavement details.

(6) Earth Dam—Earth dams are huge structures in which soil is used as a construction material (Fig. 1.7). The earth dams are built for creating water reservoirs. Since the failure of an earth dam may cause widespread catastrophe, extreme care is taken in its design and construction. It requires a thorough knowledge of soil engineering.

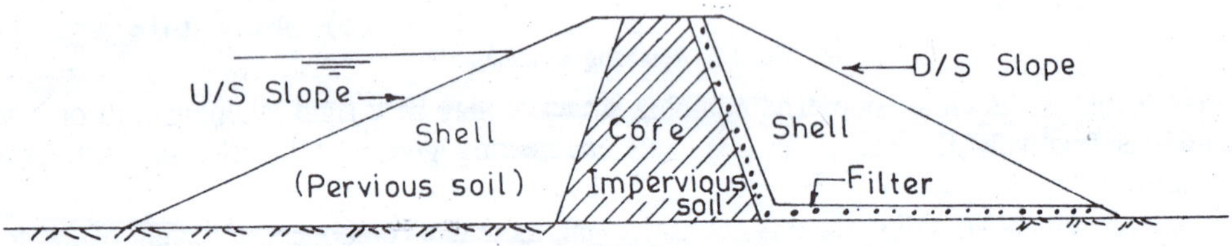

Fig. 1.7. Earth Dam.

(7) Miscellaneous Soil Problems—The geotechnical engineer has sometimes to tackle miscellaneous problems related with soil, such as soil heave, soil subsidence, frost heave, shrinkage and swelling of soils. Soil engineering provides an in-depth study of such problems.

1.5. ORIGIN OF SOILS

Soils are formed by weathering of rocks due to mechanical disintegration or chemical decomposition. When a rock surface gets exposed to atmosphere for an appreciable time, it disintegrates or decomposes into small particles and thus the soils are formed.

Soil may be considered as an incidental material obtained from the geologic cycle which goes on continuously in nature. The geologic cycle consists of erosion, transportation, deposition and upheaval of soil (Fig. 1.8). Exposed rocks are eroded and degraded by various physical and chemical processes. The products of erosion are picked up by agencies of transportation, such as water and wind, and are carried to new locations where they are deposited. This shifting of the material disturbs

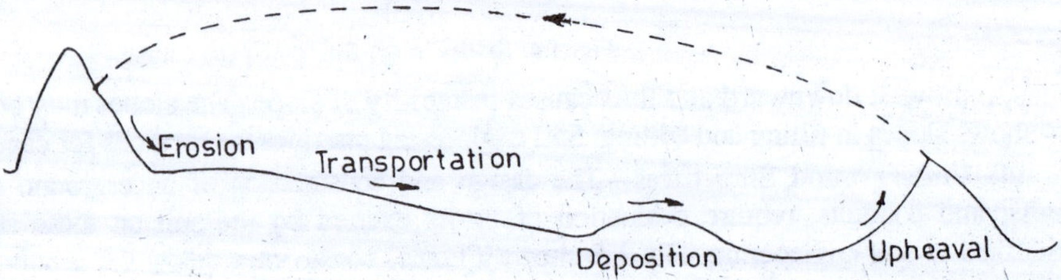

Fig. 1.8. Geologic Cycle.

the equilibrium of forces on the earth and causes large scale earth movements and upheavals. This process results in further exposure of rocks and the geologic cycle gets repeated.

If the soil stays at the place of its formation just above the parent rock, it is known as *residual soil or sedentary soil.* When the soil has been deposited at a place away from the place of its origin, it is called a *transported soil.* The engineering properties of residual soils vary considerably from the top layer to the bottom layer. Residual soils have a gradual transition from relatively fine material near the surface to large fragments of stones at greater depth. The properties of the bottom layer resemble that of the parent rock in many respects. The thickness of the residual soil formation is generally limited to a few metres.

The engineering properties of transported soils are entirely different from the properties of the rock at the place of deposition. Deposits of transported soils are quite thick and are usually uniform. Most of the soil deposits with which a geotechnical engineer has to deal are transported soils.

1.6. FORMATION OF SOILS

As mentioned above, soils are formed by either (*A*) physical disintergration or (*B*) chemical decomposition of rocks.

A. Physical Disintegration—Physical disintegration or mechanical weathering of rocks occurs due to the following physical processes :

(1) Temperature changes—Different minerals of a rock have different coefficients of thermal expansion. Unequal expansion and contraction of these minerals occur due to temperature changes. When the stresses induced due to such changes are repeated many times, the particles get detached from the rocks and the soils are formed.

(2) Wedging action of Ice—Water in the pores and minute cracks of rocks gets frozen in very cold climates. As the volume of ice formed is more than that of water, expansion occurs. Rocks get broken into pieces when large stresses develop in the cracks due to wedging action of the ice formed.

(3) Spreading of roots of plants—As the roots of trees and shrubs grow in the cracks and fissures of the rocks, forces act on the rock. The segments of the rock are forced apart and disintegration of rocks occurs.

(4) Abrasion—As water, wind and glaciers move over the surface of rock, abrasion and scouring takes place. It results in the formation of soil.

In all the processes of physical disintegration, there is no change in the chemical composition. The soil formed has the properties of the parent rock. Coarse grained soils, such as gravel and sand, are formed by the process of physical disintegration.

B. Chemical Decomposition—When chemical decomposition or chemical weathering of rocks takes place, original rock minerals are transformed into new minerals by chemical reactions. The soils formed do not have the properties of the parent rock. The following chemical processes generally occur in nature.

(1) Hydration—In hydration, water combines with the rock minerals and results in the formation of a new chemical compound. The chemical reaction causes a change in volume and decomposition of rock into small particles.

(2) Carbonation—It is a type of chemical decomposition in which carbon dioxide in the atmosphere combines with water to form carbonic acid. The carbonic acid reacts chemically with rocks and causes their decomposition.

(3) Oxidation—Oxidation occurs when oxygen ions combine with minerals in rocks. Oxidation results in decomposition of rocks. Oxidation of rocks is somewhat similar to rusting of steel.

(4) Solution—Some of the rock minerals form a solution with water when they get dissolved in water. Chemical reaction takes place in the solution and the soils are formed.

(5) Hydrolysis—It is a chemical process in which water gets dissociated into H^+ and OH^- ions. The hydrogen cations replace the metallic ions such as calcium, sodium and potassium in rock minerals and soils are formed with a new chemical decomposition.

Chemical decomposition of rocks results in formation of clay minerals. These clay minerals impart plastic properties to soils. Clayey soils are formed by chemical decomposition.

1.7. TRANSPORTATION OF SOILS

The soils formed at a place may be transported to other places be agents of transportion, such as water, wind, ice and gravity.

(1) Water transported Soils— Flowing water is one of the most important agents of transportation of soils. Swift running water carries a large quantity of soil either in suspension or by rolling along the bed. Water erodes the hills and deposits the soils in the valleys.

The size of the soil particles carried by water depends upon the velocity. The swift water can carry the particles of large size such as boulders and gravels. With a decrease in velocity, the coarse particles get deposited. The finer particles are carried further downstream and are deposited when the velocity reduces. A delta is formed when the velocity slows down to almost zero at the confluence with a receiving body of still water, such as a lake, a sea or an ocean (Fig. 1.9).

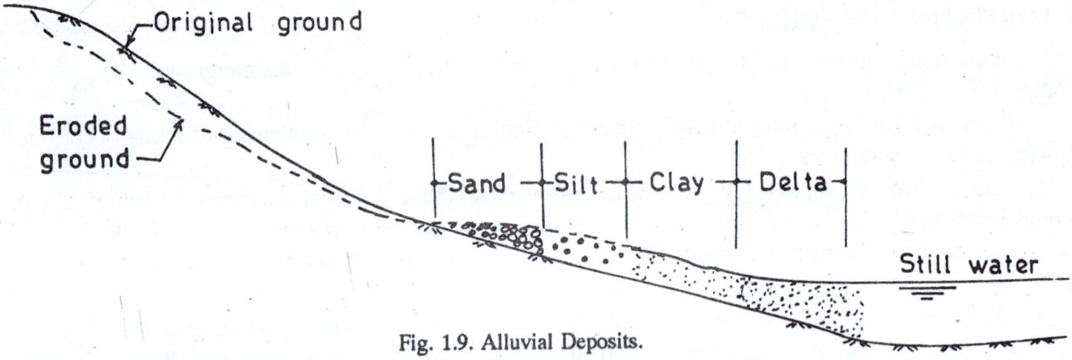

Fig. 1.9. Alluvial Deposits.

All type of soils carried and deposited by water are known as *alluvial deposits.* Deposits made in lakes are called *lacustrine deposits.* Such deposits are laminated or varved in layers. *Marine deposits* are formed when the flowing water carries soils to ocean or sea.

(2) Wind transported Soils—Soil particles are transported by winds. The particle size of the soil depends upon the velocity of wind. The finer particles are carried far away from the place of the formation. A dust storm gives a visual evidence of the soil particles carried by wind. Soils deposited by wind are known as *aeolian deposits.*

Large sand dunes are formed by winds. Sand dunes occur in arid regions and on the leeward side of sea with sandy beaches.

Loess is a silt deposit made by wind. These deposits have low density and high compressibility. The bearing capacity of such soils is very low. The permeability in the vertical direction is large.

(3) Glacier-Deposited Soils—Glaciers are large masses of ice formed by the compaction of snow. As the glaciers grow and move, they carry with them soils varying in size from fine grained to huge boulders. Soils get mixed with the ice and are transported far away from their original position. *Drift* is a general term used for the deposits made by glaciers directly or indirectly. Deposits directly made by melting of glaciers are called *till.*

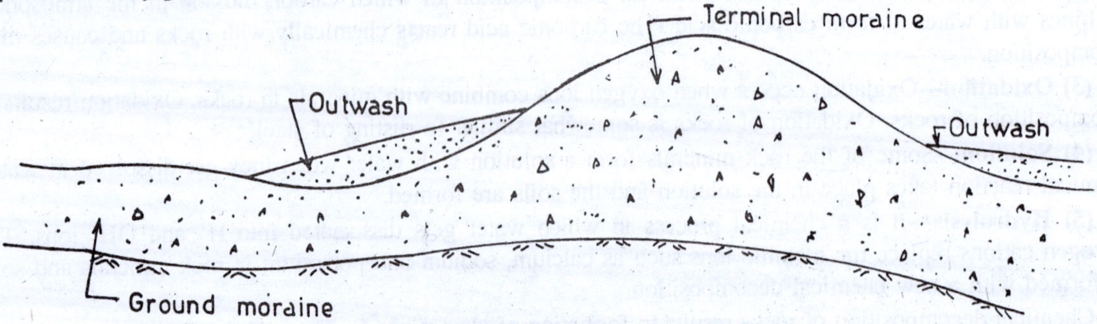

Fig. 1.10. Glacier Deposited Soils.

During their advancement, glaciers transport soils. At the terminus, a melting glacier drops the material in the form of ridges, known as terminal moraine (Fig. 1.10). The land which was once covered by glaciers and on which till has been deposited after melting is called ground moraine. The soil carried by the melting water from the front of a glacier is termed *out-wash.*

Glaciofluvial deposits are formed by glaciers. The material is moved by glaciers and subsequently deposited by streams of melting water. These deposits have stratification.

Deposits of glacial till are generally well-graded and can be compacted to a high dry density. These have generally high shearing strength.

(4) Gravity-deposited soils—Soils can be transported through short distances under the action of gravity. Rock fragments and soil masses collected at the foot of the cliffs or steep slopes had fallen from higher elevation under the action of the gravitational force. Colluvial soils, such as talus, have been deposited by the gravity.

Talus consists of irregular, coarse particles. It is a good source of broken rock pieces and coarse-grained soils for many engineering works.

(5) Soils transported by combined action—Sometimes, two or more agents of transportation act jointly and transport the soil. For example, a soil particle may fall under gravity and may be carried by wind to a far off place. It might by picked up again by flowing water and deposited. A glacier may carry it still further.

1.8. MAJOR SOIL DEPOSITS OF INDIA

The soil deposits of India may be classified in the following five major groups :

(1) Alluvial Deposits—A large part of north India is covered with alluvial deposits. The thickness of alluvium in the Indo-Gangetic and Brahmputra flood plains varies from a few metres to more than one hundred metres. Even in the peninsular India, alluvial deposits occur at some places.

The distinct characteristics of alluvial deposits is the existence of alternating layers of sand, silt and clay. The thickness of each layer depends upon the local terrain and the nature of floods in the rivers causing deposition. The deposits are generally of low density and are liable to liquefaction in earthquake-prone areas.

(2) Black Cotton Soils—A large part of central India and a portion of South India is covered with black cotton soils. These soils are residual deposits formed from basalt or trap rocks. The soils are quite suitable for growing cotton.

Black cotton soils are clays of high plasticity. They contain essentially the clay mineral montmorillonite. The soils have high shrinkage and swelling characteristics. The shearing strength of the soils is extremely low. The soils are highly compressible and have very low bearing capacity. It is extremely difficult to work with such soils.

(3) Lateritic Soils—Lateritic soils are formed by decomposition of rock, removal of bases and silica, and accumulation of iron oxide and aluminium oxide. The presence of iron oxide gives these soils the characteristic red or pink colour. These are residual soils, formed from basalt. Lateritic soils exist in the central, southern and eastern India.

The lateritic soils are soft and can be cut with a chisel when wet. However, these harden with time. A hard crust of gravel size particles, known as laterite, exists near the ground surface. The plasticity of the lateritic soils decreases with depth as they approach the parent rock. These soils, especially those which contain iron oxide, have relatively high specific gravity.

(4) Desert Soils—A large part of Rajasthan and adjoining states is covered with sand dunes. In this area, arid conditions exist, with practically little rainfall.

Dune sand is uniform in gradation. The size of the particles is in the range of fine sand. The sand is non-plastic and highly pervious. As the sand is generally in loose condition, it requires densification to increase its strength.

(5) Marine Deposits—Marine deposits are mainly confined along a narrow belt near the coast. In the south-west coast of India, there are thick layers of sand above deep deposits of soft marine clays.

The marine deposits have very low shearing strength and are highly compressible. They contain a large amount of organic matter. The marine clays are soft and highly plastic.

1.9. COMPARISON OF SOILS WITH OTHER MATERIALS

Soil is a highly complex material. It differs from conventional structural materials, such as steel and concrete.

(1) Steel is a manufactured material the properties of which are accurately controlled. The properties of concrete are also controlled to some extent during its preparation. Soil is a material which has been subjected to vagaries of nature, without any control. Consequently, soil is a highly heterogeneous and unpredictable material.

(2) The properties of a soil change not only from one place to the other but also at the place with depth. The properties also change with a change in the environmental, loading and drainage conditions. The properties of a soil depend not only on its type but also on the conditions under which it exists.

(3) The main engineering properties of steel and concrete are modulus of elasticity and tensile and compressive strength. Most of the design work can be done if these properties are known or determined. However, the engineering properties of soils depend upon a number of factors and it is not possible to characterise them by two or three parameters. Elaborate testing is required to determine the characteristics of the soil before design can be done.

(4) Because of huge quantities of soils involved, it is not economically feasible to transport the soils from other places like steel or concrete. Soils are generally used in the conditions in which they exist.

(5) Whereas steel and concrete can be inspected before use, soils for foundations are at great depth and not open to inspection. The samples of the soil taken from the bore holes are generally disturbed and do not represent the true in-situ conditions.

1.10. LIMITATIONS OF SOIL ENGINEERING

Soil engineering is not an exact science. Because of the nature and the variability of soils, sweeping assumptions are made in the derivation of equations. The solution obtained in most cases are for an idealised, hypothetical material, which may not truly represent the actual soil. A good engineering judgment is required for the interpretation of the results. In fact, each problem in soil engineering is a unique problem because the soils at two places are seldom identical.

The following limitations must be kept in mind when tackling problems related with soils.

(1) As the soil does not possess a linear or unique stress-strain relationship, the solutions of the theory of elasticity cannot be directly applied.

(2) The behaviour and the strength of soils depend upon pressure, drainage, environment and many other factors. These changes must be considered when dealing with soils.

(3) As the soil at every location is different, the results and experience from one project to the other should be transferred with caution.

(4) Since the soils are sensitive to disturbance, the results of tests conducted on soil samples should be interpreted carefully.

(5) The most of soil is underground and cannot be inspected. Adequate soil exploration should be done to determine the profile of soil strata.

(6) The methods of construction may have to be modified as the work progresses and the properties of the soil begin to unfold. Occasional observations have to be made during and even after the completion of work to check whether the assumptions made were correct.

(7) It may not be of much use to apply highly mathematical, rigorous solution to a material like soil whose properties cannot be determined to the same accuracy.

(8) The soil is a particulate material in which the particles are relatively free to move with respect to one another. The behaviour of the soil changes as the particles get shifted.

(9) The soil is a multiphase system, consisting of solid, water and air phases. The behaviour of a soil depends upon the relative proportion of the three phases.

(10) Soil mechanics is a relatively new science. It is essential to keep abreast of the latest developments in the field.

1.11. TERMINOLOGY OF DIFFERENT TYPES OF SOILS

A geotechnical engineer should be well versed with the nomenclature and terminology of different types of soils. The following list gives the names and salient characteristics of different types of soils, arranged in alphabetical order.

(1) **Bentonite**—It is a type of clay with a very high percentage of clay mineral montmorillonite. It is a highly plastic clay, resulting from the decomposition of volcanic ash. It is highly water absorbent and has high shrinkage and swelling characteristics.

(2) **Black Cotton Soil**—It is a residual soil containing a high percentage of the clay mineral montmorillonite. It has very low bearing capacity and high swelling and shrinkage properties.

(3) **Boulders**—Boulders are rock fragments of large size, more than 300 mm in size.

(4) **Calcareous soils**—These soils contain a large quantity of calcium carbonate. Such soils effervesce when tested with weak hydrochloric acid.

(5) **Caliche**—It is a type of soil which contains gravel, sand and silt. The particles are cemented by calcium carbonate.

(6) **Clay**—It consists of microscopic and sub-microscopic particles derived from the chemical decomposition of rocks. It contains a large quantity of clay minerals. It can be made plastic by adjusting the water content. It exhibits considerable strength when dry. Clay is a fine-grained soil. It is a choesive soil. The particle size is less than 0.002 mm.

Organic clay contains finely divided organic matter and is usually dark grey or black in colour. It has a conspicuous odour. Organic clay is highly compressible and its strength is very high when dry.

(7) **Cobbles**—Cobbles are large size particles in the range of 80 mm to 300 mm.

(8) **Diatomaceous earth**—Diatoms are minute unicellular marine organisms. Diatomaceous earth is a fine, light grey, soft sedimentary deposit of the silicious remains of skeletons of diatoms.

(9) **Dispersive clays**—These are special type of clays which defloculate in still water. Such soils erode if exposed to low- velocity water.

Susceptibility to dispersion depends upon the cations in the soil pore water.

(10) **Dune sands**—These are wind-transported soils. There are composed of relatively uniform particles of fine to medium sand.

(11) **Expansive clays**—These are prone to large volume changes as the water content is changed.

These soil contain the mineral montmorillonite.

(12) **Fills**—All man-made deposits of soil and waste-materials are called fills. These are the soil embankments raised above the ground surface. Engineering properties of fills depend upon the type of soil, its water content and the degree of compaction.

(13) **Gravel**—Gravel is a type of coarse-grained soil. The particle size ranges from 4.75 mm to 80 mm. It is a cohesionless material.

(14) **Hardpans**—Hardpans are types of soils that offer great resistance to the penetration of drilling tools during soil exploration. The soils are designated hardpans regardless of their particle size. These are generally dense, well-graded, cohesive aggregates of mineral particles. Hardpans do not disintegrate when submerged in water.

(15) **Humus**—It is a dark brown, organic amorphous earth of the topsoil. It consists of partly decomposed vegetal matter. It is not suitable for engineering works.

(16) **Kankar**—It is an impure form of lime stone. It contains calcium carbonate mixed with some silicious material.

(17) **Laterites**—Laterites are residual soils formed in tropical regions. Laterites are very soft when freshly cut but become hard after long exposure. Hardness is due to cementing action of iron oxide and aluminium oxide. These soils are also called lateritic soils.

(18) **Loam**—It is a mixture of sand, silt and clay. The term is generally used in agronomy. The soil is well suited to tilling operations.

(19) Loess—It is a wind blown deposit of silt. It is generally of uniform gradation, with the particle size between 0.01 to 0.05 mm. It consists of quartz and feldspar particles, cemented with calcium carbonate or iron oxide. When wet, it becomes soft and compressible because cementing action is lost. A loess deposit has a loose structure with numerous root holes which produce vertical cleavage. The permeability in the vertical direction is generally much greater than that in the horizontal direction.

(20) Marl—It is a stiff, marine calcareous clay of greenish colour.

(21) Moorum—The word *moorum* is derived from a Tamil word, meaning powdered rock. It consists of small pieces of disintegrated rock or shale, with or without boulders.

(22) Muck—It denotes a mixture of fine soil particles and highly decomposed organic matter. It is black in colour and of extremely soft consistency. It cannot be used for engineering works. The organic matter is in an advanced stage of decomposition.

(23) Peat—It is an organic soil having fibrous aggregates of macroscopic and microscopic particles. It is formed from vegetal matter under conditions of excess moisture, such as in swamps. It is highly compressible and not suitable for foundations.

(24) Sand—It is a coarse-grained soil, having particle size between 0.075 mm to 4.75 mm. The particles are visible to naked eye. The soil is cohesionless and pervious.

(25) Silt—It is a fine-grained soil, with particle size between 0.002 mm and 0.075 mm. The particles are *not visible* to naked eyes.

Inorganic silt consists of bulky, equidimensional grains of quartz. It has little or no plasticity, and is cohesionless.

Organic silt contains an admixture of organic matter. It is a plastic soil and is cohesive.

(26) Till—It is an unstratified deposit formed by melting of a glacier. The deposit consists of particles of different sizes, ranging from boulders to clay. The soil is generally well-graded. It can be easily densified by compaction. Till is also known as *boulder-clay.*

(27) Top soils—Top soils are surface soils that support plants. They contain a large quantity of organic matter and are not suitable for foundations.

(28) Tuff—It is a fine-grained soil composed of very small particles ejected from volcanoes during its explosion and deposited by wind or water.

(29) Tundra—It is a mat of peat and shrubby vegetation that covers clayey subsoil in arctic regions. The deeper layers are permanently frozen and are called *permafrost.* The surface deposit is the active layer which alternately freezes and thaws.

(30) Varved clays—These are sedimentary deposits consisting of alternate thin layers of silt and clay. The thickness of each layer seldom exceeds 1 cm. These clays are the results of deposition in lakes during periods of alternately high and low waters.

[**Note.** For glossary of technical terms, see APPENDIX A].

1.12. COHESIVE AND COHESIONLESS SOILS

Soils in which the adsorbed water and particle attraction act such that it deforms plastically at varying water contents are known as cohesive soils or clays. This cohesive property is due to presence of clay minerals in soils. Therefore, the term cohesive soil is used synonymously for clayey soils.

The soils composed of bulky grains are cohesionless regardless of the fineness of the particles. The rock flour is cohesionless even when it has the particle size smaller than 2μ size. Non-plastic silts and coarse-grained soils are cohesionless.

[**Note.** $1\ \mu = 1$ micron $= 10^{-6}$ m $= 10^{-3}$ mm].

Many soils are mixture of bulky grains and clay minerals and exhibit some degree of plasticity with varying water content. Such soils are termed cohesive if the plasticity effect is significant; otherwise, cohesionless.

Obviously, there is no sharp dividing line between cohesionless and cohesive soils. However, it is sometimes convenient to divide the soil into above two groups.

The term cohesive-soil is used for clays and plastic silt, and the term cohesionless-soil, for non-plastic silts, sands and gravel.

1.13. BRIEF HISTORY OF SOIL ENGINEERING

According to the autnor, the history of soil engineering can be divided into three periods, as described below:

(1) Ancient to Medieval period—Man's first contact with soil was when he placed his foot on the earth. In ancient times, soil was used as a construction material for building huge earth mounds for religious purposes, burial places and dwellings. Caves were built in soil to live in.

Excellent pavements were constructed in Egypt and India much before the Christian era. Some earth dams have been storing water in India for more than 2000 years. Remnants of various underground water structures, such as aqueducts, tunnels and large drains, found in the excavation at the sites of early civilisation at Mohenjodaro and Harrappa in the Indian subcontinent indicate the use of soil as foundation and construction material. Egyptian used caissons for deep foundations even 2000 B.C. Hanging garden at Babylon (Iraq) was also built during that period. The city of Babylon was built on fills above the adjoining flood plains.

During Roman times, heavy structures, such as bridges, aqueducts, harbours and buildings, were built. Some of these works are in existence even today. After the collapse of the Roman Empire, the construction activities declined. However, some heavy city walls and forts were built from the strategic considerations. Cathedrals, castles and campaniles (bell towers) were also constructed. The famous tower of Pisa, known as the leaning tower of Pisa, was also built. The tower has leaned on one side because of the differential settlement of its base.

The famous Rialto Bridge was constructed in Venice (Italy) in the seventeenth century. Leonardo da Vinci constructed a number of structures in France during the same period. The famous London Bridge in England was also built. The mausoleum Taj Mahal at Agra (India) was constructed by the emperor Shah Jehan to commemorate his favourite wife Mumtaz Mahal. It is built on masonry cylindrical wells sunk into the soil at close intervals.

It is certain that early builders, while constructing such huge structures, encountered and successfully tackled many challenging problems. However, no record in available about the methods adopted. No scientific study seems to have been made. The builders were guided by the knowledge and experience passed down from generation to generation.

(2) Period of Early Developments—The eighteenth century can be considered as the real beginning of soil engineering when early developments in soil engineering took place. In 1773, a French engineer Coulomb gave a theory of earth pressure on retaining walls. The theory is used by the geotechnical engineers even today (chapter 19). Coulomb also introduced the concept that the shearing resistance of soil consists of two components, namely, the cohesion component and the friction component (chapter 13). Culmann gave a general graphical solution for the earth pressure in 1866. Rankine, in 1857, published a theory on earth pressure considering the plastic equilibrium of the earth mass. In 1874, Rehbann gave a graphical method for computation of earth pressure based on Coulomb's theory.

Darcy gave the law of the permeability of soils in 1856. Darcy's law is used for the computation of seepage through soils (chapters 8 and 9). In the same year, Stokes gave the law for the velocity of fall of solid particles through fluids. The law is used for determining the particle size, as discussed in chapter 3.

O-Mohr gave the rupture theory for soils in 1871. He also gave a graphical method of representation of stresses, popularly known as Mohr's circle. It is extremely useful for determination of stresses on inclined planes (chapter 13).

Boussinesq, in 1885, gave the theory of stress distribution in a semi-infinite, homogeneous, isotropic, elastic medium due to an externally applied load. The theory is used for determination of stresses in soils due to loads, as discussed in chapter 11.

In 1908, Marston gave the theory for the load carried by underground conduits (chapter 22).

Atterberg, in 1911, suggested some simple tests for characterizing consistency of cohesive soils. The limits, commonly known as Atterberg's limits, are useful for identification and classification of soils, as discussed in chapters 4 and 5.

Swedish Geotechnical Commission of the State Railways of Sweden appointed a committee headed by Prof. Fellenius in 1913 to study the stability of slopes. The committee gave the Swedish circle method for checking the stability of slopes, described in chapter 18. In 1916, Petterson gave the friction circle method for the stability of slopes.

(3) Modern Era—The modern era of Soil Engineering began in 1925, with the publication of the book *Erdbaumechanic* by Karl Terzaghi. The contribution made by Terzaghi in the development of soil engineering is immense. He is fittingly called the father of soil mechanics. For the first time, he adopted a scientific approach in the study of soil mechanics. His theory of consolidation of soils (chapter 12) and the effective stress principle (chapter 10) gave a new direction.

Proctor did pioneering work on compaction of soils in 1933, as discussed in chapter 14.

Taylor made major contributions on consolidation of soils, shear strength of clays and the stability of slopes.

Casagrande made significant contributions on classification of soils, seepage through earth masses and consolidation.

Skempton did pioneering work on the pore pressures, effective stress, bearing capacity and the stability of slopes.

Meyerhof gave the theories for the bearing capacity of shallow and deep foundations.

Hvorslev did commendable work on subsurface exploration and on shear strength of remoulded clays.

The above list is far from complete. Many other distinguished geotechnical engineers have made a mark on the development of soil engineering. Because of space limitation, their mention could not be made in the above list.

PROBLEMS

A. Descriptive

1.1. Define the terms 'soil', 'soil mechanics' and soil engineering. What are limitations of soil engineering ?

1.2. What is geologic cycle ? Explain the phenomena of formation and transportation of soils.

1.3. What are the major soil deposits of India ? Explain their characteristics.

1.4. Write a brief history of soil engineering.

B. Multiple-Choice Questions

1. Colluvial soils (talus) are transported by:
(*a*) Water (*b*) Wind
(*c*) Gravity (*d*) Ice

2. Water-transported soils are termed:
(*a*) Aeoline (*b*) Alluvial
(*c*) Colluvial (*d*) Till

3. Glacier-deposited soils are called:
(*a*) Talus (*b*) Loess
(*c*) Drift (*d*) None of above

4. Cohesionless soils are formed due to:
(*a*) Oxidation (*b*) Hydration
(*c*) Physical disintegration (*d*) Chemical decomposition

5. When the products of rock weathering are not transported but remain at the place of formation, the soil is called:
(*a*) Alluvial soil (*b*) Talus
(*c*) Residual soil (*d*) Aeolian soil

6. The following type of soil is not glacier-deposited.:
(*a*) Drift (*b*) Till
(*c*) Outwash (*d*) Bentonite.

[Ans. 1 (*c*), 2 (*b*), 3 (*c*), 4 (*c*), 5 (*c*), 6 (*d*)**]**

2

Basic Definitions and Simple Tests

2.1. INTRODUCTION

A soil mass consists of solid particles which form a porous structure. The voids in the soil mass may be filled with air, with water or partly with air and partly with water. In general, a soil mass consists of solid particles, water and air. The three constituents are blended together to form a complex material (Fig. 2.1. *a*). However, for convenience, all the solid particles are segregated and placed in the lower layer of the three-phase diagram (Fig. 2.1*b*). Likewise, water and air particles are placed separately, as shown. The 3-phase diagram is also known as *Block diagram.*

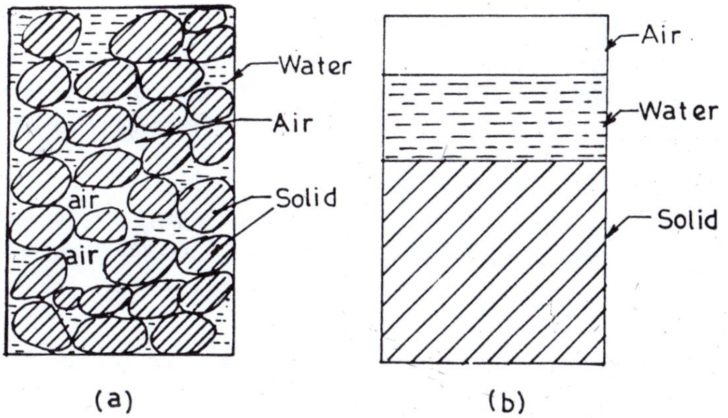

Fig. 2.1. Constituents of Soil.

It may be noted that the three constituents cannot be actually segregated, as shown. A 3-phase diagram is an artifice used for easy understanding and convenience in calculation.

Although the soil is a three-phase system, it becomes a two-phase system in the following two cases: (1) When the soil is absolutely dry, the water phase disappears (Fig. 2.2 *a*). (2) When the soil is fully saturated, there is no air phase (Fig. 2.2*b*). It is the relative proportion of the three constituents and their interaction that governs the behavior and properties of soils. The phase diagram is a simple, diagrammetic representation of a real soil. It is extremely useful for studying the various terms used in soil engineering and their inter-relationships.

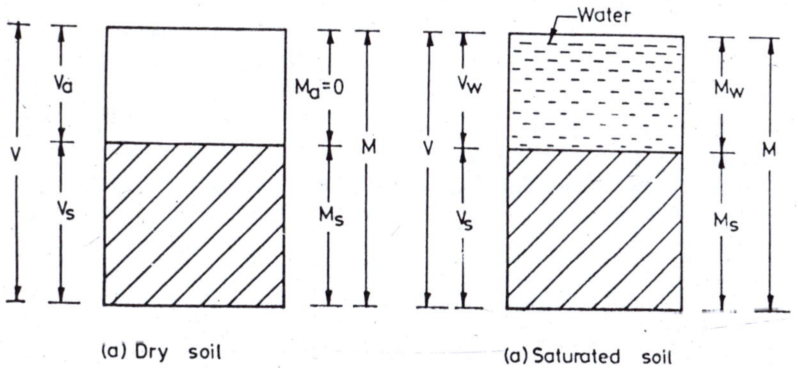

Fig. 2.2. Two-phase diagrams.

In a 3-phase diagram, it is conventional to write volumes on the left side and the mass on the right side (Fig. 2.3 *a*). The total volume of a given soil mass in designated as *V*. It is equal to the sum of the volume of solids (V_s),

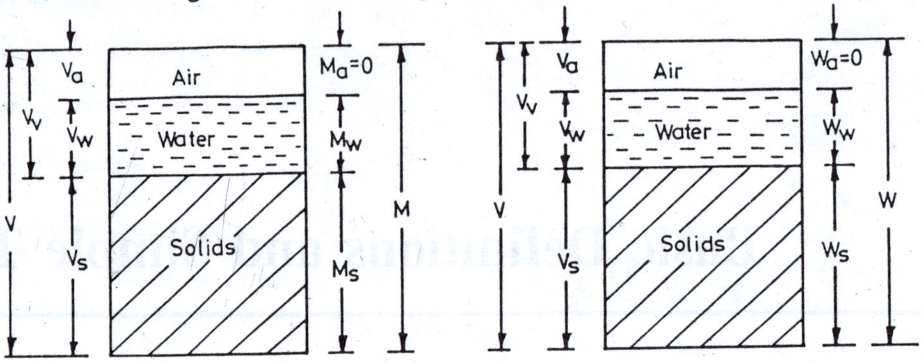

Fig. 2.3. Three-phase diagram.

the volume of water (V_w) and the volume of air (V_a). The volume of voids (V_v) is equal to the sum of the volumes of water and air.

The total mass of the soil mass is represented as *M*. The mass of air (M_a) is very small and is neglected. Therefore, the total mass of the soil is equal to the mass of solids (M_s) and the mass of water (M_w).

Fig. 2.3*b* shows the 3-phase diagram in which the weights are written on the right side.

2.2. VOLUMETRIC RELATIONSHIPS

The following five volumetric relationships are widely used in soil engineering.

(1) Void Ratio (*e*)—It is defined as the ratio of the volume of voids to the volume of solids. Thus

$$e = \frac{V_v}{V_s} \qquad \text{...(2.1)}$$

The void ratio is expressed as a decimal, such as 0.4, 0.5, etc. For coarse-grained roils, the void ratio is generally smaller than that for fine-grained soils. For some soils, it may have a value even greater than unity.

(2) Porosity (*n*)—It is defined as the ratio of the volume of voids to the total volume. Thus

$$n = \frac{V_v}{V} \qquad \text{...(2.2)}$$

Porosity is generally expressed as percentage. However, in equations, it is used as a ratio. For example, a porosity of 50% will be used as 0.5 in equations. The porosity of a soil cannot exceed 100% as it would mean V_v is greater than *V*, which is absurd. In fact, it will have a much smaller value. Porosity is also known as *percentage voids*.

Both porosity and void ratio are measures of the denseness (or looseness) of soils. As the soil becomes more and more dense, their values decrease. The term porosity is more commonly used in other disciplines such as agricultural engineering. In soil engineering, the term void ratio is more popular. It is more convenient to use void ratio than porosity. When the volume of a soil mass changes, only the numerator (*i.e.* V_v) in the void ratio changes and the denominator (*i.e.* V_s) remains constant. However, if the term porosity is used, both the numerator and the denominator change and it becomes inconvenient.

An inter-relationship can be found between the void ratio and the porosity as under.

From Eq. 2.2,
$$\frac{1}{n} = \frac{V}{V_v} = \frac{V_v + V_s}{V_v}$$

or
$$\frac{1}{n} = 1 + \frac{1}{e} = \frac{1+e}{e} \qquad \text{...}(a)$$

or
$$n = \frac{e}{1+e} \qquad \text{...(2.3)}$$

Also, from Eq. (*a*),
$$\frac{1}{e} = \frac{1}{n} - 1 = \frac{1-n}{n}$$

or
$$e = \frac{n}{1-n} \quad ...(2.4)$$

In Eqs. (2.3) and (2.4), the porosity should be expressed as a ratio (and not perentage).

(3) Degree of Saturation (S)—The degree of saturation (S) is the ratio of the volume of water to the volume of voids.

Thus
$$S = \frac{V_w}{V_v} \quad ...(2.5)$$

The degree of saturation is generally expressed as a percentage. It is equal to zero when the soil is absolutely dry and 100% when the soil is fully saturated. In expressions, the degree of saturation is used as a decimal.

In some texts, the degree of saturation is expressed as S_r.

(4) Percentage Air voids (n_a)—It is the ratio of the volume of air to the total volume.

Thus
$$n_a = \frac{V_a}{V} \quad ...(2.6)$$

As the name indicates, it is represented as a percentage.

(5) Air Content (a_c)—Air content is defined as the ratio of the volume of air to the volume of voids.

Thus
$$a_c = \frac{V_a}{V_v} \quad ...(2.7)$$

Air content is usually expressed as a percentage.

Both air content and the percentage air voids are zero when the soil is saturated ($V_a = 0$).

An inter-relationship between the percentage air voids and the air content can be obtained.

From Eq. 2.6,
$$n_a = \frac{V_a}{V} = \frac{V_a}{V_v} \times \frac{V_v}{V}$$

or
$$n_a = n\, a_c \quad ...(2.8)$$

[**Note.** In literature, the ratio V_a/V is also called air content by some authors. However, in this text, this ratio would be termed percentage air voids and not air content].

2.3. WATER CONTENT

The water content (w) is defined as the ratio of the mass of water to the mass of solids.

$$w = \frac{M_w}{M_s} \quad ...(2.9)$$

The water content is also known as the moisture content (m). It is expressed as a percentage, but used as a decimal in computation.

The water content of the fine-grained soils, such as silts and clays, is generally more than that of the coarse grained soils, such as gravels and sands. The water content of some of the fine-gained soils may be even more than 100%, which indicates that more than 50% of the total mass is that of water.

The water content of a soil is an important property. The characteristics of a soil, especially a fine-grained soil, change to a marked degree with a variation of its water content.

In geology and some other disciplines, the water content is defined as the ratio of the mass of water to the total mass. Some of the instruments, such as moisture tester, also give the water content as a ratio of the total mass. In this text water content (w) will be taken as given by Eq. 2.9, unless mentioned otherwise.

The symbol m' shall be used in this text for the water content based on the total wet mass. Thus

$$m' = \frac{M_w}{M} \times 100 \quad ...(2.10)$$

Note. Certain quantities, as defined above, are expressed as a ratio and certain other quantities, as a percentage. To avoid confusion, it is advisable to express all quantities as a ratio (or a decimal) in computations. The final result should be expressed as a percentage for the quantities which are defined as a percentage and as decimal for other quantities.

2.4. UNITS

In this text, SI units are used. In this system, mass (M), length (L) and time (T) are the basic dimensions. The mass is expressed in kilogramme (kg) units, the length in metre (M) units and the time in seconds (sec or *s*) units.

The most important derived unit is the force unit. The force is expressed in newton (N). One newton is the force which is required to give an acceleration of 1 m/sec^2 to a mass of 1 kg. Thus

$$1 \text{ N} = 1 \text{ kg} \times 1 \text{ m/sec}^2$$

In addition to kg mass and N force, the following multiples and submultiples are also frequently used.

1 milligramme (mg) $= 10^{-3}$ gram (gm or g)
1 kilogramme (kg) $= 10^3$ gm
1 megagramme (Mg) $= 10^6$ gm $= 10^3$ kg

Likewise,

1 millinewton (mN) $= 10^{-3}$ newton (N)
1 kilonewton (kN) $= 10^3$ N
1 meganewton (MN) $= 10^6$ N $= 10^3$ kN

2.5. VOLUME-MASS RELATIONSHIPS

The volume-mass relationship are in terms of mass density. The mass of soil per unit volume is known as mass density. In soil engineering, the following 5 different mass densities are used.

(1) Bulk Mass Density—The bulk mass density (ρ) is defined as the total mass (M) per unit total volume (V). Thus, from Fig. 2.3 (*a*),

$$\rho = \frac{M}{V} \qquad \text{...(2.11)}$$

The bulk mass density is also known as the wet mass density or simply bulk density or density. It is expressed in kg/m^3, gm/ml or Mg/m^3.

Obviously, 1 Mg/m^3 = 1000 kg/m^3 = 1 gm/ml

(2) Dry Mass Density—The dry mass density (ρ_d) is defined as the mass of solids per unit total volume. Thus

$$\rho_d = \frac{M_s}{V} \qquad \text{...(2.12)}$$

As the soil may shrink during drying, the mass density may not be equal to the bulk mass density of the soil in the dried condition. The total volume is measured before drying.

The dry mass density is also known as the dry density.

The dry mass density is used to express the denseness of the soil. A high value of dry mass density indicates that the soil is in a compact condition.

(3) Saturated Mass Density—The saturated mass density (ρ_{sat}) is the bulk mass density of the soil when it is fully saturated. Thus

$$\rho_{sat} = \frac{M_{sat}}{V} \qquad \text{...(2.13)}$$

(4) Submerged Mass Density—When the soil exists below water, it is in a submerged condition. When a volume *V* of soil is submerged in water, it displaces an equal volume of water. Thus the net mass of soil when submerged is reduced [Fig. 2.4 (*a*)].

The submerged mass density (ρ′) of the soil is defined as the submerged mass per unit of total volume. Thus

$$\rho' = \frac{M_{sub}}{V} \qquad ...(2.14)$$

The submerged density is also expressed as ρ_{sub} in some texts. It is also known as the buoyant mass density (ρ_b).

Fig. 2.4 (*a*) shows a soil mass submerged under water. The soil solids which have a volume of V_s are buoyed up by the water. The upthrust is equal to the mass of water diplaced by the solids.

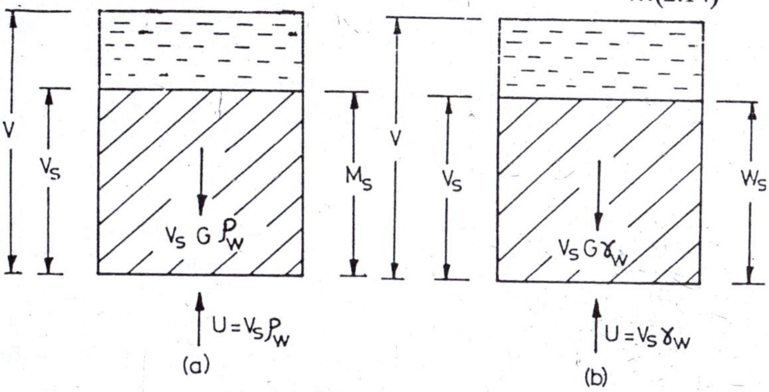

Fig. 2.4. Submerged mass.

Thus $\quad U = V_s \rho_w$

Therefore, $\quad M_{sub} = M_s - U$

$$= V_s G \rho_w - V_s \rho_w$$

From Eq. 2.14, $\quad \rho' = \dfrac{V_s \rho_w (G-1)}{V} \qquad ...(2.15)$

Alternatively, we can also consider the equilibrium of the entire volume (V). In this case, the total downward mass, including the mass of water in the voids, is given by

$$M_{sat} = M_s + V_v \rho_w$$

The total upward thrust, including that on the water in voids, is given by

$$U = V \rho_w$$

Therefore, the submerged mass is given by

$$M_{sub} = (M_s + V_v \rho_w) - V \rho_w$$

From Eq. 2.14, $\quad \rho' = \dfrac{(M_s + V_v \rho_w) - V \rho_w}{V} = \dfrac{M_{sat} - V \rho_w}{V}$

or $\quad \rho' = \dfrac{M_{sat}}{V} - \rho_w$

Using Eq. 2.13 $\quad \rho' = \rho_{sat} - \rho_w \qquad ...(2.16)$

The submerged density ρ' is roughly one-half of the saturated density.

(5) Mass Density of Solids—The mass density of solids (ρ_s) is equal to the ratio of the mass of solids to the volume of solids. Thus

$$\rho_s = \frac{M_s}{V_s} \qquad ...(2.17)$$

2.6. VOLUME-WEIGHT RELATIONSHIP

The volume-weight relationships are in terms of unit weights. The weight of soil per unit volume is known as unit weight (or specific weight). In soil engineering, the following five different unit weights are used in various computations.

(1) Bulk Unit Weight—The bulk unit weight (γ) is defined as the total weight per unit total volume [Fig. 2.3 (*b*)]. Thus

$$\gamma = \frac{W}{V} \qquad ...[2.11(a)]$$

The bulk unit weight is also known as the total unit weight (γ_t), or the wet unit weight.

In SI units, it is expressed as N/m^3 or kN/ m^3.

In some texts, the bulk unit weight is expressed as γ_b or γ_t.

(2) Dry Unit Weight—The dry unit weight (γ_d) is defined as the weight of solids per unit total volume.

Thus $$\gamma_d = \frac{W_s}{V} \qquad ...[2.12(a)]$$

(3) Saturated Unit Weight—The saturated unit weight (γ_{sat}) is the bulk unit weight when the soil is fully saturated.

Thus $$\gamma_{sat} = \frac{W_{sat}}{V} \qquad ...[2.13(a)]$$

(4) Submerged Unit Weight—When the soil exists below water, it is in a submerged condition. A buoyant force acts on the soil solids. According to Archimedes' principle, the buoyant force is equal to the weight of water displaced by the solids. The net mass of the solids is reduced. The reduced mass is known as the submerged mass or the buoyant mass.

The submerged unit weight (γ') of the soil is defined as the submerged weight per unit of total volume.

Thus $$\gamma' = \frac{W_{sub}}{V} \qquad ...[2.14(a)]$$

Fig. 2.4 (*b*) shows a soil mass submerged under water. The soil solids which have a volume of V_s are buoyed up by the water. The buoyant force (U) is equal to the weight of water displaced by the solids.

$$U = V_s \gamma_w$$

The weight of water in the voids has a zero weight in water, as the weight of water and the buoyant force just balance each other. When submerged, all voids can be assumed to be filled with water.

Therefore, $$W_{sub} = W_s - U$$
$$= V_s G \gamma_w - V_s \gamma_w = V_s \gamma_w (G - 1)$$

From Eq. 2.14, $$\gamma' = \frac{V_s \gamma_w (G-1)}{V} \qquad ...[2.15(a)]$$

We can also consider the equilibrium of the entire volume (V). The total downward force, including the wight of water in the voids, is given by

$$W_{sat} = W_s + V_v \gamma_w$$

The total upward force, including that on the water in voids, is given by $U = V \gamma_w$

Therefore, the submerged unit weight is given by

$$W_{sub} = (W_s + V_v \gamma_w) - V \gamma_w$$

From Eq. 2.14, $$\gamma' = \frac{(W_s + V_v \gamma_w) - V \gamma_w}{V} = \frac{W_{sat} - V \gamma_w}{V}$$

Using Eq. 2.13 $$\gamma' = \gamma_{sat} - \gamma_w \qquad ...[2.16(a)]$$

The submerged unit weight is roughly one-half of the saturated unit weight.

In literature, the submerged unit weight is also frequently expressed as γ_{sub}. For convenience, the submerged unit weight will be expressed as γ' in this text.

(5) Unit weight of Soil Solids—The unit weight of solids (γ_s) is equal to the ratio of the weight of solids to the volume of solids. Thus

$$\gamma_s = \frac{W_s}{V_s} \qquad ...[2.17(a)]$$

2.7. INTER-RELATION BETWEEN MASS AND WEIGHT UNITS

In Sect. 2.5, the mass-volume relationships have been developed. The corresponding weight-volume relationships are given in Sect. 2.6. The reader should carefully understand the difference between the two units and should be able to convert the mass densities to the unit weights and *vice-versa*.

The mass and weight are related by Newton's second law of motion, *viz*,

Force = Mass × Acceleration

When a force of one newton (N) is applied to a mass of one kilogramme (kg), the acceleration is 1 m/sec². The weight of 1 kg mass of material on the surface of earth is 9.81 N because the acceleration due to gravity (g) is 9.81 m/sec². Thus we can convert the mass in kg into weight in N by multiplying it by g. In otherwards, W = Mg.

Because the unit weight γ is expressed as W/V and the mass density (ρ) as M/V, the two quantities can be related as

$$\gamma = \frac{W}{V} = \frac{M \cdot g}{V} = \rho g$$

Thus unit weight in N/m³ = mass density in kg/m³ × 9.81

For example, for water ρ_w is 1000 kg/m³.

Therefore, $\gamma_w = 1000 \times 9.81 = 9810 \text{ N/m}^3 = 9.81 \text{ kN/m}^3 \sim 10 \text{ kN/m}^3$

Sometimes, the mass density is expressed in Mg/m³ or g/ml. The corresponding unit weight in kN/m³ is equal to 9.81 ρ. For example, for water ρ_w is 1 Mg/m³ or 1 g/ml. The corresponding unit weight is 9.81 kN/m³.

Likewise. mass density of 1600 kg/m³ corresponds to a unit weight of 1600 × 9.81 N/m³ = 15696 N/m³ = 15.696 kN/m³. In the reverse order, a unit weight of 18 kN/m³ corresponds to a mass density of 18000/9.81 = 1834.62 kg/m³.

It will not be out of place to give a passing reference to the MKS units still prelevant in some fields. In MKS units, the weight is expressed in kilogramme force (kgf). It is equal to the force exerted on a mass of 1 kg due to gravity. As the same force is also equal to 9.81 N, we have

$$1 \text{ kgf} = 9.81 \text{ N}$$

Thus when the force is given in kgf, it can be converted to newtons (N) and *vice versa*. For example, unit weight of water is 1000 kgf/m³ and the corresponding value in SI units is 9810 N/m³ or 9.81 kN/m³.

Likewise, $1 \text{ kgf/cm}^2 = 10^4 \text{ kgf/m}^2$

$$= 10^4 \times 9.81 \text{ N/m}^2 = 98.1 \text{ kN/m}^2 = 98.1 \text{ k Pa}$$

The unit kgf is not used in this text.

Measurement of Mass

The mass of a quantity of matter is determined with a weighing balance. The weights which have previously been used in MKS units are also used in SI units for measurement of mass. In other words, the weight of 1 kgf is called as the mass of 1 kg. The quantity of matter which weighs 1 kgf in MKS units will have a mass of 1 kg in SI units. Of course, the weight of that quantity of matter will be 9.81 N.

Thus the weighing balances and weights which were previously used for determining the weights in kgf are used to determine the mass in kg.

2.8. SPECIFIC GRAVITY OF SOLIDS

The specific gravity of solid particles (G) is defined as the ratio of the mass of a given volume of solids to the mass of an equal volume of water at 4°C. Thus, the specific gravity is given by

$$G = \frac{\rho_s}{\rho_w} \quad \text{...(2.18)}$$

The mass density of water ρ_w at 4°C is one gm/ml, 1000 kg/m³ or 1 Mg/m³.

[**Note.** In some texts, the specific gravity is represented as G_s.]

The specific gravity of solids for most natural soils falls in the general range of 2.65 to 2.80; the smaller values are for the coarse-grained soils. Table 2.1 gives the average values for different soils. It may be mentioned that the specific gravity of different particles in a soil mass may not be the same. Whenever the specific gravity of a soil mass is indicated, it is the average value of all the solid particles present in the soil mass. Specific gravity of solids is an important parameter. It is used for determination of void ratio and particle size.

Table 2.1. Typical Values of G

S.No.	*Soil Type*	*Specific Gravity*
1	Gravel	2.65—2.68
2	Sand	2.65—2.68
3	Silty Sands	2.66—2.70
4	Silt	2.66—2.70
5	Inorganic Clays	2.68—2.80
6	Organic Soils	Variable, may fall below 2.00

In addition to the standard term of specific gravity as defined, the following two terms related with the specific gravity are also occasionally used.

(1) **Mass Specific Gravity (G_m)**—It is defined as the ratio of the mass density of the soil to the mass density of water.

$$G_m = \frac{\rho}{\rho_w} \qquad \text{...(2.19)}$$

Obviously, the value of the mass specific gravity of a soil is much smaller than the value of the specific gravity of solids.

The mass specific gravity is also known as the *apparent specific gravity or the bulk specific gravity.*

(2) **Absolute Specific Gravity (G_a)**—The soil solids are not perfect solids but contain voids. Some of these voids are permeable through which water can enter, whereas others are impermeable. Since the permeable voids get filled when the soil is wet, these are in reality a part of void space in the total mass and not a part of soil solids. If both the permeable and impermeable voids are excluded from the volume of solids, the remaining volume is the true or absolute volume of the solids.

The mass density of the absolute solids $(\rho_s)_a$ is used for the determination of the absolute specific gravity of solids as under. Thus

$$G_a = \frac{(\rho_s)_a}{\rho_w} \qquad \text{...(2.20)}$$

The absolute specific gravity is not of much practical use, as it is difficult to differentiate between the permeable and impermeable voids. In most cases, the impermeable parts are taken as the part of solids. In this text, the term specific gravity of soil solids (G) is used to denote the specific gravity of soil solids inclusive of the impermeable voids. In Eq. 2.18, the soil solids therefore mean the solids with their impermeable voids.

2.9. THREE-PHASE DIAGRAM IN TERMS OF VOID RATIO

The relationships developed in the preceding sections are independent of the actual dimensions of the soil mass. For convenience, the volume of solids is taken as unity in 3-phase diagram. If the area of cross-section of the soil mass is also taken unity, the volume of solids is also equal to the height of solids. Fig. 2.5 (*a*) shows the phase diagram with volume of solids V_s equal to unity. Since the void ratio is equal to the ratio of the volume of

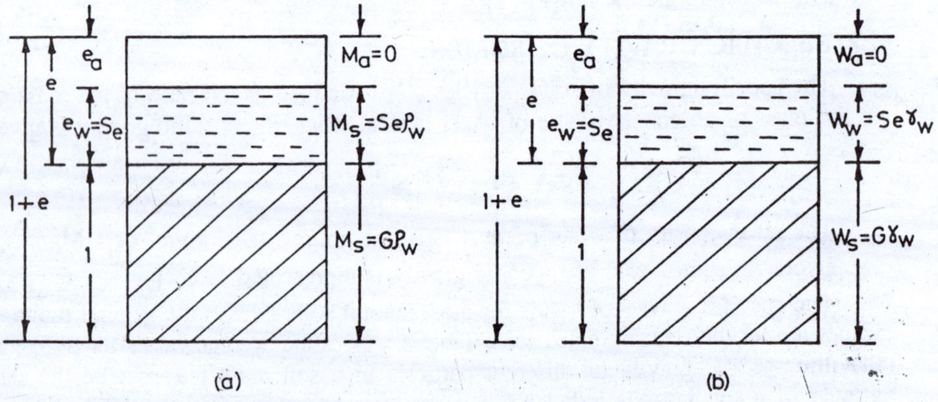

Fig. 2.5. Three-phase diagram in terms of void ratio.

voids to the volume of solids, the volume of voids in Fig. 2.5 (*a*) becomes equal to *e*. The total volume (V) is obviously equal to $(1 + e)$. The volume of air is shown by e_a and the volume of water, by e_w.

The volumes are shown on the left side and the corresponding mass on the right side in Fig. 2.5 (*a*). The volumetric relationships developed in Sect. 2.2 can be written directly in terms of void ratio as under:

Porosity, $$n = \frac{V_v}{V} = \frac{e}{1 + e}$$

Degree of saturation, $$S = \frac{V_w}{V_v} = \frac{e_w}{e}$$

The volume of water (V_w) is shown as Se in Fig. 2.5(*a*). Obviously, the volume of air (V_a) is equal to $(e - Se) = e(1 - S)$.

Therefore, percentage air voids, $$n_a = \frac{V_a}{V} = \frac{e(1 - S)}{1 + e}$$

and air content, $$a_c = \frac{V_a}{V_v} = \frac{e(1 - S)}{e} = (1 - S)$$

Various mass densities discussed in Sect. 2.4 can be expressed in terms of the void ratio from Fig. 2.5 (*a*).

From Eq. 2.11, $$\rho = \frac{M}{V} = \frac{M_s + M_w}{1 + e} = \frac{G\rho_w + Se\rho_w}{1 + e}$$

or $$\rho = \frac{(G + Se)\rho_w}{1 + e} \qquad \text{...(2.21)}$$

From Eq. 2.12 $$\rho_d = \frac{M_s}{V} = \frac{G\rho_w}{1 + e} \qquad \text{...(2.22)}$$

From Eq. 2.13, $$\rho_{sat} = \frac{M_{sat}}{V}$$

As the degree of saturation for a saturated soil is 1.0 (*i.e.* 100%), Eq. 2.21 gives

$$\rho_{sat} = \frac{(G + e)\rho_w}{1 + e} \qquad \text{...(2.23)}$$

From Eq. 2.16 $$\rho' = \rho_{sat} - \rho_w = \frac{(G + e)\rho_w}{1 + e} - \rho_w$$

or $$\rho' = \frac{(G - 1)}{1 + e}\rho_w \qquad \text{...(2.24)}$$

In case the soil is not fully saturated, the submerged mass density is given by $\rho' = \rho - \rho_w$

From Eq. 2.21 $$\rho' = \frac{(G + Se)\rho_w}{1 + e} - \rho_w$$

$$= \frac{(G + Se)\rho_w - (1 + e)\rho_w}{1 + e}$$

or $$\rho' = \frac{[(G - 1) - e(1 - S)]\rho_w}{1 + e} \qquad \text{...(2.25)}$$

Eq. 2.25 reduces to Eq. 2.24 when the soil is fully saturated ($S = 1.0$).

Equations in Weight Units

Eqs. 2.21 to 2.25 can be expressed in terms of weights. Equations can be derived considering the volume-weight phase diagrams [Fig. 2.5 (*b*)] or simply by multiplying both sides of the equations by g and remembering that $\gamma = \rho g$. Thus

Eq. 2.21 becomes $$\gamma = \frac{(G + Se)\gamma_w}{1 + e} \qquad [2.21(a)]$$

Eq. 2.22 becomes $$\gamma_d = \frac{G\gamma_w}{1 + e} \qquad ...[2.22\ (a)]$$

Eq. 2.23 becomes $$\gamma_{sat} = \frac{(G + e)\gamma_w}{1 + e} \qquad ...[2.23\ (a)]$$

Eq. 2.24 becomes $$\gamma' = \frac{(G - 1)}{1 + e}\gamma_w \qquad ...[2.24\ (a)]$$

Eq. 2.25 becomes $$\gamma' = \frac{[(G - 1) - e(1 - S)]\gamma_w}{1 + e} \qquad ...[2.25\ (a)]$$

In geotechnical engineering, unit weights are generally expressed in kN/m^3. The unit weight of water (γ_w) is 9.81 kN/m^3, which is sometimes taken as 10 kN/m^3, for convenience.

It may be mentioned once again that mass density in g/ml can be converted into unit weight in kN/m^3 by multiplying it by 9.81.

For water, $\rho_w = 1$ g/ml. $\quad \gamma_w = 9.81$ kN/m^3

For soils, if $\rho = 2$ g/ml. $\quad \gamma = 19.62$ kN/m^3

2.10. THREE-PHASE DIAGRAM IN TERMS OF POROSITY

Fig. 2.6(*a*) shows the three-phase diagram in which the total volume is taken as unity. According to the definition of porosity *n*,

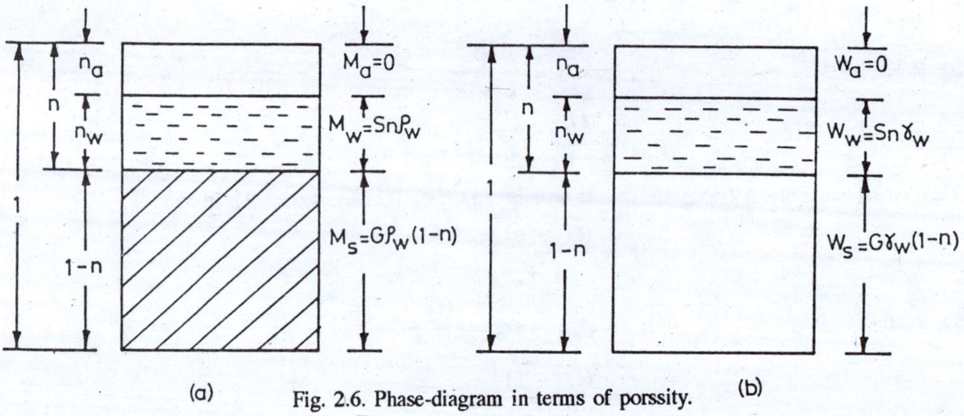

Fig. 2.6. Phase-diagram in terms of porssity.

$$n = \frac{V_v}{V} = \frac{V_v}{1} = V_v$$

Therefore, the volume of voids in shown as *n*.

Obviously, the volume of solids is $(1 - n)$.

Void ratio, $$e = \frac{n}{1 - n} \qquad \text{(same as Eq. 2.4)}$$

From Eq. 2.11, $$\rho = \frac{M}{V} = \frac{M_s + M_w\rho_w}{V} = \frac{G\rho_w(1 - n) + Sn\rho_w}{1}$$

or $$\rho = [G(1 - n) + Sn]\rho_w \qquad ...(2.26)$$

From Eq. 2.12, $$\rho_d = \frac{M_s}{V} = \frac{G\rho_w(1 - n)}{1}$$

or $$\rho_d = G\rho_w(1 - n) \qquad ...(2.27)$$

From Eq. 2.13 $$\rho_{sat} = \frac{M_{sat}}{V} = \frac{G\,\rho_w(1-n) + n\,\rho_w}{1}$$

or $$\rho_{sat} = [G\,(1-n) + n]\,\rho_w \qquad ...(2.28)$$

From Eq. 2.15 $$\rho' = \frac{V_s\,\rho_w\,(G-1)}{V} = \frac{(1-n)\,\rho_w\,(G-1)}{1}$$

or $$\rho' = (G-1)\,(1-n)\,\rho_w \qquad ...(2.29)$$

It may be mentioned that Eqs. (2.26) to (2.29) in terms of porosity can also be derived from Eqs. (2.21) to (2.25) directly by substituting $e = n/(1-n)$. This is left an exercise for the readers.

Equations in terms of Weight units

Eqs. 2.26 to 2.29 can be written in terms of unit weights as under.

Eq. 2.26 becomes $$\gamma = [G\,(1-n) + Sn]\,\gamma_w \qquad ...[2.26(a)]$$

Eq. 2.27 becomes $$\gamma_d = G\,\gamma_w\,(1-n) \qquad ...[2.27(a)]$$

Eq. 2.28 becomes $$\gamma_{sat} = [G\,(1-n) + n]\,\gamma_w \qquad ...[2.28(a)]$$

Eq. 2.28 becomes $$\gamma' = (G-1)\,(1-n)\,\gamma_w \qquad ...[2.29(a)]$$

2.11. RELATIONSHIP BETWEEN THE VOID RATIO AND THE WATER CONTENT

An extremely useful relationship between the void ratio (e) and the water content (w) can be developed as under.

Fig. 2.7 (*a*) shows the three-phase diagram.

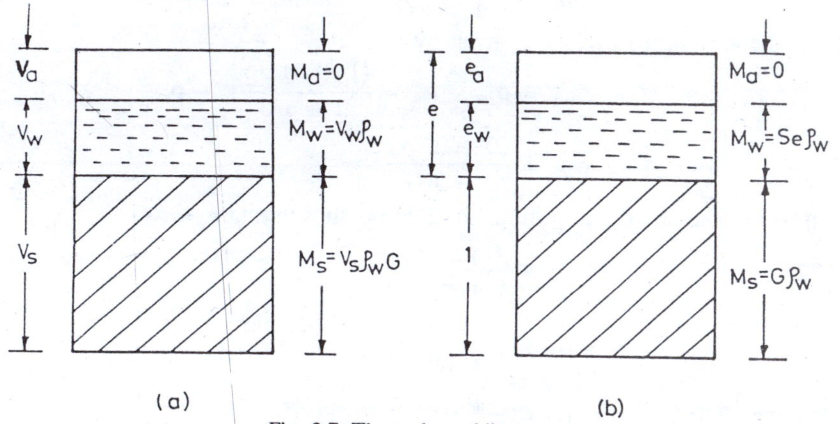

Fig. 2.7. Three-phase ddiagram.

From Eq. 2.9 $$w = \frac{M_w}{M_s}$$

or $$w = \frac{V_w\,\rho_w}{V_s\,\rho_s}$$

From Eq. 2.5, $$V_w = SV_v$$

and from Eq. 2.18, $$\frac{\rho_s}{\rho_w} = G \quad \text{or } \rho_s = G\,\rho_w$$

Therefore, $$w = \frac{SV_v}{V_s G}$$

But V_v / V_s = void ratio (e). Therefore.

$$w = \frac{Se}{G} \quad \text{or } e = \frac{wG}{S} \qquad ...(2.30)$$

For a fully saturated soil, $S = 1.0$, and

$$e = wG \quad \text{...(2.31)}$$

Alternatively Eqs. (2.30) and (2.31) can be derived using the 3-phase diagram in terms of the void ratio [Fig. 2.7 (*b*)].

From Eq. 2.9, $$w = \frac{M_w}{M_s} \quad \text{or} \quad w = \frac{Se\,\rho_w}{G\,\rho_w}$$

or $$w = \frac{Se}{G} \quad \text{or} \quad e = \frac{wG}{S} \quad \text{...(same as Eq. 2.30)}$$

It may be noted that it is more convenient to work with 3-phase diagram in terms of void ratio. The reader is advised to use 3-phase diagram in terms of void ratio as far as possible.

2.12. EXPRESSIONS FOR MASS DENSITY IN TERMS OF WATER CONTENT

The expressions for mass density can be written in terms of water content by writing the void ratio in terms of water content using Eq. 2.30.

From Eq. 2.21 $$\rho = \frac{(G + Se)\,\rho_w}{1 + e}$$

or $$\rho = \frac{(G + wG)\,\rho_w}{1 + (wG)/S} \quad \text{or} \quad \rho = \frac{(1 + w)\,G\rho_w}{1 + (wG)/S} \quad \text{...(2.32)}$$

If the soil is fully saturated, $S = 1.0$, and Eq. 2.32 becomes

$$\rho_{sat} = \frac{(1 + w)\,G\rho_w}{1 + wG} \quad \text{...(2.33)}$$

From Eq. 2.16, the submerged density is given by

$$\rho_{sub} = \rho_{sat} - \rho_w = \frac{(1 + w)\,G\rho_w}{1 + wG} - \rho_w$$

or $$\rho_{sub} = \frac{(G - 1)\,\rho_w}{1 + wG} \quad \text{...(2.34)}$$

Eq.2.34 can also be obtained directly from Eq. 2.24 by substituting $e = wG$.

From Eq. 2.22, $$\rho_d = \frac{G\,\rho_w}{1 + e}$$

or $$\rho_d = \frac{G\,\rho_w}{1 + (wG/S)} \quad \text{...(2.35)}$$

From Eqs. 2.32 and 2.35, $$\rho_d = \frac{\rho}{1 + w} \quad \text{...(2.36)}$$

Eq. 2.36 is an extremely useful equation for determination of the dry density from the bulk density and *vice versa*.

For a given water content w, a soil becomes saturated when $S = 1.0$ in Eq. 2.35. The dry density of the soil in such a condition can be represented as

$$(\rho_d)_{sat} = \frac{G\rho_w}{1 + wG} \quad \text{...(2.37)}$$

$(\rho_d)_{sat}$ is called saturated dry density.

The reader should carefully note the difference between $(\rho)_{sat}$ and $(\rho_d)_{sat}$. In the first case, the water content of a partially saturated is increased so that all the voids are filled with water, whereas in the second case, the water content is kept constant and the air voids are removed by compaction so that all the remaining voids are saturated with water. The latter condition is only hypothetical as it is not feasible to remove all the air voids.

Equations in terms of Weight Units

Eqs. 2.32 to 2.37 can be written in terms of unit weights as under.

Eq. 2.32 becomes $$\gamma = \frac{(1 + w)G\gamma_w}{1 + (wG)/S} \qquad ...[2.32(a)]$$

Eq. 2.33 becomes $$\gamma_{sat} = \frac{(1 + w)G\gamma_w}{1 + wG} \qquad ...[2.33(a)]$$

Eq. 2.34 becomes $$\gamma_{sub} = \frac{(G - 1)\gamma_w}{1 + wG} \qquad ...[2.34(a)]$$

Eq. 2.35 becomes $$\gamma_d = \frac{G\gamma_w}{1 + (wG/S)} \qquad ...[2.35(a)]$$

Eq. 2.36 becomes $$\gamma_d = \frac{\gamma}{1 + w} \qquad ...[2.36(a)]$$

Eq. 2.37 becomes $$(\gamma_d)_{sat} = \frac{G\gamma_w}{1 + wG} \qquad ...[2.37(a)]$$

2.13. RELATIONSHIP BETWEEN DRY MASS DENSITY AND PERCENTAGE AIR VOIDS

In the study of compaction of soils (Chapter 14), a relationship between the dry mass density and the percentage air voids is required. The relationship can be developed from the 3-phase diagram shown in Fig. 2.8 (*a*).

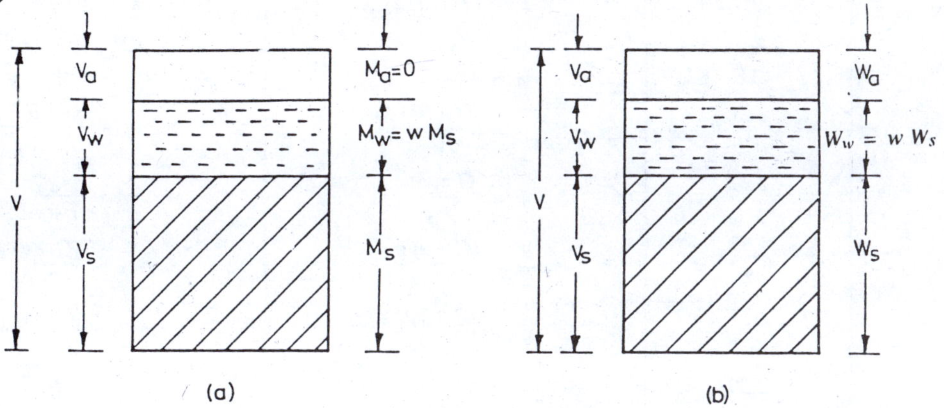

Fig. 2.8. Three-phase diagram

Now $$V = V_s + V_w + V_a$$

or $$1 = \frac{V_s}{V} + \frac{V_w}{V} + \frac{V_a}{V}$$

But $$\frac{V_a}{V} = n_a \qquad \text{(Eq. 2.6)}$$

Therefore $$1 = \frac{V_s}{V} + \frac{V_w}{V} + n_a$$

or $$(1 - n_a) = \frac{V_s}{V} + \frac{V_w}{V}$$

or $$(1 - n_a) = \frac{M_s/(G\rho_w)}{V} + \frac{M_w/\rho_w}{V}$$

$$= \frac{\rho_d}{G\rho_w} + \frac{(wM_s)/\rho_w}{V}$$

$$= \frac{\rho_d}{G\rho_w} + \frac{w\,\rho_d}{\rho_w} = \left(w + \frac{1}{G}\right)\frac{\rho_d}{\rho_w}$$

or
$$\rho_d = \frac{(1 - n_a)\,G\rho_w}{1 + wG} \qquad ...(2.38)$$

When the soil becomes fully saturated at a given water content, $n_a = 0$, and Eq. (2.38) can be written as

$$(\rho_d)_{n_a = 0} = \frac{G\rho_w}{1 + wG}$$

A little reflection will show that $(\rho_d)_{n_a = 0}$ and $(\rho_d)_{sat}$ represent the same condition.

In terms of unit weights Eq. 2.38 becomes [Fig. 2.8 (*b*)]

$$\gamma_d = \frac{(1 - n_a)\,G\,\gamma_w}{1 + e} \qquad ...[2.38\ (a)]$$

Table 2.2 gives a summary of the various relationships. The reader should make these equations as a part of his soil engineering vocabulary.

Table 2.2. Basic Relationships

S. No.	*Eq. No.*	*Relationship in Mass Density*	*Relationship in Unit weight*
1.	2.3	$n = e/(1 + e)$	$n = e/(1 + e)$
2.	2.4	$e = n/(1 - n)$	$e = n/(1 - n)$
3.	2.8	$n_a = n\,a_c$	$n_a = n\,a_c$
4.	2.21	$\rho = \frac{(G + Se)\,\rho_w}{1 + e}$	$\gamma = \frac{(G + Se)\,\gamma_w}{1 + e}$
5.	2.22	$\rho_d = \frac{G\,\rho_w}{1 + e}$	$\gamma_d = \frac{G\,\gamma_w}{1 + e}$
6.	2.23	$\rho_{sat} = \frac{(G + e)\,\rho_w}{1 + e}$	$\gamma_{sat} = \frac{(G + e)\,\gamma_w}{1 + e}$
7.	2.24	$\rho' = \frac{(G - 1)\,\rho_w}{1 + e}$	$\gamma' = \frac{(G - 1)\,\gamma_w}{1 + e}$
8.	2.30	$e = wG/S$	$e = wG/S$
9.	2.36	$\rho_d = \rho/(1 + w)$	$\gamma_d = \gamma/(1 + w)$
10.	2.38	$\rho_d = \frac{(1 - n_a)\,G\,\rho_w}{1 + wG}$	$\gamma_d = \frac{(1 - n_a)\,G\gamma_w}{1 + wG}$

[**Note.** $\rho_w = 1000\ \text{kg/m}^3 = 1.0\ \text{g/ml}$. $\gamma_w = 9810\ \text{N/m}^2 = 9.81\ \text{kN/m}^3 \approx 10\ \text{kN/m}^3$]

2.14. WATER CONTENT DETERMINATION

The water content of a soil is an important parameter that controls its behaviour. It is a quantitative measure of the wetness of a soil mass. The water content of a soil can be determined to a high degree of precision, as it involves only mass which can be determined more accurately than volumes. The water content of soil is determined as a routine matter in most of the other tests.

The water content of a soil sample can be determined by any one of the following methods:

(1) Oven Drying method (2) Torsion Balance method (3) Pycnometer method
(4) Sand Bath method (5) Alcohol method (6) Calcium Carbide method
(7) Radiation method.

(1) Oven Drying method. The oven drying method is a standard, laboratory method. This is a very accurate method.

The soil sample is taken in a small, non-corridible, airtight container. The mass of the sample and that of the container are obtained using an accurate weighing balance. According to IS : 2720 (Part II)—1973, the mass of the sample should be taken to an accuracy of 0.04 per cent. The quantity of the sample to be taken for the test depends upon the gradation and the maximum size of the particles and the degree of wetness of the soil. The drier the soil, the more shall be the quantity of the specimen. Table 2.3 gives the minimum quantity of soil specimen to be taken for the test.

The soil sample in the container is then dried in an oven at a temperature of 110° ± 5°C for 24 hours. The temperature range selected is suitable for most of the soils. The temperature lower than 110° ± 5°C may not cause complete evaporation of water and a temperature higher than this temperature may cause the breaking down of the crystalline structure of the soil particles and loss of chemically bound, structural water. However, oven-drying at 110° ± 5°C does not give reliable results for soils containing gypsum or other minerals having loosely bound water of hydration. This temperature is also not suitable for soils containing significant amount of organic matter. For all such soils, a temperature of 60° to 80°C is recommended. At higher temperature, gypsum loses its water of crystalline and the organic soils tend to decompose and get oxidized.

Table 2.3. Minimum Quantity of Soil for Water Content Determination

S. No.	*Size of Particles more than 90% passing*	*Minimum Quantity (gm)*
1.	425—micron IS sieve	25
2.	2 mm IS sieve	50
3.	4.75 mm IS sieve	200
4.	10 mm IS sieve	300
5.	20 mm IS sieve	500
6.	40 mm IS sieve	1000

The drying period of 24 hours has been recommended for normal soils, as it has been found that this period is sufficient to cause complete evaporation of water. The sample is dried till it attains a constant mass. The soil may be deemed to be dry when the difference in successive weighings of the cooled sample does not exceed about 0.1 percent of the original mass. The soils containing gypsum and organic matter may require drying for a period longer than 24 hours.

The water content of the soil sample is calculated from the following equation.

$$w = \frac{M_w}{M_s} = \frac{M_2 - M_3}{M_3 - M_1} \times 100 \qquad \text{...(2.39)}$$

where M_1 = mass of container, with lid

M_2 = mass of container, lid and wet soil

or M_3 = mass of container, lid and dry soil

The water content of the soil is reported to two significant figures.

(Refer to Chapter 30, Sect. 30.1 for the laboratory experiment)

(2) Torsion Balance Method. The infra-red lamp and torsion balance moisture meter is used for rapid and accurate determination of the water content. The equipment has two main parts : (*i*) the infra-red lamp, and (*ii*) the torsion balance. The infra-red radiation is provided by a 250 W lamp built in the torsion balance for use with an alternating current 220—230 V, 50 cycles, single- phase main supply [IS : 2720 (Part II)—1973].

As the moisture meter is generally calibrated for 25 gm of soil, the maximum size of particle in the specimen shall be less than 2 mm. The sample is kept in a suitable container so that its water content is not affected by ambient conditions. The torque is applied to one end of the torsion wire by means of a calibrated drum to balance the loss of weight of the sample as it dries out under infrared lamp. A thermometer is provided for recording the drying temperature which is kept at 110° ± 5°C. Provision is made to adjust the input voltage to the infra-red lamp to control the heat for drying of the specimen.

The weighing mechanism, known as a torsion balance, has a built- in magnetic damper which reduces pan vibrations for quick drying. The balance scale (drum) is divided in terms of moisture content (m') based on wet mass. The water content (w), based on the dry mass, can be determined from the value of m' as under.

From Eq. 2.10,
$$m' = \frac{M_w}{M} = \frac{M_w}{M_s + M_w}$$

or
$$\frac{1}{m'} = \frac{M_s + M_w}{M_w} = \frac{1}{w} + 1$$

or
$$w = \frac{m'}{1 - m'} \qquad \text{...[2.40(a)]}$$

If w and m' are expressed as percentage,

$$w = \frac{m'}{100 - m'} \times 100 \qquad \text{...[2.40(b)]}$$

The time required for the test depends upon the type of the soil and the quantity of water present. It takes about 15 to 30 minutes. Since drying and weighing occur simultaneously, the method is useful for soils which quickly re-absorb moisture after drying.

(3) Pycnometer method. A pycnometer is a glass jar of about I litre capacity and fitted with a brass conical cap by means of a screw-type cover (Fig. 2.9). The cap has a small hole of 6 mm diameter at its apex. A rubber or fibre washer is placed between the cap and the jar to prevent leakage. There is a mark on the cap and also on the jar. The cap is screwed down to the same mark such that the volume of the pycnometer used in calculations remains constant. The pycnometer method for the determination of water content can be used only if the specific gravity of solid (G) particles is known.

A sample of wet soil, about 200 to 400 g, is taken in the pycnometer and weighed. Water is then added to the soil in the pycnometer to make it about half full. The contents are thoroughly mixed using a glass rod to remove the entrapped air. More and more water is added and stirring process continued till the pycnometer is filled flush with the hole in the conical cap. The pycnometer is wiped dry and weighed.

The pycnometer is then completely emptied. It is washed thoroughly and filled with water, flush with the top hole. The pycnometer is wiped dry and weighed.

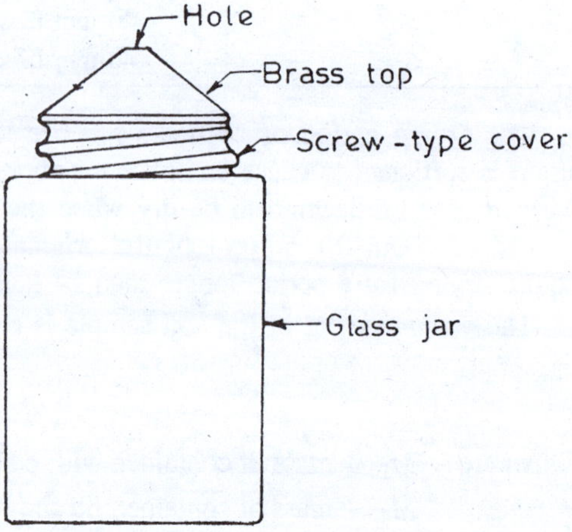

Fig. 2.9 Pycnometer.

Let M_1 = mass of pycnometer

M_2 = mass of pycnometer + wet soil

M_3 = mass of pycnometer + wet soil + water

M_4 = mass of pycnometer filled with water only.

Obviously, the mass M_4 is equal to mass M_3 minus the mass of solids M_s plus the mass of an equal volume of water (see Fig. 2.10).

Thus
$$M_4 = M_3 - M_s + \frac{M_s}{(G\,\rho_w)} \cdot \rho_w$$

or
$$M_4 = M_3 - M_s + \frac{M_s}{G}$$

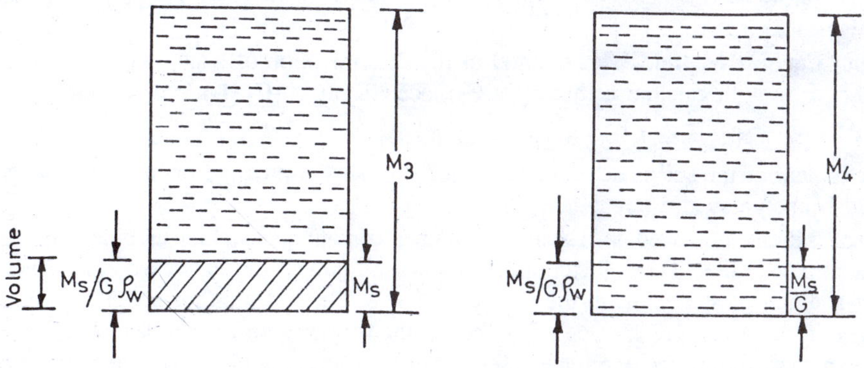

Fig. 2.10. Pycnometer Method Derivation.

$$= M_3 - M_s\left(1 - \frac{1}{G}\right)$$

or
$$M_s = (M_3 - M_4)\cdot\left(\frac{G}{G-1}\right)$$

Now, mass of wet soil = $M_2 - M_1$

Therefore, mass of water $M_w = (M_2 - M_1) - (M_3 - M_4)\left(\frac{G}{G-1}\right)$

From Eq. 2.9.
$$w = \frac{M_w}{M_s} \times 100$$

$$= \left[\frac{(M_2 - M_1)}{(M_3 - M_4)}\left(\frac{G-1}{G}\right) - 1\right] \times 100 \qquad \text{...(2.41)}$$

This method for the determination of the water content is quite suitable for coarse-grained soils from which the entrapped air can be easily removed. If a vacuum pump is available, the pycnometer can be connected to it for about 10 to 20 minutes to remove the entrapped air. The rubber tubing of the pump should be held tightly with the pycnometer to prevent leakage.

(Refer to Chapter 30, Sect. 30.2 for the laboratory experiment)

(4) Sand Bath Method. Sand bath method is a field method for the determination of water content. The method is rapid, but not very accurate. A sand bath is a large, open vessel containing sand filled to a depth of 3 cm or more.

The soil sample is taken in a tray. The sample is crumbled and placed loosely in the tray. A few pieces of white paper are also placed on the sample. The tray is weighed and the mass of wet sample is obtained.

The tray is then placed on the sand-bath. The sand bath is heated over a stove. Drying takes about 20 to 60 minutes, depending upon the type of soil. During heating, the specimen is turned with a palette knife. Overheating of soil should be avoided. The white paper turns brown when overheating occurs. The drying should be continued till the sample attains a constant mass. When drying is complete, the tray is removed from the sand bath, cooled and weighed. The water content is determined using Eq. 2.39.

(5) Alcohol Method. The soil sample is taken in an evaporating dish. Large lumps of soil, if any, should be broken and crumbled. The mass of the wet sample is taken. The sample is then mixed with methylated spirit (alcohol). The quantity of methylated spirit required is about one millilitre for every gram of soil. The methylated spirit and the soil should be turned several times, with a palette knife, to make the mixture uniform.

The methylated spirit is then ignited. The mixture is stirred with a spatula or a knife when ignition is taking place. After the methylated spirit has burnt away completely, the dish is allowed to be cooled, and the mass of the dry soil obtained. The method takes about 10 minutes.

Methylated spirit is extremely volatile. Care shall be taken to prevent fire. The method cannot be used if the soil contains a large proportion of clay, organic matter, gypsum or any other calcareous material. The method is quite rapid, but not very accurate.

(6) Calcium Carbide Method. This method of the determination of water content makes use of the fact that when water reacts with calcium carbide ($Ca\ C_2$), acetylene gas ($C_2\ H_2$) is produced.

$$CaC_2 + 2H_2O = C_2H_2 + Ca\,(OH)_2$$

The water content of the soil is determined indirectly from the pressure of the acetylene gas formed. The instrument used is known as *moisture tester.*

The wet soil sample is placed in a sealed container containing calcium carbide. The samples of sand require no special preparation. The soil sample is ground and pulverised. However, cohesive and plastic soils are tested after addition of steel balls in the pressure vessels. The test requires about 6 g of soil.

The pressure of the acetylene gas produced acts on the diaphragm of the moisture tester. The quantity of gas is indicated on a pressure gauge. From the calibrated scale of the pressure gauge, the water content (m') based on the total mass is determined. The water content (w) based on the dry mass is determined using Eq. 2.40 (a).

As calcium carbide is highly susceptible to absorption of moisture, it should not be exposed to atmosphere. The lid of the container should be firmly fixed.

(7) Radiation Method. Radio-active isotopes are used for the determination of water content of soils. A device containing a radio-active isotopes material, such as cobalt 60, is placed in a capsule. It is then lowered

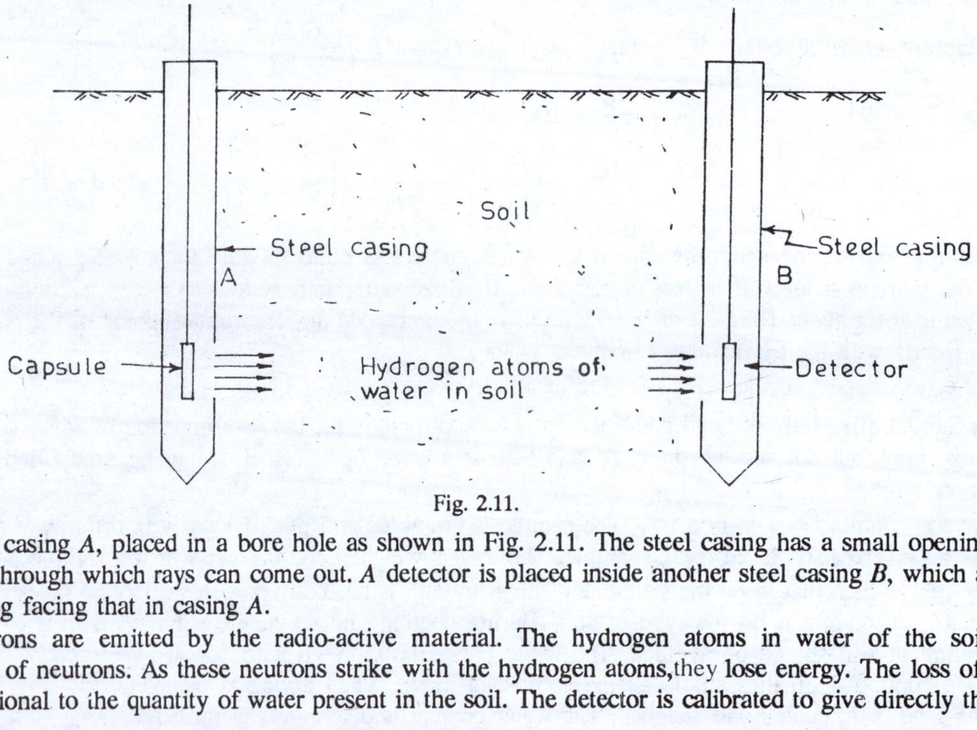

Fig. 2.11.

in a steel casing A, placed in a bore hole as shown in Fig. 2.11. The steel casing has a small opening on its one side through which rays can come out. A detector is placed inside another steel casing B, which also has an opening facing that in casing A.

Neutrons are emitted by the radio-active material. The hydrogen atoms in water of the soil cause scattering of neutrons. As these neutrons strike with the hydrogen atoms,they lose energy. The loss of energy is proportional to the quantity of water present in the soil. The detector is calibrated to give directly the water content

The method is extremely useful for the determination of water content of a soil in the in-situ conditions. The method should be very carefully used, as it may lead to radiation problems if proper shielding precautions are not taken.

2.15. SPECIFIC GRAVITY DETERMINATION

The specific gravity of solid particles is determined in the laboratory using the following methods:

(1) Density bottle method (2) Pycnometer method (3) Measuring flask method
(4) Gas jar method (5) Shrinkage limit method.

The last method of determining the specific gravity of solid particles from the shrinkage limit is discussed in Sect. 4.6.

(1) Density Bottle Method. The specific gravity of solid particles can be determined in a laboratory using a density bottle fitted with a stopper having a hole (Fig. 2.12). The density bottle of 50 ml capacity is generally used [IS : 2720 (Part II) 1980].

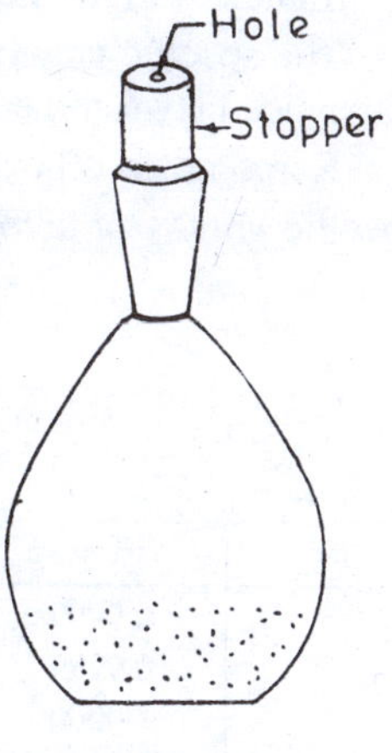

Fig. 2.12. Density bottle.

The density bottle is cleaned and dried at a temperature of 105° to 110°C and cooled. The mass of the bottle, including that of stopper, is taken. About 5-10 g of oven dry sample of soil is taken in the bottle and weighed. If the sample contains particles of large size, it shall be ground to pass a 2-mm sieve before the test.

Distilled water is then added to cover the sample. The soil is allowed to soak water for about 2 hours. More water is added until the bottle is half full. Air entrapped in the soil is expelled by applying a vacuum pressure of about 55 cm of mercury for about one hour in a vacuum dessicator. Alternatively, the entrapped air can be removed by gentle heating. More water is added to the bottle to make it full. The stopper is inserted in the bottle and its mass is taken. The temperature is also recorded.

The bottle is emptied, washed and then refilled with distilled water. The bottle must be filled to the same mark as in the previous case. The mass of the bottle filled with water is taken. The temperature should be the same as before.

Let M_1 = mass of empty bottle

M_2 = mass of bottle and dry soil

M_3 = mass of bottle, soil and water

M_4 = mass of bottle filled with water.

If the mass of solids M_s is subtracted from M_3 and replaced by the mass of water equal to the volume of solid, the mass M_4 is obtained.

Thus
$$M_4 = M_3 - M_s + \frac{M_s}{G\,\rho_w}(\rho_w)$$

or
$$M_s\left(1 - \frac{1}{G}\right) = M_3 - M_4$$

But
$$M_s = M_2 - M_1$$

Therefore
$$(M_2 - M_1)\left(1 - \frac{1}{G}\right) = M_3 - M_4$$

or
$$\frac{1}{G}(M_2 - M_1) = (M_2 - M_1) - (M_3 - M_4)$$

or
$$G = \frac{M_2 - M_1}{(M_2 - M_1) - (M_3 - M_4)} \quad \text{...(2.42)}$$

Alternatively,
$$G = \frac{M_s}{M_s + M_4 - M_3} \quad \text{...[2.42(a)]}$$

Eq. 2.42 gives the specific gravity of solids at the temperature at which the test was conducted.

Specific gravity of solids is generally reported at 27°C (IS : 2720—II) or at 4°C. The specific gravity at 27°C and 4°C can be determined from the following equations.

$$G_{27} = G_t \times \frac{\text{specific gravity of water at } t°\text{C}}{\text{specific gravity of water at 27°C}} \quad \text{...(2.43)}$$

and
$$G_4 = G_t \times \text{specific gravity of water at } t°\text{C} \quad \text{...(2.44)}$$

where G_{27} = sp sr. of particles at 27°, G_4 = sp. gr. of particles at 4°C, G_t = sp. gr. of particles at t°C

Table 2.4 gives the specific gravity of water at different temperatures.

The specific gravity of solids is reported as the average of the two tests, to the nearest 0.01, provided the difference between the two tests does not differ by 0.03.

Kerosene is a better wetting agent than water and is sometimes used in place of water. If G_k is the specific gravity of kerosene at the test temperature, Eq. 2.42 becomes

$$G = \frac{(M_2 - M_1)\, G_k}{(M_2 - M_1) - (M_3 - M_4)} \quad \text{...(2.45)}$$

Table 2.4. Specific Gravity of Water

$t°C$	*sp. gr.*	$t°C$	*sp. gr.*	$t°C$	*sp. gr.*	$t°C$	*sp. gr.*
1	0.9999	11	0.9996	21	0.9980	31	0.9954
2	0.9999	12	0.9995	22	0.9978	32	0.9951
3	1.0000	13	0.9994	23	0.9976	33	0.9947
4	1.0000	14	0.9993	24	0.9973	34	0.9944
5	1.0000	15	0.9991	25	0.9971	35	0.9941
6	1.0000	16	0.9990	26	0.9968	36	0.9937
7	0.9999	17	0.9988	27	0.9965	37	0.9934
8	0.9999	18	0.9986	28	0.9963	38	0.9930
9	0.9998	19	0.9984	29	0.9960	39	0.9926
10	0.9997	20	0.9982	30	0.9957	40	0.9922

Sometimes, other liquids, such as paraffin, alcohol and benzene, are also used.

Density bottle method is suitable for fine-grained soils, with more than 90% passing 2 mm—IS sieve. However the method can also be used for medium and coarse-grained soils if they are pulverised such that the particles pass 2 mm—IS sieve.

(See Chapter 30, Sect. 30.3 for the laboratory experiment).

(2) Pycnometer Method. The method is similar to the density bottle method. As the capacity of the pycnometer is larger, about 200—300 g of oven-dry soil is required for the test. The method can be used for all types of soils, but is more suitable for medium-grained soils, with more than 90% passing a 20 mm IS sieve and for coarse-grained soils with more than 90% passing a 40 mm IS sieve.

(See Chapter 30, Sect. 30.4 for the laboratory experiment).

(3) Measuring Flask Method. A measuring flask is of 250 ml (or 500 ml) capacity, with a graduation mark at that level. It is fitted with an adaptor for connecting it to a vacuum line for removing entrapped air. The method is similar to the density bottle method. About 80—100 g of oven dry soil is required in this case. The method is suitable for fine-grained and medium grained soils.

(4) Gas Jar Method. In this method, a gas jar of about 1 litre capacity is used. The jar is fitted with a rubber bung (Fig. 2.13). The gas jar serves as a pycnometer. The method is similar to the pycnometer method.

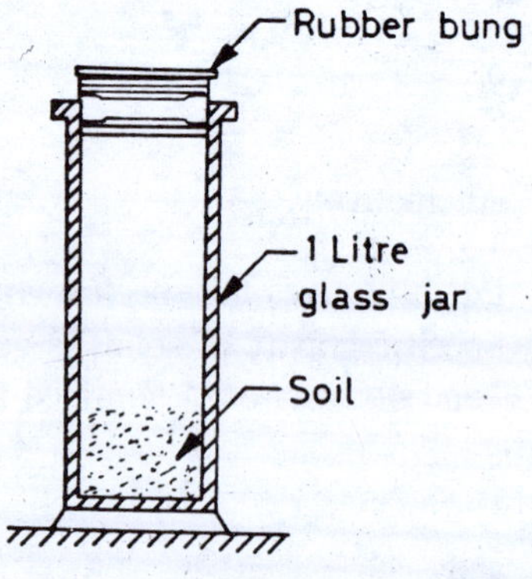

Fig. 2.13. Gas Jar.

2.16. MEASUREMENT OF MASS DENSITY

The bulk mass density of a soil sample, as per Eq. 2.11, is the mass per unit volume. Although the mass of a soil sample can be determined to a high degree of precision, it is rather difficult to determine the volume of the sample accurately. The methods discussed below basically differ in the procedure for the measurement of the volume. Once the bulk mass density has been determined, the dry mass density is found using Eq. 2.36. Thus

$$\rho = \frac{M}{V} \quad \text{and} \quad \rho_d = \frac{\rho}{1 + w}$$

The volume of the specimen used in various tests can be computed from the measured dimensions, as they have regular shapes, such as a cylinder or a cube. However, precise measurements are not possible. If the sample is made in a container of known dimensions, much more accurate measurements are possible.

The following methods are generally used for the determination of mass density.

(1) Water Displacement Method (2) Submerged mass density Method
(3) Core Cutter Method (4) Sand Replacement Method.
(5) Water Balloon Method (6) Radiation Method.

The methods are discussed below. The first two methods are laboratory methods and the rest, field methods.

(1) Water Displacement Method. The volume of the specimen is determined in this method by water displacement. As the soil mass disintegrates when it comes in contact with water, the sample is coated with paraffin wax to make it impervious. A test specimen is trimmed to more or less a regular shape and weighed. It is then coated with a thin layer of paraffin wax by dipping it in molten wax. The specimen is allowed to cool and weighed. The difference between the two observations is equal to the mass of the paraffin.

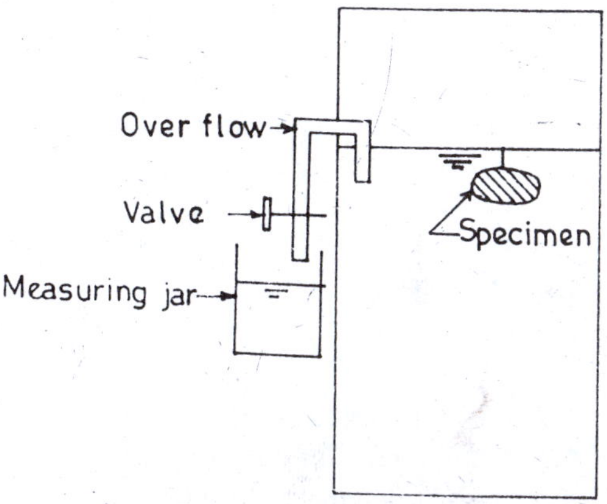

Fig. 2.14. Water displacement container.

The waxed specimen is then immersed in a water-displacement container shown in Fig. 2.14. The volume of the specimen is equal to the volume of water which comes out of the outflow tube. The actual volume of the soil specimen is less than the volume of the waxed specimen. The volume of the wax is determined from the mass of the wax peeled off from the specimen after the test and the mass density of wax.

Now
$$V = V_t - \frac{(M_t - M)}{\rho_p} \qquad \ldots(2.46)$$

where V = volume of specimen, V_t = volume of waxed specimen,
M_t = mass of waxed specimen, M = mass of specimen,
ρ_p = mass density of paraffin (approximately 0.998 gm/ml).

A representative sample of the soil is taken from the middle of specimen for the water content determination.

Once the mass, volume and the water content of the specimen have been determined, the bulk density and the dry density are found from Eqs. 2.11 and 2.36, respectively.

(See Chapter 30, Sect. 30.7 for the laboratory experiment).

(2) Submerged Mass Density method. This method is based on Archimedes' principle that when a body is submerged in water, the reduction in mass is equal to the mass of the volume of water displaced. The sample is first trimmed and weighed and then it is immersed in molten wax and again weighed, as in the water displacement method.

The specimen is then placed in the cradle of special type balance. The cradle dips in the water contained in the bucket placed just below. The apparent mass of the waxed specimen in water is determined.

The volume of the specimen is determined as below:

$$V = \frac{(M_t - M_1)}{\rho_w} - \frac{(M_t - M)}{\rho_p} \qquad \ldots(2.47)$$

where M_t = mass of waxed specimen, M = mass of specimen,

M_1 = mass of waxed specimen in water, ρ_p = mass density of wax,

ρ_w = mass density of water.

Eq. 2.47 can be derived, using Archimedes' principle.

$$M_1 = M_t - U = M_t - V_t \rho_w$$

or
$$V_t = \frac{M_t - M_1}{\rho_w}$$

Substituting this value V_t in Eq. 2.46, we get Eq. 2.47.

This method is suitable for fine-grained soils.

(3) Core Cutter Method. It is a field method for determination of mass density. A core cutter consists of an open, cylindrical barrel, with a hardened, sharp cutting edge (Fig. 2.15). A dolly is placed over the cutter and it is rammed into the soil. The dolly is required to prevent burring of the edges of the cutter. The cutter containing the soil is taken out of the ground. Any soil extruding above the edges of the cutter is removed. The mass of the cutter filled with soil is taken. A representative sample is taken for water content determination.

The volume of the soil is equal to the internal volume of the cutter, which can be determined from the dimensions of the cutter or by filling the cutter with water and finding the mass of water.

Bulk mass density, $$\rho = \frac{M_2 - M_1}{V} \quad ...(2.41)$$

where M_2 = mass of cutter, with soil,

M_1 = mass of empty cutter,

V = internal volume of cutter.

The method is quite suitable for soft, fine grained soils. It cannot be used for stoney, gravelly soils. The method is practicable only at the places where the surface of the soil is exposed and the cutter can be easily driven.

(See Chapter 30, Sect. 30.5 for the experiment).

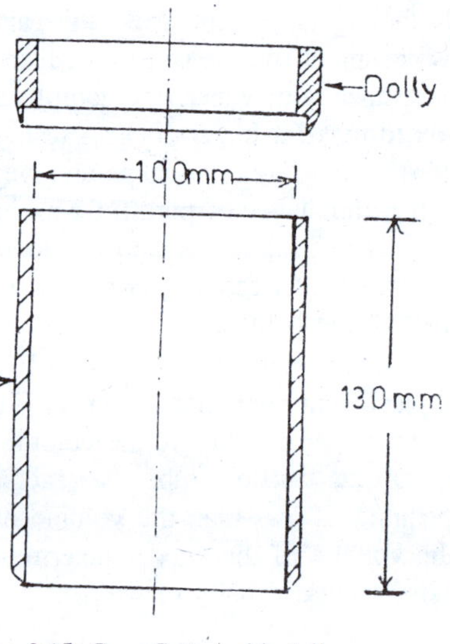

Fig. 2.15. Core Cutter with dolly.

(4) Sand Replacement Method. Fig. 2.16 shows a sand-pouring cylinder, which has a pouring cone at its base. The cylinder shown is placed with its base at the ground level. There is a shutter between the cylinder and the cone. The cylinder is first calibrated to determine the mass density of sand. For good results, the sand used should be uniform, dry and clean, passing a 600 micron sieve and retained on a 300 micron sieve.

(*a*) Calibration of apparatus—The cylinder is filled with sand and weighed. A calibrating container is then placed below the pouring cylinder and the shutter is opened. The sand fills the calibrating container and the cone. The shutter is closed, and the mass of the cylinder is again taken. The mass of the sand in the container and the cone is equal to the difference of the two observations.

The pouring cylinder is again filled to the initial mass. The sand is allowed to run out of the cylinder, equal to the volume of the calibrating container and the shutter is closed. The cylinder is then placed over a plain surface and the shutter is opened. The sand runs out of the cylinder and fills the cone. The shutter is closed when no further movement of sand takes place. The cylinder is removed and the sand filling the cone is collected and weighed (M_2).

The mass density of the sand is determined as under:

$$\rho_s = \frac{M_1 - M_2 - M_3}{V_c} \quad ...(2.49)$$

where M_1 = initial mass of cylinder with sand,

M_2 = mass of sand in cone only,

M_3 = mass of cylinder after pouring sand into the cone and the container,

V_c = Volume of the container.

Note. Mass of sand in both the container and cone is $M_1 - M_3$.

(b) Measurement of Volume of Hole—A tray with a central hole is placed on the prepared ground surface which has been cleaned and properly levelled. A hole about 100 mm diameter and 150 mm deep is excavated in the ground, using the hole in the tray as a pattern. The soil removed is carefully collected and weighed.

The sand pouring cylinder is then placed over the excavated hole as shown in Fig. 2.16. The shutter is opened and the sand is filled in the cone and the hole. When the sand stops running out, the shutter is closed. The cylinder is removed and weighed. The volume of the hole is determined from the mass of sand filled in the hole and the unit mass density of sand.

$$\text{Volume of hole} = \frac{M_1 - M_4 - M_2}{\rho_s} \qquad ...(2.50)$$

where M_1 = mass of cylinder and sand before pouring into the hole,

M_2 = mass of sand in cone only,

M_4 = mass of cylinder after pouring sand into the hole,

ρ_s = mass density of sand, as found from calibration.

The bulk mass density of the in-situ soil is determined from the mass of soil excavated and the volume of the hole.

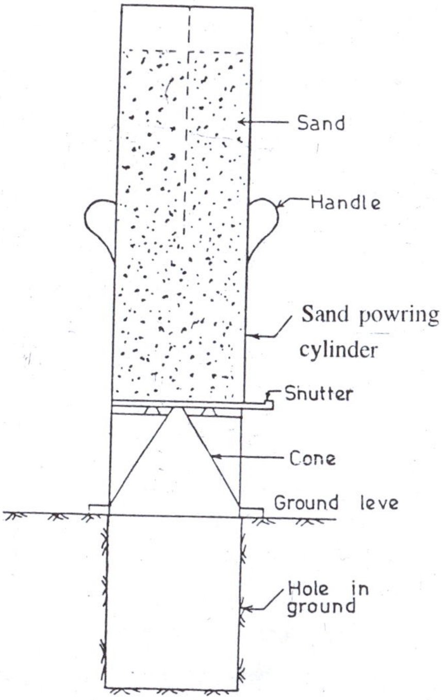

Fig. 2.16. Sand Replacement method.

The method is widely used for soils of various particle sizes, from fine-grained to coarse-grained. For accurate results, the height of sand column in the cylinder is kept approximately the same as that in the calibration test. The depth of the hole should also be equal to the depth of calibrating container.

(See Chapter 30, Sect. 30.6 for the experiment).

5. Rubber Balloon Method. The volume of the hole in this method is determined using a rubber balloon or by filling water in the hole after covering it with a plastic sheet. The rubber balloon method is explained below.

The apparatus consists of a density plate and a graduated cylinder, made of lucite, enclosed in an airtight aluminium case (Fig. 2.17). The cylinder is partly filled with water. There is an opening in the bottom of the case, which is sealed by a rubber balloon. The balloon can be pulled up into the cylinder or may be pushed down through the bottom. A pump is attached to the cylinder for this purpose. When the pressure is applied, balloon comes out the aluminium case through the hole in the density plate. When a vacuum is applied, the balloon is pulled up into the cylinder.

For determination of the volume of the hole, the density plate is placed on the levelled ground. The cylinder is then placed over the plate. The pressure is applied to the balloon. The balloon deflates against the surface of the soil. The volume of water in the cylinder is observed.

The cylinder is removed from the base plate. The soil is taken out

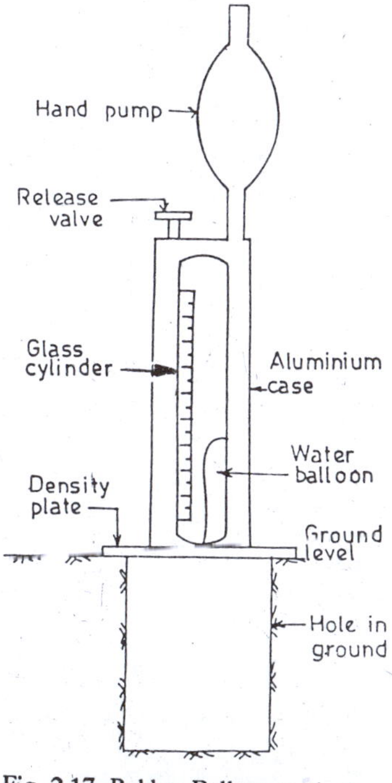

Fig. 2.17. Rubber Balloon method.

from the hole through the opening in the base plate. All loose material is removed. The soil removed is collected and weighed. The cylinder is again placed over the opening in the plate and pressure is applied to the balloon till it fills the hole. The volume of water in the cylinder is observed. The volume of the hole is found from the initial and final observation of water volume.

The method is general and is suitable for all types of soils. However, it is not so accurate, as it is difficult to fit the balloon exactly in an irregular hole.

IS : 2720 (Part XXXIV)—1972 describes the method in detail.

(6) Radiation Method. The bulk mass density of in-situ soil can be determined using the radiation method. The meter consists of two probes, one containing a radio isotope source and the other a gamma ray detector. The meter is placed on the surface which had been carefully cleaned and levelled. The probe extends to a maximum depth of 200 mm to 300 mm into the ground, and, therefore, gives an average mass density for that depth. The detector records the amount of radiation which passes through the soil from the probe attached to the meter when inserted into the ground. The denser the soil, the greater is the absorption of gamma rays, and the lesser will be the gamma rays energy at the detector. The method is known as the *direct transmission method.*

There is another method, known as the *back scatter method.* Both the source and the detector are contained in one probe. The detector records radiations which had been reflected by the soil. The bulk mass density of the soil is determined from the radiation count over a fixed time period. The mass density obtained is for the top 40 to 50 mm. The method is simpler than the direct transmission method, but it requires a greater source strength.

Radiation methods for determination of the mass density of soils are quick and convenient and are gaining popularity. However, precautions must be taken against the radiation hazard.

2.17. DETERMINATION OF VOID RATIO, POROSITY AND DEGREE OF SATURATION

The void ratio of a soil sample is a measure of its denseness. It is one of the important parameters of soils. Engineering properties of soils depend upon void ratio to a large extent. The void ratio is determined in the laboratory indirectly from the dry mass density. From Eq. 2.22,

$$e = \frac{G\rho_w}{\rho_d} - 1 \qquad \text{...(2.51)}$$

The methods for determination of the specific gravity of solids G and the dry density ρ_d have been discussed in the preceding sections.

For a saturated soil, the void ratio is determined using Eq. 2.31, $e = wG$. This method is a very convenient and accurate method, as the water content of a soil can be determined quite easily and accurately. The specific gravity of soil (G) can also be determined in the laboratory.

Once the void ratio has been determined, other volumetric relationships such as porosity and degree of saturation can be determined using Eqs. 2.3 and 2.30, respectively.

Percentage air voids are determined indirectly, using Eq. 2.38,

$$\rho_d = \frac{(1 - n_a)\, G\, \rho_w}{1 + wG}$$

or

$$n_a = 1 - \frac{\rho_d}{G\rho_w}(1 + wG) \qquad \text{...(2.52)}$$

Eq. 2.52 can be reduced to the following form

$$n_a = n - \frac{\rho_d\, w}{\rho_w} \qquad \text{...(2.53)}$$

This is left as an exercise for the readers.

Table 2.5 gives typical values of void ratio, porosity, dry density, and dry unit weight of different soils in loosest and densest conditions.

Table 2.5. Typical Values of Void Ratio and Dry Density and Dry Units Weights

S.No.	*Soil type*	*State of soil*	*Void Ratio*	*Porosity (%)*	*Dry density (kg/m^3)*	*Dry unit weight (kN/m^3)*
1.	Gravel	Loosest Densest	0.60 0.30	38 23	1600 2000	16 20
2.	Coarse sand, Medium sand	Loosest Densest	0.75 0.35	42 26	1500 1900	15 19
3.	Uniform, fine sand	Loosest Densest	0.85 0.4	46 29	1400 1900	14 19
4.	Coarse silt	Loosest Densest	1.0 0.45	50 31	1300 1800	13 18
5.	Fine silt	Softest Hardest	1.00 0.4	50 29	1300 1900	13 19
6.	Lean Clay	Softest Hardest	1.20 0.40	55 29	1300 1900	13 19
7.	Fat clay	Softest Hardest	2.20 0.40	69 29	1000 2000	10 20

ILLUSTRATIVE EXAMPLES

Illustrative Example 2.1. *The mass of a chunk of moist soil is 20 kg, and its volume is 0.011 m^3. After drying in an oven, the mass reduces to 16.5 kg. Determine the water content, the density of moist soil, the dry density, void ratio, porosity and the degree of saturation. Take G = 2.70.*

Solution. Mass of water, $M_w = 20.0 - 16.50 = 3.50$ kg

From Eq. 2.9, water content, $$w = \frac{3.50}{16.50} = 0.2121\ (21.21\%)$$

From Eq. 2.11, the wet mass density $$\rho = \frac{20}{0.011} = \mathbf{1818.18\ kg/m^3}$$

From Eq. 2.12, the dry density, $$\rho_d = \frac{16.5}{0.011} = \mathbf{1500.0\ kg/m^3}$$

From Eq. 2.22, $$1 + e = \frac{G\rho_w}{\rho_d}$$

or $$e = \frac{2.70 \times 1000}{1500} - 1 = \mathbf{0.80}$$

From Eq. 2.3, $$n = \frac{e}{1+e} = \frac{0.80}{1.80} = 0.444\ (\mathbf{44.44\%})$$

From Eq. 2.30, $$S = \frac{wG}{e} = \frac{0.2121 \times 2.70}{0.80} = \mathbf{0.7158\ (71.58\%)}$$

Illustrative Example 2.2. *A soil specimen has a water content of 10% and a wet unit weight of 20 kN/m^3. If the specific gravity of solids is 2.70, determine the dry unit weight, void ratio, and the degree of saturation. Take $\gamma_w = 10\ kN/m^3$.*

Solution. From Eq. 2.36 (*a*), $$\gamma_d = \frac{\gamma}{1+w} = \frac{20}{1+0.1} = 18.18\ kN/m^3$$

From Eq. 2.22 (*a*), $$1 + e = \frac{G\,\gamma_w}{\gamma_d} = \frac{2.70 \times 10}{18.18} = 1.49 \qquad \text{or } e = 0.49$$

From Eq. 2.30, $$S = \frac{wG}{e} = \frac{0.1 \times 2.70}{0.49} = \mathbf{0.551\ (55.1\%)}$$

Illustrative Example 2.3. *A sample of dry soil weighs 68 gm. Find the volume of voids if the total volume of the sample is 40 ml and the specific gravity of solids is 2.65. Also determine the void ratio.*

Solution. From Eq. 2.12, $\rho_d = \dfrac{M_s}{V} = \dfrac{68}{40} = 1.70$ gm/ml

Volume of solids, $V_s = \dfrac{M_s}{G\rho_w} = \dfrac{68}{2.65 \times 1.0} = 25.66$ ml

Volume of voids, $V_v = V - V_s = 40.00 - 25.66 = \mathbf{14.34}$ **ml**

From Eq. 2.1, $e = \dfrac{V_v}{V_s} = \dfrac{14.34}{25.66} = \mathbf{0.56}$

Illustrative Example 2.4. *A moist soil sample weighs 3.52 N. After drying in an oven, its weight is reduced to 2.9 N. The specific gravity of solids and the mass specific gravity are, respectively, 2.65 and 1.85. Determine the water content, void ratio, porosity and the degree of saturation. Take* $\gamma_w = 10$ *kN/m*^{3}.

Solution. Weight of water = 3.52—2.90 = 0.62 N

From Eq. 2.9, water content, $w = \dfrac{0.62}{2.90} = 0.2138$ **(21.38%)**

From Eq. 2.19, $\gamma = G_m \gamma_w = 1.85 \times 10 = 18.5$ kN/m^3

From Eq. 2.36 *a*, $\gamma_d = \dfrac{\gamma}{1 + w} = \dfrac{18.5}{1 + 0.2138} = 15.24$ kN/m^3

From Eq. 2.22 *a*, $1 + e = \dfrac{G\gamma_w}{\gamma_d} = \dfrac{2.65 \times 10}{15.24} = 1.74$

or $e = \mathbf{0.74}$

From Eq. 2.3, $n = \dfrac{e}{1 + e} = \dfrac{0.74}{1 + 0.74} = 0.4253$ (42.53%)

From Eq. 2.30, $S = \dfrac{wG}{e} = \dfrac{0.2138 \times 2.65}{0.74} = 0.7656$ (76.56%)

Illustrative Example 2.5. *A soil has a porosity of 40%, the specific gravity of solids of 2.65 and a water content of 12%. Determine the mass of water required to be added to 100 m*^{3} *of this soil for full saturation.*

Solution. Let us take unit volume of solids, *i.e.* $V_s = 1.0$ m^3.

Mass of solids, $M_s = G\rho_w = 2.65 \times 1000 = 2650$ kg

From Eq. 2.9, mass of water, $M_w = 0.12 \times 2650 = 318$ kg

Volume of water $= \dfrac{318}{1000} = 0.318$ m^3

From Eq. 2.4, $e = \dfrac{n}{1 - n} = \dfrac{0.40}{1.0 - 0.40} = 0.667$

From Eq. 2.1, volume of voids, $V_v = e\,V_s = 0.667 \times 1.0 = 0.667$ m^3

Therefore, volume of air, $= 0.667 - 0.318 = 0.349$ m^3

Volume of additional water for full saturation = 0.349 m^3

Total volume of soil, $V = V_s + V_v = 1.0 + 0.667 = 1.667$ m^3

Volume of water required for 100 m^3 of soil $= \dfrac{0.349}{1.667} \times 100 = 20.94$ m^3

Mass of water required = 20940 kg.

Illustrative Example 2.6. *A sample of saturated soil has a water content of 25 percent and a bulk unit weight of 20 kN/m*^{3}. *Determine dry density, void ratio and specific gravity of solid particles.*

What would be the bulk unit weight of the same soil at the same void ratio but at a degree of saturation of 80% ? Take $\gamma_w = 10\ kN/m^3$.

Solution. From Eq. 2.33 (*a*),

$$\gamma_{sat} = \frac{G\gamma_w}{1 + wG}(1 + w)$$

or

$$20 = \frac{G \times 10\ (1 + 0.25)}{1 + 0.25 \times G} \quad \text{or } G = \mathbf{2.67}$$

From Eq. 2.30, taking $S = 1.0$,

$$e = wG = 0.25 \times 2.67 = \mathbf{0.67}$$

From Eq. 2.22 (*a*),

$$\gamma_d = \frac{G\gamma_w}{1 + e} = \frac{2.67 \times 10}{1 + 0.67} = \mathbf{15.99\ kN/m^3}$$

In the second case, as $e = 0.67$ and $S = 0.80$, Eq. 2.21 (*a*) gives

$$\gamma = \frac{(G + Se)\gamma_w}{1 + e} = \frac{(2.67 + 0.80 \times 0.67) \times 10}{1 + 0.67} = \mathbf{19.20\ kN/m^3}$$

Illustrative Example 2.7. *A sample of clay was coated with paraffin wax and its mass, including the mass of wax, was found to be 697.5 gm. The sample was immersed in water and the volume of the water displaced was found to be 355 ml. The mass of the sample without wax was 690.0 gm, and the water content of the representative specimen was 18%.*

Determine the bulk density, dry density, void ratio and the degree of saturation. The specific gravity of the solids was 2.70 and that of the wax was 0.89.

Solution.

$$\text{Mass of wax} = 697.5 - 690.0 = 7.5 \text{ gm}$$

$$\text{Volume of wax} = \frac{7.50}{0.89 \times 1.0} = 8.43 \text{ ml}$$

$$\text{Volume of soil} = 355.0 - 8.43 = 346.57 \text{ ml}$$

$$\text{Bulk density} = \frac{690}{346.57} = 1.99 \text{ gm/ml}$$

From Eq. 2.36, dry density

$$= \frac{1.99}{1 + 0.18} = \mathbf{1.69\ gm/ml}$$

From Eq. 2.22,

$$1 + e = \frac{2.70 \times 1.0}{1.69} = 1.60 \quad \text{or } e = \mathbf{0.60}$$

From Eq. 2.30,

$$S = \frac{wG}{e} = \frac{0.18 \times 2.70}{0.60} = \mathbf{0.81\ (81\%)}$$

Illustrative Example 2.8. *(a) During a test for water content determination on a soil sample by pycnometer, the following observations were recorded.*

(1) Mass of wet soil sample = 1000 gm
(2) Mass of pycnometer with soil and filled with water = 2000 gm
(3) Mass of pycnometer filled with water only = 1480 gm
(4) Specific gravity of solids = 2.67

Determine the water content.

(b) If the bulk density of the soil is 2.05 gm/ml, determine the degree of saturation.

Solution. (*a*) From Eq. 2.41,

$$w = \left[\frac{(M_2 - M_1)}{(M_3 - M_4)} \cdot \left(\frac{G-1}{G}\right) - 1\right] \times 100$$

$$= \left[\frac{1000}{(2000 - 1480)} \times \left(\frac{2.67 - 1.0}{2.67}\right) - 1\right] \times 100 = \mathbf{20.28\%}$$

(*b*) From Eq. 2.36, $\rho_d = \dfrac{\rho}{1 + w} = \dfrac{2.05}{1 + 0.2028} = 1.70$ gm/ml

From Eq. 2.22, $1 + e = \dfrac{2.67 \times 1.0}{1.70} = 1.57$ or $e = 0.57$

Now $S = \dfrac{wG}{e} = \dfrac{0.2028 \times 2.67}{0.57} = 0.950 =$ **95.0%**

Illustrative Example 2.9. *The mass of an empty gas jar was 0.498 kg. When completely filled with water, its mass was 1.528 kg. An oven-dried sample of soil of mass 0.198 kg was placed in the jar and water was added to fill the jar and its mass was found to be 1.653 kg. Determine the specific gravity of particles.*

Solution. From Eq. 2.42, $G = \dfrac{M_2 - M_1}{(M_2 - M_1) - (M_3 - M_4)}$

or $G = \dfrac{0.198}{0.198 - (1.653 - 1.528)} = 2.71$

Illustrative Example 2.10. *In a compaction test on a soil, the mass of wet soil when compacted in the mould was 1.855 kg. The water content of the soil was 16%. If the volume of the mould was 0.945 litres, determine the dry density, void ratio, degree of saturation and percentage air voids. Take G = 2.68.*

Solution. Bulk density $\rho = \dfrac{1.855}{0.945 \times 10^{-3}} = 1962.96 \text{ kg/m}^3$

From Eq. 2.36, $\rho_d = \dfrac{\rho}{1 + w} = \dfrac{1962.96}{1 + 0.16} = $ **1692.21 kg/m³**

From Eq. 2.22, $1 + e = \dfrac{2.68 \times 1000}{1692.21} = 1.584$ or $e = 0.584$

From Eq. 2.30, $S = \dfrac{wG}{e} = \dfrac{0.16 \times 2.68}{0.584} = 0.7342 =$ **73.42%**

From Eq. 2.38, $\rho_d = \dfrac{(1 - n_a)\, G \cdot \rho_w}{1 + wG}$

or $1 - n_a = \dfrac{1692.21 \times (1 + 0.16 \times 2.68)}{2.68 \times 1000} = 0.9022$

or $n_a = 0.0978$ (**9.78%**)

Illustrative Example 2.11. *A compacted cylindrical specimen, 50 mm dia and 100 mm length, is to be prepared from oven-dry soil. If the specimen is required to have a water content of 15% and the percentage air voids of 20%, calculate the mass of the soil and water required for the preparation of the sample. Take G = 2.69.*

Solution. Let M_s be the mass of solids in kg.

Mass of water, $= wM_s = 0.15\, M_s$

Volume of solids, $V_s = \dfrac{M_s}{G \rho_w} = \dfrac{M_s}{2.69 \times 1000} = \dfrac{M_s}{2690} \text{ m}^3$

Volume of water, $V_w = \dfrac{0.15\, M_s}{1000} = 0.15 \times 10^{-3}\, M_s \cdot \text{m}^3$

Total volume of sample, $V = \pi/4 \times (0.05)^2 \times 0.10 = 196.35 \times 10^{-6} \text{ m}^3$

From Eq. 2.6, volume of air, $V_a = 0.20 \times V = 0.2 \times 196.35 \times 10^{-6} = 39.27 \times 10^{-6} \text{ m}^3$

Now $V_s + V_w + V_a = V$

or $$\frac{M_s}{2690} + 0.15 \times 10^{-3} M_s + 39.27 \times 10^{-6} = 196.35 \times 10^{-6}$$

or $$M_s = 0.301 \text{ kg}$$

$$\text{Mass of water} = 0.15 \times 0.301 = \mathbf{0.045\ kg}$$

Illustrative Example 2.12. *A borrow area soil has a natural water content of 10% and a bulk density of 1.80 Mg/m³. The soil is used for an embankment to be compacted at 18% moisture content to a dry density of 1.85 Mg/m³. Determine the amount of water to be added to 1.0 m³ of borrow soil. How many cubic metres of excavation is required for 1 m³ of compacted embankment ?*

Solution. Borrow area soil. $\rho_d = \frac{1.80}{1.10} = 1.636$ g/ml

$$\text{Dry unit weight} = 1.636 \times 9.81 = 16.05 \text{ kN m}^3$$

Let us consider 1 m³ of borrow soil.

$$W_s = \text{Dry weight/m}^3 = 16.05 \text{ kN}$$

$$W_w = \text{Weight of water/m}^3 = 1.605 \text{ kN}$$

In embankment, $W_{w1} = 0.18 \times 16.05 = 2.889$ kN

$$\text{Water to be added} = 2.889 - 1.605 = 1.284 \text{ kN}$$

$$\text{Weight of dry soil in embankment/m}^3 = 1.85 \times 9.81 = 18.15 \text{ kN}$$

$$\text{Volume of soil required/m}^3 \text{ of embankment} = \frac{18.150}{16.05} = \mathbf{1.131\ m^3}$$

Illustrative Example 2.13. *There are two borrow areas A and B which have soils with void ratios of 0.80 and 0.70, respectively. The inplace water content is 20%, and 15%, respectively. The fill at the end of construction will have a total volume of 10,000 m³, bulk density of 2 Mg/m³ and a placement water content of 22%. Determine the volume of the soil required to be excavated from both areas. G = 2.65.*

If the cost of excavation of soil and transportation is Rs. 200/- per 100 m³ for area A and Rs. 220/- per 100 m³ for area A, which of the borrow area is more economical ?

Solution. *Borrow area A.* $\rho_d = \frac{2.65 \times 1.0}{1 + 0.80} = 1.47$ g/ml (14.44 kN/m³)

Let us consider 1 m³ of borrow soil. W_s = Dry weight/m³ = 14.44 kN

In embankment. $\rho_d = \frac{2}{1 + 0.22} = 1.639$ g/ml (= 16.08 kN/m³)

$$\text{Weight of dry soil per m}^3 = 16.08 \text{ kN}$$

Volume of soil required $= \frac{16.08}{14.44} = 1.114 \text{ m}^3$

$$\text{Cost of soil} = Rs\ 200/100 \times 1.114 = \text{Rs } 2.23 \text{ per m}^3$$

Borrow area B. $\rho_d = \frac{2.65 \times 1.0}{1.70} = 1.559$ g/ml (15.29 kN/m³)

$$W_s = \text{dry weight/m}^3 = 15.29 \text{ kN}$$

In embankment, weight of dry soil = 16.08 kN

Volume of soil required $= \frac{16.08}{15.29} = 1.052 \text{ m}^3$

$$\text{Cost of soil} = \text{Rs } 220/100 \times 1.052 = \text{Rs } 2.31 \text{ per m}^3.$$

Borrow area *A* is more economical.

PROBLEMS

A. Numerical

2.1. (*a*) Define the terms void ratio, specific gravity of particles, degree of saturation and dry density.
(*b*) Develop a relationship between the void ratio, water content, specific gravity of particles and the degree of saturation.

2.2. (*a*) Describe oven-drying method for the determination of water content of a soil sample in a laboratory.
(*b*) A sample of wet soil has a volume of 0.0192 m^3 and a mass of 32 kg. When the sample is dried out in an oven, its mass reduces to 28.5 kg. Determine (*i*) Bulk density, (*ii*) Water content, (*iii*) Dry density, (*iv*) Saturated density, (*v*) Void ratio, (*vi*) Porosity, (*vii*) Degree of saturation. Take the specific gravity of solids as 2.65. **[Ans.** 1666.67 kg/m^2; 12.28% : 1484.4 kg/m^3; 1924.50 kg/m^3; 0.785; 44% and 41.4%]

2.3. (*a*) A sample of saturated soil has a water content of 25 percent and a bulk unit weight of 20 kN/m^3. Determine the dry unit weight, void ratio and the specific gravity of solids.
(*b*) What would be the bulk unit weight of the soil in (*a*) if it is compacted to the same void ratio but has a degree of saturation of 90% ? **[Ans.** 16 kN/m^3; 0.667, 2.667 19.60 kN/m^3]

2.4. A sample of soil has a volume of 65 ml and weighs 0.96 N. After complete drying, its weight reduces to 0.785 N. If the specific gravity of solid particles is 2.65, determine the degree of saturation. **[Ans.** 51%]

2.5. A saturated soil sample has a water content of 40%. If the specific gravity of solids is 2.67, determine the void ratio, saturated density, and submerged density. **[Ans.** 1.07 ; 1807 kg/m^3 ; 807 kg/m^3]

2.6. (*a*) Define the terms void ratio, dry density, submerged density and mass specific gravity.
(*b*) Derive an expression for bulk density in terms of its water content, void ratio, specific gravity of solids and density of water.

2.7. A partially saturated sample of a soil has a density of 1950 kg/m^3 and a water content of 21%. If the specific gravity of solids is 2.65, calculate the degree of saturation and void ratio.
If the sample subsequently gets saturated, determine its saturated density. **[Ans.** 86%; 0.645 ; 2003 kg/m^3]

2.8. A sample of soil has a volume of 1 litre and a weight of 17.5 N. The specific gravity of the solids is 2.68. If the dry unit weight of the soil is 14.8 kN/m^3, determine (*a*) water content, (*b*) void ratio, (*c*) porosity, (*d*) saturated unit weight, (*e*) submerged density and (*f*) degree of saturation.
[Ans. 18.2% : 0.811 : 44.8% ; 19.28 kN/m^3, 9.28 kN/m^3 and 60.2%]

2.9. A fully saturated clay sample has a mass of 130 gm and has a volume of 64 cm^3. The sample mass is 105 gm after oven drying. Assuming that the volume does not change during drying, determine the following : (*i*) specific gravity of soil solids. (*ii*) void ratio, (*iii*) porosity, (*iv*) dry density.
[Ans. 2.69 ; 0.64 ; 39% and 1.641 gm/cm^3]

2.10. Prove that the water content (*w*) of a partially saturated soil can be expressed as

$$w = \frac{1 - (G_m/G)}{(G_m/S) - 1}$$

where G_m = mass specific gravity, G = specific gravity of solids and S = degree of saturation.

2.11 (*a*) Prove that the degree of saturation of a partially saturated soil can be expressed as

$$S = \frac{w}{\frac{\rho_w}{\rho}(1 + w) - \frac{1}{G}}$$

where ρ = bulk density, G = specific gravity of solids and w water content.
(*b*) A cylindrical specimen of soil is 7.50 cm long and 3.75 cm in diameter and has a mass of 175 gm. If the water content is 18 percent and the specific gravity of solids is 2.68, determine the degree of saturation.
What would be the error in the degree of saturation if there has been an error of 1 mm in measuring the length ?
[Ans. 96.7%, 4.62%]

2.12. A pycnometer having a mass of 600 gm was used in the following measurements of three samples of soil. Sample No. 1 was ovendried; sample no. 2 was partially saturated and sample no. 3 was fully saturated. The bulk density of the sample no. 2 was 2.05 gm/ml.

Sample	No. 1	No.2	No. 3
Mass of samples (gm)	960	970	1000
Mass of sample + water + pycnometer (gm)	2080	2050	2010

If the mass of pycnometer when filled with water only was 1475 gm, determine the specific gravity of solids.

(*b*) Also determine the water content and void ratio of samples no. 2 and 3, and the degree of saturation of sample no. 2. **[Ans.** 2.70; 6.3%, 0.40; 11.70; 0.32 and 41.85%]

2.13. An undisturbed specimen of clay was tested in a laboratory and the following results were obtained.

Wet mass = 210 gm
Oven dry mass = 175 gm
Specific gravity of solids = 2.70

What was the total volume of the original undisturbed specimen assuming that the specimen was 50% saturated ?

[**Ans.** 134.8 ml]

2.14. A soil deposit to be used for construction of an earth embankment has an average dry density of 1.62 gm/ml. If the compacted embankment is to have an average dry density of 1.72 gm/ml, determine the volume of soil to be excavated for 1000 m^3 of embankment. The water content of the soil in the borrow pit is 10%.

[**Ans.** 1.061×10^3 m^3]

2.15. Determine the specific gravity of solids from the following observations:

(*i*) Mass of dry sample = 0.395 kg
(*ii*) Mass of pycnometer full of water = 1.755 kg
(*iii*) Mass of pycnometer containing soil and full of water = 2.005 kg. [**Ans.** 2.72]

2.16. A sample of clay having a mass of 675 gm was coated with paraffin wax. The combined mass of the clay and the wax was found to be 682 gm. The volume was found by immersion in water as 345 ml. The sample was then broken open and the water content and the specific gravity of solids were found to be 15% and 2.70, respectively. Calculate the bulk density of soil, its void ratio, and degree of saturation. Take specific gravity of wax as 0.89. [**Ans.** 2.002 gm/ml, 0.551 and 73.5%]

2.17. In order to determine the bulk density of a soil insitu, 4.7 kg of soil was extracted from a hole at the surface of the soil. The hole required 3.65 kg of loose dry sand for its filling. If it takes 6.75 kg of the same sand to fill a calibrating can of 4.5 litre capacity, determine the bulk density of the soil. [**Ans.** 1932 kg/m^3]

2.18. A litre capacity cutter of mass 1 kg was pushed into an embankment under construction and the mass of the cutter with soil was found to be 2.865 kg. If the sample had water content of 11%, determine the void ratio of the soil in embankment. $G = 2.67$. [**Ans.** 0.59]

B. Descriptive and Objective Types

2.19 What is a block diagram ? What is its use ?

2.20. Differentiate between :

(*a*) Percentage air voids and air content,
(*b*) Void ratio and porosity.
(*c*) Specific gravity of solids and mass specific gravity.
(*d*) Water content based on solid material and that based on total mass.
(*e*) Saturated density and bulk density.

2.21. How do you determine the void ratio of a soil ?

2.22. Discuss various methods for determination of water content in a laboratory.

2.23. Describe a method for determination of the specific-gravity of solids of fine-grained soils.

2.24. How would you determine the bulk density of a soil specimen in a laboratory ?

2.25. Discuss various methods for the determination of bulk density of a soil in field.

2.26. State whether the following statements are true or false

(*a*) The water content of a soil can be more than 100%.
(*b*) The porosity of a soil can be more than 100%
(*c*) The specific gravity of particles of coarse-grained is seldom greater than 2.70.
(*d*) The submerged density is about one-half of the saturated density.
(*e*) For determination of water content of all types of soils, the oven temperature is 100° ± 5°C.

[**Ans.** True (*a*), (*c*), (*d*)]

2.27. (*a*) Which of the following relation is not correct ?

(i) $e = \dfrac{n}{1-n}$ (ii) $n = \dfrac{e}{1-e}$

(iii) $\rho_d = \dfrac{G\rho_w}{1+e}$ (iv) $\rho' = \dfrac{(G-1)\rho_w}{1+e}$ **[Ans. (ii)]**

(b) Which of the following statements is wrong ?

(i) The void ratio of a saturated soil can be determined from its water content.

(ii) The dry density is the bulk density of soil in dried condition.

(iii) 100% saturation line and zero percent air void lines are identical. **[Ans. (ii)]**

C. Multiple-Choice Questions

1. The water content of a highly organic soil is determined in an oven at a temperature of:

(a) 105°C (b) 80°C
(c) 60°C (d) 27°C

2. Pycnometer method for water content determination is more suitable for:

(a) Clay (b) Loess
(c) Sand (d) Silt

3. The gas formed by the reaction of calcium carbide with water is:

(a) Carbon dioxide (b) Sulphur dioxide
(c) Ethane (d) Acetylene

4. The ratio of the volume of voids to the total volume of soil is:

(a) Voids ratio (b) Degree of saturation
(c) Air content (d) Porosity

5. Dry density of soil is equal to the:

(a) Mass of solids to the volume of solids.
(b) Mass of solids to the total volume of soil.
(c) Density of soil in the dried condition.
(d) None of the above.

6. The most accurate method for the determination of water content in the laboratory is:

(a) Sand bath method. (b) Oven-drying method.
(c) Pycnometer method. (d) Calcium carbide method.

7. A soil has a bulk density of 1.80 g/cm^3 at a water content of 5%. If the void ratio remains constant then the bulk density for a water content of 10% will be

(a) 2.00 g/cm^3 (b) 1.88 g/cm^3
(c) 1.82 g/cm^3 (d) 1.95 g/cm^3

8. In a wet soil mass, air occupies one-sixth of its volume and water occupies one-third of its volume. The void ratio of the soil is

(a) 0.25 (b) 0.50
(c) 1.50 (d) 1.00

9. A soil sample has a specific gravity of 2.60 and a void ratio of 0.78. The water content required to fully saturate the soil at that void ratio will be

(a) 20% (b) 30%
(c) 40% (d) 60%

[Ans. 1. (c), 2. (c), 3. (d), 4. (d), 5. (b), 6. (b), 7. (b), 8. (d), 9. (b)**]**

3

Particle Size Analysis

3.1. INTRODUCTION

(*a*) **Engineering Properties**—The main engineering properties of soils are permeability, compressibility, and shear strength. Permeability indicates the facility with which water can flow through soils. It is required for estimation of seepage discharge through earth masses. Compressibility is related with the deformations produced in soils when they are subjected to compressive loads. Compression characteristics are required for computation of the settlements of structures founded on soils. Shear strength of a soil is its ability to resist shear stresses. The shear strength determines the stability of slopes, bearing capacity of soils and the earth pressure on retaining structures. Engineering properties of soils are discussed in latter chapters.

(*b*) **Index Properties**—The tests required for determination of engineering properties are generally elaborate and time-consuming. Sometimes, the geotechnical engineer is interested to have some rough assessment of the engineering properties without conducting elaborate tests. This is possible if index properties are determined. The properties of soils which are not of primary interest to the geotechnical engineer but which are indicative of the engineering properties are called *index properties.* Simple tests which are required to determine the index properties are known as *classification* tests. The soils are classified and identified based on the index properties, as discussed in chapter 5. The main index properties of coarse-grained soils are particle size and the relative density, which are described in this chapter. For fine-grained soils, the main index properties are Atterberg's limits and the consistency (chapter 4).

The index properties are sometimes divided into two categories. (1) Properties of individual particles, and (2) Properties of the soil mass, also known as aggregate properties. The properties of individual particles can be determined from a remoulded, disturbed sample. These depend upon the individual grains and are independent of the manner of soil formation. The soil aggregate properties depend upon the mode of soil formation, soil history and soil structure. These properties should be determined from undisturbed samples or preferably from in-situ tests. The most important properties of the individual particles of coarse- grained soils are the particle size distribution and grain shape. The aggregate property of the coarse-grained soils of great practical importance is its relative density.

The index properties give some information about the engineering properties. It is tacitly assumed that soils with like index properties have identical engineering properties. However, the correlation between index properties and engineering properties is not perfect. A liberal factor of safety should be provided if the design is based only on index properties. Design of large, important structures should be done only after determination of engineering properties.

3.2. MECHANICAL ANALYSIS

The mechanical analysis, also known as *particle size analysis,* is a method of separation of soils into different fractions based on the particle size. It expresses quantitatively the proportions, by mass, of various sizes of particles present in a soil. It is shown graphically on a particle size distribution curve.

The mechanical analysis is done in two stages : (1) Sieve Analysis, (2) Sedimentation Analysis. The first analysis is meant for coarse-grained soils (particle size greater than 75 micron) which can easily pass through a set of sieves. The second analysis is used for fine-grained soils (size smaller than 75 microns). Sedimentation analysis is also known as *wet analysis.* As a soil mass may contain the particles of both types of soils, a combined analysis comprising both sieve analysis and sedimentation analysis may be required for such soils.

Particle size smaller than 0.2 micron cannot be determined by the sedimentation method. These can be determined by an electron microscope or by X-ray diffraction techniques. However, such analysis is of little practical importance in soil engineering.

3.3. SIEVE ANALYSIS

The soil is sieved through a set of sieves. Sieves are generally made of spun brass and phosphor bronz (or stainless steel) sieve cloth. According to IS : 1498—1970. the sieves are designated by the size of square opening, in mm or microns (1 micron = 10^{-6} m = 10^{-3} mm). Sieves of various sizes ranging from 80 mm to 75 microns are available. The diameter of the sieve is generally between 15 to 20 cm.

As mentioned before, the sieve analysis is done for coarse-grained soils. The coarse-grained soils can be further sub-divided into gravel fraction (size > 4.75 mm) and sand fraction (75μ < size < 4.75 mm), where Greek letter μ is used to represent micron. A set of coarse sieves, consisting of the sieves of size 80 mm, 40 mm, 20 mm, 10 mm and 4.75 mm, is required for the gravel fraction. The second set of sieves, consisting of the sieves of size 2 mm, 1 mm, 600 μ, 425 μ, 212 μ, 150 μ and 75 μ, is used for sieving minus 4.75 mm fraction. However, all the sieves may not be required for a particular soil. The selection of the required number of sieves is done to obtain a good particle size distribution curve. The sieves are stacked one over the other, with decreasing size from the top to the bottom. Thus the sieve of the largest opening is kept at the top. A lid or cover is placed at the top of the largest sieve. A receiver, known as *pan,* which has no opening, is placed at the bottom of the smallest sieve.

***(a)* Dry Sieve Analysis**—The soil sample is taken in suitable quantity, as given in Table 3.1, The larger the particle size, the greater is the quantity of soil required.

The soil should be oven-dry. It should not contain any lump. If necessary, it should be pulverized. If the soil contains organic matter, it can be taken air-dry instead of oven dry.

The sample is sieved through a 4.75 mm IS sieve. The portion retained on the sieve is the gravel fraction or plus 4.75 mm material. The gravel fraction is sieved through the set of coarse sieves manually or using a mechanical shaker. Hand sieving is normally done. The weight of soil retained on each sieve is obtained.

The minus 4.75 mm fraction is sieved through the set of fine sieves. The sample is placed in the top sieve and the set of sieves is kept on a mechanical shaker (Fig. 3.1) and the machine is started. Normally, 10 minutes of shaking is sufficient for most soils. The mass of soil retained on each sieve and on pan is obtained to the nearest 0.1 gm. The mass of the retained soil is checked against the original mass.

Dry sieve analysis is suitable for cohesionless soils, with little or no fines. If the sand is sieved in wet conditions, the surface tension may cause a slight increase in the size of the particles and the particles smaller than the aperture size may be retained on the sieve and the results would be erroneous.

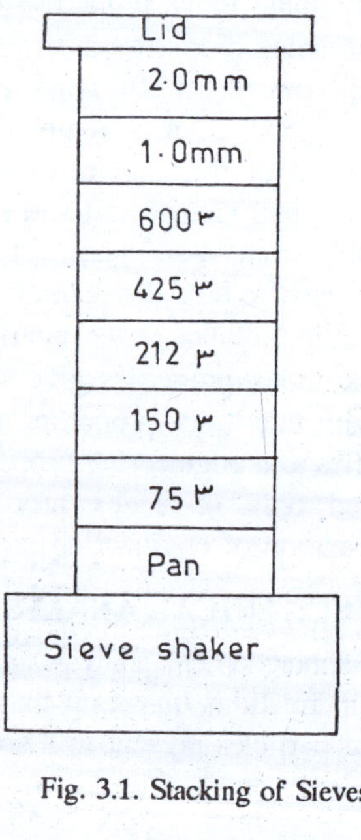

Fig. 3.1. Stacking of Sieves.

Table 3.1. Quantity of Soil for Sieve Analysis

S.No.	*Maximum Size*	*Quantity (kg)*
1.	80 mm	60
2.	20 mm	6.5
3.	4.75 mm	0.5

***(b)* Wet Sieve Analysis**—If the soil contains a substantial quantity (say, more than 5%) of fine particles, a wet sieve analysis is required. All lumps are broken into individual particles. A representative soil sample in the required quantity is taken, using a riffler, and dried in an oven. The dried sample is taken in a tray and soaked with water. If deflocculation is required, sodium hexameta-phosphate, at the rate of 2 g per litre of water, is added. The sample is stirred and left for a soaking period of at least one hour. The slurry is then sieved through a 4.75 mm IS sieve, and washed with a jet of water. The material retained on the sieve is the gravel fraction. It is dried in an oven, and sieved through set of coarse sieves.

The material passing through 4.75 mm sieve is sieved through a 75 μ sieve. The material is washed until the wash water becomes clear. The material retained on the 75 μ sieve is collected and dried in an oven. It is then sieved through the set of fine sieves of the size 2 mm, 1 mm, 600 μ, 425 μ, 212 μ, 150 μ, and 75 μ. The material retained on each sieve is collected and weighed. The material that would have been retained on pan is equal to the total mass of soil minus the sum of the masses of material retained on all sieves.

Computation of Percentage Finer

For determination of the particle size distribution curve, percentage of particles finer than a particular size is required. This can be found as under:

Let us consider the case when the sieving has been done through seven sieves, no 1 (coarsest) to no. 7 (finest). Let the mass of soil retained on these sieves be respectively, $M_1, M_2 \ldots, M_7$, and the mass of soil retained on the pan (receiver) be M_8. The sum of all these masses is, obviously, equal to the total mass of sample M.

Expressed as percentage, the materials retained on the sieves and pan are

$$p_1 = \frac{M_1}{M} \times 100 \qquad p_2 = \frac{M_2}{M} \times 100, \text{ etc.}$$

and

$$p_7 = \frac{M_7}{M} = 100 \quad \text{and} \quad p_8 = \frac{M_8}{M} \times 100$$

The cumulative percentage (C) of material retained on any sieve is equal to the sum of the percentage of soil retained on the sieve and that retained on all sieves coarser than that sieve. Therefore,

$$C_1 = p_1$$

$$C_2 = p_1 + p_2$$

$$C_7 = p_1 + p_2 + \ldots + p_7$$

The percentage finer (N) than any sieve size is obtained by subtracting the cumulative percentage retained on the sieve from 100%.

Thus, $$N_1 = 100 - C_1; \qquad N_2 = 100 - C_2, \text{ etc.}$$

and $$N_7 = 100 - C_7$$

It may be noted that the dimension of the soil particle that controls whether a particle shall pass through a sieve opening is the intermediate dimension (width) of the particle.For example, a particle with dimensions 3 mm × 2 mm × 1 mm shall pass through a sieve of size 2 mm if it is assumed that the particle is aligned such that the largest dimension is normal to the plane of sieve opening and is at right angles to the side of the square.

(See Chapter 30, Sect. 30.8 for the laboratory experiment)

3.4. STOKES' LAW

Soil particles finer than 75 μ size cannot be sieved. The particle size distribution of such soils is determined by sedimentation analysis. The analysis is based on Stokes' law, which gives the terminal velocity of a small sphere settling in a fluid of infinite extent. When a small sphere settles in a fluid, its velocity first increases under the action of gravity, but the drag force comes into action, and retards the velocity. After an initial adjustment period, steady conditions are attained and the velocity becomes constant. The velocity attained is known as *terminal velocity.* The expression for terminal velocity can be obtained from the equilibrium of the particle.

The drag force, F_D, experienced by a sphere of radius r when it falls through a fluid of viscosity η is given by

$$F_D = 6\pi\eta rv \quad \text{...(a)}$$

where v is the velocity.

The other two forces acting on the sphere are the weight (W) of the sphere and the buoyant force (U).

$$W = 4/3 \quad \pi r^3 \ \gamma_s = 4/3 \quad \pi r^3 \ (\rho_s g) \quad \text{...(b)}$$

where γ_s is the unit weight of the material of sphere

and
$$U = 4/3 \quad \pi r^3 \ \gamma_w = 4/3 \quad \pi r^3 \ (\rho_w g) \quad \text{...(c)}$$

From equilibrium of forces in vertical direction,

$$W = U + F_D$$

or
$$4/3\,\pi r^3 \gamma_s = 4/3\,\pi r^3 \gamma_w + 6\,\pi\eta rv$$

or
$$4/3\,\pi r^3 g\,\rho_s = 4/3\,\pi r^3 g\,\rho_w + 6\pi\eta rv$$

or
$$v = \frac{2}{9}\cdot\frac{r^2}{\eta}(\rho_s - \rho_w)\,g$$

or
$$v = \frac{1}{18}\cdot\frac{g D^2 (G-1)\rho_w}{\eta} \quad \text{...(3.1)}$$

where D is the diameter of the sphere, G is the specific gravity of the material of sphere, and g is the acceleration due to gravity.

If a spherical particle falls through a height H_e centimeters in t minutes,

$$v = \frac{H_e}{60\,t} \ \text{cm/sec} \quad \text{...(3.2)}$$

From Eqs. (3.1) and (3.2),

$$\frac{H_e}{60t} = \frac{1}{18}\frac{gD^2(G-1)\rho_w}{\eta} \quad \text{...(3.3)}$$

or
$$D = \sqrt{\frac{0.3\,\eta \times H_e}{g\,(G-1)\,\rho_w \times t}} \quad \text{...[3.4(a)]}$$

or
$$D = M\sqrt{\frac{H_e}{t}} \quad \text{...[3.4(b)]}$$

where M is a factor, equal to $\left[\dfrac{0.3\,\eta}{g\,(G-1)\,\rho_w}\right]^{1/2}$

in which η is the viscosity in poise (dyne- sec/cm^2), g = 981 cm/sec^2, and ρ_w is in gm/ml. D is in centimeters.

Table 3.2 gives the values of the coefficient of viscosity η for water at different temperatures.

The values of the factor M can be computed and tabulated for different temperatures. For example, for G = 2.67 and T = 20°C, and taking ρ_w = 1.0 gm/ml, and $\eta = 10.09 \times 10^{-3}$ poise, g = 981 cm/sec^2,

$$M = \left[\frac{0.3 \times 10.09 \times 10^{-3}}{981 \times 1.67 \times 1.0}\right]^{1/2} = 1.36 \times 10^{-3}$$

An approximate expression for diameter D of the particle can be obtained from Eq. 3.1.

Table 3.2 Coefficient of viscosity of water η
(Values in millipoise)

$T°C$	η	$T°C$	η	$T°C$	η	$T°C$	η
0	17.94	10	13.10	20	10.09	30	8.00
1	17.32	11	12.74	21	9.84	31	7.83
2	16.74	12	12.39	22	9.61	32	7.67
3	16.19	13	12.06	23	9.38	33	7.51
4	15.68	14	11.75	24	9.16	34	7.36
5	15.19	15	11.45	25	8.95	35	7.21
6	14.73	16	11.16	26	8.75	36	7.06
7	14.29	17	10.88	27	8.55	37	6.92
8	13.87	18	10.60	28	8.36	38	6.79
9	13.48	19	10.34	29	8.18	39	6.66

[**Note.** 1 millipoise = 0.1 mN-s/m^2]

$$v = \frac{1}{18}\,\frac{981 \times D^2\,(2.67 - 1.0)}{10.09 \times 10^{-3}} \times 1.00$$

or

$$v = 9020\,D^2 \qquad \ldots[3.5(a)]$$

where v is the velocity in cm/sec and D is the diameter in cm.

If v is expressed in mm/sec and D in mm,

$$v = 902\,D^2 \qquad \ldots[3.5(b)]$$

If v is expressed in cm/sec and D in mm,

$$v = 90.2\,D^2 \qquad \ldots[3.5(c)]$$

Table 3.3 gives the time required for the settlement of particles of different sizes through a height of 100 mm.

Table 3.3. Time of Settlement for 100 mm Height

S.No.	Diameter (mm)	Time
1.	0.075	19.72 sec
2.	0.02	4.62 min
3.	0.006	51.36 min
4.	0.002	7.70 hr
5.	0.001	30.81 hr

3.5. PREPARATION OF SUSPENSION FOR SEDIMENTATION ANALYSIS

About 50 g of oven-dried soil is weighed accurately and transferred to an evaporating dish. To have proper dispersion of soil, about 100 ml of a dispersion solution is added to the evaporating dish to cover the soil. IS : 2720-Part IV recommends the use of dispersion solution obtained after adding 33g of sodium hexameta-phosphate and 7g of sodium carbonate to distilled water to make one litre of solution. After the dispersing solution has been added to soil, the mixture is warmed gently for about 10 minutes. The contents of the evaporating dish are then transferred to the cup of a mechanical stirrer. Distilled water is added to make the cup about three-fourth full. The suspension is stirred for about 15 minutes. However, the stirring period is more for clayey soils.

The suspension is then washed through a 75 μ sieve, using jets of distilled water. The portion of the suspension which has passed through the sieve is used for sedimentation analysis. The specimen is washed into a 1000 ml jar and enough water is added to make 1000 ml of suspension.

If the soil contains organic matter and calcium compounds, it should be pretreated before adding the dispersing agent. This is done is two stages.

(1) The soil is taken in a beaker and first treated with a 20 volume hydrogen peroxide solution to remove the organic matter, at the rate of about 100 ml of hydrogen peroxide for 100 gm of soil. The mixture is

warmed to a temperature not exceeding 60°C. Hydrogen peroxide causes oxidation of organic matter and gas is liberated. When no more gas comes out, the mixture is boiled to decompose the remaining hydrogen peroxide. The mixture is then cooled.

(2) Calcium compounds in the soil are removed by adding 0.2 N hydrochloric acid at the rate of 100 ml for every 100 g of soil. When the reaction is complete, the mixture is filtered. The filtrate is washed with distilled water until it is free from the acid. The damp soil on the filter is placed in a evaporating dish and dried in an oven to constant mass.

3.6. THEORY OF SEDIMENTATION

At the commencement of the sedimentation, the soil particles are uniformly dispersed throughout the suspension, and the concentration of particles of different sizes is the same at all depths. After a time period, at a particular depth, only those particles remain which have not settled. Because all particles of the same size have the same velocity, the particles of a given size, if they exist at any level, are in the same concentration as at the beginning of sedimentation. In other words, all particles smaller than a particular size D will be present at a depth H_e in the *same degree of concentration* as at the beginning. All particles larger than the size D would have settled below that depth.

For illustration, let us assume that the soil is composed of particles of only three sizes, which have terminal velocities in the ratio of 1:2:3. The three types of particles, two at each level, are shown in the left

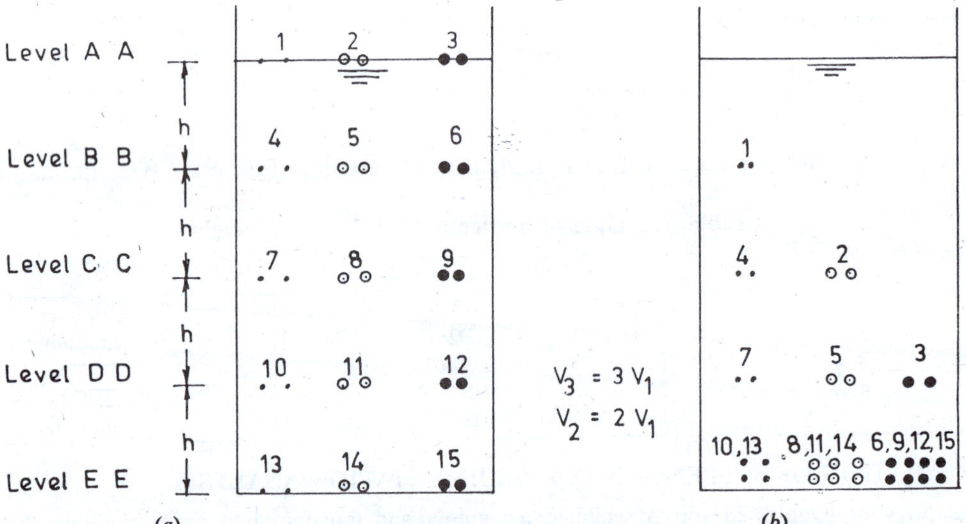

(*a*) (*b*)

Fig. 3.2. Settlement of particles.

column, middle column and the right column in Fig. 3.2 (*a*). At the beginning of the sedimentation, the concentration of particles is the same at all levels.

After some time, the particles take the position as shown in Fig. 3.2 (*b*). The particles of the smallest size have settled to a depth h, those of the intermediate size and the largest size to $2h$ and $3h$, respectively. At lever B—B, only the particles of the smallest size exist, and the concentration of these particles is the same as at the beginning, *viz.* 2 particles. At level C—C, the concentration of the particles of the smallest and intermediate sizes is the same as at beginning. Likewise, at level D—D, the particles of all the three sizes exist with the same concentration.

If m_D is the mass of particles per ml of suspension at depth H_e after time t, and m_s is the mass of particles per ml of suspension at the beginning of sedimentation, the percentage finer than the size D is given by

$$N = \frac{m_D}{m_s} \times 100 \qquad \text{...(3.6)}$$

The particle size D is determined using Eq. 3.4(*a*).

3.7. PIPETTE METHOD

In this method, 500 ml of soil suspension is required. The procedure for preparation of 1000 ml of suspension has been discussed in Sect. 3.5. All the quantities required for 1000 ml of suspension are halved to get a 500 ml of suspension. The suspension is taken in a sedimentation tube. Fig. 3.3 shows a 10 ml capacity pipette used for extraction of the sample. The pipette is fitted with a suction inlet.

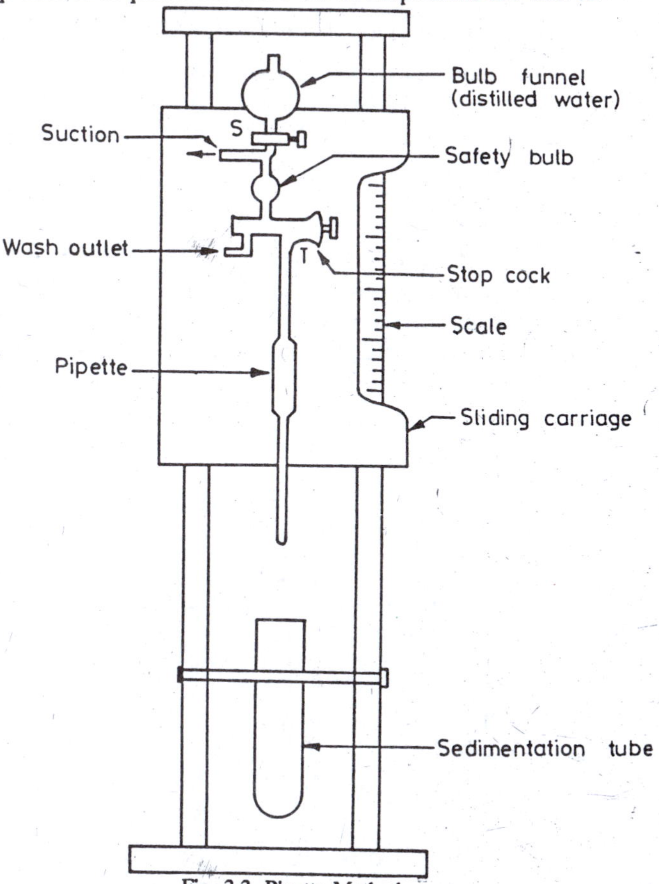

Fig. 3.3. Pipette Method

(a) Calibration of Pipette

For determination of the volume of pipette, it is calibrated before use. For calibration, the nozzle of the pipette is immersed in distilled water. The stop cock *S* is closed. The three-way stop cock *T* is opened, and the water is sucked up into the pipette until it rises in safety bulb. The stop cock *T* is then closed and the pipette is taken out. The stop cock *T* is now turned the other way round to connect it to the wash outlet to drain the excess water from the safety bulb. The stop cock is then turned the other way round to discharge the water contained in the pipette into a glass weighing bottle. The mass of water in the bottle in grams is equal to the volume of the pipette in ml.

(b) Sedimentation Test

The sedimentation tube containing the suspension is placed in a constant-temperature bath at 27°C for about one hour. The tube is then removed from the bath, a rubber bung is placed on its top to close the mouth. The tube is inverted end-over-end a number of times to cause thorough mixing. The bung is removed and the tube is again placed in the constant temperature bath kept just below the tip of the pipette. The instant when the tube is placed in the constant temperature bath is taken as the beginning of the sedimentation. The stop watch is started to record the time.

The pipette is gradually lowered into the suspension in the sedimentation tube. The samples are taken from a depth of 100 mm below the surface of the suspension. The first sample is taken after 2 minutes of the start of sedimentation. The pipette is lowered slowly about 20 seconds before the sample is due to be taken. More samples are taken after 4, 8, 15 and 30 minutes, and 1, 2, 4, 8 and 24 hours. Exact time at which the sample is taken is noted from the stop watch.

The procedure for taking samples is as follows. The stop cock *T* is opened and the suspension is drawn into the pipette until it is full of suspension. The time taken for actual sampling is about 20 seconds. The stop cock *T* is then closed, and any surplus suspension dawn up in the safety bulb is drained away through the wash outlet. The safety bulb is flushed out with distilled water stored in the bulb funnel. The stop cock is again turned and the soil sample in the pipette is transferred to a weighing bottle. Distilled water is used to transfer any solid particles adhering to the inside. The samples taken are dried in an oven at 105—110°C for 24 hours to obtain the mass of solids per ml. As the solids also contain dispersing agent, a correction in the mass of solids is required. If m is the mass of dispersing agent per ml of suspension, the weight of soil solids per ml is given by

$$m_D = m_D' - m \quad \text{...(3.7)}$$

where m_D' = mass of solids/ml as obtained from the sample.

m_D = actual mass of soil solids/ml.

The percentage finer than any size D can be obtained using Eq. 3.6.

Merits and Demerits of the Pipette Method

The pipette method is a standard laboratory method for the particle size analysis of fine-grained soils. It is a very accurate method. However, the apparatus is quite delicate and expensive. It requires a very sensitive weighing balance. For quick particle size analysis, the hydrometer method, described in the following section, is more convenient.

3.8. HYDROMETER METHOD

A hydrometer is an instrument used for the determination of the specific gravity of liquids. As the specific gravity of the soil suspension depends upon the particle size, a hydrometer can be used for the particle size analysis. A special type of hydrometer with a long stem (neck) is used. The stem is marked from top to bottom, generally in the range of 0.995 to 1.030 (Fig. 3.4). At the time of commenement of

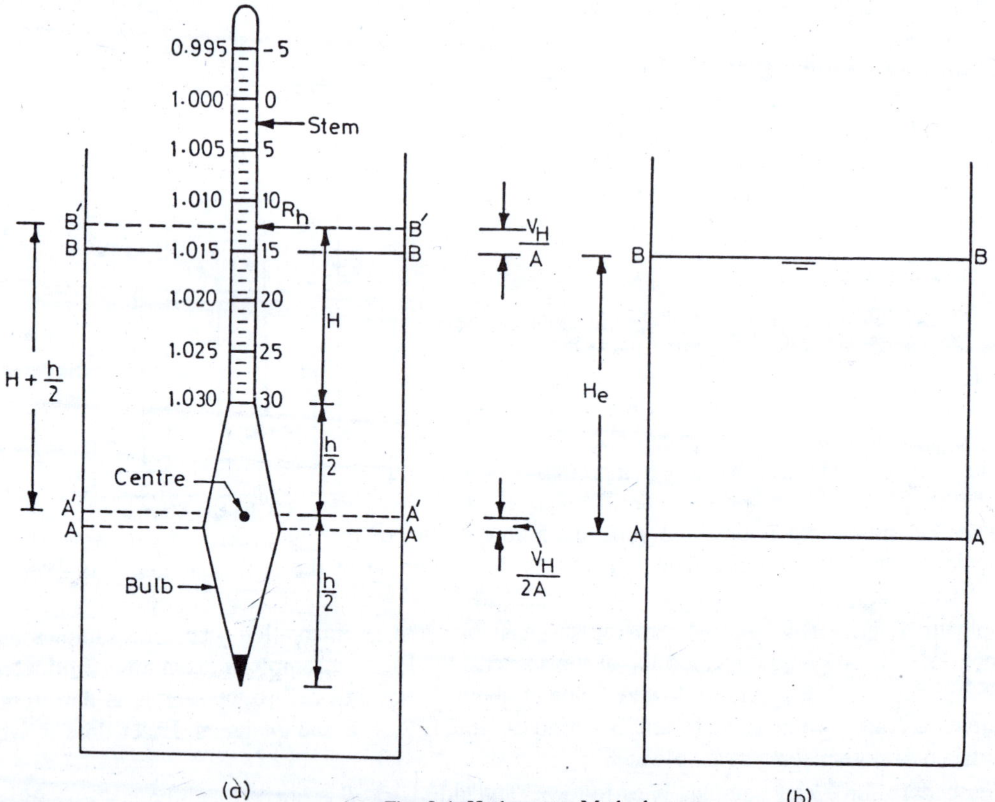

Fig. 3.4. Hydrometer Method

sedimentation, the specific gravity of suspension is uniform at all depths. When the sedimentation takes place, thee larger particles settle more deeper than the smaller ones. This results in non-uniform specific gravity of the suspension at different depths. The lower layers of the suspension have specific gravity greater than that of the upper layers.

Casagrande has shown that the hydrometer measures the specific gravity of suspension at a point indicated by the centre of the immersed volume. If the volume of the stem is neglected, the centre of the immersed volume of the hydrometer is the same as the centre of the bulb. Thus, the hydrometer gives the specific gravity of the suspension at the *centre* of the bulb.

(a) Calibration of hydrometer

To determine the depth at which the specific gravity is measured, calibration of the hydrometer is done. The volume of the hydrometer, V_H, is first determined by immersing it in a graduated cylinder partly filled with water and noting down the volume due to the rise in water level. The volume of the hydrometer can also be determined indirectly from its mass. The volume of hydrometer in ml is approximately equal to the mass of hydrometer in grams, assuming that the specific gravity of hydrometer is unity.

The depth of any layer A-A from the free surface B-B is the effective depth at which the specific gravity is measured by the hydrometer [(Fig. 3.4 (b)]. As soon as the hydrometer is inserted in the jar, the layer of suspension which was at level A- A rises to the level $A' - A'$, and that at level B- B rise to the level $B' - B'$. The effective depth H_e is given by

$$H_e = \left(H + \frac{h}{2}\right) - \frac{V_H}{A} + \frac{V_H}{2A} \quad \text{...(a)}$$

where H = depth from the free surface $B' - B'$ to the lowest mark on the stem,
h = height of bulb,
V_H = volume of hydrometer,
A = cross-sectional area of jar.

In Eq. (a), it has been assumed that the rise in suspension level from $A - A$ to $A' - A'$ at the centre of the bulb is equal to half the total rise due to the volume of the hydrometer.

Thus
$$H_e = H + \frac{1}{2}\left(h - \frac{V_H}{A}\right) \quad \text{...(3.8)}$$

The markings on the hydrometer stem give the specific gravity of the suspension at the centre of the bulb. The hydrometer readings are recorded after subtracting unity from the value of the specific gravity and multiplying the remaining digit by 1000. Thus, a specific gravity of 1.015 is represented by a hydrometer reading R_h of (1.015 – 1.000) × 1000 = 15. The graduations on the right side of the stem directly give the reading R_h. As the effective depth H_e depends upon the hydrometer reading R_h, a calibration chart can be obtained between the hydrometer reading R_h and the effective depth H_e. For determination of the effective depth H_e from Eq. (3.8), an accurate scale is used to determine the height h and the depth H to various graduations. Fig. 3.5 shows a typical calibration chart.

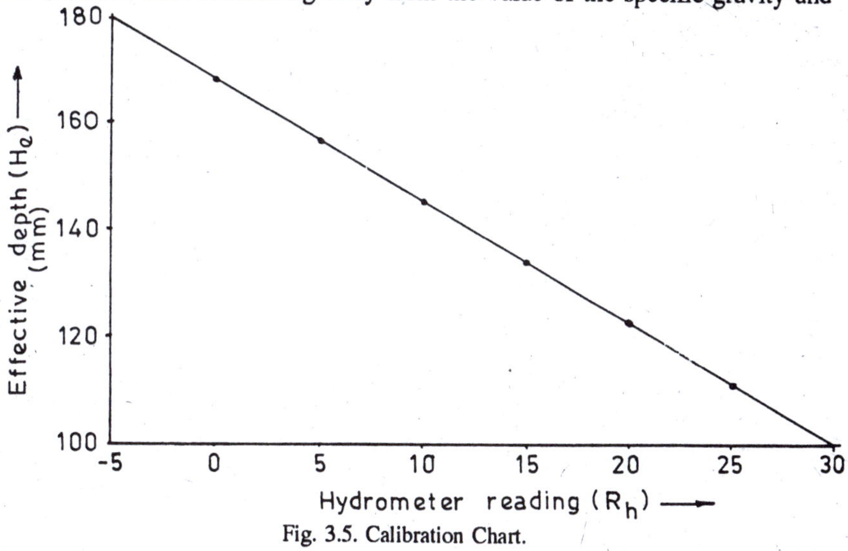

Fig. 3.5. Calibration Chart.

As the sedimentation progresses, the specific gravity of the suspension decreases and the hydrometer goes deeper and deeper, and the effective depth increases. The hydrometer reading R_h, of course, decreases (Fig. 3.6).

(b) Test Procedure

Exactly 1000 ml of suspension is prepared as explained in Sect. 3.5. After stirring, the suspension is washed into a 1000 ml jar and water is added to it to bring the level to 1000 ml mark. The suspension is mixed thoroughly by placing a bung (or the palm of a hand) on the open end of the jar and turning it upside down and back a few times. The jar is then placed on a table, and a stop watch is started.

The hydrometer is inserted in the suspension and the first reading is taken after $\frac{1}{2}$ minute of the commencement of the sedimentation. Further readings are taken after one minute, two minutes and four minutes of the commencement of the sedimentation. The hydrometer is then removed from the jar and rinsed with distilled water and floated in a comparison cylinder containing distilled water with the dispersing agent added to the same concentration as in the soil suspension.

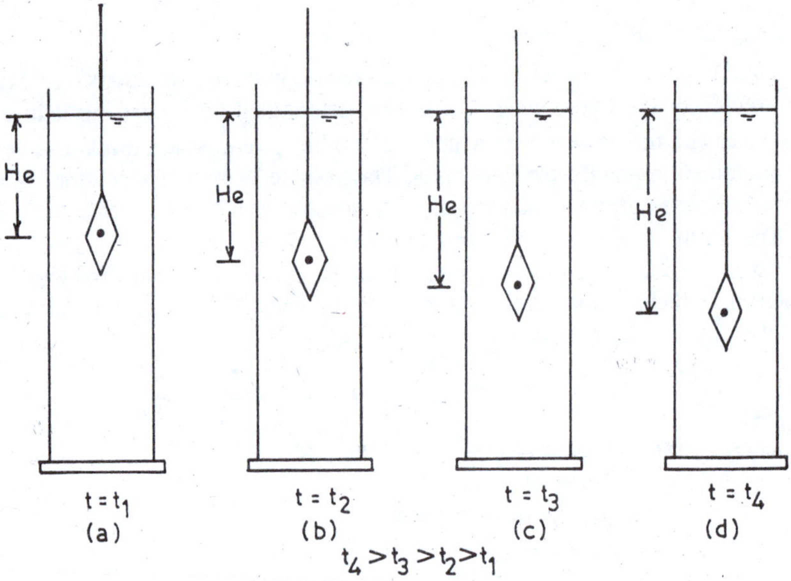

Fig. 3.6. Downward movement of hydrometer.

Further readings are taken after 8, 15 and 30 minutes and 1, 2, 4, 8 and 24 hours reckoned from the beginning of sedimentation. For each of these readings, the hydrometer is inserted about 20 seconds before the reading. The hydrometer is taken out after the reading and floated in the comparison cylinder.

(c) Corrections of Hydrometer Reading

The hydrometer readings are corrected as under:

(i) Meniscus correction—Since the suspension is opaque, the observations are taken at the top of the meniscus. The meniscus correction is equal to the reading between the top of the meniscus and the level of the suspension. As the marking on the stem increases downward, the correction is positive.

The meniscus correction (C_m) is determined from the readings at the top and bottom of meniscus in the comparison cylinder. The meniscus correction is constant for a hydrometer.

If R_h' is the hydrometer reading of the suspension at a particular time, the corrected hydrometer R_h reading is given by

$$R_h = R_h' + C_m \qquad \text{...(3.9)}$$

The corrected hydrometer reading (R_h) is required for determining the effective depth from the calibration chart (Fig. 3.5).

(ii) Temperature correction—The hydrometer is generally calibrated at 27°C. If the temperature of the suspension is different from 27°C, a temperature correction (C_t) is required for the hydrometer reading. If the temperature is more than 27°C, the suspension is lighter, and the actual reading will be less than the corrected reading. The temperature correction is positive. On the other hand, if the temperature is less than 27°C, the temperature correction is negative.

The temperature correction is obtained from the charts supplied by the manufacturer.

(iii) Dispersion agent Correction—Addition of the dispersing agent to the soil specimen causes and increase in the specific gravity of the suspension. Therefore, the dispersing agent correction is always negative. The dispersing agent correction (C_d) can be determined by noting the hydrometer reading in clear water and again in the same water after adding the dispersing agent.

Thus, the corrected reading R can be obtained from the observed reading R_h' as under.

$$R = R_h' + C_m \pm C_t - C_d \qquad ...(3.10)$$

Composite Correction—Instead of finding the corrections individually, it is convenient to find one composite correction. The composite correction (C) is the algebraic sum of all the corrections. Thus,

$$R = R_h' \pm C \qquad ...(3.11)$$

The composite correction is found directly from the readings taken in a comparison cylinder, which has distilled water and the dispersing agent in the same concentration, and has the same temperature. As the hydrometer has been calibrated at 27°C to indicate a specific gravity of 1.000, the difference between the reading taken at the top of meniscus and 1.000 is in magnitude equal to the composite correction. The negative of the hydrometer reading in the comparison cylinder is equal to the composite correction. The composite correction can be positive or negative. For example, if the hydrometer reading is +2 (*i.e.* 1.002), the correction is –2, and if the reading is –3 (*i.e.* 0.997), the correction is +3.

The composite correction is found before the start of the test and at every 30 minute interval.

3.9. RELATION BETWEEN PERCENTAGE FINER AND HYDROMETER READING

The corrected hydrometer reading R can be related to the percentage finer N than any size D as under:

Let M_s be the mass of dry soil in a suspension of volume V. At the commencement of the sedimentation, the soil-water suspension is uniform, and, therefore, the mass of solids per unit volume of suspension at any depth is M_s/V.

The initial density of suspension is given by

$$\rho_i = \frac{M_s + \text{mass of water in suspension}}{V}$$

or

$$\rho_i = \frac{M_s}{V} + \text{mass of water/volume of suspension.} \qquad ...(a)$$

The mass of water per unit volume of suspension can be determined from the volume of water per unit volume of suspension, as explained below.

Mass of solids/volume of suspension $= \dfrac{M_s}{V}$

Volume of solids/volume of suspension $= \dfrac{M_s}{V(G\,\rho_w)}$

Volume of water/volume of suspension $= 1 - \dfrac{M_s}{V(G\,\rho_w)}$

Mass of water/volume of suspension $= \left[1 - \dfrac{M_s}{V(G\,\rho_w)}\right]\rho_w$

From Eq. (a),

$$\rho_i = \frac{M_s}{V} + \left[1 - \frac{M_s}{V(G\,\rho_w)}\right]\rho_w$$

$$= \rho_w + \frac{M_s}{V}\left(1 - \frac{1}{G}\right)$$

or

$$\rho_i = \rho_w + \frac{M_s}{V}\left(\frac{G-1}{G}\right) \qquad ...(3.12)$$

If M_D is the mass of solids in volume V at that depth after time t, Eq. 3.12 gives the density of suspension at that depth as

$$\rho = \rho_w + \frac{M_D}{V}\left(\frac{G-1}{G}\right) \qquad ...(3.13)$$

From Eq. 3.6, the percentage finer N than any size is given by

$$N = \frac{m_D}{m_s} \times 100$$

or

$$m_D = \frac{N \cdot m_s}{100}$$

where $m_D = M_D/V$ and $m_s = M_s/V$

Therefore, Eq. 3.13 becomes

$$\rho = \rho_w + \frac{N\, m_s}{100}\left(\frac{G-1}{G}\right) \quad \text{...(3.14)}$$

or

$$\rho - \rho_w = \frac{N\, m_s}{100}\left(\frac{G-1}{G}\right)$$

or

$$N = \left(\frac{G}{G-1}\right)\frac{(\rho-\rho_w)}{m_s} \times 100 \quad \text{...(3.15)}$$

As the hydrometer reading R is equal to $(\rho - \rho_w) \times 1000$, Eq. 3.15 can be written as

$$N = \left(\frac{G}{G-1}\right) \cdot \frac{R}{1000} \times \frac{1}{m_s} \times 100 \quad \text{...(3.16)}$$

As $m_s = M_s/V$,

$$N = \left(\frac{G}{G-1}\right) \cdot \frac{R}{1000} \times \frac{V}{M_s} \times 100 \quad \text{...[3.17(a)]}$$

or

$$N = \left(\frac{G}{G-1}\right) \cdot \frac{R}{M_s} \times 100 \quad \text{...[3.17(b)]}$$

where M_s is the mass of the solids in a volume V of 1000 ml.

The particle size D is determined using Eq. 3.4, taking the value of effective depth H_e from the calibration curve for the hydrometer reading R_h.

(See Chapter 30, Sect. 30-9 for the laboratory experiment)

3.10. LIMITATION OF SEDIMENTATION ANALYSIS

The sedimentation analysis does not give correct values of the particle size and the percentage finer due to the following limitations.

(1) The sedimentation analysis gives the particle size in terms of equivalent diameter, which is less than the particle size given by sieve analysis. The soil particles are not spherical. The equivalent diameter is close to the thickness (smallest dimension) rather than the length or width. (The equivalent diameter is the diameter of the sphere which falls with the same velocity as the actual particle.)

(2) As the specific gravity of solids for different particles is different, the use of an average value of G in Eq. 3.17 (*b*) is a source of error. However, as the variation of the values of G is small, the error is negligible.

(3) Stokes' law is applicable only when the liquid is infinite. The presence of walls of the jar affects the results to some extent.

(4) In Stokes' law, it has been assumed that only one sphere settles, and there is no interference from other spheres. In the sedimentation analysis, as many particles settle simultaneously, there is some interference.

However, the effect of errors mentioned in paras (3) and (4) is negligibly small if the mass of dry soil used per 1000 ml of suspension is not more than 50g.

(5) The sedimentation analysis cannot be used for particles larger than 0.2 mm as turbulent conditions develop and Stokes' law is not applicable.

(6) The sedimentation method is not applicable for particles smaller than 0.2μ because Brownian movement takes place and the particles do not settle as per Stokes' law.

(7) The sedimentation method cannot be used for chalky soils, because of the removal of the calcium carbonate of chalky soils in the pretreatement by hydrochloric acid.

Despite above limitations,the sedimentation analysis is used for determination of the particle size of fine-grained soils. The particle sizes of such soils is not of much practical significance and, therefore, even approximate analysis is good enough. The index properties of such soils are plasticity characteristics and not the particle size. The main use of the sedimentation analysis is to determine the clay content (particles less than 2 μ size) in a soil mass.

3.11. COMBINED SIEVE AND SEDIMENTATION ANALYSIS

It the soil mass consists of particles of both coarse-grained and fine-grained soils, a combined analysis is done. The slurry of the soil is made as mentioned in the wet sieve analysis (Sect. 3.3). The slurry is sieved through a 4.75 mm IS sieve. The material retained on the sieve is oven-dried and a coarse-sieve analysis is done.

The material retained on a 75 μ IS sieve is also oven-dried and the sieve is analysis is done using the set of fine sieves.

The suspension passing the 75 μ sieve is mixed with a deflocculating agent, if not already done. The hydrometer test is performed on the suspension, as explained in Sect. 3.8.

The percentage finer than any size can be calculated on the basis of the *original* mass of soil taken for the combined analysis.

3.12. PARTICLE SIZE DISTRIBUTION CURVE

The particle size distribution curve, also known as a *gradation curve,* represents the distribution of particles of different sizes in the soil mass. The percentage finer N than a given size is plotted as ordinate (on natural scale) and the particle size as abscissa (on log scale). In Fig. 3.7 (a), the particle size decreases from

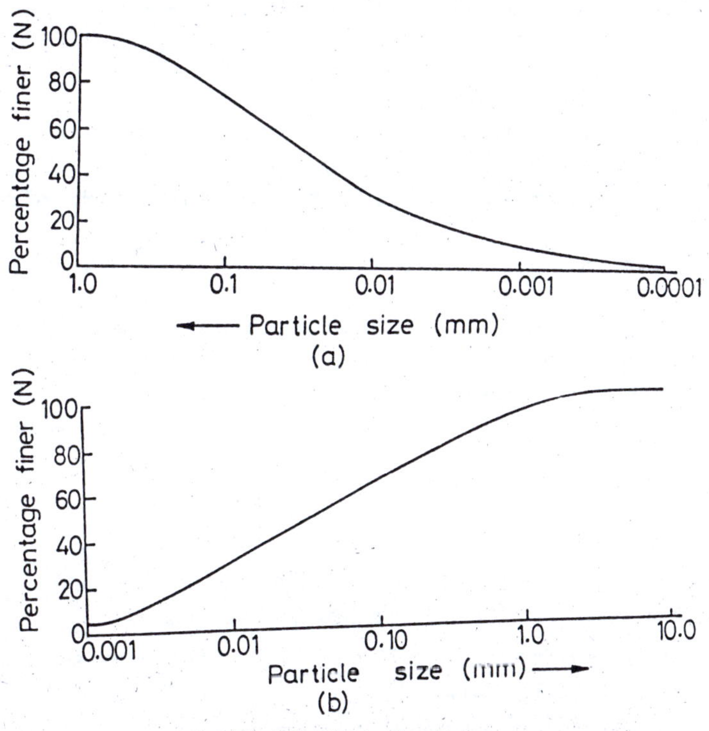

Fig. 3.7. Particle Size Curve.

left to right, whereas in Fig. 3.7 (*b*), the particle size increases from left to right. Both the methods are prelevant. The reader should carefully observe the horizontal scale of the particle size distribution curve. In this text, the particle size distribution is shown as in Fig. 3.7 (*b*), *i.e.*, the particle size increases from left to right, which is also the usual convention.

The semi-log plot for the particle size distribution, as shown in Fig. 3.7, has the following advantages over natural plots.

(1) The soils of equal uniformity exhibit the same shape, irrespective of the actual particle size.

(2) As the range of the particle sizes is very large, for better representation, a log scale is required.

Grading of Soils—The distribution of particles of different sizes in a soil mass is called grading. The grading of soils can be determined from the particle size distribution curves. Fig. 3.8 shows the particle size distribution curves of different soils.

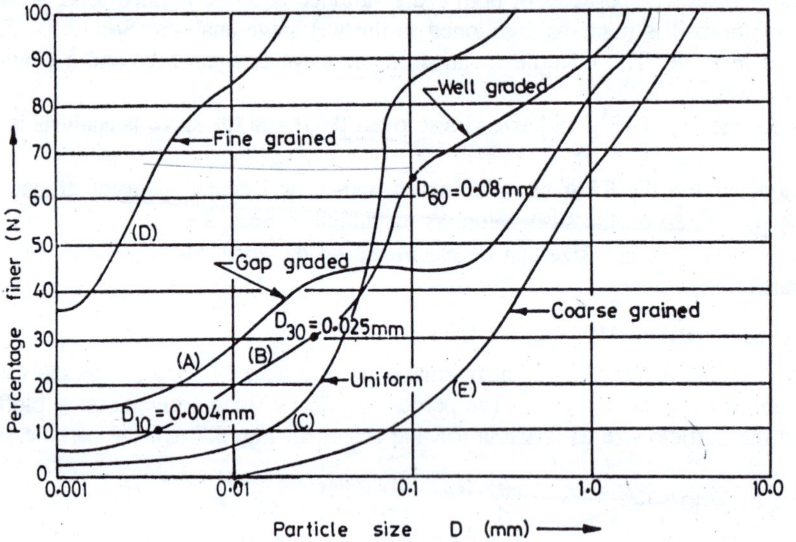

Fig. 3.8. Grading os Soils.

A curve with a hump, such as curve *A*, represents the soil in which some of the intermediate size particles are missing. Such a soils is called *gap-graded or skip-graded.*

A flat *S*-curve, such as curve *B*, represents a soil which contains the particles of different sizes in good proportion. Such a soil is called a *well-graded* (or uniformly graded) soil.

A steep curve, like *C*, indicates a soil containing the particles of almost the same size. Such soils are known as *uniform soils.*

The particle size distribution curve also reveals whether a soil is coarse-grained or fine-grained. In general, a curve situated higher up and to the left (curve *D*) indicates a relatively fine-grained soil, whereas a curve situated to the right (curve *E*) indicates a coarse-grained soil.

The uniformity of a soil is expressed qualitatively by a term known as uniformity coefficient, C_u, given by

$$C_u = \frac{D_{60}}{D_{10}} \qquad ...(3.18)$$

where D_{60} = particle size such that 60% of the soil is finer than this size, and

D_{10} = particle size such that 10% of the soil is finer than this size.

D_{10} size is also known as the *effective size.* In Fig. 3.8, D_{60} and D_{10} for the soil *B* are, respectively, 0.08 mm and 0.004 mm. Therefore, C_u = 0.08/0.004 = 20

The larger the numerical value of C_u, the more is the range of particles. Soils with a value of C_u less than 2 are uniform soils. Sands with a value of C_u of 6 or more, are well-graded. Gravels with a value of C_u of 4 or more are well-graded.

The general shape of the particle size distribution curve is described by another coefficient known as the coefficient of curvature (C_c) or the coefficient of gradation (C_g).

$$C_c = \frac{(D_{30})^2}{D_{60} \times D_{10}} \quad \text{...(3.19)}$$

where D_{30} is the particle size corresponding to 30% finer.

For a well-graded soil, the value of the coefficient of curvature lies between 1 and 3.

It may be noted that the gap grading of the soil cannot be detected by C_u only. The value of C_c is also required to detect it.

For the soil shown by curve *B* in Fig. 3.9, the particle size D_{30} is 0.025 mm. Therefore,

$$C_c = \frac{(0.025)^2}{0.08 \times 0.004} = 1.95$$

3.13. USES OF PARTICLE SIZE DISTRIBUTION CURVE

The particle size distribution curve is extremely useful for coarse-grained soils. As the behavior of fine-grained soils (minus 75 μ) depends upon the plasticity characteristic and not on the particle size, its use for fine- grained soils is limited.

(1) The particle size distribution curve is used in the classification of coarse-grained soils (see chapter 5).

(2) The coefficient of permeability of a coarse-grained soil depends to a large extent on the size of the particles. An approximate value of the coefficient of permeability can be determined from the particle size as discussed in chapter 8.

(3) The particle size is used to know the susceptibility of a soil to frost action.

(4) The particle size distribution curve is required for the design of drainage filters.

(5) The particle size distribution provides an index to the shear strength of the soil. Generally, a well-graded, compacted sand has high shear strength.

(6) The compressibility of a soil can also be judged from its particle size distribution curve. A uniform soil is more compressible than a well-graded soil.

(7) The particle size distribution curve is useful in soil stabilisation and for the design of pavements.

(8) The particle size distribution curve may indicate the mode of deposition of a soil. For example, a gap-graded soil indicates deposition by two different agencies.

(9) The particle size distribution curve of a residual soil may indicate the age of the soil deposit. With increasing age, the average particle size decreases because of weathering. The particle size distribution curve which is initially wavy becomes smooth and regular with age.

3.14. SHAPE OF PARTICLES

The engineering properties of soils, especially coarse-grained soils, depend upon the shape of particles. As it is more difficult to measure the shape than the size, the shape of the particles does not get the required attention.

When the length, width and thickness of the particles are of same order of magnitude, the particles are known to have a bulky shape. Cohesionless soils have bulky particles. As stated in chapter 1, bulky particles are formed by physical disintegration of rocks. Rock flour, which has the size of the particles in the range of fine-grained soils, behaves like cohesionless soils because its particles are bulky. Soils containing bulky grains behave like a heap of loose bricks or broken stone pieces. Such soils can support heavy loads in static conditions. However, when vibration takes place, large settlements can occur.

Cohesive, clayey soils have particles which are thin and flaky, like a sheet of paper. Soils composed of flaky particles are highly compressible. These soils deform easily under static loads, like dry leaves or loose papers in a basket subjected to a pressure. However, such soils are relatively more stable when subjected to vibrations.

The shape of the coarse-grained soils can be described in terms of sphericity, flatness or angularity. Sphericity (S) of the particle is defined as

$$S = D_e/L$$

where D_e is equivalent diameter of the particle assuming it to be a sphere, given by $D_e = (6V/\pi)^{1/3}$, where V is the volume of the particle and L is the length of the particle.

The particles with a high value of sphericity (more roundness) are easy to manipulate in construction and their tendency to fracture is low.

Flatness (F) and elongation (E) are defined as

as $$F = B/T \quad \text{and} \quad E = L/B$$

where L, B and T are, respectively, length, width and thickness.

The higher the value of the flatness or the elongation, the more is the tendency of the soil to fracture.

The angularity (R) of a particle is defined as

$$R = \frac{\text{average radius of corners and edges}}{\text{radius of maximum inscribed circle}}$$

Depending upon angularity, the particles are qualitatively divided into 5 shapes (Fig. 3.9).

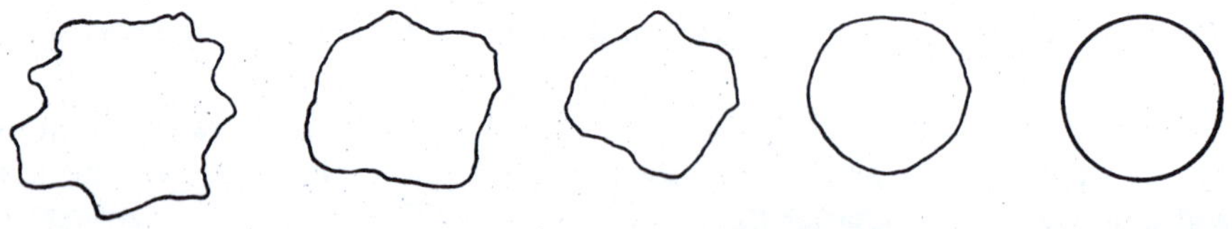

Fi.g 3.9. Different shapes of particles

The angularity of particles has great influence on the behavior of coarse-grained soils. The particles with a high value of angularity tend to resist the displacement, but have more tendency for fracturing. On the other hand, the particles with low angularity (more roundness) do not crush easily under loads, but have low resistance to displacements as they have a tendency to roll. In general, the angular particles have good engineering properties, such as shear strength.

3.15. RELATIVE DENSITY

The most important index aggregate property of a cohesionless soil is its relative density. The engineering properties of a mass of cohesionless soil depend to a large extent on its relative density (D_r), also known as *density index* (I_D). The relative density is defined as

$$D_r = \frac{e_{max} - e}{e_{max} - e_{min}} \times 100 \qquad \text{...(3.20)}$$

where e_{max} = maximum void ratio of the soil in the loosest condition.

e_{min} = minimum void ratio of the soil in the densest condition.

e = void ratio in the natural state.

The relative density of a soil gives a more clear idea of the denseness than does the void ratio. Two types of sands having the same void ratio may have entirely different state of denseness and engineering properties. However, if the two sands have the same relative density, they usually behave in identical manner.

The relative density of a soil indicates how it would behave under loads. If the deposit is dense, it can take heavy loads with very little settlements. Depending upon the relative density, the soils are generally divided into 5 categories (Table 3.3).

Table 3.3. Denseness of Soils

Denseness	*Very Loose*	*Loose*	*Medium Dense*	*Dense*	*Very Dense*
D_r (%)	< 15	15 to 35	35 to 65	65 to 85	85 to 100

3.16. DETERMINATION OF RELATIVE DENSITY

Figs. 3.10 (*a*), (*b*) and (*c*) show the soil in the densest, natural and loosest states. As it is difficult to measure the void ratio directly, Eq. 3.20 cannot be used. However, it is convenient to express the void ratio in terms of dry density (ρ_d),

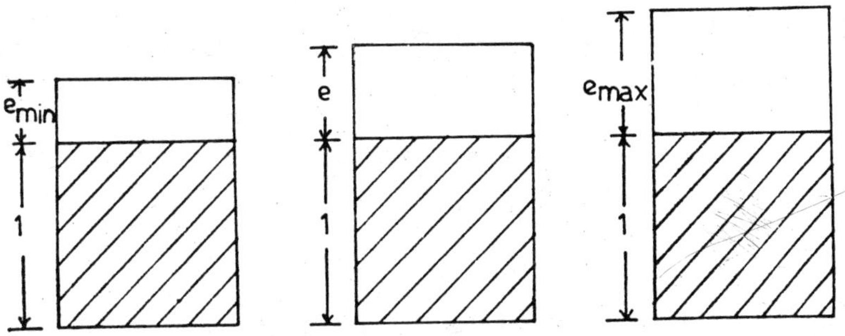

Fig. 3.10

$$e = \frac{G\rho_w}{\rho_d} - 1$$

Representing the dry density in the loosest, densest and natural conditions as ρ_{min}, ρ_{max} and ρ_d , Eq. 3.20 becomes

$$D_r = \frac{\left(\dfrac{G\rho_w}{\rho_{min}} - 1\right) - \left(\dfrac{G\rho_w}{\rho_d} - 1\right)}{\left(\dfrac{G\rho_w}{\rho_{min}} - 1\right) - \left(\dfrac{G\rho_w}{\rho_{max}} - 1\right)}$$

or

$$D_r = \frac{\rho_{max}}{\rho_d}\left(\frac{\rho_d - \rho_{min}}{\rho_{max} - \rho_{min}}\right) \qquad \text{...(3.21)}$$

Eq. 3.21 is used to determine the relative density of an in-situ deposit. The methods for the determination of the dry density (ρ_d) have been discussed in chapter 2. The methods for the determination of ρ_{max} and ρ_{min} are described below.

For determination of the minimum dry density, a representative, oven dry sample of soil is taken. The sample is then pulverised and sieved through the required sieve. The minimum dry density is found by pouring the dry soil in a mould using a pouring device (IS : 2720—Part XIV). The spout of the pouring device is so adjusted that the height of free fall is always 25mm. The mass and the volume of the soil deposited are found, and the dry density of the soil is determined as under.

$$\rho_{min} = \frac{M_{min}}{V_m} \qquad \text{...(3.22)}$$

where M_{min} = mass of dry soil.

V_m = volume of soil deposited in the mould.

The maximum dry density is determined either by the dry method or the wet method. In the dry method, the mould is filled with thoroughly mixed oven-dry soil. A surcharge load is placed on the soil surface, and the mould is fixed to a vibrator deck. The specimen is vibrated for 8 minutes. The mass and volume of the soil in the compacted state are found. The maximum dry density is given by

$$\rho_{max} = \frac{M_{max}}{V_m}$$

where M_{max} = mass of dry soil and

V_m = volume of mould.

The maximum dry density of a soil can be determined also in the saturated state. In this method, the mould is filled with wet soil and water is added till a small quantity of free water accumulates on the free surface of the soil. During and just after filling the mould, vibration is done for a total of six minutes. Water appearing on the surface of soil is removed. A surcharge mass is placed on the soil and the mould is vibrated again for 8 minutes. The volume (V_m) of the soil is determined. The mass M_{max} of the soil is determined after oven drying the sample.

Note. If the sand is vibrated under more severe conditions, it may have a relative density of more than 100%.

ILLUSTRATIVE EXAMPLES

Illustrative Example 3.1. *The results of a sieve analysis of a soil are given below:*
Total mass of sample = 900 gm.

IS Sieve	20 *mm*	10 *mm*	4.75 *mm*	2 *mm*	1.0 *mm*	0.6 *mm*	4.25 μ	212 μ	150 μ	75 μ	*Pan*
Mass of soil retained (gm)	35	40	80	150	150	140	115	55	35	25	75

Draw the particle size distribution curve and hence determine the uniformity coefficient and the coefficient of curvature.

Solution. The calculations for percentage finer N than different sizes are shown (Table 3.1).

Table E-3.1

IS Sieve (1)	*Mass retained* (2)	*Percentage retained* $= \frac{(2)}{(900)} \times 100$ (3)	*Cumulative percentage retained* (4)	*Pereentage Finer (N)* = 100 − (4) (5)
20 mm	35 gm	3.89	3.89	96.11
10	40	4.44	8.33	91.67
4.75	80	8.89	17.22	82.78
2.0	150	16.67	33.89	66.11
1.0	150	16.67	50.56	49.44
0.6	140	15.56	66.12	33.88
425 μ	115	12.78	78.90	21.10
212 μ	55	6.11	85.01	14.99
150 μ	35	3.89	88.90	11.10
75 μ	25	2.78	91.68	8.32
Pan	75	8.32	100.00	

Σ = 900.0 gm

The particle size distribution curve is shown in Fig. E-3.1.

From plot, D_{60} = 1.55 mm; D_{30} = 0.53 mm; D_{10} = 0.115 mm

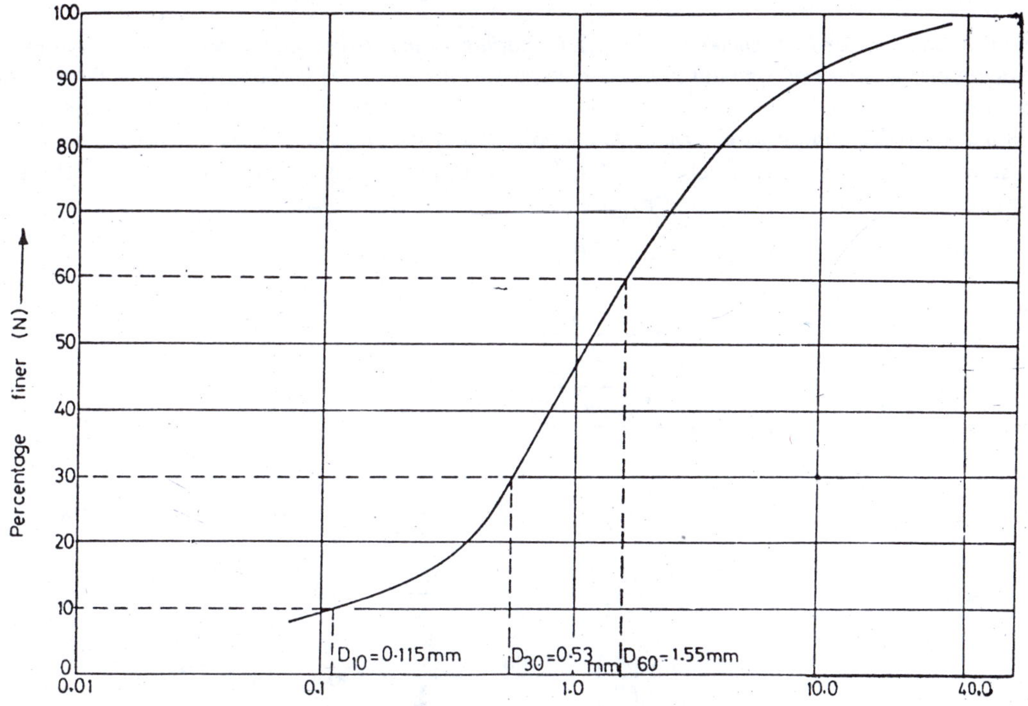

Fig. E-3.1 Particle size (mm)

From Eq. 3.18 $$C_u = \frac{D_{60}}{D_{10}} = \frac{1.55}{0.115} = \mathbf{13.48}$$

From Eq. 3.19, $$C_c = \frac{(D_{30})^2}{D_{60} \times D_{10}} = \frac{(0.53)^2}{1.55 \times 0.115} = \mathbf{1.58}$$

Illustrative Example 3.2. *The following observations were taken during a pipette analysis for the determination of particle size distribution of a soil sample.*

(a) Depth below the water surface at which the sample was taken = 100 mm

(b) Capacity of pipette = 10 ml

(c) Mass of sample when dried = 0.3 gm

(d) Time of taking sample = 7 minutes after the start.

(e) Volume of soil suspension in the sedimentation tube = 500 ml.

(f) Dry mass of soil used in making suspension = 25 gm.

Determine the coordinate of the point on the particle size distribution curve corresponding to above observations.

Take $G = 2.70$ *and* $\eta = 10.09$ *millipoise.* $\rho_w = 1$ *gm/ml.*

Solution. From Eq. 3.4 (*a*), $$D = \sqrt{\frac{0.30\ \eta H_e}{g\,(G-1)\,\rho_w \times t}}$$

$$D = \sqrt{\frac{0.30 \times 10.09 \times 10^{-3} \times 10}{981 \times (2.70-1) \times 7}} = 0.00161 \text{ cm}$$

From Eq. 3.6, $$N = \frac{m_D}{m_s} \times 100 = \frac{0.3/10}{25/500} \times 100 = 60\%$$

The coordinates of the point on the particle-size distribution curve are (0.0161 mm. 60%).

Illustrative Example 3.3. *A dry sample of mass 50 gm is mixed with distilled water to prepare a* *suspension of 1000 ml for hydrometer analysis. The reading of the hydrometer taken after 5 minutes was 25 and the depth of the centre of the bulb below the water surface when the hydrometer was in the jar was 150 mm. The volume of the hydrometer was 62 ml and the area of cross-section of the jar was 55 cm². Assuming G = 2.68 and η = 9.81 millipoise, determine the coordinates of the point corresponding to above observation.*

Solution. The depth between levels $B' - B'$ and $A' - A'$ in Fig. 3.4, is given to be 150 mm. The effective depth between B – B and A – A is given by,

$$H_e = 15.0 - \frac{V_H}{A} + \frac{V_H}{2A} = 15.0 - \frac{V_H}{2A}$$

$$= 15.0 - \frac{62}{2 \times 55} = 14.436 \text{ cm}$$

From Eq. 3.4 (*a*),
$$D = \sqrt{\frac{0.3 \times \eta H_e}{g(G-1)t}}$$

$$= \sqrt{\frac{0.3 \times 9.81 \times 10^{-3} \times 14.436}{981 \times (2.68 - 1.0) \times 5}} = 0.0023 \text{ cm} = \mathbf{0.023 \text{ mm}}$$

From Eq. 3.17 (*a*),
$$N = \left(\frac{G}{G-1}\right) \times \left(\frac{V}{M_s}\right)\left(\frac{R}{1000}\right) \times 100$$

$$= \left(\frac{2.68}{1.68}\right) \times \frac{1000}{50} \times \frac{25}{1000} \times 100 \quad \text{or } N = \mathbf{79.76}$$

The coordinates of the point on the particle-size distribution curve are (0.023 mm, 79.76%).

Illustrative Example 3.4. *A soil has a dry density of 1.816 gm/ml in the natural condition. When 410 gm of the soil was poured in a vessel in a very loose state, its volume was 290 ml. The same soil when vibrated and compacted was found to have a volume of 215 ml. Determine the relative density.*

Solution. From Eq. 3.22,
$$\rho_{min} = \frac{M_{min}}{V_m} = \frac{410}{290} = 1.414 \text{ gm/ml}$$

From Eq. 3.23,
$$\rho_{max} = \frac{M_{max}}{V_m} = \frac{410}{215} = 1.907 \text{ gm/ml}$$

From Eq. 3.21,
$$D_r = \frac{\rho_{max}}{\rho_d}\left(\frac{\rho_d - \rho_{min}}{\rho_{max} - \rho_{min}}\right) \times 100$$

or
$$= \frac{1.907}{1.816}\left(\frac{1.816 - 1.414}{1.907 - 1414}\right) \times 100$$

or
$$D_r = \mathbf{85.63\ \%}$$

Illustrative Example 3.5. *A test for the relative density of soil in place was performed by digging a small hole in the soil. The volume of the hole was 400 ml and the moist weight of the excavated soil was 9 N. After oven drying, the weight was 7.8 N. Of the dried soil, 4 N was poured into a vessel in a very loose state, and its volume was found to be 270 ml. The same weight of soil when vibrated and tamped had a volume of 200 ml. Determine the relative density.*

Solution.
$$w = \frac{9.0 - 7.80}{7.80} = 0.1538$$

$$\gamma_d = \frac{7.8}{400} = 0.0195 \text{ N/ml} = 19.5 \text{ kN/m}^3$$

$$(\gamma_d)_{min} = \frac{4.0}{270} = 0.0148 \text{ N/ml} = 14.8 \text{ kN/m}^3$$

$$(\gamma_d)_{max} = \frac{4.0}{200} = 0.02 \text{ N/ml} = 20.0 \text{ kN/m}^3$$

From Eq. 3.21, substituting γ for ρ throughout,

$$D_r = \frac{20.0}{19.5}\left(\frac{19.5 - 14.8}{20.0 - 14.8}\right) \times 100 = \mathbf{92.70\%}$$

Illustrative Example 3.6. *A sample of sand has a volume of 1000 ml in its natural state. Its minimum volume when compacted is 840 ml. When gently poured in a measuring cylinder, its maximum volume is 1370 ml. Determine the relative density.*

Solution. Let M be the dry mass (in gm) of the sample.

Therefore, $\rho_{max} = \dfrac{M}{840}$; $\rho_{min} = \dfrac{M}{1370}$; $\rho_d = \dfrac{M}{1000}$

From Eq. 3.21, $D_r = \dfrac{\rho_{max}}{\rho_d}\left(\dfrac{\rho_d - \rho_{min}}{\rho_{max} - \rho_{min}}\right)$

$$= \frac{M/840}{M/1000}\left(\frac{M/1000 - M/1370}{M/840 - M/1370}\right) \times 100$$

$$= 0.6981 \times 100 = \mathbf{69.81\%.}$$

Illustrative Example 3.7. *In order to find the relative density of a sand, a mould of volume 1000 ml was used. When the sand was dynamically compacted in the mould, its mass was 2.10 kg, whereas when the sand was poured in loosely, its mass was 1.635 kg. If the in-situ density of the soil was 1.50 Mg/m^3, calculate the relative density. G = 2.70. Assume that the sand is saturated.*

Solution. $(\rho_{max})_{sat} = \dfrac{2.10 \times 10^3}{1000} = 2.1 \text{ g/ml}$

$$(\rho_{min})_{sat} = \frac{1.635 \times 10^3}{1000} = 1.635 \text{ g/ml}$$

As $\rho_d = 1.50 \text{ Mg/m}^3 = 1.50 \text{ g/ml}$, $e = \dfrac{G\rho_w}{\rho_d} - 1 = \dfrac{2.70}{1.5} - 1 = 0.80$

Now $(\rho_{max})_{sat} = \left(\dfrac{G + e_{min}}{1 + e_{min}}\right)\rho_w$

or $2.1 = \left(\dfrac{2.70 + e_{min}}{1 + e_{min}}\right) \times 1.0$ or $e_{min} = 0.545$

Likewise $(\rho_{min})_{sat} = \left(\dfrac{G + e_{max}}{1 + e_{max}}\right)\rho_w$

or $1.635 = \left(\dfrac{2.70 + e_{max}}{1 + e_{max}}\right) \times 1.0$ or $e_{max} = 1.677$

From Eq. 3.20 $D_r = \dfrac{e_{max} - e}{e_{max} - e_{min}} \times 100$

$$= \frac{1.677 - 0.80}{1.677 - 0.545} \times 100 = 77.47\%$$

PROBLEMS

A. Numerical

3.1. One kg of soil was sieved through a set of 8 sieves, with the size 4.75 mm, 2.0 mm, 600 μ, 425 μ, 300μ, 212μ, 150μ and 75μ. The mass of soil retained on these sieves was found to be 50, 78, 90, 150, 160, 132, 148 and 179 gm, respectively. Determine the percentage finer than the corresponding sizes.

[**Ans.** 99.5, 87.5, 78.2, 63.2, 47.2, 34.0, 19.2 and 1.3]

3.2. Prove that the particle diameter and the terminal velocity of particle are related as

$$v = 9020\, D^2$$

where v = velocity in cm/sec,

D = diameter in cm

Clearly state the various assumptions made.

3.3. Determine the maximum void ratio for a sand composed of grains of spherical shapes.

[**Hint** Consider a cubical box of size $2d$, where d is the diameter of sphere. The number of grains in the box is 8]

[**Ans.** 0.91]

3.4. The minimum and the maximum dry density of a sand were found to be 1.50 and 1.70 gm/ml. Calculate the dry density corresponding to relative densities of 50% and 75%. [**Ans.** 1.594 gm/l; 1.645 gm/ml]

3.5. An undisturbed sample of fine sand has a dry unit weight of 18 kN/m^3. At the maximum density, the void ratio is 0.35, and that at the minimum density, 0.90. Determine the relative density of the undisturbed soil. G = 2.65.

[**Ans.** 77.82%]

3.6. A coarse-grained soil is compacted to a wet density of 2Mg/m^3 at a water content of 15%. Determine the relative density of the compacted sand, given $e_{max} = 0.85$ and $e_{min} = 0.40$ and G = 2.67.

[**Ans.** 70%]

3.7. How long would it take for a particle of soil 0.002 cm in diameter to settle from the surface to the bottom of the pond 15 m deep ? Take G = 2.60 and $\eta = 1.0 \times 10^{-5}$ gmf-sec/cm^2 (9.81 mP). [**Ans.** 11.72 hours]

3.8 A sample of soil of mass 40 gm is dispersed in 1000 ml of water. How long after the commencement of sedimentation should the hydrometer reading be taken in order to estimate the percentage of particles less than 0.002 mm effective diameter ? The centre of the bulb is at an effective depth of 20 cm below the surface of water. Take G = 2.70, η = 0.01 poise. [**Ans.** 14.99 hours]

3.9. In a sedimentation test, 25 gm of soil was dispersed in 1000 ml of water (η = 0.01 poise). One hour after the commencement of sedimentation, 25 ml of the suspension was taken by means of a pipette from a depth of 10 cm. The mass of solid particles obtained on drying was 0.09 g. Determine

(*a*) The largest size of the particle remaining in suspension at a depth of 10 cm after one hour of the beginning of sedimentation.

(*b*) The percentage of particles finer than this size in the original suspension.

(*c*) The time interval from the commencement, after which the largest particle remaining in suspension at 10 cm depth is one-half of this size.

(**Hint.** Volume of suspension = 1009.3 ml) [**Ans.** 0.0055 mm; 14.53%; 4 hours]

3.10 The results of a sedimentation test of a sample passing 75μ sieve are given below. Determine the grain size distribution. Use approximate formula $v = 9100\ D^2$.

Observavion No.	*Time*	*Depth*	*Mass of soil in 25 ml sample*
1.	Zero	all depth	25 gm
2.	60 seconds	10 cm	15 gm
3.	5 minutes	10 cm	10 gm
4.	10 minutes	5 cm	5 gm
5.	5 hours	5 cm	0.5 gm

[**Ans.** Percentage finer than 0.075 mm, 0.0428 mm, 0.0191 mm 0.0095 mm and 0.0017 mm, respectively, 100%, 60%, 40%, 20% and 2%].

3.11 In a test 10 gm of fine-grained soil of specific gravity 2.70 was dispersed to make 500 ml of suspension. A sample of volume 10 ml was taken by means of a pipette at a depth of 100 mm, 50 minutes after the commencement of sedimentation. The sample was dried in an oven. If the dry mass of the soil was 0.03 gm, calculate the largest size of the particle remaining in the suspension at a depth of 100 mm and the percentage of particles finer than this size in the original soil. η = 0.01 poise. **[Ans.** 0.006 mm; 15%]

3.12. During a sedimentation test for grain size analysis, the corrected hydrometer reading in a 1000 ml uniform soil suspension at the commencement of sedimentation is 1.028. After 30 minutes, the corrected hydrometer reading is 1.012, and the corresponding effective depth is 10.5 cm. Determine (*i*) the total mass of solids dispersed in 1000 ml of suspension, (*ii*) the particle size corresponding to the 30 minute reading, and (*iii*) the percentage finer than this size. Take G = 2.67 and η = 0.01 poise. **[Ans.** 44.77 gm; 0.00796 mm; 42.86%]

3.13. A dry soil sample is 49 gm in mass. It is composed of the following:

Particle size (mm)	0.05	0.02	0.01	0.001
Mass (gm)	7	20	18	4

The sample is mixed with enough water to make a uniform suspension of 1000 ml. Determine

(*i*) The largest particle size at a depth of 10 cm after 5 minutes of the commencement of sedimentation and the specific gravity of the suspension at that time at that depth.

(*ii*) The time required for all the particles to settle below 10 cm depth. Take G = 2.70 and η = 9.81 millipoise. **[Ans.** 0.01 mm; 1.014; 1.06×10^{-5} seconds]

3.14. An air-dry soil sample weighing 25 kg was sieved in a laboratory. The results are given below.

IS Sieve (mm)	2.0	1.0	0.6	0.425	0.212	0.150	0.075	Pan
Mass retained (kg)	0	2.02	3.51	7.53	8.15	2.81	0.90	0.08

Draw the grain size distribution curve and determine the coefficient of curvature and the uniformity coefficient. **[Ans.** 1.15 ; 2.59]

3.15. A 1000 ml suspension containing 30 gm of dry soil was prepared for a hydrometer analysis. If the temperature is the same as that at which it was calibrated, what whould be the hydrometer reading at the instant of commencement of sedimentation ? Take G = 2.70. **[Ans.** 1.019]

B. Descriptive and Objective Type

3.16. What do you understand about index properties ? What is their importance?

3.17. How would you determine the percentage finer than different sieve sizes in the laboratory ?

3.18. What are the main index properties of a coarse-grained soil? How are these determined?

3.19. Differentiate between the dry sieve analysis and the wet sieve analysis. Why the wet sieve analysis is required?

3.20. State Stokes' law. What is its use in the sedimentation method of analysis? What are its limitations ?

3.21. Compare the pipette method and the hydrometer method. Why the hydrometer method is more popular?

3.22. State the various corrections required for a hydrometer reading. How these corrections are determined?

3.23. What is particle size distribution curve? What is its use in soil engineering?

3.24. What is relative density? How is it determined ? What is its importance for a coarse-grained soil?

3.25. What do you understand by calibration of a hydrometer? How is it done ?

3.26. State whether the following statements are true or false.

(*a*) The silt size particles can be seen by unaided (naked) eye.

(*b*) The sieve analysis gives the largest dimension of the soil particle.

(*c*) The wet sieve analysis gives slightly larger size than that by the dry sieve analysis.

(*d*) The readings on a hydrometer increase in upward direction.

(*e*) The sedimentation analysis is useful for all soil particles smaller than 75μ size.

(*f*) The rock flour even of clay size particles is non- plastic.

(*g*) A gap-graded soil is also called a uniform soil.

(*h*) A well-graded soil contains particles of one size.

[Ans. True (*c*), (*f*)]

C. Multiple Choice Questions

1. In Stokes' law, the terminal velocity of the particle is
   (*a*) Proportional to the radius of the particle.
   (*b*) Proportional to the square of the radius of particle.
   (*c*) Inversely proportional to the square of the radius of particle.
   (*d*) None of the above.
2. Stoke's law does not hold good if the size of particles is
   (*a*) Greater than 0.2 mm (*b*) less than 0.2 μm
   (*c*) Neither (*a*) Nor (*b*) (*d*) Both (*a*) and (*b*)
3. Pretreatement of soil to remove the organic matter by oxidation is done with
   (*a*) Sodium hexametaphosphate (*b*) Oxygen
   (*c*) Hydrogen peroxide (*d*) Hydrochloric acid
4. The particle-size distribution curve with a hump is obtained for a
   (*a*) Uniform soil (*b*) Well-graded soil
   (*c*) Gap-graded soil (*d*) Poorly-graded soil
5. For a well-graded sand, the coefficient of curvature should be
   (*a*) More than 3 (*b*) Between 1 and 3
   (*c*) Less than 1 (*d*) None of above
6. For a dense sand, the relative density is
   (*a*) Between 35 and 65 (*b*) Between 65 and 85
   (*c*) Between 85 and 100 (*d*) Greater than 100
7. A well-graded sand should have
   (*a*) Cu ≥ 4.00 (*b*) Cu ≥ 6.00
   (*c*) Cu ≥ 1.00 (*d*) Cu ≥ 3
8. In hydrometer analysis for a soil mass
   (*a*) Both meniscus correction and dispersing agent correction are negative
   (*b*) Both meniscus correction and dispersing agent correction are positive
   (*c*) Meniscus correction is positive while dispersing agent correction is negative
   (*d*) Meniscus correction is negative while dispersing agent correction is positive
9. For a particle of diameter 0.075 mm, the terminal velocity will be about
   (*a*) 0.05 cm/s (*b*) 0.50 cm/s
   (*c*) 1.0 cm/s (*d*) 1.50 cm/s

[Ans. 1. (*b*), 2. (*d*), 3. (*c*), 4. (*c*), 5. (*b*), 6. (*b*), 7. (*b*), 8. (*c*), 9. (*b*)**]**

4

Plasticity Characteristics of Soils

4.1. PLASTICITY OF SOILS

The plasticity of a soil is its ability to undergo deformation without cractking or fracturing. A plastic soil can be moulded into various shapes when it is wet. Plasticity is an important index property of fine-grained soils, especially clayey soils.

Plasticity in soils is due to presence of clay minerals. The clay particles carry a negative charge on their surfaces, as discussed in chapter 6. The water molecules are dipolar (dipoles) and are attracted towards the clay surface. The phenomenon is known as *adsorption* (not absorption) of water, and the water so attracted to the clay surface is called adsorbed water. Plasticity of the soil is due to adsorbed water.

The clay particles are separated by layers of adsorbed water which allow them to slip over one another. When the soil is subjected to deformations, the particles do not return to their original positions, with the result that the deformations are plastic (irreversible). As the water content of the soil is reduced, the plasticity of the soil is reduced. Ultimately, the soil becomes dry when the particles are cemented together as a solid mass.

The presence of adsorbed water is necessary to impart plasticity characteristics to a soil. The soil does not become plastic when it is mixed with a non-polarizing liquid, such as kerosene or paraffin oil. These liquids do not have electromagnetic properties to react with clay minerals.

The soil becomes plastic only when it has clay minerals. If the soils contains only non-clay minerals, such as quartz, it would not become plastic whatever may be the fineness of soil. When such soils are ground to very fine size, these cannot be rolled into threads. Rock flour, which contains very fine particles of non-clay particles, does not become plastic.

This chapter deals with plasticity characteristics and consistency of fine-grained soils.

4.2. CONSISTENCY LIMITS

The consistency of a fine-grained soil is the physical state in which it exists. It is used to denote the degree of firmness of a soil. Consistency of a soil is indicated by such terms as soft, firm or hard.

In 1911, a Swedish agriculture engineer Atterberg mentioned that a fine-grained soil can exist in four states, namely, liquid, plastic, semi-solid or solid state. The water contents at which the soil changes from one state to the other are known as *consistency limits or Atterberg's limits.*

The water content alone is not an adequate index property of a soil. At the same water content, one soil may be relatively soft, whereas another soil may be hard. However, the soils with the same consistency limits behave somewhat in a similar manner. Thus consistency limits are very important index properties of fine-grained soils

A soil containing high water content is in a liquid state. It offers no shearing resistance and can flow like liquids. It has no resistance to shear deformation and, therefore, the shear strength is equal to zero. As the water content is reduced, the soil becomes stiffer and starts developing resistance to shear deformation. At some particular water content, the soil becomes plastic (Fig. 4.1). The water content at which the soil changes from the liquid state to the plastic state is known as liquid limit (*LL*, w_l). In other words, the liquid limit is the water content at which the soil ceases to be liquid.

The soil in the plastic state can be moulded into various shapes. As the water content is reduced, the plasticity of the soil decreases. Ultimately, the soil passes from the plastic state to the semi-solid state when it stops behaving as a plastic. It cracks when moulded. The water content at which the soil becomes semi-solid is known as the plastic limit (*PL*, w_p). In other words the plastic limit is the water content at which the soil just fails to behave plastically.

The numerical difference between the liquid limit and the plastic limit is known as plasticity index (*PI*, I_p).

Thus $$PI = LL - PL$$

The soil remains plastic when the water content is between the liquid limit and the plastic limit. The plasticity index is an important index property of fine-grained soils.

When the water content is reduced below the plastic limit, the soil attains a semi-solid state. The soil cracks when moulded. In the semi-solid state, the volume of the soil decreases with a decrease in water content till a stage is reached when further reduction of the water content does not cause any reduction in the volume of the soil. The soil is said to have reached a solid state. (In solids, no appreciable change in volume is observed with a change in water content). The water content at which the soil changes from the semi-solid state to the solid state is known as the shrinkage limit (*SL*, w_s).

Below the shrinkage limit, the soil does not remain saturated. Air enters the voids of the soil. However, because of capillary tension developed, the volume of the soil does not change. Thus, the shrinkage limit is the water content at which the soil stops shrinking further and attaines a constant volume. The shrinkage limit may also be defined as the lowest water content at which the soil is fully saturated.

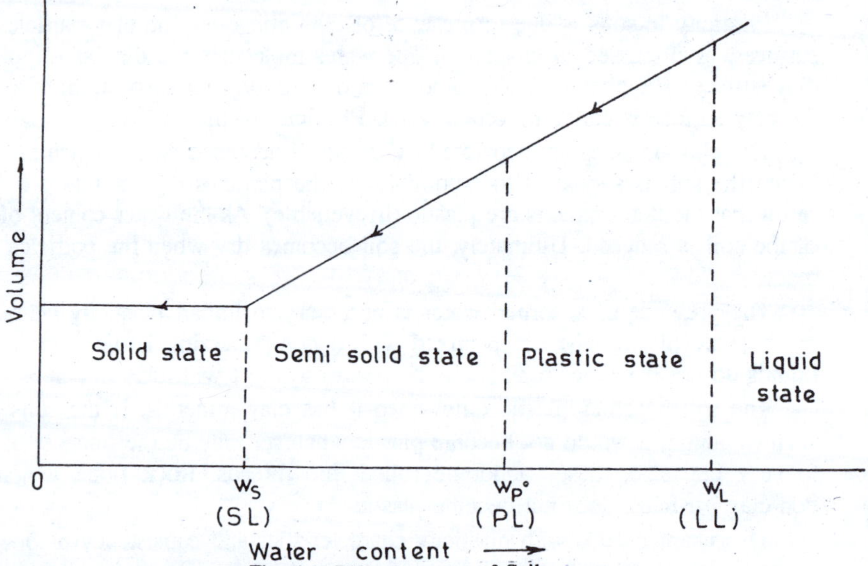

Fig. 4.1. Different states of Soil.

Fig. 4.1 shows sudden changes in the states of the soil at different consistency limits. Actually, the transition between different states is gradual. The consistency limits are determined rather arbitrarily, as explained in the following sections.

[**Note.** In liquid state, the soil is like soup; in plastic state, like soft butter; in semi-solid state, like cheese; and in solid state like hard candy.]

4.3. LIQUID LIMIT

As defined above, the liquid limit is the water content at which the soil changes from the liquid state to the plastic state. At the liquid limit, the clay is practically like a liquid, but possesses a small shearing strength. The shearing strength at that stage is the smallest value that can be measured in the laboratory. The liquid limit of soil depends upon the clay mineral present. The stronger the surface charge and the thinner the particle, the greater will be the amount of adsorbed water and, therefore, the higher will be the liquid limit.

The liquid limit is determined in the laboratory either by Casagrande's apparatus or by cone penetration method. The latter is discussed in Sect. 4.4. The device used in Casagrande's method consists of a brass cup which drops through a height of 1 cm on a hard base when operated by the handle (Fig. 4.2). The device is operated by turning the handle which raises the cup and lets it drop on the rubber base. The height of drop is adjusted with the help of adjusting screws.

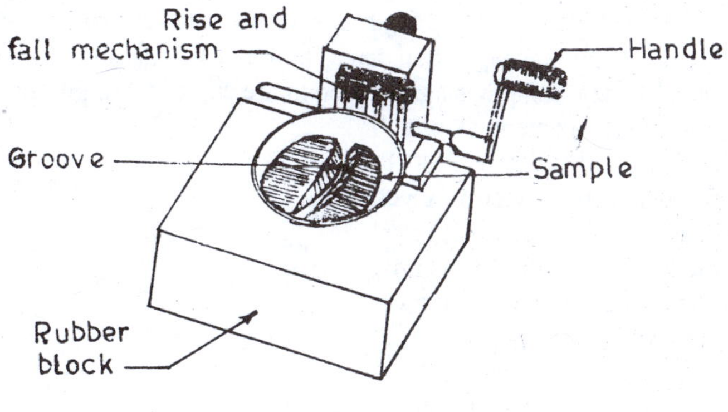

Fig. 4.2. Liquid Limit Apparatus.

About 120 gm of an air-dried sample passing through 425 μ IS sieve is taken in a dish and mixed with distilled water to form a uniform paste. A portion of this paste is placed in the cup of the liquid limit device, and the surface is smoothened and a levelled with a spatula to a maximum depth of 1 cm. A groove is cut through the sample along the symmetrical axis of the cup, preferably in one stroke, using a standard grooving tool. IS : 2720—Part V recommends two types of grooving tools : (1) Casagrande tool, (2) ASTM tool. The Casagrande tool cuts a groove of width 2 mm at the bottom, 11 mm at the top and 8 mm deep. The ASTM tool cuts a groove of width 2 mm at the bottom, 13.6 mm at the top and 10 mm deep (Fig. 4.3). The Casagrande tool is recommended for normal fine-grained soils, whereas the ASTM tool is recommended for sandy, fine grained soils, in which the Casagrande tool tends to tear the soil in the groove.

After the soil pat has been cut by a proper grooving tool, the handle is turned at a rate of 2 revolutions per second until the two parts of the soil sample come into contact at the bottom of the groove along a distance of 12 mm. The groove should close by a flow of the soil, and not by slippage between the soil and the cup. When the groove closes by a flow, it indicates the failure of slopes formed on the two sides of the groove.

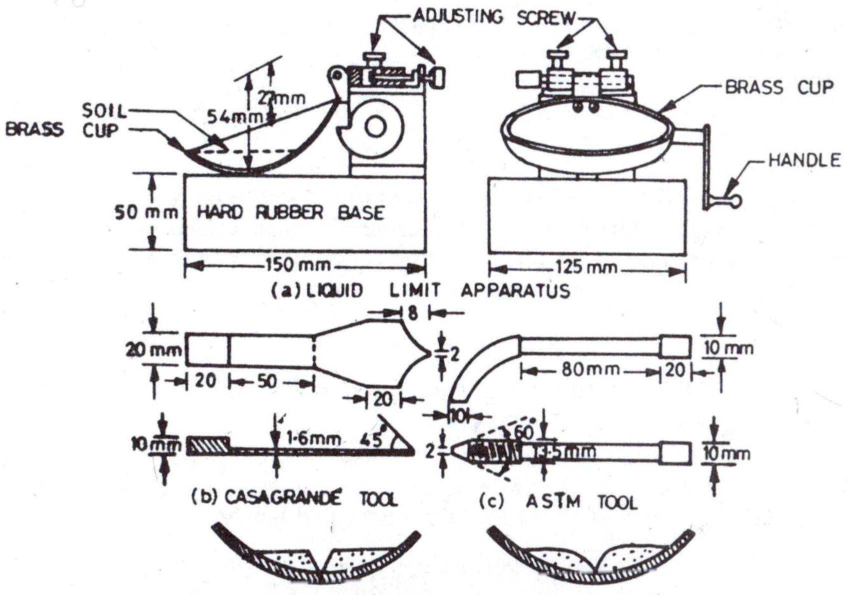

Fig. 4.3. Details of Apparatus and Tools.

The soil in the cup is again mixed, and the test is repeated until two consecutive tests give the same number of blows. About 15 gm of soil near the closed groove is taken for water content determination.

The soil in the cup is transferred to the dish containing the soil paste and mixed thoroughly after adding more water. The soil sample is again taken in the cup of the liquid limit device and the test is repeated. The liquid limit is the water content at which the soil is sufficiently fluid to flow when the device is given 25 blows. As it is difficult to get exactly 25 blows for the sample to flow, the test is conducted at different water contents so as to get blows in the range of 10 to 40. A plot is made between the water content as ordinate and the number of blows on log scale as abscissa. The plot is approximately a straight line. The plot is known as *flow curve.* The liquid limit is obtained, from the plot, corresponding to 25 blows (Fig. 4.4). The liquid limit is expressed as the nearest whole number.

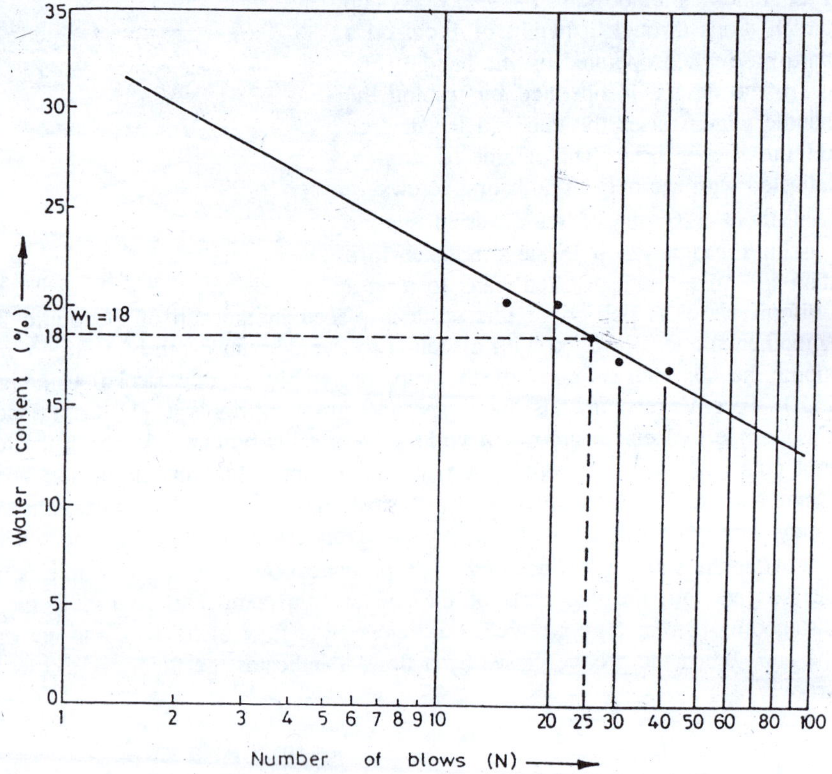

Fig. 4.4. Flow Curve.

The rappings in the liquid limit device cause small shearing stresses on the sample. The liquid limit is arbitrarily taken as the water content when the soil has shear strength just sufficient to withstand the shearing stresses induced in 25 blows. The shear strength of the soil at liquid limit is about 2.7 kN/m^2.

One-point Method

The above procedure for determining the liquid limit requires the test to be repeated at least 4—5 times at different water content and plotting the results. The procedure is inconvenient and time-consuming. It is possible to obtain an approximate value of the liquid limit by conducting only one test, provided the number of blows is in the limited range. The method is based on the premise that the flow curve is a straight line. The liquid limit is given by

$$w_l = w_N (N/25)^n \qquad \text{...[4.1(a)]}$$

where w_N = water content of the soil when the groove closes in N blows.

n = an index, as given below.

According to IS : 2720—V, for soils with liquid limit less than 50%, the value of n is equal to 0.092 and for soils with liquid limit greater than 50%, the value of n = 0.12. The accepted range for N is 15 to 35 for soils with liquid limit less than 50% and 20 to 30 for soils with liquid limit more than 50%.

Alternatively,

$$w_l = \frac{w_N}{1.3215 - 0.23 \log_{10} N} \qquad \text{...[4.1(b)]}$$

Eq. 4.1 (*a*) can be written as

$$w_l = C\, w_N \qquad \text{...(4.2)}$$

where C is the correction factor.

The value of the factor is approximately 0.98 for $N = 20$ and 1.02 for $N = 30$.

(See Chapter 30, Sect. 30.10 for the laboratory experiment)

4.4. CONE PENETROMETER METHOD

The liquid limit of a soil can also be determined by Cone Penetrometer (IS : 2720-V). It consists of a stainless steel cone having an apex angle of 30° ± 1° and a length of 35 mm. The cone is fixed at the lower end of a sliding rod which is fitted with a disc at its top (Fig. 4.5). The total mass of the cone, sliding rod and the disc is 80 g ± 0.05 g.

The soil sample is prepared as in the case of the Casagrande method. The soil pat is placed in a cup of 50 mm internal diameter and 50 mm height. The cup is filled with the sample, taking care so as not to entrap air. Excess soil is removed and the surface of the soil is levelled up.

The cup is placed below the cone, and the cone is gradually lowered so as to just touch the surface of the soil in the cup. The graduated scale is adjusted to zero. The cone is released, and allowed to penetrate the soil for 30 seconds. The water content at which the penetration is 25 mm is the liquid limit. Since it is difficult to obtain the penetration of 25 mm exactly, liquid limit is determined from the equation given below.

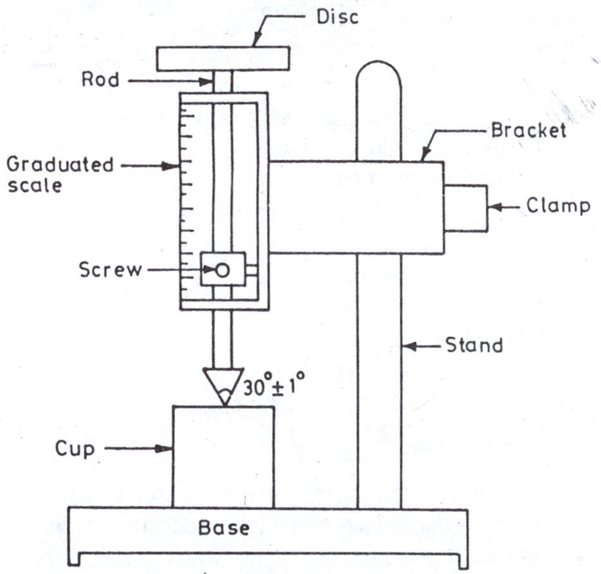

Fig. 4.5. Cone Penetrometer

$$w_l = w_y + 0.01\,(25 - y)\,(w_y + 15) \quad ...(4.3)$$

where y (in mm) is the penetration when the water content is w_y, and w_l = liquid limit.

Eq. 4.3 is applicable provided the depth of penetration y is between 20 to 30 mm. If the penetration is not in this range, the soil in the cup is taken out, and the water content adjusted to get the required penetration.

A chart can also be drawn for direct determination for the liquid limit from the observed value of y and w_y.

The shear strength of soil at liquid limit, as determined by this method, is about 1.76 kN/m^2 which occurs when the penetration is 25 mm.

The cone penetrometer method has several advantages over the Casagrande method.

(1) It is easier to perform.

(2) The method is applicable to a wide range of soils.

(3) The results are reliable, and do not depend upon the judgment of the operator.

4.5. PLASTIC LIMIT

Plastic limit is the water content below which the soil stops behaving as a plastic material. It begins to crumble when rolled into a thread of soil of 3 mm diameter. At this water content, the soil loses its plasticity and passes to a semi-solid state.

For determination of the plastic limit of a soil, it is air-dried and sieved through a 425μ IS sieve. About 30 gm of soil is taken in an evaporating dish. It is mixed thoroughly with distilled water till it becomes plastic and can be easily moulded with fingers.

About 10 gm of the plastic soil mass is taken in one hand and a ball is formed. The ball is rolled with fingers on a glass plate to form a soil thread of uniform diameter (Fig. 4.6). The rate of rolling is kept about 80 to 90 strokes per minute. If the diameter of thread becomes smaller than 3 mm, without crack formation, it shows that the water content is more than the plastic limit. The soil is kneaded further. This results in the reduction of the water content, as some water is evaporated due to the heat of the hand. The soil is re-rolled and the procedure repeated till the thread crumbles. The water content at which the soil can be rolled into a thread of approximately 3 mm in diameter without crumbling is known as the plastic limit (*PL* or w_p).

The test is repeated, taking a fresh sample each time. The plastic limit is taken as the average of three values. The plastic limit is reported to the nearest whole number.

The shear strength at the plastic limit is about 100 times that at the liquid limit.

(See Chapter 30, Sect. 30.11 for the laboratory experiment)

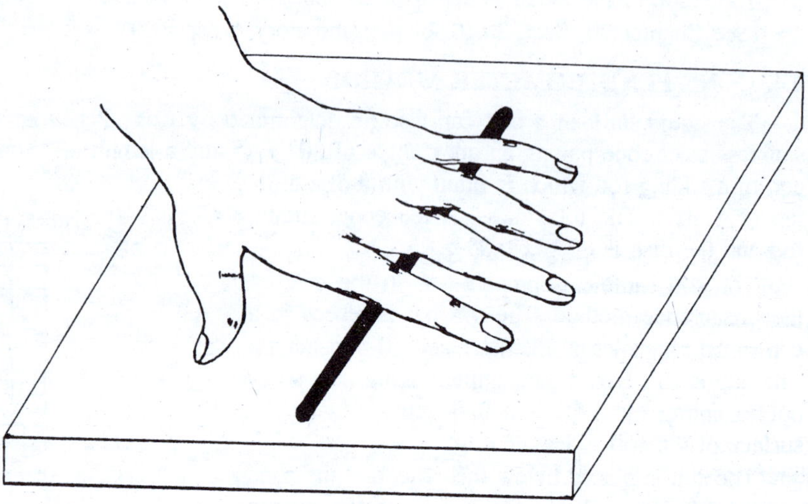

Fig. 4.6. Determination of Plastic limit.

4.6. SHRINKAGE LIMIT

Shrinkage limit is the smallest water content at which the soil is saturated. It is also defined as the maximum water content at which a reduction of water content will not cause a decrease in the volume of the soil mass. In other words, at this water content, the shrinkage ceases. An expression for the shrinkage limit can be obtained as given below

Fig. 4.7 (*a*) shows the block diagram of a soil sample when it is fully saturated and has the water content greater than the expected shrinkage limit. Fig. 4.7 (*b*) shows the sample at shrinkage limit. Fig. 4.7 (*c*) depicts

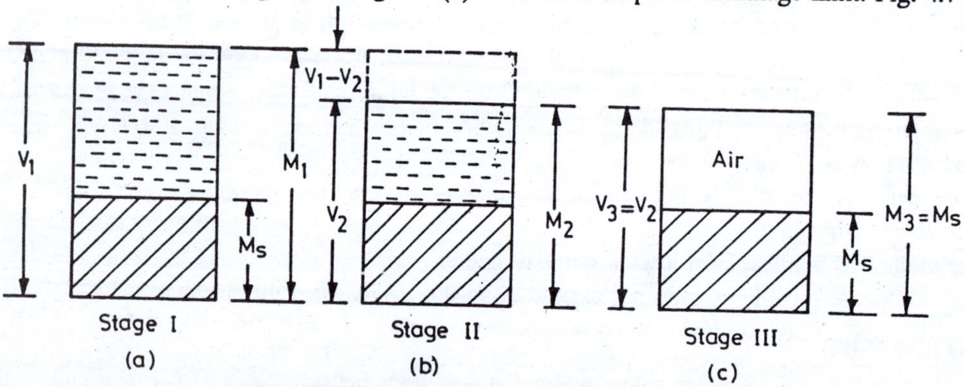

Fig. 4.7. Stages for Derivation of Shrinkage Limit.

the condition when the soil sample has been ovendried. The total volume V_3 in Fig. 4.7 (*c*) is the same as the total volume V_2 in Fig. 4.7 (*b*). The three figures indicate, respectively, stage I, II and III.

Let M_s be the mass of solids.

Mass of water in stage I $= M_1 - M_s$

Loss of mass of water from stage I to stage II $= (V_1 - V_2)\,\rho_w$

Mass of water in stage II $= (M_1 - M_s) - (V_1 - V_2)\,\rho_w$

From definition, shrinkage limit = water content in stage II

or

$$w_s = \frac{(M_1 - M_s) - (V_1 - V_2)\,\rho_w}{M_s} \qquad \text{...(4.5)}$$

or $$w_s = w_1 - \frac{(V_1 - V_2)}{M_s}\,\rho_w \qquad ...(4.6)$$

where w_1 represents the water content in stage I.

For determination of the shrinkage limit in the laboratory, about 50 gm of soil passing a 425 μ sieve is taken and mixed with distilled water to make a creamy paste. The water content (w_1) of the soil is kept greater then the liquid limit.

A circular shrinkage dish, made of porcelain or stainless steel and having a diameter 30 to 40 mm and a height of 15 mm, is taken. The shrinkage dish has a flat bottom and has its internal corners well rounded. The capacity of the shrinkage dish is first determined by filling it with mercury. The shrinkage dish is placed in a large porcelain evaporating dish and filled with mercury. Excess mercury is removed by pressing a plain glass plate firmly over the top of the shrinkage dish. The mass of mercury is the shrinkage dish is obtained by transferring the mercury into a mercury weighing dish. The capacity of the shrinkage dish in ml is equal to the mass of mercury in gm divided by the specific gravity of mercury (usually, taken as 13.6).

The inside surface of the empty shrinkage dish is coated with a thin layer of vaseline or silicon grease. The mass of empty shrinkage dish is obtained accurately. The soil sample is placed in the shrinkage dish, about one-third its capacity. The dish is tapped on a firm surface to ensure that no air is entrapped. More soil is added and the tapping continued till the dish is completely filled with soil. The excess soil is removed by striking off the top surface with a straight edge. The mass of the shrinkage dish with soil is taken to obtain the mass (M_1) of the soil. The volume of the soil V_1 is equal to the capacity of the dish.

The soil in the shrinkage dish is allowed to dry in air until the colour of the soil pat turns light. It is then dried in a oven. The mass of the shrinkage dish with dry soil is taken to obtain the mass of dry soil M_s.

For determination of the volume of the dry pat, a glass cup, about 50 mm diameter and 25 mm height, is taken and placed in a large dish. The cup is filled with mercury. The excess mercury is removed by pressing a glass plate with three prongs firmly over the top of the cup. Any mercury adhering on the side of the cup is wiped off, and the cup full of mercury is transferred to another large dish.

The dry pat of the soil is removed from the shrinkage dish, and placed on the surface of the mercury in the cup and submerged into it by pressing it with the glass plate having prongs (Fig. 4.8). The mercury displaced by the soil pat is transferred to a mercury weighing dish and weighed. The volume of the mercury is determined from its mass and specific gravity. The volume of the dry pat V_d is equal to the volume of the mercury displaced. Of course, the volume V_2 in stage II is also equal to V_d.

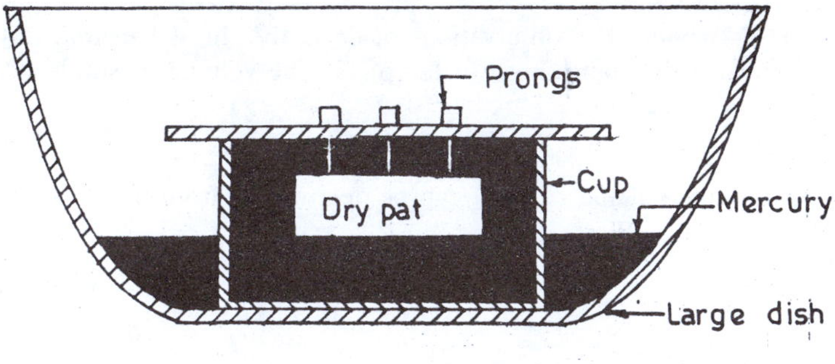

Fig. 4.8. Determination of Volume of dry pat.

The shrinkage limit of the soil is determined, using Eq. 4.5, from the measured values of V_1, V_2, M_1 and M_s.

(See Chapter 30, Sect. 30.12 for the laboratory experiment).

4.7. ALTERNATIVE METHOD FOR DETERMINATION OF SHRINKAGE LIMIT

The shrinkage limit of a soil can be determined by an alternative method if the specific gravity of solid particles (G) is known or is determined separately. An expression for shrinkage limit in terms of the specific gravity of solids can be developed from Fig. 4.7 (*b*). At that stage, the water content is at the shrinkage limit, given by,

$$w_s = \frac{(V_2 - V_s)\,\rho_w}{M_s} \qquad ...(4.7)$$

where V_s is the volume of solids.

Eq. 4.7 can be written as $$w_s = \left[\frac{V_2}{M_s} - \frac{V_s}{G\,\rho_w\,(V_s)}\right]\rho_w$$

or $$w_s = \left[\frac{V_2\,\rho_w}{M_s} - \frac{1}{G}\right] \qquad ...(4.8)$$

Now, from the definition of the dry mass density, $\rho_d = \dfrac{M_s}{V_2}$

Therefore, $$w_s = \left(\frac{\rho_w}{\rho_d} - \frac{1}{G}\right) \qquad ...(4.9)$$

Eq. 4.8 can be used for the determination of the shrinkage limit, as explained below.

A smooth, round-edged pat of wet soil is made in a shrinkage dish. It is then dried in an oven and cooled in a dessicator. Any dust on the sample is brushed off. The dry mass M_s of the sample is determined.

The volume V_2 of the dry soil pat is obtained by placing it in a glass cup and determining the displacement of mercury, as discussed in Sect. 4.6.

Determination of Specific Gravity of Solids from Shrinkage Limit

I. Method—The specific gravity of solids (G) can be determined using Eq. 4.8 if the shrinkage limit has already been determined.

From Eq. 4.8, $$G = \frac{1}{(V_2\,\rho_w/M_s) - w_s} \qquad ...[4.10(a)]$$

Sometimes, Eq. 4.9 is written in terms of mass specific gravity (G_m) in dried state. Taking $G_m = \rho_d/\rho_w$,

$$G = \frac{1}{1/(G_m) - w_s} \qquad ...[4.10(b)]$$

II. Method—The observations made in the shrinkage limit test, as described in Section 4.6, can be used to determine the approximate value of G. The volume of solids (V_s) is stage III (Fig. 4.7(c)]

$$V_s = \frac{M_s}{G\,\rho_w} \qquad ...(a)$$

Also, the volume of solid can be determined from the volume V_1 in Fig. 4.7(a) (stage I) as

$$V_s = V_1 - \text{volume of water}$$

$$V_s = V_1 - \frac{(M_1 - M_s)}{\rho_w} \qquad ...(b)$$

From Eqs. (a) and (b), $$\frac{M_s}{G\,\rho_w} = V_1 - \frac{(M_1 - M_s)}{\rho_w}$$

or $$\frac{M_s}{G} = V_1\,\rho_w - (M_1 - M_s)$$

or $$G = \frac{M_s}{V_1\,\rho_w - (M_1 - M_s)} \qquad ...(4.11)$$

The methods for determination of V_1, M_1 and M_s have already been discussed in Sect. 4.6.

4.8. SHRINKAGE PARAMETERS

The following parameters related with shrinkage limit are frequently used in soil engineering.

(1) Shrinkage Index—The shrinkage index (I_s) is the numerical difference between the liquid limit (w_l) and the shrinkage limit (w_s).

$$I_s = w_l - w_s \qquad ...(4.12)$$

(2) Shrinkage Ratio—The shrikage ratio (SR) is defined as the ratio of a given volume change, expressed as a percentage of dry volume, to the corresponding change in water content.

$$SR = \frac{(V_1 - V_2)/V_d}{w_1 - w_2} \times 100 \qquad \ldots(4.13)$$

where V_1 = volume of soil mass at water content w_1

V_2 = volume of soil mass at water content w_2

V_d = volume of dry soil mass.

When the volume V_2 is at the shrinkage limit,

$$SR = \frac{(V_1 - V_d)/V_d}{w_1 - w_s} \times 100 \qquad \ldots(4.14)$$

Another expression for shrinkage ratio (SR) can be found from Eq. 4.13, by expressing the water content as

$$w_1 - w_2 = \frac{(V_1 - V_2)\,\rho_w}{M_s}$$

Therefore,
$$SR = \frac{M_s}{V_d\,\rho_w}$$

or
$$SR = \frac{\rho_d}{\rho_w} = G_m \qquad \ldots(4.15)$$

Thus the shrikage ratio is equal to the mass gravity of the soil in dry state (G_m).

From Eqs. 4.9 and 4.15, the shrinkage limit,

$$w_s = \left(\frac{1}{\text{S.R.}} - \frac{1}{G}\right) \qquad \ldots[4.15(a)]$$

(3) Volumetric Shrinkage—The volumetric shrinkage (*VS*), or volumetric change, is defined as the change in volume expressed as a percentage of the dry volume when the water content is reduced from a given value to the shrinkage limit. Thus

$$VS = \left(\frac{V_1 - V_d}{V_d}\right) \times 100 \qquad \ldots(4.16)$$

But From Eq. 4.14,

$$[(V_1 - V_d)/V_d] \times 100 = SR\,(w_1 - w_s)$$

Therefore,
$$VS = SR\,(w_1 - w_s) \qquad \ldots(4.17)$$

(4) Linear Shrinkage—Linear shrinkage (*LS*) is defined as the change in length divided by the initial length when the water content is reduced to the shrinkage limit. It is expressed as a percentage, and reported to the nearest whole number.

Thus
$$LS = \left(\frac{\text{Initial length} - \text{final length}}{\text{Initial length}}\right) \times 100 \qquad \ldots(4.18)$$

The linear shrinkage can be determined in a laboratory (IS : 2720-Part XX). A soil sample about 150 gm in mass and passing through a 425μ sieve is taken in a dish. It is mixed with distilled water to form a smooth paste at a water content greater than the liquid limit. The sample is placed in a brass mould, 140 mm long and with a semi-circular section of 25 mm diameter.

The sample is allowed to dry slowly first in air and then in an oven. The sample is cooled and its final length measured. The linear shrinkage is calculated using the following equation.

$$LS = \left[1 - \frac{\text{Length of oven–dry sample}}{\text{Initial length of specimen}}\right] \times 100 \qquad \ldots(4.19)$$

In Eq. 4.19, it has been assumed that the length of the specimen in oven-dried state is the same as that at the shrinkage limit.

The linear shrinkage may also be obtained from the volumetric shrinkage (*VS*) as under.

$$LS = 100\left[1 - \left(\frac{100}{VS + 100}\right)^{1/3}\right] \quad ...(4.20)$$

The linear shrinkage is related to the plasticity index (I_p), as under:

$$LS = 2.13 \times (LS) \quad ...(4.21)$$

4.9. PLASTICITY, LIQUIDITY AND CONSISTENCY INDEXES

(1) Plasticity Index—Plasticity index (I_p or PI) is the range of water content over which the soil remains in the plastic state. It is equal to the difference between the liquid (w_l) and the plastic limit (w_p). Thus,

$$I_p = w_l - w_p \quad ...(4.22)$$

When either w_l or w_p cannot be determined, the soil is non-plastic (*NP*). When the plastic limit is greater than the liquid limit, the plasticity index is reported as zero (and not negative).

(2) Liquidity Index—Liquidity index (I_l or LI) is defined as

$$I_l = \frac{w - w_p}{I_p} \times 100 \quad ...(4.23)$$

where w = water content of the soil in natural condition.

The liquidity index of a soil indicates the nearness of its water content to its liquid limit. When the soil is at its liquid limit, its liquidity index is 100% and it behaves as a liquid. When the soil is at the plastic limit, its liquidity index is zero. Negative values of the liquidity index indicate a water content smaller than the plastic limit. The soil is then in a hard (dessicated) state.

The liquidity index is also known as *Water-Plasticity ratio.*

(3) Consistency Index—Consistency index (I_c, *CI*) is defined as

$$I_c = \frac{w_l - w}{I_p} \times 100 \quad ...(4.24)$$

where w = water contents of the soil in natural condition.

The consistency index indicates the consistency (firmness) of a soil. It shows the nearness of the water content of the soil to its plastic limit. A soil with a consistency index of zero is at the liquid limit. It is extremely soft and has negligible shear strength. On the other hand, a soil at a water content equal to the plastic limit has a consistency index of 100%, indicating that the soil is relatively firm. A consistency index of greater than 100% shows that the soil is relatively strong, as it is the semi-solid state. A negative value of consistency index is also possible, which indicates that the water content is greater than the liquid limit.

The consistency index is also known are *relative consistency.*

It is worth noting that the sum total of the liquidity index and the consistency index is always equal to 100%, indicating that a soil having a high value of liquidity index has a low value of consistency index and vice-versa.

4.10. FLOW INDEX

Flow index (I_f) is the slope of the flow curve obtained between the number of blows and the water content in Casagrande's method of determination of the liquid limit (Fig. 4.4). Thus

$$I_f = \frac{w_1 - w_2}{\log_{10}(N_2/N_1)} \quad ...[4.25(a)]$$

where N_1 = number of blows required at water content of w_1.

and N_2 = number of blow required at water content of w_2.

Eq. 4.25 (*a*) can be written in the general form

$$w = -I_f \log_{10}(N) + C \quad ...[4.25(b)]$$

The flow index can be determined from the flow curve from any two points. For convenience, the number of blows N_3 and N_1 are taken corresponding to one log cycle, *i.e.* $N_3/N_1 = 10$. In that case,

$$I_f = w_1 - w_3 \qquad \ldots[4.25(c)]$$

It may be mentioned that the number of blows actually observed in tests are in a narrow range, normally in the range of 20 to 30 and the ratio of $N_3/N_1 = 10$ can be obtained only after extrapolation of the plot.

The flow index is the rate at which a soil mass loses its shear strength with an increase in water content. Fig. 4.9 shows the flow curves of two soils (1) and (2). The soil—(2) with a greater value of flow index has a steeper slope and possesses lower shear strength as compared to soil (1)—with a flatter slope. In order to decrease the water content by the same amount, the soil with a steeper slope takes a smaller number of blows, and, therefore, has lower shear strength.

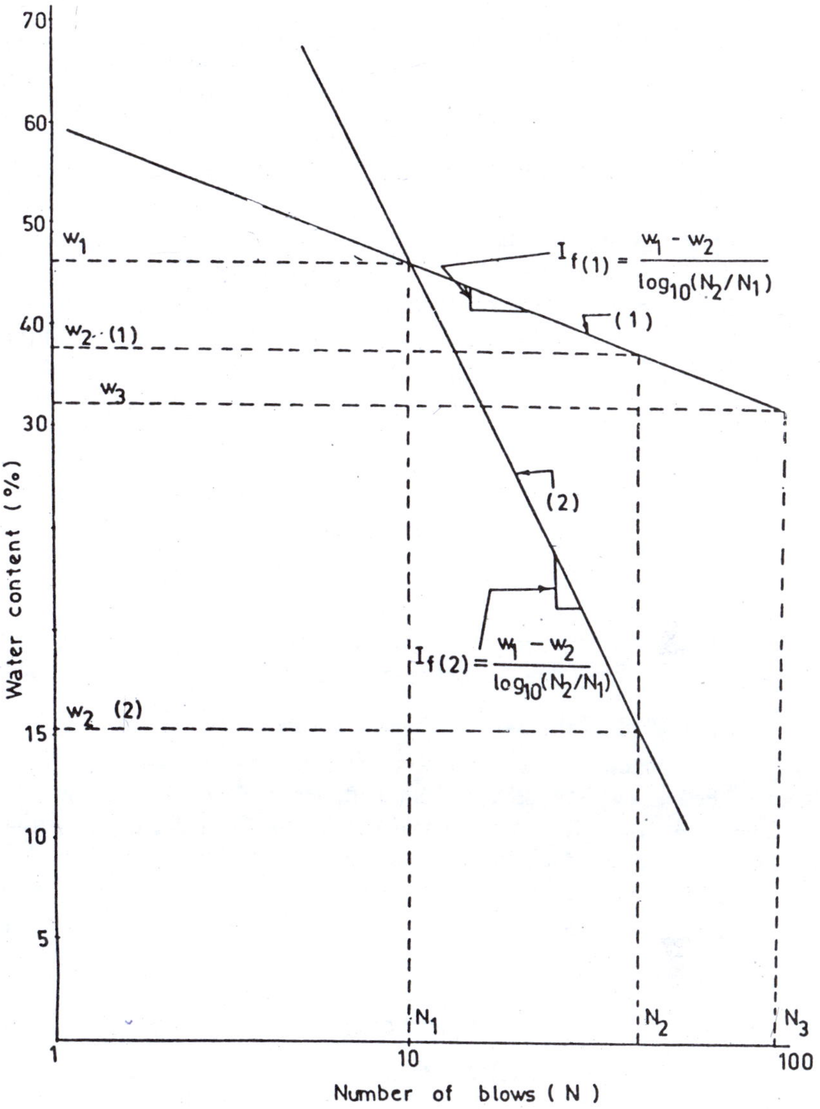

Fig. 4.9. Flow Indexes

4.11. TOUGHNESS INDEX

Toughness index (I_t) of a soil is defined as the ratio of the plasticity index (I_p) to the flow index (I_f.)

Thus
$$I_t = \frac{I_p}{I_f} \qquad \ldots(4.26)$$

Toughness index of a soil is a measure of the shearing strength of the soil at the plastic limit. This can be proved as under:

Let us assume that the flow curve is a straight line between the liquid limit and the plastic limit. As the shearing resistance of the soil is directly proportional to the number of blows in Casagrande's device,

$$k\ S_l = N_l \qquad \ldots(a)$$

and
$$k\ S_p = N_p \qquad \ldots(b)$$

where N_l = number of blows at the liquid limit when the shear strength is S_l

N_p = number of blows at the plastic limit when the shear strength is S_p

k = constant.

From Eq. 4.25 (*a*), taking $w_2 = w_l$ and $N_1 = 1.0$,

$$I_f = \frac{w_1 - w_l}{\log_{10}(N_l/1)} = \frac{w_1 - w_l}{\log_{10} N_l}$$

$$w_l = w_1 - I_f \log_{10} N_l \qquad ...(c)$$

Likewise, for the plastic limit.

$$w_p = w_1 - I_f \log_{10} N_p \qquad ...(d)$$

From Eqs. (*c*) and (*d*),

$$w_l - w_p = - I_f \log_{10}(N_l/N_p)$$

Substituting the value of (N_l/N_p) from Eqs. (*a*) and (*b*),

$$w_l - w_p = -I_f \log_{10}(S_l/S_p) = I_f \log(S_p/S_l) \qquad ...(e)$$

Since the shearing strength of all soils at the liquid limit is almost constant and equal to 2.7 kN/m², Eq. (*e*) can be written as, taking $w_l - w_p = I_p$,

$$I_p = I_f \log(S_p/2.7)$$

or

$$I_t = \log_{10}(S_p) - \log_{10}^{(2.7)}$$

or

$$\log_{10}(S_p) = I_t + C \qquad ...(4.27)$$

where C is constant equal to $\log_{10}^{(2.7)}$ (= 0.431).

Eq. (4.27) proves that the shear strength at plastic limit depends upon the toughness index.

The value of the toughness index of most soils lies between 0 to 3.0. A value of toughness index less than unity indicates that the soil is friable at the plastic limit.

4.12. MEASUREMENT OF CONSISTENCY

Consistency of a soil, as defined earlier, is its resistance to deformation. Consistency is conventionally described as very soft, soft, medium, stiff, very stiff and hard. These terms are relative and may have different interpretation to different geotechnical engineers. For quantitative measurement of consistency, it is related to the shear strength or compressive strength.

The unconfined compressive strength (q_u) of a soil is equal to the failure load per unit area when a standard, cylindrical specimen is tested in an unconfined compression testing machine (chapter 13). As the unconfined compressive strength is twice the shear strength (s), it can be obtained from the vane shear test also.

Table 4.1 gives the unconfined compressive strength of soils of different consistency.

Table 4.1. Consistency in Terms of Consistency Index and Unconfined Compressive Strength (q_u)

S. NO.	*Consistency*	*Consistency index (%)*	*Unconfined compressive strength (q_u) (kN/m²)*	*Characteristics of soil*
1.	Very soft	0—25	< 25 kN/m²	Fist can be pressed into soil
2.	Soft	25—50	25—50	Thumb can be pressed into soil
3.	Medium (Firm)	50—75	50—100	Thumb can be pressed with pressure
4..	Stiff	75—100	100—200	Thumb can be pressed with great difficulty
5.	Very stiff	> 100	200—400	The soil can be readily indented with thumb nail
6.	Hard	> 100	> 400	The soil can be indented with difficulty by thumb nail

4.13. SENSITIVITY

A cohesive soil in its natural state of occurrence has a certain structure (see chapter 6). When the structure is disturbed, the soil becomes remoulded, and its engineering properties change considerably. Sensitivity (S_t) of a soil indicates its weakening due to remoulding. It is defined as the ratio of the undisturbed strength to the remoulded strength at the same water content.

$$S_t = \frac{(q_u)_u}{(q_u)_r} \qquad \text{...(4.28)}$$

where $(q_u)_u$ = unconfined compressive strength of undisturbed clay

$(q_u)_r$ = unconfined compressive strength of remoulded clay.

Depending upon sensitivity, the soils can be classified into six types, as given in Table 4.2.

Table 4.2. Classification of Soils based on Sensitivity

S.No.	*Sensitivity*	*Soil Type*
1.	< 1.00	Insensitive
2.	1.0—2.0	Little sensitive
3.	2.0—4.00	Moderately sensitive
4.	4.0—8.00	Sensitive
5.	8.0—16.0	Extra sensitive
6.	> 16.0	Quick

For most clays, sensitivity lies between 2 and 4. Clays considered sensitive have S_t values between 4 and 8. In case of sensitive clays, remoulding causes a large reduction in strength. Quick clays are unstable. These turn into slurry when remoulded.

High sensitivity in clays is due to a well-developed flocculent structure which is disturbed when the soil is remoulded. High sensitivity may also be due to leaching of soft glacial clays deposited in salt water and subsequently uplifted.

Extra-sensitive clay, such as clays of Mexico city, are generally derived from the decomposition of volcanic ash.

4.13. THIXOTROPY

The word *Thixotropy* is derived from two words : *thixis* meaning touch, and *tropo*, meaning to change. Therefore, thixotropy means any change that occurs by touch.

The loss of strength of a soil due to remoulding is partly due to change in the soil structure and partly due to disturbance caused to water molecules in the adsorbed layer. Some of these changes are reversible. If a remoulded soil is allowed to stand, without loss of water, it may regain some of its lost strength. In soil engineering, this gain in strength of the soil with passage of time after it has been remoulded is called *thixotropy*. It is mainly due to a gradual reorientation of molecules of water in the adsorbed water layer and due to re-establishment of chemical equilibrium.

Thixotropy of soils is of great practical importance in soil engineering. For example, when a pile is driven into ground, the loss of strength occurs due to disturbance caused. Thixotropy indicates how much shear strength will be regained after the pile has been driven and left in place for some time.

4.14. ACTIVITY OF SOILS

Activity (A) of a soil is the ratio of the plasticity index and the percentage of clay fraction (minus 2μ size). Thus

$$A = \frac{I_p}{F} \qquad \text{...(4.29)}$$

where I_p = plasticity index, F = clay fraction.

The clay fraction F is percentage finer than 2μ size.

The amount of water is a soil mass depends upon the type of clay mineral present. Activity is a measure of the water-holding capacity of clayey soils. The changes in the volume of a clayey soil during swelling or shrinkage depend upon the activity.

A number of samples of a particular soil are taken and their plasticity index and clay fraction determined. If a plot is obtained between the clay fraction (as abscissa) and the plasticity index (as ordinate), it is observed that all the points for a particular soil lie on a straight line (Fig. 4.10).

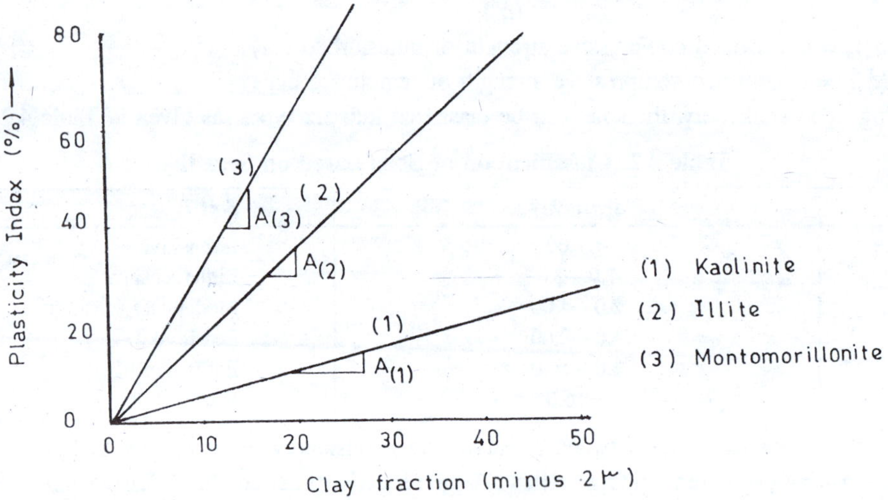

Fig. 4.10. Activity of Soils.

The slope of the line gives the activity of soil. The steeper the slope, the greater is the activity. The lines with different slopes are obtained for different soils.

The soils containing the clay mineral montmorillonite have very high activity ($A > 4$). The soil containing the mineral kaolinite are least active ($A < 1$), whereas the soils containing the mineral illite are moderately active ($A = 1$ to 2). Depending upon activity, the soils are classified into three types (Table 4.3).

Table 4.3 Classification of Soils Based on Activity

S. No.	*Activity*	*Soil type*
1.	$A < 0.75$	Inactive
2.	$A = 0.75$ to 1.25	Normal
3.	$A > 1.25$	Active

Activity gives information about the type and effect of clay mineral in a soil. The following two points are worth noting:

(1) For a soil of specific origin, the activity is constant. The plasticity index increases as the amount of clay fraction increases.

(2) Highly active minerals, such as montmorillonite, can produce a large increase in the plasticity index even when present in small quantity.

4.15. USES OF CONSISTENCY LIMITS

The consistency limits are determined for remoulded soils. However the shrinkage limit can also be obtained for the undisturbed sample. Since the actual behavior of a soil depends upon its natural structure, the consistency limits do not give complete information about the in-situ soils. They give at best a rough estimate about the behaviour of in-situ soils.

Although it is not possible to interprete the consistency limits and other plasticity characteristics in fundamental terms, yet these parameters are of great practical use as index properties of fine-grained soils. The engineering properties of such soils can be empirically related to these index properties as under.

(1) It has been found that both the liquid and plastic limits depend upon the type and amount of clay in

a soil. However, the plasticity index depends mainly on the amount of clay. The plasticity index of a soil is a measure of the amount of clay in soil.

(2) As the particle size decreases, both the liquid and plastic limits increase, but the former increases at a greater rate. Therefore, the plasticity index increases at a rapid rate (Fig. 4.11).

When silt is added to clay, it becomes leaner. Its liquid limit and plastic limit both decrease, but the former at a faster rate. The net effect is that the plasticity index decreases.

Plasticity index is, therefore a measure of the fineness of the particles.

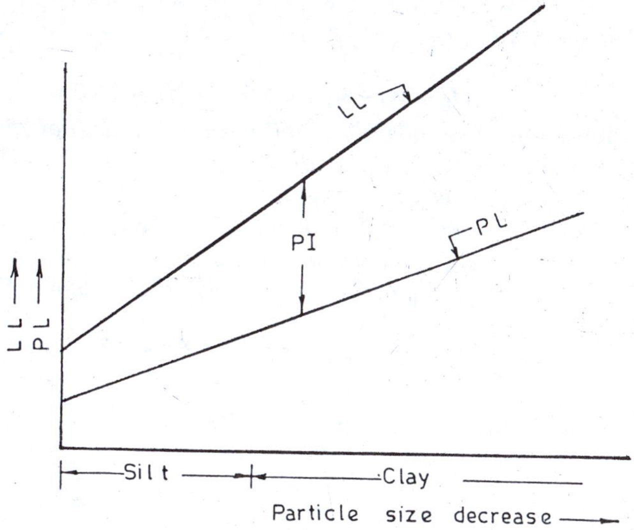

Fig. 4.11. Variation of LL and PL with particle size

(3) The study of plasticity index, in combination with liquid limit, gives information about the type of clay. Plasticity chart, which is a plot between the plasticity index and liquid limit, is extremely useful for classification of fine-grained soils, as discussed in chapter 5. In fact, the main use of consistency limits is in classification of soils.

(4) Sandy soils change from the liquid state to the semi-solid rather abruptly. These soils do not possess plasticity and are classified as non-plastic (*NP*).

Soils with the liquid limit less than 20% are generally sands.

(5) The plastic limit of a soil increases if organic matter is added, without any significant increase in the liquid limit. Therefore, soils with high organic content have low plasticity index.

(6) The liquid limit of a soil is an indicator of the compressibility of a soil. The compressibility of a soil generally increases with an increase in liquid limit (see chapter 12).

(7) The shrinkage index is directly proportional to the percentage of clay-size fraction present in the soil. It can be used as an indicator for the amount of clay.

(8) The toughness index is a measure of the shearing strength of the soil at the plastic limit.

A high value of toughness index indicates a high percentage of colloidal clay containing mineral montmorillonite.

(9) When comparing the properties of two soils with equal values of plasticy index, it is found that as the liquid limit increases, the dry strength and toughness decrease, whereas compressibility and permeability increase.

(10) When comparing the properties of two soils with equal liquid limits, it is found that as the plasticity index increases, the dry strength and toughness increase, whereas the permeability decreases. However, the compressibility remains almost the same.

ILLUSTRATIVE EXAMPLES

Illustrative Example 4.1. *A test for the determination for the liquid limit was carried on a soil sample. The following sets of observations were taken. Plot the flow curve and determine the liquid limit and the flow index.*

No. of Blows (N)	38	27	20	13
Wateer content (w) %	47.5	49.5	51.9	53.9

Solution. Fig. E-4.1 shows the plot.

For $N = 25$, $w_l = 49\%$.

From Eq. 4.25 (*a*), for $\frac{N_2}{N_1} = \frac{100}{10} = 10.$

$$I_f = w_1 - w_3 = 55.0 - 42.0 = \mathbf{13.0\%}$$

Illustrative Example 4.2. *A soil has a liquid limit of 25% and a flow index of 12.5%. If the plastic limit is 15%, determine the plasticity index and the toughness index.*

If the water content of the soil in its natural condition in the field is 20%, find the liquidity index and the relative consistency.

Solution. From Eq. 4.22, $I_p = w_l - w_p = 25 - 15 = 10\%$

From Eq. 4.26, $I_t = \frac{I_p}{I_f} = \frac{10}{12.5} = 0.8\ (\mathbf{80\%})$

From E. 4.23, $I_l = \frac{w - w_p}{I_p} \times 100$

$$= \frac{0.20 - 0.15}{0.10} \times 100 = \mathbf{50\%}$$

From Eq. 4.24, $I_c = \frac{w_l - w}{I_p} \times 100$

$$= \frac{0.25 - 0.20}{0.1} \times 100 = \mathbf{50\%}$$

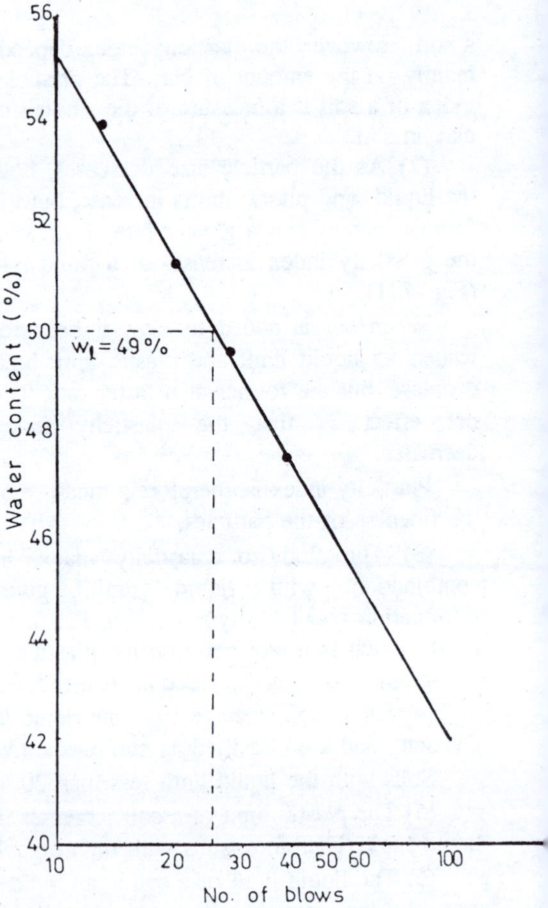

Fig. E-4.1.

Illustrative Example 4.3. *A cone penetrometer test was conducted on a sample of soil for the determination of the liquid limit, and the following observations were recorded.*

Cone penetration (mm)	19.1	20.5	22.2	25.2	26.3
Water content (w)%	51.5	53.2	55.2	58.1	59.5

Determine the liquid limit.

Solution. Fig. E-4.3 shows the plot between the cone penetration and the water content. From the plot, the water content corresponding to cone penetration of 25 mm is 58%.

Thus $w_l = 58\%$.

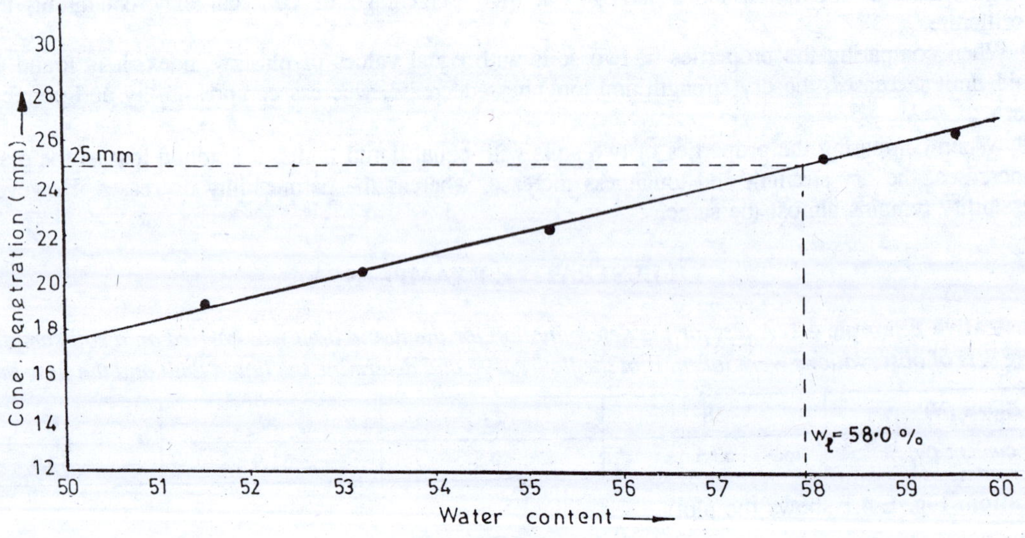

Fig. E-4.3.

Illustrative Example 4.4. *A sample of clay has the liquid limit and the shrinkage limit of, respectively, 60% and 25%. If the sample has a volume of 10 ml at the liquid limit, and a volume of 6.40 ml at the shrinkage limit, determine the specific gravity of solids.*

Solution. Let M_s be the mass of solids, in gm.

Therefore, mass of water at the liquid limit = $0.6\ M_s$

and mass of water at the shrinkage limit = $0.25\ M_s$

Mass of water lost between the liquid limit and the shrinkage limit = $(0.6 - 0.25)\ M_s = 0.35\ M_s$

Reduction in volume = $0.35\ M_s$ ml

But actual reduction in volume = 10.0 − 6.40 = 3.60 ml

Therefore, $0.35\ M_s = 3.60$ or $M_s = 10.29$ gm

Thus, the mass of water at the shrinkage limit

$$= 0.25 \times 10.29 = 2.57 \text{ gm}$$

Volume of water at the shrinkage limit = 2.57 ml

Volume of solid particles, $V_s = 6.40 - 2.57 = 3.83$ m

Therefore, specific gravity of solids, $G = \dfrac{M_s}{\rho_w V_s} = \dfrac{10.29}{3.83} = \mathbf{2.69}$

Alternatively, directly from Eq. 4.10 (*a*),

$$G = \frac{1}{(V_2\,\rho_w/M_s) - w_s} = \frac{1}{(6.40 \times 1.0/10.29) - 0.25} = 2.69$$

Illustrative Example 4.5. *In an experiment for the determination of the shrinkage limit, the following observations were taken.*

(*a*) *Volume of saturated soil* = *9.75 ml*
(*b*) *Mass of saturated soil* = *16.5 gm*
(*c*) *Volume of dry soil after shrinkage* = *5.40 ml*
(*d*) *Mass of dry soil after shrinkage* = *10.9 gm*

Compute the shrinkage limit and the specific gravity of solids.

Solution. Given values are $V_1 = 9.75$ ml, $V_2 = 5.40$ ml, $M_1 = 16.5$ gm and $M_s = 10.9$ gm.

From Eq. 4.5,
$$w_s = \frac{(M_1 - M_s) - (V_1 - V_2)\,\rho_w}{M_s}$$

Therefore,
$$w_s = \frac{(16.5 - 10.9) - (9.75 - 5.40) \times 1.0}{10.9}$$
$$= \frac{5.6 - 4.35}{10.9} = 0.1147\ (\mathbf{11.47\%})$$

From Eq. 4.11,
$$G = \frac{M_s}{V_1\,\rho_w - (M_1 - M_s)}$$
$$= \frac{10.9}{9.75 \times 1.0 - (16.5 - 10.9)} = \mathbf{2.63}$$

Illustrative Example 4.6. *A soil has liquid limit and plastic limit of 47% and 33%, respectively. If the volumetric shrinkages at the liquid limit and plastic limit are 44% and 29%, determine the shrinkage limit.*

Solution. From Eq. 4.16,
$$VS = \frac{V_1 - V_d}{V_d} \times 100$$

At liquid limit,
$$VS = \frac{V_l - V_d}{V_d} \times 100 = 44$$

or
$$\frac{V_l}{V_d} = 0.44 + 1.0 = 1.44$$

or $$V_d = 0.694\ V_l \quad \text{...(a)}$$

At plastic limit, $$VS = \frac{V_p - V_d}{V_d} \times 100 = 29$$

or $$\frac{V_p}{V_d} = 0.29 + 1.0 = 1.29$$

or $$V_p = 1.29\ V_d \quad \text{...(b)}$$

Let the volume at liquid limit, V_l, be 1.0 ml.

From Eq. (*a*), V_d = volume at shrinkage limit = 0.694 ml

From Eq. (*b*), V_p = volume at plastic limit = 0.895 ml

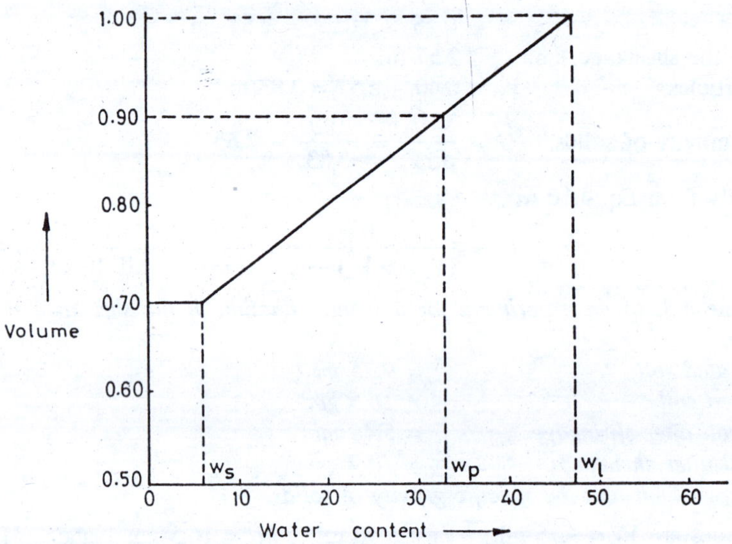

Fig. E-4.6

From Fig. E 4.6 by proportion,

$$\frac{w_l - w_s}{1.0 - 0.694} = \frac{w_p - w_s}{0.895 - 0.694}$$

or $$\frac{0.47 - w_s}{0.306} = \frac{0.33 - w_s}{0.201}$$

or $$w_s = \mathbf{0.06\ (6.0\%)}$$

Illustrative Example 4.7. *The following index properties were determined for two soils A and B.*

Index property	*A*	*B*
Liquid limit	65	35
Plastic limit	25	20
Water content	35	25
Sp. gr. of solids	2.70	2.65
Degree of saturation	100%	100%

Which of he two soils (i) contains more clay particles, (ii) has a greater bulk density. (iii)) has a greater dry density, (iv) has a greater void ratio ?

Solution. (see Table below)

S. No.		*SOIL A*	*SOIL B*
1.	Plasticity index $PI = w_l - w_p$	65—25 = 40%	35—20 = 15%
2.	Void ratio $e = w\,G$	0.35 × 2.7 = **0.945**	0.25 × 2.65 = 0.663
3.	Dry density $\rho_d = \dfrac{G\,\rho_w}{1+e}$	$\dfrac{2.7 \times 1.0}{1.945}$ = **1.388 g/ml**	$\dfrac{2.65 \times 1.0}{1.663}$ = 1.594 g/ml
4.	Bulk density $\rho = \rho_d(1+w)$	1.388 × 1.35 = 1.874 g/ml	1.594 × 1.25 = **1.992 g/ml**

As the plasticity index of soil *A* is more than that of soil *B*, it has more clay particles.

PROBLEMS

A. Numerical

4.1. The consistency limits of a soil sample are:
Liquid limit = 52%
Plastic limit = 32%
Shrinkage limit = 17%
If the specimen of this soil shrinks from a volume of 10 cm^3 at liquid limit to 6.01 cm^3 at the shrinkage limit, calculate the specific gravity of solids. **[Ans.** 2.80]

4.2. A cone penetration test was carried out on a sample of soil with the following results:

Cone penetration (mm)	16.1	17.6	19.3	21.3	22.6
Moisture content (%)	50	52.1	54.1	57.0	58.2

Determine the liquid limit of the soil. **[Ans.** 60%]

4.3. In a shrinkage limit test, a dish with volume of 10.5 ml was filled with saturated clay. The mass of the saturated clay was 18.75 gm. The clay was dried gradually first in atmosphere and then in an oven. The mass of the dry clay was 12.15 gm and its volume 5.95 ml. Determine the shrinkage limit. **[Ans.** 16.9%]

4.4. A sample of clay has a void ratio of 0.70 in the undisturbed state and of 0.50 in a remoulded state. If the specific gravity of solids is 2.65, determine the shrinkage limit in each case. **[Ans.** 26.4%, 18.9%]

4.5. A fully saturated clay has a water content of 40% and a mass specific gravity of 1.85. After oven-drying, the mass specific gravity reduces to 1.75. Determine the specific gravity of solids and the shrinkage limit. **[Ans.** 2.80, 21.4%]

4.6. The Atterberg limits of a clay are : *LL* = 60%, *PL* = 45%, and *SL* = 25%. The specific gravity of soil solids is 2.70 and the natural moisture content is 50%.

(*i*) What is its state of consistency in nature ?

(*ii*) Calculate the volume to be expected in the sample when moisture content is reduced by evaporation to 20%. Its volume at liquid limit is 10 cm^3. **[Ans.** consistency index = 66.7%, 6.40 cm^3]

B. Descriptive and Objective Type

4.7. Discuss the importance of Atterberg's limits in soil engineering.

4.8. What are the main index properties of a fine-grained soil ? How are these determined in a laboratory ?

4.9. What do you understand by consistency of soil ? How is it determined ?

4.10. What are the different methods for determination of the liquid limit of a soil? What are their relative merits and demerits ?

4.11. Describe the method for determination of shrinkage limit of a soil.

4.12. Discuss various shrinkage parameters. How would you determine linear shrinkage ?

4.13. What are uses of consistency limits? What are their limitations ?

4.14. Differentiate between:

(*a*) Liquidity index and consistency index.

(*b*) Flow index and toughness index.

(*c*) Plasticity and consistency.

(*d*) Activity and sensitivity.

4.15. State whether the following statements are true of false.

(*a*) All the consistency limits are determined for the soil in disturbed condition.

(*b*) The liquidity index cannot be more than 100%.

(*c*) The consistency index can be negative.

(*d*) Plastic limit is the water content of soil which represents the boundary between the plastic state and the semi-solid state.

(*e*) At shrinkage limit, the soil is fully saturated.

(*f*) The activity of a clay mineral is a constant.

(*g*) The soils with soft consistency have high strength.

(*h*) The soils with a dispersed structure have a high sensitivity. **[Ans.** True, (*c*), (*d*), (*e*), (*f*)]

C. Multiple-Choice Questions

1. At shrinkage limit, the soil is

(*a*) Dry (*b*) Partially saturated

(*c*) Saturated (*d*) None of above

2. The shrinkage index is equal to

(*a*) Liquid limit minus plastic limit.

(*b*) Liquid limit minus shrinkage limit.

(*c*) Plastic limit minus shrinkage limit.

(*d*) None of above.

3. Toughness index of a soil is the ratio of

(*a*) Plasticity index to the flow index.

(*b*) Liquidity index to the flow index.

(*c*) Consistency index to the flow index.

(*d*) Shrinkage index to the flow index.

4. A stiff clay has a consistency index of

(*a*) 50—75 (*b*) 75—100

(*c*) Greater than 100 (*d*) Less than 50

5. The plasticity index of a highly plastic soil is about

(*a*) 10—20 (*b*) 20—40

(*c*) Grater than 40 (*d*) Less than 10

6. The activity of the mineral montmorillonite is

(*a*) Less than 0.75 (*b*) Between 0.75 and 1.25

(*c*) Between 1.25 and 4 (*d*) Greater than 4

7. A soil sample has L_L = 45%, P_L = 25% and S_L = 15%.

For a natural water content of 30%, the consistency index will be

(*a*) 75% (*b*) 50%

(*c*) 40% (*d*) 25%

8. For the soil with L_L = 45%, P_L = 25% and S_L = 15%, the plasticity index is

(*a*) 50% (*b*) 20%

(*c*) 60% (*d*) 40%

[Ans. 1. (*c*), 2. (*c*), 3. (*a*), 4. (*b*), 5. (*b*), 6. (*d*), 7. (*a*), 8. (*b*)]

5

Soil Classification

5.1. INTRODUCTION

Soil classification is the arrangement of soils into different groups such that the soils in a particular group have similar behaviour. It is a sort of labelling of soils with different labels. As there is a wide variety of soils covering earth, it is desirable to systematize or classify the soils into broad groups of similar behaviour. It is more convenient to study the behaviour of groups than that of individual soils. Classification of various commodities and species is also common in many other disciplines. For example, a chemist classifies the chemicals into various groups, and a zoologist classifies the species into a number of groups. Likewise, a geotechnical engineer classifies the soils into various groups.

For a soil classification system to be useful to the geotechnical engineers, it should have the following basic requirements:

(1) It should have a limited number of groups.

(2) It should be based on the engineering properties which are most relevant for the purpose for which the classification has been made.

(3) It should be simple and should use the terms which are easily understood.

Most of the classification systems developed satisfy the above requirements.

A geotechnical engineer is interested to know the suitability or otherwise of a soil as a foundation or a construction material. For complete knowledge, all the engineering properties are determined after conducting a large number of tests. However, an approximate assessment of the engineering properties can be obtained from the index properties after conducting only classification tests, as discussed in chapters 3 and 4. A soil is classified according to index properties, such as particle size and plasticity characteristics.

If the classification of a soil has been done according to some standard classification system, its properties and behaviour can be estimated based on the experience gained from similar soils elsewhere. A classification system thus provides a common language between engineers dealing with soils. It is useful in exchange of information and experience between the geotechnical engineers. For example, if a soil has been classified as *SW* according to Unified Soil Classification system, the geotechnical engineer anywhere in the world would know that the soil is well graded sand, is quite pervious, has low compressibility and high shear strength. All this information is exchanged only in two letters *SW*.

It may be mentioned that soil classification is no substitute for exact analysis based on engineering properties. For final design of large structures, the engineering properties should be determined by conducting elaborate tests on undisturbed samples.

[**Note.** The soil classification system can be likened to classification of human beings into 12 zodiac signs done by an astrologer. Although general behaviour of a human being under a particular zodiac sign can be estimated from his zodiac sign, for complete prediction, his detailed horoscope is required].

5.2. PARTICLE SIZE CLASSIFICATION

The size of individual particles has an important influence on the behaviour of soils. It is not surprising that the first classification of soils was based on the particle size. It is a general practice to classify the soils into four broad groups, namely, gravel, sand, silt size and clay size. While classifying the fine grained soils on the basis of particle size, it is a good practice to write silt size and clay size and not just silt and clay. In general usage, the terms silt and clay are used to denote the soils that exhibit plasticity and cohesion over a wide range of water content. The soil with clay-size particles may not exhibit the properties associated with clays. For example, rockflour has the particles of the size of the clay particles but does not possess plasticity. It is classified as clay-size and not just clay in the particle size classification systems.

Any system of classification based only on particle size may be misleading for fine-grained soils. The behaviour of such soils depends on the plasticity characteristics and not on the particle size. However, classification based on particle size is of immense value in the case of coarse-grained soils, since the behaviour of such soils depends mainly on the particle size.

Some of the classification system based on particle size *alone* are discussed below.

(1) MIT System—MIT system of classification of soils was developed by Prof. G. Gilboy at Massachusettes Institute of Technology in USA. In this system, the soil is divided into four groups (Fig. 5.1 *a*).

(*i*) Gravel, particle size greater than 2 mm.

(*ii*) Sand, particle size between 0.06 mm to 2.0 mm.

(*iii*) Silt size, particle size between 0.002 mm to 0.06 mm.

(*iv*) Clay size, particle size smaller than 0.002 mm (2μ).

Boundaries between different types of soils correspond to limits when important changes occur in the soil properties. The particles less than 2μ size are generally colloidal fraction and behave as clay. The soils with particle size smaller than 2μ are classified as clay size.

The naked eye can see the the particle size of about 0.06 mm and larger. The soils with particle size smaller than 0.06 mm but larger than 2μ are classified as silt-size. Important changes in the behaviour of soil occur if particle size is larger than 0.06 mm when it behaves as cohesionless soil.

The boundary between gravel and sand is abritrarily kept as 2 mm. This is about the size of lead in the pencil.

The soils in sand and silt-size-range are further subdivided into three categories : coarse (*C*), medium (*M*) and fine (*F*), as shown in the figure. It may be noted that MIT system uses only two integers 2 and 6, and is easy to remember.

(2) International Classification System—The International Classification System was proposed for general use at the International Soils Congress held as Washington in 1927. This classification system was

Clay (size)	Silt (size)			Sand			Gravel
	F	M	C	F	M	C	

0.002 (2μ) 0.006 0.02 0.06 0.2 0.6 2.0 mm

(a) MIT System

Legend
F = Fine M = Medium
C = Coarse

Ultra Clay	Clay		Silt		Mo		Sand				Gravel
	F	C	F	C	F	C	F	M	C	VC	

0.2 μ 0.6 μ 2 μ 0.006 0.02 0.05 0.1 0.2 0.5 1.0 2.0 mm

(b) International Classification

VC = Very coarse

Clay (size)	Silt (size)	Sand				Fine Gravel	Gravel
		Very fine	Fine	Medium	Coarse		

0.005 0.05 0.10 0.25 0.5 1.0 2.0

(c) U.S. Bureau of Soils Classification

Fig. 5.1. Classification Systems.

known as the Swedish classification system before it was adopted as International system. However, the system was not adopted by the United States.

In this system [Fig. 5.1 (*b*)], in addition to sand, silt, and clay, a term *mo* has been used for soil particles in the size range between sand and silt.

(3) U.S. Bureau of Soils Classification—This is one of the earliest classification systems developed in 1895 by U.S. Bureau of Soils [Fig. 5.1 (*c*)]. In this system, the soils below the size 0.005 mm are classified as clay size in contrast to 0.002 mm size in other systems. The soils between 0.005 mm and 0.05 mm size are classified as silt size. Sandy soils between the size 0.05 mm and 1.0 mm are subdivided into four categories as very fine, fine, medium and coarse sands. Fine gravels are in the size range of 1.0 to 2.0 mm.

5.3. TEXTURAL CLASSIFICATION

Texture means visual appearance of the surface of a material such as fabric or cloth. The visual appearance of a soil is called its texture. The texture depends upon the particle size, shape of particles and gradation of particles. The textural classification incorporates only the particle size, as it is difficult to incorporate the other two parameters.

In fact, all the classification systems based on the particle size, as discussed in Sect. 5.2, are textural classification systems. However, in soil engineering, the term textural classification is used rather in a restricted sense. The triangular classification system suggested by U.S. Bureau of Public Roads in commonly known as the textural classification system (Fig. 5.2). The term texture is used to express the percentage of the three constituents of soils, namely, sand, silt and clay.

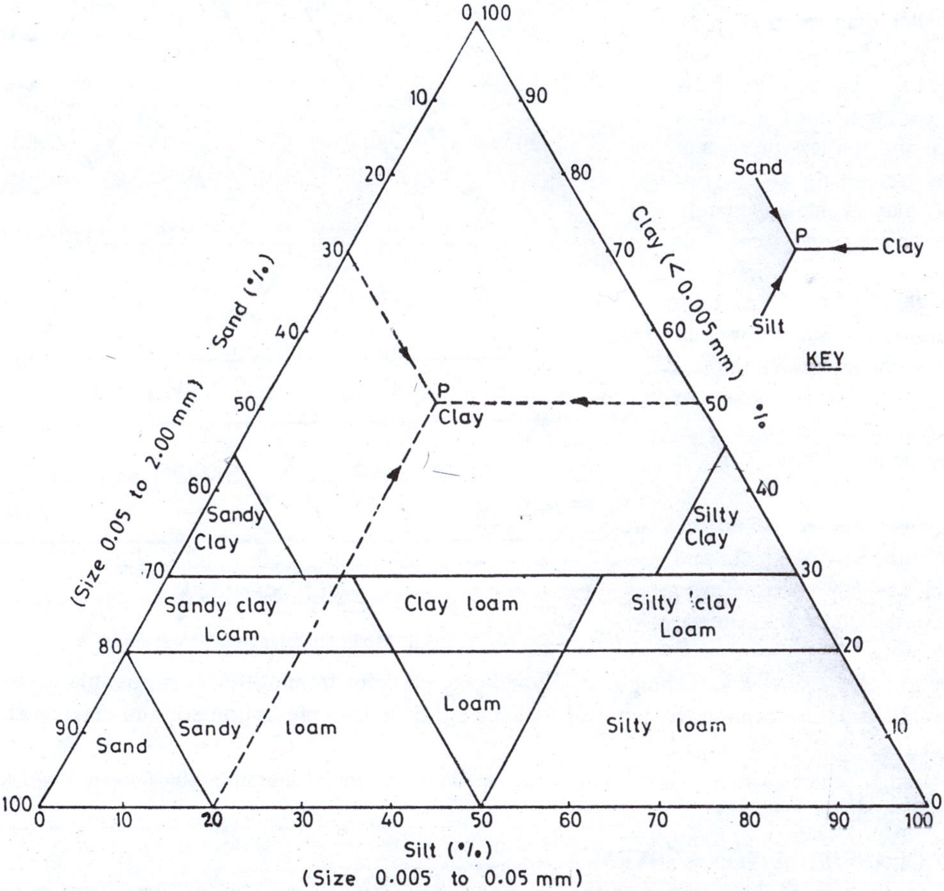

Fig. 5.2. Textural classification System.

According to the textural classification system, the percentages of sand (size 0.05 to 2.0 mm), silt (size 0.005 to 0.05 mm) and clay (size less than 0.005 mm) are plotted along the three sides of an equilateral triangle. The equilateral triangle is divided into 10 zones, each zone indicates a type of soil. The soil can be classified by determining the zone in which it lies. A key is given that indicates the directions in which the lines are to be drawn to locate the point. For example if a soil contains 30% sand and 20% silt and 50% clay, it is shown by point (*P*) in the figure. The point falls in the zone labelled clay. Therefore, the soil is classified as clay.

The textural classification system is useful for classifying soils consisting of different constituents. The system assumes that the soil does not contain particles larger than 2.0 mm size. However, if the soil contains a certain percentage of soil particles larger than 2.0 mm, a correction is required in which the sum of the percentages of sand, silt and clay is increased to 100%. For example, if a soil contains 20% particles of size lager than 2 mm size, the actual sum of the percentages of sand, silt and clay particles is 80%. Let these be respectively 12, 24 and 44%. The corrected percentages would be obtained by multiplying with a factor of 100/80. Therefore, the corrected percentages are 15, 30 and 55%. The textural classification of the soil would be done based on these corrected percentages.

In this system, the term loam is used to describe a mixture of sand, silt and clay particles in various proportions. The term loam originated in agricultural engineering where the suitability of a soil is judged for crops. The term is not used in soil engineering. In order to eliminate the term loam, the Mississipi River Commission (USA) proposed a modified triangular diagram (Fig. 5.3). The term loam is replaced by soil engineering terms such as silty clay. The principal component of a soil is taken as a noun and the less prominent component as an adjective. For example, silty clay contains mainly particles of a clay, but some silt particles are also present. It must be noted that the primary soil type with respect to behaviour is not necessarily the soil type that constitutes the largest part of the sample. For example, the general character of a mixed soil is determined by clay fraction if it exceeds 30%.

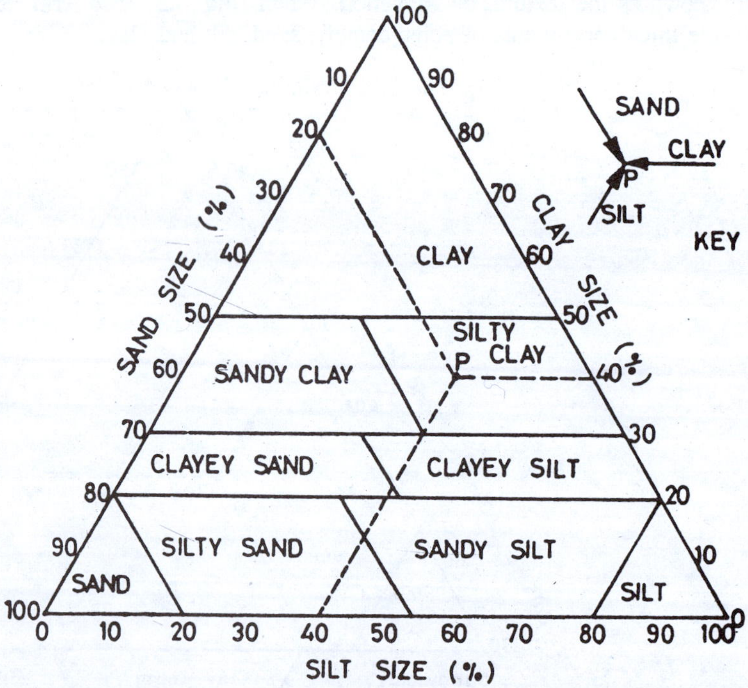

Fig. 5.3. Modified Triangular Diagram.

Right Triangle Chart. Since the sum of the percentages of sand, silt and clay size particles is 100%, there is no need to plot all the three percentage. The percentage of sand particles can be found by deduction of the sum of percentages of silt and clay particles from 100%. It is possible to determine the textural classification by locating the point of intersection of lines representing silt and clay, as shown in right-triangle chart (Fig. 5.4).

The right-triangle chart is more convenient than the conventional triangular chart as it involves only orthogonal arrangement of grid lines.

5.4. AASHTO CLASSIFICATION SYSTEM

American Association of State Highway and Transportation Official (AASHTO) Classification system is useful for classifying soils for highways. The particle size analysis and the plasticity characteristics are required to classify a soil. The classification system is a complete system which classifies both coarse-grained and fine-grained soils. In this system, the soils are divided into 7 types, designated as *A*-1 to *A*-7. The soils *A*-1 and *A*-7 are further subdivided into two categories, and the soil *A*-2, into four categories, as shown in Table 5.1.

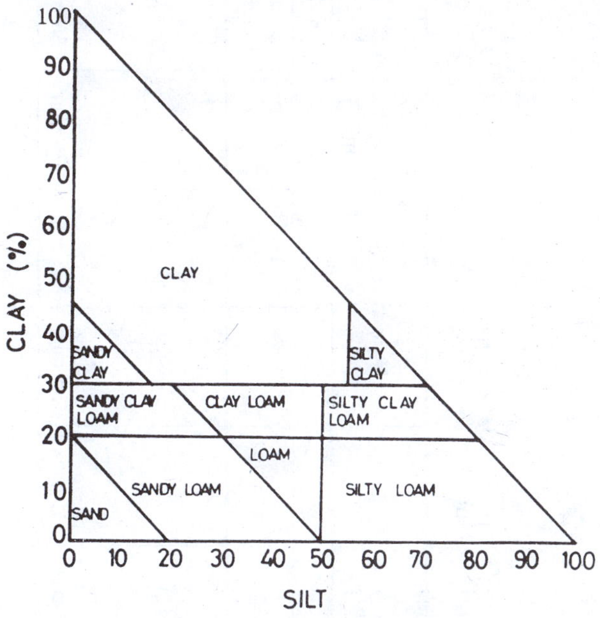

Fig. 5.4. Right Triangle chart.

To classify a soil, its particle size analysis is done, and the plasticity index and liquid limit are determined. With the values of these parameters known, one examines the first (extreme left) column of Table 5.1 and ascertains whether the known parameters satisfy the limiting values in that column. If these satisfy the requirements, the soil is classified as *A*-1-*a*. If these do not satisfy, one enters the second column (from the left) and determines whether these satisfy the limiting values in that column. The procedure is repeated for the next column until the column is reached when the known parameters satisfy the requirements. The soil is classified as per nomenclature given at the top of that column.

The soil with the lowest number, *A*-1, is the most suitable as a highway material or subgrade. In general, the lower is the number of soil, the more suitable is the soil. For example, the soil *A*-4 is better than the soil *A*-5. In Table 5.1, the column for soil *A*-3 is to the left of the column for soil *A*-2. This arrangement is only to determine the classification of the soil. This does not indicate that soil *A*-3 is more suitable for highways than *A*-2 soil.

Fine-grained soils are further rated for their suitability for highways by the group index (*GI*), determined as follows:

$$GI = (F - 35)\,[0.2 + 0.005\,(w_l - 40)] + 0.01(F - 15)\,(I_p - 10) \qquad \ldots(5.1)$$

where F = percentage by mass passing American Sieve no. 200 (size 0.075 mm), expressed as a whole number.

w_l = liquid limit (%) expressed as a whole number.

I_p = plasticity index (%), expressed as a whole number.

Eq. 5.1 is sometimes expressed as

$$GI = 0.2\,(F-35) + 0.005\,(F-35)\,(w_l - 40) + 0.01\,(F-15)\,(I_p-10)$$

While calculating *GI* from the above equation, if any term in the parentheses becomes negative, it is dropped, and not given a negative value. The maximum values of (F–35) and (F–15) are taken as 40 and that of (w_l – 40) and (I_p – 10) as 20.

The group index is rounded off to the nearest whole number. If the computed value is negative, the group index is reported as zero. The group index is appended to the soil type determined from Table 5.1. For example *A*-6 (15) indicates the soil type *A*-6, having a group index of 15. The smaller the value of the group index, the better is the soil in that category. A group index of zero indicates a good subgrade, whereas a group index of 20 or greater shows a very poor subgrade. The group index must be mentioned even when it is zero to indicate that the soil has been classified as per AASHTO system.

Table 5.1. AASHTO Classification System

General Classification	*Granular materials (35% or less passing No. 200 Sieve (0.075 mm)*							*Silt-clay Materials More than 35% passing No. 200 Sieve (0.075 mm)*			
Group Classification	*A—1*		*A—3*	*A—2*				*A—4*	*A—5*	*A—6*	*A—7*
	A—1—a	*A—1—b*		*A—2—4*	*A—2—5*	*A—2—6*	*A—2—7*				*A—7—5* *A—7—6*
(*a*) Sieve Analysis: Percent Passing											
(*i*) 2.00 mm (No. 10)	50 max										
(*ii*) 0.425 mm (No. 40)	30 max	50 max	51 min								
(*iii*) 0.075 mm (No. 200)	15 max	25 max	10 max	35 max	35 max	35 max	35 max	36 min	36 min	36 min	36 min
(*b*) Characteristics of fraction passing 0.425 mm (No. 40)											
(*i*) Liquid limit				40 max	41 min	40 max	41 min	40 max	41 min	40 max	41 min
(*ii*) Plasticity index	6 max		N.P.	10 max	10 max	11 min	11 min	10 max	10 max	11 min	11 min*
(*c*) Usual types of significant Constituent materials	Stone Fragments Gravel and sand		Fine Sand	Silty or Clayey Gravel Sand				Silty Soils		Clayey Soils	
(*d*) General rating as subgrade.	Excellent to Good							Fair to Poor			

* If plasticity index is equal to or less than (liquid Limit—30), the soil is A—7—5 (*i.e.* PL > 30%)
If plasticity index is greater than (Liquid Limit—30), the soil is A—7—6 (*i.e.* PL < 30%)

5.5. UNIFIED SOIL CLASSIFICATION SYSTEM

The Unified Soil Classification System (USC) was first developed by Casagrande in 1948. and later, in 1952, was modified by the Bureau of Reclamation and the Corps of Engineers of the United States of America. The system has also been adopted by American Society of Testing Materials (ASTM). The system is the most popular system for use in all types of engineering problems involving soils. The various symbols used are given in Table 5.2.

Table 5.2. Symbols used in USC System

	Symbols	*Description*
Primary	G	Gravel
	S	Sand
	M	Silt (Symbol M is obtained from the Swedish word 'mo')
	C	Clay
	O	Organic
	Pt	Peat
Secondary	W	Well-graded
	P	Poorly graded
	M	Non-plastic fines
	C	Plastic fines
	L	Low Plasticity
	H	High plasticity

The system uses both the particle size analysis and plasticity characteristics of soils, like AASHTO system. In this system, the soils are classified into 15 groups (Table 5.3). The soils are first classified into two categories.

(1) Coarse-grained soils—If more than 50% of the soil is retained on No. 200 (0.075 mm) sieve, it is designated as coarse-grained soil. There are 8 groups of coarse-grained soils.

(2) Fine-grained soils—If more than 50% of the soil passes No. 200 sieve, it is called fine-grained soil. There are 6 groups of fine-grained soils.

1. Coarse-grained Soils—The coarse-grained soils are designated as gravel (*G*) if 50% or more of coarse fraction (plus 0.075 mm) is retained on No. 4 (4.75 mm) sieve; otherwise it is termed sand (*S*).

If the coarse-grained soils contains less than 5% fines and are well-graded (*W*), they are given the symbols *GW* and *SW*, and if poorly graded (*P*), symbols *GP* and *SP*. The criteria for well-grading are given in Table 5.3. If the coarse-grained soils contain more than 12% fines, these are designated as *GM, GC, SM* or *SC*, as per criteria given. If the percentage of fines is between 5 to 12% dual symbols such as *GW-GM, SP-SM,* are used.

2. Fine-grained Soils—Fine-grained soils are further divided into two types : (1) Soils of low compressibility (*L*) if the liquid limit is 50% or less. These are given the symbols *ML, CL* and *OL.* (2) Soils of high compressibility (*H*) if the liquid limit is more than 50%. These are given the symbols *MH, CH* and *OH*. The exact type of the soil is determined from the plasticity chart (Fig. 5.5). The *A*-line has the equation $I_p = 0.73\ (w_l - 20)$. It separates the clays from silts. When the plasticity index and the liquid limit plot in the hatched portion of the plasticity chart, the soil is given double symbol *CL- ML.*

The inorganic soil *ML* and *MH* and the organic soils *OL, OH* plot in the same zones of the plasticity chart. The distinction between the inorganic and organic soils is made by oven-drying. If oven drying decreases the liquid limit by 30% or more, the soil is classified organic (*OL* or *OH*); otherwise, inorganic (*ML* or *MH*)

Highly Organic Soils—Highly organic soils are identified by visual inspection. These soils are termed peat (P_t).

5.6. COMPARISON OF AASHTO AND USC SYSTEMS

AASHTO system is for finding out the suitability or otherwise of soils as subgrade for highways only.

Table 5.3. Unified Soil Classification System

Major Division			*Group Symbols*	*Typical names*	*Classification criteria*		
Coarse-Grained Soils. [More than 50% retained on No. 200 sieve (0.075 mm)]	Gravel [50% or more of coarse fraction retained on No. 4 sieve (4.75 mm)]	Clean Gravels	GW	Well graded gravels	Percentage of fines (*a*) Less than 5% passing No. 200, GW, GP, SW, SP (*b*) more than 12% passing No. 200, GM, GC, SM, SC (*c*) 5 to 12% passing No. 200 use of dual symbols as GW—GM, SP—SC.	$C_u > 4$ $C_c = 1$ to 3	
			GP	Poorly graded gravels		Not meeting both criteria for GW	
		Gravels with fines	GM	Silty gravels		Atterberg Limits below A-line or plasticity index less than 4	Atterberg Limits in hatched area GM—GC
			GC	Clayey gravels		Atterberg Limits above A-line and plasticity index greater than 7	
	Sand [more than 50% of coarse faction passing No. 4 sieve (4.75 mm)]	Clean Sands	SW	Well-graded sands		$C_u > 6$ $C_c = 1$ to 3	
			SP	Poorly graded sands		Not meeting both criteria for SW	
		Sands with fines	SM	Silty sands		Atterberg Limits below A-line or plasticity index less than 4	Atterberg Limits in hatched area SM—SC
			SC	Clayey sands		Atterberg Limits above A-line and plasticity index greater than 7	
Fine grained soils [50% or more passing No. 200 sieve (0.075 mm)]	Silts and clays Liquid Limit 50% or less		ML	Inorganic silts of low plasticity	See Plasticity Chart (Fig. 5.5)		
			CL	Inorganic clays of low to medium plasticity			
			OL	Organic silts of low plasticity			
	Silts and clays Liquid Limit greater than 50%		MH	Inorganic silts of high plasticity			
			CH	Inorganic clays of high plasticity			
			OH	Organic clays of medium of high plasticity			
Highty organic Soils			Pt	Peat, muck and other highly organic soils	Visual-manual identification		

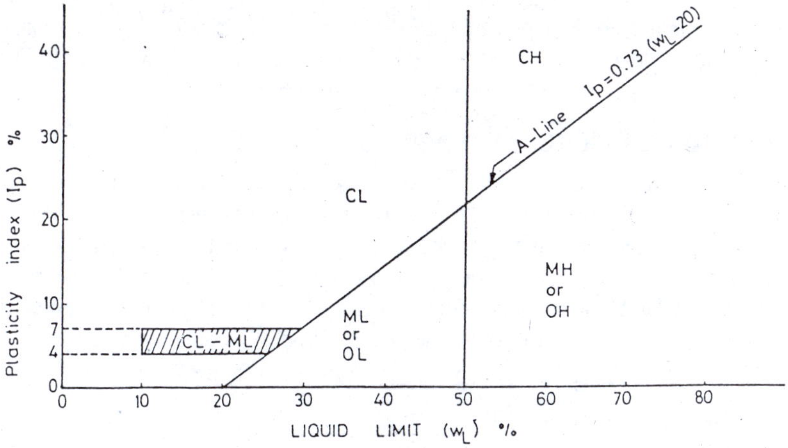

Fig. 5.5. Plasticity chart (USC).

USC system is for determining the suitability of soils for general use. Both the systems, however, have the same basis. They classify the soils according to the particle size analysis and the plasticity characteristics. Both the systems divide the soils into two major categories, namely, coarse-grained and fine-grained soils.

The following differences between the two systems are worth mentioning.

(1) According to AASHTO system, a soil is termed fine-grained if more than 35% passes No. 200 (0.075 mm) sieve, whereas in the USC system, if more than 50% passes that sieve. In this respect, the AASHTO system is somewhat better because the soil behaves as fine-grained when the percentage of fines is 35%, and the limit of 50% in USC system is somewhat higher.

(2) In AASHTO system, sieve No. 10 (2.0 mm size) is used to divide the soil into gravel and sand, whereas in USC system, sieve No. 4(4.75 mm size) is used.

(3) In USC system, the gravelly and sandy soils are clearly separated, whereas in AASHTO system, clear demarcation is not done. The soil A-2 in the latter system contains a large variety of soils.

(4) Symbols used in USC system are more descriptive and are more easily remembered than those in AASHTO system.

(5) Organic soils are also classified as *OL* and *OH* and as peat (P_t) if highly organic in USC system. In AASHTO, there is no place for organic soils.

(6) USC system is more convenient to use than the AASHTO system. In the latter, the process of elimination is required which is time-consuming.

Table 5.4. Approximate Equivalence Between AASHTO and USC System

AASHTO System	*USC system (most probable)*
A-1-a	GW, GP
A-1-b	SW, SM, GM, SP
A-2-4	GM, SM
A-2-5	GM, SM
A-2-6	GC, SC
A-2-7	GM, GC, SM, SC
A-3	SP
A-4	ML, OL, MH, OH
A-5	MH, OH, ML, OH
A-6	CL
A-7-5	OH, MH, CL, OL
A-7-6	CH, CL, OH

Table 5-4 gives approximate equivalence in both the systems. If the soil has been classified according to one system, its classification according to the other can be determined. However, the equivalence is only approximate. For exact classification, the corresponding procedure should be used.

5.7. INDIAN STANDARD CLASSIFICATION SYSTEM

Indian Standard Classification (ISC) system adopted by Bureau of Indian Standards is in many respects similar to the Unified Soil Classification (USC) system. However, there is one basic difference in the classification of fine-grained soils. The fine- grained soils in ISC system are subdivided into three categories of low, medium and high compressibility instead of two categories of low and high compressibility in USC system. A brief outline of Classification and Identification of Soils for general engineering purposes (IS : 1498-1970) is given below. For complete details, the reader should consult the code.

ISC system classifies the soils into 18 groups as per Tables 5.6 and 5.7.

Soils are divided into three broad divisions:

(1) Coarse-grained soils, when 50% or more of the total material by weight is retained on 75 micron IS sieve.

(2) Fine-grained soils, when more than 50% of the total material passes 75 micron IS sieve.

(3) If the soil is highly organic and contains a large percentage of organic matter and particles of decomposed vegetation, it is kept in a separate category marked as peat (P_t).

In all, there are 18 groups of soils : 8 groups of coarse- grained, 9 groups of fine-grained and one of peat. Basic soil components are given in Table 5.5. Symbols used are the same as in USC system (Table 5.2).

Table 5.5. Basic Soil Components in ISC System

Soil	*Soil components*	*Symbol*	*Particle size range and dercription*
(*i*) Coarse-grained components	Boulder	None	Rounded to angular, bulky, hard, rock, particle; average diameter more than 300 mm
	Cobble	None	Rounded to angular, bulky, hard, rock particle; average diameter smaller than 300 mm but ratained on 80 mm IS sieve
	Gravel	G	Rounded to angular, bulky, hard, rock particles; passing 80 mm IS sieve but retained on 4.75 mm IS sieve **Coarse :** 80 mm to 20 mm IS sieve **Fine :** 20 mm to 4.75 mm IS sieve
	Sand	S	Rounded to angular, bulky, hard, rock particle; passing 4.75 mm IS sieve, but retained on 75 micron sieve **Coarse :** 4.75 mm to 2.0 mm IS sieve **Medium :** 2.0 mm to 425 micron IS sieve **Fine :** 425 micron to 65 micron IS sieve
(*ii*) Fine-grained components	Silt	M	Particles smaller than 75 micron IS sieve; identified by behaviour, that is, slightly plastic or non-plastic regardless of moisture and exhibits little or no strength when air-dried.
	Clay	C	Particles smaller than 75 micron IS sieve : identified by behaviour, that is, it can be made to exhibit plastic properties within a certain considerable strength when air dried.
	Organic matter	O	Organic matter in various sizes and stages of decomposition.

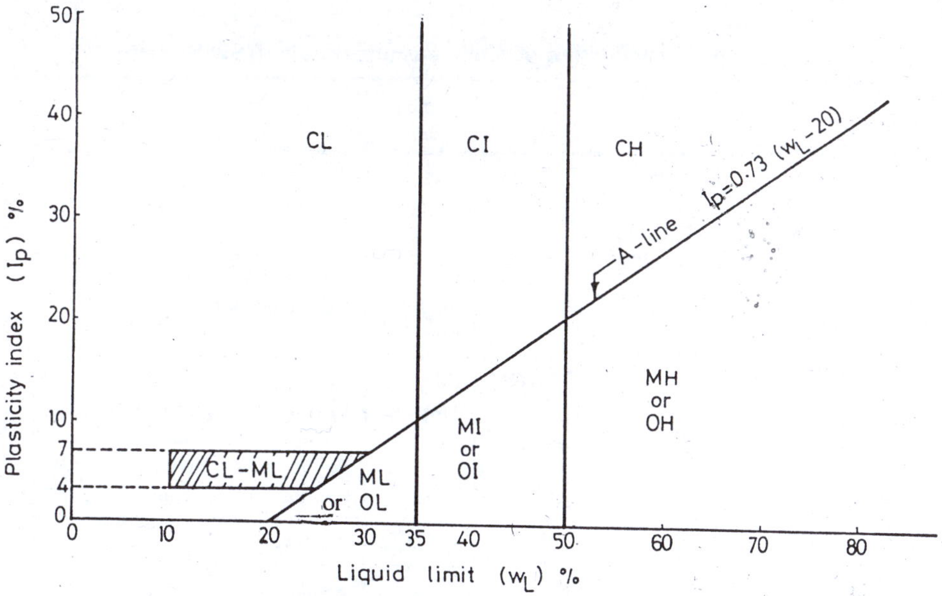

Fig. 5.6. Plasticity Chart (ISC)

1. Coarse-grained Soils—Coarse-grained soils are subdivided into gravel and sand. The soil is termed gravel (G) when more than 50% of coarse fraction (plus 75μ) is retained on 4.75 mm IS sieve, and termed sand (S) if more than 50% of the coarse fraction is smaller than 4.75 mm IS sieve. Coarse-grained soils are further subdivided as given in Table 5.6 into 8 groups.

2. Fine-grained Soils—The fine-grained soils are further divided into three subdivisions, depending upon the values of the liquid limit:

(*a*) Silts and clays of low compressibility—These soils have a liquid limit less than 35 (represented by symbol L).

(*b*) Silts and clays of medium compressibility—These soils have a liquid limit greater than 35 but less than 50 (represented by symbol I).

(*c*) Silts and clays of high compressibility—These soils have a liquid limit greater than 50 (represented by symbol *H*).

Fine-grained soils are further subdivided, in 9 groups as given in Table 5.7.

5.8. BOUNDARY CLASSIFICATIONS

Sometimes, it is not possible to classify a soil into any one of 18 groups discussed above. A soil may possess characteristics of two groups, either in particle size distribution or in plasticity. For such cases, boundary classifications occur and dual symbols are used.

(*a*) Boundary classification for coarse-grained soils

The following boundary classification can occur:

(1) Boundary classifications within gravel group or sand group can occur. The following classification are common.

GW—GP, GM—GC, GW—GM, GW—GC, GP—GM
SW—SP, SM—SC, SW—SM, SW—SC, SP—SM

While giving dual symbols, first write a coarser soil then a finer soil.

(2) Boundary classification can occur between the gravel and sand groups such as

GW—SW, GP—SP, GM—SM, and *GC—SC*

The rule for correct classification is to favour the non-plastic classification. For example, a gravel with 10% fines, C_u = 20 and C_c = 2.0 and I_p = 6 will be classified as *GW—GM,* and not *GW—GC.*

Table 5.6. Classification of Coarse-grained Soils (ISC System)

Division	*Subdivision*		*Group symbol*	*Typical names*	*Laboratory Criteria*		*Remark*
(1) Coarse-grained soils (More than half of material is larger than 75-micron IS sieve size)	Gravel (G) (more than half of coarse fraction is larger than 4.75 mm IS sieve)	Clean gravels (Fines less than 5%)	(1) GW	Well graded gravels	C_u grater than 4 C_c between 1 and 3		When fines are between 5% to 12%, border line cases requiring dual symbols such as GP—GM, SW—SC, etc.
			(2) GP	Poorly graded gravels	Not meeting all gradation requirements for GW		
		Gravels with appreciable amount of fines (Fines more than 12%)	(3) GM	Silty gravels	Atterberg Limits below A-line or I_p less than 4	Atterberg Limits plotting above A-line with I_p betwen 4 and 7 are border-line cases requiring use of dual symbol GM—GC	
			(4) GC	Clayey gravels	Atterberg limits above A-line and I_p grater than 7		
	Sand (S) (More than half of coarse fraction is smaller than 4.75 mm IS sieve)	Clean sands (Fines less than 5%)	(5) SW	Well-graded sands	C_u greeater than 6 C_c between 1 and 3		
			(6) SP	Poorly-graded sands	Not meeting all gradation requirements for SW		
		Sands with appreciable amout of fines (Fines more than 12%)	(7) SM	silty sands	Atterberg Limits below A-line or I_p less than 4	Atteraerg's Limits plotting above A-line with I_p between 4 and 7 are border-line cases requiring use of double symbols SM—SC	
			(8) SC	Clayey sands	Atterberg limits above A line with I_p greater than 7		

(*b*) **Boundary classification for fine-grained soils.**

(1) Within the same compressibility subdivision, such as
ML—CL, ML—OL, CL—OL; CI—MI, MI—OI, CI—OI; MH—CH, MH—OH; CH—OH.

First write a coarser soil when there is a choice and then a finer soil.

(2) Between low and medium compressibility, such as
ML—MI, CL—CI, OL—OI

(3) Between medium and high compressibility
ML—MH, CI—CH, OI—OH

(*c*) **Boundary Classification between coarse-grained and fine-grained soils.**

Boundary classification can occur between a coarse-grained soil and a fine-grained soil, such as
SM—ML, SC—CL

Table 5.7. Classification of Fine-grained Soils (ISC System)

Division	*Subdivision*	*Group Symbols*	*Typical names*	*Laboratory Criteria (see Fig 5.6)*		*Remarks*
2) Fine-grained soils (more than 50% pass 75μ, IS sieve)	Low-compressibility (*L*) (Liquid Limit less than 35%)	(1) ML	Inorganic silts with none to low plasticity	Atterberg limits plot below A-line or I_p less than 7	Atterberg limits pioting above A-line with I_p between 4 to 7 (hatched zone) ML-CL	(1) Organic and inorganic soils plotted in the same zone in plasticity chart are distinguished by odour and colour or liquid limit test after oven-drying. A reduction in liquid limit after oven- drying to a value less than three- fourth of the liquid limit before oven- drying is positive identification of organic soils.
		(2) CL	Inorganic clays of low plasticity	Atterberg limits plot above A-line and I_p greater than 7		
		(3) OL	Organic silts of low plasticity	Atterberg limits plot below A-line		
	Intermediate compressibility (I) (Liquid limit greater than 35% but less than 50%	(4) MI	Inorganic silts of medium plasticity	Atterberg limits plot below A-line		
		(5) CI	Inorganic clays of medium plasticity	Atterberg limits plot above A-line		(2) Black cotton soils of India lie along a band partly above the A-line and partly below tha A line
		(6) OI	Orgaic silts of medium plasticity	Atterberg limits plot below A-line		
	High compressibiliy (*H*) (Liquid limit greater than 50%)	(7) MH	Inorganic silts of high compressibility	Atterberg limits plot below A-line		
		(8) CH	Inorganic clays of high plasticity	Atterberg limits plot above A-line		See, plasticity chart (Fig. 5.6)
		(9) OH	Organic clays of medium to high plasticity	Atterberg limits plot below A-line		
(3) Highly organic soil		Pt	Peat and other highly organic soils	Readily identified by colour, odour, spongy feel and fibrous texture		

5.9. FIELD IDENTIFICATION OF SOILS

The soils can be identified in the field by conducting the following simple tests.

The sample is first spread on a flat surface. If more than 50% of the particles are visible to the naked eye (unaided eye), the soil is coarse-grained; otherwise, it is fine grained. The fine- grained particles are smaller than 75μ size and are not visible to unaided eye. The fraction of soil smaller than 75μ size, that is, the clay and silt fraction, is referred to as *fines*.

(1) Coarse-grained Soils—If the soil is coarse-grained, it is further identified by estimating the percentage of (*a*) gravel size particles (4.75 mm to 80 mm), (*b*) sand size particles, (75μ to 4.75 mm) and (*c*) silts and clay size particles (smaller than 75μ size). Gravel particles are larger than 4.75 mm size and can be identified visually.

If the percentage of gravel is greater than that of sand, the soil is a gravel; otherwise, it is sand.

Gravels and sands are further classified as clean if they contain fines less than 5% and as dirty if they contain fines more than 12%. Gravels and sands containing 5 to 12% fines are given boundary classification. The fine fraction of the coarse-grained soils is identified using the procedure given below for fine- grained soils to determine whether it is silty or clayey.

To differentiate fine sand from silt, dispersion test is adopted. When a spoonful of soil is poured in a jar full of water, fine sand settles in a minute or so, whereas silt takes 15 minutes or more.

(2) Fine-grained soils—If the soil is fine-grained, the following tests are conducted for identification on the fraction of the soil finer than the 425-micron IS sieve to differentiate silt from clay.

(*a*) Dilatancy (reaction to shaking) test—A small pat of moist soil of about 5 ml in volume is prepared. Water is added to make the soil soft but not sticky. The pat is placed in the open palm of one-hand and shaken horizontally, striking against the other hand several times during shaking. If the soil gives a positive reaction, the water appears on its surface which changes to a lively consistency and appears glossy. When the pat is squeezed between the fingers, the water and gloss disappear from the surface. It becomes stiff and ultimately crumbles.

The rapidity with which water appears on the surface during shaking and it disappears during squeezing is used in the identification of fine-grained soils (Tables 5.8). The larger the size of the particles, the quicker is the reaction. The reaction is called quick if water appears and disappears quickly. The reaction is termed slow if water appears and disappears slowly. For no reaction, the water does not appear at the surface.

(*b*) Toughness test—The pat used in the dilatancy test is dried by working and remoulding until it has the consistency of putty. The time required to dry the pat depends upon the plasticity of the soil.

The pat is rolled on a smooth surface or between the palms into a threads of about 3 mm in diameter. The thread is folded and re-rolled to reduce the water is soil, due to evaporation by heat of hand, until the 3 mm diameter thread just crumbles. The water content at that stage is equal to the plastic limit and the resistance to moulding at that stage is called the toughness.

After the thread crumbles, the pieces of the sample are lumped together and subjected to kneading until the lump also crumbles. The tougher the thread at the plastic limit and the stiffer the kneaded lump just before it crumbles, the higher is the toughness of the soil. The toughness is low if the thread is weak and the soil mass cannot be lumped together when drier than plastic limit. The toughness is high when the lump can be moulded drier than plastic limit and high pressure is required to roll the thread.

The toughness depends upon the potency of the colloidal clay.

Table 5.8. Field Identification Tests

Test	*ML*	*CL*	*OL*	*MI*	*CI*	*OI*	*MH*	*CH*	*OH*
(*a*) Dilatancy	Quick	None to very slow	Slow	Quick to slow	None	Slow	Slow to none	None	None to very slow
(*b*) Toughness	None	Medium	Low	None	Medium	Low	Low to medium	High	Low to medium
(*c*) Dry strength	None of low	Medium	Low	Low	Medium to high	Low to medium	Low to medium	High to very high	Medium to high

(*c*) Dry strength test—The pat of the soil is completely dried by air drying, sun drying or oven-drying.

The dry strength is determined by breaking the dried pat and crumbling it between fingers. The dry strength is a measure of plasticity of the soil. The dry strength depends upon the colloidal fraction of the soil.

The strength is termed high if the dried pat cannot be powdered at all; medium, if considerable pressure is required; and low, if the dry pat can be easily powdered.

Table 5.8 can be used for the field identification of different soils.

5.10. GENERAL CHARACTERISTICS OF SOILS OF DIFFERENT GROUPS

General characteristics of the soils of various groups as classified by ISC system and USC system are given in Table 5.9. The information given in the table should be considered as a rough guidance about the engineering properties of soils. For complete information, the tests should be conducted and the engineering properties determined.

Table 5.9. General Propeties of Soils

Soil Group	*Permeability*	*Compressibility*	*Shear Strength*	*Workability*
(*a*) *Gravels*				
GW	Pervious	Negligible	Excellent	Excellent
GP	Very pervious	Negligible	Good	Good
GM	Semi-pervious to impervious	Negligible	Good	Good
GC	Impervious	Very low	Good to fair	Good
(*b*) Sands				
SW	Pervious	Negligible	Excellent	Excellent
SP	Pervious	Very low	Good	Fair
SM	Semi-pervious to impervious	Low	Good	Fair
SC	Impervious	Low	Good to fair	Good
(*c*) Low & medium Plasticity silt & clays				
ML, MI	Semi-pervious to impervious	Medium	Fair	Fair
CL, CI	Impervious	Medium	Fair	Good to fair
OL, OI	Semi-pervious to impervious	Medium	Fair	Fair
(*d*) High- plasticity silts & clays				
MH	Semi-pervious to impervious	High	Fair to poor	Poor
CH	Impervious	High	Poor	Poor
OH	Impervious	High	Poor	Poor

Note. Highly organic soils [Peats] are not used in engineering works.

ILLUSTRATIVE EXAMPLES

Illustrative Example 5.1. A sample of soil was tested in a laboratory, and the following observations were recorded:

Liquid Limit = 45%, *Plastic Limit* = 16%

U.S. Sieve No.	*No. 4*	*No. 10 (2.0 mm)*	*No. 40 (0.425 mm)*	**No. 200 (0.075 mm)**
Percentage Passing	100	91.5	80.0	60.0

Classify the soil according to AASHTO system.

Solution. Plasticity Index = 45 – 16 = 29%

Referring to Table 5.1, and proceeding from the extreme left column to right, the first column in which the properties fit is A-7.

To ascertain whether the soil is A-7-5 or A-7-6, the value of $(w_l - 30)$ is required. In this case,

$$w_l - 30 = 45 - 30 = 15\%$$

As $I_p > (w_l - 30)$, the soil A-7-6.

From Eq. 5.1, taking $F = 60$,

$$\text{Group Index} = (F - 35)\,[0.2 + 0.005\,(w_l - 40)] + 0.01\,(F - 15)(I_p - 10)$$

or

$$GI = (60 - 35)[0.2 + 0.005 \times 5)] + 0.01\,(40)(19)$$

$$= 13.3, \text{ say } 13.$$

The soil is classified as A-7-6 (13).

Illustrative Example 5.2. *Classify the soils A and B, with the properties as shown below, according to USC system.*

Soil	w_l *(%)*	I_p *(%)*	*% passing No. 4 sieve*	*% passing No. 200 sieve*
A	45	29	100	59
B	55	15	100	85

Solution. (*a*) *Soil A*. As more than 50% passes No. 200 sieve, the soil is fine-grained.

As w_l is less than 50%, the soil is of low plasticity. The Atterberg limits plot above the A-line in Fig. 5.5. The soil is classified as *CL*.

(*b*) *Soil B*. The soil is fine-grained. As the liquid limit is greater than 50%, the soil is of high compressibility. The Atterberg limits plot below A-line. It can be either *MH* or *OH*. If the soil is *OH*, its liquid limit will decrease considerably on oven-drying.

Illustrative Example 5.3. *Classify the soil with the following properties according to ISC system.*

Liquid Limit	*Plassicity index*	*% passing 4.75 mm sieve*	*% passing 75 μ sieve*
40%	10%	60%	45%

Solution. As more than 50% is retained on 75μ IS sieve, the soil is coarse-grained.

Coarse fraction = 55%; Gravel fraction = 40%; Sand fraction = 15%

As more than half the coarse-fraction is larger than 4.75 mm IS sieve, the soil is gravel.

The soil has more than 12% fines, it can be either *GM* or *GC*.

As the Atterberg limits plot below A-line (Fig. 5.6), the soil is *GM*.

Illustrative Example 5.4. *Fig E 5.4 shows the grain size distribution curves for two soils A and B. The plasticity characteristics of the soils are given below.*

Soil A *Liquid Limit = 40%;* *Plasticity Index = 10%;*

Soil B *Liquid Limit = 28%;* *Plasticity Index = 12%*

Classify the soils according to IS classification and comment on their shear strength.

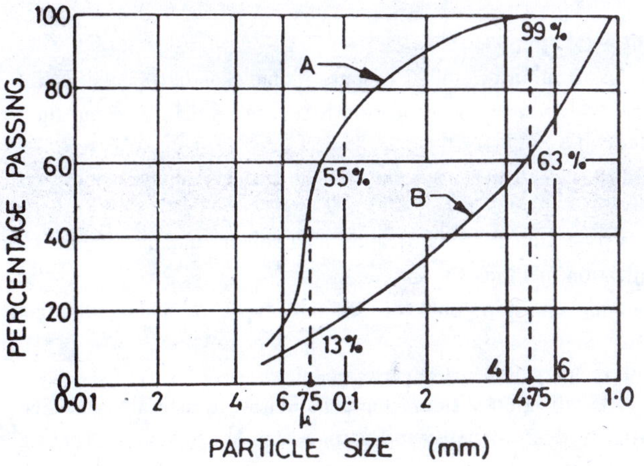

Fig. E-5.4.

Solution. (*a*) *Soil A.* As more than 50% passes 75μ sieve, the soil is fine-grained. The Atterberg limits plot below A-line (Fig. 5.6) in the zone of intermediate compressibility. It can be either *MI* or *OI*. If the liquid limit reduces to three-fourth of the original value or more on oven drying, it *OI*; otherwise *MI*.

(*b*) *Soil B.* As more than 50% of the total material is larger than 75 μ sieve, the soil is coarse- grained.

Coarse fraction = 87%,
Gravel fraction = 37%;
Sand fraction = 50%.

As more than half of coarse fraction is smaller than 4.75 mm sieve, the soil is sand. As fines are more than 12%, the soil can be *SM* or *SC*. As the Atterberg limits plot above A- line (Fig. 5.6), the soil is *SC*.

PROBLEMS

A. Numerical

5.1 Atterberg limit tests were carried out on a soil sample, with the following results:
Liquid limit = 40%; Plastic limit = 25%.
Classify the soil according to Unified Classification system and the Indian Standard classification system.
[**Ans.** *CL; CI*]

5.2. The following results were obtained from the classification tests of a soil.
Percentage passing 75μ sieve = 40%
Liquid limit = 35%; Plastic Limit = 15%
Calculate the group index of the soil and classify it according to AASHTO system. [**Ans.** 4; A-6(4)]

5.3. The sieve analysis of a soil gave the following results :
% passing 75μ sieve = 4; % retained on 4.75 mm sieve = 50
Coefficient of curvature = 2; Uniformity coefficient = 5
Classify the soil according to *ISC* system. [**Ans.** *GW*]

5.4. The sieve analysis of a soil gave the following results:
%passing 75μ sieve = 8; % retained on 4.75 mm sieve = 35
Coefficient of curvature = 2.5; Uniformity coefficient = 7
The fine fraction gave the following results :
Plasticity index = 3; Liquid Limit = 15.
Classify the soil according to *ISC* system. [**Ans.** *SW—SM*]

5.5. A soil has the following characteristics :
% passing 75μ sieve = 58%; Liquid Limit = 40%
Plasticity Index = 10%; Liquid Limit of oven-dried sample = 25%
Classify the soil according to *ISC* system. [**Ans.** *OI*]

B. Descriptive and Objective Type

5.6. What is the use of classification of soils ? Discuss Indian Standard Classification system.

5.7. What is the difference between the classification based on particle size and the textural classification ? Discuss the limitations of the two systems.

5.8. Compare the AASHTO classification system and Unified soil classification system. Why the latter system is more commonly used?

5.9. Give the step-by-step procedure for classification of a soil by Indian standard classification system.

5.10. Discuss field identification methods for soils.

5.11. Give the general engineering properties of different types of soils classified according to Indian standard classification system.

5.12. State whether the following statements are true or false :

(*a*) According to all classification systems, the soils below 2μ size are classified as clay size.

(*b*) According to MIT system, the soils with particle size larger than 4.75 mm are classified as gravel.

(*c*) The texture of a soil depends only on the particle size.

(*d*) In AASHTO system, the soil A—2 is better than the soil A—3.

(*e*) The group index of zero indicates a soil of very poor quality.

(*f*) The group index of a soil can be negative.

(*g*) According to USC system, the fine-grained soils are of 9 types.

(*h*) The soil with particles size less than 2μ are clayey soils.

(*i*) A coarse-grained sand is well-graded if C_c = 1 to 3 and C_u is greater than 6.

(*j*) Oven-drying reduces the liquid limit of an inorganic soil considerably.

(*k*) The fine-grained soils with a high percentage of colloidal fraction have high dry strength.

[Ans. True, (*d*), (*i*), (*k*)]

C. Multiple-Choice Questions

1. IS classification of soil is in many respects similar to
(*a*) AASHTO classification (*b*) Textural classification
(*c*) Unified soil classification (*d*) MIT classification

2. The maximum size of particles of silt is
(*a*) 75 μ (*b*) 60 μ
(*c*) 2 μ (*d*) 0.2 μ

3. The maximum size of particles of clay is
(*a*) 0.2 mm (*b*) 0.02 mm
(*c*) 0.002 mm (*d*) 0.0002 mm

4. According to IS classification system, the soils can be classified into
(*a*) 15 groups (*b*) 18 groups
(*c*) 3 groups (*d*) 7 groups

5. The soils which plot above the A line in the plasticity chart are
(*a*) clays (*b*) silts
(*c*) sands (*d*) organic soils

6. A silty soil gives a positive reaction in
(*a*) Toughness test (*b*) Dilatancy test
(*c*) Dry strength test (*d*) None of above

7. A soil has the liquid limit of 30. The corresponding plasticity index given by the A-line is
(*a*) 7.3 (*b*) 7.5
(*c*) 9.0 (*d*) 9.5

8. The maximum value of the term (F-15) in the group index is taken as
(*a*) 20 (*b*) 30
(*c*) 40 (*d*) 60

[Ans. 1. (*c*), 2. (*a*), 3. (*c*), 4. (*b*), 5. (*a*), 6. (*b*), 7. (*a*), 8. (*c*)]

6

Clay Mineralogy and Soil Structure

6.1. INTRODUCTION

The coarse-grained soils generally contain the minerals quartz and feldspar. These minerals are strong and electrically inert. The behaviour of such soils does not depend upon the nature of the mineral present. The behavior of fine-grained soils, on the other hand, depends to a large extent on the nature and characteristics of the minerals present. The most significant properties of clay depend upon the type of mineral. The crystalline minerals whose surface activity is high are clay minerals. These clay minerals impart cohesion and plasticity. The study of clay minerals is essential for understanding the behaviour of clayey soils. Clay mineralogy is the the science dealing with the structure of clay minerals on microscopic, molecular and atomic scale. It also includes the study of the mineralogical composition and electrical properties of the clay particles. The study of clay minerals is important for particles smaller than about 2 micron size.

Soil structure means the geometrical arrangement of soil particles in a soil mass. It is concerned with the shape, size and orientation of particles. If the individual particles are packed very close to one another, the void ratio is low and the soil is dense and strong. If the particles are so arranged that there are more voids, the soil is loose and weak. Engineering properties and behaviour of both coarse-grained and fine-grained depend upon the structure.

This chapter is mainly devoted to clay mineralogy. The soil structure is considered in the last section. In fact, clay mineralogy also discusses the structure of clayey soils not as a whole mass but at a particle level.

6.2. GRAVITATIONAL AND SURFACE FORCES

The gravitational force in a soil particles is proportional to its mass. As the specific gravity of particles is approximately constant, the gravitational force is proportional to the volume of the particle. The volume depends upon the particle size . Thus, the gravitational force on a particle is related to the particle size. In other words, the larger the particle size, the greater would be the gravitational force.

Bonding or surface forces between particles depend upon the surface area of the particles and not upon the volume. The surface area also depends upon the particle size.. However, the surface forces become more important only when the paticle size is small. As the particle size decreases, the effect of surface forces on a particle becomes more predominant than the gravitational force.

The relative magnitude of volume and the surface area can be judged if we consider, say, a cube whose each side is 10 mm (volume = $10^3 mm^3$). When the cube is subdivided into smaller cubes, the ratio of the surface area to the volume increases, as shown in Table 6.1. The ratio increases ten thousand times when the side of the smaller cube becomes 1 micron. The magnitude of the surface area per unit volume (or mass) is known as *specific surface.*

The particles of coarse-grained soils are larger than 0.075 mm size. For such soils, the ratio of surface area to the volume is relatively small. These soils do not possess plasticity and cohesion which are predominant only when the surface forces are large. In fine-grained soils, the gravity forces are relatively insignificant compared

Table 6.1. Ratio of Surface area to Volume

S. N.	*Side Length*	*Number of cubes*	*Surface area*	*Volume*	*Surface area volume (mm^2/mm^3)*
1.	10 mm	1	600 mm^2	10^3 mm^3	0.60
2.	1 mm	10^3	$6 \times 10^3 \times 1$	,,	6.0
3.	0.1 mm	10^6	$6 \times 10^6 \times 0.01$	,,	60.0
4.	0.01 mm	10^9	$6 \times 10^9 \times (0.01)^2$	,,	600.0
5.	0.001 mm (1μ)	10^{12}	$6 \times 10^{12} \times (0.001)^2$	,,	6000.0

with the surface forces. The fine-grained soils possess the plasticity characteristics depending upon the surface area, the type of minerals and the nature of environment present around the soil particle.

A material in which the surface forces are predominant is known as a *colloid*. The term colloid has been derived from Greek words *kolla* and *oidos*, meaning a gluey material and alike. For colloids, the ratio of the surface area to the volume is very large. It varies between 600 to 10^5 mm^2/mm^3.The clayey soils with particles smaller than 2 micron size are generally colloidal in nature. The colloids have very large specific surface.

6.3. PRIMARY VALENCE BONDS

Primary valence bonds hold together the atoms of a molecule. These are of two types:

(1) Ionic bond, (2) Covalent bond.

1. Ionic bond—In an atom, the electrons carrying a negative charge revolve about the nucleus. Some elements have an excess or a deficiency of the electrons in the outer shell. One atom joins another atom by adding some of the electrons to its outer shell or by losing some of electrons from its outer shell. For example, an atom of sodium has an excess electron in its outer shell and an atom of chlorine has one deficient electron in its outer shell. A molecule of sodium chloride is formed by ionic bond when an atom of sodium combines with an atom of chlorine. The atom which loses an ion becomes a positive ion (cation) and that which gains an ion becomes a negative ion (anion). In ionic bonds, the forces bind the positive ions and negative ions.

$$Na + Cl = Na^{1+}\ Cl^{1-}$$

The number of electrons required to complete the first six shells individually are respectively, 2, 8, 8, 18, 18 and 32. The total number of electrons required to complete are, therefore , 2, 10, 18, 36, 54 and 86. The deficiency or excess of electrons in a particular shell of an element is determined from the number of electrons available and that required to complete the outershell. For example, aluminium has 13 electrons. It has an excess of 3 electrons over the second shell (total 10 electrons). Likewise, oxygen which has 8 electrons, lack 2 electrons in the second shell (total 10 electrons). An atom of hydrogen has equal excess and deficiency. It has only one electron which can be considered either as one deficient in the first shell or one excess electron. Likewise, the atom of silicon has 14 electrons which has equal excess and deficiency of 4 each. It has an excess of 4 over the second shell or a deficiency of 4 in the third shell (total 18 electrons). See Table 6.2 for ionic structure of various elements.

The atoms of two different elements combine to satisfy their individual deficiency or excess. For example, when aluminium and oxygen combine two atoms of aluminium (excess 6) combine with 3 atoms of oxygen (deficiency 6) to form aluminium oxide (Fig. 6.1).

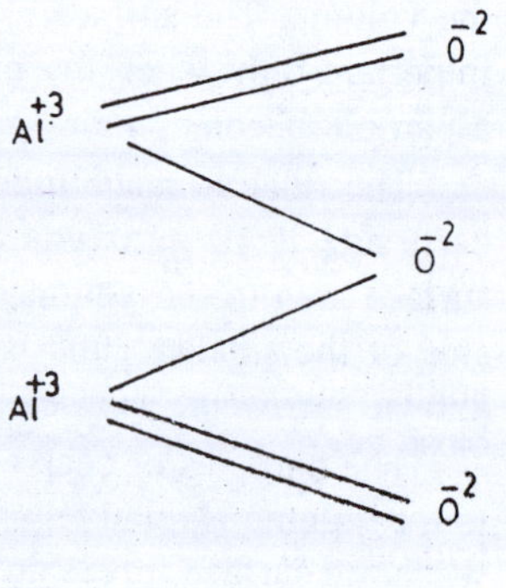

Fig. 6.1. Aluminium oxide

Table 6.2. Ionic Structure of Various Atoms

S. No.	*Element*	*Symbol*	*Number of Electron*	*Defficiency in outer shell*	*Excess in outer shell*	*Remark*
1.	Hydrogen	H	1	− 1	+ 1	Can either lose or gain one ion
2.	Oxygen	O	8	− 2	—	
3.	Silicon	Si	14	− 4	+ 4	Can either lose or gain 4 ions
4.	Aluminium	Al	13	—	+ 3	
5.	Ferrous	Fe	26	—	+ 8	
6.	Calcium	Ca	20	—	+ 2	
7.	Sodium	Na	11	—	+ 1	
8.	Potassium	K	19	—	+ 1	
9.	Magnesium	Mg	12	—	+ 2	
10.	Chlorine	Cl	17	− 1	—	

2. Covalent Bond—Covalent bond develops between two atoms by sharing of electrons in their outer shell. Two atoms, each lacking one electron, may combine by sharing of a pair of electrons. Likewise, two atoms, each lacking two electrons, may combine by sharing four electrons. For example, the bond between two atoms of oxygen in a oxygen molecule is a covalent bond. Each atom lacks 2 electrons in the outer shell. The two atoms bond by sharing 4 electrons in their outer shells. In other words, a covalent bond occurs when there is sharing of electrons by atoms of like valence. The covalent bond occurs generally in elements of negative valences or in non-electrolytes, such as carbon. (A non-electrolyte does not form ions).

Primary valence bonds are very strong. These do not break in normal soil engineering applications. Therefore, primary valence bonds are not of much relevance in soil engineering. However, the study of ionic structure is useful in understanding the behaviours of various atoms.

6.4. HYDROGEN BOND

The hydrogen atom has only one electron. The number of electrons required to fill the first shell is 2. The atom can be considered either as a cation (with one excess electron) or an anion (with one electron deficiency). The bond between the hydrogen cation (H^+) and anions of two atoms of another element is called the hydrogen bond. The hydrogen atom is attracted by two atoms instead of only one atom as suggested by its ionic structure. The hydrogen atom cannot decide to which of the two atoms it should bond itself, and, therefore, it shares its bond with both atoms. The most common example of the hydrogen bond is the bond between the hydrogen atoms and oxygen atoms in a water molecule. The hydrogen atom links one molecule of water to the other (Fig. 6.2).

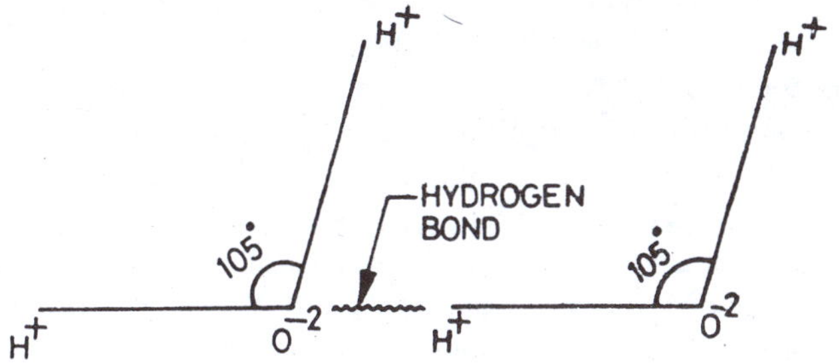

Fig. 6.2. Hydrogen Bond.

A hydrogen atom can bond only two atoms of oxygen, as it is small in size and can fit in only two anions which are of large size (Fig. 6.3). In other words, only two anions can approach the hydrogen cation close enough to form a hydrogen bond.

Only the strong electro-negative atoms, such as oxygen and chlorine, can join with hydrogen to form a hydrogen bond. The hydrogen bond is considerably weaker than primary valence bonds. However, it is fairly strong and cannot be broken during normal soil engineering problems.

6.5. SECONDARY VALENCE BONDS

Secondary valence bonds are intermolecular bonds which develop between atoms in one molecule to atoms in another molecule. A molecule is electrically neutral, *i.e.*, it has no charge. However, the construction of the molecule may be such that the centres of the negative and positive charges do not exactly coincide. The molecule may behave like a small bar magnet, with two electrical poles. Consequently, an electrical moment is developed inside the molecule. A molecule with such a structure is called a *dipole*. In nature, two dipolar molecules orient themselves in such a way that net attraction occurs. The attractive forces so developed are known as *Vander Waal Forces*, after Vander Waal who postulated the existence of a common attractive forces between molecules of all matters in 1873.

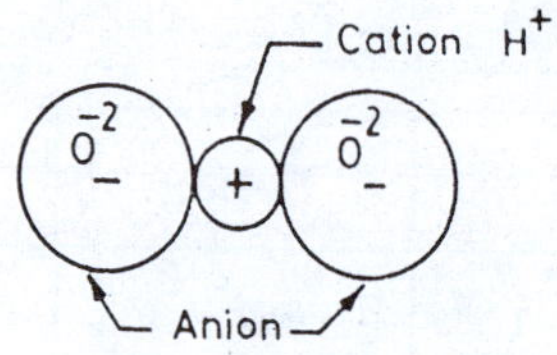

Fig. 6.3. A caation joining two anions.

Vander Waal forces develop due to any one of the following three effects.

(1) Orientation effect—This effect occurs between the oppositely charged ends of permanent dipoles, as shown in Fig. 6.4.

(2) Induction effect—Even in a non-polar molecule, a pole can be induced. When a non-polar molecule is placed in an electric field, it gets polarised and starts behaving as a dipole. Induction effect occurs between an induced pole and another dipole.

(3) Dispersion effect—As all electrons oscillate, the centre of negative charges goes on changing periodically. This results in the formation of a temporary, fluctuating pole. Dispersion effect occurs between a fluctuating pole and another dipole.

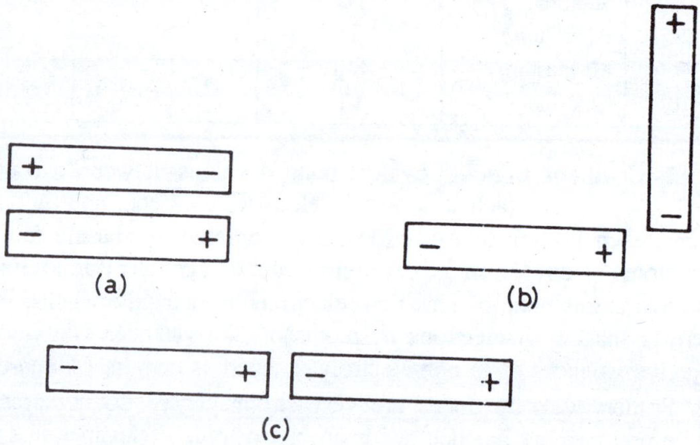

Fig. 6.4. Orientation Effect.

As all molecules behave as permanent or induced or fluctuating dipoles, Vander Waal forces are always present in molecules. These exist in all matters. The relative magnitude of orientation, induction and dispersion effects in a water molecule are 77%, 4% and 19% respectively. Thus the orientation effect is the most predominant effect.

A common example of secondary valence bond is the attractive force between molecules of water. The water molecules act as a bar magnet because the positive and negative charges are not centrally located. It may be noted that all liquids are not dipoles. Some of the liquids, such as kerosene and carbon tetrachloride, are non-polar, as shown by construction in Fig. 6.5.

Vander Waal forces also develop between the surfaces of two parallel particles of clay mineral, separated by water. The magnitude of the forces depend upon the distance between the clay particles, structure of the minerals and the characteristics of water.

The secondary valence bonds are relatively weak and are easily broken. The Vander Waal forces play an important part in the behaviour of clayey soils.

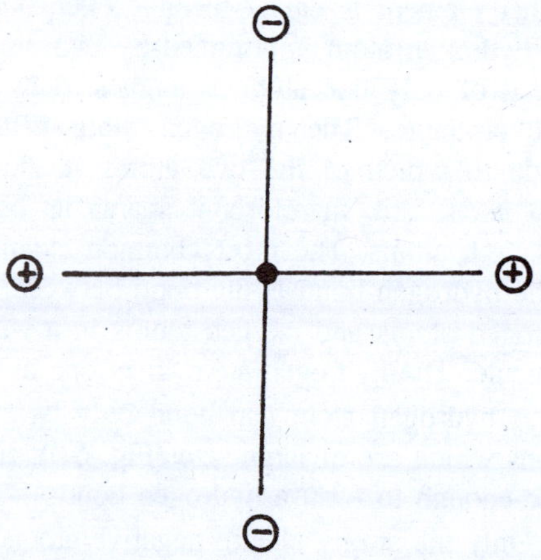

Fig. 6.5. Non-polar System

6.6. BASIC STRUCTURAL UNITS OF CLAY MINERALS

Clay minerals are composed of two basic structural units: (1) Tetrahedral unit, (2) Octahedral unit.

1. Tetrahedral Unit--A tetrahedral unit consists of a silicon atom (Si^{4+}) surrounded by four oxygen atoms (O^{2-}), forming the shape of a tetrahedron. Oxygen atoms are at the tips of the tetrahedron, whereas the silicon atom is at its centre (Fig. 6.6). There is a net negative charge of 4. An individual terahedron unit cannot exist in nature.

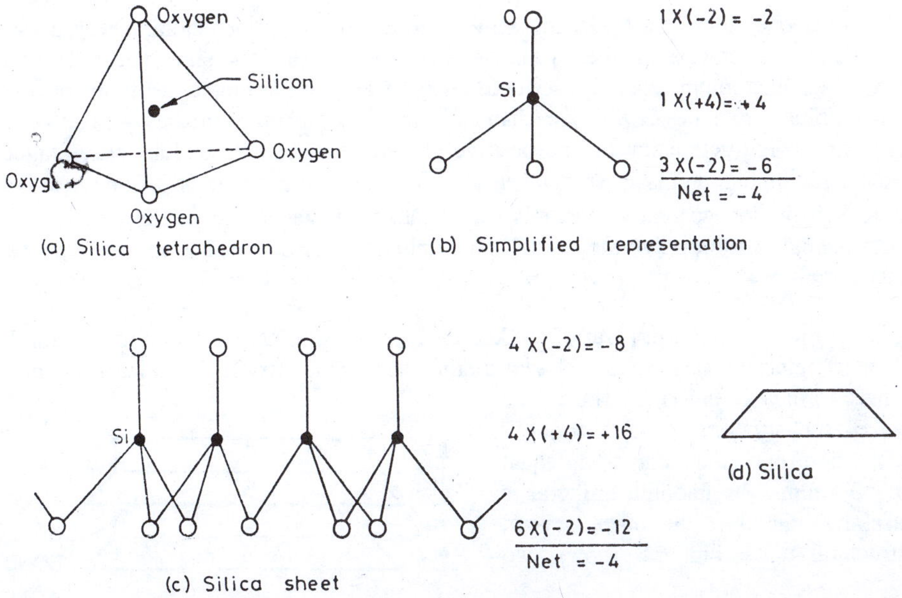

Fig. 6.6. Tetrahedral Unit.

A number of tetrahedral unit combine to form a sheet, with oxygen atoms at the base of all tetrahedra in a common plane, and all the tips pointing in the same direction. Each oxygen atom at the base is shared by two tetrahedra. A silica sheet is formed by terahedral units. The three oxygen atoms at the base being common to two tetrahedra get their negative charge shared and the tip oxygen atom has two negative charges. Thus, there are 5 negative charges and 4 positive charges, leaving a net negative charge of one per tetrahedron. Fig. 6.6 (*c*) shows 4 tetrahedra units combined and having a net negative charge of 4. Fig. 6.6 (*d*) gives a simple representation of *silica sheet,* commonly used in clay minerals.

2. Octahedral Unit—An octahedral unit consists of six hydroxyls (OH^{-1}) forming a configuration of an

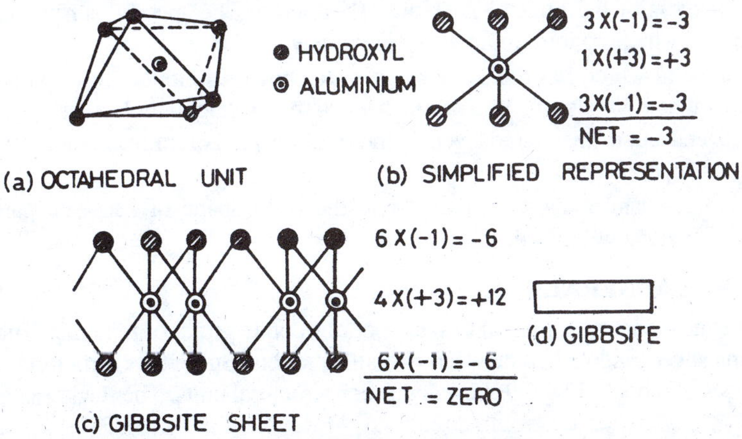

Fig. 6.7. Octahedral Unit.

octahedron and having one aluminium atom at the centre (Fig. 6.7). As the aluminium (Al^{+3}) has three positive charges, an octahedral unit has 3 negative charges. Because of net negative charge, an octahedral unit cannot exist in isolation.

Several octahedral units combine to form a gibbsite sheet. Fig. 6.7 (*c*) shows a *gibbsite sheet* formed by four octahedral units. The sheet is electrically neutral. Fig. 6.7 (*d*) shows a simple representation.

6.7. ISOMORPHOUS SUBSTITUTION

It is possible that one atom in a basic unit may be replaced by another atom. The process is known as isomorphous substitution (isomorphous means same form). For example, one silicon atom in a tetrahedral unit may be substituted by aluminium atom. This would occur if aluminium atoms are more readily available in water. As an aluminium atom has 3 positive charges whereas a silicon atom has 4 positive charges, there would be a net unit charge deficiency of one positive charge per substitution. Likewise, magnesium (Mg^{+2}) atoms may replace aluminium atoms in an octahedral unit and cause a reduction of one positive charge.

Isomorphous substitution generally increases the negative charge on the particle, owing to reduction of positive charges. A slight distortion of the crystal lattice also usually occurs due to isomorphous substitution.

6.8. KAOLINITE MINERAL

Kaolinite is the most common mineral of the kaolinite group of minerals. Its basic structural unit consists of an alumina sheet (gibbsite) (G) combined with a silica sheet (S). Tips of the silica sheet and one base of the alumina sheet form a common interface. The total thickness of the structural unit is about 7 Angstorm (A°), where one Angstorm A° is equal to 10^{-10} m or 10^{-7} mm. The kaolinite mineral is formed by stacking, one over the other, several such basic structural units. Fig. 6.8 shows two such units.

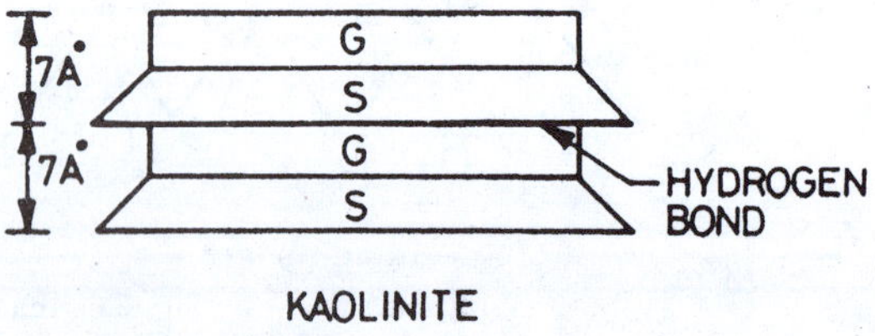

Fig. 6.8. Kaolinite Mineral.

The structural units join together by hydrogen bond, which develops between the oxygen of silica sheet and the hydroxyls of alumina sheet. As the bond is fairly strong, the mineral is stable. Moreover, water cannot easily enter between the structural units and cause expansion.

The kaolinite mineral is electrically neutral. However, in the presence of water, some hydroxyle ions dissociate and lose hydrogen and leave the crystal with a small residual negative charge. The flat surfaces of the mineral attract positive ions (cations) and water. A thick layer of adsorbed water is formed on the surface.

The kaolinite minerals generally have a hexagonal shape in plan, with the side of the hexagon between 0.5 to 1.0 micron. The thickness of the mineral is about 0.05 micron. The specific surface is about 15 m^2/g. The most common example of the kaolinite mineral is China clay.

Halloysite is a clay mineral which has the same basic structure as Kaolinite, but in which the successive structural units are more randomly packed, and are separated by a single molecular layer of water. The properties of halloysite depend upon this water layer. If the water layer is removed by drying, the properties of the mineral drastically change.

Halloysite particles are tubular in shape, in contrast to the platy shape of kaolinite particles. The soils containing halloysite have a very low mass density.

6.9. MONTMORILLONITE MINERAL

Montmorillonite is the most common mineral of the montmorillonite group of minerals. The basic structural unit consists of an alumina sheet sandwiched between two silica sheets. Successive structural units are stacked one over another, like leaves of a book. Fig. 6.9 shows two such structural units. The thickness of each structural unit is about 10 A°.

The two successive structural units are joined together by a link between oxygen ions of the two silica sheets. The link is due to natural attraction for the cations in the intervening space and due to Vander Waal forces. The negatively charged surfaces of the silica sheet attract water in the space between two structural units. This results in an expansion of the mineral. It may also cause dissociation of the mineral into individual structural units of thickness 10 A°. The soil containing a large amount of the mineral montmorillonite exhibits high shrinkage and high swelling characteristics. The water in the interventing space can be removed by heating at 200° to 300°C.

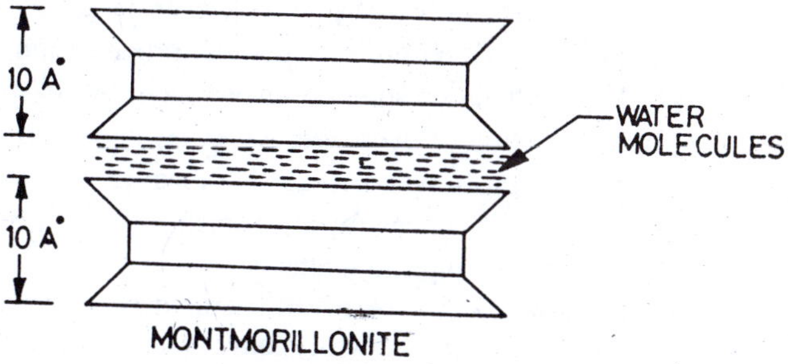

Fig. 6.9. Montmorillonite mineral.

Montmorillonite minerals have lateral dimensions of 0.1 μ to 0.5 μ and the thickness of 0.001 μ to 0.005 μ. The specific surface is about 800 m^2/gm.

The gibbsite sheet in a montmorillonite mineral may contain iron or magnesium instead of aluminium. Some of the silicon atoms in the silica sheet may also have isomorphous substitution. This results in giving the mineral a residual negative charge. It attracts the soil water to form an adsorbed layer, which gives plasticity characteristics to the soil.

6.10. ILLITE MINERAL

Illite is the main mineral of the illite group. The basic structural unit is similar to that of the mineral montmorillonite. However, the mineral has properties different from montmorillonite due to following reasons.

(1) There is always a substantial amount of isomorphous substitution of silicon by aluminium in silica sheet. Consequently, the mineral has a larger negative charge than that in montmorillonite.

(2) The link between different structural units is through non- exchangeable potassium (K^+) and not through water. This bonds the units more firmly than in montmorillonite (Fig. 6.10).

(3) The lattice of illite is stronger than that of montmorillonite, and is, therefore, less susceptible to cleavage.

(4) Illite swells less than montmorillonite. However, swelling is more than in kaolinite.

(5) The space between different structural units is much smaller than in montmorillonite, as the potassium ions just fit in between the silica sheet surfaces.

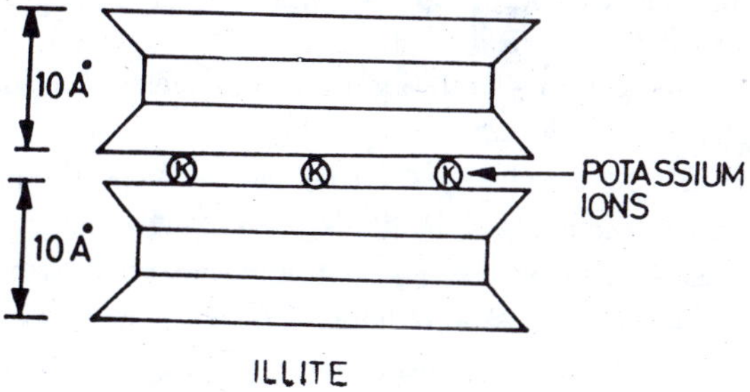

Fig. 6.10. Illite Mineral.

The properties of the mineral illite are somewhat intermediate between that of kaolinite and montmorillonite. The bond between the non-exchangeable K^+ ions, though stronger than that in montmorillonite, is considerably weaker than hydrogen bond of kaolinite. The swelling of illite is more than that of kaolinite, but less than that of montmorillonite.

The lateral dimensions of the mineral illite are the same as that of the mineral montmorillonite, equal to 0.1 μ to 0.5 μ. However, the thickness is much greater than that of montmorillonite and is between 0.005 μ and 0.05 μ. The specific surface is about 80 m^2/gm.

6.11. ELECTRICAL CHARGES ON CLAY MINERALS

As mentioned before, the particles of clay carry an electric charge. This fact can be proved by inserting two electrodes in a beaker containing clay mixed with water. When the electrodes are connected to an electrical circuit containing a battery and an ammeter, there is a deflection of the needle of the ammeter. This proves that there is a flow of current through the medium. Theoretically, a soil particle can carry either a negative charge or positive charge. However, in actual tests, only negative charges have been measured.

The net negative charge may be due to one or more of the following reasons.

(1) Isomorphous substitution of one atom by another of lower valency.

(2) Dissociation of hydroxyle ion (OH) into hydrogen ions.

(3) Adsorption of anions (negative ions) on clay surface.

(4) Absence of cations (positive ions) in the lattice of the crystal.

(5) Presence of organic matter.

The magnitude of the electrical charge depends on the surface area of the particle. It is very high in small particles, such as colloids, which have very large surface area. A soil particle attracts the cations in the environment to neutralise the negative charge. The phenomenon is known as *adsorption.*

6.12. BASE EXCHANGE CAPACITY

The cations attracted to the negatively charged surface of the soil particles are not strongly attached. These cations can be replaced by other ions and are, therefore, known as exchangeable ions. The soil particle and the exchangeable ions make the system neutral.

The phenomenon of replacement of cations is called cation exchange or base exchange. The net negative charge on the mineral which can be satisfied by exchangeable cations is termed *cation-exchange capacity or base-exchange capacity.* In other words, base-exchange capacity is the capacity of the clay particles to change the cation adsorbed on the surface.

Base-exchange capacity is expressed in terms of the total number of positive charges adsorbed per 100 gm of soil. It is measured in milliequivalent (meq), which is equal to 6×10^{20} electronic charges. Thus, one meq per 100 gm means that 100 gm of material can exchange 6×10^{20} electronic charges if the exchangeable ions are univalent, such as Na^+. However, if the exchangeable ions are divalent, such as Ca^{2+}, 100 gm of material will replace 3×10^{20} calcium ions.

According to another definition, one milliequivalent (meq) is also equal to one milligram of hydrogen or its equivalent other material which will replace one milligram of hydrogen. For example, calcium has a molecular weight of 40, whereas that of hydrogen is unity. However, calcium is divalent in contrast to hydrogen which is univalent. Therefore, one mg of hydrogen is equivalent to 20 mg of calcium in base exchange capacity. If 100 gm of a dry material adsorbs 60 mg of calcium, the base exchange capacity of the material is 60/20 *i.e.* 3 meq/100 gm.

The base-exchange capacity of clay depends upon the P_H value of the water in the environment. If the water is acidic ($P_H < 7$), the base-exchange capacity is reduced.

Some cations are more strongly adsorbed than others. The adsorbed cations commonly found in soils, arranged in a series in terms of their affinity for attraction are as follows:

$$Al^{3+} > Ca^{2+} > Mg^{+2} > NH_4^+ > H^+ > Na^+ > Li^+$$

For example, Al^{3+} cations are more strongly attracted than Ca^{2+} cations. Thus Al^{3+} ions can replace Ca^{2+} ions. Likewise, Ca^{2+} ions can replace Na^+ ions.

The base formula of the clay mineral is altered by base exchange. For example, if calcium chloride is added to a soil containing sodium chloride, there would be an exchange of Ca^{2+} ions for Na^+ ions, and the sodium clay would turn into the calcium clay. Thus

$$\text{Sodium clay} + Cl_2 = \text{Calcium clay} + NaCl.$$

The properties of the clay therefore change due to base exchange.

The base exchange capacity of the montmorillonite mineral is about 70—100 meq per 100 gm. However, that of kaolinite and illite are respectively 4.0 and 40.0 meq per 100 g.

6.13. DIFFUSE DOUBLE LAYER

The faces of clay minerals carry a net negative charge. The edges of the mineral may have either positive charges or negative charges. The charges in clay minerals are due to molecular grouping and arrangement of ions. The electrical charges in the minerals are responsible for their behaviour when they come in contact with other particles and with water present in the soil. Clay deposits, because of their sedimentary nature, always exist in the presence of water.

Because of the net negative charge on the surface, the clay particles attract cations, such as potassium, calcium and sodium, from moisture present in the soil to reach an electrically balanced equilibrium. These cations, in turn, attract particles with negative charges and water dipoles.

(The engineering behaviour of coarse particles is not affected by surface electrical charges, because of their low ratio of surface area to volume. In such soils, the gravitational forces are more important).

The plasticity characteristics of clays are because of the unusual molecular structure of water in soil deposits. Experiments conducted with clays using nonpolar liquid, such as kerosene, in place of water, has shown that plasticity does not occur, and the soil behaves as a coarse-grained sands soil.

The water molecule is a dipole, since the hydrogen atoms are not symmetrically oriented around the oxygen atoms. The molecule acts as a bar magnet (Fig. 6.11). As the faces of clay particles carry a negative

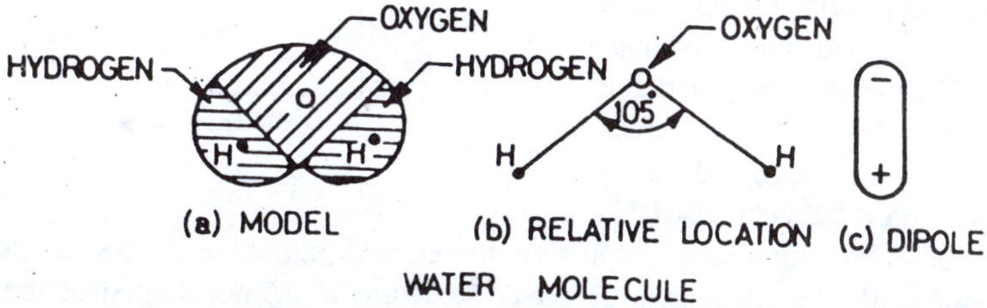

Fig. 6.11. Structure of a water molecule. (*a*) Model, (*b*) Relative location, (*c*) Dipole water molecule

charge, there is attraction between the negatively charged faces and the positive ends of dipoles [Fig. 6.12 (*a*)]. The second mode of attraction between the water dipoles and the clay surface is through cations [Fig. 6.12 (*b*)]. Cations are attracted to the soil surface and water dipoles are attached to these cations through their

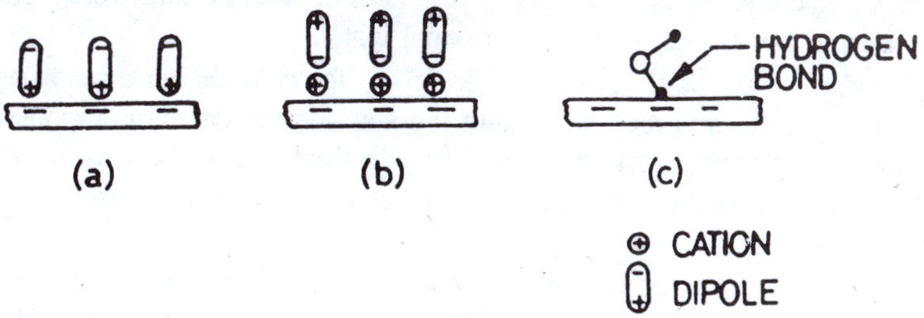

Fig. 6.12. Attraction of water molecules to soil surface.

negative charged ends. The third possible mode by which the attraction between the water and the clay surface occurs is by sharing of the hydrogen atom in the water molecule by hydrogen bonding between the oxygen atoms in the clay particles and the oxygen atoms in the water molecules [Fig. 6.12 (*c*)].

The cations attracted to a clay mineral surface also try to move away from the surface because of their thermal energy. The net effect of the forces due to attraction and that due to repulsion is that the forces of attraction decrease exponentially with an increase in distance from the clay particles surface. The layer extending from the clay particle surface to the limit of attraction is known as the *diffuse double layer* (Fig. 6.13).

It is believed that immediately surrounding the particle, there is a thin, very tightly held layer of water (called stern layer) about 10 A° thick. Beyond this thickness there is a second layer, in which water is more mobile. This second layer extends to the limit of attraction, and is known as diffuse-double layer (Fig. 6.13). The water held in the diffuse-double layer is known as adsorbed water or oriented water. Outside the diffuse double layer, the water is normal, non-oriented. The total thickness of the diffuse-double layer is about 400 A°.

For a given soil particle, the thickness of the cation layer depends mainly on the valency and concentration of cations. The mono-valent cations, such as Na^+, lead to a thicker layer compared to that by divalent cations, such as Ca^{2+}. The number of monovalent cations required is twice the number of divalent cations. Increasing the concentration of cations close to the surface, reduces the thickness of the cation layer required to neutralise the negative charge.

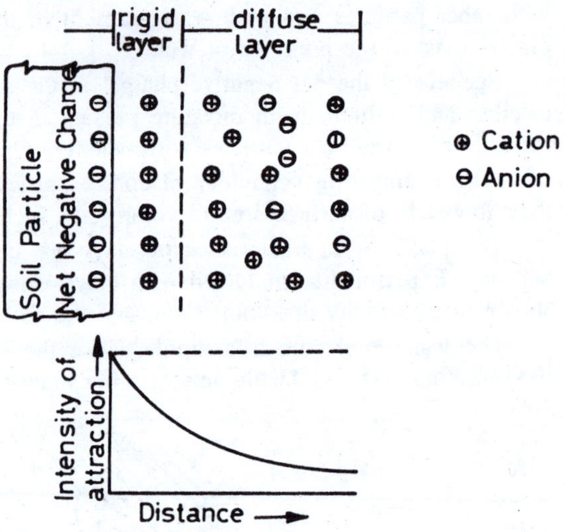

Fig. 6.13. Diffuse Double Layer.

Repulsion occurs between the like charges of the two double layers of two particles. The forces of repulsion between the two particles depend upon the characteristics of the double layers. An increase in cation valency or concentration results in a decrease of repulsive forces. However, Vander Waal forces of attraction do not depend upon the characteristics of the double layer. These forces decrease rapidly with an increasing distance between particles. The net force between particles will depend upon the relative magnitudes of repulsive forces and attractive forces. Depending upon the relative magnitudes, the soil will take either as a dispersed structure or a flocculent structure, as discussed in Section 6.15.

6.14. ADSORBED WATER

The water held by electro-chemical forces existing on the soil surface is adsorbed water. As the adsorbed water is under the influence of electrical forces, its properties are different from that of normal water. It is much more viscous, and its surface tension is also greater. It is heavier than normal water. The boiling point is higher, but the freezing point is lower than that of the normal water.

The thickness of the adsorbed water layer is about 10 to 15 A° for colloids but may be upto 200 A° for silts. The attractive forces between the adsorbed water and the soil surface decrease exponentially with the distance until the double layer merges into normal water. The adsorbed water exists in an almost solidified state. The pressure required to pull away the adsorbed water layer from the soil surface is very high; it may be as high as 10,000 atmospheres.

Adsorbed water imparts plasticity characteristics to soils. The adsorbed water depends upon the clay minerals present in the soil. The presence of highly active clay minerals is necessary to give the soil plasticity. The fine-grained soils without clay minerals may develop cohesion if the particle size is very small, but these soils cannot be moulded into small threads as these are not plastic.

6.15. SOIL STRUCTURES

The geometrical arrangement of soil particles with respect to one another is known as soil structure. The soils in nature have different structures depending upon the particle size and the mode of formation. The following types of structures are usually found. The first two types are for coarse-grained soils and types (3) and (4) for clays. Types (5) and (6) are for mixed soils.

(1) Single-grained Structure—Cohesionless soils, such as gravel and sand, are composed of bulky grains in which the gravitational forces are more predominant than surface forces. When deposition of these soils occurs, the particles settle under gravitational forces and take an equilibrium position as shown in Fig. 6.14 (*a*). Each particle is in contact with those surrounding it. The soil structure so formed is known as single-grained structure. The arrangement is somewhat similar to the stacking of oranges on a grocer's counter or to that of marble pieces at a toys' shop.

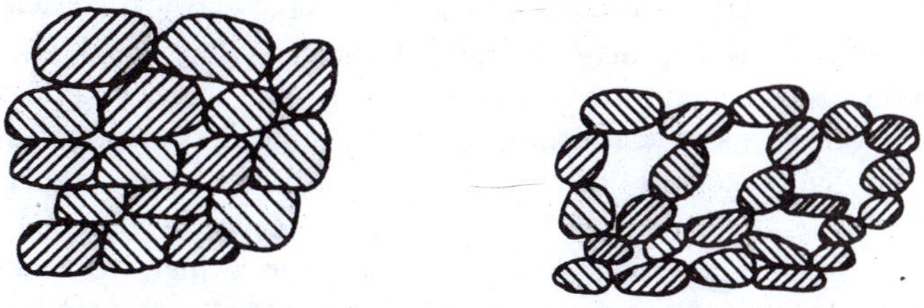

(a) Single Grained Structure (b) Honey-comb Structure

Fig. 6.14. Soil structure in sands and silts. (*a*) Single Grained Structure, (*b*) Honey-comb Structure

Depending upon the relative position of the particles, the soil may have a loose structure or a dense structure. Fig. 6.15 shows spherical particles in the loosest and those in the densest condition. It can be proved that for the loosest condition, the void ratio is 0.90 and that for the densest state, is 0.35. In actual sand deposits, as the particles are not exactly spherical, the void ratio between the loosest and densest conditions varies between 0.90 and 0.35.

As mentioned in chapter 3, the engineering properties of sands improve considerably with a decrease in void ratio or an increase in relative density. In general, the smaller the void ratio, the higher the shear strength, and the lower the compressibility and permeability. Loose sands are inherently more unstable. When subjected to shocks and vibrations, the particles move into a more dense state. Dense sands are quite stable, as they are not affected by shock and vibrations.

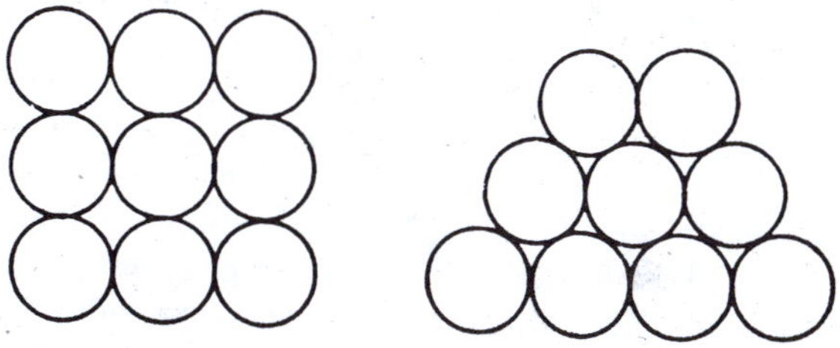

(a) LOOSE (b) DENSE

Fig. 6.15. Spheres in loosest and densest states.

(2) Honey-Comb Structure—It is possible for fine sands or silts to get deposited such that the particles when settling develop a particle-to-particle contact that bridges over large voids in the soil mass [Fig. 6.14(*b*)]. The particles wedge between one another into a stable condition and form a skeleton like an arch to carry the weight of the overlying material. The structure so formed is known as honey-comb structure. The honey-comb structure usually develops when the particle size is between 0.002 mm and 0.02 mm.

Honey-comb structure occurs in soils having small granular particles which have cohesion because of their fineness. The particles are held in position by mutual attraction due to cohesion. The particles, however, do not possess plasticity characteristics associated with clayey soils.

Soils in honey-comb structure are loose. They can support loads only under static conditions. Under vibrations and shocks, the structure collapses and large deformations take place. In nature, honey-comb structure usually occurs in small pockets, and can be easily detected.

Honey-comb structure can also develop when fine sand is dumped into a filling without densification of or when water is added to dry fine sand. The phenomenon is known as *bulking* of sand (see chap. 7).

(3) Flocculated Structure—Flocculated structure occurs in clays. The clay particles have large surface area and, therefore, the electrical forces are important in such soils. The clay particles have a negative charge on the surface and a positive charge on the edges. Interparticle contact develops between the positively charged edges and the negatively charged faces. This results in a flocculated structure [Fig. 6.16(*a*)].

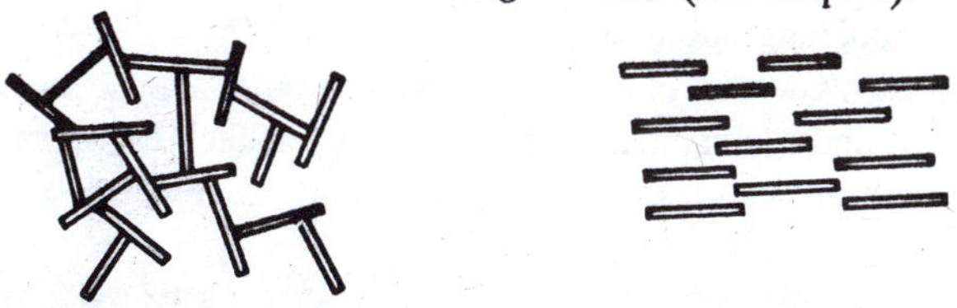

(a) Flocculated Structure (b) Dispersed Structure

Fig. 6.16. Soil structure in clay (*a*) Plocculated Structure, (*b*) Dispersed Structure

Flocculent structure is formed when there is a net attractive force between particles.

When clay particles settle in water, deposits formed have a flocculated structure. The degree of flocculation of a clay deposit depends upon the type and concentration of clay particles, and the presence of salts in water. Clays settling out in a salt water solution have a more flocculent structure than those settling out in a fresh water solution. Salt water acts as an electrolyte and reduces the repulsive forces between the particles.

Soils with a flocculent structure are light in weight and have a high void ratio and water content. However, these soils are quite strong and can resist external forces because of a strong bond due to attraction between particles. The soils are insensitive to vibrations. In general, the soils in a flocculated structure have a low compressibility, a high permeability and a high shear strength.

(4) Dispersed Structure—Dispersed structure develops in clays that have been reworked or remoulded. The particles develop more or less a parallel orientation [Fig. 6.16 (*b*)]. Clay deposits with a flocculent structure when transported to other places by nature or man get remoulded. Remoulding converts the edge-to-face orientation to face-to-face orientation. The dispersed structure is formed in nature when there is a net repulsive force between particles.

The soils in dispersed structure generally have a low shear strength, high compressibility and low permeability. Remoulding causes a loss of strength in a cohesive soil. With the passage of time, however, the soil may regain some of its lost strength. Due to remoulding, the chemical equilibrium of the particles and associated adsorbed ions and water molecules within the double layer is disturbed. The soil regains strength as a result of re- establishing a degree of chemical equilibrium. This phenomenon of regain of strength with the passage of time, with no change in water content, is known as thixotropy, as already discussed in chapter 4.

(5) Coarse-grained Skeleton—A coarse-grained skeleton is a composite structure which is formed when the soil contains particles of different types. When the amount of bulky, cohesionless particles is large compared with that of fine-grained clayey particles, the bulky grains are in particle-to-particle contact. These particles form a framework or skeleton [Fig. 6.17 (*a*)]. The space between the bulky grains is occupied by clayey particles, known as binders. In nature, the bulky grains are deposited first during sedimentation and the *binder* is subsequently deposited.

Fig. 6.17. Composite Structure (*a*) coarse Graind Skeltion, (*b*) Clay Matrix

As long as the soil structure is not disturbed, a coarse-grained skeleton can take heavy loads without much deformations. However, when the structure is disturbed, the load is transferred from the coarse-grained particles to clayey particles, and the supporting power and the stability of the soil is considerably reduced.

(6) Clay-Matrix Structure—Clay-matrix structure is also a composite structure formed by soils of different types. However, in this case, the amount of clay particles is very large as compared with bulky, coarse- grained particles [Fig. 6.17 (*b*)]. The clay forms a matrix in which bulky grains appear floating without touching one another.

The soils with a clay-matrix structure have almost the same properties as clay. Their behaviour is similar to that of an ordinary clay deposit. However, they are more stable, as disturbance has very little effect on the soil formation with a clay-matrix structure.

PROBLEMS

A. Numerical

6.1. A dry mineral has a mass of 100 gm and adsorbs 50 mg of calcium. Determine its base exchange capacity.

[**Ans.** 2.5 meg per 100 mg]

B. Descriptive and Objective type

6.2. Differentiate between gravitational forces and surface forces. What is the effect of increased surface area on the properties of soils.

6.3. What are primary valence bonds ? What is their importance in soil engineering ?

6.4. What do you understand by hydrogen bond ? Give examples.

6.5. What are secondary valence bonds ? Write a short note on Vander Waal forces.

6.6. Describe the constitution of the two basic structural units required in the formation of clay minerals. Are these electrically neutral ?

6.7. Discuss the characteristics and the construction of Kaolinite, Montmorillonite and Illite mineral groups.

6.8. Write short notes on:

(*i*) Base exchange capacity. (*ii*) Isomorphous substitution.

(*iii*) Electrical double layer (*iv*) Adsorbed water

6.9. What are different types of soil structures which can occur in nature ? Describe is brief.

6.10. State whether the following statements are true or false.

(*a*) The mineral quartz is electrically active.

(*b*) The clay minerals are responsible for plasticty characteristics of soils.

(*c*) The hydrogen bond is stronger than secondary valence bonds.

(*d*) Isomorphous substitution does not change the electrical charge.

(*e*) The soils containing the mineral halloysite have a high unit weight.

(*f*) The mineral montmorillonite causes excessive swelling and shrinkage.

(*g*) The adsorbed water imparts plasticity to soils.

(*h*) Honey-comb structure occurs in clayey soils.

(*i*) Remoulded fine-grained soils have a flocculated structure. [**Ans.** True, (*b*), (*c*), (*f*), (*g*)]

C. Multiple-Choice Questions.

1. The behaviour of clay is governed by

(*a*) Mass energy (*b*) Surface energy

(*c*) Both (*a*) and (*b*) (*d*) Neither (*a*) and (*b*)

2. Honey-combed structure is found in

(*a*) Gravels (*b*) Coarse sands

(*c*) Fine sands and silts (*d*) clay

3. The weakest bond in soils is

(*a*) Ionic bond (*b*) Covalent bond

(*c*) Hydrogen bond (*d*) Secondary valance bond

4. An octahedral unit has

(*a*) Four negative charges (*b*) Three negative charges

(*c*) One negative (*d*) No negative charge

5. In illite mineral, the bond between structural units is

(*a*) Hydrogen bond (*b*) Potassium ion bond

(*c*) Water molecules bond (*d*) Covalent bond

6. The plasticity characteristics of clays are due to

(*a*) Adsorbed water (*b*) Free water

(*c*) Capillary water (*d*) None of above

7. In fine sands and silts, the most common type structure is

(*a*) Single grained (*b*) Honey comb

(*c*) Flocculated (*d*) Dispersed

8. The base exchange capacity of the mineral montmorrillonite is about

(*a*) 70 meq/100 g (*b*) 700 meq/100 g

(*c*) 7 meq/100 g (*d*) 40 meq/100 g

[**Ans.** 1. (*b*), 2. (*c*), 3. (*d*), 4. (*b*), 5. (*b*) 6. (*a*), 7 (*b*), 8 (*a*)]

7

Capillary Water

7.1. TYPES OF SOIL WATER

The soil water is broadly classified into two categories: (1) Free water, and (2) Held water. Free water moves in the pores of the soil under the influence of gravity. The held water is retained in the pores of the soil, and it cannot move under the influence of gravitational force.

Free water flows from one point to the other wherever there is a difference of total head. The rate at which the head is lost along the flow passage is equal to the hydraulic gradient. The flow of free water in soils is just like laminar flow in pipes. Because of very small flow passages in the soil, the velocity head is generally neglected, and the total head is taken equal to the sum of the elevation head and the pressure head. Free water is discussed in chapter 8.

Held water is further divided into three types : (1) Structural water, (2) Adsorbed water, and (3) Capillary water. The structural water is chemically combined water in the crystal structure of the mineral of the soil. This water cannot be removed without breaking the structure of the mineral. A temperature of more than 300 °C is required for removing the structural water. In soil engineering, the structural water is considered as an integral part of the soil solid.

The water held by electrochemical forces existing on the soil surface is known as adsorbed water, as discussed in chapter 6. The quantity of adsorbed water depends upon the colloidal fraction in the soil, the chemical composition of the clay mineral and the environment surrounding the particle. The adsorbed water is important only for clayey soils. For coarse- grained soils, its amount is negligible or zero.

The adsorbed water is also sometimes called *hygroscopic water*. The amount of water in an air-dried soil is defined as hygroscopic water. Since air drying removes capillary water and free water, the remaining water is approximately equal to the adsorbed water. Hygroscopic water depends upon the humidity and temperature of air. It is assumed that oven drying removes all the hygroscopic water. The amount of water in an air-dried sample, expressed as a percentage of the dry mass, is known as hygroscopic water content.

The water held in the interstices of soils due to capillary forces is called capillary water. This chapter discusses mainly the capillary water and its effect on soils.

7.2. SURFACE TENSION

To understand surface tension, let us consider a molecule of water surrounded by other molecules in the body of water, as shown in Fig. 7.1 (*a*). The forces due to molecular attraction act all around, and the molecule is in equilibrium. However, at the free surface, as shown in Fig. 7.1 (*b*), the pull from the air above is smaller than the pull from the water molecules below and the equilibrium is disturbed.

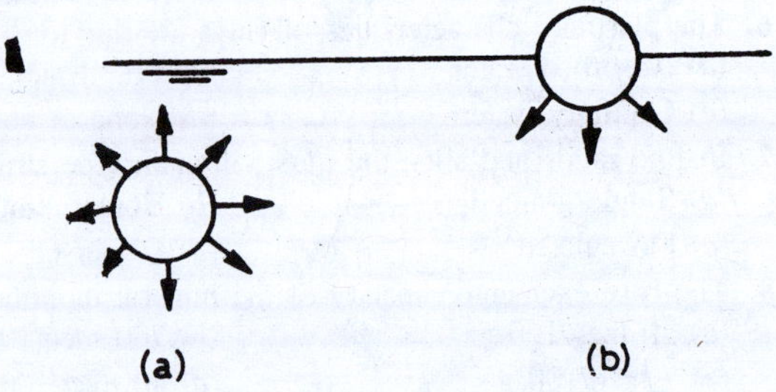

Fig. 7.1. Effect of Surface tension.

The forces tend to reduce the surface area of the air-liquid surface to a minimum. The surface assumes a curved shape to maintain equilibrium. The interface behaves like a stretched membrane or a skin. The surface tension exists at the interface. Surface tension is defined as the force per unit length of a line drawn on the surface. It acts in the direction normal to that line. The surface tension of water at normal temperature is about 0.073 N/m at 20°C. It decreases with an increase in temperature.

It is because of surface tension that a small needle can float on water, and insects can walk on it.

Capillary water exists in soils so long as there is an air-water interface. As soon as the soil is submerged under water, the interface is destroyed, and the capillary water becomes normal, free water. The capillary water is always under tension (negative pressure). However, the properties of the capillary water are the same as that of normal, free water.

7.3. CAPILLARY RISE IN SMALL DIAMETER TUBES

Water rises in small diameter, capillary tubes, because of adhesion and cohesion. Adhesion occurs because water adheres or sticks to the solid walls of the tube. Cohesion is due to mutual attraction of water molecules. If the effect of cohesion is less significant than the effect of adhesion, the liquid wets the surface and the liquid rises at the point of contact. However, if the effect of cohesion is more predominant than adhesion, the liquid level is depressed at the point of contact, as in the case of mercury.

If a glass tube of small diameter, open at both ends, is lowered into water, the water level rises in the tube, as the water wets the tube. Let θ be the angle of contact between the water and the wall of the tube [Fig. 7.2 (*a*)].

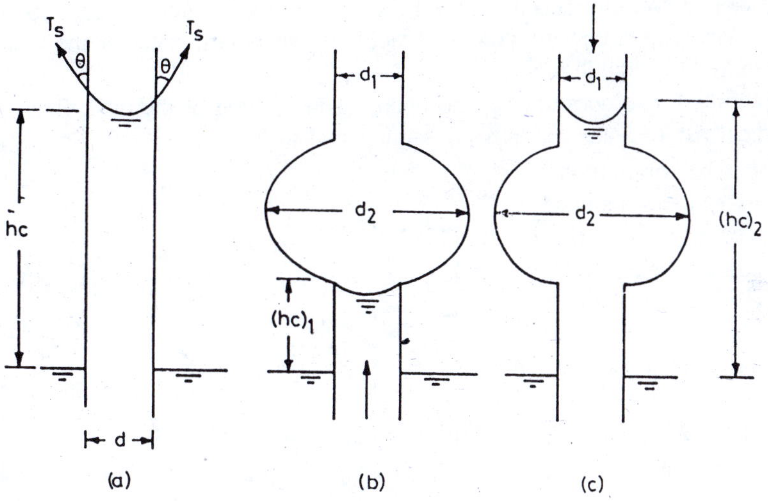

Fig. 7.2. Capillary Rise

F_u = Upward pull due to surface tension = $(T_s \cos\theta)\,\pi d$

where T_s = surface tension and d diameter of the tube.

F_d = Downward force due to mass of water in the tube

$= \gamma_w (\pi/4\, d^2) \times h_c$

where h_c = height of capillary rise.

For equilibrium, $$F_u = F_d$$

or $$(T_s \cos\theta)\,\pi d = \gamma_w (\pi/4\, d^2)\, h_c$$

or $$h_c = \frac{4\,T_s \cos\theta}{\gamma_w d} = \frac{4\,T_s \cos\theta}{g\,\rho_w d} \quad \text{...(7.1)}$$

For a clean glass tube and pure water, the meniscus is approximately hemispherical, *i.e.* θ = 0. Therefore,

$$h_c = \frac{4\,T_s}{\gamma_w\, d} \qquad \text{...(7.2)}$$

Taking $T_s = 0.073 \text{ N/m}$, $\gamma_w = 9810 \text{ N/m}^3$,

$$h_c = \frac{4 \times 0.073}{9810\, d} = \frac{3 \times 10^{-5}}{d} \text{ metres}$$

where d is in metres.

If d is in centimeters, $h_c = \dfrac{3 \times 10^{-3}}{d}$ metres

If h_c and d both are in cm, $h_c = \dfrac{0.3}{d}$ cm ...(7.3)

Capillary rise in tubes of non-uniform diameter depends upon the direction of flow of water. If a tube with a large bulb is dipped in water, the water is lifted due to capillary action, but it may not rise past the bulb where the diameter is d_2 [Fig. 7.2 (*b*)]. The capillary rise is limited to a height of $(h_c)_1$ because water cannot maintain equilibrium at a large diameter d_2.

If the same tube, with a large bulb is filled by pouring water from above or by lowering the tube below the water level and then raising when filled, an equilibrium is maintained at a height $(h_c)_2$ [Fig. 7.2 (*c*)]. The water is able to maintain equilibrium at the diameter d_1 above the bulb.

Thus the capillary rise in tubes of non-uniform diameter is more if the flow is downward than when it is upward. The capillary rise does not depend upon the shape and the diameter of the tube below the meniscus when the flow is downward. In upward flow, the capillary rise is terminated if the diameter of the tube is greater than that required for equilibrium.

The height of capillary rise does not depend upon the inclination of the tube. Even if the capillary tube is inclined, the vertical rise of water remains the same, equal to h_c.

In a capillary tube of uniform diameter, no water can be retained when lifted. The upward forces (F_u) due to surface tension are balanced by downward forces (F_d) at the lower end [Fig. 7.3 (*a*)]. However, if the tube is necked, with smaller diameter at top, the upward force (F_u) is greater than the downward force (F_d), and some water can be retained in the tube [Fig. 7.3 (*b*)].

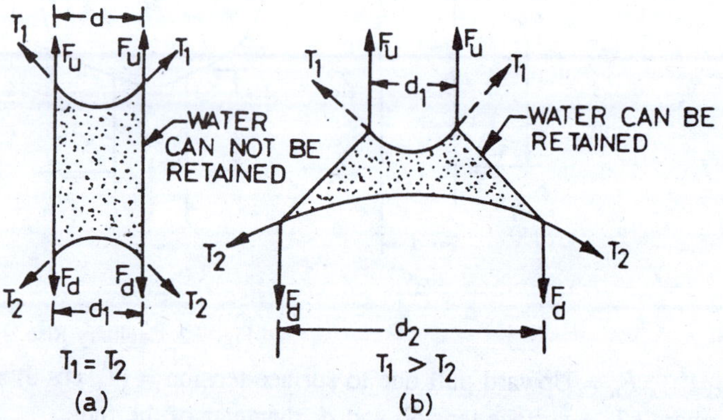

Fig. 7.3. Capillary Tube when lifted

7.4. CAPILLARY TENSION

The water in a capillary tube is under a negative pressure, commonly called *tension*. The pressure variation in the capillary tube of Fig. 7.4 (*a*) can be determined as follows. The pressure at point *D* at the free surface is atmospheric *i.e.*, equal to zero gauge pressure. (In soil engineering, generally gauge pressures are used). The pressure at point *C*, which is at the same level as point *D*, is also zero. From laws of hydrostatic, the pressure at point *B*, which is at a height of h_c above the free surface, is given by

$$p_B = -\,\gamma_w\, h_c \qquad \text{...(7.4)}$$

The pressure is negative because it is less than atmospheric pressure. In other words, the water at point *B* is under tension.

The capillary rise at any point *E* is *h*, and therefore the pressure is given by

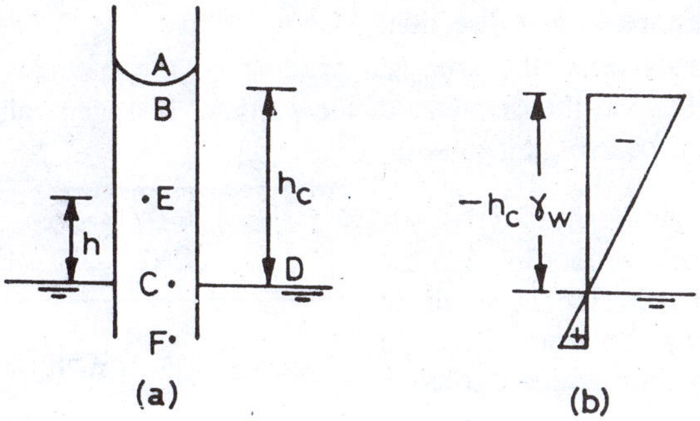

Fig. 7.4. Pressure Variation.

$$p_E = -\gamma_w h$$

The capillary tension, therefore, varies linearly with the height of point above the water surface, as shown in Fig. 7.4 (*b*). The pressure at point *F* below the water surface is, of course, positive (hydrostatic).

As the capillary tube is open to atmosphere, the pressure at point *A* above the meniscus is atmospheric, *i.e.* zero. Therefore, the pressure difference across the two sides of the meniscus is equal to $\gamma_w h_c$. The pressure difference is also known as *pressure deficiency* (p'').

Thus $$p'' = \gamma_w h_c$$

Substituting the value of h_c from Eq. 7.2,

$$p'' = \gamma_w \left(\frac{4T_s}{\gamma_w d}\right) = \frac{4T_s}{d} \quad \text{...(7.5)}$$

If the meniscus is not hemi-spherical and it has diameters d_1 and d_2 in two orthogonal directions, it can be shown that

$$p'' = T_s\left(\frac{2}{d_1} + \frac{2}{d_2}\right) \quad \text{...(7.6)}$$

Capillary water can be likened to hanging of a weight to the inside walls of a chimney. The walls of the chimney support the load and transfer it as reaction to the base. The weight causes compressive stresses in the walls of the chimney. In a similar manner, the capillary water causes compression in the walls of the glass tube. The compressive force (*F*) is equal to the weight of suspended column of water.

$$F = \left(\frac{\pi}{4} d^2 h_c\right) \gamma_w \quad \text{...(7.7)}$$

The compressive stress in the wall of the tube can be determined from the contact area and the compressive force. The compressive stress is *constant* in the entire height h_c of the tube.

7.5. CAPILLARY RISE IN SOILS

The water which falls on the ground as rain flows under gravity and passes through the soil and reaches a surface known as *ground water table*. The soil is saturated below the ground water table. The level to which underground water rises in an observation well is called ground water table or simply water table. The ground water table is also called as the ground *phreatic surface*, a term derived from the Greek word *phreos*, meaning a well. Ground water, which is a form of free water, is not static. It is a moving stream which flows under gravitational force. The water table is not horizontal. It takes the shape according to the topography. The water is drawn above the water table (abbreviated at W.T.) due to capillary action.

A soil mass consists of a number of interconnected interstices which act as capillary tubes of varying diameters. Although the channels formed by interconnected interstices are not circular in cross-section, the results of capillary rise in circular tubes are useful for understanding the phenomenon of capillary rise in soils. The channels formed in the soil are a sort of capillary tubes of varying diameter but not necessarily vertical. These capillary tubes may be inclined in any direction.

Capillary rise in soils depends upon the size and grading of the particles. The diameter (d) of the channels in pore passage depends upon the diameter of the particle. It is generally taken as one-fifth of the effective diameter (D_{10}) in case of coarse-grained soils.

Thus, $d = 0.2\ D_{10}$

As the capillary rise is inversely proportional to the diameter of the tube, the capillary rise is small in coarse-grained soils, but it may be very large in fine-grained soils. In some very fine-grained soils, it may be even more than 30 m.

The space above the water table can be divided into two regions : (1) Zone of capillary saturation, in which the soil is fully saturated. (2) Zone of aeration, in which the soil is not saturated (Fig. 7.5). The height to which capillary water rises in soils is known as *capillary fringe*. It includes the zone of capillary saturation and a part of the zone of aeration in which the capillary water exists in interconnected channels.

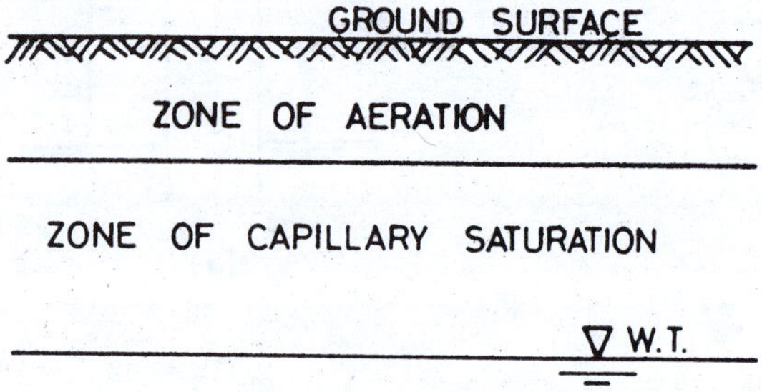

Fig. 7.5. Capillary zone.

The soil above the capillary fringe may contain water in the form of contact water (Fig. 7.6). In this form, water forms a meniscus around the point of contact. Surface tension holds the water in contact with soil. Because of the tension in the capillary water, there is an equal and opposite force induced at the points of contacts which presses the particles together. The contact pressure depends upon the water content, particle size, angle of contact and density of packing. The contact pressure decreases as the water content increases because of an increase of radius of meniscus. Eventually, a stage is reached when the contact pressure becomes zero as soon as the soil becomes fully saturated.

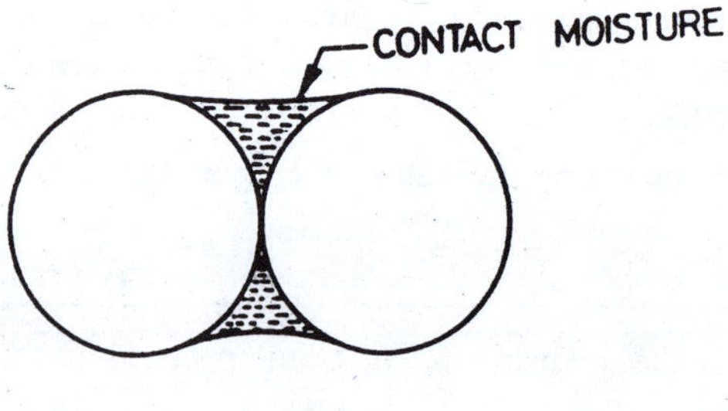

Fig. 7.6.

Terzaghi and Peck (1948) gave a relationship between the maximum height of capillary fringe and the effective size, as

$$(h_c)_{max} = \frac{C}{eD_{10}} \quad \ldots(7.8)$$

where C = constant, depending upon the shape of the grain and impurities.

e = void ratio.

D_{10} = effective diameter, the size corresponding to 10% percentage finer.

If D_{10} is in mm, the value of C varies between 10 to 50 mm^2, and the height $(h)_{max}$ is also given in mm. If D_{10} and $((h_c)_{max}$ are in centimeters, C = 0.1 to 0.5 cm^2.

Table 7.1 gives representative heights of capillary rise in different soils.

Table 7.1. Representative Heights of Capillary Rise

S.No.	*Soil Type*	*Capillary rise (m)*
1.	Fine gravel	0.02 to 0.10
2.	Coarse sand	0.10 to 0.15
3.	Fine sand	0.30 to 1.00
4.	Silt	1.0 to 10.0
5.	Clay	10.0 to 30.0
6.	Colloid	more than 30.0

7.6. SOIL SUCTION

As in the case of a capillary tube, the water in the soil above the water table has a negative pressure. The soil is in a state of reduced pressure, known as soil suction. The soil suction is measured in term of the height of water column suspended in soil. Generally, it is expressed in terms of the common logarithm of the height in centimeters, and is known as P_F value. For example, a soil suction of 100 cm of water column can be represented as P_F equal to 2, because,

$$100 = 10^2 \text{cm} \qquad \text{and } \log_{10}^{100} = 2$$

A P_F value of zero corresponds to a soils suction of 1 cm, as $\log_{10}^{1} = 0$.

Although the soil suction represents a negative pressure, it is customary to omit the negative sign.

The soil suction can also be represented in pressure unit, using the relation,

$$1 \text{ cm of water column} = 0.0098 \text{ N/cm}^2$$

7.7. CAPILLARY POTENTIAL

The tenacity with which the soil holds the capillary water is measured in terms of the capillary potential or matrix potential. The capillary potential (ψ) is defined as the work done to take away a unit mass of water from a unit mass of soil. It is numerically equal to the tension (negative pressure) in the soil water but it is of opposite sign. Therefore,

$$\psi = -p \qquad \text{...(7.9)}$$

where p is tension in soil.

[**Note.** Some authors express capillary potential as energy per unit mass in kJ/kg.

For example, if $p = 1 \text{ bar} = 100 \text{ kN/m}^2$,

$$\text{height of water column} = \frac{100}{9.81} = 10.2 \text{ m and } \Psi = 10.2 \times 9.81 = 100 \text{ kJ/kg}]$$

It is worth noting that the capillary potential is always negative. The maximum possible value of ψ is equal to zero when the soil tension is zero, which occurs when the water is at atmospheric pressure. As the water content in the soil decreases, the tension increases. This causes a decrease in capillary potential. The capillary potential is minimum when the water content is minimum.

Water in the capillary fringe is seldom under equilibrium. It moves from a region of high potential (more water content) to a region of low potential (less water content). The water starts moving as soon as the suction equilibrium is disturbed either due to evaporation of water or due to an increase in water content. The velocity of the capillary water is given by

$$v = k_u \cdot i_s \qquad \text{...(7.10)}$$

where k_u = coefficient of unsaturated permeability,

i_s = suction gradient, which is equal to the potential difference per unit length.

7.8. CAPILLARY TENSION DURING DRYING AND WETTING OF SOILS

Capillary tension develops not only in the soils above the water table but also in a soil when its water content is reduced. When the water content of a saturated soil is reduced by drying, the water recedes into the interstices of the soil and forms menisci. As the water content is reduced further, the menisci recede. The radii of curvature decrease, and there is a corresponding increase in soil suction.

Fig. 7.7 shows the relationship between the soil suction and the water content of a soil. The suction at a particular water content is more when the soil is drying than when the soil is wetting, and a hysterisis loop is formed. The reason for the difference in soil suction is that during drying the release of water from the larger pores is controlled by the surrounding smaller pores, whereas during wetting it is not controlled by the smaller pores. The phenomenon is somewhat similar to the flow of capillary water in tubes of non- uniform diameter discussed in Sect 7.3. The process of drying is analogous to the flow of water in the downward direction, in which the capillary rise does not depend upon the larger diameter of the bulb.

The increase in soil suction with decreasing water content is continuous over the entire range of water content. Its value is zero when the soil is saturated and is very high when the soil is oven dry. When a dry soil is submerged under water, the meniscus is destroyed and the soil suction is reduced to zero. The capillary water changes to free water.

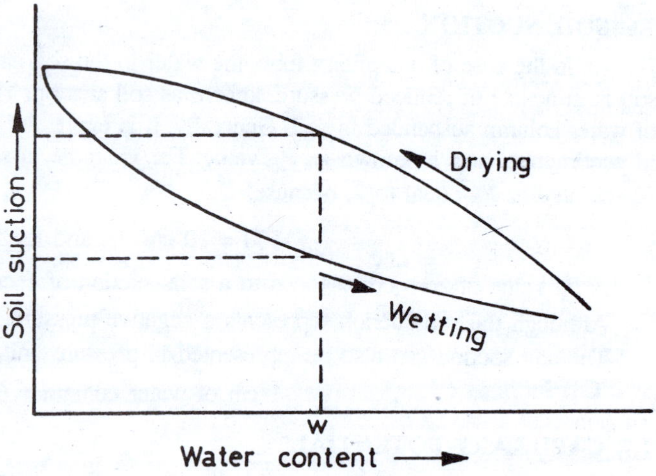

Fig. 7.7. Drying and Wetting Cycle.

7.9. FACTORS AFFECTING SOIL SUCTION

The suction in soils depends mainly on the following factors:

(1) **Particle size**—In general the smaller the particle size, the greater is the soil suction. The soils with fine particles have a large number of small pores with small radii of menisci. It results in a large capillary rise and hence greater suction.

(2) **Water content**—As the water content of a soil decreases, the soil suction increases and it attains the maximum value when the soil is dry. As discussed above, with a decrease in water content, water recedes into smaller pores resulting in the decrease of the radius of curvature of the meniscus.

(3) **History of drying or wetting cycle**—As discussed in the preceding section, for the same water content, the soil suction is greater during the drying cycle than in the wetting cycle.

(4) **Soil Structure**—The soil structure governs the size of interstices in the soils. As the soil suction depends upon the size of interstices, a change in the soil structure affects the soil suction.

(5) **Temperature**—A rise in temperature causes a reduction in surface tension (T_s) of the water. Consequently, the soil suction decreases as the temperature increases.

(6) **Denseness of soil**—As the denseness of a soil increases, generally soil suction increases. When the soil is loose, with a low density, the pores are of large radius and the soil suction is low.

(7) **Angle of contact**—The angle of contact between water and soil particles depends upon the mineralogical composition of soils. As the angle of contact (θ) increases, the soil suction decreases. The soil suction is maximum when the angle of contact is zero.

(8) **Dissolved salts**—The surface tension of water increases with an increase in impurities, such as salt. Therefore, the dissolved salts cause an increase in soil suction.

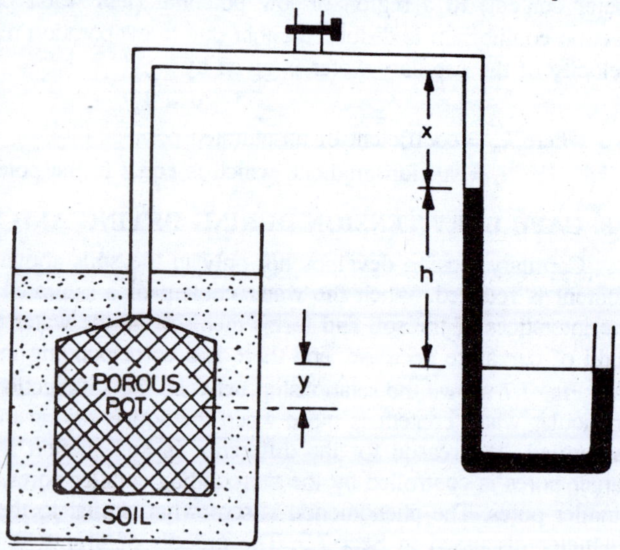

Fig. 7.8. Tensiometer..

7.10. MEASUREMENT OF SOIL SUCTION

Suction in a soil mass can be measured using the following methods:

(1) **Tensiometer Method**—A tensiometer consists of a porous pot filled with water. The top of the porous pot is connected to a U- tube containing mercury. The pot is placed in the soil

whose suction is to be determined (Fig. 7.8).

The soil draws water from the porous pot. The process continues till an equilibrium is attained, when the suction inside and that outside the pot are equal. The suction (p'') inside the porous pot can be calculated, using the manometer equation (see any text on Fluid Mechanics), as

$$0 - 13.6\,h \times 9.81 + (h + y) \times 9.81 = p''$$

or $$p'' = -(12.6\,h + y) \times 9.81 \qquad \text{...(7.11)}$$

where h = deflection of mercury in manometer in metres,

y = vertical intercept between the mercury level in the right limb of the manometer and the centre of the pot.

The soil suction (p'') can be calculated using Eq. 7.11 once the values of h and y have been determined. This method is suitable for determination of soil suction upto 0.8 bar or 80 kN/m^2 or 800 cm of water.

(2) Suction Plate Method—In this method, the soil sample is placed over a porous plate known as suction plate. The suction plate is in contact with water in the reservoir (Fig. 7.9). The water reservoir is connected to a pipe. A mercury manometer is attached to the pipe as shown. The other end of the pipe is connected to a vacuum pump.

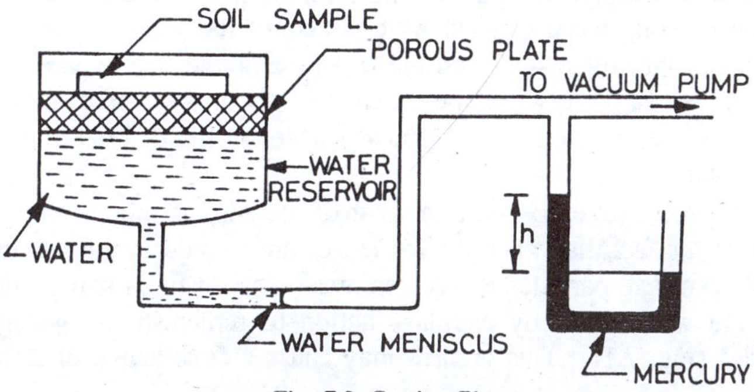

Fig. 7.9. Suction Plate.

The soil sample takes water from the reservoir through the porous plate. The meniscus in the pipe has a tendency to move towards left, which indicates that the water is being drawn into the soil. The water meniscus is kept stationary by means of the vacuum pump. The soil suction is equal to the reduction in the pressure as shown by the deflection (h) of mercury in the manometer.

This method of measurement is suitable for a suction upto 1 bar or 100 kN/m^2 or 10 m of water.

(3) Centrifuge Method—In this method, the centrifuging action occurs on the soil due to its rotation. The soil sample is placed on a porous pot as shown in Fig. 7.10. The pot is enclosed in a brass case which can be rotated about the centre of rotation.

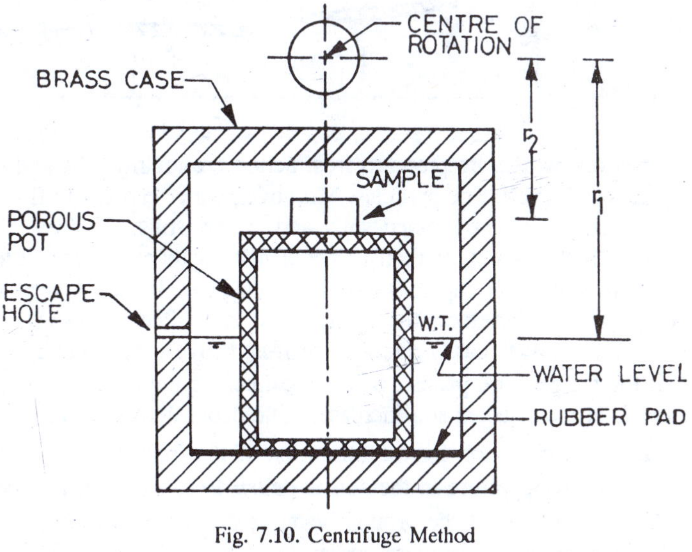

Fig. 7.10. Centrifuge Method

As the rotation occurs, water from the soil comes out and travels through the walls of the porous pot to the water reservoir in the brass case marked with water level as the water table (W.T.). The level of the water table is kept constant, as the excess water passes through the escape hole provided at that level. The migration of the water from the soil to the water table continues till the suction of the water left in the soil is just equal to that required for equilibrium. The soil suction can be determined as

$$h = \frac{\omega^2}{2g}\,(r_1^2 - r_2^2) \qquad \text{...(7.12)}$$

where h = soil suction, expressed in terms of the height of water column ($\log h = P_F$).

ω = rotational speed (radians per second)

r_1 = radial distance from the centre of rotation to the water table

r_2 = radial distance from the centre of rotation to the middle of the soil sample.

The test is conducted at various speeds to obtain a relationship between the water content and the soil suction.

The centrifuge method can be used for determination of very high suctions, of the order of several thousands of kN/m^2. For accurate results, thin samples shall be used. If the sample is relatively thick, it is subjected to an additional overburden pressure due to its own weight and erroneous results are obtained.

7.11. FROST HEAVE

The water which migrates upward from the water table to the capillary fringe may freeze if the atmospheric temperature falls to the freezing point, and the ice is formed. This results in an increase in the volume of soil, because when water is converted into ice, there is about 9% increase in its volume. If the porosity of the soil is 45% and the soil is saturated, the expansion of the soil would be (0.09 × 45) = 4.05%. In other words, there would be a heave of about 4 cm in every one metre thickness of the soil deposit. Due to frost heave, the soil at the ground surface is lifted. This may cause the lifting of light structures built on the ground.

The frost heave observed in most of the soils is much more than a heave of about 4 cm per metre. This is due to the fact that when the ice lenses are formed in the soil due to freezing of water, the water film from the adjacent soil particles is also removed. This disturbs soil suction equilibrium and more water is drawn up from the water table by capillary action to replenish the water deprived by the ice lenses from the soil particles (Fig. 7.11). This process may cause a frost heave of 20 to 30% of the soil depth.

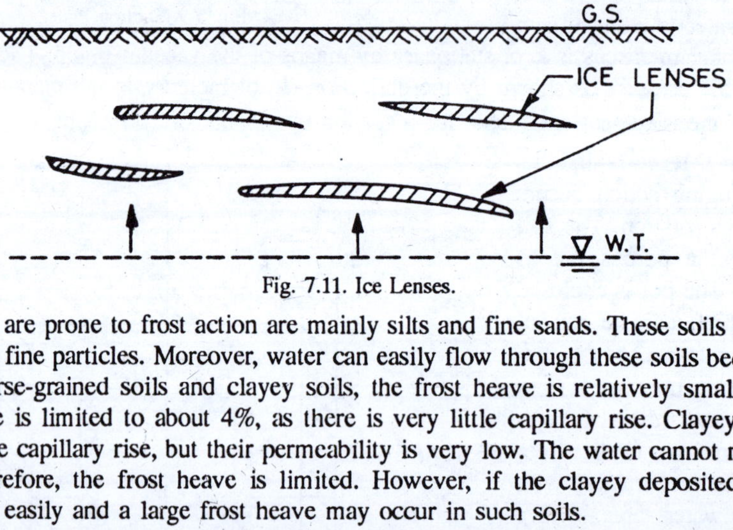

Fig. 7.11. Ice Lenses.

The soils which are prone to frost action are mainly silts and fine sands. These soils have large capillary rise due to relatively fine particles. Moreover, water can easily flow through these soils because of fairly good permeability. In coarse-grained soils and clayey soils, the frost heave is relatively small. In coarse-grained soils, the frost heave is limited to about 4%, as there is very little capillary rise. Clayey soils, on the other hand, have very large capillary rise, but their permeability is very low. The water cannot move easily through these soils and, therefore, the frost heave is limited. However, if the clayey deposited have fissures and cracks, water moves easily and a large frost heave may occur in such soils.

If the temperature persists below the freezing point for a long period, frost penetrates the soil further, and the depth of the affected soil increases. The depth upto which the water may freeze is known as the *frost line.*

The basic condition for the formation of the frost heave may be summarised as under:

(1) The temperature in the soil is below freezing point and persists for a long period.

(2) A reservoir of the ground water is available sufficiently close to the frost line to feed the growing ice lenses by capillary action.

(3) The soil is saturated at the beginning and during the freezing period.

(4) The soil has sufficiently high capillary potential to lift the water above the ground water table.

(5) The soil has good permeability so that water moves quickly through it.

The cracks and fissure also permit rapid movement of water.

(6) The soil particles of size about 0.02 mm are most prone to frost heave.

If a uniform soil contains more than 10% particles of the size 0.02 mm or if a well-graded soil contains more than 3% particles of this size, the soil is prone to frost heave.

The foundations of structures should be carried below the frost depth to avoid possible frost heave after the completion of the structure. However, highways and runways have limited depth below the ground surface and cannot be constructed below the frost line. In such cases, other measures are taken to reduce frost heave, as discussed in Sect. 7.13.

7.12. FROST BOIL

After the occurrence of frost heave, if the temperature rises, the frozen soil thaws and free water is liberated. Thawing process starts from the upper layer and moves downwards. The liberated water is trapped in the upper layer while the lower layers are still frozen. The strength of the soil in the upper layer is reduced due to its softening caused by an increases in water content. The process of softening of soil due to liberation of water during thawing is known as *frost boil.*

Frost boil affects the structures resting on the ground surface. The effect is more pronounced on highway pavements. A hole is generally formed in the pavement due to extrusion of soft soil and water under the action of wheel loads. In extreme cases, the pavement breaks under traffic, and there is ejection of subgrade soil in a soft and soapy condition.

Coarse-grained soils are not affected much by frost boil, as the quantity of liberated water is small, and that too is drained away quickly. The soils most prone to the softening effect are silty soils. These soils have low plasticity index and become very soft with a small increase in water content. Clayey soils are not affected as much as silty soils since the quantity of liberated water is small and the plasticity index is high.

7.13. PREVENTION OF FROST ACTION

The frost heave and frost boil cause great difficulties in the maintenance of highways and runways, as discussed above. The following measures are usually taken to mitigate the ill effects of the frost action.

(1) The most effective method of prevention of frost action is to replace the frost-susceptible soil by coarse-grained soils such as gravels or coarse sands. In most cases, the method is not economically feasible owing to large quantities of soils involved.

(2) The frost action can be prevented by providing an insulating blanket between the water table and the ground surface. The insulating blanket consists of gravel, and has a thickness of 15 to 30 cm. The blanket reduces the capillary action and hence the migration of water and the formation of ice lenses (Fig. 7.12).

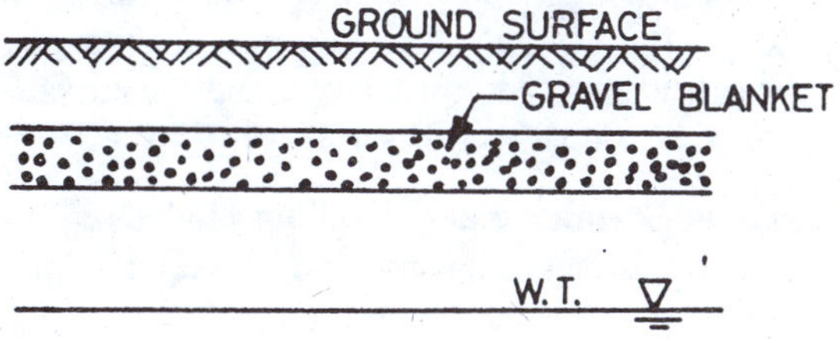

Fig. 7.12. Insulating Blanket.

(3) A good drainage system prevents the frost action in two ways: (*i*) It lowers the water table and thus increases the distance between the ground surface and the water table, (*ii*) The water liberated during thawing is drained away quickly by the drainage system.

(4) Sometimes additives are used to reduce frost action. Dispersion agents, such as sodium polyphosphate, when mixed with soil, decrease the permeability of the soil.

(5) Water proofing materials and other chemicals are also used to change the adsorbed cations on the clay minerals to reduce the tendency of soils to attract the water dipole.

7.14. SHRINKAGE AND SWELLING OF SOILS

Shrinkage A clayey soil shrinks when water evaporates from it. If water is added to such soils, swelling takes place. Shrinkage and swelling are characteristics of clayey soil. The coarse-grained soils have very little shrinkage and swelling.

Shrinkage is due to tension in soil water. When tension (negative pressure) develops in water, compressive forces act on the solid particle. The compressive forces induced in the solid particles are similar to those induced in the walls of the capillary tube discussed in Sect. 7.4. When the water content of a soil mass reduces due to evaporation, the meniscus retreats. This causes compression of the solid particles and hence a reduction in the volume of the soil mass.

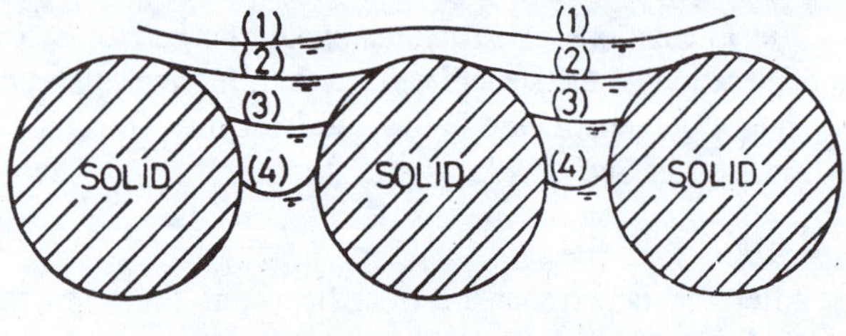

Fig. 7.13. Retreating of Meniscus.

The stresses in pore water during shrinkage can be studied from the capillary tube analogy (Sect. 7.4). Let us consider a soil mass consisting of spherical, solid particles, shown in Fig. 7.13. When the capillary spaces between the particles are completely filled with water, the meniscus forms a plane surface, as indicated by 1-1. The tension in water is zero. As evaporation takes place, water is removed from the free surface and the meniscus retreats to the position 2-2. This process causes tension in the water and corresponding compressive forces on the solid grains. The tension developed depends upon the radius of the meniscus.

With further evaporation, the meniscus retreats to position 3-3 and the radius decreases. This increases the compressive forces acting on the solid particles. Eventually, when the meniscus attains the minimum radius, shown by position 4-4, it is fully developed and the compressive forces induced are maximum. Further recession of the meniscus does not increase the compressive forces, as there are no pores of smaller radius.

The lower limit of the volume occurs at the shrinkage limit. At the shrinkage limit, the soil is still saturated, but there is no free water at the soil surface. Further drying does not cause a reduction in its volume as the soil resistance exceeds the compressive forces. As soon as the shrinkage limit is reached, the surface becomes dry. It is indicated by a change in the colour of the soil surface to a lighter shade.

There may be a small additional shrinkage after the shrinkage limit, but this is usually ignored.

Swelling When water is added to a clayey soil which had shrunk by evaporation of the pore water, the menisci are destroyed. The tension in soil water becomes zero. The compressive forces between the solid particles reduce considerably, and elastic expansion of the soil mass occurs and this causes some *swelling*. However swelling mainly occurs due to attraction of dipolar molecules of water to the negatively charged soil particles. The swelling also depends upon a number of other factors, such as mutual repulsion of clay particles and their adsorbed layers and the expansion of entrapped air. The mechanism of swelling is much more complex than that of shrinkage.

Effects of Shrinkage and Swelling of Soils

Shrinkage and swelling create many problems, as discussed below.

(1) Shrinkage and swelling cause the deformations and stresses in the structures resting on or in the soil.

(2) High swelling pressures develop if the soil has an access to water, but is prevented from swelling. The light structures may be lifted if the swelling pressure is excessive.

(3) In semi-arid regions, the clay near the ground surface is subjected to shrinkage during dry periods and the cracks are formed. During wet periods, the clay swells and the cracks are closed. This process of the formation and closing of the cracks may cause the development of fissures in soils.

(4) If silt particles drop into the shrinkage cracks formed behind the retaining wall, particles later swell and force the retaining wall out of the plumb. It may cause the failure of the wall if it had not been properly designed to resist the pressure so developed.

(5) If the soil below the pavements has high shrinkage and swelling properties, it creates the problems in the maintenance of highways and runways.

7.15. SLAKING OF CLAY

When a clay that had been dried well below the shrinkage limit is suddenly immersed in water, it disintegrates into a soft, wet mass. The process is known as slaking of clay. Slaking can be explained as below.

When the soil dries to a water content lower than the shrinkage limit, some of the voids get filled with air (Fig. 7.14). Water enters these air-filled voids when the soil is immersed in water. This causes an explosion of the voids, and therefore disintegration of soil occurs.

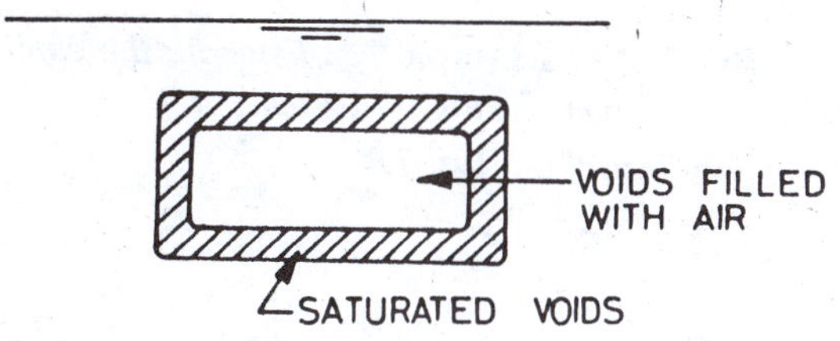

Fig. 7.14. Slaking of Clay.

According to another interpretation, when water enters the pores, it forms menisci which react against the air in the void. The entrapped air is subjected to very high pressure and the soil mass disintegrates.

7.16. BULKING OF SAND

If a damp sand is loosely deposited, its volume is much more than that when the same sand is deposited in a loose, dry state. The phenomenon of increase in volume of sand due to dampness is known as bulking of sand.

In damped state, cohesion develops between the particles due to capillary water. The cohesion prevents the particles from taking a stable position. A sort of honey-comb structure is formed. The effect is predominant when the water content is between 4 to 5%. The increase in volume due to bulking is between 20 to 30% for most sands.

If the damp sand is saturated by adding more water, the effect of capillary action is eliminated and the volume of the sand mass is decreased.

7.13. CAPILLARY SIPHONING

In an earth dam with an impervious core, capillary siphoning may occur (Fig. 7.15). The water rises in the outer shell due to capillary action. If the crest (top level) of the impervious core is in the reach of

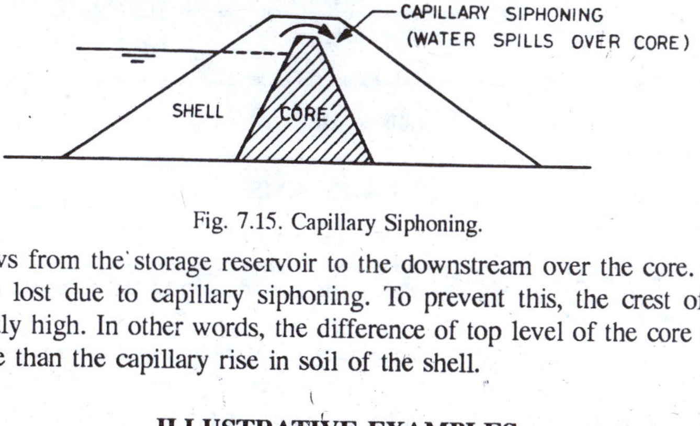

Fig. 7.15. Capillary Siphoning.

capillary rise, water flows from the storage reservoir to the downstream over the core. Considerable quantity of stored water may be lost due to capillary siphoning. To prevent this, the crest of the impervious core should be kept sufficiently high. In other words, the difference of top level of the core and water level in the reservoir should be more than the capillary rise in soil of the shell.

ILLUSTRATIVE EXAMPLES

Illustrative Example 7.1. *What is the negative pressure in the water just below the meniscus in a capillary tube of diameter 0.1 mm filled with water. The surface tension is 0.075 N/m and wetting angle is 10 degrees.*

Solution. From Eq. 7.1,

$$h_c = \frac{4\,T_s \cos\theta}{g\,\rho_w\,d} = \frac{4 \times 0.075 \times 0.9848}{9.81 \times 1000 \times 0.1 \times 10^{-3}} = 0.301 \text{ m}$$

$$\text{Negative pressure} = \gamma_w h_c = 9.81 \times 1000 \times 0.301$$

$$= 2952.81 \text{ N/m}^2 = 2.953 \text{ kN/m}^2$$

Illustrative Example 7.2. *Estimate the capillary rise in a soil with a void ratio of 0.60 and an effective size of 0.01 mm. Take C = 15 mm².*

Solution. From Eq. 7.8,

$$h_c = \frac{C}{e\,D_{10}} = \frac{15}{0.6 \times 0.01} = 2500 \text{ mm} = \mathbf{2.5\ m}$$

Illustrative Example 7.3. *The p_F of a soil is 2.50. Determine the capillary potential of the soil.*

Solution.

$$\text{Soil suction} = (10)^{2.5} = 316.23 \text{ cm} = 3.1623 \text{ m}$$

$$\text{Capillary potential} = -3.1623 \times 9.81 \times 10^3 \text{ N/m}^2$$

$$= -31.02 \text{ kN/m}^2$$

Illustrative Example 7.4. *The capillary rise in a soil A with an effective size of 0.02 mm was 60 cm. Estimate the capillary rise in a similar soil B with an effective size of 0.04 mm.*

Solution. From Eq. 7.8,

$$\frac{(h_c)_1}{(h_c)_2} = \frac{(D_{10})_2}{(D_{10})_1}$$

or

$$\frac{60}{(h_c)_2} = \frac{0.04}{0.02} = 2 \quad \text{or} \quad (h_c)_2 = \mathbf{30\ cm}$$

Illustrative Example 7.5. *The capillary rise in silt is 50 cm and that in fine sand is 30 cm. What is the difference in the pore size of the two soils ?*

Solution. From Eq. 7.3,

$$h_c = \frac{0.30}{d} \text{ cm}$$

For silt,

$$(h_c)_1 = 50 = \frac{0.30}{d_1} \quad \text{or} \quad d_1 = 6.0 \times 10^{-3} \text{ cm}$$

For fine sand,

$$(h_c)_2 = 30 = \frac{0.30}{d_2} \quad \text{or} \quad d_2 = 10.0 \times 10^{-3} \text{ cm}$$

Difference in pore size

$$= (10.00 - 6.0) \times 10^{-3}$$

$$= \mathbf{4.00 \times 10^{-3}\ cm}$$

PROBLEMS

A. Numerical

7.1 Find the capillary rise in a sandy soil which has a void ratio of 0.65 and the effective size of particle is 0.07 mm. Take $C = 0.1$ cm². **[Ans. 21.98 cm]**

7.2. The effective size of a soil is 0.015 mm. Estimate the height of capillary rise. Take surface tension as 0.074 N/m. **[Ans. 10 m]**

7.3. Compute the maximum capillary tension for a capillary tube of 0.1 mm diameter. Take surface tension as 0.075 N/m. **[Ans. 3.0 kN/m²]**

7.4. The glass vessel shown in Fig. P 7.4 is filled with water. It has two holes of diameter 0.01 cm and 0.03 cm as shown. If a fully- developed meniscus is formed in the upper hole, determine the height h of the wall of the vessel. **[Ans. 20.27 cm]**

7.5. In Prob. 7.4, if both the holes are of the same diameter, equal to 0.01 cm, determine the contact angle in the lower hole if that in the upper hole is zero and h = 20.27 cm. **[Ans. 70.54°]**

B. Descriptive and Objective Type

7.6. What are different categories of soil water ? Describe in brief.

7.7. Discuss the phenomenon of capillary rise in soils. What are the factors that effect the height of capillary zone ?

7.8. What is soil suction ? How is it measured ? What are the factors that affect soil suction ?

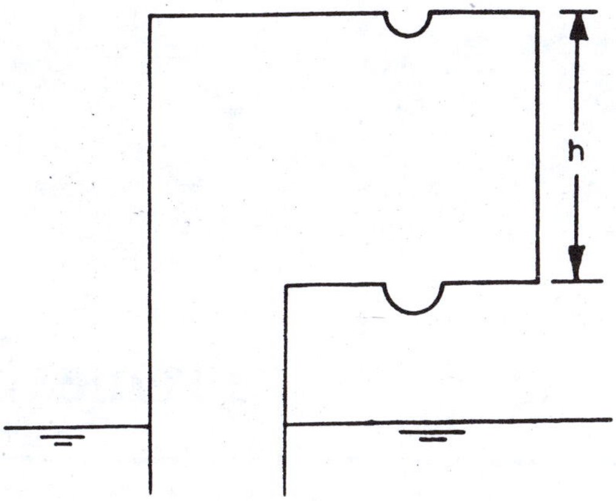

Fig. P. 7.4

7.9. Differentiate between frost heave and frost boil. What is their effect on soils ? How frost action can be prevented ?

7.10. Write a note on shrinkage and swelling of soils.

7.11. Discuss the phenomena of slaking and bulking.

C. Multiple Choice Questions

1. Capillary rise in a small tube is due to
(*a*) Cohesion (*b*) Adhesion
(*c*) Both cohesion and adhesion (*d*) Neither (*a*) nor (*b*)

2. The surface tension of water at normal temperatures is about
(*a*) 0.73 dynes/m (*b*) 0.73 N/m
(*c*) 0.073 N/m (*d*) 0.073 kN/m

3. The capillary rise in clay is usually between
(*a*) 0.10 and 0.15 m (*b*) 0.3 and 1.0 m
(*c*) 1.0 and 10.0 m (*d*) greater than 10 m

4. A pF value of zero corresponds to a soil section of
(*a*) 1 m (*b*) zero metre
(*c*) 1 cm (*d*) 10 cm

5. The frost heave in the following type of soils is generally high
(*a*) Coarse sands (*b*) clays
(*c*) Fine sands and silts (*d*) gravels

6. Bulking of sands is usually
(*a*) Less than 10% (*b*) Between 20 to 30%
(*c*) Greater than 30% (*d*) Between 10 to 20%

7. The frost heave depth as percentage of the soil depth in fine sands and silts is about
(*a*) 4 to 5% (*b*) 5 to 10%
(*c*) 10 to 15% (*d*) 20 to 30%

8. A tension of 1kN/m^2 corresponds to a capillary potential of
(*a*) 1 kJ/kg (*b*) 10 kJ/kg
(*c*) 100 kJ/kg (*d*) 1000 kJ/kg

[**Ans.** 1. (*c*), 2. (*c*), 3. (*d*), 4. (*c*), 5. (*c*), 6. (*b*), 7. (*d*), 8. (*a*)]

8

Premeability of Soils

8.1. INTRODUCTION

A material is porous if it contains interstices. The porous material is permeable if the interstices are interconnected or continuous. A liquid can flow through a permeable material. Electron photomicrographs of even very fine clays indicate that the interstices are interconnected. However, the size, cross- section, and orientation of the interstices in different soils are highly variable. In general, all the soils are permeable.

The property of a soil which permits flow of water (or any other liquid) through it, is called the permeability. In other words, *the permeability is the ease with which water can flow through it.* A soil is highly pervious when water can flow through it easily. In an impervious soil, the permeability is very low and water cannot easily flow through it. A completely impervious soil does not permit the water to flow through it. However, such completely impervious soils do not exist in nature, as all the soils are pervious to some degree. A soil is termed impervious when the permeability is extremely low.

Permeability is a very important engineering property of soils. A knowledge of permeability is essential in a number of soil engineering problems, such as settlement of buildings, yield of wells, seepage through and below the earth structures. It controls the hydraulic stability of soil masses. The permeability of soils is also required in the design of filters used to prevent piping in hydraulic structures.

As mentioned in chapter 7, free water or gravitational water flows through soils under the influence of gravity. Flow of free water depends upon the permeability of the soil and the head causing flow. This chapter deals with Darcy's law for flow of water, the methods for the determination of permeability and the factors affecting the permeability of soils. Further details of flow of water and seepage problems are discussed in the next chapter.

8.2. HYDRAULIC HEAD

The total head at any point in a flowing fluid is equal to the sum of the elevation (or datum) head, the pressure head and the velocity head. The elevation head (z) is equal to the vertical distance of the point above the datum. The pressure head (p/γ_w) is equal to the head indicated by a piezometer with its tip at that point. The velocity head is equal to $v^2/2g$. However, for flow of water through soils, as the velocity (v) is extremely small, the velocity head is neglected. Therefore, the total head of water in soil engineering problems is equal to the sum of the elevation head and the pressure head. For flow problems in soils, the downstream water level is generally taken as datum. The piezometric level is the water level shown by a piezometer inserted at that point. The line joining the piezometric levels at various points is called a piezometric surface. The piezometric surface also represents the hydraulic gradient line (HGL). The sum of the pressure head and the elevation head is known as the piezometric head.

Fig. 8.1 shows two vessels A and B containing water at different levels and connected by a small tube containing soil sample. Let the length of the tube be L. The flow takes place from the vessel A with a high head to the vessel B with a low head through the tube. With datum at the water level in the vessel B, the

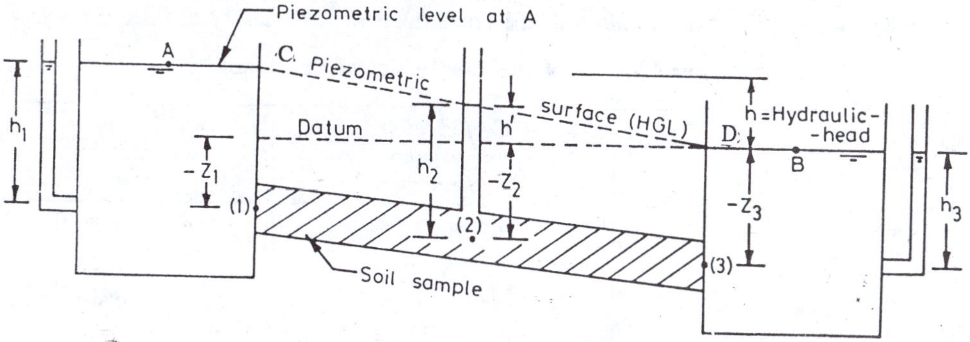

Point	Elevation head	Pressure head	Total head
1	$-Z_1$	h_1	h
2	$-Z_2$	h_2	h'
3	$-Z_3$	h_3	0

Fig. 8.1. Various Heads.

elevation head, the pressure head and the total head at three points 1, 2 and 3 are also shown in the figure. The total head at point 1 is h and that at point 3 is zero. The head h is known as the *hydraulic head.* It is equal to the difference in the elevations of water levels at the entry and exit points in a soil mass. Obviously, it is equal to the loss of head through the soil. The hydraulic head is also known as the *effective head.*

The loss of head per unit length of flow through the soil is equal to the hydraulic gradient (i),

$$i = h/L \qquad \text{...(8.1)}$$

where h hydraulic head, and L = length of the soil specimen.

The variation of head at various points is represented by the line CD, known as the hydraulic gradient line (H.G.L.) or pressure gradient line. If a piezometer is inserted at any intermediate point 2, the water will rise upto the level of the hydraulic gradient line at that point. The line CD, therefore, represents a piezometric surface. It is generally assumed that the loss of head over the length of the soil sample is uniform and, therefore, the variation of head is linear.

8.3. DARCY'S LAW

The flow of free water through soil is governed by Darcy's law. In 1856, Darcy demonstrated experimentally that for laminar flow in a homogeneous soil, the velocity of flow (v) is given by

$$v = k\,i \qquad \text{...(8.2)}$$

where k = coefficient of permeability, i = hydraulic gradient.

The velocity of flow is also known as the *discharge velocity or the superficial velocity.*

Eq. 8.2 is known as Darcy's law, which is one of the corner stones of soil engineering. The discharge q is obtained by multiplying the velocity of flow (v) by the total cross- sectional area of soil (A) normal to the direction of flow. Thus

$$q = vA = k\,i\,A \qquad \text{...(8.3)}$$

The area A includes both the solids and the voids.

The coefficient of permeability can be defined using Eq. 8.2. If the hydraulic gradient is unity, the coefficient of permeability is equal to the velocity of flow. In other words, the coefficient of permeability is defined as the velocity of flow which would occur under unit hydraulic gradient. The coefficient of permeability has the dimensions of velocity [L/T]. It is measured in mm/sec, cm/sec, m/sec, m/day or other velocity units. The coefficient of permeability depends upon the particle size and upon many other factors as explained later. Table 8.1 gives the typical values of the coefficient of permeability of different soils.

Table 8.1. Typical Values of the Coefficient of Permeability

S. No.	*Soil Type*	*Coefficient of permeability (mm/sec)*	*Drainage properties*
1.	Clean gravel	10^{+1} to 10^{+2}	Very good
2.	Coarse and medium sands	10^{-2} to 10^{+1}	Good
3.	Fine sands, loose silt	10^{-4} to 10^{-2}	Fair
4.	Dense silt, clayey silts	10^{-5} to 10^{-4}	Poor
5.	Silty clay, clay	10^{-8} to 10^{-5}	Very poor

According to USBR, the soils having the coefficient of permeability greater than 10^{-3} mm/sec are classified as pervious and those with a value less than 10^{-5} mm/sec as impervious. The soils with the coefficient of permeability between 10^{-5} to 10^{-3} mm/sec are designated as semi-pervious.

8.4. VALIDITY OF DARCY'S LAW

Darcy's law is valid if the flow through soils is laminar. The flow of water through soils depends upon the dimension of interstices, which, in turn, depend upon the particle size. In fine-grained soils, the dimensions of the interstices are very small and the flow is necessarily laminar. In coarse-grained soils, the flow is also generally laminar. However, in very coarse-grained soils, such as coarse gravels, the flow may be turbulent.

For flow of water through pipes, the flow is laminar when the Reynolds number is less than 2000. For flow through soils, it has been found that the flow is laminar if the Reynolds number is less than unity. For flow through soils, the characteristic length in the Reynolds number is taken as the average particle diameter (D).

Thus $$\frac{\rho v D}{\eta} \leq 1$$

where ρ = is the mass density and η is the coefficient of viscosity.

Using Allen Hazen's equation (Eq. 8.30) for velocity, it can be shown that the maximum diameter of the particle for the flow to be laminar is about 0.50 mm.

The value of the critical Reynolds number of unity is, however, conservative. It has been demonstrated that the flow remains laminar even upto the Reynolds number of 75. It has been observed that Darcy's law is valid for flow in clays, silts and fine sands. In coarse sands, gravels and boulders, the flow may be turbulent and Darcy's law may not be applicable. It is difficult to predict the exact range of the validity of Darcy's law. The best method to ascertain the range is to conduct experiments and determine the actual relationship between the velocity v and the hydraulic gradient. For Darcy's law to be valid, this relationship should be approximately linear.

For flow through coarse sands, gravels and boulders, the actual relationship between the velocity and the hydraulic gradient is non-linear. Hough gave the following equation for the velocity when the flow is turbulent.

$$v = k\,(i)^n \qquad \text{...(8.4)}$$

where n = exponent, with a value of 0.65

In extremely fine-grained soils, such as a colloidal clay, the interstices are very small. The velocity is therefore very small. In such soils also, the Darcy law is not valid.

8.5. DETERMINATION OF COEFFICIENT OF PERMEABILITY

The coefficient of permeability of a soil can be determined using the following methods.

(a) Laboratory Methods. The coefficient of permeability of a soil sample can be determined by the following methods :

(i) Constant-head permeability test.

(ii) Variable-head permeability test.

The instruments used are known as *permeameters.* The former test is suitable for relatively more pervious soils, and the latter for less pervious soils.

(b) Field Methods. The coefficient of permeability of a soil deposit in-situ conditions can be determined by the following fields methods :

(i) Pumping-out tests.

(ii) Pumping-in tests.

The pumping-out tests influence a large area around the pumping well and give an overall value of the coefficient of permeability of the soil deposit. The pumping-in test influences a small area around the hole and therefore gives a value of the coefficient of permeability of the soil surrounding the hole.

(c) Indirect Methods. The coefficient of permeability of the soil can also be determined indirectly from the soil parameters by

(i) Computation from the particle size or its specific surface.

(ii) Computation from the consolidation test data.

The first method is used if the particle size is known. The second method is used when the coefficient of volume change has been determined from the consolidation test on the soil.

(d) Capillarity-Permeability test. The coefficient of permeability of an unsaturated soil can be determined by the capillarity—permeability test (Sect. 8.16).

8.6. CONSTANT HEAD PERMEABILITY TEST

The coefficient of permeability of a relatively more permeable soil can be determined in a laboratory by the constant-head permeability test. The test is conducted in an instrument known as constant-head permeameter. It consists of a metallic mould, 100 mm internal diameter, 127.3 mm effective height and 1000 ml capacity according to IS : 2720 (Part XVII). The mould is provided with a detachable extension collar, 100 mm diameter and 60 mm high, required during compaction of soil. The mould is provided with a drainage base plate with a recess for porous stone. The mould is fitted with a drainage cap having an inlet valve and an air release valve. The drainage base and cap have fittings for clamping to the mould.

Fig. 8.2 shows a schematic sketch. The soil sample is placed inside the mould between two porous discs. The porous discs should be at least ten times more permeable than the soil. The porous discs should be deaired before these are placed in the mould. The water tubes should also be deaired. The sample can also be prepared in the permeameter by pouring the soil into it and tamping it to obtain the required density. The base is provided with a dummy plate, 12 mm thick and 108 mm in diameter, which is used when the sample is compacted in the mould.

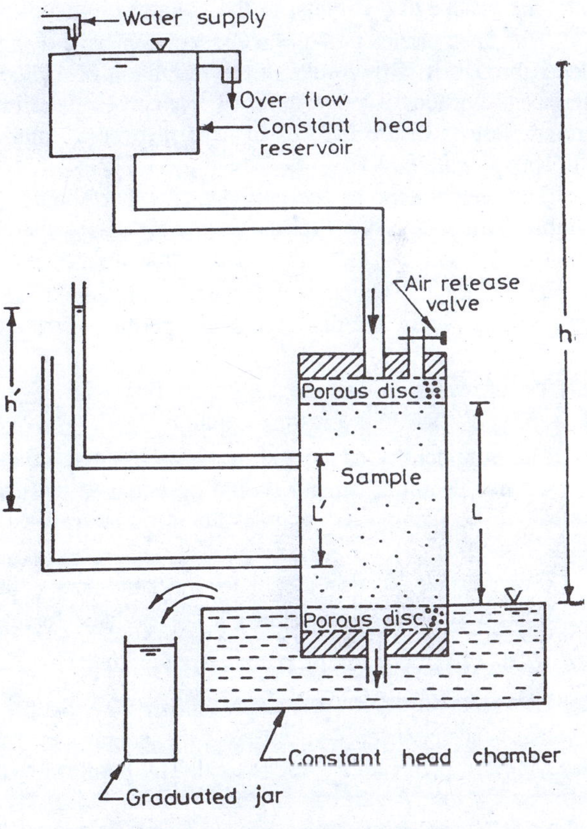

Fig. 8.2. Constant Head Permeameter

It is essential that the sample is fully saturated. This is done by one of the following three methods.

(*i*) By pouring the soil in the permeameter filled with water and thus depositing the soil under water.

(*ii*) By allowing water to flow upward from the base to the top after the soil has been placed in the mould. This is done by attaching the constant-head reservoir to the drainage base. The upward flow is maintained for sufficient time till all the air has been expelled out.

(*iii*) By applying a vacuum pressure of about 700 mm of mercury through the drainage cap for about 15 minutes after closing the drainage valve. Then the soil is saturated by allowing deaired water to enter from the drainage base. The air-release valve is kept open during saturation process.

After the soil sample has been saturated, the constant-head reservoir is connected to the drainage cap. Water is allowed to flow out from the drainage base for some time till a steady-state is established. The water level in the constant-head chamber in which the mould is placed is kept constant. The chamber is filled to the brim at the start of the experiment. The water which enters the chamber after flowing through the sample spills over the chamber and is collected in a graduated jar for a convenient period. The head causing flow (h) is equal to the difference in water levels between the constant-head reservoir and the constant-head chamber.

If the cross-sectional area of the specimen is A, the discharge is given by (Eq. 8.3)

$$q = k\,i\,A$$

or

$$q = k\frac{h}{L}A$$

or

$$k = \frac{qL}{Ah} \qquad ...(8.5)$$

where L = length of specimen, h = head causing flow.

The discharge q is equal to the volume of water collected divided by time.

The finer particles of the soil specimen have a tendency to migrate towards the end faces when water flows through it. This results in the formation of a ***filter skin*** at the ends. The coefficient of permeability of these end portions is quite different from that of the middle portion. For more accurate results, it would be preferable to measure the loss of head h' over a length L' in the middle to determine the hydraulic gradient (i). Thus $i = h'/L'$.

The temperature of the permeating water should be preferably somewhat higher than that of the soil sample. This will prevent release of the air during the test. It also helps in removing the entrapped air in the pores of the soil. As the water cools, it has a tendency to absorb air.

To reduce the chances of formation of large voids at points where the particles of the soil touch the permeameter walls, the diameter of the permeameter is kept at least 15 to 20 times the particles size.

To increase the rate of flow for the soils of low permeability, a gas under pressure is applied to the surface of water in the constant-head reservoir. The total head causing flow in that case increases to $(h + p/\gamma_w)$, where p is pressure applied.

The bulk density of the soil in the mould may be determined from the mass of the soil in the mould and its volume. The bulk density should be equal to that in the field. The undisturbed sample can also be used instead of the compacted sample. For accurate results, the specimen should have the same structure as in natural conditions.

(See Chapter 30, Sect. 30.13 for the laboratory experiment).

The constant had permeability test is suitable for clean sand and gravel with $k > 10^{-2}$ mm/sec.

8.7. VARIABLE-HEAD PERMEABILITY TEST

For relatively less permeable soils, the quantity of water collected in the graduated jar of the constant-head permeability test is very small and cannot be measured accurately. For such soils, the variable-head permeability test is used. The permeameter mould is the same as that used in the constant-head permeability test. A vertical, graduated stand pipe of known diameter is fitted to the top of permeameter. The sample is placed between two porous discs. The whole assembly is placed in a constant head chamber filled with water to the brim at the start of the test. Fig. 8.3 shows a schematic sketch. The porous discs and water tubes should be de-aired before the sample is placed. If in-situ, undisturbed sample is available, the same can be used; otherwise the soil is taken in the mould and compacted to the required density.

The valve at the drainage base (not shown in figure) is closed and a vacuum pressure is applied slowly through the drainage cap to remove air from the soil. The vacuum pressure is increased to 700 mm of mercury and maintained for about 15 minutes. The sample is saturated by allowing deaired water to flow upward from the drainage base when under vacuum. When the soil is saturated, both the top and bottom outlets are closed. The standpipe is filled with water to the required height.

The test is started by allowing the water in the stand pipe to flow through the sample to the constant-head chamber from which it overflows and spills out. As the water flows through the soil, the water level in the standpipe falls. The time required for the water level to fall from a known initial head (h_1) to a known final head (h_2) is determined. The head is measured with reference to the level of water in the constant- head chamber.

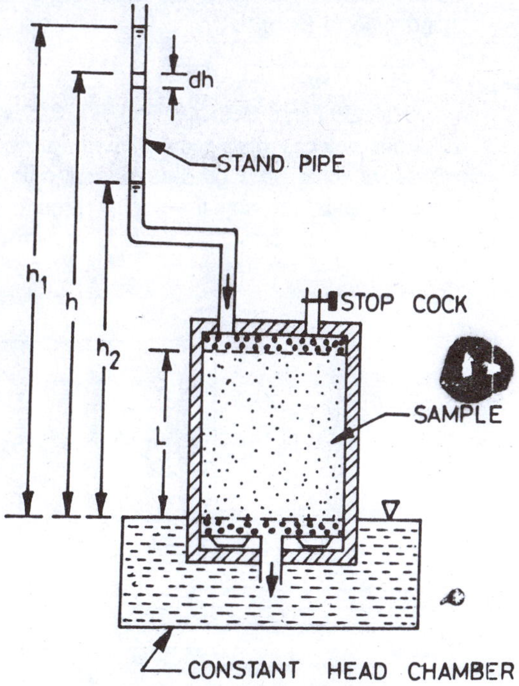

Fig. 8.3. Variable Head Permeameter.

Let us consider the instant when the head is h. For the infinitesimal small time dt, the head falls by dh. Let the discharge through the sample be q. From continuity of flow,

$$a\,dh = -q\,dt$$

where a is cross-sectional area of the standpipe.

or
$$a\,dh = -(A \times k \times i) \times dt$$

or
$$a\,dh = -A\,k \times \frac{h}{L} \times dt$$

or
$$\frac{A\,k\,dt}{a\,L} = \frac{-dh}{h}$$

Integrating,
$$\frac{A\,k}{a\,L}\int_{t_1}^{t_2} dt = -\int_{h_1}^{h_2} \frac{dh}{h}$$

or
$$\frac{A\,k}{a\,L}(t_2 - t_1) = \log_e (h_1/h_2)$$

or
$$k = \frac{a\,L}{A\,t} \log_e (h_1/h_2) \qquad \ldots(8.6)$$

where $t = (t_2 - t_1)$, the time interval during which the head reduces from h_1 to h_2.

Eq. 8.6 is sometimes written as

$$k = \frac{2.30\,a\,L}{A\,t} \log_{10} (h_1/h_2) \qquad \ldots(8.7)$$

The rate of fall of water level in the stand pipe and the rate of flow can be adjusted by changing the area of the cross-section of the standpipe. The smaller diameter pipes are required for less pervious soils.

The coefficient of permeability is reported at 27°C as per IS : 2720 (Part XVII). The void ratio of the soil is also generally determined.

The variable head permeameter is suitable for very fine sand and silt with $k = 10^{-2}$ to 10^{-5} mm/sec.

(See Chapter 30, Sect. 30.14 for the laboratory experiment).

Sometimes, the permeability test is conducted using the consolidometer instead of the permeameter mould (see chapter 12). The fixed-ring consolidometer is used as a variable-head permeameter by attaching a stand pipe to its base.

8.8. SEEPAGE VELOCITY

The discharge velocity v in Eq. 8.2 is not the actual velocity through the interstices of the soil. It is a fictitious velocity obtained by dividing the total discharge (q) by the total cross-sectional area (A). The total cross- sectional area consists of not only the voids but also the solids. As the flow can take place only through voids, the actual velocity through the voids is much greater than the discharge velocity. The actual velocity on a macroscopic scale is known as the seepage velocity (v_s).

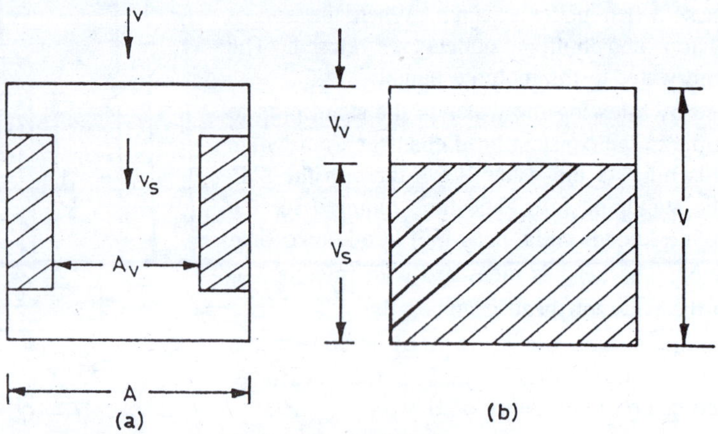

Fig. 8.4. Seepage Velocity

Fig. 8.4 (*a*) shows the longitudinal section through a soil sample in which the voids and the solid particles are segregated. However, it must be clearly understood that the voids and solids in actual soils form a complex system and it is not possible to segregate them. From the continuity of flow.

$$q = vA = v_s A_v \qquad \text{...(a)}$$

where A_v is the area of flow through voids and v_s is the actual seepage velocity.

From Eq. (*a*), $v_s = v \times (A/A_v)$

Multiplying the numerator and denominator by the length (L) of the specimen,

$$v_s = v \times \left(\frac{A \times L}{A_v \times L}\right) \qquad \text{...(b)}$$

The product ($A \times L$) is equal to the total volume V and the product ($A_v \times L$), equal to the volume of voids (V_v) [Fig. 8.4 (*b*)].

Therefore, $$v_s = v \times \frac{V}{V_v} \qquad \text{...(c)}$$

As the ratio V_v/V is equal to the porosity,

$$v_s = \frac{v}{n} \qquad \text{...(8.8)}$$

In other words, the seepage velocity is equal to the discharge velocity divided by porosity.

Using Eq. 8.2, $$v_s = \frac{k\,i}{n}$$

or $$v_s = k_p \times i \qquad \text{...(8.9)}$$

where $$k_p = k/n \qquad \text{...(8.10)}$$

The coefficient k_p is known as the *coefficient of percolation.* Its value is always greater than the coefficient of permeability (k).

Strictly speaking, the seepage velocity is not be absolute velocity through the interstices. The interstices are tortuous and irregular in cross-section and cannot be represented as shown in Fig. 8.4 (*a*). The absolute velocity varies from point to point. Its direction may also change and, at times, it may be directly opposite to the general direction of flow. In fact, the problem is so complex that an analysis based on the absolute velocity is not possible. Although on the microscopic scale, the flow path is tortuous, on a macroscopic scale, it can be considered as a straight line. The seepage velocity can be taken as the macroscopic velocity at which the line of wetting progresses in the direction of flow. Obviously, it is not equal to the absolute velocity as the water flows not in a straight line but it detours around solid particles. Fortunately, the absolute velocity is not of much practical use in soil engineering. The geotechnical engineer is interested in the macroscopic behaviour of the soil and not in its microscopic behaviour.

The total discharge can be computed using either the discharge velocity (v) or the seepage velocity (v_s). The discharge velocity is more convenient and is commonly used in soil engineering. In this text, when the term velocity is used without any qualification, it means discharge velocity.

8.9. GENERAL EXPRESSION FOR LAMINAR FLOW

For understanding the flow of water through soils, let us first consider the laminar flow through pipes. Fig. 8.5 shows a horizontal pipe of circular cross-section of radius R. Let us take a small cylindrical fluid element of radius r and length l, as shown in figure. The shear srress τ on the surface of the fluid element is given by Newton's law of viscosity as

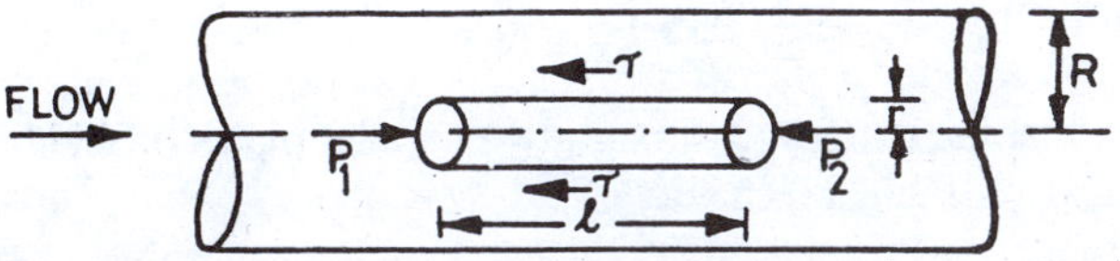

Fig. 8.5. Laminar flow in a pipe.

$$\tau = -\mu \left(\frac{dv}{dr} \right) \qquad \ldots(a)$$

where μ = coefficient of viscosity and $\frac{dv}{dr}$ = velocity gradient.

For steady flow, the net force acting on the element in the horizontal direction is zero. Therefore,

$$(p_1 - p_2)\, \pi r^2 - (2\, \pi\, r l)\, \tau = 0$$

Substituting the value of τ from Eq. (*a*), and simplifying,

$$\frac{dv}{dr} = \frac{-r\,(p_1 - p_2)}{2\, \mu\, l}$$

The pressure p_1 and p_2 can be expressed in terms of piezometric heads h_1 and h_2 as

$$p_1 = \gamma_w h_1 \qquad \text{and} \qquad p_2 = \gamma_w h_2$$

Thus
$$\frac{dv}{dr} = \frac{-r\, \gamma_w\, (h_1 - h_2)}{2\, \mu\, l}$$

Representing $(h_1 - h_2)/l$ by the hydraulic gradient (i),

$$\frac{dv}{dr} = \frac{-r\, \gamma_w\, i}{2\mu}$$

Integrating,
$$v = \frac{-\gamma_w\, i}{2\, \mu} \left(\frac{r^2}{2} \right) + C$$

The constant of integration C can be obtained from the condition of no slip (*i.e.*, $v = 0$) at the boundary (*i.e.*, $r = R$). Thus

$$C = \frac{\gamma_w i R^2}{4\mu}$$

Therefore $$v = \frac{-\gamma_w i}{4\mu} r^2 + \frac{\gamma_w i R^2}{4\mu}$$

or $$v = \frac{\gamma_w i}{4\mu}(R^2 - r^2) \qquad \text{...(8.11)}$$

Eq. 8.11 indicates that the velocity distribution in a pipe is of the shape of paraboloid of revolution, with the maximum velocity at its centre. The equation is known as *Hagen—Poisseuille equation* for laminar flow through pipes. The equation can be used to determine the discharge q in the pipe as under.

Discharge through a small circular ring of radius r and thickness dr is given by

$$dq = (2\pi r\, dr)\, v = 2\pi r dr \frac{\gamma_w i}{4\mu}(R^2 - r^2)$$

Integrating, $$q = \int_0^R 2\pi r \left(\frac{\gamma_w i}{4\mu}\right)(R^2 - r^2)\, dr = \frac{\pi \gamma_w i R^4}{8\mu}$$

Writing the radius R in terms of the hydraulic radius R_H (*i.e* $R_H = D/4 = R/2$) and the area A for πR^2,

$$q = \frac{1}{2}\frac{\gamma_w i}{\mu} R_H^2 \times A \qquad \text{...(8.12)}$$

Likewise, it can be shown that the discharge through two parallel plates of width B and placed at distance d apart is given by (see any text on Fluid Mechanics),

$$q = \frac{1}{3}\frac{\gamma_w i}{\mu}(2 B d_2)$$

Substituting $A = 2Bd$ and $R_H = \frac{2Bd}{2B} = d$,

$$q = \frac{1}{3}\frac{\gamma_w i}{\mu} R_H^2 \times A \qquad \text{...(8.13)}$$

Comparing Eqs. 8.12 and 8.13, it is observed that the general form of the equation for laminar flow through passages of different shapes is the same. The difference is only in the numerical value of the constants. The general equation for discharge in a conduit of any shape can be written as

$$q = C_s \left(\frac{\gamma_w i}{\mu}\right) R_H^2 A \qquad \text{...(8.14)}$$

where C_s is a constant which depends on the shape of the conduit.

Eq. 8.14 is sometimes called the *generalized Hagen- Poisseuille equation*. This equation can be used in a modified form for the flow through soils as explained in the next section.

8.10. LAMINAR FLOW THROUGH POROUS MEDIA

Since the flow through porous media is laminar, Eq. 8.14 can be used. However, the area of flow passage in the case of porous media is equal to the porosity times the total cross-sectional area and, therefore, Eq. 8.14 becomes

$$q = C_s \left(\frac{\gamma_w i}{\mu}\right) R_H^2 (nA) \qquad \text{...(a)}$$

when n is the porosity of soil, represented as ratio.

The hydraulic radius R_H for a porous medium can be written as

$$R_H = \frac{\text{area of flow}}{\text{wetted perimeter}} = \frac{A_v}{P_v}$$

Multiplying the numerator and the denominator by the length of the passage (L).

$$R_H = \frac{A_v \times L}{P_v \times L} = \frac{\text{volume of flow channel}}{\text{surface area of flow channel}}$$

The volume of flow channel may be taken as the volume of voids (V_v), which is equal to $e\,V_s$, where e is the void ratio and V_s is the volume of solids. The surface area of the flow channel may be worked out on the basis of a hypothetical spherical grain of diameter D and having the same volume/area ratio as the entire mass. Thus

$$R_H = \frac{V_v}{A_s} = \frac{e\,V_s}{A_s} = e\,\frac{\pi D^3/6}{\pi D^2} = \frac{eD}{6}$$

Substituting the above value of R_H in Eq. (a) and taking $n = e/1 + e$,

$$q = C_s \left(\frac{\gamma_w i}{\mu}\right)\left(\frac{e\,D}{6}\right)^2 \left(\frac{e}{1+e}\right) A$$

or

$$q = \frac{C_s}{36}\left(\frac{\gamma_w}{\mu}\right)\left(\frac{e^3}{1+e}\right) D^2 i A$$

Replacing $C_s/36$ by another coefficient C,

$$q = C\left(\frac{\gamma_w}{\mu}\right)\left(\frac{e^3}{1+e}\right) D^2 i A$$

or

$$\frac{q}{A} = C\left(\frac{\gamma_w}{\mu}\right)\left(\frac{e^3}{1+e}\right) D^2 i$$

Using Eqs.. (8.2) and (8.3), the above equation can be written as

$$v = C\left(\frac{\gamma_w}{\mu}\right)\left(\frac{e^3}{1+e}\right) D^2 i = k\,i$$

where

$$k = C\left(\frac{\gamma_w}{\mu}\right)\left(\frac{e^3}{1+e}\right) D^2 \qquad \text{...(8.15)}$$

Eq. 8.15 gives a general expression for the coefficient of permeability of soil.

8.11. FACTORS AFFECTING PERMEABILITY OF SOILS

The following factors affect the permeability of soils.

(1) Particle size. As it is evident from Eq. 8.15, the coefficient of permeability of a soil is proportional to the square of the particle size (D). The permeability of coarse-grained soils is very large as compared to that of fine- grained soils. The permeability of coarse sand may be more than one million times as much that of clay.

(2) Structure of soil mass. The coefficient C in Eq. 8.15 takes into account the shape of the flow passage. The size of the flow passage depends upon the structural arrangement. For the same void ratio, the permeability is more in the case of flocculated structure as compared to that in the dispersed structure.

Stratified soil deposits have greater permeability parallel to the plane of stratification than that perpendicular to this plane. Permeability of a soil deposit also depends upon shrinkage cracks, joints, fissures and shear zones. Loess deposits have greater permeability in the vertical direction than in the horizontal direction.

The permeability of a natural soil deposit should be determined in undisturbed condition. The disturbance caused during sampling may destroy the original structure and affect the permeability. The effect of disturbance is more pronounced in the case of fine- grained soils than in the case of coarse-grained soils.

(3) Shape of Particles. The permeability of a soil depends upon the shape of particles. Angular particles have greater specific surface area as compared with the rounded particles. For the same void ratio, the soils with angular particles are less permeable than those with rounded particles, as the permeability is inversely proportional to the specific surface. However, in a natural deposit, the void ratio for a soil with angular particles may be greater than that for rounded particles, and the soil with angular particles may be actually more permeable.

(4) Void Ratio. Eq. 8.15 indicates that the coefficient of permeability varies as $e^3/(1+e)$. For a given soil, the greater the void ratio, the higher is the value of the coefficient of permeability.

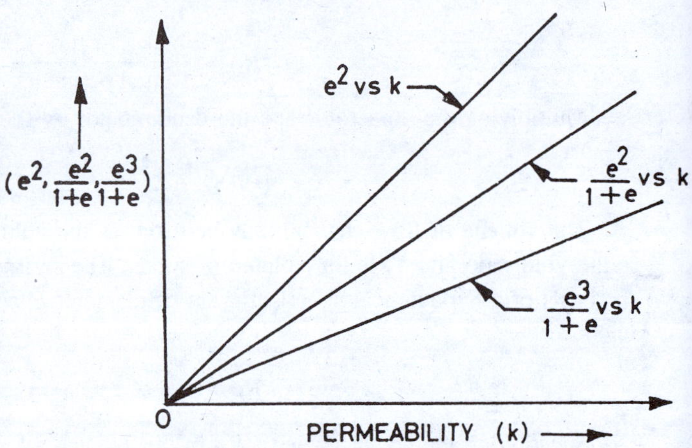

Fig. 8.6. Variation of k with e^2, $\frac{e^2}{1+e}$ and $\frac{e^3}{1+e}$

Based on other concepts, it has been established that the permeability of a soil varies as e^2 or $e^2/(1+e)$ (Fig. 8.6). Whatever may be the exact relationship, all soils have e versus log k plot as a straight line (Fig. 8.7).

It must be noted that each plot in Fig. 8.7 is for a given soil. The permeability of a soil at a given void ratio may not have any relationship with that of another soil at the same void ratio. Paradoxically, the soils with the largest void ratio (*i.e.* clays) are the least pervious. This is due to the fact that the individual void passages in clays are extremely small through which water cannot flow easily.

If the permeability of a soil at a void ratio of 0.85 is known, its value at another void ratio of e can be determined using the following equation given by Casagrande :

$$k = 1.4\, k_{0.85}\, e^2 \qquad ...(8.16)$$

where $k_{0.85}$ = permeability at a void ratio of 0.85, k = permeability at a void ratio of e.

(5) Properties of water. As indicated in Eq. 8.15, the coefficient of permeability is directly proportional to the unit weight of water (γ_w) and is inversely proportional to its viscosity (μ). The unit weight of water does not vary much over the range of temperature ordinarily encountered in soil engineering problems. However, there is a large variation in the value of the coefficient of viscosity (μ). The

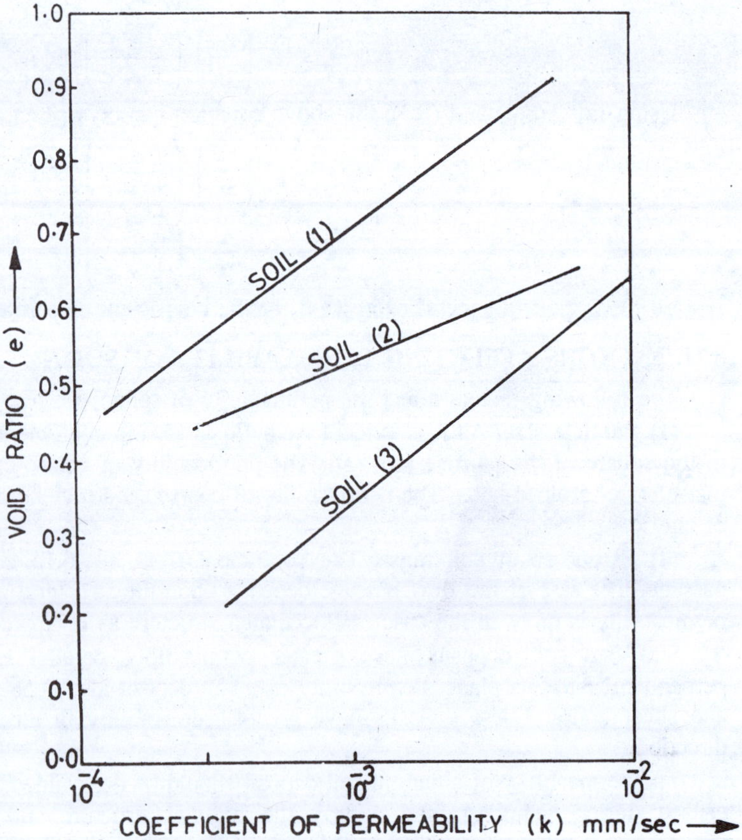

Fig. 8.7. Variation of log k with e.

coefficient of permeability increases with an increase in temperature due to reduction in the viscosity.

It is usual practice (IS : 2720 Part XVII) to report the coefficient of permeability at 27°C. The following equation can be used for conversion of the permeability to 27°C.

$$k_{27} = k_t \frac{\mu_t}{\mu_{27}} \qquad \text{...(8.17)}$$

where k_{27} = coefficient of permeability at 27°C when viscosity is μ_{27}.

and k_t = coefficient of permeability of t °C when viscosity is μ_t.

Eq. 8.17 can be written as

$$k_{27} = C_t\, k_t \qquad \text{...(8.18)}$$

where C_t is the correction factor, equal to (μ_t/μ_{27}).

The correction factor C_t can be determined from the values of the coefficient of viscosity given in Table 3.2.

(6) Degree of Saturation. If the soil is not fully saturated, it contains air pockets formed due to entrapped air or due to air liberated from percolating water. Whatever may be the cause of the presence of air in soils, the permeability is reduced due to presence of air which causes blockage of passage. Consequently, the permeability of a partially saturated soil is considerably smaller than that of a fully saturated soil. In fact, Darcy's law is not strictly applicable to such soils.

The permeability of a partially saturated soil is measured in the laboratory by the capillarity—permeability test (Sect. 8.16).

(7) Adsorbed water. The fine-grained soils have a layer of adsorbed water strongly attached to their surface. This adsorbed water layer is not free to move under gravity. It causes an obstruction to flow of water in the pores and hence reduces the permeability of soils.

It is difficult to estimate the voids occupied by the adsorbed water. According to one estimate, the void ratio occupied by adsorbed water is about 0.10. The effective void ratio available for flow of water is thus about $(e - 0.1)$ and not e. In some cases, at very low hydraulic gradients, the coefficient of permeability of fine-grained soils becomes negligibly small due to presence of adsorbed water.

(8) Impurities in water. Any foreign matter in water has a tendency to plug the flow passage and reduce the effective voids and hence the permeability of soils.

8.12. COEFFICIENT OF ABSOLUTE PERMEABILITY

As discussed above, the coefficient of permeability of a soil depends not only on the properties of soil but also on the properties of permeant (water). Attempts have been made to separate the effect of properties of permeant from the effect of the properties of soil. Another coefficient, known as the coefficient of absolute permeability (K) (also called intrinsic permeability), has been introduced. It is related to the coefficient of permeability (k) as under:

$$K = k(\mu/\gamma_w) \qquad \text{...(8.19)}$$

Using Eq. 8.15,

$$K = C\left(\frac{e^3}{1+e}\right)D^2$$

Therefore, the coefficient of absolute permeability (K) is independent of the properties of water. It depends only on the characteristics of soils.

The dimensions of the coefficient of absolute permeability can be determined from Eq. 8.19 as

$$[K] = \left[\frac{L}{T}\right]\left[\frac{FT}{L^2}\right]\left[\frac{L^3}{F}\right] = [L^2]$$

It has the dimension of area.

The units of K are mm^2, cm^2, m^2 or darcy.

$$1 \text{ darcy} = 0.987 \times 10^{-8} \text{ cm}^2$$

The coefficient of absolute permeability for a soil with a given void ratio and structure is constant. It has the same value whatever may be the fluid.

8.13. PUMPING-OUT TESTS

The laboratory methods for the determination of the coefficient of permeability, as discussed before, do not give correct results. The samples used are generally disturbed and do not represents the true in-situ structure. For more accurate, representative values, the field tests are conducted. The field tests may be in the form of pumping out test wherein the water is pumped out from the wells drilled for this purpose. The other type of the field tests are pumping-in tests, wherein the water is pumped into the drilled holes, as discussed in the following article.

For large engineering projects, it is the usual practice to measure the permeability of soils by pumping-out tests. The method is extremely useful for a homogeneous, coarse grained deposits for which it is difficult to obtain undisturbed samples. In a pumping out test, the soil deposit over a large area is influenced, and therefore the results represent an overall coefficient of permeability of a large mass of soil. However, the tests are very costly and can be justified only for large projects.

Ground water occurs in pervious soil deposits known as *aquifers*. The aquifers are reservoirs of ground water that can be easily drained or pumped out. An aquiclude is a soil deposit which is impervious. If an aquifer does not have an aquiclude at its top and the water table is in the acquifer itself, it is called an unconfined aquifer. If the acquifer is confined between two aquicludes, one at its top and the other at its bottom, it is known as confined aquifer. The coefficient of permeability of the soil can be found using the equations developed below separately for unconfined aquifer and confined aquifer.

(*a*) **Unconfined Aquifer** In an unconfined aquifer, a tube well is drilled as shown in Fig. 8.8. The well reaches the underlying impervious stratum. The tube used for the well is perforated so that water can enter the well. The tube is surrounded by a screen called strainer to check the flow of soil particles into the well. Water is pumped out of the tube well till a steady state is reached. At that stage, the discharge becomes

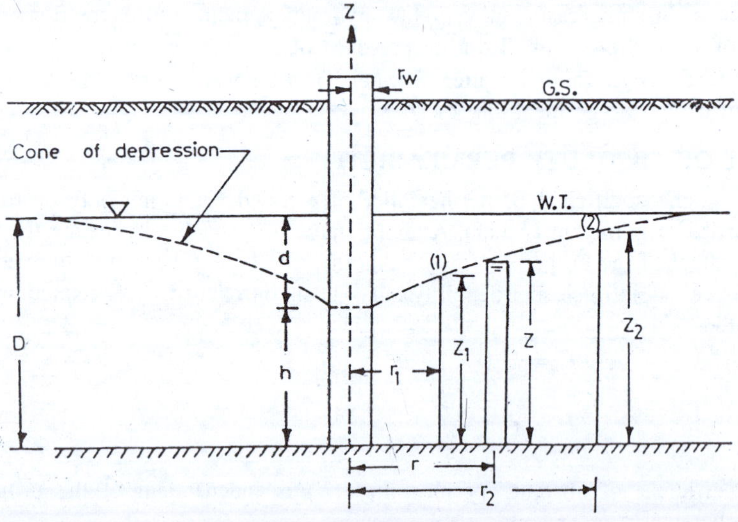

Fig. 8.8. Unconfined Aquifer

constant and the water level in the well does not change. The water table, which was originally horizontal before the pumping was started, is depressed near the well. The water table near the well forms an inverted cone, known as the *cone of depression*. The maximum depression of the water table is known as the draw-down (*d*).

The expression for the coefficient of permeability can be derived making the following assumptions, known as Dupuit's assumption.

(1) The flow is laminar and Darcy's law is valid.
(2) The soil mass is isotropic and homogeneous.
(3) The well penetrates the entire thickness of aquifer.
(4) The flow is steady.
(5) The coefficient of permeability remains constant throughout.
(6) The flow towards the well is radial and horizontal.
(7) Natural ground water regime remains constant.
(8) The slope of the hydraulic gradient line is small, and can be taken as the tangent of the angle in place of the sine of angle, *i.e.*

$$i = \frac{dz}{dr} \qquad \text{...(8.20)}$$

Let us consider the flow through a cylindrical surface of height z at a radial distance of r from the centre of the well. From Darcy's law,

$$q = k\,i\,A$$

Substituting the value of i from Eq. (8.20) and taking A equal to $2\pi r z$,

$$q = k\left(\frac{dz}{dr}\right)(2\pi rz)$$

or

$$\frac{dr}{r} = \frac{2\pi k z\,dz}{q}$$

Integrating,

$$\int_1^2 \frac{dr}{r} = \frac{2\pi k}{q}\int_1^2 z\,dz$$

or

$$\log_e (r_2/r_1) = \frac{2\pi k}{q}\,\frac{(z_2^2 - z_1^2)}{2}$$

or

$$k = \frac{q}{\pi(z_2^2 - z_1^2)}\log_e (r_2/r_1) \qquad \text{...(8.21)}$$

or

$$k = \frac{2.30\,q}{\pi(z_2^2 - z_1^2)}\log_{10} (r_2/r_1) \qquad \text{...(8.22)}$$

Near the test well, there is a rapid drop in head and the slope of the hydraulic gradient is steep, and assumption (8) is not satisfied. The observation wells 1 and 2 should be drilled at considerable distance from the well for accurate measurements. The radial distance r_1 of the well should be at least equal to the thickness of aquifer (D). The observation wells are usually arranged in two orthogonal lines, one along the general direction of flow of the ground water and the other at right angle to this direction.

An approximate value of the coefficient of permeability can be determined if the radius of influence (R) is known or is estimated. The circle of influence, over which the effect of pumping is observed, extends to a very large area. In fact, it gradually merges asymptotically to the water table. The radius of influence varies between 150 to 300 m. According to Sichardt, it can be found using the relation

$$R = 3000\,d\sqrt{k}$$

where R = radius of influence (m), d = drawdown (m)
and k = coefficient of permeability (m/sec)

According to Kozeny (1933), the radius of influence,

$$R = [(12\,t/n)\,(qk/\pi)^{1/2}]^{1/2}$$

where t is the time required to establish steady conditions, and n is the porosity.

Eq. 8.21 can be written as

$$k = \frac{q}{\pi (D^2 - h^2)} \log_e (R/r_w) \qquad \ldots(8.23)$$

where r_w = radius of test well, R = radius of influence,

D = depth of aquifer measured below the water table, h = depth of water in the well.

Eq. 8.23 gives an approximate value of the coefficient of permeability, because the slope of the water surface near the well is steep and Dupuit's assumption is not justified. Further, the value of the radius of influence (R) is also approximate.

(*b*) **Confined Aquifer.** Fig. 8.9 shows a confined aquifer of thickness b and lying between the two aquicludes. The piezometric surface is above the top of the aquifer. In confined aquifer, the water pressure is indicated by the piezometric surface (PS).

Thus the piezometric surface is the water table equivalent for a confined aquifer.

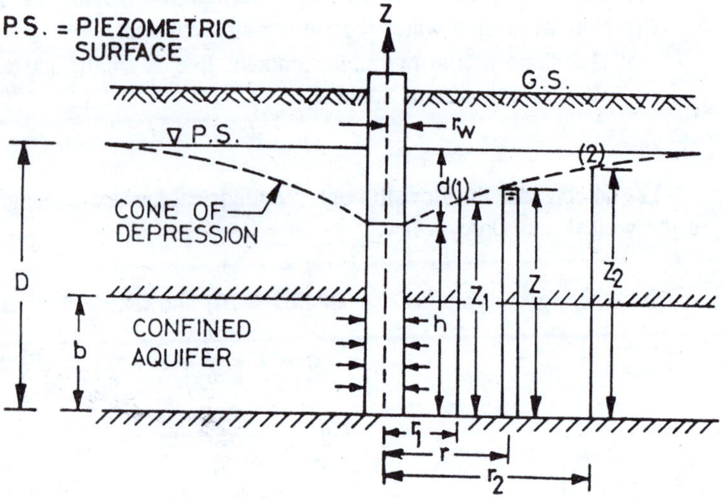

Fig. 8.9. Confined Aquifer.

Initially, the piezometric surface is horizontal. When the pumping is started from the well, it is depressed and a cone of depression is formed. The expression for the coefficient of permeability can be derived making the same assumptions as in the case of unconfined aquifer. Let us consider the discharge through a cylindrical surface at a radial distance r from the centre and of height z. From Darcy's law,

$$q = k\,i\,A$$

or
$$q = k\left(\frac{dz}{dr}\right)(2\,\pi r b) \qquad \ldots(a)$$

Integrating,
$$[\log_e (r)]_1^2 = \frac{2\,\pi\,k\,b}{q}\,[\,z\,]_1^2$$

or
$$k = \frac{q \log_e (r_2/r_1)}{2\,\pi\,b\,(z_2 - z_1)} \qquad \ldots(8.24)$$

or
$$k = \frac{2.30\,q \log_{10} (r_2/r_1)}{2\,\pi\,b\,(z_2 - z_1)} \qquad \ldots(8.25)$$

where z_1 = height of water level in observation well (1) at a radial distance of r_1 and

z_2 = height of water level in observation well (2) at a radial distance of r_2.

As in the case of an unconfined aquifer, an approximate value of k can be determined if the radius of influence R is known or estimated. In this case,

$$k = \frac{q \log_e (R/r_w)}{2\,\pi\,b\,(D - h)} \qquad \ldots(8.26)$$

8.14. PUMPING-IN TESTS

Pumping-in tests are conducted to determine the coefficient of permeability of an individual stratum through which a hole is drilled. These tests are more economical than the pumping-out test. However, the pumping-out tests give more reliable values than that given by pumping-in tests. The pumping-in tests give the value of the coefficient of permeability of stratum just close to the hole, whereas the pumping-out tests give the value for a large-area around the hole.

There are basically two types of pumping-in tests : (1) Open-end tests, (2) Packer tests. In an open-end tests, the water flows out of the test hole through its bottom end, whereas in packer tests, the water flows out through the sides of the section of a hole enclosed between packers. The value of the coefficient of permeability is obtained from the quantity of water accepted by the hole. The water pumped-in should be clean, as the impurities, such as silt, clay or any other foreign matter, may cause plugging of the flow passages. If the water available is turbid, it should be clarified in a settling tank or by using a filter. The temperature of the water pumped in should be slightly higher than the temperature of the ground water to preclude the formation of air bubbles in stratum.

(1) Open-end Tests. A pipe casing is inserted into the bore hole to the desired depth and it is cleaned out. The hole is kept filled with water during cleaning if it extends below the water table. This is necessary to avoid squeezing of the soil into the bottom of the pipe casing when the driving tool is withdrawn.

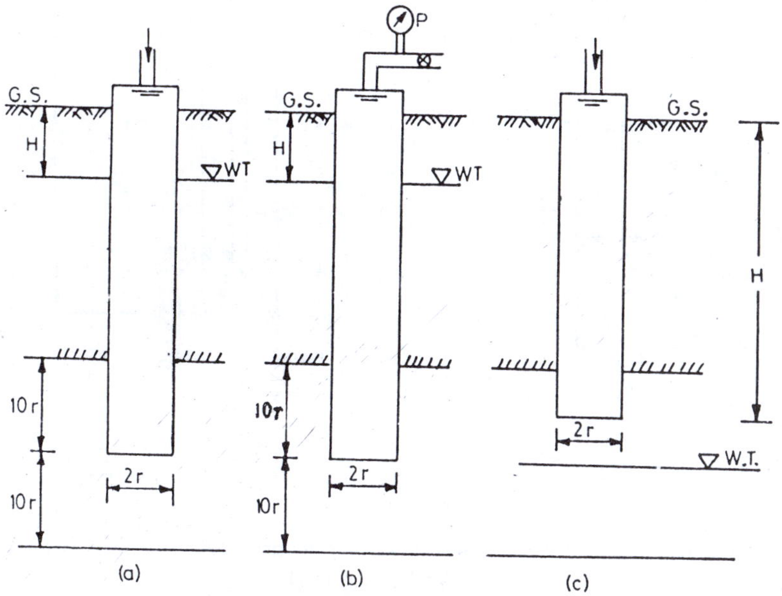

Fig. 8.10. Open-end tests.

After the hole has been cleaned out, water is added to the hole through a metring system. The constant rate of flow (q) is determined at which the steady conditions are established. The coefficient of permeability is determined by the following equation (USBR, 1961).

$$k = \frac{q}{5.5\, rH} \quad \text{...(8.27)}$$

where r = inside radius of the casing,

H = difference of levels between the inlet to the casing and the water table [Fig. 8.10 (*a*)],

q = discharge

If required, the discharge can be increased by pumping-in water under a pressure p [Fig. 8.10 (*b*)]. In this case, the value of H becomes equal to $(H + p/\gamma_w)$.

For accurate results, the lower end of the pipe should be at a distance of not less than 10 r from the top as well as from the bottom of the stratum.

The open-end test can also be conducted above the water table [Fig. 8.10 (*c*)]. In this case, however, it is difficult to maintain a constant water level in the casing and some surging of this level has to be tolerated. Eq. 8.27 can also be used in this case. However, in this case H is equal to the difference of inlet level and the bottom end of the pipe. If required, the rate of flow (q) can be increased by pumping-in water under a pressure p, with a total head of $(H + p/\gamma_w)$.

(2) Packer Tests. The packer tests are performed in an uncased portion of the pipe casing. The packer tests are more commonly used for testing of rocks. The tests are occasionally used for testing of soils if the bore hole can stay open without any casing.

(*a*) Single packer tests. If the hole cannot stand without a casing, single-packer test is used. The packer is placed as shown in Fig. 8.11 (*a*). Water is pumped into the hole. It comes out of the sides of uncased portion of the hole below the packer. If the casing is used for the full depth, it should have perforations in the portion of the stratum being tested. The lower end of the casing is plugged.

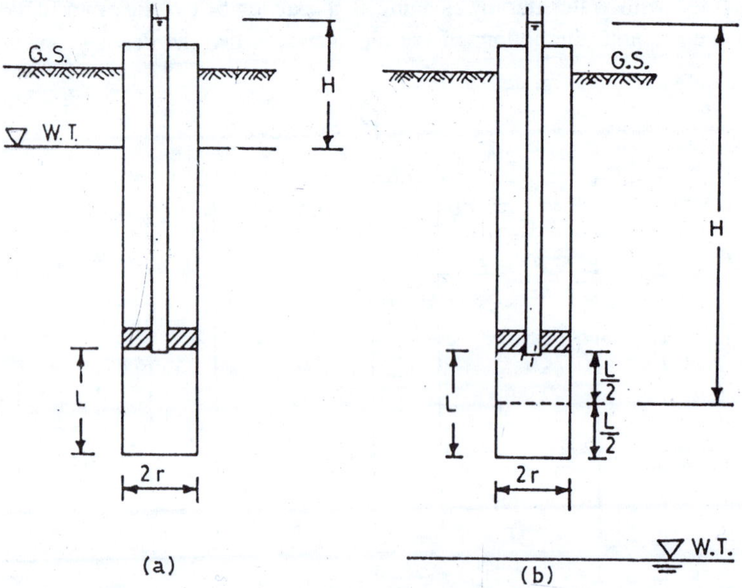

Fig. 8.11. Packer tests.

When the steady conditions are attained, the constant rate of flow (q) is determined. The value of the coefficient of permeability is found by the following equation (USBR, 1961).

$$k = \frac{q}{2\pi LH} \log_e (L/r) \qquad \text{if } L \geq 10r \qquad \ldots(8.28)$$

or

$$k = \frac{q}{2\pi LH} \sinh^{-1} (L/2r) \qquad \text{if } 10r > L \geq r \qquad \ldots(8.29)$$

where r = inside radius of hole, L = length of the hole tested,

H = difference of water levels at the entry and the ground water table for the hole tested below the water table.

$\sinh^{-1}$ = arc hyperbolic sine.

For the holes tested above the water table, H is equal to the difference of levels of water at the entry and the middle of the test section [Fig. 8.11 (*b*)].

If the water is applied under pressure (p), the value of H becomes $(H + p/\gamma_w)$, as in the case of open-end tests.

After the test is complete, the packer is removed. If required, the hole is made deeper and again a packer is placed and the test repeated for that portion.

(*b*) **Double-packer test.** If the hole can stand without a casing, double-packer test can be used. The hole is drilled to the final depth. It is filled with water, surged and bailed out. Two packers are fitted to a small diameter pipe, as shown in Fig. 8.12. The bottom of the pipe fitted with packers is plugged. Fig. 8.12 (*a*) shows the conditions when the test section is below the ground water table and Fig. 8.12 (*b*), when above the ground water table. The value of the coefficient of permeability is determined using Eq. 8.28 or Eq. 8.29, depending upon the value of L and r as specified.

The double-packer test is conducted first in the lowest portion near the bottom of the hole and later repeated for the upper layers.

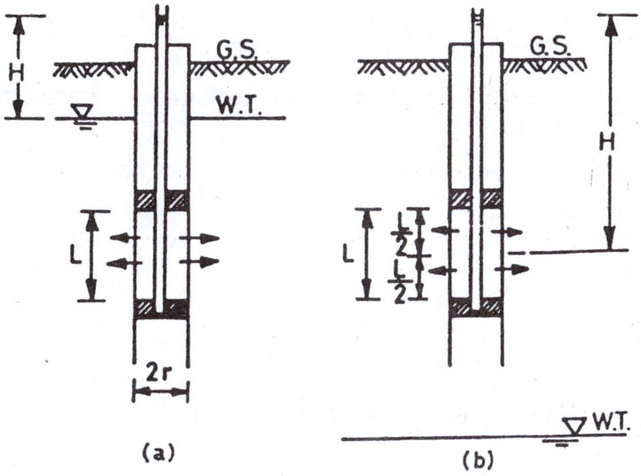

Fig. 8.12. Double-packer test.

The packer tests give better results when conducted below the water table than when above the water table. For reliable results, the thickness of the stratum should be at least five times the length (L) of the hole tested.

8.15. COEFFICIENT OF PERMEABILITY BY INDIRECT METHODS

The value of the coefficient of permeability of a soil can be estimated by using indirect methods, without conducting laboratory tests or field tests. The following methods are commonly used.

(1) Allen Hazen's formula. Allen Hazen conducted a large number of tests on filter sands of particle size between 0.1 mm and 3.0 mm, having a coefficient of uniformity of less than 5, and gave the following relation :

$$k = C D_{10}^2 \qquad \text{...(8.30)}$$

where k = coefficient of permeability (cm/sec), D_{10} = effective size (cm),

C = constant, with a value between 100 and 150

If k and D_{10} are taken in mm/sec and mm, respectively, the value of the constant C lies between 10 and 15.

Although Allen Hazen formula was derived for uniform sands in a loose state of compaction, it is frequently used used for other soils. However, the computed values may be in error by ± 100%.

(2) Kozeny—Carman equation. The coefficient of the permeability of a soil can be estimated using the Kozeny—Carman equation:

$$k = \frac{g\,\rho_w}{(C_s\,\mu\,S^2)\,T^2} \cdot \frac{e^3}{1+e} \qquad \text{...(8.31)}$$

where k = coefficient of permeability (cm/sec), ρ_w = mass density of water (gm/ml),

C_s = shape factor, which can be taken as 2.5 for granular soils,

μ = coefficient of viscosity (poise), e = void ratio, g = 981 cm/sec^2,

T = tortuosity, with a value of $\sqrt{2}$ for granular soils and

S = surface area per unit volume of soil solids, known as specific surface (cm^2/cm^3).

The Kozeny-Carman equation gives good results for coarse-grained soils such as sands and some silts. However, when the equation is used for clayey soils, serious discrepancies are observed. The accuracy for coarse-grained soils is about 20%.

For computation of k from Eq. 8.31, the value of specific surface S is required. The specific surface (S) of a particle is equal to the surface area of the particle per unit volume of the particle. It depends upon the shape and size of the particle. For a spherical particle of diameter D, specific surface (S) is given by

$$S = \frac{(\pi/D^2)}{(\pi D^3/6)} = \frac{6}{D} \qquad ...(8.32)$$

The specific surface of spheres uniformly distributed in size between the mesh size a and b, is given by

$$S = 6/\sqrt{ab} \qquad ...(8.33)$$

For accurate results, the ratio a/b should not be greater than 2.

If the particles are of irregular shape, the specific surface can be determined indirectly from a comparison with the specific surface of uniform sphere of the same size, and using a factor known as angularity factor (f).

$$f = \frac{\text{Specific surface of the actual sand particle}}{\text{Specific surface of spheres of the same size}}$$

The value of f depends upon the angularity of the particles. Its value is usually taken as 1.1 for rounded sands, 1.25 for sands of medium angularity and 1.4 for angular sands.

If $M_1, M_2 \ldots M_n$ are the percentage of the total soil sample retained on different sieves, the overall specific surface of the total sample is given by

$$S = f(M_1 S_1 + M_2 S_2 + \ldots\ldots M_n S_n) \qquad ...(8.34)$$

where $S_1, S_2 \ldots\ldots S_n$ are the specific surface of spheres uniformly distributed within the corresponding sieves.

(3) Loudon's Formula. Loudon gave the following empirical formula.

$$\log_{10}(k S^2) = a + bn \qquad ...(8.35)$$

where k = coefficient of permeability (cm/sec), S = specific surface (cm^2/cm^3),
n = porosity, expressed as a ratio, a = constant, with an average value of 1.365 at 10°C,
b = constant, with an average value of 5.15 at 10°C.

The Loudon formula is much more convenient to use than the Kozeny-Carman equation and gives approximately the same accuracy.

(4) Consolidation test data. The coefficient of permeability of fine-grained soils can be determined indirectly from the data obtained from a consolidation test conducted on the sample (see chapter 12). It is given by

$$k = C_v \gamma_w m_v = C_v \rho_w g m_v \qquad ...(8.36)$$

where k = coefficient of permeability (m/sec), C_v = coefficient of consolidation (m^2/sec),
ρ_w = density of water (kg/m^3), g = 9.81 m/sec^2,
m_v = coefficient of volume compressibility (cm^2/N), γ_w = unit weight of water (N/m^3).

This method is suitable for very fine-grained soils ($k < 10^{-5}$ mm/sec) for which permeability test cannot be easily conducted in the laboratory.

8.16. CAPILLARITY—PERMEABILITY TEST

The coefficient of permeability of soil in unsaturated condition can be determined from the capillarity-permeability test. The apparatus consists of a transparent tube made of lucite or glass, about 35 cm long and 4 cm diameter (Fig. 8.13). The sample of the dry soil in powdered form is placed in the tube and screens are fixed at both ends. One end of the transparent tube is connected to high level water reservoirs and the other end is open to atmosphere through an air-vent pipe. The air-vent pipe is connected to the screen at that end with a spring.

The valve D connecting to the higher reservoir is initially closed. When the valve C connecting to the lower reservoir is opened, capillary action in soil occurs and it draws water into it. The wetted surface starts advancing towards the open end. Let us consider the stage when the wetted surface has advanced by a distance of x. Let the negative capillary head be h_c, as shown by an imaginary manometer in figure. (The manometer is imaginary and in actual tests, no manometer is used. It has been shown in the figure just to

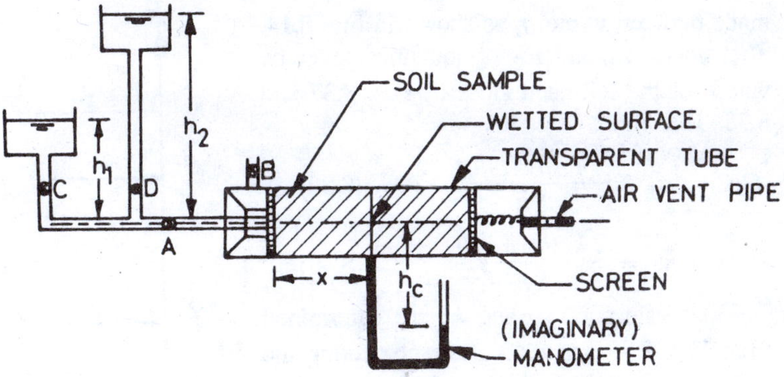

Fig. 8.13. Capillarity—Permeability test.

indicate the negative head). The total head causing flow is increased because of the negative head (h_c) and is given by

$$h = h_1 + h_c$$

Assuming a uniform hydraulic gradient over the entire length x, the velocity is given by Darcy's law,

$$v = k\,i = k\,\frac{(h_1 + h_c)}{x} \qquad ...(a)$$

The wetted surface moves (on a macroscopic scale) with a seepage velocity (v_s), given by Eq. 8.8 as

$$v_s = v/n$$

Therefore, the seepage velocity is given by

$$v_s = \frac{k}{n} \cdot \frac{(h_1 + h_c)}{x}$$

For partially saturated soils, the above equation is modified taking actual saturated porosity as $S \times n$. Thus

$$v_s = \frac{k_u}{S\,n}\,\frac{(h_1 + h_c)}{x}$$

where k_u = coefficient of permeability in unsaturated condition

S = degree of saturation, expressed as a ratio.

Substituting $v_s = dx/dt$, we have

$$\frac{dx}{dt} = \frac{k_u}{S\,n}\,\frac{(h_1 + h_c)}{x}$$

or

$$x\,dx = \frac{k_u (h_1 + h_c)}{S\,n}\,dt$$

Integrating,

$$\int_1^2 x\,dx = \frac{k_u (h_1 + h_c)}{S\,n} \int_1^2 dt$$

or

$$\frac{x_2^2 - x_1^2}{2} = \frac{k_u\,(h_1 + h_c)}{S\,n}\,(t_2 - t_1)$$

or

$$\frac{x_2^2 - x_1^2}{(t_2 - t_1)} = \frac{2\,k_u\,(h_1 + h_c)}{S\,n} \qquad ...(8.37)$$

Eq. 8.37 can be used to determine the coefficient of permeability (k_u) if all other variables are given. As the capillary head (h_c) is also not known, there are two unknowns (k_u and h_c) on the right-hand side of the equation. Therefore, one more equation is required.

The second equation can be derived if the head is changed from h_1 to h_2 when the water surface has advanced to about half the length of the transparent tube by closing the valve C and opening the valve D. Let x_2 and x_3 by the distances measured from the left end at the time t_2 and t_3. Eq. 8.37 becomes, for this case, as

$$\frac{x_3^2 - x_2^2}{(t_3 - t_2)} = \frac{2\,k_u\,(h_2 + h_c)}{S\,n} \qquad ...(8.38)$$

The values of the unknown k_u and h_c can be obtained analytically from Eqs. 8.37 and 8.38. A plot is made between x^2 and t, as shown in Fig. 8.14. The slopes m_1 and m_2 of the lines give the values of the left-hand sides of Eqs. 8.37 and 8.38. Therefore,

$$m_1 = \frac{2\,k_u\,(h_1 + h_c)}{S\,n} \quad ...(8.39)$$

$$m_2 = \frac{2\,k_u\,(h_2 + h_c)}{S\,n} \quad ...(8.40)$$

The values of k_u and h_c are determined from Eqs. 8.39 and 8.40 after substituting the values of m_1 and m_2 obtained from the plot.

The porosity n of the soil sample is determined from its dry density, as discussed in chapter 2.

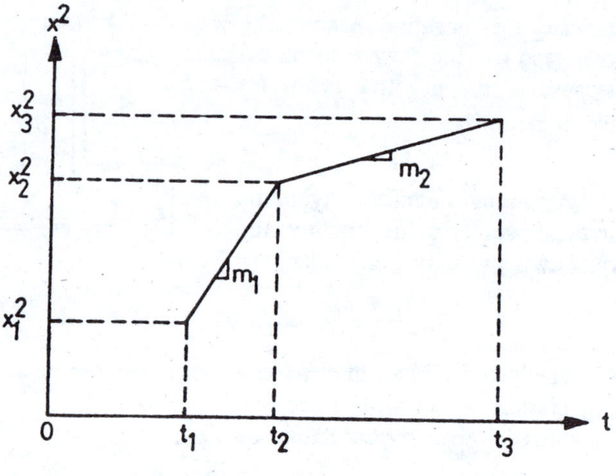

Fig. 8.14. Plot of t and x^2.

$$\rho_d = \frac{M_s}{V} = \frac{G\,\rho_w}{1 + e}$$

or
$$e = \frac{G\,\rho_w}{\rho_d} - 1$$

and
$$n = \frac{e}{1 + e}$$

The degree of saturation (S) is obtained from the water content of the soil determined after the test, using the equations developed in chapter 2.

$$S = wG/e$$

For accurate results, the capillary head (h_c) should be maintained constant along the vertical wetting surface. It is done by slowly revolving the tube about its axis.

8.17. PERMEABILITY OF STRATIFIED SOIL DEPOSITS

A stratified soil deposit consists of a number of soil layers having different permeabilities. The average permeability of the deposit as a whole parallel to the planes of stratification and that normal of the planes of stratification can be determined as explained below.

(a) **Flow Parallel to Planes of Stratification.** Let us consider a deposit consisting of two horizontal layers of soil of thickness H_1 and H_2 as shown in Fig. 8.15.

For flow parallel to the planes of stratification, the loss of head (h) over a length L is the same for both the layers. Therefore, the hydraulic gradient (i) for each layer is equal to the hydraulic gradient of the entire deposit. The system is analogous to the two resistances in parallel in an electrical circuit, wherein the potential drop is the same in both the resistances.

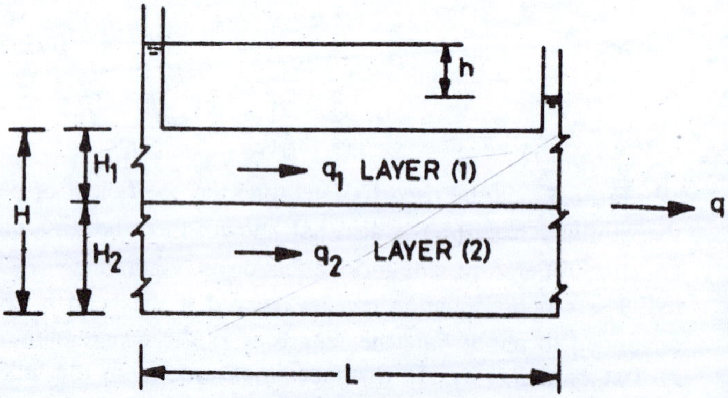

Fig. 8.15

From the continuity equation, the total discharge (q) per unit width is equal to the sum of the discharges in the individual layers, *i.e.*,

$$q = q_1 + q_2 \quad ...(a)$$

Let $(k_h)_1$ and $(k_h)_2$ be the permeability of the layers 1 and 2 respectively, parallel to the plane of stratification and (k_h) be the overall permeability in that direction. From Eq. (*a*), using Darcy's law,

$$k_h \times i \times (H_1 + H_2) = (k_h)_1 \times i \times H_1 + (k_h)_2 \times i \times H_2$$

or
$$k_h = \frac{(k_h)_1 \times H_1 + (k_h)_2 \times H_2}{H_1 + H_2}$$

If there are *n* layers instead of two.

$$k_h = \frac{(k_h)_1 \times H_1 + (k_h)_2 \times H_2 + \ldots + (k_h)_n \times H_n}{H_1 + H_2 + \ldots + H_n} \quad \ldots(8.41)$$

(*b*) Flow normal to the plane of stratification. Let us consider a soil deposit consisting of two layers of thickness H_1 and H_2 in which the flow occurs normal to the plane of stratification (Fig. 8.16).

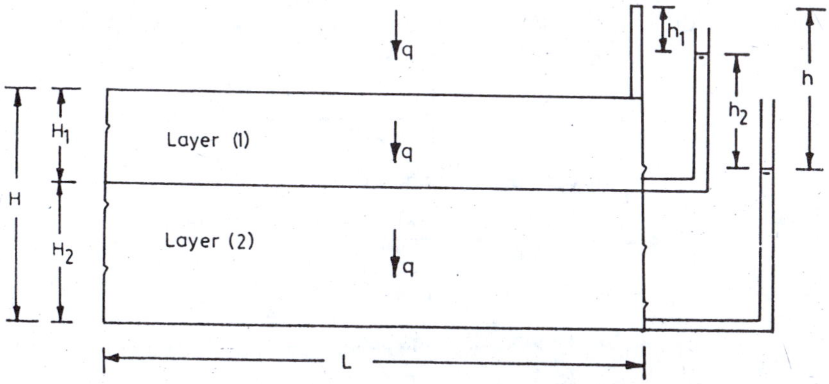

Fig. 8.16. Flow normal to plane of stratification.

Let $(k_v)_1$ and $(k_v)_2$ be the coefficient of permeability of the layers 1 and 2 in the direction perpendicular to the plane of stratification, and k_v be the average coefficient of permeability of the entire deposit in that direction. In this case, the discharge per unit width is the same for each layer and is equal to the discharge in the entire deposit. The case is analogous to the resistances in series in an electrical circuit, wherein the current is the same for all resistances.

Therefore, $$q = q_1 = q_2 \quad \ldots(a)$$

Using Darcy's law, considering unit area perpendicular to flow,

$$h_v \times i_n \times 1 = (k_v)_1 \times (i_v)_1 \times 1 = (k_v)_2 \times (i_v)_2 \times 1 \quad \ldots(b)$$

where i_v = overall hydraulic gradient, $(i_v)_1$ = hydraulic gradient in layer 1,
$(i_v)_2$ = hydraulic gradient in layer 2

From Eq. (*b*), $$(i_v)_1 = [(k_v)/(k_v)_1] \times i_v \quad \ldots(c)$$

and $$(i_v)_2 = [(k_v)/(k_v)_2] \times i_v \quad \ldots(d)$$

As the total loss of head (*h*) over the entire deposit is equal to the sum of the loss of heads in the individual layers,

$$h = h_1 + h_2$$

Writing in terms of hydraulic grandient (*i*) and the distance of flow, remembering $h = i \times L$,

$$i_v \times H = (i_v)_1 \times H_1 + (i_v)_2 \times H_2$$

Using Eqs. (*c*) and (*d*),

$$i_v \times H = \frac{(k_v)}{(k_v)_1} \times i_v \times H_1 + \frac{(k_v)}{(k_v)_2} \times i_v \times H_2$$

or $$k_v \left[\frac{H_1}{(k_v)_1} + \frac{H_2}{(k_v)_2} \right] = H = H_1 + H_2$$

or $$k_v = \frac{H_1 + H_2}{\dfrac{H_1}{(k_v)_1} + \dfrac{H_2}{(k_v)_2}}$$

In general, when there are n such layers,

$$k_v = \frac{H_1 + H_2 + \ldots + H_n}{\dfrac{H_1}{(k_v)_1} + \dfrac{H_2}{(k_v)_2} + \ldots + \dfrac{H_n}{(k_v)_n}} \qquad \ldots(8.42)$$

Evan (1962) proved that for isotropic ($k_v = k_h$) and homogeneous layers, the average permeability of the entire deposit parallel to the plane of stratification is always greater than that normal to this plane. For illustration, let us consider a deposit consisting of two layers of thickness 1 m and 2 m, having the coefficient of permeability of 1×10^{-2} cm/sec and 1×10^{-4} cm/sec, respectively.

From Eq. 8.41, $$k_h = \frac{1 \times 10^{-2} \times 100 + 1 \times 10^{-4} \times 200}{100 + 200}$$

$$= 0.34 \times 10^{-2} \text{ cm/sec}$$

From Eq. 8.42, $$k_v = \frac{100 + 200}{\dfrac{100}{1 \times 10^{-2}} + \dfrac{200}{1 \times 10^{-4}}}$$

$$= 1.49 \times 10^{-4} \text{ cm/sec}$$

Thus $$k_h > k_v$$

It may be noted that the average permeability parallel to the plane of stratification depends mainly on the permeability of the most permeable layer and its value is close to the permeability of that layer. On the other hand, the average permeability normal to the plane of stratification is close to that for the most impermeable layer. In other words, the average flow parallel to the plane of stratification is governed by the most permeable layer and that peıpendicular to the plane of stratification by the least permeable layer.

ILLUSTRATIVE EXAMPLES

Illustrative Example 8.1. In a constant head permeameter test, the following observations were taken.

Distance between piezometer tappings = *100 mm*
Difference of water levels in piezometers = *60 mm*
Diameter of the test sample = *100 mm*
Quantity of water collected = *350 ml*
Duration of the test = *270 seconds*

Determine the coefficient of permeability of the soil.

Solution. From Eq. 8.5, $$k = \frac{qL}{Ah}$$

In this case, $$q = 350/270 = 1.296 \text{ ml/sec}$$

Therefore, $$k = \frac{1.296 \times 10.0}{(\pi/4) \times (10)^2 \times 6.0} = \mathbf{0.0275 \text{ cm/sec.}}$$

Illustrative Example 8.2. *The falling-head permeability test was conducted on a soil sample of 4 cm diameter and 18 cm length. The head fell from 1.0 m to 0.40 m in 20 minutes. If the cross- sectional area of the stand pipe was 1 cm², determine the coefficient of permeability.*

Solution. From Eq. 8.6, $k = \frac{aL}{At} \log_e (h_1/h_2)$

$$= \frac{1.0 \times 18.0}{(\pi/4) \times (4.0)^2 \times 20 \times 60} \log_e (1.0/0.40)$$

$$= \mathbf{1.09 \times 10^{-3}\ cm/sec.}$$

Illustrative Example 8.3. *A soil has the coefficient of permeability of 4.75 × 10^{-2} mm/sec at 30°C. Determine its value at 27°C. Take the coefficient of viscosity at 30°C and 27°C as 8.0 milli poise and 8.5 milli poise, respectively.*

Solution. From Eq. 8.17, $k_{27} = k_t \frac{\mu_t}{\mu_{27}}$

$$= 4.75 \times 10^{-2} \times 8.0/8.5 = \mathbf{4.48 \times 10^{2}\ mm/sec.}$$

Illustrative Example 8.4. *Estimate the value of the coefficient of permeability of a soil with an effective diameter of 0.2 mm.*

Solution. From Eq. 8.30, $k = CD_{10}^2$

Taking $C = 125$, $k = 125 \times (0.02)^2 = \mathbf{0.05\ cm/sec.}$

Illustrative Example 8.5. *The coefficient of permeability of a soil at a void raio of 0.7 is 4 × 10^{-4} cm/sec. Estimate its value at a void ratio of 0.50.*

Solution. From Eq. 8.15, $k = C\left(\frac{\gamma_w}{\mu}\right)\left(\frac{e^3}{1+e}\right) D^2$

As all the parameters remain constant, except e,

$$\frac{k_{0.7}}{k_{0.5}} = \frac{(0.70)^3}{(1+0.70)} \times \frac{(1+0.50)}{(0.50)^3}$$

or $$\frac{4 \times 10^{-4}}{k_{0.5}} = 2.421$$

or $$k_{0.5} = \mathbf{1.65 \times 10^{-4}\ cm/sec.}$$

Alternative Method

From Eq. 8.16, $k = 1.4\, k_{0.85}\, e^2$

or $4 \times 10^{-4} = 1.4\, k_{0.85} \times (0.7)^2$

or $k_{0.85} = 5.83 \times 10^{-4}$

For $e = 0.50$, $k = 1.40 \times 5.83 \times 10^{-4}\ (0.5)^2$

$$= \mathbf{2.04 \times 10^{-4}\ cm/sec}$$

Illustrative Example 8.6. *A sandy layer 10 m thick overlies an impervious stratum. The water table is in the sandy layer at a depth of 1.5 m below the ground surface. Water is pumped out from a well at the rate of 100 litres per second and the drawdown of the water table at radial distances of 3.0 m and 25.0 m is 3.0 and 0.50 m, respectively. Determine the coefficient of permeability.*

Solution. From Eq. 8.21, $k = \frac{q}{\pi (z_2^2 - z_1^2)} \log_e (r_2/r_1)$

In this case, $z_2 = 8.50 - 0.50 = 8.0$ m and $z_1 = 8.50 - 3.0 = 5.50$ m

Therefore, $$k = \frac{100 \times 10^{-3}}{\pi [(8)^2 - (5.50)^2]} \log_e (25/3)$$

$$= 0.002\ m/sec = \mathbf{2\ mm/sec.}$$

Illustrative Example 8.7. *Determine the coefficient of permeability of a confined aquifer 5 m thick which gives a steady discharge of 20 litres/sec through a well of 0.3 m radius. The height of water in the well which was 10 m above the base before pumping dropped to 8 m. Take the radius of influence as 300 m.*

Solution. From Eq. 8.26,
$$k = \frac{q \log_e (R/r)}{2\pi b (D-h)}$$

or
$$k = \frac{0.020 \log_e (300/0.3)}{2\pi \times 5 \times (10-8)} = \mathbf{0.0022\ m/sec.}$$

Illustrative Example 8.8. *Determine the average coefficient of permeability in the horizontal and vertical directions for a deposit consisting of three layers of thickness 5 m, 1 m and 2.5 m and having the coefficients of permeability of 3×10^{-2} mm/sec, 3×10^{-5} mm/sec, and 4×10^{-2} mm/sec, respectively. Assume the layers are isotropic.*

Solution. From Eq. 8.41, taking $n = 3$,

$$k_h = \frac{(k_h)_1 \times H_1 + (k_h)_2 \times H_2 + (k_h)_3 \times H_3}{H_1 + H_2 + H_3}$$

$$= \frac{3 \times 10^{-2} \times 5 \times 10^3 + 3 \times 10^{-5} \times 1 \times 10^3 + 4 \times 10^{-2} \times 2.5 \times 10^3}{(5+1+2.5) \times 10^3}$$

$$= \frac{0.15 + 0.00003 + 0.10}{8.50} = \mathbf{0.0294\ mm/sec.}$$

From Eq. 8.42,
$$k_v = \frac{H_1 + H_2 + H_3}{\dfrac{H_1}{(k_v)_1} + \dfrac{H_2}{(k_v)_2} + \dfrac{H_3}{(k_v)_3}}$$

or
$$k_v = \frac{(5+1+2.5)\, 10^3}{\left(\dfrac{5 \times 10^3}{3 \times 10^{-2}} + \dfrac{1 \times 10^3}{3 \times 10^{-5}} + \dfrac{2.5 \times 10^3}{4 \times 10^{-2}}\right)} = \mathbf{2.5 \times 10^{-4}\ mm/s}$$

Illustrative Example 8.9. *Fig. E 8.9 shows an aquifer inclined at 10° to the horizontal. The difference of water levels in two observation wells at a horizontal distance of 60 m is 5 m. Determine the discharge through the aquifer per unit width if k = 0.7 mm/sec. The depth of aquifer normal to the direction of flow is 2.951 m.*

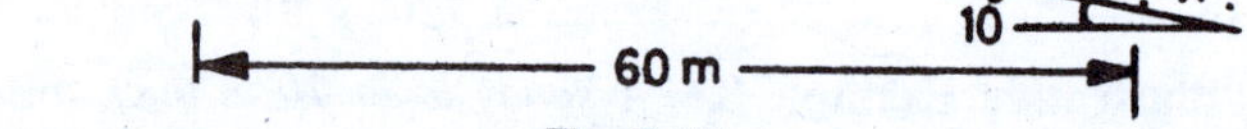

Fig. E-8.9

Solution. Length of aquifer between two observation wells

$$= 60/\cos 10^0 = 60.926 \text{ m}$$

Hydraulic gradient $= h/L = \dfrac{5.0}{60.926} = 0.082$

From Darcy's law, discharge per unit width,

$$q = k\,i\,A$$

$$= 0.7 \times 10^{-3} \times 0.082 \times (2.95 \times 1) = 0.169 \times 10^{-3} \text{ m}^3/\text{sec}$$

$$= \mathbf{0.169\ lit/sec.}$$

Illustrative Example 8.10. *Fig. E-8.10 shows an experimental set- up. If 30% of the effective head is lost in the soil A, determine the total head and the piezometric head at points 1 and 2.*

Determine the coefficient of permeability of the soil B if that of the soil A is 0.5 mm/sec.

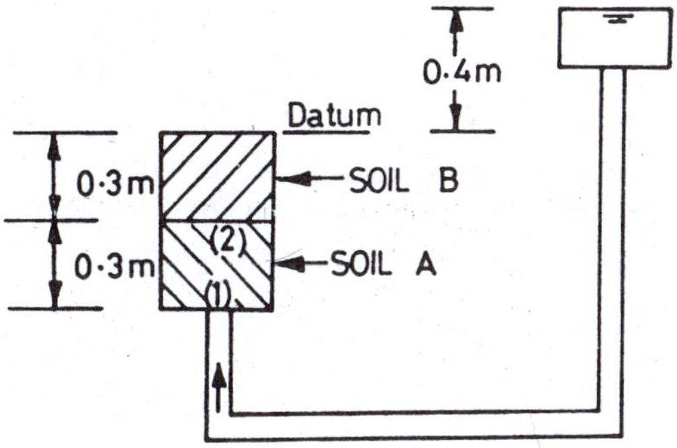

Fig. E-8.10.

Solution.

Piezometric head at (1) = 0.3 + 0.3 + 0.4 = 1.0 m
Datum head at (1) = – 0.60 m
Total head at (1) = 1.0 – 0.6 = 0.4 m
Head lost in Soil A = 0.3 × 0.4 = 0.12 m
Total head at (2) = 0.40 – 0.12 = 0.28 m
Datum head at (2) = –0.30 m
Piezometric head at (2) = 0.28 – (–0.30) = 0.58 m
Loss of head in soil B = 0.7 × 0.4 = 0.28 m
Let k_B be the coefficient of permeability of soil B.

Since the discharge and area are the same in both the soils,

$$k_B \times i_B = k_A \times i_A$$

or $$k_B \times 0.28/0.30 = 0.5 \times 0.12/0.30$$

or $$k_B = \mathbf{0.214\ mm/sec.}$$

Illustrative Example 8.11. *Determine the discharge per unit width of the slot in Fig. E-8 11 if the drawdown is 2.3 m. The coefficient of permeability of the soil is 1 × 10^{-2} mm/sec.*

Also determine the elevation of the water surface at a distance of 30 m from the centre of the slot.

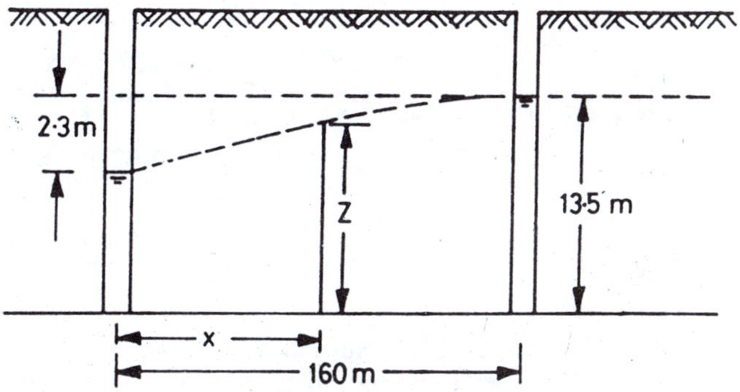

Fig. E-8.11.

Solution. Let us consider flow at a distance of x from the centre of the slot. Using Darcy's law, discharge per unit width

$$q = k\left(\frac{dz}{dx}\right) \times (z \times 1)$$

or $$q\,dx = kz\,dz$$

Integrating, $$q\int_1^2 dx = \int_1^2 k\,zdz$$

or $$q = \frac{k(z_2^2 - z_1^2)}{2(x_2 - x_1)} \qquad \ldots(a)$$

Substituting the values given,

$$q = \frac{1 \times 10^{-2} \times 10^{-3}\,(13.5^2 - 11.2^2)}{2\,(160 - 0)}\ \text{m}^3/\text{sec} = \mathbf{0.001775\ lit/sec}$$

The water surface z at x = 30 m can be determined using Eq. (a).

$$0.1775 \times 10^{-5} = \frac{1 \times 10^{-2} \times 10^{-3}\,(z^2 - 11.2^2)}{2\,(30 - 0)}$$

or $$z^2 - 125.44 = 2 \times 30 \times 0.1775 = 10.65$$

or $$z = \mathbf{11.666\ m.}$$

Illustrative Example 8.12. *A capillarity—permeability test was conducted in two stages under a head of 50 cm and 200 cm at the end of entry of water. In the first stage, the wetted surface advanced from its initial position of 2 cm to 8 cm in 6 minutes. In the second stage it advanced from 8 cm to 20 cm in 20 minutes. If the degree of saturation at the end of the test was found to be 90% and the porosity was 30%, determine the capillarity head and the coefficient of permeability.*

Solution. From Eq. 8.37, $$\frac{x_2^2 - x_1^2}{t_2 - t_1} = \frac{2\,k_u\,(h_1 + h_c)}{S \times n}$$

or $$\frac{(8)^2 - (2)^2}{6 \times 60} = \frac{2\,k_u\,(50 + h_c)}{0.9 \times 0.30}$$

or $$k_u\,(50 + h_c) = 0.0225 \qquad \ldots(1)$$

From Eq. 8.38, $$\frac{x_3^2 - x_2^2}{t_3 - t_2} = \frac{2\,k_u\,(h_2 + h_c)}{S \times n}$$

or $$\frac{(20)^2 - (8)^2}{20 \times 60} = \frac{2\,k_u\,(200 + h_c)}{0.9 \times 0.3}$$

or $$k_u\,(200 + h_c) = 0.0378 \qquad \ldots(2)$$

From Eqs. (1) and (2),

$$\frac{200 + h_c}{50 + h_c} = \frac{0.0378}{0.0225} = 1.68$$

or $$h_c = \mathbf{170.59\ cm}$$

From Eq. (1), $$k_u\,(50 + 170.59) = 0.0225$$

or $$k_u = \mathbf{1.02 \times 10^{-4}\ cm/sec.}$$

PROBLEMS

A Numericals

8.1. (*a*) A constant-head permeability test was run on a sand sample 30 cm in length and 20 cm^2 in area. When a loss of head was 60 cm, the quantity of water collected in 2 minutes was 250 ml. Determine the coefficient of permeability of the soil.

(*b*) If the specific gravity of grains was 2.65, and dry mass of the sample, 1.1 kg, find the void ratio of the sample. **[Ans.** 0.052 cm/sec; 0.445**]**

8.2. A falling-head permeability test was performed on a sample of clean, uniform sand. One minute was required for the initial head of 100 cm to fall to 50 cm in the stand pipe of the cross- sectsonal area 1.50 cm^2. If the sample was 4 cm in diameter and 30 cm long, calculate the coefficient of permeability of the sand. **[Ans.** 0.0275 cm/sec**]**

8.3. During a pumping test, a well was sunk through a stratum of dense sand 10 m deep overlying an impervious stratum. Observation holes were drilled at 15 m and 6.75 m from the well. Initially, the water level in the well was 2.50 m below the ground surface. After pumping until steady conditions had been achieved, the water levels in the observation wells had dropped 1.95 m and 0.50m, respectively. If the steady discharge was 5 litres/sec, determine the coefficient of permeability. **[Ans.** 0.698 × 10^{-2} cm/sec**]**

8.4. A permeameter of diameter 82.5 mm contains a column of fine sand 460 mm long. When water flows through it under a constant head at a rate of 191 ml/minute, the loss of head between two points 250 mm apart is 380 mm, calculate the coefficient of permeability.

If falling head test is made on the same sample using a stand pipe of diameter 30 mm, in what time will the water level in stand pipe fall from 1560 mm to 1066 mm above outflow level.

[Ans. 3.92 × 10^{-1} mm/sec; 59.1 sec**]**

8.5. Calculate the coefficient of permeability of a soil sample 8 cm in height and cross-sectional area 60 cm^2. It is observed that in 12 minutes, 600 ml of water passed down under an effective constant head of 50 cm.

On oven drying, the test specimen weighs 750 gm. Taking 2.70 as specific gravity of soil, calculate the seepage velocity of water during the test. [**Ans.** 2.22×10^{-3} cm/sec; 0.33 cm/sec.]

8.6. Fig. P-8.6 shows a cross-section through the strata underlying a site. Calculate the equivalent permeability of the layered system in the vertical and horizontal direction. Assume that each layer is isotropic.

[**Ans.** 1.41×10^{-6} cm/sec; 0.081 cm/sec]

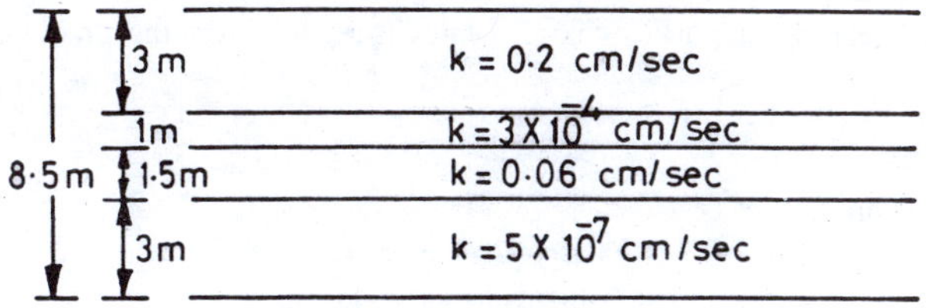

Fig. P-8.6.

8.7. A glacial clay deposit contains a series of silt partings in it at an average vertical spacing of 2 m. If the silt layers are about 5 mm in thickness and have a permeability of one hundred times that of the clay, determine the ratio of the horizontal and vertical permeabilities. [**Ans.** 1.244]

8.8. In a falling-head permeameter if the time intervals for drop in levels from h_1 to h_2 and h_2 to h_3 are equal, prove that

$$h_2 = \sqrt{h_1 \times h_3}$$

8.9. If the effective grain size of the soil is 0.3 mm, estimate the coefficient of permeability. Take Hazen's $C = 10$.

[**Ans.** 0.9 mm/sec]

8.10. A soil has a coefficient of permeability of 0.5×10^{-4} cm/sec at 20°C. Determine its value when the temperature rises to 35°C. ($\mu_{20°} = 10.09 \times 10^{-3}$ poise and $\mu_{35} = 7.21 \times 10^{-3}$ poise). [**Ans.** 0.7×10^{-4} cm/sec]

8.11. A drainage pipe beneath a dam has become clogged with sand whose coefficient of permeability is 10 m/day. It has been observed that the flow through the pipe is 0.15 m^3/day when the difference of water levels on the upstream and downstream is 20 m. If the cross sectional area of the pipe was 200 cm^2, what length of the pipe was filled with sand ? [**Ans.** 26.67 m]

8.12. A soil has the coefficient of permeability of 0.4×10^{-4} cm/sec at a void ratio of 0.65 and a temperature of 30°C. Determine the coefficient of permeability at the same void ratio and a temperature of 20°C. At 20°C, $\rho_w = 0.998$ gm/ml and $\mu = 0.0101$ poise and at 30°C, $\rho_w = 0.996$ gm/ml and $\mu = 0.008$ poise.

What would be the coefficient of permeability at a void ratio of 0.75 and a temperature of 20°C ?

[**Ans.** 0.317×10^{-4} cm/sec; 0.422×10^{-4} cm/sec]

B. Descriptive and Objective type

8.13. What is Darcy's law ? What are its limitations ?

8.14. What are different methods for determination of the coefficient of permeability in a laboratory ? Discuss their limitations.

8.15. Describe pumping-out methods for the determination of the coefficient of permeability in the field. What are their advantages and disadvantages ? What are Dupuit's assumptions ?

8.16. Discuss open-end and packers methods for the determination of the coefficient of permeability. Compare these methods with the pumping-out methods.

8.17. What is Allen Hazen's formula for the coefficient of permeability ? What is its use ? Compare this with Kozeny—Carman equation and Loudon's formula.

8.18. Describe in brief the capillarity-permeability test ? Why the values of the coefficient of permeability obtained from this test differ from those obtained from other tests ?

8.19. How would you determine the average permeability of a soil deposit consisting of a number of layers ? What is its use in soil engineering ?

8.20. Write whether the following statements are true or false.

(*a*) The coefficient of permeability of a soil increases with an increase in temperature.

(*b*) The soils with a higher void ratio have always greater permeability than soils with a smaller void ratio.

(*c*) The coefficient of permeability decreases with an increase in the specific surface.

(*d*) For a given soil, the coefficient of permeability increases with an increase in void ratio.

(*e*) For a soil deposit consisting of isotropic layers, the coefficient of permeability parallel to the plane of stratification is always greater than that normal to this plane.

(*f*) The variable-head permeability test is used for fine- grained soils.

(*g*) The line joining the piezometric surfaces is also known as the hydraulic gradient line.

[**Ans.** True (*a*), (*c*), (*d*), (*e*), (*f*), (*g*)]

C. Multiple-Choice Questions

1. The permeability of soil varies

(*a*) inversely as square of grain size (*b*) as square of grain size

(*c*) as grain size (*d*) inversely as void ratio.

2. The maximum particle size for which Darcy's law is applicable is

(*a*) 0.2 mm (*b*) 0.5 mm

(*c*) 1.0 mm (*d*) 2.0 mm

3. According to U.S.B.R., a soil with a coefficient of permeability of 10^{-4} mm/sec will be classified as

(*a*) Pervious (*b*) Impervious

(*c*) Semi-pervious (*c*) Highly pervious

4. The coefficient of permeability of clay is generally.

(*a*) Between 10^{-4} and 10^{-2} mm/s (*b*) Between 10^{-5} and 10^{-4} mm/s

(*c*) Between 10^{-5} and 10^{-8} mm/s (*d*) Less then 10^{-8} mm/s

5. A constant-head permeameter is used for

(*a*) Coarse-grained soils (*b*) Silty soils

(*c*) Clayey soils (*d*)Organic soils

6. The coefficient of permeability of a soil

(*a*) increases with a increase in temperature.

(*b*) increases with a decrease in temperature.

(*c*) increases with a decrease in unit weight of water.

(*d*) decreases with an increase in void ratio.

7. A soil has a discharge velocity of 6×10^{-7} m/s and a void ratio of 0.50. Its seepage velocity is

(*a*) 18×10^{-7} m/s (*b*) 12×10^{-7} m/s

(*c*) 24×10^{-7} m/s (*d*) 36×10^{-7} m/s

8. In a pumping-out test, the drawdown is 5m. If the coefficient of permeability of the soil is 10^{-4}m/s, the radius of influence will be about

(*a*) 250 m (*b*) 300 m

(*c*) 150 m (*d*) 200 m

9. For a sphere of 0.5 mm diameter, the specific surface is

(*a*) 12 mm^{-1} (*b*) 6 mm^{-1}

(*c*) 8 mm^{-1} (*d*) 9 mm^{-1}

[**Ans.** 1. (*b*), 2. (*b*), 3. (*c*), 4. (*c*), 5. (*a*), 6. (*a*), 7. (*a*), 8. (*c*), 9. (*a*)]

9

Seepage Analysis

1. INTRODUCTION

Seepage is the flow of water under gravitational forces in a permeable medium. Flow of water takes place from a point of high head to a point of low head. The flow is generally laminar.

The path taken by a water particle is represented by a flow line. Although an infinite number of flow lines can be drawn, for convenience, only a few are drawn. At certain points on different flow lines, the total head will be the same. The lines connecting points of equal total head can be drawn. These lines are known as equipotential lines. As flow always takes place along the steepest hydraulic gradient, the equipotential lines cross flow lines at right angles. The flow lines and equipotential lines together form a flow net. The flow net gives a pictorial representation of the path taken by water particles and the head variation along that path.

Fig. 9.1 (*a*) shows a glass cylinder containing a soil sample of length *L*. A steady flow occurs vertically downward through the soil sample under a head of *h*. The elevation head, the pressure head and the total head

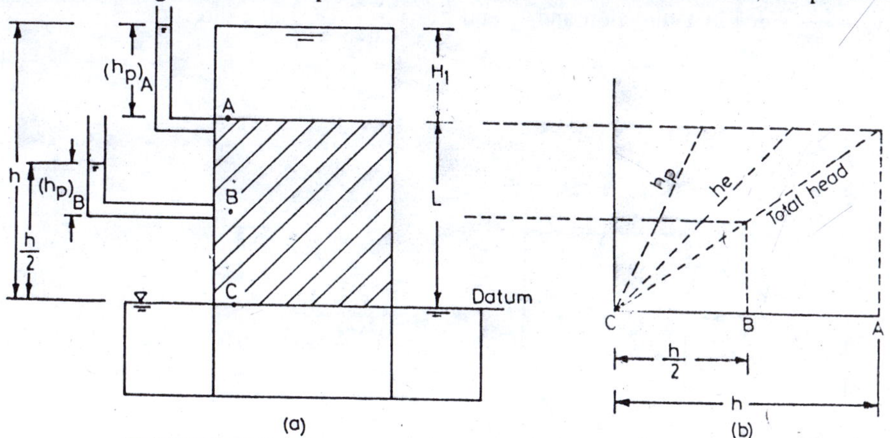

Point	Elevation head (he)	Pressure head (h_p)	Total head (H)
A	L	H_1	$L + H_1 = h$
B	0.5 L	$0.5L + H_1 - 0.5h = 0.5h - 0.5L$	$L + H_1 - 0.5h = 0.5h$
C	0	0	0

(c)

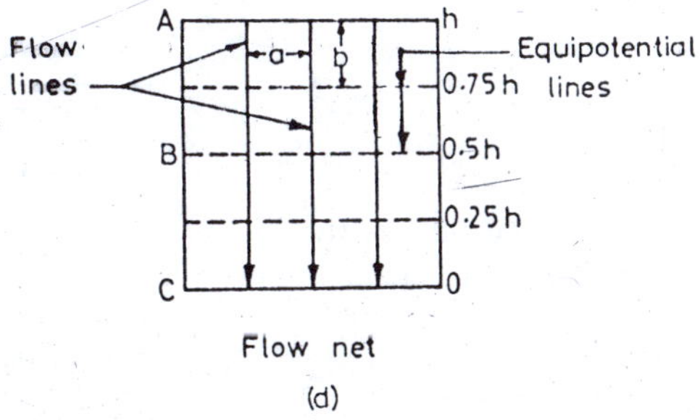

Fig. 9.1. Vertical flow through soil.

at points. A, B and C can be worked out as shown in Fig. 9.1 (b) and 9.1 (c). The point B is at a height of 0.5 L above the datum. As the rate of loss of head is linear, the loss of head upto point B is $h/2$. Therefore, the total head at point B is $h/2$. Fig. 9.1 (d) shows a simple flow net, in which five flow lines and an equal number of equipotential lines are drawn. The equipotential lines are horizontal and the flow lines are vertical in this case. If a dye is inserted at a few points on the top of the soil sample, the paths taken by the dye represent the flow lines. The flow nets in actual soil engineering problems are not as simple as shown in the figure.

In this chapter, the methods for construction of flow net and their uses are discussed. The forces associated with seepage and their effect on the stresses are dealt in the following chapter.

9.2. LAPLACE'S EQUATION

The simple method of construction of flow net as explained above cannot be used for soil engineering problems in which the flow is generally two-dimensional. The Laplace equation is used in the construction of the flow net in such cases.

The following assumptions are made in the derivation of the Laplace equation:

(1) The flow is two-dimensional.

(2) Water and soil are incompressible.

(3) Soil is isotropic and homogeneous.

(4) The soil is fully saturated.

(5) The flow is steady, *i.e.*, flow conditions do not change with time.

(6) Darcy's law is valid.

Let us consider an element of soil of size dx by dz through which flow is taking place (Fig. 9.2). The third dimension along y-axis is large. For convenience, it is taken as unity. Let the velocity at the inlet and outlet faces be v_x and $\left(v_x + \frac{\partial v_x}{\partial x} \cdot dx\right)$ in x-direction and v_z and $\left(v_z + \frac{\partial v_z}{\partial z} \cdot dz\right)$ in z-direction.

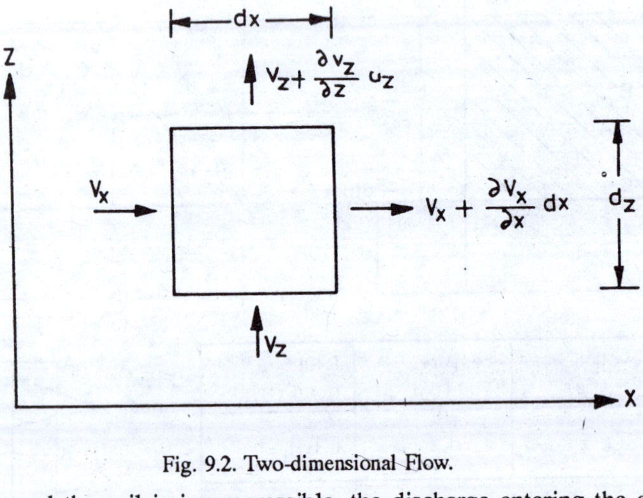

Fig. 9.2. Two-dimensional Flow.

As the flow is steady and the soil is incompressible, the discharge entering the element is equal to that leaving the element.

Thus $$v_x\,dz + v_z\,dx = \left(v_x + \frac{\partial v_x}{\partial x} \cdot dx\right) dz + \left(v_z + \frac{\partial v_z}{\partial z} \cdot dz\right) dx$$

or $$\left(\frac{\partial v_x}{\partial x} + \frac{\partial v_z}{\partial z}\right) dx\,dz = 0$$

or $$\left(\frac{\partial v_x}{\partial x} + \frac{\partial v_z}{\partial z}\right) = 0 \qquad ...(9.1)$$

Eq. 9.1 is the continuity equation for two-dimensional flow.

Let h be the total head at any point. The horizontal and vertical components of the hydraulic gradient are, respectively,

$$i_x = -\frac{\partial h}{\partial x}, \text{ and } i_z = -\frac{\partial h}{\partial z}$$

The minus indicates that the head decreases in the direction of flow.

From Darcy's law, $$v_x = -k_x \frac{\partial h}{\partial x}, \qquad v_z = -k_z \frac{\partial h}{\partial z}$$

Substituting in Eq. 9.1, $$-k_x \frac{\partial^2 h}{\partial x^2} - k_z \frac{\partial^2 h}{\partial z^2} = 0$$

or $$k_x \frac{\partial^2 h}{\partial x^2} + k_z \frac{\partial^2 h}{\partial z^2} = 0$$

As the soil is isotropic, $k_x = k_z$. Therefore,

$$\frac{\partial^2 h}{\partial x^2} + \frac{\partial^2 h}{\partial z^2} = 0 \qquad ...(9.2)$$

Eq. 9.2 is the *Laplace equation* in terms of head h.

Sometimes, the Laplace equation is represented in terms of velocity potential ϕ, given by

$$\phi = -kh$$

Therefore, $$\frac{\partial \phi}{\partial x} = v_x = -k\frac{\partial h}{\partial x}$$

and $$\frac{\partial \phi}{\partial z} = v_z = -k\frac{\partial h}{\partial z}$$

Substituting the values of v_x and v_z in Eq. 9.1,

$$\frac{\partial^2 \phi}{\partial x^2} + \frac{\partial^2 \phi}{\partial z^2} = 0 \qquad ...(9.3)$$

Eq. 9.3 is the Laplace equation in terms of velocity potential.

Laplace's equation can be solved if the boundary conditions at the inlet and exit are known. The equation represents two families of curves which are orthogonal to each other. One family represents the flow lines along which the flow takes place. The other family represents the equipotential lines along which the potential (ϕ) or total head (h) is constant. The graphical representation of the Laplace equations is, therefore, a flow net.

9.3. STREAM AND POTENTIAL FUNCTIONS

Stream function (ψ) is a scalar function of the coordinates x, z such that its partial derivatives satisfy the following equations:

$$\frac{\partial \psi}{\partial x} = -v_z \qquad ..[9.4(a)]$$

and $$\frac{\partial \psi}{\partial z} = v_x \qquad ...[9.4(b)]$$

As a stream function is a continuous function, its total differential is given by

$$\partial \psi = \frac{\partial \psi}{\partial x} \cdot dx + \frac{\partial \psi}{\partial z} \cdot dz$$

Substituting the values of $\frac{\partial \psi}{\partial x}$ and $\frac{\partial \psi}{\partial z}$ from Eq. 9.4,

$$d\psi = -v_z\, dx + v_x\, dz$$

If the stream function is constant along a curve, $d\psi = 0$. Therefore, $-v_z\, dx + v_x\, dz = 0$

or
$$\left(\frac{dz}{dx}\right)_\psi = \frac{v_z}{v_x} \qquad \text{...(9.5)}$$

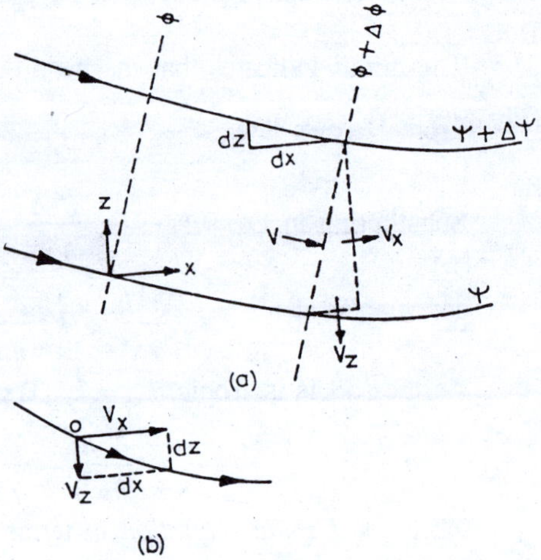

Fig. 9.3. Flow Curves.

The tangent at any point on the ψ-curve gives the directions of the resultant velocity (v) (Fig. 9.3). Hence, the ψ-curve represents the flow line. The curves with constant values $\psi_1, \psi_2 ..., \psi_n$ are the flow lines.

Velocity potential (φ) is a scalar function of x and z such that it derivatives satisfy the following equations (Refer Sect. 9.2).

$$\frac{\partial \phi}{\partial x} = v_x = -k\frac{\partial h}{\partial x} \qquad \text{...[9.6(a)]}$$

and
$$\frac{\partial \phi}{\partial z} = v_z = -k\frac{\partial h}{\partial z} \qquad \text{...[9.6(b)]}$$

Integrating, Eqs. (*a*) and (*b*), $\phi(x, z) = -kh(x, z) + f(z)$

and $\varphi(x, z) = -kh(x, z) + g(x)$

Since x and z can be varied independently,

$$f(z) = g(x) = \text{constant, say } C$$

Therefore, $\phi(x, z) = -kh(x, z) + C$

If the total head h is taken as a constant, it represents a curve for which φ has a constant value. This is an equipotential line. Assigning different values to h such as $h_1, h_2, ... h_n$, we get a number of equipotential lines $\phi_1, \phi_2, ... \phi_n$.

The slope along an equipotential line φ can be determined as under. The total differential is given by

$$d\phi = \frac{\partial \varphi}{\partial x} \cdot dx + \frac{\partial \varphi}{\partial z} \cdot dz$$

If φ is a constant along a curve, $d\varphi = 0$

Hence,
$$0 = \frac{\partial \varphi}{dx} \cdot dx + \frac{\partial \phi}{\partial z}\, dz$$

or
$$\left(\frac{dz}{dx}\right)_\phi = -\frac{\partial \phi/\partial x}{\partial \varphi/\partial z} = -\frac{v_x}{v_z} \qquad \text{...(9.7)}$$

From Eqs. (9.5) and (9.7),

$$\left(\frac{dz}{dx}\right)_\psi \times \left(\frac{dz}{dx}\right)_\phi = -\frac{v_z}{v_x} \times \frac{v_x}{v_z} = -1$$

Thus, the stream function and the potential function are orthogonal to each other.

From Eqs. (9.4) and (9.6),
$$\frac{\partial \phi}{\partial x} = \frac{\partial \psi}{\partial z}$$

or
$$\frac{\partial^2 \phi}{\partial x \partial z} = \frac{\partial^2 \psi}{\partial^2 z}$$

and
$$-\frac{\partial \phi}{\partial z} = \frac{\partial \psi}{\partial x}$$

or $$-\frac{\partial^2 \phi}{\partial x \partial z} = \frac{\partial^2 \psi}{\partial x^2}$$

Therefore, $$\frac{\partial^2 \psi}{\partial x^2} + \frac{\partial^2 \psi}{\partial z^2} = -\frac{\partial^2 \phi}{\partial x \partial z} + \frac{\partial^2 \phi}{\partial x \partial z} = 0$$

Thus, the stream function (ψ) also satisfies the Laplace equation.

Determination of Discharge

The discharge Δq between two adjacent flow lines ψ and $(\psi + \Delta\psi)$ can be determined as follows [Fig. 9.3 (*a*)].

The discharge is equal to the resultant velocity v multiplied by the normal distance (Δn) between ψ and $(\psi + d\psi)$, Obviously,

$$\text{discharge} = -v_z\,dx + v_x\,dz$$

Therefore, $$\Delta q = \int_{\psi}^{\psi + \Delta\psi} (-v_z\,dx + v_x\,dz)$$

Substituting the value of v_x and v_z from Eq. 9.4,

$$\Delta q = \int_{\psi}^{\psi + \Delta\psi} \left(\frac{\partial\psi}{\partial x} \cdot dx + \frac{\partial\psi}{\partial z} \cdot dz\right) = \int_{\psi}^{\psi + \Delta\psi} d\psi = \Delta\psi \qquad ...(9.8)$$

In other words, the flow between two adjacent flow lines is constant and is equal to the difference of stream functions of the two lines.

9.4. CHARACTERISTICS OF FLOW NET

Fig. 9.4 shows a flow field formed between two adjacent flow lines and equipotential lines. If v_s is the velocity along the stream line represented by ψ,

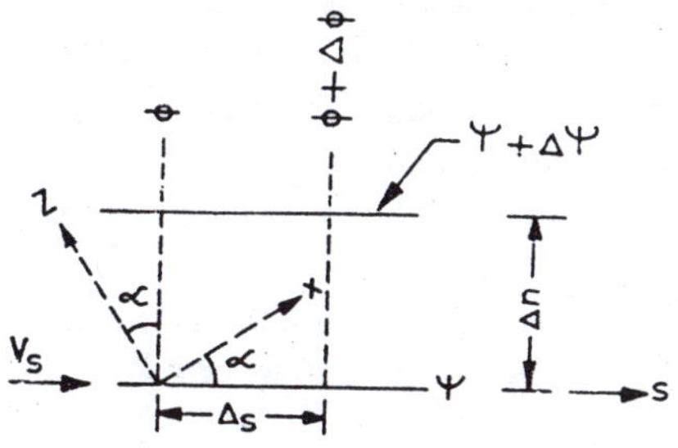

Fig. 9.4. Flow field.

$$v_x = v_s \cos\alpha \qquad ...[9.9(a)]$$

and $$v_z = -v_s \sin\alpha \qquad ...[9.9(b)]$$

The potential function φ can be written as

$$\frac{\partial\phi}{\partial s} = \frac{\partial\phi}{\partial x} \cdot \frac{dx}{ds} + \frac{\partial\phi}{\partial z} \cdot \frac{dz}{ds}$$

Using Eqs. 9.6, $$\frac{\partial\phi}{\partial s} = v_x \cos\alpha + v_z(-\sin\alpha)$$

Substituting the values of v_x and v_z from Eqs. 9.9,

$$\frac{\partial\phi}{\partial s} = v_s \cos^2\alpha + v_s \sin^2\alpha = v_s \qquad ...(9.10)$$

Likewise, $$\frac{\partial\psi}{\partial n} = \frac{\partial\psi}{\partial x}\cdot\frac{dx}{\partial n} + \frac{\partial\psi}{dz}\cdot\frac{dz}{dn}$$

or $$\frac{\partial\psi}{dn} = -v_z \sin\alpha + v_x \cos\alpha$$

or $$\frac{\partial\psi}{dn} = v_s \sin^2\alpha + v_s \cos^2\alpha = v_s \qquad \text{...(9.11)}$$

From Eqs.. 9.10 and 9.11, $$\frac{\partial\phi}{\partial s} = \frac{\partial\psi}{\partial n}$$

or $$\frac{\Delta\phi}{\Delta s} = \frac{\Delta\psi}{\Delta n} \qquad \text{...(9.12)}$$

The flow net must satisfy Eq. 9.12.

It is convenient to construct the flow net such that the change in stream function ($\Delta\psi$) between two adjacent flow lines and the change in potential function ($\Delta\phi$) between two adjacent equipotential lines are constant.

Therefore $\Delta\phi/\Delta\psi$ = constant

From Eq. 9.12, $\Delta s/\Delta n$ = constant

Although any fixed ratio of $\Delta s/\Delta n$ can be used, for convenience, $\Delta s/\Delta n$ is kept unity. Therefore, in actual practice, the flow net consists of approximate squares,

$$\Delta s = \Delta n \qquad \text{...(9.13)}$$

Thus the distance between two adjacent flow lines is equal to the distance between two adjacent equipotential lines.

The characteristics of flow net can be summarised as under:

(1) The fundamental condition that is to be satisfied is that every intersection between a flow line and an equipotential line should be at right angles.

(2) The second condition to be satisfied is that the discharge (Δq) between any two adjacent flow lines is constant and the drop of head (Δh) between the two adjacent equipotential lines is constant.

(3) The ratio of the length and width of each field ($\Delta s/\Delta n$) is constant. The ratio is generally taken as unity for convenience. In other words, the flow net consists of approximate squares.

The flow net can be obtained by any one of the following methods, as discussed in the following sections.

(1) Graphical method, (2) Electrical analogy method,
(3) Soil Models, (4) Plastic models,
(5) Solution of Laplace's equation.

It will be assumed that the flow is two-dimensional. In many of soil engineering problems, such as flow through a long earth dam, seepage under a long sheet pile and seepage below long gravity dams, the flow is actually two-dimensional. In all such cases, vertical sections at different points along the length are identical. The velocity has components only in two orthogonal directions (x, z), the component in the third direction (y-direction) is zero. However, if the length of the soil mass in the third direction (y-direction) is small, the end effects are important and the flow is not truly two-dimensional and Laplace's equation, as derived above, does not apply.

9.5. GRAPHICAL METHOD

The graphical method of flow net construction is the most commonly used method. The hydraulic boundary conditions which define the limiting flow lines and equipotential lines should be first identified and established. A reasonably good flow net can be drawn by the graphical method even by a novice with some practice. However, for getting a good flow net, a lot of practice and patience is required. Fortunately, the accuracy of the computation of hydraulic quantities, such as discharge and pore water pressure, does not depend much on the exactness of the flow net. A reasonably good estimate of hydraulic quantities can be made even from a rough flow net.

The following points should be kept in mind while sketching the flow net.

(1) Too many flow channels distract the attention from the essential features. Normally, three to five flow channels are sufficient. (The space between two flow lines is called a flow channel).

(2) The appearance of the entire flow net should be watched and not that of a part of it. Small details can be adjusted after the entire flow net has been roughly drawn.

(3) The curves should be roughly elliptical or parabolic in shape.

(4) All transitions should be smooth.

(5) The flow lines and equipotential lines should be orthogonal and form approximate squares.

(6) The size of the square in a flow channel should change gradually from the upstream to the downstream.

The procedure for drawing the flow net can be divided into the following steps:

(1) First identify the hydraulic boundary conditions. In Fig. 9.5, the upstream bed level *GDAK* represents 100% potential line and the downstream bed level *CFJ*, 0% potential line. The first flow line *KLM* hugs the

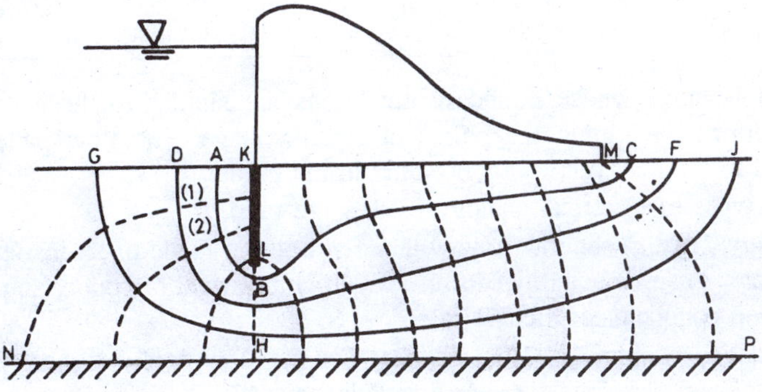

Fig. 9.5. Flow Net.

hydraulic structure and is formed by the flow of water on the upstream of the sheet pile, the downstream of the sheet pile and at the interface of the base of the dam and the soil surface. The last flow line is indicated by the impervious stratum *NP*.

(2) Draw a trial flow line *ABC* adjacent to the boundary line. The line must be at right angles to the upstream and downstream beds.

The location of the first trial line is determined from experience. An experienced person will make a good estimate of the first trial line and subsequent work would be reduced.

(3) Starting from the upstream end, divide the first flow channel into approximate squares by equipotential lines. The size of the square should change gradually.

Some of the squares may, however, be quite irregular. Such squares are called singular squares.

(4) Extend downward the equipotential lines forming the sides of the squares. These extensions point out approximate width of the squares, such as squares marked (1) and (2).

Other sides of the squares are set equal to the widths as determined above. Irregularities are smoothened out, and the next flow line *DF* is drawn joining these bases. While sketching the flow line, care should be taken to make flow fields as approximate squares throughout.

(5) The equipotential lines are further extended downward, and one more flow line *GHJ* is drawn, repeating the step (4).

(6) If the flow fields in the last flow channel are inconsistent with the actual boundary conditions, the whole procedure is repeated after taking a new trial flow line.

It is not necessary that the last flow channel should make complete squares. The flow fields in the last channel may be approximate rectangles with the same length to width ratio. In this case, the number of flow channels would not be full integer. In fact, the flow channels will be an integer only by chance.

9.6. ELECTRICAL ANALOGY METHOD

According to Darcy's law, the discharge in a soil mass is proportional to the hydraulic head (h). According to Ohm's law, the current in an electrical conductor is proportional to the voltage (E). An analogy exists between the two types of flow. The analogous quantities in the two systems are given in Table 9.1.

Table 9.1 Analogous Quantities

S.No.	*Flow of water*	*Flow of Current*
1.	Law : $q = k\frac{h}{L}A$	Law : $I = K \cdot \frac{E}{L} \cdot A$
2.	Discharge, q	Current, I
3.	Head, h	Voltage, E
4.	Length, L	Length, L
5.	Area, A	Area, A
6.	Permeability, k	Conductivity, K

An electrical model is made whose boundary conditions are similar to those of the soil model. The equipotential lines are drawn by joining the points of equal voltage. The flow pattern obtained from the electrical model are used in the construction of flow net in the model.

The following three types of electrical analogy models are used.

(1) Electrical Analogy Tray. A shallow tray, with a flat bottom, made of an insulating material is taken. The tray is filled with water. A small quantity of salt or hydrochloric acid or copper sulphate solution is added to water to make it a good conductor of electricity.

The hydraulic boundaries are simulated on the tray. For the flow below a sheet pile shown in Fig. 9.6 (*a*), the boundary flow lines are *ABC* and *FG*. An insulating material, such as ebonite or perspex, is used to simulate the boundary flow lines. The insulating material is fixed to the tray by means of some non-conducting adhesive, such as plasticene or bee wax.

The boundary equipotential lines *DA* and *CE* are simulated by some good conductor of electricity such as copper bars.

For obtaining the flow pattern, an electrical potential difference of 20 V is applied to the two electrodes *DA* and *CE*. A voltage dividing variable resistor, known as potential divider, is connected in parallel to the alternating current source to vary the voltage in the range of 0 to 20 V. A galvanometer (or any other null indicator) and a probe are connected to the variable potential arm [Fig. 9.6 (*b*)].

The position of the equipotential lines is determined by locating the points of constant potential (voltage). To trace the equipotential line corresponding to a given percentage of total potential (say 10%), the voltage divider is set at that potential (2V). The probe is moved in the tray till the galvanometer shows no current flow. That position of the probe gives the point corresponding to 2V potential. By moving the probe, other points corresponding to that potential are obtained. A graph sheet is generally placed below the transparent plate to determine the coordinates of the points. A line joining all these points gives the equipotential line corresponding to 10% of the total head. Likewise, the equipotential line corresponding to 20% of the total head is obtained by changing the setting on the voltage divider to 4V and repeating the procedure. Other equipotential lines can be drawn in the same manner.

After the equipotential lines have been drawn, flow lines can be sketched manually. The flow lines should be orthogonal to the equipotential lines and must satisfy the actual hydraulic boundary conditions. Alternatively, the flow lines can be drawn electrically by interchanging the boundaries. The copper strips are used for impermeable boundaries *ABC* and *FG* and insulating strips for *DA* and *CE*. The voltage difference is applied across the new positions of copper strips. The new equipotential lines, which are actually flow lines, are traced by locating the points with the help of probe.

(2) Conducting Paper Method. A conducting paper is made by introducing graphite during its manufacture. One side of the graphite paper is coated with a non-conducting material and the other side with a positive aluminum coating. The paper is cut to the shape of the hydraulic structure for which the flow net is required. The boundary equipotential lines, such as *DA* and *CE* in Fig. 9.6 (*a*), are given a coating of silver paint. When the paint has dried, the connecting wires are spaced out along the boundary strips in individual strands and are stapled in position.

Direct current (D.C.) supply can be used as there are no polarization effects. A 2-V accumulator is used for feeding the circuit. The lines of equal potential are traced, as in the electrical analogy tray.

The conducting paper method is quicker and more convenient than the tray method. However, the accuracy is low. As the transverse resistance of the paper is generally greater than the longitudinal resistance, it causes error. The scales of the model in the longitudinal and transverse directions are sometimes kept different to account for difference in resistances. This makes the method more complex.

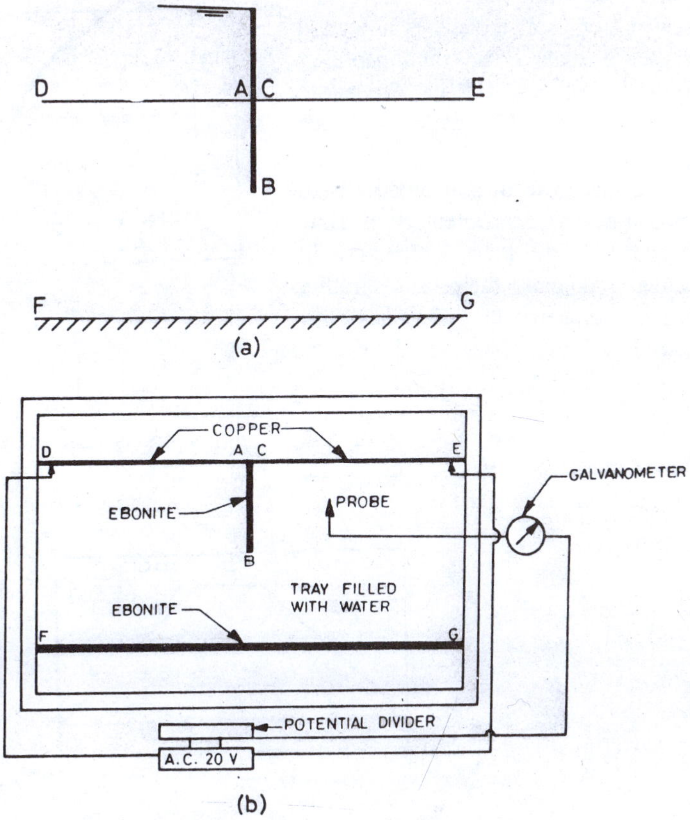

Fig. 9.6. Electrical Analogy Tray.

(3) Potential Analyser Method. A potential analyser is made in the form of a mesh of resistances (usually, of 100 ohms), separated at each node by pins of negligible resistance. The mesh is cut to the required shape. It is well insulated against temperature and humidity.

A direct current with a voltage difference of 1 V is applied to the appropriate boundaries of the model. The potential at any nodal point can be read with a high degree of accuracy. The equipotential lines are then drawn through the points of equal potential..

The method is quite convenient and gives fairly accurate results.

9.7. SOIL MODELS

Flow net can be obtained from a small scale soil model of the hydraulic structure. The soil model is placed between two transparent plates, about 100 mm apart. Fig 9.7 shows a soil model of an earth dam, with a horizontal filter at its toe.

The flow lines are traced directly by introducing a dye at suitable points on the upstream face of the dam. The equipotential lines can be drawn by connecting the points with the same piezometric levels. For this purpose, tiny piezometers are installed in the model at suitable points (not shown in figure). However, it is more convenient to draw equipotential lines manually after the flow lines have been drawn.

The accuracy of the flow net obtained from soil models is not good because of scale effects and capillary effects. Sometimes, viscous fluids are used in place of water to reduce capillary effects.

The main use of soil models is to demonstrate the fundamentals of flow net and seepage in a laboratory. In practical problems, their use is rather limited, because of the time and effort required in the construction of these models.

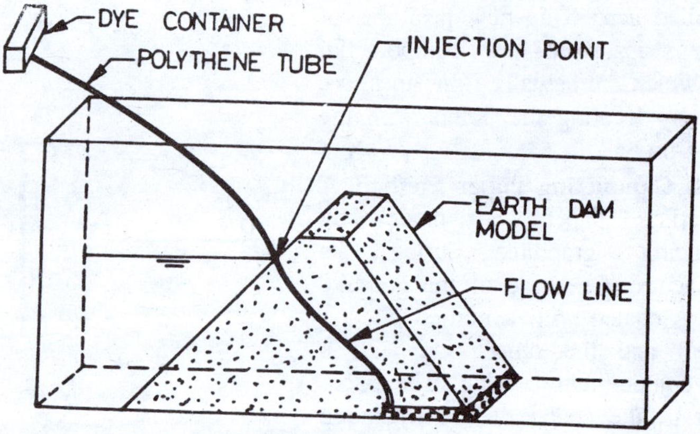

Fig. 9.7. Soil model.

9.8. PLASTIC MODELS

A seepage flume of width of a few centimeters is used in this method. A model made of plastic is fastened to one side wall of the flume, leaving a small space of 2.5 mm or less between the model and the other side wall (Fig. 9.8).

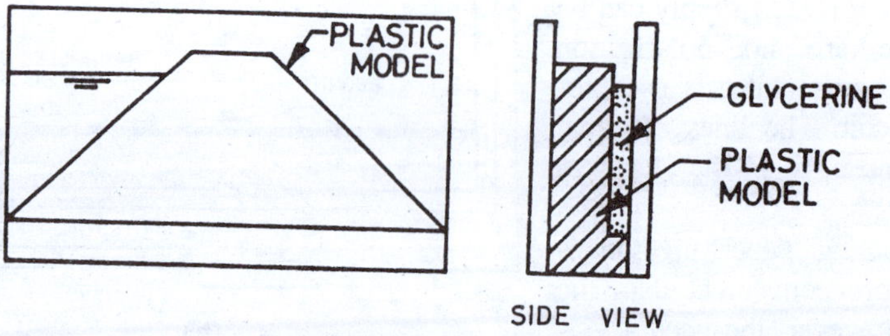

Fig. 9.8. Plastic model.

A highly viscous fluid, such as glycerine, is made to seep through the small space between the model and the side wall. The flow is laminar. As the fluid flows, it gives an accurate representation of seepage through soil. The flow lines can be observed directly by injecting a dye at suitable points.

Plastic models can be constructed more quickly than soil models. The flow lines in such models are also better defined. Consequently, the flow net obtained is more accurate than that obtained from soil models. Different permeabilities of the soil can be accounted for by varying the space between the model and the wall. Anisotropic soils can be represented by a zig-zag face.

9.9. FLOW NET BY SOLUTION OF LAPLACE'S EQUATION

Laplace's equation can be solved by numerical techniques, such as finite difference method. Relaxation method is generally used to find the potentials at various points. Once the potentials have been determined at different nodal points, the equipotential lines are drawn by joining the points of equal potentials. Potentials can be obtained very quickly if a high-speed digital computer is available.

The Laplace equation (Eq. 9.3) can be written in finite difference form, as

$$\phi_1 + \phi_2 + \phi_3 + \phi_4 - 4\phi_0 = 0 \qquad \text{...(9.14)}$$

where ϕ_1, ϕ_2, ϕ_3 and ϕ_4 are the potentials at the four adjoining points around the central point O with the potential ϕ_0 (Fig. 9.9).

The cross-section of the earth structure, for which the flow net is required, is covered with a square grid with a number of nodes. The values of the potential (ϕ) at various nodal points are assumed, satisfying the hydraulic boundary conditions. As the assumed values are not correct, there would be a residual R_0 at point O, given by the equation,

$$\phi_1 + \phi_2 + \phi_3 + \phi_4 - 4\phi_0 = R_0 \qquad ...(9.15)$$

Each node is considered as a central node in turn and the residual determined. The object of the relaxation method is to reduce these residuals to zero. It must be borne in mind that the potentials at different nodes are inter-related and any change in potential at one node has an effect on the residuals at the adjacent nodes. The process is, therefore, quite tedius and time-consuming. However, special relaxation techniques have been devised to reduce the effort.

The final correct value of ϕ give the true picture of the variation of potential. The equipotential lines are drawn through the points of equal potentials. The flow lines are then drawn orthogonal to equipotential lines.

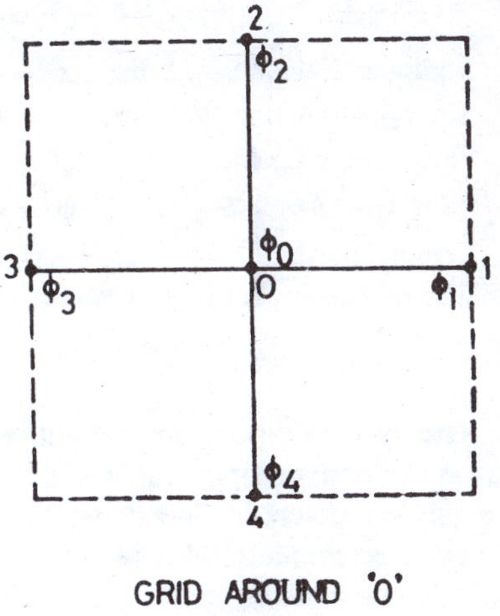

Fig. 9.9. Finite Difference Grid

9.10. FLOW NET IN EARTH DAMS WITH A HORIZONTAL FILTER

The methods of drawing a flow net discussed in the preceding sections are used when the boundary flow lines and equipotential lines are given. Seepage through an earth dam is a case of *unconfined seepage* in which the upper boundary of flow net is not known. In such cases, it becomes necessary to first locate the upper boundary before a flow net can be drawn.

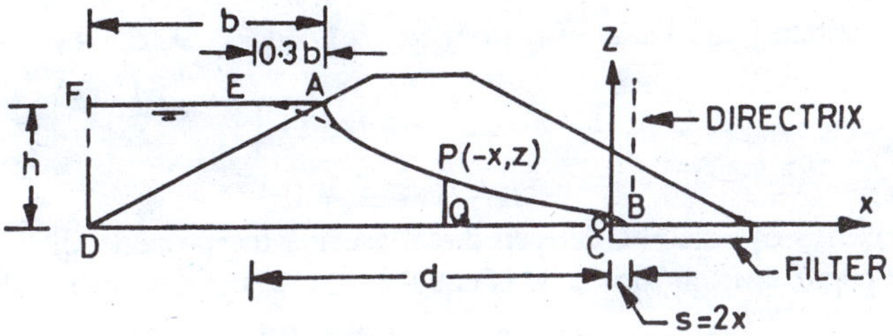

Fig. 9.10. Earth Dam with a horizontal filter

Let us consider the case of a homogeneous earth dam on an impervious foundation and having a horizontal filter at the downstream end (Fig. 9.10). The horizontal filter starts at point C. The impermeable boundary CD is a flow line which forms the lower boundary of the flow net. The upstream face AD is an equipotential line as the total head at every point on this face is equal to h. The discharge face CB is the equipotential line of zero potential. Thus, three hydraulic boundary conditions are known.

The fourth boundary of the flow net is the top flow line AB, which is not known in the beginning. Below the line AB, the soil is saturated and the pressure every where on the AB is atmospheric. The line AB is known as *phreatic line or seepage line.* As the pressure head is zero on the phreatic line, the total head is equal to the elevation head. Consequently, there are equal vertical intercepts between the points of intersection of successive equipotential lines and the phreatic line. Once the phreatic line has been located, the flow net can be drawn by the usual methods.

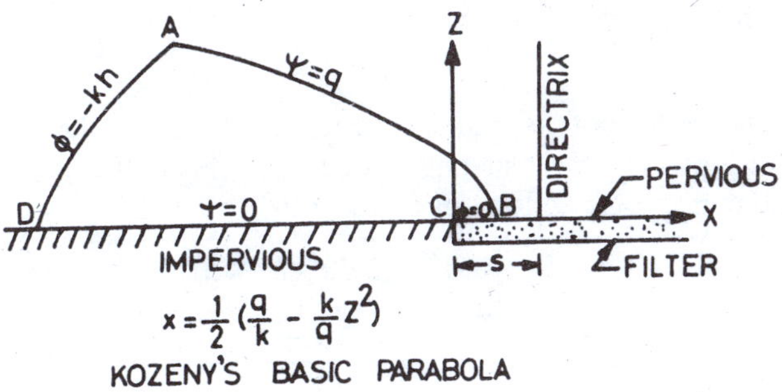

Fig. 9.11. Kozeny's Solution.

Kozeny studies the problem using the method of conformal transformation. The boundary conditions for the flow region $ABCD$ are as under (Fig. 9.11).

Equipotential line, AD, has $\phi = -kh$
Equipotential line, BC, has $\phi = 0$
Flow line, DC, has $\psi = 0$
Flow line, AB, has $\psi = q$

Kozeny's solution represents a family of confocal parabolas of flow lines and equipotential lines. The equation of Kozeny's basic parabola AB, with C as focus as well as origin, is

$$x = \frac{1}{2}\left(\frac{q}{k} - \frac{k}{q}z^2\right) \quad \text{...(9.16)}$$

Kozeny's conditions are not entirely fulfilled by any practical earth dam. However, an earth dam with a horizontal drainage approximates the conditions at exit. An inconsistency occurs due to the fact that the upstream equipotential line in an actual earth dam is a plane surface and not a parabola as assumed by Kozeny. Casagrande (1940) recommended that the seepage line in actual dams can also be taken as basic parabola, provided the starting point for the parabola is taken at point E, such that $AE = 0.3\ AF$ (Fig. 9.10). The distance AF is the projection of the upstream slope on the water surface. The coordinates of the phreatic line can be determined using Eq. 9.16. The origin is at C, which is also the focus.

Substituting $z = 0$ in Eq. 9.16, the value of x is given by

$$x_0 = \frac{1}{2}\left(\frac{q}{k}\right) = \frac{q}{2k} \quad \text{or} \quad q = 2k\,x_0$$

The distance $2x_0$ between the focus and the directrix is known as focal distance (s). Thus

$$q = k\,s \quad \text{...(9.17)}$$

Substituting the value of q from Eq. 9.17 in Eq. 9.16,

$$x = \frac{1}{2}\left(\frac{ks}{k} - \frac{k}{ks}z^2\right) = \frac{s}{2} - \frac{z^2}{2s}$$

or $$s^2 - 2xs - z^2 = 0 \quad \text{...(9.18)}$$

Eq. 9.18 can also be derived directly using the property of the parabola that the distance of any point P on the parabola from the focus is equal to the distance from the directrix. (Fig. 9.12). Thus

$$FP = PB$$

or $$\sqrt{x^2 + z^2} = s - x$$

By squaring, $$x^2 + z^2 = s^2 + x^2 - 2sx$$

or $$s^2 - 2xs - z^2 = 0$$

If x is taken positive towards left of F, the above equation becomes

$$s^2 + 2xs - z^2 = 0 \quad \text{...(9.19)}$$

The value of s can be determined using the coordinates of the starting point E (Fig. 9.10). Substituting $x = d$ and $z = h$ in Eq. 9.19.

$$s^2 + 2\,d\,s - h^2 = 0$$

or $$s = \frac{-2\,d \pm \sqrt{4d^2 + 4h^2}}{2}$$

Taking positive sign, $s = \sqrt{(d^2 + h^2)} - d$ ...(9.20)

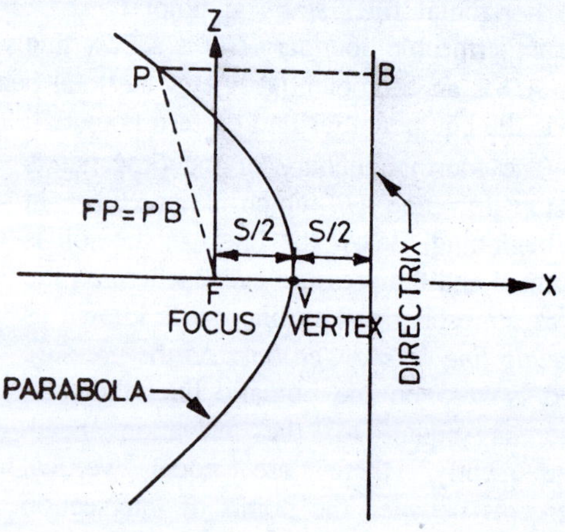

Fig. 9.12. Properties of Parabola.

Once the value of s has been determined, Eq. 9.19 can be used to determine the coordinates of the various points on the phreatic line. For different value of x, the corresponding z coordinates are computed and plotted.

An entrance correction is required for the phreatic line obtained by the above procedure. The actual flow line must start at point *A* and not point *E*. Further, the flow line must be normal to the upstream face which is an equipotential line. The entry correction is made by eye judgment as shown in Fig. 9.10. The actual phreatic line is shown in solid line. Fig. 9.13 shows the entry correction when there is pervious gravel on the upstream. The phreatic line in this case is horizontal at the entry, as it cannot rise above for being normal to the inclined d/s face of gravel.

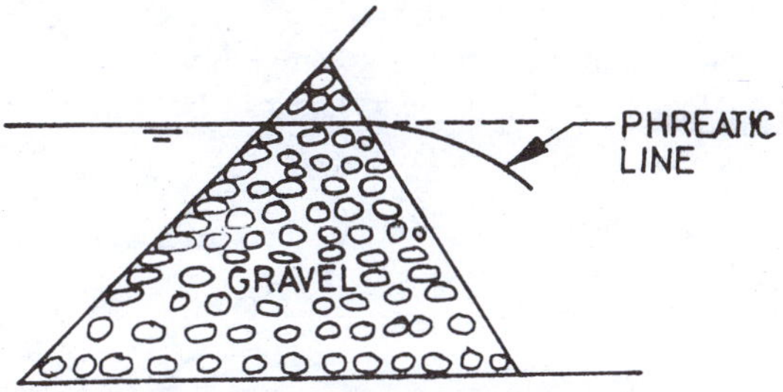

Fig. 9.13. Entry Correction for an u/s face with gravel.

Once the phreatic line has been drawn, the flow net can be completed using the methods already discussed. Fig. 9.14 shows a typical flow net.

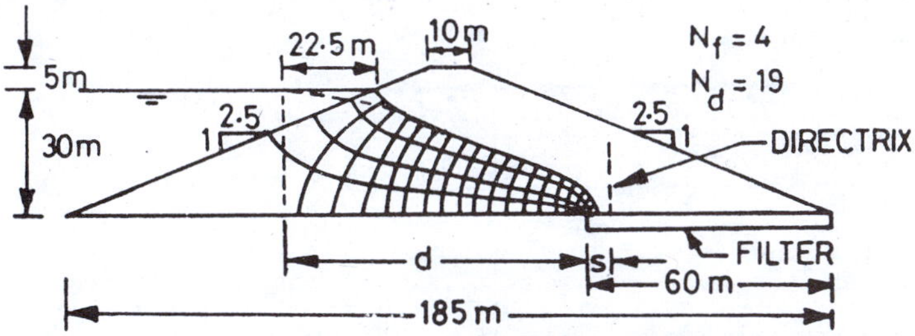

Fig. 9.14. Flow Net in an Earth dam.

Discharge through the body of the dam

To determine the discharge through the body of the earth dam, let us consider the flow passing through the section *PQ* (Fig. 9.10). From Darcy's law, discharge per unit length is given by

$$q = kiA$$

or $$q = k \cdot \frac{dz}{dx} \cdot (z \times 1) \qquad \ldots(a)$$

From Eq. 9.19, $$z = (2xs + s^2)^{1/2}$$

or $$\frac{dz}{dx} = \frac{s}{(2xs + s^2)^{1/2}}$$

Therefore, Eq. (*a*) gives, $$q = k \frac{s}{(2xs + s^2)^{1/2}} (2xs + s^2)^{1/2}$$

or $$q = ks \qquad \ldots(9.21)$$

Eq. 9.21 is a simple equation which gives approximate discharge through the body of the dam.

The discharge can also be obtained from the flow net, as explained later (Sect. 9.14).

9.11. SEEPAGE THROUGH EARTH DAM WITH SLOPING DISCHARGE FACE

Fig. 9.15 shows an earth dam without any filter on the downstream side. The downstream face through which water escapes is inclined to the horizontal. In this case, the phreatic line cuts the downstream face. It is normally not permitted in earth dams as it may cause the failure of downstream slope due to sloughing action. The

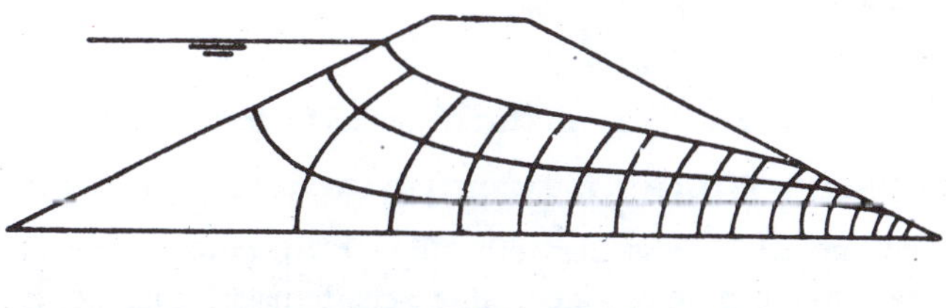
Fig. 9.15. Flow Net for earth dam without filter.

down stream face of the dam acts as the discharge face.

Fig. 9.16 (*a*) shows the downstream face when the phreatic line cuts the downstream face. The downstream face makes an angle β with the horizontal. The angle is measured clockwise from the horizontal. In this case, the phreatic line can be drawn as in the case of the dam with a horizontal filter (Sect. 9.10),

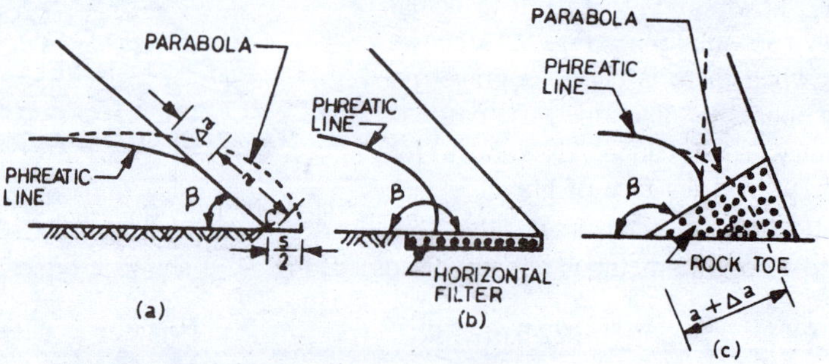

β	30°	60°	90°	120°	150°	180°
$\frac{\Delta a}{a+\Delta a}$	0·36	0·32	0·26	0·18	0·10	0·0

(d)

Fig. 9.16.

taking the point *C* as the focus and also the origin. The phreatic line is given the entry correction as before. An additional correction at exit is required in this case, as the basic parabola goes outside the downstream face, which is impossible. The actual seepage line meets the discharge face tangentially for β < 90°. The seepage line has been shown by full line, whereas the theoretical basic parabola is shown by dotted line.

In the case of horizontal filter, the angle β is 180° [Fig. 9.16 (*b*)]. For a rock toe [Fig. 9.16 (*c*)], the angle β is greater than 90°. The phreatic line drops vertically in this case.

Casagrande gave the charts for the exit correction. The basic parabola is shifted by distance Δa to locate the point where the actual seepage line cuts the discharge face. The value of Δa is obtained from the value of $\Delta a/(a + \Delta a)$ after the distance $(a + \Delta a)$ is obtained from the basic parabola. The value of $\Delta a/(a + \Delta a)$ depends upon the angle β, given in Fig. 9.16 (*d*). The value is also available in the form of a curve (Fig. 9.17). It is worth noting that the correction is zero when the angle β is 180. That is the reason why exit correction was not applied in the case of horizontal filter. The chart is applicable for β ≥ 30°.

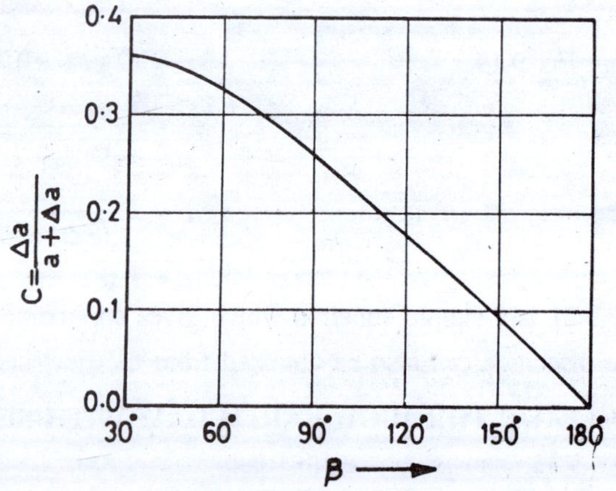

Fig. 9.17. Casagrand's Chart.

Obviously, $\Delta a = C(a + \Delta a)$

where *C* is the correction factor obtained from the chart (Fig. 9.17)

9.12. SEEPAGE THROUGH EARTH DAM WITH DISCHARGE ANGLE LESS THAN 30°

If the angle β is less than 30° (Fig. 9.18), point *S* at where the seepage line becomes tangential to downstream face can be obtained using Schaffernack's method. It is assumed that part *CS* of the seepage line is a straight line. A tangent at point *S* coincides over the length *CS* with the seepage line.

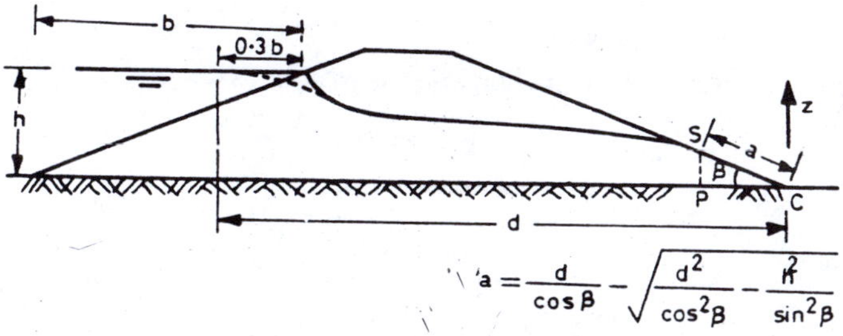

Fig. 9.18

The discharge is given by, $q = kz \cdot \frac{dz}{dx}$ ...(9.22)

But $\frac{dz}{dx} = i = \tan\beta$

and z = distance $SP = a \sin\beta$, where $SC = a$

Therefore, $q = k\,(a \sin\beta) \tan\beta$ ...(9.23)

From Eqs. 9.22 and 9.23,

$$kz \frac{dz}{dx} = k\,a \sin\beta \tan\beta$$

or $$z\,dz = a \sin\beta \tan\beta\, dx$$

Integrating between $x = a\cos\beta$ to $x = d$, and between $z = a \sin\beta$ to h,

$$\int_{a\sin\beta}^{h} z\,dz = a\,\beta \tan\beta \int_{a\cos\beta}^{d} dx$$

or $$\frac{1}{2}(h^2 - a^2 \sin^2\beta) = a \sin\beta \tan\beta\,(d - a\cos\beta)$$

or $$h^2 - a^2 \sin^2\beta = 2\,a \frac{\sin^2\beta}{\cos\beta}(d - a\cos\beta)$$

or $$\frac{h^2 \cos\beta}{\sin^2\beta} - a^2\cos\beta = 2\,ad - 2\,a^2\cos^2\beta$$

or $$a^2 \cos\beta - 2\,ad + \frac{h^2\cos\beta}{\sin^2\beta} = 0$$

or $$a = \frac{+\,2d \pm \sqrt{4\,d^2 - 4\,(h^2\cos\beta/\sin^2\beta)\cos\beta}}{2\cos\beta}$$

or $$a = \frac{d}{\cos\beta} - \sqrt{\frac{d^2}{\cos^2\beta} - \frac{h^2}{\sin^2\beta}}$$ ...(9.24)

Once the value of a has been determined from Eq. 9.24, the discharge can be found using Eq. 9.23.

9.13. SEEPAGE THROUGH EARTH DAM WITH DISCHARGE ANGLE GREATER THAN 30° BUT LESS THAN 60°.

Eq. 9.24 was obtained on the basis of Dupuit's assumption that the hydraulic gradient is equal to dz/dx, Casagrande suggested that the actual hydraulic gradient for discharge angle greater than 30°is given by

$$i = \frac{dz}{ds}$$

where distance s is measured along the curve.

Based on this assumption, the discharge expression can be written as

$$q = k\left(\frac{dz}{ds}\right) z \qquad ...(9.25)$$

Referring to Fig. 9.19, z = distance $SP = a \sin \beta$

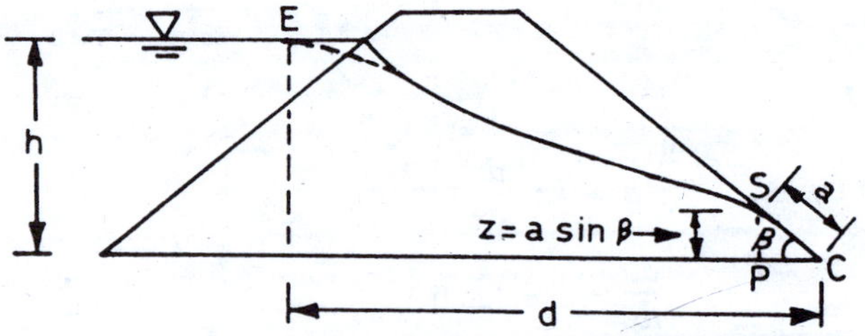

Fi.g 9.19. Earth Dam with discharge angle greater than 30°.

and $$\frac{dz}{ds} = \sin \beta$$

Therefore, Eq. (9.25) becomes $$q = k\,a \sin^2 \beta \qquad ...(9.26)$$

From Eqs. 9.25 and 9.26, $$k \frac{dz}{ds} z = k\,a \sin^2 \beta$$

or $$z\,dz = a \sin^2 \beta\, ds$$

Integrating, $$\int_{a \sin \beta}^{h} z\,dz = a \int_{a}^{S} (\sin^2 \beta)\, ds$$

or $$\frac{1}{2}(h^2 - a^2 \sin^2 \beta) = a \sin^2 \beta\,(S - a)$$

or $$h^2 - a^2 \sin^2 \beta = 2aS \sin^2 \beta - 2a^2 \sin^2 \beta$$

or $$a^2 - 2aS + \frac{h^2}{\sin^2 \beta} = 0$$

or $$a = \frac{+2S \pm \sqrt{4S^2 - 4h^2/\sin^2 \beta}}{2}$$

or $$a = S - \sqrt{S^2 - h^2/\sin^2 \beta} \qquad ...(9.27)$$

The approximate length S of the straight line CE can be determined as

$$S = \sqrt{d^2 + h^2} \qquad ...(9.28)$$

Therefore, $$a = \sqrt{d^2 + h^2} - \sqrt{d^2 - h^2 \cot^2 \beta} \qquad ...(9.29)$$

Once the value of a has been determined, the discharge can be obtained from Eq. 9.26.

For angle $\beta > 60°$, the error introduced due to approximation in Eq. 9.28 becomes large and this method is not normally used.

9.14. USES OF FLOW NET

The flow net can be used for a number of purposes as explained below :

(1) Discharge. The space between two adjacent flow lines is called a flow channel. Let N_f be the number of flow channels. The difference between two adjacent equipotential lines is called equipotential drop. Let N_d be the number of equipotential drops. In Fig. 9.20, there are 5 flow channels and 10 equipotential drops.

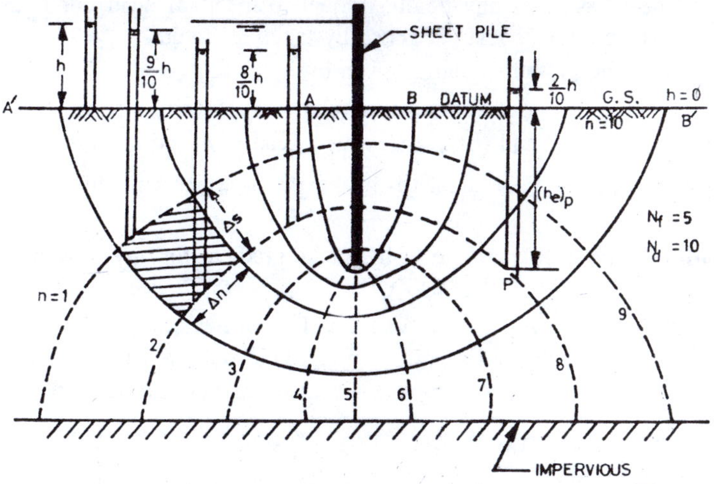

Fig. 9.20. Uses of Flow Net.

Let us consider the flow through the flow field shown hatched. From Darcy's law, the discharge through the flow field per unit length,

$$\Delta q = k \cdot \left(\frac{\Delta h}{\Delta s}\right)(\Delta n \times 1) \qquad \ldots(a)$$

where Δh is equipotential drop in the flow field,

and Δs and Δn are dimensions of the flow field.

Substituting $\Delta h = \dfrac{h}{N_d}$ in Eq. (*a*),

$$\Delta q = k \cdot \frac{h}{N_d} \cdot \left(\frac{\Delta n}{\Delta s}\right)$$

Total discharge,

$$q = N_f \quad \Delta q = k.h. \frac{N_f}{N_d} \cdot \left(\frac{\Delta n}{\Delta s}\right)$$

Taking $\Delta s / \Delta n$ = unity,

$$q = k.h. \frac{N_f}{N_d} \qquad \ldots(9.30)$$

In Fig. 9.20,

$$q = k \times h \times \frac{5}{10} = 0.5\, kh$$

The ratio (N_f/N_d) is a characteristic of the flow net. It is known as *shape factor* (ʃ). It is independent of the permeability (k) of the soil. It depends only on the configuration or the shape of the soil mass.

It is not necessary that N_f and N_d be always full integer. The last flow channel may consist of rectangles. However, in the last flow channel, the length/breadth ($\Delta s / \Delta n$) ratio should be approximately the same for all flow fields.

(2) Total head. The loss of head (Δh) from one equipotential line to the next is h/N_d. The total head at any point (P) can be determined as under.

$$h_p = h - n \times (h/N_d) \qquad \ldots(9.31)$$

where n is the number of the equipotential drops upto point P.

In Fig. 9.20, n = 8 for point P. Therefore, total head at P is

$$h_p = h - 8 \times (h/10) = 0.2\,h$$

It may be noted than if piezometers were placed at different points on the same equipotential line, water would rise in these piezometers to the same elevation.

(3) Pressure head. The pressure at any point is equal to the total head minus the elevation head. As mentioned above, the downstream water level is generally taken as datum.

For example, for point P, the pressure head is given by

$$(h_p)_P = h_p - (-(he)_p) = h_p + (he)_p \qquad ...(9.32)$$

where $(h_p)_P$ = pressure head at P and $(he)_p$ = elevation head at P and h_p is the total head.

Obviously, the pressure head at P is equal to the height of water column in the piezometers at P, as shown in the figure.

(4) Hydraulic gradient. The average value of hydraulic gradient for any flow field is given by

$$i = \Delta h / \Delta s \qquad ...(9.33)$$

where Δs is the length of the flow field and Δh is the loss of head.

The hydraulic gradient is generally maximum at the exit near point B where the length Δs is a minimum. As the velocity depends upon the hydraulic gradient, it is also maximum at the exit.

9.15. FLOW NET FOR ANISOTROPIC SOIL

The coefficient of permeability of stratified soil deposits parallel to the plane of stratification is generally greater than that normal to this plane. Such soils are anisotropic in permeability. Let us take the axes $x - x$ and $z - z$ parallel and perpendicular to the plane of stratification, respectively. Therefore $k_x > k_z$. From Darcy's law,

$$v_x = k_x\, i_x = -k_x \frac{\partial h}{\partial x} \qquad ...(a)$$

and

$$v_z = k_z\, i_z = -k_z \frac{\partial h}{\partial z} \qquad ...(b)$$

Substituting the values of v_x and v_z in the continuity equation (Eq. 9.1),

$$-k_x \frac{\partial^2 h}{\partial x^2} - k_z \frac{\partial^2 h}{\partial z^2} = 0$$

or

$$k_x \frac{\partial^2 h}{\partial x^2} + k_z \frac{\partial^2 h}{\partial z^2} = 0 \qquad ...(9.34)$$

As Eq. 9.34 is not Laplace's equation, the principles of flow net construction, as described in the preceding sections, are not applicable to anisotropic soils.

Eq. 9.34 can however be converted to Laplace's equation by transformation. Let the x coordinate be transformed to the new coordinate x_t by the transformation (Fig. 9.21).

$$x_t = x\sqrt{k_z/k_x} \qquad ...(9.35)$$

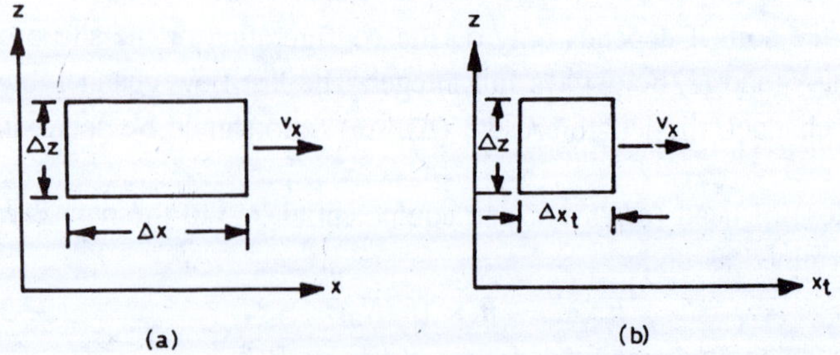

Fig. 9.21. Transformation of Coordinates.

Eq. 9.34 can be written as

$$\left(\frac{k_x}{k_z}\right)\frac{\partial^2 h}{\partial x^2} + \frac{\partial^2 h}{\partial z^2} = 0$$

or

$$\frac{\partial^2 h}{\partial x_t^2} + \frac{\partial^2 h}{\partial z^2} = 0 \qquad \ldots(9.36)$$

Eq. 9.36 is the Laplace equation in x_t and z. Therefore, the principles of flow net construction can be used for anisotropic soils after transformation.

The cross-section of the soil mass whose flow net is required is redrawn keeping the z-scale unchanged but reducing the x- scale by the ratio $\sqrt{k_z/k_x}$. The flow net is constructed for the transformed section by usual methods [Fig. 9.22 (*b*)]. The flow net for the actual section is obtained by transferring back the flow net to the natural section by increasing the x-scale in the ratio $\sqrt{k_x/k_z}$. Obviously, the flow net for the natural section does not have the flow lines and the equipotential lines orthogonal to each other [Fig. 9.22 (*a*)].

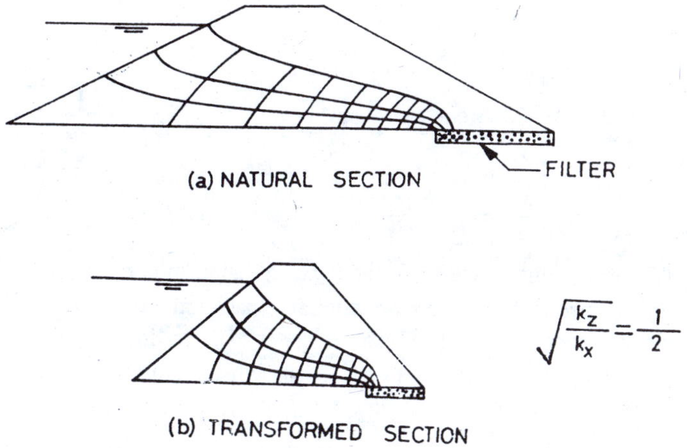

Fig. 9.22. Flownet for anisotropic soils.

The discharge through an anisotropic soil mass can be obtained from an equation similar to Eq. 9.30,

$$q = k' h \cdot (N_f/N_d) \qquad \ldots(9.37)$$

where k' is the modified coefficient of permeability as determined below.

Discharge through a flow channel on the transformed scale per unit width is given by

$$\Delta q = k' (\Delta h/\Delta x_t)\ \Delta z \qquad \ldots(a)$$

Discharge through the same flow channel on the natural scale per unit width is given by

$$\Delta q = k_x (\Delta h/\Delta x)\ \Delta z \qquad \ldots(b)$$

Since the discharge is the same in both the channels,

$$k' (\Delta h/\Delta x_t) \cdot \Delta z = k_x \cdot (\Delta h/\Delta x) \cdot \Delta z$$

or $$k' = k_x \cdot (\Delta x_t/\Delta x)$$

Using Eq. 9.35, $$k' = k_x \cdot \sqrt{(x_z/k_x)}$$

or $$k' = \sqrt{k_x k_z} \qquad \ldots(9.38)$$

The discharge q is determined using Eq. 9.37 with a value of k' obtained from Eq. 9.38.

9.16. COEFFICIENT OF PERMEABILITY IN AN INCLINED DIRECTION

Let k_x and k_z be the coefficients of permeability along x and z directions respectively and k_s be the coefficient of permeability in inclined s-direction (Fig. 9.23).

By partial differentiation,

$$\frac{\partial h}{\partial s} = \frac{\partial h}{\partial x} \cdot \frac{\partial x}{\partial s} + \frac{\partial h}{\partial z} \cdot \frac{\partial z}{\partial s} \qquad \ldots(a)$$

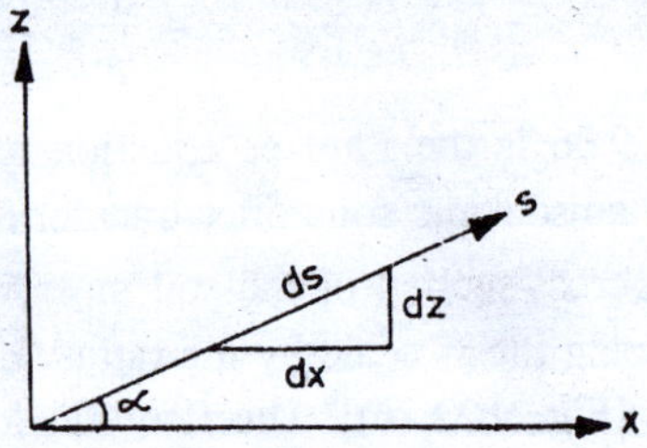

Fig. 9.23. Permeability in an inclined direction

Using the relations,

$$v_x = -k_x \frac{\partial h}{\partial x}, \quad v_z = -k_z \frac{\partial h}{\partial z}$$

and $$v_s = -k_s \frac{\partial h}{\partial s}$$, Eq. (*a*) becomes

$$-\frac{v_s}{k_s} = -\frac{v_x}{k_x} \cdot \frac{\partial x}{\partial s} - \frac{v_z}{k_s} \cdot \frac{\partial z}{\partial s} \qquad \ldots(b)$$

Now $\quad v_x = v_s \cos \alpha \quad$ and $v_z = v_s \sin \alpha$

and $$\frac{dx}{ds} = \cos \alpha \quad \text{and} \quad \frac{dz}{ds} = \sin \alpha$$

Eq. (*b*) can be written as

$$\frac{1}{k_s} = \frac{\cos^2 \alpha}{k_x} + \frac{\sin^2 \alpha}{k_z} \qquad \ldots(9.39)$$

or

$$\frac{s^2}{k_s} = \frac{x^2}{k_x} + \frac{z^2}{k_z} \qquad \ldots(9.40)$$

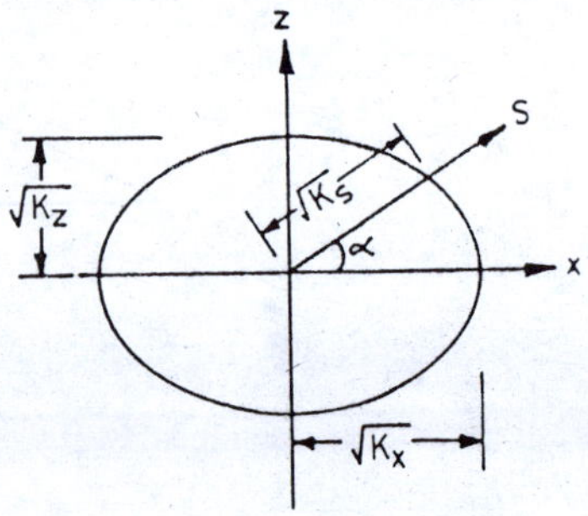

Fig. 9.24. Directional variation of permeability.

Eq. 9.40 is equation of an ellipse with $\sqrt{k_x}$ and $\sqrt{k_z}$ as semi-major and semi-minor axes, respectively. The directional variation of permeability can be determined from the ellipse (Fig. 9.24). A line making an angle *a* with *x*-axis gives the intercept $\sqrt{k_s}$ as shown in figure. Thus k_s can be found.

9.17. FLOW NET IN A NON-HOMOGENEOUS SOIL MASS

Sometimes, two different soils are used in a soil mass, thus making it non-homogeneous. The flow lines and equipotential lines get deflected at the interface. The flow net thus gets modified.

Let the coefficients of the permeability of the two soils be k_1 and k_2. We shall consider separately the two cases when (1) $k_1 > k_2$ and (2) $k_1 < k_2$.

Case 1. $k_1 > k_2$. Fig. 9.25 (*a*) shows the case when the soil—1 has permeability more than the soil—2. The flow lines get deflected towards the normal after crossing the interface. The phenomenon of deflection of the flow lines is somewhat similar to refraction of light rays from a sparse medium to a dense medium.

Let α_1 be the angle which the flow line makes with the normal in soil—1 and α_2 be the angle, in soil—2. Let ϕ_1 and ϕ_2 be the two equipotential lines. The discharge through the flow channel between the two flow lines in two soils is given by

$$\Delta q_1 = k_1 (\Delta h/\Delta s_1)\, \Delta n_1$$

and $$\Delta q_2 = k_2 \cdot (\Delta h/\Delta s_2) \cdot \Delta n_2$$

For continuity of flow across the interface, the discharge through the flow channel remains the same. Therefore,

$$\Delta q_1 = \Delta q_2$$

or $$k_1 \cdot (\Delta h/\Delta s_1) \cdot \Delta n_1 = k_2 \cdot (\Delta h/s_2) \cdot \Delta n_2$$

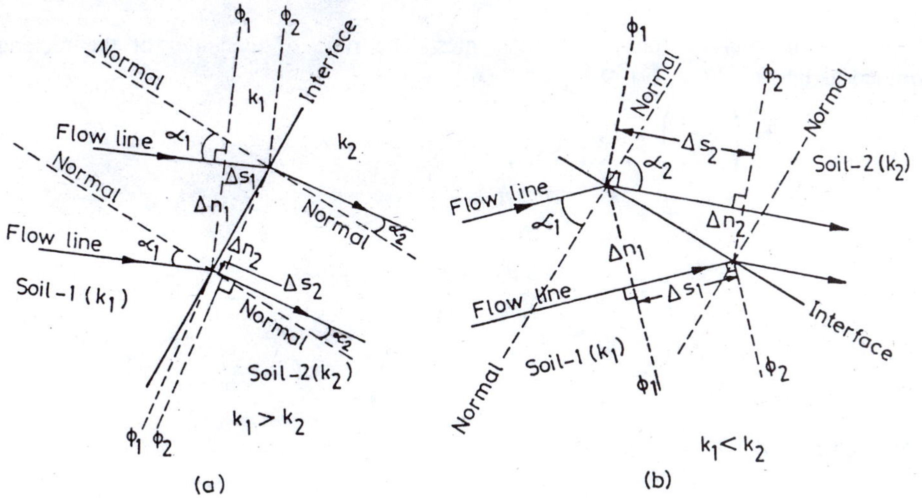

Fig. 9.25. Flownet in a non-homogeneous soil.

or $$k_1 \cdot (\Delta n_1/\Delta s_1) = k_2 \cdot (\Delta n_2/\Delta s_2) \quad ...(9.41)$$

or $$\frac{k_1}{\tan \alpha_1} = \frac{k_2}{\tan \alpha_2}$$

or $$\frac{k_1}{k_2} = \frac{\tan \alpha_1}{\tan \alpha_2} \quad ...(9.42)$$

Eq. 9.41 must be satisfied at the interface by every flow line crossing it.

Case (2) $k_1 < k_2$. Fig. 9.25 (*b*) shows the case when the flow takes place from a soil of low permeability to that of high permeability. At the interface, the flow line is deflected away from the normal. Using a procedure similar to that for the first case, it can be shown that

$$\frac{k_1}{\tan \alpha_1} = \frac{k_2}{\tan \alpha_2}$$

or $$\frac{k_1}{k_2} = \frac{\tan \alpha_1}{\tan \alpha_2} \quad \text{(same as Eq. 9.42)}$$

As $k_2 > k_1$, the angle α_2 is greater than angle α_1 and the flow line deflects away from the normal.

Flow Net for Non-Homogeneous Section

Fig. 9.26 shows the flow net for an earth dam consisting of two soils of different permeability. The flow net is drawn using the following concepts.

(1) The flow net consists of squares in soil—1.

(2) The flow lines deflect at the interface, according ot Eq. 9.42.

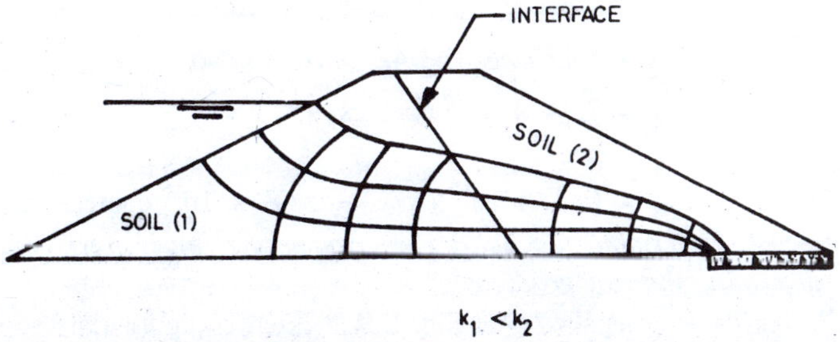

Fig. 9.26. Non-homogeneous section.

(3) The flow net in soils—2 consists of rectangles. The ratio of the sides of the rectangle can be determined as under : From Eq. 9.41,

$$k_1\left(\frac{\Delta n_1}{\Delta s_1}\right) = k_2\left(\frac{\Delta n_2}{\Delta s_2}\right)$$

or $$\frac{\Delta n_2}{\Delta s_2} = \frac{k_1}{k_2}\cdot\left(\frac{\Delta n_1}{\Delta s_1}\right)$$

As $$\frac{\Delta n_1}{\Delta s} = 1, \qquad \frac{\Delta n_2}{\Delta s_2} = \frac{k_1}{k_2}$$

or $$\frac{\Delta s_2}{\Delta n_2} = \frac{k_2}{k_1}$$

In Fig. 9.26, as $k_2 > k_1$, $$\frac{\Delta s_2}{\Delta n_2} > 1.0$$

If the ratio of permeability is greater than 10, flow net in the soil of higher permeability need not be drawn. The loss of head in the soil of higher permeability is neglected. For example, in Fig. 9.26, if $k_1 > 10k_2$, the flow net in soil—1 is neglected and it is assumed that the flow lines in soil—1 are horizontal. The flow net will be constructed only for soil—2, taking the interface as the upstream face. On the other hand, if $k_2 > 10\,k_1$, the flow net will be drawn only for soil—1. In this latter case, the interface will act as a discharge face.

ILLUSTRATIVE EXAMPLES

Illustrative Example 9.1. *Determine the coordinates of the phreatic line for the earth dam shown in Fig. 9.14, Find the discharge through the earth dam from the flow net and also analytically. Take k = 4.5 × 10^{-4} cm/sec.*

Solution. From Eq. 9.20, taking d = 72.5 m and h = 30 m,

$$s = \sqrt{(d^2 + h^2)} - d$$

$$= \sqrt{(30^2 + 72.5^2)} - 72.5 = 5.96 \text{ m}$$

The coordinates of the phreatic line are determined from Eq. 9.19.

$$s^2 + 2xs - z^2 = 0$$

or $$(5.96)^2 + 2x(5.96) - z^2 = 0$$

or $$35.52 + 11.92x - z^2 = 0$$

The z-coordinates are determined for different values of x as under.

x	+ 10	+ 20	+ 30	+ 40	+ 50	+ 60	+72.5 m
z	12.44	16.25	19.83	22.63	25.13	27.40	30 m

From Eq. 9.30, $$q = kh\,(N_f/N_d) = 4.5 \times 10^{-6} \times 30 \times (4/19)$$

or $$q = 2.84 \times 10^{-5} \text{ cumecs/m}$$

Analytically, from Eq. 9.21, $$q = ks$$

or $$q = (4.5 \times 10^{-6})(5.96) = 2.68 \times 10^{-5} \text{ cumecs/m}$$

Illustrative Example 9.2. *Determine the uplift pressure on the impervious concrete floor of the weir shown in Fig. 9.2. Also determine the exit gradient.*

Solution. Flow net in constructed as shown in Fig. E 9.2. Each equipotential drop Δh is 7.5/15 = 0.5 m, as n_d = 15.

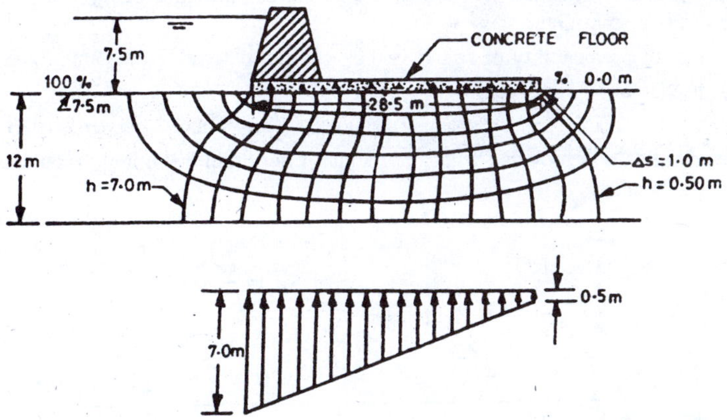

Fig. E-9.2

The total head at the two extremities of the floor are 7.0 m and 0.5 m. These are also equal to the pressure heads, as the underface of the floor is at the datum (d/s level).

Total uplift force $U = \frac{1}{2}(h_1 + h_2)\,\gamma_w \times \text{area}$

$$= \frac{1}{2}(7.0 + 0.50) \times 9.81 \times (28.5 \times 1)$$

or $U =$ **1048.4 kN per metre length**

The length (Δs) of the last flow field near toe is 1.0 m.

Therefore, exit gradient (i) $= \Delta h/\Delta s = 0.5/1.00 =$ **0.50**

PROBLEMS

9.1. Determine the seepage discharge through the foundation of an earth dam if the flow net has 10 equipotential drops and 3.5 flow channels. The length of the dam is 300 m and the coefficient of permeability of the soil is 2.5×10^{-4} cm/sec. The level of water above the base of the dam is 12 m on upstream and 4 m on downstream.

[**Ans.** $66.23 \times 10^3\ m^3$/ year]

9.2. In the experimental set up shown in Fig. P 9.2, flow takes place under a constant head through the soils *A* and *B*.

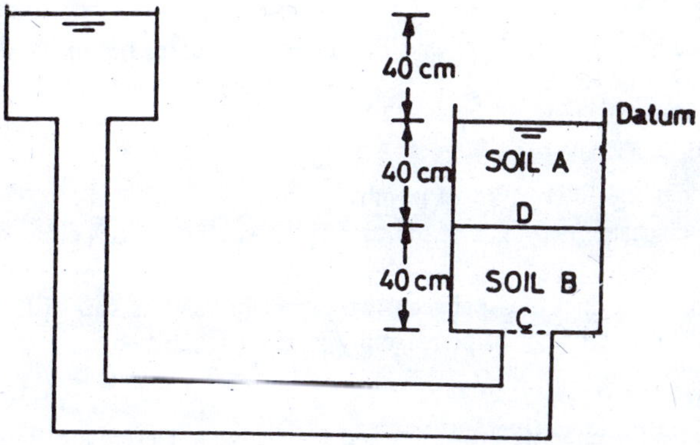

Fig. P 9.2.

(*i*) Determine the piezometric head at point *C*.

(*ii*) If 40% of the excess hydrostatic pressure is lost in flowing through soil *B*, what are the hydraulic head and piezometric head at point *D*.

(*iii*) If the coefficient of permeability of soil *B* is 0.05 cm/sec, determine the same for soil *A*.

(*iv*) What is the discharge per unit area ?

[Ans. (*i*) 120 cm, (*ii*) 24 cm, 64 cm, (*iii*) 0.033 cm/sec (*iv*) 0.02 ml/sec.

9.3. A homogeneous earth dam is provided with a horizontal filter drain 30 m long at its toe, as shown in Fig. P 9.3. Determine the focal length.

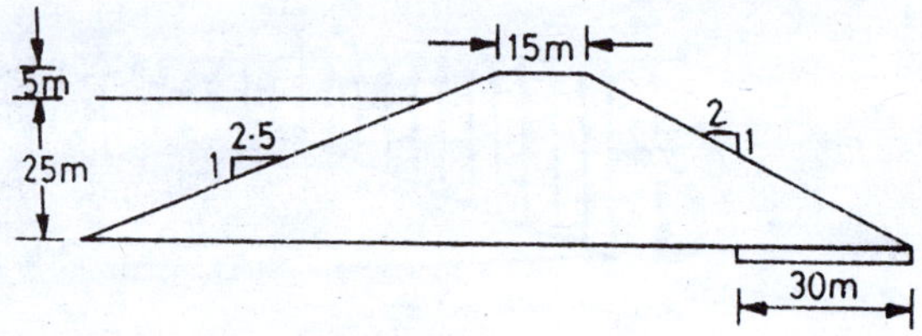

Fig. P 9.3.

Also determine the seepage discharge per unit length if the coefficient of permeability is 40 m/day.

[Ans. s = 3.99 m, q = 159.6 m^3/day]

9.4. A sandy stratum 5 m thick has a slope of 1 in 10 and lies between two impervious strata (Fig. P 9.4). If the piezometers inserted at two points 20 m apart indicate a pressure difference of 3.5m and the coefficient of permeability is 1.91 × 10^{-4} cm/sec, determine the seepage discharge. **[Ans.** 5.96 litres/hour]

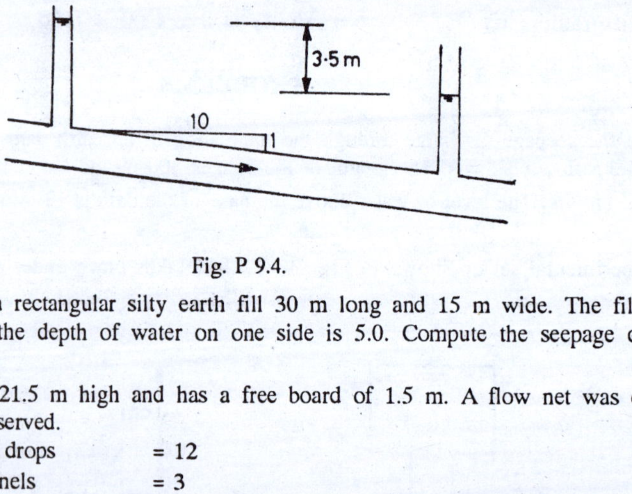

Fig. P 9.4.

9.5. Water percolates across a rectangular silty earth fill 30 m long and 15 m wide. The fill is founded on an impervious stratum and the depth of water on one side is 5.0. Compute the seepage discharge. k = 0.15 cm/minute.. **[Ans.** 108 m^3/day]

9.6. A homogeneous dam is 21.5 m high and has a free board of 1.5 m. A flow net was constructed and the following results were observed.

No. of potential drops = 12

No. of flow chanels = 3

The dam has a horizontal filter of 15 m length

Calculate the discharge/m length of the dam if the coefficient of permeability of the dam material is 2.7 × 10^{-6} m/sec. **[Ans.** 1.35 × 10^{-5} cumecs/m]

9.7. A silty-clay deposit 4 m thick lies between two layers of sand as shown in Fig. P 9.7. Calculate the seepage through the clay if the coefficient of permeability is 1 × 10^{-6} cm/sec. **[Ans.** 1.296 × 10^{-3} m^3/day/m^2]

B. Descriptive and Objective Type

9.8. What is a flow net ? Describe its properties and applications. Describe different methods used to construct the flow net.

9.9. Explain the uses of a flow net.

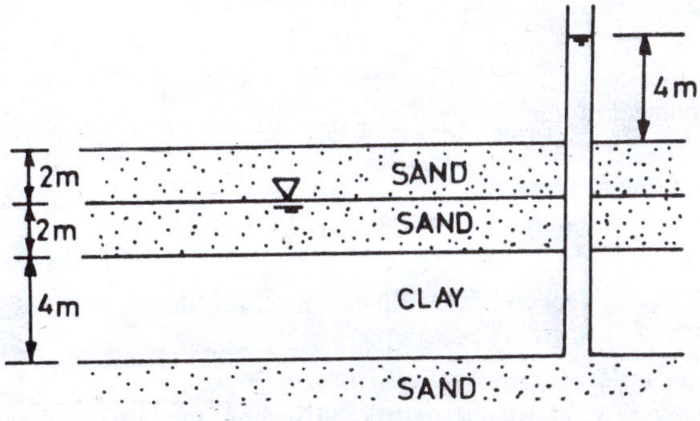

Fig. P 9.7.

9.10. Describe the electrical analogy method of flow net construction.

9.11. Prove that the discharge per unit width of an earth dam with a horizontal filter at its toe is equal to the coefficient of permeability times the focal length.

9.12. Prove that the discharge through an earth mass is given by

$$q = k \cdot \frac{h}{N_d} \cdot N_f$$

where k = coefficient of permeability, h = head, N_f = number of flow channels,
N_d = number of equipotential drops.

9.13. How would you draw the flow net for a homogeneous earth dam without any filter ?

9.14. What is entry correction of the flow net ? How is it done ?

9.15. How would you construct the flow net when the soil is anisotropic ?

9.16. Explain the method of constructing the flow net in an earth dam consisting of two different zones.

9.17. Mention whether the following statements are true or false.

(*a*) The flow lines and equipotential lines are orthogonal for an isotropic soil.
(*b*) The number of equipotential lines and flow lines is always a full integer.
(*c*) In two-dimensional flow, the velocity in the third direction is zero.
(*d*) The velocity potential is equal to the total head.
(*e*) The flow net for anisotropic soil can be obtained from Laplace's equation.
(*f*) The electrical analogy method can be used to obtain directly flow lines.
(*g*) Relaxation method is used to determine the potentials at various points.
(*h*) The upstream face of an earth dam is an equipotential line.
(*i*) The shape factor depends upon the type of soil.
(*j*) When the flow passes from a soil of high permeability to that of low permeability, the flow lines are deflected away from the normal.
(*k*) The equipotential lines make equal vertical intercepts on the phreatic line.
(*l*) The phreatic line of a homogeneous section always cuts the downstream face.
(*m*) The phreatic line at the entrance may rise upward.
(*n*) For an earth dam with a horizontal filter at its downstream toe, the Casagrande exit correction is zero.

[**Ans.** True, (*a*), (*c*), (*f*), (*g*), (*h*), (*k*), (*l*), (*n*)]

C. Multiple-Choice Questions

1. The phreatic line in a homogeneous dam is

(*a*) Circular (*b*) Elliptical
(*c*) Hyperbolic (*d*) Parabolic

2. If there is flow from a soil of permeability k_1 to that of k_2, the angles θ_1 and θ_2 which the flow line makes with the normal to the interface are related as

(a) $\frac{\sin \theta_1}{\sin \theta_2} = \frac{k_1}{k_2}$ (b) $\frac{\tan \theta_1}{\tan \theta_2} = \frac{k_1}{k_2}$

(c) $\frac{\cos \theta_1}{\cos \theta_2} = \frac{k_1}{k_2}$ (b) $\frac{\cot \theta_1}{\cot \theta_2} = \frac{k_1}{k_2}$

3. The pressure on a phreatic line is
(a) equal to atmospheric pressure.
(b) greater than atmospheric pressure.
(c) less than atmospheric pressure.
(d) not related to the atmospheric pressure.

4. A flow net has 4 flow channels and 20 equipotential drops, the shape factor is
(a) 1/5 (b) 5
(a) 80 (d) None of above

5. For an isotopic soil, k_x/k_z = 9. For the transposed section, the horizontal scale should be
(a) 1/9 (b) 1/3
(c) Three times (d) Nine times

6. The starting point of the horizontal drainage is usually taken as . . . . of parabola
(a) Focus (b) Origin
(c) Vertex (d) Both (a) and (b)

7. If the flow net of a cofferdam foundation has
h = 6m, N_f = 6 and N_d = 18, $k = 4 \times 10^{-5}$ m/min, then the seepage discharge (in m^3/d) per m length is
(a) 0.2304 (b) 0.1152
(c) 1.0368 (d) 2.304

8. A flow net can be used to determine
(a) Seepage, coefficient of permeability and uplift pressure
(b) Seepage, coefficient of permeability and exit gradient
(c) Seepage, exit gradient and uplift pressure
(d) Seepage and exit gradient only

9. For an anisotropic soil with kx = 4kz, the value of the modified coefficient of permeability k′ is
(a) 2 kx (b) 4 kx
(c) 0.5 kx (d) 0.25 kx

10. An isotropic soil has the vertical permeability of k and the horizontal permeability of 4 k, the equivalent permeability of the transformed section is
(a) 2 k (b) 8 k
(c) 0.5 k (d) 1.5 k

[**Ans.** 1. (d), 2. (b), 3. (a), 4. (a) 5. (b), 6. (d), 7. (b), 8. (c), 9. (a), 10. (a)]

10

Effective Stress Principle

10.1. INTRODUCTION

The effective stress principle enunciated by Karl Terzaghi in 1936 forms an extremely useful basis of the most important theories in soil engineering. The effective stress principle consists of two parts :

1. Definition of the effective stress.
2. Importance of the effective stress in engineering behaviour of soil.

This chapter is devoted mainly to the first part. The second part dealing with the importance of effective stress is discussed briefly in the following article. The role of effective stress on compression characteristics and shear strength is dealt in detail in chapters 12 and 13, respectively.

The methods for determination of effective stress in soils for hydrostatic conditions and for steady seepage conditions are discussed separately. The effect of seepage pressure on the stability of the soil masses in described. Piping failures and the methods for its prevention are also discussed.

10.2. EFFECTIVE STRESS PRINCIPLE

(1) Definition of Effective Stress

Fig. 10.1 shows a soil mass which is fully saturated. Let us consider a prism of soil with a cross-sectional area *A*. The weight *P* of the soil in the prism is given by

$$P = \gamma_{sat}\, h\, A \qquad \text{...}(a)$$

where γ_{sat} is the saturated weight of the soil, and h is the height of the prism.

Total stress (σ) on the base of the prism is equal to the force per unit area. Thus

$$\sigma = \frac{P}{A} = \gamma_{sat}\, h \qquad \text{...(10.1)}$$

While dealing with stresses, it is more convenient to work in terms of unit weights rather than density. As discussed in chapter 2,

$$\gamma = \rho \cdot g$$

where γ is in N/m^3 and ρ is in kg/m^3, g = 9.81 m/sec^2

Thus, $$\gamma_{sat} = \rho_{sat} \times g = 9.81\, \rho_{sat}$$

Generally, the unit weights are expressed in kN/m^3 and the mass density in kg/m^3. In that case,

$$\gamma_{sat} = \frac{\rho_{sat} \times g}{1000} = 9.81 \times 10^{-3}\, \rho_{sat} \qquad \text{...}(a)$$

For example, if ρ_{sat} = 2000 kg/m^3,

$$\gamma_{sat} = 9.81 \times 10^{3} \times 2000 = 19.62 \text{ kN/m}^3$$

Sometimes, Eq. (*a*) is approximated as

$$\gamma_{sat} \approx 0.01\ \rho_{sat} \quad \text{...(b)}$$

In that case $\gamma_{sat} = 0.01 \times 2000 = 20.00\ \text{kN/m}^3$

For convenience, Eq. (*b*) is sometimes used.

Pore water pressure (u) is the pressure due to pore water filling the voids of the soil. Thus

$$u = \gamma_w h \quad \text{...(10.2)}$$

Pore water pressure is also known as *neutral pressure* or *neutral stress,* because it cannot resist shear stresses.

Pore water pressure is taken as zero when it is equal to atmospheric pressure, because in soil engineering the pressures used are generally gauge pressure and not absolute pressures.

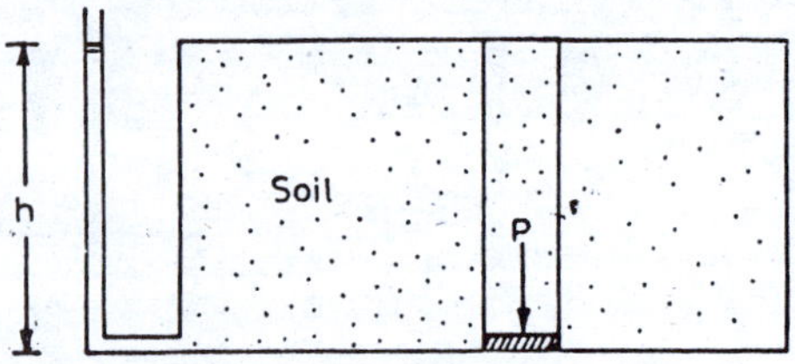

Fig. 10.1. Saturated soil mass.

The effective stress ($\bar{\sigma}$) at a point in the soil mass is equal to the total stress minus the pore water pressure. Thus

$$\bar{\sigma} = \sigma - u \quad \text{...(10.3)}$$

For saturated soils, it is obtained as

$$\bar{\sigma} = \gamma_{sat}\ h - \gamma_w h$$

or $$\bar{\sigma} = (\gamma_{sat} - \gamma_w)\ h$$

or $$\bar{\sigma} = \gamma'\ h$$

where γ' is the submerged unit weight.

The effective stress is also represented by σ' in some texts.

It may be noted that the effective stress is an abstract quantity, as it cannot be measured directly in the laboratory. It is deduced from two physical, measurable quantities σ and u. Thus the effective stress is a mathematical concept and not a physical quantity.

(2) Importance of Effective Stress

The effective stress controls the engineering properties of soils. Compression and shear strength of a soil are dependent on the effective stress. Thus

compression $= f(\bar{\sigma})$

and shear strength $= \varphi(\bar{\sigma})$

where f and φ represent some functions.

As the effective stress in a soil increases, the compression of the soil occurs. This causes settlement of structures built on soils.

The shear strength of a soil depends on its effective stress. As the effective stress is changed, the shear strength changes. The stability of slopes, the earth pressures against retaining structure and the bearing capacity of soils depend upon the shear strength of the soil and hence, the effective stress. The importance of shear strength in soil engineering problem cannot be over-emphasised. It is one of the most important properties of soils.

As discussed in chapter 8, the permeability of a soil depends upon the void ratio. With a change in effective stress, the void ratio of the soil changes. Therefore, to some extent, the permeability of a soil is also governed by the effective stress.

10.3. NATURE OF EFFECTIVE STRESS

Let us consider a physical model of a soil mass, fully saturated, as shown in Fig. 10.2(*a*). Let us take a wavy plane *X*- *X* passing through the points of contact of solid particles. On the macroscopic scale, the wavy plane cannot be distinguished from a true horizontal plane as the individual particles are of relatively small size. Therefore, for all practical purposes, the plane *X*-*X* can be assumed as horizontal.

The total normal force P acting on the soil model is resisted partly by the interparticle forces at the points of contact (P_m) and partly by the pore water pressure force (P_w) [Fig. 10.2 (*b*)].

Thus $$P = P_m + P_w \quad ...(10.5)$$

At every point of contact, the interparticle force F can be resolved into the normal component (N) and the tangential component (T) to the plane $X—X$ [Fig. 10.2 (*c*)]. The interparticle forces are random in both

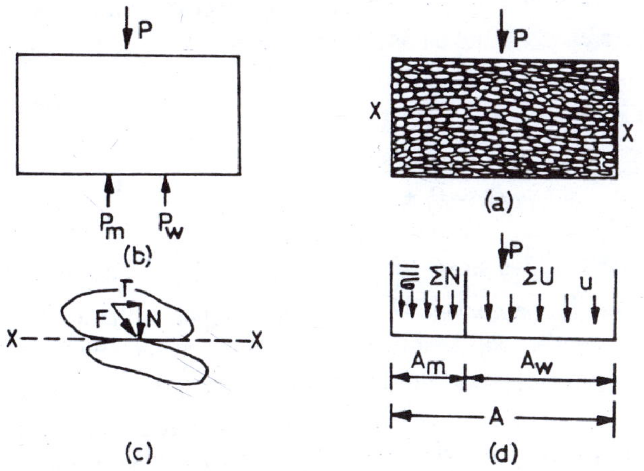

Fig. 10.2. Physical model of a soil mass.

magnitude and direction throughout the soil mass. The tangential components, however, neutralise one another and the resultant of all the normal components is downward.

The effective stress is the nominal stress transmitted through the solid particles, and is given by

$$\bar{\sigma} = \frac{\text{sum of all N–components}}{\text{area of X—X plane}}$$

or $$\bar{\sigma} = \frac{\Sigma N}{A} \quad ...(10.6)$$

Let the area of cross-section occupied by the solid particles (minerals) be A_m and that occupied by water be A_w. [Fig. 10.2 (*d*)].

Therefore, $$A = A_m + A_w$$

or $$A_w = A - A_m \quad ...(10.7)$$

Let u be the pore water pressure. From Eq. 10.5,

$$P = P_m + P_w = \Sigma N + \Sigma U$$

or $$\sigma A = \bar{\bar{\sigma}} \ A_m + u A_w \quad ...(10.8)$$

where $\bar{\bar{\sigma}}$ is the actual normal stress transmitted at the points of contact of the solid particles, and σ is the total stress (Eq. 10.1).

Eq. 10.8 can be written as

$$\sigma = \bar{\bar{\sigma}} \ (A_m/A) + u \ (A_w/A)$$

Using Eq. 10.7, $$\sigma = \bar{\bar{\sigma}} \ (A_m/A) + u \ (1 - A_m/A)$$

or $$\sigma = \bar{\bar{\sigma}} \ a_m + u \ (1 - a_m) \quad ...(10.9)$$

where $a_m = A_m/A$.

The geotechnical engineer is interested in the effective stress ($\bar{\sigma}$) not in the actual contact stress ($\bar{\bar{\sigma}}$). Let us again consider the equilibrium in the vertical direction [Fig. 10.2 (*d*)]. We have

$$P = \Sigma N + u A_w$$

or

$$\sigma A = \Sigma N + u A_w$$

or

$$\sigma = \Sigma N/A + u\,(A_w/A) \qquad \text{...(10.10)}$$

In Eq. 10.7, as the area occupied by the interparticle contact (mineral to mineral) A_m is very small (about 3% for granular soils), the area A_w be taken approximately equal to the total area A. In other words,

$$A_w \approx A.$$

Therefore, Eq. 10.10 becomes $\sigma = \Sigma N/A + u$

Designating $\Sigma N/A$ by the nominal effective stress, $\bar{\sigma}$,

$$\sigma = \bar{\sigma} + u$$

or

$$\bar{\sigma} = \sigma - u \qquad \text{(same as Eq. 10.3)}$$

It must be noted that the effective stress ($\bar{\sigma}$) depends upon the normal force (ΣN) transmitted at the points of contact, but it is not equal to the contact stress ($\bar{\bar{\sigma}}$). It is equal to the total normal force N transmitted at the points of contact divided by the total area A, including that occupied by water. It has no physical meaning and, therefore, cannot be directly measured. It is much smaller than the actual contact stress $\bar{\bar{\sigma}}$.

The pore water pressure due to water in voids acts equally in all directions (Pascal's law). It does not resist any shear stress. and, therefore, is also called the neutral stress. However, it is very important as the effective stress depends upon the pore water pressure.

In clayey soils, there may not be direct contact between the minerals due to the surrounding adsorbed water layers. However, it has been established by actual experiments that the interparticle contact forces can be transmitted even through the highly viscous adsorbed water. The above equations which have been developed assuming the soil as coarse-grained may be used for clayey soils as well.

For surface active minerals, Eq. 10.3 is modified as

$$\bar{\sigma} = \sigma - u + (A' - R') \qquad \text{...[10.3 (a)]}$$

where A' and R' are respectively the attractive and repulsive forces per unit area.

10.4. EFFECT OF WATER TABLE FLUCTUATIONS ON EFFECTIVE STRESS

Let us consider a soil mass shown in Fig. 10.3. The depth of the water table (W.T.) is H_1 below the ground surface. The soil above the water table is assumed to be wet, with a bulk unit weight of γ. The soil below the water table is saturated, with a saturated weight of γ_{sat}.

Total down ward force (P) at section X—X is equal to the weight (W) of the soil. Thus,

$$P = W = \gamma H_1 A + \gamma_{sat} H_2 A$$

where A is the area of cross-section of the soil mass.

Dividing by A throughout,

$$\frac{P}{A} = \gamma H_1 + \gamma_{sat} H_2$$

The left-hand side is equal to the total stress (Eq. 10.1).

Therefore,

$$\sigma = \gamma H_1 + \gamma_{sat} H_2$$

The pore water pressure (u) is given by Eq. 10.2 as

$$u = \gamma_w H_2$$

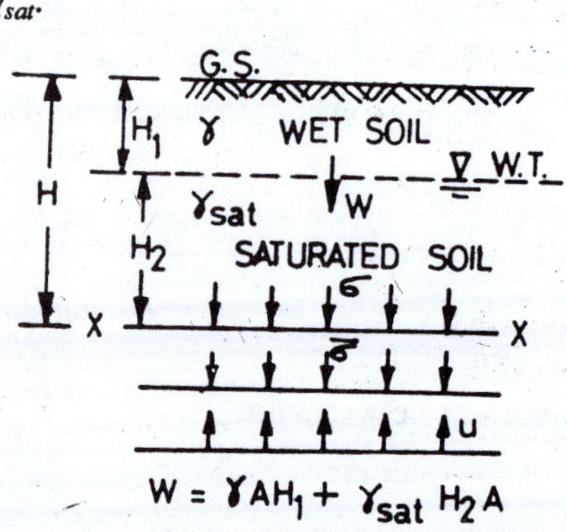

Fig. 10.3

From Eq. 10.3,

$$\bar{\sigma} = \sigma - u = (\gamma H_1 + \gamma_{sat} H_2) - \gamma_w H_2$$

or

$$\bar{\sigma} = \gamma H_1 + (\gamma_{sat} - \gamma_w)\, H_2$$

or $$\bar{\sigma} = \gamma H_1 + \gamma' H_2 \qquad \text{...(10.11)}$$

Eq. 10.11 gives the effective stress at section *X—X*. Fig. 10.3 also shows the directions of $\bar{\sigma}$ and u at *X—X*.

(*a*) If the water table rises to the ground surface, the whole of the soil is saturated, and

$$\bar{\sigma} = \gamma' (H_1 + H_2) = \gamma' H \qquad \text{...(10.12)}$$

As $\gamma' < \gamma$, the effective stress is reduced due to rise of water table.

(*b*) If the water table is depressed below the section *X—X*,

$$\bar{\sigma} = \gamma H \qquad \text{...(10.13)}$$

In this case, the effective stress is increased.

Thus, it is observed that the fluctuations in water table level cause changes in the pore water pressure and the corresponding changes in the effective stress.

10.5. EFFECTIVE STRESS IN A SOIL MASS UNDER HYDROSTATIC CONDITIONS

Fig. 10.4 (*a*) shows a soil mass under hydrostatic conditions, wherein the water level remains constant. As the interstices in the soil mass are interconnected, water rises to the same elevation in different piezometers fixed to the soil mass. The effective stress at various sections can be determined using Eq. 10.3.

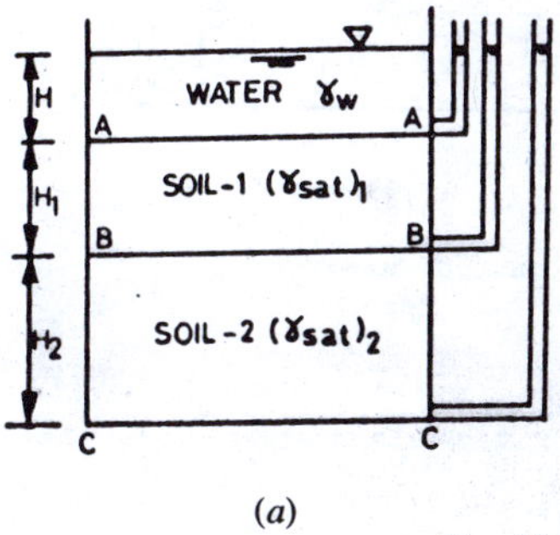

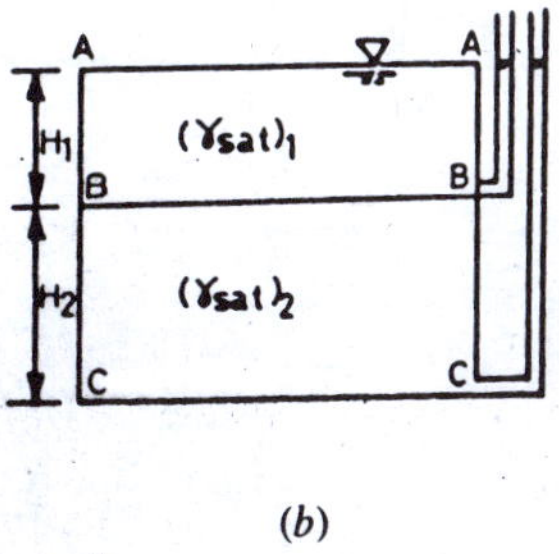

Fig. 10.4. Water Table above soil

(1) Water Table above the soil surface A—A:

Section A—A $\sigma = \gamma_w H, \quad u = \gamma_w H$

Therefore, $\bar{\sigma} = \sigma - u = 0$ ...(*a*)

Section B—B $\sigma = \gamma_w H + (\gamma_{sat})_1 H_1$, and $u = \gamma_w (H + H_1)$

Therefore, $\bar{\sigma} = [(\gamma_{sat})_1 - \gamma_w] H_1 = \gamma_1' H_1$ ...(*b*)

where γ_1' is the submerged unit of soil—1.

Section C—C $\sigma = \gamma_w H + (\gamma_{sat})_1 H_1 + (\gamma_{sat})_2 H_2$

$u = \gamma_w (H + H_1 + H_2)$

Therefore $\bar{\sigma} = \gamma_1' H_1 + \gamma_2' H_2$ ...(*c*)

where γ_2' is the submerged unit weight of soil-2.

(2) Water Table at the soil surface A—A

Fig. 10.4 (*b*) shows the condition when the depth *H* of water above the section A—A is reduced to zero. In this case, the effective stresses at various sections are determined as under:

Section A—A $\sigma = u = \bar{\sigma} = 0$ ...(*d*)

Section B—B $\sigma = (\gamma_{sat})_1 H_1$, and $u = \gamma_w H_1$

and $\bar{\sigma} = \gamma_1' H_1$ ...(*e*)

Section C—C $\sigma = (\gamma_{sat})_1 H_1 + (\gamma_{sat})_2 H_2$

$u = \gamma_w (H_1 + H_2)$

Therefore, $\bar{\sigma} = \gamma_1' H_1 + \gamma_2' H_2$ ...(*f*)

Comparing Eqs. (*a*), (*b*), (*c*) with (*d*), (*e*) and (*f*), it is observed that the depth H of water above the soil surface does not contribute to the effective stress at all. In other words, the effective stress in a soil mass is independent of the depth of water above the soil surface. It should therefore not be surprising that the marine soil deposits, which are under a very large depth of water, have a low effective stress and correspondingly low shear strength.

(3) Water Table in Soil—1

Fig. 10.5 (*a*) shows the case when the water table is at D—D in the Soil—1 at depth H_1'. The effective stresses at various sections are determined as follows.

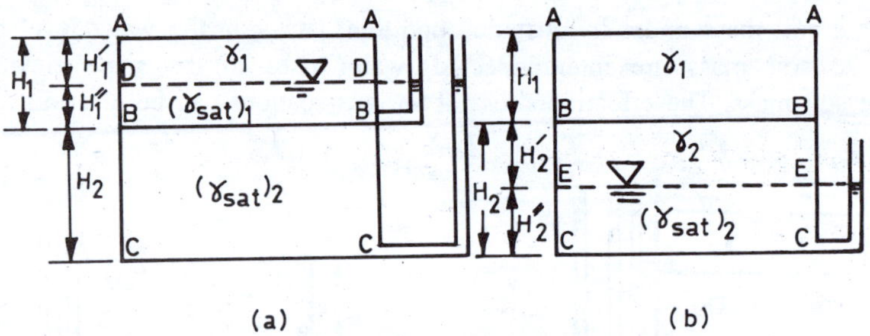

Fig. 10.5. Water Table in (*a*) Soil-1 and (*b*) Soil-2.

Section A—A $\sigma = u = \bar{\sigma} = 0$

Section D—D $\sigma = \gamma_1 H_1'$,

where γ_1 is the unit weight of soil above D—D. $u = 0$

Therefore, $\bar{\sigma} = \gamma_1 H_1'$

Section B—B $\sigma = \gamma_1 H_1' + (\gamma_{sat})_1 H_1''$ (Note. $H_1' + H_1'' = H_1$)

$u = \gamma_w H_1''$

Therefore, $\bar{\sigma} = \gamma_1 H_1' + [(\gamma_{sat})_1 - \gamma_w] H_1'' = \gamma_1 H_1' + \gamma_1' H_1''$

Section C—C $\sigma = \gamma_1 H_1' + (\gamma_{sat})_1 H_1'' + (\gamma_{sat})_2 H_2$

$u = \gamma_w (H_2 + H_1'')$

Therefore, $\bar{\sigma} = \gamma_1 H_1' + \gamma_1' H_1'' + \gamma_2' H_2$

When $H_1'' = 0$, $\bar{\sigma} = \gamma_1 H_1' + \gamma_2' H_2$

(4) Water Table in Soil—2

Fig. 10.5 (*b*) shows the condition when the water table is at *EE* in Soil—2 at depth H_2'. The effective stresses at various sections are as under:

Section A—A $\sigma = u = \bar{\sigma} = 0$

Section B—B $\sigma = \gamma_1 H_1, \quad u = 0, \quad \bar{\sigma} = \gamma_1 H_1$

Section E—E $\sigma = \gamma_1 H_1 + \gamma_2 H_2'$, $u = 0$ (Note. $H_2' + H_2'' = H_2$)

$\bar{\sigma} = \gamma_1 H_1 + \gamma_2 H_2'$

Section C—C $\sigma = \gamma_1 H_1 + \gamma_2 H_2' + (\gamma_{sat})_2 H_2''$

$u = \gamma_w H_2''$

$\bar{\sigma} = \gamma_1 H_1 + \gamma_2 H_2' + \gamma_2' H_2''$

where γ_2' is submerged unit weight of Soil-2.

(5) Water Table below C—C

Fig. 10.6 shows the condition when the water table is below C—C. As the pore water pressure is zero everywhere, the effective stresses are also equal to the total stresses.

Section B—B $\sigma = \bar{\sigma} = \gamma_1 H_1$

Section C—C $\sigma = \bar{\sigma} = \gamma_1 H_1 + \gamma_2 H_2$

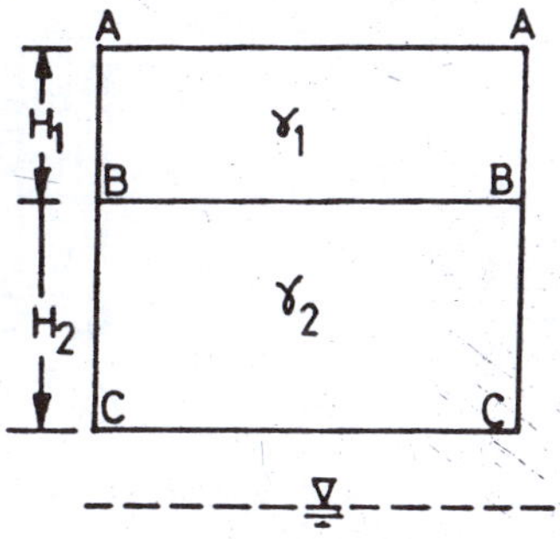

Fig. 10.6. Water Table below soil.

The following points are worth noting in the five cases studied above.

(1) The effective stress at any section goes on increasing as the water table goes down.

(2) The effective stress depends upon the bulk unit weight above the water table and the submerged unit weight below the water level.

(3) The effective stresses in a soil mass can be determined from the basic definitions, without memorising any formula.

10.6. INCREASE IN EFFECTIVE STRESSES DUE TO SURCHARGE

Let us consider the case when the soil surface is subjected to a surcharge load of intensity q per unit area. Let us assume that the water table is at level B—B (Fig. 10.7). The stresses at various sections are determined as under.

Section A—A $\sigma = q, \quad u = 0, \quad \bar{\sigma} = q$

i.e., all the points on the soil surface are subjected to an effective stress equal to q.

Section B—B $\sigma = q + \gamma_1 H_1, \quad u = 0$

Therefore, $\bar{\sigma} = q + \gamma_1 H_1$

Section C—C $\sigma = q + \gamma_1 H_1 + (\gamma_{sat})_2 H_2$

$u = \gamma_w H_2$

Therefore, $\bar{\sigma} = q + \gamma_1 H_1 + \gamma_2' H_2$

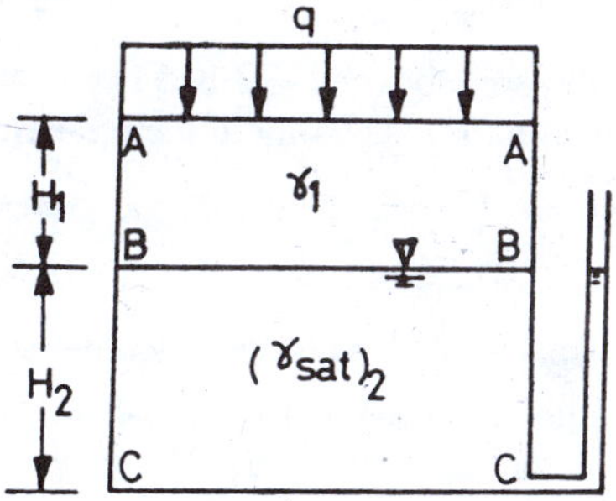

Fig. 10.7. Effect of Surcharge.

From the above illustrations, it is clear that the effective stress throughout the depth is greater than the case with no surcharge discussed in the preceding section. The difference is equal to the intensity q. In other words, the effective stress is increased by q throughout.

10.7. EFFECTIVE STRESSES IN SOILS SATURATED BY CAPILLARY ACTION

If the soil above the water table is saturated by capillary action, the effective stresses can be determined using Eq. 10.3. However, in this case the pore water pressure above the water table is negative [Fig. 10.8 (*a*)]. The water table is at level B—B. Let us consider two cases:

(1) Soil saturated upto surface level A—A [Fig. 10.8 (*a*)]

(2) Soil saturated upto level D—D [Fig. 10.8 (*b*)]

(1) Soil saturated upto surface level A—A [Fig. 10.8 (*a*)]

The pore pressure diagram is drawn on the right side.

The stresses at various sections are determined as under.

Section A—A $\sigma = 0, \quad u = -\gamma_w H_1$

Therefore, $\bar{\sigma} = 0 - (-\gamma_w H_1) = \gamma_w H_1$

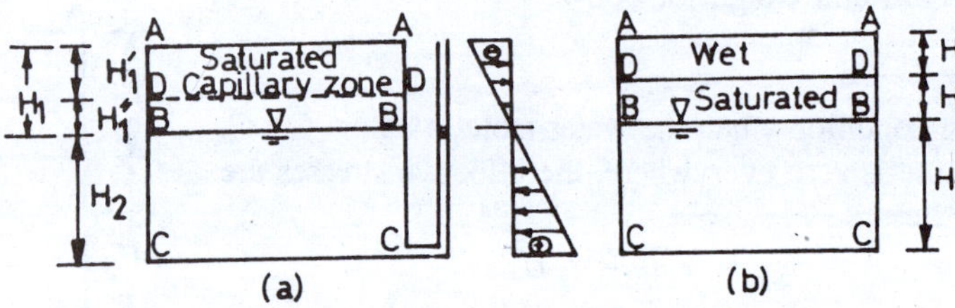

Fig. 10.8. Effect of capillary rise.

If the soil was not saturated with capillary action, the effective stress at section A—A would have been zero. Thus, the capillary action has increased the effective stress by $\gamma_w H_1$. In other words, the negative pressure acts like a surcharge (q).

Section D—D $\sigma = (\sigma_{sat})_1 \; H_1'$ (Note. $H_1' + H_1'' = H_1$)

$u = -\gamma_w H_1'' = -\gamma_w (H_1 - H_1')$

Therefore, $\bar{\sigma} = [(\gamma_{sat})_1 H_1' - \gamma_w H_1'] + \gamma_w H_1$

or $\bar{\sigma} = \gamma_1' \; H_1' + \gamma_w H_1$

If the soil had been saturated due to rise in water-table to A—A, the effective stress at section D—D would have been $\gamma_1' H_1'$. Thus, the effective stress is increased by $\gamma_w H_1$ due to capillary action.

Section B—B $\sigma = (\gamma_{sat})_1 H_1$, $u = 0$

Therefore, $\bar{\sigma} = (\gamma_{sat})_1 H_1 = \gamma_1' H_1 + \gamma_w H_1$

If the soil above B—B had been saturated due to rise in water table to A—A, the effective stress would have been $\gamma_1' H_1$. Thus the effective stress is increased by $\gamma_w H_1$ by capillary action

Section C—C $\sigma = (\gamma_{sat})_1 H_1 + (\gamma_{sat})_2 H_2$, $u = \gamma_w H_2$

Therefore, $\bar{\sigma} = (\gamma_{sat})_1 H_1 + \gamma_2' H_2 = \gamma_1' H_1 + \gamma_2' H_2 + \gamma_w H_1$

At this section also, the effective stress has also increased by $\gamma_w H_1$.

It may be noted that the effective stress at all levels below the plane of saturation A—A, due to capillary water, is increased by $\gamma_w H_1$. The capillary water pressure $\gamma_w H_1$ acts as if a surcharge. The effect is somewhat similar to the constant compressive stresses induced in the walls of the capillary tubes discussed in chapter 7.

(2) Soil saturated upto level D—D [Fig. 10.8 (*b*)].

Let us now consider the case when the soil above the water table B—B is saturated only upto level D—D upto a height H_1''. The soil above level D—D is wet and has a unit weight of γ.

The capillary rise in this case is H_1''.

The stresses at various sections can be determined as under.

Section A—A $\sigma = u = \bar{\sigma} = 0$

There is no effect of capillary water.

Section D—D $\sigma = \gamma_1 H_1'$, $u = -\gamma_w H_1''$

Therefore, $\bar{\sigma} = \gamma_1 H_1' + \gamma_w H_1''$

The effective stress due to capillary pressure is increased by $\gamma_w H_1''$.

Section B—B $\sigma = \gamma H_1' + (\gamma_{sat})_1 \; H_1'' = \gamma_1' H_1' + \gamma_1' H_1'' + \gamma_w H_1''$

$u = 0$

Therefore, $\bar{\sigma} = (\gamma_1 H_1' + \gamma_1 H_1'') + \gamma_w H_1''$

The effective stress is increased by $\gamma_w H_1''$ due to capillary action.

Likewise, it can be shown that the effective stress is increased by $\gamma_w H_1''$ at section C—C also.

The following points may be noted from the study of both cases :

(1) The capillary water above the water table causes a negative pressure $\gamma_w H$, where H is the capillary rise. This negative pressure causes an increase in the effective stresses at all levels below the saturation level. The increase is equal to $\gamma_w H$. The capillary action is equivalent to a surcharge $q = \gamma_w H$.

(2) If the soil is saturated due to rise in water table, the effective stress depends upon the submerged unit weight; whereas for the soil saturated with capillary water, the effective stress depends upon the saturated unit weight. In the latter case, the water does not contribute to hydrostatic pressure.

(3) If the water table rises to the top soil surface, the meniscus is destroyed and the capillary water changes to the free water, and the effective stress is reduced throughout.

(4) Eq. 10.3 is applicable in all cases. However, it should be remembered that the pore water pressure in the capillary zone is negative.

10.8. SEEPAGE PRESSURE

As the water flows through a soil, it exerts a force on the soil. The force acts in the direction of flow in the case of isotropic soils. The force is known as the *drag force or seepage force.* The pressure induced in the soil is termed *seepage pressure.*

Let us consider the upward flow of water in a soil sample of length L and cross-sectional area A under a hydraulic head of h [Fig. 10.9 (*a*)]. The expression for seepage force and seepage pressure can be derived considering the boundary water pressures u_1 and u_2 acting on the top and bottom of the soil sample, as shown in Fig. 10.9 (*b*)(*i*). The boundary water pressure can be resolved into two components, namely, the hydrostatic pressure and the hydrodynamic pressure as shown in Fig. 10.9(*b*)(*ii*) and 10.9(*b*)(*iii*).

(1) The hydrostatic pressures $u_1(s)$ and $u_2(s)$ are the components which would occur if there were no flow. If the samples were submerged under water to a depth of H_1, these pressures would have occurred.

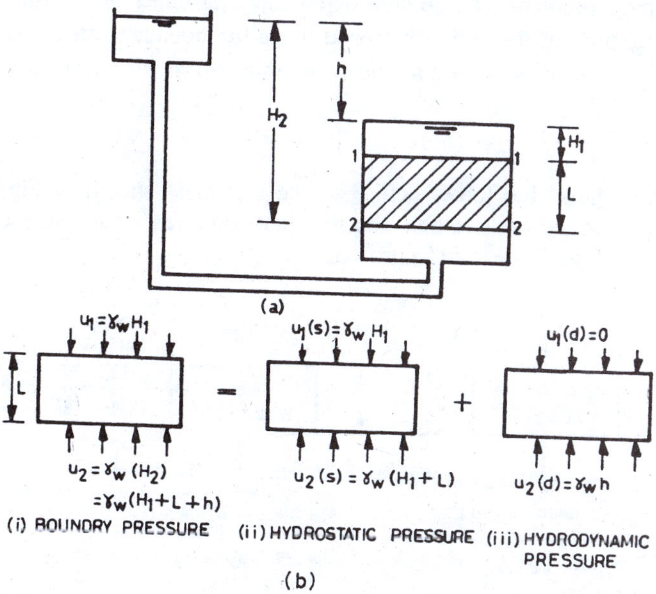

Fig. 10.9. Vertically upward flow.

(2) The hydrodynamic pressures $u_1(d)$ and $u_2(d)$ are the components which are responsible for flow of water. This pressure is spent as the water flows through the soil. These components cause the seepage pressure.

At the top of the sample, $u_1 = u_1(s) + u_1(d)$

or $\gamma_w H_1 = \gamma_w H_1 + 0$

At the bottom of the sample, $u_2 = u_2(s) + u_2(d)$

or $\gamma_w (H_1 + L + h) = \gamma_w (H_1 + L) + \gamma_w h$

The hydrodynamic pressure is due to hydraulic head h. The seepage force (J) acts on the soil skeleton due to flowing water through frictional drag. It is given by

$$J = \gamma_w h A \qquad \text{...(10.14)}$$

The seepage pressure (p_s) is the seepage force per unit area,

$$p_s = J/A = \gamma_w h \qquad \text{...(10.15)}$$

The seepage pressure (p_s) can be expressed in terms of the hydraulic gradient. From 10.15,

$$p_s = \gamma_w h = \gamma_w \cdot (h/L) \cdot L$$

or
$$p_s = i \gamma_w L \qquad \text{...(10.16)}$$

The seepage force (J) can be expressed as the force per unit volume (j), as

$$j = \frac{J}{A \times L} = \frac{\gamma_w h A}{AL} = \gamma_w \frac{h}{L}$$

or
$$j = i \gamma_w \qquad \text{...(10.17)}$$

Thus, the seepage force per unit volume is equal to the product of the hydraulic gradient (i) and the unit weight of water. As the hydraulic gradient is dimensionless, the seepage force per unit volume has the dimensions of the unit weight (*i.e.*)[F/L^3]. It has the units of N/m^3. For isotropic soils, the seepage force acts in the direction of flow.

10.9. FORCE EQUILIBRIUM IN SEEPAGE PROBLEMS

Force equilibrium in seepage problems can be considered adopting either of the following approaches.

(1) Considering the equilibrium of the entire mass and using the boundary pressures.

(2) Considering the equilibrium of the solid particle or the mineral skeleton, and using the hydrodynamic pressures.

(*a*) Vertical Flow

(*i*) *Upwards*. Fig. 10.10(*a*) shows the forces acting on the soil mass shown in Fig. 10.9 (*a*). The unit weight of the soil used is the saturated unit weight. The resultant force (R) on the soil mass considering the equilibrium of the entire mass, adopting the first approach,

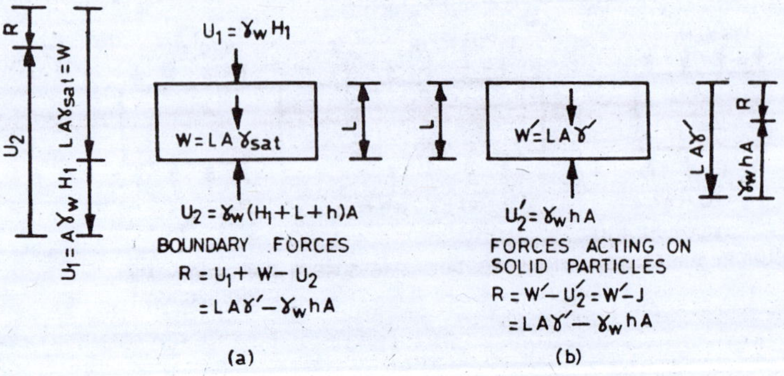

Fig. 10.10. Force equilibirum in vertical flow.

$$R = (W + U_1) - U_2$$

or $$R = (LA\,\gamma_{sat} + \gamma_w AH_1) - \gamma_w\,A\,(H_1 + L + h)$$

or $$R = LA\,\gamma' - \gamma_w h A \qquad ...(10.18)$$

The figure on the left-hand side shows the force diagram. The resultant force R acts downwards. For stability of the mass, R must act downwards.

Fig. 10.10(b) shows the forces acting on the solid particles, adopting the second approach. The unit weight of the soil used in this approach is the submerged unit weight. The resultant force (R) on the soil skeleton is given by

$$R = W' - U_2'$$

or $$R = LA\,\gamma' - \gamma_w h A \qquad \text{(same as Eq. 10.18)}$$

In Eq. 10.18, the first term gives the submerged unit weight and the second term, the seepage force (Eq. 10.14). It must be noted that in the first approach, the seepage force (J) is not considered separately. It is automatically accounted for in the boundary forces.

(ii) *Downwards*. Adopting a similar procedure, it can be shown that the resultant force when the flow is downward is given by

$$R = LA\,\gamma' + \gamma_w\,h A \qquad ...(10.19)$$

(b) Inclined Flow

Fig. 10.11 (a) shown the flow through an inclined soil specimen. In this case also, the resultant force R can also be determined by adopting either of the two approaches discussed above for the vertical flow. As for the vertical flow, in the first approach, the resultant force R is the vectorial sum of the saturated weight

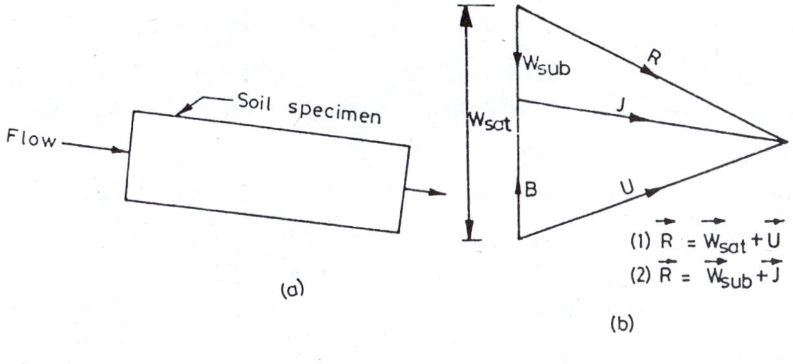

Fig. 10.11. Force equilibrium in inclined flow.

(W_{sat}) and the boundary forces U. In the second approach, the resultant force R is the vectorial sum of the submerged weight (W_{sub}) and the seepage force (J). The force triangles are shown in Fig. 10.11(b). Thus

(1) $$\overline{R} = \overline{W_{sat}} + \overline{U} \qquad ...(10.20)$$

(2) $$\overline{R} = \overline{W_{sub}} + \overline{J} \qquad ...(10.21)$$

Thus the two approaches give the same resultant force R.

In Fig. 10.11 (b), the buoyant force is represented by B. Obviously,

$$B = W_{sat} - W_{sub} \qquad ...(10.22)$$

The two approaches give identical results. However, the first approach is more popular. It is more convenient to determine the boundary forces than to determine the seepage forces. The seepage forces depend upon the direction of flow and change from point to point. It becomes difficult to determine the seepage forces, especially in two-dimensional flow.

10.10. EFFECTIVE STRESSES UNDER STEADY SEEPAGE CONDITIONS

As the water flows through the soil, it exerts a seepage force on the soil particles. The seepage force affects the interparticle forces and hence the effective stresses. The effective stress is increased when the flow is downward, as the seepage force increases the interparticle forces. On the other hand, when the flow is upward, the effective stress is decreased as the seepage force decreases the interparticle forces. The two cases are discussed separately below.

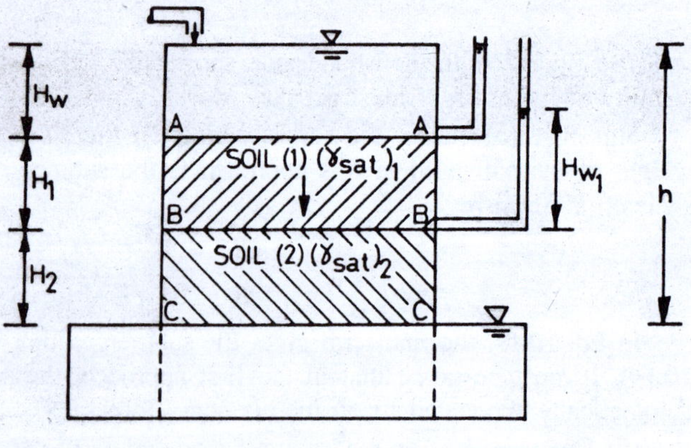

Fig. 10.12. Downward Flow.

***(a)* Downward Flow.** Let us consider the case when the flow is downward (Fig. 10.12). The head causing flow in h. The pore water pressure at sections A—A and B—B are indicated by the piezometers. The effective stresses at various sections are determined using Eq. 10.3.

Section A—A $\sigma = \gamma_w H_w, \quad u = \gamma_w H_w$

Therefore, $\bar{\sigma} = 0$

Section B—B $\sigma = (\gamma_{sat})_1 H_1 + \gamma_w H_w, \quad u = \gamma_w H_{w1}$

Therefore, $\bar{\sigma} = (\gamma_{sat})_1 H_1 + \gamma_w H_w - \gamma_w H_{w1}$

or $\bar{\sigma} = (\gamma_1' H_1 + \gamma_w H_1) + \gamma_w H_w - \gamma_w H_{w1}$

or $\bar{\sigma} = \gamma_1' H_1 + \gamma_w (H_w + H_1 - H_{w1})$

For hydrostatic conditions, the effective stress is $\gamma_1' H_1$, as discussed in Sect. 10.5. The second term indicates the effect due to flow. As $(H_w + H_1) > H_{w1}$, the effective stress is increased due to downward flow.

Section C—C $\sigma = \gamma_w H_w + (\gamma_{sat})_1 H_1 + (\gamma_{sat})_2 H_2, \quad u = 0$

Therefore, $\bar{\sigma} = \gamma_w H_w + (\gamma_1' + \gamma_w) H_1 + (\gamma_2' + \gamma_w) H_2$

or $\bar{\sigma} = \gamma_1' H_1 + \gamma_2' H_2 + \gamma_w (H_1 + H_w + H_2)$

or $\bar{\sigma} = \gamma_1' H_1 + \gamma_2' H_2 + \gamma_w h$

A comparison with the effective stresses corresponding to hydrostatic conditions shows that the effective stress is *increased* by $\gamma_w h$.

The conclusion that the effective stress is increased due to downward flow can also be drawn from intitutive feeling that as the water flows downward, it exerts a drag force in the downward direction and causes an increase in the interparticle forces.

***(b)* Upward Flow.** Fig. 10.13 shows the case when the flow is upward. The piezometers at various elevations indicate the pore water pressure.

Section A—A $\sigma = \gamma_w H_w, \quad u = \gamma_w H_w$

Therefore, $\bar{\sigma} = 0$

Section B—B $\sigma = \gamma_w H_w + (\gamma_{sat})_1 H_1, \quad u = \gamma_w H_{w1}$

Therefore, $\bar{\sigma} = (\gamma_{sat})_1 H_1 + \gamma_w H_w - \gamma_w H_{w1}$

$= (\gamma_1' + \gamma_w) H_1 + \gamma_w H_w - \gamma_w H_{w1}$

or $\bar{\sigma} = \gamma_1' H_1 + \gamma_w (H_1 + H_w - H_{w1})$

As $H_{w1} > (H_1 + H_w)$, the term $\gamma_w (H_1 + H_w - H_{w1})$ is negative, and the effective stress is less than that from the corresponding hydrostatic conditions

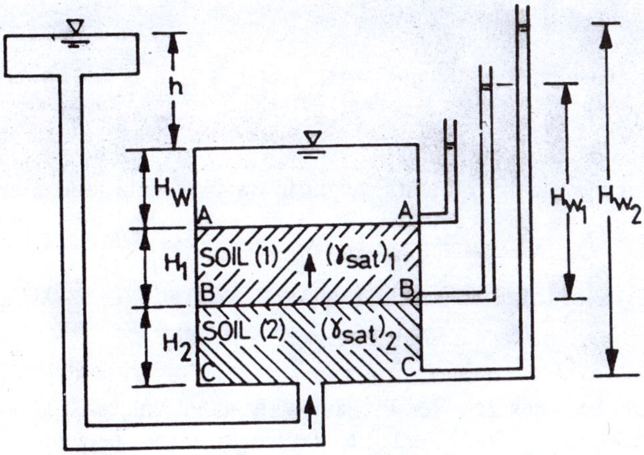

Fig. 10.13. Upward Flow.

Section C—C $\sigma = \gamma_w H_w + (\gamma_{sat})_1 H_1 + (\gamma_{sat})_2 H_2, \quad u = \gamma_w H_{w2}$

or $\bar{\sigma} = \gamma_w H_w + (\gamma_1' + \gamma_w) H_1 + (\gamma_2' + \gamma_w) H_2 - \gamma_w H_{w2}$

or $\bar{\sigma} = \gamma_1' H_1 + \gamma_2' H_2 - (H_{w2} - H_w - H_1 - H_2) \gamma_w$

or $\bar{\sigma} = \gamma_1' H_1 + \gamma_2' H_2 - \gamma_w h$

Thus, the effective stress is *reduced* by $\gamma_w h$ from the corresponding hydrostatic conditions.

10.11. QUICK SAND CONDITIONS

As discussed above, the effective stress is reduced due to upward flow of water. When the head causing upward flow is increased, a stage is eventually reached when the effective stress is reduced to zero. The condition so developed is known as quick sand condition. Fig. 10.14 shows a soil specimen of length L subjected to an upward pressure. Let us consider the stresses developed at section C—C.

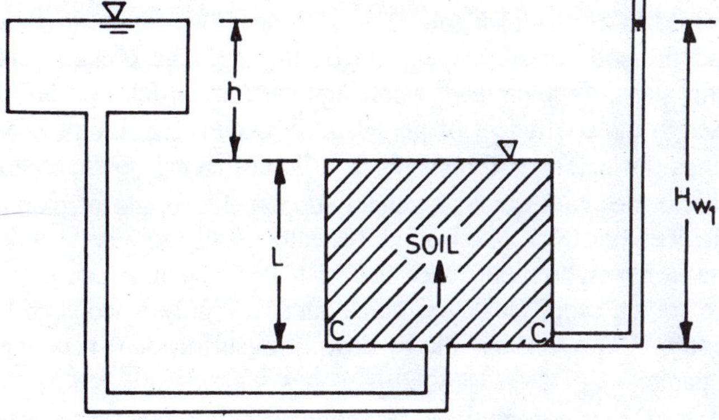

Fig. 10.14. Quick sand conditions.

$$\sigma = \gamma_{sat} L = (\gamma' + \gamma_w) L, \quad u = \gamma_w H_{w1} = \gamma_w (L + h)$$

Therefore, $\bar{\sigma} = (\gamma' + \gamma_w) L - \gamma_w (L + h)$

or $\bar{\sigma} = \gamma' L - \gamma_w h$

The second term can be written in terms of the hydraulic gradient as under.

$$\gamma_w h = \gamma_w \cdot (h/L) \cdot L = \gamma_w \cdot i \cdot L$$

Therefore, $\bar{\sigma} = \gamma' L - \gamma_w i L$

The effective stress becomes zero if $\gamma' L = \gamma_w i L$

or $$i = \gamma'/\gamma_w \qquad \text{...[10.23(a)]}$$

The hydraulic gradient at which the effective stress becomes zero is known as the critical gradient (i_c). Thus

$$i_c = \gamma'/\gamma_w \qquad \text{...(10.23)}$$

Substituting the value of the submerged unit weight in terms of void ratio from Eq. 2.24 (*a*),

$$i_c = \frac{(G - 1)}{1 + e} \qquad \text{...(10.24)}$$

Taking the specific gravity of solids (*G*) as 2.67, and the void ratio (*e*) as 0.67,

$$i_c = \frac{2.67 - 1}{1 + 1.67} = 1.0$$

Thus the effective stress becomes zero for the soil with above values of *G* and *e* when the hydraulic gradient is unity *i.e.* the head causing flow is equal to the length of the specimen.

Alternative method

The above expression for the critical gradient can also be obtained from the equilibrium of forces. When the quick condition develops, the upward force is equal to the downward weight. Thus

$$\gamma_{sat} \, (L \times A) = (h + L) A \, \gamma_w$$

or $$(\gamma_{sat} - \gamma_w) LA = A \, h \, \gamma_w$$

or $$L\gamma' = h \, \gamma_w \qquad \text{...[10.23(b)]}$$

or $$h/L = \gamma'/\gamma_w$$

or $$i = \gamma'/\gamma_w \qquad \text{...[same as Eq. 10.23 (a)]}$$

The shear strength of a cohesionless soil depends upon the effective stress (see chapter 13). The shear strength is given by

$$s = \bar{\sigma} \tan \phi$$

where ϕ is the angle of shearing resistance. The shear strength becomes zero when the effective stress $\bar{\sigma}$ is zero. The soil is then said to be in quick (alive) condition. If the critical gradient is exceeded, the soil particles move upward, and the soil surface appears to be boiling. The quick condition is also known as *boiling condition.* During this stage, a violent and visible agitation of particles occurs. The discharge suddenly increases due to an increases in the coefficient of permeability occurred in the process. If a weight is placed on the surface of soil, it sinks down. The soil behaves as a liquid having no shear strength.

When a natural soil deposit becomes quick, it cannot support the weight of man or animal. But contrary to common belief, the soil does not suck the victims beneath its surface. As a matter of fact, quick sand behaves like a liquid with a unit weight about twice that of water. A person can easily float in it with about one-third of his body out of quick sand. However, quick sand is highly viscous and movement in it would require a great effort and energy. A person may die by drowning (suffocation) if he gets tired and let his head fall into the quick sand in panic.

If a person is caught in quick sand conditions, he should keep his head high above the soil surface and move slowly towards the bank. He should try to catch some tree on the bank and try to pull himself out of the quick sand.

It is to be emphasized that quick sand is not a special type of sand. It is a condition which occurs in a soil when the effective stress is zero. Any cohesionless soil can become quick when the upward seepage force is large enough to carry the soil particles upward. The quick sand condition may also develop in gravel when the hydraulic gradient exceeds the critical gradient. However, the discharge required to maintain quick condition in gravels is very large, which may not be available. The required discharge depends upon the permeability of the soil.

The shear strength of cohesive soils is given by (see chapter 13),

$$s = c + \bar{\sigma} \tan \phi$$

The soil has shear strength equal to the cohesion intercept (c) even when the effective stress is reduced to zero. The cohesive soils, therefore, do not become quick as soon as the effective stress is reduced to zero.

The quick sand conditions may be summarised as under:

(1) Quick sand is not a special type of soil. It is a hydraulic condition.

(2) A cohesionless soil becomes quick when the effective stress is equal to zero.

(3) The critical gradient at which a cohesionless soil becomes quick is about unity.

(4) The discharge required to maintain a quick condition in a soil increases as the permeability of the soil increases.

(5) The cohesive soil does not become quick when the effective stress is equal to zero, as it still possesses some strength equal to the cohesion intercept.

(6) A quick condition is most likely to occur in silt and fine sand.

10.12. SEEPAGE PRESSURE APPROACH FOR QUICK CONDITIONS

The expression for critical gradient in the proceeding section has been developed using the first approach mentioned in Sect. 10.9, *i.e.*, considering the boundary pressures. The same expression can be developed using the second approach, *i.e.*, considering the seepage pressure.

Let us consider the equilibrium of the soil skeleton in Fig. 10.15. The downward force is given by

$$W_{sub} = (A \times L)\, \gamma'$$

The upward force is equal to the seepage force (Eq. 10.14).

$$J = i\gamma_w (A \times L)$$

When the soil becomes quick, the resultant force is zero.

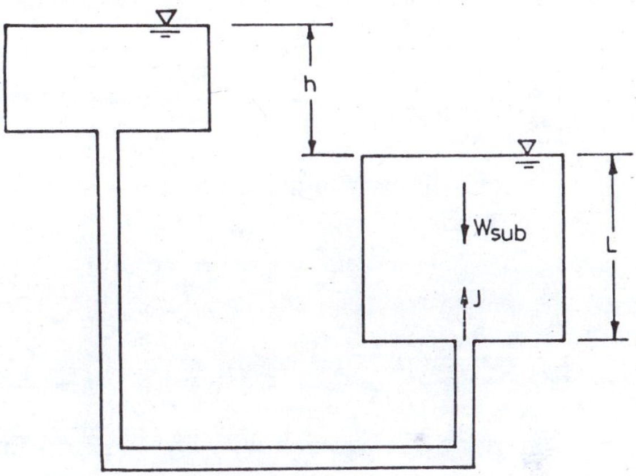

Fig. 10.15. Seepage Pressure Approach

Therefore, $W_{sub} = J$

or $AL\gamma' = i\gamma_w (A \times L)$

or $i = \dfrac{\gamma'}{\gamma_w}$ ...(some as Eq. 10.23)

Alternatively, one can work with pressures instead of forces. The downward pressure due to the submerged weight of soil is given by

$$\bar{\sigma} = \gamma' L$$

The seepage pressure is given by Eq. 10.16 as

$$p_s = i\gamma_w L$$

The net effective stress would be zero when $\bar{\sigma} = p_s$.

Thus $iL\gamma_w = \gamma' L$

or $i = \gamma'/\gamma_w$ ...(same as Eq. 10.23)

10.13. EFFECT OF SURCHARGE AND SUBMERGENCE ON QUICK CONDITIONS

Fig. 10.16 shows a soil specimen submerged under water and subjected to surcharge load of intensity q. Let us consider the stresses at section C—C.

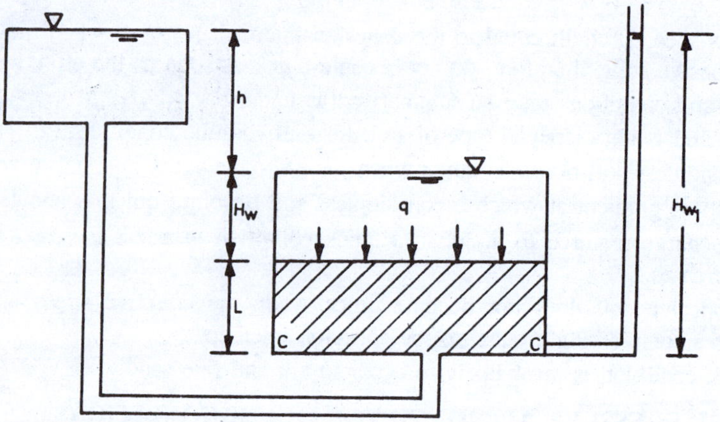

Fig. 10.16. Effect of Surcharge.

$$\sigma = \gamma_w H_w + q + \gamma_{sat} L$$

and $$u = \gamma_w H_{w1} \quad \text{or } u = \gamma_w (h + H_w + L)$$

Therefore, $$\bar{\sigma} = (\gamma_w H_w + q + \gamma_{sat} L) - \gamma_w (h + H_w + L)$$

The soil will become quick when $\bar{\sigma} = 0$. Thus

$$\gamma_w (h + H_w + L) = \gamma_w H_w + q + \gamma_{sat} L$$

$$\gamma_w h + \gamma_w H_w + \gamma_w L = \gamma_w H_w + q + \gamma_{sat} L$$

or $$\gamma_w h = \gamma' L + q$$

or $$h = \frac{\gamma' L + q}{\gamma_w} \qquad ...(10.25)$$

Comparing this equation with Eq. 10.23 (*b*) for the case when there is no surcharge as discussed in the preceding sections, it is observed that the head required to cause quick condition is increased by q/γ_w. In other words, the downward weight increases the stability against quick conditions. The critical gradient is, however, not affected by the depth of water (H_w) over the soil surface. Substituting $q = 0$ in Eq. 10.25,

$$h = \frac{\gamma' L}{\gamma_w}$$

or $$h/L = \gamma'/\gamma_w$$

or $$i = \gamma'/\gamma_w \qquad \text{...(same as Eq. 10.23)}$$

10.14. FAILURES OF HYDRAULIC STRUCTURES BY PIPING

Hydraulic structures, such as weirs and dams, built on pervious foundations sometimes fail by formation of a pipe-shaped channel in its foundation, known as *piping failure.* The failure occurs when water flowing through the foundation has a very high hydraulic gradient and it carries soil particles with it. There are two types of such failures :

(1) Backward-erosion piping failure.

(2) Heave-piping failure.

(1) Backward Erosion Piping. This type of piping begins with the removal of soil particles by flowing water near the exit points. A scour hole forms near the exit when the hydraulic gradient is high. The hole extends upstream along the foundation. The failure occurs as soon as the scour hole approaches the upstream reservoir. The failure mechanism of backward erosion piping is not amenable to theoretical solution.

Backward erosion of soil is caused by the percolating water, and the piping begins when the hydraulic gradient at exit, known as *exit gradient,* exceeds the critical gradient (i_c), given by Eq. 10.24. The soil at the exit is removed by the percolating water. When the soil near the exit has been removed, the flow net gets modified. There is more concentration of the flow lines in the remaining soil mass, resulting in an increase

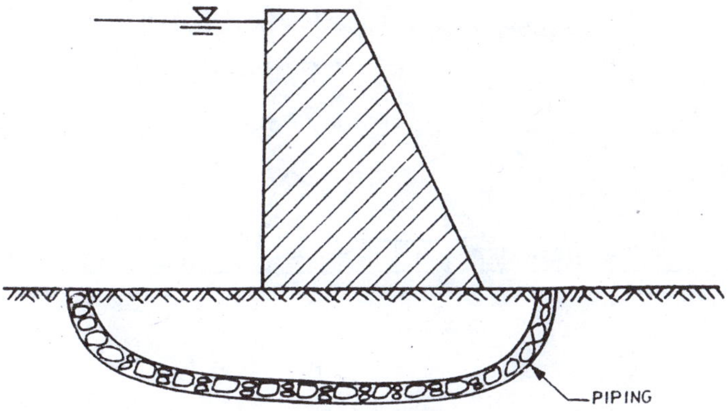

Fig. 10.17. Backward erosion Piping.

of the exit gradient. This causes further removal of the soil. This process of backward erosion continues towards the upstream reservoir and a sort of pipe is formed (Fig. 10.17). As soon as the channel approaches the reservoir, a large amount of water rushes through the channel so formed and the hydraulic structure fails.

Backward erosion piping may also occur in the body of earth structure, such as an earth dam. This takes place when the phreatic line cuts the downstream face of the dam and the seepage pressure is high. It is indicated by a progressive sloughing of the downstream face. Such failures can occur even when the exit gradient is low. If the downstream face has the slope angle equal to the angle of internal friction of the cohesionless soil, the critical gradient at which failure occurs is approximately equal to zero. In other words, the failure may occur even when the seepage is almost horizontal towards the downstream face.

Backward erosion piping may also occur along any weak bedding plane in the foundation, or along the periphery of a conduit embedded in the earth dam when the seepage pressure is high.

Generally, backward erosion piping failure occurs when the exit gradient is greater than the critical gradient. But, in exceptional cases, it may occur even when the overall downward submerged weight of the soil is greater than the upward force due to seepage. In such a case, some of the fine particles of the soil are carried by the percolating water even though the most of soil particles are restrained. Thereafter, the seepage concentrates in the loosened soil and results in piping failure.

(2) Heave Piping. Failure by heave piping occurs in the form of a rise or a heave of a large mass of soil due to seepage pressure. When the seepage force due to upward flow of water at any level is greater than the submerged weight of the soil above that level, the entire soil mass in that zone heaves up and is blown out by the percolating water. This type of failure is known as heave piping failure.

Heave piping may occur on the downstream of a sheet pile cutoff wall of a hydraulic structure (Fig. 10.18). According to Terzaghi, heave piping occurs within a distance of $D/2$ on the downstream of the sheet pile, where D is the depth of pile below the ground surface. It occurs in the zone marked *a b c d* when the upward seepage force is greater than the submerged weight of the soil in this zone. The seepage force can be determined from the flow net.

In Fig. 10.18, the equipotential line of potential 0.4 h passes through d and that of 0.3 h through c. The average excess hydrostatic pressure on the base $c\ d$ of the prism $a\ b\ c\ d$ is 0.35 h. Therefore, upward seepage force, $U = \gamma_w\ (0.35\ h)\ (D/2 \times 1)$ per unit length and downward force due to submerged weight,

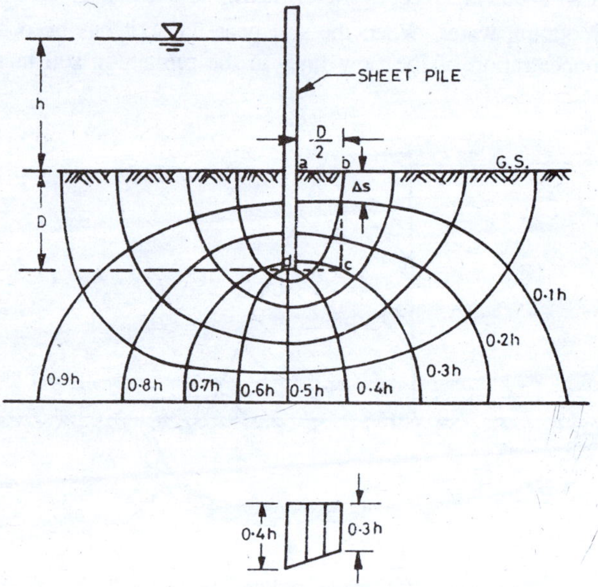

Fig. 10.18. Heave Piping.

$$W' = \gamma' \times (D/2 \times D)$$

Heave piping would occur when $U \geq W'$. The failure is associated with an expansion of the soil which results in an increase in the permeability of the soil. The flow suddenly increases and ultimately leads to failure.

The factor of safety with respect to heave piping can be obtained from the following equation.

$$F = \frac{W'}{U} = \frac{\gamma' (D^2/2)}{\gamma_w (h_a)(D/2)}$$

or

$$F = \frac{\gamma' D}{\gamma_w h_a} \qquad \text{...(10.26)}$$

where h_a is the average excess hydrostatic pressure at the base of the prism *a b c d*.

10.15. PREVENTION OF PIPING FAILURES

The occurrence of piping in and below a hydraulic structure such as an earth dam is disastrous. This may cause catastrophe. The following measures are generally adopted to prevent piping failures.

(1) Increasing the path of percolation. The hydraulic gradient (i) depends upon the path of percolation (L). If the length of the path is increased, the exit gradient will decrease to a safe value. The length of the path of percolation can be increased by adopting the following methods.

(*a*) Increasing the base width of the hydraulic structure.

(*b*) Providing vertical cut off walls below the hydraulic structure.

(*c*) Providing an upstream impervious blanket, as shown in Fig. 10.19.

(2) Reducing Seepage. With a reduction of seepage through the dam, the chances of piping failure through the body of the dam are considerably reduced. The quantity of seepage discharge is reduced by providing an impervious core, as shown in Fig. 10.19.

(3) Providing drainage filter. A drainage filter changes the direction of flow away from the downstream face. It prevents the movement of soil particles along with water. The drainage filter is properly graded, as discussed in the following section.

The drainage filter may be horizontal or in the form of a rock toe (Sect. 9.11). It may also be in the form

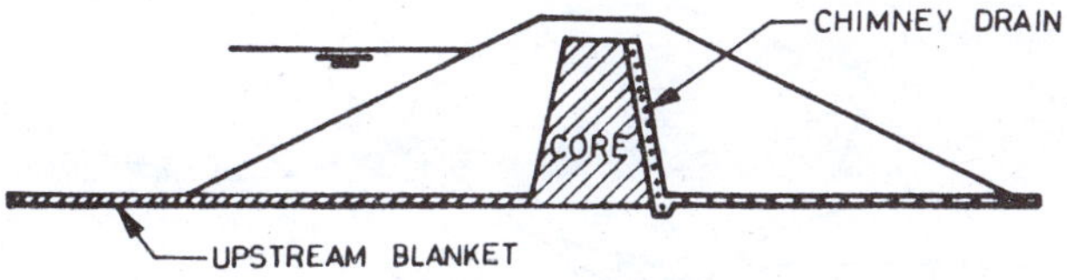

Fig. 10.19. Prevention of Piping.

of a chimney drain, as shown in Fig. 10.19. A chimney drain is effective for stratified soil deposits in which the horizontal permeability is greater than the vertical permeability.

(4) Loaded Filter. A loaded filter consists of graded sand and gravels. The function of the loaded filter is to increase the downward force without increasing the upward seepage force.

The loaded filter is placed at the exit point where the water emerges from the foundation. For the sheet pile wall, the filter is placed over the affected zone *a b c d* in Fig. 10.18. The loaded filter increases the factor of safety against heave piping. The factor of safety (F) is given by

$$F = \frac{W' + W}{U} \quad \text{...(10.27)}$$

where W is the weight of filter.

The loaded filter should be of pervious material such that it does not increase the hydrostatic pressure. It should only increase the downward force.

10.16. DESIGN OF A GRADED FILTER

A graded filter consists of layers of pervious material which permit flow of water but prevent the movement of soil particles. The soil particles in a particular layer are coarser than that in the preceding layer. However, the difference of sizes of the particles in the two layers should not be excessive otherwise the particles of the preceding layer will be carried into the next layer. The particle sizes of different layers are fixed according to the design criteria given below :

(1) The filter material should be coarse enough so that the percolating water moves easily without any build up of water pressure in the filter.

For the filter to provide free drainage, its coefficient of permeability should be 25 times or more than the coefficient of permeability of the soil to be protected, known as a *base material.* As the coefficient of permeability varies as the square of the particles size, the ratio of the particle diameters should be at least 5. Therefore, the first criterion of design of graded filter is that D_{15} of filter material is greater than $5D_{15}$ of base material.

or $$D_{15}\ (f) > 5\, D_{15}\ (b) \quad \text{...(10.28)}$$

where (f) stands for filter and (b) for base material.

(2) The filter material should be fine enough that the soil particles of the base material are not washed through the filter.

It would not be necessary to screen out all the particles in the base material.. If the filter openings restrain the coarsest 15% *i.e.* D_{85} size of base material, the soil particles are checked due to formation of a skeleton. The coarser 15% particles collect over the openings in the filter material and form smaller openings to trap the smaller particles, as shown in Fig. 10.20. Therefore, the size of the openings formed in the filter must be less than D_{85} of the soil. It has been established that the diameter of the openings is about 1/5 of the D_{15} size of the filter. Therefore, the second criterion is

$$\frac{1}{5} D_{15} \text{ of filter material} < D_{85} \text{ of base of material}$$

or $$D_{15}\ (f) < 5\, D_{85}\ (b) \quad \text{...(10.29)}$$

Combining both the criteria (1) and (2),

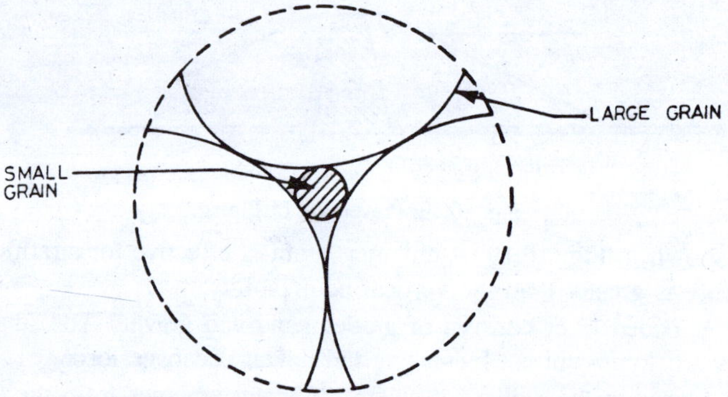

Fig. 10.20. Openings in filter

$$\frac{D_{15}(f)}{D_{85}(b)} < 5 < \frac{D_{15}(f)}{D_{15}(b)} \quad \text{...(10.30)}$$

The U.S. Corps of Engineers have recommended that

$$\frac{D_{50}(f)}{D_{50}(b)} \le 25 \quad \text{...[10.30 (a)]}$$

In a graded filter, each layer is designed considering it as a filter and the preceding layer as a base material. The particle sizes of the layers increase in the direction of flow. (Fig. 10.21).

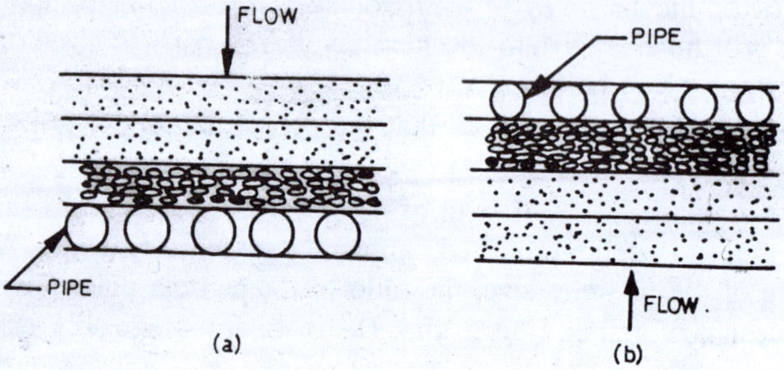

Fig. 10.21. Graded filter.

(4) The material of the last layer should be coarse enough not to be carried away through the openings of the perforated drainage pipes, if provided.

If the D_{85} size of the last layer satisfies the following criterion, the chances of washing of the filter material into the pipes are reduced.

For circular holes in the pipe,

$$\frac{D_{85} \text{ of filter material}}{\text{Diameter of the hole}} > 1.2 \quad \text{...(10.31)}$$

For slotted openings,

$$\frac{D_{85} \text{ of filter material}}{\text{Width of slot}} > 1.4 \quad \text{...(10.32)}$$

Generally, $\dfrac{D_{85} \text{ of filter material}}{\text{size of opening}}$ for both types of the openings is kept equal to or grater than 2.0.

(5) The grain size curve of the filter material should be roughly parallel to that of the base material.

(6) To avoid segregation, filter should not contain the particles of size larger than 75 mm.

(7) For proper working, the filter material should not contain more than 5% of the fines passing 75μ IS sieve.

(8) The thickness and area of the filter should be sufficient to carry the seepage discharge safely.

If the filter has to work as a loaded filter, the total thickness should be large enough to provide adequate weight.

10.17. EFFECTIVE STRESS IN PARTIALLY SATURATED SOILS

In partially saturated soils, air is also present along with water. In the discussion given below, it is assumed that air is in sufficient quantity such that there is continuity in both the air phase and the water phase. Because of meniscus formation, the air pressure is greater than the water pressure. It is assumed that the air pressure and water pressure are constant throughout the void spaces. Thus, there are three measurable stresses in a partially saturated, namely, total stress σ, pore water pressure u_w and air pressure u_a.

Let us consider the forces acting on the wavy plane *X—X*, shown in Fig. 10.22 (*a*). The wavy plane passes through the points of contact of solid particles. The wavy plane can be taken as a plane as already mentioned. Fig. 10.22 (*b*) shows the forces acting on the plane. From equilibrium in the vertical direction.

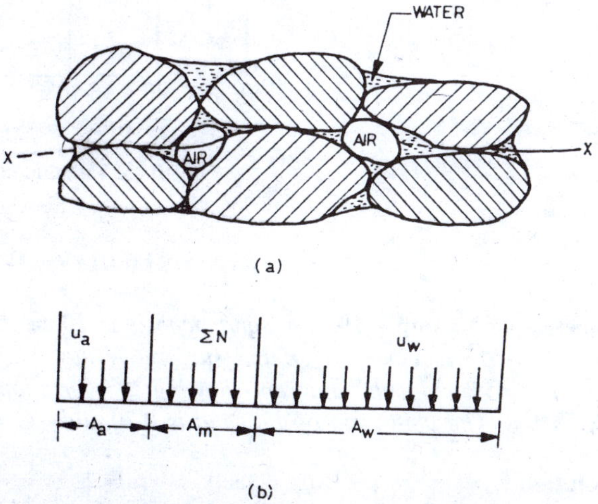

Fig. 10.22. Partially saturated soil.

$$\sigma A = \Sigma N + u_w A_w + u_a A_a \qquad \ldots(a)$$

where σ = total stress, *A* total area of the plane, A_w = area of the plane passing through water,
A_a = area of the plane passing through air, and
ΣN = summation of normal forces acting at the particle to particle contact points.

From Eq. (*a*),
$$\sigma = \frac{\Sigma N}{A} + u_w \frac{A_w}{A} + u_a \frac{A_a}{A}$$

or
$$\sigma = \bar{\sigma} + u_w a_w + u_a (1 - a_w) \qquad \ldots(b)$$

where $\bar{\sigma}$ = effective stress (= $\Sigma N/A$), $a_w = A_w/A$, and $a_a = A_a/A = 1 - a_w$

The area at the points of contact is neglected as it is very small as compared to the area through the water and that through the air. Eq. (*b*) can be written as

$$\bar{\sigma} = \sigma - u_w a_w - u_a (1 - a_w) \qquad \ldots(10.32)$$

Eq. 10.32 cannot be verified experimentally as it is difficult to measure the area a_w. Bishop et al (1960) conducted a large number of tests and gave the following equation for the effective stress in partially saturated soils.

$$\bar{\sigma} = \sigma - u_a + \chi (u_a - u_w) \qquad \ldots(10.33)$$

where χ (pronounced as *Chi*) represents the fraction of the area of the soil occupied by water. It depends mainly on the degree of saturation *S* (Fig. 10.23). Its value is zero for dry soil and is unity for fully saturated soil. The value of χ also depends upon the soil structure, the cycle of wetting and drying, and stress changes.

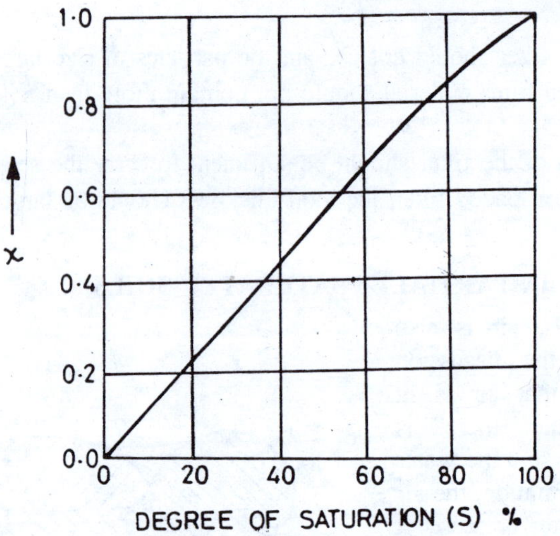

Fig. 10.23. Variation of χ with S.

ILLUSTRATIVE EXAMPLES

Illustrative Example 10.1. *A sand deposit is 10 m thick and overlies a bed of soft clay. The ground water table is 3 m below the ground surface. If the sand above the ground water table has a degree of saturation of 45%, plot the diagram showing the variation of the total stress, pore water pressure and the effective stress. The void ratio of the sand is 0.70. Take G = 2.65.*

Solution Bulk density, $\rho = \left(\frac{G + Se}{1 + e}\right) \rho_w$

or $\rho = \left(\frac{2.65 + 0.45 \times 0.70}{1 + 0.70}\right) \times 1000 = 1744.12 \text{ kg/m}^3$

or $\gamma = 1744.12 \times 9.81 \times 10^{-3} = 17.11 \text{ kN/m}^3$

For saturated soils, S = 1.0, and

$$\rho = \left(\frac{2.65 + 0.70}{1 + 0.70}\right) \times 1000 = 1970.59 \text{ kg/m}^2$$

or $\gamma = 1970.59 \times 9.81 \times 10^{-3} = 19.33 \text{ kN/m}^3$

Fig. E 10.1 shows the soil profile. The stresses at section B-B and C-C are as under:

Section B—B $\sigma = 17.11 \times 3 = 51.33 \text{ kN/m}^2, \quad u = 0$

$\bar{\sigma} = 51.33 \text{ kN/m}^2$

Section C—C $\sigma = 17.11 \times 3 + 19.33 \times 7 = 186.64 \text{ kN/m}^2$

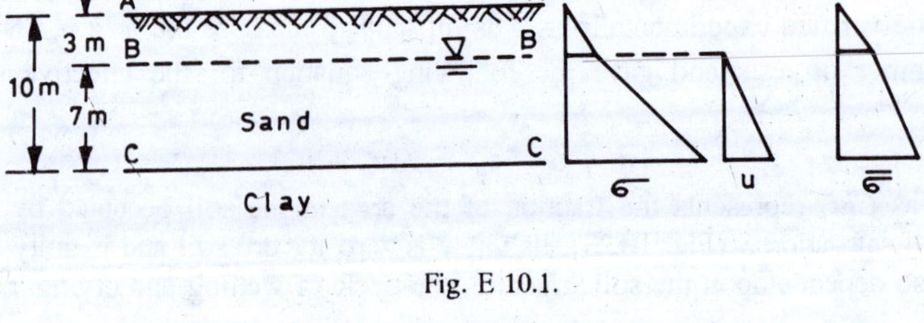

Fig. E 10.1.

$u = 7 \times 9.81 = 68.67\ \text{kN/m}^2,\quad \bar{\sigma} = 186.64 - 68.67 = 117.97\ \text{kN/m}^2$

Fig. E 10.1 shows the variation of stresses.

Illustrative Example 10.2. *Determine the increase in the effective stress at section C—C in Illustrative Example 10.1 when the water table is lowered by 5 m. Assume that the soil above the water table has the degree of saturation of 45% throughout.*

Solution. $\sigma = 8 \times 17.11 + 2 \times 19.33 = 175.54\ \text{kN/m}^2$

$u = 2 \times 9.81 = 19.62\ \text{kN/m}^2,\quad \bar{\sigma} = 175.54 - 19.62 = 155.92\ \text{kN/m}^2$

Increase in effective stress $= 155.92 - 117.97$

$= \mathbf{37.95\ kN/m^2}$

Illustrative Example 10.3. *A soil profile consists of a surface layer of clay 4 m thick ($\gamma = 19.5\ kN/m^3$) and a sand layer 2 m thick ($\gamma = 18.5\ kN/m^3$) overlying an impermeable rock. The water table is at the ground surface. If the water level in a standpipe driven into the sand layer rises 2 m above the ground surface, draw the plot showing the variation of σ, u and $\bar{\sigma}$. Take $\gamma_w = 10\ kN/m^3$.*

(*b*) *Determine the increase in effective stress at the top of the rock when the artesian head in the sand is reduced by 1 m.*

Solution. Fig. E 10.3 shows the profile.

Section B—B (Clay), $\sigma = 19.5 \times 4 = 78.0\ \text{kN/m}^2,\quad u = 4 \times 10.0 = 40.0\ \text{kN/m}^3$

$\bar{\sigma} = 78.0 - 40.0 = 38.0\ \text{kN/m}^2$

Section B—B (sand) $\sigma = 19.5 \times 4 = 78.0\ \text{kN/m}^2$

$u = 6 \times 10.0 = 60.0\ \text{kN/m}^2,\quad \bar{\sigma} = 78.0 - 60.0 = 18.0\ \text{kN/m}^2$

Section C—C $\sigma = 4 \times 19.5 + 2 \times 18.5 = 115.0\ \text{kN/m}^2$

$u = 8 \times 10.0 = 80.0\ \text{kN/m}^2,\quad \bar{\sigma} = 115.0 - 80.0 = 35.0\ \text{kN/m}^2$

Fig. E 10.3 shows the variation of stresses. It is to be noted that there is a sudden change in u and $\bar{\sigma}$ at level B—B.

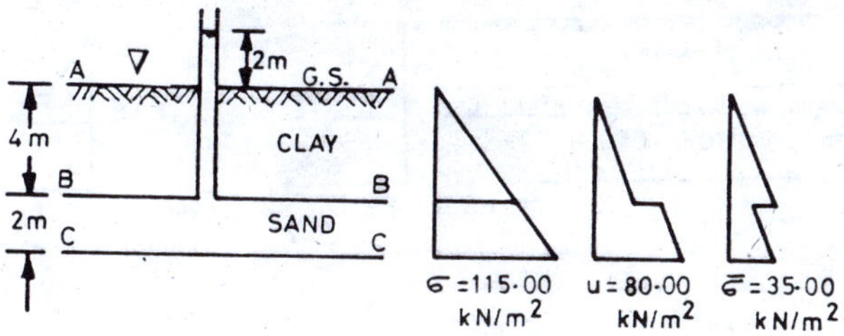

Fig. E 10.3.

(*b*) When artesian pressure is reduced by 1 m

Section C—C $\sigma = 4 \times 19.5 + 2 \times 18.5 = 115.0\ \text{kN/m}^2$

$u = 7 \times 10.0 = 70.0\ \text{kN/m}^2,\quad \bar{\sigma} = 115.0 - 70.00 = 45.0\ \text{kN/m}^2$

Increase in stress $= 45.0 - 35.0 = \mathbf{10.0\ kN/m^2}$.

Illustrative Example 10.4. *A soil profile consists of a surface layer of sand 3.5 m thick ($\rho = 1.65\ Mg/m^3$), an intermediate layer of clay 3 m thick ($\rho = 1.95\ Mg/m^3$) and the bottom layer of gravel 3.5 m thick ($\rho = 1.925\ Mg/m^3$). The water table is at the upper surface of the clay layer. Determine the effective pressure at various levels immediately after placement of a surcharge load of 58.86 kN/m² to the ground surface.*

Solution. Fig. E 10.4 shows the soil profile.

Section A—A $\sigma = 58.86\ \text{kN/m}^2,\quad u = 0,\quad \bar{\sigma} = 58.86\ \text{kN/m}^2$

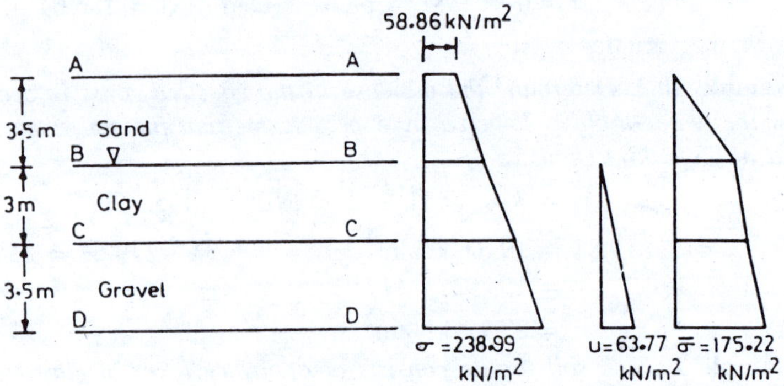

Fig. E 10.4

Section B—B $\sigma = 58.86 + 3.5\,(1.65 \times 9.81) = 115.51$

$u = 0,\ \bar{\sigma} = 115.51\ \text{kN/m}^2$

Section C—C $\sigma = 115.51 + 3 \times (1.95 \times 9.81) = 172.90$

$u = 3 \times 9.81 = 29.43,\ \bar{\sigma} = 172.90 - 29.43 = \mathbf{143.47\ kN/m^2}$

Section D—D $\sigma = 172.90 + 3.5\,(1.925 \times 9.81) = 238.99$

$u = 6.5 \times 9.81 = 63.77,\ \bar{\sigma} = 238.99 - 63.77 = \mathbf{175.22\ kN/m^2}.$

Illustrative Example 10.5. *A sand deposit consists of two layers. The top layer is 2.5 m thick (ρ = 1709.67 kg/m^3) and the bottom layer is 3.5 m thick (ρ_{sat} = 2064.52 kg/m^3). The water table is at a depth of 3.5 m from the surface and the zone of capillary saturation is 1 m above the water table. Draw the diagrams, showing the variation of total stress, neutral stress and effective stress.*

Solution. Fig. E 10.5 shows the soil profile.

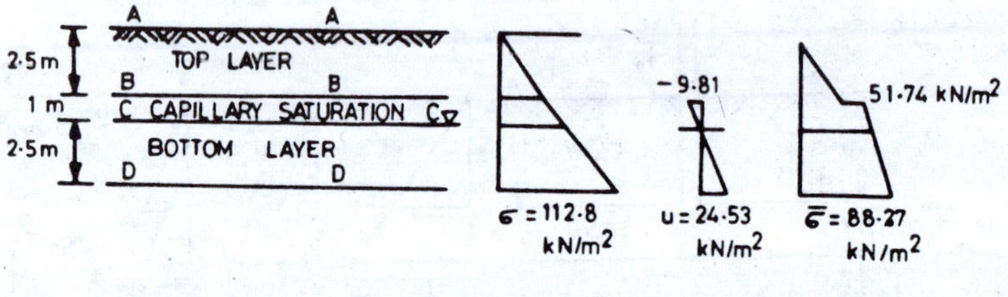

Fig. E 10.5

$$\gamma = 1709.67 \times 9.81 \times 10^{-3} = 16.77\ \text{kN/m}^3$$

$$\gamma_{sat} = 2064.52 \times 9.81 \times 10^{-3} = 20.25\ \text{kN/m}^3$$

Level A—A $\sigma = u = \bar{\sigma} = 0$

Level B—B (top layer) $\sigma = 16.77 \times 2.5 = 41.93\ \text{kN/m}^2\ u = 0$

$\bar{\sigma} = \mathbf{41.93\ kN/m^2}$

Level B—B (Bottom layer) $\sigma = 16.77 \times 2.5 = 41.93\ \text{kN/m}^2$

$u = -1 \times 9.81 = -9.81\ \text{kN/m}^2$

$\bar{\sigma} = 41.93 - (-9.81) = \mathbf{51.74\ kN/m^2}$

Level C—C $\sigma = 16.77 \times 2.5 + 1 \times 20.25 = 62.18\ \text{kN/m}^2$

$u = 0;\ \bar{\sigma} = 62.18 - 0 = \mathbf{62.18\ kN/m^2}$

Level D—D $\sigma = 16.77 \times 2.5 + 20.25 \times 3.5 = 112.8 \text{ kN/m}^2$

$$u = 2.5 \times 9.81 = 24.53 \text{ kN/m}^2, \ \bar{\sigma} = 112.8 - 24.53 = \mathbf{88.27 \text{ kN/m}^2}$$

Illustrative Example 10.6. *A 8 m thick layer of stiff saturated clay (γ = 19.0 kN/m³) is underlain by a layer of sand. The sand is under an artesian pressure of 5 m. Calculate the maximum depth of the cut that can be made without causing a heave.*

Solution. Fig. E 10.6 shows the cut. Let H be the required depth of the cut. Heave will occur when the effective stress $\bar{\sigma}$ becomes zero at level A—A.

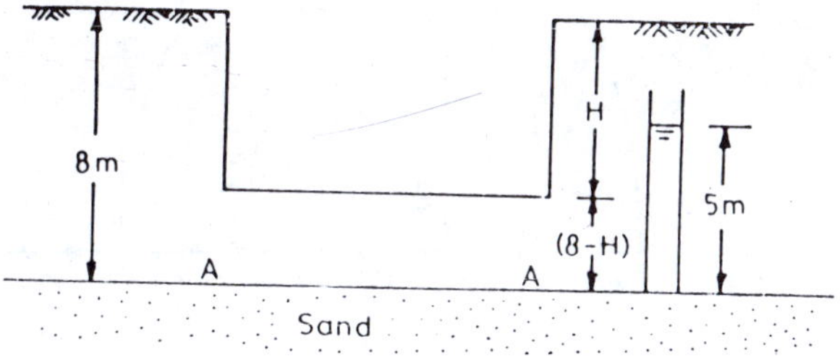

Fig. E 10.6.

$$\sigma = 19.0 \times (8 - H) = 152.0 - 19H, \ u = 5 \times 9.81 = 49.05$$

$$\bar{\sigma} = 152.0 - 19H - 49.05 = 102.95 - 19H = 0$$

or $\quad H = \mathbf{5.42 \text{ m.}}$

Illustrative Example 10.7. *A 10 m thick layer of silty clay (ρ = 1864.64 kg/m³) overlies a gravel layer. The gravel is under an artesian pressure of 12.5 m. It is proposed to excavate a foundation trench 2 m deep. Determine the factor of safety against heaving.*

(b) What would be the factor of safety against heaving when a uniform pressure of 98.1 kN/m² is applied to the footing constructed in the above trench ?

Solution. Factor of safety $= \dfrac{\text{Down–ward force at the top of gravel layer}}{\text{Upward force at the top of gravel layers}}$

$$F = \frac{1864.64\,(9.81 \times 10^{-3}) \times (10-2)}{9.81 \times 12.5} = \mathbf{1.19}$$

(b) After Construction. The downward force is increased due to the uniform pressure of 98.1 kN/m².

$$F = \frac{1864.64\,(9.81 \times 10^{-3}) \times (10-2) + 98.1}{9.81 \times 12.5} = \mathbf{1.99}$$

Illustrative Example 10.8. *Determine the factor of safety against heave failure in the hydraulic structure shown in Fig. Ex. 10.8. ρ = 1850 kg/m³.*

Solution. Average pressure on the base of soil prism,

$$h_a = 0.42\,h = 0.42 \times 8 = 3.36 \text{ m}$$

$$\gamma' = 1850 \times 9.81 \times 10^{-3} - 9.81 = 8.34 \text{ kN/m}^2$$

From Eq. 10.26, factor of safety, $F = \dfrac{\gamma' D}{\gamma_w\, h_a}$

or $\quad F = \dfrac{8.34 \times 5}{9.81 \times 3.36} = \mathbf{1.27}$

Illustrative Example 10.9. *Determine the approximate limits of the filter material required for the soil of the base material which has D_{15} = 0.01 mm and D_{85} = 0.10 mm, and the grading curve as shown in Fig. E 10.9.*

Solution. From Eq. 10.28, $\quad D_{15\,(f)} > 5\,D_{15}\,(b)$

or $\quad D_{15\,(f)} > 5 \times 0.01 > 0.05 \text{ mm}$

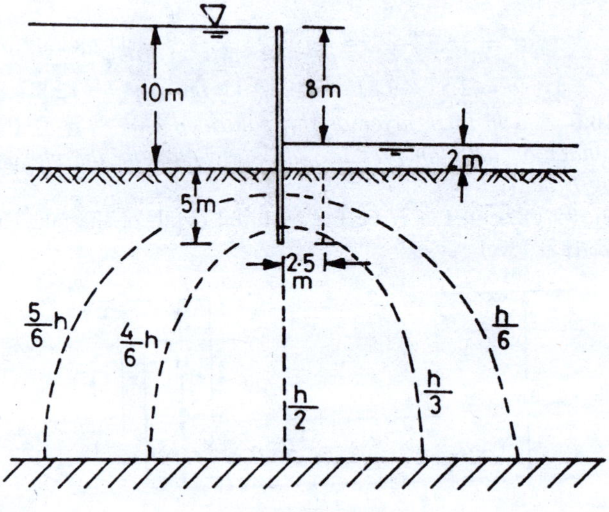

Fig. E 10.8.

From Eq. 10.29, $D_{15(f)} < 5\ D_{85}\ (b)$

or $D_{15(f)} < 5 \times 0.10 < 0.5$ mm

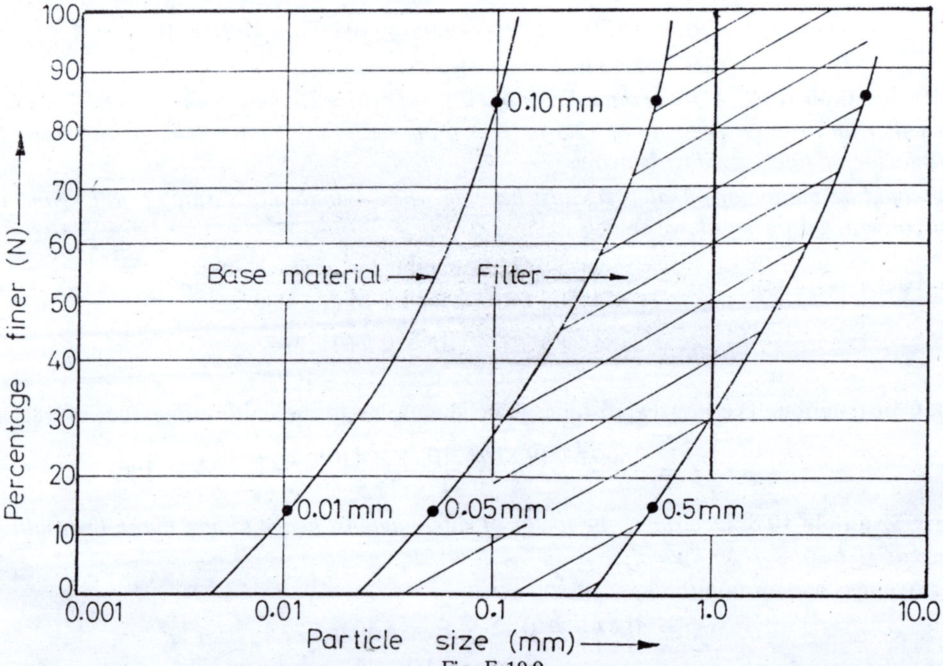

Fig. E 10.9

Therefore, D_{15} of filter material should lie between 0.05 mm and 0.5 mm. As the gradation curve of the filter material should be roughly parallel to that of the base material, the hatched portion indicates the limits of the material suitable as filter.

Illustrative Example 10.10. *If excavation is carried out in a soil with a porosity of 0.40 and the specific gravity of solids of 2.65, determine the critical gradient. A 1.50 m layer of the soil is subjected to an upward seepage head of 1.95 m. What depth of coarse sand would be required above the soil to provide a factor of safety of 2.50 ? Assume that sand has the same porosity and sp. gr. of solids as the soil.*

Solution.
$$e = \frac{n}{1-n} = \frac{0.40}{0.60} = 0.667$$

$$\text{Critical gradient} = \frac{G-1}{1+e} = \frac{2.65 - 1}{1 + 0.667} = 0.99$$

$$\text{Saturated density} \quad \rho_{sat} = \left(\frac{G + e}{1 + e}\right) \rho_w$$

$$= \left(\frac{2.65 + 0.667}{1 + 0.667}\right) \times 1.0 = 1.99 \text{ g/ml } (19.52 \text{ kN/m}^3)$$

Let x be the depth of sand layer.

$$\text{Effective pressure at bottom} = (1.50 + x)(19.52 - 9.81)$$

$$= (1.50 + x)\ 9.71$$

$$\text{Upward pressure at bottom} = 1.95 \times 9.81$$

$$\text{Now factor of safety, F.S.} = \frac{(1.50 + x) \times 9.71}{1.95 \times 9.81} = 2.50$$

or $x = \mathbf{3.42\ m}$

PROBLEMS

A. Numerical

10.1. Determine the total, neutral and effective stresses at the bottom of the deposit shown in Fig. P 10.1.

[**Ans.** 199.14, 83.39, 115.75 kN/m^2]

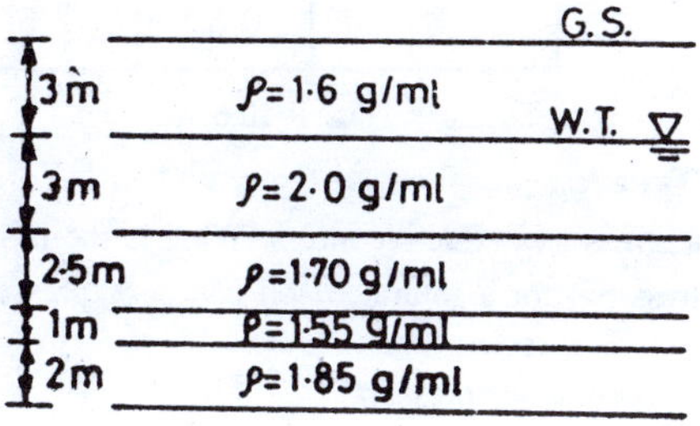

Fig. P 10.1

10.2. The water table in a deposit of uniform sand is located at 2 m below the ground surface. Assuming the soil above the water table is dry, determine the effective stress at a depth of 5 m below the ground surface. The void ratio is 0.75 and the specific gravity of solids is 2.65.

(*b*) If the soil above the water table is saturated by capillary action, what is the effective stress at the that depth?

[**Ans.** 57.43 kN/m^2; 65.83 kN/m^2]

10.3. A deposit of fine sand has a void ratio of 0.54 and the specific gravity of solid particles is 2.67. Compute the safe exit gradient, with a factor of safety of 4.

[**Ans.** 0.271]

10.4. A deposit of silty clay lies between two layers of sand, as shown in Fig. P 10.4. The lower sand layer is under artesian pressure of 4 m and the water level in the upper sand layer is 2.0 m below the ground surface. Determine the effective stress at the bottom.

(*b*) Also, determine the head above G.S. that would cause heaving at the base of the clay.

[**Ans.** 32 kN/m^2, 7.2 m]

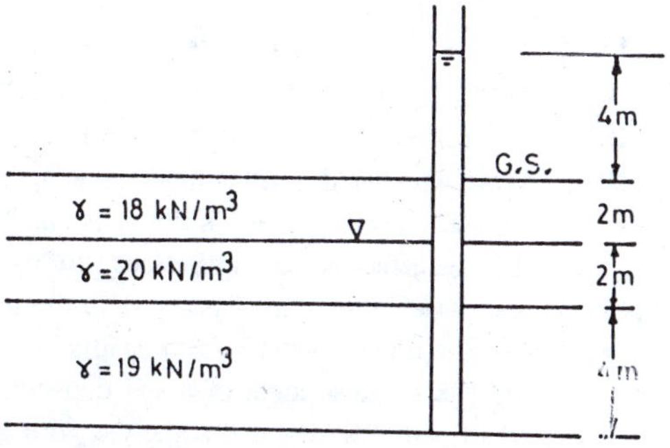

Fig. P 10.4.

10.5. The porosity of a sample of sand in the loose state was 54% and in dense state, 38%. Find out the critical hydraulic gradient in both the states if the specific gravity of the soil grain was 2.60. Also find out the saturated densities in kg/m^3. [**Ans.** 0.736 ; 0.992; 1737.33 kg/m^3; 1992.56 kg/m^3]

10.6. If 40% of the excess hydrostatic pressure is lost in flowing through soil *B* which has a coefficient of permeability of 0.05 cm/sec (Fig. P 10.6), determine the discharge velocity and seepage velocity through each soil and the hydraulic head at which instability occurs.

[**Ans.** Soil *A* : 0.02 cm/sec, 0.06 cm/sec; Soil *B* : 0.02 cm/sec, 0.053 cm/sec; 73.33 cm]

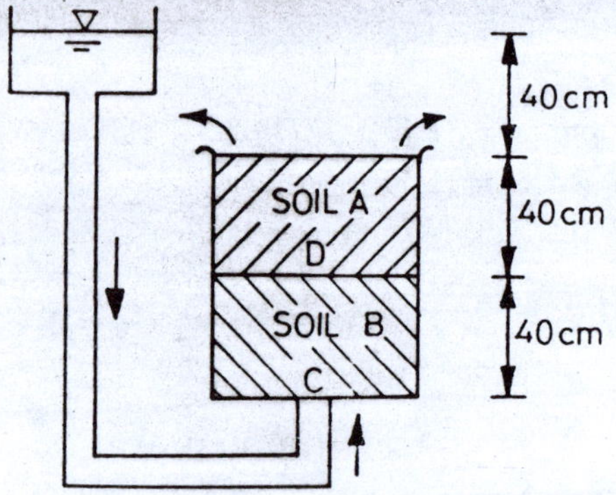

SOIL TYPE	e	G
A	0·50	2·65
B	0·60	2·67

Fig. P 10.6.

B. Descriptive and Objective Type

10.7. Define total stress, neutral stress and effective stress. What is the importance of the effective stress ?

10.8. Prove that the effective stress ($\overline{\sigma}$) for a saturated soil can be expressed as

$$\overline{\sigma} = \sigma - u$$

where σ = total stress, u = pore water pressure.

10.9. What is the effect of surcharge and the capillary action on the effective stress ?

10.10. Prove that the seepage force per unit volume is equal to the product of the hydraulic gradient and the unit weight of water.

10.11. Discuss two different approaches of considering the equilibrium in seepage problems. Which approach is more convenient and why ?

10.12. What is the effect of the seepage pressure on the effective stress ? Give examples.

10.13. What is quick sand ? How would you calculate the hydraulic gradient required to create quick sand conditions in a sample of sand ?

10.14. Explain the mechanics of piping in hydraulic structures. What methods are used to increase the factor of safety against piping ?

10.15. Why a filter is used on the downstream of an earth dam ? How would you design a filter ?

10.16. What are two different types of piping failures ? Explain with the help of sketches.

10.17. What is effective stress principle ?

19.18. Write whether the following statements are correct.

(*a*) The effective stress is the stress at the points of contact of the soil particles.

(*b*) The effective stress stress can be measured directly in the field.

(*c*) The effective stress is equal to the total stress minus the pore water pressure.

(*d*) The rise of water due to capillary action reduces the effective stress.

(*e*) The shear strength of a soil depends upon its effective stress.

(*f*) In partially saturated soils, the pore air pressure is more than the pore water pressure.

(*g*) Quick sand is a type of sand.

(*h*) Quick sand conditions can develop even in gravel.
(*i*) Backward—erosion piping is amenable to the theoretical solution.
(*j*) Heave piping occurs on the downstream of the pile for a distance of half the depth of the pile.
(*k*) The piping can be checked by providing a loaded filter.
(*l*) The graded filter consists of a layer of pervious materials which is well-graded.
(*m*) The diameter of the openings between particles is equal to about D_{15} size of the particle.
(*n*) The graded filter checks the flow of water.
(*o*) Piping is the same as quick sand.
(*p*) The effective stress at various points increases due to rise in the water table.
(*q*) The effective stress in a fully submerged soil depends upon its submerged unit weight.

[Ans. True, (*c*), (*e*), (*f*), (*h*), (*j*), (*k*), (*q*)]

C. Multiple Choice Questions

1. The critical gradient of a soil increases
(*a*) with increase in void ratio (*b*) decrease in void ratio
(*c*) decrease in specific gravity (*d*) None of above

2. The exit gradient is equal to the ratio of
(*a*) total head to total length (*b*) slope of flow line
(*c*) slope of equipotential line (*d*) head loss to length of flow field at exit.

3. The effective stress is
(*a*) actual contact stress (*b*) an abstract quantity
(*c*) equal to total stress (*d*) None of above

4. The effective stress controls the following properties of soils :
(*a*) Shear strength (*b*) Compressibility
(*c*) Permeability (*d*) All the above

5. Quick sand is
(*a*) a type of sand
(*b*) a condition in which a cohesionless soil loses its strength because of upward flow of water.
(*c*) a condition in which a cohesive soil loses its strength.
(*d*) none of above.

6. The seepage pressure is proportional to
(*a*) hydraulic gradient (*b*) unit weight of water
(*c*) length of the specimen (*d*) all the above.

7. A deposit of fine sand has a porosity *n* and specific gravity of solid is G. The critical gradient is equal to
(*a*) (G-1) (1 + n) (*b*) (G-1) (1-n)
(*c*) (G-1)/(1+n) (*d*) (G-1)/(1-n)

8. For a void ratio of 0.60, the relationship between the specific gravity of solids (G) and the hydraulic gradient (*i*) for the quick sand condition is
(*a*) G = 0.6 *i* + 1 (*b*) G = *i* + 0.6
(*c*) G = 1.6 *i* + 1 (*d*) G = 1.6 *i* -1

9. For a soil deposit having *n* = 33% and G = 2.60, the critical gradient is
(*a*) 1.0 (*b*) 1.05
(*c*) 1.07 (*d*) 1.10

[Ans. 1. (*b*), 2. (*d*), 3. (*b*), 4. (*d*), 5. (*b*), 6. (*d*), 7. (*b*), 8. (*c*), 9. (*c*)]

11

Stressees Due to Applied Loads

11.1. INTRODUCTION

Stresses are induced in a soil mass due to weight of overlying soil and due to the applied loads. These stresses are required for the stability analysis of the soil mass, the settlement analysis of foundations and the determination of the earth pressures. The stresses due to self weight of soil have been discussed in chapter 10. These stresses are summarised in Section 11.3. The rest of the chapter is devoted to the determination of stresses due to applied loads.

The stresses induced in soil due to applied loads depend upon its stress-strain characteristic. The stress-strain behaviour of soils is extremely complex and it depends upon a large number of factors, such as drainage conditions, water content, void ratio, rate of loading, the load level, and the stress path. However, simplifying assumptions are generally made in the analysis to obtain stresses. It is generally assumed that the soil mass is homogeneous and isotropic. The stress-strain relationship is assumed to be linear. The theory of elasticity is used to determine the stresses in the soil mass. It involves considerable simplification of real soil behaviour and the stresses computed are approximate ones. Fortunately, the results are good enough for soil problems usually encountered in practice. For more accurate results, realistic stress-strain characteristics should be used. However, the procedure becomes complex and numerical techniques and a high speed computer are required.

11.2. STRESS-STRAIN PARAMETERS

The main stress-strain parameters required for the application of elastic theories are modulus of elasticity (E) and Poisson's ratio (ν). The modulus of elasticity can be determined in the laboratory by conducting a triaxial compression test (see chapter 13). The stress-strain curve is plotted between the deviator stress ($\sigma_1 - \sigma_3$), and the axial strain (ε_1). An unconsolidated-undrained (*UU*) or an unconfined compression test can be performed for saturated, cohesive soils. A consolidated drained (*CD*) is usually conducted for cohesionless soils. The value of modulus is generally taken as the secant modulus at 1/2 to 1/3 of the peak stress (Fig. 11.1). Sometimes, instead of the secant modulus, the initial tangent modulus or the tangent modulus at 1/2 to 1/3 of the peak stress is also used.

The value of Poisson's ratio (ν) for an elastic material varies from 0.0 to 0.50. For undrained conditions, the value of Poisson's ratio is 0.50. For drained conditions, the Poisson's ratio is less than 0.50. As the soil is not a purely elastic material, the value of ν outside the elastic range of 0.0 to 0.50 is also occasionally encountered. It is difficult to ascertain the exact value of Poisson's ratio. Fortunately, the effect of

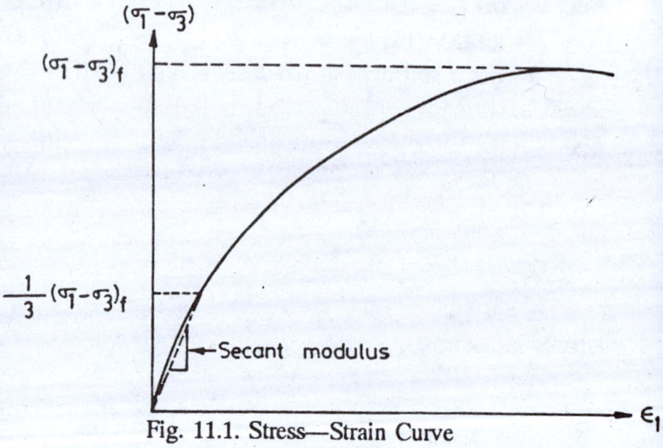

Fig. 11.1. Stress—Strain Curve

Poisson's ratio on the computed stresses is not significant and an approximate value can be used without much error.

Tables 11.1 and 11.2 give typical range of values of modulus of elasticity and Poisson's ratio, respectively, for some soils.

Table 11.1. Typical Values of E

S. No.	*Type of Soil*	E	
		MN/m^2	kN/m^2
1.	Soft Clay	1.5—4.0	1500—4000
2.	Hard clay	6.0—15.0	6000—15000
3.	Silty Sand	6.0—20.0	6000—20000
4.	Loose Sand	10.0—25.0	10000—25000
5.	Dense Sand	40.0—80.0	40000—80000
6.	Dense gravel	100—200	1×10^5 to 2×10^5

Table 11.2. Typical Values of Poisson's Ratio (ν)

S. No.	*Type of Soil*	ν
1.	Saturated clay	0.4—0.5
2.	Unsaturated clay	0.1—0.3
3.	Silt	0.3—0.35
4.	Loose sand	0.30—0.50
5.	Dense sand	0.20—0.30

11.3. GEOSTATIC STRESSES

The method for the determination of total vertical stresses due to self weight of the soil have been discussed in chapter 10. The stresses due to self weight of soils are generally large in comparison with those induced due to imposed loads. This is unlike many other civil engineering structures, such as steel bridges, wherein the stresses due to self weight are relatively small. In soil engineering problems, the stresses due to self weight are significant. In many cases the stresses due to self weight are a large proportion of the total stresses and may govern the design.

When the ground surface is horizontal and the properties of the soil do not change along a horizontal plane, the stresses due to self weight are known as *geostatic stresses.* Such a condition generally exists in sedimentary soil deposits. In such a case, the stresses are normal to the horizontal and vertical planes, and there are no shearing stresses on these planes. In other words, these planes are principal planes. The vertical and horizontal stresses can be determined as under.

***(a)* Vertical stresses.** The vertical stresses are determined using the methods described in chapter 10. Let us consider the horizontal plane A—A at a depth z below the ground surface [Fig. 11.2 *(a)*]. Let the area of cross-section of the prism be A. If the unit weight of soil (γ) is constant, the vertical stress (σ_z) is equal to the weight of soil in prism divided by the area of base. Thus

$$\sigma_z = \frac{\text{Weight of soil in prism}}{\text{Area of base}} = \frac{\gamma\,(z \times A)}{A}$$

or

$$\sigma_z = \gamma z \qquad \text{...(11.1)}$$

If the soil is stratified, having n layers of thickness $z_1, z_2, \ldots z_n$, with unit weight $\gamma_1, \gamma_2, \ldots \gamma_n$, the vertical stress is given by

$$\sigma_z = \sum_{i=1}^{n} \gamma_i z_i \qquad \text{...(11.2)}$$

In natural deposits, generally the density of the soil increases with an increase in depth due to the weight of soil above. In such a case, the unit weight of soil cannot be taken as constant. In this case, the weight of soil in the prism is given by

$$W = \int_0^z \gamma A \; dz$$

where dz is the thickness of a small strip of soil at depth z.

Therefore, the vertical stress is given by

$$\sigma_z = \frac{W}{A} = \frac{\int_0^z \gamma A \, dz}{A}$$

or
$$\sigma_z = \int_0^z \gamma \, dz \quad \ldots(11.3)$$

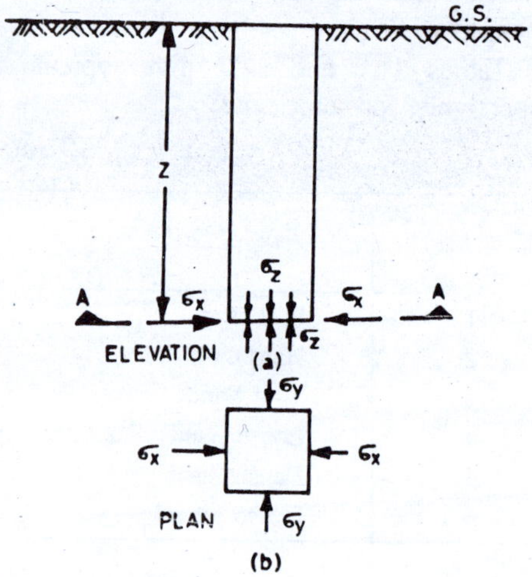

Fig. 11.2. Geostatic Stress.

If the soil is stratified and also has a variable unit weight, the vertical stress is given by

$$\sigma_z = \int_0^{z_1} \gamma_1 \; dz + \int_0^{z_2} \gamma_2 \; dz + \ldots + \int_0^{z_n} \gamma_n \, dz \quad \ldots(11.4)$$

(*b*) **Horizontal stresses.** The horizontal stresses (σ_x and σ_y) act on vertical planes, as shown in Fig. 11.2 (*b*). The horizontal stresses at a point in a soil mass are highly variable. These depend not only upon the vertical stresses, but also on the type of the soil and on the conditions whether the soil is stretched or compressed laterally. In the treatment that follows it would be assumed that $\sigma_x = \sigma_y$.

The ratio of the horizontal stress (σ_x) to the vertical stress (σ_z) is known as the coefficient of lateral stress or lateral stress ratio (K). Thus
$$K = \frac{\sigma_x}{\sigma_z}$$

or
$$\sigma_x = K \sigma_z \quad \ldots(11.5)$$

In natural deposits, generally there is no lateral strain. The lateral stress coefficient for this case is known as the *coefficient of lateral pressure at rest* (K_0). The value of its coefficient can be obtained from the theory of elasticity, as explained below. In retaining structures (chapter 19), there is either stretching or contraction of soils and the value of K is different.

The strain in x-direction is given by (see any text on theory of elasticity or mechanics of materials)

$$\varepsilon_x = \frac{1}{E} \; [\sigma_x - \nu \, (\sigma_y + \sigma_z)]$$

For conditions of no lateral strain, $\varepsilon_x = 0$.

Thus
$$\sigma_x = \nu \, (\sigma_y + \sigma_z)$$

Taking $\sigma_x = \sigma_y$ and simplifying,

or
$$\sigma_x (1 - \nu) = \nu \; \sigma_z$$

or
$$\sigma_x = \left(\frac{\nu}{1 - \nu}\right) \sigma_z$$

or
$$\sigma_x = K_0 \, \sigma_z \quad \ldots(11.6)$$

where
$$K_0 = \frac{\nu}{1 - \nu} \quad \ldots(11.7)$$

The value of K_0 can be obtained if the Poisson's ratio ν is known or estimated. Eq. 11.7 is not of much practical use as the soil is not a purely elastic material and it is difficult to estimate the Poisson ratio.

The value of K_0 is determined from actual measurement of soil pressure or from experience. For a sedimentary sand deposit, its value varies from 0.30 to 0.6, and for a normally consolidated clay, its value generally lies between 0.5 and 1.10. Table 11.3 gives the average values of K_0 for different types of soils.

Jaky's formula is commonly used, according to which

$$K_0 = 1 - \sin \varphi'$$

where ϕ' is the angle of shearing resistance.

Table 11.3. Values of Lateral Pressure Coefficient at Rest (K_0)

S. No.	*Type of Soil*	K_0
1.	*Loose sand*	0.5—0.60
2.	*Dense sand*	0.3—0.50
3.	*Clay (drained)*	0.5—0.60
4.	*Clay (undrained)*	0.80—1.1
5.	*Over-consolidated clay*	1.0—3.0

11.4. VERTICAL STRESSES DUE TO A CONCENTRATED LOAD

Boussinesq (pronounced as Boo-si-nesk) gave the theoretical solutions for the stress distribution in an elastic medium subjected to a concentrated load on its surface. The solutions are commonly used to obtain the stresses in a soil mass due to externally applied loads. The following assumptions are made.

(1) The soil mass is an elastic continuum, having a constant value of modulus of elasticity (E), *i.e.*, the ratio between the stress and strain is constant.

(2) The soil is homogeneous, *i.e.*, it has identical properties at different points.

(3) The soil is isotropic, *i.e.*, it has identical properties in all directions.

(4) The soil mass is semi-infinite, *i.e.*, it extends to infinity in the downward direction and lateral directions. In other words, it is limited on its top by a horizontal plane and extends to infinity in all other directions.

(5) The soil is weightless and is free from residual stresses before the application of the load.

[**Note.** The stresses due to self weight are computed separately as explained in the preceding section].

Fig. 11.3 shows a horizontal surface of the elastic continuum subjected to a point load Q at point O. The origin of the coordinates is taken at O. Using logarithmic stress function for the solution of elasticity problem, Boussinesq proved that the polar stress σ_R at point P (x, y, z) is given by

$$\sigma_R = \frac{3}{2\pi} \frac{Q \cos \beta}{R^2} \quad ...(11.8)$$

where R = polar distance between the origin O and point P.

β = angle which the line OP makes with the vertical.

Obviously, $R = \sqrt{x^2 + y^2 + z^2}$

or $R = \sqrt{r^2 + z^2}$ where $r^2 = x^2 + y^2$

and $\sin \beta = r/R$ and $\cos \beta = z/R$

The vertical stress (σ_z) at point P is given by

$$\sigma_z = \sigma_R \cos^2 \beta$$

$$= \frac{3}{2\pi}\left(\frac{Q \cos \beta}{R^2}\right)\cos^2 \beta$$

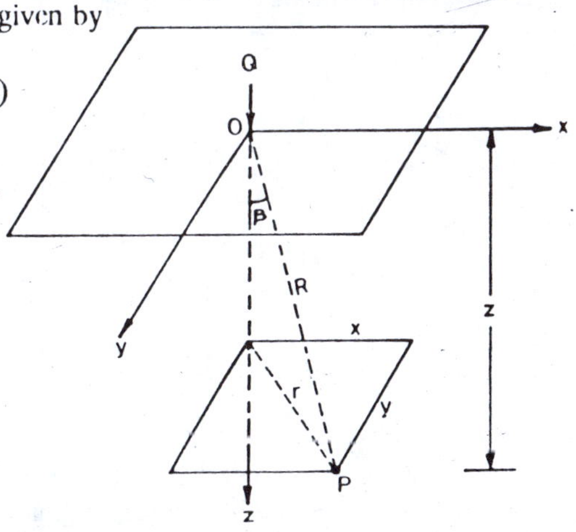

Fig. 11.3. Stresses due to a concentrated load

or $$\sigma_z = \frac{3\,Q}{2\,\pi} \cdot \frac{\cos^3 \beta}{R^2}$$

or $$\sigma_z = \frac{3Q}{2\pi} \cdot \frac{(z/R)^3}{R^2} = \frac{3Q}{2\pi} \cdot \frac{z^3}{R^5}$$

or $$\sigma_z = \frac{3Q}{2\pi} \cdot \frac{1}{z^2} \cdot \frac{z^5}{R^5}$$

or $$\sigma_z = \frac{3Q}{2\pi} \cdot \frac{1}{z^2} \cdot \left[\frac{z^5}{(r^2 + z^2)^{5/2}}\right]$$

or $$\sigma_z = \frac{3Q}{2\pi} \cdot \frac{1}{z^2} \cdot \frac{1}{[1 + (r/z)^2]^{5/2}} \quad \text{...(11.9)}$$

or $$\sigma_Z = I_B \cdot \frac{Q}{z^2} \quad \text{...(11.10)}$$

where $$I_B = \frac{3}{2\,\pi\,[1 + (r/z)^2\,]^{5/2}} \quad \text{...(11.11)}$$

The coefficient I_B is known as the Boussinesq influence coefficient for the vertical stress. The value of I_B can be determined for the given value of r/z from Eq. 11.11. The computed values are tabulated in Table 11.4.

Table 11.4. Values of Boussinesq's Coefficient (I_B)

r/z	I_B	r/z	I_B	r/z	I_B	r/z	I_B
0.00	0.4775	1.05	0.0745	2.05	0.0077	3.25	0.0011
0.05	0.4745	1.10	0.0658	2.10	0.0070	3.50	0.0008
0.10	0.4657	1.15	0.0581	2.15	0.0064	3.75	0.0005
0.15	0.4516	1.20	0.0513	2.20	0.0058	4.00	0.0004
0.20	0.4329	1.25	0.0454	2.25	0.0053	4.25	0.0003
0.25	0.4103	1.30	0.0402	2.30	0.0048	4.50	0.0002
0.30	0.3849	1.35	0.0357	2.35	0.0044	4.75	0.0002
0.35	0.3577	1.40	0.0317	2.40	0.0040	5.00	0.0001
0.40	0.3295	1.45	0.0282	2.45	0.0037	10.00	0.0000
0.45	0.3011	1.50	0.0251	2.50	0.0034		
0.50	0.2733	1.55	0.0224	2.55	0.0031		
0.55	0.2466	1.60	0.0200	2.60	0.0029		
0.60	0.2214	1.65	0.0179	2.65	0.0026		
0.65	0.1978	1.70	0.0160	2.70	0.0024		
0.70	0.1762	1.75	0.0144	2.75	0.0022		
0.75	0.1565	1.80	0.0129	2.80	0.0021		
0.80	0.1386	1.85	0.0116	2.85	0.0019		
0.85	0.1226	1.90	0.0105	2.90	0.0018		
0.90	0.1083	1.95	0.0095	2.95	0.0016		
0.95	0.0956	2.00	0.0085	3.00	0.0015		
1.00	0.0844						

The following points are worth noting when using Eq. 11.10.

(1) The vertical stress does not depend upon the modulus of elasticity (E) and the Poisson ratio (ν). But the solution has been derived assuming that the soil is linearly elastic. The stress distribution will be the same in all linearly elastic materials.

(2) The intensity of vertical stress just below the load point is given by

$$\sigma_Z = 0.4775 \frac{Q}{z^2} \qquad \ldots(11.12)$$

(3) At the surface ($z = 0$), the vertical stress just below the load is theoretically infinite. However, in an actual case, the soil under the load yields due to very high stresses. The load point spreads over a small but finite area and, therefore, only finite stresses develop.

(4) The vertical stress (σ_z) decreases rapidly with an increase in r/z ratio. Theoretically, the vertical stress would be zero only at an infinite distance from the load point. Actually, at $r/z = 5.0$ or more, the vertical stress becomes extremely small and is neglected.

(5) In actual practice, foundation loads are not applied directly on the ground surface. However, it has been established that the Boussinesq solution can be applied conservatively to field problems concerning loads at shallow depths, provided the distance z is measured from the point of application of the load.

(6) Boussinesq's solution can even by used for negative (upward) loads. For example, if the vertical stress decrease due to an excavation is required, the negative load is equal to the weight of the soil removed. However, as the soil is not fully elastic, the stresses determined are necessarily approximate.

(7) The field measurements indicate that the actual stresses are generally smaller than the theoretical values given by Boussinesq's solution, especially at shallow depths. Thus, the Boussinesq solution gives conservative values and is commonly used in soil engineering problems.

Limitations of Boussinesq's Solution. The solution was initially obtained for determination of stresses in elastic solids. Its application to soils may be questioned, as the soils are far from purely elastic solids. However. experience indicates that the results obtained are satisfactory.

The application of Boussinesq's solution can be justified when the stress changes are such that only a stress increase occurs in the soil. The real requirement for use of the solution is not that the soil be elastic (*i.e.*, fully recoverable), but it should have a constant ratio between stress and strain. When the stress decrease occurs, the relation between stress and strain is not linear and, therefore, the solution is not strictly applicable. If the stresses induced in the soil are small in comparison with the shear strength of the soil, the soil behaves somewhat elastically and the Boussinesq solution can be used.

For practical cases, the Boussinesq solution can be safely used for homogeneous deposits of clay, man-made fills and for limited thickness of uniform sand deposits. In deep sand deposits, the modulus of elasticity increases with an increase in depth and, therefore, the Boussinesq solution will not give satisfactory results. In this case, the assumption of proportionality between stress and strain cannot be justified. For such a case, non-linear elastic solutions or elastic-plastic solutions are required.

The point loads applied below ground surface cause somewhat smaller stresses than are caused by surface loads, and, therefore, the Boussinesq solution is not strictly applicable. However, the solution is frequently used for shallow footings, in which z is measured below the base of the footing.

11.5. HORIZONTAL AND SHEAR STRESSES DUE TO A CONCENTRATED LOAD

The method for determination of the vertical stress (σ_z) has been discussed in the preceding section. In most soil engineering problems, only the vertical stresses are required. Occasionally, other stress components (σ_x, σ_y ,τ_{xy}, and τ_{yz}) are also required. These components can be determined as follows :

Fig. 11.4 shows an elementary stress block, indicating all the stress components. In all there are 9 stress components. namely, σ_x, σ_y, σ_z, τ_{xy}, τ_{xz}, τ_{yx}, τ_{yz}, τ_{zx} and τ_{zy}. However, the moment equation gives the following relations.

$$\tau_{xy} = \tau_{yx}\,; \quad \tau_{yz} = \tau_{zy}\,; \quad \tau_{zx} = \tau_{xz}.$$

and, therefore independent unknown components are only six σ_x, σ_y, σ_z, τ_{xy}, τ_{yz}, τ_{zx}. The equations for determination, of σ_z have already been given. The corresponding equations for other components are :

$$\sigma_x = \frac{3Q}{2\pi}\left[\frac{x^2 z}{R^5} + \frac{(1-2\nu)}{3}\left\{\frac{1}{R(R+z)} - \frac{(2R+z)x^2}{R^3(R+z)^2} - \frac{z}{R^3}\right\}\right]$$

$$\sigma_y = \frac{3Q}{2\pi}\left[\frac{y^2 z}{R^5} + \frac{(1-2\nu)}{2}\left\{\frac{1}{R(R+z)} - \frac{(2R+z)\,y^2}{R^3(R+z)^2} - \frac{z}{R^3}\right\}\right]$$

$$\tau_{xy} = \frac{3Q}{2\pi}\left[\frac{xyz}{R^5} - \frac{(1-2\nu)}{3}\,\frac{(2R+z)xy}{R^3(R+z)^2}\right]$$

$$\tau_{yz} = \frac{3Q}{2\pi}\cdot\frac{yz^2}{R^5} \qquad \ldots(11.13)$$

$$\tau_{zx} = \frac{3Q}{2\pi}\cdot\frac{xz^2}{R^5}$$

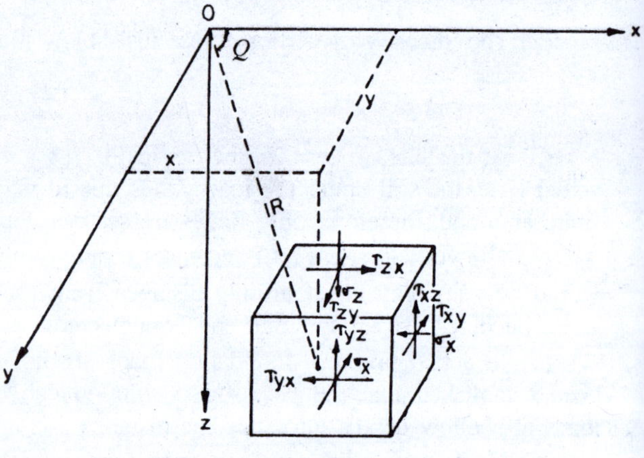

Fig. 11.4. Different Stress Components.

It may be noted that σ_x, σ_y and τ_{xy} depend upon Poisson's ratio.

Cylindrical Coordinates. Sometimes, it is more convenient to use cylindrical coordinates (r, θ, z) instead of cartesian coordinates (x, y, z). The Boussinesq solution in terms of cylindrical coordinates is as under (Fig. 11.5).

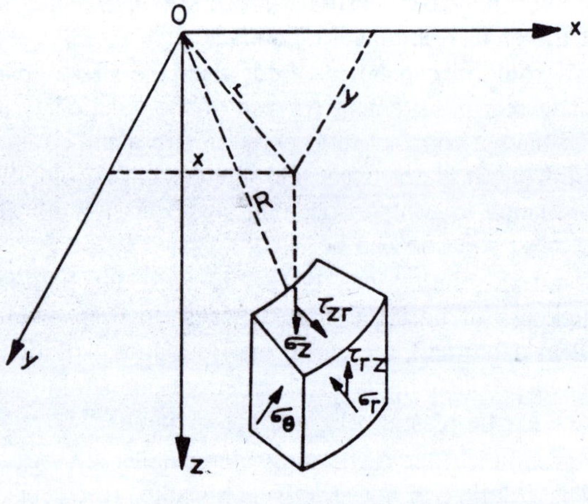

Fig. 11.5. Cylindrical Coordinates.

Vertical stress, $\sigma_z = \dfrac{3Q}{2\pi}\cdot\dfrac{z^3}{R^5}$

Radial stress, $\sigma_r = \dfrac{Q}{2\pi}\left[\dfrac{3zr^2}{R^5} - \dfrac{(1-2\nu)}{R(R+z)}\right]$ ...(11.14)

Tangential stress, $\sigma_\theta = \dfrac{Q}{2\pi}(1-2\nu)\left[\dfrac{1}{R(R+z)} - \dfrac{z}{R^3}\right]$

Shear stress, $\tau_{rz} = \dfrac{3Q}{2\pi}\cdot\dfrac{rz^2}{R^5}$

Shear stresses $\tau_{r\theta} = \tau_{z\theta} = 0$

where $R = \sqrt{r^2 + z^2}$, as before.

11.6. ISOBAR DIAGRAM

An isobar is a curve joining the points of equal stress intensity. In other words, an isobar is a contour of equal stress. An isobar is a spatial curved surface of the shape of an electrical bulb or an onion. The curved surface is symmetrical about the vertical axis passing through the load point.

The isobar of a particular intensity can be obtained using Eq. 11.10. The calculations are shown in a tabular form. Table 11.5 gives calculations for an isobar of intensity 0.1 Q per unit area.

From Eq. 11.10,
$$\sigma_z = I_B \frac{Q}{z^2}$$

Taking $\sigma_z = 0.1\,Q$,
$$0.1\,Q = I_B \cdot \frac{Q}{z^2}$$

or
$$I_B = 0.1\,z^2 \quad \ldots(a)$$

For different depths z, the value of I_B is computed from Eq. (a), as shown in the second row of Table 11.5. The values of r/z for computed values of I_B are obtained from Eq. 11.11 or Table 11.4. Once the values of r/z have been determined, the radial distance r can be obtained as shown in table. It may be observed that r is zero at the load point, and it attains a maximum value at r/z = 0.75 and again decreases. As the isobar is symmetrical about the load axis, the other half can be drawn from symmetry. The shape of an isobar approaches a leminscate curve (not circle). Fig. 11.6 shows the pressure bulb of intensity 0.1 Q.

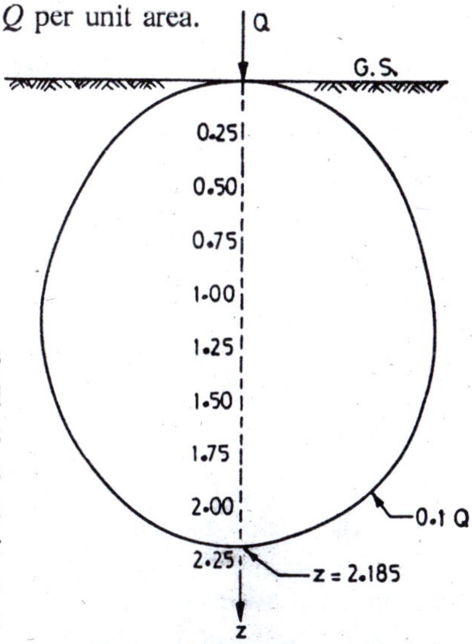

Fig. 11.6. Isobar of 0.1 Q

In the same manner, the isobars of other intensity 0.2 Q, 0.3 Q, etc. can be drawn. Obviously, the isobars of higher intensity shall lie within the isobar of 0.1 Q.

Table 11.5. Calculations for Isobar of 0.1 Q

Depth z	0.25	0.50	0.75	1.0	1.25	1.50	1.75	2.00	2.185
I_B	0.00625	0.25	0.05625	0.10	0.1562	0.225	0.3062	0.400	0.4775
r/z	2.16	1.50	1.16	0.93	0.75	0.59	0.44	0.27	0.000
r	0.54	0.75	0.87	0.93	0.938	0.885	0.770	0.540	0.000

Isobars are useful for determining the effect of the load on the vertical stresses at various points. The zone within which the stresses have a significant effect on the settlement of structures is known as the *pressure bulb*. It is generally assumed that an isobar of 0.1 Q forms a pressure bulb. The area outside the pressure bulb is assumed to have negligible stresses.

11.7. VERTICAL STRESS DISTRIBUTION ON A HORIZONTAL PLANE

The vertical stresses at various points on a horizontal plane at a particular depth z can be obtained using Eq. 11.10. Let us determine the stresses at a depth of 2 m. Therefore,

$$\sigma_z = I_B \cdot (Q/z^2) = I_B \cdot (Q/4) = 0.25\, I_B\, Q.$$

The value of σ_z are computed (see Table 11.6) for different values of r/z, after obtaining I_B from Eq. 11.11 or Table 11.4.

Table 11.6. Calculation for vertical stress at z = 2m

r	0	0.50	1.00	1.50	2.00	2.50	3.00	4.00
r/z	0	0.25	0.50	0.75	1.00	1.25	1.50	2.00
I_B	0.4775	0.4103	0.2733	0.1565	0.0844	0.0454	0.0251	0.0085
σ_z	0.1194 Q	0.1026 Q	0.0683 Q	0.0390 Q	0.0211 Q	0.0113 Q	0.0063 Q	0.0021 Q

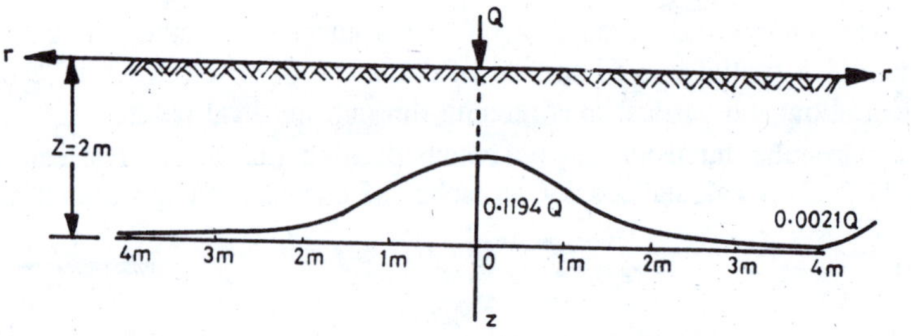

Fig. 11.7. Vertical Stress on a horizontal plane.

Fig. 11.7 shows the vertical stress distribution diagram. The diagram is symmetrical about the vertical axis. The maximum stress occurs just below the load (r = 0), and it decreases rapidly as the distance r increases.

The vertical stress distribution diagram on a horizontal plane can also be obtained graphically if the isobars of different intensity are available. The horizontal plane is drawn on the isobars diagram. The points of intersection of the horizontal plane with the isobar of a particular intensity give that vertical stress.

11.8. INFLUENCE DIAGRAMS

An influence diagram is the vertical stress distribution diagram on a horizontal plane at a given depth, due to a unit concentrated load. In Fig. 11.7, if the concentrated load Q is taken as unity, the diagram becomes an influence diagram. The influence diagrams are useful for determination of the vertical stress at any point on that horizontal plane due to a number of concentrated loads applied at the ground surface.

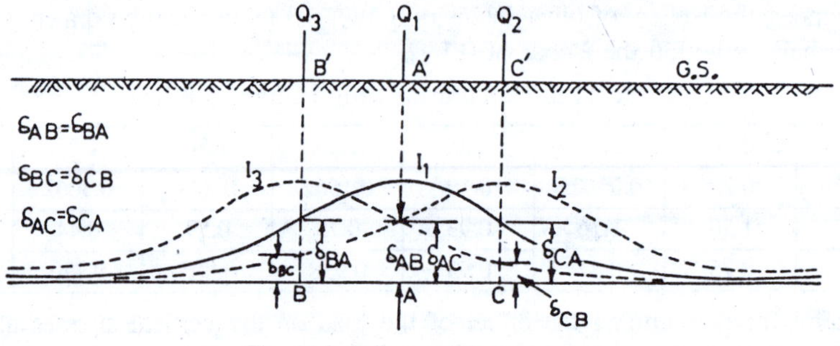

Fig. 11.8. Infiuence Diagrams.

Fig. 11.8 shows three influence diagrams, marked I_1, I_2 and I_3, due to unit loads applied at three points A', C' and B' on the ground surface. The stress at any point A on the horizontal plane at depth z due to three loads Q_1, Q_2 and Q_3 is given by

$$(\sigma_z)_A = Q_1 \sigma_{AA} + Q_2 \sigma_{AB} + Q_3 \sigma_{AC} \quad \text{...(11.15)}$$

where σ_{AA} = vertical stress at A due to unit load at A'

σ_{AB} = vertical stress at A due to unit load at B'

and σ_{AC} = vertical stress at A due to unit load at C'

The values of σ_{AA}, σ_{AB} and σ_{AC} can be abtained from the influence diagram I_1, I_3 and I_2.

The computation work is considerably simplified using the reciprocal theorem, according to which

$$\sigma_{AB} = \sigma_{BA}, \quad \sigma_{AC} = \sigma_{CA} \text{ and } \sigma_{BC} = \sigma_{CB}$$

where the first suffix denotes the point where the stress is required and the second suffix gives the point above which the load is applied. Accordingly, Eq. 11.15 can be written as

$$(\sigma_z)_A = Q_1 \sigma_{AA} + Q_2 \sigma_{BA} + Q_3 \sigma_{CA} \quad \text{...(11.16)}$$

where σ_{AA} = vertical stress at A due to unit load at A',
σ_{BA} = vertical stress at B due to unit load at A',
and σ_{CA} = vertical stress at C due to unit load at A'.

Therefore, there is no need of drawing three influence diagrams in this case. Only one influence diagram (I_1) with unit load at A' is sufficient. The values of σ_{BA} and σ_{CA} are determined from I_1 diagram below the load points B' and C'.

If the stresses at any other point, say point B, are required, then the influence line for load above that point (B' in this case) would be drawn. Alternatively, the influence line diagram I_1 can be traced on a paper and placed in such a way that its axis of symmetry passes through the point B'.

11.9. VERTICAL STRESS DISTRIBUTION ON A VERTICAL PLANE

The vertical stress distribution on a vertical plane at a radial distance of r can be obtained using Eq. 11.10. In this case, the radial distance r is constant and the depth changes. The values of r/z are obtained for different depths z. The values of I_B are obtained from Eq. 11.11 or Table 11.4 and the stresses computed as $\sigma_z = (I_B/z^2)\,Q$. Table 11.7 shows the calculations for vertical stresses on a vertical plane at r = 1 m.

Table 11.7. Calculations of Vertical stresses at r = 1m

z	0.25	0.50	1.00	1.50	2.00	2.50	5.00
r/z	4.0	2.0	1.00	0.667	0.50	0.40	0.20
I_B	0.0004	0.0085	0.0844	0.1904	0.2733	0.3294	0.4329
σ_z	0.0064 Q	0.0340 Q	0.0844 Q	0.0845 Q	0.0683 Q	0.0527 Q	0.017 Q

Fig. 11.9 shows the variation of vertical stresses on a vertical plane at r = 1 m. The vertical stresses are plotted horizontally along r-axis, and the depth, parallel to the z-axis. It may be noted that the vertical stress first increases and then decreases. The maximum vertical stress occurs at r/z = 0.817. This corresponds to the point of intersection of the vertical plane with the line drawn at 39°15′ to the vertical axis of the load.

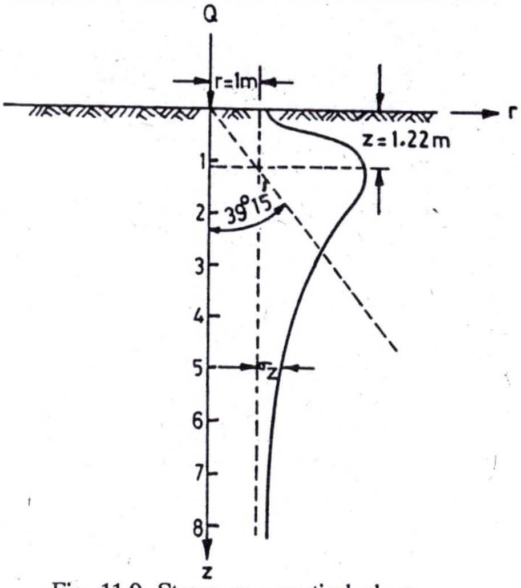

Fig. 11.9. Stress on a vertical plane

11.10. VERTICAL STRESSES DUE TO A LINE LOAD

The vertical stresses in a soil mass due to a vertical line load can be obtained using Boussinesq's solution. Let the vertical line load be of intensity q' per unit length, along the y-axis, acting on the surface of a semi-infinite soil mass, as shown in Fig. 11.10.

Let us consider the load acting on a small length δy. The load can be taken as a point load of $q'\,\delta y$ and Boussinesq's solution can be applied to determine the vertical stress at $P\,(x, y, z)$. From Eq. 11.9,

$$\Delta\sigma_z = \frac{3(q'\,\delta y)}{2\pi} \cdot \frac{z^3}{(r^2 + z^2)^{5/2}} \qquad \ldots(a)$$

The vertical stress at P due to the line load extending from $-\infty$ to $+\infty$ is obtained by integration,

$$\sigma_z = \frac{3q'\,z^3}{2\pi} \int_{-\infty}^{+\infty} \frac{dy}{(r^2 + z^2)^{5/2}}$$

or

$$\sigma_z = \frac{3\,q'\,z^3}{2\,\pi} \int_{-\infty}^{+\infty} \frac{dy}{(x^2 + y^2 + z^2)^{5/2}} \qquad \ldots(b)$$

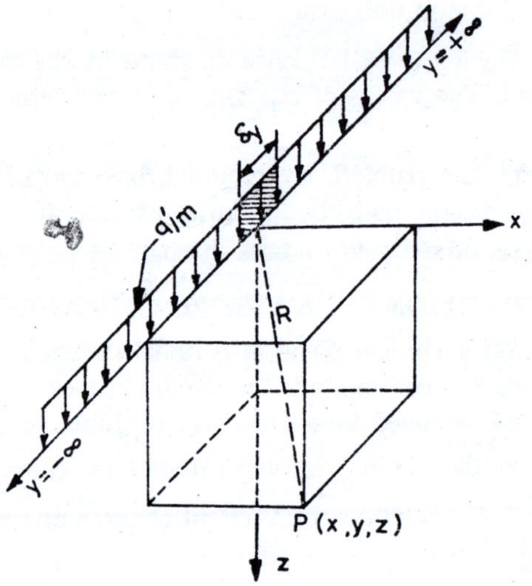

Fig. 11.10

Substituting $x^2 + z^2 = u^2$ in Eq. (*b*),

$$\sigma_z = \frac{3\,q'\,z^3}{2\,\pi} \int_{-\infty}^{+\infty} \frac{dy}{(u^2 + y^2)^{5/2}} \qquad \text{...(c)}$$

Let $y = u \tan\theta$. Therefore, $dy = u \sec^2\theta \, d\theta$.

Eq. (*c*) can be written as

$$\sigma_z = \frac{3\,q'\,z^3}{2\,\pi} \int_0^{\pi/2} \frac{u \sec^2\theta}{u^5 \sec^5\theta} \, d\theta$$

or

$$\sigma_z = \frac{3\,q'\,z^3}{2\,\pi u^4} \int_0^{\pi/2} \cos^3\theta \, d\theta \qquad \text{...(d)}$$

Let $\sin\theta = t$. Therefore, $\cos\theta \, d\theta = dt$

Eq. (*d*) becomes

$$\sigma_z = \frac{3\,q'\,z^3}{\pi u^4} \int_0^1 (1 - t^2)\, dt$$

or

$$\sigma_z = \frac{3\,q'\,z^3}{\pi u^4} \left[t - \frac{1}{3} t^3 \right]_0^1$$

or

$$\sigma_z = \frac{3\,q'\,z^3}{\pi u^4} \times \frac{2}{3} = \frac{2\,q'\,z^3}{\pi\,(x^2 + z^2)^2}$$

or

$$\sigma_z = \frac{2\,q'}{\pi z} \left[\frac{1}{1 + (x/z)^2} \right]^2 \qquad \text{...(11.17)}$$

Eq. 11.17 can be used to determine the vertical stress at point P.

When the point P lies vertically below the line load, $x = 0$.

Therefore,

$$\sigma_z = \frac{2\,q'}{\pi z} \qquad \text{...(11.18)}$$

The expressions for the stresses σ_x and τ_{xz} can be obtained in a similar manner, starting from Eq. 11.13.

$$\sigma_x = \frac{2q'}{\pi} \frac{x^2 z}{(x^2 + z^2)^2} \quad ...(11.19)$$

and

$$\tau_{xz} = \frac{2q'}{\pi} \cdot \frac{xz^2}{(x^2 + z^2)^2} \quad ...(11.20)$$

11.11. VERTICAL STRESS UNDER A STRIP LOAD

The expression for vertical stress at any point P under a strip load can be developed from Eq. 11.17 of the line load. The expression will depend upon whether the pint P lies below the centre of the strip load or not.

Note. The length of the strip is very long. For convenience, unit length is considered.

(1) Point P below the centre of the strip

Fig. 11.11 shows a strip load of width B (= $2b$) and intensity q. Let us consider the load acting on a small elementary width dx at a distance x from the centre of the load. This small load of $q\,dx$ can be considered as a line load of intensity q'. From Eq. 11.17,

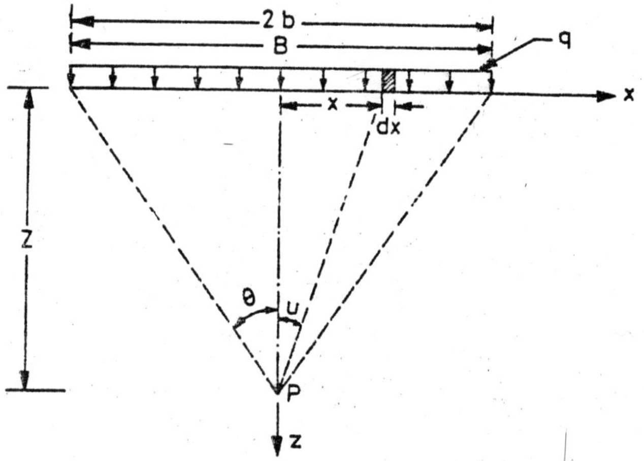

Fig. 11.11. Strip Load, point P below centre.

$$\Delta \sigma_z = \frac{2q\,dx}{\pi z} \left[\frac{1}{1 + (x/z)^2} \right]^2$$

The stress due to entire strip load is obtained as

$$\sigma_z = \frac{2q}{\pi z} \int_{-b}^{+b} \frac{1}{[1 + (x/z)^2]^2} dx \quad ...(a)$$

Let $x/z = \tan u$. Therefore, $dx = z \sec^2 u\, du$

Substituting in Eq. (a), $\sigma_z = \dfrac{2q}{\pi z} \times 2 \displaystyle\int_0^{\theta} \frac{z \sec^2 u}{(1 + \tan^2 u)^2} du$

where $\theta = \tan^{-1}(b/z)$ = angle made by extemities of the strip at P.

or

$$\sigma_z = \frac{4q}{\pi} \int_0^{\theta} \cos^2 u\, du$$

or

$$\sigma_z = \frac{4q}{\pi} \int_0^{\theta} \left(\frac{1 + \cos 2u}{2} \right) du$$

or $$\sigma_z = \frac{q}{\pi} (2\theta + \sin 2\theta) \quad ...(11.21)$$

(2) Point *P* not below the centre of the strip

Fig. 11.12 shows the case when the point *P* is not below the centre of the strip. The extremities of the strip make angles of β_1 and β_2 at *P*. As in the previous case, the load $q\ dx$ acting on a small length dx can be considered as a line load. The vertical stress at *P* is given by Eq. 11.17 as

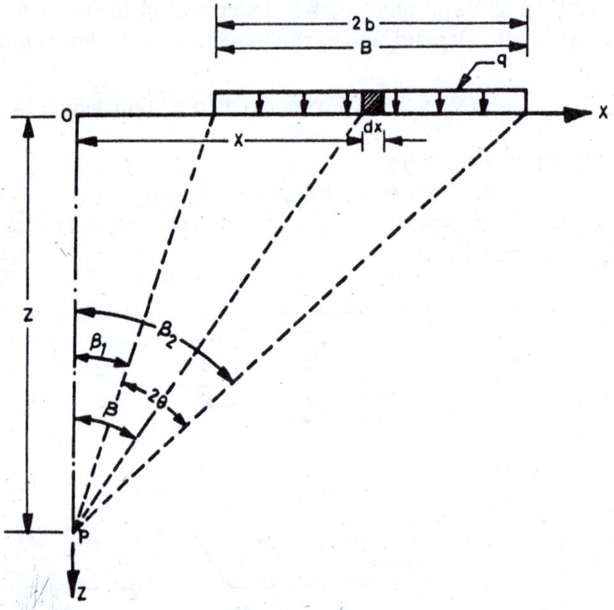

Fig. 11.12. Strip load, point *P* not at the centre.

$$\Delta\sigma_z = \frac{2q\,dx}{\pi z} \left[\frac{1}{1 + (x/z)^2} \right]^2 \quad ...(a)$$

Eq. (*a*) is simplified making the following substitution,

$$x = z \tan \beta \qquad \text{or} \qquad dx = z \sec^2 \beta\, d\beta$$

Therefore $$\Delta \sigma_z = \frac{2q\,(z \sec^2 \beta)\, d\beta}{\pi z} \left[\frac{1}{1 + \tan^2 \beta} \right]^2$$

or $$\Delta\sigma_z = \frac{2q}{\pi} \cos^2 \beta\, d\beta$$

Integrating, $$\sigma_z = \frac{q}{\pi} \int_{\beta 1}^{\beta 2} (1 + \cos 2\beta)\, d\beta$$

$$= \frac{q}{\pi} \left[\beta + \frac{1}{2} \sin 2\beta \right]_{\beta 1}^{\beta 2}$$

or $$\sigma_z = \frac{q}{\pi} \left[(\beta_2 - \beta_1) + (\sin \beta_2 \cos \beta_2 - \sin \beta_1 \cos \beta_1) \right]$$

Substituting $\beta_2 - \beta_1 = 2\,\theta$,

$$\sigma_z = \frac{q}{\pi} \left[2\,\theta + (\sin \beta_2 \cos \beta_2 - \sin\beta_1 \cos \beta_1) \right] \quad ...(b)$$

If $(\beta_1 + \beta_2) = 2\varphi$, it can be shown that

$$\sin \beta_2 \cos \beta_2 - \sin \beta_1 \cos \beta_1 = \sin 2\theta \cos 2\varphi$$

Therefore, Eq. (*b*) becomes

$$\sigma_z = \frac{q}{\pi} [2\theta + \sin 2\theta \cos 2\varphi] \quad \text{...(11.22)}$$

The expressions for σ_x and τ_{xz} can be likewise derived.

$$\sigma_x = \frac{q}{\pi} [2\theta - \sin 2\theta \cos 2\varphi] \quad \text{...(11.23)}$$

and

$$\tau_{xz} = \frac{q}{\pi} [\sin 2\theta \sin 2\varphi] \quad \text{...(11.24)}$$

It may be mentioned that Eqs. 11.22 to 11.24 are general equations which can be used even for the case when the point *P* is below the centre of the load.

In this case, $\beta_2 = -\beta_1 = \theta$

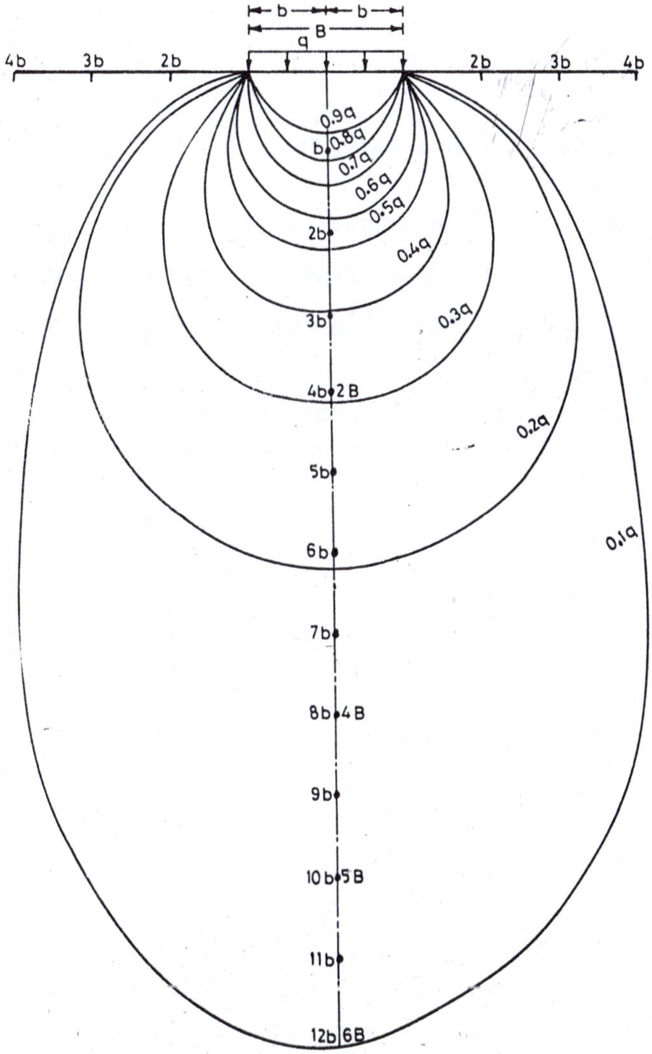

Fig. 11.13. Isobars of strip load.

and $\beta_1 + \beta_2 = 0$ or $\varphi = 0$

Therefore, Eq. 11.22 gives

$$\sigma_z = \frac{q}{\pi}(2\theta + \sin 2\theta) \qquad \text{...(same as Eq. 11.21)}$$

Eq. 11.22 can be used to determine isobars of different intensity due to strip load. Fig. 11.13 shows the isobars. The isobar of load intensity 0.1 q is at a depth of about 6 B below the load. Fig. 11.13 can be used for determination of vertical stresses at various points.

11.12. MAXIMUM SHEAR STRESSES AT POINTS UNDER A STRIP LOAD

The shear stress at any point P below a strip load is given by Eq. 11.24 as $\tau_{xz} = \frac{q}{\pi} \sin 2\theta \sin 2\varphi$. The planes on which the shear stresses are zero are known as principal planes. Therefore for principal planes, $\tau_{xz} = 0$.

or $$\frac{q}{\pi} \sin 2\theta \sin 2\varphi = 0$$

As q and θ cannot be zero, τ_{xz} will be zero when

$\sin 2\phi = 0$ or $2\varphi = 0$

The principal stresses are obtained from Eqs. 11.22 and 11.23, substituting $2\varphi = 0$.

$$\sigma_1 = \sigma_z = \frac{q}{\pi}(2\theta + \sin 2\theta) \qquad \text{...(11.25)}$$

$$\sigma_2 = \sigma_x = \frac{q}{\pi}(2\theta - \sin 2\theta) \qquad \text{...(11.26)}$$

The maximum shear stress is equal to half the difference of the principal stresses. Thus

$$\tau_{max} = \frac{1}{2}(\sigma_1 - \sigma_3) = \frac{q}{\pi} \sin 2\theta \qquad \text{...(11.27)}$$

Eq. 11.27 indicates that the maximum shear stress at P depends upon the angle 2θ subtended by the strip load at the point P. Obviously, the maximum shear stress will remain constant if the angle 2θ does not change. Let us draw a circle with the centre O obtained by the intersection of lines OA and OB making angles

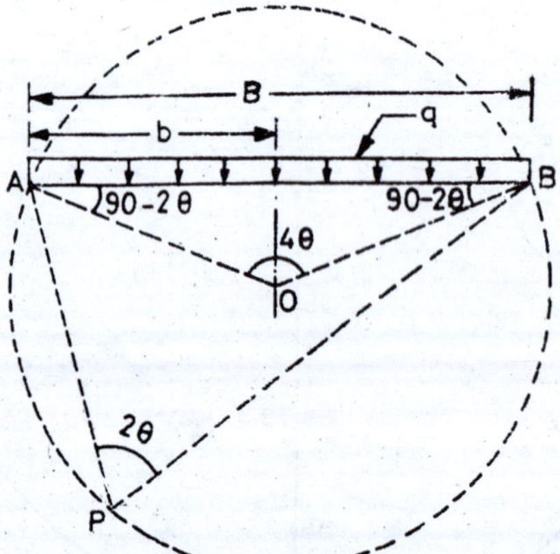

Fig. 11.14. Maximum shear stresses

(90–2θ) with the ends of the load, as shown in Fig. 11.14. As the angle subtended at the centre of circle is twice that at the circumference, the point P makes an angle 2θ. All the points on this circle will subtend an angle 2θ.

From Eq. 11.27, as the maximum shear stress depends on the angle 2θ, the circle is the locus of all points with shear stress equal to τ_{max}. The absolute maximum value of shear stress, $(\tau_{max})_{max}$ will occur when $\sin 2\theta = 1$ in Eq. 11.27. Thus

$$(\tau_{max})_{max} = \frac{q}{\pi}$$

The locus of $(\tau_{max})_{max}$ is a semi circle, which has the width of the loaded strip, B, as its diameter, In this case,

$$\sin 2\theta = 1 \quad \text{or} \quad 2\theta = 90°.$$

11.13. VERTICAL STRESSES UNDER A CIRCULAR AREA

The loads applied to soil surface by footings are not concentrated loads. These are usually spread over a finite area of the footing. It is generally assumed that the footing is flexible and the contact pressure is uniform. In other words, the load is assumed to by uniformly distributed over the area of the base of footings.

Let us determine the vertical stress at the point P at depth z below the centre of a uniformly loaded circular area (Fig. 11.15). Let the intensity of the load be q per unit area and R be the radius of the loaded area. Boussinesq's solution can be used to determine σ_z. The load on the elementary ring of radius r and width dr is equal to q $(2\pi r)\,dr$. The load acts at a constant radial distance r from the point P. From Eq. 11.9,

$$\Delta\sigma_z = \frac{3\,(q \times 2\pi r dr)}{2\pi} \cdot \frac{1}{z^2} \cdot \frac{1}{[1 + (r/z)^2]^{5/2}}$$

The vertical stress due to entire load is given by

$$\sigma_z = 3qz^3 \int_0^R \frac{r\,dr}{(r^2 + z^2)^{5/2}} \qquad \ldots(a)$$

Let $r^2 + z^2 = u$. Therefore, $2r\,dr = du$

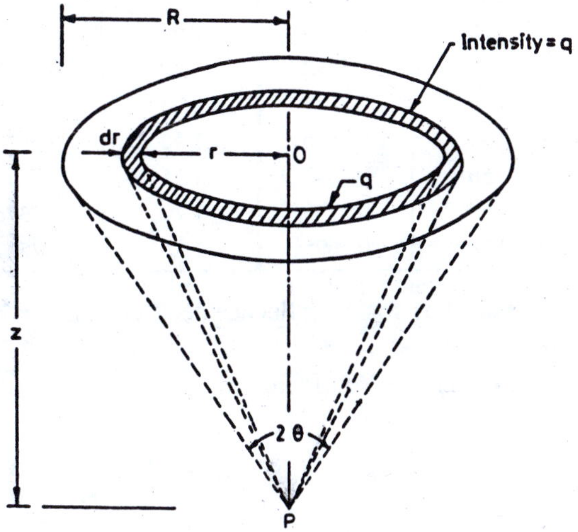

Fig. 11.15. Circular Load.

Eq. (a) becomes
$$\sigma_z = 3qz^3 \int_{z^2}^{(R^2 + z^2)} \frac{du}{2u^{5/2}}$$

$$= \frac{3}{2}\,qz^3 \left(-\frac{2}{3}\right)\left[u^{-3/2}\right]_{z^2}^{R^2 + z^2}$$

$$= -qz^3 \left[\frac{1}{(R^2 + z^2)^{3/2}} - \frac{1}{(z^2)^{3/2}}\right]$$

$$= qz^3 \left[\frac{1}{z^3} - \frac{1}{(R^2 + z^2)^{3/2}}\right]$$

or
$$\sigma_z = q\left[1 - \left\{\frac{1}{1 + (R/z)^2}\right\}^{3/2}\right] \qquad \ldots(11.29)$$

or
$$\sigma_z = I_c \cdot q \qquad \ldots(11.30)$$

where I_c is the influence coefficient for the circular area, and is given by

$$I_c = \left[1 - \left\{\frac{1}{1 + (R/z)^2}\right\}^{3/2}\right] \qquad ...(11.31)$$

Table 11.8 gives the value of the influence coefficient I_c for different values of R/z.

Table 11.8. Influence Coefficients I_c for the Circular Area

R/z	I_c	R/z	I_c	R/z	I_c	R/z	I_c
0.00	0.0000	0.65	0.4109	1.30	0.7734	1.95	0.9050
0.05	0.0037	0.70	0.4502	1.35	0.7891	2.00	0.9106
0.10	0.0148	0.75	0.4880	1.40	0.8036	2.50	0.9488
0.15	0.0328	0.80	0.5239	1.45	0.8170	3.00	0.9684
0.20	0.0571	0.85	0.5577	1.50	0.8293	3.50	0.9793
0.25	0.0869	0.90	0.5893	1.55	0.8407	4.00	0.9857
0.30	0.1286	0.95	0.6189	1.60	0.8511	5.00	0.9925
0.35	0.1592	1.00	0.6465	1.65	0.8608	6.00	0.9956
0.40	0.2079	1.05	0.6720	1.70	0.8697	7.00	0.9972
0.45	0.2416	1.10	0.6956	1.75	0.8779	8.00	0.9981
0.50	0.2845	1.15	0.7175	1.80	0.8855	9.00	0.9987
0.55	0.3273	1.20	0.7376	1.85	0.8925	10.00	0.9990
0.60	0.3695	1.25	0.7562	1.90	0.8990	∞	1.0000

Eq. 11.31 for the influence coefficient I_c can be written in terms of the angle 2θ subtended at point P by the load.

Let $\tan\theta = R/z$. Therefore,

$$I_c = \left[1 - \left\{\frac{1}{1 + \tan^2\theta}\right\}^{3/2}\right]$$

or

$$I_c = 1 - (\cos^2\theta)^{3/2} = 1 - \cos^3\theta \qquad ...(11.32)$$

Eq. 11.32 indicates that as θ tends to 90°, the value of I_c approaches unity. In other words, when a uniformly loaded area tends to by very large in comparison with the depth z, the vertical stress at the point P is approximately equal to q.

When the point P is not below the centre of the load, analysis becomes complicated and is outside the scope of this text. In that case, the isobars shown in Fig. 11.16 can be used to determine the vertical stress at any point. It may be noted that the isobar of $0.1q$ cuts the axis of the load at a depth of about $4R$ (= $2D$) below the loaded area. The zone within which the stresses is indicated by this isobar, as mentioned above, is known as the *bulb of pressure*. The reader should compare this pressure bulb with that below the strip load, which is much deeper.

11.14. VERTICAL STRESS UNDER A CORNER OF A RECTANGULAR AREA

The vertical stress under a corner of a rectangular area (Fig. 11.17) with a uniformly distributed load of intensity q can be obtained from Boussinesq's solution. From Eq. 11.9, the stress at depth z is given by, taking $dQ = q\,dA = q\,dx\,dy$,

$$\Delta\sigma_z = \frac{3\,(q\,dx\,dy)\,z^3}{2\pi}\,\frac{1}{(x^2 + y^2 + z^2)^{5/2}}$$

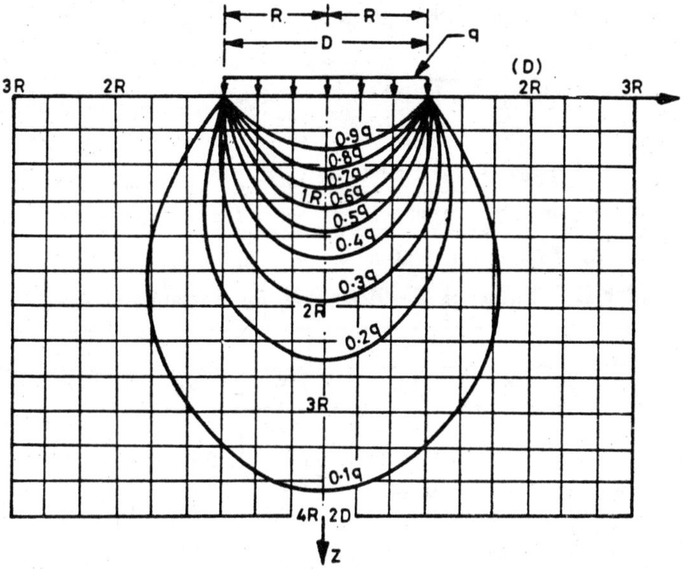

Fig. 11.16. Isobars for circular loaded area.

By integration, $$\sigma_z = \frac{3\,qz^3}{2\pi}\int_0^L\int_0^B \frac{q\,dx\,dy}{(x^2+y^2+z^2)^{5/2}}$$

Although the integral is quite complicated, Newmark was able to perform it The results were presented as follows:

$$\sigma_z = \frac{q}{2\pi}\left[\frac{mn}{\sqrt{m^2+n^2+1}}\cdot\frac{m^2+n^2+2}{m^2+n^2+m^2n^2+1}+\sin^{-1}\left(\frac{mn}{\sqrt{m^2+n^2+m^2n^2+1}}\right)\right] \quad \text{...(11.33)}$$

where $m = B/z$ and $n = L/z$

The values of m and n can be interchanged without any effect on the values of σ_z. **Eq. 11.33 can be expressed as**

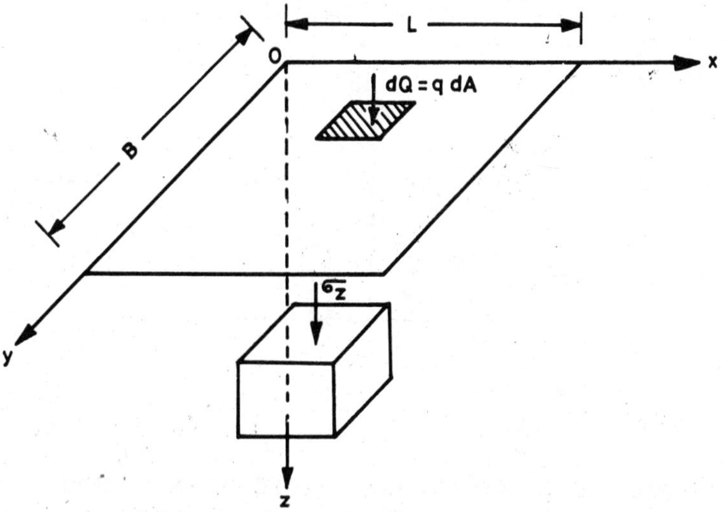

Fig. 11.17. Vertical stress under corner.

$$\sigma_z = I_N q \qquad \ldots(11.34)$$

where I_N is Newmark's influence coefficient, given by

$$I_N = \frac{1}{2\pi}\left[\frac{mn}{\sqrt{m^2+n^2+1}} \cdot \frac{m^2+n^2+2}{m^2+n^2+m^2n^2+1} + \sin^{-1}\left(\frac{mn}{\sqrt{m^2+n^2+m^2n^2+1}}\right)\right]$$

Alternatively,

$$I_N = \frac{1}{4\pi}\left[\frac{2mn\,(m^2+n^2+1)^{1/2}}{m^2+n^2+m^2n^2+1} \cdot \frac{m^2+n^2+2}{m^2+n^2+1} + \tan^{-1}\frac{2mn\,(m^2+n^2+1)^{1/2}}{m^2+n^2+1-m^2n^2}\right]$$

Table 11.9 gives the values of I_N for different values of m and n.

Table 11.9. Influence Coefficients I_N for Rectangular Area

m	n								
	0.2	0.4	0.6	0.8	1.0	2.0	3.0	5.0	10.0
0.2	0.0179	0.0328	0.0435	0.0504	0.0547	0.0610	0.0619	0.0620	0.0620
0.4	0.0328	0.0602	0.0801	0.0931	0.1013	0.1134	0.1150	0.1154	0.1154
0.6	0.0435	0.0801	0.1069	0.1247	0.1361	0.1533	0.1555	0.1561	0.1562
0.8	0.0504	0.0931	0.1247	0.1461	0.1598	0.1812	0.1841	0.1849	0.1850
1.0	0.0547	0.1013	0.1361	0.1598	0.1752	0.1999	0.2034	0.2044	0.2046
2.0	0.0610	0.1134	0.1533	0.1812	0.1999	0.2325	0.2378	0.2395	0.2399
3.0	0.0618	0.1150	0.1555	0.1841	0.2034	0.2378	0.2439	0.2461	0.2465
5.0	0.0620	0.1154	0.1561	0.1849	0.2044	0.2395	0.2461	0.2486	0.2491
10.0	0.0620	0.1154	0.1562	0.1850	0.2046	0.2399	0.2465	0.2491	0.2498

Fadum gave charts for determination of the influence factor I_N (Fig. 11.18). These charts can be used in a design office. The charts can also be used for determination of the vertical stress under a strip load, in which case the length tends to infinity and the curve for $n = \infty$ can be used.

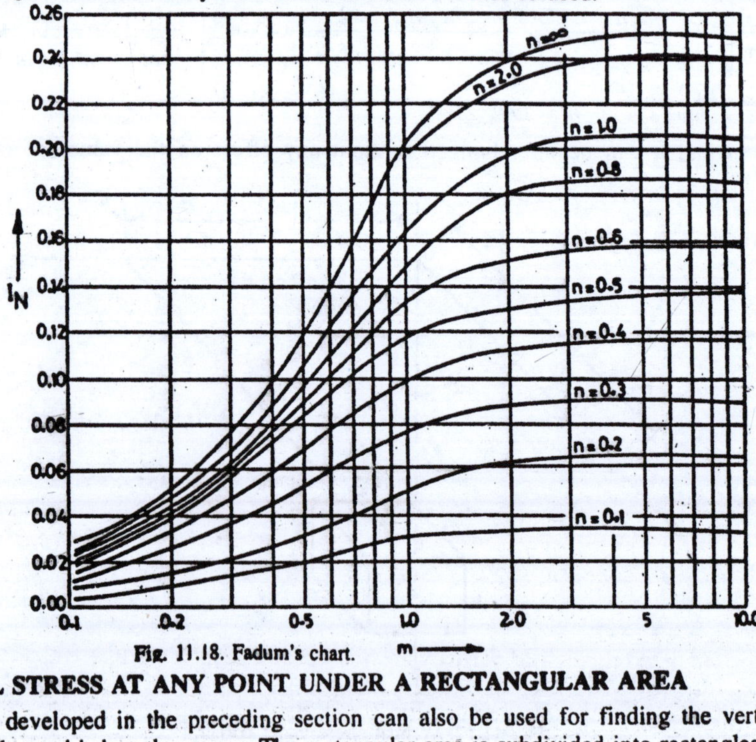

Fig. 11.18. Fadum's chart.

11.15. VERTICAL STRESS AT ANY POINT UNDER A RECTANGULAR AREA

The equations developed in the preceding section can also be used for finding the vertical stress at a point which is not located below the corner. The rectangular area is subdivided into rectangles such that each

11.15. VERTICAL STRESS AT ANY POINT UNDER A RECTANGULAR AREA

The equations developed in the preceding section can also be used for finding the vertical stress at a point which is not located below the corner. The rectangular area is subdivided into rectangles such that each rectangles has a corner at the point where the vertical stress is required. The vertical stress is determined using the principle of superposition. The following three cases can occur.

(1) Point anywhere below the rectangular area. Fig. 11.19 (*a*) shows the location of the point *P* below the rectangular area *ABCD*. The given rectangle is subdivided into 4 small rectangles *AEPH, EBFP, HPGD* and *PFCG,* each having one corner at *P.* The vertical stress at *P* due to the given rectangular load is equal to that from the four small rectangles. Therefore, using Eq. 11.34,

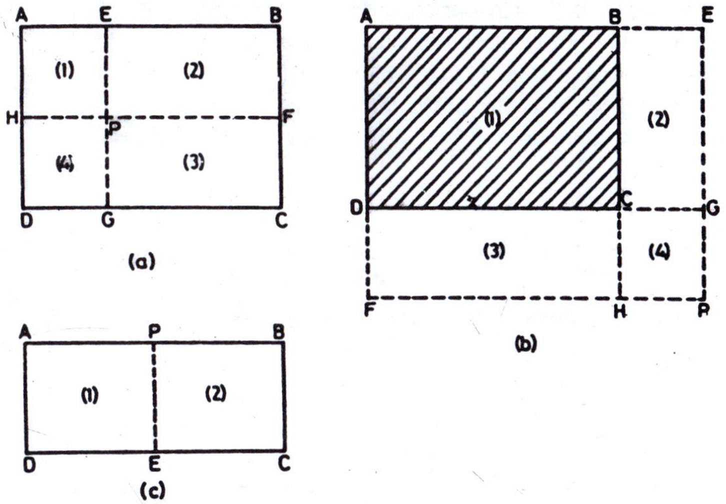

Fig. 11.19. Vertical stress under a rectangular area.

$$\sigma_Z = q\,[(I_N)_1 + (I_N)_2 + (I_N)_3 + (I_N)_4] \qquad ...(11.35)$$

where $(I_N)_1$, $(I_N)_2$, $(I_N)_3$ and $(I_N)_4$ are Newmark's influence factors obtained from Table 11.9 for the four rectangles marked (1), (2), (3) and (4).

For the special case, when the point *P* is at the centre of the rectangle *ABCD*, all the four small rectangles are equal, and Eq. 11.35 becomes

$$\sigma_Z = 4\, I_N \qquad ...(11.36)$$

where I_N is the influence factor for the small rectangle.

(2) Point outside the loaded area. Fig. 11.19 (*b*) shows the point *P* outside the loaded area *ABCD*. In this case, a large rectangle *AEPF* is drawn with its one corner at *P.*

Now rectangle *ABCD* = rectangle *AEPF* – rectangle *BEPH* – rectangle *DGPF* + rectangle *CGPH*

The last rectangle *CGPH* is given plus sign because this area has been deducted twice, once in rectangle *BEPH* and once in *DGPF.*

Therefore, the stress at *P* due to a load on rectangle *ABCD* is given by

$$\sigma_z = q\,[(I_N)_1 - (I_N)_2 - (I_N)_3 + (I_N)_4] \qquad ...(11.37)$$

where $(I_N)_1$, $(I_N)_2$, $(I_N)_3$ and $(I_N)_4$ are the influence coefficients for the rectangles *AEPF, BEPH, DGPF* and *CGPH,* respectively.

(3) Point below the edge of the loaded area. If the point *P* is below the edge of the loaded area *ABCD* (Fig. 11.19 *c*), the given rectangle is divided into two small rectangles *APED* and PBCE. In this case,

$$\sigma_z = q\,[(I_{N1}) + (I_N)_2] \qquad ...(11.38)$$

where $(I_N)_1$ and $(I_N)_2$ are influence coefficients for rectangles *APED* and *PBCE*, respectively.

11.16. NEWMARK'S INFLUENCE CHARTS

The methods for the determination of vertical stresses under a strip, a circular and a rectangular area have been discussed in the preceding sections. In practice, sometimes one has to find the vertical stresses under a uniformly loaded areas of other shapes. For such cases, Newmark's influence charts are extremely useful.

Newmark's chart is based on the concept of the vertical stress below the centre of the circular area, discussed in Sect. 11.13. Let us consider a uniformly loaded circular area of radius R_1, divided into 20 equal sectors (Fig. 11.20). The vertical stress at point P at depth z just below the centre of the loaded area due to load on one sector (hatched area (1)) will be (1/20) of that due to load on full circle. From Eq. 11.29,

$$\sigma_z = \frac{1}{20} q \left[1 - \left\{ \frac{1}{1 + (R_1/z)^2} \right\}^{3/2} \right] \quad ...(a)$$

If the vertical stress (σ_z) is given an arbitrary fixed value, say $0.005q$, Eq. (a) becomes

$$0.005q = \frac{q}{20} \left[1 - \left\{ \frac{1}{1 + (R_1/z)^2} \right\}^{3/2} \right] \quad ...(b)$$

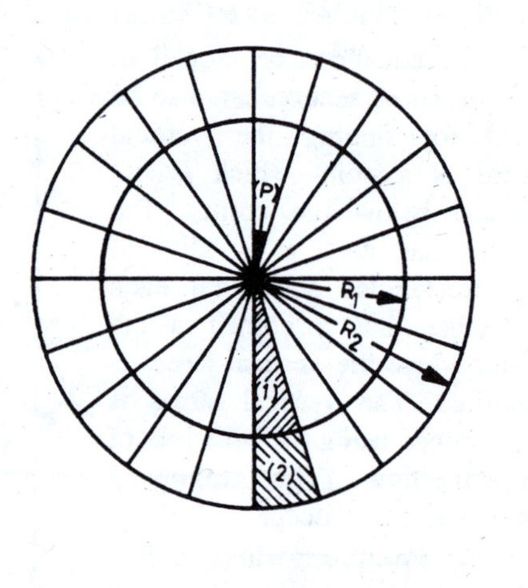

Fig. 11.20. Concentric circles for R_1 and R_2.

Solving Eq. (b), $R_1/z = 0.270$ ...(11.39)

Thus every one-twentieth sector of the circle, with a radius R_1 equal to 0.270 z, would give a vertical stress of 0.005 q at its centre.

Let us now consider another concentric circle of radius R_2 and divide it again into 20 equal sectors. Each larger sector is divided into two sub-areas. If the small area (marked 2) exerts a stress of 0.005 q at P, the vertical stress due to both area (1) and (2) would be equal to 2 × 0.005 q. Thus,

$$2 \times 0.005\, q = \frac{q}{20} \left[1 - \left\{ \frac{1}{1 + (R_2/z)^2} \right\}^{3/2} \right] \quad ...(c)$$

Solving, $R_2/z = 0.40$.

In other words, the radius of the second circle should be equal to 0.40 z.

Likewise, the radii of the third to the ninth circles can be determined. The values obtained are 0.52 z, 0.64 z, 0.77 z, 0.92 z, 1.11 z, 1.39 z and 1.91 z. The radius of $9\frac{1}{2}$ circle is 2.54 z. The radius for the tenth circle R_{10} is given by

$$10 \times 0.005\, q = \frac{q}{20} \left[1 - \left\{ \frac{1}{1 + (R_{10}/z)^2} \right\}^{3/2} \right]$$

or $R_{10} = \infty$

Therefore, the tenth circle cannot be drawn.

Fig. 11.21 shows the complete Newmark's influence chart, in which only $9\frac{1}{2}$ circles have been drawn for z equal to the distance AB marked on the chart.

Use of Newmark's Chart. The chart can be used to determine the vertical stress at point P below the loaded area. A plan of the loaded area is drawn on a tracing paper to a scale such that the length AB(= 2 cm in this case) is equal to the depth (z) of the point P below the surface. For example, if the pressure is required at a depth of 1 m, the plan should be drawn to a scale of 2 cm = 1 m or R.F. = 1/50. The traced plan of the loaded area is placed over the Newmark chart such that the point P at which the pressure is required coincides with the centre of the chart. The vertical stress at point P is given by

$$\sigma_z = I \times n \times q \quad ...(11.40)$$

where I = influence coefficient (= 0.005 in this case),

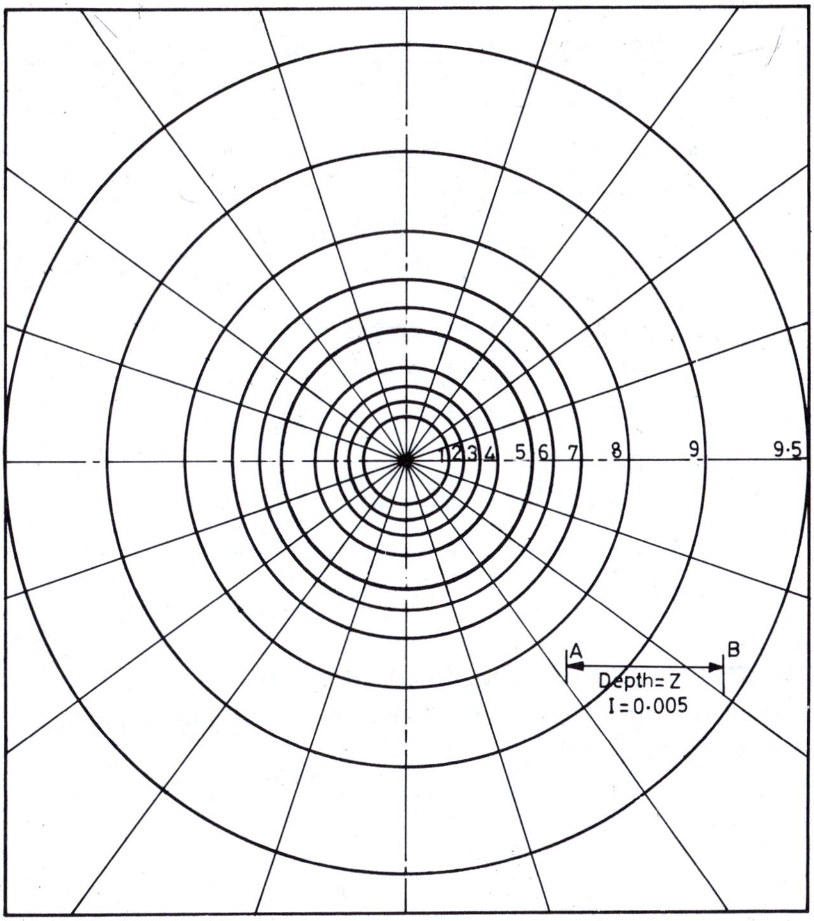

Fig. 11.21. Newmark's Charts.

n = number of small area units covered by the plan. Each area between two successive radial lines and two successive concentric circle is taken as one unit.

q = intensity of load.

The following points are worth noting:

(1) The fractions of the unit areas should also be counted and properly accounted for.

(2) If the plan of the loaded area extends beyonds the $9\frac{1}{2}$th circle. it may be assumed to approach the 10th circle for the purpose of counting the unit areas.

(3) The point P at which the vertical stress is required may be anywhere within or outside the loaded area.

(4) If the depth at which the stress is required is changed, a fresh plan is required such that the new depth is equal to the distance AB on the chart.

11.17. COMPARISON OF STRESSES DUE TO LOADS ON AREAS OF DIFFERENT SHAPES

The variation of vertical stress with depth depends upon the shape and size of the loaded area. Fig. 11.22 shows the variation of the vertical stress with depth below the centre of circular, square and strip loads.

The vertical stresses are equal to the load intensity at the surface and decrease rapidly with an increase in depth (z). In the case of circular and square loads, the vertical stress is about 10% of the surface load (q) at a depth of about $2B$. However, in the case of strip loads, the stresses are much greater. Even at $z = 3B$, the vertical stress is about 20% of the surface load (q). In other words, the pressure bulb in this case is much deeper.

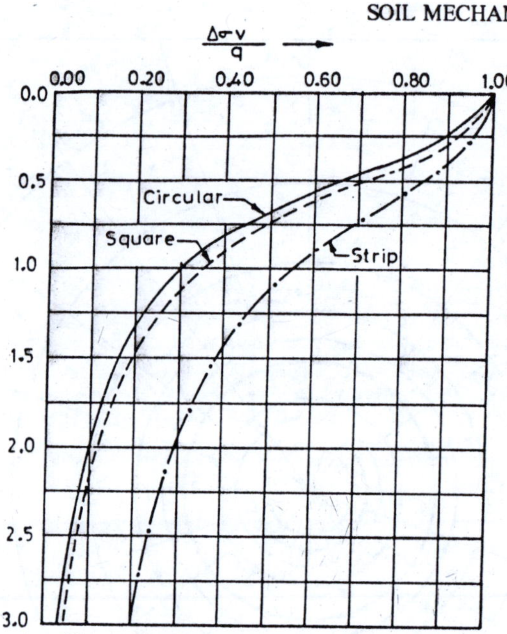

Z/B	Δσv/q Circular	Square	Strip
0.5	0.65	0.70	0.85
1.0	0.28	0.34	0.55
1.5	0.14	0.18	0.39
2.0	0.08	0.14	0.31
2.5	0.06	0.07	0.25
3.0	0.04	0.05	0.20

B = Width or diameter

Fig. 11.22. Comparison of circular, square and strip loads.

11.18. VERTICAL STRESSES UNDER TRIANGULAR LOADS

Fig. 11.23 shows a triangular load with a width of $2b$ and the intensity varying from 0 to q. The vertical stress (σ_z) at a point $P(x, z)$ is given by

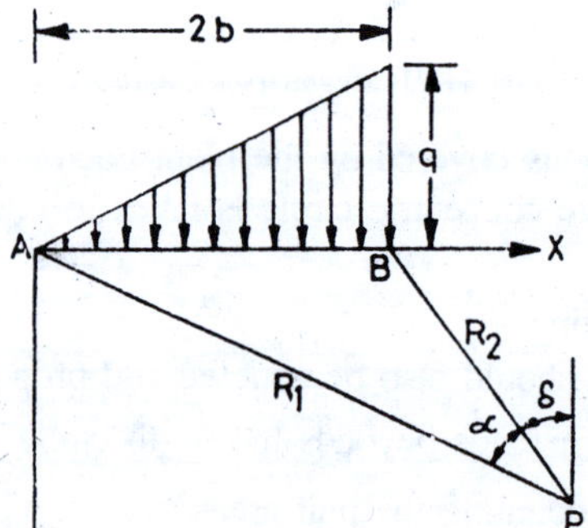

Fig. 11.23. Triangular load.

$$\sigma_z = \frac{q}{2\pi}\left[\frac{x}{b}\alpha - \sin 2\delta\right] \quad \text{...(11.41)}$$

where δ is the angle which the line PB marks with vertical, and α is the angle subtended by PA and PB at P.

If the point P is exactly below the end B, $x = 2b$ and $\delta = 0$. Therefore,

$$\sigma_z = \frac{q}{2\pi}\left(\frac{2b}{b}\alpha\right) = \frac{q\alpha}{\pi} \quad \text{...(11.42)}$$

The above equations can also be applied to the case when the intensity of the load increases linearly from zero at one end to a maximum q and then decreases to zero (Fig. 11.24).

For the load shown in Fig. 11.24 (a),

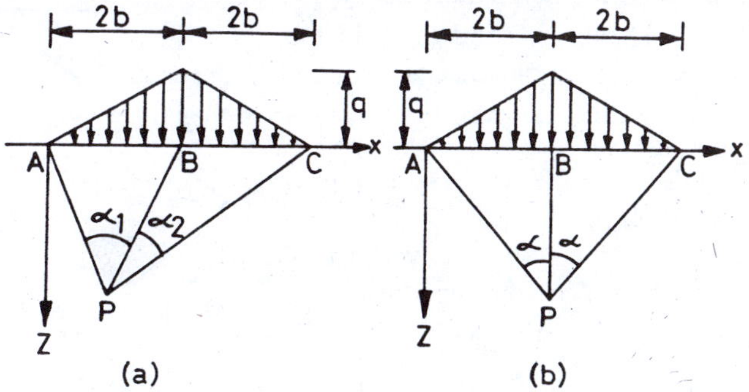

Fig. 11.24. Triangular load with maximum intensity at centre.

$$\sigma_z = \frac{q}{2\pi b}[2b(\alpha_1 + \alpha_2) + x(\alpha_1 - \alpha_2)] \qquad ...(11.43)$$

When the point P is exactly below the point B, $\alpha_1 = \alpha_2 = \alpha$ and $x = 2b$. [Fig. 11.24 (b)]. Therefore,

$$\sigma_z = \frac{q}{2b\pi}[2b \times 2\alpha + 2b(\alpha - \alpha)]$$

or

$$\sigma_z = \frac{2q\alpha}{\pi} \qquad ...(11.44)$$

11.19. VERTICAL STRESSES UNDER TRAPEZOIDAL LOADS

Fig. 11.25 (a) shows a trapezoidal load due to an embankment, which consists of a triangular load over width a and a uniform load of intensity q over width b. The vertical stress at point P is given by

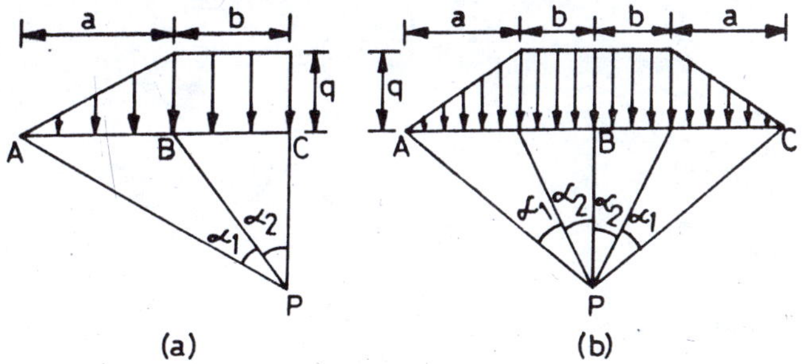

Fig. 11.25. Trapezoidal load.

$$\sigma_z = \frac{q}{\pi}\left[\left(\frac{a+b}{a}\right)(\alpha_1 + \alpha_2) - \frac{b}{a}\alpha_2\right]$$

or

$$\sigma_z = \frac{q}{\pi}\left[(\alpha_1 + \alpha_2) + \frac{b}{a}\alpha_1\right]$$

or

$$\sigma_z = \frac{q}{\pi a}[a(\alpha_1 + \alpha_2) + b\alpha_1] \qquad ...(11.45)$$

Obviously, for the trapezoidal load shown in Fig. 11.25 (b), the vertical stress at P,

$$\sigma_z = \frac{2q}{\pi a}\left[a(\alpha_1 + \alpha_2) + b\alpha_1\right] \qquad \ldots(11.46)$$

11.20. STRESSES DUE TO HORIZONTAL LOAD

(*a*) **Line Load Q_1** Fig. 11.26 shows a horizontal load Q_1 per unit run acting on the soil surface shown by solid lines. In this case, it is more convenient to determine the radial stress (σ_r) and the tangential stress (σ_θ). The radial stress is given by

$$\sigma_r = \frac{2\,Q_1 \sin\theta}{r\,(2\alpha - \sin 2\alpha)} \qquad \ldots(11.47)$$

where θ = angle made by radial line r with the vertical

α = angle made by soil surface with the vertical

For horizontal ground surface, $\alpha = \pi/2$. Therefore,

$$\sigma_r = \frac{2Q_1 \sin\theta}{r\,(\pi - 0)}$$

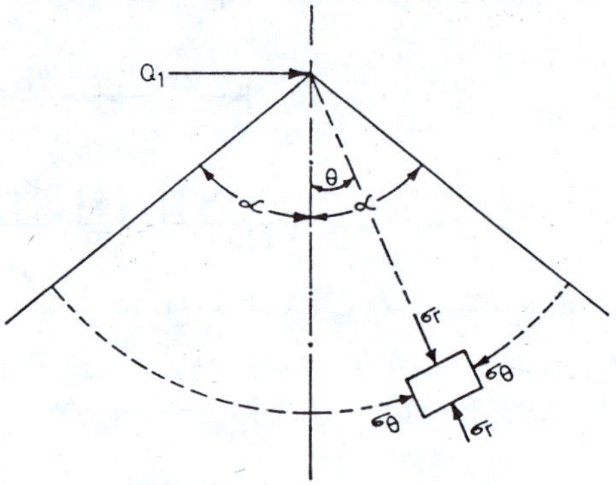

Fig. 11.26. Horizontal load.

or

$$\sigma_x = \frac{2\,Q_1 \sin\theta}{\pi r} \qquad \ldots(11.48)$$

(*b*) **Concentrated Load Q.** In cartesian coordinates, the stresses due to a horizontal load Q can be written as

$$\sigma_z = \frac{3\,Q\,xz^2}{2\,\pi\,R^5}$$

$$\sigma_x = \frac{Q}{2\pi}\cdot\frac{x}{R^3}\times\left[\frac{3x^2}{R^2} - (1-2\nu) + \frac{(1-2\nu)\,R^2}{(R+z)^2}\left\{3 - \frac{x^2\,(3\,R+z)}{R^2\,(R+z)}\right\}\right]$$

$$\sigma_y = \frac{Q}{2\pi}\cdot\frac{x}{R^3}\times\left[\frac{3y^2}{R^2} - (1-2\nu) + \frac{(1-2\nu)\,R^2}{(R+z)^2}\left\{3 - \frac{y^2\,(3\,R+z)}{R^2\,(R+z)}\right\}\right]$$

$$\tau_{xy} = \frac{Q}{2\pi}\cdot\frac{y}{R^3}\times\left[\frac{3x^2}{R^2} + \frac{(1-2\nu)\,R^2}{(R+z)^2}\left\{1 - \frac{x^2\,(3\,R+z)}{R^2\,(R+z)}\right\}\right]$$

$$\tau_{xz} = \frac{3Q}{2\pi}\cdot\frac{x^2 z}{R^5}$$

$$\tau_{yz} = \frac{3Q}{2\pi}\cdot\frac{xyz}{R^5} \qquad \ldots(11.49)$$

where x,y,z are coordinates of point P and R is the distance OP, as shown in Fig. 11.4.

This case is generally referred to as *Cerutti's problem.*

11.21. STRESSES DUE TO INCLINED LOAD

Fig. 11.27 shows an inclined load Q_2 per unit run acting on the soil surface. The radial stress at a point at an angle θ is given by

$$\sigma_r = \frac{2Q_2}{r}\left(\frac{\cos\beta\cos\theta}{2\alpha + \sin 2\alpha} + \frac{\sin\beta\sin\theta}{2\alpha - \sin 2\alpha}\right) \qquad \ldots(11.50)$$

where β = angle which the load Q_2 makes with vertical,

α = angle the soil surface makes with the vertical.

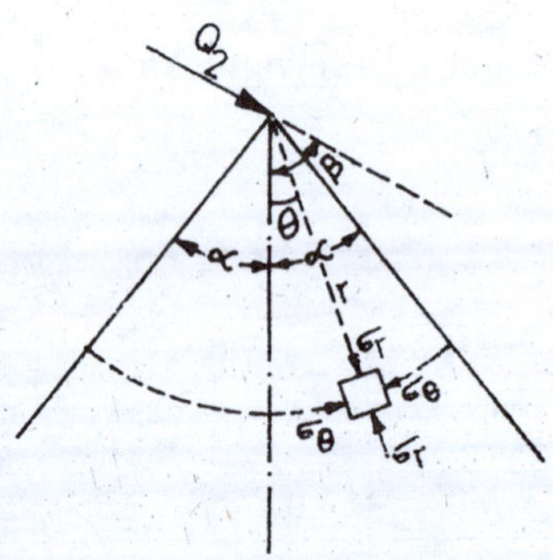

Fig. 11.27. Inclined load.

For the horizontal ground surface, $\alpha = \pi/2$. Thus

$$\sigma_r = \frac{2Q_2}{r}\left[\frac{\cos\beta\cos\theta}{\pi} + \frac{\sin\beta\sin\theta}{\pi}\right]$$

or

$$\sigma_r = \frac{2Q_2}{\pi r}\cos(\theta - \beta) \qquad \text{...(11.51)}$$

When the load is vertical, $\beta = 0$.

$$\sigma_r = \frac{2Q_2}{\pi r}\cos\theta \qquad \text{...[(11.51 (a)]}$$

11.22. WESTERGAARD'S SOLUTION

Boussinesq' solution assumes that the soil deposit is isotropic. Actual sedimentary deposits are generally anisotropic. There are generally thin layers of sand embedded in homogeneous clay strata. Westergaard's solution assumes that there are thin sheets of rigid materials sand-wiched in a homogeneous soil mass. These thin sheets are closely spaced and are of infinite rigidity and are, therefore, incompressible. These permit only downward displacement of the soil mass as a whole without any lateral displacement. Therefore, Westergaard's solution represents more closely the actual sedimentary deposits.

According to Westergaard, the vertical stress at a point P at a depth z below the concentrated load Q is given by

$$\sigma_z = \frac{c/2\pi}{[c^2 + (r/cz)^2]^{3/2}} \cdot \frac{Q}{z^2} \qquad \text{...(11.52)}$$

where c depends upon the Poisson ratio (ν) and is given by

$$c = \sqrt{(1 - 2\nu)/(2 - 2\nu)}$$

For an elastic material, the value of ν varies between 0.0 to 0.50. For the case when ν is zero, Eq. 11.52 is simplified considerably, taking $c = 1/\sqrt{2}$,

$$\sigma_z = \frac{1}{\pi[1 + 2(r/z)^2]^{3/2}} \cdot \frac{Q}{z^2} \qquad \text{...(11.53)}$$

or

$$\sigma_z = I_w \frac{Q}{z^2} \qquad \text{...(11.54)}$$

where I_w is known as *Westergaard influence coefficient.*

$$I_w = \frac{1}{\pi[1 + 2(r/z)^2]^{3/2}} \qquad \text{...(11.55)}$$

The values of I_w are considerably smaller than the Boussinesq influence factor (I_B). Table 11.10 gives the values of I_w. The values of I_B are also given for comparison.

Table 11.10. Comparison of I_W and I_B

r/z	0.0	0.1	0.2	0.3	0.4	0.5	0.6
I_B	0.4775	0.4657	0.4329	0.3849	0.3295	0.2733	0.2214
I_W	0.3183	0.3090	0.2836	0.2483	0.2099	0.1733	0.1411
r/z	0.7	0.8	0.9	1.0	2.0	3.0	6.0
I_B	0.1762	0.1386	0.1083	0.0844	0.0085	0.0015	0.0001
I_W	0.1142	0.0925	0.0751	0.0613	0.0118	0.0038	0.0005

Fig. 11.28 shows the variation of I_B and I_W with r/z. The Westergaard influence factor is about 2/3

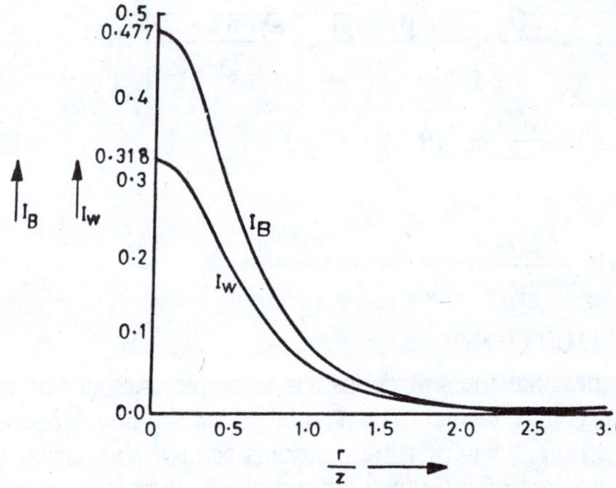

Fig. 11.28. Comparison of I_B and I_W.

of the Boussinesq values for small values of r/z. But for r/z more than 2.0, the Westergaard values are slightly greater. The effect of the load is negligibly small in both the cases when r/z is greater than about 2.0.

11.23. FENSKE'S CHARTS

Just like Newmark's Charts which are based on Boussinesq's solution, Fenske's Charts are based on Westergaard's solution. The Fenske chart can be prepared using Eq. 11.52.

$$\sigma_z = \frac{Q}{2\pi} \cdot \frac{1}{(cz)^2 [1 + (r/cz)^2]^{3/2}}$$

The above equation can be integrated to obtain the vertical stress (σ_z) below the centre of a uniform circular load of intensity q and radius R as was done for the Boussinesq solution for derivation of Eq. 11.29. In this case,

$$\sigma_z = q \left[1 - \left\{\frac{1}{1 + (R/cz)^2}\right\}^{1/2}\right] \quad ...(11.56)$$

If instead of the full circle, only 1/8th sector of the circle is considered, the stress is given by

$$\sigma_z = \frac{q}{8} \left[1 - \left\{\frac{1}{1 + (R/cz)^2}\right\}^{1/2}\right] \quad ...(11.57)$$

Eq. 11.57 is similar to the equation used for Newmark's chart, with one difference that the depth used here is the modified depth cz.

The radius R_1 of the first circle can be determined for a constant value of σ_z (say, 0.001 q). Thus

$$0.001\, q = \frac{q}{8} \left[1 - \left\{\frac{1}{1 + (R_1/cz)^2}\right\}^{1/2}\right]$$

or $\quad \dfrac{R_1}{cz} = 0.127$

or $\quad R_1 = 0.127\,(cz)$

The modified depth cz is marked as the distance AB in Fig. 11.29.

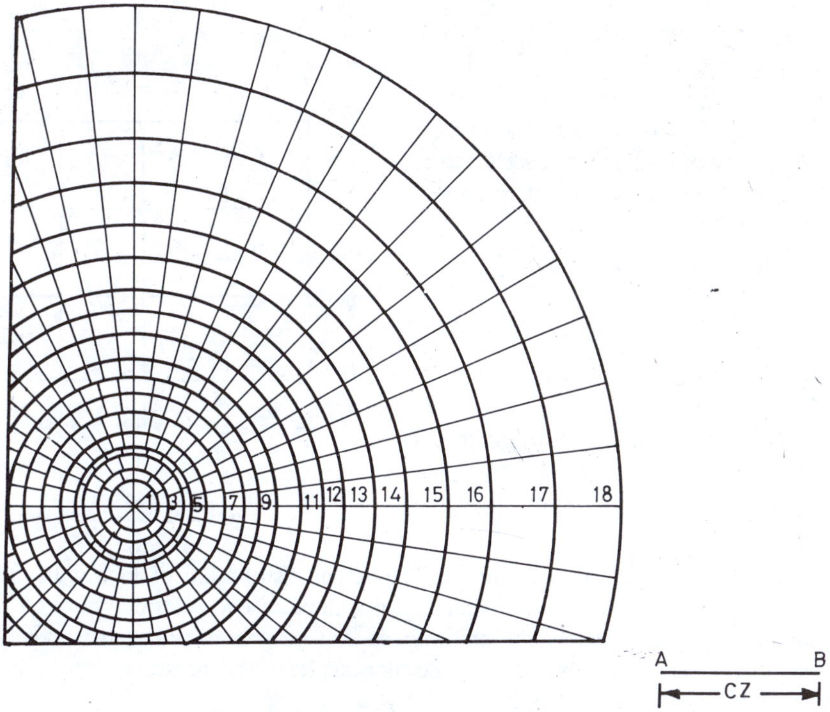

Fig. 11.29. Fenske's Chart (I = 0.001).

Likewise, the radii of other circles are determined. Unlike the Newmark chart, the radial divisions are also changed in Fenske's chart. There are 8 radial divisions for the first circle and 48 radial divisions for the 18th circle. The radii of the circular arcs and the number of radial divisions are so chosen that each influence area unit is approximately a square. Table 11.11 gives the values of $R/(cz)$ for different circles and their corresponding number of division.

The method of using the Fenske chart is similar to that for the Newmark chart. However, in this case the distance AB represents the modified depth cz. The plan of the loaded area is drawn on a tracing paper to a scale such that the distance AB is equal to c times the depth z of the point P at which the stress is required. For Poisson's ratio of zero, the value of c is equal to 0.707.

Table 11.11. Values of R/cz for Fenske's Chart

Circle No.	1	2	3	4	5	6	7	8	9
R/ (cz)	0.127	0.204	0.292	0.376	0.472	0.560	0.664	0.772	0.900
Divisions	8	12	20	24	32	32	40	40	48

Circle No.	10	11	12	13	14	15	16	17	18	19
R/(cz)	1.032	1.176	1.332	1.512	1.712	1.952	2.236	2.592	3.044	4.420
Divisions	48	48	48	48	48	48	48	48	48	48

11.24. APPROXIMATE METHODS

The methods discussed in the preceding sections are relatively more accurate, but are time- consuming. Sometimes, the engineer is interested to estimate the vertical stresses approximately for preliminary designs. The following methods can be used.

(1) Equivalent Point-Load Method. The vertical stress at a point under a loaded area of any shape can be determined by dividing the loaded area into small areas and replacing the distributed load on each small area by an equivalent point load acting at the centroid of the area (Fig. 11.30); *e.g.* in Fig. 11.30, $Q = qa^2$ for each area. The total load is thus converted into a number of point loads. The vertical stress at any point below or outside the loaded area is equal to the sum of the vertical stresses due to these equivalent point loads. Using Eq. 11.10,

$$\sigma_z = \frac{[Q_1 (I_B)_1 + Q_2 (I_B)_2 + \ldots + Q_n (I_B)_n]}{z^2}$$

$$\text{or } \sigma_z = \frac{1}{z^2} \sum_{i=1}^{n} Q_i (I_B)_i \qquad \ldots(11.58)$$

Eq. 11.58 gives fairly accurate results if the side a of the small area unit is equal to or less than one-third of the depth z of point P at which the vertical stress is required.

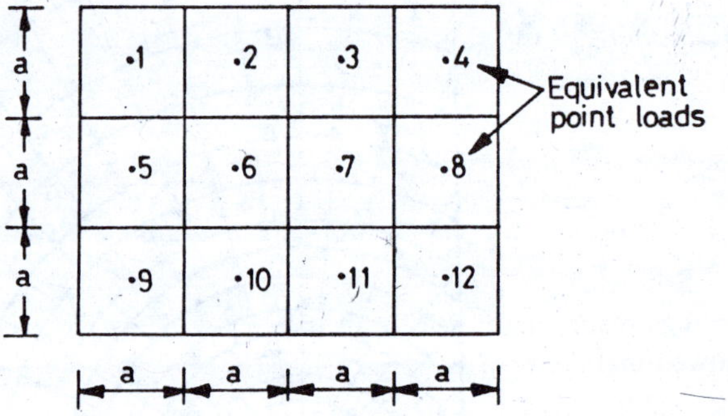

Fig. 11.30. Equivalent Point loads.

(2) Two-to-one Load Distribution Method. The actual distribution of load with the depth is complex. However, it can be assumed to spread approximately at a slope of two (vertical) to one (horizontal). Thus the vertical pressure at any depth z below the soil surface can be determined approximately by constructing a frustum of pyramid (or cone) of depth z and side slopes (2:1). The pressure distribution is assumed to be uniform on a horizontal plane at that depth.

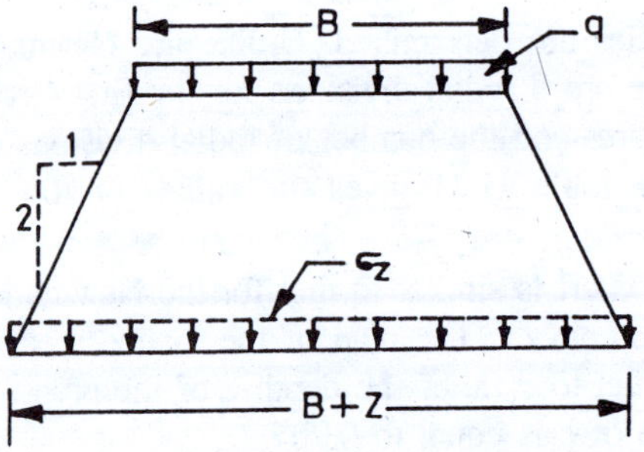

Fig. 11.31. Two-to-One Distribution.

The average vertical stress σ_z depends upon the shape of the loaded area, as given below (see Fig. 11.31).

(1) Square Area (B × B), $$\sigma_z = \frac{qB^2}{(B + z)^2} \qquad \ldots(11.59)$$

(2) Rectangular Area (B × L), $$\sigma_z = \frac{q(B \times L)}{(B + z)(L + z)} \qquad \ldots(11.60)$$

(3) Strip Area (width B, unit length), $$q_z = \frac{q \times (B \times 1)}{(B + z) \times 1} \qquad \ldots(11.61)$$

(4) Circular Area (diameter D), $$\sigma_z = \frac{qD^2}{(D + z)^2} \qquad \ldots(11.62)$$

The above method gives fairly accurate values of the average vertical stress if the depth z is less than 2.5 times the width of the loaded area. The maximum stress is generally taken as 1.5 times the average stress determined above.

(3) Sixty Degree Distribution. This method is similar to the preceding method. In this case, the pressure distribution is assumed along lines making an angle of 60° with the horizontal instead of $63\frac{1}{2}°$ (2 : 1). The method gives approximately the same results.

11.25. CONTACT PRESSURE DISTRIBUTION

The upward pressure due to soil on the underside of the footing is termed contact pressure. In the derivations of the preceding sections, it has been assumed that the footing is flexible and the contact pressure distribution is uniform and equal to q. Actual footings are not flexible as assumed. The actual distribution of the contact pressure depends upon a number of factors such as the elastic properties of the footing material and soil, the thickness of footings. In fact, it is a *soil- structure interaction* problem.

Borowicka (1936, 1938) studied the contact pressure distribution of uniformly loaded strips and circular footings resting on a semi-infinite elastic mass, assuming the base of the footing as frictionless. The analysis showed that the contact pressure distribution depends upon the relative rigidity (K_r) of the footing-soil system. The relative rigidity is defined as

$$K_r = \frac{1}{6}\frac{(1-\nu_s^2)}{(1-\nu_f^2)}\left(\frac{E_f}{E_s}\right)\cdot\left(\frac{t}{b}\right)^3 \qquad \text{...(11.63)}$$

where ν_s, ν_f = Poisson's ratios for soil and footing material, respectively,

E_s, E_f = Moduli of elasticity for soil and footing material, respectively.

$2b$ = width (or diameter) of footing, t = thickness of footing.

Fig. 11.32 shows the contact pressure distribution of circular and strip footings for different values of relative stiffness. For a perfectly rigid footing ($K_r = \infty$), the contact pressure is minimum at the centre, with

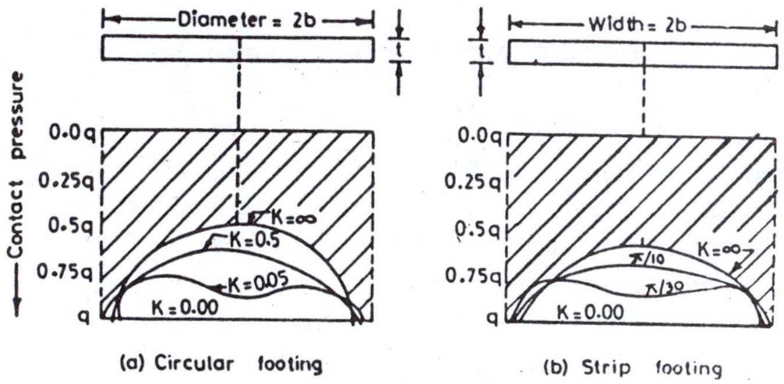

Fig. 11.32. Contact Pressure (Borowicka).

a value of about 0.5 q for the circular footing and 0.67 q for the strip footing. The contact pressure is very large at the edges. In fact, it tends to infinity. For purely flexible footings ($K_r = 0$), the contact pressure is uniform and equal to q. (In Fig. 11.32, K_r is represented as K).

Borowicka's results can be used to determine the contact pressure on a cohesive soil which behaves like an elastic soil mass. In a cohesionless soil, modulus of elasticity increases with depth due to an increase in confining pressure. Such soils are non-homogeneous.

Contact pressure on saturated clay. Fig. 11.33 shows the qualitative contact pressure distribution under flexible and rigid footings resting on a saturated clay and subjected to a uniformly distributed load q. When the footing is flexible, it deforms into the shape of a bowel, with the maximum deflection at the centre. The contact pressure distribution is uniform.

If the footing is rigid, the settlement is uniform. The contact pressure distribution is minimum at the centre and the maximum at the edges. The stresses at the edges in real soils cannot be infinite as theoretically determined for an elastic mass. In real soils, beyond a certain limiting value of stress, the plastic flow occurs and the pressure becomes finite.

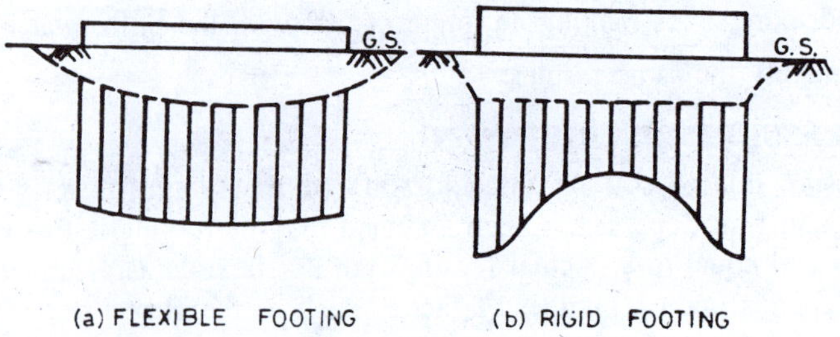

Fig. 11.33. Contact pressure on saturated clay.

Contact Pressure on sand. Fig. 11.34 shows the qualitative contact pressure distribution under flexible and rigid footings resting on a sandy soil and subjected to a uniformly distributed load *q*. In this case, the edges of the flexible footing undergo a larger settlement than at the centre. The soil at the centre is confined and, therefore, has a high modulus of elasticity and deflects less for the same contact pressure. The contact pressure is uniform.

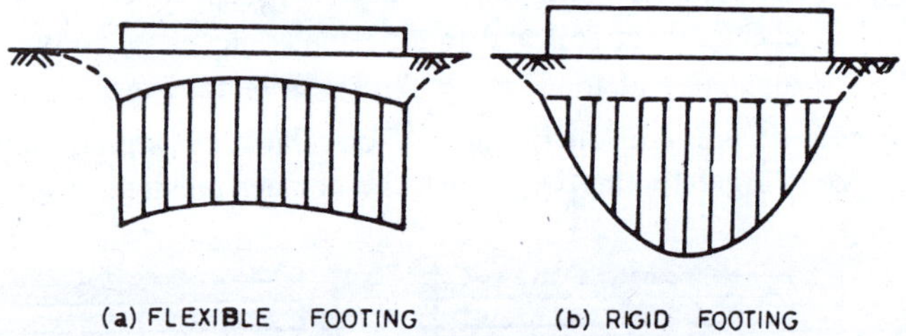

Fig. 11.34. Contact Pressure on Sand.

If the footing is rigid, the settlement is uniform. The contact pressure increases from zero at the edges to a maximum at the centre. The soil, being unconfined at edges, has low modulus of elasticity. However, if the footing is embedded, there would be finite contact pressure at edges.

Usual Assumption. As discussed above, the contact pressure distribution for flexible footings is uniform for both clay and sand. The contact pressure for rigid footing is maximum at the edges for footings on clay, but for the rigid footings on sand, it is minimum at the edges. For convenience, the contact pressure is assumed to be uniform for all types of footings and all types of soils (Fig. 11.35) if load is symmetric.

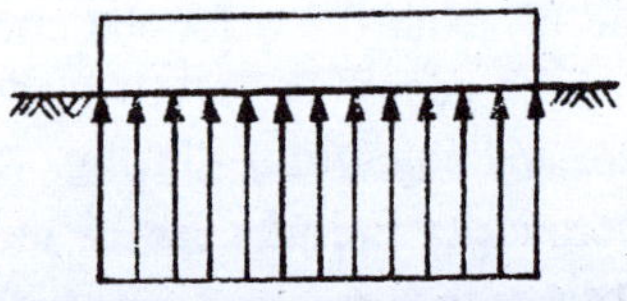

Fig. 11.35. Uniform contact Pressure.

The above assumption of uniform pressure distribution will result in a slightly unsafe design for rigid footing on clays, as the maximum bending moment at the centre is underestimated. It will give a conservative design for rigid footings on sandy (cohesionless) soils, as the maximum bending moment is overestimated. However, at the ultimate stage just before the failure, the soil behaves as an elasto-plastic material (and not an elastic material) and the contact pressure is uniform, and the assumption is justified at the ultimate stage.

11.26. LIMITATIONS OF ELASTIC THEORIES

Both Boussinesq's and Westergaard's theories are applicable to elastic materials. Actual soils do not behave in the manner as assumed in the analysis. The results obtained are necessarily approximate. The theories have the following limitations.

(1) The soil mass is never truly isotropic and homogeneous.

(2) The soil mass is not elastic as the particles do not return to the original position when the load is removed.

(3) The stress-strain ratio for most soils is not constant.

However, for most soils the stress-strain ratio is approximately constant provided the stresses are well below the failure stresses, and no unloading occurs.

Although the applicability of elastic theories to soil problems is questionable, yet the results are generally not much different from the observed values. A difference of 20 to 30% between the theoretical and the measured values may occur. This difference is generally ignored considering many complexities of the problem. The elastic theories are used as better theories are not yet available which can be used in a design office.

ILLUSTRATIVE EXAMPLES

Illustrative Example 11.1. *A concentrated load of 2000 kN is applied at the ground surface. Determine the vertical stress at a point P which is 6 m directly below the load. Also calculate the vertical stress at a point R which is at a depth of 6 m but at a horizontal distance of 5 m form the axis of the load.*

Solution. From Eq. 11.9, $$\sigma_z = \frac{3Q}{2\pi z^2} \cdot \frac{1}{[1 + (r/z)^2]^{5/2}}$$

Point P, $r/z = 0$, $$\sigma_z = \frac{3 \times 2000}{2\pi(6)^2} \cdot \frac{1}{[1 + 0]^{5/2}} = 26.53 \text{ kN/m}^2$$

Point R, $r/z = 5/6$, $$\sigma_z = \frac{3 \times 2000}{2\pi(6)^2} \cdot \frac{1}{[1 + (5/6)^2]^{5/2}} = 7.1 \text{ kN/m}^2$$

Illustrative Example 11.2. *A long strip footing of width 2 m carries a load of 400 kN/m. Calculate the maximum stress at a depth of 5 m below the centre line of the footing. Compare the results with 2 : 1 distribution method.*

Solution. From Eq. 11.21, $$\sigma_z = \frac{q}{\pi}(2\theta + \sin 2\theta)$$

In this case, $b = 1$ m and $z = 5$ m.

$\tan\theta = 1/5 = 0.2$ and $2\theta = 0.395$ radians

Taking $q = 400/2 = 200$ kN/m^2,

$$\sigma_z = \frac{200}{\pi}(0.395 + 0.385) = 49.6 \text{ kN/m}^2$$

2 : 1 Distribution method. From Eq. 11.61,

$$\sigma_z = \frac{q \times B}{B + z} = \frac{200 \times 2}{2 + 5} = \mathbf{57.1\ kN/m^2}$$

$$\text{Percentage error} = \frac{57.1 - 49.6}{49.6} \times 100 = \mathbf{15.2\%}$$

Illustrative Example 11.3. *There is a line load of 120 kN/m acting on the ground surface along y-axis. Determine the vertical stress at a point P which has x and z coordinates as 2 m and 3.5 m, respectively.*

Solution. From Eq. 11.17, $$\sigma_z = \frac{2q'}{\pi z}\left[\frac{1}{1 + (x/z)^2}\right]^2$$

At point P, $$\sigma_z = \frac{2 \times 120}{\pi \times 3.5}\left[\frac{1}{1 + (2/3.5)^2}\right]^2 = \mathbf{12.40\ kN/m^2}$$

Illustrative Example 11.4. *The unit weight of the soil in a uniform deposit of loose sand (K_0 = 0.50) is 16.5 kN/m³. Determine the geostatic stresses at a depth of 2 m.*

Solution. From Eq. 11.1,

$$\sigma_z = \gamma z = 16.5 \times 2.0 = \mathbf{33.0\ kN/m^2}$$

From Eq. 11.6, $\sigma_x = K_0\,\sigma_z = 0.5 \times 33.0 = \mathbf{16.5\ kN/m^2}$

Illustrative Example 11.5. *Determine the vertical stress at a point P which is 3 m below and at a radial distance of 3 m from the vertical load of 100 kN. Use Westergaard's solution (ν = 0.0).*

Solution. From Eq. 11.53, $\sigma_z = \dfrac{1}{\pi\,[1 + 2\,(r/z)^2]^{3/2}} \cdot \dfrac{Q}{z^2}$

or $$\sigma_z = \frac{1}{\pi\,[1 + 2(3/3)^2]^{3/2}} \cdot \frac{100}{(3)^2} = 0.681\ \text{kN/m}^2$$

Alternatively Using Eq. 11.54, $\sigma_z = I_w \cdot \dfrac{Q}{z^2}$

Taking I_w from Table 11.10 as 0.0613,

$$\sigma_z = 0.0613 \times \frac{100}{(3)^2} = \mathbf{0.681\ kN/m^2}$$

Illustrative Example 11.6. *Calculate the vertical stress at a point P at a depth of 2.5 m directly under the centre of the circular area of radius 2 m and subjected to a load 100 kN/m². Also calculate the vertical stress at a point Q which is at the same depth of 2.5 m but 2.5 m away from the centre of the loaded area.*

Solution. From Eq. 11.29 $\sigma_z = q\left[1 - \left\{\dfrac{1}{1 + (R/z)^2}\right\}^{3/2}\right]$

$$\sigma_z \text{ at } P = 100\left[1 - \left\{\frac{1}{1 + (2.0/2.5)^2}\right\}^{3/2}\right] = 52.39\ \text{kN/m}^2$$

From Fig. 11.16, the vertical stress at z = 1.25 R and r = 1.25 R is about 0.2 q.

Therefore, σ_z at Q = 0.2 × 100 = 20 kN/m²

Illustrative Example 11.7. *An L—shaped building in plan (Fig. E 11.7) exerts a pressure of 75 kN/m² on the soil. Determine the vertical stress increment at a depth of 5 m below the interior corner P.*

Solution. The loaded area is subdivided into three small areas such that each small area has one corner at P.

From Eq. 11.35, $\sigma_z = q\,[(I_N)_1 + (I_N)_2 + (I_N)_3]$

For area A_1 $m = n = 10/5 = 2.0$

From Table 11.9, $(I_N)_1 = 0.2325$

For area A_2 $m = 15/5 = 3,\ n = 10/5 = 2.0$

From Table 11.9, $(I_N)_2 = 0.2378$

For area A_3 $m = 20/5 = 4,\ n = 15/5 = 3$

From Table 11.9, $(I_N)_3 = 0.2450$

Therefore, σ_z = 75 [0.2325 + 0.2378 + 0.2450] = **53.65 kN/m²**.

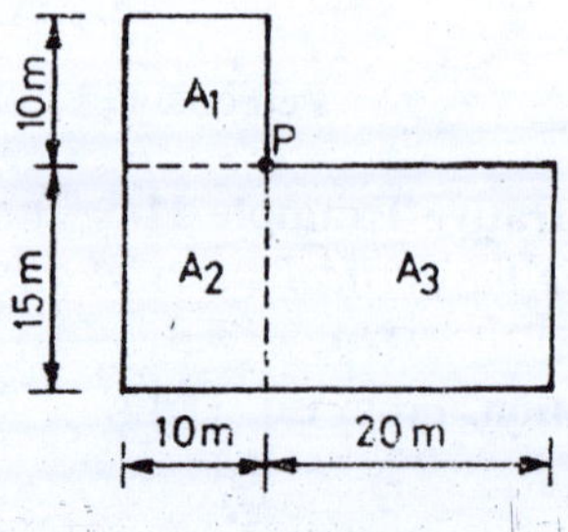

Fig. E 11.7.

Illustrative Example 11.8. *A rectangular foundation 4 m by 5 m carries a uniformly distributed load of*

200 kN/m². Determine the vertical stress at a point P located as shown in Fig. E 11.8 and at a depth of 2.5 m.

Solution. From Eq. 11.35, $\sigma_z = q[(I_N)_1 + (I_N)_2 + (I_N)_3 + (I_N)_4]$

In this case For A_1 and A_2, $m = 2/2.5 = 0.80$, $n = 2/2.5 = 0.80$

$(I_N)_1 = 0.1461$,

For A_3 and A_4, $m = 3/2.5 = 1.20$, $n = 2/2.5 = 0.80$

$(I_N)_3 = 0.1684$

Therefore, $\sigma_z = 200\,[0.1461 + 0.1461 + 0.1684 + 0.1684]$

$= \mathbf{125.8\ kN/m^2}$

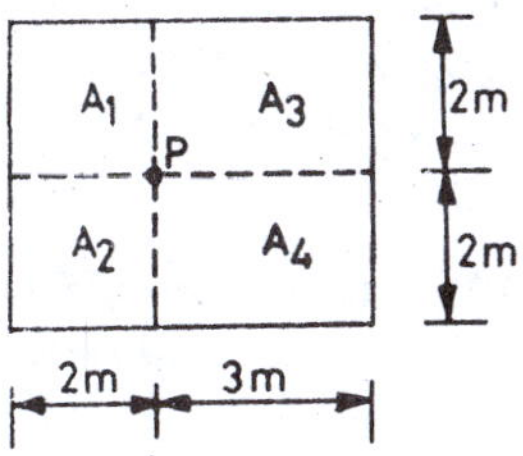

Fig. Ex. 11.8

Illustrative Example 11.9. *A T-shaped foundation (Fig. E 11.9) is loaded with a uniform load of 120 kN/m². Determine the vertical stress at point P at a depth of 5.0 m. Use Newmark's influence chart. Compare the answer by exact method.*

Solution. The foundation plan is drawn on a tracing paper with a scale such that the distance AB in Fig. 11.21 represents 5.0 m. The plan is placed on the Newmark chart such that point P is at the centre of the chart.

Number of area units occupied by plan = 63

From Eq. 11.40, $\sigma_z = I \times n \times q$

or $\sigma_z = 0.005 \times 63 \times 120 = 37.8\ \text{kN/m}^2$

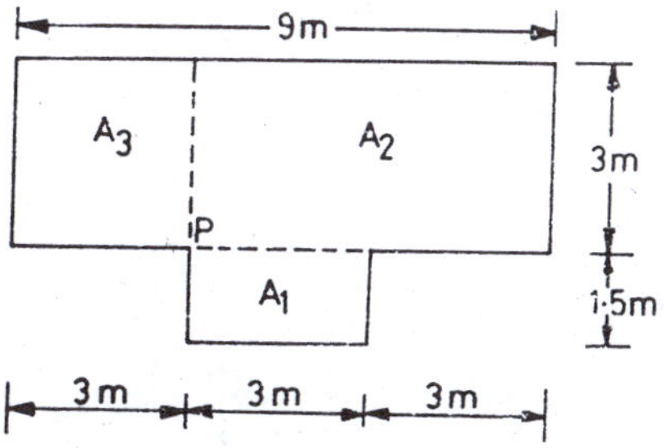

Fig. E 11.9

Exact method. The loaded area is divided into 3 areas, such that they have one corner at P.

Area A_1	$m = 3/5.00 = 0.60$; $n = 1.5/5.00 = 0.30$,	$(I_N)_1 = 0.0629$
Area A_2	$m = 3/5.00 = 0.60$; $n = 6/5.00 = 1.20$,	$(I_N)_2 = 0.1431$
Area A_3	$m = 3/5.00 = 0.60$; $n = 3/5.00 = 0.60$,	$(I_N)_2 = 0.1069$

From Eq. 11.35, $\sigma_z = q\,[(I_N)_1 + (I_N)_2 + (I_N)_3\,]$

or $\sigma_z = 120\,[0.0629 + 0.1431 + 0.1069] = 37.55\ \text{kN/m}^2$

Illustrative Example 11.10. *A rectangular loaded area 2 m × 2.5 m carries a load of 80 kN/m² (Fig. E 11.10). Determine the vertical stress at point P located outside the loaded area at a depth of 2.5 m.*

Solution. From Eq. 11.37, $\sigma_z = [(I_N)_1 - (I_N)_2 - (I_N)_3 + (I_N)_4)]$

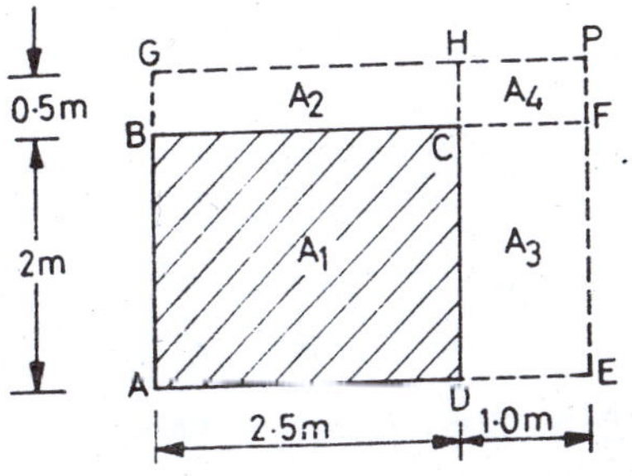

Fig. E 11.10.

For large rectangle *AEPG*, $m = 3.50/2.50 = 1.40$, $n = 2.50/2.50 = 1.00$

and $(I_N)_1 = 0.1914$

Area A_2 $m = 3.5/2.5 = 1.40$, $n = 0.5/2.5 = 0.20$, $(I_N)_2 = 0.0589$

Area A_3 $m = 2.5/2.5 = 1.0$, $n = 1.0/2.50 = 0.40$, $(I_N)_3 = 0.1013$

Area A_4 $m = 0.5/2.5 = 0.2$, $n = 1.0/2.5 = 0.40$, $(I_N)_4 = 0.0328$

Therefore,
$$\sigma_z = 80\,[0.1914 - 0.059 - 0.1013 + 0.0328]$$
$$= \mathbf{5.12\ kN/m^2}$$

Illustrative Example 11.11. *A rectangular foundation 3.0 × 1.50 m carries a uniform load of 40 kN/m². Determine the vertical stress at P which is 3 m below the ground surface (Fig. E 11.11). Use equivalent point load method.*

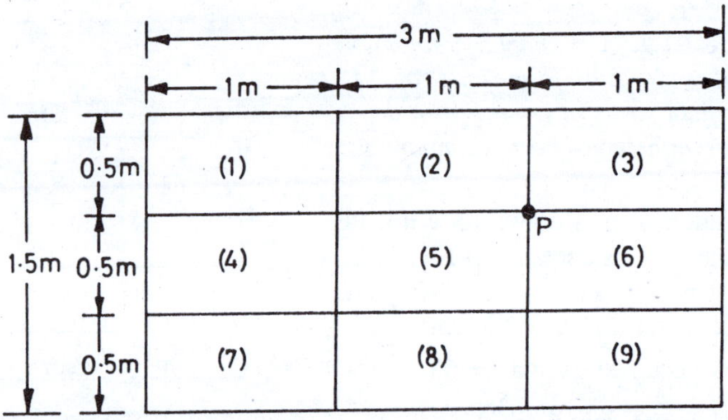

Fig. E 11.11.

Solution. Let us divide the loaded area into 9 small areas of size 0.5 m × 1.0 m.

Load on each area $= 40 \times (1.0 \times 0.5) = 20\text{kN}$

The stresses at point P are determined due to 9 point loads, using Boussinesq's solution (Eq. 11.9).

For loads (1) and (4), $r = \sqrt{1.5^2 + (0.25)^2} = 1.521$, $r/z = 0.507$

For loads (2), (3), (5), (6), $r = \sqrt{0.5^2 + 0.25^2} = 0.559$, $r/z = 0.186$

For loads (8) and (9), $r = \sqrt{(0.75)^2 + (0.5)^2} = 0.901$; $r/z = 0.300$

For load (7), $r = \sqrt{(1.5)^2 + (0.75)^2} = 1.677$; $r/z = 0.559$

Therefore,
$$\sigma_z = \Sigma \frac{3Q}{2\pi (z^2)} \times \frac{1}{[1 + (r/z)^2]^{5/2}}$$

In this case,
$$\sigma_z = \frac{3 \times 20}{2\pi (3)^2} \times \left[\frac{2}{[1 + (0.507)^2]^{5/2}} + \frac{4}{[1 + (0.186)^2]^{5/2}} + \frac{2}{[1 + (0.30)^2]^{5/2}} + \frac{1}{[1 + (0.559)^2]^{5/2}}\right]$$

or
$$\sigma_z = 1.061\ [1.129 + 3.674 + 1.612 + 0.507] = \mathbf{7.34\ kN/m^2}$$

Illustrative Example 11.12. *Determine the vertical stress at a point P which is 3 m below the ground surface and is on the centre line of the embankment shown in Fig. E 11.12. Take γ = 18 kN/m³.*

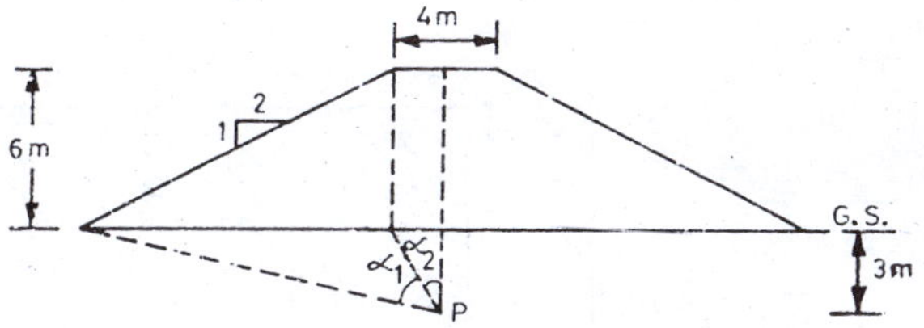

Fig. E 11.12.

Solution. From Eq. 11.46,

$$\sigma_z = \frac{2q}{\pi a}\left[a(\alpha_1 + \alpha_2) + b\,\alpha_1\right]$$

In this case, a = 12 m and b = 2 m, q = 6 × 18 = 108 kN/m^2

$$\tan \alpha_2 = 2/3.0 = 0.667; \qquad \alpha_2 = 0.588 \text{ radians}$$

$$\tan(\alpha_1 + \alpha_2) = \frac{14}{3.00} = 4.667; \quad (\alpha_1 + \alpha_2) = 1.359 \text{ radians}$$

Therefore,
$$\sigma_z = \frac{2 \times 108}{\pi \times 12}\left[12(1.359) + 2 \times (1.359 - 0.588)\right] = \mathbf{102.3\ kN/m^2}$$

PROBLEMS

A. Numerical

11.1. A monument weighing 15 MN is erected on the ground surface. Considering the load as a concentrated one, determine the vertical pressure directly under the monument at a depth of 8 m below the ground surface.

[**Ans.** 111.9 kN/m^2]

11.2. A concentrated load of 50 kN acts on the surface of a homogeneous soil mass of large extent. Determine the stress intensity at a depth of 5 m, directly under the load, and at a horizontal distance of 2.5 m.

[**Ans.** 0.955 kN/m^2; 0.55 kN/m^2]

11.3. Two columns A and B are situated 6 m apart. Column A transfers a load of 500 kN and column B, a load of 250 kN. Determine the resultant vertical stress on a horizontal plane 20 m below the ground surface at points vertically below the points A and B. [**Ans.** 59.8 kN/m^2 ; 29.9 kN/m^2]

11.4. An excavation 3 m × 6 m for foundation is to be made to a depth of 2.5 m below ground level in a soil of bulk unit weight 20 kN/m^3. What effect this excavation will have on the vertical pressure at a depth of 6 m measured from the ground surface vertically below the centre of foundation ? I_N for m = 0.43 and n = 0.86 is 0.10.

[**Ans.** decrease 20 kN/m^2]

11.5. A square foundation (5 m × 5 m) is to carry a load of 4000 kN. Calculate the vertical stress at a depth of 5 m below the centre of the foundation. I_N = 0.084 for m = n = 0.50.

(*ii*) Also determine the vertical stress using 1 : 2 distribution. [**Ans.** 53.76 kN/m^2 ; 40 kN/m^2]

11.6. A water tower has a circular foundation of 10 m diameter. If the total weight of the tower, including the foundation, is 2 × 10^4 kN, calculate the vertical stress at a depth of 2.5 m below the foundation level.

[**Ans.** 231.9 kN/m^2]

11.7. A rectangular foundation, 3 m × 2.1 m, is perfectly flexible and carries a load of 300 kN/m^2. Determine the vertical pressure at a depth of 5 m below a point P shown in Fig. P 11.7. [**Ans.** 31.8 kN/m^2]

11.8. The contact pressure for a square footing 2 m × 2 m is 400 kN/m^2. Using 1 : 2 distribution, determine the depth at which the contact pressure is 100 kN/m^2. [**Ans.** 2 m]

11.9. A rectangular foundation 20 m × 10 m subjects the subgrade to a contact pressure of 250 kN/m^2. Determine the vertical stress at a point P located at a depth of 5 m (Fig. P 11.9).

(Use Table for I_N values) [**Ans.** 3.375 kN/m^2]

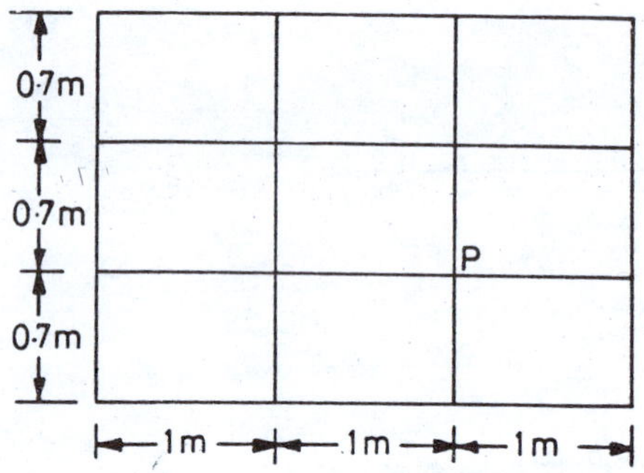

Fig. P 11.7.

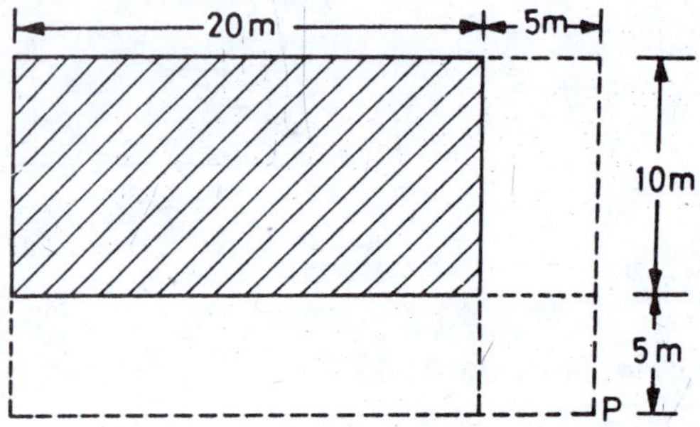

Fig. P 11.9.

11.10. A 1000 kN load is uniformly distributed on a surface area of 3 m × 2.5 m. Find the approximate value of vertical stress at a depth of 2 m, using,
(*i*) 2 : 1 distribution
(*ii*) 60° distribution. **[Ans.** 44.4 kN/m^2 ; 39.2 kN/m^2]

11.11. A concentrated load of 1000 kN acts vertically at the ground surface. Determine the vertical stress at a point which is at
(*i*) a depth of 2.5 m and a horizontal distance of 4.0 m
(*ii*) at a depth of 5.0 and a radial distance of 2.5 m. **[Ans.** 3.2 kN/m^2; 10.93 kN/m^2]

B. Descriptive and Objective Type

11.12. State the assumptions made in computing stresses below the ground surface due to a point load acting on it. Discuss their validity in practice.

11.13. Derive an expression for the vertical stress at a point due to a point load, using Boussinesq's theory.

11.14. What do you understand by geostatic stresses ? How are these determined ?

11.15. What is an influence diagram ? What is its use in practice ?

11.16. Derive an expression for the vertical stress at a point due to a line load. Give examples of a line load.

11.17. How would you determine the stresses at a point due to a (*a*) strip load (*b*) circular load. Compare the zones of influence due to the two types of loads.

11.18. Describe the method of calculating the stress at a point below the corner of a rectangular load. How is this method used for finding the stresses at points other than that below the corner ?

11.19. Discuss the basis of the construction of Newmark's influence chart. How is it used ?

11.20. Explain Westergaard's theory for the determination of the vertical stress at a point. How is it different from Boussinesq's solution ?

11.21. What is Fenske's chart ? Explain its construction and use.

11.22. Discuss various approximate methods for the determination of the vertical stress at a point What are their limitations ?

11.23. What do you understand by contact pressure ? What are the factors that affect the contact pressure distributions? Draw the contact pressure distribution diagram for flexible and rigid footings on sand and clayey soils.

11.24. Mention whether the following statements are true or false.
(*a*) The vertical stress due to a point load depends upon modulus of elasticity.
(*b*) For determination of the deformation, the secant modulus at the peak stress is used.
(*c*) The Poisson ratio for most of the soils is zero.
(*d*) The horizontal stress can be more than the vertical stress.
(*e*) The Boussinesq influence coefficient just below the point load is zero.
(*f*) The maximum shear stress due to a strip load is constant at all points.
(*g*) The zone of influence due to a circular load is deeper than that due to a strip load.
(*h*) While determining Newmark's influence coefficient, the constant *m* and *n* can be interchanged.

(*i*) The Boussinesq solution always gives stresses greater than the Westergaard solution.

(*j*) The equivalent point load gives reliable results if the dimension of the area is greater than three times the depth.

(*k*) Two-to-one load distribution and sixty-degree distribution give approximately the same stresses.

(*l*) In actual design, the contact pressure distribution is generally taken as uniform.

[**Ans.** True, (*d*), (*h*), (*k*), (*l*)]

C. Multiple-Choice Questions

1. The stress developed at a point in the soil exactly below a point load at the surface is
(*a*) proportional to the depth of point.
(*b*) proportional to the square of the depth of point.
(*c*) inversely proportional to the depth of point.
(*d*) inversely proportional to the square of the depth of point.

2. An isobar is a curve which
(*a*) joins points of equal horizontal stress.
(*b*) joins points of equal vertical stress.
(*c*) joins points of zero vertical stress.
(*d*) joins points of maximum vertical stress.

3. If the entire semi-infinite soil mass is loaded with a load intensity of q at the surface, the vertical stress at any depth is equal to
(*a*) q (*b*) 0.5 q
(*c*) zero (*d*) infinity

4. For a strip of width B subjected to a load intensity of q at the surface, the pressure bulb of intensity 0.2 q extends to a depth of
(*a*) 3B (*b*) 6 B
(*c*) 1.5 B (*d*) B

5. Newmark's influence chart can be used for the determination of vertical stress under
(*a*) circular load area only (*b*) rectangular loaded area only
(*c*) strip load only (*d*) Any shape of loaded area

6. The Westergaard analysis is used for
(*a*) homogeneous soils (*b*) cohesive soils
(*c*) sandy soils (*d*) stratified soils

7. A concentrated load of 1000 kN acts vertically at a point on the soil surface. According to Boussinesq's equation the ratio of the vertical stresses at depths of 3m and 5m is
(*a*) 0.35 (*b*) 0.70
(*c*) 1.75 (*d*) 2.78

8. A load of 2000 kN is uniformly distributed over an area of 3 m × 2m. The average vertical stress at a depth of 2m using 2 : 1 distribution is
(*a*) 160 kN/m^2 (*b*) 100 kN/m^2
(*c*) 48 kN/m^2 (*d*) 37 kN/m^2

[**Ans.** 1. (*d*), 2. (*b*), 3. (*a*), 4. (*a*), 5. (*d*), 6. (*d*), 7. (*d*), 8. (*b*)]

12

Consolidation of Soils

12.1. INTRODUCTION

When a soil mass is subjected to a compressive force, like all other materials, its volume decreases. The property of the soil due to which a decrease in volume occurs under compressive forces is known as the compressibility of soil. The compression of soils can occur due to one or more of the following causes.

(1) Compression of solid particles and water in the voids.

(2) Compression and expulsion of air in the voids.

(3) Expulsion of water in the voids.

Compression of solid particles is negligibly small. Compression of water in the voids is also extremely small, as the water is almost incompressible in the range of stresses involved in soil engineering. Therefore, the compression due to the first cause is not much significant.

Air exists only in partially saturated soils and dry soils. The compression of the air is rapid as it is highly compressible. Further, air is expelled quickly as soon as the load is applied. However, the compression due to the second cause is not relevant for saturated soils.

When the soil is fully saturated, compression of soil occurs mainly due to the third cause, namely, expulsion of water. As this chapter is mainly concerned with saturated soils, only this cause is relevant.

The compression of a saturated soil under a steady static pressure is known as *consolidation.* It is entirely due to *expulsion of water from the voids.* It is similar to the action of squeezing of water from a saturated sponge under pressure. The soil behaves as a saturated sponge. As the consolidation of soils occurs, the water escapes. The solid particles shift from one position to the other by rolling and sliding and thus attain a closer packing. It is worth noting that the decrease in volume of soil occurs not due to compression of solids or water but due to the shifting of positions of the particles as the water escapes. Small volume changes may occur due to bending, distortion and fracture of the soild particles, but such changes are insignificant in the ordinary range of stresses involved in soil engineering problems. However, bending, distortion and fracture are indirectly responsible for a further decrease in volume due to shifting of particles.

Settlement of a structure is its vertical, downward movement due to a volume decrease of the soil on which it is built. In other words, the settlement is the gradual sinking of a structure due to compression of the soil below. A study of consolidation characteristics is extremely useful for forecasting the magnitude and time of the settlement of the structure.

The compression of soils due to expulsion of air due to dynamic methods, such as rolling and tamping, is known as compaction (see chapter 14).

This chapter deals mainly with consolidation of fully saturated soils.

(Note. In geology, consolidation means hardening of soils due to solidification. It should not be confused with the word consolidation used in soil engineering).

12.2. INITIAL, PRIMARY AND SECONDARY CONSOLIDATION

The consolidation of a soil deposit can be divided into 3 stages :

(1) Initial Consolidation. When a load is applied to a partially saturated soil, a decrease in volume occurs due to expulsion and compression of air in the voids. A small decrease in volume also occurs due to compression of solid particles. The reduction in volume of the soil just after the application of the load is known as initial consolidation or initial compression. For saturated soils, the initial consolidation is mainly due to compression of solid particles.

(2) Primary Consolidation. After initial consolidation, further reduction in volume occurs due to expulsion of water from voids. When a saturated soil is subjected to a pressure, initially all the applied pressure is taken up by water as an excess pore water pressure, as water is almost incompressible as compared with solid particles. A hydraulic gradient develops and the water starts flowing out and a decrease in volume occurs. The decrease depends upon the permeability of the soil and is, therefore, time dependent. This reduction in volume is called primary consolidation.

In fine-grained soils, the primary consolidation occurs over a long time. On the other hand, in coarse-grained soils, the primary consolidation occurs rather quickly due to high permeability. As water escapes from the soil, the applied pressure is gradually transferred from the water in the voids to the soild particles. Thus, the effective stress is increased.

(3) Secondary Consolidation. The reduction in volume continues at a very slow rate even after the excess hydrostatic pressure developed by the applied pressure is fully dissipated and the primary consolidation is complete. This additional reduction in the volume is called secondary consolidation. The causes for secondary consolidation are not fully established. It is attributed to the plastic readjustument of the solid particles and the adsorbed water to the new stress system. In most inorganic soils, it is generally small.

In the discussions that follow the word consolidation means primary consolidations unless otherwise stated. The primary consolidation is the most important component of the total consolidation and forms the main subject matter in this chapter.

12.3. SPRING ANALOGY FOR PRIMARY CONSOLIDATION

The process of primary consolidation can be explained with the help of the spring analogy given by Terzaghi. Fig. 12.1 (*a*) shows a cylinder fitted with a tight-fitting piston having a valve. The cylinder is filled with water and contains a spring of specified stiffness. Let the initial length of the spring be 100 mm and the stiffness of spring be 10 mm/N. Let us assume that the piston is weightless and the spring and water are initially free of stress.

When a load *P* (say, 1 N) is applied to the piston, with its valve closed, the entire load is taken by water [Fig. 12.1 (*b*)]. The stiffness of the spring is negligible compared with that of water, and consequently, no load is taken by spring. From equilibrium,

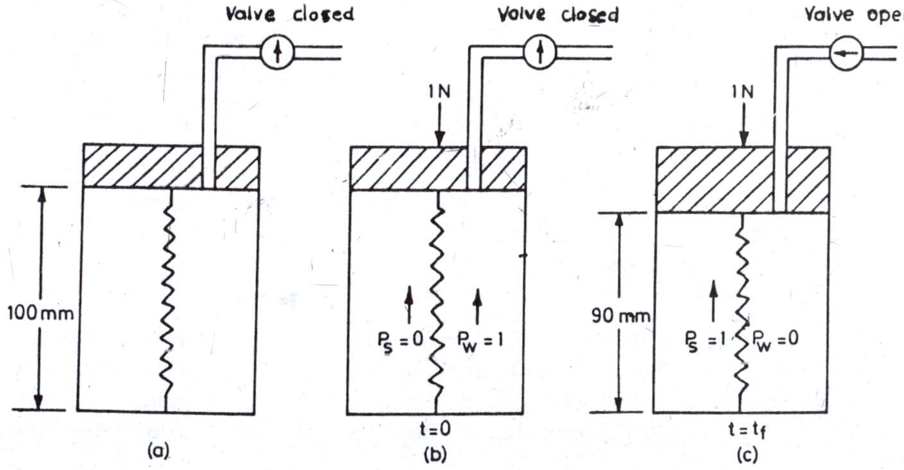

Fig. 12.1. Spring Analogy.

$$P_w + P_s = P \qquad ...(12.1)$$

where P_w = load taken by water, P_s = load taken by spring, and P = total load.

For P = 1 N, Eq. 12.1 becomes

$$P_w + P_s = 1.0 \qquad ...(12.2)$$

Initially (t = 0) when valve is closed, P_s = 0.00. Therefore, $P_w = 1.0$

If the valve is now gradually opened, water starts escaping from the cylinder. The spring starts sharing some load and a decrease in its length occurs. When a portion (ΔP) of the load is transferred from the water to the spring, Eq. 12.2 becomes

$$\Delta P + (1.0 - \Delta P) = 1.0$$

As more and more water escapes, the load carried by the spring increases. Fig. 12.2 shows the transfer of the load from the water to the spring. Eventually, when the steady conditions are established, the water stops escaping. Finally, at time $t = t_f$, the entire load is taken by spring. Thus, $P_w = 0$ and $P_s = 1.00$.

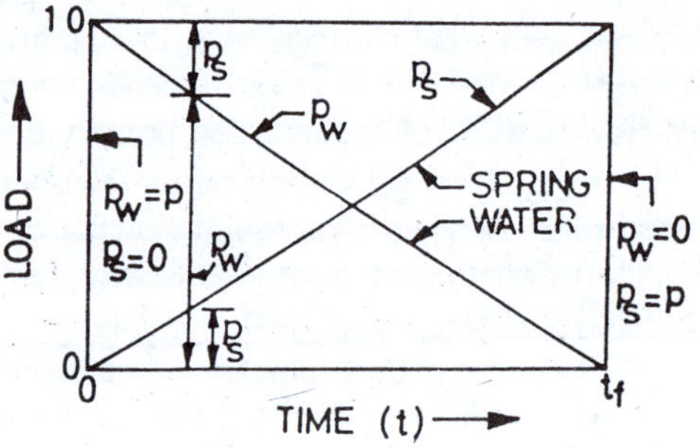

Fig. 12.2. Load Sharing between spring and water.

This load causes a decreases in length of the spring by 10 mm. The final length is 90 mm, as shown in Fig. 12.1 (*c*).

As the load carried by water is zero, it is again free of excess pressure. Now if the valve is closed and the load P is increased to 2N, the process of transfer of load repeats and finally the spring takes the complete load and its length becomes 80 mm. Likewise, the process is repeated till the final increment of the load has been applied.

12.4. BEHAVIOUR OF SATURATED SOILS UNDER PRESSURE

The behaviour of saturated soils when subjected to a steady, static pressure is similar to that of the spring analogy model. The solid particles in the saturated soil behave like springs, while the water in the voids behaves like water in the cylinder. The permeability of the soil controls the flow of water and it can be likened to the valve in the piston. The pore water pressure ($\bar{u}$) in the soil is analogous to the pressure carried by water in the cylinder. Finally the stress developed in the spring is analogous to the effective stress ($\bar{\sigma}$) developed in the soil.

When a pressure $\Delta\sigma_1$ is applied to a saturated soil sample of unit cross-sectional area, the pressure is shared by the solid particles and water as

$$\Delta\bar{\sigma} + \bar{u} = \Delta\sigma_1 \qquad ...(12.3)$$

Initially, just after the application of pressure, the entire load is taken by water. The pressure developed in water, also known as *excess hydrostatic pressure* or *hydrodynamic pressure* ($\bar{u}$), is equal to the applied pressure. The pressure taken by the solid particles, represented as the effective stress, $\Delta\bar{\sigma}$, is zero. Thus

$$0 + (\bar{u}_1) = \Delta\sigma_1 \qquad ...(12.4)$$

The excess hydrostatic pressure developed after the application of the load sets up a hydraulic gradient, and the water starts escaping from the voids. As the water escapes, the applied pressure is transferred from the water to the solids. Eventually, the whole of the pressure is transferred to the soil solids as the effective stress, and the excess water pressure becomes zero. Thus

$$\Delta\bar{\sigma} = \Delta\sigma_1 \qquad ...(12.5)$$

As the effective stress increases, the volume of the soil decreases. The decrease in volume is generally expressed as change in void ratio. Fig. 12.3 shows decrease in void ratio with time, as the effective stress increases due to transfer of pressure to the solid particles. Initially, just after the application of the pressure (t = 0), the void ratio is e_0. Finally, when the pressure has been fully transferred to the solid particles ($t = t_f$), the void ratio is $(e_f)_1$. It must be noted that the curve shown in Fig. 12.3 is drawn for application of one pressure increment $\Delta\sigma_1$.

If the applied pressure is now increased to $\Delta\sigma_2$, the process of load transfer repeats and the soil attains eventually a different final void ratio $(e_f)_2$ when the entire load is transferred to the solid particles. A curve can be dawn between the final void ratios and the corresponding effective stresses for different load increments. (Fig. 12.4). It may be noted that as the effective stress increases, the final void ratio decreases, and, therefore, the volume of the soil decreases. The reduction in volume is due to expulsion of water from the voids under excess hydrostatic pressure and is, therefore, primary consolidation.

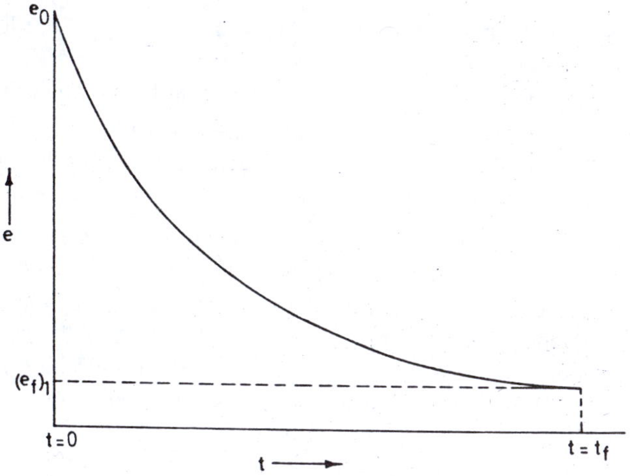

Fig. 12.3. Variation of void ratio with time

12.5. CONSOLIDATION TEST

The consolidation test is conducted in a laboratory to study the compressibility of a soil. The test is performed in the consolidation test apparatus, known as the *consolidometer* or *an oedometer* [(Fig. 12.5 (*a*)]. It consists of a loading device and a cylindrical container called consolidation cell. The soil specimen is placed in the cell between top and bottom porous stones. The consolidation cells are of two types : (1) Floating or free ring cell [Fig. 12.5 (*b*)] in which both the top and bottom porous stones are free to move. The top porous stone can move downward and the bottom stone can move upward as the sample consolidates. (2) Fixed ring cell [Fig. 12.5 (*c*)] in which the bottom porous stone cannot move. Only the top porous stone can move downward as the specimen consolidates. The fixed ring cell can also be used as a variable-head permeability test apparatus. For this purpose, a piezometer is attached to the base of the cell.

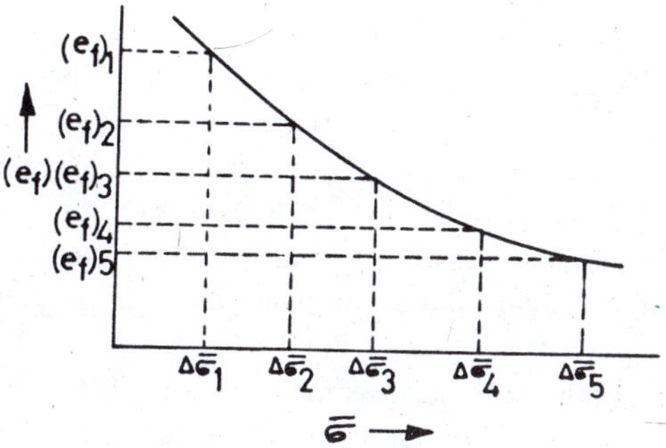

Fig. 12.4. Plot between e_f and $\bar{\sigma}$.

The inside surface of the ring should be smooth and polished to reduce friction. The ring imposes a condition of zero lateral strain on the soil sample. The internal diameter of the cell is usually 60 mm, but the

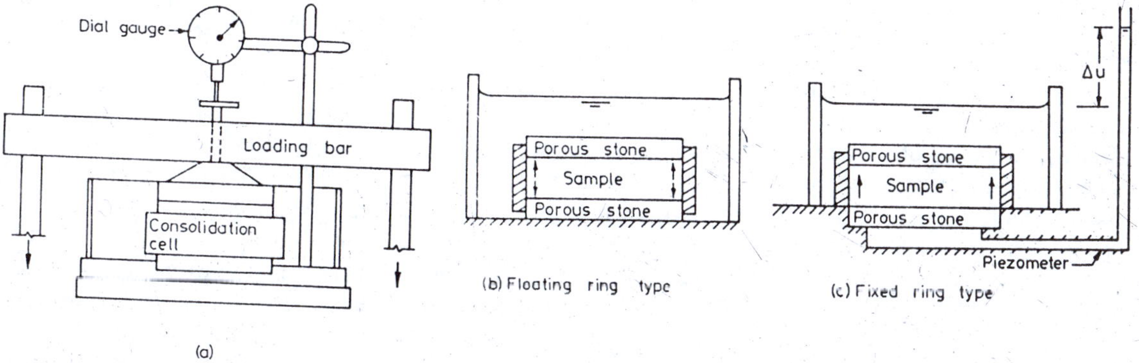

Fig. 12.5. Consolidation Test.

cells with a diameter upto 100 mm are also available. The thickness of the sample is fixed from the following considerations:

(1) The thickness of the sample should be as small as possible to reduce side friction, but a minimum thickness of 20 mm is usually required to get uniform distribution of pressure on the sample.

(2) The diameter to the thickness ratio should be a minimum of 3.

(3) The thickness of the specimen should not be less than 10 times the maximum size of the particle.

The thickness of the sample for a 60 mm diameter cell is usually taken as 20 mm. The specimens of diameter 50, 70 and 100 mm may be used in special cases.

The consolidometer has arrangements for the application of the desired load increment, saturation of sample and measurement of change in thickness of the sample at every stage of consolidation process. The consolidation cell is placed in a water jacket or water trough so that water has free access into and out of the sample. The cell is provided with a perforated pressure pad at its top for the application of load. The load is applied either by suspending weights from a hanger resting at the centre of the pressure pad or by a lever arrangement. The arrangement for saturation of the sample consists of a small water reservoir connected to the cell with a plastic tube (not shown in figure). A dial gauge is used to measure the change in thickness as the consolidation takes place. The sample is kept submerged under water to prevent evaporation from its surface.

Before conducting the test, the porous stones are saturated either by boiling them in distilled water for about 15 minutes or by keeping them submerged under water for 4 to 8 hours. The bottom porous stone is first placed in the consolidation cell and a filter paper is fixed on the porous stone. The ring containing the sample is then placed on the bottom porous stone. Another filter paper is kept on the top of the sample and then top porous stone is placed. The loading pad is placed on the top porous stone. The bolts are tightened so as to hold the entire assembly, and then the consolidation cell is kept under the loading unit. It should be centred carefully so that the load is applied axially. The dial gauge is mounted and adjusted. The mould assembly is connected to the water reservoir to saturate the sample. The level of water in the reservoir should be approximately same as that of the sample.

An initial setting pressure of about 5.0 kN/m^2 (for very soft soils, 2.5 kN/m^2) is applied to the sample. The initial setting pressure is chosen such that there is no swelling. The load is allowed to stand till there is no change in the dial gauge reading or 24 hours whichever is less. The final dial gauge reading under the initial setting pressure is noted.

The first increment of load to give a pressure of 10 kN/m^2 is then applied to the specimen. The dial gauge readings are taken after 0.25, 1.0, 2.25, 4.0, 6.25, 9.0, 12.25, 16.0, 20.25, 25, 36, 49, 64, 81, 100, 121, 144, 169, 196, 225, 289, 324, 400, 500, 600 and 1440 minutes (24 hours). Sometimes, after 49 minutes, readings are taken at 1, 2, 4, 8, 10 and 24 hours. The primary consolidation in the sample is usually complete within 24 hours.

The second increment of the load is then applied. It is usual practice to double the previous load in each increment. The successive pressures usually applied are 20, 40, 80, 160, 320 and 640 kN/m^2, etc., till the desired maximum required load intensity is reached. The maximum load intensity is governed by the actual loading on the soil in the field after the construction of the structure.

After the consolidation under the final load increment is complete, the load is reduced to one-fourth of the final load (160 kN/m^2 in above case) and allowed to stand for 24 hours. The sample takes water and swells. The reading of the dial gauge is taken when the swelling is complete. The load is further reduced to one-fourth intensity (40 kN/m^2) and the swelling recorded after 24 hours. The load is then reduced to 10 kN/m^2 and the swelling is noted. The load is finally reduced to the initial setting load and kept for 24 hours and the final dial gauge reading taken. Throughout the test, the container gutter should be kept filled with water.

Immediately after complete unloading, the ring with the sample is taken out. The excess surface water is dried using a blotting paper. The weight of the ring and the sample is taken. The sample is then dried in an oven (maintained at 110°C) for 24 hours and its dry mass M_s and the water content are determined.

(See Chapter 30, Sect. 30.15 for the laboratory experiment).

12.6. DETERMINATION OF VOID RATIO AT VARIOUS LOAD INCREMENTS

The results of a consolidation test are plotted in the form of a plot between the void ratio and the effective stress. It is, therefore, required to determine the void ratio at various load increments. There are two methods :

(1) Height of solids method, (2) Change in void ratio method.

The first method is a general method applicable to both saturated and unsaturated soils. The second method is applicable only to saturated soils.

(1) Height of Solids method. In this method, equivalent height of solids is determined from the dry mass of the soil. The height of solids is given by

$$H_s = \frac{V_s}{A} = \left(\frac{M_s}{G\rho_w}\right) \cdot \frac{1}{A} \qquad ...(12.6)$$

where H_s = height of solids, V_s = volume of solids,

M_s = dry mass of sample, G = specific gravity of solids, A = cross-sectional area of specimen.

From definition of void ratio,

$$e = \frac{\text{volume of voids}}{\text{volume of solids}} = \frac{V - V_s}{V_s} \qquad ...(a)$$

Eq. (*a*) can be written as,

$$e = \frac{(A \times H) - (A \times H_s)}{(A \times H_s)} = \frac{H - H_s}{H_s} \qquad ...(12.7)$$

where H is the total height (total thickness).

Thus, the void ratio is determined from the total height (H) and the height of solids. The total thickness of the sample is measured at least once during the test, usually either before the start of the test or at the end of the test. At other stages of loading, the thickness H is worked out from the measured thickness and the difference in dial gauge readings.

Therefore, $$H = H_0 \pm \Sigma \Delta H \qquad ...(12.8)$$

where H_0 = initial height and ΔH = change in height.

See Table 12.1 for the illustration of the method.

Table 12.1. Computation of Void ratios by Height of Solids Method

Given data H_0 = 25 mm, A = 50 cm^2, Volume = 125 ml,

M_s = 190.24 gm, G = 2.67, w_f = 24.94%

Least count of dial gauge = 0.01 mm

Observations			*Calculations*		
Applied pressure (kN/m²)	*Dial gauge reading*	*Change in thickness ΔH (mm)*	$H = H_0 \pm \Sigma\Delta H$	$H - H_s$	*e fom Eq. (a)* $= (H - H_s)/H_s$
0.0	490		25.00	10.75	0.754
10.0	482	− 0.08	24.92	10.67	0.748
20.0	470	− 0.12	24.80	10.55	0.740
40.0	431	− 0.39	24.41	10.16	0.713
80.0	390	− 0.41	24.00	9.75	0.684
160.0	343	− 0.47	23.53	9.28	0.651
320.0	295	− 0.48	23.05	8.80	0.617
640.0	249	− 0.46	22.59	8.34	0.585
0.0	364	+ 1.15	23.74	9.49	0.666

From Eq. 12.6, $H_s = \frac{190.24}{2.67 \times 1.0} \times \frac{1}{50} = 1.425 \text{ cm} = 14.25 \text{ mm}$

From Eq. 12.7, $e = \frac{H - H_s}{H_s} = \frac{H - 14.25}{14.25}$ ...(a)

Obviously, the initial void ratio (e_0) at the start of the test is given by

$$e = \frac{H_0 - H_s}{H_s} \quad ...(12.9)$$

For an intermediate stage, $e = \frac{H - H_s}{H_s}$ ...(12.10)

After determination of the void ratio and the water content at the beginning and at the end of the test, the corresponding degree of saturation can be found from the relation, $S = wG/e$.

From the calculated void ratios, a plot of 'e' versus log $\overline{\sigma}$ can be made, as shown in Fig. 12.8.

(2) Change in Void ratio method. In this method, the final void ratio (e_l), corresponding to complete swelling conditions after the load has been removed, is determined from its water content, using the equation,

$$e_l = wG$$

assuming that the soil is fully saturated.

The void ratio corresponding to intermediate loading stages is determined as explained below. From the definition of void ratio,

$$e = \frac{V - V_s}{V_s} = \frac{V}{V_s} - 1 \quad ...(a)$$

where V = total volume and V_s is the volume of solids. Eq. (a) can be written as

$$V = V_s\,(1 + e)$$

or $A \times H = V_s\,(1 + e)$ ...(b)

where A is the cross-sectional area of the specimen and H is its total height.

By partial differentiation of (b), $A\,dH = V_s\,de$ ...(c)

From Eqs. (b) and (c), $\frac{dH}{H} = \frac{de}{1 + e}$

or $\Delta e = \frac{(1 + e)}{H}\,\Delta H$ ...(12.11)

Eq. 12.11 can also be derived directly, taking the volume of solids as unity and the cross-sectional area also as unity. In this case.

Original volume $= H = 1 + e$

Change in volume $= \Delta e$, and change in height $= \Delta H$

Therefore, $\frac{\Delta e}{1 + e} = \frac{\Delta H}{H}$

or $\Delta e = \frac{(1 + e)}{H}\,\Delta H$ ...(same as 12.11)

As the void ratio e_l and the total height H of the sample are known at the end of the test, the void ratio at any other stage can be determined from the change in thickness ΔH measured by the dial gauge. Thus, the change in void ratio (Δe) under each pressure increment is calculated from Eq. 12.11 by working backwards from the known value of void ratio e_l at the end of the test after swelling.

Thus $\Delta e = \frac{(1 + e_l)}{H_l}\,\Delta H$ ...[12.11(a)]

where ΔH is the change in thickness, as measured with respect to the thickness H_l at the end of the test. See Table 12.2 for the illustration of the method.

Table 12.2. Computation of Void Ratios by Change in Void Ratio Method

Given Data $H_0 = 25$ mm, $A = 50\ cm^2$, Volume = 125 ml, $M_s = 190.24$ gm,

$G = 2.67$, $w_l = 24.94\%$, $H_l = 23.74$

$e_l = w_l \times G = 0.2494 \times 2.67 = 0.666.$

From Eq. 12.11 (*a*),
$$\Delta e = \frac{(1 + 0.666)}{23.74} \times \Delta H = 0.0702\, \Delta H \quad ...(d)$$

Least count of dial gauge = 0.01 mm.

Observations			*Calculations*		
Applied pressuree (kN/m^2)	*Dial gauge reading*	*Change in thickness* ΔH *(mm)*	$H = H_0 \pm \Sigma \Delta H$	Δe *(from Eq. (d)*	e
0.0	490	–0.08	25.00		0.754
10.0	482		24.92	– 0.006	0.748
20.0	470	– 0.12	24.80	– 0.008	0.740
40.0	431	– 0.39	24.41	– 0.027	0.713
80.0	390	– 0.41	24.00	– 0.029	0.684
160.0	343	– 0.47	23.53	– 0.033	0.651
320.0	295	– 0.48	23.05	– 0.034	0.617
640.0	249	– 0.46	22.59	– 0.032	0.585
0.0	364	+ 1.15	23.74	+ 0.081	0.666 ↑

12.7. CONSOLIDATION TEST RESULTS

(1) Dial gauge reading-time plot. Fig. 12.6 (*a*) shows the plot between the dial gauge reading and time for a typical load increment for clay and sand samples. The thickness just after the application of the load increment ($t = t_0$), is a maximum which decreases as the time increases. The decrease is rapid initially but it

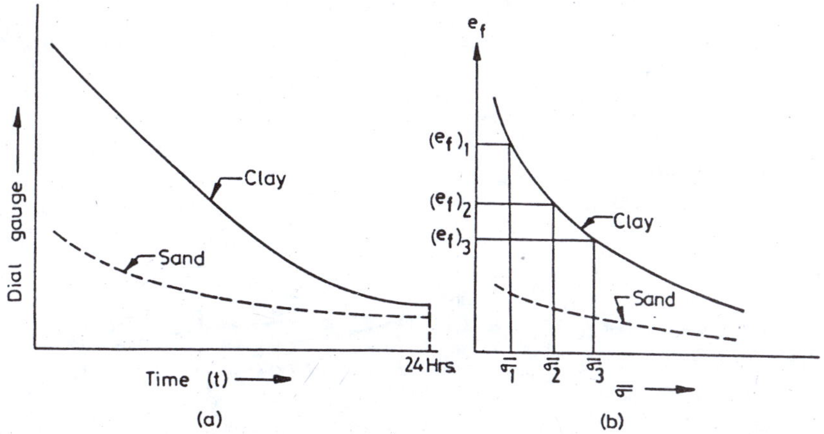

Fig. 12.6. (*a*) Dial gauge reading-time plot. (*b*) Final void ratio–$\bar{\sigma}$ plot.

slows down as the time passes. There is practically no change in thickness after 24 hours. The consolidation at that load increment is considered to the complete at 24 hours. For sand, the change in thickness occurs very quickly and stops after a few minutes. This is due to high permeability of the sand which permits easy flow of water.

The plot between the dial gauge reading and time is required for determining the coefficient of consolidation, which is useful for obtaining the *rate* of consolidation in the field.

(2) Final void ratio-effective stress plot. The thickness of the specimen after 24 hours of application of the load increment is taken as the final thickness for that increment. The final void ratio (e_f) corresponding to the final thickness for each increment is determined using the methods discussed in the preceding section.

Fig. 12.6 (*b*) shows the plot between the final void ratios, $(e_f)_1$, $(e_f)_2$, $(e_f)_3$ ... etc. and the corresponding effective stresses $\bar{\sigma}_1$, $\bar{\sigma}_2$ $\bar{\sigma}_3$, ...for load increments 1, 2, 3, ... etc. As the sand is relatively less compressible, the change in void ratio is small. The plot between the final void ratio and the effective stress is required for determination of the *magnitude* of the consolidation settlement in the field.

The reader must carefully note the difference between Fig. 12.6 (*a*) and Fig. 12.6 (*b*). The former shows the process of consolidation under a particular increment. For each load increment, a plot like Fig. 12.6 (*a*) can be plotted. The latter shows the plot between the final void ratios reached under different load increments and the corresponding effective stresses under those increments.

(3) Final void ratio—log σ plot. Fig. 12.7 (*a*) shows a plot between the final void ratio and the effective stress, which is similar to one in Fig. 12.6 (*b*). For convenience the suffix *f* has been dropped. The curve has

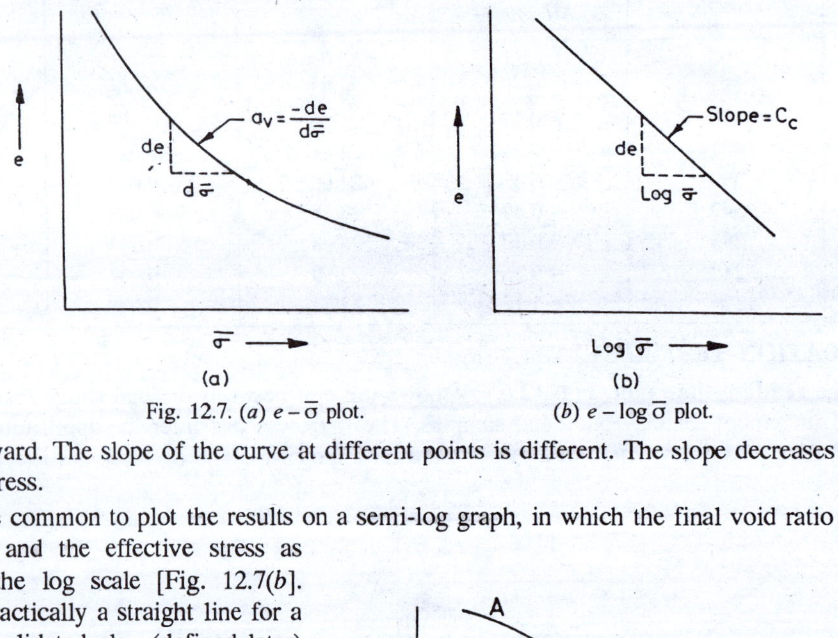

Fig. 12.7. (*a*) $e-\bar{\sigma}$ plot. (*b*) $e-\log\bar{\sigma}$ plot.

concavity upward. The slope of the curve at different points is different. The slope decreases with an increase in effective stress.

It is more common to plot the results on a semi-log graph, in which the final void ratio is plotted on the natural scale and the effective stress as abscissa on the log scale [Fig. 12.7(*b*)]. The plot is practically a straight line for a normally consolidated clay (defined later) within the range of pressure usually encountered in practice.

(4) Unloading and Reloading plot. In Fig. 12.8, the curve *AB* indicates the decrease in void ratio with an increase in the effective stress. The curve is similar to one shown in Fig. 12.7 (*b*). It is the loading curve.

After the sample has reached equilibrium at the effective stress of $\bar{\sigma}_2$, as shown by point *B*, the pressure is reduced and the sample is allowed to take up water and swell. The curve *BEC* is obtained in

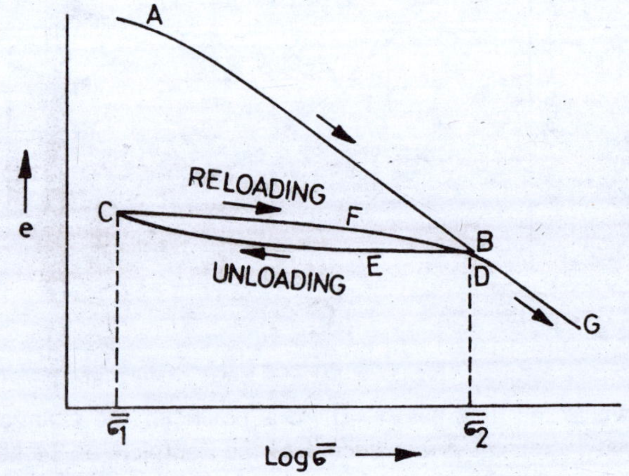

Fig. 12.8. Loading, unloading and reloading plot.

unloading. This is known as the expansion curve or swelling curve. It may be noted that the soil cannot attain the void ratio existing before the start of the test, and there is always some permanent set or residual deformation.

If the specimen which has swelled to the point C is reloaded, the recompression curve CFD is obtained. As the load approaches the maximum value of the load previously applied corresponding to point B, there is reversal of curvature of the curve and then the plot DG continues as an extension of the first loading curve AB. However, the reloaded specimen remains at a slightly lower void ratio at point D than that attained at B during the initial compression for the same load.

12.8. BASIC DEFINITIONS

The following basic definitions related to consolidation are of paramount importance.

(1) Coefficient of Compressibility. The coefficient of compressibility (a_v) is defined as decrease in void ratio per unit increase in effective stress. It is equal to the slope of the $e - \bar{\sigma}$ curve at the point under consideration [Fig. 12.7 (*a*)].

Thus
$$a_v = \frac{-de}{d\bar{\sigma}} = \frac{-\Delta e}{\Delta\bar{\sigma}} \quad \text{...(12.12)}$$

As the effective stress increases, the void ratio decreases, and therefore, the ratio $de/d\bar{\sigma}$ is negative. The minus sign makes a_v positive. For convenience, the coefficient of compressibility a_v is reported as positive.

As the value of a_v is different at various effective stresses, while reporting its value, the effective stress to which that value corresponds must be mentioned. The coefficient of compressibility decreases with an increase in the effective stress. In other words, the soil becomes stiffer (less compressible) as the effective stress is increased and the curve becomes flatter.

The coefficient of compressibility (a_v) has the dimensions of $[L^2/F]$. The units are m^2/kN. It may be noted that the units are inverse of that for pressure.

(2) Coefficient of Volume change. The coefficient of volume change (or the coefficient of volume compressibility) is defined as the volumetric strain per unit increase in effective stress. Thus

$$m_v = \frac{-\Delta V/V_0}{\Delta\bar{\sigma}} \quad \text{...(12.13)}$$

where m_v = coefficient of volume change, V_0 = initial volume,

ΔV = change in volume, and $\Delta\bar{\sigma}$ = change in effective stress.

The reader should note that the coefficient of volume change is inverse of the bulk modulus used in solid mechanics and fluid mechanics. For most clays, $m_v = 1 \times 10^{-3}$ to $1 \times 10^{-4} m^2/kN$.

The volumetric strain ($\Delta V/V_0$) can be expressed in terms of either void ratio or the coefficient the thickness of the specimen as explained under:

(*a*) Let e_0 be the initial void ratio. Let the volume of solids be unity. Therefore, the initial volume V_0 is equal to $(1 + e_0)$. If Δe is the change in void ratio due to change in volume ΔV, we have $\Delta V = \Delta e$. Thus

$$\frac{\Delta V}{V_0} = \frac{\Delta e}{1 + e_0}$$

Therefore, Eq. 12.13 becomes
$$m_v = \frac{-\Delta e/(1 + e_0)}{\Delta\bar{\sigma}} \quad \text{...(12.14)}$$

(*b*) As the area of cross-section of the sample in the consolidometer remains constant, the change in volume is also proportional to the change in height. Thus $\Delta V = \Delta H$

Therefore,
$$\frac{\Delta V}{V_0} = \frac{\Delta H}{H_0}$$

where H_0 = initial height.

Therefore, Eq. 12.13. becomes $$m_v = \frac{-\Delta H/H_0}{\Delta \bar{\sigma}} \qquad ...(12.15)$$

or $$\Delta H = -m_v H_0 \Delta \bar{\sigma} \qquad ...[12.15(a)]$$

The relationship between a_v and m_v can be obtained from Eqs. 12.12 and 12.14 as

$$m_v = \frac{a_v}{1 + e_0} \qquad ...(12.16)$$

Like a_v , the coefficient of volume change m_v depends upon the effective stress at which it is determined. Its value also decreases with an increase in the effective stress. The unit of m_v is the same as that of a_v. However, the coefficient of volume change m_v is more commonly used in practice than the coefficient of compressibility a_v.

(3) Compression Index. The compression index (C_c) is equal to the slope of the linear portion of the void ratio versus log $\bar{\sigma}$ plot [Fig. 12.7 (*b*)]. Thus

$$C_c = \frac{-\Delta e}{\log_{10} (\bar{\sigma}/\bar{\sigma}_0)} \qquad ...(12.17)$$

where $\bar{\sigma}_0$ = initial effective stress, $\bar{\sigma}$ = final effective stress, Δe = change in void ratio

Sometimes, Eq. 12.17 is expressed as

$$C_c = \frac{-\Delta e}{\log_{10}\left(\dfrac{\bar{\sigma}_0 + \Delta \bar{\sigma}}{\bar{\sigma}_0}\right)} \qquad ...(12.18)$$

where $\Delta\bar{\sigma}$ is the change in effective stress.

The numerical value of C_c can be easily determined from the difference in void ratio corresponding to one log cycle. Thus

$$C_c = \frac{-\Delta e}{1} = -\Delta e \qquad ...(12.19)$$

The compression index is extremely useful for determination of the settlement in the field.

The compression index of a clay is related to its index properties, especially the liquid limit. Terzaghi and Peck gave the following empirical relationship for clays of low to medium sensitivity (($S_t \le 4$).

(*a*) For undisturbed soils, $$C_c = 0.009 (w_L - 10) \qquad ...(12.20\ a)$$

(*b*) For remoulded soils, $$C_c = 0.007 (w_L - 10) \qquad ...(12.20\ b)$$

where w_L = liquit limit (%).

The value of C_c normally varies between 0.30 for highly plastic clays and 0.075 for low plastic clays.

The compression index is also related to the insitu void ratio e_0 or water content (w_0) as under

$$C_c = 0.54 (e_0 - 0.35) \qquad ...(12.21)$$

$$C_c = 0.0054 (2.6 w_0 - 35) \qquad ...(12.22)$$

The coefficient of compressibility a_v may be calculated from the compression index as under :

$$a_v = 0.435 \frac{C_c}{\bar{\sigma}_a}$$

where $\bar{\sigma}_a$ is the average pressure for the increment.

(4) Expansion Index. The expansion index or swelling index (C_e) is the slope of the *e*-log $\bar{\sigma}$ plot obtained during unloading (BEC in Fig. 12.8).

$$C_e = \frac{\Delta e}{\log_{10}\left(\dfrac{\bar{\sigma} + \Delta \bar{\sigma}}{\bar{\sigma}}\right)} \qquad ...12.23)$$

As it is evident, the expansion index is much smaller than the compression index.

(5) Recompression Index. Recompression is the compression of a soil which had already been loaded and unloaded. The load during recompression is less than the load to which the soil has been subjected previously. The slope of the recompression curve obtained during reloading (CFD in Fig. 12.8) when plotted as e-log $\bar{\sigma}$, is equal to the recompression index (C_r). Thus

$$C_r = \frac{-\Delta e}{\log\left(\frac{\bar{\sigma} + \Delta \bar{\sigma}}{\bar{\sigma}}\right)} \quad ...(12.24)$$

The recompression index is appreciably smaller than the compression index C_c. It is usually in the range of 1/10 to 1/5 of the compression index.

(6) Normally consolidated and Over-consolidated clays. A normally consolidated soil is one which had not been subjected to a pressure greater than the present existing pressure. A soil is said to be over-consolidated if it had been subjected in the past to a pressure in excess of the present pressure.

The portion AB of the curve in Fig. 12.8 represents the soil in normally consolidated condition. The curve in this range is also called the *virgin compression curve.* The soil in the range CD when it is recompressed represents overconsolidated condition, as the soil had been previously subjected to a pressure $\bar{\sigma}_2$, which is greater than the pressure in the range CD.

The maximum pressure to which an over-consolidated soil had been subjected in the past divided by the present pressure is known as the overconsolidation ratio (O.C.R.). For example, the soil indicated by the condition at point C has an over-consolidation ratio of $\bar{\sigma}_2/\bar{\sigma}_1$.

It may be emphasized that normally consolidated soils and overconsolidated soils are not different types of soils but these are conditions in which a soil exists. The same type of soil can behave as normally consolidated in a certain pressure range and an over-consolidated in some other pressure range. For example, in Fig. 12.8, the soil which behaves as overconsolidated in the range CD would again behave as normally consolidated in the range DG.

The liquidity index of a normally consolidated clay is generally between 0.6 and 1.00, whereas that for an over-consolidated clay between 0.0 and 0.60.

As the recompression index (C_r) is very small as compared with the compression index (C_c), the soils in the overconsolidated state have smaller compressibility. The settlements of the structures built an over-consolidated clays are small.

(7) Underconsolidated clays. If the clay deposit has not reached equilibrium under the applied overburden loads, it is said to be underconsolidated. This normally occurs in areas of recent land fill.

12.9. TERZAGHI'S THEORY OF CONSOLIDATION

(1) Assumptions. Terzaghi (1925) gave the theory for the determination of the rate of consolidation of a saturated soil mass subjected to a static, steady load. The theory is based on the following assumptions :

(1) The soil is homogeneous and isotropic.

(2) The soil is fully saturated.

(3) The solid particles and water in the voids are incompressible. The consolidation occurs due to expulsion of water from the voids.

(4) The coefficient of permeability of the soil has the same value at all points, and it remains constant during the entire period of consolidation.

(5) Darcy's law is valid throughout the consolidation process.

(6) Soil is laterally confined, and the consolidation takes place only in axial direction. Drainage of water also occurs only in the vertical direction.

(7) The time lag in consolidation is due entirely to the low permeability of the soil.

(8) There is a unique relationship between the void ratio and the effective stress, and this relationship

remains constant during the load increment. In other words, the coefficient of compressibility and the coefficient of volume change are constant.

Comments on the assumptions

The assumptions made by Terzaghi are not fully satisfied in actual field problems. The results obtained from the use of the theory to practical problems are approximate. However, considering complexity of the problem, the theory gives reasonably accurate estimate of the time rate of settlement of a structure built on the soil. A brief comment on the various assumptions and their effect is given below.

Assumptions 1 to 3 are generally satisfied for fully saturated, clay deposits. However, the presence of air may affect the accuracy.

Assumptions 4 and 5 are not fully satisfied. In fact, the coefficient of permeability varies at different points in the deposit. Its value decreases as the consolidation progresses due to an increase in the effective stress. Further, at very low hydraulic gradients, the Darcy law in not strictly applicable. Fortunately, the errors introduced due to these assumptions are small.

The largest error is probably due to the assumption 6. In the field, the consolidation is usually 3-dimensional and not one- dimensional. However, in the case of deposits having large areas compared with their thicknesses have essentially one-dimensional consolidation and the error is not much.

Assumption 7 is not fully justified, as some secondary consolidation does occur along with the primary consolidation. However, for most inorganic soils, the secondary consolidation is small and does not introduce much error.

The actual relationship between the void ratio and the effective stress is not linear, contrary to the assumption 8. However, if a large number of samples are taken from the same stratum and an average value of the coefficient of volume change (m_v) is taken for the appropriate range of the effective stress, the error introduced due to this assumption is not unduly high. The only justification for making this assumption is to get a relatively simple expression. The theory becomes more complex when actual relationship is used.

(2) Derivation of Differential Equation. The basic differential equation of one-dimensional consolidation can be derived as under:

Let us consider a saturated clay layer of thickness $2d\,(=H)$ sandwiched between two layers of sand (Fig. 12.9). When a uniform pressure of $\Delta\sigma$ is applied on the surface of the top sand layer, the total stress developed at all points in the clay layer is increased by $\Delta\sigma$.

As explained in the spring analogy model (Sect. 12.3), initially the whole of the pressure is taken up by water, and the hydrostatic excess pressure of $\Delta\sigma/\gamma_w$ develops. Fig. 12.9 shows the excess hydrostatic pressure diagram on the right side. It is assumed that various points along the thickness of the clay layer are connected by flexible tubes to the piezometers. At time $t = 0$, just after the application of the load, the excess hydrostatic pressure $\bar{u}_i$ is equal to $\Delta\sigma/\gamma_w$ throughout the clay layer. This is represented by the horizontal line *AB*. The

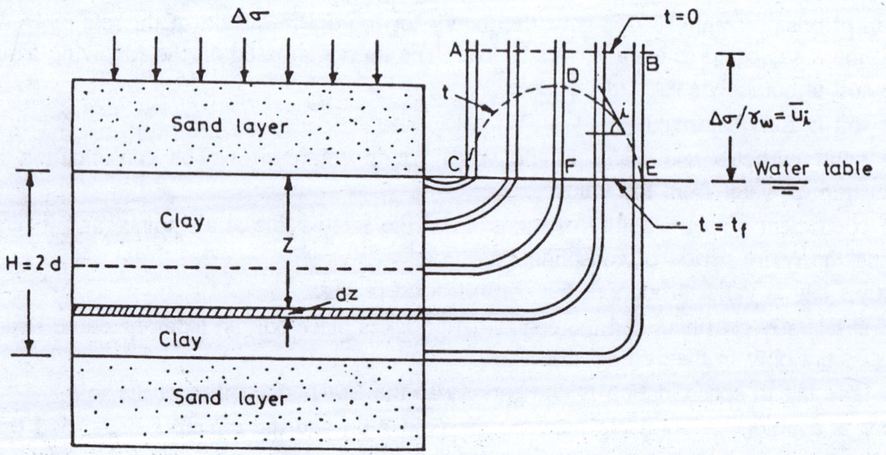

Fig. 12.9. Excess Hydrostatic Pressure.

excess hydrostatic pressure is independent of the position of the water table. For convenience, the water table is assumed at the level of the surface of the clay layer.

Water starts escaping towards the upper and the lower sand layers due to excess hydrostatic pressure developed. The hydrostatic excess pressure at the top and the bottom of the clay layer, indicated by points *C* and *E* in the pressure diagram, drops to zero. However, the excess hydrostatic pressure in the middle portion of the clay layer at *D* remains high. The curves indicating the distribution of excess hydrostatic pressure are known as *isochrones*. The isochrone *CDE* indicates the distribution of excess hydrostatic pressure at time *t*.

As the consolidation progresses, the excess hydrostatic pressure in the middle of the clay layer decreases. Finally at time $t = t_f$, the whole of the excess hydrostatic pressure has been dissipated, and the pressure distribution is indicated by the horizontal isochrone *CFE*.

Let us consider the equilibrium of an element of the clay at a depth of z from its top at time t. The consolidation pressure $\Delta\sigma$ is partly carried by water and partly by solid particles as

$$\Delta\sigma = \Delta\bar{\sigma} + \bar{u} \qquad \ldots(12.25)$$

where $\Delta\bar{\sigma}$ is the pressure carried by solid particles,

and $\bar{u}$ is the excess hydro-static pressure (pressure units)

The hydraulic gradient (i) at that depth is equal to the slope of the isochrone *CDE* at a horizontal distance z from the point *C* in the pressure diagram.

Thus,
$$i = \frac{\partial h}{\partial z} = \frac{\partial(\bar{u}/\gamma_w)}{\partial z} = \frac{1}{\gamma_w}\left(\frac{\partial \bar{u}}{\partial z}\right) \qquad \ldots(a)$$

where $\bar{u}$ is the excess hydrostatic pressure at depth z.

The expression for the hydraulic gradient i can also be obtained as under. Let us consider a thin slice of clay layer Δz at depth z (Fig. 12.10). The pressure difference ($\Delta\bar{u}$) across this thickness is given by

$$\Delta\bar{u} = \left(\bar{u} + \frac{\partial \bar{u}}{\partial z}\cdot dz\right) - \bar{u} = \frac{\partial \bar{u}}{\partial z}\cdot dz$$

The unbalanced head across the thickness is given by

$$\Delta h = \frac{\Delta\bar{u}}{\gamma_w} = \frac{1}{\gamma_w}\left(\frac{\partial \bar{u}}{\partial z}\cdot dz\right)$$

The hydraulic gradient becomes

$$i = \frac{\Delta h}{dz} = \frac{1}{\gamma_w}\left(\frac{\partial \bar{u}}{\partial z}\right) \qquad \ldots[\text{same as } (a)]$$

From Darcy's law, the velocity of flow at depth z is given by

$$v = ki = k\cdot\frac{1}{\gamma_w}\left(\frac{\partial \bar{u}}{\partial z}\right)$$

The velocity of flow at the bottom of the element of thickness Δz can be written as

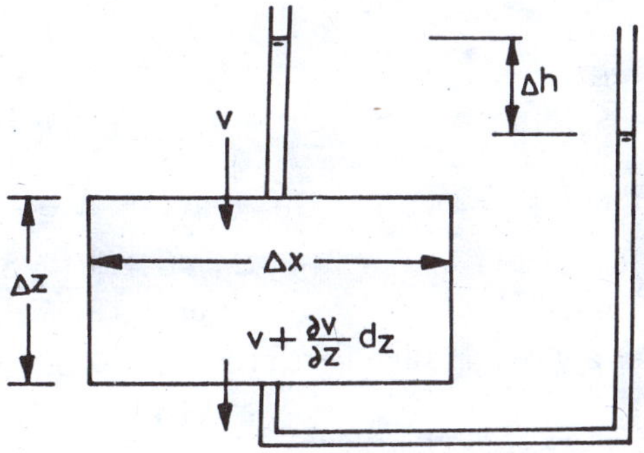

Fig. 12.10. Pressure Difference on a Thin Slice.

$$v + \frac{\partial v}{\partial z}\cdot dz = v + \frac{\partial}{\partial z}\left[\frac{k}{\gamma_w}\left(\frac{\partial \bar{u}}{\partial z}\right)\right] dz$$

or
$$v + \frac{\partial v}{\partial z}\,dz = v + \frac{k}{\gamma_w}\left(\frac{\partial^2 \bar{u}}{\partial z^2}\right) dz$$

Therefore,
$$\frac{\partial v}{\partial z} = \frac{k}{\gamma_w}\cdot\frac{\partial^2 \bar{u}}{\partial z^2} \qquad \ldots(12.26)$$

The discharge entering the element Q_{in} is

$$Q_{in} = v(\Delta x \times \Delta y)$$

where Δx and Δy are the dimensions of the element is plan.

The discharge leaving the element Q_{out} is

$$Q_{out} = \left(v + \frac{\partial v}{\partial z} \cdot dz \right) (\Delta x \times \Delta y)$$

Therefore, the net discharge squeezed out of the element is given by

$$\Delta Q = Q_{out} - Q_{in}$$

or
$$\Delta Q = \left[\left(v + \frac{\partial v}{\partial z} \cdot dz \right) - v \right] \Delta x \times \Delta y$$

or
$$\Delta Q = \frac{\partial v}{\partial z} (\Delta x \times \Delta y \times \Delta z) \qquad \text{...}(b)$$

As the water is squeezed out, the effective stress increases and the volume of the soil mass decreases. From Eq. 12.13,

$$\Delta V = -m_v V_0 \Delta \bar{\sigma}$$

where V_0 = initial volume of soil mass (= $\Delta x \, \Delta y \, \Delta z$)

and $\Delta \bar{\sigma}$ = increase in effective stress.

The decrease in volume of soil per unit time

$$\frac{\partial (\Delta V)}{\partial t} = -m_v (\Delta x \; \Delta y \; \Delta z) \frac{\partial (\Delta \bar{\sigma})}{\partial t} \qquad \text{...}(c)$$

As the decrease in volume of soil mass per unit time is equal to the volume of water squeezed out per unit time, Eqs. (*b*) and (*c*) give

$$\frac{\partial v}{\partial z} (\Delta x \; \Delta y \; \Delta z) = -m_v (\Delta x \; \Delta y \; \Delta z) \frac{\partial (\Delta \bar{\sigma})}{\partial t}$$

or
$$\frac{\partial v}{\partial z} = -m_v \frac{\partial (\Delta \bar{\sigma})}{\partial t} \qquad \text{...}(d)$$

From Eq. 12.25,
$$\Delta \bar{\sigma} = \Delta \sigma - \bar{u}$$

or
$$\frac{\partial \Delta \bar{\sigma}}{\partial t} = \frac{\partial (\Delta \bar{\sigma})}{\partial t} - \frac{\partial \bar{u}}{\partial t}$$

For a given pressure increment, $\partial \Delta \sigma = 0$. Therefore,

$$\frac{\partial (\Delta \bar{\sigma})}{\partial t} = -\frac{\partial \bar{u}}{\partial t}$$

Therefore, Eq. (*d*) becomes
$$\frac{\partial v}{\partial z} = -m_v \left(-\frac{\partial \bar{u}}{\partial t} \right) = m_v \left(\frac{\partial \bar{u}}{\partial t} \right) \qquad \text{...}(12.27)$$

Equating two values of $\partial v / \partial z$ from Eqs. 12.26 and 12.67,

$$\frac{k}{\gamma_w} \frac{\partial^2 \bar{u}}{\partial z^2} = m_v \left(\frac{\partial \bar{u}}{\partial t} \right)$$

or
$$c_v \frac{\partial^2 \bar{u}}{\partial z^2} = \frac{\partial \bar{u}}{\partial t} \qquad \text{...}(12.28)$$

where c_v is the coefficient of consolidation and is given by

$$c_v = \frac{k}{\gamma_w m_v} = \frac{k}{g \rho_w m_v} \quad ...(12.29)$$

Eq. 12.28 is the basic differential equation of one-dimensional consolidation. It gives the distribution of hydrostatic excess pressure $\bar{u}$ with depth z and time t.

12.10. SOLUTION OF BASIC DIFFERENTIAL EQUATION OF CONSOLIDATION

The solution of the basic differential equation of one- dimensional consolidation (Eq. 12.28) can be obtained using Fourier series. Let us express hydrotatic excess pressure $\bar{u}$ as

$$\bar{u} = f_1(z) \cdot f_2(t) \quad ...(12.30)$$

where $f_1(z)$ and $f_2(t)$ indicate some function of z and t, respectively.

Substituting the above value of $\bar{u}$ in Eq. 12.28,

$$c_v \left[f_2(t) \frac{\partial^2}{\partial z^2} [f_1(z)] \right] = f_1(z) \frac{\partial [f_2(t)]}{\partial t}$$

or

$$\frac{\frac{\partial^2}{\partial z^2} [f_1(z)]}{f_1(z)} = \frac{\frac{\partial}{\partial t} [f_2(t)]}{c_v f_2(t)}$$

The left-hand side of the above equation is a function of z only and the right-hand side is a function of t only. In other words, if the left-hand side is equal to some constant (say, $-A^2$) when t is taken as a variable and the right-hand side is equal to the same constant when z is considered as a variable.

Thus,

$$\frac{\partial^2}{\partial z^2} [f_1(z)] = -A^2 f_1(z) \quad ...(a)$$

and

$$\frac{\partial^2}{\partial z^2} [f_2(t)] = -A^2 c_v f_2(t) \quad ...(b)$$

Eq. (*a*) has the solution given by

$$f_1(z) = C_1 \cos Az + C_2 \sin Az \quad ...(c)$$

where C_1 and C_2 are constants of integration.

Eq. (*b*) has the solution given by

$$f_2(t) = C_3 e^{-A^2 c_v t} \quad ...(d)$$

where C_3 is a constant of integration and e is the base of the hyperbolic or Napierian logarithm.

Substituting the above solutions in Eq. 12.30,

$$\bar{u} = [C_1 \cos Az + C_2 \sin Az] \, C_3 \, e^{-A^2 c_v t}$$

or

$$\bar{u} = [C_4 \cos Az + C_5 \sin Az] \, e^{-A^2 c_v t} \quad ...(12.31)$$

where C_4 and C_5 are other constants, such that

$$C_4 = C_1 C_3 \text{ and } C_5 = C_2 C_3$$

The constants C_4 and C_5 can be determined from the boundary conditions :

(*i*) $t = 0$, $\bar{u} = \bar{u}_i$, for any value of z

where $\bar{u}_i$ is initial hydrostatic pressure.

(*ii*) $t = \infty$, $\bar{u} = 0$, for any value of z

(*iii*) $z = 0$, $\bar{u} = 0$, for any value of t

(*iv*) $z = H\ (= 2d)$, $\bar{u} = 0$, for any value of t

For the boundary condition (*iii*), Eq. 12.31 gives $C_4 = 0$.

Therefore, Eq. 12.31 becomes $\bar{u} = C_5 \sin(Az)\, e^{-A^2 c_v t}$ ...(*e*)

For the boundary condition (*iv*), $\bar{u} = 0$ at $z = H$.

Therefore, $C_5 \sin(AH)\, e^{-A^2 c_v t} = 0$

The above equation is satisfied if $AH = n\pi$, where n is any integer. The equation can be written in the following form:

$$\bar{u} = B_1 \sin(\pi z/H)\, e^{-(\pi^2/H^2) c_v t} + B_2 \sin(2\pi z/H)\, e^{-(4\pi^2/H^2) c_v t} + \ldots + B_n \sin(n\pi z/H)\, e^{-(n^2\pi^2/H^2) c_v t} + \ldots$$

or

$$\bar{u} = \sum_{n=1}^{n=\infty} B_n \sin(n\pi z/H)\, e^{-(n^2\pi^2/H^2) c_v t} \qquad \ldots(12.32)$$

where $B_1, B_2 \ldots B_n$ are constants.

From boundary condition (*i*), when $t = 0$, $\bar{u} = \bar{u}_i$. Therefore,

or

$$\bar{u}_i = \sum_{n=1}^{n=\infty} B_n \sin(n\,\pi\, z/H) \qquad \ldots[12.32\,(a)]$$

If m and n are two unequal integers, the following identities hold good.

$$\int_0^{\pi} \sin mx \sin nx\, dx = 0$$

and

$$\int_0^{\pi} \sin^2 nx\, dx = \pi/2$$

In the above identities if $(\pi z/H)$ is substituted for x, the differential dx changes to $(\pi/H)\, dz$ and the limits of integration change to 0 to H.

Therefore,

$$\int_0^{H} \sin(m\,\pi\, z/H) \sin(n\,\pi\, z/H)\, dz = 0 \qquad \ldots(f)$$

and

$$\int_0^{H} \sin^2(n\,\pi\, z/H)\, dz = H/2 \qquad \ldots(g)$$

Multiplying both sides of Eq. 12.32 (*a*) by $\sin(n\,\pi\, z/H)$ and integrating between the limits 0 to H,

$$\int_0^{H} \bar{u}_i \sin(n\,\pi\, z/H)\, dz = \sum_{\substack{m=1 \\ (m \neq n)}}^{\infty} \left[B_m \int_0^{H} \sin(m\,\pi\, z/H) \sin(n\,\pi\, z/H)\, dz \right] + \sum_{n=1}^{\infty} \left[B_n \cdot \int_0^{H} \sin^2(n\,\pi\, z/H)\, dz \right]$$

The right-hand side of the above equation has been split into two parts: (*i*) when $m \neq n$ (*ii*) when $m = n$. Using identities (*f*) and (*g*), the above equation becomes

$$\int_0^{H} \bar{u}_i \sin(n\,\pi\, z/H)\, dz = B_n \cdot (H/2)$$

Therefore,

$$B_n = \frac{2}{H} \int_0^{H} \bar{u}_i \sin(n\,\pi\, z/H)\, dz \qquad \ldots(h)$$

Substituting the above value of B_n in Eq. 12.32,

$$\bar{u} = \sum_{n=1}^{n=\infty} \left[\frac{2\bar{u}_i}{H} \int_0^H \sin(n\pi z/H)\, dz \right] \sin(n\pi z/H)\, e^{-(n^2\pi^2/H^2)\, c_v t}$$

or
$$\bar{u} = \sum_{n=1}^{n=\infty} \frac{2\,\bar{u}_i}{n\pi} (1 - \cos n\pi)(\sin n\pi z/H)\, e^{-(n^2\pi^2/H^2)\, c_v t} \qquad ...(k)$$

Only odd integers n are relevant, because for even integer $1 - \cos n\pi = 0$

and for odd integers $1 - \cos n\pi = 2$

Substituting $n = 2N + 1$, where N is an integer, Eq. (k) becomes

$$\bar{u} = \frac{4}{\pi}\bar{u}_i \sum_{N=0}^{N=\infty} \frac{1}{(2N+1)} \left[\sin \frac{(2N+1)\pi z}{H} \right] e^{-((2N+1)^2\pi^2/H^2)\, c_v t} \qquad ...(12.33)$$

Eq. 12.33 is the required solution of the basic differential equation of one-dimensional consolidation. It gives the variation of hydrostatic excess pressure $\bar{u}$ with depth z at any time t in terms of the initial hydrostatic excess pressure $\bar{u}_i$ (equal to $\Delta\sigma/\gamma_w$).

Substituting $H = 2d$, where d is the length of the drainage path, in Eq. 12.33,

$$\bar{u} = \frac{4\bar{u}_i}{\pi} \sum_{N=0}^{\infty} \frac{1}{(2N+1)} \left[\sin \frac{(2N+1)\pi z}{2d} \right] e^{-((2N+1)^2\pi^2/4d^2)\, c_v t} \qquad ...(12.34)$$

Let us introduce a non-dimensional parameter, known as *time factor* (T_v), given by

$$T_v = \frac{c_v t}{d^2} \qquad ...(12.35)$$

Eq. 12.34 becomes,
$$\bar{u} = \frac{4\bar{u}_i}{\pi} \sum_{N=0}^{\infty} \frac{1}{(2N+1)} \times \left[\sin \frac{(2N+1)\pi z}{2d} \right] e^{-(2N+1)^2\pi^2 T_v/4} \qquad ...(12.36)$$

A series of isochrones indicating the variation of $\bar{u}$ with z can be plotted for different values of T_v. The shape of the isochrones depends upon the initial distribution of excess pore water pressure $\bar{u}_i$ and the drainage conditions at the boundaries of the clay layer. If both the upper and lower boundaries are free draining, the clay layer is known as *open layer.* If only one boundary of the clay layer is free draining, the layer is called *half-closed layer.*

Fig. 12.11 (a) shows the *isochrones* for an open layer of thickness $2d$ when the initial distribution of pressure is uniform. The upper half of the diagram is also applicable for a half-closed layer of thickness d [Fig. 12.11 (b)]. The hydraulic gradient at any point is equal to the slope of the isochrone at that point.

The progress of consolidation at any point depends upon the pore water pressure $\bar{u}$ at that point. The degree of consolidation (U_z) at any point at depth z is equal to the ratio of the dissipated excess pore water pressure to the initial excess pore water pressure, *i.e.*,

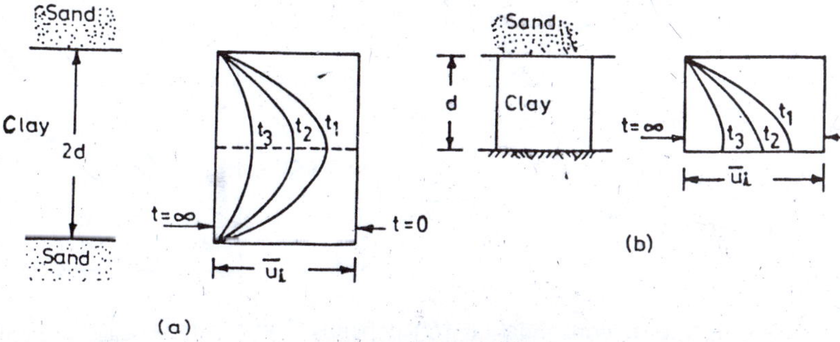

Fig. 12.11. Isochrones.

$$U_z = \frac{\bar{u}_i - \bar{u}}{\bar{u}_i} = 1 - \frac{\bar{u}}{\bar{u}_i} \qquad \text{...(12.37)}$$

Substituting the value of $\bar{u}/\bar{u}_i$ from Eq. 12.36,

$$U_z = 1 - \sum_{N=0}^{\infty} \frac{2}{M} \sin\left(\frac{Mz}{d}\right) e^{-M^2 T_v} \qquad \text{...(12.38)}$$

where $$M = \frac{\pi}{2}(2N + 1)$$

Eq. 12.38 gives the degree of consolidation at a point. In practical problems, the main interest is to know the *average degree of consolidation* of the whole layer. The average degree of consolidation (U) is defined as

$$U = \frac{U_i - U_t}{U_i} \qquad \text{...(12.39)}$$

where U_i is initial excess hydrostatic pressure over the entire depth,

$$U_i = \frac{1}{2d}\int_0^{2d} \bar{u}_i \, dz$$

and U_t is the average excess hydrostatic pressure after time t over the entire depth,

$$U_t = \frac{1}{2d}\int_0^{2d} \bar{u} \, dz$$

Eq. 12.39 can be written as, $$U = 1 - \frac{\frac{1}{2d}\int_0^{2d} \bar{u} \, dz}{\frac{1}{2d}\int_0^{2d} \bar{u}_i \, dz}$$

or $$U = 1 - \frac{\int_0^{2d} \bar{u} \, dz}{\int_0^{2d} \bar{u}_i \, dz} \qquad \text{...(12.40)}$$

For constant initial excess pore water pressure $\bar{u}_i$, Eq. 12.40 becomes

$$U = 1 - \frac{1}{2d\bar{u}_i}\int_0^{2d} \bar{u} \, dz \qquad \text{...(12.41)}$$

It may be noted that the average degree of consolidation U is equal to the area of the hatched portion of the rectangle shown in Fig. 12.12.

Substituting the value of $\bar{u}$ from Eq. 12.36 in Eq. 12.41,

$$U = 1 - \frac{1}{2d\bar{u}_i}\int_0^{2d} \sum_{N=0}^{\infty} \frac{2\bar{u}_i}{M} \sin\left(\frac{Mz}{d}\right) e^{-M^2 T_v}$$

where $M = \frac{\pi}{2}(2N + 1)$ as before.

Thus $$U = 1 - \sum_{N=0}^{N=\infty} \frac{2}{M^2} e^{-M^2 T_v} \qquad \text{...(12.42)}$$

or $$U = f(T_v) \qquad \text{...(12.43)}$$

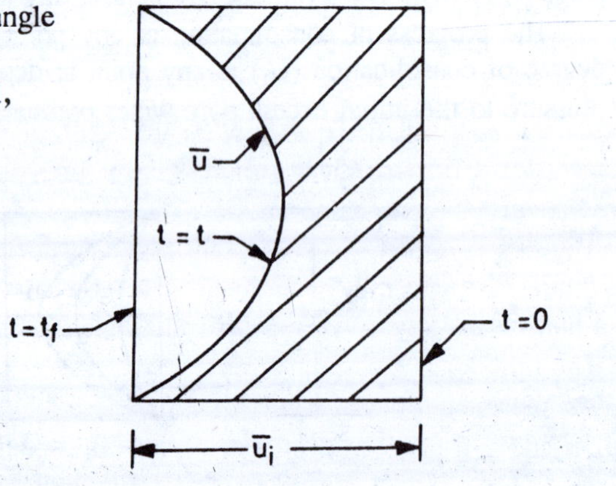

Fig. 12.12. Average Degreee of Consolidation

Therefore the average degree of consolidation (U) depends

upon the non-dimensional time factor T_v. The curves can be obtained between U and T_v for different drainage conditions and the pressure distribution.

Table 12.3 gives the values of T_v for different values of the average degree of consolidation (U). Case (1) is applicable for uniform pressure distribution for both open layer and half-closed layer. This is also applicable for triangular (linear) distribution of pressure in the case of open layers. Case (2) is applicable for triangular distribution of pressure for half-closed layer with maximum pressure near the drainage face. Case (3) is also applicable for triangular distribution of pressure for half-closed layer but with zero pressure near the drainage face. Fig. 12.13 gives the curve for the three cases considered.

Table 12.3. Variation of U with T_v.

Pressure distribution	 *Curve (1)* *Case (1)*	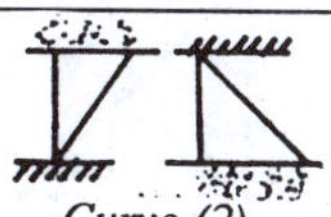 *Curve (2)* *Case (2)*	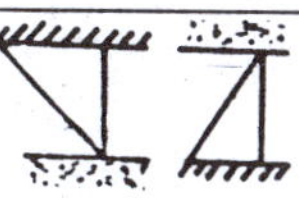 *Curve (3)* *Case (3)*
U	T_v	T_v	T_v
0	0.0	0.0	0.0
0.1	0.008	0.003	0.047
0.2	0.031	0.009	0.100
0.3	0.071	0.024	0.158
0.4	0.126	0.048	0.221
0.5	0.196	0.092	0.294
0.6	0.287	0.160	0.383
0.7	0.403	0.270	0.500
0.8	0.567	0.440	0.665
0.9	0.848	0.720	0.940
1.0	∞	∞	∞

The time factor T_v depends upon the coefficient of consolidation (c_v), time t and the drainage path d (Eq. 12.35). The coefficient of consolidation represents the combined effect of the coefficient of permeability (k) and the coefficient of volume change (m_v) as indicated by Eq. 12.29.

Thus $$T_v = \frac{c_v t}{d^2} = \left(\frac{k}{m_v \gamma_w}\right)\frac{t}{d^2} = \left(\frac{k}{m_v g \rho_w}\right)\frac{t}{d^2} \qquad \text{...(12.44)}$$

As the codsolidation progresses, both the coefficient of permeability (k) and the coefficient of volume change (m_v) decrease, but the ratio (k/m_v) remains almost constant over a considerable range of pressure. Consequently, the coefficient of consolidation remains almost constant.

The drainage path (d) represents the maximum distance that the water has to travel before reaching the free-drainage boundary. For an open layer, the value of d is equal to half the thickness of the layer, whereas for a half-closed layer, it is equal to the thickness of the layer.

The time (t) is measured from the instant the load is applied to the layer.

As indicated by Eq. 12.44, the time factor T_v and hence the degree of consolidation depends upon k, m_v, d and t. It also depends upon the distribution of pressure across the thickness.

The curve (1) in Fig. 12.13 is parabolic. The relationship can be represented by the following empirical equations.

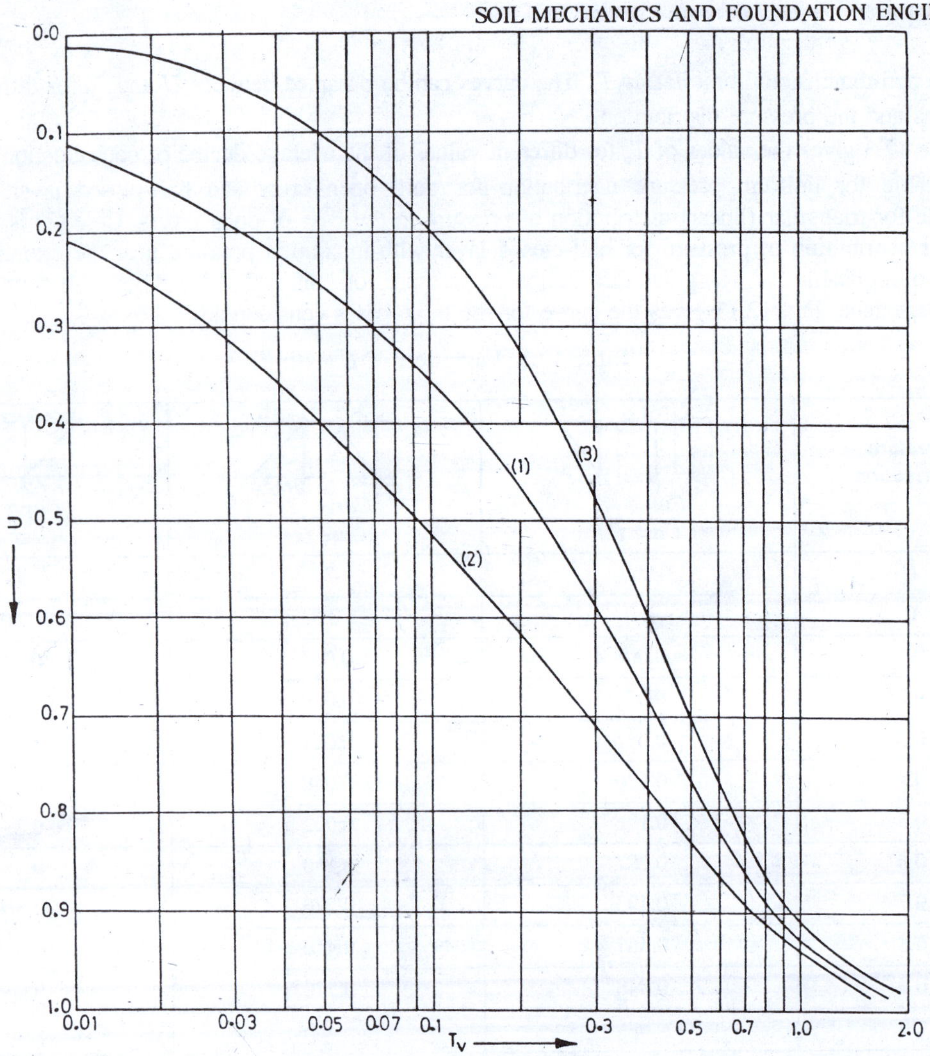

Fig. 12.13. U—T_v Curves.

$$T_v = \frac{\pi}{4} U^2 \quad \text{(for } U < 0.60\text{)} \qquad \text{...(12.45)}$$

and

$$T_v = -0.933 \log_{10} (1 - U) - 0.085 \quad \text{(for } U > 0.60\text{)} \qquad \text{...(12.46)}$$

where U is expressed as a ratio (not percentage).

Eq. 12.46 can also be written as,

$$T_v = 1.781 - 0.933 \log_{10} (100 - U\%) \qquad \text{...[12.46 (a)]}$$

These approximate relations are extremely useful when the curves between U and T_v are not available.

Limitation of the Consolidation Theory. One-dimensional consolidation theory is based on a number of assumptions which are not realised in practice. The equation has the following limitations :

(1) The value of the coefficient of consolidation (c_v) has been assumed to be constant. In reality, it changes with a change in the consolidation pressure. For accurate predictions of the time-rate of consolidation in the field, its value should be determined for the expected pressure range.

(2) The distance d of the drainage path cannot be measured accurately in the field. The thickness of the deposit is generally variable, and an average value has to be estimated.

(3) There is sometimes difficulty in locating the drainage face. Sometimes thin pervious seams which

can act as good drainage faces are missed in the boring operations. On the other hand, sometimes isolated sand pockets are wrongly taken as drainage faces.

(4) The equation is based on the assumption that the consolidation is one-dimensional. In field, the consolidation is generally 3- dimensional. The lateral drainage may have a significant effect on the time rate of consolidation.

(5) The initial consolidation and the secondary consolidation have been neglected. Sometimes, these form an important part of the total consolidation.

(6) In the field, the load is seldom applied instantaneously. The effect of the loading period has to be considered, as explained in Sect. 12.15.

(7) In actual practice, the pressure distribution may be far from linear or uniform. The theory becomes complicated when correct distribution is considered.

Notwithstanding the above limitations, the consolidation theory is used to predict the time rate of settlement of the structures built on a soil. The results are fairly accurate if the theory is applied with caution, keeping above limitations in mind.

12.11. DETERMINATION OF COEFFICIENT OF CONSOLIDATION

The curve between dial gauge reading and time t obtained in the laboratory by testing the soil sample is similar in shape to the theoretical curve between U and T_v obtained from the consolidation theory. This similarity between the laboratory curve and the theoretical curve is used for the determination of the coefficient of consolidation (c_v) of the soil. The methods are known as the *fitting methods*. The following two methods are commonly used.

(1) Square-root of time method. The method, devised by Taylor, utilizes the theoretical relationship between U and $\sqrt{T_v}$. The relationship is linear up to the value of U equal to about 60% (Eq. 12.45). It has been further established that at U = 90%, the value of $\sqrt{T_v}$ is 1.15 times the value obtained by the extension of the initial straight line portion [Fig. 12.14 (*a*)].

The sample of the soil whose coefficient of consolidation is required is tested as explained in Sect. 12.5 *For a given load increment*, the dial gauge readings are taken for different time intervals. A curve is plotted between the dial gauge reading (R), as ordinate, and the $\sqrt{t}$, as abscissa [Fig. 12.14 (*b*)]. The curve $ABCDE$ shows the plot. The curve begins at the dial gauge reading R_0 at time t_0, indicated by point A.

As the load increment is applied, there is an initial compression. It is obtained by producing back the

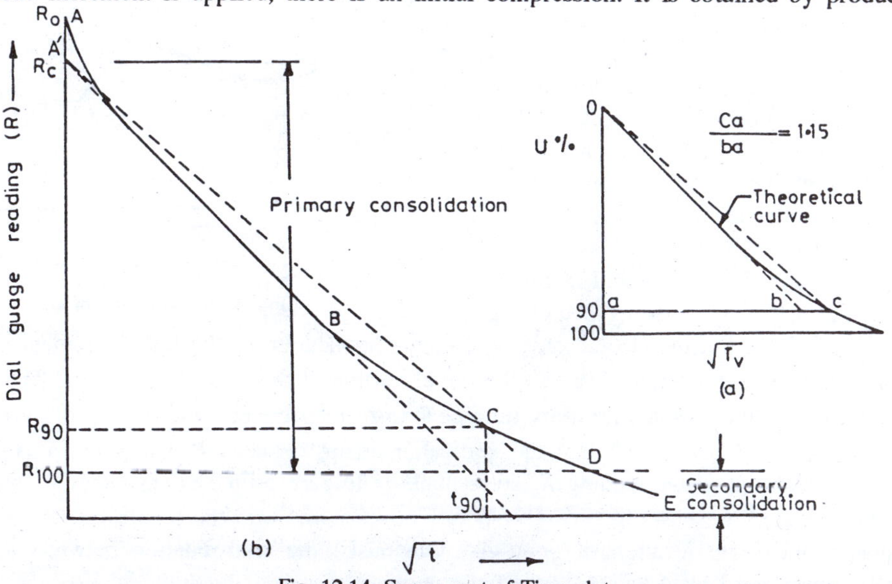

Fig. 12.14. Square-root of Time plot.

initial linear part of the curve to intersect the dial-gauge reading axis at point A'. This corresponds to the corrected zero reading (R_c). The consolidation between the dial gauge reading R_0 and R_c is the *initial compression.* The Terzaghi theory of consolidation is not applicable in this range.

From the corrected zero reading point A', a line $A'C$ is drawn such that its abscissa is 1.15 times that of the initial linear portion $A'B$ of the curve. The intersection of this line with the curve at point C indicates 90% of U. The dial gauge reading corresponding to C is shown as R_{90} and the corresponding abscissa as $\sqrt{t_{90}}$.

The point D for 100% primary consolidation can be obtained from R_{90} as,

$$R_c - R_{100} = \frac{10}{9}(R_c - R_{90})$$

The consolidation after 100% of primary consolidation, in the range DE, is the secondary consolidation.

The value of the coefficient of consolidation of the soil for that load increment is obtained from the value of $\sqrt{t_{90}}$ obtained from that plot. From Table 12.3, for U = 90%, the value of T_v = 0.848. Therefore, using Eq. 12.35,

$$c_v = \frac{T_v\, d^2}{t} = \frac{0.848\, d^2}{(\sqrt{t_{90}})^2} = \frac{0.848\, d^2}{t_{90}} \quad \text{...(12.47)}$$

The distance of the drainage path d is half the total thickness. The total thickness may be taken as the average of the initial thickness (H_i) and final thickness (H_f) of the sample.

Thus $$d = \frac{H}{2} = \frac{1}{2}\left[\frac{H_i + H_f}{2}\right] \quad \text{...[12.48(a)]}$$

For single drainage, $$d = H = \left(\frac{H_i + H_f}{2}\right) \quad \text{...[12.48(b)]}$$

The test is repeated for different load increments and an average value of c_v obtained, as shown in Fig. 12.15.

(2) Logarithm of time method. The method given by Casagrande uses the theoretical curve between U and Log T_v, as shown in Fig. 12.16 (*a*). The curve consists of three parts : (*i*) an initial portion which is parabolic in shape, (*ii*) a middle portion which is almost linear, and (*iii*) the last portion to which the horizontal axis is an asymptote. It is observed that the point of intersection of the tangent drawn at the point of inflexion on the curve and the asymptote of the lower portion gives the value of 100% consolidation.

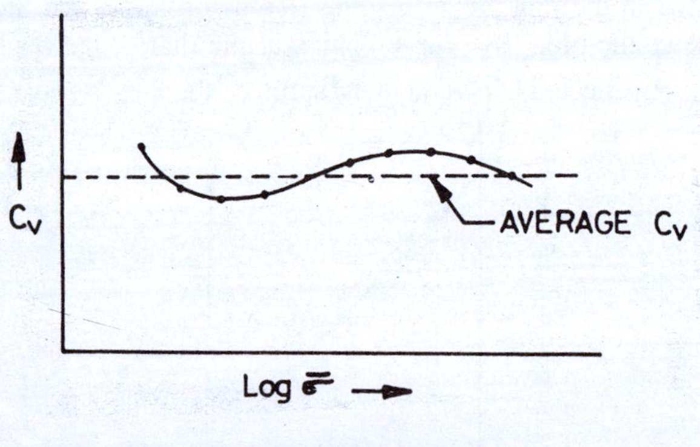

Fig. 12.15. Variation of C_v.

The sample of the soil is tested as explained in Section 12.5. For a given load increment, a curve is plotted between the dial gauge reading R and log t [Fig. 12.16 (*b*)]. Let R_0 be the initial dial gauge before the application of the load increment. The corrected zero reading (R_c) is obtained using the fact that the initial portion of the curve is parabolic. Two points B and C are selected corresponding to some arbitrary time t_1 and $4t_1$, respectively, and having the vertical intercept a, as shown. Point A' is located such that the vertical intercept between B and A' is also equal to a. It represents the corrected dial gauge reading R_c corresponding to zero primary consolidation. As a check, the procedure can be repeated by selecting two other points (not shown) with the time ratio 1 : 4. It should also give approximately the same location of point A'. Obviously, the consolidation between the dial gauge reading R_0 to R_c, represented by A and A', is initial compression.

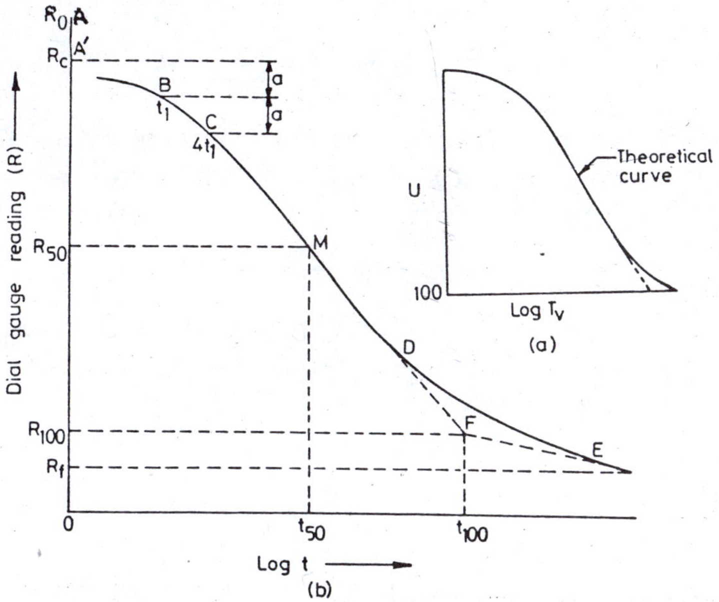

Fig. 12.16. Logarithm of time Plot.

The final portion of the experimental curve is linear. The point F corresponding to 100% consolidation is obtained from the intersection of the two linear parts, as shown. The values of R_{100} and t_{100} are obtained corresponding to point F. The compression between the dial gauge readings R_c and R_{100} is the primary consolidation, and that between R_{100} and R_f is the secondary consolidation.

The point M corresponding to 50% primary consolidation (R_{50}) is located midway between R_c and R_{100} and the value of time t_{50} is obtained.

Thus $$R_c - R_{50} = \frac{1}{2}(R_c - R_{100})$$

From Table 12.3, for $U = 50\%$, the value of T_v is equal to 0.196. From Eq. 12.35,

$$c_v = \frac{0.196\, d^2}{t_{50}} \quad ...(12.49)$$

The distance d of the drainage path is determined using Eq. 12.48, as in the first method.

The test is repeated for different load increments and an average value of c_v for the desired load range is determined, as shown in Fig. 12.15.

Comparison of the two methods. The two methods for determination of the coefficient of consolidation give comparable results for most of the soils. However, the following points must be carefully noted.

(1) For some soils, the square-root of time plot does not give a straight line for the initial portion and, therefore, to locate the corrected zero R_c becomes difficult. For such soils, the log-of-time method is better.

(2) The square-root of time method is more suitable for soils exhibiting high secondary consolidation. In such soils the log t–plot does not show the characteristic shape required to locate the point corresponding to 100% consolidation.

(3) The square-root of time method is more convenient for a general case, as it requires dial gauge readings covering a much shorter period of time compared with the log-time method. The latter method requires accurate plotting of the secondary consolidation curve in order to locate the asymptote.

Compression Ratios. The following definitions for different compression ratios are used.

(*i*) Initial compression ratio (r_i). It is the ratio of the initial compression to the total compession. In terms of dial gauge readings, it is expressed as

$$r_i = \frac{R_0 - R_c}{R_0 - R_f} \quad \text{...(12.50)}$$

where R_0 = zero dial gauge reading, R_c = corrected zero reading, and R_f = final dial gauge reading.

***(ii)* Primary compression ratio (r_p).** It is ratio of the primary compression to the total compression. In terms of dial gauge readings, it is given by

$$r_p = \frac{R_c - R_{100}}{R_0 - R_f} \quad \text{...12.51)}$$

where R_{100} = dial gauge reading corresponding to 100% primary consolidation.

***(iii)* Secondary compression ratio (r_s).** It is the ratio of the secondary compression to the total compression. In terms of dial gauge readings,

$$r_s = \frac{R_{100} - R_f}{R_0 - R_f} \quad \text{...(12.52)}$$

It can also be written as $r_s = 1 - (r_i + r_p)$ ...(12.53)

because $r_s + r_i + r_p = 1$

12.12. PRECONSOLIDATION PRESSURE

The maximum pressure to which an overconsolidated soil had been subjected in the past is known as the preconsolidation pressure or over-consolidation pressure ($\bar{\sigma}_c$). When a soil specimen is taken from a natural deposit, the weight of the overlying material (over-burden) is removed. This causes an expansion of the soil due to a reduction in pressure. Thus the specimen is generally preconsolidated or over-consolidated. When the specimen is loaded in the consolidation test, the initial portion *AB* of the compression curve *ABCD* (Fig. 12.17) is actually a recompression curve. Consequently, the initial portion *AB* is relatively flat. It is followed by a straight line *CD* with a steep slope which indicates the compression of a virgin (normally consolidated) soil.

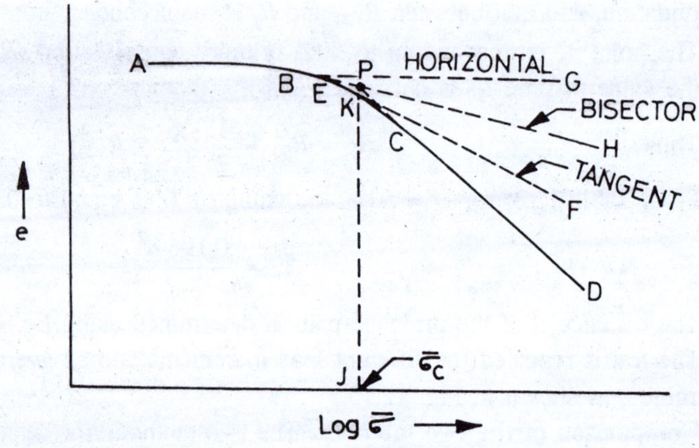

Fig. 12.17. Determination of $\bar{\sigma}_c$.

In the transition range *BC*, the slope gradually changes. The preconsolidation pressure ($\bar{\sigma}_c$) falls in this range. It can be obtained using the method given by Casagrande.

The procedure consists of the following steps :

(1) Determine the point *E* on the curve where the curvature is maximum, *i.e.*, the radius of curvature is minimum.

(2) Draw the tangent *EF* to the curve at *E*.

(3) Draw a horizontal line *EG* at *E*.

(4) Bisect the angle between the tangent *EF* and the horizontal *EG*, and draw the bisector *EH*.

(5) Produce back the straight line portion *CD* of the curve and determine the point of intersection *P* of the bisector *EH* and the backward extension of *CD*.

(6) Draw the vertical *PJ* through *P* which cuts the log $\bar{\sigma}$-axis at *J*. The point *J* indicates the preconsolidation pressure $\bar{\sigma}_c$.

(7) The vertical *PJ* cuts the curve at point *K*. The portion *ABK* of the curve represents the recompression curve and the portion *KCD* as the virgin compression curve.

12.13. CAUSES OF PRECONSOLIDATION IN SOILS

Preconsolidation in a soil deposit may be due to one or more of the causes mentioned below:

(1) Preconsolidation may be due to the overburden which had been later removed by erosion.

(2) It may be due to loads of buildings and other structures which had been demolished.

(3) It may be due to melting of glaciers which covered the soil deposit in the past.

(4) Preconsolidation may be due to capillary pressure which acted on the soil in the past but was later destroyed due to a rise in water table.

(5) Preconsolidation may be due to dessication of the clay deposit. During drying due to rise in temperature, tension develops in pore water and the effective stress increases. After dessication, the soil remained preconsolidated.

(6) Sustained downward seepage forces cause an increase in the effective stress. When the seepage later stops, the soil becomes preconsolidated due to reduction in the effective stress.

(7) Pre-consolidation may be due to tectonic forces caused by the movement of earth's crust which later became less severe.

12.14. FINAL SETTLEMENT OF A SOIL DEPOSIT IN THE FIELD

Computation of settlement of a soil deposit in the field consists of two parts :

(*i*) Computation of magnitude of final settlement.

(*ii*) Determination of the time-rate of settlement.

For the computation of final settlement, the coefficient of volume change or the compression index is required, which is based on the plot between void ratio and the effective stress. For the time-rate of computation, the Terzaghi theory is used. It requires the coefficient of consolidation and is discussed in Sect. 12.10.

(1) Final Settlement Using Coefficient of Volume Change

Let us consider a small element of thickness Δz at a depth z in the clay deposit of total thickness H_0 (Fig. 12.18). Let the effective pressure increment causing the settlement be $\Delta\bar{\sigma}$. From Eq. 12.15,

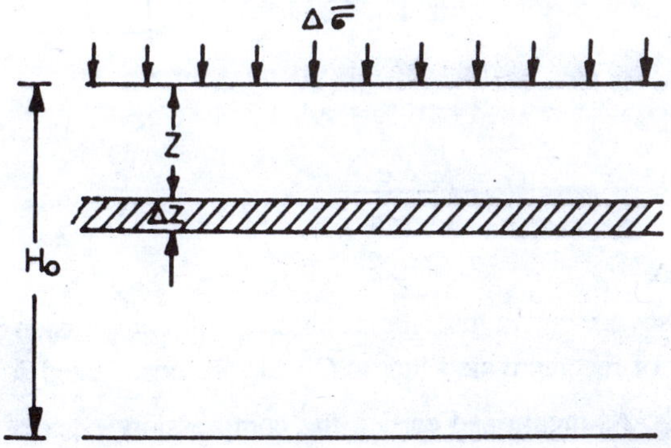

Fig. 12.18. Layer Subjected to $\Delta\bar{\sigma}$.

$$\Delta H = m_v H_0 (\Delta\bar{\sigma})$$

Representing the final settlement as Δs_f and taking $H_0 = \Delta z$,

$$\Delta s_f = m_v \Delta z (\Delta\bar{\sigma})$$

Total settlement of the complete layer,

$$s_f = \int_0^{H_0} \Delta s_f = \int_0^{H_0} m_v \, \Delta \bar{\sigma} \, dz$$

If both m_v and $\Delta\bar{\sigma}$ are constant,

$$s_f = m_v \, \Delta \bar{\sigma} \, H_0 \qquad \text{...(12.54)}$$

In an actual case, as the pressure intensity $\Delta\bar{\sigma}$ and the coefficient of volume change m_v decrease with depth z, it is convenient to divide the stratum into a number (n) of small layers and to assume $\Delta\bar{\sigma}$ and m_v as constant in each layer. The final settlement is given by

$$s_f = \sum_{i=1}^{i=n} (m_v)_i \, (\Delta \bar{\sigma})_i \, (\Delta z)_i \qquad \text{...(12.55)}$$

The settlement s_f may also be obtained by the method of graphical integration. The variation of $\Delta\bar{\sigma}$ and m_v with the depth z is shown in Fig. 12.19 (*a*), and Fig. 12.19 (*b*) respectively. Fig. 12.19 (*c*) shows the variation of the product ($\Delta\bar{\sigma} \times m_v$) with the depth z. The final settlement is equal to the area of the diagram in Fig. 12.19 (*c*).

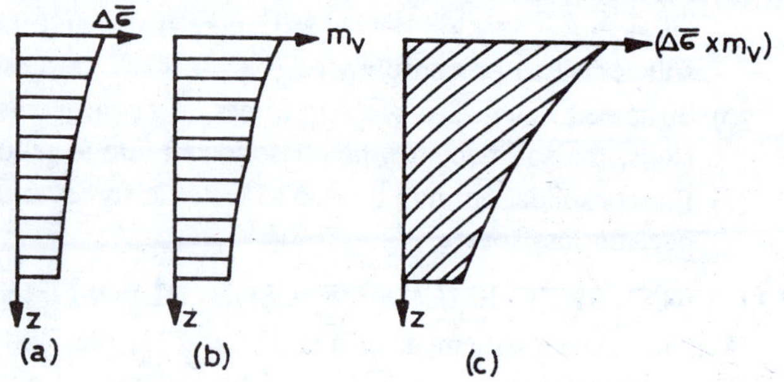

Fig. 12.19. Graphical Integration Method.

However, if the thickness of the clay layer is relatively small, the variations of $\Delta\bar{\sigma}$ and m_v with depth z may be assumed as linear (Fig. 12.20). The values at the mid-depth may be taken as representative values, and Eq. 12.15 may be used as

$$s_f = (m_v)_m \, (\Delta\bar{\sigma})_m \, H_0 \qquad \text{...(12.56)}$$

where $(m_v)_m$ and $(\Delta\bar{\sigma})_m$ are the values at mid-depth.

(2) Final settlement using Void Ratio

If $e - \bar{\sigma}$ plot for the soil is available, it can be used to determine the final settlement. The value of Δe corresponding to the given load increment is read off from the plot and substituted in Eq. 12.11.

$$\Delta H = H_0 \left(\frac{\Delta e}{1 + e_0} \right)$$

or

$$s_f = H_0 \cdot \left(\frac{\Delta e}{1 + e_0} \right) \qquad \text{...(12.57)}$$

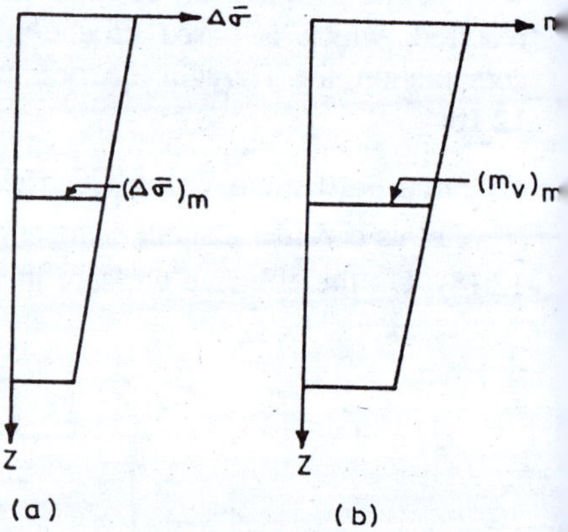

Fig. 12.20

where e_0 is the initial void ratio.

The usual practice is not to use Δe but to use the coefficient of compression index (C_c) for normally consolidated soils and the coefficient of recompression index (C_r) for preconsolidated soils as explained below.

***(a)* Normally consolidated soils.** As mentioned earlier, the compression index of a normally consolidated soil is constant. For such soils, it is more convenient to use compression index. From Eq. 12.18,

$$C_c = \frac{-\Delta e}{\log_{10} (\bar{\sigma}_0 + \Delta\bar{\sigma})/\bar{\sigma}_0}$$

or

$$-\Delta e = C_c \log_{10} (\bar{\sigma}_0 + \Delta\bar{\sigma})/\bar{\sigma}_0$$

where Δe is change in void ratio when the effective pressure is increased from $\bar{\sigma}_0$ to ($\bar{\sigma}_0 + \Delta\bar{\sigma}$). Substituting this value of Δe in Eq. 12.57,

$$s_f = \frac{C_c}{1 + e_0} \cdot H_0 \cdot \log_{10}\left(\frac{\bar{\sigma}_0 + \Delta\bar{\sigma}}{\bar{\sigma}_0}\right) \qquad ...(12.58)$$

(b) Preconsolidated soils. The final settlements are small in the case of preconsolidated soils as the recompression index C_r is considerably smaller than the compression index. From Eq. 12.24,

$$-\Delta e = C_r \log\left(\frac{\bar{\sigma}_0 + \Delta\bar{\sigma}}{\bar{\sigma}_0}\right)$$

Therefore, Eq. 12.57 gives

$$s_f = \frac{C_r}{1 + e_0} \cdot H_0 \cdot \log\left(\frac{\bar{\sigma}_0 + \Delta\bar{\sigma}}{\bar{\sigma}_0}\right) \qquad ...(12.59)$$

The above equation is applicable when $(\bar{\sigma}_0 + \Delta\bar{\sigma})$ is smaller than the perconsolidation pressure $\bar{\sigma}_c$.

If the perconsolidation pressure $\bar{\sigma}_c$ is greater than $\bar{\sigma}_0$ but less than $(\bar{\sigma}_0 + \Delta\bar{\sigma})$, the settlement is computed in two parts:

(i) Settlement for pressure $\bar{\sigma}_0$ to $\bar{\sigma}_c$.

(ii) Settlement for pressure $\bar{\sigma}_c$ to $(\bar{\sigma}_0 + \Delta\bar{\sigma})$

For the first, part, the recompression index is applicable, whereas for the second part, the compression index is used. Thus

$$s_f = \frac{C_r}{1 + e_0} \cdot H_0 \log(\bar{\sigma}_c / \bar{\sigma}_0) + \frac{C_c H_0}{1 + e_0} \log\left(\frac{\bar{\sigma}_0 + \Delta\bar{\sigma}}{\bar{\sigma}_c}\right) \qquad ...(12.60)$$

In this case, the first part is relatively small and is sometimes neglected.

12.15. TIME-SETTLEMENT CURVE

The load is not applied to a soil deposit in the field instantaneously as assumed in the consolidation theory. First excavation is done for the footing. It causes a decrease in the stresses. After the excavation, the footing is constructed and the load of the super-structure is applied gradually, as shown in Fig. 12.21 (*a*). The net load on the soil becomes zero, say at time $t = t_0$. It then gradually increases to full load P at $t = t_p$. Thus, there is expansion of soil due to removal of load in the beginning and it is followed by recompression. It is generally assumed that the net compression of the soil during the period $t = 0$ to $t = t_0$ is negligible. The actual loading period is from time $t = t_0$ to $t = t_p$. During this period, the loading is assumed to vary linearly from zero to full load P.

The time-settlement curve in the field is obtained based on the assumption that the settlement at the end of construction period (t_p) is the same as that would have occurred in half as much time had the entire load been applied instantaneously. In other words, the actual settlement at time t_p is equal to that at $t_p/2$ due to instantaneous loading. In Fig. 12.21 (*b*), the curve *OB* is the load settlement curve obtained using the Terzaghi theory, assuming that the full load P is applied instantaneously at time t_0. The corrected curve is obtained from the instantaneous loading curve. The point *C* on the corrected curve has the settlement *FC* at time t_p. This settlement is equal to the settlement *AH* at time $t_p/2$ of the instantaneous curve. A horizontal line *AC* is drawn from point *A* to cut the vertical *FC* at time t_p at *C*.

The settlement on the corrected curve at any other time is also obtained from the instantaneous curve. The settlement *LG* at time t is obtained from the settlement *KD* at time $t/2$ of the instantaneous curve, but a correction is applied. At time t, as the full load is not acting, the settlement is not exactly equal to *KD*. The actual load acting at t is equal to $P \times (t/t_p)$, *i.e.* the load acting at time t is (t/t_p) times the full load P. The correction to the settlement is made graphically, as described below. A horizontal line *DE* is drawn to cut the vertical at time t_p at *E*. The diagonal *OE* is drawn. It intersects the vertical *LG* at time t at point *G*. The actual settlement at time t is given by *LG*. Obviously, this is equal to $FE \times (t/t_p)$ or $KD \times (t/t_p)$. Therefore, the correction factor is (t/t_p).

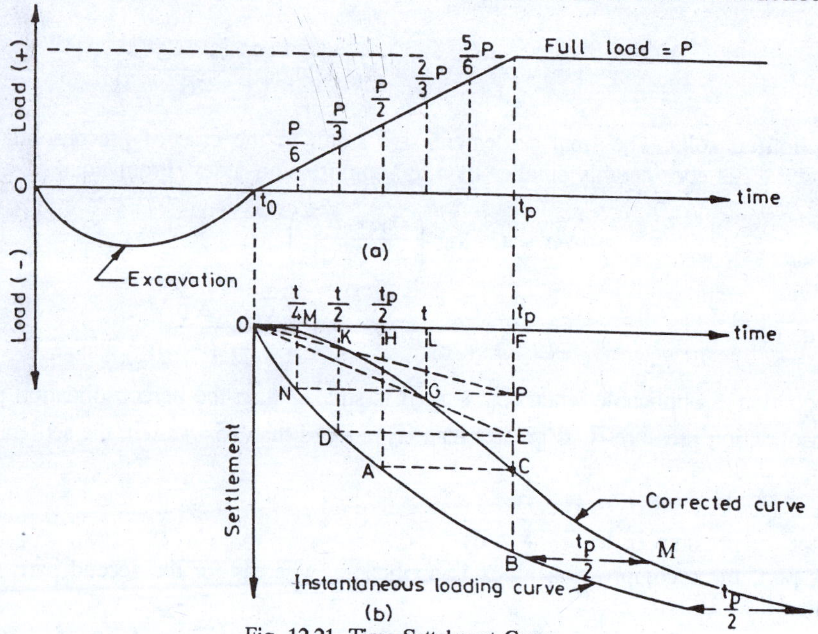

Fig. 12.21. Time-Settelment Curve.

Likewise, the settlement at any other time can be obtained. Fig. 12.21 (*b*) also indicates the settlement at time *t*/4 of the instaneous curve. A smooth curve is drawn through all points so obtained. Thus the corrected curve *OC* is drawn. The curve *CM* beyond the loading period is extended by making offset *BM* equal to *AC*. The load after time t_p is equal to the full load *P*. The offset, therefore, remains equal to the one-half of the loading period ($t_p/2$). In other words, the horizontal offset between the corrected curve and the instantaneous curve after the loading period is constant, and equal to $t_p/2$.

Alternative method.

The corrected curve can be obtained easily if the rate of settlement is not of interest during the construction period (t_p). In this case, the corrected curve can be taken as the curve for instaneous loading in which the origin is taken at $t = t_p/2$. In other words, the whole load *P* is assumed to be applied instantaneously at half the loading period ($t_p/2$).

12.16. FIELD CONSOLIDATION CURVE

The compression characteristics of in-situ soils are different from those obtained from the tests conducted on the soil samples. Even the so-called undisturbed samples are also slightly disturbed when these are taken out. The disturbance causes a slight decrease in the slope of the compression curve obtained in the laboratory after conducting a consolidation test. Consequently, the slope of the curve for in-situ soils is expected to be greater than that obtained from the test. The methods for obtaining the field consolidation curve from the laboratory consolidation curve are discussed separately for normally consolidated, over-consolidated and under-consolidated soils.

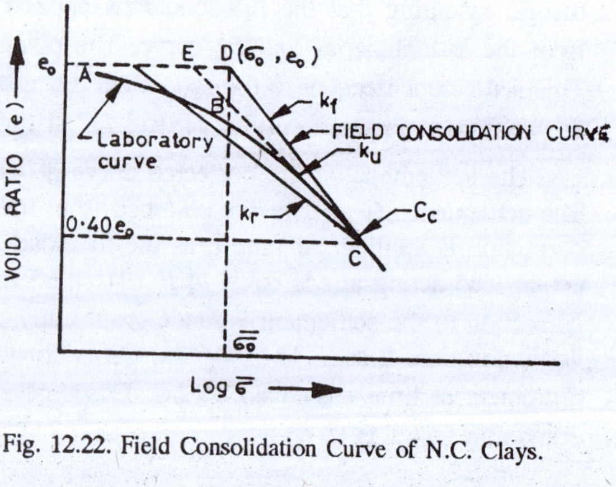

Fig. 12.22. Field Consolidation Curve of N.C. Clays.

(*a*) **Normally consolidated soil.** Schmertmann established that the laboratory virgin curve intersects the field consolidation line at a void ratio of 0.40 e_0, where e_0 is the initial void ratio. The initial void ratio (e_0) can be taken as the void ratio at the start of the laboratory test. Thus the field consolidation line must pass through point

C corresponding to a void ratio of 0.40 e_0 (Fig. 12.22). [Note. In some texts, it is taken as 0.42 e_0]. The coordinates of point D represent the natural void ratio (e_0) and the effective overburden pressure ($\bar{\sigma}_0$) before the sample was extracted. When the sample has been taken out, the overburden pressure reduces to zero, but the water content (w) and hence void ratio remain essentially the same. The process is represented by line $D\,e_0$.

When the undisturbed sample is tested in the laboratory, the compression curve (k_u), represented by the curve ABC is obtained. The portion AB of this curve represents the recompression and the portion BC as the virgin compression. The upward extension of the linear portion BC intersects the horizontal line through e_0 at point E. For a normally consolidated soil, the point E always lies towards the left of the point D. Once it has been established that the soil is normally consolidated, the field consolidation line or in-situ consolidation line (k_f) is drawn joining the points D and C. The slope of this line DC is equal to the compression index of the in- situ soil.

If the sample is remoulded and again tested, the compression curve (k_r) is obtained. The slope of the k_r-curve is somewhat smaller than that of the k_u-curve. However, the downward extension of the k_r-line also intersects the horizontal line drawn from 0.4 e_0 at point C.

(*b*) **Preconsolidated soil.** In Fig. 12.23, the curve ABC is the laboratory compression curve (k_u) for the undisturbed sample. The point C corresponds to a void ratio of 0.4 e_0. The backward extension of the linear portion of the curve ABC meets the horizontal line e_0 D at point E. In case of preconsolidated soils, the point E lies towards the right of point D representing the in-situ condition ($\bar{\sigma}_0$, e_0). It is obvious that there is recompression of the soil from a pressure of $\bar{\sigma}_0$ to the preconsolidation pressure ($\bar{\sigma}_c$). A smooth curve DG is drawn from point D parallel to the recompression curve. Its intersection with the vertical passing through $\bar{\sigma}_c$ is denoted by the point G. The point G is joined to the point C by a straight line. The curve DGC represents the field compression curve (k_f). For pressure between $\bar{\sigma}_0$ to $\bar{\sigma}_c$, the relevant index is the recompression index C_r and for the pressure more than $\bar{\sigma}_c$, it is the compression index (C_c) given by the k_f-line.

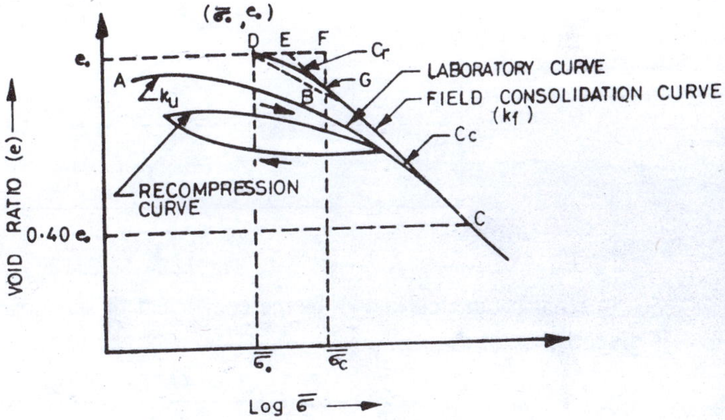

Fig. 12.23. Field Consolidation Curve for O.C. Clays.

(*c*) **Underconsolidated soils.** In areas of recent landfill, sometimes the soil is underconsolidated. It has not reached equilibrium under the effective overburden pressure ($\bar{\sigma}_0$). For such soils, the preconsolidation pressure ($\bar{\sigma}_c$) is less than the overburden pressure.

Thus $$\bar{\sigma}_c < \bar{\sigma}_0 \quad \text{or} \quad \text{O.C.R.} < 1$$

The total compression for such soils when subjected to external loads is equal to the sum of the compression due to overburden pressure till equilibrium is reached and the additional compression due to external loads. Thus

$$\Delta e = \Delta e_1 + \Delta e_2$$

where Δe_1 = decrease in void ratio due to $\bar{\sigma}_0$

Δe_2 = decrease in void ratio due to applied pressure, as found in normally consolidated soils.

12.17. SECONDARY CONSOLIDATION

According to Terzaghi's theory of consolidation, the primary consolidation stops when the excess pore water pressure becomes zero. In actual practice, it has been observed that some consolidation continues even after the full dissipation of the excess pore water pressure. This additional consolidation is known as secondary consolidation. In other words, the secondary consolidation is the consolidation which occurs after the completion of the primary consolidation.

The causes of secondary consolidation are not well understood. This is probably due to the gradual readjustment of the soil skeleton which occurs after the stresses caused during primary consolidation. There is plastic readjustment of the soil particles to the new stress. (In this respect, secondary consolidation is somewhat analogous to the creep in steel when it is overstressed and is in the plastic state.) Secondary consolidation may also be due to progressive fracture of the interparticle bonds and the particles themselves.

The rate of secondary consolidation is given by the secondary compression index (C_t), as defined below.

$$C_t = \frac{-\Delta e}{\log_{10}(t_2/t_1)} \qquad \text{...(12.61)}$$

where Δe is change in void ratio between time t_1 and t_2

It is equal to the slope of the final linear part of the curve drawn between the void ratio and log time (Fig. 12.24).

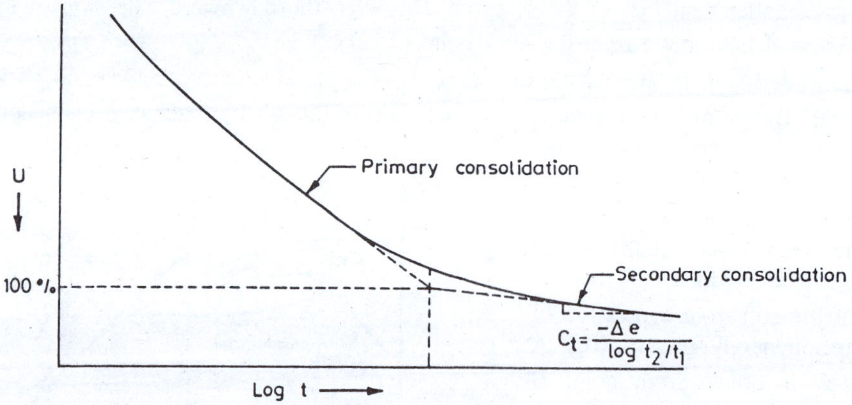

Fig. 12.24. Secondary Consolidation.

Another parameter, known as the coefficient of secondary consolidation (C_α), is more commonly used. It is given by

$$C_\alpha = \frac{C_t}{1+e_p} = \frac{\Delta e}{1+e_p} \cdot \frac{1}{\log_{10}(t_2/t_1)} \qquad \text{...(12.62)}$$

where e_p = void ratio at the end of primary consolidation

and Δe = change in void ratio between time t_1 and t_2.

The magnitude of the secondary consolidation is given by

$$s_s = C_\alpha \times H \log_{10}(t_2/t_1) \qquad \text{...[12.62(a)]}$$

Time t_1 corresponds to the end of primary consolidation.

The rate of secondary consolidation depends upon the plastic characteristics of the soil. It is controlled by highly-viscous, adsorbed water layer surrounding the clay minerals. As the secondary consolidation is highly erratic, its estimate using the above equations is seldom accurate.

For a particular soil, the rate of secondary consolidation increases as the ratio of the pressure increment to the existing pressure is decreased. For standard consolidation test, the ratio is kept one. The rate also increases with a decrease in the thickness of the specimen used in the test. There are a number of other factors which control the rate of secondary consolidation. In general, the value of C_α for normally consolidated soils varies with the compressibility and hence with the natural water content.

The rate depends upon the length of time the preload may have acted on the soil. It also depends upon the shear stresses and on the degree of disturbance of the sample.

Secondary consolidation is important only for highly plastic clays and organic soils. In some organic soils, the secondary consolidation may even be more than the primary consolidation. In over-consolidated, inorganic clays, the secondary consolidation is usually small and hence it is neglected.

12.18. 3-D CONSOLIDATION EQUATION IN CARTESIAN COORDINATES

Terzaghi's theory of one-dimensional consolidation discussed above is based on the assumption that the soil is laterally confined and the consolidation takes place only in the vertical direction. In field, as the layers are not laterally confined, the consolidation takes place in all the three-dimensions. In general, the consolidation in the horizontal direction is small and, therefore, neglected. However, in some special cases, such as in sand drains, there is significant radial drainage, in addition to the vertical drainage. For such cases, three-dimensional consolidation equation is required to determine the rate of consolidation. The equation for 3-D consolidation is derived below, making the following assumptions :

(1) The soil mass is homogenous.

(2) The soil is completely saturated.

(3) The soil particles as well as the water in the voids are incompressible. The consolidation takes place due to reduction in voids caused by flow water.

(4) Darcy's law, in generalised form, is applicable to anisotropic soils.

(5) Pressure increment $\Delta\bar{\sigma}$ is applied instantaneously to develop an initial excess pore water pressure u_i.

Fig. 12.25 shows a parallelepiped of soil mass with sides dx, dy, dz with its centre at point $P(x, y, z)$. Let the velocity components at the point P by v_x, v_y and v_z. The velocities on the six faces are obtained using the partial derivatives and are shown in the figure.

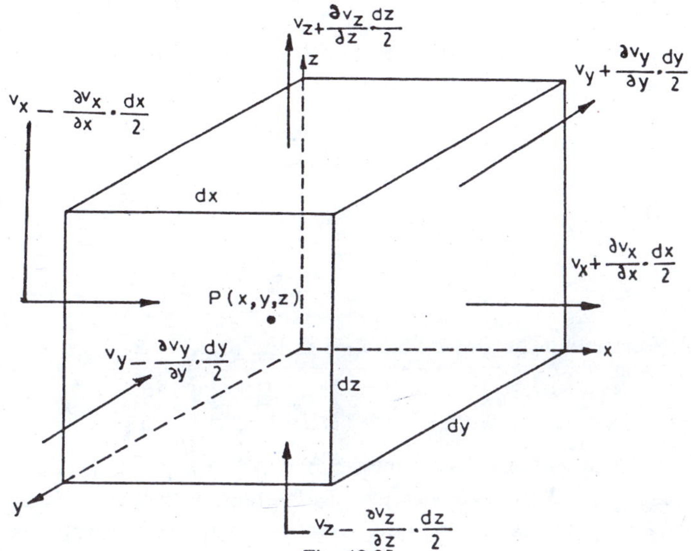

Fig. 12.25

The volume of water entering the parallelepiped per unit time (Q_i) is obtained from the products of the relevant velocities and areas. Thus

$$Q_i = \left(v_x - \frac{\partial v_x}{2} \cdot \frac{dx}{2}\right) dy\, dz + \left(v_y - \frac{\partial v_y}{\partial z} \cdot \frac{dy}{2}\right) dx\, dz + \left(v_z - \frac{\partial v_z}{\partial z} \cdot \frac{dz}{2}\right) \times dx\, dy$$

Likewise, the volume of water going out per unit time (Q_0) is given by

$$Q_0 = \left(v_x + \frac{\partial v_x}{2} \cdot \frac{dx}{2}\right) dy\, dz + \left(v_y + \frac{\partial v_y}{\partial z} \cdot \frac{dy}{2}\right) dx\, dz + \left(v_z + \frac{\partial v_z}{\partial z} \cdot \frac{dz}{2}\right) \times dx\, dy$$

Therefore, the volume of water squeezed out of the parallelepiped per unit time is given by

$$dQ = Q_0 - Q_i$$

or

$$dQ = \left(\frac{\partial v_x}{\partial x} + \frac{\partial v_y}{\partial y} + \frac{\partial v_z}{\partial z}\right) dx\, dy\, dz \quad \ldots(a)$$

The volume of the parallelepiped V is equal to dx , dy , dz. It is also equal to $V_s(1 + e)$, where V_s is the volume of the solids and e is the void ratio. Thus

$$V_s = \frac{V}{1 + e} = \frac{dx\,dy\,dz}{1 + e}$$

or
$$V = V_s\,(1 + e)$$

or
$$\frac{\partial V}{\partial t} = \frac{\partial}{\partial t}\left[\,V_s\,(1 + e)\right] = V_s\,\frac{\partial e}{\partial t}$$

or
$$\frac{\partial V}{\partial t} = \frac{dx\,dy\,dz}{1 + e}\cdot\frac{\partial e}{\partial t} \qquad \ldots(b)$$

Obviously the volume of water squeezed out per unit time is equal to the change in volume of parallelepiped per unit time. From Eqs. (*a*) and (*b*),

$$\frac{dx\,dy\,dz}{1 + e}\cdot\frac{\partial e}{\partial t} = \left(\frac{\partial v_x}{\partial x} + \frac{\partial v_y}{\partial y} + \frac{\partial v_z}{\partial z}\right) dx\,dy\,dz$$

or
$$\frac{\partial e}{\partial t} = (1 + e)\left(\frac{\partial v_x}{\partial x} + \frac{\partial v_y}{\partial y} + \frac{\partial v_z}{\partial z}\right) \qquad \ldots(c)$$

If $\bar{u}$ is the excess hydrostatic pressure, the velocities in *x*, *y*, and *z* directions are obtained from Darcy's law as

$$v_x = k_x\,i_x = k_x\,\frac{\partial h}{\partial x} = k_x\,\frac{1}{\gamma_w}\cdot\frac{\partial \bar{u}}{\partial x}$$

$$v_y = k_y\,i_y = k_y\,\frac{\partial h}{\partial y} = k_y\,\frac{1}{\gamma_w}\cdot\frac{\partial \bar{u}}{\partial y}$$

$$v_z = k_z\,i_z = k_z\,\frac{\partial h}{\partial z} = k_z\,\frac{1}{\gamma_w}\cdot\frac{\partial \bar{u}}{\partial z}$$

Substituting the above velocities in Eq. (*c*),

$$\frac{\partial e}{\partial t} = \frac{1 + e}{\gamma_w}\left(k_x\,\frac{\partial^2 \bar{u}}{\partial x^2} + k_y\,\frac{\partial^2 \bar{u}}{\partial y^2} + k_z\,\frac{\partial^2 \bar{u}}{\partial z^2}\right) \qquad \ldots(d)$$

As soon as the pressure increment ($\Delta\,\sigma$) is applied, the pore water pressure develops. Initially, the load is entirely taken by pore water, but as the time passes, water is squeezed out. The excess pore water pressure gradually decreases and the effective stress increases, as in the one-dimensional consolidation. Thus

$$\Delta\,\sigma = \Delta\,\bar{\sigma} + \Delta \bar{u}$$

where $\Delta\,\bar{\sigma}$ = effective stress, and $\Delta\,\bar{u}$ = pore water pressure.

As any increase in effective stress ($\bar{\sigma}$) is equal to a decrease in excess hydrostatic pressure $\bar{u}$,

$$\Delta\,\bar{\sigma} = -\,\Delta\,\bar{u}$$

Therefore,
$$\frac{\partial e}{\partial \bar{\sigma}} = -\,\frac{\partial e}{\partial \bar{u}}$$

But $\frac{\partial e}{\partial \bar{\sigma}}$ = coefficient of compressibility a_v (Eq. 12.12).

Therefore,
$$\frac{\partial e}{\partial \bar{u}} = -\,a_v$$

From the rule of partial differentiation.

$$\frac{\partial e}{\partial t} = \frac{\partial e}{\partial \bar{u}}\cdot\frac{\partial \bar{u}}{\partial t} = a_v\cdot\frac{\partial \bar{u}}{\partial t} \qquad \ldots(e)$$

From Eqs. (*d*) and (*e*),

$$a_v\,\frac{\partial \bar{u}}{\partial t} = \frac{1 + e}{\gamma_w}\left(k_x\,\frac{\partial^2 \bar{u}}{\partial x^2} + k_v\,\frac{\partial^2 \bar{u}}{\partial y^2} + k_z\,\frac{\partial^2 \bar{u}}{\partial z^2}\right)$$

or $$\frac{\partial \bar{u}}{\partial t} = \frac{(1+e)}{a_v} \cdot \frac{1}{\gamma_w}\left(k_x \frac{\partial^2 \bar{u}}{\partial x^2} + k_y \frac{\partial^2 \bar{u}}{\partial y^2} + k_z \frac{\partial^2 \bar{u}}{\partial z^2}\right)$$

The above equation can be written in terms of the coefficient of volume change m_v.

From Eq. 12.16, $$m_v = \frac{a_v}{1+e}$$

Thus $$\frac{\partial \bar{u}}{\partial t} = \frac{1}{m_v \gamma_w}\left(k_x \frac{\partial^2 \bar{u}}{\partial x^2} + k_y \frac{\partial^2 \bar{u}}{\partial y^2} + k_z \frac{\partial^2 \bar{u}}{\partial z^2}\right)$$

The equation can be written in terms of coefficients of consolidation c_{vx}, c_{vy} and c_{vz} in the three directions using Eq. 12.29.

Therefore, $$\frac{\partial \bar{u}}{\partial t} = c_{vx} \frac{\partial^2 \bar{u}}{\partial x^2} + c_{vy} \frac{\partial^2 \bar{u}}{\partial y^2} + c_{vz} \frac{\partial^2 \bar{u}}{\partial z^2} \qquad \text{...(12.63)}$$

Eq. 12.63 is the general equation for three-dimensional consolidation (3-D consolidation).

12.19. 3-D CONSOLIDATION EQUATION IN CYLINDRICAL COORDINATES

Three-dimensional consolidation equation, obtained in the proceeding section, can be transformed into cylindrical coordinates (r, θ, z) by making the following substitution:

$$x = r\cos\theta, \qquad y = r\sin\theta \quad \text{and} \quad z = z$$

where r = radial distance (polar distance),

and θ = angle made by the radius with the pole.

Thus, $$\tan\theta = y/x \quad \text{or} \quad \theta = \text{arc}\tan(y/x) \qquad \text{...}(a)$$

and $$r^2 = x^2 + y^2 \qquad \text{...}(b)$$

Differentiating Eq. (b), $$2r\frac{\partial r}{\partial x} = 2x \quad \text{or} \quad \frac{\partial r}{\partial x} = \frac{x}{r} = \cos\theta$$

Likewise, $$\frac{\partial r}{\partial y} = \frac{y}{r} = \sin\theta$$

From Eq. (a), $$\frac{\partial \theta}{\partial x} = -\frac{y}{x^2}\frac{1}{\sec^2\theta}$$

or $$\sec^2\theta \frac{\partial \theta}{\partial x} = -\frac{y}{x^2}$$

or $$\frac{\partial \theta}{\partial x} = \frac{-y}{x^2(1+y^2/x^2)} = \frac{-y}{r^2} = -\frac{\sin\theta}{r}$$

Likewise, $$\frac{\partial \theta}{\partial y} = \frac{x}{r^2} = \frac{\cos\theta}{r}$$

Therefore, $$\frac{\partial \bar{u}}{\partial x} = \frac{\partial \bar{u}}{\partial r}\cdot\frac{\partial r}{\partial x} + \frac{\partial \bar{u}}{\partial \theta}\cdot\frac{\partial \theta}{\partial x}$$

or $$\frac{\partial \bar{u}}{\partial x} = \frac{\partial \bar{u}}{\partial r}\cdot\cos\theta - \frac{\partial \bar{u}}{\partial \theta}\cdot\left(\frac{\sin\theta}{r}\right)$$

and $$\frac{\partial^2 \bar{u}}{\partial x^2} = \left(\frac{\partial}{\partial r}\cos\theta \quad \frac{1}{r}\sin\theta\frac{\partial}{\partial r}\right)\left(\frac{\partial \bar{u}}{\partial r}\cos\theta - \frac{1}{r}\frac{\partial \bar{u}}{\partial \theta}\sin\theta\right)$$

$$= \frac{\partial^2 \bar{u}}{\partial r^2}\cos^2\theta - 2\frac{\partial^2 \bar{u}}{\partial\theta\,\partial r}\cdot\frac{\sin\theta\cos\theta}{r} + \frac{\partial \bar{u}}{\partial r}\cdot\frac{\sin^2\theta}{r}$$

$$+ 2\frac{\partial \bar{u}}{\partial \theta} \cdot \frac{\sin\theta\cos\theta}{r^2} + \frac{\partial^2 \bar{u}}{\partial \theta^2} \cdot \frac{\sin^2\theta}{r^2} \qquad \text{...(c)}$$

Likewise,

$$\frac{\partial^2 \bar{u}}{\partial y^2} = \frac{\partial^2 \bar{u}}{\partial r^2}\sin^2\theta + 2\frac{\partial^2 \bar{u}}{\partial \theta\,\partial r} \cdot \frac{\sin\theta\cos\theta}{r}$$

$$+ \frac{\partial \bar{u}}{\partial r} \cdot \frac{\cos^2\theta}{r} \frac{-2\,\partial \bar{u}}{\partial \theta} \cdot \frac{\sin\theta\cos\theta}{r} + \frac{\partial^2 \bar{u}}{\partial \theta^2} \cdot \frac{\cos^2\theta}{r^2} \qquad \text{...(d)}$$

From Eqs. (*c*) and (*d*),

$$\frac{\partial^2 \bar{u}}{\partial x^2} + \frac{\partial^2 \bar{u}}{\partial y^2} = \frac{\partial^2 \bar{u}}{\partial r^2} + \frac{1}{r}\frac{\partial \bar{u}}{\partial r} + \frac{1}{r^2}\frac{\partial^2 \bar{u}}{\partial \theta^2} \qquad \text{...(e)}$$

In the case of axial symmetry, $c_{vx} = c_{vy} = c_{vr}$ (say)

From Eq. 12.63,

$$\frac{\partial \bar{u}}{\partial t} = c_{vx}\frac{\partial^2 \bar{u}}{\partial x^2} + c_{vy}\frac{\partial^2 \bar{u}}{\partial y^2} + c_{vz}\frac{\partial^2 \bar{u}}{\partial z^2}$$

or

$$\frac{\partial \bar{u}}{\partial t} = c_{vr}\left(\frac{\partial^2 \bar{u}}{\partial x^2} + \frac{\partial^2 \bar{u}}{\partial y^2}\right) + c_{vz}\frac{\partial^2 \bar{u}}{\partial z^2}$$

Using Eq. (*e*),

$$\frac{\partial \bar{u}}{\partial t} = c_{vr}\left(\frac{\partial^2 \bar{u}}{\partial r^2} + \frac{1}{r}\frac{\partial \bar{u}}{\partial r} + \frac{1}{r^2}\frac{\partial^2 \bar{u}}{\partial \theta^2}\right) + c_{vz}\frac{\partial^2 \bar{u}}{\partial z^2}$$

In the case of axial symmetry, $\dfrac{\partial^2 \bar{u}}{\partial \theta^2} = 0$

Thus,

$$\frac{\partial \bar{u}}{\partial t} = c_{vr}\left(\frac{\partial^2 \bar{u}}{\partial r^2} + \frac{1}{r}\frac{\partial \bar{u}}{\partial r}\right) + c_{vz}\frac{\partial^2 \bar{u}}{\partial z^2} \qquad \text{...(12.64)}$$

Eq. 12.64 is the governing equation for consolidation in three- dimensions in terms of cylindrical coordinates for the case of axial symmetry. The equation can be split into two parts :

(*i*) Radial Flow :

$$c_{vr}\left(\frac{\partial^2 \bar{u}}{\partial r^2} + \frac{1}{r}\frac{\partial \bar{u}}{\partial t}\right) = \frac{\partial \bar{u}}{\partial t} \qquad \text{...(12.65)}$$

(*ii*) Vertical flow :

$$c_{vz}\left(\frac{\partial^2 \bar{u}}{\partial z^2}\right) = \frac{\partial \bar{u}}{\partial t} \qquad \text{...(12.66)}$$

where $c_{vr} = k_r/(m_v\,\gamma_w)$ and $c_{vz} = k_z/(m_v\,\gamma_w)$.

If U_r and U_v are the average degrees of consolidation in radial and vertical directions, respectively, using Eq. 12.43,

$$U_r = f(t_r) \qquad \text{...(12.67)}$$

and

$$U_v = f(T_v) \qquad \text{...(12.68)}$$

where T_r and T_v are time factors in radial and vertical directions, respectively, given by (using Eq. 12.44),

$$T_r = c_{vr}\,t/(4R^2) \qquad \text{...(12.69)}$$

and

$$T_v = c_{vz}\,t/d^2 \qquad \text{...(12.70)}$$

where R is the radius of the drainage area,

d is the drainage length in the vertical direction.

It has been shown that the overall average degree of consolidation U under combined radial and vertical directions can be expressed as

$$(1 - U) = (1 - U_v)\,(1 - U_r) \qquad \text{...(12.71)}$$

The value of U_v can be obtained using the theory of one-dimensional consolidation as already discussed. The value of U_r is obtained as explained in the following section for the sand drains.

12.20. SAND DRAINS

The main application of the radial consolidation is in the design of sand drains used to increase the rate of drainage in the embankment. Sand drains are constructed by driving a casing (or a hollow mandrel) into the embankment and making vertical bore holes. The holes are backfilled with a suitably graded sand. The casing is withdrawn after the sand has been filled. The drains are generally laid either in a square pattern [Fig. 12.26 (*a*)] or a triangular pattern [Fig. 12.26 (*b*)]. The spacing (S) of the drains is kept smaller than the thickness of the embankment ($2H$) in order to reduce the length of the radial drainage path. The zone of influence of each drain in a triangular pattern is hexagonal in plan, which can be approximated by an equivalent circle of radius R, where $R = 0.525\ S$. In case of a square pattern, the radius of circle of influence R is equal to $0.554S$. The radius of the sand drain is represented by r_w.

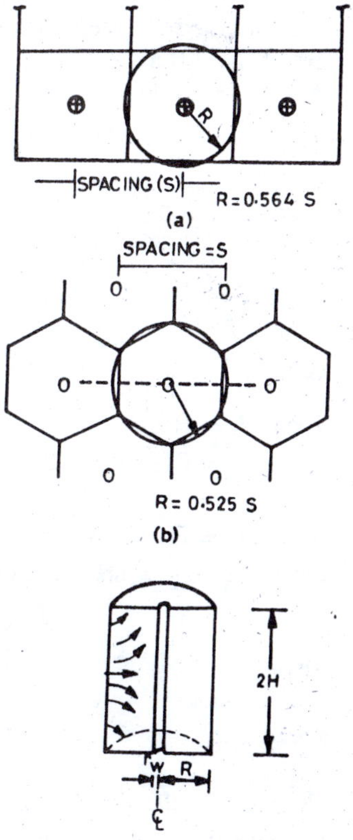

Fig. 12.26. Patterns of Sand Drains.

Fig. 12.27 shows the sand drains installed in position. A sand blanket is placed over the top of the sand drains to connect all the sand drains. To accelerate the drainage, a surcharge load is placed on the sand blanket. The surcharge is generally in the form of dumped soil. Due to surcharge load, the pore water pressure increases in the embankment. The drainage occurs in the vertical and horizontal directions. The horizontal drainage occurs because of sand drains. The sand drains accelerate the process of dissipation of excess pore water created by the surcharge.

The theory of sand drains was given by Rendulic (1935) and Barron (1948). Later, Richart (1959) summarised the theories. Depending upon the type of strain, there are two cases.

(1) Free strain case. (2) Equal strain case.

(1) Free Strain Case. If the surcharge load placed over the sand blanket is flexible, free strain case occurs. In this case, there is uniform distribution of surface loads, but the settlements at the surface are uneven. The basic differential equation for radial drainage, given by Eq. 12.65, is

$$\frac{\partial \bar{u}}{\partial t} = c_{vr}\left(\frac{\partial^2 \bar{u}}{\partial r^2} + \frac{1}{r}\frac{\partial \bar{u}}{\partial r}\right)$$

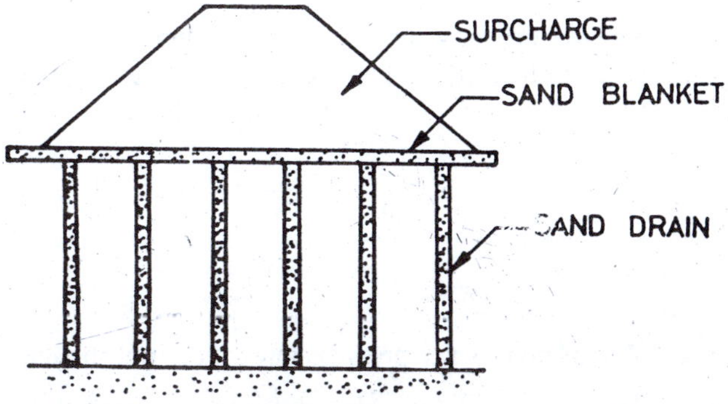

Fig. 12.27. Sand Drains Installation.

In the case of free strain case, the boundary conditions are as under

At time $t = 0$, $\bar{u} = \bar{u}_i$. At time $t > 0$, $\bar{u} = 0$ at $r = r_w$. and at $r = R$, $\frac{\partial \bar{u}}{\partial r} = 0$

The solution for excess pore water pressure $\bar{u}$ at any time t and at a radial distance r is obtained by the solution of the differential equation as

$$\bar{u} = \sum_{\alpha_1, \alpha_2 \ldots}^{\alpha = \infty} \frac{-2\, U_1(\alpha)\, U_0(\alpha_r / r_w)}{\alpha\,[n^2 U_0^2(\alpha n) - U_1^2(\alpha)]}\, e^{(-4\,\alpha^2 n^2 T_r)}$$

where $n = R/r_w$ and

$$U_1(\alpha) = J_1(\alpha)\, Y_0(\alpha) - Y_1(\alpha)\, J_0(\alpha)$$

$$U_0(\alpha n) = J_0(\alpha n)\, Y_0(\alpha) - Y_0(\alpha n)\, J_0(a)$$

and $$U_0(\alpha r/r_w) = J_0(\alpha r/r_w)\, Y_0(\alpha) - Y_0\left(\frac{\alpha r}{r_w}\right) J_0(\alpha)$$

where J_0 = Bessel function of first kind of zero order.

J_1 = Bessel function of first kind of first order.

Y_0 = Bessel function of second kind of zero order.

Y_1 = Bessel function of second kind of first order.

and $\alpha_1, \alpha_2 \ldots$ are roots of Bessel function which satisfy the equation

$$J_1(\alpha n)\, Y_0(\alpha) - Y_1(\alpha n)\, J_0(\alpha) = 0$$

Also $$T_r = C_{vr}\, t/(2R)^2$$

where $$c_{vr} = \frac{k_r}{m_v \gamma_w} = \frac{k_h}{m_v \gamma_w}$$

in which k_h is coefficient of permeability in horizontal direction.

The average pore water pressure u_{av} throughout the soil mass may be written as

$$u_{av} = u_i \sum_{\alpha_1, \alpha_2 \ldots}^{\alpha = \infty} \frac{4 U_1^2(\alpha)}{\alpha^2 (n^2 - 1)\, [n^2 U_0^2(\alpha n) - U_1^2(\alpha)]}\, e^{(-4\,\alpha^2 n^2 T_r)}$$

The average degree of radial consolidation U_r can be determined from the equation

$$U_r = 1 - \frac{u_{av}}{u_i} \qquad \ldots(12.72)$$

Fig. 12.28 shows the variation of U_r with the time factor T_r by dotted lines for different values of n, where $n = R/r_w$.

(2) Equal Strain Case. This case occurs when the surcharge applied is rigid, such as heavy steel plates. In this case, the settlements are uniform, but the distribution of pressure is non-uniform. The problem was solved by Barron, who gave the expression for excess pore water pressure $\bar{u}$ as

$$\bar{u} = \frac{4\, u_{av}}{(2R)^2\, F(n)} \left[R^2 \log_e (r/r_w) - \frac{(r^2 - r_w^2)}{2} \right] \qquad \ldots(12.73)$$

where $$F(n) = \frac{n^2}{n^2 - 1} \log_e^{(n)} - \frac{(3n^2 - 1)}{4 n^2}$$

u_{av} = average value of pore water pressure throughout out the embankment

or $u_{av} = \bar{u}_i\, e^{\lambda}$ in which $\lambda = \frac{-8T_r}{F(n)}$

The average degree of consolidation in radial direction is given by

$$U_r = 1 - e^{(-8T_r/F(n))} \quad \ldots(12.74)$$

Fig. 12.28 shows the variation of U_r with T_r by firm lines for 3 values of n.

Tr (Equal strain case)

Ur (%)	n = 5	n = 10	n = 40
10	0.012	0.021	0.039
20	0.026	0.046	0.082
30	0.042	0.070	0.131
40	0.060	0.101	0.188
50	0.081	0.137	0.255
60	0.107	0.180	0.337
70	0.137	0.231	0.431
80	0.188	0.317	0.592
90	0.270	0.455	0.847
95	0.351	0.590	1.102
99	0.539	0.907	1.693

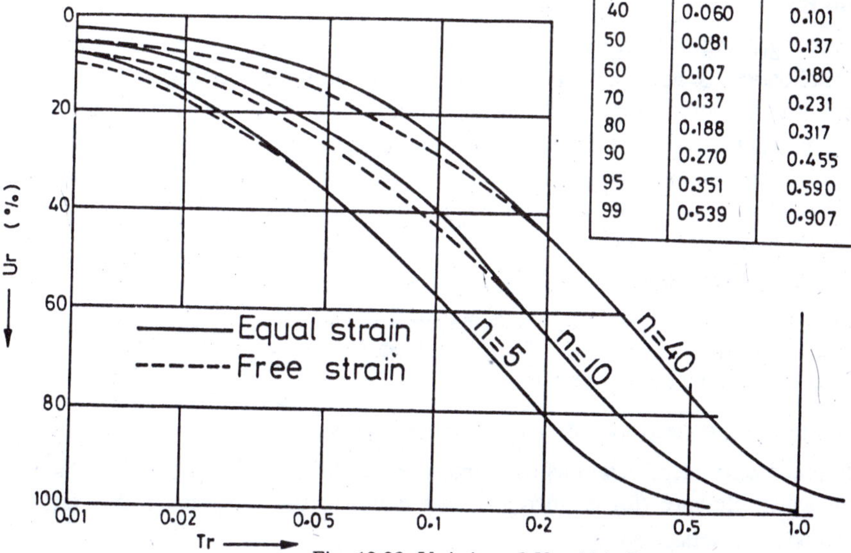

Fig. 12.28. Variation of U_r which T_r.

It may be observed that the curves for free strain and equal strain are not much different and they give approximately the same results. Equal strains case is generally preferred as it is more convenient. Fig. 12.28 also gives the value of T_r in a tabular form for the equal strain case.

Effect of Smear Zone. A smear zone is formed around a sand drain due to the remouldining of clay caused during its construction. A decrease in the coefficient of permeability in the radial direction occurs due to remoulding. Barron extended the analysis of the equal-strain case taking into account the effect of smear zone.

The analysis is based on the assumption that the clay in the smear zone has zero excess pore water pressure on the inner boundary and the time-dependent excess pore water pressure on the outer boundary. Fig. 12.29 shows a cross-section through a sand drain having a smear zone. The radial distance from the centre line of the drain well to the farthest point on the smear zone is equal to the radius of smear zone r_s.

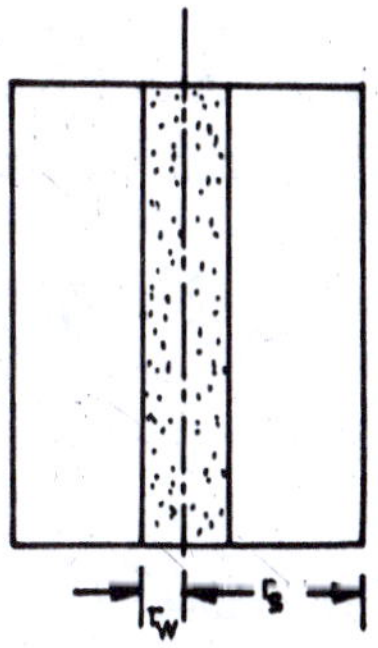

Fig. 12.29. Smear Zone

The excess pore water pressure $\bar{u}$ in this case in given by

$$\bar{u} = \frac{1}{m} u_{av} \left[\log_e (r/R) - \frac{(r^2 - r_s^2)}{2R^2} + \frac{k_h}{k_s} \left(\frac{(n^2 - B^2)}{n^2} \right) \log_e B \right] \quad \ldots(12.75)$$

where k_h = coefficient of permeability of the soil in horizontal direction

k_s = coefficient of permeability of smear zone,

$B = r_s/r_w$ and $n = R/r_w$, where R = radius of influence

$$m = \frac{n^2}{n^2 - B^2} \log_e (n/B) - \frac{3}{4} + \frac{B^2}{4n^2} + \frac{k_h}{k_s} \left(\frac{n^2 - B^2}{n^2} \right) \log_e B$$

and $$u_{av} = u_i\, e^{-(8T_r/m)}$$

The average degree of consolidation is given by

$$U_r = 1 - \frac{u_{av}}{u_i} = 1 - e^{-(8T_r/m)} \quad ...(12.76)$$

The solutions for values of m are also available in the form of curves for different values of k_h/k_s and B and n. It may be noted that for no smear zone, B = 1.0 and the results of equal-strain case apply.

The net effect of the smear zone on the consolidation is that the influence area of the drain is reduced. As an approximation, the effect of smear zone is sometimes taken indirectly into account by reducing the radius of influence R to 0.5 R.

The following points regarding sand drains may be noted.

(1) Secondary consolidation is not taken into account in the design of sand drains. In fact, the sand drains are ineffective in controlling the secondary consolidation for highly plastic and organic soils.

(2) Sand drains tend to act as weak piles and reduce the stresses in the clay. Consequently, the excess pore water pressure developed is generally less as compared with that in the case when there are no sand drains. This factor is not taken into account in equation given above.

(3) The typical design parameters for the sand drains are as under :

(*a*) Radius of sand drains well, r_w = 0.2 to 0.3 m
(*b*) Spacing of sand drains, S = 2 to 5 m
(*c*) Depth of sand drains, $2H$ = 3 to 35 m
(*d*) Thickness of sand blanket = 0.6 to 1 m.

12.21. EFFECT OF LATERAL STRAINS ON CONSOLIDATION

In field, the condition of zero lateral strain as assumed in Terzaghi's consolidation theory is not satisfied. The initial excess pore water pressure (u_i) is, therefore, not equal to the increase in the vertical stress as assumed in one- dimensional consolidation.

Skempton and Bjerrum developed a semi-empirical method of calculating settlement, taking into account the effect of lateral strain. The final settlement is expressed by them as

$$s_f = \mu\, s_{oed} \quad ...(12.77)$$

where s_{oed} is the final settlement based on the assumption of no lateral strain using one-dimensional consolidation theory, as explained in Sec. 12.14. The correction factor μ is given by

$$\mu = \frac{\int_0^H m_v\, \Delta\sigma_1 \left\{ A + \frac{\Delta\sigma_3}{\Delta\sigma_1}(1 - A) \right\} dz}{\int_0^H m_v\, \Delta\sigma_1\, dz} \quad ...(12.78)$$

where A is the pore pressure coefficient as discussed in chapter 13.

If m_v and A are assumed to be constant, Eq. 12.78 reduces to

$$\mu = A + (1 - A)\,\alpha \quad ...(12.79)$$

where $$\alpha = \frac{\int_0^H \Delta\sigma_3\, dz}{\int_0^H \Delta\sigma_1\, dz}$$

The value of α depends upon the shape of the loaded area and the thickness D of the clay stratum in relation to the dimension of the loaded area. It can be calculated using the theory of elasticity. The value of α for the circular and strip footings are given in Table 12.4.

Table 12.4. Values of coefficient α

D/B	0.0	0.25	0.50	1.0	2.0	4.0	10.0	∞
α *(circular footing)*	1.0	0.67	0.50	0.38	0.30	0.28	0.26	0.25
α *(strip footing)*	1.0	0.80	0.63	0.53	0.45	0.40	0.36	0.25

Note : (1) B is equal to the width of strip footing or diameter of the circular footing.

(2) For square footings, use the values of α for the circular footing of the same area.

The values of the pore pressure coefficient A depend upon the type of the clay.

The typical values of the correction factor μ are given in Table 12.5. Fig. 12.30 gives the values of μ for different values of A and D/B ratio, where D is the thickness of soil layer.

Table 12.5. Values of μ.

Type of Soil	μ
Heavily over-consolidated clay	0.25 to 0.4
Moderately overconsolidated clay	0.4 to 0.7
Normally consolidated clay	0.6 to 1.00
Soft, sensitive clay	1.0 to 1.20

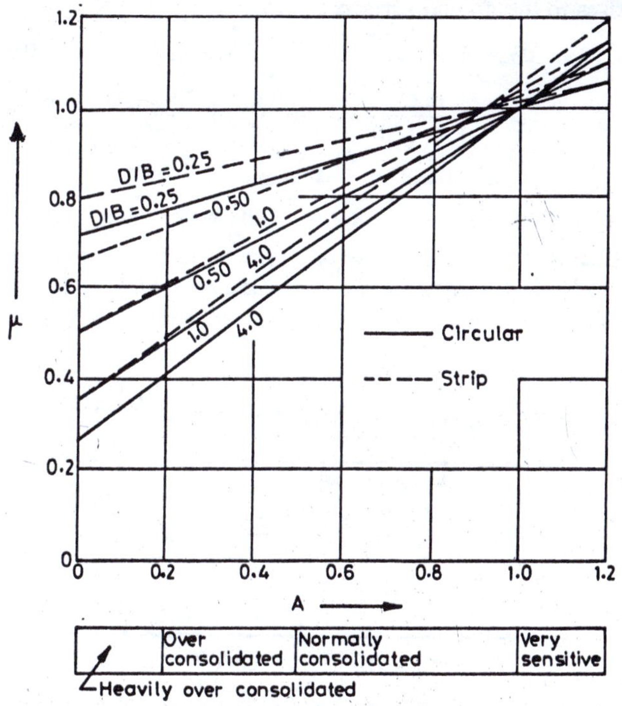

Fig. 12.30. Plot for value of μ.

ILLUSTRATIVE EXAMPLES

Illustrative Example 12.1. *Calculate the final settlement of the clay layer shown in Fig. E 12.1 due to an increase of pressure of 30 kN/m² at mid-height of the layer. Take* γ_w = 10*kN/m*³.

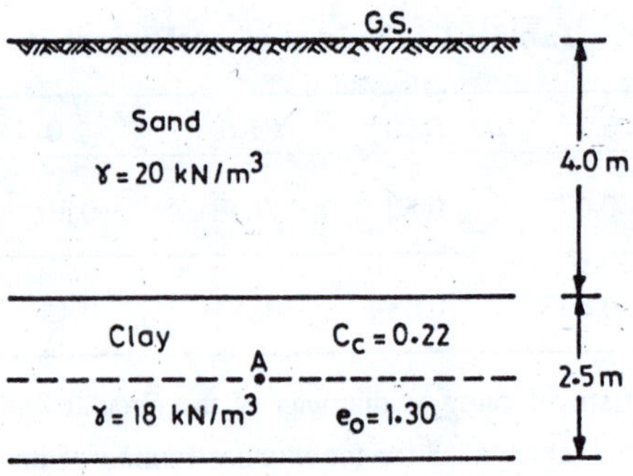

Fig. E 12.1

Also calculate the settlement when the water table rises to the ground surface.

Solution. Initial pressure at the centre of the clay layer.

$$\bar{\sigma}_0 = 4 \times 20 + 1.25 \times 18 = 102.5 \text{ kN/m}^2.$$

From Eq. 12.58,

$$s_f = \frac{C_c}{1 + e_0} \cdot H_0 \log_{10} \left(\frac{\bar{\sigma}_0 + \Delta \bar{\sigma}}{\bar{\sigma}_0} \right)$$

$$= \frac{0.22}{1 + 1.30} \times 2.50 \log_{10} \left(\frac{102.5 + 30.0}{102.5} \right)$$

$$= 0.0263 \text{ m} = 2.63 \text{ cm}$$

When the water table rises to the ground surface,

$$\bar{\sigma}_0 = 4 \times (20 - 10) + 1.25 \times (18 - 10) = 50 \text{ kN/m}^2$$

Therefore,

$$s_f = \frac{0.22}{1 + 1.30} \times 2.50 \log_{10} \left(\frac{50 + 30}{50} \right)$$

$$= 0.0488 \text{ m} = \mathbf{4.88 \text{ cm}}$$

As expected, the settlement increases due to the rise of the water table.

Illustrative Example 12.2. *A footing has a size of 3.0 m by 1.50 m and it causes a pressure increment of 200 kN/m² at its base (Fig. E 12.2). Determine the consolidation settlement at the middle of the clay layer. Assume 2 : 1 pressure distribution and consider the variation of pressure across the depth of the clay layer.* $\gamma_w = 10$ *kN/m³.*

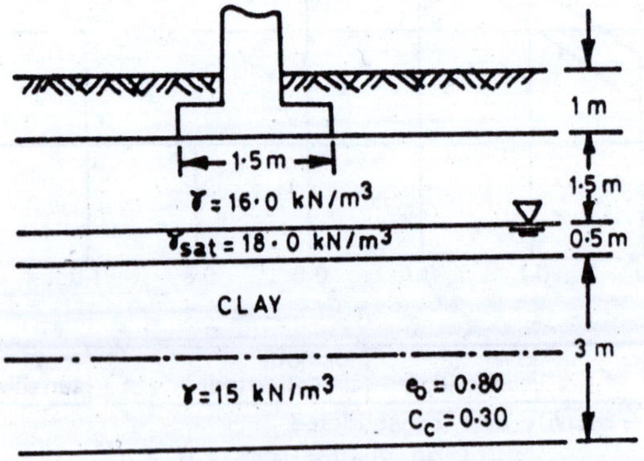

Fig. E 12.2.

Solution. Initial pressure at the centre of the clay layer,

$$\bar{\sigma}_0 = 2.5 \times 16 + 0.5\ (18 - 10) + 1.5 \times (15 - 10.0) = 51.5 \text{ kN/m}^2$$

The pressure increase at the top, middle and the bottom of the clay layer are found as follows (Eq. 11.60).

$$(\Delta\sigma)_t = \frac{200 \times (3 \times 1.50)}{(3.0 + 2.0)\,(1.5 + 2.0)} = 51.4 \text{ kN/m}^2$$

$$(\Delta\sigma)_m = \frac{200 \times (3 \times 1.50)}{(3.0 + 3.5)\,(1.5 + 3.5)} = 27.7 \text{ kN/m}^2$$

$$(\Delta\sigma)_b = \frac{200 \times (3 \times 1.50)}{(3.0 + 5.0)\,(1.5 + 5.0)} = 17.3 \text{ kN/m}^2$$

The average pressure can be found from the following equation (Simpson's rule).

$$\Delta\sigma = \frac{1}{6}\,[(\Delta\sigma)_t + 4\,(\Delta\sigma)_m + (\Delta\sigma)_b]$$

or $$\Delta\sigma = \frac{1}{6}\,[51.4 + 4 \times 27.7 + 17.3] = 29.9 \text{ kN/m}^2$$

Therefore, from Eq. 12.58, $$s_f = \frac{0.30}{1 + 0.80} \times 3.0 \log_{10}\left(\frac{51.5 + 29.9}{51.5}\right)$$

or $$s_f = 0.09941 \text{ m} = \mathbf{99.41\ mm}$$

Illustrative Example 12.3. *A stratum of clay is 2 m thick and has an initial overburden pressure of 50 kN/m² at its middle. Determine the final settlement due to an increase in pressure of 40 kN/m² at the middle of the clay layer. The clay is over-consolidated, with a preconsolidation pressure of 75 kN/m². The values of the coefficients of recompression and compression index are 0.05 and 0.25, respectively. Take initial void ratio as 1.40.*

Solution. From Eq. 12.60,

$$s_f = \frac{C_r}{1 + e_0} H_0 \log_{10}\left(\frac{\bar{\sigma}_c}{\bar{\sigma}_0}\right) + \frac{C_c}{1 + e_0} H_0 \log_{10}\left(\frac{\bar{\sigma}_0 + \Delta\bar{\sigma}}{\bar{\sigma}_c}\right)$$

or $$s_f = \frac{0.05 \times 2.0}{1 + 1.40} \log_{10}\left(\frac{75}{50}\right) + \frac{0.25 \times 2}{1 + 1.40} \log_{10}\left(\frac{50 + 40}{75}\right)$$

or $$s_f = 7.34 \times 10^{-3} + 16.50 \times 10^{-3} \text{ m}$$

or $$s_f = 23.84 \times 10^{-3} \text{ m} = \mathbf{23.84\ mm}$$

Illustrative Example 12.4. *A consolidation test was conducted on a sample of a normally consolidated clay, with an initial void ratio of 1.55, and the following results were obtained.*

$\bar{\sigma}$ (kN/m²)	80	160	320	640	1280
e	1.35	1.28	1.14	0.96	0.78

Plot e – log $\bar{\sigma}$ curve. If the initial overburden pressure is 150 kN/m², draw the field consolidation line and hence determine the coefficient of compression.

(b) If the thickness of the clay layer in the field is 4 m and the increase in the pressure due to loading is 50 kN/m², compute the settlement.

Solution. Fig. E 12.4 shows the required plot. The field consolidation line is drawn between the point (σ_0, e_0) and the point D where the plot cuts the horizontal line through $e = 0.40\, e_0 = 0.4 \times 1.55 = 0.62$.

From the figure, $$C_c = \frac{-\Delta e}{\log_{10}(\sigma_2/\sigma_1)} = \frac{1.47 - 0.88}{\log_{10}(1000/200)} = 0.844$$

From Eq. 12.58, $$s_f = \frac{C_c}{1 + e_0} H_0 \log_{10}\left(\frac{\sigma_0 + \Delta\bar{\sigma}}{\bar{\sigma}_0}\right)$$

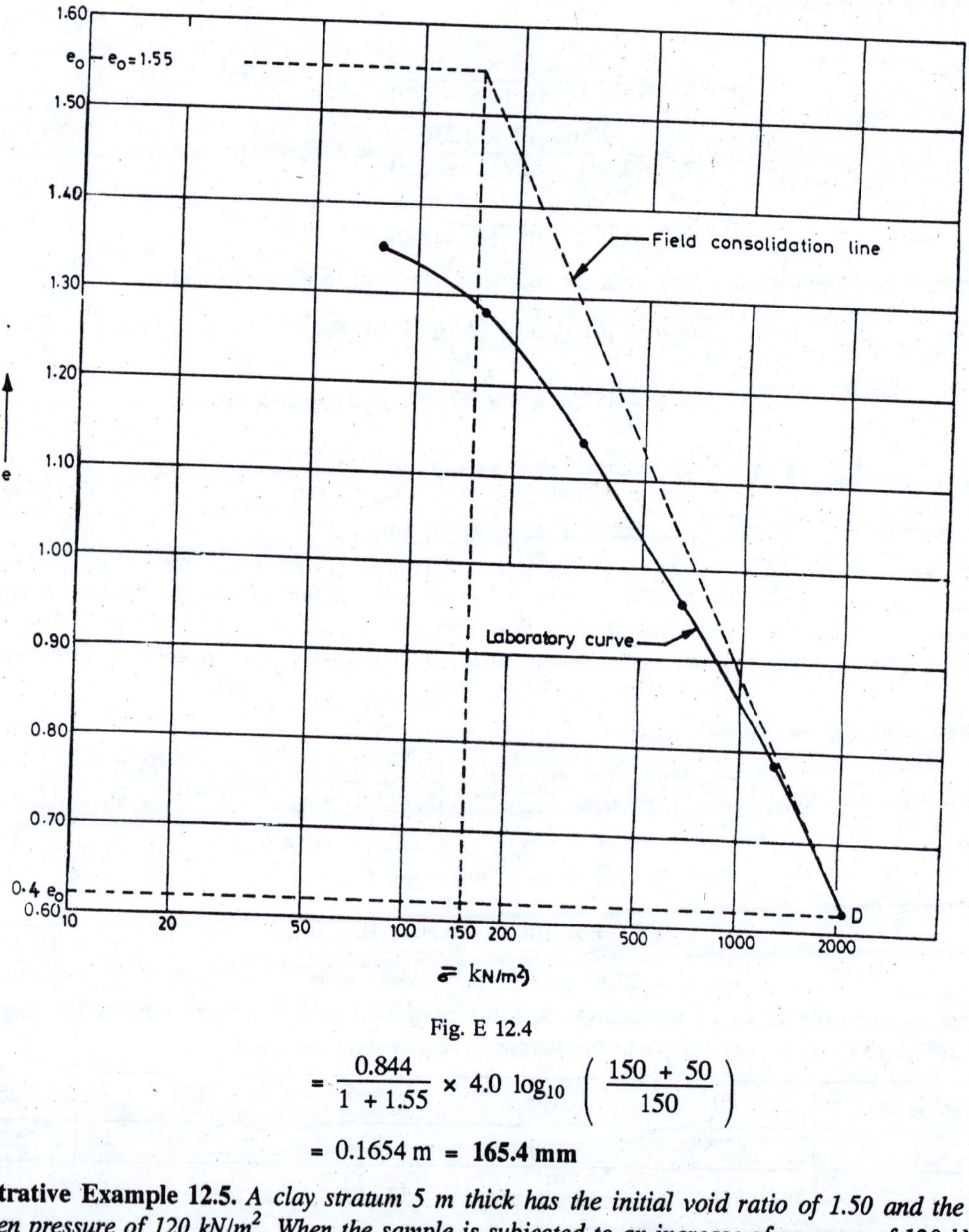

Fig. E 12.4

$$= \frac{0.844}{1 + 1.55} \times 4.0 \log_{10}\left(\frac{150 + 50}{150}\right)$$

$$= 0.1654 \text{ m} = \mathbf{165.4 \text{ mm}}$$

Illustrative Example 12.5. *A clay stratum 5 m thick has the initial void ratio of 1.50 and the effective overburden pressure of 120 kN/m². When the sample is subjected to an increase of pressure of 120 kN/m², the void ratio reduces to 1.44. Determine the coefficient of the volume compressibility and the final settlement of the str^tum.*

Solution From Eq. 12.14, $$m_v = \frac{-\Delta e/(1 + e_0)}{\Delta\bar{\sigma}}$$

or $$m_v = \frac{(1.50 - 1.44)/(1 + 1.50)}{120} = 2 \times 10^{-4} \text{ m}^2/\text{kN}$$

From Eq. 12.54, $$s_f = m_v \, \Delta \, \bar{\sigma} \, H_0$$

or $$s_f = 2 \times 10^{-4} \times 120 \times 5 \times 10^3 \text{ mm} = \mathbf{120 \text{ mm}}$$

Illustrative Example 12.6. *Determine the coefficient of consolidation of soil whose test data is plotted in Fig. E 12.6. The sample was 2.0 cm thick and had double drainage.*

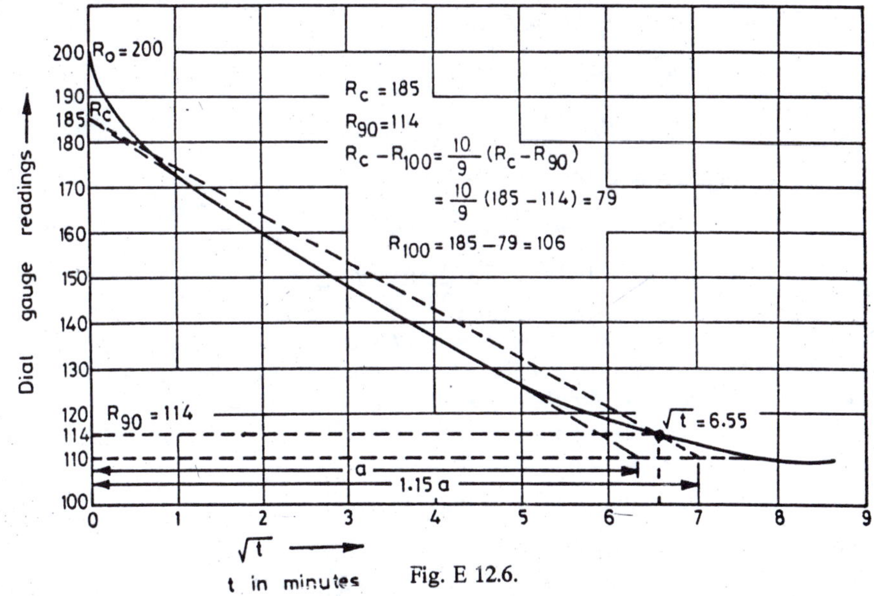

Fig. E 12.6.

Solution. The construction lines are shown dotted.

From the figure, $\sqrt{t_{90}} = 6.55$ *or* $t_{90} = 42.90$ minutes

From Eq. 12.47, $c_v = \dfrac{0.848\, d^2}{t_{90}} = \dfrac{0.848 \times (1.0)^2}{42.90} = \mathbf{0.0198\ cm^2/min.}$

Illustrative Example 12.7. *Determine the coefficient of consolidation of a soil whose data is plotted in Fig. E 12.7. The sample was 20 mm thick and had double drainage.*

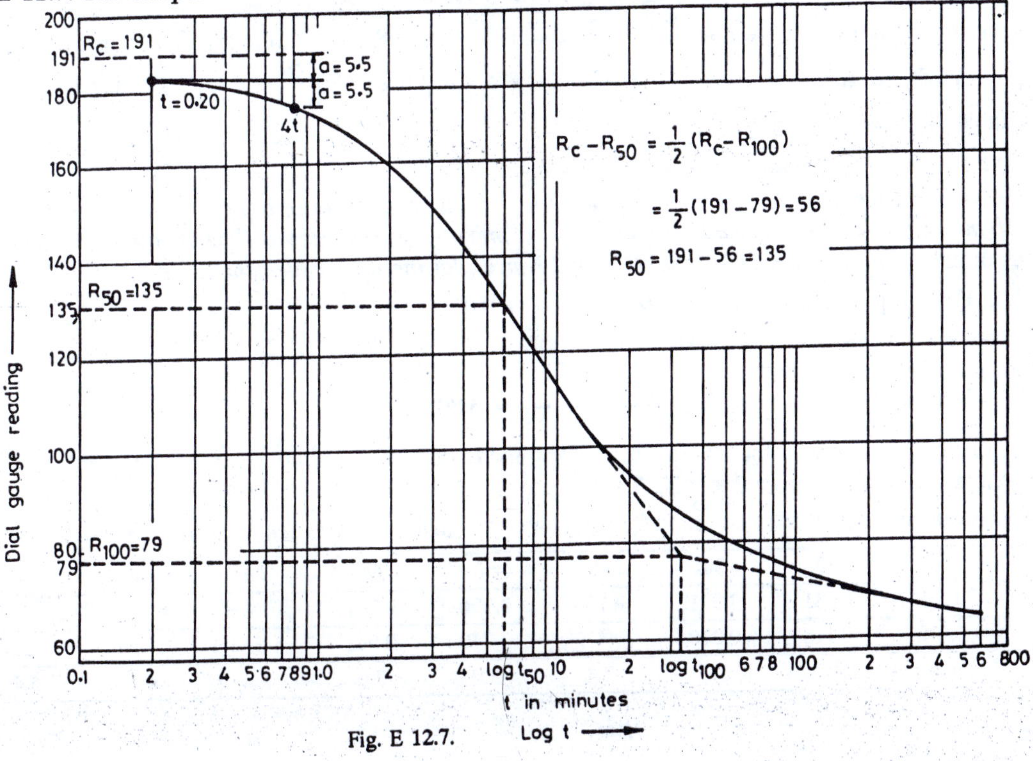

Fig. E 12.7.

Solution. The construction lines are shown in Fig. E 12.7. R_{100} is obtained from the intersection of the two straight lines drawn. The point 50% consolidation (R_{50}) and the corresponding t_{50} are determined.

$$R_c - R_{50} = \frac{1}{2}(R_c - R_{100})$$

In this case, $R_c = 191,\ R_{100} = 79$

Therefore, $R_{50} = R_c - \frac{1}{2}(R_c - R_{100}) = 191 - \frac{1}{2}(191 - 79) = 135$

From the plot, $\log_{10} t_{50} = 0.58$

or $t_{50} = 3.802 \text{ minutes} = 288.12\ s$

From Eq. 12.49, $c_v = \dfrac{0.196\, d^2}{t_{50}} = \dfrac{0.196 \times (10)^2}{228.12} = 0.086\ \text{mm}^2/\text{sec}$

Illustrative Example 12.8. *A 3 m thick clay layer beneath a building is overlain by a permeable stratum and is underlain by an impervious rock. The coefficient of consolidation of the clay was found to be 0.025 cm²/minute. The final expected settlement for the layer is 8 cm. (a) How much time will it take for 80% of the total settlement to take place ? (c) Determine the time required for a settlement of 2.5 cm to occur. (d) Compute the settlement that would occur in one year.*

Solution. (*a*) Length of drainage path, $d = 3$ m = 300 cm.

From Eq. 12.35, $C_v = T_v d^2/t$

Substituting the values, $0.025 = T_v (300)^2/t$ or $t = 3.6 \times 10^6\ T_v$ ...(*a*)

From Table 12.3, when $U = 80\%,\quad T_v = 0.567$

Therefore, from Eq. (*a*), $t = 3.6 \times 10^6 \times 0.567 = 2.041 \times 10^6$ minutes = 3.883 years.

(*b*) When the settlement is 2.5 cm, $U = \dfrac{2.5}{8.0} \times 100 = 31.25\%.$

From Table 12.3, when $U = 31.25\%,\quad T_v = 0.078$

Therefore, from Eq. (*a*), $t = 3.6 \times 10^6 \times 0.078 = 0.2808 \times 10^6$ minutes = 195 days

(*c*) From Eq. (*a*), $1 \times 365 \times 24 \times 60 = 3.6 \times 10^6\ T_v$ or $T_v = 0.146$

From Table 12.3, when $T_v = 0.146,\quad U = 0.429.$

Therefore, settlement $= 0.429 \times 8 = 3.432$ cm

Illustrative Example 12.9. *A clay layer 4 m thick has a final settlement of 6.0 cm. The layer has double drainage. If the coefficient of consolidation is 0.02 cm²/minute, determine the time required for different percentages of consolidation from 10% upto 90% and hence plot the time-settlement curve.*

Solution. From Eq. 12.35, $T_v = C_v t/d^2$

or $T_v = 0.02 \times t/(200)^2$ or $t = 2 \times 10^6\ T_v$

If time is taken in years, $t = \dfrac{2 \times 10^5\ T_v}{60 \times 24 \times 365} = 3.805\ T_v$

The calculations are given in the tabular from below for U of 10, 20, 30, 40, 50, 60, 70, 80 and 90% and the corresponding T_v obtained from Table 12.3.

The settlements are calculated from the values of U and the final settlement (s_f) *i.e.* $s = U \times s_f$ or $s = 6U$

U (%)	10	20	30	40	50	60	70	80	90
T_v	0.008	0.031	0.071	0.126	0.196	0.287	0.403	0.567	0.848
t (year)	0.030	0.118	0.270	0.479	0.746	0.092	1.533	2.157	3.227
s (cm)	0.6	1.2	1.8	2.4	3.0	3.6	4.2	4.8	5.4

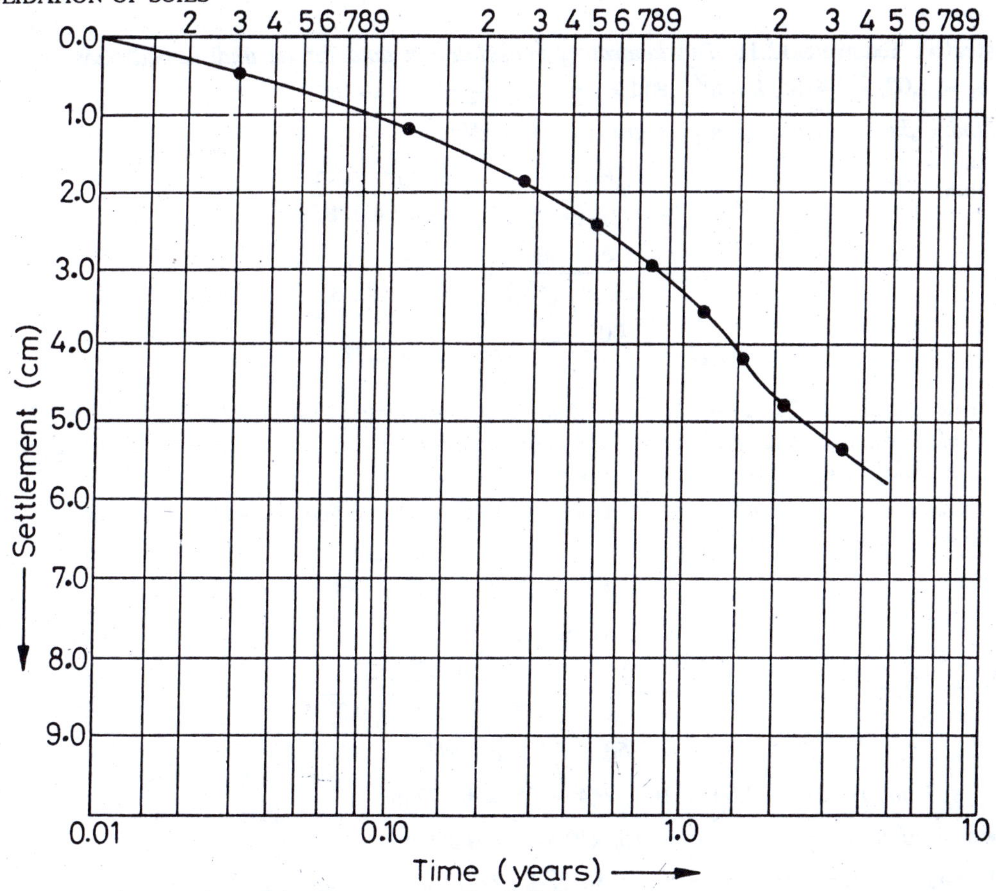

Fig. E-12.9.

Fig. E. 12.9 shows the settlement-log t curve.

Illustrative Example 12.10. *An area is underlain by a stratum of cloy layer 6 m thick. The layer is doubly drained and has the coefficient of consolidation of 0.3m²/month.*

(a) How long would it take for a surcharge load to cause a settlement of 40 cm if the same load causes a final settlement of 60 cm ?

(b) If the sand drains (S = 3 m and r_w = 0.30) are used, determine the time required for 90% consolidation. Take c_{vr} = 2.0 m²/month. Assume the triangular layout of drains. Neglect vertical consolidation in this case.

Solution. (*a*) $U = 40/60 = 0.6667$ *i.e.* 66.67%

From Table 12.3, for U = 66.67%, $T_v = 0.364$

From Eq. 12.35, $c_v = T_v d^2/t$

or $t = 0.364 \times (3)^2/0.30 = 10.92$ months.

(*b*) For triangular layout of drains, $R = 0.525 \times S$

or $R = 0.525 \times 3 = 1.575$ m

$n = R/r_w = 1.575/0.30 = 5.25$

From Fig. 12.28, for U_r = 90% and n = 5.25, we have $T_r \doteq 0.270$

From Eq. 12.69, $T_r = c_{vr}\, t / 4R^2$

or $$0.270 = \frac{2.0 \times t}{4 \times (1.575)^2}$$ or t = 1.34 months.

Illustrative Example 12.11. *The laboratory consolidation data for an undisturbed clay sample are as follows.* $e_1 = 1.00$, $\bar{\sigma}_1 = 85$ *kN/m*^{2}, and $e_2 = 0.80$, $\bar{\sigma}_2 = 465$ *kN/m*^{2}.

Determine the void ratio for a pressure $\bar{\sigma}_3$ *of 600 kN/m*^{2}.

Solution.
$$C_c = \frac{e_1 - e_2}{\log_{10}(\bar{\sigma}_2/\bar{\sigma}_1)} = \frac{1.00 - 0.80}{\log_{10}(465/85)} = 0.271$$

Now
$$C_c = \frac{e_1 - e_3}{\log_{10}(\bar{\sigma}_3/\bar{\sigma}_1)} = \frac{e_1 - e_3}{\log_{10}(600/85)}$$

or
$$0.271 = \frac{1.00 - e_3}{0.849} \quad \text{or} \quad e_3 = \mathbf{0.77}$$

Illustrative Example 12.12. A *clay layer 4 m thick is subjected to a pressure of 55 kN/m*^{2}. *If the layer has a double drainage and undergoes 50% consolidation in one year, determine the coefficient of consolidation. Take* $T_v = 0.196$.

If the coefficient of permeability is 0.020 m/yr, determine the settlement in one year and rate of flow of water per unit area in one year.

Solution.
$$c_v = T_v d^2/t = 0.196 \times (2.0)^2/(1)$$

or
$$c_v = 0.784 \text{ m}^2/\text{yr}$$

From Eq. 12.29,
$$m_v = \frac{k}{c_v \gamma_w} = \frac{0.020 \times 1000}{0.784 \times 1000 \times 9.81}$$
$$= 2.60 \times 10^{-3} \text{ m}^2/\text{kN}$$

$$s_f = m_v H_0 \Delta\bar{\sigma} = 2.60 \times 10^{-3} \times 4 \times 55 = 0.572 \text{ m}$$

Settlement after one year $= 0.5 \times 0.572 = 0.286$ m.

Settlement rate Since U is proportional to $\sqrt{t}$ for $U < 0.60$, the settlement (s) is also proportional to $\sqrt{t}$. Thus

$$s^2 \propto t \quad or \quad t = Cs^2$$

When $t = 1$ year, $s = 0.286$ m. Therefore,

$$C = \frac{1}{(0.286)^2} = 12.226$$

Thus
$$t = 12.226\, s^2$$

or
$$\frac{ds}{dt} = \frac{1}{2 \times 12.226\, s} = \frac{1}{24.452\, s} = \mathbf{0.143\ m/yr}$$

Discharge per unit area per surface $= 0.143/2 = 0.072$ m^3/yr/m^2

PROBLEMS

A. Numerical

12.1. A saturated soil stratum 4 m thick lies above an impervious stratum and below a pervious stratum. It has a void ratio of 1.50 at an initial pressure of 150 kN/m^2.

(*i*) Compute the change in void ratio due to to an increase of stress of 50 kN/m^2. Take $C_c = 0.20$.

(*ii*) Also compute the final settlement of the soil stratum due to above increase in stress.

(*iii*) What would be the time required for 50 percent consolidation ? Take $T_v = 0.20$ and $k = 3.0 \times 10^{-4}$ cm/sec.

[**Ans.** 0.025; 4 cm; 34.18 minutes]

12.2. In a laboratory, the consolidation test was performed on a specimen of clay 3 cm thick. The sample was drained at top and bottom. The time required for 50% consolidation of the sample was observed to be 15 minutes. Determine the coefficient of consolidation of clay.

Calculate time required for 50% and 90% consolidation for this clay deposit in the field 3 m thick and drained at both ends. **[Ans.** 4.93×10^{-4} cm^2/sec; 104.17 days; 450.69 days]

12.3. There is a layer of soft clay 4 m thick under a newly constructed building. The overburden pressure over the centre of the clay layer is 300 kN/m^2. Compute the settlement if there is an increase in pressure due to construction of 100 kN/m^2. Take $C_c = 0.50$, $G = 2.70$. The water content of the deposit was found to be 50%. **[Ans.** 10.63 cm]

12.4. In a consolidation test, an increase of 100 kN/m^2 in the vertical pressure was applied to a saturated clay sample initially 2.5 cm thick. The thickness of the sample reduced to 2.46 cm after 24 hours. The sample was then relieved of pressure and allowed to take up water. The final thickness was 2.465 cm and the moisture content was 30%. Assuming that the sample was saturated throughout the test, calculate the following :

(*i*) the initial void ratio. (Take $G = 2.68$)

(*ii*) the void ratio after consolidation.

(*iii*) the void ratio after expansion

(*iv*) the coefficient of compressibility. **[Ans.** 0.83, 0.80, 0.804, 1.6×10^{-4} m^2/kN]

12.5. In a consolidation test on a soil, the void ratio of the sample decreased from 1.25 to 1.10 when the pressure is increased from 200 kN/m^2 to 400 kN/m^2. Calculate the coefficient of consolidation if the coefficient of permeability is 8×10^{-8} cm/sec. **[Ans.** 7.55 m^2/year]

12.6. The time required to reach 60% consolidation for a sample 1 cm thick tested in consolidometer under conditions of double drainage was found to be 35 seconds. Determine the time required for a layer 10 m thick to reach the same degree of consolidation, if it has drainage only on one side. **[Ans.** 1.4×10^8 seconds]

12.7. A clay stratum 2.5 m thick lies over a sandy stratum and has drainage on both sides. Calculate the values of the pore water pressure and the effective stress at the middle of the clay stratum when 60% of consolidation has taken place under an increment of load from 500 to 2500 kN/m^2. **[Ans.** 800 kN/m^2; 1700 kN/m^2]

12.8. In a consolidation test, a fully saturated clay sample was subjected to a load of 500 kN/m^2. After 12 hours, the average pore pressure was found to be 200 kN/m^2. Find out the time required for 50% consolidation to take place. Assume $T_v = (\pi/4)\,U^2$. **[Ans.** 8.33 hours]

12.9. There is clay layer 8 m thick with a layer of sand on either side. An undisturbed sample 2.5 cm thick of the clay when tested in the laboratory required 25 minutes to reach 50% consolidation ($T_v = 0.20$). It is proposed to construct a building at the above site. Estimate the time required for 90% consolidation to take place. ($T_v = 0.85$). **[Ans.** 7555.6 days]

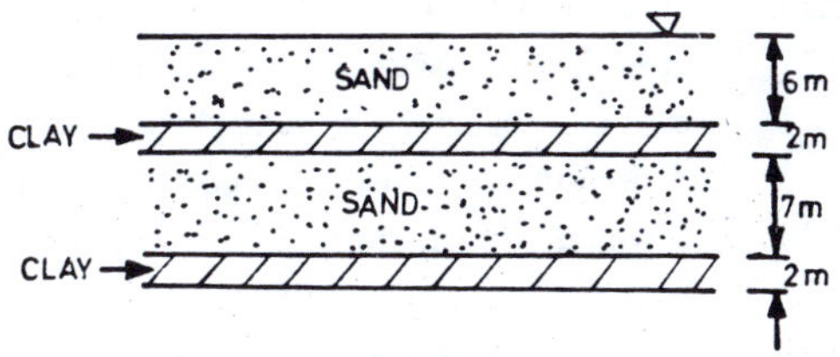

Fig. P 12.10.

12.10. Fig. P-12.10 shows a bore hole log obtained during sub-surface exploration. There are two layers of clay of thickness 2 m each, located 6 m and 15 m below the ground surface. The bulk densities of sand and clay were found to be 2.0 gm/ml and 1.82 gm/ml, respectively. Compute the total settlement of each clay layer under a uniformly distributed load of 400 kN/m^2 spread over a large area on ground surface. $C_c = 0.40$, $e_0 = 1.08$. **[Ans.** 32.45 cm, 21.57 cm, total = 54.02 cm]

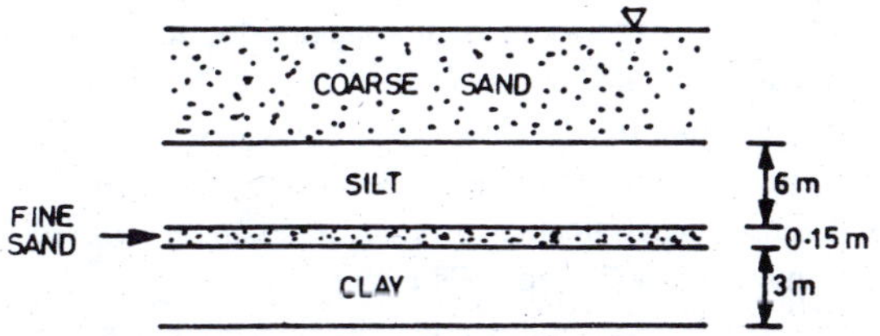

Fig. P-12.11.

12.11. A 6 m thick fine silt stratum [$C_v = 7.2 \times 10^{-3}$ cm^2/sec] underlies a coarse sand deposit and overlies a 0.15 cm thick fine sand layer under which there is a clay stratum 3 m thick (Fig. P-12.11) ($c_v = 9 \times 10^{-5}$ cm^2/sec). A pervious stratum lies below the clay layer. When the clay layer reaches a degree of consolidation of 20%, what would be the degree of consolidation of the silt layer ? ($T_v = 0.03$ for 20% U and U = 82% for $T_v = 0.6$).

[Ans. 82%]

12.12. A clay layer of 10 m thickness underlies a sand stratum of 10 m and overlies a pervious layer. The sand layer carries a point load of 10 MN. Assume $e = 0.70$, and $G = 2.72$, L.L. = 60% and $c_v = 25 \times 10^{-4}$ cm^2/sec. The water table is located 5 m above the top of the clay layer. Find how long would the clay take to settle 4.7 cm.

[Ans. 36.55 hours]

12.13. A 15 m thick hydraulically isotropic clay stratum overlies an impervious stratum. If the coefficient of consolidation is 5×10^{-4} cm^2/sec, find the time required for 50% and 90% consolidation (T_v = 0.20 and 0.85, respectively).

(*ii*) In order to accelerate the rate of settlement, vertical sand drains of 1 m diameter were installed at 5 m centre to centre. Calculate how much the settlement is accelerated due to provision of sand drains. For $n = 5$, T_r = 0.081 for U_r = 50% and T_r = 0.27 for U_r = 90%.

[Ans. 9.0×10^8 seconds ; 38.25×10^8 seconds.
(*ii*) 0.405×10^8 seconds ; 1.35×10^8 seconds]

B. Descriptive and Objective type

12.14. Describe the consolidometer test. Show how the results of this test are used to predict the rate of settlement and the magnitude of settlement.

12.15. Define the following terms :
(*i*) Coefficient of compressibility, (*ii*) Coefficient of volume change.
(*iii*) Compression index, (*iv*) Expansion index
(*v*) Recompression index.

12.16. Discuss the spring analogy for primary consolidation. What are its uses ?

12.17. Differentiate between primary consolidation and secondary consolidation.

12.18. Differentiate between normally consolidated and the overconsolidated soils. How would you determine the overconsolidation pressure ?

12.19. Discuss Terzaghi's theory of consolidation, stating the various assumptions and their validity.

12.20. What is the coefficient of consolidation ? What is its use in the settlement analysis ? How is it determined ?

12.21. What is the time factor ? How is it related to the average degree of consolidation ?

12.22. Discuss the limitations of Terzaghi's theory of consolidation. Why is theory used despite its limitations ?

12.23. What are different causes of preconsolidation in soils ? What is the effect of preconsolidation on the settlement?

12.24. How would you determine the time-settlement curve in the field ?

12.25. What is field consolidation curve ? How is it obtained ?

12.26. Explain the phenomenon of secondary consolidation. Differentiate between the secondary consolidation index and the coefficient of secondary consolidation.

12.27. Explain the theory of 3-dimensional consolidation. What is its practical use ?

12.28. Describe sand drains. How are these designed ? Discuss their uses. What is the effect of smear ?

12.29. Discuss Skempton-Bjerrum's theory for calculating settlement, taking into account the effect of lateral strains.

12.30. Write whether the following statements are correct.

(*a*) When the soil is fully saturated, the compression occurs mainly due to compression of water.
(*b*) The initial consolidation of a fully saturated soil is zero.
(*c*) The secondary consolidation is neglible in organic soils.
(*d*) The friction in the fixed ring cell is more than that in the floating-ring cell.
(*e*) The height of solids method for the determination of void ratio cannot be used for partially saturated soils.
(*f*) The compression index of normally consolidated soils is constant.
(*g*) The expansion index and the recompression index are approximately equal.
(*h*) The coefficient of consolidation varies with the change in pressure.
(*i*) The time taken for a half-closed layer to attain a particular degree of consolidation is twice of that for an equivalent open layer.

(j) The field consolidation curve is generally steeper than the laboratory curve.

(k) The actual settlement is always greater than that given by Terzaghi's theory if lateral strains occur.

[Ans. True, (b), (d), (f), (g), (h), (j)]

C. Multiple Choice Questions

1. The coefficient of compressibility is the ratio of

(a) Change in void ratio to change in effective stress

(b) Volumetric strain to change in effective stress.

(c) Change in thickness to change in effective stress.

(d) Stress to strain.

2. With an increase in the liquid limit, compression index

(a) decreases (b) increases

(c) remain constant (d) may increase or decrease

3. The recompression index is about . . . . of the compression index

(a) 5 times (b) 1/5

(c) 1/2 (d) 1/20

4. When consolidation of a saturated soil sample occurs, the degree of saturation

(a) increases (b) decreases

(c) remains constant (d) may increase or decrease

5. Consolidation time of a soil sample

(a) increases with an increase permeability.

(b) increases with a decreases in compressibility.

(c) increases with a decrease in unit weight of water.

(d) increases with a decrease in permeability.

6. The ultimate settlement of a soil deposit increases with

(a) an increase in the compression index.

(b) an increase in the initial void ratio.

(c) a decrease in thickness of the stratum.

(d) an increase in time.

7. Under load, the void ratio of a submerged, saturated clay decreases from 1.0 to 0.92. The ultimate settlement of a layer 2m thick layer will be

(a) 2.0 cm (b) 4.0 cm

(c) 8.0 cm (d) 16.0 cm

8. A normally consolidated clay layer settles 2 cm when the effective stress is increased from 80 to 160 kN/m^2. When the effective stress is further increased to 320 kN/m^2, the further settlement will be

(a) 2 cm (b) 4 cm

(c) 1 cm (d) 8 cm

9. A fully saturated clay specimen is subjected to a pressure of 200 kN/m^2 in the consolidation test. After a period of time when the average pore pressure is 60 kN/m^2, the degree of consolidation is

(a) 60 (b) 70

(c) 30 (d) 50

[Ans. 1. (a), 2. (b), 3. (b), 4. (c), 5. (d), 6. (a), 7. (c), 8. (a), 9. (b)]

13

Shear Strength

13.1. INTRODUCTION

The shear strength of a soil is its maximum resistance to shear stresses just before the failure. Soils are seldom subjected to direct shear. However, the shear stresses develop when the soil is subjected to direct compression. Although shear stresses may also develop when the soil is subjected to direct tension, but these shear stresses are not relevant, as the soil in this case fails in tension and does not fail in shear. In field, soils are seldom subjected to tension, as it causes opening of the cracks and fissures. These cracks are not only undesirable, but are also detrimental to the stability of the soil masses. Thus, the shear failure of a soil mass occurs when the shear stresses induced due to the applied compressive loads exceed the shear strength of the soil. It may be noted that the failure in soil occurs by relative movements of the particles and not by breaking of the particles.

Shear strength is the principal engineering property which controls the stability of a soil mass under loads. It governs the bearing capacity of soils, the stability of slopes in soils, the earth pressure against retaining structures and many other problems, as explained in later chapters. All the problems of soil engineering are related in one way or the other with the shear strength of the soil. Unfortunately, the shear strength is one of the most complex engineering properties of the soil. The current research is giving new concepts and theories. This chapter presents the basic concepts and the accepted theories of the shear strength.

13.2. STRESS-SYSTEM WITH PRINCIPAL PLANES PARALLEL TO THE COORDINATE AXES

In general, a soil mass is subjected to a three—dimensional stress system. However, in many soil engineering problems, the stresses in the third direction are not relevant and the stress system is simplified as two-dimensional. The plane strain conditions are generally assumed, in which the strain in the third (longitudinal) direction is zero. Such conditions exist, for example, under a strip footing of a long retaining wall.

At every point in a stressed body, there are three planes on which the shear stresses are zero. These planes are known as *principal planes.* The plane with the maximum compressive stress (σ_1) is called the major principal plane, and that with the minimum compressive (σ_3) as the minor principal plane. The third principal plane is subjected to a stress which has the value intermediate between σ_1 and σ_3, and is known as the intermediate principal plane. Generally, the stresses on a plane perpendicular to the intermediate principal plane are required in the analysis. Therefore, the stresses on the intermediate principal plane are not much relevant. Only the major principal stress (σ_1) and the minor principal stress (σ_3) are generally important.

In solid mechanics, the tensile stresses are taken as positive. In soil engineering problems, tensile stresses rarely occur. To avoid many negative signs, compressive stresses are taken as positive and the tensile stresses as negative in soil engineering.

Fig. 13.1 shows a plane which is perpendicular to the intermediate principal plane. The major and minor principal stresses act on this plane. The major principal plane is horizontal and the minor principal plane is vertical. Let us consider plane AB which is inclined at an angle θ (measured counterclockwise) to the major principal plane AC.

Resolving the forces acting on the wedge ABC in the horizontal (x-direction),

$$\sigma_3 BC = \sigma AB \sin\theta - \tau AB \cos\theta$$

where σ = normal stress on AB, τ = shear stress on AB.

In above expression, the length perpendicular to the plane of the paper has been taken as unity, which is the general practice for two dimensional stress system. The above equation can be simplified as

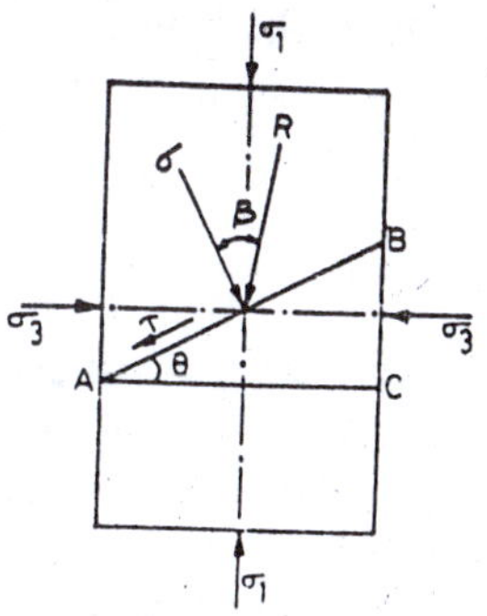

Fig. 13.1.

$$\sigma_3 \frac{BC}{AB} = \sigma \sin\theta - \tau\cos\theta$$

or $$\sigma_3 \sin\theta = \sigma\sin\theta - \tau\cos\theta \quad ...(a)$$

Likewise, resolving the forces in the vertical (y- direction),

$$\sigma_1 AC = \sigma AB\cos\theta + \tau AB \sin\theta$$

or $$\sigma_1\cos\theta = \sigma\cos\theta + \tau\sin\theta \quad ...(b)$$

Multiplying Eq. (a) by $\cos\theta$ and Eq. (b) by $\sin\theta$, and subtracting,

$$(\sigma_1 - \sigma_3)\sin\theta\cos\theta = \sigma(\cos\theta\sin\theta - \cos\theta\sin\theta) + \tau(\sin^2\theta + \cos^2\theta)$$

or $$(\sigma_1 - \sigma_3)\sin\theta\cos\theta = \tau$$

or $$\tau = \frac{1}{2}(\sigma_1 - \sigma_3)\sin 2\theta \quad ...(13.1)$$

Substituting the above value of τ in Eq. (a),

$$\sigma_3\sin\theta = \sigma\sin\theta - \frac{(\sigma_1-\sigma_3)}{2}\sin 2\theta\cos\theta$$

or $$\sigma_3 = \sigma - (\sigma_1 - \sigma_3)\cos^2\theta$$

or $$\sigma = \sigma_3 + (\sigma_1 - \sigma_3)\frac{(1 + \cos 2\theta)}{2}$$

or $$\sigma = \frac{(\sigma_1 + \sigma_3)}{2} + \frac{(\sigma_1 - \sigma_3)}{2}\cos 2\theta \quad ...(13.2)$$

Eqs. 13.1 and 13.2 give the stresses on the inclined plane AB, making an angle θ (measured counterclockwise) with the major principal plane AC.

13.3. MOHR'S CIRCLE

Otto Mohr, a German scientist, devised a graphical method for the determination of stresses on a plane inclined to the principal planes. The graphical construction is known as *Mohr's circle* and is extremely useful. In this method, an origin O is selected and the normal stresses are plotted along horizontal axis and the shear stresses on the vertical axis. As the compressive stresses are taken positive in soil engineering, these are plotted towards the right of the origin, *i.e.* along positive x-axis. The shear stress is generally taken as positive if it causes a counterclockwise couple at a point inside the wedge ABC in Fig. 13.1. Thus the shear stress marked on the plane AB is positive. The positive shear stresses are plotted upward from the origin, *i.e.*, along positive y-axis. Let us draw the Mohr circle for the stresses shown in Fig. 13.1.

In Fig. 13.2, the point E represents the minor principal stress σ_3 and the point F, the major principal stress σ_1. The point C is the middle point with the normal stress coordinate equal to $(\sigma_1 + \sigma_3)/2$. The circle is drawn with C as the centre and EF as diameter. The circle is known as Mohr's circle. Each point on the circle gives the stresses σ and τ on a particular plane. It can be shown that the point D on the circle gives the stresses on the plane AB inclined at an angle θ to the major principal plane. The line DE makes an angle θ with σ-axis. The angle DCF subtended at the centre is obviously twice the angle DEC.

From the figure, $OH = OC + CH$

or $$OH = \frac{\sigma_1 + \sigma_2}{2} + \frac{(\sigma_1 - \sigma_3)}{2} \cos 2\theta = \sigma$$

and $$DH = CD \sin 2\theta = \frac{\sigma_1 - \sigma_3}{2} \sin 2\theta = \tau$$

Comparing the above equations with Eqs. 13.2 and 13.1, it is obvious that the coordinates of the point *D* represent the stresses on the inclined plane *AB* in Fig. 13.1.

The point *E* is a unique point, which is known as *the pole (P) or the origin of planes (OP)*. If a line is drawn from any point (say *D*) on the Mohr circle parallel to the plane (say, *AB*) whose stresses are represented by that point, it will intersect the circle at the pole *P*. When the major principal plane is horizontal (parallel to *x*-axis), the minor principal plane is vertical and the pole lies at the point *E* which indicates the minor principal stress. Once the pole has been located, the stresses any other plane making an angle α with the major principal plane can be determined graphically by drawing a line through the pole and making an angle α with σ-axis. The coordinates of the point obtained by the intersection of this line with the circle give the stresses on that plane.

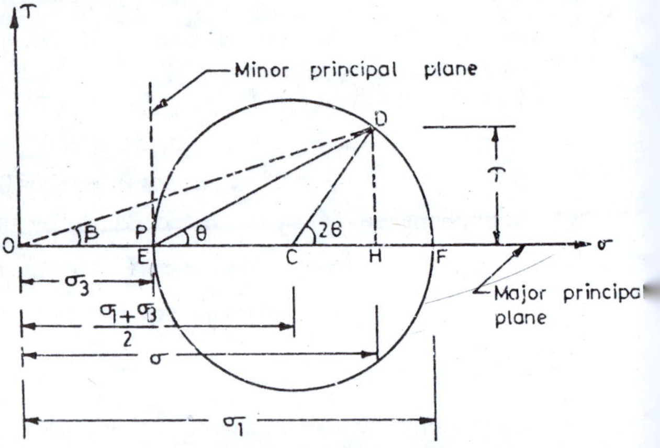

Fig. 13.2.

The line *OD* represents the magnitude of the resultant stress on the inclined plane *AB*. The angle of the obliquity of the resultant with the normal of the plane *AB* is equal to the angle β.

It may be mentioned that the equations developed in Sect. 13.2 and the Mohr circle discussed in this section are based on the principles of mechanics.These are valid for all materials irrespective of their stress-strain characteristics. The same relations are also applicable in solid mechanics.

13.4. STRESS-SYSTEM WITH PRINCIPAL PLANES INCLINED TO THE COORDINATE AXIS

Fig. 13.3 (*b*) shows a stressed element in which the principal planes are inclined to the coordinate axes. In other words, the principal planes are not vertical and horizontal. The stresses on a plane inclined at an angle θ to the major principal plane can be determined using Eqs. 13.1 and 13.2 or using the Mohr circle, as

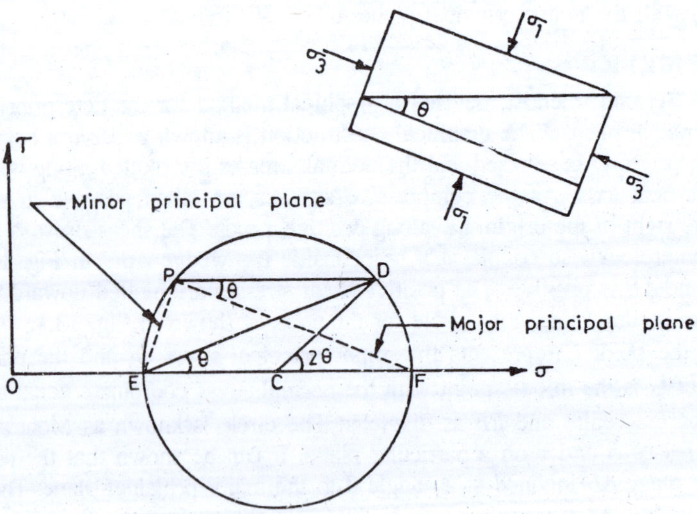

Fig. 13.3. Inclined Principal Planes.

these relations are independent of the inclination of the principal planes. However, the orientation of the principal planes is different from that in Fig. 13.2.

In Fig. 13.3 (*a*), the points *E* and *F* are located and the Mohr circle is drawn, as in Fig. 13.2. The pole *P* is located by drawing a line *EP* from *E* parallel to minor principal plane *BC* or by drawing a line *FP* from *F* parallel to the major principal plane *AC*. The stresses on the plane *AB* inclined at an angle θ to the major principal plane can be determined by drawing a line *PD* through the pole *P* and parallel to *AB*. The coordinates of point *D* give the stresses σ and τ on the plane *AB*. It may be noted that the line *ED* also makes an angle θ with σ-axis.

13.5. STRESS-SYSTEM WITH VERTICAL AND HORIZONTAL PLANES NOT PRINCIPAL PLANES

Fig. 13.4 (*a*) shows an element *ABC* in which the vertical and horizontal planes *BC* and *AC* are subjected to shear stresses in addition to the normal stresses. Hence, these planes are not the principal planes. The stresses on the plane *AB* inclined at an angle θ to plane *AC* can be determined from the equilibrium of forces.

Resolving the forces in *x*-direction [Fig. 13.4 (*b*)],

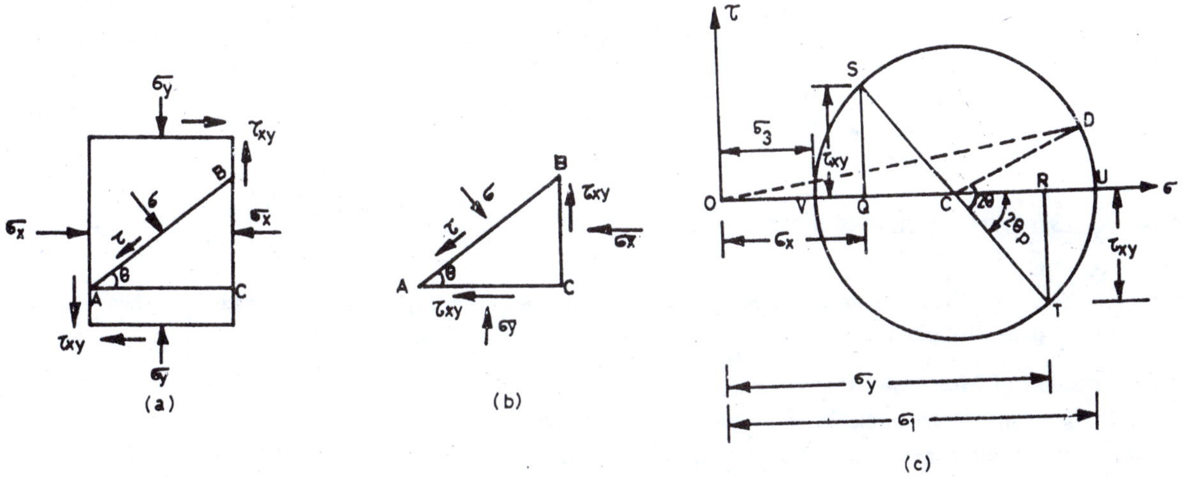

Fig. 13.4. Vertical Planes not Principal Planes.

$$\sigma\, AB \sin\theta = \sigma_x BC + \tau_{xy} AC + \tau\, AB\cos\theta$$

or $$\sigma \sin\theta = \sigma_x \sin\theta + \tau_{xy}\cos\theta + \tau\cos\theta \quad \ldots(a)$$

Resolving the forces in *y*-direction,

$$\sigma AB\cos\theta + \tau\, AB\sin\theta = \tau_{xy} BC + \sigma_y AC$$

or $$\sigma\cos\theta + \tau\sin\theta = \tau_{xy}\sin\theta + \sigma_y\cos\theta$$

or $$\sigma\cos\theta = \tau_{xy}\sin\theta + \sigma_y\cos\theta - \tau\sin\theta \quad \ldots(b)$$

Multiplying Eq. (*a*) by sin θ and Eq. (*b*) by cos θ and adding

$$\sigma = \sigma_x\sin^2\theta + \tau_{xy}\sin\theta\cos\theta + \tau\sin\theta\cos\theta + \tau_{xy}\sin\theta\cos\theta + \sigma_y\cos^2\theta - \tau\sin\theta\cos\theta$$

$$\sigma = \sigma_x\sin^2\theta + \sigma_y\cos^2\theta + 2\tau_{xy}\sin\theta\cos\theta$$

or $$\sigma = \sigma_x\left(\frac{1-\cos 2\theta}{2}\right) + \sigma_y\left(\frac{1+\cos 2\theta}{2}\right) + \tau_{xy}\sin 2\theta$$

or $$\sigma = \frac{\sigma_x+\sigma_y}{2} + \frac{(\sigma_y-\sigma_x)}{2}\cos 2\theta + \tau_{xy}\sin 2\theta \quad \ldots(13.3)$$

Eq. 13.3 gives the normal stress σ on the plane *AB*.

Substituting the above value of σ in Eq. (*a*) and simplifying,

$$\tau = \frac{\sigma_y - \sigma_x}{2} \sin 2\theta - \tau_{xy} \cos 2\theta \qquad \ldots(13.4)$$

Eq. 13.4 gives the stress τ on the plane *AB*.

Eq. 13.3. can be written as,

$$\sigma - \frac{(\sigma_x + \sigma_y)}{2} = \frac{(\sigma_y - \sigma_x)}{2} \cos 2\theta + \tau_{xy} \sin 2\theta \qquad \ldots[13.3(a)]$$

Mohr's circle Squaring Eqs. 13.3 (*a*) and 13.4 and adding we get

$$\left[\sigma - \left(\frac{\sigma_x + \sigma_y}{2}\right)\right]^2 + \tau^2 = \left[\left(\frac{\sigma_y - \sigma_x}{2}\right)^2 \cos^2 2\theta + \tau_{xy}^2 \sin^2 2\theta + 2\left(\frac{\sigma_y - \sigma_x}{2}\right) \cos 2\theta \; \tau_{xy} \; \sin 2\theta\right]$$

$$+ \left[\left(\frac{\sigma_y - \sigma_x}{2}\right)^2 \sin^2 2\theta + \tau_{xy}^2 \cos^2 2\theta - 2\frac{(\sigma_y - \sigma_x)}{2} \sin 2\theta \, (\tau_{xy} \cos 2\theta)\right]$$

or

$$\left[\sigma - \left(\frac{\sigma_x + \sigma_y}{2}\right)\right]^2 + \tau^2 = \left(\frac{\sigma_y - \sigma_x}{2}\right)^2 + \tau_{xy}^2 \qquad \ldots(13.5)$$

Eq. 13.5 is the equation of a circle whose centre has the coordinates $\left[\left(\frac{\sigma_x + \sigma_y}{2}\right), 0\right]$ and whose radius is $\left[\left(\frac{\sigma_y - \sigma_x}{2}\right) + \tau_{xy}^2\right]^{1/2}$. The coordinates of any point on the circle represent the normal and shear stresses (σ, τ) on a particular plane. The circle is Mohr's circle.

Construction of Mohr Circle

To draw the Mohr circle in this case, the normal stresses σ_x and σ_y are marked as points *Q* and *R* respectively on the σ-axis [Fig. 13.4 (*c*)]. At point *Q*, a perpendicular *QS* is drawn, such that $QS = \tau_{xy}$. The shear stress τ_{xy} is positive on the plane *BC* as it causes a counterclockwise moment at a point inside the wedge. Likewise, the perpendicular *RT* is equal to τ_{xy}. However, this is negative because the shear stress on the plane *AC* causes a clockwise moment.

Point *C* is at the middle point of *QR* and has the coordinates $\left[\left(\frac{\sigma_x + \sigma_y}{2}\right), 0\right]$. It also lies on the line drawn through point *S* and *T*. *A* circle is drawn with its centre at *C* and its radius equal to $\left[\left(\frac{\sigma_y - \sigma_x}{2}\right) + \tau_{xy}^2\right]^{1/2}$. The circle passes through points *S* and *T*. It may be noted that the point *T* in the Mohr circle represents the stresses on the plane *AC* and the point *S*, on the plane *BC*. The line *CD* drawn at an angle of 2θ to the line *CT* intersects the Mohr circle at *D*. Therefore, the point *D* gives the stresses on the inclined plane *AB*.

Principal Planes. As the principal planes are the planes with zero shear stresses, from Eq. 13.4,

$$0 = \frac{\sigma_y - \sigma_x}{2} \sin 2\theta_p - \tau_{xy} \cos 2\theta_p$$

or

$$\tan 2\theta_p = \frac{\tau_{xy}}{(\sigma_y - \sigma_x)/2} \qquad \ldots(13.6)$$

where θ_p is the angle which the principal plane makes with the plane *AC* in Fig. 13.4 (*b*). As the angle subtended at the centre is twice this angle, the line *CU* indicates one of the principal planes and the point *U* represents the major principal stress. This can also be proved from the above value of $\tan 2\theta_p$. We have

$$\sin 2\theta_p = \pm \frac{\tau_{xy}}{\sqrt{\left(\frac{\sigma_y - \sigma_x}{2}\right)^2 + \tau_{xy}^2}}$$

$$\cos 2\theta_p = \pm \frac{(\sigma_y - \sigma_x)/2}{\sqrt{\left(\frac{\sigma_y - \sigma_x}{2}\right)^2 + \tau_{xy}^2}}$$

Substituting these values of $\sin 2\theta_p$ and the $\cos 2\theta_p$ in Eq. 13.3,

$$\sigma = \frac{\sigma_x + \sigma_y}{2} \pm \left(\frac{\sigma_y - \sigma_x}{2}\right) \times \frac{(\sigma_y - \sigma_x)/2}{\sqrt{\left(\frac{\sigma_y - \sigma_x}{2}\right)^2 + \tau_{xy}^2}} \pm \frac{\tau_{xy} \times \tau_{xy}}{\sqrt{\left(\frac{\sigma_y - \sigma_x}{2}\right)^2 + \tau_{xy}^2}}$$

or

$$\sigma = \frac{\sigma_x + \sigma_y}{2} \pm \sqrt{\left(\frac{\sigma_y - \sigma_x}{2}\right)^2 + \tau_{xy}^2}$$

Therefore, the two principal stresses are as under.

Major principal stress,

$$\sigma_1 = \frac{\sigma_x + \sigma_y}{2} + \sqrt{\left(\frac{\sigma_y - \sigma_x}{2}\right)^2 + \tau_{xy}^2} \quad \text{...(13.7)}$$

The point U gives the major principal stress (σ_1).

Minor principal stress,

$$\sigma_3 = \frac{\sigma_x + \sigma_y}{2} - \sqrt{\left(\frac{\sigma_y - \sigma_x}{2}\right)^2 + \tau_{xy}^2} \quad \text{...(13.8)}$$

The point V gives the minor principal stress (σ_3)

Also, because $\tan 2\theta_p = \tan (2\theta_p + 180°)$, the second principal plane is indicated by the line CV.

13.6. IMPORTANT CHARACTERISTICS OF MOHR'S CIRCLE

The following important characteristics of Mohr's circle should be carefully noted, as these are required for further study.

(1) The maximum shear stress τ_{max} is numerically equal to $(\sigma_1 - \sigma_3)/2$ and it occurs on a plane inclined at 45° to the principal planes (Fig. 13.5).

(2) Point D on the Mohr circle represents the stresses (σ, τ) on a plane make an angle θ with the major principal plane.

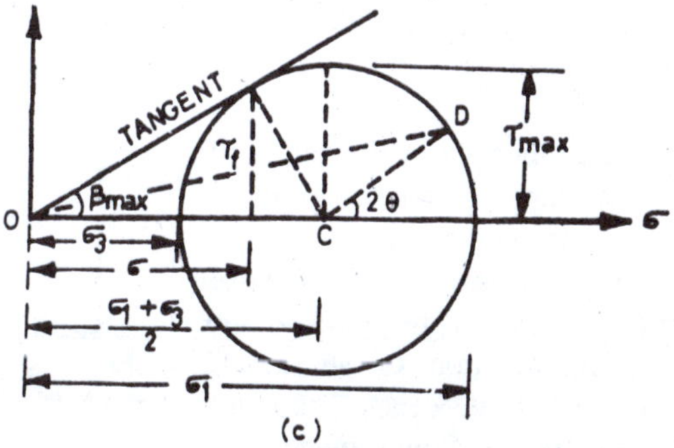

Fig. 13.5. Characteristics of Mohr's Circle.

The resultant stress on that plane is equal to $\sqrt{\sigma^2 + \tau^2}$ and its angle of obliquity with the normal of the plane is equal to angle β, given by

$$\beta = \tan^{-1}(\tau/\sigma) \qquad \text{...(13.9)}$$

(3) The maximum angle of obliquity β_{max} is obtained by drawing a tangent to the circle from the origin *O*.

$$\beta_{max} = \sin^{-1}\frac{(\sigma_1 - \sigma_3)/2}{(\sigma_1 + \sigma_2)/2} = \sin^{-1}\left(\frac{\sigma_1 - \sigma_3}{\sigma_1 + \sigma_3}\right) \qquad \text{...(13.9)}$$

(4) The shear stress τ_f on the plane of the maximum obliquity is less than the maximum shear stress τ_{max}.

(5) Shear stresses on planes at right angles to each other are numerically equal but are of opposite signs, as shown in Fig. 13.4 (*c*).

(6) As the Mohr circle is symmetrical about σ-axis, it is usual practice to draw only the top half circle for convenience.

(7) There is no need to be rigid about sign convention for plotting the shear stresses in Mohr's circle. These can be plotted either upward or downward. Although the sign convention is required for locating the orientation of the planes, the numerical results are not affected.

13.7. MOHR-COULOMB THEORY

The soil is a particulate material. The shear failure occurs in soils by slippage of particles due to shear stresses. The failure is essentially by shear, but shear stresses at failure depend upon the normal stresses on the potential failure plane. According to Mohr, the failure is caused by a critical combination of the normal and shear stresses.

The soil fails when the shear stress (τ_f) on the failure plane at failure is a unique function of the normal stress (σ) acting on that plane.

$$\tau_f = f(\sigma)$$

Since the shear stress on the failure plane at failure is defined as the shear strength (*s*), the above equation can be written as

$$s = f(\sigma) \qquad \text{...(13.11)}$$

The Mohr theory is concerned with the shear stress at failure plane at failure. A plot can be made between the shear stress τ and the normal stress σ at failure. The curve defined by Eq. 13.11 is known as the Mohr envelope [Fig. 13.6 (*a*)]. There is a unique failure envelope for each material.

Failure of the material occurs when the Mohr circle of the stresses touches the Mohr envelope. As discussed in the preceding sections, the Mohr circle represents all possible combinations of shear and normal stresses at the stressed point. At the point of contact (*D*) of the failure envelope and the Mohr circle, the critical combination of shear and normal stresses is reached and the failure occurs. The plane indicated by the line *PD* is, therefore, the failure plane. Any Mohr's circle which does not cross the failure envelope and

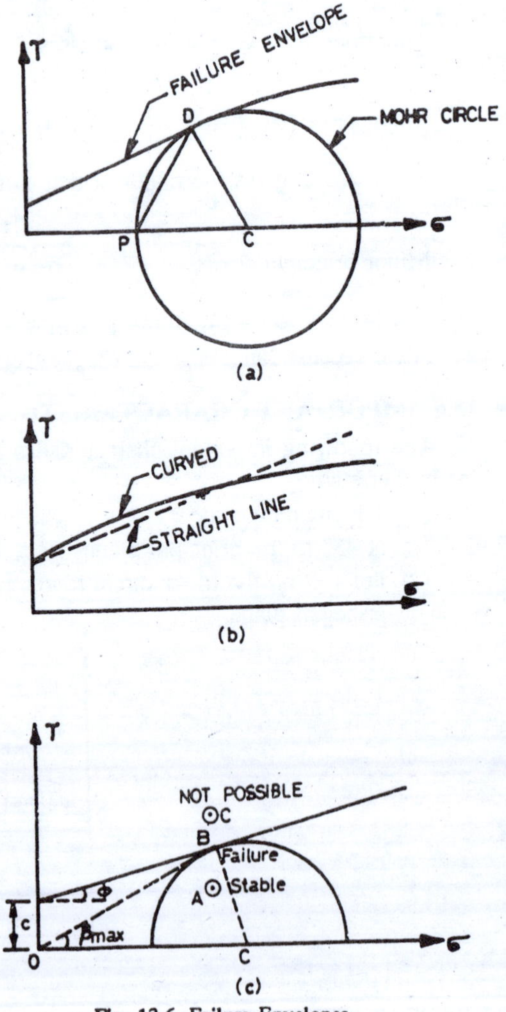

Fig. 13.6. Failure Envelopes.

lies below the envelope represents a (non-failure) stable condition. The Mohr circle cannot cross the Mohr envelope, as the failure would have already occurred as soon as the Mohr circle touched the envelope.

The shear strength (s) of a soil at a point on a particular plane was expressed by Coulomb as a linear function of the normal stress on that plane, as

$$s = c + \sigma \tan \phi \quad ...(13.12)$$

In other words, the Mohr envelope is replaced by a straight line by Coulomb as shown in Fig. 13.6 (*b*).

In Eq. 13.12, c is equal to the intercept on τ-axis and ϕ is the angle which the envelope makes with σ-axis [Fig. 13.6 (*c*)]. The component c of the shear strength is known as *cohesion.* Cohesion holds the particles of the soil together in a soil mass, and is independent of the normal stress. The angle ϕ is called the *angle of internal friction.* It represents the frictional resistance between the particles, which is directly proportional to the normal stress.

As mentioned before, the failure occurs when the stresses are such that the Mohr circle just touches the failure envelope, as shown by point B in Fig. 13.6 (*c*). In other words, shear failure occurs if the stresses σ and τ on the failure plane plot as point B. If the stresses plot as point A below the failure envelope, it represents a stable, non-failure condition. On the other hand, a state of stress represented by point C above the failure envelope is not possible. It may be noted that a material fails along a plane when the critical combination of the stresses σ and τ gives the resultant with a maximum obliquity (β_{max}), in which case the resultant just touches the Mohr circle.

13.8. REVISED MOHR-COULOMB EQUATION

Later research showed that the parameters c and ϕ in Eq. 13.12 are not necessarily fundamental properties of the soil as was originally assumed by Coulomb. These parameters depend upon a number of factors, such as the water content, drainage conditions, conditions of testing. The current practice is to consider c and ϕ as mathematical parameters which represent the failure conditions for a particular soil under given conditions. That is the reason why c and ϕ are now called cohesion intercept and the angle of shearing resistance. These indicate the intercept and the slope of the failure envelope, respectively.

Terzaghi established that the normal stresses which control the shear strength of a soil are the effective stresses and not the total stresses. In terms of effective stresses, Eq. 13.12 is written as

$$s = c' + \bar{\sigma} \tan \phi' \quad ...(13.13)$$

where c' and ϕ' are the cohesion intercept and the angle of shearing resistance in terms of the effective stresses.

Eq. 13.13 is known as the *Revised Mohr—Coulomb* equation for the shear strength of the soil. The equation has replaced the original equation (Eq. 13.12). It is one of the most important equations of soil engineering.

The Mohr—Coulomb theory shows a reasonably good agreement with the observed failures in the field and in the laboratory. The theory is ideally suited for studying the behaviour of soils at failure. The theory is used for estimation of the shear strength of soils. However, even this theory is not perfect. It has the following main limitations :

(1) It neglects the effect of the intermediate principal stress (σ_2),

(2) It approximates the curved failure envelope by a straight line, which may not give correct results.

(3) When the Mohr envelope is curved, the actual obliquity of the failure plane is slightly smaller than the maximum obliquity. Therefore, the angle of the failure plane, as found, is not correct.

(4) For some clayey soils, there is no fixed relationship between the normal and shear stresses on the plane of failure. The theory cannot be used for such soils.

13.9. DIFFERENT TYPES OF TESTS AND DRAINAGE CONDITIONS

The following tests are used to measure the shear strength of a soil.

(1) Direct shear test (2) Triaxial Compression test

(3) Unconfined Compression test (4) Shear Vane test.

The shear test must be conducted under appropriate drainage conditions that simulate the actual field problem. In shear tests, there are two stages :

(1) Consolidation stage in which the normal stress (or confining pressure) is applied to the specimen and it is allowed to consolidate.

(2) Shear stage in which the shear stress (or deviator stress) is applied to the specimen to shear it.

Depending upon the drainage conditions, there are three types of tests as explained below :

(1) Unconsolidated—Undrained Condition. In this type of test, no drainage is permitted during the consolidation stage. The drainage is also not permitted in the shear stage.

As no time is allowed for consolidation or dissipation of excess pore water pressure, the test can be conducted quickly in a few minutes. The test is known as unconsolidated—undrained test (*UU* test) or quick test (*Q*-test).

(2) Consolidated—Undrained Condition. In a consolidated—undrained test, the specimen is allowed to consolidate in the first stage. The drainage is permitted until the consolidation is complete.

In the second stage when the specimen is sheared, no drainage is permitted. The test is known as consolidated—undrained test (*CU* test) It is also called a *R*-test, as the alphabet *R* falls between the alphabet *Q* used for quick test, and the alphabet *S* used for slow test.

The pore water pressure can be measured in the second stage if the facilities for its measurement are available. In that case, the test is known as $\overline{CU}$ test.

(3) Consolidated—Drained Condition. In a consolidated—drained test, the drainage of the specimen is permitted in both the stages. The sample is allowed to consolidate in the first stage. When the consolidation is complete, it is sheared at a very slow rate to ensure that fully drained conditions exist and the excess pore water is zero.

The test is known as a consolidated—drained test (*CD* test) or drained test. It is also known as the slow test (*S*—test).

13.10. MODE OF APPLICATION OF SHEAR FORCE

The shear force in a shear test is applied either by increasing the shear displacement at a given rate or by increasing the shearing force at a given rate. Accordingly, the shear tests are either strain—controlled or stress-controlled.

(1) Strain controlled tests. In a strain-controlled test, the test is conducted in such a way that the shearing strain increases at a given rate. Generally, the rate of increase of the shearing strain is kept constant, and the specimen is sheared at a uniform strain rate.

The shear force acting on the specimen is measured indirectly using a proving ring. The rate of shearing strain is controlled manually or by a gear system attached to an electric motor.

Most of the shear tests are conducted as strain—controlled. The stress—strain characteristic are easily obtained in these tests, as the shape of the stress—strain curve beyond the peak point can be observed only in a strain—controlled test. A strain—controlled test is easier to perform than a stress- controlled test.

(2) Stress—Controlled tests. In a stress—controlled test, the shear force is increased at a given rate. Usually, the rate of increase of the shear force is maintained constant. The shear load is increased such that the shear stresses increase at a uniform rate. The resulting shear displacements are obtained by means of a dial gauge.

Stress—controlled tests are preferred for conducting shear tests at a very low rate, because an applied load can easily be kept constant for any given period of time. Further, the loads can be conveniently applied and removed. The stress-controlled test represents the field conditions more closely.

13.11 DIRECT SHEAR TEST

(*a*) Apparatus. A direct shear test is conducted on a soil specimen in a shear box which is split into two halves along a horizontal plane at its middle (Fig. 13.7). The shear box is made of brass or gunmetal. It is

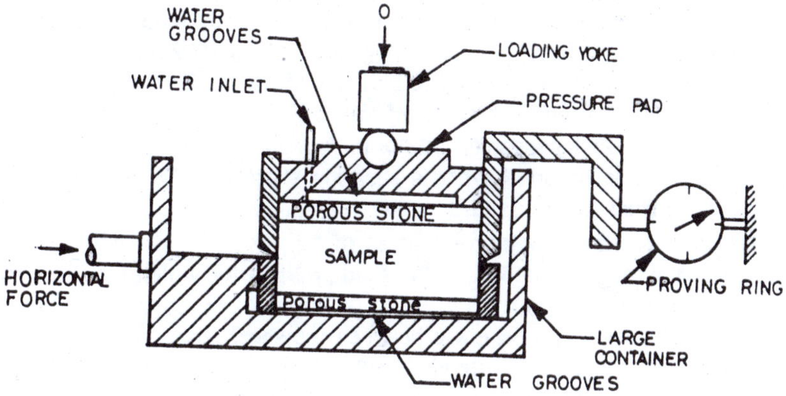

Fig. 13.7. Direct Shear Test.

either square or circular in plan. A square box of size 60 × 60 × 50mm is commonly used. The box is divided horizontally such that the dividing plane passes through the centre. The two halves of the box are held together by locking pins. Suitable spacing screws to separate the two halves are also provided. The spacing screws are fixed to the upper half and they butt against the top of the lower half.

The box is provided with the gripper or the grid plates which are toothed and fitted inside it. The gripper plates are plain (without perforations) for undrained tests and perforated for drained tests. Porous stones are placed at the top and the bottom of the specimen in drained tests. A pressure pad of brass or gun metal is fitted into the box at its top to transmit the normal load to the sample. The normal load from the loading yoke is applied on the top of the specimen through a steel ball bearing upon the pressure pad.

The lower half of the box is fixed to the base plate which is rigidly held in position in a large container. The large container is supported on rollers (rollers not shown). The container can be pushed forward at a constant rate by a geared jack which works as a strain-controlled device. The jack may be operated manually or by an electric motor.

A loading frame is used to support the large container. It has the arrangement of a loading yoke and a lever system for applying the normal load.

A proving ring is fitted to the upper half of the box to measure the shear force.The proving ring butts against a fixed support. As the box moves, the proving ring records the shear force. The shear displacement is measured with a dial gauge fitted to the container. Another dial gauge is fitted to the top of the pressure pad to measure the change in the thickness of the specimen.

***(b)* Test.** A soil specimen of size 60 × 60 × 25 mm is taken. It may be either an undisturbed sample or made from compacted and remoulded soil. The specimen may be prepared directly in the box and compacted. The base plate is attached to the lower half of the box. A porous stone is placed in the box. For undrained tests, a plain grid is kept on the porous stone, keeping its segregations at right angles to the direction of shear. For drained tests, perforated grids are used instead of plain grids. The mass of the base plate, porous stone and grid is taken. The specimen if made separately is transferred to the box and its mass taken.

The upper grid, porous stone and the pressure pad are placed on the specimen. The box is placed inside the large container and mounted on the loading frame. The upper half of the box is brought in contact with the proving ring. The loading yoke is mounted on the steel ball placed on the pressure pad. The dial gauge is fitted to the container to give the shear displacement. The other dial gauge is mounted on the loading yoke to record the vertical movement.

The locking pins are removed and the upper half box is slightly raised with the help of spacing screws. The space between the two halves is adjusted, depending upon the maximum particle size. The space should be such that the top half of the box does not ride on soil grains which come between the edges.

The normal load is applied to give a normal stress of 25 kN/m^2. Shear load is then applied at a constant rate of strain. For undrained tests, the rate is generally between 1.0 mm to 2.00 mm per minute. For drained tests, the strain rate depends upon the type of soil. For sandy soils, it may be taken as 0.2 mm/minute; whereas for clayey soils, it is generally between 0.005 to 0.02 mm/min. The sample shears along the horizontal plane between the two halves. The readings of the proving-ring and the dial gauges are taken every 30 seconds. The test is continued till the specimen fails. The failure is indicated when the proving ring dial gauge begins to recede after having reached the maximum. For the soils which do not give a peak point, the failure is assumed to have occurred when a shearing strain of 20% is reached. At the end of the test, the specimen is removed from the box and its water content found.

The test is repeated under the normal stress of 50, 100, 200 and 400 kN/m^2. The range of the normal stress should cover the range of loading in the field problem for which the shear parameters are required. The shear stress at any stage during shear is equal to the shear force indicated by the proving ring divided by the area of the specimen. A plot can be made between the shear stress and the shear strain. The shear strain is equal to the shear displacement (ΔH) divided by the length of the specimen (L). The shear stress is obtained from the shear load indicated by the proving ring and the cross-sectional area.

Direct shear tests can be conducted for any one of the three drainage conditions. For *U-U* test, plain grids are used and the sample is sheared rapidly. For *CU* test, perforated grids are used. The sample is consolidated under the normal load and after the completion of consolidation, it is sheared rapidly in about 5—10 minutes. In a *CD* test, the sample is consolidated under the normal load and then sheared slowly so that excess pore water pressure is dissipated. A *CD* test may take a few hours for cohesionless soils. For cohesive soils, it may take 2 to 5 days.

The direct shear test is generally conducted on cohesionless soils as *CD* test. It is convenient to perform and it gives good results for the strength parameters. It is occasionally used to determine the strength parameters of silt and clay under unconsolidated—undrained and consolidated drained conditions, but it does not offer the flexibility of a triaxial compression test, as explained later.

13.12. PRESENTATION OF RESULTS OF DIRECT SHEAR TEST

(*a*) **Stress-Strain Curve.** A stress-strain curve is a plot between the shear stress τ and the shear displacement ($\Delta H/L$) [Fig. 13.8 (*a*)]. In case of dense sand (and also over-consolidated clays), the shear stress attains a ***peak value*** at a small strain. With further increase in strain, the shear stress decreases slightly and becomes more or less constant, known as ultimate stress. In case of loose sands (and normally consolidated clays), the shear stress increases gradually and finally attains a constant value, known as the *ultimate stress* or residual strength. It has been observed that the ultimate shear stress attained by both dense and loose sands tested under similar conditions is approximately the same. The figure also shows the stress-strain curve of a medium dense sand.

Generally, the failure strain is 2 to 4% for dense sand and 12 to 16% for loose sand.

Fig. 13.8 (*b*) shows the volume changes with an increase in shear strain for *CD* tests. Since the cross-sectional area of the specimen remains unchanged, the volume change is proportional to the change in thickness measured by the dial gauge. In case of dense sands (and over-consolidated clays), the volume first decreases slightly,

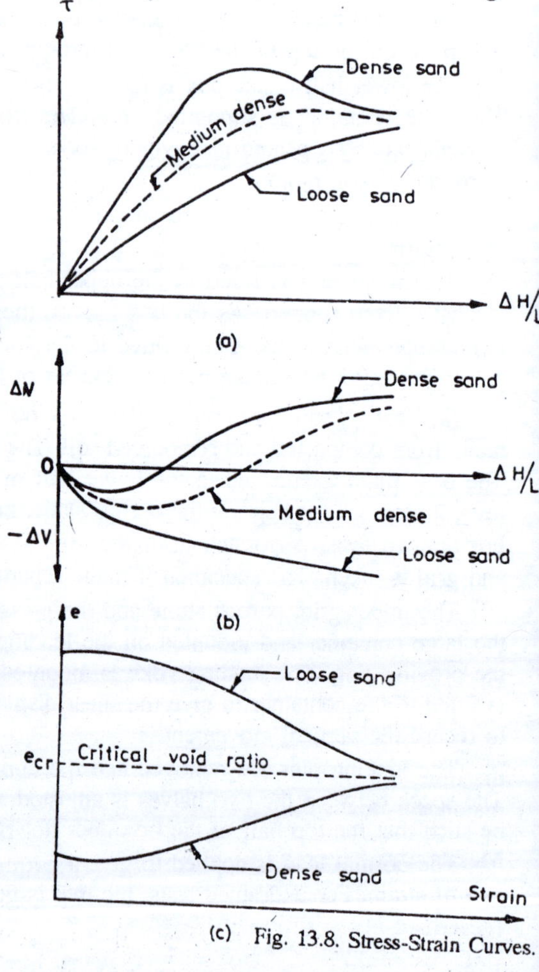

(c) Fig. 13.8. Stress-Strain Curves.

but it increases with further increase in strain. In the case of loose sands (and normally consolidated clays), the volume decreases with an increase in shear strain. The figure also shows the curve for medium dense sand.

It may be observed that the void ratio of an initial loose sand decreases with an increase in shear strain, whereas that for the initially dense sand increases with an increase in strain [Fig. 13.8 (*c*)]. The void ratio at which there is no change in it with an increase in strain is known as the *critical void ratio.* If the sand initially is at the critical void ratio, there would be practically no change in volume with an increase in shear strain.

(*b*) **Failure Envelope.** For obtaining a failure envelope, a number of identical specimens are tested under different normal stresses. The shear stress required to cause failure is determined for each normal stress. The failure envelope is obtained by plotting the points corresponding to shear strength at different normal stresses and joining them by a straight line [Fig. 13.9 (*a*)]. The inclination of the failure envelope to

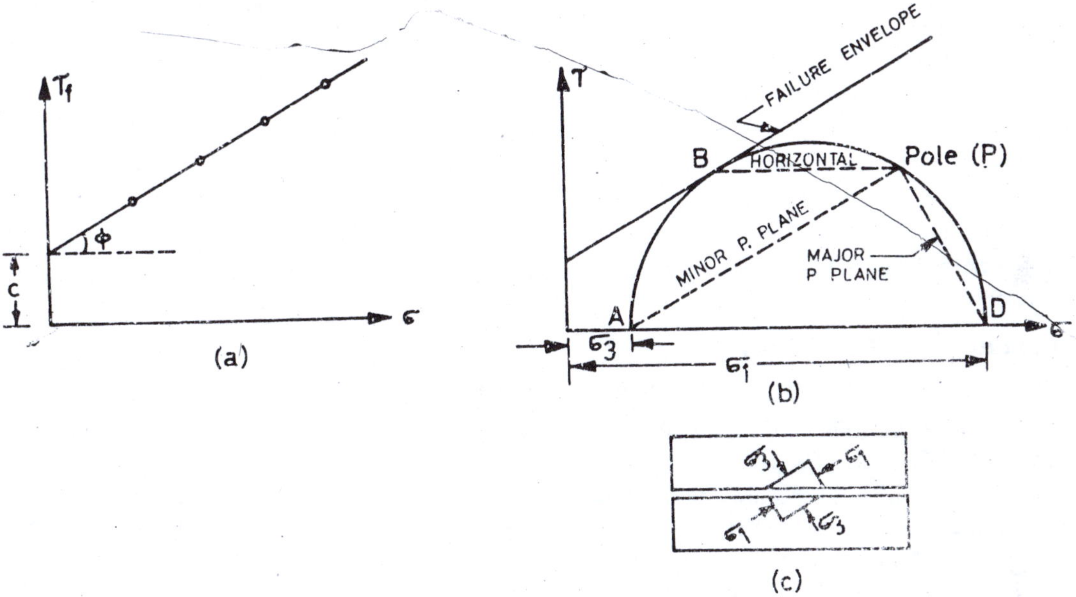

Fig. 13.9. Failure Envelope.

the horizontal gives the angle of shearing resistance ϕ and its intercept on the vertical axis is equal to the cohesion intercept *c*.

For dense sands, the failure envelope can be drawn either for peak stress or for ultimate stress. The values of the parameters ϕ and *c* for the two envelopes will be different. For loose sands, the failure envelope is drawn for ultimate stress, which is usually taken as the shear stress at 20% shear strain.

(c) Mohr-Circle. In a direct shear test, the stresses on planes other than the horizontal plane are not known. It is, therefore, not possible to draw Mohr stress circle at different shear loads. However, the Mohr circle can be drawn at the failure condition assuming that the failure plane is horizontal.

In Fig. 13.9 (*b*), the point *B* represents the failure condition for a particular normal stress. The Mohr circle at failure is drawn such that it is tangential to the failure envelope at *B*. The horizontal line *BP* gives the direction of the failure plane. The point *P* is the pole. The lines *PD* and *PA* give the directions of the major and minor principal planes, respectively. The principal planes are also shown in Fig. 13.9 (*c*).

Merits and Demerits of Direct Shear Test

The direct shear test has the following merits and demerits as compared to the triaxial compression test (described in the following section).

Merits.

(1) The sample preparation is easy. The test is simple and convenient.

(2) As the thickness of the sample is relatively small, the drainage is quick and the pore pressure dissipates very rapidly. Consequently, the consolidated-drained and the consolidated- undrained tests take relatively small period.

(3) It is ideally suited for conducting drained tests on cohesionless soils.

(4) The apparatus is relatively cheap.

Demerits.

(1) The stress conditions are known only at failure. The conditions prior to failure are indeterminate and, therefore, the Mohr circle cannot be drawn.

(2) The stress distribution on the failure plane (horizontal plane) is not uniform. The stresses are more at the edges and lead to the progressive failure, like tearing of a paper. Consequently, the full strength of the soil is not mobilised simultaneously on the entire failure plane.

(3) The area under shear gradually decreases as the test progresses. But the corrected area cannot be determined and, therefore, the original area is taken for the computation of stresses.

(4) The orientation of the failure plane is fixed. This plane may not be the weakest plane.

(5) Control on the drainage conditions is very difficult. Consequently, only drained tests can be conducted on highly permeable soils.

(6) The measurement of pore water pressure is not possible.

(7) The side walls of the shear box cause lateral restraint on the specimen and do not allow it to deform laterally.

13.13. DIFFERENT TYPES OF SOILS

On the basis of shear strength, soils can be divided into three types.

(1) Cohesionless soils. (2) Purely cohesive soils and

(3) Cohesive-frictional soils.

1. Cohesionless soils. These are the soils which do not have cohesion *i.e.*, $c' = 0$. These soils derive the shear strength from the intergranular friction. These soils are also called *frictional soils.* For example, sands and gravels.

2. Purely cohesive soils. These are the soils which exhibit cohesion but the angle of shearing resistance $\phi = 0$. For example, saturated clays and silts under undrained conditions. These soils are also called $\phi_u = 0$ soils.

3. Cohesive-frictional soils. These are composite soils having both c' and ϕ'. These are also called c—ϕ soils. For example, clayey sand, silty sand, sandy clay, etc.

[**Note.** Sometimes, cohesive-frictional soils are also called cohesive soils. Thus any soil having a value of c' is called a cohesive soil.]

13.14. TRIAXIAL COMPRESSION TEST APPARATUS

The triaxial compression test, or simply triaxial test, is used for the determination of shear characteristics of all types of soils under different drainage conditions. In this test, a cylindrical specimen is stressed under conditions of axial symmetry, as shown in Fg. 13.10. In the first stage of the test, the specimen is subjected to an all round confining pressure (σ_c) on the sides and at the top and the bottom. This stage is known as the consolidation stage.

In the second stage of the test, called the shearing stage, an additional axial stress, known as the deviator stress (σ_d), is applied on the top of the specimen through a ram. Thus, the total stress in the axial direction at the

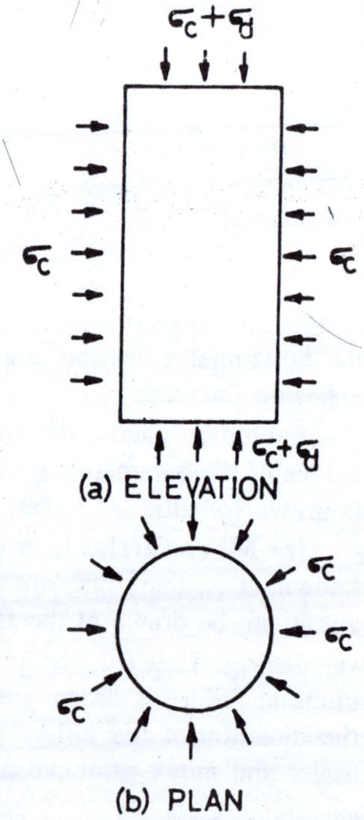

Fig. 13.10.

time of shearing is equal to $(\sigma_c + \sigma_d)$. It may be noted that when the axial stress is increased, the shear stresses develop on inclined planes due to compressive stresses on the top.

The vertical sides of the specimen are principal planes, as there are no shear stresses on the sides. The confining pressure σ_c is equal to the minor principal stress (σ_3). The top and bottom planes are the major principal planes. The total axial stress which is equal to the sum of the confining pressure and the deviator stress, is the major principal stress (σ_1). Because of axial symmetry, the intermediate principal stress (σ_2) is also equal to the confining pressure (σ_c).

[**Note.** The above interpretation of the stress conditions in the triaxial test is not strictly correct according to the theory of elasticity. In the case of cylindrical specimens, the three principal stresses are the axial, radial and the circumferential stresses. The state of stress is statically indeterminate throughout the specimen. For convenience, in the triaxial test, the circumferential stress is taken equal to the radial stress and the principal stresses σ_2 and σ_3 are assumed to be equal].

The main features of a triaxial test apparatus are shown in Fig. 13.11. It consists of a circular base that has a central pedestal. The pedestal has one or two holes which are used for the drainage of the specimen in a drained test or for the pore pressure measurement in an undrained test. A triaxial cell is fitted to the top of the base plate with the help of 3 wing nuts (not shown in the figure) after the specimen has been placed on the pedestal. The triaxial cell is a perspex cylinder which is permanently fixed to the top cap and the bottom brass collar. There are three tie rods which support the cell. The top cap is a bronze casting with its central boss forming a bush through which a stainless steel ram can slide. The ram is so designed that it has minimum of friction and at the same time does not permit any leakage. There is an air-release valve in the top cap which is kept open when the cell is filled with water (or glycerine) for applying the confining pressure. An oil valve is also provided in the top cap to fill light machine oil in the cell to reduce the leakage of water past the ram in long duration tests. The apparatus is mounted on a loading frame. The deviator stress is applied to the specimen from a strain-controlled loading machine. The loading system consists of either a screw jack operated by an electric motor and gear box or a hydraulic ram operated by a pump.

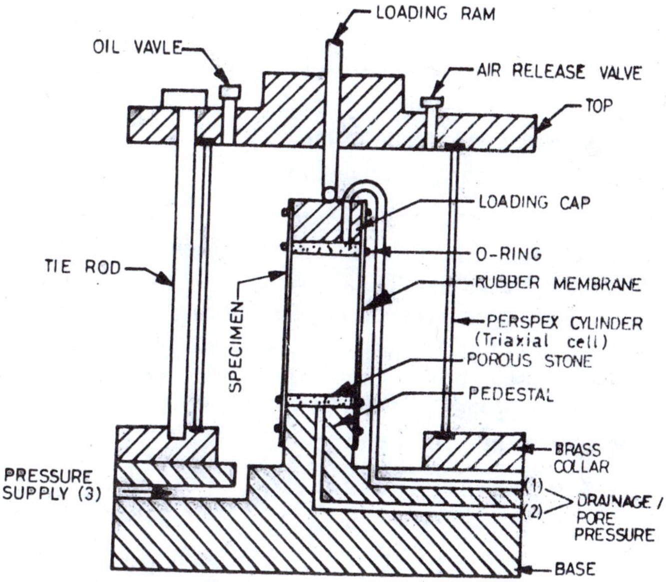

Fig. 13.11. Triaxial Tes Apparatus.

The triaxial test apparatus has the following special attachments.

1. Mercury Control System. The cell pressure is a triaxial test in maintained constant with a self-compensating mercury control system, developed by Bishop and Henkel. It consists of two limbs of a water-mercury manometer (Fig. 13.12). The pressure in the water of the triaxial cell develops due to the difference in levels of the mercury in the two pots. The water pressure at the centre of the specimen in the triaxial cell, at a height of h_3 above the datum, can be calculated using the theory of manometers. As the mercury surface in the upper pot is open to atmosphere, the (gauge) pressure there is zero. From the manometer equation,

$$0 + \gamma_m h_1 - \gamma_m h_2 - (h_3 - h_2)\,\gamma_w = \sigma_c$$

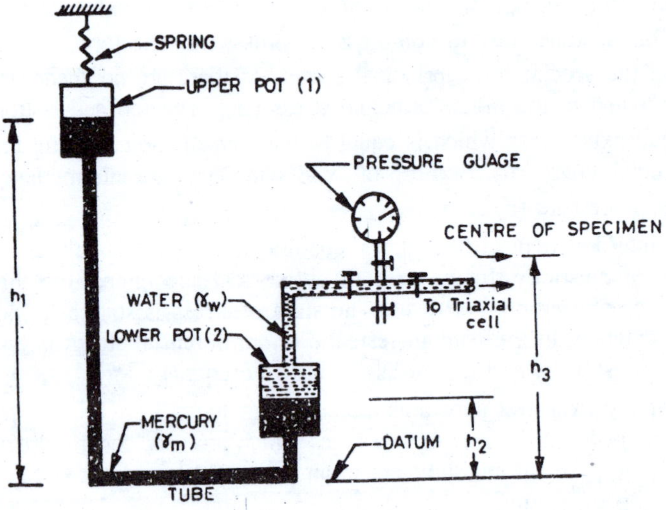

Fig. 13.12. Mercury Control System.

where σ_c is the cell pressure at the centre of the specimen,

γ_w is the unit weight of water, and

γ_m is the unit weight of mercury.

The above equation can be simplified as

$$\sigma_c = \gamma_m (h_1 - h_2) + (h_2 - h_3) \gamma_w \quad ...(13.14)$$

The upper pot is supported by a spring. When the volume of the specimen decreases due to consolidation or when the water leaks past the ram, water flows from the lower pot to the cell and the mercury level in the lower pot rises by a small amount Δh. The mercury level in the upper pot would also fall by the same amount if the two pots are of the same cross-sectional area. However, the difference of mercury levels in the two pots is maintained constant by the spring. The stiffness (k) of the spring is selected such that it reduces in length and causes a rise of the upper pot as soon as its weight decreases due to flow of mercury. The stiffness of the spring is given by

$$k = A \gamma_m \left[\frac{1}{2 - (\gamma_w/\gamma_m)} \right] - W \quad ...(13.15)$$

where A = cross-sectional area of the mercury pot,

and W = weight of unit length of the tube filled with mercury which is also lifted above the floor.

2. Pore water Pressure Measurement Device. The pore water pressure in the triaxial specimen is measured by attaching it to the device shown in Fig. 13.13. It consists of a null indicator in which no- flow condition is maintained. For accurate measurement, *no- flow condition is essential because* the flow of water from the sample to the gauge would modify the actual magnitude of the pore water pressure. Further, the flow of water leads to a time lag in the attainment of a steady state in samples of cohesive soils because of low permeability.

The null indicator is essentially a U-tube partly filled with mercury. One limb of the null indicator is connected to the specimen in the triaxial cell and the other limb is connected to a pressure gauge. A control cylinder, which is filled with water, is attached to the system. The water can be displaced by a screw-controlled piston in the control cylinder. The whole system is filled with deaired water. The tubes connecting the specimen and the null-indicator should be such that these undergo negligible volume changes under pressure and are free from leakage.

Any change in the pore-water pressure in the specimen tends to cause a movement of the mercury level **in the null-indicator. However, the no-flow condition is maintained by making a corresponding change in the**

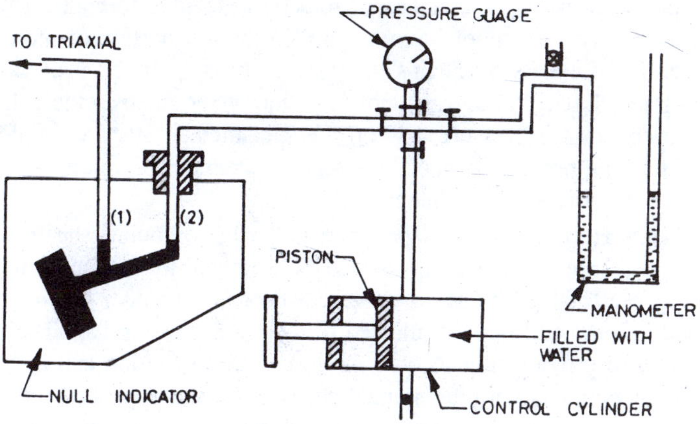

Fig. 13.13. Pore Water Pressure Measurement Device.

other limbs by means of the control cylinder. Thus the mercury levels in the two limbs remain constant. The pressure applied by the control cylinder is recorded by pressure gauge or the manometer.

If the specimen is partially saturated, a special fine, porous ceramic disc is placed below the sample in the triaxial cell. The ceramic disc permits only pore water to flow, provided the difference between the pore air pressure and pore water pressure is below a certain value, known as *the air-entry value* of the ceramic disc. Under undrained conditions, the ceramic disc will remain fully saturated, provided the air-entry value is high. It may be mentioned that if the required ceramic disc is not used and instead the usual coarse, porous disc is used, the device would measure air pressure and not water pressure in a partially saturated soil.

In modern equipment, sometimes the pore water pressure is measured by means of a transducer and not by conventional null indicator.

3. Volume Changes Measurement. Volume changes in a drained test and during consolidation stage of a consolidated undrained test are measured by means of a burette connected to the specimen in the triaxial cell. For accurate measurements, the water level in the burette should be approximately at the level of the centre of the specimen (Fig. 13.14).

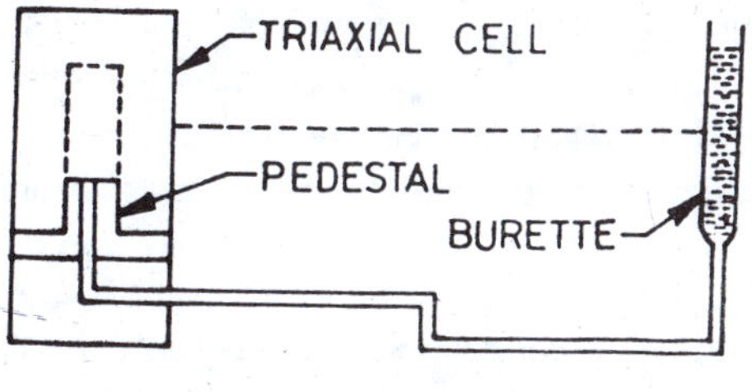

Fig. 13.14. Volume Change Measurement.

During consolidation stage, the volume of the specimen decreases and the water level in the burette rises. The change in the volume of the specimen is equal to the volume of the water increased in the burette. During shearing of specimens of dense sand when the volume of the sample increases, the water flows from the burette to the specimen. The increase in volume of the specimen is equal to the volume of water decreased in the burette.

13.15. TRIAXIAL TESTS ON COHESIVE SOILS

The following procedure is used for conducting the triaxial tests an cohesive soils.

***(a)* Consolidated-undrained test.** A deaired, coarse porous disc or stone is placed on the top of the pedestal in the triaxial test apparatus. A filter paper disc is kept over the porous stone. The specimen of the cohesive soil is then placed over the filter paper disc. The usual size of the specimen is about 37.5 mm diameter and 75.0 mm height. A porous stone is also placed on the top of the specimen. Deaired vertical filter strips are placed at regular spacing around the entire periphery such that these touch both the porous stones. The sample in then enclosed in a rubber membrane, which is slid over the specimen with the help of a membrane stretcher. The membrane is sealed to the specimen with O-rings.

The triaxial cell is placed over the base and fixed to it by tightening the nuts. The cell is then filled with water by connecting it to the pressure supply. Some space in the top portion of the cell is filled by injecting oil through the oil valve. When excess oil begins to spill out through the air-vent valve, both the valves (oil valve and air-vent valve) are closed. Pressure is applied to the water filled in the cell by connecting it to the mercury-pot system. As soon as the pressure acts on the specimen, it starts consolidating. The specimen is connected to the burette through pressure connections for measurement of volume changes. The consolidation is complete when there is no more volume change.

When the consolidation is complete, the specimen is ready for being sheared. The drainage valve is closed. The pore water pressure measurement device is attached to the specimen through the pressure connections. The proving ring dial gauge is set to zero. Using the manual control provided in the loading frame, the ram is pushed into the cell but not allowed to touch the loading cap. The loading machine is then run at the selected speed. The proving ring records the force due to friction and the upward thrust acting on the ram. The machine is stopped, and with the manual control, the ram is pushed further into the cell bringing it in contact with the loading cap. The dial gauge for the measuring axial deformation of the specimen is set to zero.

The sample is sheared by applying the deviator stress by the loading machine. The proving ring readings are generally taken corresponding to axial strains of 1/3%, 2/3%, 1%, 2%, 3%, 4%, 5%, ...until failure or 20% axial strain.

Upon completion of the test, the loading is shut off. Using the manual control, all additional axial stress is removed. The cell pressure is then reduced to zero, and the cell is emptied. The triaxial cell is unscrewed and removed from the base. O-rings are taken out, and the membrane is removed. The specimen is then recovered after removing the loading cap and the top porous stone. The filter paper strips are peeled off. The post- shear mass and length are determined. The water content of the specimen is also found.

(*b*) Unconsolidated Undrained test. The procedure is similar to that for a consolidated-undrained test, with one basic difference that the specimen is not allowed to consolidate in the first stage. The drainage valve during the test is kept closed. However, the specimen can be connected to the pore-water pressure measurement device if required.

Shearing of the specimen is started just after the application of the cell pressure. The second stage is exactly the same as in the consolidated-undrained test described above.

(*c*) Consolidated Drained test. The procedure is similar to that for a consolidated-undrained test, with one basic difference that the specimen is sheared slowly in the second stage. After the consolidation of the specimen in the first stage, the drainage valve is not closed. It remains connected to the burette throughout the test. The volume changes during the shearing stage are measured with the help of the burette. As the permeability of cohesive soils is very low, it takes 4-5 days for the consolidated drained test.

13.16. TRIAXIAL TESTS ON COHESIONLESS SOILS

Triaxial tests on specimens of cohesionless soils can be conducted using the procedure as described for cohesive soils. As the samples of cohesionless soils cannot stand of their own, a special procedure is used for preparation of the sample as described below.

A metal former, which is a split mould of about 38.5 mm internal diameter, is used for the preparation of the sample (Fig. 13.15). A coarse porous stone is placed on the top of the pedestal of the triaxial base, and the pressure connection is attached to a burette (not shown). One end of a membrane is sealed to the pedestal by O-rings. The metal former is clamped to the base. The upper metal ring of the former is kept inside the top end of the rubber membrane and is held with the help of a clamp before placing the funnel and the rubber bung in position as shown in figure.

The membrane and the funnel are filled with deaired water. The cohesionless soil which is to be tested is saturated by mixing it with enough water in a beaker. The mixture is boiled to remove the entrapped air. The saturated soil is deposited in the funnel, with a stopper in position, in the required quantity. The glass rod is then removed and the sample builds up by a continuous rapid flow of saturated soil in the former. The funnel is then removed. The sample may be compacted if required. The surface of the sample is leveled and a porous stone is placed on its top. The loading cap is placed gently on the top porous stone. O-rings are fixed over the top of the rubber membrane.

A small negative pressure is applied to the sample by lowering the burette. The negative pressure gives rigidity to the sample and it can stand without any lateral support. For sample of 37.5 mm diameter, a negative pressure of 20 cm of water (or 2 kN/m^2) is sufficient. As soon as the negative pressure is applied, the consolidation of the sample occurs and it slightly shortens. The diameter of the upper porous stone should be slightly smaller than that of the specimen so that it can go inside when the sample shortens; otherwise, a neck is formed.

The split mould is then removed, and the diameter and the height of the sample are measured. The thickness of the membrane is deducted from the total diameter to get the net diameter of the sample. The cell is then placed over the base and clamped to the base. It is then filled with water.

The rest of the procedure is the same as for cohesive soils.

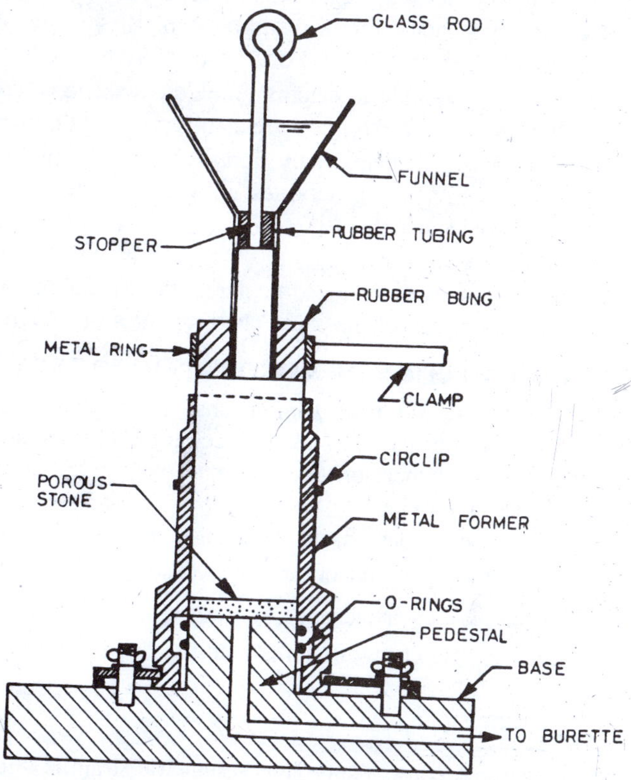

Fig. 13.15. Preparation of Sample of Cohesionless Soil.

13.17. MERITS AND DEMERITS OF TRIAXIAL TEST

The triaxial test has the following merits and demerits.

Merits.

(1) There is complete control over the drainage conditions. Tests can be easily conducted for all three types of drainage conditions.

(2) Pore pressure changes and the volumetric changes can be measured directly.

(3) The stress distribution on the failure plane is uniform.

(4) The specimen is free to fail on the weakest plane.

(5) The state of stress at all intermediate stages upto failure is known. The Mohr circle can be drawn at any stage of shear.

(6) The test is suitable for accurate research work. The apparatus is adaptable to special requirements such as extension test and tests for different stress paths.

Demerits.

(1) The apparatus is elaborate, costly and bulky.

(2) The drained test takes a longer period as compared with that in a direct shear test.

(3) The strain condition in the specimen are not uniform due to frictional restraint produced by the loading cap and the pedestal disc. This leads to the formation of the dead zones at each end of the specimen.

The non-uniform distribution of stresses can be largely eliminated by lubrication of end surfaces. However, non-uniform distribution of stresses has practically no effect on the measured strength if length/diameter ratio is equal to or more than 2.0.

(4) It is not possible to determine the cross-sectional area of the specimen accurately at large strains, as the assumption that the specimen remains cylindrical does not hold good.

(5) The test simulates only axis-symmetrical problems. In the field, the problem is generally 3-dimensional. A general test in which all the three stresses are varied would be more useful.

(6) The consolidation of the specimen in the test is isotropic; whereas in the field, the consolidation is generally anisotropic.

Despite the above-mentioned demerits, the triaxial test is extremely useful. It is the only reliable test for accurate determination of the shear characteristics of all types of soils and under all the drainage conditions.

13.18. COMPUTATION OF VARIOUS PARAMETERS

(a) **Post-Consolidation Dimensions.** In consolidated-drained and consolidated-undrained tests, the consolidation of the specimen takes place during the first stage. As the volume of the specimen decreases, its post-consolidation dimensions are different from the initial dimensions. The post consolidation dimensions can be determined approximately assuming that the sample remains cylindrical and it behaves isotropically. Let L_i, D_i, and V_i be the length, diameter and the volume of the specimen before consolidation. Let L_0, D_0 and V_0 be the corresponding quantities after consolidation.

Therefore, volumetric change, $\Delta V_i = V_i - V_0$

The volumetric change (ΔV_i) is measured with the help of burette.

Volumetric strain, $$\varepsilon_v = \frac{\Delta V_i}{V_i}$$

For isotropic consolidation, the volumetric strain is three times the linear strain (ε_l), Thus

$$\varepsilon_l = \varepsilon_v/3$$

Thus $$L_0 = L_i - \Delta L_i = L_i - L_i \times \varepsilon_l$$

or $$L_0 = L_i(1 - \varepsilon_l) = L_i\,(1 - \varepsilon_v/3) \qquad \text{...(13.16)}$$

Likewise, $$D_0 = D_i\,(1 - \varepsilon_v/3)$$

The post consolidation diameter D_0 can also be computed after L_0 has been determined from the relation,

$$(\pi/4 \cdot D_0^2) \times L_0 = V_0$$

or $$D_0 = \sqrt{\frac{V_0}{(\pi/4) \times L_0}} \qquad \text{...(13.17)}$$

(b) **Cross-sectional Area During Shear Stage.** As the sample is sheared, its length decreases and the diameter increases. The cross-sectional area A at any stage during shear can be determined assuming that the sample remains cylindrical in shape. Let ΔL_0 be the change in length and ΔV_0 be the change in volume. The volume of the specimen at any stage is given by $V_0 \pm \Delta V_0$.

Therefore, $$A\,(L_0 - \Delta L_0) = V_0 \pm \Delta V_0$$

or $$A = \frac{V_0 \pm \Delta V_0}{L_0 - \Delta L_0} = \frac{V_0\left(1 \pm \dfrac{\Delta V_0}{V_0}\right)}{L_0\left(1 - \dfrac{\Delta L_0}{L_0}\right)} \qquad \text{...(13.18)}$$

Eq. 13.18 is the general equation which gives the cross-sectional area of the specimen.

The above equation can be written as

$$A = \frac{A_0 L_0 \left(1 \pm \frac{\Delta V_0}{V_0}\right)}{L_0 (1 - \varepsilon_1)} = \frac{A_0 \left(1 \pm \frac{\Delta V_0}{V_0}\right)}{(1 - \varepsilon_1)} \quad \ldots(13.19)$$

where ε_1 is the axial strain in the sample.

For an undrained test, as the volumetric change (ΔV_0) is zero, Eq. 13.19 becomes

$$A = \frac{A_0}{1 - \varepsilon_1} \quad \ldots(13.20)$$

The stresses in the specimen at various stages of shear should be calculated using the cross-sectional area A as found above.

The correction so applied is known as *area correction.*

(c) Stresses *(i) Deviator stress.* The deviator stress (σ_d) acting on the specimen when the axial load applied by the loading machine is P can be obtained as

$$\sigma_d = \frac{P}{A} \quad \ldots(13.21)$$

The deviator stress (σ_d) is equal ($\sigma_1 - \sigma_3$)

It may be noted that the load indicated by the proving ring is slightly more than P because of friction on the ram and the upward thrust on the ram due to pressure of the water in the cell. The correction can be determined separately.

A more convenient procedure is to lift the ram above the specimen when the cell pressure has been applied. The machine is started keeping the strain rate the same as to be used in the actual test. The proving ring records the load. To account for correction, the dial gauge on the proving ring is set to zero to indicate zero load. This automatically compensates the ram friction and the upward thrust on the ram due to cell pressure. Thus the load indicated by the proving ring during shear would be equal to the load P applied to the specimen.

(ii) Principal stresses. The minor principal stress (σ_3) is equal to the cell pressure (σ_c). The major principal stress (σ_1) is equal to the sum of the cell pressure and the deviator stress.

Thus $\sigma_1 = \sigma_3 + (\sigma_1 - \sigma_3)$

or $\sigma_1 = \sigma_3 + \sigma_d \quad \ldots(13.22)$

(iii) Compressive strength. The deviator stress at failure, $(\sigma_1 - \sigma_3)_f$, is known as the compressive strength of the soil.

13.19. PRESENTATION OF RESULTS OF TRIAXIAL TESTS

(a) **Stress—Strain Curves.** *Drained Test.* Fig. 13.16 *(a)* shows the stress—strain curve for a drained test. The y-axis shows the deviator stress ($\sigma_1 - \sigma_3$) and the x-axis, the axial strain (ε_1). For dense sand (and over-consolidated clay), the deviator stress reaches a peak value and then it decreases and becomes almost constant, equal to the ultimate stress, at large strains. For loose sand (and normally consolidated clay), the deviator stress increases gradually till the ultimate stress is reached.

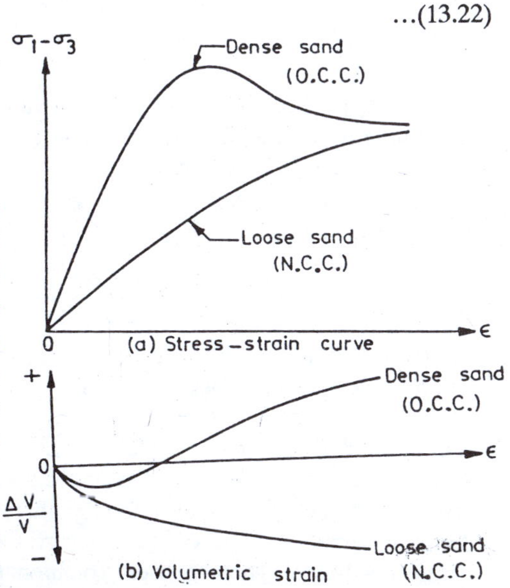

Fig. 13.16. Stress-strain Curve for Drained Test.

The volumetric strain is shown in Fig. 13.16 *(b)*. In dense sand (and over-consolidated clay), there is a decrease in the volume at low strains, but at large strains, there is an increase in the volume. In loose sand (and normally consolidated clay),

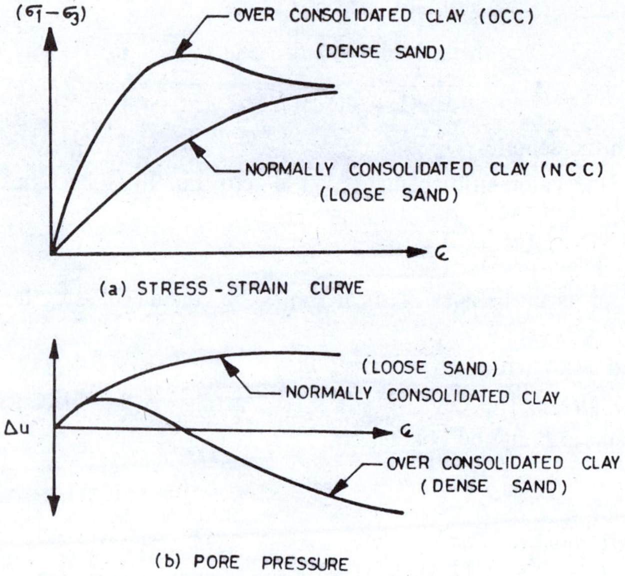

Fig. 13.17. Sterss-Strain Curve, *CU* Test.

the volume decreases at all strains. (For some loose sands, there is a slight tendency to increase in the volume at large strains).

Consolidated—undrained test. Fig. 13.17 (*a*) shows the stress-strain curve for a consolidated—undrained test. The shape of the curves is similar to that obtained in a consolidated-drained test. In a consolidated—undrained test, there is an increase in the pore water pressure throughout for loose sand (and normally consolidated clay), as shown in Fig. 13.17 (*b*). However, in the case of dense sands (and over- consolidated clay), the pore water pressure increases at low strains but at large strains it becomes negative (below atmospheric pressure).

(*b*) **Mohr Envelopes.** For drawing the failure envelopes, it is necessary to test at least three samples at three different cell pressures in the stress range of interest. For dense sands and over-consolidated clays, the failure envelope can be drawn either for the peak stress or for the ultimate stress. For loose sands and normally consolidated clays, the failure envelope is drawn for the ultimate stress, which is usually taken at 20% strain. Further, the failure envelope can be drawn either in terms of effective stresses or in terms of total stress. Of course, the two envelopes will give different values of strength parameters (c and φ).

Britlleness I_B is expressed as

$$I_B = \frac{\text{Peak strength} - \text{Ultimate strength}}{\text{Ultimate strength}}$$

(*i*) *Effective Stresses.* Fig. 13.18 (*a*) shows the failure envelope for a normally consolidated clay in terms of effective stresses obtained from a *consolidated drained test.* The failure envelope has an angle of shearing resistance of ϕ and passes through origin. First the Mohr circles for the three tests are drawn in terms of effective stresses corresponding to failure conditions. Then the best common tangent is drawn to the three circles. The common tangent is the

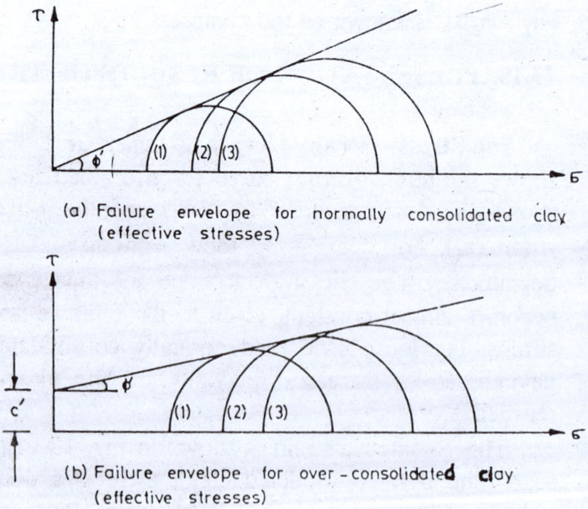

Fig. 13.18. Failure Envelopes (effective stress).

failure envelope. As each circle represents a failure, there must be at least one point on it which gives the stresses satisfying the failure criterion. Obviously, the common tangent joins all such points of the three circles.

Thus for normally consolidated clays, shear strength is

$$s = \bar{\sigma} \tan \varphi'.$$

Fig. 13.18 (*b*) shows the failure envelope for overconsolidated clay in terms of effective stresses. The failure envelope is slightly curved in the initial portion, but, for convenience, it is approximated as a straight line. The failure envelope has an intercept c' on the τ- axis. The angle of shearing resistance is ϕ'. In the case of over-consolidated clays, shear strength is

$$s = c' + \bar{\sigma} \tan \varphi'$$

The failure envelopes in terms of effective stresses can also be drawn from the results of a consolidated-undrained test ($\overline{CU}$ test) when the pore water pressure measurements are also taken. The shear strength parameters c' and φ' obtained from the consolidated-undrained tests and that from consolidated drained tests are approximately equal. Drained tests on dense sands and over-consolidated clays give slightly higher values of ϕ' due to extra work required during dilation (increase in volume), but the difference is small, and, therefore, usually neglected.

(*ii*) *Total stresses*. The failure envelope in terms of total stresses can be drawn from the test results of a *consolidated-undrained test.*

[**Note.** In consolidated-drained tests, the total stresses are also equal to the effective stresses, as the pore water pressure throughout is zero].

The failure envelopes are similar in shape to that in terms of effective stresses but the values of the strength parameters are quite different. Fig. 13.19 shows the failure envelopes for effective stresses and also for total stresses for a normally consolidated clay. The angle of shearing resistance in terms of total stresses (ϕ_{cu}) is much smaller than that for the effective stresses (ϕ').

In the case of normally consolidated clays, shear strength is

$$s = \sigma \tan \phi_{cu}$$

Fig. 13.20 shows the failure envelope for an overconsolidated clay in terms of total stresses. The angle of shearing resistance (ϕ_{cu}) is much smaller than the angle ϕ' obtained in terms of effective stresses. In the case of overconsolidated clays, shear strength is

$$s = c_{cu} + \sigma \tan \phi_{cu}$$

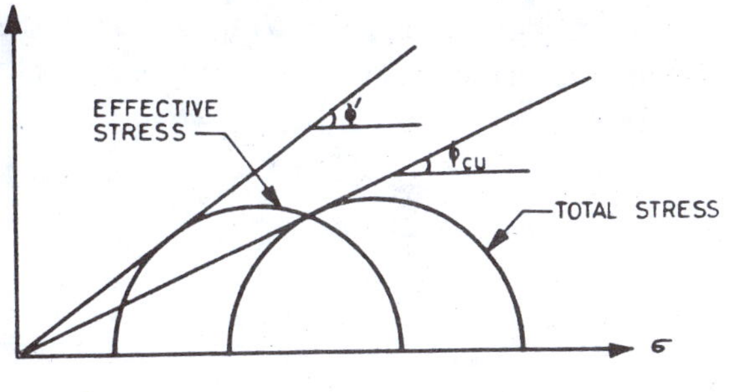

Fig. 13.19.

The angle of shearing resistance ϕ_{cu} obtained from the total stress envelopes is also known as apparent angle of shearing resistance.

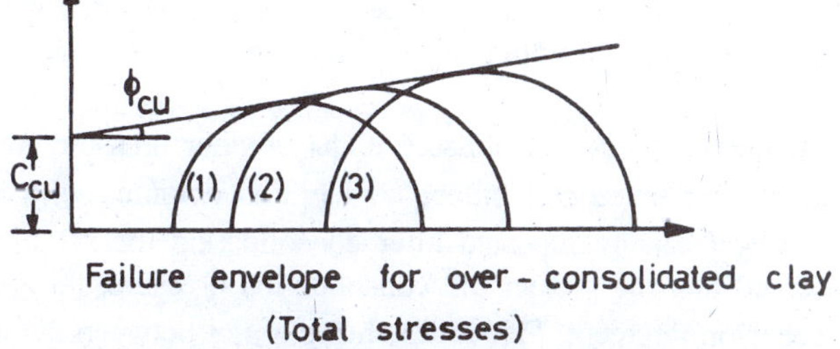

Fig. 13.20. Strength Envelope.

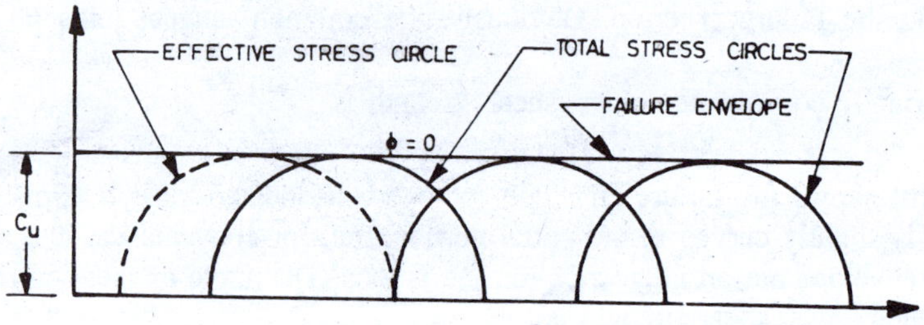

Fig. 13.21.

Fig. 13.21 shows the failure envelope in terms of total stress obtained from an *unconsolidated-undrained test* on a normally consolidated clay. The failure envelope is horizontal ($\phi = 0$), and has a cohesion intercept of c_u. In this case, shear strength is $s = c_u$. The failure envelope for an over-consolidated clay is also horizontal, but the value of c_u will be more, depending upon the degree of overconsolidation.

For an unconsolidated-undrained test, the failure envelope cannot be drawn in terms of effective stresses. In all the tests conducted at different confining pressures, the effective stress remains the same. This is due to the fact that an increase in confining pressure results in an equal increase in pore water pressure for a saturated soil under undrained conditions. Thus only one Mohr circle (shown dotted) in terms of effective stresses is obtained from all the three tests. It may be noted that the deviator stress at failure is the same for all specimens.

13.20 EFFECT OF CONSOLIDATION PRESSURE ON UNDRAINED STRENGTH

As discussed in the preceding section, the shear strength of a cohesive soil under undrained conditions depends upon the consolidation pressure. If a remoulded specimen of saturated clay is first consolidated under a confining pressure, say $(\sigma_c)_1$, and then sheared under undrained conditions, with different confining pressures, the deviator stress is independent of the cell pressure. The failure envelope is horizontal and the cohesion intercept is c_{u1} [Fig. 13.22 (*a*)].

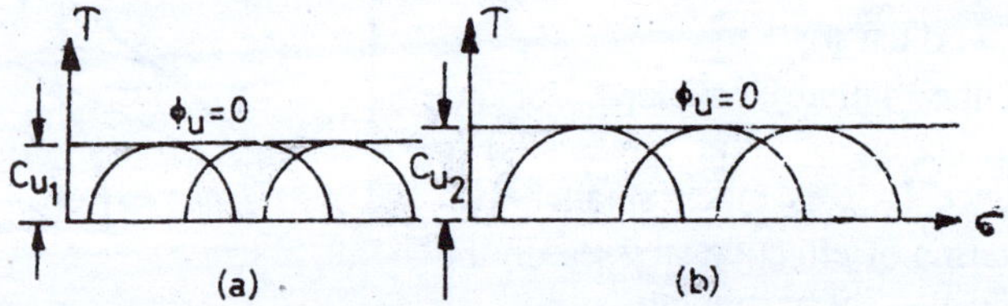

Fig. 13.22. Effect of Consolidation Pressure on C_u

The reader should carefully note the difference between the conventional consolidated-undrained test discussed earlier and the test described in this section. In the former, each specimen is consolidated under a certain pressure, equal to confining pressure, and then sheared under undrained conditions with the same cell pressure. Thus, only one Mohr circle at failure is obtained for one confining pressure. In the latter, a set of specimens are consolidated under the same confining pressure and then sheared with different cell pressures. Thus, a number of Mohr circles at failure are obtained in which the deviator stress at failure is the same for all specimens.

If another set of identical specimens is consolidated under another pressure, say $(\sigma_c)_2$, and then sheared under different cell pressure, another horizontal failure envelope is obtained with the cohesion intercept c_{u2} [Fig. 13.22 (*b*)]. Likewise, the test can be repeated after consolidating the set of specimens under another pressure, say $(\sigma_c)_3$. It is observed that the greater the consolidation pressure, the greater the undrained shear strength represented by the cohesion intercept. Fig. 13.23 shows a plot between the consolidation pressure and the undrained strength. The plot is a straight line for normally consolidated clays.

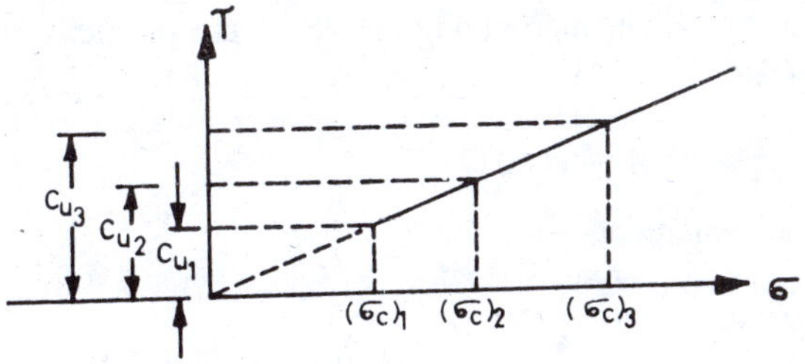

Fig. 13.23. C_u – σ Plot for Normally consolidated Clay.

For over-consolidated clays, there is a discontinuity in the plot between c_u and σ at a pressure equal to the preconsolidation pressure ($\overline{\sigma}_c$), as shown in Fig. 13.24.

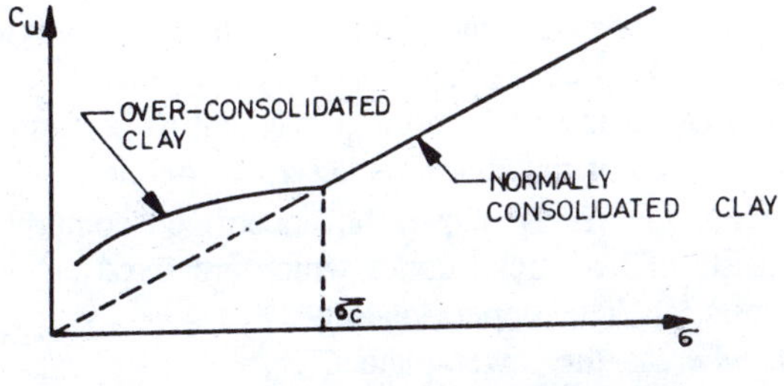

Fig. 13.24. C_u – σ Plot for Over-consolidated Clay

13.21. RELATIONSHIP BETWEEN UNDRAINED SHEAR STRENGTH AND EFFECTIVE OVERBURDEN PRESSURE

A relationship can be obtained between the undrained strength (c_u) and the effective overburden pressure $\overline{\sigma}_3$ for a normally consolidated natural deposit. Fig. 13.25 shows the Mohr circle in terms of total stresses obtained from a conventional *CU* test conducted at a confining pressure of $\overline{\sigma}_3$ on a normally consolidated clay.

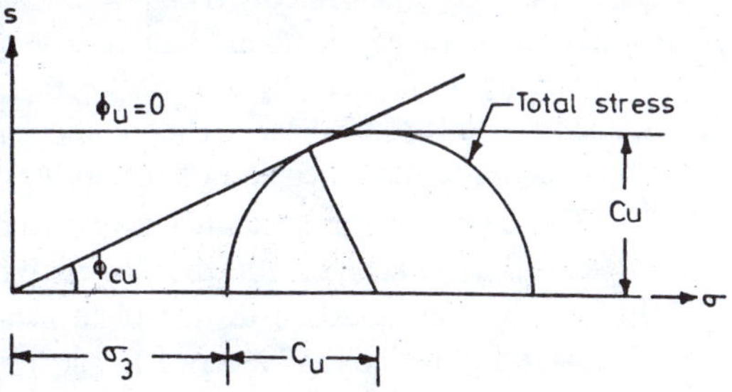

Fig. 13.25. Relationship between C_u, σ_3 and ϕ_{cu}.

By geometry, $$\frac{c_u}{\sigma_3 + c_u} = \sin \phi_{cu}$$

or $$\frac{\sigma_3 + c_u}{c_u} = \frac{1}{\sin \phi_{cu}}$$

or $$\frac{\sigma_3}{c_u} = \frac{1}{\phi_{cu}} - 1 = \frac{1 - \sin \phi_{cu}}{\sin \phi_{cu}}$$

or $$\frac{c_u}{\sigma_3} = \frac{\sin \phi_{cu}}{1 - \sin \phi_{cu}}$$

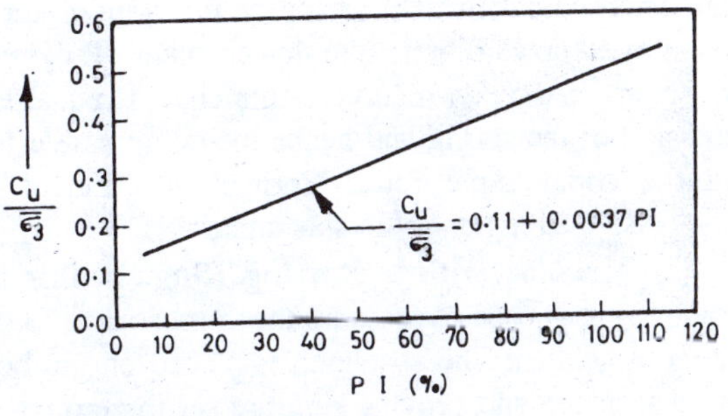

Fig. 13.26. Plot Between $C_u/\overline{\sigma}_3$ and PI.

The ratio c_u/σ_3 is a constant for a given clay. Skempton suggested that a similar constant ratio exists between the undrained shear strength of normally consolidated natural deposits and the effective overburden pressure. It has been established that the ratio $(c_u/\overline{\sigma})$ is constant provided the plasticity index

(PI) of the soil remains constant (Fig. 13.26). An approximate value of the undrained shear strength of a normally consolidated deposit can be obtained from Fig. 13.26, if the plasticity index has been determined. The relationship is expressed as (Skempton, 1957).

$$\frac{c_u}{\bar{\sigma}} = 0.11 + 0.0037\, PI$$

where c_u = undrained cohesion intercept,

$\bar{\sigma}$ = effective over-burden pressure

PI = plasticity index (%)

The value of the ratio ($c_u / \bar{\sigma}$) determined in a consolidated-undrained test on undisturbed samples is generally greater than actual value because of anisotropic consolidation in the field. The actual value is best determined by in-situ shear vane test, as explained later.

13.22. UNCONFINED COMPRESSION TEST

The unconfined compression test is a special form of a triaxial test in which the confining pressure is zero. The test can be conducted only on clayey soils which can stand without confinement. The test is generally performed on intact (non- fissured), saturated clay specimens. Although the test can be conducted in a triaxial test apparatus as a *U—U* test, it is more convenient to perform it in an unconfined compression testing machine. There are two types of machines, as described below.

(1) Machine with a spring. Fig. 13.27 shows the unconfined compression testing machine in which a loaded spring is used. It consists of two metal cones which are fixed on horizontal loading plates *B* and *C* supported on the vertical posts *D*. The upper loading plate *B* is fixed in position, whereas the lower plate *C* can slide on the vertical posts The soil specimen is placed between the two metal cones.

When the handle is turned, the plate *A* is lifted upward. As the plate *A* is attached to the plate *C*, the latter plate is also lifted. When the handle is turned slowly, at a speed of about half a turn per second, a compressive force acts on the specimen. Eventually, the specimen fails in shear. The compressive load is proportional to the extension of the spring.

The strain in the specimen is indicated on a chart fixed to the machine. As the lower plate *C* moves upward, the pen attached to this plate swings sideways. The lateral movement of the pen (in arc) is proportional to the strain in the specimen.

The chart plate is attached to the yoke *Y*. As the yoke moves upward when the handle is rotated, the chart plate moves upward. The pivot of the arm of the pen also moves upward with the lower plate. The vertical movement of the pen relative to the chart is equal to the extension of the spring and hence the compressive force. Thus the chart gives a plot between the deformation and the compressive force. Springs of different stiffnesses can be used depending upon the expected compressive strength of the specimen.

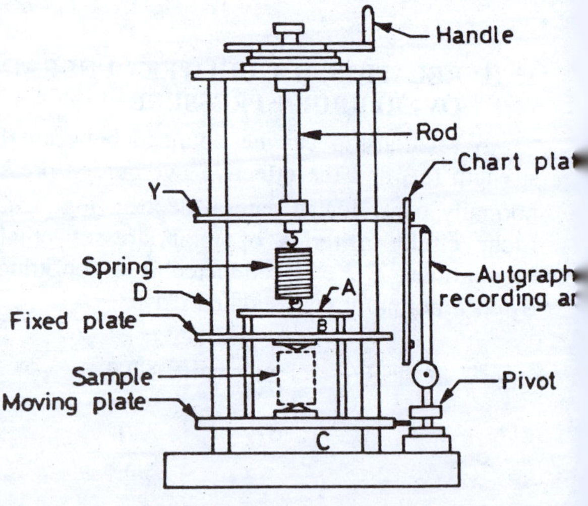

Fig. 13.27. Unconfined Compression Testing Machine (Spring Type).

(2) Machine with a Proving Ring. In this type of the unconfined compression testing machine, a proving ring is used to measure the compressive force (Fig. 13.28). There are two plates, having cone seatings for the specimen. The specimen is placed on the bottom plate so that it makes contact with the upper plate. The dial gauge and proving ring are set to zero.

The compressive load is applied to the specimen by turning the handle. As the handle is turned, the upper plate moves downward and causes compression. (In some machines, the upper plate is fixed and the compressive load is applied by raising the lower plate). The handle is turned gradually so as to produce an axial strain of 1/2% to 2% per minute. The shearing is continued till the specimen fails or till 20% of the axial strain occurs, whichever is earlier.

The compressive force is determined from the proving ring reading, and the axial strain is found from the dial gauge reading.

Presentation of Results. In an unconfined compression test, the minor principal stress (σ_3) is zero. The major principal stress (σ_1) is equal to the deviator stress, and is found from Eq. 13.21.

$$\sigma_1 = P/A$$

where P = axial load,

and A = area of cross-section.

The axial stress at which the specimen fails is known as the unconfined compressive strength (q_u). The stress- strain curve can be plotted between the axial stress and the axial strain at different stages before failure.

While calculating the axial stress, the area of cross-section of the specimen at that axial strain should be used. The corrected area can be obtained from Eq. 13.20 as

$$A = A_0/(1 - \varepsilon)$$

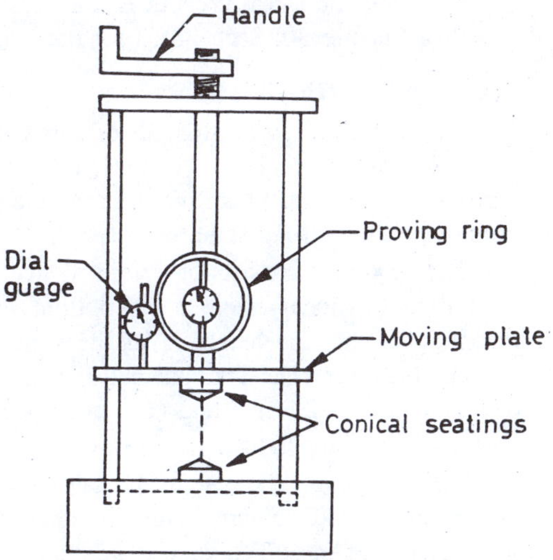

Fig. 13.28. Unconfined Compression Testing Machine (Proving Ring Type).

The Mohr circle can be drawn for stress conditions at failure. As the minor principal stress is zero, the Mohr circle passes through the origin (Fig. 13.29). The failure envelope is horizontal (ϕ_u = 0). The cohesion intercept is equal to the radius of the circle, *i.e.*

$$s = c_u = \frac{\sigma_1}{2} = \frac{q_u}{2} \quad \text{...(13.25)}$$

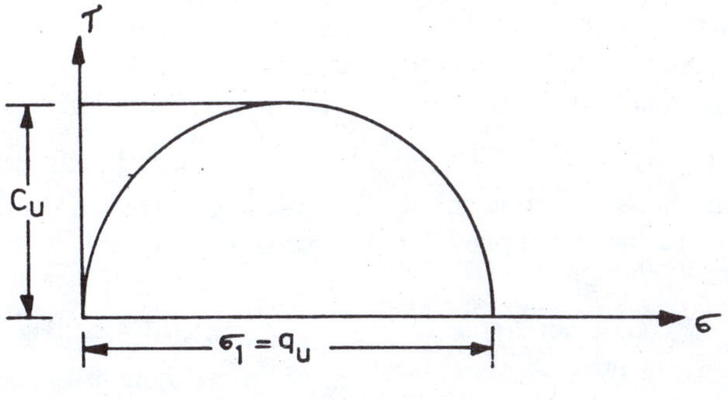

Fig. 13.29. Mohr Circle for Unconfined Compression Test.

Merits and Demerits of the test

Merits

(1) The test is convenient, simple and quick.

(2) It is ideally suited for measuring the unconsolidated-undrained shear strength of intact, saturated clays.

(3) The sensitivity of the soil may be easily determined by conducting the test on an undisturbed sample and then on the remoulded sample.

Demerits

(1) The test cannot be conducted on fissured clays.

(2) The test may be misleading for soils for which the angle of shearing resistance is not zero. For such soils, the shear strength is not equal to half the compressive strength.

(See Chapter 30, Sect. 30.17 for the laboratory experiment).

13.23. VANE SHEAR TEST

The undrained shear strength of soft clays can be determined in a laboratory by a vane shear test. The test can also be conducted in the field on the soil at the bottom of a bore hole. The field test can be performed even without drilling a bore hole by direct penetration of the vane from the ground surface if it is provided with a strong shoe to protect it.

The apparatus consists of a vertical steel rod having four thin stainless steel blades (vanes) fixed at its bottom end. IS : 2720—XXX—1980 recommends that the height H of the vane should be equal to twice the overall diameter D. The diameter and the length of the rod are recommended as 2.5 mm and 60 mm respectively. Fig. 13.30 (*a*) shows a vane shear test apparatus.

For conducting the test in the laboratory, a specimen of the size 38 mm diameter and 75 mm height is taken in a container which is fixed securely to the base. The vane is gradually lowered into the specimen till the top of the vane is at a depth of 10 to 20 mm below the top of the specimen. The readings of the strain indicator and torque indicator are taken.

Torque is applied gradually to the upper end of the rod at the rate of about 6° per minute (*i.e.* 0.1° per second). The torque acting on the specimen is indicated by a pointer fixed to the spring. The torque is continued till the soil fails in shear. The shear strength of the soil is determined using the formula derived below.

Derivation of Formula. In the deviation of the formula, it is assumed that the shear strength (s) of the soil is constant on the cylindrical sheared surface and at the top and bottom faces of the sheared cylinder. The torque applied (T) must be equal to the sum of the resisting torque at the sides (T_1) and that at the top and bottom (T_2). Thus,

$$T = T_1 + T_2 \qquad \ldots(a)$$

The resisting torque on the sides is equal to the resisting force developed on the cylindrical surface multiplied by the radial distance. Thus,

$$T_1 = (s\pi DH) \times D/2 \qquad \ldots(b)$$

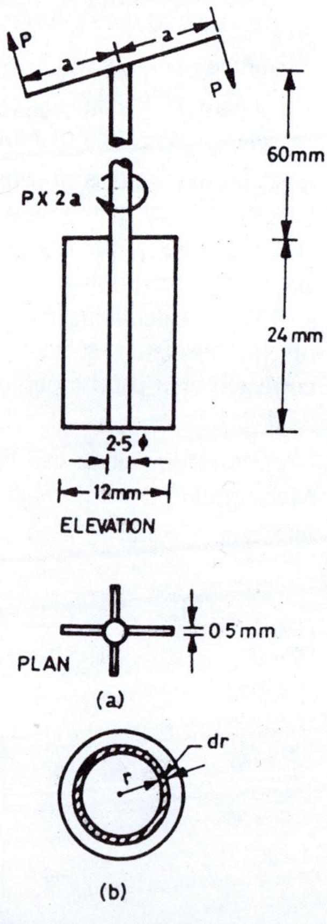

Fig. 13.30. Vane Shear Test.

The resisting torque T_2 due to the resisting forces at the top and bottom of the sheared cylinder can be determined by the integration of the torque developed on a circular ring of radius r and width dr [Fig. 13.30 (*b*)]. Thus.

$$T_2 = 2\int_0^{D/2} [s\,(2\pi r)\,dr]\,r = 4\pi s\left[\frac{r^3}{3}\right]_0^{D/2}$$

or

$$T_2 = \pi s\,\frac{D^3}{6} \qquad \ldots(c)$$

From Eqs. (*a*), (*b*) and (*c*), $T = \pi\, s\,[D^2 H/2 + D^3/6]$

or
$$s = \frac{T}{\pi (D^2 H/2 + D^3/6)} \quad \ldots(13.27)$$

For example, if D = 1.2 cm, and H = 2.4 cm, $s = 0.158\, T$

where T is in N-cm and s in N/cm^2.

Eq. 13.27 is modified if the top of the vane is above the soil surface and the depth of the vane inside the sample is H_1. In such a case.

$$s = \frac{T}{\pi (D^2 H_1/2 + D^3/12)} \quad \ldots(13.28)$$

The shear strength of the soil under undrained conditions is equal to the apparent cohesion c_u.

The vane shear test can be used to determine the sensitivity of the soil. After the initial test, the vane is rotated rapidly through several revolutions such that the soil becomes remoulded. The test is repeated on the remoulded soils and the shear strength in remoulded state is determined. Thus,

$$\text{Sensitivity } (S_t) = \frac{(s) \text{ undisturbed}}{(s) \text{ remoulded}}$$

Merits and Demerits of Shear Vane Test

Merits.

(1) The test is simple and quick.

(2) It is ideally suited for the determination of the in-situ undrained shear strength of non-fissured, fully saturated clay.

(3) The test can be conveniently used to determine the sensitivity of the soil.

Demerits.

(1) The test cannot be conducted on the fissured clay or the clay containing sand or silt laminations.

(2) The test does not give accurate results when the failure envelope is not horizontal.

13.24. PORE PRESSURE PARAMETERS

A knowledge of the pore water pressure is essential for the determination of effective stresses from the total stresses. The pore water pressure is usually measured in the field by installing piezometers. However, in some cases, it becomes difficult and impractical to install the piezometers and measure the pore water pressure directly in the field. For such cases, a theoretical method for the determination of the pore water pressure is useful. Skempton gave the pore pressure parameters which express the response of pore pressure due to changes in the total stresses under undrained conditions. These parameters are used to predict pore water pressure in the field under similar conditions. The expressions for pore pressure parameters are derived separately for isotropic consolidation, for deviatoric stress and for the combined effect.

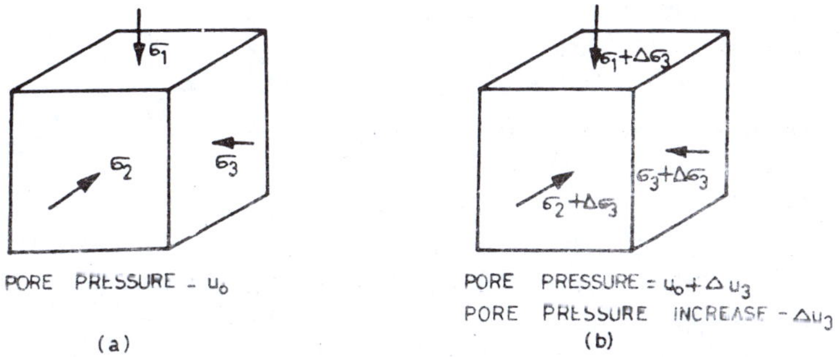

Fig. 13.31. Pore Pressure Under Isotropic Consolidation.

(1) Pore Pressure Under Isotropic Consolidations. Let us consider a small element of a saturated soil mass which is in equilibrium under three principal stresses σ_1, σ_2 and σ_3 [Fig. 13.3. (*a*)]. Let the initial pore pressure be u_0. When the element is subjected to an equal increase $\Delta\sigma_3$ in all the three directions, let the increase in the pore pressure be Δu_3 [Fig. 13.31 (b)]. Consequently, the increase in the effective stress in each direction would be equal to $(\Delta\sigma_3 - \Delta u_3)$.

Let the initial volume be V_0 and the porosity be n. As the soil is saturated, the volume of water would be equal to nV_0. If C_s is the coefficient of compressibility of the soil skeleton, the reduction in the volume of the soil mass due to an increase in the effective stress $(\Delta\sigma_3 - \Delta u_3)$ is given by

$$\Delta V_0 = C_s V_0(\Delta \sigma_3 - \Delta u_3) \qquad ...(a)$$

If C_v is the coefficient of volume compressibility of pore fluid under isotropic conditions, the reduction in the volume of voids is given by

$$\Delta V_v = C_v (n V_0) \Delta u_3 \qquad ...(b)$$

Assuming that the solid particles are incompressible, the reduction in the volume of the soil mass is equal to the reduction in the volume of voids. Therefore, from Eqs. (*a*) and (*b*),

$$C_s V_0 (\Delta \sigma_3 - \Delta u_3) = C_v (n V_0) \Delta u_3$$

or $$\Delta u_3 [nC_v + C_s]V_0 = C_s V_0 \Delta \sigma_3$$

or $$\Delta u_3 = \frac{C_s \Delta \sigma_3}{nC_v + C_s}$$

or $$\Delta u_3 = B \Delta \sigma_3 \qquad ...(13.30)$$

where B is known as the pore pressure parameter for the isotropic consolidation, and is given by

$$B = \frac{1}{1 + n (C_v/C_s)} \qquad ...(13.31)$$

In a fully saturated soil, the compressibility of the pore water (C_v) is negligible compared with the compressibility of the soil mass (C_s). Therefore, the ratio (C_v/V_s) tends to zero and the coefficient B becomes equal to unity. In other words, the change in pore water pressure is equal to the change in the total stress, *i.e.*

$$\Delta u_3 = \Delta \sigma_3$$

In a partially saturated soil, the compressibility of the air in the voids is high. The ratio (C_v/V_s) has a value greater than unity, and, therefore, the pore pressure coefficient B has a value of less than unity. The pore pressure coefficient B increases with the degree of saturation, but the variation is not linear (Fig. 13.32). There is a steep rise in the value of B at S = 80% or so.

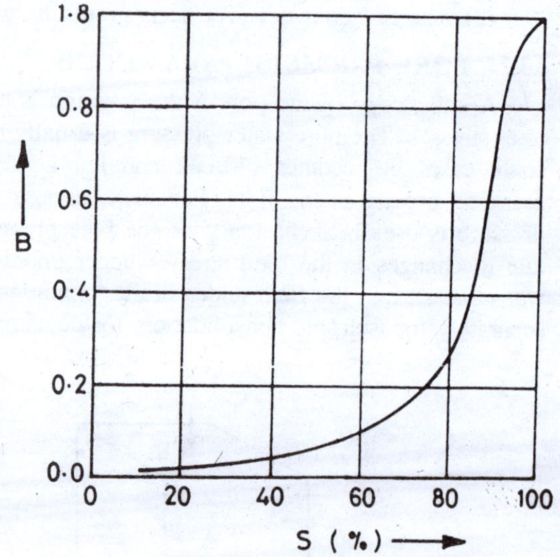

Fig. 13.32. Variation of B with S.

(2) Pore Pressure Under Deviator Stress. Let us consider the element of a saturated soil which is in equilibrium under three principal stresses σ_1, σ_2 and σ_3 (Fig. 13.33). When the element is subjected to a deviator stress $\sigma_d (= \sigma_1 - \sigma_3)$, let the increase in pore water pressure be Δ_{ud}. The changes in the effective stresses in the three directions are given by

$$\Delta \bar{\sigma}_1 = \Delta \sigma_d - \Delta u_d \qquad ...(a)$$

$$\Delta \bar{\sigma}_2 = - \Delta u_d \qquad ...(b)$$

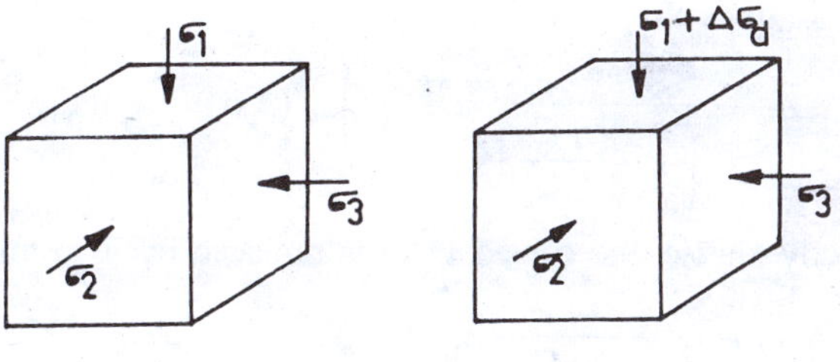

(a) (b)

Fig. 13.33. Pore Pressure Under Deviator Stress.

and $$\Delta \bar{\sigma}_3 = -\Delta u_d \qquad \ldots(c)$$

In an elastic material, the volumetric strain $\Delta V_0/V_0$ is equal to the sum of the linear strains in three directions, and is given by

$$\frac{\Delta V_0}{V_0} = \Delta \varepsilon_1 + \Delta \varepsilon_2 + \Delta \varepsilon_3$$

where ε_1 = strain in the direction-1 $= \dfrac{\Delta \sigma_1}{E} - \nu \left(\dfrac{\Delta \sigma_2}{E} + \dfrac{\Delta \sigma_3}{E} \right)$

ε_2 = strain in the direction-2 $= \dfrac{\Delta \sigma_2}{E} - \nu \left(\dfrac{\Delta \sigma_1}{E} + \dfrac{\Delta \sigma_3}{E} \right)$

and ε_3 = strain in the direction-3 $= \dfrac{\Delta \sigma_3}{E} - \nu \left(\dfrac{\Delta \sigma_1}{E} + \dfrac{\Delta \sigma_2}{E} \right)$

Therefore, $$\frac{\Delta V_0}{V_0} = \frac{1 - 2\nu}{E} (\Delta \sigma_1 + \Delta \sigma_2 + \Delta \sigma_3)$$

or $$\frac{\Delta V_0}{V_0} = \frac{3(1 - 2\nu)}{E} \left(\frac{\Delta \sigma_1 + \Delta \sigma_2 + \Delta \sigma_3}{3} \right) \qquad \ldots(d)$$

Because the soil is not a purely elastic material, Eq. (*d*) for soils is modified as

$$\frac{\Delta V_0}{V_0} = C_s \left(\frac{\Delta \sigma_1 + \Delta \sigma_2 + \Delta \sigma_3}{3} \right) \qquad \ldots(e)$$

where C_s is the coefficient of volume compressibility of the soil.

Substituting the values of $\Delta\bar{\sigma}_1$, $\Delta\bar{\sigma}_2$ and $\Delta\bar{\sigma}_3$ from Eqs. (*a*), (*b*) and (*c*) in Eq. (*e*),

$$\frac{\Delta V_0}{V_0} = C_s \left(\frac{\Delta \sigma_d - \Delta u_d - \Delta u_d - \Delta u_d}{3} \right)$$

or $$\frac{\Delta V_0}{V_0} = \frac{C_s}{3} (\Delta \sigma_d - 3 \Delta u_d) \qquad \ldots(f)$$

As in the case of isotropic consolidation, the reduction in the volume of fluid in voids is given by

$$\Delta V_v = C_v (n V_0) \Delta u_d \qquad \ldots(g)$$

As the change in the volume of the soil mass is equal to the reduction in the volume of voids,

$$\frac{C_s}{3} (\Delta \sigma_d - 3 \Delta u_d) V_0 = C_v (n V_0) \Delta u_d$$

or $$\Delta u_d (n C_v + C_s) V_0 = C_s V_0 (\Delta \sigma_d/3)$$

$$\Delta u_d = \left(\frac{C_s}{n C_v + C_s} \right) \left(\frac{\Delta \sigma_d}{3} \right)$$

or
$$\Delta u_d = \frac{1}{\left(1 + \frac{n\,C_v}{C_s}\right)} \times \left(\frac{1}{3}\right) \cdot (\Delta \sigma_d)$$

Because a soil is not perfectly elastic, the constant 1/3 is replaced by A in the above expression. Thus

$$\Delta u_d = \frac{A}{\left(1 + \frac{n\,C_v}{C_s}\right)} \times (\Delta \sigma_d)$$

Using Eq. 13.31, the above expression is written as

$$\Delta u_d = AB\,\Delta\sigma_d \quad \text{...(13.32)}$$

Eq. 13.32 can also be written as

$$\Delta u_d = \bar{A}\,\Delta\sigma_d \quad \text{...(13.33)}$$

where
$$\bar{A} = A \times B \quad \text{...(13.34)}$$

For a fully saturated soils, $\bar{A}$ is also equal to A, as B is unity.

The value of the pore pressure parameter A can be determined experimentally in a triaxial test. A saturated soil specimen is set up under a certain confining pressure and the pore water pressure is measured. A deviator stress, $\Delta\sigma_d$, is then applied to the specimen and the change in pore water pressure Δu_d is measured under undrained conditions. The value of the coefficient $\bar{A}$, can be determined using Eq. 13.33. The coefficient A is also equal to $\bar{A}$, as B is equal to unity. Although the value of the parameter A can be determined at any stage of the loading, its value at failure (A_f) is of main interest. The parameter A_f is required for the determination of the shear strength which depends upon the effective stress at failure. The value of A_f depends upon such factors as the degree of saturation and over-consolidation ratio.

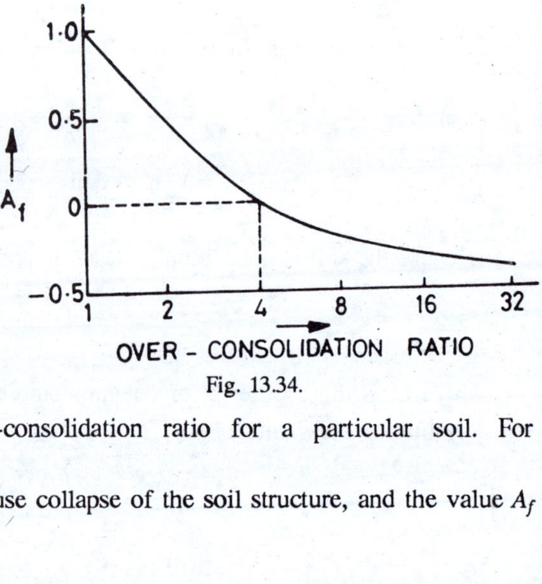

Fig. 13.34.

For a heavily overconsolidated clay, there is a tendency for the soil to increase in volume (dilate) when the deviator stress is applied. As no water can be drawn into the soil to cause an increase in volume, a negative pore pressure develops. This gives a negative value of A_f. Fig. 13.34 shows the variation of A_f with over-consolidation ratio for a particular soil. For overconsolidation ratio of greater than 4.0, A_f is negative.

For a highly sensitive clay, the deviator stress may cause collapse of the soil structure, and the value A_f may be even greater than unity.

Table 13.1 gives the values of A_f for different soils.

Table 13.1. Values of A_f

S.No.	Type of soil	A_f
1.	Sensitive clay	1.5—2.5
2.	Normally consolidated clay	0.5—1.3
3.	Lightly overconsolidateed clay	0..3—0.7
4.	Highly overconsolidated clay	–0.5—0.0
5.	Very loose fine sand	2.0—3.0
6.	Medium fine sand	0.0—1.00
7.	Dense fine sand	—0.3—0.0

(3) Pore pressure due to both Isotropic Consolidation and Deviator stress.

In a conventional triaxial test, the specimen is first isotropically consolidated under a pressure of $\Delta\sigma_3$, and then it is sheared under a deviator stress of $(\Delta\sigma_1 - \Delta\sigma_3)$. The pore pressure developed can be obtained by combining Eqs. 13.30 and 13.32, as

$$\Delta u = \Delta u_3 + \Delta u_d$$

or $$\Delta u = B\,\Delta\sigma_3 + AB\;(\Delta\sigma_1 - \Delta\sigma_3)$$

or $$\Delta u = B\,[\Delta\sigma_3 + A\;(\Delta\sigma_1 - \Delta\sigma_3)] \quad \ldots(13.35)$$

or $$\Delta u = B\,\Delta\sigma_3 + \bar{A}\;(\Delta\sigma_1 - \Delta\sigma_3) \quad \ldots(13.36)$$

where $$\bar{A} = A \times B$$

For a fully saturated soil, as $B = 1.0$,

$$\Delta u = \Delta\sigma_3 + A\;(\Delta\sigma_1 - \Delta\sigma_3) \quad \ldots(13.37)$$

The above equations can be used for determination of the pore water pressure in the field when the values of the parameters A and B have been determined.

13.25. MOHR-COULOMB FAILURE CRITERION

The Mohr-Coulomb equation was discussed in Sect. 13.6. The Mohr- Coulomb failure criterion can be written in terms of principal stresses. Fig. 13.35 shows the failure envelope of a soil. It is tangent to the

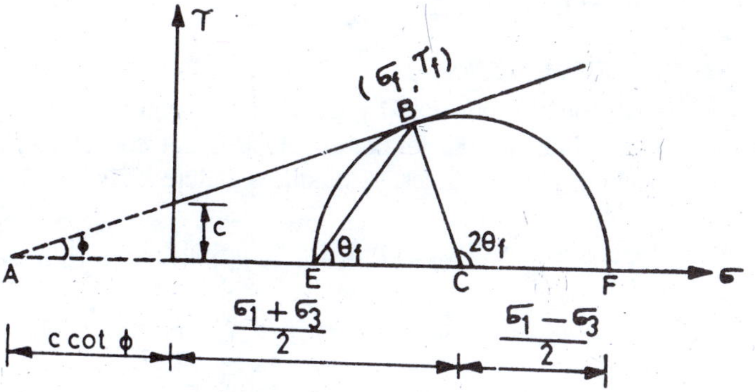

Fig. 13.35. Mohr-Coulomb Failure Criterion.

Mohr-circle at failure. The stress on the failure plane at failure is indicated by the point B on the Mohr circle. The failure plane is indicated by the line BE. The failure plane makes an angle θ_f with the major principal plane. The failure envelope has an intercept c on the τ-axis and it cuts the σ-axis (when extended back) at point A.

From the triangle ABC, $$\sin\phi = \frac{BC}{AC}$$

or $$\sin\varphi = \frac{1/2\,(\sigma_1 - \sigma_3)}{c\cot\varphi + \left(\dfrac{\sigma_1 + \sigma_3}{2}\right)} \quad \ldots[13.38(a)]$$

or $$(\sigma_1 - \sigma_3) = 2c\cos\varphi + (\sigma_1 + \sigma_3)\sin\varphi \quad \ldots[13.38(b)]$$

In terms of effective stresses, the above equation becomes

$$(\bar{\sigma}_1 - \bar{\sigma}_3) = 2c'\cos\varphi' + (\bar{\sigma}_1 + \bar{\sigma}_3)\sin\varphi' \quad \ldots(13.39)$$

where c' and ϕ' are shear strength parameters in terms of effective stresses.

Eq. 13.39 may be written in a slightly modified form as

$$\bar{\sigma}_1 - \bar{\sigma}_1\sin\varphi = \bar{\sigma}_3\sin\varphi + \bar{\sigma}_3 + 2c'\cos\varphi'$$

or $$\bar{\sigma}_1 = \bar{\sigma}_3 \frac{(1 + \sin \varphi')}{(1 - \sin \varphi')} + \frac{2c' \cos \varphi'}{1 - \sin \phi'}$$

or $$\bar{\sigma}_1 = \bar{\sigma}_3 \tan^2 (45° + \phi'/2) + 2c' \tan (45° + \phi'/2) \quad ...(13.40)$$

$\text{Tan}^2 (45° + \phi/2)$ is called *flow ratio* (N_ϕ).

Thus $$\bar{\sigma}_1 = \bar{\sigma}_3 N_\phi + 2c' \sqrt{N_\phi}$$

For cohesionless soil, $c' = 0$.

Therefore, $$\bar{\sigma}_1 = \bar{\sigma}_3 \tan^2 (45° + \phi'/2) \quad ...(13.41)$$

or $$\bar{\sigma}_1 = \bar{\sigma}_3 \left(\frac{1 + \sin \phi'}{1 - \sin \phi'} \right) \quad ...(13.42)$$

When the stresses in a soil mass satisfy the above failure criterion, the failure is imminent. The soil mass is said to have reached a state of *plastic equilibrium*. The failure criterion is extensively used for the determination of earth pressure on retaining walls and in many other applications.

The angle θ_f can be determined from the angle of shearing resistance ϕ'. As the failure depends upon the effective stresses and not the total stresses, the angle of shearing resistance affecting the orientation of the failure plane is ϕ' and not the apparent angle ϕ. From the triangle *ABC*.

$$2\theta_f = 90 + \phi'$$

or $$\theta_f = (45° + \phi'/2) \quad ...(13.43)$$

13.26. MODIFIED FAILURE ENVELOPE

Mohr's failure envelope is generally obtained by drawing a common tangent to Mohr circles at failure [Fig. 13.36 (*a*)]. This method of obtaining the failure envelope is not convenient, as it becomes difficult to draw the required tangent touching all the circles. A modified failure envelope as explained below is more convenient.

A modified failure envelope is a plot between *p* and *q* values at failure [Fig. 13.36 (*b*)], where

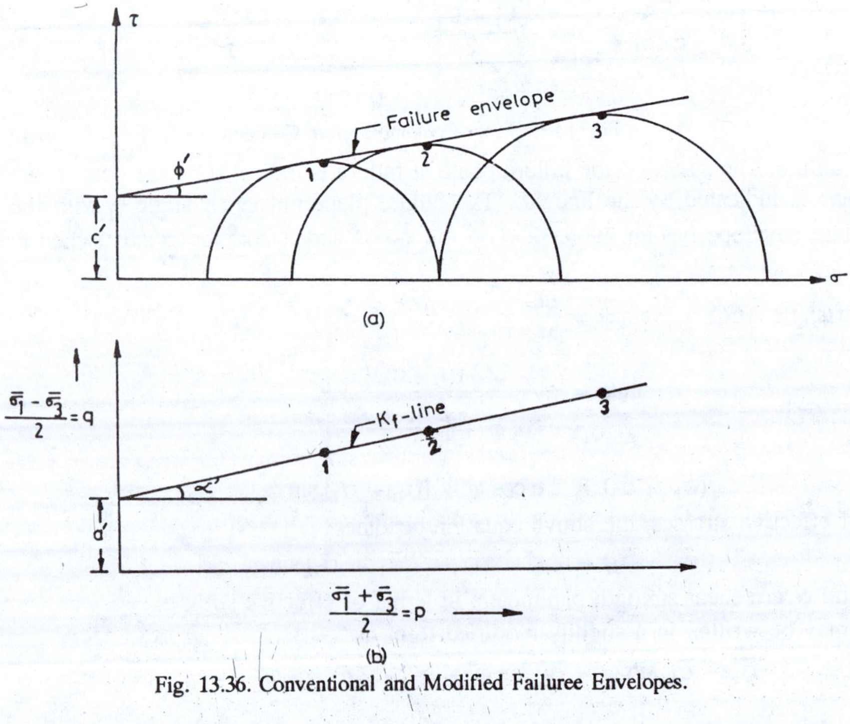

Fig. 13.36. Conventional and Modified Failuree Envelopes.

$$p = \frac{\sigma_1 + \sigma_3}{2} \quad \text{...[13.44 (a)]}$$

$$q = \frac{\sigma_1 - \sigma_3}{2} \quad \text{...[13.44 (b)]}$$

As discussed in Sect. 13.6, the coordinates of the top point of the Mohr circle corresponding to the maximum shear stress are $(\sigma_1 + \sigma_2)/2$ and $(\sigma_1 - \sigma_3)/2$ and are, therefore, equal to p and q.

In Fig. 13.36 (*a*), the points 1, 2 and 3 give the maximum shear stresses reached in the three tests at the time of failure. These points are transferred to p-q plot in Fig. 3.36 (*b*), and a line is drawn through these points. The line makes an angle α' with the p-axis and has an intercept a' on the q-axis. This line is known as the *modified failure envelope*, and has the following equation in terms of effective stresses.

$$q = p \tan \alpha' + a'$$

or $$\left(\frac{\bar{\sigma}_1 - \bar{\sigma}_3}{2}\right) = \left(\frac{\bar{\sigma}_1 + \bar{\sigma}_3}{2}\right) \tan \alpha' + a' \quad \text{...(13.45)}$$

or $$(\bar{\sigma}_1 - \bar{\sigma}_3) = (\bar{\sigma}_1 + \bar{\sigma}_3) \tan \alpha' + 2a' \quad \text{...(13.46)}$$

A relationship between the shear strength parameters c' and ϕ' and the parameters of the modified envelope α' and a' can be obtained by comparing Eqs. 13.39 and 13.46,

$$\sin \varphi' = \tan \alpha' \quad \text{...(13.47)}$$

and $$c' \cos \varphi' = a' \quad \text{or} \quad c' = a' \sec \phi' \quad \text{...(13.48)}$$

The values of parameters c' and ϕ' are obtained from the intercept a' and the slope α', using above equations.

It may be noted that points 1, 2, 3, etc. represent the maximum shear stresses which are greater than the shear stresses on the failure plane. The Mohr envelope gives the shear stresses on the failure planes, which are represented by the points of tangency; whereas the modified failure envelope joins the points of maximum shear stresses at the time of failure.

The main advantage of the modified failure envelope is that the stress conditions at failure are represented by one point instead of a Mohr circle. As the averaging of scattered points is easier than drawing a common tangent to a number of circles, it is more convenient than the Mohr-Coulomb failure envelope. Despite this, Mohr-Coulomb failure envelope is used more commonly than the modified failure envelope. A Mohr envelope is more popular because it not only gives the shear strength parameters, but also the principal stresses at the time of failure and directions of the failure plane. It gives a better insight into the whole phenomenon.

The modified failure envelope is also known as the K_f-line. It can be drawn either in terms of effective stresses or in terms of total stresses.

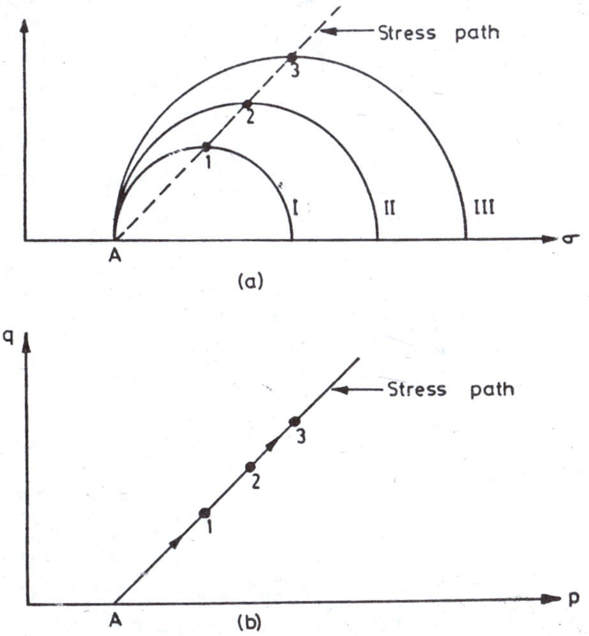

Fig. 13.37. Stress Path.

13.27. STRESS PATH

A stress path is a curve which shows the changes in stresses as the load acting on the soil specimen changes. Lambe's stress path is a commonly used stress path. It is a line drawn through the points representing the maximum shear stresses acting on the specimen as the load is changed. In other words, it is the locus of points of maximum shear stresses experienced by a specimen as the load changes take place.

Fig. 13.37 (*a*) shows the Mohr circle I, II, III which represent the stress conditions of a specimen as the load on the specimen is increased such that the minor principal stress remains constant. The line joining points 1, 2, 3 at the tops of the Mohr circles is the stress path. It is more convenient to draw the stress path on a *p-q* plot, as shown in Fig. 13.37 (*b*). There is no need to draw the complete stress circles. Only the stress points 1, 2, 3, etc. corresponding to maximum shear stresses are plotted. The direction of arrow on the stress path indicates the direction of the stress changes.

There are basically three types of stress paths.

(1) **Effective Stress Path (ESP).** It is plotted between effective stresses $(\bar{\sigma}_1 + \bar{\sigma}_3)/2$ and $(\bar{\sigma}_1 - \bar{\sigma}_3)/2$

(2) **Total Stress Path (TSP).** It is plotted between total stresses $(\sigma_1 + \sigma_3)/2$ and $(\sigma_1 - \sigma_3)/2$.

(3) **Total Stress minus static pore pressure path (TSSP).** It is a plot between $[(\sigma_1 + \sigma_3)/2 - u_s]$ and $(\sigma_1 - \sigma_3)/2$, where u_s is the static pore water pressure.

It may be noted that the vertical axis for all the stress paths is the same, as

$$\frac{\sigma_1 - \sigma_3}{2} = \frac{\bar{\sigma}_1 - \bar{\sigma}_3}{2} = q$$

If the principal planes are horizontal and vertical, the vertical stress (σ_v) and horizontal, stress (σ_h) are the principal stresses. In that case, the above three basic stress paths can be drawn between

(1) $\dfrac{\bar{\sigma}_v + \bar{\sigma}_h}{2}, \dfrac{\bar{\sigma}_v - \bar{\sigma}_h}{2}$; (2) $\dfrac{\sigma_v + \sigma_h}{2}, \dfrac{\sigma_v - \sigma_h}{2}$

(3) $\left(\dfrac{\sigma_v + \sigma_h}{2} - u_s\right)$ and $\dfrac{\sigma_v - \sigma_h}{2}$, respectively.

In the normal triaxial test, as the static pore water pressure (u_s) is zero, the stress paths (2) and (3) coincide. However, if a back pressure is used in testing, the two stress paths will be different. In field, the static pore water pressure depends upon the position of the water table with respect to the point.

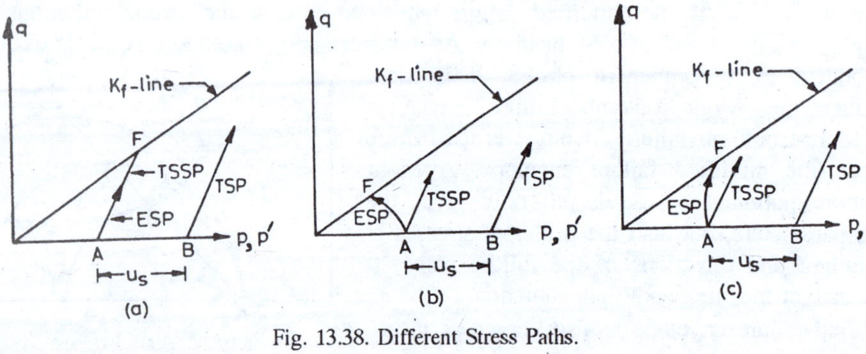

Fig. 13.38. Different Stress Paths.

Fig. 13.38 (*a*) shows the stress paths for a drained test. The point *A* corresponds to the stress condition when only the confining pressure acts on the specimen. (As $\sigma_1 = \sigma_3, p = 0$). The point *F* represents the failure. Fig. 13.38 (*b*) shows the stress paths for a consolidated undrained test on a normally consolidated clay and Fig. 13.38 (*c*), on an over-consolidated clay.

Fig. 13.39 shows the effective stress paths *A* – 1 for conventional triaxial compression test (σ_h = constant) *i.e.* $\Delta \sigma_v$ = positive and $\Delta\sigma_h = 0$. The typical example is the case of a footing subjected to the vertical loads.

The stress path *A* – 2 is for the unloading case, when $\Delta\sigma_h$ = negative and $\Delta\sigma_v = 0$. The typical example is the case of active earth pressure on the retaining walls.

The stress path *A* – 4 is for the loading case in extension test when $\Delta\sigma_v = 0$ and $\Delta\sigma_h$ = positive. It represents the passive earth pressure case.

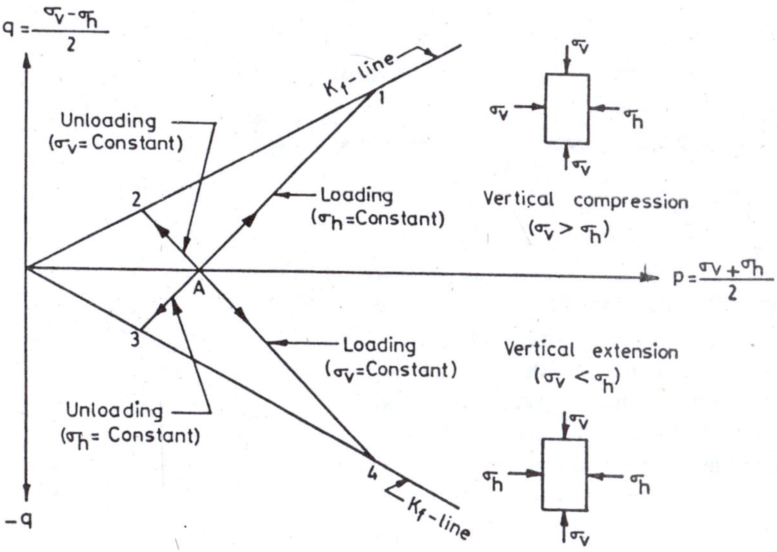

Fig. 13.39. Loading and Unloading Stress Paths.

The stress path $A - 3$ is for the unloading case in the extension test, when $\Delta\sigma_h = 0$ and $\Delta\sigma_v$ = negative. It represents the case of the foundation excavation.

The following points about the stress path are worth noting.

(1) A negative value of q indicates that the horizontal stress is greater than the vertical stress; and a positive value, that the vertical stress is greater than the horizontal.

(2) *TSP* to the right of *ESP* indicates a positive pore water pressure and *TSP* to the left of *ESP*, a negative pore water pressure.

(3) *TSSP* to the right of *ESP* indicates a positive excess pore pressure and *TSSP* to the left of *ESP*, a negative excess pressure. When both coincide, excess pore pressure is zero.

(4) Lines from any location on a stress path drawn at 45° to the axis cut the horizontal axis at a stress equal to the horizontal stress.

Note. The stress path can also be drawn using the space diagonal (Rendulic stress path), which is outside the scope of the text.

13.28. SHEAR STRENGTH OF PARTIALLY SATURATED SOILS

The effective stress in a partially saturated soil is determined using Bishop's equation (see chapter 10).

$$\bar{\sigma} = \sigma - u_a - \chi(u_w - u_a) \qquad ...(13.45)$$

The parameter χ depends upon the degree of saturation. As it is difficult to determine accurately the effective stresses in partially saturated soils, it is common practice to draw the failure envelope in terms of total stresses.

Unconsolidated-undrained triaxial tests are conducted on a number of samples with a given degree of initial saturation (S). The Mohr circles in terms of total stresses are drawn at failure and the failure envelope obtained (Fig. 13.40). As an increase in confining pressure causes an increase in the solubility of air in voids, the degree of saturation increases, and the failure envelope tends to the horizontal. For design purposes, the curved envelope is approximated as a straight line.

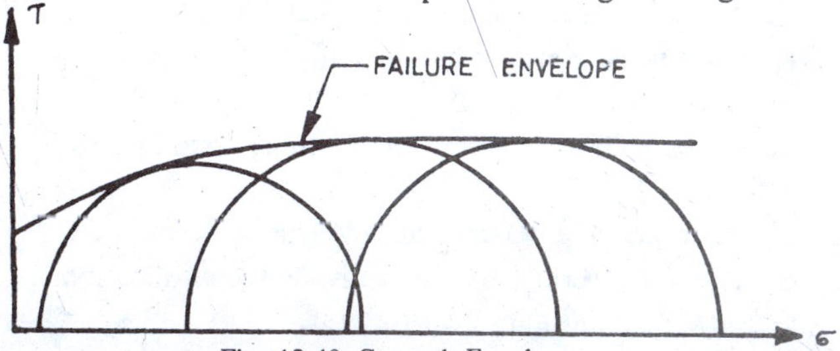

Fig. 13.40. Strength Envelope.

$$s = c_{us} + \sigma \tan \phi_{us} \quad ...(13.50)$$

where s = shear strength, c_{us} = cohesion intercept,

ϕ_{us} = angle of shearing resistance.

As it is difficult to draw the correct failure envelope, the values obtained are necessarily approximate.

As the initial degree of saturation is increased, the shear stress to cause failure is decreased (Fig. 13.41). For a fully saturated soil, the total stress failure envelope is horizontal, as already discussed for fully saturated soils.

The tests are conducted simulating the field conditions. However, an actual soil deposit in field may not remain unsaturated if it has an access to water. For such a case, the soil sample is saturated in the laboratory and the undrained shear strength in saturated conditions should be determined.

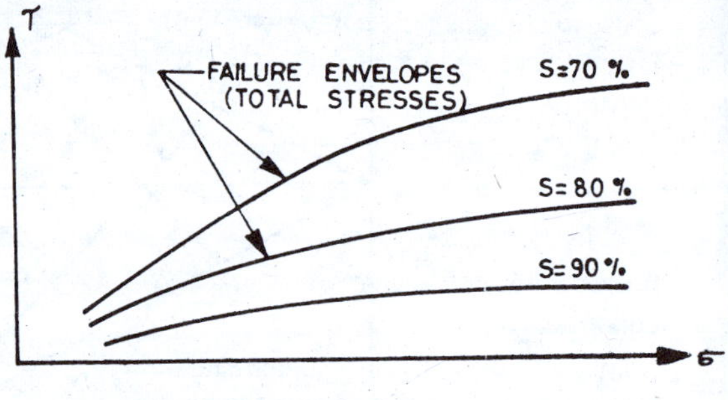

Fig. 13.41. Failure Envelopes for Different Degree of Saturation.

Partially saturated soils are encountered in the field in artificial fills or residual soils. A compacted soils is ordinarily placed at optimum water content and it is therefore partially saturated. Its behaviour depends upon the amount of air present in the pore space. If the volume of air is relatively small, the soil may get saturated under stresses. If the air content is very large, the soil remains unsaturated and undergoes large volume changes even in undrained conditions. The shear strength of partially saturated soils is not yet fully understood. It is still in the active research stage.

13.29. HVORSLEV'S STRENGTH THEORY

According to Hvorslev's hypothesis, the shear strength of remoulded, saturated clays is given by

$$s = c_e + \bar{\sigma} \tan \phi_e \quad ...(13.51)$$

where c_e = true cohesion, ϕ_e true angle of internal friction, $\bar{\sigma}$ = effective stress on the failure plane at failure.

The constants c_e and ϕ_e are known as the *Hvorslev shear strength parameters.*

The true cohesion c_e depends upon the water content (and hence void ratio) of the soil at failure, whereas the true angle of internal friction ϕ_e is approximately constant for a soil. It depends upon the plasticity index of the soil. The angle ϕ_e decreases with an increase of plasticity index of the soil.

At a given water content, the true cohesion c_e is directly proportional to the equivalent consolidation pressure $\bar{\sigma}_e$, and may be written as

$$c_e = K \bar{\sigma}_e$$

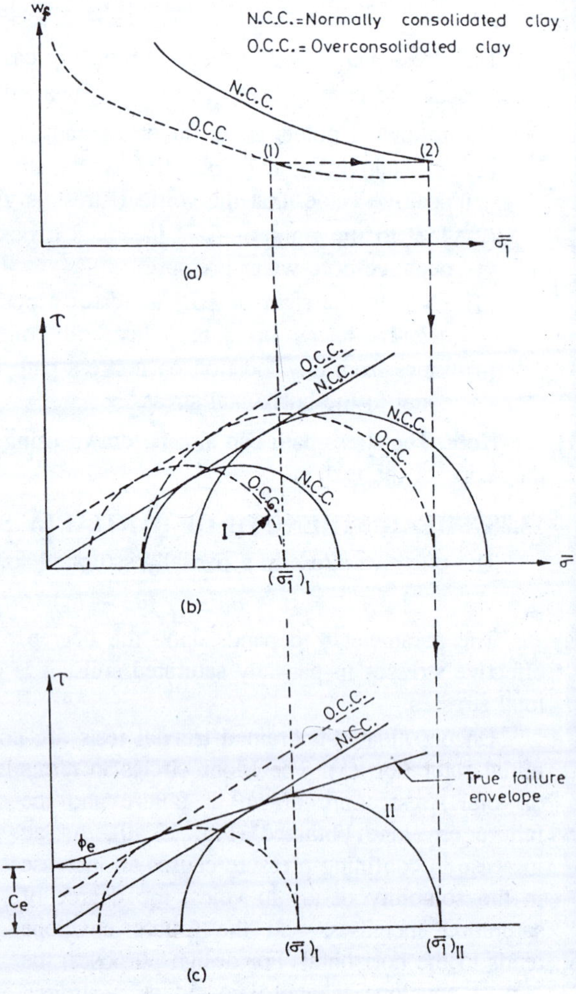

Fig. 13.42.

where K is known as Hvorslev coefficient of cohesion. Accordingly, the shear strength can be expressed as

$$s = K\bar{\sigma}_e + \bar{\sigma}\tan\varphi_e \quad ...(13.52)$$

Bishop and Henkel (1962) suggested a method for determination of c_c and ϕ_e from a series of consolidated-undrained triaxial tests on normally consolidated and over-consolidated specimens. The two failure envelopes are obtained as usual and are shown in Fig. 13.42 (*b*). The water content at failure for the two types of specimens is plotted against the maximum principal stress as shown in Fig. 13.42 (*a*).

For determination of the true failure envelope, any circle (say left circle I) for the over-consolidated clay in Fig. 13.42 (*b*) is chosen. The point corresponding to its maximum stress $(\bar{\sigma}_1)_I$ is projected upward to the $\bar{\sigma}_1 - w_f$ curve in Fig. 13.42 (*a*) to get the point 1 on the curve for over-consolidated clay. The point 1 is projected horizontally across at constant water content to obtain point 2 on the curve for the normally consolidated clay. The point 2 is projected downward to obtain the point $(\bar{\sigma}_1)_{II}$ in Fig. 13.42 (*c*). Through this point, a Mohr circle II is drawn to touch the failure envelope for normally consolidated clay. In Fig. 13.42 (*c*), the left circle I is the same as the circle I in Fig. 13.42 (*b*). The common tangent to the circle I and II in Fig. 13.42 (*c*) is the true failure envelope. The parameters c_e and ϕ_e are obtained from this envelope.

The true failure envelope has been obtained using the concept that two samples can exist at the same water content, one as normally consolidated and one as over-consolidated. As the water contents at points 1 and 2 are equal, the true cohesion is the same and the difference between the shear strength of the two samples is due to the internal friction only.

The fundamental properties of soils can be studied in terms of Hvorslev shear strength parameter. However, the theory is generally used only for research purposes. For practical use in engineering problems, the Mohr-Coulomb theory is commonly used.

13.30. LIQUEFACTION OF SANDS

As discussed earlier, the shear strength of sandy soils is given by the Mohr-Coulomb equation (Eq. 13.13), taking the cohesion intercept as zero.

Thus $$s = \bar{\sigma}\tan\phi' \quad ....(13.53)$$

If the sand deposit is at a depth of z below the ground and the water table is at the ground surface, the effective stress is given by (see Chapter 10),

$$\bar{\sigma} = \gamma_{sat}\, z - \gamma_w z = \gamma' z$$

Therefore, $$s = \gamma' z \tan\phi'$$

If the sand deposit is shaken due to an earth-quake or any other oscillatory load, extra pore water pressure (u') develops, and the strength equation becomes

$$s = (\gamma' z - u')\tan\phi'$$

It can also be expressed in the term of extra pore pressure head h, where $u' = \gamma_w h$. Thus

$$s = (\gamma' z - \gamma_w h)\tan\phi' \quad ...(13.54)$$

As indicated by Eq. 13.54, the shear strength of sand decreases as the pore water increases. Ultimately, a stage is reached when the soil loses all its strength. In which case,

$$\gamma' z - \gamma_w h = 0$$

or $$\frac{h}{z} = \frac{\gamma'}{\gamma_w}$$

Expressing h/z as critical gradient,

$$i_{cr} = \frac{(G-1)\gamma_w}{1+e} \cdot \frac{1}{\gamma_w}$$

or $$i_{cr} = \frac{G-1}{1+e} \quad ...(13.55)$$

The phenomenon when the sand loses its shear strength due to oscillatory motion is known as *liquefaction of sand.* The structures resting on such soils sink. In the case of partial liquefaction, the structure may undergo excessive settlement and the complete failure may not occur.

The soils most susceptible to liquefaction are the saturated, fine and medium sands of uniform particle size. When such deposits have a void ratio greater than the critical void ratio and are subjected to a sudden shearing stresses, these decrease in volume and the pore pressure u' increases. The soil momentarily liquefies and behaves as a dense fluid. Extreme care shall be taken while constructing structures on such soils. If the deposits are compacted to a void ratio smaller than the critical void ratio, the chances of liquefaction are reduced. (See Chapter 32 for more details on liquefaction of sand.)

13.31. SHEAR CHARACTERISTICS OF COHESIONLESS SOILS

The shear characteristics of cohesionless soils can be summarized as given below.

The shear strength of cohesionless soils, such as sands and non- plastic silts, is mainly due to friction between particles. In dense sands, interlocking between particles also contributes significantly to the strength.

The stress-strain curve for dense sands exhibits a relatively high initial tangent modulus. The stress reaches a maximum value at its peak at a comparatively low strain and then decreases rapidly with an increasing strain and eventually becomes more or less constant, as discussed earlier. The stress-strain curve for loose sands exhibits a relatively low initial tangent modulus. At large strains, the stress becomes more or less constant.

The dense sand shows initially a volume decrease in a drained test, but as the strain increases, the volume starts increasing. The loose sand shows a volume decrease throughout.

In the case of loose sand, the specimen bulges and ultimately fails by sliding simultaneously on numerous planes. The failure is known as the *plastic failure* [Fig. 13.43 (*a*)]. In the case of dense sand, the specimen shows a clear failure plane and the failure is known as the *brittle failure* [Fig. 13.43 (*b*)].

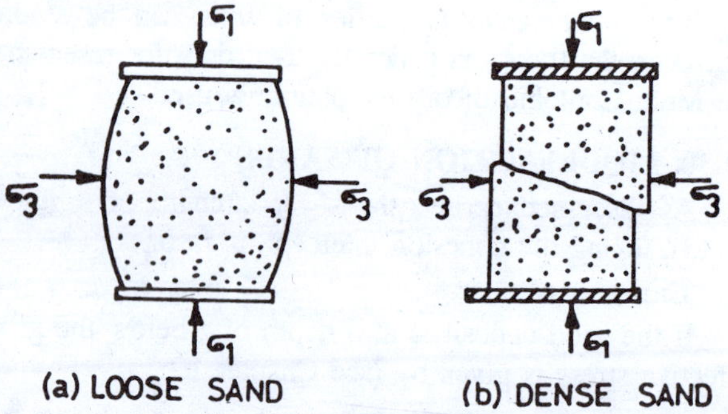

Fig. 13.43. Types of Failure.

The failure envelope for dense sand can be drawn either for the peak stresses or for the ultimate stresses. The value of the angle of shearing resistance (ϕ') for the failure envelope for peak stresses is considerably greater than that for the ultimate stresses. In the case of loose sands, as the peak stress and the ultimate stress are identical, there is only one failure envelope. The angle of shearing resistance in very loose state is approximately equal to the angle of repose. The angle of repose is the angle at which a heap of dry sand stands without any support. It has been established that air-dry sand gives approximately the same value of ϕ' as the saturated sand. As it is easier to perform tests on dry sand, tests can be performed on dry sand instead of saturated sand.

If the failure envelope is slightly non-linear, a straight line may be drawn for the given pressure range and the angle of shearing resistance is taken as the slope of this line. The cohesion intercept, if any, is usually neglected.

The angle of shearing resistance of sands in the field can be determined indirectly by conducting in-situ tests, such as the standard penetration test (SPT) as explained in chapter 17.

The factors that affect the shear strength of cohesionless soils are summarized below:

(1) Shape of particles. The shearing strength of sands with angular particles having sharp edges is greater than that with rounded particles, other parameters being identical.

(2) Gradation. A well-graded sand exhibits greater shear strength than a uniform sand.

(3) Denseness. The degree of interlocking increases with an increase in density. Consequently, the greater the denseness, the greater the strength. The value of ϕ' is related to the relative density (D_r) as $\phi' = 26° + 0.2\,D_r$. However, the ultimate value of ϕ' is not affected by denseness.

(4) Confining pressure. The shear strength increases with an increase in confining pressure. However, for the range of pressures in the common field problems, the effect of confining pressure on the angle of shearing resistance is not significant.

(5) Deviator stress. The angle ϕ' decreases under very high stresses. As the maximum deviator stress is increased from 500 to 5000 kN/m^2, the value of ϕ' decreases by about 10%. This is due to the crushing of particles.

(6) Intermediate principal stress. The intermediate principal stress affects the shear strength to a small extent. The friction angle for dense sands in the plane strain case is about 2° to 4° greater than that obtained from a standard triaxial test. However, for loose sand, there is practically no difference in the two values.

(7) Loading. The angle of shearing resistance of sand is independent of the rate of loading. The increase in the value of ϕ' from the slowest to the fastest possible rate of loading is only about 1 to 2%.

The angle of shearing resistance in loading is approximately equal to that in unloading.

(8) Vibrations and Repeated loading. Repeated loading can cause significant changes. A stress much smaller than the static failure stress if repeated a large number of times can cause a very large strain and hence the failure.

(9) Type of minerals. If the sand contains mica, it will have a large void ratio and a lower value of ϕ'. However, it makes no difference whether the sand is composed of quartz or feldspar minerals.

(10) Capillary moisture. The sand may have apparent cohesion due to capillary moisture. The apparent cohesion is destroyed as soon as the sand becomes saturated.

A person can easily walk on damp sand near the sea beach because it possesses strength due to capillary moisture. On the same sand in saturated conditions, it becomes difficult to walk as the capillary action is destroyed.

Table 13.2 gives the representative values of ϕ' for different types of cohesionless soils.

Table 13.2. Representative Values of ϕ' for Sands and Silts

S. No.	Soil	ϕ'
1.	Sand, round grains, uniform	27° to 34°
2.	Sand, angular, well-graded	33° to 45°
3.	Sandy gravels	35° to 50°
4.	Silty sand	27° to 34°
5.	Inorganic silt	27° to 35°

Note. Smaller values are for loose conditions and larger values are for dense conditions.

13.32. SHEAR CHARACTERISTICS OF COHESIVE SOILS

The shear characteristics of cohesive soils are summarized below :

The shear characteristics of a cohesive soil depend upon whether a soil is normally consolidated or over-consolidated. The stress- strain curve of an over-consolidated clay is similar to that of a dense sand and that of a normally consolidated clay is identical to that of a loose sand. However, the strain required to reach peak stress are generally greater in clay than in sand. The high strength at the peak point in an over-consolidated clay is due to structural strength; whereas in the dense sand, it is mainly due to interlocking. In over-consolidated clay, strong structural bonds develop between the particles. Loose sands tend to increase in volume at large strains whereas normally consolidated clays show no tendency to expand after a decrease in volume.

The effective stress parameters (c', ϕ') for an overconsolidated clay are determined from the failure envelope,

$$s = c' + \bar{\sigma} \tan \phi'$$

However, for a normally consolidated clay, the failure envelope passes through the origin and hence $c' = 0$.

For clay, when tested in unconsolidated-underained test, the failure envelope in terms of total stresses is horizontal ($\phi_u = 0$), and the shear strength is given by

$$s = c_u$$

For heavily over-consolidated clays, negative pore pressure develops during shear, and the undrained strength is more than the drained strength. The $\phi_u = 0$ concept, therefore, leads to the results on the unsafe side. Further, the continuity of heavily over-consolidated clays is commonly disrupted by a network of hair cracks and the average pressure is reduced. This results in a substantial decrease in its shear strength.

The general behaviour of compacted (partially saturated) clays is similar to a moderately over-consolidated clay. A total stress analysis is used for such soils.

The factors affecting the shear strength of cohesive soils may be summarised as under:

(1) Structure of clay. The clay exhibits a definite structure. Even a normally-consolidated clay exhibits a small peak due to structural strength. In case of over-consolidated clays, the structural strength is predominant.

(2) Clay content. The ultimate friction angle ϕ' of the cohesive soil depends upon the clay content. As the clay content increases, the angle decreases. Further, the difference between the ϕ' at peak and that at ultimate condition increases with an increase in clay content, because the clay particles do not reach a fully oriented face-to-face alignment at peak.

(3) Drainage conditions. As the cohesive soils have low permeability, the shear strength will depend whether it is in drained condition or in undrained conditions. The cohesive soils have very low strength just after the application of the load when undrained conditions exist.

(4) Rate of strain. In the case of normally consolidated clays, the effect of rate of strain upon the angle of shearing resistance is relatively small. The value of ϕ' may decrease by about 10% if the strain rate is reduced by a factor of 10. However, in some cases, the angle ϕ' is found to increase with a decrease in rate of strain. In the case of over-consolidated clays, some of the shear strength is always lost when the rate of strain is decreased.

(5) Intermediate principal stress. The values of c' and ϕ' are affected very little by the magnitude of the intermediate principal stress.

(6) Repeated loading. For clays tested at constant water content, the shear strength is increased due to a large number of repetitions of the stress. However, if the stress intensities are too high, the cumulative deformations may result in failure.

(7) Confining pressure. The shear strength of clays increases with an increase in the confining pressure, provided there is enough time available for the pore water pressure to dissipate.

(8) Plasticity index. The value of ϕ' decreases with an increase in plasticity index of the clay. The following relation is commonly used.

$$\sin \phi' = 0.814 - 0.234 \log_{10} I_P$$

(9) Stress history. The values of strength parameters depend upon the stress history.

(10) Disturbance. The shear strength of disturbed sample is less than that of the undisturbed samples.

Table 13.3. gives the representative values of c_u for different types of cohesive soils for undrained conditions.

Table 13.3. Representative Values of c_u for clay

S. No.	*Soil*	*Cohesion c_u (kN/m^2)*
1.	Very soft clay	< 12
2.	Soft to medium clay	12 — 25
3.	Stiff clay	50 — 100
4.	Very stiff	100 — 200
5.	Hard	> 200

13.33. CHOICE OF TEST CONDITIONS AND SHEAR PARAMETERS

Test conditions and shear parameters should be chosen to represent the field conditions as closely as possible.

In case of coarse grained soils, drainage takes place quickly during the application of the load. Tests on these soils are generally carried out under drained conditions, and therefore consolidated drained tests are more common. However, in some special cases when large masses of saturated, fine sands are subjected to quick loadings due to earthquakes or other causes, the undrained conditions are more relevant.

In case of fine-grained soils, drainage takes place very slowly. Immediately after the application of the load, undrained conditions are suitable. However, for final stability problems, drained tests are relevant, as the water content of such soils adjusts itself to the new conditions in due course of time. Consolidated-undrained tests are required for the cases when the soil gets consolidated under a certain loading and then additional load is applied; for example, in the case of earth dams, wherein the soil gets consolidated under self weight before the reservoir is filled and the water pressure causes additional stresses. Fig. 13.44 shows the shear strength envelopes obtained for a consolidated drained (*CD*), consolidated-undrained (*CU*) and unconsolidated undrained (*UU*) tests for a particular soil.

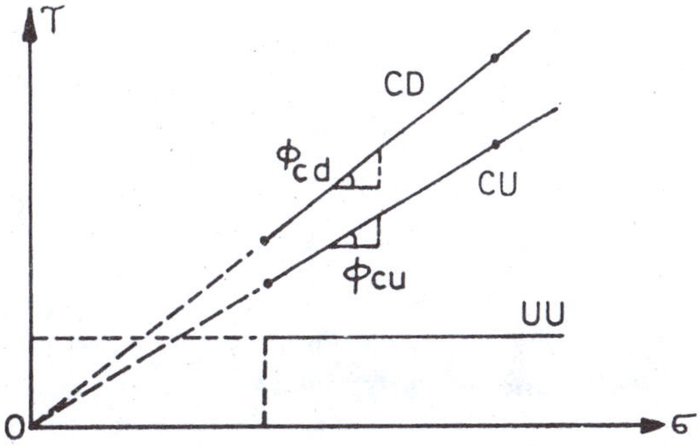

Fig. 13.44. Envelopes for Different Types of Tests.

The choice between the effective stress analysis (*ESA*) using the effective stress parameters c' and ϕ' and the total stress analysis (TSA) using the apparent parameters c and ϕ depends upon the condition whether the pore water pressure can be estimated or not. If the pore water pressure can be estimated (or measured), the effective stress analysis should be done, as it is more rational. It is based on well-established, unique functional relationship between the shear strength and the effective stress on the failure plane at failure.

In case the pore water pressure cannot be accurately estimated or measured, the total stress analysis is more convenient. However, it gives no indication of the real factor of safety. It is not known whether the analysis would give error on the side of safety or on the unsafe side. The laboratory tests are conducted such that the total stresses in the field are simulated. A tacit assumption is made that the pore water pressure that develops in the field would be the same as in the laboratory. The assumption may not be realised in practice. Further, it is not known whether the pore water pressure in the field is more than or less than in the laboratory.

It may be emphasized that shear strength parameters do not have unique values for a given soil. These parameters are meaningless unless drainage conditions, type of strength envelope, the normal stress range, etc. are mentioned. Selection of these parameters will depend upon the field conditions and the type of analysis.

ILLUSTRATIVE EXAMPLES

Illustrative Example 13.1. *A series of direct shear tests was conducted on a soil, each test was carried out till the sample failed. The following results were obtained.*

Sample No.	*Normal stress* (kN/m^2)	*Shear stresses* (kN/m^2)
1	15	18
2	30	25
3	45	32

Determine the cohesion intercept and the angle of shearing resistance.

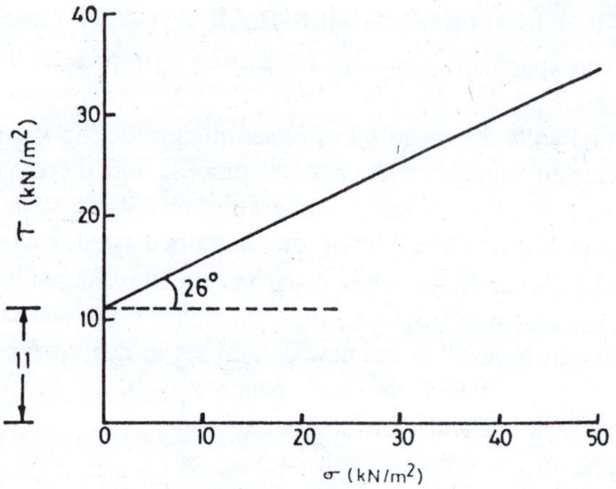

Fig. E-13.1.

Solution. Fig. Ex. 13.1 shows the plot between the shear stresses and normal stresses at failure.

From the plot, $c = 11\ \text{kN/m}^2$, $\phi = 26°$

Illustrative Example 13.2. *The following results were obtained from a series of consolidated undrained tests on a soil, in which the pore water pressure was not determined. Determine the cohesion intercept and the angle of shearing resistance.*

Sample No.	*Confining preessure (kN/m^2)*	*Deviator stress at failure (kN/m^2)*
1	100	600
2	200	750
3	300	870

Solution. The major principal stresses in the three test are 700, 950 and 1170 kN/m², respectively. Fig. Ex. 13.2 shows the Mohr circle at failure in terms of total stresses.

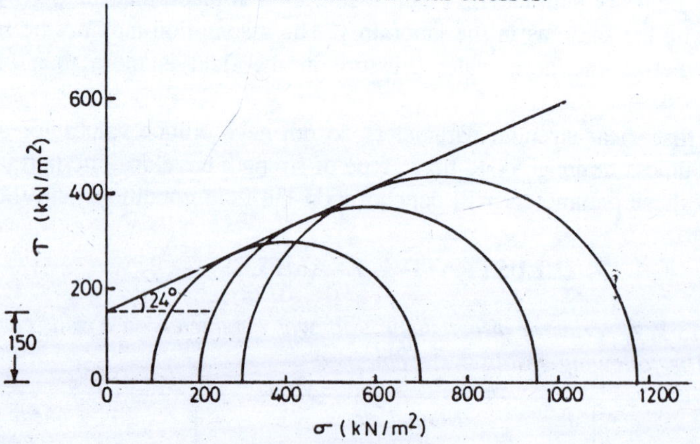

Fig. Ex. 13.2.

A common tangent is drawn to the Mohr circle.

From the plot, $c = 150\ \text{kN/m}^2$, $\phi = 24^0$.

Illustrative Example 13.3. *A series of consolidated-undrained ($C\bar{U}$) triaxial tests was conducted on an over-consolidated clay and the following results were obtained.*

Sample No.	*Cell pressure (kN/m^2)*	*Deviator stress (kN/m^2)*	*Pore-water pressure (kN/m^2)*
1	125	510	– 70
2	250	620	– 10
3	500	850	+ 120

Plot the strength envelopes in terms of total stresses and effective stresses, and hence determine the strength parameters.

If the soil was preconsolidated to a pressure of 1000 kN/m^2, plot the variation of the pore pressure parameter A_f with the over-consolidation ratio.

Solution. The table below shows the required calculations.

Sample No.	σ_3	σ_1	u	$\bar{\sigma}_3 = \sigma_3 - u$	$\bar{\sigma}_1 = \sigma_1 - u$	$A_f = u/\sigma_d$	*O.C.R.*
1	125	635	– 70	195 (= 125 + 70)	705 (= 635 + 70)	– 0.137 (= –70/510)	8.0 (= 1000/125)
2	250	870	– 10	260	880	– 0.016	4.0
3	500	1350	+ 120	380	1230	+ 0.141	2.0

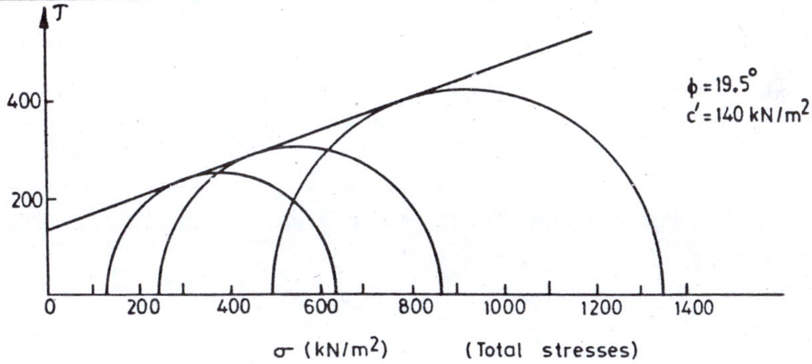

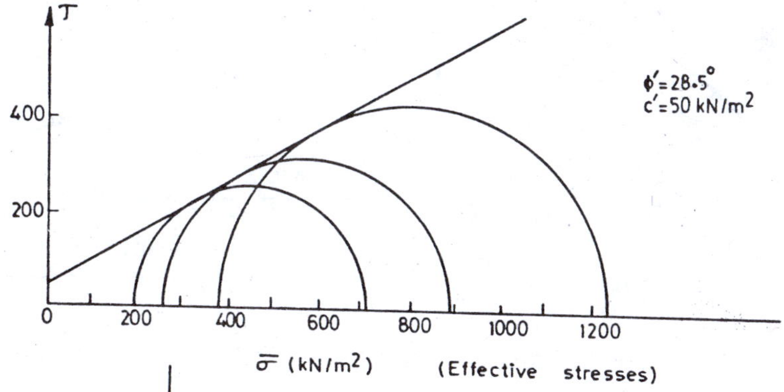

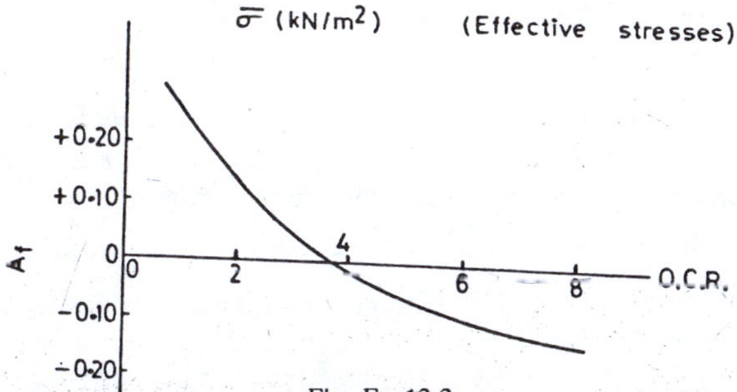

Fig. Ex-13.3.

Fig. Ex. 13.3 (*a*) shows the strength envelope in terms of total stresses. From the plot,

$$\phi = 19.5°, \; c = 140 \text{ kN/m}^2$$

Fig. Ex. 13.3 (b) shows the strength envelope in terms of effective stresses. From the plot,

$$\phi' = 28.5°, \; c' = 50 \text{ kN/m}^2$$

Fig. Ex. 13.3 (*c*) shows the plot between A_f and *O.C.R.*

Illustrative Example 13.4. *The following results were obtained from a consolidated—undrained* ($\overline{CU}$) *test on a normally consolidated clay. Plot the strength envelope in terms of total stresses and effective stresses and determine the strength parameters.*

Sample No.	*Cell pressure (kN/m²)*	*Deviator stress (kN/m²)*	*Pore water pressure (kN/m²)*
1	250	152	120
2	500	300	250
3	750	455	350

Solution. The table below shows the calculations for principal total stresses and effective stresses.

Sample No.	σ_3	σ_1	u	$\bar{\sigma}_3$	$\bar{\sigma}_1$
1	250	402	120	130	282
2	500	800	250	250	550
3	750	1205	350	400	855

Fig. Ex. 13.4 (*a*) and Fig. Ex. 13.4 (*b*) show the plot in terms of effective stresses and total stresses, respectively.

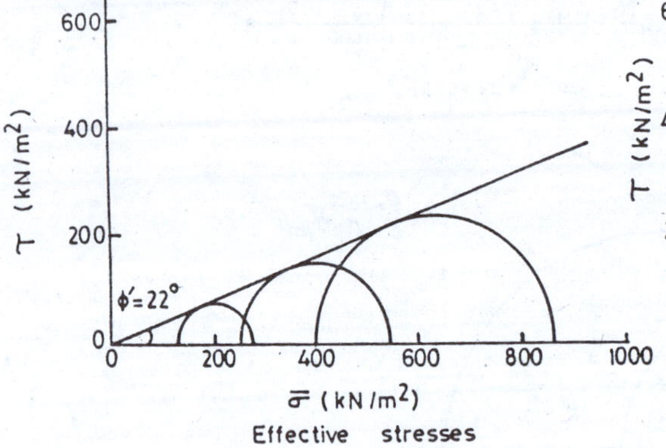

Fig. Ex. 13.4 (*a*)

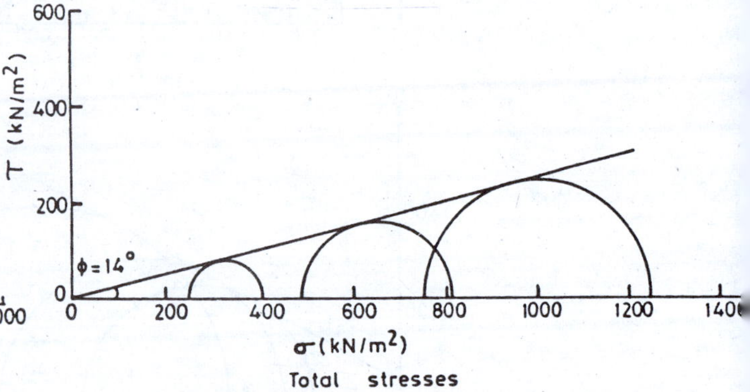

Fig. Ex. 13.4 (*b*)

From the plot in terms of total stresses, $\phi = 14°$

From the plot in terms of effective stresses, $\phi' = 22°$

Illustrative Example 13.5. *A sample of a soil failed in a triaxial test under a deviator stress of 200 kN/m² when the confining pressure was 100 kN/m². If for the same sample, the confining pressure had been 200 kN/m², what would have been the deviator stress at failure? Assume the soil has (a) c = 0 and (b) φ = 0.*

Solution. The stress σ_1 in the first test = 200 + 100 = 300.

(*a*) Fig. E 13.5 (*a*) shows the plot for $c = 0$. The envelope has been drawn tangential to the first circle. The second circle has been drawn such that it passes through $\sigma_3 = 200$ and is also tangential to the envelope. From the plot, $\phi = 30°$.

Alternative method

From Eq. 13.38 (*a*), $\sin\phi = \dfrac{(300 - 100)/2}{(300 + 100)/2} = 0.5$

In the second test, ϕ remains equal to 30°.

Therefore, from Eq. 13.38 (*a*), $\sin\phi = \dfrac{(\sigma_1 - \sigma_3)/2}{(\sigma_1 + \sigma_3)/2}$

or $0.50 = \dfrac{\sigma_1 - 200}{\sigma_1 + 200}$ or $\sigma_1 = 600$ kN/m^2

Therefore, $\sigma_1 - \sigma_3 = 600 - 200 = 400$ kN/m^2

(*b*) From the plot when $\phi' = 0$,

$\sigma_1 - \sigma_3 = 200$, as shown in Fig. Ex. 13.5 (*b*)

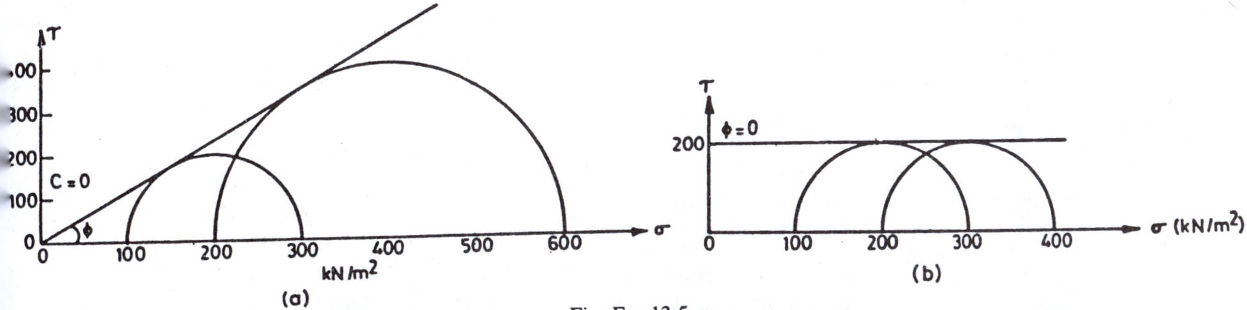

Fig. Ex. 13.5.

Illustrative Example 13.6. *A sample of dry cohesionless soil was tested in a triaxial machine. If the angle of shearing resistance was 36° and the confining pressure, 100 kN/m², determine the deviator stress at which the sample failed.*

Solution. For dry cohesionless, the cohesion intercept is zero.

From Eq. 13.38 (*a*), $\sin\phi = \dfrac{(\sigma_1 - \sigma_3)/2}{(\sigma_1 + \sigma_3)/2}$

or $\sin 36° = \dfrac{\sigma_1 - 100}{\sigma_1 + 100}$ or $\sigma_1 = 385.4$

Deviator stress at failure, $\sigma_d = 385.4 - 100 = 285.4$ kN/m^2

Illustrative Example 13.7. *The stresses on a failure plane in a drained test on a cohesionless soil are as under :*

Normal stress $(\sigma) = 100$ *kN/m²*.
Shear stress $(\tau) = 40$ *kN/m²*.

(a) Determine the angle of shearing resistance and the angle which the failure plane makes with the major principal plane.

(b) Find the major and minor principal stresses.

Solution. Fig. Ex. 13.7 shows the Coulomb failure line passing through the origin and the point *A* with coordinates (100, 40).

$$\tan\phi' = \frac{40}{100} = 0.4 \quad \text{or} \quad \phi' = 21.80°$$

From Eq. 13.43, the angle which the failure plane makes with the major principal plane,

$$\theta = 45 + \phi'/2 = 45 + 10.9 = 55.9°$$

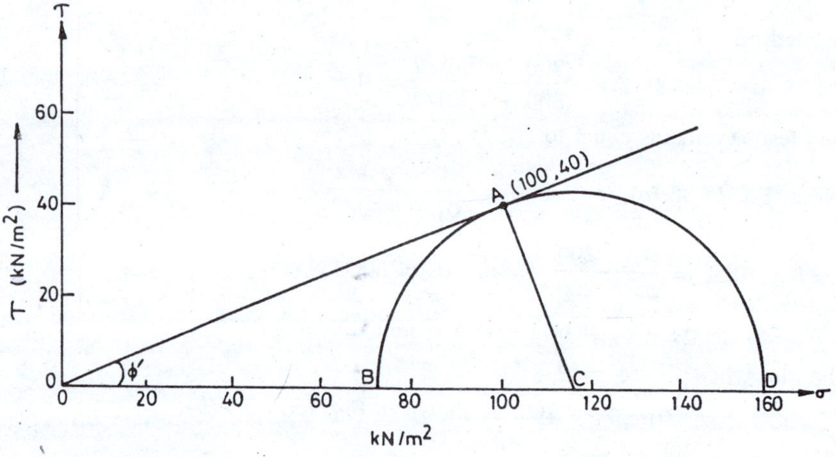

Fig. Ex. 13.7.

(*b*) The centre *C* of the Mohr circle is located by drawing a normal *AC* to line *OA* at *A*. Mohr's circle is drawn through point *A*, with its centre at *C*. The circle cuts the σ-axis at points *B* and *D*.

From the figure, $\sigma_3 = 73\ \text{kN/m}^2,\ \sigma_1 = 159\ \text{kN/m}^2$

The stresses σ_3 and σ_1 can also be obtained analytically as follows.

Length $OA = \sqrt{100^2 + 40^4} = 107.7$

$$AC = OA \tan \phi' = 107.7 \times \tan 21.8° = 43\ \text{kN/m}^2$$

$$OC = OA \sec \phi' = 107.7 \times \sec 21.8° = 116\ \text{kN/m}^2$$

$$OD = OC + AC = 116 + 43 = 159\ \text{kN/m}^2 = \sigma_1 .$$

$$OB = OC - AC = 116 - 43 = 73\ \text{kN/m}^2 = \sigma_3 .$$

Illustrative Example 13.8. *An undrained triaxial compression test was conducted on a sample of compacted clay. Pore-water pressure was measured after the application of the cell pressure and also at failure, as given below. Find the pore pressure coefficients A and B.*

(a) Consolidation stage. *Change in cell pressure* $= 0$ *to* $100\ kN/m^2$

Change in pore water pressure $= -60$ *to* $+10\ kN/m^2$

(b) Shearing stage. *Deviator stress at failure* $= 500\ kN/m^2$

Pore pressure at failure $= -70\ kN/m^2$.

Solution. From Eq. 13.30, $\Delta u_3 = B\, \Delta \sigma_3$

or $[10 - (-60)] = B\,[100 - 0]$ or $B = \mathbf{0.70}$

From Eq. 13.32, $\Delta u_d = AB\, \Delta \sigma_d$

or $(-70 - 10) = A \times 0.7\,(500)$ or $A = \mathbf{-0.23}$.

Illustrative Example 13.9. *A shear vane of 7.5 cm diameter and 11.0 cm length was used to measure the shear strength of a soft clay. If a torque of 600 N-m was required to shear the soil, calculate the shear strength.*

The vane was then rotated rapidly to cause remoulding of the soil. The torque required in the remoulded state was 200 N-m. Determine the sensitivity of the soil.

Solution. From Eq. 13.27,

$$s = \frac{T}{\pi\,(D^2 H/2 + D^3/6)} = \frac{600 \times 10^{-3}}{\pi\,[(7.5)^2 \times 11.0/2 + (7.5)^3/6] \times 10^{-6}}$$

or $s = 503\ \text{kN/m}^2$

In the remoulded state,

$$s_{rem} = \frac{200 \times 10^{-3}}{\pi [(7.5)^2 \times 11.0/2 + (7.5)^3/6] \times 10^{-6}} = 168 \text{ kN/m}^2$$

From Eq. 13.29, sensitivity $= \frac{503}{168} = 3.0$

Illustrative Example 13.10. *A series of triaxial tests was conducted on samples of a cohesionless soil and the following readings were taken for the deviator stress (kN/m²) at different strains. Draw the stress–strain curves and hence obtain the values of the secant moduli at one-half the ultimate stress.*

Sample No.	*Strain (%)* / σ_3 *(kN/m²)*	1	2	3	4	5	6	7	8	9	10
1	50	15	27	36	43	49	54	57	61	63	64
2	100	22	37	49	58	65	72	76	79	81	82
3	150	32	52	65	75	89	88	96	96	98	99

Solution. Fig. Ex. 13.10 shows the stress-strain curves for the soil at three different confining pressures. From the plot, the values of secant moduli are obtained.

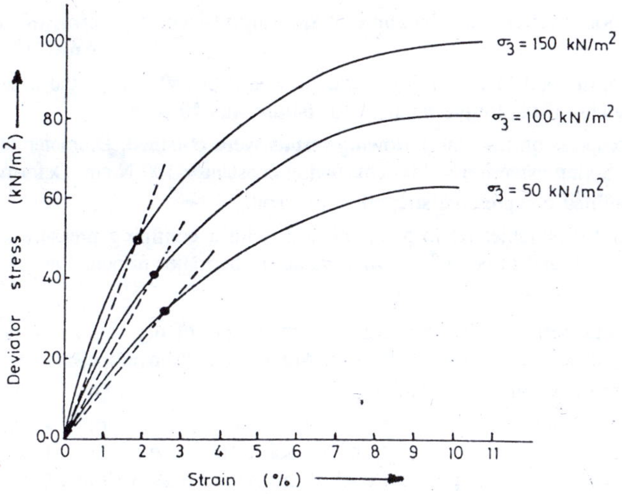

Fig. E-13.10.

$(E_s)_1$ for $\sigma_c = 50$ kN/m² is 1250 kN/m²

$(E_s)_2$ for $\sigma_c = 100$ kN/m² is 1750 kN/m²

$(E_s)_3$ for $\sigma_c = 150$ kN/m² is 2700 kN/m²

PROBLEMS

A. Numerical

13.1. The principal stresses at a point in a material are 80 kN/m² and 40 kN/m². Determine the normal, shear and resultant stress on a plane inclined at 30° to the major principal plane.
Find also, for this plane, the maximum value of obliquity. [**Ans.** 70; 17.3; 72.1 kN/m²; 19.47°]

13.2. On a failure plane in a cohesionless soil sample, the normal and shear stresses are found as 10 kN/m² and 4 kN/m². Determine the resultant stress on the plane of failure, the angle of shearing resistance and the inclination of failure plane to the major principal plane. [**Ans.** 10.8 kN/m² ; 21.8° ; 55.9°]

13.3. A consolidated-undrained triaxial test was conducted on a saturated clay. When the confining pressure was 200 kN/m^2, the sample failed at a deviator stress of 500 kN/m^2. The pore water pressure was 150 kN/m^2. The failure plane occurred at an angle of 60° to the horizontal. Determine the normal and shear stresses on the failure plane at failure. Also calculate the the maximum shear stress.

(*ii*) If the same specimen were tested in a drained test with a confining pressure of 200 kN/m^2, what would have been the deviator stress at failure? c = 44 kN/m^2. **[Ans.** 226.5; 108; 125; 552.4 kN/m^2]

11.4. A cylindrical soil sample failed at an axial load of 140 kN/m^2 in an unconfined compression test. The failure plane makes an angle of 54° with horizontal. Determine the soil properties. **[Ans.** φ = 18°]

13.5. The following results were obtained from an undrained shear box test on a soil.

Normal Load (N)	250	500	750
Failure load (N)	320	460	610

Determine the strength parameters in terms of total stresses. The cross-sectional area of the shear box was 36 cm^2. **[Ans.** ϕ = 29°, c = 55 kN/m^2]

13.6. The results of a $C\overline{U}$ test on a compacted soil are given below:

Sample No.	σ_3 (kN/m^2)	$(\sigma_1 - \sigma_3)$ (kN/m^2)	u (kN//m^2)
1	70	230	– 20
2	350	550	+ 90

Determine the cohesion intercept and the angle of shearing resistance in terms of (*a*) total stresses (*b*) effective stresses. **[Ans.** 21°, 55 kN/m^2 ; 29°, 20 kN/m^2]

13.7. A shear vane, 7.5 cm dia and 11.25 cm long, was pressed into soft clay at the bottom of a bore hole. Find the shear strength of the clay if the torque required for failure was 40 N-m. **[Ans.** 33 kN/m^2]

13.8. In an unconfined compression test, the following results were obtained. Diameter of sample = 3.75 cm. Length of sample = 7.5 cm. Spring extension = 3.00 cm. Spring constant = 100 N/cm. Deformation of sample = 12 mm. Determine the unconfined compressive strength of the soil. **[Ans.** 228.1 kN/m^2]

13.9. A sample of dry sand was subjected to a triaxial test, with a confining pressure of 250 kN/m^2. The angle of shearing resistance was found to be 36°. At what value of the major principal stress, the sample is likely to fail? **[Ans.** 963 kN/m^2]

13.10. A direct shear test was performed on a 6 cm × 6 cm sample of dry sand. The normal load was 360 N. The failure occurred at a shear load of 180 N. Plot the Mohr strength envelope, and determine ϕ. Assume c = 0. Also determine the principal stresses at failure. **[Ans.** 26.57°; 69.1 kN/m^2; 181.1 kN/m^2]

13.11. An embankment is constructed of soil, with c' = 50 kN/m^2, ϕ' = 20° and γ = 16 kN/m^2. Determine the pore water pressure, effective stress, shear strength of the soil at the base of embankment just after the fill has been raised from 3 m to 6 m. Take pore pressure coefficients A and B as 0.50 and 0.80, respectively, and the lateral pressure as one-half of the vertical pressure. **[Ans.** 28.8; 67.2; 74.5 kN/m^2]

13.12. A series of shear tests was performed on a soil. Each test was carried out until the soil sample sheared and the principal stresses for each test are as follows.

Test	σ_3 (*kN/m*^{2})	σ_1 (*kN/m*^{2})
1	300	875
2	400	1160
3	500	1460

Plot the Mohr circle of stress and determine the strength envelope and angle of internal friction of the soil. **[Ans.** 29°]

B. Descriptive and Objective Type

13.13. What is Mohr's strength theory for soils? Sketch typical strength envelopes for a clean sand.

13.14. Describe direct shear test. What are its merits and demerits?

13.15. Describe the triaxial shear test. What are the advantages of triaxial shear test over the direct shear test?

13.16. What is unconfined compression test? Sketch the apparatus used. What is its advantage over a triaxial test?

13.17. Define slow, quick and consolidated quick triaxial shear test, illustrating their use by at least one field example.

13.18. What is Mohr's circle? Discuss its important characteristics.

13.19. Write revised Mohr-Coulomb equation. How does it differ from the original equation?

13.20. Differentiate between the stress-controlled and the strain-controlled tests. Why the latter tests are more commonly conducted?

13.21. Discuss the shear characteristics of cohesionless soils and cohesive soils.

13.22. Explain the working of the following accessories of a triaxial test apparatus.

(*a*) Pore pressure measurement device.

(*b*) Mercury control system.

13.23. Derive a relationship between the principal stresses at failure using Mohr-Coulomb failure criterion.

13.24. Discuss modified failure envelope. What are its advantages and disadvantages over the standard failure envelope?

13.25. What is stress path? Sketch different types of stress paths that can be obtained in a triaxial test.

13.26. Discuss the shear characteristics of partially saturated soils.

13.27. What is Hvorslev' strength theory ? How does it differ from the the Mohr-Coulomb theory ?

13.28. What is liquefaction of sands ? How can it be prevented ?

13.29. Discuss how the shear test conditions are decided. Where would you use the effective stress analysis and where, the total stress analysis ?

13.30. Write whether the following statements are true or false.

(*a*) The origin and the pole are the same point in a Mohr circle.

(*b*) On the failure plane, the shear stress is maximum.

(*c*) The Mohr circle can be drawn for all intermediate loadings in a shear box test.

(*d*) According to Mohr's theory, the failure envelope is a straight line.

(*e*) The shear strength of a soil depends upon the total stresses.

(*f*) The Mohr circle for unconfined compression test passes through the origin.

(*g*) In a stress-controlled test, the stress-strain curve after the peak can easily be obtained.

(*h*) Consolidated drained test is also known as slow test.

(*i*) Consolidated undrained test on sand can be conducted easily in a direct shear machine.

(*j*) The dense sand increases in volume during shear.

(*k*) At critical void ratio, the volume change during shear is maximum.

(*l*) The pore pressure during a shear test on a normally consolidated clay is negative.

(*m*) The effective stress failure envelope of a normally consolidated clay passes through origin.

(*n*) The failure envelope for a normally consolidated clay in terms of total stresses obtained from an unconsolidated undrained test is horizontal.

(*o*) The dense sand has a brittle fracture.

(*p*) Vane shear test can be conducted on all types of soil.

(*q*) The unconfined compression test can be conducted on a cohesionless soil.

(*r*) The pore pressure parameter B is equal to unity for saturated soils.

(*s*) The failure plane makes an angle of $(45° + \phi/2)$ with the major principal plane, where ϕ is the angle of shearing resistance in terms of total stresses.

(*t*) It is easier to draw the modified failure envelope than the Mohr-Coulomb envelope.

(*u*) Dense sands are liable to liquefaction.

[**Ans.** True, (*f*), (*h*), (*j*), (*m*), (*n*), (*o*), (*r*), (*t*)]

C. Multiple Choice Questions

1. The shear strength of plastic undrained clay depends upon

(*a*) Internal friction (*b*) Cohesion

(*c*) Both (*a*) and (*b*) (*d*) Neither (*a*) and (*b*)

2. When drainage is permitted throughout the triaxial test, the test is known as

(*a*) Quick test (*b*) Drained test

(*c*) Consolidated undrained test (*d*) None of (*a*), (*b*) and (*c*)

3. The shear strength of a cohesionless soil is
(*a*) proportional to the angle of shearing resistance.
(*b*) inversely proportional to the angle of shearing resistance.
(*c*) proportional to the tangent of the angle of shearing resistance.
(*d*) None of above.

4. The angle of the failure plane with the major principal plane is given by
(*a*) $45° + \phi'$ (*b*) $45° + \phi'/2$
(*c*) $45° - \phi'/2$ (*d*) $45 - \phi'$
where ϕ' is the angle of shearing resistance.

5. For a heavily over-consolidated clay, the pore pressure coefficient A_f is in the range of
(*a*) 0.7 to 1.3 (*b*) 0.3 to 0.7
(*c*) – 0.5 to 0.0 (*d*) –1.0 to –0.50

6. If the angle of the modified failure envelope is α', the angle of shearing resistance ϕ' is given by
(*a*) $\cos \varphi' = \tan \alpha'$ (*b*) $\sin \varphi' = \tan \alpha'$
(*c*) $\tan \phi' = \sin \alpha'$ (*d*) None of above

7. Coulomb's equation for shear strength can be represented as
(*a*) $c = s + \sigma \tan \phi$ (*b*) $c = s - \sigma \tan \phi$
(*c*) $s = \sigma + c \tan \phi$ (*d*) $s = c - \sigma \tan \phi$

8. For saturated, normally consolidated soils, Skempton's pore pressure coefficients can be represented as
(*a*) $A < 1, B = 1$ (*b*) $A > 1, B > 1$
(*c*) $A > 1, B < 1$ (*d*) $A < 1, B > 1$

9. In an undrained triaxial compression test, the sample failed at a deviator stress of 200 kN/m^2 when the cell pressure was 100 kN/m^2. The cohesion intercept is
(*a*) 200 kN/m^2 (*b*) 100 kN/m^2
(*c*) 300 kN/m^2 (*d*) 50 kN/m^2

10. A dry sand specimen was tested in a triaxial machine with the cell pressure of 50 kPa. If the deviator stress at failure was 100 kPa, the angle of shearing resistance is
(*a*) 30° (*b*) 15°
(*c*) 45° (*d*) 60°

[**Ans.** 1. (*b*), 2. (*b*), 3. (*c*), 4. (*b*), 5. (*c*), 6. (*b*), 7. (*b*), 8. (*a*), 9. (*b*), 10. (*a*)]

14

Compaction of Soils

14.1. INTRODUCTION

Compaction means pressing the soil particles close to each other by mechanical methods. Air during compaction is expelled from the void space in the soil mass and, therefore, the mass density is increased. Compaction of a soil mass is done to improve its engineering properties. Compaction generally increases the shear strength of the soil, and hence the stability and bearing capacity. It is also useful in reducing the compressibility and permeability of the soil.

Compaction is an entirely different process than consolidation discussed in chapter 12. It is important to note the following basic differences between the two processes, even though both the processes cause a reduction in the volume.

(1) Consolidation is a gradual process of reduction of volume under sustained, static loading; whereas compaction is a rapid process of reduction of volume by mechanical means such as rolling, tamping and vibration.

(2) Consolidation causes a reduction in volume of a saturated soil due to squeezing out of water from the soil; whereas in compaction, the volume of a partially saturated soil decreases because of expulsion of air from the voids at the unaltered water content (Fig. 14.1).

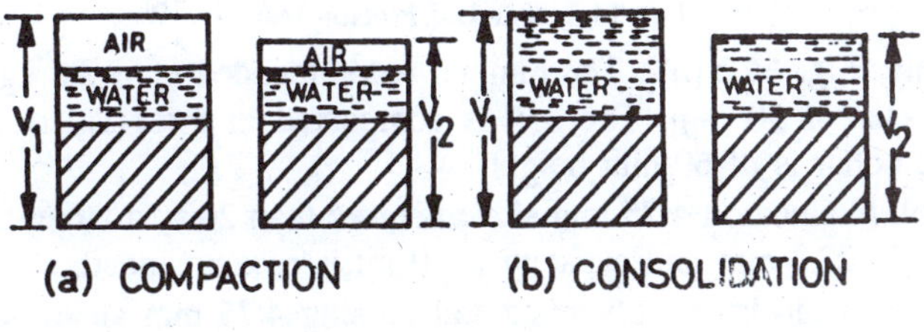

Fig. 14.1.

(3) Consolidation is a process which occurs in nature when the saturated soil deposits are subjected to static loads caused by the weight of the buildings and other structures. In contrast, compaction is an artificial process which is done to increase the density (unit weight) of the soil to improve its properties before it is put to any use.

Compaction of soil is required for the construction of earth dams, canal embankments, highways, runways and in many other engineering applications. This chapter deals with various methods of compaction and their effects on the engineering properties of the soil. Various other methods of site improvement are also discussed.

Stabilisation of soils is discussed in chapter 15. For ground improvement, see chapter 33.

14.2. STANDARD PROCTOR TEST

To assess the amount of compaction and the water content required in the field, compaction tests are done on the same soil in the laboratory. The tests provide a relationship between the water content and the dry density. The water content at which the maximum dry density is attained is obtained from the relationships provided by the tests.

Proctor (1933) used a standard mould of 4 inches internal diameter and an effective height of 4.6 inches, with a capacity of 1/30 cubic foot. The mould had a detachable base plate, and a removable collar of 2 inches height at its top. The soil was compacted in the mould in 3 equal layers, each layer was given 25 blows of 5.5 pounds rammer falling through a height of 12 inches. A curve was obtained between the dry density and the water content.

IS : 2720 (Part VII) recommends essentially the same specifications as in Standard Proctor test, with some minor modifications and metrification. The mould recommended is of 100 mm diameter, 127.3 mm

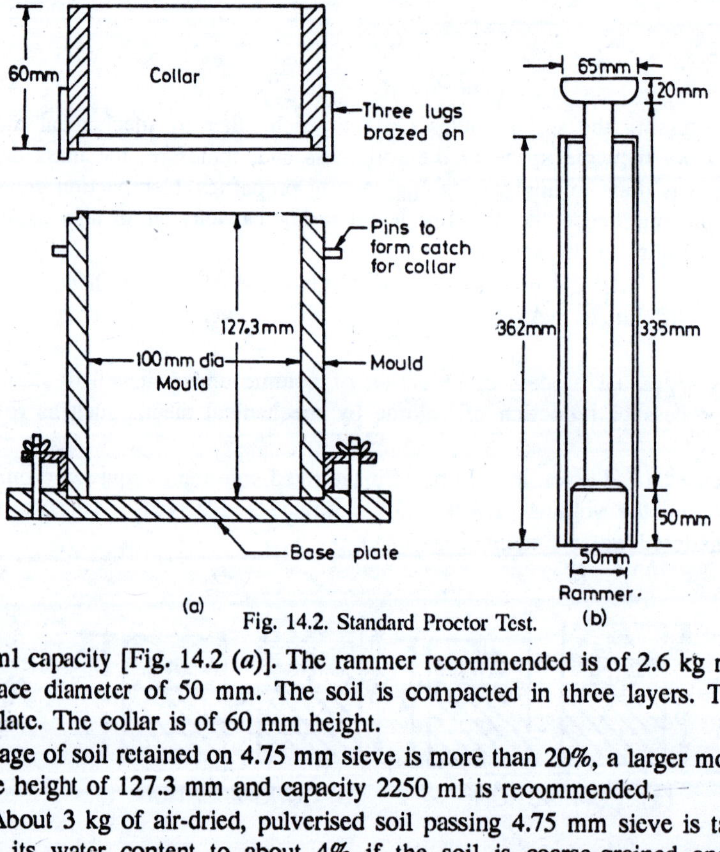

Fig. 14.2. Standard Proctor Test.

height and 1000 ml capacity [Fig. 14.2 (*a*)]. The rammer recommended is of 2.6 kg mass with a free drop of 310 mm and a face diameter of 50 mm. The soil is compacted in three layers. The mould is fixed to a detachable base plate. The collar is of 60 mm height.

If the percentage of soil retained on 4.75 mm sieve is more than 20%, a larger mould of internal diameter 150 mm, effective height of 127.3 mm and capacity 2250 ml is recommended.

Procedure. About 3 kg of air-dried, pulverised soil passing 4.75 mm sieve is taken. Water is added to the soil to bring its water content to about 4% if the soil is coarse-grained and to about 8% if it is fine-grained. The water content should be much less than the expected optimum water content (Table 14.1). The soil is mixed thoroughly and covered with a wet cloth and left for maturing for about 15 to 30 minutes.

Table 14.1. Range of Optimum Water Content

Sand	*Sandy silt or silty sand*	*Silt*	*Clay*
6 to 10%	8 to 12%	12 to 16%	14 to 20%

The mould is cleaned, dried and greased lightly. The mass of the empty mould with the base plate, but without collar, is taken. The collar is then fitted to the mould. The mould is placed on a solid base and filled with fully matured soil to about one-third its height. The soil is compacted by 25 blows of the rammer, with a free fall of 310 mm. (The number of blows required for the bigger mould of 2250 ml capacity is 56 instead of 25). The blows are evenly distributed over the surface. The soil surface is scratched with a spatula before the second layer is placed. The mould is filled to about two-thirds height with the soil and compacted again by 25 blows. Likewise, the third layer is placed and compacted. The third layer should project above the top of the mould into the collar by not more than 6 mm.

The collar is rotated to break the bond between the soil in the mould and that in collar. The collar is then removed, and the soil is trimmed off flush with the top of the mould. The mass of the mould, base plate and the compacted soil is taken, and thus the mass of the compacted soil is determined. The bulk density of the soil is computed from the mass of the compacted soil and the volume of the mould.

Representative soil samples are taken from the bottom, middle and top of the mould for determining the water content. The dry density is computed from the bulk density and the water content.

Bulk mass density, $$\rho = \frac{M}{V} \text{ gm/ml} \quad \ldots(14.1)$$

where M = mass of compacted soil (gm), V = volume of the mould (ml).

Dry density, $$\rho_d = \frac{\rho}{1 + w} \quad \ldots(14.2)$$

where w is the water content.

The soil removed from the mould is broken with hand. More water is added to the soil so as to increase the water content by 2 to 3%. It is thoroughly mixed and allowed to mature. The test is repeated and the dry density and the water content are determined.

Compaction Curve. A compaction curve is plotted between the water content as abscissa and the corresponding dry density as ordinate (Fig. 14.3). It is observed that the dry density initially increases with an increase in water content till the maximum density $(\rho_d)_{max}$ is attained. With further increase in water content, the dry density decreases. The water content corresponding to the maximum dry density is known as the optimum water content (O.W.C.) or the optimum moisture content (O.M.C.).

At a water content lower than the optimum, the soil is rather stiff and has lot of void spaces and, therefore, the dry density is low. As the water content is increased, the soil particles get lubricated and slip over each other, and move into densely packed positions and the dry density is increased. However, at a water content more than the optimum, the additional water reduces the dry density, as it occupies the space that might have been occupied by solid particles, as further explained in Sect. 14.8.

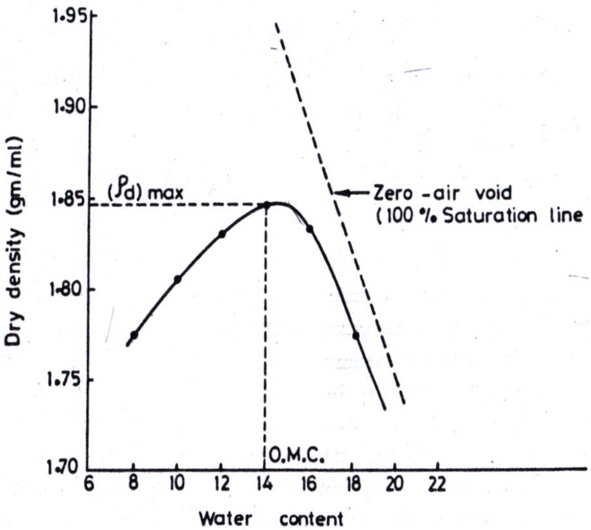

Fig. 14.3. Compaction Curve.

For a *given water content,* theoretical maximum density, $(\rho_d)_{theomax}$, is obtained corresponding to the condition when there are no air voids (*i.e.* degree of saturation is equal to 100%). The theoretical maximum dry density is also known as saturated dry density $(\rho_d)_{sat}$. In this condition, the soil becomes saturated by reduction in air voids to zero but with no change in water content. The soil could also become saturated by increasing the water content such that all air voids are filled. As we are interested in the dry density at a given water content, the latter case is not relevant here. An expression for the theoretical maximum density is developed below.

From the equations developed in chapter 2, the dry density (ρ_d) is expressed as

$$\rho_d = \frac{G\,\rho_w}{1 + e}$$

As $e = w\,G/S$, $$\rho_d = \frac{G\,\rho_w}{1 + (w\,G/S)}$$

The theoretical maximum dry density occurs when $S = 100\%$.

Thus $$(\rho_d)_{theomax} = \frac{G\,\rho_w}{1 + w\,G} \qquad \text{...(14.4)}$$

It may be mentioned that compaction methods cannot remove all the air voids, and, therefore, the soil never becomes fully saturated. Thus, the theoretical maximum dry density is only hypothetical. It can be calculated from Eq. 14.4 for any value of w if the value of G is known. The line indicating the theoretical maximum dry density can be plotted along with the compaction curve, as shown in Fig. 14.3. It is also known as *zero air void line or 100% saturation line.*

Likewise, the lines for other degrees of saturation, say 80%, 90% etc. can be drawn. For example, for $S = 90\%$, Eq. 14.3 becomes

$$\rho_d = \frac{G\,\rho_w}{1 + (w\,G/0.90)} \qquad \text{...(14.5)}$$

Instead of drawing lines corresponding to different degrees of saturation, it is sometimes more convenient to draw lines corresponding to different percentage air voids (n_a). From equations developed in chapter 2,

$$\rho_d = \frac{(1 - n_a)\,G\,\rho_w}{1 + w\,G} \qquad \text{...(14.6)}$$

For theoretical maximum density, $n_a = 0$. Therefore,

$$(\rho_d)_{theomax} = \frac{G\,\rho_w}{1 + w\,G} \qquad \text{(same as Eq. 14.4)}$$

Thus, the zero-air void line and 100% saturation line are identical.

The lines for other percentages of air voids, such as 10%, 20% etc. can be drawn. For example, for 10% air voids, Eq. 14.6 gives

$$\rho_d = \frac{0.90 \times G\,\rho_w}{1 + w\,G} \qquad \text{...(14.7)}$$

It may be noted that 10% air-void line and 90% saturation line are *not identical.*

The water content at which the soil is compacted in the field is controlled by the value of the optimum water content determined by the laboratory compaction test. The amount of compaction in the field should be approximately equal to that in the laboratory. The standard Proctor test described above is adequate to represent the compaction of fills behind retaining walls and in highways and earth dams where light rollers are used. In such cases, the optimum water content obtained from the standard Proctor test can be used as a control criterion. However, in situations where heavier compaction is required, for example in modern highways and runways, the standard Proctor test does not represent the equivalent compaction in the laboratory. For such conditions, the modified Proctor test, as described in the following section, is used to represent the compaction in the field.

(See Chapter 30, Sect. 30.18 for the laboratory experiment).

14.3. MODIFIED PROCTOR TEST

The modified Proctor test was developed to represent heavier compaction than that in the standard Proctor test. The test is used to simulate the field conditions where heavy rollers are used. The test was standardised by the American Association of State Highway Officials and is, therefore, also known as modified AASHO-test. The Indian Standard Code IS : 2720 (Part VIII) gives the specifications for heavy compaction based on this test.

In the modified Proctor test, the mould used is the same as in the standard Proctor test. However, the rammer used is much heavier and has a greater drop than that in the standard Proctor test. Its mass is 4.89 kg and the free drop is 450 mm. The face diameter is 50 mm as in the standard Proctor test. The soil is compacted in five equal layers, each layer is given 25 blows. The compactive effort in the modified Proctor test, measured in kJ/m^3 of soil, is about 4.56 times that in the standard Proctor test. Thus, a much heavier compaction is attained.

(Compactive effort in modified Pactor test = 2700 kJ/m^3; in standard Proctor test = 592 kJ/m^3).

If the percentage of soil retained on a 4.75 mm sieve is more than 20%, the larger mould of 150 mm internal diameter, effective height of 127.3 mm and capacity 2250 ml is used. In this case, 56 blows are required for each layer. The rest of the procedure is similar to that in the standard Proctor test.

The dry densities are obtained for different water contents and the compaction curve is drawn. Fig. 14.4 shows the compaction curve for the modified Proctor test (curve No. 2). The curve is higher than and to the left of that obtained from a standard Proctor test (curve no. 1). The heavier compaction increases the maximum dry density but decreases the optimum water content. The percentage increase of the dry density is between 3 to 18% for most soils; the percentage increase is more for clayey soils than for the sandy soils.

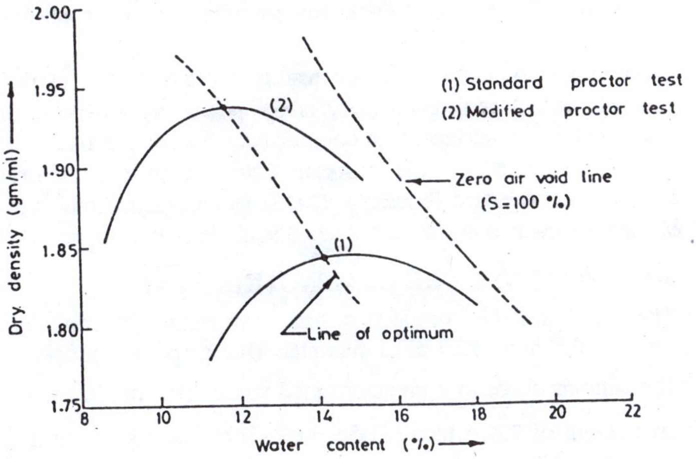

Fig. 14.4. Compaction Curves of Standard Proctor Test and Modified Proctor Test.

Fig. 14.4 also shows the zero air-void line. It may be noted that the maximum dry density attained even in the modified Proctor test is lower than the theoretical maximum dry density indicated by the zero air-void line. The *line of optimums* shown in the figure joins the points indicating the maximum dry density. It is roughly parallel to the zero air-void line.

14.4. COMPACTION OF SANDS

The compaction curves shown in Fig. 14.3 and 14.4 are obtained for soils which contain at least some percentage of cohesive soils, In case of pure sandy soils, the effect of water content on the dry density is not well defined when the water content is below the optimum value. There is a large scattering of the points on the compaction curve. Generally, the dry density decreases with an increase in the water content in this range (Fig. 14.5). The dry density decreases due to capillary tension in pore water. The capillary tension resists the tendency of soil particles to take a dense state and hence the volume increases. The phenomenon is known as the *bulking of sand*. The maximum bulking occurs at a water content of about 4 to 5%. With further increase in the water content,

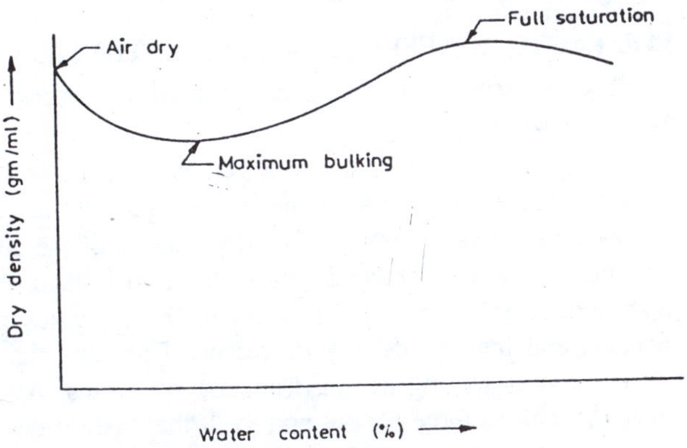

Fig. 14.5. Campaction Curvee for Cohesionless Soil.

the dry density increases as the meniscus is destroyed and the particles are able to shift and take a closer packing. The maximum dry density occurs when the soil is fully saturated. If the water content is increased beyond this point, the dry density again decreases. The coarse- grained soils do not adsorb water and are not amenable to lubrication. These do not display a distinct optimum water content.

For sandy soils, the compaction curve is of little practical use. For such soils, the relative density, as discussed in chapter 3, is used as a criterion for measurement of compactness (or denseness). The dry density of the sand is measured in the embankment and its relative density is determined if the dry densities in the loosest and densest states are known.

14.5. JODHPUR MINI COMPACTOR TEST

The Jodhpur Mini Compactor test was developed by Prof. Alam Singh (1965). A small mould of internal diameter 79.8 mm (cross- sectional area = 5000 mm^2), effective height 60 mm and a capacity of 300ml is used. The rammer used is of 2.5 kg mass and is known as the dynamic ramming tool (*DRT*). The mass slides down a stem through a height of 250 mm and falls over a foot of 40 mm diameter and 75 mm height and compacts the soil. The test is suitable for both fine-grained soils and coarse-grained soils (minus 4.75 mm sieve).

The procedure for conducting the test is similar to that in the standard Proctor test, but the soil is compacted only in 2 layers. Each layer is compacted by 15 blows of the dynamic ramming tool uniformly distributed over the soil surface. The compactive effort is approximately equal to that obtained in the standard Proctor test. It is claimed that the optimum water content and the dry density obtained in the test are almost equal to that in the standard Proctor test. It is recommended that, for fine- grained soils, a fresh soil sample shall be taken for each test after allowing a suitable maturing time.

14.6. HARVARD MINIATURE COMPACTION TEST

In Harvard miniature compaction test, compaction is done by the kneading action of a cylindrical tamping foot of 0.5 inch (12.7 mm) diameter. The tamping foot operates through a pre-set compression spring to give the tamping force to a predetermined value. The mould used is of $1\frac{5}{16}$ inch (33.34 mm) diameter and of effective height of 2.816 inch (71.53 mm). The capacity of the mould is 1/456 cubic foot (= 62.4 ml).

The number of layers, the tamping force and the number of tamps per layer are selected depending upon the type of the soil and the amount of compaction required.

14.7. ABBOT COMPACTION TEST

In the Abbot compaction test, a metal cylinder (mould) of 5.2 cm internal diameter and 40 cm effective height is used. The cylinder is clamped to the base. The soil is taken in the cylinder and compacted by a 2.5 kg rammer having a circular face of 5 cm diameter. The rammer is lifted up and dropped inside the cylinder through a height of 35 cm above the base.

14.8. FACTORS AFFECTING COMPACTION

The dry density of the soil is increased by compaction. The increase in the dry density depends upon the following factors :

(1) Water Content. At low water content, the soil is stiff and offers more resistance to compaction. As the water content is increased, the soil particles get lubricated. The soil mass becomes more workable and the particles have closer packing. The dry density of the soil increases with an increase in the water content till the optimum water content is reached. At that stage, the air voids attain approximately a constant volume. With further increase in water content, the air voids do not decrease, but the total voids (air plus water) increase and the dry density decreases. Thus the higher dry density is achieved upto the optimum water content due to forcing air out from the soil voids. After the optimum water content is reached, it becomes more difficult to force air out and to further reduce the air voids.

The effect of water content on the dry density of the soil can also be explained with the help of electrical double layer theory (chapter 6). At low water content, the forces of attraction in the adsorbed water layer are large, and there is more resistance to movements of the particles. As the water content is increased, the electrical double layer expands and the interparticle repulsive forces increase. The particles easily slide over one another and are closely packed. This results in higher dry density.

(2) Amount of Compaction. As discussed earlier, the effect of increasing the amount of compactive effort is to increase the maximum dry density and to decrease the optimum water content (Fig. 14.4). At a water content less than the optimum, the effect of increased compaction is more predominant. At a water content more than the optimum, the volume of air voids becomes almost constant and the effect of increased compaction is not significant.

It may be mentioned that the maximum dry density does not go on increasing with an increase in the compactive effort. For a certain increase in the compactive effort, the increase in the dry density becomes smaller and smaller. Finally, a stage is reached beyond which there is no further increase in the dry density with an increase in the compactive effort.

The line of optimums which joins the peaks of the compaction curves of different compactive efforts follows the general trend of the zero-air void line. This line corresponds to air voids of about 5%.

(3) Type of Soil. The dry density achieved depends upon the type of soil. The maximum dry density and the optimum water content for different soils are shown in Fig. 14.6. In general, coarse-grained soils can be compacted to higher dry density than fine- grained soils. With the addition of even a small quantity of fines to a coarse-grained soil, the soil attains a much higher dry density for the same compactive effort. However, if the quantity of fines is increased to a value more than that required to fill the voids of the coarse-grained soils, the maximum dry density decreases. A well graded sand attains a much higher dry density than a poortly graded soil.

Cohesive soils have high air voids. These soils attain a relatively lower maximum dry density as compared with the cohesionless soils. Such soils require more water than cohesionless soils and, therefore, the optimum water content is high. Heavy clays of very high plasticity have very low dry density and a very high optimum water content.

(4) Method of Compaction. The dry density achieved depends not only upon the amount of compactive effort but also on the method of compaction. For the same amount of compactive effort, the dry density will depend upon whether the method of compaction utilizes kneading action, dynamic action or static action. For example, in Harvard Miniature compaction test, the soil is compacted by the kneading action, and, therefore, the

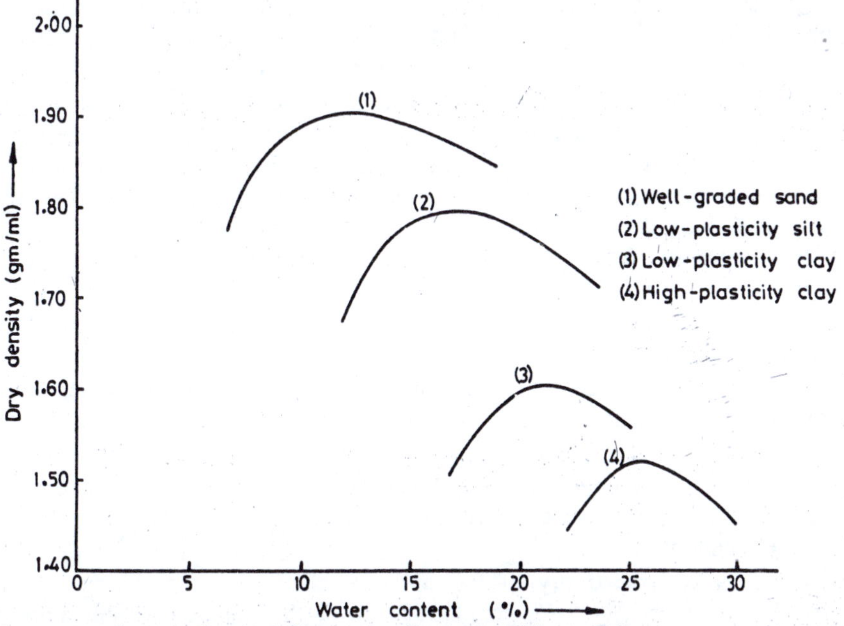

Fig. 14.6. Compaction Curves for Different Soils.

compaction curve obtained is different from that obtained from the other conventional tests in which an equal compactive effort is applied.

Different methods of compaction give their own compaction curves. Consequently, the lines of optimums are also different.

(5) Admixture. The compaction characteristics of the soils are improved by adding other materials, known as admixtures. The most commonly used admixtures are lime, cement and bitumen, as discussed in chapter 15. The dry density achieved depends upon the type and amount of admixtures.

14.9. EFFECT OF COMPACTION ON PROPERTIES OF SOILS

The engineering properties of soils are improved by compaction. The desirable properties are achieved by proper selection of the soil type, the mode of placement and the method of compaction. The effect of compaction on various soil properties is discussed below. In the following discussions, the dry of optimum means when the water content is less than the optimum, and the wet of optimum means when the water content is more than the optimum.

(1) Soil structure. The water content at which the soil is compacted plays an important role in the engineering properties of the soil. Soils compacted at a water content less than the optimum water content generally have a flocculated structure, regardless of the method of compaction. Soils compacted at a water content more than the optimum water content usually have a dispersed structure if the compaction induces large shear strains and a flocculated structure if the shear strains are relatively small.

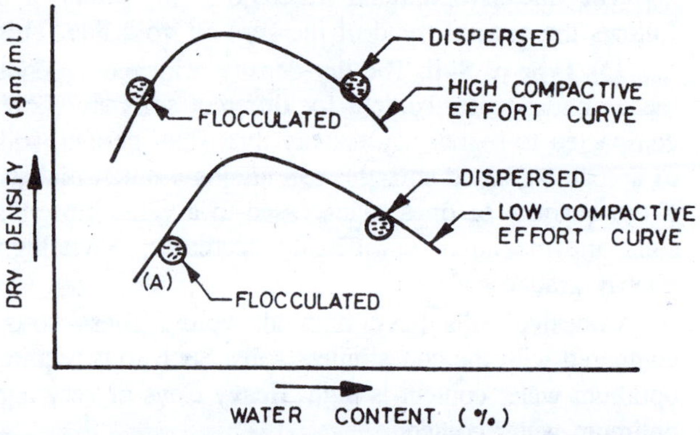

Fig. 14.7. Soil Structure in Compacted Soils.

In Fig. 14.7, at point *A* on the dry side of the optimum, the water content is so low that the attractive forces are more predominant than the repulsive forces. This results in a flocculated structure. As the water content is increased beyond the optimum, the repulsive forces increase and the particles get oriented into a dispersed structure. If the compactive effort is increased, there is a corresponding increase in the orientation of the particles and higher dry densities are obtained, as shown by the upper curve.

(2) Permeability. The permeability of a soil depends upon the size of voids, as discussed in chapter 8. The permeability of a soil decreases with an increase in water content on the dry side of the optimum water content. There is an improved orientation of the particles and a corresponding reduction in the size of voids which cause a decrease in permeability. The minimum permeability occurs at or slightly above the optimum water content. After that stage, the permeability slightly increases, but it always remains much less than that on the dry side of the optimum. The slight increase in the dry density is more pronounced than the effect of improved orientation.

If the compactive effort is increased, the permeability of the soil decreases due to increased dry density and better orientation of particles.

(3) Swelling. A soil compacted dry of the optimum water content has high water deficiency and more random orientation of particles. Consequently, it imbibes more water than the sample compacted wet of the optimum, and has, therefore, more swelling.

(4) Pore water pressure. A sample compacted dry of the optimum has low water content. The pore water pressure developed for the soil compacted dry of the optimum is therefore less than that for the same soil compacted wet of the optimum.

(5) Shrinkage. Soils compacted dry of the optimum shrink less on drying compared with those compacted wet of the optimum. The soils compacted wet of the optimum shrink more because the soil particles in the dispersed structure have nearly parallel orientation of particles and can pack more efficiently.

(6) Compressibility. The flocculated structure developed on the dry side of the optimum offers greater resistance to compression than the dispersed structure on the wet side. Consequently, the soils on the dry side are less compressible.

However, the compressibility of the soil depends upon a number of other factors. It increases with an increase in the degree of saturation. The compressibility of a soil compacted on the wet side of the optimum is also influenced by the method of compaction. If the compaction is of kneading or impact type, it creates a more dispersed structure with a corresponding increase in the compressibility. If the compaction causes very large stresses, the compressibility increases due to breakdown of the structure and greater orientation of the particles.

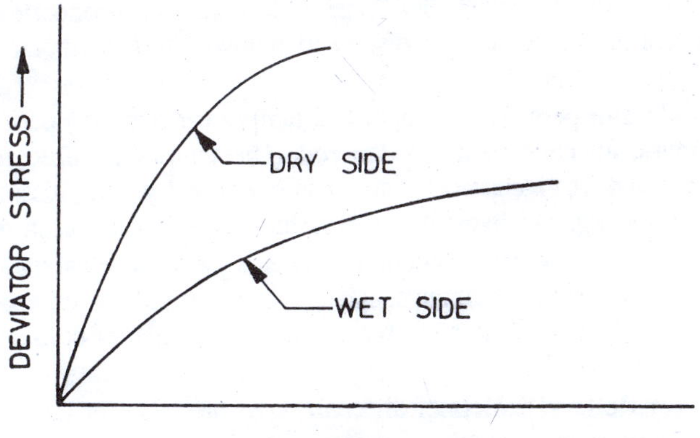

Fig. 14.8. Stress-Strain Curves.

(7) Stress-Strain relationship. The soils compacted dry of the optimum have a steeper stress-strain curve than those on the wet side (Fig. 14.8). The modulus of elasticity for the soils compacted dry of the optimum is therefore high. Such soils have brittle failure like dense sands or over-consolidated clays. The soils compacted wet of the optimum have relatively flatter stress-strain curve and a corresponding lower value of the modulus of elasticity. The failure in this case occurs at a large strain and is of plastic type.

(8) Shear Strength. In general, at a given water content, the shear strength of the soil increases with an increase in the compactive effort till a critical degree of saturation is reached. With further increase in the compactive effort, the shear strength decreases. The shear strength of the compacted soils depends upon the soil type, the moulded water content, drainage conditions, the method of compaction, etc. The shear strength of the compacted silts and clays at the moulded water content and at a water content when fully saturated are quite different, as discussed below.

(*a*) *Shear strength at moulded water content.* Two samples are compacted to the same dry density, one dry of the optimum and the other wet of the optimum, and tested for shear strength. Fig. 14.9 shows the Mohr-Coulomb failure envelopes. The soils compacted dry of the optimum have a higher shear strength at low strains. However, at large strains the flocculated structure for the soil on the dry side is broken and the ultimate strength is approximately equal for both the samples.

On the wet wide, the shear strength is further reduced if the compaction is. by kneading action. It causes a greater orientation towards a dispersed structure than that by static compaction methods.

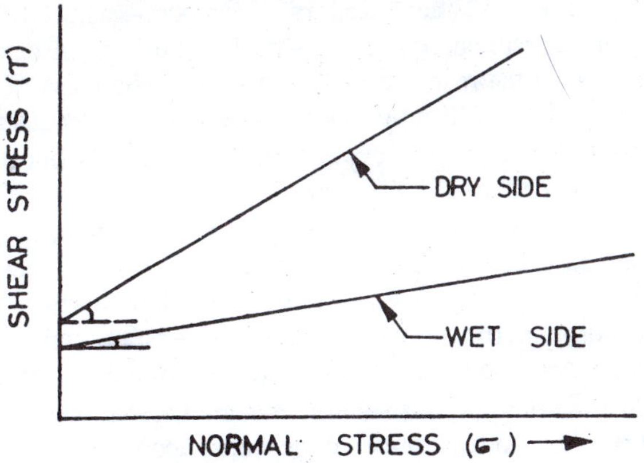

Fig. 14.9. Failure Envelopes.

(*b*) *Shear strength after saturation.* Two samples are compacted to the same dry density, one dry of the optimum and the other wet of the optimum, and then soaked in water, *without any volume change,* to have full saturation. The samples are then tested for shear strength. The samples compacted dry of the optimum show greater strength. However, the difference in the strength of the two samples is much smaller than that prior to saturation. The difference in the water deficiency of the two samples and the consequent pore water tension is greatly reduced after saturation.

If swelling is permitted during saturation, the difference in strength of the two samples is further reduced. In some cases, the sample compacted on the wet side may exhibit even more strength.

The drained shear strength of the two samples is almost equal.

14.10. METHODS OF COMPACTION USED IN FIELD

Several methods are used for compaction of soil in field. The choice of the method will depend upon the soil type, the maximum dry density required, and economic consideration. Some of the more commonly used conventional methods are discussed below. Other methods of compaction, such as vibroflotation, pounding, are discussed later.

(1) Tampers. A hand-operated tamper (or rammer) consists of a block of iron (or stone), about 3 to 5 kg in mass, attached to a wooden rod. The tamper is lifted for about 0.30 m and dropped on the soil to be compacted. A mechanical rammer is operated by compressed air or gasoline power. It is much heavier, about 30 to 150 kg. Mechanical rammers have been used upto a mass of 1000 kg in some special cases.

Tampers are used to compact soils adjacent to existing structures or confined areas, such as trenches and behind the bridge abutments, where other methods of compaction cannot be used. Owing to very low output, tampers are not economical where large quantities of soils are involved. Tampers can be used for all types of soils.

(2) Rollers. Rollers of different types are used for compaction of soils. The compaction depends upon the following factors.

(*i*) **Contact Pressure.** In general, the compaction increases with an increase in the contact pressure. For a smooth-wheel roller, the contact pressure depends upon the load per unit width and the diameter of the roller.

(*ii*) **Number of passes.** The compaction of a soil increases with an increase in the number of passes made. However, beyond a certain limit, the increase in the density with an increase in the number of passes in not appreciable. From economy consideration, the number of passes is generally restricted to a reasonable limit between 5 to 15.

(*iii*) **Layer thickness.** The compaction of a soil increases with a decrease in the thickness of the layer. However, for economy consideration, the thickness is rarely kept less than 15 cm.

(*iv*) **Speed of roller.** The compaction depends upon the speed of the roller. The speed should be so adjusted that the maximum effect is achieved.

Types of Rollers

(*a*) Smooth—Wheel Rollers. A smooth—wheel roller generally consists of three wheels; two large wheels in the rear and one small wheel in the front. A tandem type smooth—wheel roller consists of only two drums; one in the rear and one in the front. The mass of a smooth—wheel roller generally varies between 2 to 15 Mg. These rollers are operated by internal combustion engines.

[**Note.** Some authors express 1000 kg mass as one tonne (1t). As tonne is not a standard SI unit, it is better to express 1000 kg as 1 Mg].

Smooth-wheel rollers are useful for finishing operations after compaction of fills and for compacting granular base courses of highways. These are not effective for compaction of deep layers of soils, as the resulting compaction pressures induced are low. Further, these rollers also cause stratification in deep layers due to non-uniform compaction. These rollers are generally used to "seal" the surface of the fill at the end of day's work to provide a smooth surface to quickly drain off any rain water.

(*b*) Pneumatic-tyred rollers. Pneumatic-tyred rollers use compressed air to develop the required inflation pressure. The contact pressure depends upon the area of contact and the inflation pressure. The roller generally consists of 9 to 11 wheels fixed on two axles, with the pneumatic tyres so spaced that a complete coverage is obtained with each pass of the roller. The rollers are available in a wide range of load sizes. The gross mass of the roller varies between 5 to 200 Mg. However, the rollers with mass more than 50 Mg are rarely used. The inflation pressure varies between 200 to 1000 kN/m^2. The rollers are available as a self-propelled unit as well as a towed unit.

The roller compacts the soil primarily by kneading action. These rollers are effective for compacting cohesive as well as cohesionless soils. These rollers are the best type of equipment for general use. Light rollers (mass upto 20 Mg) are effective for compacting soil layers of small thickness upto 15 cm, whereas heavy rollers are useful for layers of thickness upto 30 cm.

Sometimes, the rollers are designed to produce a wobble effect, due to which a slightly weaving path is tracked. This improves the compaction of the soil. Pneumatic-tyred rollers are generally provided with a weight box or ballast box. The box can be filled with ballast to increase the weight of the roller.

(c) Sheep-foot rollers. In ancient time before the advent of the rollers, it was usual practice to pass a flock of sheep on the newly formed soil fill to cause its compaction. The same principle is used in the design of sheep-foot rollers.

The sheep-foot roller consists of a hollow drum with a large number of small projections (known as feet) on its surface. These projections penetrate the soil layers during the rolling operations and cause compaction. The drums are mounted on a steel frame. The drum can be filled with water or ballast to increase the mass. Sheep-foot rollers are available both as a self- propelled unit and a towed unit. As rolling is done, most of the weight of the roller is imposed through the projections on the soil. The contact pressure is generally between 700 to 4200 kN/m^2. The roller may sink into the soil if the contact pressure is more than the bearing capacity of the soil.

The sheep-foot rollers are ideally suited for compaction of *cohesive soils.* The rollers compact the soil by a combination of tamping and kneading action. When the roller is passed for the first time, the projections penetrate the soil layer and the lower portion of the layer is compacted. In successive passes, compaction is obtained in the middle and the top portion of the layer. This continually rising effect of the compaction is called *walking-out* of the roller.

The depth of layer that can be compacted depends upon the length of the projections and the weight of the roller. Small rollers can compact layers of 15 cm thickness, whereas heavy rollers can compact layers of 30 cm thickness. In general, the thickness of the layer compacted is kept not more than 5 cm greater than the length of the projection.

(3) Vibratory compactors. In vibratory compactors, vibrations are induced in the soil during compaction. The compactors are available in a variety of forms. When the vibrator is mounted on a drum, it is called a *vibratory roller.* These rollers are available both as pneumatic type and the smooth-wheel type. In a smooth-wheel type, a separate motor drives an arrangement of eccentric weights to create high frequency, low amplitude, up- and-down oscillations of the drum. These rollers are suitable for compacting *granular soils,* with no fines, in layers upto 1 m thickness. However, if there is appreciable percentage of fines, the thickness has to be reduced. In a pneumatic-tyred vibratory compactor, a separate vibrating unit is attached to the wheel axle. The ballast box is suspended separately from the axle so that it does not vibrate. These compactors are suitable for compacting granular soils with thickness of layer of about 30 cm.

Another form of a vibratory compactor is a *vibrating-plate compactor.* In this system, there are a number of small plates, each plate is operated by a separate, vibrating unit. Hand- operated vibrating plates are also available. The effect of the vibrating plates is limited to small depths. Their main use is to compact granular base courses for highways and runways where the thickness of layers is small.

Vibratory compactors can compact the granular soils to a very high maximum dry density.

14.11. PLACEMENT WATER CONTENT

As the methods used for compaction in the field are different from that for compaction in the laboratory, the optimum water content in the field may not be necessarily be the same as in the laboratory. The laboratory value may be taken as a rough guide for placement water content in the field. The ideal placement water content when the pneumatic-tyred rollers are used is approximately equal to the optimum water content as obtained from a standard Proctor test. The placement water content when the sheep-foot rollers, smooth-wheel rollers and vibratory rollers are used, is of the order of the optimum water content obtained in the modified Proctor test.

For important works, a full-scale test is conducted in the field to determine the placement water content, the thickness of layer, mass and speed of roller and the number of passes. Sometimes, in case of small, unimportant works, the placement water content is taken equal to the optimum water content of the standard Proctor test for light compaction and equal to that of the modified Proctor test for heavy compaction. However, the field water content is sometimes kept intentionally different from the optimum water content in order to achieve or to improve a specific engineering property of the soil.

To avoid large expansions and swelling pressure under pavements and the floors, cohesive soils in such cases are generally compacted at a water content more than the optimum water content with the resulting dry density less than the maximum dry density. The clayey soil in the impervious core of an earth dam ia also compacted on the wet side of the optimum to reduce swelling pressure. On the other hand, the highway embankments of cohesive soils are generally compacted at a water content somewhat lower than the optimum water content in order to achieve high shear strength and low compressibility. Likewise, the soil in the outer shells of earth dams is compacted dry of the optimum to obtain high shear strength, high permeability and low pore pressure.

As discussed earlier, cohesionless soils do not exhibit a well- defined optimum water content. For such soils, the maximum dry density is achieved either in completely dry condition or in completely saturated condition. In the field, completely saturated condition is preferred for practical reasons to achieve the maximum compaction.

If the water content of the soil in the borrow area is less than the required placement water content, water is sprinkled over the area. On the other hand, if it is more than the desired value, the soil is excavated from the borrow pit, spread and allowed to dry. However, in wet weather, it becomes rather difficult to decrease the water content and the work has to be stopped.

14.12. RELATIVE COMPACTION

The dry density achieved in the field is compared with the maximum dry density obtained in the standard Proctor test or that in the modified Proctor test. The ratio of the dry density in the field to the maximum dry density is known as the relative compaction or percent compaction. Thus

$$\text{Relative compaction} = \frac{\rho_d \text{ in the field}}{(\rho_d)_{max} \text{ in the laboratory}} \times 100 \qquad \text{...(14.8)}$$

For cohesive soils, the dry density of the order of 95% of the maximum dry density of the standard Proctor test (*i.e.* 95% relative compaction of the standard Proctor test) can be achieved using a sheep-foot roller or a pneumatic-tyred roller. However, if the soil is very heavy clay, only sheep-foot rollers are effective. For moderately cohesive soils, the dry density of the order of 95% of that in the modified Proctor test can be achieved using pneumatic tyred roller with an inflation pressure of 600 kN/m^2 or more.

For cohesionless soils, the dry density of the order of 100% or even more of that in the modified Proctor test can be obtained using pneumatic-tyred rollers, vibratory rollers and other vibratory equipment.

14.13. COMPACTION CONTROL

The laboratory compaction tests give the optimum water content and the maximum dry density. In the field, during the compaction of the soils, it is essential to check the dry density and the water content in order to effect proper quality control. The geotechnical engineer has to ensure that the specified amount of compaction and the desired dry densities are achieved.

Compaction control is done by measuring the dry density and the water content of the compacted soil in the field.

(1) Dry Density. The dry density is measured using the methods discussed in chapter 2. The core-cutter method and the sand replacement method are commonly used. The nuclear methods are occasionally used as these are non-destructive and require no physical or chemical processing of the soil and are very convenient.

(2) Water Content. The oven-drying method of the determination of the water content takes 24 hours. This method, though very accurate, cannot be used for controlling construction, as the soil layer from which the sample was taken would be buried by the time the water content is known. Therefore, the basic requirement is that the method used be such that it gives quick results. In the field, the water content is generally determined using the sand-bath method, alcohol method or the calcium carbide method, as discussed in chapter 2. The nuclear methods are also being used increasingly.

The water content can also be determined indirectly using a *Proctor needle* (also known as plasticity needle). The Proctor needle consists of a rod attached to a spring-loaded plunger (Fig. 14.10). The stem of the plunger is marked to read the resistance in newton. A sliding ring on the stem indicates the maximum resistance recorded during the test. The needle-shank has graduations to indicate the depth of penetration. The equipment is provided with a series of needle points of different cross-sectional areas (0.25, 0.50, 1.0 and 2.5 cm^2) to obtain a wide range of the penetration resistance. For cohesive soils, the needle points of larger cross-sectional areas are required and for cohesionless soil, those of smaller cross- sectional areas are used. The needle point used should be such that it is neither too small for accurate measurement nor too large.

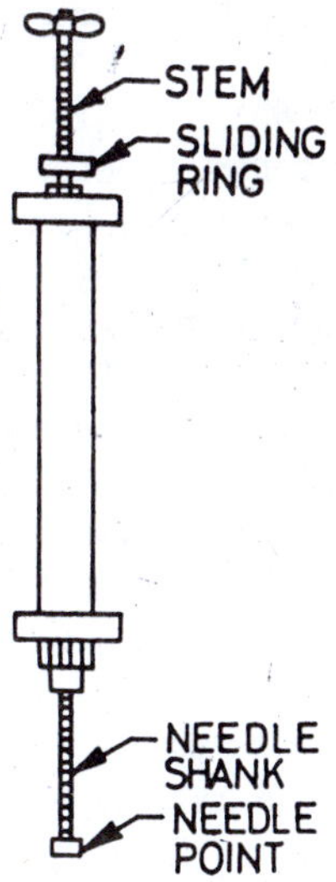

Fig. 14.10. Proctor Needle.

A suitable needle point is selected and screwed to the needle shank. After the soil has been compacted at a given water content in the compaction test in the laboratory, the Proctor needle is forced 7.5 cm into it at the rate of about 1.25 cm/sec. The maximum force used is found from the compression of the spring. From the known area of the needle point, the penetration resistance per unit area is computed. A number of such measurements are made in the laboratory during the compaction test, and a calibration curve is obtained between the penetration resistance (*R*) and the water content, as shown in Fig. 14.11. It is found that for a given degree of compaction, the penetration resistance decreases with an increase in water content.

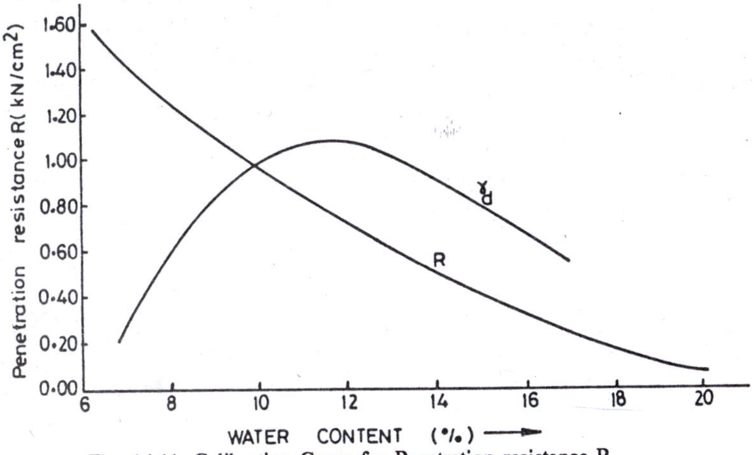

Fig. 14.11. Calibration Curve for Penetration resistance R.

To determine the water content of the compacted soil in the field, the soil is compacted in the standard compaction mould in the field in the same manner as was used during the calibration of the needle. The penetration resistance of the compacted soil is measured. The moisture content is then obtained form the calibration curve.

This method of the determination of the water content is quite rapid and reliable for fine-grained soils. However, it does not give accurate results for cohesionless soils and for soils having a large percentage of gravels and stone pieces.

14.14. VIBROFLOTATION METHOD

Vibroflotation is used for compacting thick deposits of loose, sandy soils upto 30 m depth. A vibroflot consists of a cylindrical tube, about 2 m diameter, fitted with water jets at the top and the bottom. It contains a rotating eccentric mass which develops a horizontal vibratory motion.

The vibroflot is sunk into the loose soil upto the desired depth using the lower water jet [Fig. 14.12 (*a*)]. As water comes out of the jet, it creates a momentary quick condition ahead of the vibroflot due to which the shear strength of the soil is reduced. The vibroflot settles due to its own mass. When the desired depth has been reached, the vibrator is activated. The vibroflot then vibrates laterally and causes the compaction of the soil in the horizontal direction to a radius of about 1.5. m.

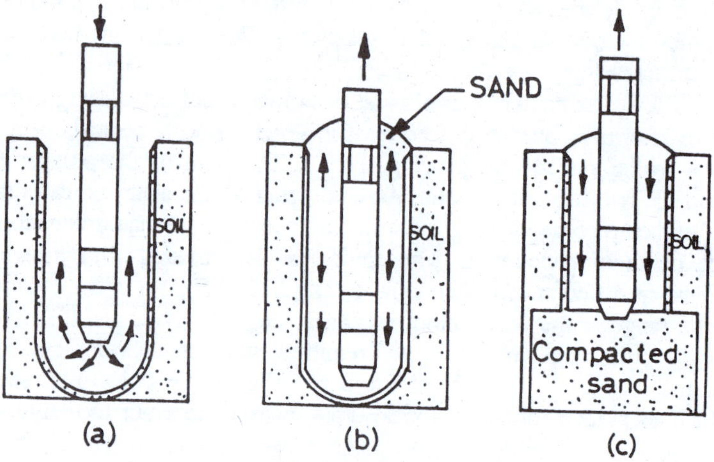

Fig. 14.12. Vibroflotation.

The water from the lower jet is transferred to the top jet and the pressure is reduced so that it is just enough to carry the sand poured at the top to the bottom of the hole [Fig. 14.12 (*b*)]. Vibration continues as the vibroflot is slowly raised to the surface. Additional sand is continually dropped into the space (crator) around the vibroflot. By raising the vibroflot in stages and simultaneously backfilling, the entire depth of the soil is compacted [Fig. 14.12 (*c*)].

The spacing of the holes is usually kept between 2 to 3 m on a grid pattern. The relative density (density index) achieved for the sandy soils is 70% or more. In soft, cohesive soils, vibroflotation is not effective. For cohesive soils, it can be used to form a sand pile to reinforce the deposit and to accelerate consolidation and thus improve its engineering properties.

14.15. TERRA PROBE METHOD

Terra probe method in many respects is similar to the vibroflotation method. The terra probe consists of an open-ended pipe, about 75 cm diameter. It is provided with a vibratory pile drive. The vibratory pile driver when activated gives vertical vibrations to the terra probe and it goes down. After reaching the desired depth, the terra probe is gradually raised upward while the vibrodriver continues to operate. Thus, the soil within and around the terra probe is densified.

The terra probe method has been successfully used upto depth of 20 m. The spacing of the holes is usually kept about 1.5 m. Saturated soil conditions are ideal for the success of the method. For the sites where the water table is deep, water jets are fitted to the terra probe to assist the penetration and densification of the soil.

The terra probe method is considerably faster than the vibroflotation method. As it does not require backfilling of sand, it can even be used at offshore locations. However, the method is less effective than vibroflotation method. In the terra probe, the zone of influence is considerably smaller and the relative density achieved is also lower.

14.16. COMPACTION BY POUNDING

To densify large deposits of loose, sandy soils, the pounding method has also been recently used. The method is also known as heavy tamping, dynamic compaction or high-energy compaction. Pounding is done by dropping a heavy mass (2 to 50 Mg) from a large height (7 to 35 m) on the ground surface. The actual mass and the height are selected depending upon the crane available and the depth of the soil deposit. A closely spaced grid pattern is selected for the pounding locations. At each location, 5 to 10 poundings are given.

The pounding method is used to compact the soil deposits to a great depth. It is very effective for densifying loose sandy deposits. Recently, the method has been successfully used to compact fine-grained soil deposits as well. The depth (D) in metres upto which the method is effective can be determined from the following relation:

$$D = C\sqrt{MH} \qquad ...(14.9)$$

where C = coefficient (0.5 to 0.75), M = mass (Mg), H = height of drop (m).

While using the pounding method, care shall be taken that harmful vibrations are not transferred to the adjacent buildings. The radius of influence (R) in metres beyond which no harmful vibrations are transmitted can be determined from the relation.

$$R = 130\sqrt{MH} \qquad ...(14.10)$$

where M = mass (Mg), and H = height of drop (m).

14.17. COMPACTION BY EXPLOSIVES

Buried explosives are sometimes used to density cohesionless soils. The shock wave and vibrations produced by explosives are somewhat similar to that produced by vibratory, compaction equipment. The method is quite effective when the cohesionless soil is fully saturated. The shock waves cause liquefaction of sand, which is followed by densification. In partially saturated cohesionless soils, compressive stresses develop due to capillary action and prevent the soil particles from taking closer positions. The method is not effective for partially saturated soils.

The depth upto which the blast is effective is limited to about 25 m. The uppermost zone of the soil upto a depth of about 1 m gets displaced in a random manner and is, therefore, not properly densified. This zone should be compacted using the conventional methods by rollers.

Explosive charges usually consist of about 60% dynamite and 30% special gelatin dynamite and ammonite. The charges are placed at two-thirds the thickness of the stratum to be densified. The spacing of the explosive points is kept between 3 to 8 m. Three to five blasts are generally required at each location.

The radius of influence (R) of compaction can be determined using the relation

$$R = (M/C)^{1/3} \qquad ...(14.11)$$

where R = radius of influence (m), M = mass of charge (kg), C constant (= 0.04 for 60% dynamite)

14.18. PRECOMPRESSION

As discussed in chapter 12, precompression improves the properties of the cohesive soils. In this method, the soil is preloaded before the application of the design loads. Preloading causes settlement before actual construction begins. The preload is generally is the form of an earth fill which is left in place for a long time so as to induce the required settlement. After the required compression has been achieved, the preload is removed prior to the construction. A monitoring system consisting of settlement plates and piezometers may be used to check the progress of settlement.

The precompression method is effective for compaction of silts, clays, organic soils and sanitary land fills. The preload must be carefully selected so as not to cause shear failures in the soil. The stability of the soil deposit under preload should be checked. Sufficient soil data should be collected to predict the rate and magnitude of the settlement. Sometimes, vertical sand drains are used to decrease the time of settlement.

14.19. COMPACTION PILES

Cohesionless soils can be densified by constructing compaction piles. A capped, pipe pile is driven into the soil. The soil surrounding the pile is compacted due to vibrations caused during driving. The pile is then extracted and the hole formed is backfilled with sand. Thus the compaction pile is formed. (For more details, see chapter 25.)

14.20. SUITABILITY OF VARIOUS METHODS OF COMPACTION

The suitability criteria of various methods of compaction can be summarised as under:

(1) *Cohesionless Soils only.* Smooth-wheel rollers are suitable for compacting layers of small thickness

in base courses. Vibratory rollers are effective for compacting deposits of large thickness.

(2) *Cohesive Soils only.* Sheep-foot rollers are suitable for compaction of cohesive soils.

(3) *Both cohesionless and cohesive soils.* The following methods are universal. These can be used for both cohesionless soils and cohesive soils.

(*i*) Tampers are effective for compacting soils in a confined space of all types.

(*ii*) Pneumatic-tyred rollers are extremely useful for compacting all types of soils.

ILLUSTRATIVE EXAMPLES

Illustrative Example 14.1. *A sample of soil was prepared by mixing a quantity of dry soil with 10% by mass of water. Find the mass of this wet mixture required to produce a cylindrical, compacted specimen of 15 cm diameter and 12.5 cm deep and having 6% air content. Find also the void ratio and the dry density of the specimen if G = 2.68.*

Solution. Air content, $a_c = V_a/V_v = 0.06$

or $V_a = 0.06\, V_v$. Hence $V_w = 0.94 V_v$

Thus $$V_a = 0.06\left(\frac{V_w}{0.94}\right) = 0.0638\, V_w$$

Volume of specimen $(V) = \pi/4 \times (15)^2 \times (12.5) = 2208.9$ ml

Now, with usual notations, $V = V_s + V_w + V_a$

or $2208.9 = V_s + V_w + 0.0638\, V_w = V_s + 1.0638\, V_w$

Writing volumes in terms of mass,

$$2208.9 = \frac{M_s}{(2.68 \times 1.0)} + 1.0638\left(\frac{M_w}{1.0}\right)$$

Substituting $M_w = 0.10\, M_s$,

$$2208.9 = \frac{M_s}{(2.68)} + 1.0638 \times 0.1\, M_s$$

or $M_s = 4606.54$ gm. $M_w = 460.65$ gm

Mass of wet soil, $M = M_s + M_w = 4606.54 + 460.65 = 5067.19$

Bulk density, $$\rho = \frac{M}{V} = \frac{5067.19}{2208.9} = 2.294 \text{ gm/ml}$$

Dry density, $$\rho_d = \frac{\rho}{1 + w} = \frac{2.294}{1 + 0.10} = 2.085 \text{ gm/ml}$$

Therefore, $$e = \frac{G\rho_w}{\rho_d} - 1$$

$$= \frac{2.68 \times 1.0}{2.085} - 1 = \mathbf{0.285}$$

Illustrative Example 14.2. *The following results were obtained from a standard compaction test on a sample of soil.*

Water content (%)	*0.12*	*0.14*	*0.16*	*0.18*	*0.20*	*0.22*
Mass of wet soil (kg)	*1.68*	*1.85*	*1.91*	*1.87*	*1.87*	*1.85*

The volume of the mould used was 950 ml. Make necessary calculations and plot the compaction curve and obtain the maximum dry density and the optimum water content. Also calculate the void ratio, the degree of saturation and the theoretical maximum dry density (G = 2.70).

Solution. Calculations are shown in tabular form.

Water content (w)	0.12	0.14	0.16	0.18	0.20	0.22
Mass of wet soil (M) (kg)	1.68	1.85	1.91	1.87	1.87	1.85
Bulk density $\rho = M/V = \frac{V}{0.950}$	1.77	1.95	2.01	1.97	1.97	1.95
Dry density $\rho_d = \rho/(1+w)$	1.58	1.71	1.73	1.67	1.64	1.60
Void ratio $e = \frac{G\rho_w}{\rho_d} - 1$	0.71	0.58	0.56	0.62	0.65	0.69
Degree of saturation $S = \frac{wG}{e}$	0.46	0.65	0.77	0.78	0.83	0.86
Theoretical mmaximum dry density $(\rho_d)\text{theomax} = \frac{G\rho_w}{1+Gw}$	2.04	1.96	1.89	1.82	1.75	1.69

Fig. Ex. 14.2 shows the compaction curve.

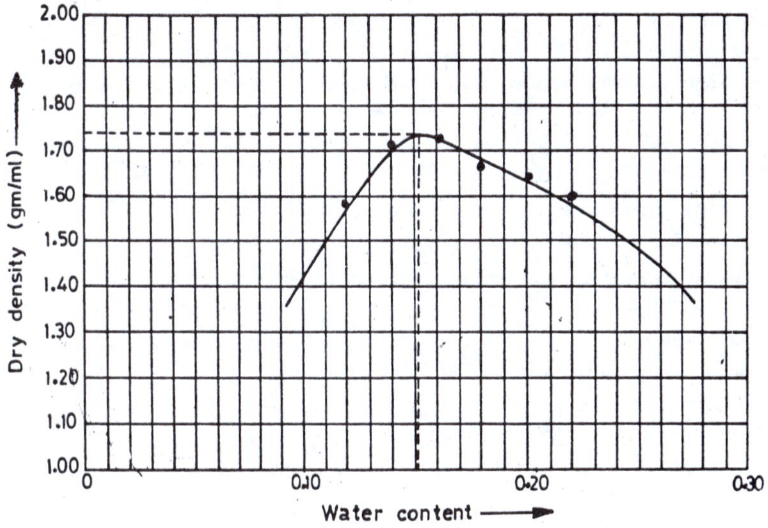

Fig. E-14.2.

From the plot. $(\rho_d)_{max}$ = 1.74 gm/ml, O.W.C. = 15.2%,

Illustrative Example 14.3. *The maximum dry density of a sample by the light compaction test is 1.78 g/ml at an optimum water content of 15%. Find the air voids and the degree of saturation. G = 2.67.*

What would be the corresponding value of dry density on the zero air void line at O.W.C. ?

Solution.

$$\rho_d = \frac{G\rho_w}{1+e} = \frac{G\rho_w}{1+(wG/S)}$$

or
$$1.78 = \frac{2.67 \times 1.0}{1 + (0.15 \times 2.67/S)}$$

$$1.78 + \frac{0.713}{S} = 2.67 \quad \text{or } S = 0.801 \quad \text{or } \mathbf{80.1\%}$$

Now
$$\rho_d = \frac{(1-n_a)\,G\,\rho_w}{1 + wG} = \frac{(1-n_a) \times 2.67 \times 1.0}{1 + 0.15 \times 2.67} = 1.78$$

or
$$n_a = 0.066 \quad \text{or } \mathbf{6.6\%}$$

$$(\gamma_d)_{theomax} = \frac{G\,\rho_w}{1 + wG} = \frac{2.67 \times 1.0}{1 + 0.15 \times 2.67} = \mathbf{1.91\ g/ml}$$

PROBLEMS

A. Numerical

14.1. A cylindrical specimen of a cohesive soil of 10 cm diameter and 20 cm length was prepared by compaction in a mould. If the wet mass of the specimen was 3.25 kg and its water content was 15%, determine the dry density and the void ratio.

If the specific gravity of the particles was 2.70, find the degree of saturation. [**Ans.** 1.80 gm/ml; 0.50; 81%]

14.2. The following are the results of a standard compaction test performed on a sample of soil.

Water Content (%)	7.7	11.5	14.6	17.5	19.7	21.2
Mass of wet soil (kg)	1.7	1.89	2.03	1.99	1.96	1.92

If the volume of the mould used was 950 c.c. and the specific gravity of soil grains was 2.65, make necessary calculations and plot the water content-dry density curve and obtain the optimum water content and the maximum dry density. [**Ans.** 15%; 1.83 gm/ml]

14.3. An earthen embankment of 10^6 m^3 volume is to be constructed with a soil having a void ratio of 0.80 after compaction. There are three borrow pits marked A, B and C, having soils with void ratios of 0.90, 1.50 and 1.80, respectively. The cost of excavation and transporting the soil is Rs. 0.25, Rs. 0.23 and Rs. 0.18 per m^3, respectively. Calculate the volume of soil to be excavated from each pit. Which borrow bit is the most economical ? (G = 2.65). [**Ans.** 1.055×10^6; 1.389×10^6; 1.555×10^6 m^3; *A*]

B. Descriptive and Objective Type

14.4. Differentiate between consolidation and compaction. Give examples.

14.5. Describe Standard Proctor test and the modified Proctor test. How would you decide the type of the test to be conducted in the laboratory ?

14.6. What is a compaction curve ? Give its salient features. What is a zero-air void line ?

14.7. What are the factors that affect compaction ? Discuss in brief.

14.8. What is the effect of compaction on the engineering properties of the soil ? How would you decide whether the soil should be compacted the dry of optimum or the wet of optimum ?

14.9. What are the different methods of compaction adopted in the field? How would you select the type of roller to be used?

14.10. Write short notes on

(*a*) Placement water content
(*b*) Relative compaction
(*c*) Compaction control
(*d*) Terra probe
(*e*) Vibroflotation
(*f*) Compaction by pounding
(*g*) Precompression
(*h*) Compaction by explosives

14.11. Write whether the following statement are correct or not.

(*a*) Compaction occurs under a sustained, static load on a saturated soil.
(*b*) The theoretical maximum dry density can be attained in the laboratory.
(*c*) The zero-air void line and 100% saturation line are identical.

(*d*) As the compaction is increased, the optimum water content increases.
(*e*) The modern highways and runways have compaction equal to that attained in a standard Proctor test.
(*f*) Vibroflotation is effective for highly cohesive soil.
(*g*) The Proctor needle can be used to determine the dry density achieved in the field.
(*h*) The relative compaction is the same as the relative density.
(*i*) The pneumatic-tyred rollers can be used for both cohesionless and cohesive soils.
(*j*) The water content of the soil in the field is always kept equal to the optimum water content.
(*k*) The soils compacted dry of the optimum have higher modulus of elasticity than those on the wet side.
(*l*) The core in an earth dam is generally compacted wet side of the optimum.
(*m*) The permeability of the soil decreases by compaction.
(*n*) The Jodhpur mini-compactor test gives lower dry density than the standard Proctor test.
(*o*) In the Jodhpur mini-compactor test, kneading action takes place.
(*p*) The number of passes made by a roller is generally more than ten.
(*q*) The shear strength of a soil always increases with an increase in compaction.
(*r*) In pneumatic tyres, the contact pressure depends upon the inflation pressure.
(*s*) The thickness of the layer during compaction is kept about 10 cm.
(*t*) The smooth-wheel rollers can be used for compaction of deep fills.

[**Ans.** True, (*c*), (*i*), (*k*), (*l*), (*m*), (*r*)]

C. Multiple Choice Questions

1. Pneumatic-tyred rollers are useful for compacting
   (*a*) Cohesive soils (*b*) Cohesionless soils
   (*c*) Both (*a*) and (*b*) (*d*) For soils in confined space
2. Vibratory rollers are best suited for compacting
   (*a*) Coarse sand and gravels (*b*) Silts
   (*c*) Clays (*d*) organic soils
3. Precompression method is useful for compacting
   (*a*) Silts (*b*) Clays
   (*c*) Organic soils (*d*) All the above
4. The line of optimums generally corresponds to percentage air voids of about
   (*a*) zero percent (*b*) 5 percent
   (*c*) 10 percent (*d*) 20 percent
5. The range of optimum water contents for the standard proctor test for clayey soils is
   (*a*) 6 to 10% (*b*) 8 to 12%
   (*c*) 12 to 15% (*d*) 14 to 20%
6. Soil compacted dry of the optimum as compared to that wet of the optimum
   (*a*) has less permeability (*b*) swells less
   (*c*) shrinks less (*d*) has less resistance to compression
7. For a Standard Compaction test, the mass of hammer and the drop of hammer are as follows :
   (*a*) 2.6 kg and 450 mm (*b*) 2.60 kg and 310 mm
   (*c*) 4.8 kg and 310 mm (*d*) 4.89 kg and 450 mm
8. Select the correct statement
   (*a*) Relative compaction is the same as relative density
   (*b*) Vibratory rollers are effective in the case of highly cohesive soils
   (*c*) 'Zero air void line' and 100% saturation line are identical

[**Ans.** 1. (*c*), 2. (*a*), 3. (*d*), 4. (*b*), 5. (*d*), 6. (*c*), 7. (*b*), 8. (*c*)]

15

Soil Stabilisation

15.1. INTRODUCTION

Soil stabilisation is the process of improving the engineering properties of the soil and thus making it more stable. It is required when the soil available for construction is not suitable for the intended purpose. In its broadest senses, stabilisation includes compaction, preconsolidation, drainage and many other such processes. However, the term stabilisation is generally restricted to the processes which *alter the soil material itself* for improvement of its properties. A cementing material or a chemical is added to a natural soil for the purpose of stabilisation.

Soil stabilisation is used to reduce the permeability and compressibility of the soil mass in earth structures and to increase its shear strength. Soil stabilisation is required to increase the bearing capacity of foundation soils. However, the main use of stabilisation is to improve the natural soils for the construction of highways and airfields. The principles of soil stabilisation are used for controlling the grading of soils and aggregates in the construction of bases and sub-bases of the highways and airfields.

Soil stabilisation is also used to make an area trafficable within a short period of time for military and other emergency purposes. Sometimes, soil stabilisation is used for city and suburban streets to make them more noise-absorbing.

This chapter deals with the various methods of soil stabilisation and their effects on the engineering properties of the soils.

15.2. MECHANICAL STABILISATION

Mechanical stabilisation is the process of improving the properties of the soil by changing its gradation. Two or more types of natural soils are mixed to obtain a composite material which is superior to any of its components. To achieve the desired grading, sometimes the soils with coarse particles are added or the soils with fine particles are removed.

Mechanical stabilisation is also known as *granular stabilisation.*

For the purpose of mechanical stabilisation, the soils are subdivided into two categories:

(1) **Aggregates :** These are the soils which have a granular bearing skeleton and have particles of the size larger than 75μ.

(2) **Binders :** These are the soils which have particles smaller than 75μ size. They do not possess a bearing skeleton.

The aggregates consist of strong, well-graded, angular particles of sand and gravel which provide internal friction and incompressibility to a soil. The binders provide cohesion and imperviousness to a soil. These are composed of silt and clay. The quantity of binder should be sufficient to provide plasticity to the soil, but it should not cause swelling.

Proper blending of aggregates and binders is done in order to achieve required gradation of the mixed soil. The blended soil should possess both internal friction and cohesion. The material should be workable during placement. When properly placed and compacted, the blended material becomes mechanically stable. The load-carrying capacity is increased. The resistance against the temperature and moisture changes is also improved.

The mechanical stability of the mixed soil depends upon the following factors.

(1) Mechanical strength of the aggregate. The mixed soil is stable if the aggregates used have high strength. However, if the mixture is properly designed and compacted, even the aggregates of relatively low strength can provide good mechanical stability.

(2) Mineral composition. The mechanical stability of the mixed soil depends upon the composition of the minerals in it. The minerals should be weather-resistant.

Sodium sulphates and sodium carbonates cause large volume changes due to their hydration and dehydration. These are detrimental to the stability.

(3) Gradation. The gradation of the mixed soil should be such that the voids of the coarser particles are filled with finer particles so that a high density is obtained. According to Fuller (1907), the maximum density is achieved if the particle size distribution of the mixture satisfies the following criterion:

$$p = (d/D)^{0.50} \times 100 \qquad \ldots(15.1)$$

where p = percentage of the soil mixture passing sieve of size d,

D = maximum particle size.

The U.S. Bureau of Public Roads recommends that the value of the exponent in Eq. 15.1 should be taken as 0.45 instead of 0.50.

It is found by experience that to obtain sufficient cohesion in the mixture, it is necessary to have a greater proportion of the material passing 75μ sieve than that given by Eq. 15.1. Generally, an ideal mixture would include about 25% binder. The basic requirement is that the mixture shall contain a fair proportion of different size particles. It should contain sufficient quantity of fines to provide cohesion, especially in the wearing surface.

(4) Plasticity characteristics. Soils with high liquid limit and plasticity index are suitable as binders for soils used for surfacing. Such soils possess greater cohesion and better moisture retention capacity. They provide a better seal against the downward movement of surface water.

For soils used in base courses, the requirement of plasticity characteristics are quite different from those for surfacing. The soil in base course should have low plasticity to avoid excessive accumulation of water and the resulting loss of strength.

As the soil available at site may seldom meet both the gradation and the plasticity characteristics, it usually becomes necessary to mix soils from different sources to obtain the desired mixture. This is normally done by trial mixes. As far as possible, the maximum use of locally available soils should be made for economy.

(5) Compaction. The mechanical stability of the stabilised soil mass depends upon the degree of compaction attained in the field. Normally, the compaction is done at or near the optimum water content.

Uses of Mechanical Stabilisation. It is the simplest method of soil stabilisation. It is generally used to improve the subgrades of low bearing capacity. It is extensively used in the construction of bases, sub-bases and surfacing of roads.

15.3. CEMENT STABILISATION

Cement Stabilisation is done by mixing pulverised soil and portland cement with water and compacting the mix to attain a strong material. The material obtained by mixing soil and cement is known as *soil-cement.* The soil-cement becomes a hard and durable structural material as the cement hydrates and develops strength.

Types of Soil-cement

Mitchell and Freitag (1959) have divided the soil-cement into 3 categories.

(1) Normal Soil-Cement. It consists of 5 to 14% of cement by volume. The quantity of cement mixed with soil is sufficient to produce a hard and durable construction material. The quantity of water used should be just sufficient to satisfy hydration requirements of the cement and to make the mixture workable.

The normal soil-cement is quite weather-resistant and strong. It is commonly used for stabilising sandy and other low plasticity soils.

(2) Plastic Soil-Cement. This type of soil-cement also contains cement 5 to 14% by volume, but it has more quantity of water to have wet consistency similar to that of plastering mortar at the time of placement.

The plastic soil-cement can be placed on steep or irregular slopes where it is difficult to use normal road-making equipment. It has also been successfully used for water-proof lining of canals and reservoirs. The plastic soil-cement can be used for protection of steep slopes against erosive action of water.

(3) Cement-Modified soil. It is a type of soil-cement that contains less than 5% of cement by volume. It is a semi-hardened product of soil and cement. It is quite inferior to the other two types.

As the quantity of cement used is small, it is not able to bind all the soil particles into a coherent mass. However, it interacts with the silt and clay fractions and reduces their affinity for water. It reduces the swelling characteristics of the soil. The use of cement-modified soils is limited.

The following discussion is confined to the first two types of soil-cement.

Factors affecting cement stabilisation

The factors affecting cement stabilisation can be summarised as under:

(1) Type of soil. Granular soils with sufficient fines are ideally suited for cement stabilisation. Such soils can be easily pulverized and mixed. They require the least amount of cement.

Granular soils with deficient fines, such as beach sands and wind-blown sands, can also be stabilised but these soils require more cement. As it is difficult to move road-making equipment over such soils when dry, it is desirable to keep them wet for better traction.

Silty and clayey soils can produce satisfactory soil-cement but those with a high clay-content are difficult to pulverise. Moreover, the quantity of cement increases with an increase in clay content. The quality of soil-cement in this case is not good, as it may have high shrinkage properties.

Organic matter, if present in colloidal form, interferes with the hydration of cement and causes a reduction in the strength of soil-cement. The trouble is more common in sandy soils than in clayey soils. The soil should be treated with calcium chloride to remedy the situation. Sodium hydroxide is also effective in correcting the ill effects due to organic matter. Sometimes, addition of a small quantity of silt or clay to a sandy soil may aid in the cement reaction.

(2) Quantity of cement. A well-graded soil requires about 5% cement, whereas a poorly graded, uniform sand may require about 9% cement. Non-plastic silts require about 10% cement, whereas plastic clays may need about 13% cement.

The actual quantity of cement required for a particular soil is ascertained by laboratory tests. For base courses, samples are subjected to durability tests for determination of the quantity of the cement required. It consists of 12 cycles of freezing and thawing or 12 cycles of wetting and drying. The maximum volume change (swelling plus shrinkage) of 2% is generally permitted.

Sometimes, the quantity of cement is determined according to the minimum unconfined compressive strength. Generally, a minimum strength of about 1500 kN/m^2 for clayey soils and of about 5500 kN/m^2 for sandy soils is specified. High strength is obtained by decreasing the water-cement ratio. This is done by increasing the cement content for the same water content.

As a rough guide, the cement content can be taken as 6% for sandy soils and 15% for clayey soils.

(3) Quantity of Water. The quantity of water used must be sufficient for hydration of cement and silt-clay cement and for making the mix workable. Generally, the amount of water ascertained from compaction consideration is adequate for hydration as well.

Water used should be clean and free from harmful salts, alkalies, acids or organic matter. In general, the water which is potable is also satisfactory for soil-cement.

(4) Mixing, Compaction and Curing. The mixture of the soil, cement and water should be thoroughly mixed, as the success of cement stabilisation depends mainly on thorough mixing. If it is not properly mixed, it may result in a non-homogeneous, weak product. However, the mixing should not be continued after the cement has started hydrating, as it would result in a loss of strength.

Soil-cement should be properly compacted. Compaction is generally done as for soil alone (Chapter 14). For good results, fine- grained soils should be compacted wet of optimum, and coarse- grained soils, dry of optimum. After compaction, the surface is finished by a rubber-tyred roller.

Soil-cement should be protected against loss of moisture by providing a thin bituminous coating. Sometimes, other materials, such as water-proof paper, moist straw or dirt, are also used.

(5) Admixture. To increase the effectiveness of cement as stabiliser, admixtures are sometimes added to soil cement. Admixtures may permit a reduction in the amount of cement required. These may also help stabilisation of soils which are not responsive to cement alone.

Lime and calcium chloride have been used as admixtures for clays and soils containing harmful organic matter to make them more responsive to cement. Fly ash acts as a pozzolana and is effective for stabilisation of dune sand. Sodium carbonate and sodium sulphates have also been used as admixture.

Construction Methods

The construction of soil-cement bases and sub-bases is done using the following methods.

(1) Mix-in-place method. In this method of construction, mixing of soil-cement is done at the place where it would be finally placed. It consists of the following steps:

(*i*) The subgrade is cleared of all undesirable materials such as boulders, debris, stumps. It is then levelled to the required formation level.

(*ii*) The levelled subgrade is scarified to a depth equal to the proposed thickness of the soil cement.

(*iii*) The scarified soil is then pulverised till at least 80% of the soil passes 4.75 mm IS sieve. It can be done either manually or with the help of a machine.
Pulverisation of highly plastic soil can be done easily if about 4% lime is added to it.

(*iv*) The pulverised soil is properly shaped to the required grade and the required quantity of cement is spread uniformly over the surface. It is then intimately mixed dry with rotary tillers or special soil mixers.

(*v*) The required quantity of water is sprinkled over the surface and wet mixing is done till the mixture has a uniform colour. The operation should not last longer than 3 hours.
The surface is then properly graded using towed graders.

(*vi*) Compaction is done using suitable methods. The thickness of the layer should not be more than 15 cm. Compaction should not take more than 2 hours.
After compaction, the surface is properly finished.

(*vii*) The compacted soil-cement is moist cured for at least 7 days by providing a bituminous primary coat. Alternatively, it is kept damp by frequent application of a light spray of water.

The mix-in-place method of construction is quite simple, cheap and easily adaptable to different field conditions. The main disadvantage is that the mixing is not uniform and high strength cannot be achieved.

(2) Plant-mix method

There are two types of plants used in the plant-mix method of construction.

(*a*) Stationary plant. In this method, the excavated soil is transported to a stationary plant located at a suitable place. The required quantity of cement is added to the soil in the plant. Mixing is done after adding water. The time required to obtain a uniform mixture depends upon the type of soil. The mixed material is then discharged into dumper trucks and transported back to the subgrade. It is spread and properly compacted.

The stationary plant is useful for obtaining a uniform mix. In this method, the depth of treatment can be better controlled. However, the method is quite expensive as compared with mix-in- place method. The material has to be compacted as delivered and not as a complete section of the road. A further disadvantage is that the work may have to be stopped even after a minor breakdown in the plant.

(*b*) Travelling Plant. A travelling plant can move along the road under construction. The soil, after placement of cement over it, is lifted up by an elevator and discharged into the hopper of the mixer of the travelling plant. Water is added and proper mixing is done. The mix is then discharged on the subgrade and spread by a grader. It is then properly compacted.

The travelling plant method, like stationary plant, is useful for accurate proportioning and uniform mixing. The depth of treatment is also properly controlled and a uniform subgrade surface is attained. However, the initial cost is very high.

15.4. LIME STABILISATION

Lime stabilisation is done by adding lime to a soil. It is useful for stabilisation of clayey soils. When lime reacts with soil, there is exchange of cations in the adsorbed water layer and a decrease in plasticity of the soil occurs. The resulting material is more friable than the original clay, and is, therefore, more suitable as subgrade.

Lime is produced by burning of lime stone in kilns. The quality of lime obtained depends upon the parent material and the production process. There are basically 5 types of limes.

(*i*) High calcium, quick lime (CaO)

(*ii*) Hydrated, high calcium lime [$Ca(OH)_2$]

(*iii*) Dolomitic lime ($CaO + MgO$)

(*iv*) Normal, hydrated dolomitic lime [$Ca(OH)_2 + MgO$]

(*v*) Pressure, hydrated dolomitic lime [$Ca(OH)_2 + MgO_2$].

The quick lime is more effective as stabiliser than the hydrated lime; but the latter is more safe and convenient to handle. Generally, the hydrated lime is used. It is also known as *slaked lime.* The higher the magnesium content of the lime, the less is the affinity for water and the less is the heat generated during mixing.

The amount of lime required for stabilisation varies between 2 to 10% of the soil. However, if the lime is used only to modify some of the physico—chemical characteristics of the soil, the amount of lime is about 1 to 3%. The following amount may be used as a rough guide.

(*i*) 2 to 5% for clay gravel material having less than 50% of silt- clay fraction.

(*ii*) 5 to 10% for soils with more than 50% of silt-clay fraction.

(*iii*) For soils having particle size intermediate between (*i*) and (*ii*) above, the quantity of lime required is between 3 to 7%.

(*iv*) About 10% for heavy clays used as bases and sub-bases.

Lime stabilisation is not effective for sandy soils. However, these soils can be stabilised in combination with clay, fly ash or other pozzolanic materials, which serve as hydraulically reactive ingredients. The ratio of fly ash to lime generally varies between 3 to 5. The fly ash used is about 10 to 20% of the soil weight.

Chemical and Physical Changes in lime stabilisation

When lime reacts with wet soil, it alters the nature of the adsorbed layer by base exchange. Calcium ions replace the sodium or hydrogen ions. The double layer is usually depressed due to an increase in the cation concentration. However, sometimes the double layer may expand due to high P_H value of lime.

Lime reacts chemically with available silica and alumina in soils. A natural cement composed of calcium alumino silicate complexes is formed, which causes a cementing action. The reaction depends upon the effective concentration of the reactants and temperature.

In lime stabilisation, the liquid limit of the soil generally decreases but the plastic limit increases. Thus, the plasticity index of the soil decreases. The soil becomes more friable and workable. The strength of the lime-stabilised soil is generally improved. It is partly due to a decrease in the plastic properties of the soil and partly due to formation of cementing material. Increase in the unconfined compressive strength is sometimes as high as 60 times. The modulus of elasticity of the soil also increases substantially.

Addition of lime causes a high concentration of calcium ions in the double layer. It causes a decrease in the tendency of attraction of water. Consequently, the resistance of the soil to water adsorption, capillary rise and volume changes on wetting or drying is substantially increased. The lime-stabilised bases or sub-bases form a water resistant barrier which stops penetration of rain water. There is an increase in the optimum water content and a reduction in the maximum dry density. In swampy areas where the water content is above the optimum, application of lime to soil helps in drying of the soil.

Cyclic freezing and thawing can cause a temporary loss of strength, but because of subsequent healing action, there is no loss of strength in long run.

Construction Methods

Construction methods used in lime stabilisation are similar to those used in cement stabilisation. However, the following differences should be carefully noted.

(*i*) As the reaction in the case of lime is slow, there is no maximum time limit between the addition of lime to the soil and the completion of compaction. However, care should be taken to avoid carbonation of lime in the process.

(*ii*) Lime may be added in the form of slurry instead of dry powder.

(*iii*) A rest period of 1 to 4 days is generally required after spreading lime over a heavy clay before final mixing is done. This facilitates proper mixing of lime and soil.

(*iv*) The soil-lime is compacted to the required maximum dry density.

After compaction, the surface is kept moist for 7 days and then covered with a suitable wearing coat. Sometimes, the wearing coat is applied soon after compaction to help hold the moisture.

15.5. BITUMINOUS STABILISATION

Bitumens are non-aqueous systems of hydrocarbons that are soluble in carbon di-sulphide. Tars are obtained by the destructive distillation of organic materials such as coal. Asphalts are materials in which the primary components are natural or refined petroleum bitumens.

Bituminous stabilisation is generally done with asphalt as binder. As asphalts are normally too viscous to be used directly, these are used as cut-back with some solvent, such as gasoline. These are also used as emulsions, but in this form they require a longer drying period.

Any inorganic soil which can be mixed with asphalt is suitable for bituminous stabilisation. In cohesionless soils, asphalt binds the soil particles together and thus serves as a bonding or cementing agent. In cohesive soils, asphalt protects the soil by plugging its voids and water proofing it. It helps the cohesive soil to maintain low moisture content and to increase the bearing capacity.

The amount of bitumen required generally varies between 4 to 7% by weight. The actual amount is determined by trial.

Types of soil-bitumen

According to the Highway Research Board of USA, there are four types of soil-bitumen.

(1) Soil-bitumen (proper). This is a water-proof, cohesive soil system. The best results are obtained if the soil satisfies the following criteria.

(*a*) Passing No. 4 (4.76 mm) Sieve 50%.

(*b*) Passing No. 40 (0.425 mm) Sieve 35 to 100%.

(*c*) Passing No. 200 (0.074 mm) Sieve 10 to 50%.

(*d*) Plastic limit less than 18%.

(*e*) Liquid limit less than 40%.

(*f*) The maximum size of the particle should not be greater than one- third the compacted thickness of the soil-bitumen.

The quantity of bitumen varies from 4 to 7% of the dry weight.

(2) Sand-bitumen. This is a bitumen stabilised cohesionless soil system. The sand should be free from vegetal matter or lumps of clay. The sand may require filler for its mechanical stability. However, it should not contain more than 25% minus No. 200 sieve material (*i.e.* the material finer than No. 200 Sieve) for dune sands and not more than 12% in case in other types of sand.

The amount of bitumen required varies from 4 to 10%.

(3) Water-Proofed Clay Concrete. A soil possessing a good gradation is water proofed by a uniform distribution of 1 to 3% of bitumen in this system. Soils of three different gradations have been recommended. For the three gradations, the percentage passing No. 200 sieve varies between (*i*) 8 to 12; (*ii*) 10 to 16 and (*iii*) 13 to 30.

(4) Oiled earth. In this system, a soil surface consisting of silt-clay material is made water proof by spraying bitumen in two or three applications. Slow or medium curing bitumen or emulsions are used. The bitumen penetrates only a short depth into the soil. The amount of bitumen required is about 5 litres per square metre of the soil surface.

Factors affecting bituminous stabilisation

(1) Type of soil. Bituminous stabilisation is very effective in stabilising sandy soils having little or no fines. If a cohesive soil has the plastic limit less than about 20% and the liquid limit less than 40%, it can be effectively stabilised. However, plastic clays cannot be properly treated because of the mixing problems and large quantity of asphalt required. Fine-grained soils of the arid regions which have high p_H value and contain dissolved salts do not respond well.

(2) Amount of asphalt. The quality of the bitumen-stabilised soil improves with the amount of asphalt upto a certain limit. However, if the amount of the asphalt is excessive, it results in a highly fluid mixture that cannot be properly compacted.

(3) Mixing. The quality of the product improves with more thorough mixing.

(4) Compaction. The dry density of the bitumen-soil depends on the amount and type of compaction. It also depends upon the *volatile content.* In modified AASHO test, the maximum dry density occurs at a volatile content of about 8%. For samples cured and then immersed in water, the maximum strength occurs at a moulding volatile content corresponding to the maximum compacted density.

Construction Methods

Construction methods for bituminous stabilisation are similar to those used for soil-cement stabilisation. However, the following points should be noted.

(1) The optimum volatile content for compaction is generally much greater than that for stability. The volatile content required for thorough mixing may be even greater, especially for clayey soils. It is, therefore, necessary to aerate the mix between mixing and compaction and between compaction and application.

(2) To obtain a high stability, the layer method of construction is preferred. Each layer is kept about 5 cm thick. When the lower layer has dried up, the subsequent layer is laid. The total thickness for bases is kept between 10 to 20 cm.

(3) In the mix-in-place method, the bitumen is sprayed in several passes. Each layer is partially mixed before the next pass. This method prevents the saturation of the surface of the subgrade.

(4) Climatic conditions influence the amount of bitumen that can be applied, as the amount of fluid (moisture) already present is the soil depends upon the climatic conditions.

15.6. CHEMICAL STABILISATION

In chemical stabilisation, soils are stabilised by adding different chemicals. The main advantage of chemical stabilisation is that setting time and curing time can be controlled. Chemical stabilisation is however generally more expensive than other types of stabilisation.

The following chemicals have been successfully used.

(1) Calcium Chloride. When calcium chloride is added to soil, it causes colloidal reaction and alters the characteristics of soil water. As calcium chloride is deliquescent and hygroscopic, it reduces the loss of moisture from the soil. It also reduces the chances of frost heave, as the freezing point of water is lowered. Calcium chloride is very effective as dust palliative. As the soils treated with calcium chloride do not easily pick up water, the method is effective for stabilisation of silty and clayey soils which lose strength with an increase in water content.

Calcium chloride causes a slight increase in the maximum dry density. However, the optimum water content is slightly lower than that for the untreated soil. It causes a small decrease in the strength of the soil. However, if the compacted soil is put to water imbibition, water pick up is reduced and the strength of the treated soil is greater than that of the untreated soil.

It may be noted that most of the benefits of stabilisation require the presence of the chemical in the pore fluid. As soon as the chemical is leached out, the benefits are lost. The performance of treated soils depends to a large extent on the ground-water movement.

The construction methods are similar to those used for lime stabilisation. The quantity of calcium chloride required is about $\frac{1}{2}$% of the weight of the soil.

(2) Sodium Chloride. The action of sodium chloride is similar to that of calcium chloride in many respects. However, the tendency for attraction of moisture is somewhat lesser than that of calcium chloride. When sodium chloride is added to the soil, crystallisation occurs in the pores of the soil and it forms a dense hard mat with the stabilised surface. The pores in the soil get filled up and retard further evaporation of water. Sodium chloride also checks the tendency for the formation of shrinkage cracks.

Sodium chloride is mixed with the soil either by the mix-in-place method or by the plant-mix method. It should not be applied directly to the surface.

The quantity of sodium chloride required is about 1% of the soil weight.

(3) Sodium Silicate. Sodium silicates, as well as other alkali silicates, have been successfully used for soil stabilisation. The chemical is used as solution in water, known as *water glass.* The chemical is injected into the soil. Sodium silicate gives strength to soil when it reacts with it. It also makes the soil impervious. It also acts as a dispersing agent. The maximum compacted density is increased. The quantity of the chemical required varies between 0.1 to 0.2% of the weight of the soil.

This method of stabilisation is relatively inexpensive, but its long-term stability is doubtful. The treated soil may lose strength when exposed to air or to ground water.

(4) Polymers. Polymers are long-chained molecules formed by polymerising of certain organic chemicals called *monomers.* Polymers may be natural or synthetic. Resins are natural polymers. Calcium acrylate is a commonly used synthetic polymer. When a polymer is added to a soil, reaction takes place. Sometimes, the monomers are added with a catalyst to the soil. In that case, polymerisation occurs along with the reaction.

(5) Chrome Lignin. The chemical lignin is obtained as a by- product during the manufacture of paper from wood. Chrome lignin is formed from black liquor obtained during sulphite paper manufacture. Sodium bicarbonate or potassium bicarbonate is added to sulphite liquor to form chrome lignin. It slowly polymerises into a brown gel. When the chemical is added to the soil, it slowly reacts to cause bonding of particles. The quantity of lignin required varies from 5 to 20% by weight.

As lignin is soluble in water, its stabilising effect is not permanent.

(6) Other Chemicals. (*i*) Some water proofers such as alkyl chloro silanes, siliconates amines and quaternary ammonium salts, have been used for water proofing of soils.

(*ii*) Coagulating chemicals, such as calcium chloride and ferric chloride, have been used to increase the electrical attraction and to form flocculated structure in order to improve the permeability of the soil.

(*iii*) Dispersant, such as sodium hexa-metaphosphate, are used to increase electrical repulsion and to cause dispersed structure. The compacted density of the soil is increased.

(*iv*) Phosphoric acid combined with a wetting agent can be used for stabilisation of cohesive soils. It reacts with clay minerals and forms an insoluble aluminum phosphate.

15.7. THERMAL STABILISATION

Thermal change causes a marked improvement in the properties of the soil. Thermal stabilisation is done either by heating the soil or by cooling it.

(*a*) Heating. As the soil is heated, its water content decreases. Electric repulsion between clay particles is decreased and the strength of the soil is increased. When the temperature is increased to more than 100°C, the adsorbed water is driven off and the strength is further increased.

When the soil is heated to temperature of 400°C to 600°C, some irreversible changes occur which make the soil non-plastic and non-expansive. The clay clods are converted into aggregates.

With further increase in temperature, there is some fusion and vitrification, and a brick-like material is obtained which can be used as an artificial aggregate for mechanical stabilisation.

This method of stabilisation is quite expensive because of large heat input. It is rarely used in practice.

(*b*) **Freezing.** Cooling causes a small loss of strength of clayey soils due to an increase in interparticle repulsion. However, if the temperature is reduced to the freezing point, the pore water freezes and the soil is stabilised. Ice so formed acts as a cementing agent.

Water in cohesionless soils freezes at about 0°C. However, in cohesive soils, water may freeze at a much lower temperature. The strength of the soil increases as more and more water freezes.. This method of stabilisation is very costly. This method is used only in some special cases. It has been successfully used to solidify soils beneath foundations. The method is commonly used when advancing tunnels or shafts through loose silt or fine sand.

Freezing may cause serious trouble to adjacent structures if the freezing front penetrates these areas. It may cause excessive heaving. The method should be used after considering the above aspects.

15.8. ELECTRICAL STABILISATION

Electrical stabilisation of clayey soils is done by a process known as *electro-osmosis*. As a direct current (D.C.) is passed through a clayey soil, pore water migrates to the negative electrode (cathode). It occurs because of the attraction of positive ions (cations) that are present in water towards cathode. The strength of the soil is considerably increased due to removal of water. (For further details of electro-osmosis see chapter 16).

Electro-osmosis is an expensive method, and is mainly used for drainage of cohesive soils. Incidentally, the properties of the soil are also improved.

15.9. STABILISATION BY GROUTING

In this method of stabilisation, stabilisers are introduced by injection into the soil. As the grouting is always done under pressure, the stabilisers with high viscosity are suitable only for soils with high permeability. This method is not suitable for stabilising clays because of their very low permeability.

The grouting method is costlier as compared with direct blending methods. The method is suitable for stabilising buried zones of relatively limited extent, such as a pervious stratum below a dam. The method is used to improve the soil that cannot be disturbed. An area close to an existing building can be stabilised by this method.

Types of Grouting

Depending upon the stabiliser used, grouting techniques can be classified as under:

(1) Cement Grouting. A cement grout consists of a mixture of cement and water. If the hole drilled in the soil is smooth, the water-cement ratio is kept low. Sometimes, chemicals are added to grout to increase it fluidity so that it can be injected into the soil.

Cement grouting is quite effective for stabilising rocks with fissures, gravel and coarse sand.

(2) Clay Grouting. In this method, the grout used is composed of a very fine-grained soil (bentonite clay) and water. The bentonite clay readily adsorbs water on its surface. The viscosity, strength and flow characteristics of the grout can be adjusted according to the site conditions. Clay grouting is suitable for stabilising sandy soils.

Sometimes, other chemicals are added to clay grout. Clay cement grout is a mixture of clay, bentonite and cement. Clay-cement grout is a mixture of clay, bentonite and cement. Clay-chemical grout is a mixture of clay and sodium silicate. It is effective for medium and fine sands.

(3) Chemical Grouting. The grout used consists of a solution of sodium silicate in water, known as *water glass*. The solution contains both free sodium hydroxide and colloidal silicic acid. An insoluble silica gel is formed. As the reaction is slow, calcium chloride is generally added to accelerate the reaction.

The method is suitable for medium and fine sands. However, the effect of chemical grouting is not permanent.

(4) Chrome-lignin grouting. The grout used is made of lignosulphates and a hexavalant chromium compound. When it is combined with an acid, the chromium ion changes valence and thereby oxidises the lignosulphates into a gel.

The method can be used to stabilise fine sand and coarse silt.

(5) Polymer grouting. Various polymers have been successfully used in grouting of fine sands and silts.

(6) Bituminous grouting. Sandy and silty soils have been grouted successfully using emulsified asphalt. Slow-setting emulsions are generally preferred, as these can travel a large distance into the stratum.

15.10. STABILISATION BY GEOTEXTILE AND FABRICS

The soil can be stabilised by introducing geotextiles and fabrics which are made of synthetic materials, such as polythelene, polyester, nylon. The geotextile sheets are manufactured in different thicknesses ranging from 10 to 300 mils (1 mil = 0.0254 mm = 25.4 μ). The width of the sheet can be up to 10 m. These are available in rolls of length upto about 600 m. Geotextiles are manufactured in different patterns, such as woven, non-woven, grid, and hybrid. The woven geotextiles are made from continuous mono-filament or slit-film fibers. The non-woven geotextiles are made by the use of thermal or chemical bonding of continuous fibres and then pressed through rollers into relatively thin sheets. The grids of geotextiles are made from a sheet of polymer by punching it and then elongating it in at least one direction. The hybrid geotextile are nothing but combinations of woven, non-woven and grid.

The geotextiles are quite permeable. Their permeability is comparable to that of fine sand to coarse sand. These ar quite strong and durable. These are not affected by even hostile soil environment. The use of geotextiles in geotechnical and construction engineering has increased considerably in the last 2 decades. Geotextiles are being increasingly used for the site improvement, soil stabilisation and various other related works. While selecting geotextiles for a particular job, due importance should be given to the major function that the geotextile has to perform, as explained below.

1. Geotextiles as separators. Geotextiles are commonly used as separators between two layers of soils having a large difference in particle sizes to prevent migration of small-size particles into the voids of large-size particles. The main use as separators is in the construction of highways on clayey soils. As the particle size of granular base course of the highway is much larger than that of the subgrade (clayey soil in this case), it is the usual practice to provide an intervening soil layer of a soil containing grain-sizes intermediate between that of the subgrade and the base course to prevent migration of clay particles into the base course. Instead of the intervening soil layer, geotextile can be provided to serve the same purpose. The size of perforations should be according to the requirement. Thus a geotextile sheet is used between the subgrade and the base course (Fig. 15.1).

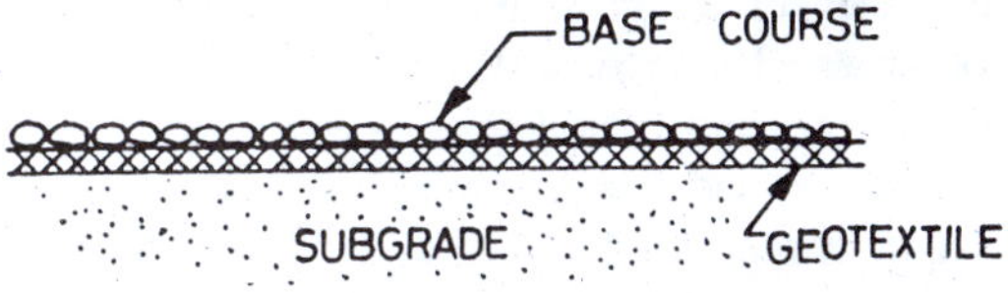

Fig. 15.1.

2. Geotextile as Filter. It is the usual practice to provide a properly graded filter to prevent the movement of soil particles due to seepage forces The filter is so designed that the particle size of the filter is small enough to hold the protected material in place, as discussed in chapter 10. If the filter material is not properly selected, the particles of the soil move into the pores of the filter and may prevent proper functioning of the drainage. It may also lead to piping.

Geotextiles can be used as filters instead of conventional filter. When the silt-laden turbid water passes through the geotextile, the silt particles are prevented from movement by the geotextile. The modification in the soil and void of the geotextile occurs, and

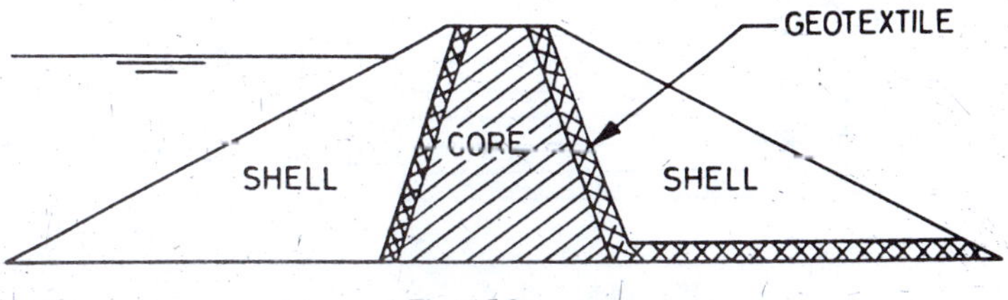

Fig. 15.2.

after some time an equilibrium stage is attached. For relatively thin geotextile sheet, most of the filtration occurs within the soil just upstream of the geotextile fabric. Fig. 15.2 shows the use of geotextile as filter on the upstream and downstream of the core of a zoned earth dam. It prevents the migration of the particles of the core into the shells.

3. Geotextile as Drain. A drain is used to convey water safely from one place to the other. As the geotextiles are pervious, they themselves function as a drain. They have a relatively higher water-carrying capacity as compared to that of the surrounding soil.

Drainage occurs either perpendicular to the plane of the sheet or in-plane of the sheet. In the first case, it functions primarily as a filter. In the latter case, it acts as a water carrier, and a relatively bulky geotextile or a composite system of geotextile is required. Fig. 15.3 shows a typical application where geotextile is used for drainage behind a retaining wall.

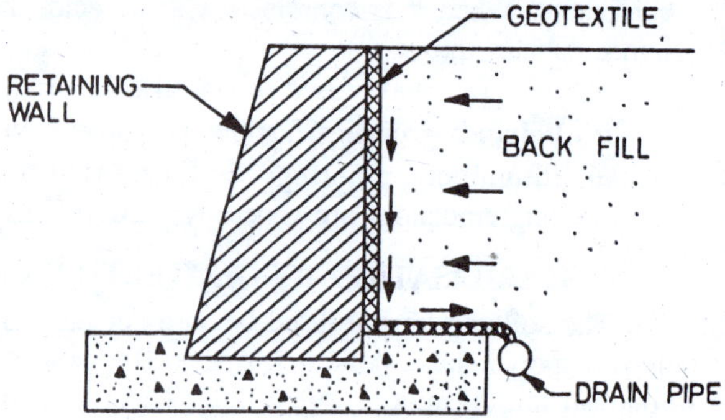

Fig. 15.3.

In all the above applications of the geotextile, the following advantages are generally achieved.

(1) The installation is generally easier and faster.

(2) The system has greater stability.

(3) The quantity of soil to be excavated and disposed of is less.

(4) The load on the structure is less.

4. Geotextile as Reinforcement for strengthening soil. Geotextiles have a high tensile strength. These can be used to increase the load-carrying capacity of the soil. Geotextiles are used as reinforcement in the soil, which is poor in tension but good in compression. The action is somewhat similar to that of steel bars in a reinforced concrete slab.

Geotextiles when used as reinforcement for soils have solved many construction problems on soft and compressible soils. Fig. 15.4 shows the reinforcement of an embankment with geotextiles. The geotextiles have been used in the construction of unpaved roads over soft soils. These are laid over the soil and the base course of the road is placed directly over it. When the vehicles pass over the road, the geotextile deforms and its strength is mobilised.

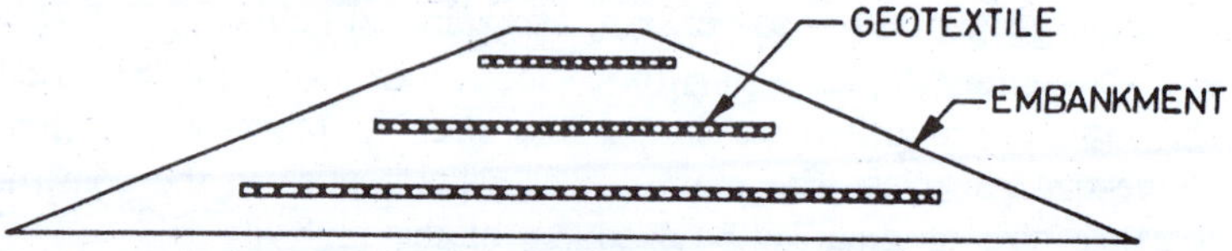

Fig. 15.4.

5. Geotextiles used as reinforcement in retaining walls. Geotextile can be used as reinforcement in the construction of earth-retaining structures. Geotextiles are used to form the facing of the retaining wall as well as reinforcement. Such retaining walls are also called fabric reinforced retaining walls (Fig. 15.5).

Fig. 15.5.

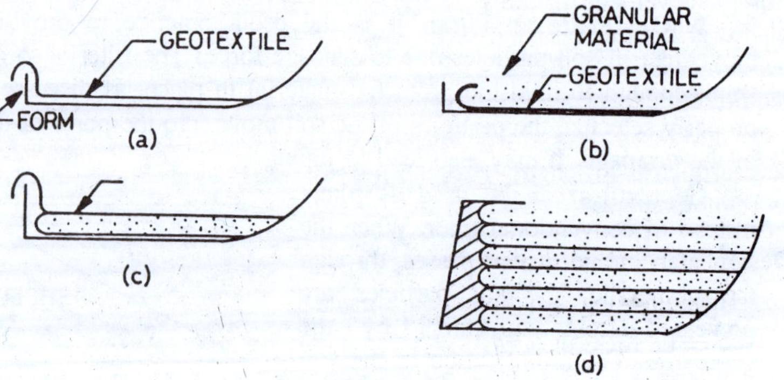

Fig. 15.6.

The following procedure is used for the construction of the fabric-reinforced wall.

(*i*) First the ground surface is levelled and the first geotextile sheet of the required width is laid over the surface such that about 1.5 m to 2 m of the sheet at the wall surface is draped over temporary wooden form (Fig. 15.6 *a*).

(*ii*) Granular material is placed over the geotextile sheet and compacted with a roller of suitable weight.

(*iii*) After compaction, the sheet is folded as shown in Fig. 15.6 (*b*).

(*iv*) The second geotextile sheet is placed over the compacted layer over the granular material and draped over the wooden form as shown in Fig. 15.6 (*c*), and the process is repeated.

(*v*) The front face of the wall is protected by the use of shotcrete or gunite. Shotcrete is the cement concrete with a low water content. It is sprayed over the soil surface at a high pressure. Fig. 15.6 (*d*) shows the completed wall.

The design of fabric reinforced retaining walls is similar in principle to that of reinforced earth discussed in the following section.

15.11. REINFORCED EARTH

The soil can be stabilised by introducing thin strips in it. In reinforced earth, thin metal strips or strips of wire or geosynthetics are used as reinforcement to reinforce the soil. The essential feature of the reinforced earth is that friction develops between the reinforcement and the soil. By means of friction, the soil transfers the forces built up in the earth mass to the reinforcement. Thus tension develops in the reinforcement when the soil mass is subjected to shear stresses under loads.

The main application of the reinforced earth is in the reinforced earth wall. The wall consists of a facing element, reinforcement and the back fill (Fig. 15.7). At the exposed vertical surface of the earth mass, facing elements are used to provide a sort of barrier so that the soil is contained. The facing units are generally prefabricated from units which are small and light so that they can be easily transported and placed in position. These are usually made of steel, aluminum, reinforced concrete or plastic. These should be strong enough to hold back the back fill. Moreover, these should be such that the reinforcement can be easily fastened to them. The facing units generally require a small plain concrete footing at the bottom so that they can be easily built.

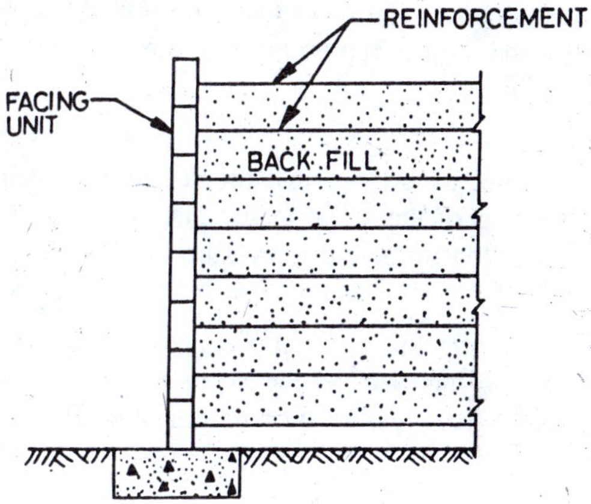

Fig. 15.7.

The reinforcement is connected to the facing element and extended back into the backfill zone. The friction developed in the reinforcement restrains the facing element. First a layer of reinforcement strips is placed at the level ground surface and the backfilling is done with a granular soil. The soil with less than 15 percent passing No. 200 sieve is used. The entire process of laying strips and backfilling is continued till the required height of the reinforced earth wall is attained.

Galvanised steel strips are commonly used as reinforcement. Each strip is about 50-100 mm wide and several metres in length. The thickness is upto 9 mm. Sometimes metal rods, wires and geotextiles are used as reinforcement.

Design of reinforced earth wall

The following assumptions are made :

1. The backfill is horizontal, without any surcharge.
2. The earth pressure acting on the facing element is the same as that acts on a rigid vertical face retaining concrete wall.

3. Rankine's earth pressure theory for active pressure (discussed in chapter 19) holds good.
4. The failure plane makes an angle of (45° + ϕ/2) with the horizontal, where ϕ is the angle of shearing resistance of the backfill material.

Let us consider the design of a retaining wall AB of height H (Fig. 15.8). When the wall rotates about the point A away from the backfill, a failure plane AC forms which makes an angle of (45° + ϕ/2) with the horizontal. The active pressure acting on the wall at any depth Z below the soil surface is given by

$$p_z = \gamma Z K_A \quad \text{...(15.1)}$$

The pressure variation is linear. The total pressure per unit length of the wall is given by

$$P_a = \frac{1}{2}\gamma H^2 K_A \quad \text{...(15.2)}$$

where K_A is Rankine's active earth pressure coefficient, equal to $\tan^2(45° + \phi/2)$, as discussed in chapter 19.

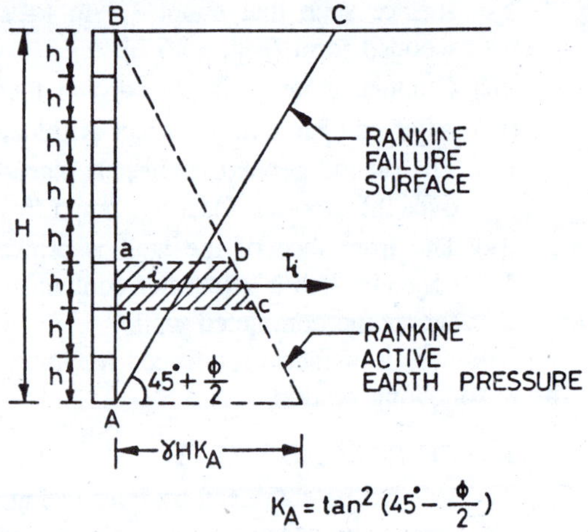

Fig. 15.8.

Let us divide the height H of the wall into small equal heights h as shown in Fig. 15.8. Let Z_i be the depth of any reinforcing strip i. The total earth pressure acting on the strip is represented by the area abcd of the pressure diagram. The average pressure p_i on the strip is given by

$$p_i = \gamma Z_i K_A \quad \text{...(15.3)}$$

Let us assume that the reinforcing strips are placed at intervals of h in the vertical direction. Let the spacing of the strips in the direction perpendicular to the plane of paper be s. Thus one reinforcing strip is subjected to the earth pressure on the area of ($h \times s$). Therefore, the tension in the strip i is given by

$$T_i = p_i A = (\gamma Z_i K_A)(h \times s) \quad \text{...(15.4)}$$

Using the same procedure, the tension in other reinforcing strips can be determined. Of course, the tension increases as the depth increases. The sum of the tension in all the reinforcing strips is equal to the total earth pressure on a length of s. Thus

$$\sum_{i=1}^{n} T_i = s P_a \quad \text{...(15.5)}$$

Length of reinforcing strip. The reinforcing strips should extend well beyond the active zone into the backfill to have proper grip length. The length of the strip lying between the wall AB and the failure plane AC is not effective for computing the grip length. The effective grip length lying on the right-hand side of the failure plane AC should be able to provide a suitable factor of safety against failure.

Let us again consider the reinforcing strip at depth Z_i. If F_i is the frictional resistance on the reinforcing strip of length L_e, then

$$F_i = F_s T_i \quad \text{...(15.6)}$$

where F_s is the factor of safety (usually taken as 2) and T_i is the tension in the strip.

If δ is the angle of surface friction, then

$$F_i = (\gamma Z_i) \tan\delta \times \text{(surface area of strip)}$$

The surface area of the strip of width b is taken equal to ($2\,bL_e$) as the resistance develops on both faces of the strip.

Thus
$$F_i = (\gamma Z_i) \tan\delta\,(2\,b L_e) \quad \text{...(15.7)}$$

Substituting the value of F_i in Eq. 15.6 and simplifying

$$L_e = \frac{F_s T_i}{2\gamma Z_i b \tan\delta} \quad \text{...(15.8)}$$

Eqs. 15.7 and 15.8 are applicable to rectangular strips. If round bars are used,

$$L_e = \frac{F_s T_i}{\pi d \gamma Z_i \tan \delta} \qquad \text{...[15.8 (a)]}$$

where d is the diameter of the bar.

If continuous reinforcing sheets are used,

$$L_e = \frac{F_s T_i}{2 \gamma Z_i \tan \delta} \qquad \text{...[15.8 (b)]}$$

The angle of surface friction δ depends upon the density and type of the backfill material. It also depends upon the roughness of the reinforcing strip. The value of δ usually varies between 0.5 ϕ and ϕ, where ϕ is the angle of shearing resistance of the backfill material.

Cross-sectional area of the reinforcing strip

The cross-sectional area of the reinforcing strip is determined if the allowable tensile stress (f_s) is known. The cross-sectional area is obviously equal to the tensile force divided by the allowable tensile stress. Thus

$$A_s = \frac{F_i}{f_s} = \frac{F_s T_i}{f_s}$$

or

$$A_s = \frac{F_s (\gamma Z_i K_A) h s}{f_s} \qquad \text{...(15.9)}$$

Eqs. 15.8 and 15.9 give different lengths L_e and the cross-sectional areas A_s for different reinforcing strips. For convenience, it is the usual practice to adopt the same length and the cross-section for all the reinforcing strips. Thus the value corresponding to the maximum tension at the base of wall is used for all strips. Hence $Z_i = H$ for all strips.

PROBLEMS

A. Descriptive and Objective type Questions

15.1. What is soil stabilisation ? What are its uses ?

15.2. What is mechanical stabilisation ? What are the factors that affect the mechanical stability of a mixed soil?

15.3. Describe in brief cement stabilisation. What are the factors that affect the stability of soil cement? Discuss construction methods.

15.4. Discuss the use of lime in stabilisation of soils. What are the chemical and physical changes which take place in lime stabilisation ?

15.5. Write a short note on bituminous stabilisation. What are different types of soil bitumen? Describe the factors affecting bituminous stabilisation.

15.6. What are different types of chemicals used in stabilisation of soils?

15.7. Write short notes on:

(*i*) Thermal stabilisation (*ii*) Electrical stabilisation.

(*iii*) Grouting (*iv*) Geotextile.

15.8. Write whether the following statements are true or false:

(*a*) Mechanical stabilisation requires addition of chemicals to soils.

(*b*) Cement stabilisation is more suitable for fine grained soils than coarse-grained soils.

(*c*) Lime stabilisation is suitable for coarse-grained soils.

(*d*) Bituminous stabilisation can be used for both coarse- grained and fine-grained soils.

(*e*) The effect of lignin as a stabilisation agent is permanent.

(*f*) Thermal stabilisation is quite inexpensive.

(*g*) Electro-osmosis is used for stabilising highly cohesive soils.

(*h*) A geotextile is embedded in the soil to give it stability.

(*i*) Cement grouting can be used for clayey soils. [**Ans.** True, (*d*), (*g*), (*h*)]

B. Multiple Choice Questions

1. Cement stabilisation is generally used for stabilising
(*a*) Sands (*b*) Silts
(*c*) Clays (*d*) All the above

2. Lime stabilisation is generally used for stabilising
(*a*) Gravels (*b*) Sands
(*c*) Clays (*d*) All the above

3. Bituminons stabilisation is generally used for stabilising
(*a*) Sands (*b*) Silts
(*c*) Clays (*d*) All the above

4. The following methods of mixing cement in the stabilisation of the soil is generally the best:
(*a*) Mix-in-place method (*b*) Stationary plant method
(*c*) Travelling plant method (*d*) None of above.

5. For stabilisation of heavy clays, the following method is generally most effective
(*a*) Mechanical stabilisation (*b*) Thermal stabilisation
(*c*) Chemical stabilisation (*d*) Electrical stabilisation.

6. Chemical grouting is generally used for
(*a*) fine sands and coarse sits (*b*) medium and fine sands
(*c*) Coarse sands (*d*) clays.

7. Lime stabilisation of clayey soils generally leads to
(*a*) Decrease in shrinkage limit (*b*) Decrease in plastic limit
(*c*) Increase in liquid limit (*d*) Flocculation of particles

8. For the maximum dry density the percentage of particles passing 75μ size is about
(*a*) 40% (*b*) 60%
(*c*) 20% (*d*) 80%

9. The material used for manufacture of geotextile is
(*a*) Polythene (*b*) Nylon
(*c*) Polyster (*d*) All the above

10. For the design of the reinforced earth wall, the following assumption is not made:
(*a*) The earth pressure distribution is the same as in a rigid retaining wall
(*b*) The Rankine theory is applicable
(*c*) The failure plane makes an angle of (45° – ϕ/2) with the horizontal
(*d*) There is no surcharge on the backfill

[**Ans.** 1. (*a*), 2. (*c*), 3. (*d*), 4. (*c*), 5. (*d*), 6. (*b*), 7. (*d*), 8. (*c*), 9. (*d*), 10. (*c*)]

16

Drainage, Dewatering and Wells

16.1. INTRODUCTION

Drainage is the process of removal of gravity water (free water) from a soil mass in order to keep it in a stable condition. Drainage may be classified into two categories : (*i*) Surface Drainage, (*ii*) Sub-surface Drainage. Surface drainage is the method of collection and diversion of the surface run off. Subsurface drainage consists of collection and disposal of the ground water. Subsurface drainage is also known as *dewatering*. It is process of removal of water from a foundation pit when it is situated below the ground water table or when it is surrounded by a coffer dam.

The purpose of dewatering is to keep the excavation dry so that concreting can be done. Dewatering is temporary if it is done at the time of construction. It is followed by restoration to its original water table after the structure has been completed. Permanent dewatering is required for removing subsurface gravitational water throughout the life of structure. It may be necessary to keep the water away from the structure to check dampness or other ill effects.

Subsurface drainage not only facilitates construction, but it also helps in improving the properties of the soil. This helps in the stabilisation of soils. In case of fine-grained soils, although the quantity of water removed is not much, improvement in the properties of the soil is significant. Subsurface drainage also helps in reducing the hydrostatic pressure acting on the base of the structure.

The method of sub-surface drainage to be adopted at a particular site would depend upon the characteristics of the soil, the position of the water table and the time period the system has to operate. Various methods of sub-surface drainage are discussed in this chapter. Theory of wells is also dealt with. The methods of surface drainage are outside the scope of this text.

16.2. INTERCEPTOR DITCHES

Interceptor ditches are used for excavation of limited depth made in a coarse soil. These ditches are constructed around the area to be dewatered. The ditches must penetrate deeper than the level of the work area (Fig. 16.1). At suitable locations, sump pits are constructed along the ditch for installation of the pump to remove the water collected.

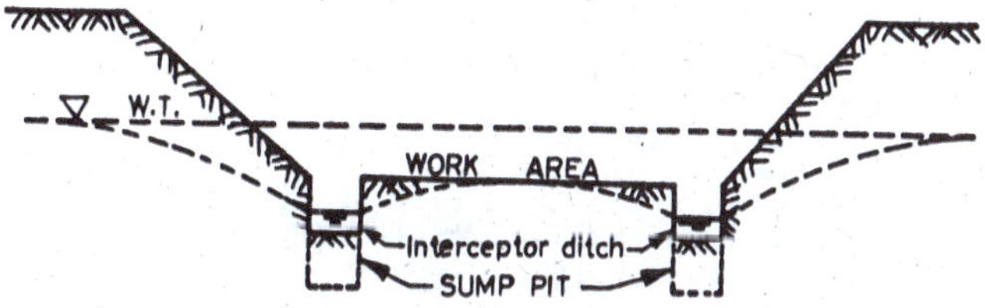

Fig. 16.1. Interceptor Ditch.

If the soil is fine sand of low permeability, boiling may occur in sumps and ditches. This may be prevented by placing filter layers on the sides and at the bottom of the ditches and sumps.

Interceptor ditches are most economical for carrying away the water which emerges on the slopes and near the bottom of the foundation pit. The method can be effectively used for rock formation, gravel and coarse sand. In fine sands and silts, there may be sloughing, erosion or quick conditions. For such soils, the method is confined to a depth of 1 to 2 m.

16.3. SINGLE-STAGE WELL POINTS

A well point is a perforated pipe about 1 m long and 5 cm in diameter. The perforations are covered with a screen to prevent clogging. A jetting nozzle is provided at its lower end. A conical steel drive point is fixed to the lower end of the well point to facilitate installation (Fig. 16.2). A ball valve is also provided near the lower end which permits flow of water only in the downward direction during installation. The well point is connected to the bottom of the *riser pipe* of the same diameter. Risers of different well points are connected to a horizontal pipe of 15 to 30 cm diameter, known as *header*. The header is connected to a specially designed pumping unit. The spacing of the well points depends upon the type of soil and the depth of water. Generally, it varies between 1 to 3 m.

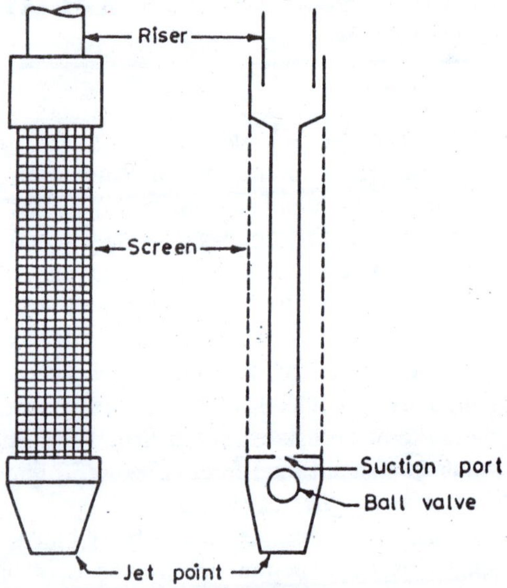

Fig. 16.2. Well Points.

Well points can be installed in a drilled hole, but generally these are installed by jetting. Water is pumped through the riser pipe in the downward direction. As it discharges through the nozzle, it displaces the soil below the tip. Jetting is continued till the required penetration of the tip is achieved. The advantage of installation by jetting is that the water under pressure washes away soil fines near the tip and leaves a relatively coarse material. It forms a natural filter around the tip. The hole formed near the tip is filled with coarse sand.

After the well points have been installed around the area to be dewatered, pumping is started. Each well point lowers the water table around it and forms a small cone of depression (Fig. 16.3). Various cones of depression join, and a common drawdown curve is obtained. The water table is thus lowered (Fig. 16.4.)

Well points are suitable for lowering the water table by 5 to 6 m in soils with a coefficient of permeability between 1×10^{-4} to 1×10^{-6} m/sec. The screen normally provided with the well points can prevent medium sand and coarse sand from entering the well point. If the stratum to be dewatered consists of

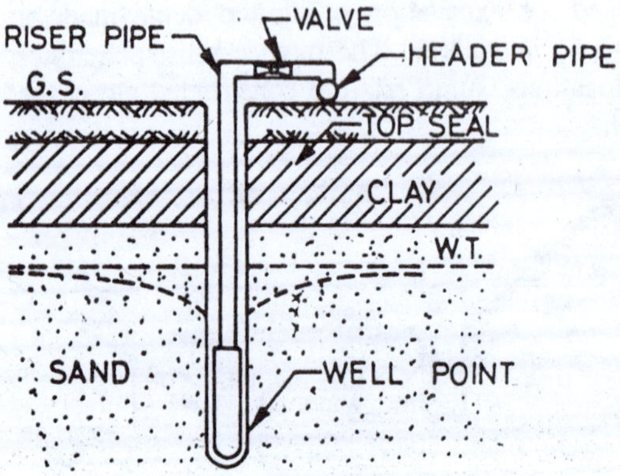

Fig. 16.3. Well-point Installation.

finer soils, a sand filter has to be provided around the well point.

It is essential to continue pumping once it has been started until the excavation is complete. If it is stopped in between, it may prove to be disastrous.

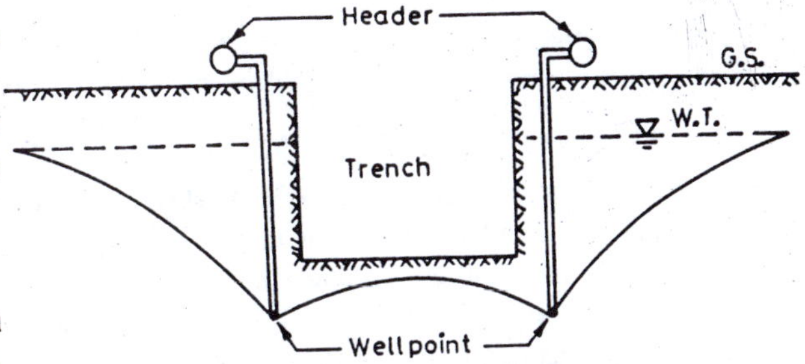

Fig. 16.4. Effect of Well Points.

16.4. MULTI-STAGE WELL POINTS

When the water table is to be lowered for a depth greater than 6 m, multi-stage well points are required. In this method, two or more rows of well points are installed at different elevations. Fig. 16.5 shows a two-stage well point system.

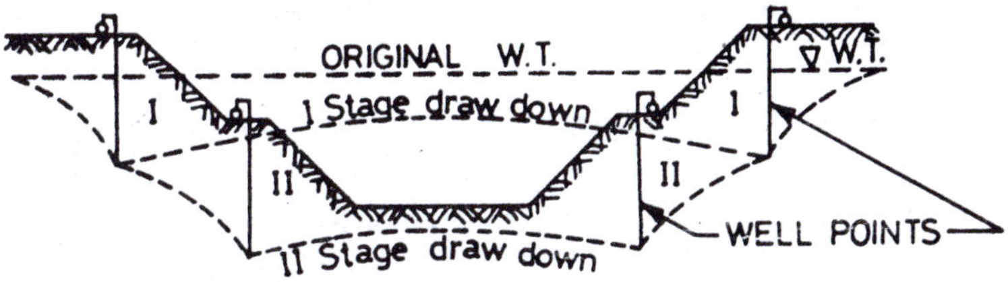

Fig. 16.5. Multi-stage Well Points.

The installation of well points is done in stages. The first stage well points (marked I) are located near the perimeter of the area, as in a single-stage well point system. These are put into operation and the water table is lowered by about 5 m and the area is excavated. The well points of the second stage (marked II) are then installed within the area already excavated. Water table is further lowered by about 5 m and the excavation of further 5 m depth is done. Thus the total depth of excavation becomes about 10 . If required, the third stage of well points can also be installed to further lower the water table.

The method is useful for excavations upto 15 m depth. Excavations exceeding 15 m depth are generally dewatered by a deep well system (Sect. 16.7).

16.5. VACUUM WELL POINTS

Well points cannot be used successfully for draining silty sands and other fine sands with an effective size less than about 0.05 mm. The coefficient of permeability of such soils is generally between 1×10^{-5} to 1×10^{-7} m/sec. These soils can be effectively drained by using vacuum well points.

For installation of a vacuum well points, a hole of about 25 cm diameter is formed around the well point and the riser pipe by jetting water under pressure. When water is still flowing, medium to coarse sand is filled into the hole upto about 1 m from the top. The top 1 m portion of the hole is then filled by tamping clay into it. It forms a sort of *seal* (Fig. 16.6). Any other impervious material can also be used instead of clay to form a seal. Well point spacing is generally closer than that in a conventional system.

When the header is connected to a vacuum pump, it creates a vacuum in the sand filter around the well point. As the pressure on the water table is equal to the atmospheric pressure, the head causing flow is increased by an amount equal to the vacuum pressure. The hydraulic gradient increases and it overcomes the flow resistance in the soil pores. The ground water flows to the region of vacuum in the well points and drainage occurs.

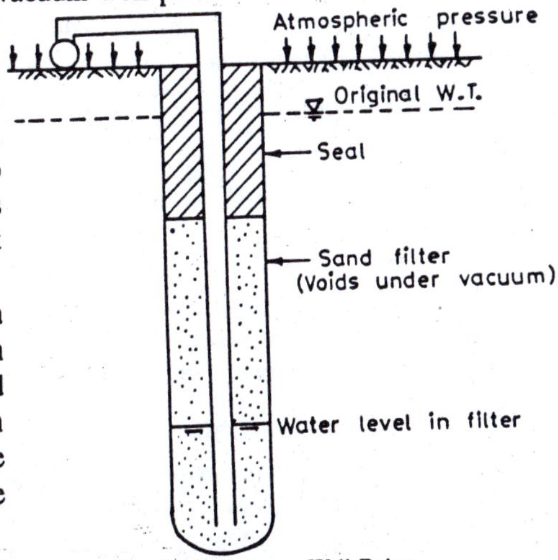

Fig. 16.6. Vaccum Well Point.

As the effective pressure on the soil is increased,

consolidation takes place. It makes the soil stiff. However, the process is slow and it may take several weeks for the soil to become stiff enough for carrying out the excavation work.

16.6. SHALLOW WELL SYSTEM

In a shallow well system, a hole of about 30 cm diameter is first bored into the ground, using a casing. A filter tube of about 15 cm diameter, covered with a special wire mesh, is then lowered into the casing. The casing is gradually withdrawn and suitable filter material is added in to the annular space between the casing and the filter tube. This forms a *filter well.* A suction pipe is lowered into the filter well. A number of such wells may be installed. The suction pipes of all these wells are connected to a common header A pumping unit is attached to the header. As the pumping is started, the drainage occurs. The suction lift of the well should not be more than 10 m for its proper working.

Shallow well system is rarely used in practice. Well-point systems, as discussed earlier, are more economical upto a depth of 10 m than a shallow well system.

16.7. DEEP WELL SYSTEM

A deep well is about 30 to 60 cm in diameter, bored to a depth of 15 to 30 m . It is provided with a casing which is perforated one in the pervious zones penetrated. Coarse filter material is placed in the annular space between the casing and the walls of the hole. The spacing of deep wells varies between 10 to 30 m, depending upon thee area to be dewatered and the location of the water table.

A submersible pump is placed inside the casing near the bottom. The pump is driven by a motor mounted on the top of the casing through a vertical shaft (Fig. 16.7). However, if the motor is submersible, it can be directly attached to the pump. As the pump is placed at the bottom, there is no restriction on the height to which the water can be lifted, unlike a shallow well system.

Deep wells are located on the outer periphery of the area to be excavated. A row of well point is also installed at the toe of side slopes of the deep excavation to intercept seepage between the deep wells and to prevent sloughing of the slopes near the toe.

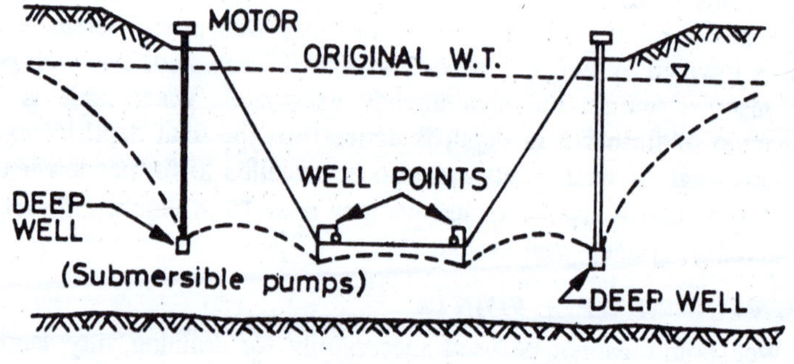

Fig. 16.7. Deep Well System.

As in the case of well point systems, pumping once started must be continued until the entire excavation work is completed. If the pumping is stopped in between, there would be a rapid development of hydrostatic and seepage pressure in the excavation, which may prove to be disastrous.

16.8. HORIZONTAL WELLS

Horizontal wells of about 5 cm to 8 cm diameter have been used for drainage of hill sides. These wells are drilled into the hill at a slightly upward slope. A perforated casing is installed in the well to collect and discharge water. Horizontal wells have been successfully installed for horizontal length of 60 m.

Large horizontal tunnels have also been successfully used to tap deep aquifers beneath hill sides.

A combined system of vertical wells and horizontal wells can be used to drain stratified soil deposits. In this system, the vertical wells intercept the aquifer and discharge the water into the horizontal well. The water collected by the horizontal well is discharged at a suitable point.

16.9. ELECTRO-OSMOSIS

Electro-osmosis is a method of drainage of cohesive soils in which a direct current (D.C.) is used. When a direct current is passed through a saturated soil between a positive electrode (anode) and a negative electrode (cathode), pore water migrates to the cathode. The cathode is a well point which collects the water drained from the soil. The water collected is discharged, as in a conventional well-point system.

The phenomenon of electro-osmosis can be explained with the help of the electrical double layer. Cations (positive ions) are formed in pore water when the dissolved minerals go into solution. These cations move towards the negatively charged surface of clay minerals to satisfy the electrical charge. As the water molecules act as dipoles, the cations also attract the negative end of dipoles. When the cations move to the cathode, they take with them the attached water molecules. In fact, the entire outer part of the diffuse double layer which is loosely adsorbed to the soil particles gets sheared along a plane. Only the inner most part of the layer about 10 A° thick remains attached to the particle.

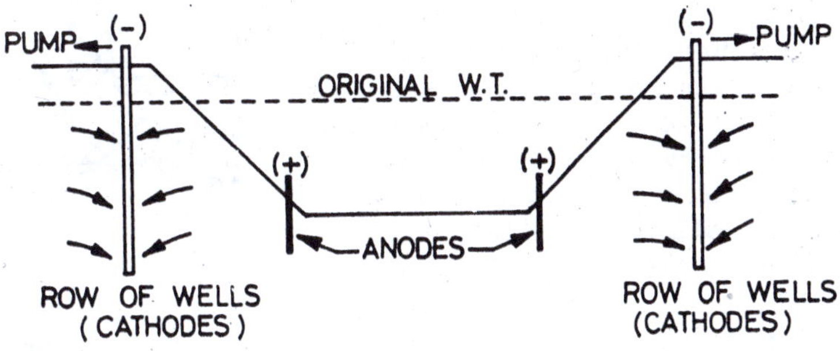

Fig. 16.8.

Fig. 16.8 shows an installation in which electro-osmosis is used. Anodes are in the form of steel rods located near the toe of the slope of the excavation. Cathodes are in the form of perforated pipes, resembling well points, installed in the soil mass about 4 to 5 m away from the slope of the cut. The electrodes are so arranged that the natural direction of flow of water is reversed and is directed away from the excavation. This arrangement is required to prevent sloughing of the slopes. In many cases, mere reversing of the direction of flow helps in increasing the stability of the slope even if there is no significant decrease in the water content of the soil.

The system requires about 20 to 30 amperes of electricity per well at a voltage of 40 to 180. The consumption of energy is between 0.5 to 10 kWh/m^3 of soil drained. Because of specialised equipment and high electricity consumption, drainage by electro-osmosis is expensive compared with other methods. The method should be used only in exceptional cases when other methods cannot be used. It is normally used to drain water in a cohesive soil of low permeability ($k = 1 \times 10^{-5}$ to 1×10^{-8} m/sec).

Electro-osmosis also helps in increasing the shear strength of the cohesive soil.

16.10. PERMANENT DRAINAGE AFTER CONSTRUCTION

When a usable part of a structure lies below the ground water table, it should be made water proof. Suitable construction techniques should be used. As far as possible, basement walls and floor should be cast monolithically. If the joints are provided, these should be as few as possible. All joints shall be provided with water stops.

Permanent drainage is done by providing foundation drains and blanket drains, as explained below.

***(a)* Foundation Drains.** When ground water flows towards the structure, provisions are made to carry the water away from the foundation. Fig. 16.9 shows an arrangement of foundation drain in which perforated drains are used. The drains are surrounded by a filter. The drains should be placed at an elevation higher than the bottom of the footing to avoid the possibility of carrying away fine soil particles.

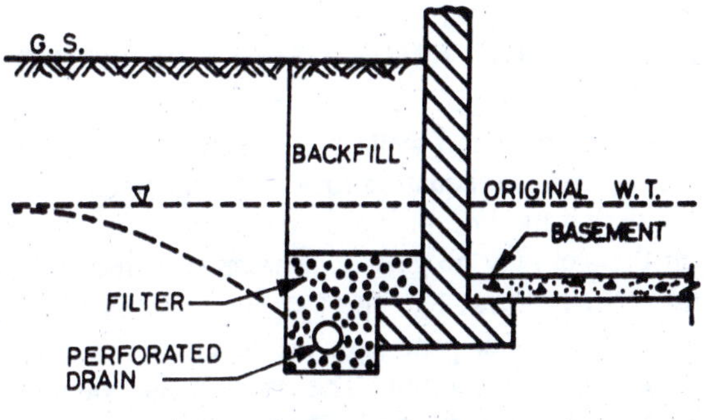

Fig. 16.9. Foundation Drain.

The water collected in the drain is disposed of by gravity to a nearby storm drain, nullah or any other drainage facility located at a lower level. If no such drain is available, the water is collected in a specially constructed sump well and pumped out. The drainage system should be provided with some periodical cleaning arrangement.

Foundation drains are effective when the depth below the water table is not too much. When the water table is very high, suitable interceptor drains are installed at some distance away from the structure to lower the water table in stages.

(*b*) **Blanket Drains.** The blanket drains are provided beneath the floor slab (Fig. 16.10) A blanket drain consists of coarse sand, gravel and crushed stones. It provides a highly pervious drainage path. The water coming out of the blanket drain is collected and drained away by gravity. Alternatively, it can be collected in a sump pit and then pumped out.

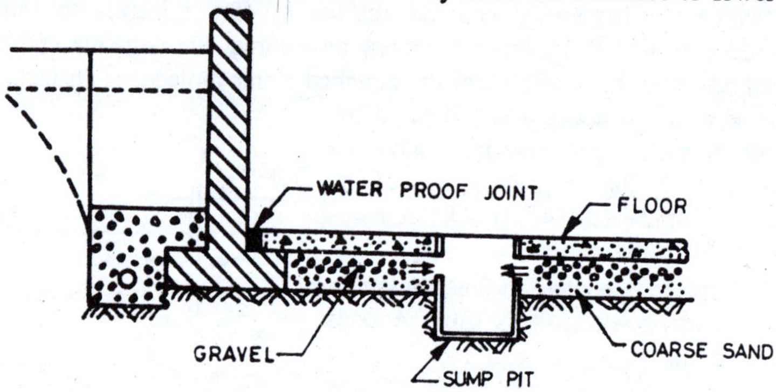

Fig. 16.10. Blanket Drain.

Blanket drains are quite effective in reducing the uplift pressure on the floor. The possibility of upward seepage flow through the basement floor is also considerably reduced.

16.11. DESIGN OF DEWATERING SYSTEMS

The design of a dewatering system consists of the determination of the number, size, spacing and penetration of the well points or wells. These parameters depend upon the expected rate of discharge, the type of the soil and the drawdown in the wells. Collectors and pumps should have sufficient capacity to serve the intended purpose.

It is extremely important to establish a fundamental relationship between the discharge and the corresponding drawdown. The rate of discharge is computed using Darcy's law. It is assumed that the stratum is homogeneous and isotropic. In case the stratum is anisotropic, it is assumed to have been transformed into an equivalent isotropic stratum using the method discussed in Chapter 9. It is further assumed that the flow is continuous and steady. Equations for the rate of discharge and corresponding drawdown for different types of wells are developed in the following sections.

The well may be either gravity well or artesian well. A gravity well penetrates a homogeneous, pervious stratum aquifer in which the water table is located. An artesian well penetrates a homogeneous, pervious stratum which is bounded by impervious strata above and below and in which the piezometric surface is above the top of the pervious stratum. A combined artesian- gravity type of flow occurs when the water table is an artesian well falls below the top of the pervious stratum.

The equations developed may also be used for the determination of discharge from wells for irrigation and other purposes. However such wells are constructed for supplying water and not drainage. In this text, the use of wells for drainage is of main concern.

16.12. DISCHARGE FROM A FULLY PENETRATING SLOT

When the drainage wells are closely spaced in a straight line, an approximate equation for the discharge can be obtained by considering the line of wells equivalent to a long slot. The flow in the slot may be gravity, artesian or combined.

(*a*) **Gravity Flow.** Fig. 16.11 shows a slot in which gravity flow occurs. The slot penetrates the pervious stratum in which the water table is located. The stratum is bounded by an impervious stratum at its base. The gravity flow is also known as *unconfined flow.*

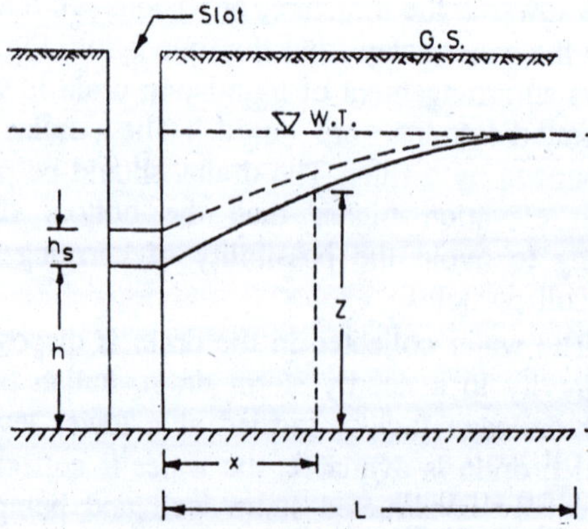

Fig. 16.11. Gravity Flow in a Slot.

The equation for the discharge is derived based on Dupuit-Forchheimer assumptions (see chapter 8). According to one of these assumptions, the hydraulic gradient at any point is taken equal to the tangent of the angle with the horizontal which the drawdown curve makes at that point. It is further assumed that the stratum is isotropic, homogeneous and Darcy's low is applicable.

Let us consider the flow through a vertical section of height z located at a distance x from the slot. From Darcy's law,

$$q = k\,i\,A$$

or
$$q = k \cdot \frac{dz}{dx} \cdot (z \cdot y) \qquad \text{...}(a)$$

where y is the length of the slot perpendicular to the plane of paper.

Transposing Eq. (a), $$z\,dz = \frac{q}{ky}\,dx$$

Integrating and substituting the boundary conditions, $x = 0$, $z = h$ and $x = L$, $z = H$,

$$\left[\frac{z^2}{2}\right]_h^H = \frac{q}{ky}\,[x]_0^L \qquad \text{...}(b)$$

or
$$\left[\frac{H^2 - h^2}{2}\right] = \frac{qL}{ky}$$

or
$$q = k\left(\frac{H^2 - h^2}{2L}\right) \cdot y \qquad \text{...(16.1)}$$

The drawdown level at a vertical section at a distance x can be obtained from Eq. (b).

$$\left[\frac{z^2}{2}\right]_z^H = \frac{q}{ky}\,[x]_x^L$$

or
$$H^2 - z^2 = \frac{2q}{ky}\,(L - x) \qquad \text{...(16.2)}$$

Substituting the value of q from Eq. (16.1),

$$H^2 - z^2 = \frac{2(L-x)}{ky} \cdot \frac{k\,(H^2 - h^2)}{2\,L} \cdot y$$

or
$$H^2 - z^2 = \frac{(L-x)}{L} \cdot (H^2 - h^2) \qquad \text{...(16.3)}$$

As the flow at the face of the slot is almost vertical, Dupuit Forchheimer assumption is not strictly valid. The actual drawdown level is given by the following equation

$$H^2 - z^2 = \frac{(L-x)}{L}\,[H^2 - (h + h_s)^2] \qquad \text{...(16.4)}$$

The value of the height h_s depends upon the ratios h/H and L/H.

(b) **Artesian Flow.** Fig. 16.12 shows artesian flow in the slot. The artesian flow is also known as *confined flow.* In field, such a condition occurs when closely spaced wells are installed near and parallel to the bank of a river such that the pervious stratum is exposed.

The equation for discharge can be derived as in the case of gravity flow. However, in this case, the area of flow is equal to ($t \times y$) and not ($z \times y$), where t is the thickness of the pervious stratum. Therefore, Eq. (a) becomes

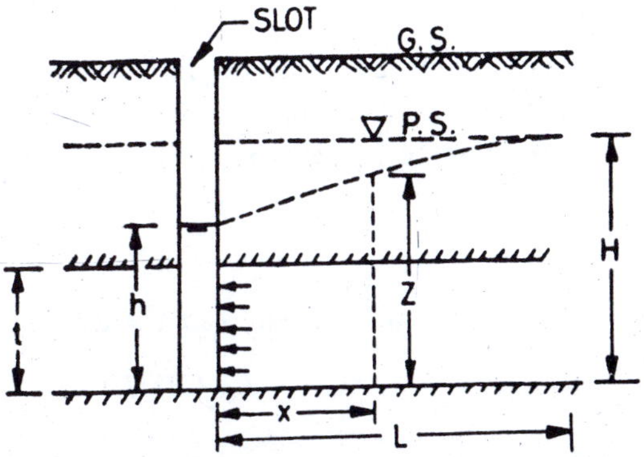

Fig 16.12. Artesian Flow in a Slot.

$$q = k \cdot \left(\frac{dz}{dx}\right) \cdot (t \times y)$$

or
$$dz = \frac{q}{kty} \cdot dx$$

Integrating and substituting the boundary conditions,

$$[z]_h^H = \frac{q}{kty}\ [x]_0^L \quad \text{...(c)}$$

or
$$(H - h) = \frac{q}{kty}\ (L - 0)$$

or
$$q = \frac{kty}{L}\ (H - h) \quad \text{...(16.5)}$$

The drawdown level at a distance x from the slot can be obtained from Eq. (c), as

$$[z]_z^H = \frac{q}{kty}\ [x]_x^L$$

or
$$(H - z) = \frac{q}{kty}\ (L - x)$$

Substituting the value of q from Eq. 16.5,

$$H - z = \frac{(H - h)(L - x)}{L} \quad \text{...(16.6)}$$

(c) **Artesian-Gravity Flow.** Fig. 16.13 shows an artesian gravity type of flow. The flow near the slot is gravity flow, whereas that away from the slot is artesian flow. Let L_G be the distance of the point at which

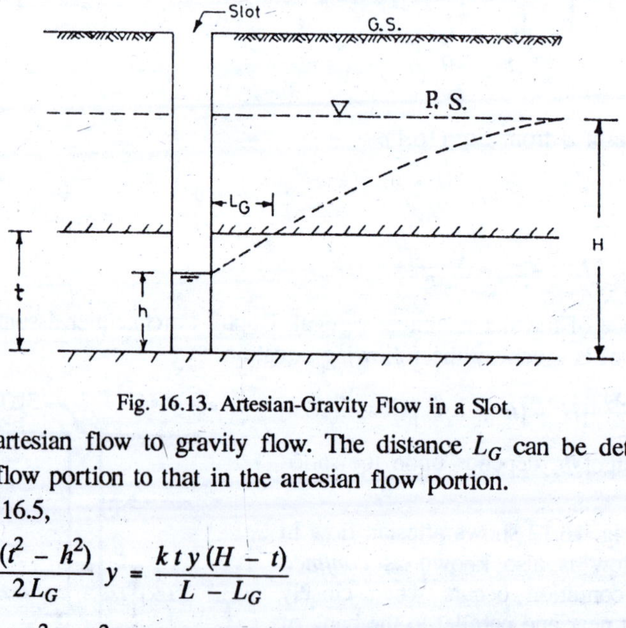

Fig. 16.13. Artesian-Gravity Flow in a Slot.

the flow changes from artesian flow to gravity flow. The distance L_G can be determined by equating the discharge in the gravity flow portion to that in the artesian flow portion.

From Eqs. 16.1 and 16.5,

$$\frac{k(t^2 - h^2)}{2L_G}\ y = \frac{kty(H - t)}{L - L_G}$$

or
$$(L - L_G)(t^2 - h^2) = 2L_G\, t\,(H - t)$$

or
$$L_G\,[2t(H - t) + (t^2 - h^2)] = L\,(t^2 - h^2)$$

or
$$L_G = \frac{L\,(t^2 - h^2)}{2tH - t^2 - h^2} \quad \text{...(16.7)}$$

From Eq. 16.1,
$$q = \frac{k(t^2 - h^2)}{2L_G} \cdot y$$

Substituting the value of L_G from Eq. 16.7,
$$q = \frac{k(t^2 - h^2)}{2L(t^2 - h^2)} y \cdot (2tH - t^2 - h^2)$$

or
$$q = \frac{ky(2tH - t^2 - h^2)}{2L} \quad \text{...(16.8)}$$

Eq. 16.8 can be used for the determination of the discharge.

The drawdown levels can be obtained as under.

(*i*) Gravity flow portion ($x \le L_G$)

From Eq. 16.3,
$$t^2 - z^2 = \frac{(L_G - x)}{L_G}(t^2 - h^2)$$

or
$$L_G(t^2 - z^2) = (L_G - x)(t^2 - h^2)$$

or
$$L_G z^2 = x(t^2 - h^2) + L_G h^2$$

or
$$z = \sqrt{\frac{x}{L_G}(t^2 - h^2) + h^2} \quad \text{...(16.9)}$$

(*ii*) Artesian flow portion ($x \ge L_G$)

From Eq. 16.6,
$$H - z = \frac{(H - t)[L - L_G - (x - L_G)]}{(L - L_G)}$$

or
$$z = \frac{(x - L)(H - t)}{(L - L_G)} + H \quad \text{...(16.10)}$$

16.13. DISCHARGE FROM A PARTIALLY PENETRATING SLOT

When the thickness of the stratum is too large for full penetration of the slot, a partially penetrating slot is used. The flow in a partially penetrating slot can be gravity flow or artesian flow.

(*a*) **Gravity flow.** (Fig. 16.14). The discharge q_p can be computed using the results of model studies conducted by Chapman (1956).

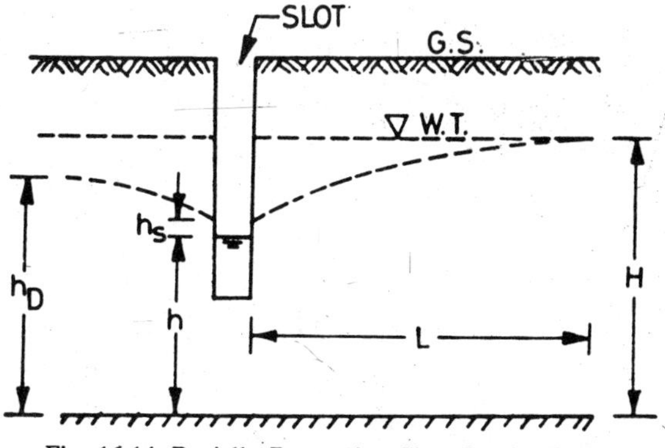

Fig. 16.14. Partially Penetrating Slot (Gravity Flow).

$$q_p = \left[0.73 + 0.27\frac{(H - h)}{H}\right]\frac{ky}{2L}(H^2 - h^2) \quad \text{...(16.11)}$$

The symbols are given in Fig. 16.14. As before, y is the length of the slot perpendicular to the plane of paper.

The maximum residual head (h_D) downstream of the slot is given by

$$h_D = h\left[\frac{1.48}{L}(H - h) + 1\right] \quad \ldots(16.12)$$

Eqs. (16.11) and (16.12) are valid for the ratio L/H equal to or greater than 3. It may be noted that h_D is greater than h.

(*b*) **Artesian flow.** Fig. 16.15 shows artesian flow in a partially penetrating slot of depth W. The discharge q_p is given by (Barron, 1952).

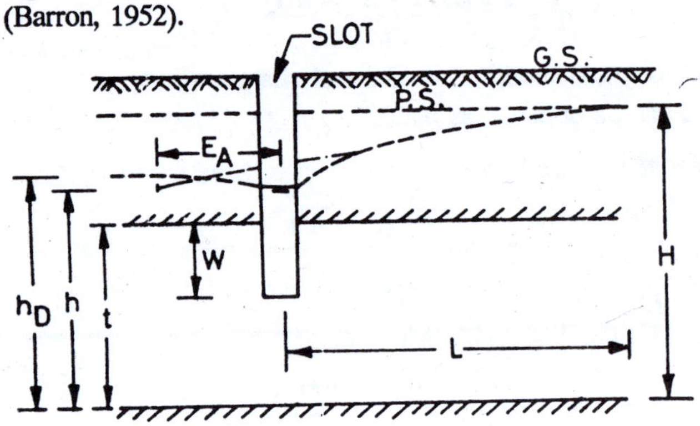

Fig. 16.15. Partially Penetrating Slot (Atresian Flow.)

$$q_p = \frac{k\,t\,y\,(H - h)}{L + E_A} \quad \ldots(16.13)$$

where E_A is the *extra-length factor,* which depends upon the ratio W/t and L/t (Fig. 16.16).

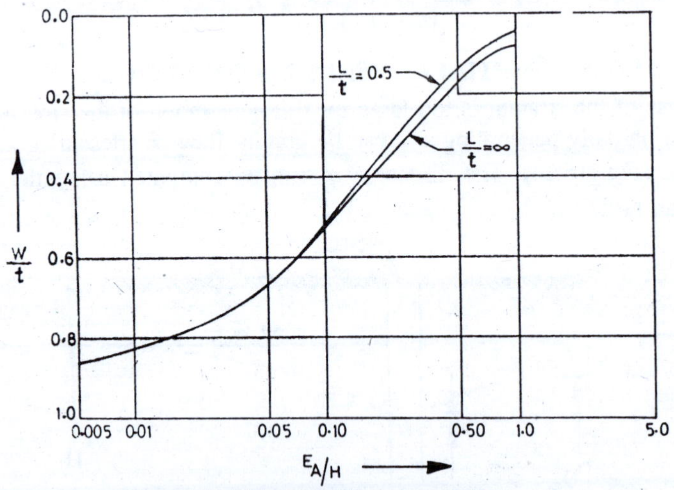

Fig. 16.16. Variation of Extra-Length Factor.

The maximum residual head h_D on the downstream of the slot is given by

$$h_D = \frac{E_A(H - h)}{L + E_A} + h \quad \ldots(16.14)$$

As in the case of gravity flow,the head h_D is greater than that at the slot.

16.14. DISCHARGE IN A SLOT FROM BOTH SIDES

In most of the practical cases, the flow towards a slot is from both sides and not only from one side. The equations for discharge from a slot from both sides are given below.

(a) Fully Penetrating Slot. The discharge from a fully penetrating slot from both sides is twice the discharge from one side. Equations developed in Sect. 16.12 can be used.

For gravity flow, Eq. 16.1 gives the discharge from one side. Therefore, the discharge from both sides is given by

$$q = k\left(\frac{H^2 - h^2}{L}\right) \cdot y \qquad \text{...(16.15)}$$

For artesian flow, Eq. 16.5 can be used to give the discharge from both sides as

$$q = \frac{2\,k\,t\,y\,(H - h)}{L} \qquad \text{...(16.16)}$$

(b) Partially Penetrating Slot. Chapman (1956) gave the following equations for the discharge of a partially penetrating slot from both sides.

For gravity flow,
$$q_p = \left[0.73 + 0.27\,\frac{(H - h)}{H}\right]\frac{ky}{L}\,(H^2 - h^2) \qquad \text{...(16.17)}$$

In case of artesian flow, the discharge is given by the equation,

$$q_p = \frac{2\,k\,t\,y\,(H - h)}{L + \lambda t} \qquad \text{...(16.18)}$$

where λ is a factor, which depends upon the ratio (W/t), as given in Fig. 16.17,
W = depth of slot in the aquifer, and t = thickness of aquifer.

All other notations are the same as in Sect. 16.12.

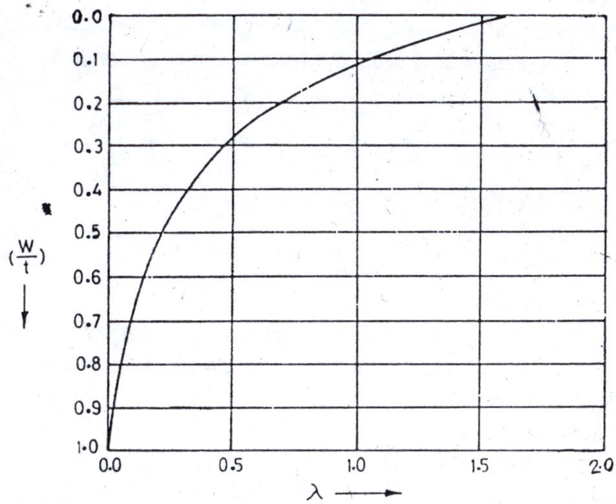

Fig. 16.17. Variation of λ.

16.15. WELL HYDRAULICS

Wells and commonly used for the dewatering of ground water. A well is a circular hole of a suitable diameter made in an aquifer. As the pumping is done from the well, a cone of depression is created all around the well. The equations for discharge can be developed using Darcy's law. However, the main use of wells is to supply water for drinking, irrigation and other purposes. Ground water is an important source of water. It is exploited through open wells, tube wells, springs and horizontal galleries.

Ground water is the water that falls as precipitation and then infiltrates the soil below the water table. The ground water reservoir is formed in the voids of the water-bearing strata, known as *aquifers*. These aquifers act as conduits for transmission of ground water. As the discharge from wells depends mainly on the permeability of the soil, pumping out tests as discussed below are frequently used to determine the coefficient of permeability of the soil, as explained in chapter 8.

As already explained, aquifers are mainly of two types. (*i*) Unconfined aquifers, (*ii*) Confined aquifers. As in the case of slots, the flow in a well may be gravity, or artesian or artesian-gravity. The equations for discharge are developed in the following sections.

As *aquiclude* is a soil formation such as clay which contains water, but which is not capable of transmitting or supplying adequate quantity of water. An *aquifuge* is a mass of rock or an impervious formation which neither transmits nor stores any water.

16.16. TERMS USED IN WELL HYDRAULICS

Various technical terms used in well hydraulics are defined below.

(1) Specific Yield. The specific yield (S_Y) of an aquifer is defined as the ratio of the volume of water drained by gravity to the total volume of the aquifer. It is usually expressed as a percentage. Thus

$$S_Y = \frac{\text{Volume of water drained by gravity}}{\text{Total volume of the aquifer}} \times 100$$

or $$S_Y = \frac{V_{WY}}{V} \times 100 \quad \text{...(16.19)}$$

The total volume of water in an aquifer depends upon porosity. However, a high value of porosity does not necessarily indicate that the aquifer will yield large volume of water. The quantity of water which can be obtained from the aquifer is that which flows under gravity. Therefore, it depends upon the permeability and hence specific yield. The specific yield is always less than porosity. It is sometimes called effective porosity. Specific yield of most of the aquifers, such as sand and gravel, varies between 15 to 30%.

(2) Specific Retention. The specific retention (S_R) of an aquifer is the ratio of the water retained in the soil after drainage to the total volume of aquifer. It is also expressed as a percentage. Thus

$$S_R = \frac{\text{Volume of water retained}}{\text{Total volume of the aquifer}} \times 100$$

or $$S_R = \frac{V_{WR}}{V} \times 100 \quad \text{...(16.20)}$$

When a saturated mass of soil is subjected to drainage, some water is not drained as it is retained in the pores of the soil due to molecular and capillary forces. The amount of water retained depends on grain size, grain shape and distribution of pores. The specific retention is high for soils with small pores, such as clayey soils.

For a saturated soil, the total volume of water V_w is equal to porosity (n) times the total volume of the aquifer (V). Thus

$$V_w = n\,V$$

But $$V_w = V_{wY} + V_{wR}$$

Therefore $$n = \frac{V_w}{V} = \frac{V_{wY} + W_{wR}}{V}$$

or $$n = S_Y + S_R \quad \text{...(16.21)}$$

Thus, the porosity of the stratum is equal to the sum of the specific yield and the specific retention.

(3) Storage Coefficient. The storage coefficient is defined as the volume of water released (or stored) by an aquifer per unit surface area per unit change in the component of the head normal to the surface.

In an unconfined aquifer, the storage coefficient corresponds to its specific yield. The storage coefficient for a unit thickness of the aquifer is equal to the specific yield, provided gravity drainage is complete. In a confined aquifer, the storage coefficient also depends upon the compression of the aquifer and the expansion of the contained water when the pressure is decreased during pumping. Typical values of storage coefficient for unconfined aquifers range from 0.02 to 0.03 and that for confined aquifers range from 0.00005 to 0.005. The actual values can be obtained from the pumping out tests.

(4) Transmissibility Coefficient. The coefficient of transmissibility (T) is defined as the rate of flow of water through a vertical strip of aquifer of unit width and extending to the fully saturation height under unit hydraulic gradient. Obviously, the coefficient of transmissibility in a confined aquifer is equal to the product of the coefficient of permeability and the thickness t of aquifer.

Thus $$T = k \times t \quad \text{...(16.22)}$$

Its units are m^2/sec or cm^2/sec.

The coefficient of transmissibility of a well in an unconfined aquifer is equal to the product of the coefficient of permeability and the average saturated thickness t_a. Thus

$$T = k \times t_a \qquad \text{...(16.23)}$$

when $t_a = (H + h)/2$ where H is the height of the original water table above the impervious stratum and h is the height of water in the well after drawdown.

16.17. DISCHARGE FROM A FULLY PENETRATING WELL

The water table is initially horizontal. When the pumping is started, the water table is lowered near the well and a cone of depression is formed. The drawdown at any point is equal to the vertical intercept between the original water level and the depressed water level. The flow may be gravity, or artesian, or artesian-gravity. The equations for discharge for all these types of flows are given below.

(1) Gravity Flow. Gravity flow occurs in an unconfined aquifer (see Fig. 8.8). This type of flow has already been discussed in chapter 8 when describing the methods for the determination of the coefficient of permeability in the field. However, for completeness of the treatment, the equations are repeated here.

Eq. 8.23 can be written as
$$q = \frac{\pi k (H^2 - h^2)}{\log_e (R/r_w)} \qquad \text{...(16.24)}$$

where H = depth of aquifer measured below the water table, h = depth of water in the well,
R = radius of influence, r_w = radius of well, k = coefficient of permeability.

The elevation z of the drawdown curve at a radial distance r from the well can be computed from the equation

$$z^2 = \frac{q \log_e (r/r_w)}{\pi k} + h^2 \qquad \text{...(16.25)}$$

The drawdown at a radial distance r can be computed from the equation

$$H^2 - z^2 = \frac{q}{\pi k} \log_e (R/r) \qquad \text{...(16.26)}$$

If there are two observation wells at radial distances r_1 and r_2, and if the depths of water in these wells are h_1 and h_2, respectively, Eq. 16.24 can be written as

$$q = \frac{\pi k (h_2^2 - h_1^2)}{\log_e (r_2/r_1)} \qquad \text{...(16.27)}$$

Eq. 16.24 can be written in terms of the effective length of the strainer (L). Let d be the drawdown at the well, *i.e.*, $H - h = d$

Thus
$$H + h = d + 2h$$

Therefore,
$$q = \frac{\pi k (H + h)(H - h)}{\log_e (R/r_w)}$$

or
$$q = \frac{\pi k d (d + 2h)}{\log_e (R/r_w)}$$

As the effective length L of the strainer in the well is equal to h,

$$q = \frac{\pi k d (d + 2L)}{\log_e (R/r_w)} \qquad \text{...(16.28)}$$

(2) Artesian Flow. The artesian flow occurs in a confined aquifer (see Fig. 8.9). Eq. 8.26 developed in chapter 8 can be written as

$$q = \frac{2 \pi k t (H - h)}{\log_e (R/r_w)} \qquad \text{...(16.29)}$$

where, k = coefficient of permeability, t = thickness of aquifer,
H = depth of the base of the aquifer below the piezometric surface,
h = depth of water in the well, R = radius of influence, r_w = radius of well.

The elevation z of the drawdown curve at a radial distance r from the well is given by

$$z = \frac{q}{2\pi kt} \log_e (r/r_w) + h \qquad ...(16.30)$$

The drawdown at that distance is given by

$$H - z = \frac{q}{2\pi kt} \log_e (R/r) \qquad ...(16.31)$$

Sometimes, Eq. 16.29 is written in terms of the coefficient of transmissibility (T) and the drawdown (d). Using Eq. 16.22 and taking ($H - h$) equal to d,

$$q = \frac{2\pi Td}{\log_e (R/r_w)} \qquad ...(16.32)$$

If h_1 and h_2 are the depths of water in two observation wells situated at distances r_1 and r_2, Eq. 16.29 becomes

$$q = \frac{2\pi kt\,(h_2 - h_1)}{\log_e (r_2/r_1)} \qquad ...(16.33)$$

(3) Artesian-Gravity Flow. Fig. 16.18 shows a well in which artesian-gravity flow occurs. As already explained, the flow is gravity type near the well and artesian type away from the well.

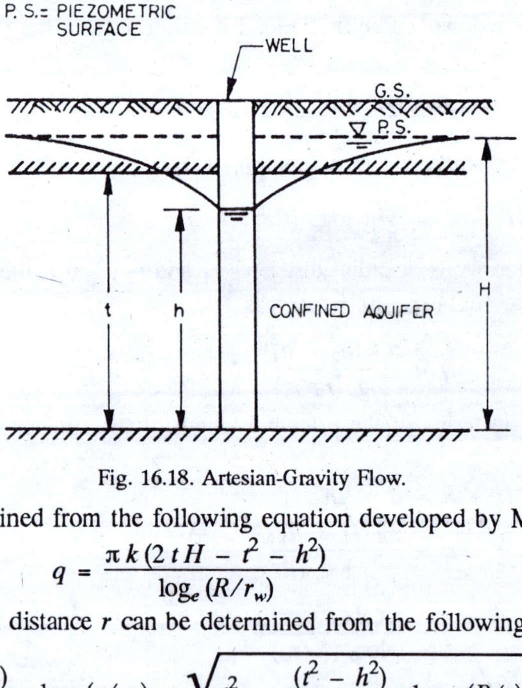

Fig. 16.18. Artesian-Gravity Flow.

The discharge can be obtained from the following equation developed by Muskat (1937),

$$q = \frac{\pi k\,(2tH - t^2 - h^2)}{\log_e (R/r_w)} \qquad ...(16.34)$$

The elevation z at a radial distance r can be determined from the following equation.

$$z = \frac{(H - t)}{\log_e (R/r_w)} \log_e (r/r_w) + \sqrt{t^2 - \frac{(t^2 - h^2)}{\log_e (R/r_w)} \log_e (R/r)} \qquad ...(16.35)$$

The radial distance R_G of the point at which the flow changes from artesian type to gravity type is given by

$$\log_e (R_G) = \frac{(t^2 - h^2)\log_e (R) + 2t\,(H - t)\log_e (r_w)}{2tH - t^2 - h^2} \qquad ...(16.36)$$

Eqs. 16.34 and 16.35 have been based on the assumption that the head at the well is at the same elevation as the water surface in the well. This is not true if the drawdown is relatively large. For such cases, the head at and in the close vicinity of the well is greater than the water depth in the well.

16.18. DISCHARGE FROM A PARTIALLY PENETRATING WELL

The discharge from a partially penetrating well depends upon the depth of penetration in the aquifer. Like a fully penetrating well, the flow can be gravity flow or artesian flow.

(*a*) **Gravity Flow.** Fig. 16.19 shows a partially penetrating well with gravity flow. The discharge is given by the following equation.

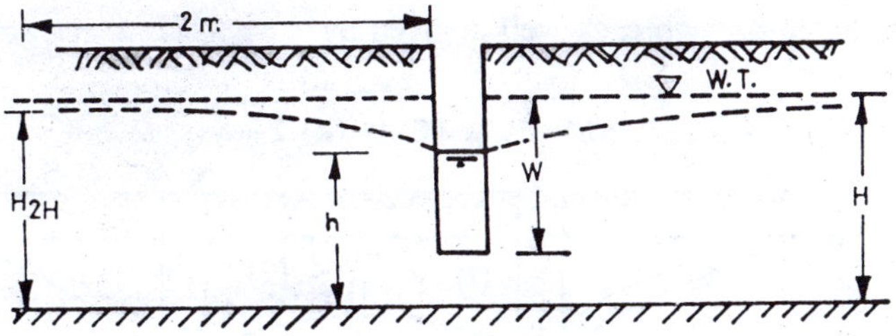

Fig. 16.19. Partially Penetrating Well (Gravity Flow).

$$q_p = \frac{2\pi k (H_{2H} - h)}{\left[\frac{1}{W} \log_e (\pi W/2r_w) + \frac{0.1}{H}\right]} \qquad ...(16.37)$$

where H_{2H} is the height of the drawdown curve at a distance of $2H$ from the well and W = depth of penetration of the well in the aquifer. The other notations are the same as in the preceding sections.

(*b*) **Artesian Flow.** Fig. 16.20 shows a partially penetrating well with artesian flow. The discharge is given by the following equation.

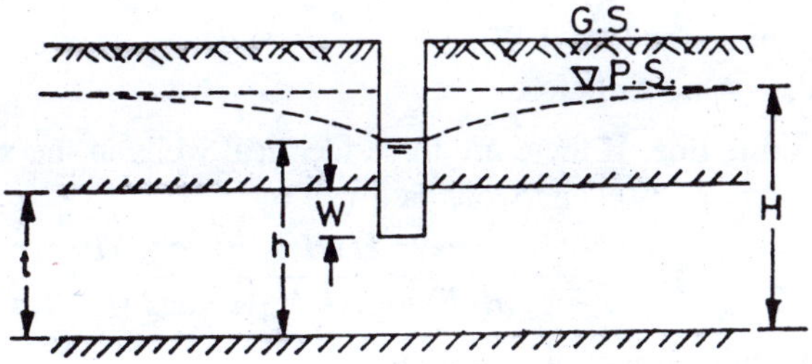

Fig. 16.20. Partially Penetrating Well (Artesian Flow).

$$q_p = \frac{2\pi kt (H - h) C}{\log_e (R/r_w)} \qquad ...(16.38)$$

where C is the correction factor for partial penetration. It is equal to the ratio of the discharge from the partially penetrating well to that for a fully penetrating well for the same drawdown.

The value of C can be obtained from the following equation given by Kozeny (1933).

$$C = \frac{W}{t}\left(1 + 7\sqrt{\frac{r_w}{2W}} \cos \frac{\pi W}{2t} \times \frac{180}{\pi}\right) \qquad ...16.39)$$

where W = depth of penetrations in the confined aquifer.

The other notations are the same as in the preceding sections.

16.19. INTERFERENCE AMONG WELLS

When two or more wells are located close to one another, their drawdown curves intersect within their zones of influence. Consequently, the discharge in the individual well is decreased. The phenomenon is known as interference among wells. The discharge from an individual well depends upon the spacing of the wells and their orientation.

(*a*) **Artesian Wells.**

(*i*) **Two wells.** Fig. 16.21 shows two artesian wells having a spacing of B. The dotted lines [curves (1) and (2)] show their individual cones of influences. The combined cone of influence is shown by the curve-3. The discharge through each well is given by

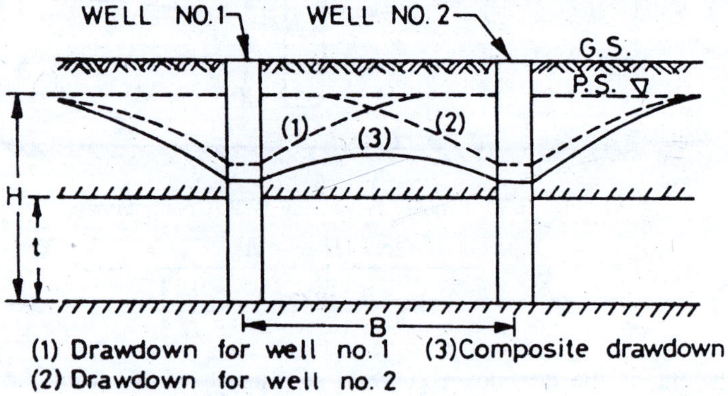

Fig. 16.21. Interference among Wells.

$$q = \frac{2\pi\, kt\,(H - h)}{\log_e \left(\dfrac{R^2}{Br_w}\right)} \qquad \ldots(16.40)$$

where R is the radius of influence ($R >> B$).

All other notations are the same as before.

(*ii*) **Three wells in the same line.** If there are three identical wells in the same line having a spacing of B, the discharge through the two end wells is given by

$$q = \frac{2\pi\, kt\,(H - h)\,\log_e (B/r_w)}{2\log_e (R/B)\log_e (B/r_w) + \log_e (B/2r_w)\log_e (R/r_w)} \qquad \ldots(16.41)$$

The discharge through the middle well is given by

$$q = \frac{2\,\pi\, kt\,(H - h)\,\log_e (B/2r_w)}{2\log_e (R/B)\log_e (B/r_w) + \log_e (B/2r_w)\log_e (R/r_w)} \qquad \ldots(16.42)$$

(*iii*) **Three wells forming a triangle.** If there are three identical wells located at the apexes of an equilateral triangle, with side B, the discharge from each well is given by

$$q = \frac{2\,\pi\, kt\,(H - h)}{\log_e (R^3/B^2\, r_w)} \qquad \ldots(16.43)$$

(*b*) **Gravity Wells.** The discharge from gravity wells can be obtained by using the equations given above for the artesian wells and making the following substitution :

$(H^2/2t)$ for H and $(h^2/2t)$ for h.

For example, the discharge from each well when the two wells are spaced B apart is obtained from Eq. 16.40 as

$$q = \frac{2\,\pi\, kt\,(H^2/2t - h^2/2t)}{\log_e (R^2/Br_w)}$$

or

$$q = \frac{\pi\, k\,(H^2 - h^2)}{\log_e (R^2/r_w B)} \qquad \ldots(16.44)$$

16.20. SPHERICAL FLOW IN A WELL

The equations developed in the preceding sections are for radial flow. Fig. 16.22 shows a well which penetrates upto the top surface of the confined aquifer. The flow in the well is spherical. In this case, the equations given in the preceding sections do not apply as the flow is not radial and the Dupuit assumption is not valid.

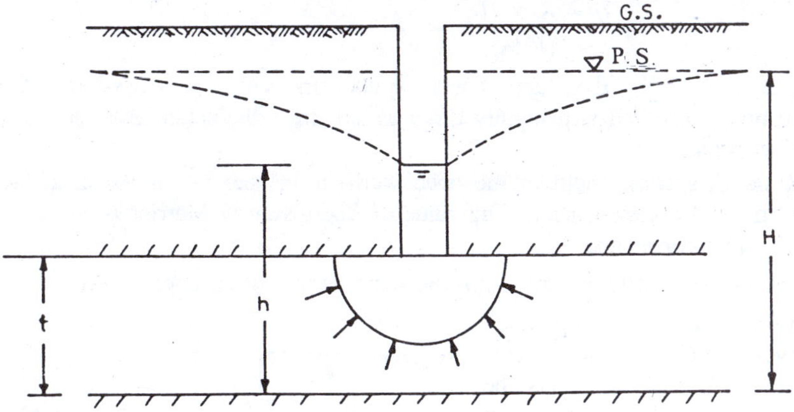

Fig. 16.22. Spherical Flow.

The discharge in the case of spherical flow is given by

$$q = 2\pi k r_w (H - h) \qquad \ldots(16.45)$$

where r_w is the radius of the well.

The discharge in case of spherical flow is much less as compared to that in a fully penetrating well. As the wells with spherical flow are not very effective, these are rarely used in practice.

16.21. DISCHARGE FROM AN OPEN WELL

An open well is a vertical hole of a large diameter 2 to 10 m. It is of shallow depth, and it draws water only from one pervious stratum. The well penetrates to such a depth below water table that there is at least a water depth of 3 to 4 m even in a dry season. Generally, the depth of an open well is limited to 30 m. The sides of the well may be lined with bricks or stones. In case of an unlined well, the discharge is from the bottom and sides. However, in the case of a lined well, it is mainly from the bottom of the well. The discharge capacity of an open well is generally limited to 0.004 cumecs, because the well can be excavated to a limited depth. Moreover, the water can be withdrawn only at a small velocity. If the velocity is greater than the critical velocity, the soil grains may be dislodged and subsidence may occur.

Open wells may be dug, bored, drilled or driven. Large diameter open wells are generally dug manually or mechanically.

Safe yield of an open well is the rate of flow at which the water percolates into the well under safe maximum working head. As the water is pumped out, the water level in the well falls. The difference between the water level in the well and the original water table is the depression head. The critical depression head occurs when the velocity is so high as to dislodge the soil particles. The safe maximum working head is usually taken as one- third of the critical depression head.

The yield from an open well is determined by the following two methods : (1) Pumping test at constant level, (2) Recuperation test.

1. Constant level Pumping test. In this method, the water level in the well is depressed by pumping out water till the maximum safe depression head is reached. The rate of pumping is so adjusted that the water level in

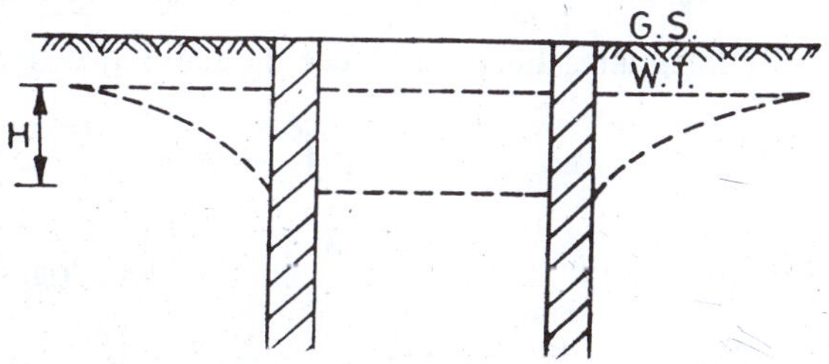

Fig. 16.23. Pumping Test.

the well remains constant (Fig 16.23). At that stage, the rate of withdrawal is equal to the percolation rate. The rate of pumping per hour gives the yield per hour of the well at a particular drawdown.

The discharge is generally expressed as

$$q = K \times H$$

$$q = (K/A) \times A \times H \qquad \text{...(16.46)}$$

where q = discharge (m^3/hr), K = Constant, depending upon the soil, A = Cross-sectional area of the cavity of flow at the well bottom (m^2). It is generally taken as 4/3 times the actual cross-sectional area of the well bottom, H depression head.

The ratio K/A is the specific yield of the open well (m^3/hr) per m^2 of the area through which water percolates under 1 m of depression head. [The value of K/A given by Marriot is equal to 0.25 for clay, 0.5 for fine sand and 1.0 for coarse sand.]

[**Note.** The specific yield of tube well is not the same as that of an open well.]

2. Recuperation Test. In a recuperation test, the water level is depressed to a level below the normal level by pumping. As the pumping is stopped, water level in the well starts rising. The time required for the water level to recuperate is noted. The equation for the discharge is derived below.

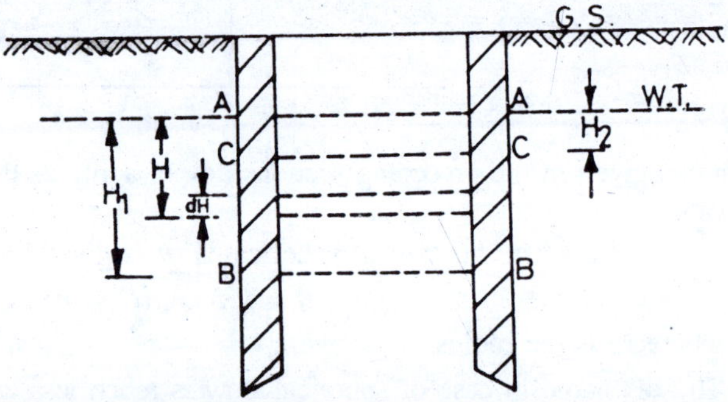

Fig. 16.24. Recuperation Test.

Let A–A be the water level before the pumping and B–B be the water level when the pumping is stopped (Fig. 16.24). Let H_1 be the depression head when the pumping is stopped, and H be the depression head at a time t after the pumping is stopped. Let H_2 be the depression head at a time T after the pumping is stopped.

Volume of water entering the well when the head recuperates by dH is given by

$$dV = A \cdot dH \qquad \text{...(a)}$$

where A = Cross-sectional area of the well bottom.

The volume of water entering the well in time dt during which head recuperates by dH is also given by

$$dV = q\, dt \qquad \text{...(b)}$$

where q = rate of discharge in the well

As the rate of discharge depends upon the depression head, it can also be written as

$$q = K \cdot H$$

Thus $$dV = KH\, dt \qquad \text{...(c)}$$

From Eqs. (*a*) and (*c*),

$$KH\, dt = -A\, dH \qquad \text{...(d)}$$

The minus sign indicates that the head decreases as time t increases.

Integrating Eq. (*d*) between the limits ($t = 0$, $H = H_1$ and $t = T$, $H = H_2$),

$$\frac{K}{A}\int_0^T dt = -\int_1^2 \frac{H_2}{H_1}\frac{dH}{H}$$

or $$\frac{K}{A} \cdot T = [\log_e H]_{H_2}^{H_1} = \log_e (H_1/H_2)$$

or $$\frac{K}{A} = \frac{1}{T}\log_e (H_1/H_2) \qquad \text{...(16.47)}$$

Knowing the values of H_1, H_2 and T, the value of the specific yield (K/A) can be computed. Once the value of K/A has been determined, the discharge can be obtained from Eq. 16.46 for any other depression head H as

$$q = (K/A) \times A \times H$$

Although recuperation test is not as reliable as the constant level pumping method, it is used when it is difficult to regulate the rate of pumping to attain a constant water level in the well required in the constant level pumping test.

16.22. ADVERSE EFFECTS OF DRAINAGE

As the water is drained out from the soil, the neutral stress decreases and the effective stress is increased. The increased effective stress may cause consolidation settlement. Therefore, the structures already existing in the zone of influence may experience undesirable settlement. This undesirable effect can be minimized by artificial recharge *i.e.* by pumping water into the ground near the existing structures through a system of well points. Thus a constant ground water level is maintained near the structures.

Another adverse effect of the drainage is that cavities are formed in the soil. These cavities may collapse after some time and may cause undulation in the ground. This adverse effect is minimised by providing suitably designed filters to check flow of soil along with water.

ILLUSTRATIVE EXAMPLES

Illustrative Example 16.1. A *slot is made in an unconfined aquifer to drain water. The flow to the slot occurs from both sides. If the water table is at a height of 12 m above the base and the drawdown is 4 m, find the discharge per metre length, assuming that the distance of the slot from both sides is 100 m. Take* $k = 5 \times 10^{-4}$ *m/sec.*

Solution. From Eq. 16.15,

$$q = k\left(\frac{H^2 - h^2}{L}\right) \cdot y$$

$$= 5 \times 10^{-4}\left(\frac{12^2 - 8^2}{100}\right) \times 1 = \mathbf{4 \times 10^{-4}\ m^3/sec}$$

Illustrative Example 16.2. *A well fully penetrates an unconfined aquifer having a saturated thickness of 10 m. If the radius of the well is 10 cm, and the drawdown is 3 m, determine the discharge. Take the radius of influence as 300 m and the coefficient of permeability as* 8×10^{-4} *m/sec.*

Solution. From Eq. 16.24,

$$q = \frac{\pi k (H^2 - h^2)}{\log_e (R/r_w)}$$

or

$$q = \frac{\pi \times 8 \times 10^{-4}(10^2 - 7^2)}{\log_e (300/0.10)} = \mathbf{0.016\ m^3/sec}$$

Illustrative Example 16.3. *Calculate the discharge from a fully- penetrating tube well, having the following particular :*

(i) Thickness of confined aquifer	= *25 m*
(ii) Tube well diameter	= *30 cm*
(iii) Drawdown	= *3 m*
(iv) Radius of influence	= *150 m*
(v) Coefficient of permeability	= *30 m/day*

Solution. From Eq. 16.29,

$$q = \frac{2\pi kt (H - h)}{\log_e (R/r_w)}$$

$$= \frac{2\pi \times 30 \times 25 \times (3)}{\log_e (150/0.15)}$$

$$= 2046.5 \text{ m}^3/\text{day} = \mathbf{0.0237 \ m^3/sec.}$$

Illustrative Example 16.4. *A tube well is driven in a confined aquifer of 25 m thickness located 20 m below the ground surface. The water table is 15 m below the ground surface. If the discharge of the well is 0.05 m^3/sec when the depression head is 10 m, find the diameter of the tube well. Take the radius of influence as 300 m and k = 3 × 10^{-4} m/sec.*

Solution. From Eq. 16.29, $$q = \frac{2\pi kt (H - h)}{\log_e (R/r_w)}$$

$$\log_e (R/r_w) = \frac{2\pi kt (H - h)}{q}$$

$$\log_e (R/r_w) = \frac{2\pi \times 3 \times 10^{-4} \times 25 \times 10}{0.05} = 9.42$$

$$R/r_w = 12332.58$$

or $$r_w = 300/12332.58 = \mathbf{0.0243 \ m}$$

Illustrative Example 16.5. *A fully penetrating well draws water from a confined aquifer of thickness 12 m and of permeability 1.5 × 10^{-3} m/sec. If the discharge is 0.03 m^3/sec, compute the drawdown at 30 m from the centre of the well. Take R/r_w = 2000.*

Solution. Eq. 16.31, $$H - z = \frac{q}{2\pi kt} \log_e (R/r_w)$$

$$= \frac{0.03}{2\pi \times 1.5 \times 10^{-3} \times 12} \log_e (2000) = \mathbf{2.017 \ m}$$

Illustrative Example 16.6. *Fig. E-16.6 shows an artesian-gravity well with a radius of influence of 300 m. If the diameter of the well is 30 cm, compute the discharge. Take k = 5 × 10^{-4} m/sec.*

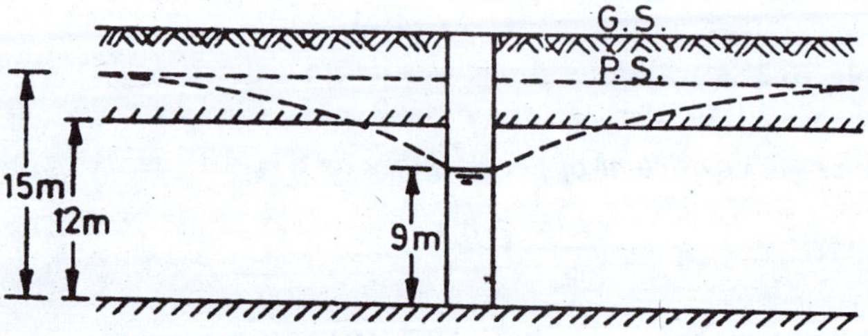

Fig. E-16.6.

Solution. From Eq. 16.34, $$q = \frac{\pi k (2tH - t^2 - h^2)}{\log_e (R/r_w)}$$

or $$q = \frac{\pi \times 5 \times 10^{-4} (2 \times 12 \times 15 - 12^2 - 9^2)}{\log_e (300/0.15)} = \mathbf{0.0279 \ m^3/sec.}$$

Illustrative Example 16.7. *A well 25 cm diameter partially penetrates a confined aquifer by 9 m. If the total thickness of the aquifer is 20 m and the discharge is 0.04 m^3/sec, compute drawdown at the well. Take R = 200 m, and k = 4 × 10^{-3} m/sec.*

Solution. From Eq. 16.38, $$q_p = \frac{2\pi kt (H - h) C}{\log_e (R/r_w)}$$

where C is given by Eq. 16.39 as

$$C = \frac{W}{t}\left(1 + 7\sqrt{\frac{r_w}{2W}} \cos \frac{\pi W}{2t} \times \frac{180}{\pi}\right)$$

$$= \frac{9}{20}\left(1 + 7\sqrt{\frac{0.125}{2 \times 9}} \cos \frac{\pi \times 9 \times 180}{2 \times 20 \times \pi}\right) = \mathbf{0.649.}$$

Therefore, $$q_p = \frac{2\pi \times 4 \times 10^{-3} \times 20\,(H - h) \times 0.649}{\log_e(200/0.125)}$$

or $$H - h = \frac{0.04 \times 7.378}{0.326} = \mathbf{0.905\ m}$$

Illustrative Example 16.8. *Two tube wells fully penetrate 12 m thick confined aquifer and are 150 m apart. The wells have following particulars :—*

(*i*) *Diameter* = *30 cm*
(ii) Depression head = 4.5 m
(*iii*) *Radius of influence* = *300 m*
(*iv*) *Coefficient of permeability* = 1.5×10^{-3} *m/sec.*

Determine the discharge from each well and compare it with for no interference.

Solution. From Eq. 16.40, $$q = \frac{2\pi kt(H - h)}{\log_e(R^2/Br_w)}$$

$$= \frac{2\pi \times 1.5 \times 10^{-3} \times 12 \times 4.50}{\log_e(300^2/150 \times 0.15)} = 0.0613\ \text{m}^3/\text{sec}$$

For no interference, Eq. 16.29 gives

$$q = \frac{2\pi\, kt(H - h)}{\log_e(R/r_w)}$$

$$= \frac{2\pi \times 1.5 \times 10^{-3} \times 12 \times 4.50}{\log_e(300/0.15)} = 0.0669\ \text{m}^3/\text{sec}$$

Percentage reduction is discharge

$$= \frac{(0.0669 - 0.0613)}{0.0669} \times 100 = \mathbf{8.37\%}$$

Illustrative Example 16.9. *Determine the discharge from a well with spherical flow and having the following particulars :*

(*i*) *Thickness of confined aquifer* = *10 m*
(*ii*) *Radius of influence* = *100 m*
(*iii*) *Radius of well* = *0.1 m*
(*iv*) *Coefficient of permeability* = 1×10^{-4} *m/sec*
(*v*) *Drawdown* = *3 m*

What would have been the discharge if the well had penetrated the confined aquifer fully ?

Solution. From Eq. 16.45, $$q = 2\pi k r_w (H - h)$$

or $$q = 2\pi \times 10^{-4} \times (0.1 \times 3) = 1.88 \times 10^{-4}\ \text{m}^3/\text{sec}$$

If the well had penetrated fully, the discharge would have been given by Eq. 16.29 as,

$$q = \frac{2\pi kt\,(H - h)}{\log_e(100/0.1)}$$

$$q = \frac{2\pi \times 10^{-4} \times 10 \times 3}{\log_e(100/0.1)} = 27.29 \times 10^{-4}\ \text{m}^3/\text{sec}$$

The ratio of the discharge in spherical flow to that in a fully penetrating well is given by

$$r = \frac{1.88}{27.27} \approx \frac{1}{14.5} = 7\%$$

i.e. the discharge in spherical flow is about 7% of that in radial flow.

Illustrative Example 16.10. *Design an open well in coarse sand (specific yield = 1.0 $m^3/hr/m^2$ under unit drawdown) for a yield of 0.003 m^2/s, when operated under a depression head of 3 m.*

Solution. From Eq. 16.46, $q = (K/A) \times A \times H$

or $0.003 \times 60 \times 60 = 1.0 \times A \times 3$ or $A = 3.6$ m

If d_w is the diameter of the well.

$$(\pi/4) \times (d_w)^2 = 3.60 \quad \text{or} \quad d_w = \mathbf{2.14\ m}$$

Illustrative Example 16.11. *During a recuperation test, the water level in an open well was depressed by 2.4 m which recuperated by an amount of 1.5 m in 60 minutes. Determine the yield from the well of 3 m diameter under a depression head of 3.0 m.*

Solution. From Eq. 16.47, $\frac{K}{A} = \frac{1}{T} \log_e (H_1/H_2)$

or $\frac{K}{A} = \frac{1}{1} \log_e (2.40/0.90) = 0.981$

From Eq. 16.46, $q = (K/A) \times A \times H$

$$= 0.981 \times \pi/4 \times (3)^2 \times 3.0 = \mathbf{20.80\ m^3/hour}$$

PROBLEMS

A Numerical

16.1. A slot is made in a confined aquifer 2 m thick to drain water. The flow to the slot occurs from both sides. If the water table is at a height of 10 m above the base and the drawdown is 3 m, find the discharge per metre length. The distance of the slot from both sides is 120 m. Take $k = 4 \times 10^{-4}$ m/sec. [**Ans.** 0.04 lit/sec]

16.2. A 30 cm diameter well penetrates 20 m below the water table. The draw down at 100 m is 0.50 m and that at 30 m is 1.10 m. If the discharge is 100 lit/sec, determine the transmissibility of the aquifer.

[**Ans.** 0.032 m^2/sec.]

16.3. A well penetrates an unconfined aquifer having a saturated depth of 90 m. When the drawdown is 10 m, the discharge is 4 lit/sec. Determine the discharge when the drawdown is 15 m. Take the radius of influence the same in both the cases. [**Ans.** 5.82 lit/sec]

16.4. A well is sunk through a layer of sand of thickness 12 m lying over an impervious stratum. When the water was pumped at a constant rate of 5 m^3/minute, the water levels in the observation wells situated at a distance of 15 m and 30 m from the pump well were 2.9 m and 2.7 m below the ground water table, respectively. Find the permeability of the sand. [**Ans.** 5 mm/sec]

16.5. Find the coefficient of permeability of an unconfined aquifer of thickness 18 m when a well of diameter 20 cm discharges 50 lit/sec under a depression head of 4.5 m. Take the radius of influence as 300 m.

[**Ans.** 9×10^{-4} m/sec]

16.6. Calculate the discharge through a fully penetrating tube well of the following particulars :

(*i*) Thickness of confined aquifer	= 25m
(*ii*) Diameter of the well	= 25 cm
(*iii*) Drawdown	= 3 m
(*iv*) Coefficient of permeability	= 0.29×10^{-3} m/sec.
(*v*) Radius of influence	= 200 m

[**Ans.** 18.51 lit/sec.]

16.7. A fully penetrating well of diameter 30 cm draws water from a confined aquifer of permeability 2×10^{-3} m/sec and thickness 12 m. If the steady discharge is 0.03 cumecs, compute the drawdown at 10 m from the centre of the well. Take radius of influence as 500 m. [**Ans.** 0.778 m]

16.8. A fully penetrating well of diameter 30 cm draws water from a 2.5 m thick confined aquifer. The steady state drawdown at 10 m and 50 m were observed to 2.5 m and 0.50 respectively. Determine the steady-stage discharge. Take $k = 1.5 \times 10^{-3}$ m/sec. [**Ans.** 29.26 lit/sec]

16.9. Design an open well in a fine sandy soil (specific yield = 0.5 m^3/hour/m^2 under unit depression head) to yield 0.004 cumecs under a depression head of 3 m. [**Ans.** 3.5 m dia]

16.10. Calculate the average yield of an open well of 3 m diameter from the recuperation test wherein the water level is depressed to the extent of 2.0 m and recuperation rate is 1.0 m per hour. The allowable depression head in the well in 3 m. [**Ans.** 14.69 m^3/hr]

B. Descriptive and Objective Type

16.11. What do you know about drainage of soils? What are its uses ? What are its ill effects?

16.12. Explain the working of a single-stage well point system. What are its limitation?

12.13. Write short notes on:

(*a*) Multi-stage well point system.

(*b*) Vacuum well points.

(*c*) Deep well system.

16.14. What is electro-osmosis? What are its advantages and disadvantages as compared with the conventional drainage systems?

16.15. What is the function of permanent drainage systems installed after construction? Describe in brief the foundation drains and blanket drains.

16.16. Differentiate between

(*a*) A slot and a well

(*b*) Gravity flow and artesian flow

(*c*) Fully penetrating and partially penetrating slots.

(*d*) Coefficient of permeability and transmissibility coefficient.

(*e*) An open well and a tube well

(*f*) Specific yield and specific retention.

16.17. Derive the expression for discharge from a fully-penetrating slot when the flow is

(*a*) gravity flow (*b*) artesian flow (*c*) artesian-gravity flow.

16.18. Derive the expression for discharge from a fully-penetrating well in

(*a*) a confined aquifer (*b*) an unconfined aquifer.

16.19. What do you understand by spherical flow? Why this type of flow is not used in practice?

16.20. What is mutual interference of wells? How can this be avoided?

16.21. How would you determine the yield of an open well?

16.22. Which of the following statements are true?

(*a*) Electro-osmosis is generally used for coarse-grained soils.

(*b*) The discharge from a slot is doubled if the flow can occur from both sides.

(*c*) In a confined aquifer, when the well fully penetrates the aquifer, the flow is artesian.

(*d*) A double-stage well point can be used for lowering the water table by 10 m.

(*e*) Foundation drains are more effective than blanket drains if the depth below the water table is not much.

(*f*) An open well gives more discharge than a tube well.

[**Ans.** True, (*b*), (*c*), (*d*), (*e*)]

C. Multiple Choice Questions

1. For lowering the water table by about 10 m, the following method is generally the most suitable.

(*a*) Well-point method (*b*) Shallow well system

(*c*) Deep well points (*d*) Elector-osmosis

2. Vacuum well points are generally used for draining

(*a*) Coarse sands (*b*) Fine sands and silty sands

(*c*) Silts (*d*) Clays

3. In a shallow-well system, the suction lift is usually limited to

(*a*) 5 m (*b*) 10 m
(*c*) 15 m (*d*) 20 m

4. The units of transmissibility coefficient are
(*a*) cm/sec (*b*) cm^2/sec
(*c*) cm^3/sec (*d*) None of above

5. The radius of influence of the gravity wells is generally assumed as
(*a*) 100 m (*b*) 300 m
(*c*) 1000 m (*d*) 50 m

6. The specific yield of an open well in coarse sand is about
(*a*) 1.00 h_r^{-1} (*b*) 0.5 h_r^{-1}
(*c*) 0.25 h_r^{-1} (*d*) 5.0 h_r^{-1}

7. Select the incorrect statement
(*a*) The confined aquifer is bounded at top and bottom by impervious strata.
(*b*) The pressure of water in the confined aquifer is greater than atmospheric pressure
(*c*) The aquifer is a fully saturated stratum
(*d*) A tube well starts flowing by itself in a confined aquifer

8. In a tube well driven in a confined aquifer if the drawdown in doubled, the discharge increases
(*a*) 4 times (*b*) 2 times
(*c*) 8 times (*d*) 16 times

9. The formation which contains water but is not able to transmit it is called
(*a*) an aquifer (*b*) An aquiclude
(*c*) an aquifuge (*d*) none of above

10. Electro-osmosis for a clayey soil generally leads to
(*a*) Decrease in shear strength (*b*) Increase in shear strength
(*c*) Increase in water content (*d*) Increase in plasticity

[**Ans.** 1. (*a*), 2. (*b*), 3. (*b*), 4. (*b*), 5. (*b*), 6. (*a*), 7. (*d*), 8. (*b*), 9. (*b*), 10. (*b*)]

PART—II

EARTH-RETAINING STRUCTURES

AND

FOUNDATION ENGINEERING

17

Site Investigations

17.1. INTRODUCTION

Site investigations or subsurface explorations are done for obtaining the information about subsurface conditions at the site of proposed construction. Site investigations in one form or the other is generally required for every big engineering project. Information about the surface and sub-surface features is essential for the design of structures and for planning construction techniques.

Site investigations consist of determining the profile of the natural soil deposits at the site, taking the soil samples and determining the engineering properties of the soils. It also includes in-situ testing of the soils.

Site investigations are generally done to obtain the information that is useful for one or more of the following purposes.

(1) To select the type and depth of foundation for a given structure.

(2) To determine the bearing capacity of the soil.

(3) To estimate the probable maximum and differential settlements.

(4) To establish the ground water level and to determine the properties of water.

(5) To predict the lateral earth pressure against retaining walls and abutments.

(6) To select suitable construction techniques.

(7) To predict and to solve potential foundation problems.

(8) To ascertain the suitability of the soil as a construction material.

(9) To investigate the safety of the existing structures and to suggest the remedial measures.

The relevant information is obtained by drilling holes, taking the soil samples and determining the index and engineering properties of the soil. In-situ tests are also conducted to determine the properties of the soils in natural conditions. This chapter discusses various methods of sub-surface explorations and in-situ testing.

17.2. PLANNING A SUBSURFACE EXPLORATION PROGRAMME

A sub-surface exploration programme depends upon the type of the structure to be built and upon the variability of the strata at the proposed site. The extent of sub-surface exploration is closely related to the relative cost of the investigations and that of the entire project for which it is undertaken. In general, the more detailed the investigations are done, the more is known about the sub-surface conditions. As a result, the greater economy can be achieved in the construction of the project because the element of uncertainty is considerably reduced. However, a limit is reached when the cost of investigations outweighs any saving in the cost of the project, and it increases the overall cost. It would not be economical to have investigations beyond that limit.

The extent of investigations would also depend upon the location of the project. A small house in an already built-up area would not require much exploration. On the other hand, if the house is to be built in a newly developed area, a detailed investigation would be required to ascertain the location of different soil strata and their physical characteristics. If a multi-storeyed building is to be constructed, extensive sub-surface explorations would be necessary. These buildings impose very heavy loads and the zone of influence is also very deep. It would, therefore, be more desirable to invest some amount on sub-surface exploration than to overdesign the building and make it costlier.

Planning of a sub-surface exploration programme is a difficult task. Besides a thorough knowledge of soil engineering, it requires experience and engineering judgment. Sometimes, the exploration programme has to be changed as the investigations progress. As the variability of the soil strata is found to increase, the extent of investigations is also increased. On the other hand, if the site is found to be underlain by uniform deposits, the extent of investigations is decreased.

In general, the aim of the investigations should be to get the maximum information that is useful in the design and construction of the project at a minimum cost. The cost of site investigations generally varies between 0.05 to 0.2% of the total cost of the entire structure. In some unusual conditions, the cost may be even upto 1%.

17.3. STAGES IN SUB-SURFACE EXPLORATIONS

Sub-surface explorations are generally carried out in three stages:

(1) Reconnaissance. Site reconnaissance is the first step in a sub-surface exploration programme. It includes a visit to the site and to study the maps and other relevant records. It helps in deciding future programme of site investigations, scope of work, methods of exploration to be adopted, types of samples to be taken and the laboratory testing and in-situ testing.

(2) Preliminary Explorations. The aim of a preliminary exploration is to determine the depth, thickness, extent and composition of each soil stratum at the site. The depth of the bed rock and the ground water table is also determined.

The preliminary explorations are generally in the form of a few borings or test pits. Tests are conducted with cone penetrometers and sounding rods to obtain information about the strength and compressibility of soils.

Geophysical methods are also used in preliminary explorations for locating the boundaries of different strata.

(3) Detailed Explorations. The purpose of the detailed explorations is to determine the engineering properties of the soils in different strata. It includes an extensive boring programme, sampling and testing of the samples in a laboratory.

Field tests, such as vane shear tests, plate load tests and permeability tests, are conducted to determine the properties of the soils in natural state. The tests for the determination of dynamic properties are also carried out, if required.

For complex projects involving heavy structures, such as bridges, dams, multi-storey buildings, it is essential to have detailed explorations. However, for small projects, especially at sites where the strata are uniform, detailed investigations may not be require. The design of such projects is generally based on the data collected during reconnaissance and preliminary explorations.

17.4. RECONNAISSANCE

The geotechnical engineer makes a visit to the site for a careful visual inspection in reconnaissance. The information about the following features is obtained in reconnaissance.

(1) The general topography of the the site, the existence of drainage ditches and dumps of debris and sanitary fills.

(2) Existence of settlement cracks in the structure already built near the site.

(3) The evidence of land slides, creep of slopes and the shrinkage cracks.

(4) The stratification of soils as observed from deep cuts near the site.

(5) The location of high flood marks on the nearby building and bridges.

(6) The depth of ground water table as observed in the wells.

(7) Existence of springs, swamps, etc. at the site.

(8) The drainage pattern existing at the site.

(9) Type of vegetation existing at the site. The type of vegetation gives a clue to the nature of the soil.

(10) Existence of underground water mains, power conduit, etc. at the site.

In addition to making site visits, the geotechnical engineer should study geological maps, aerial photographs, toposheet, soil maps and the blue prints of the existing buildings. Maps and publications of various agencies give a lot of information about the geologic character of the area.

The geotechnical engineer should also get information about the type of structure to be built and its proposed use. In the case of a multi-storeyed building, the information about the column loads and their approximate locations should be obtained. In the case of bridges, the span length and the load carried by the piers and abutments should be ascertained. In the case of a dam, the geotechnical engineer should get information about the type of the dam, its height, base width and other salient characteristics.

The information obtained during reconnaissance is helpful in evolving a suitable sub-surface investigation programme.

17.5. DEPTH OF EXPLORATION

The depth of exploration required at a particular site depends upon the degree of variation of the subsurface data in the horizontal and vertical directions. It is not possible to fix the number, disposition and depth of borings without making a few preliminary borings or soundings at the site. The geotechnical engineer having a long experience and good engineering judgment may give some guidelines.

The depth of exploration is governed by the depth of the influence zone. The depth of the influence zone depends upon the type of the structure, intensity of loading, shape and disposition of the loaded area, the soil profile, and the physical characteristics of the soil. The depth upto which the stress increment due to

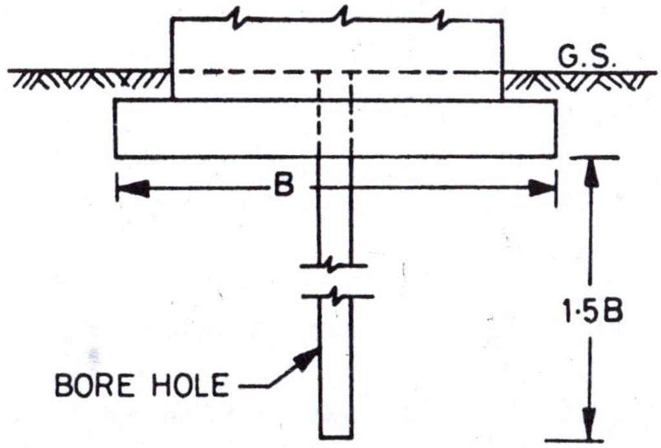

Fig. 17.1. Depth of Exploration.

superimposed loads can produce significant settlement and shear stresses is known as the *significant depth.* The depth of exploration should be at least equal to the significant depth.

The significant depth can be determined using the method discussed in chapter 11. The significant depth is generally taken as the depth at which the vertical stress is 20% of the load intensity. According to the above criterion, the depth of exploration should be about 1.5 times the width of the square footing (Fig. 17.1) and about

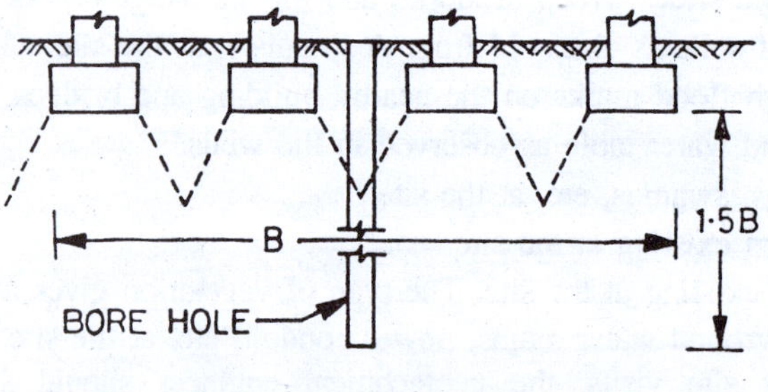

Fig. 17.2. Depth of Exploration for Closely-spaced Footings.

3.0 times the with of the strip footing. However, if the footings are closely spaced, the whole of the loaded area acts as a raft foundation. In that case, the depth of boring should be at least 1.5 times the width of the entire loaded area (Fig. 17.2). In the case of pile foundation, the depth of exploration below the tip of bearing piles is kept at least 1.5 times the width of the pile group. However, in the case of friction piles, the depth of exploration is taken 1.5 times the width of the pile group measured from the lower third point (Fig. 17.3).

It is more logical to relate the increase in stress to the in-situ stress. The depth of exploration is usually taken upto the level at which the increase in stress is 1/20th of the in-situ stress before the application of the load.

When the foundations are taken up to rock, it should be ensured that large boulders are not mistaken as bed rock. The minimum depth of core boring into the bed rock should be 3m to establish it as a rock.

In case of multi-storeyed buildings, the depth of exploration can be taken from the following formula (Sowers and Sowers, 1970),

$$D = C(S)^{0.7} \qquad ...(17.1)$$

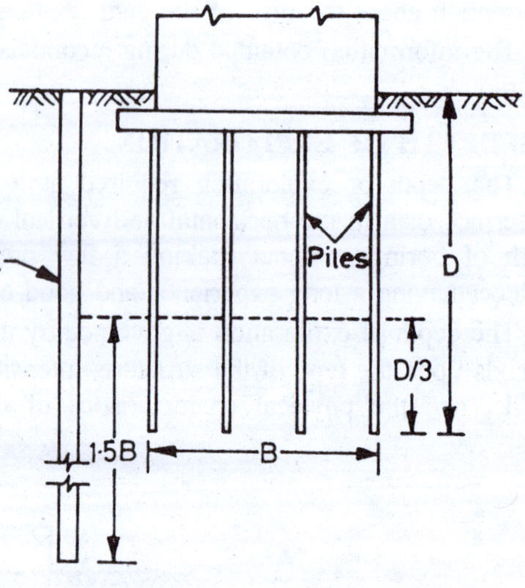

Fig. 17.3. Depth of Exploration for Friction Piles.

where D = depth of exploration (m), C = constant, equal to 3 for light steel buildings and narrow concrete buildings. It is equal to 6 for heavy steel buildings and wide concrete buildings. S = number of storeys.

If loose soil or recently deposited soil or a weak stratum is encountered, it should be explored thoroughly. Explorations should be carried to a depth at which the net increase in the vertical stress is less than the allowable bearing pressure of the soil.

For two adjacent footings, each of size $B \times L$, spaced at a clear spacing A, IS: 1892—1972 suggests that the minimum depth of boring should be 1.5 B when $A \geq 4B$; and it should be 1.5 L when $A < 2B$. For adjacent rows of such footings, the minimum recommended depth of exploration is 4.5B when $A < 2B$; it is 3.5 B when $A > 2B$ and it is 1.5 B when $A \geq 4B$.

For explorations of deep excavations, the depth of exploration below the proposed excavation level should be at least 1.5 times the depth of excavation. In case of road cuts, it is taken at least equal to the width of the cut.

In case of road fills, the minimum depth of boring is 2m below the ground surface or equal to the height of the fill, whichever is greater.

In case of gravity dams, the minimum depth of boring is twice the height of the dam.

17.6. LATERAL EXTENT OF EXPLORATIONS

The lateral extent of exploration and the spacing of bore holes depend mainly on the variation of the strata in the horizontal direction. The exploration should be extensive so as to reveal major changes in the properties of the sub-surface strata.

For small and less important buildings, even one bore hole or a trial pit in the centre may suffice. But for compact buildings, covering an area of about 0.4 hectares, there should be at least 5 bore holes, one at the centre and four near the corners (Fig. 17.4).

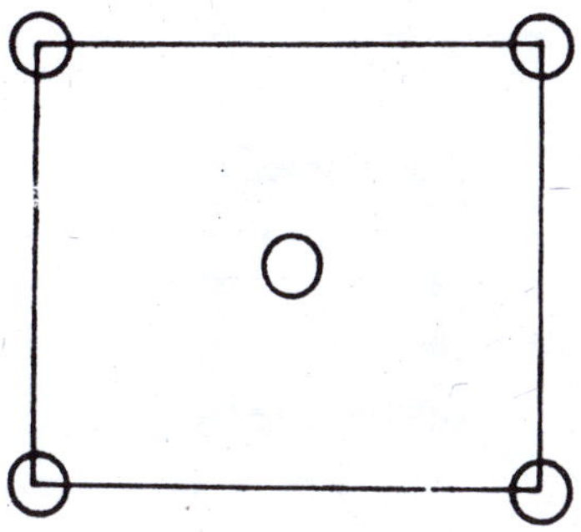

Fig. 17.4.

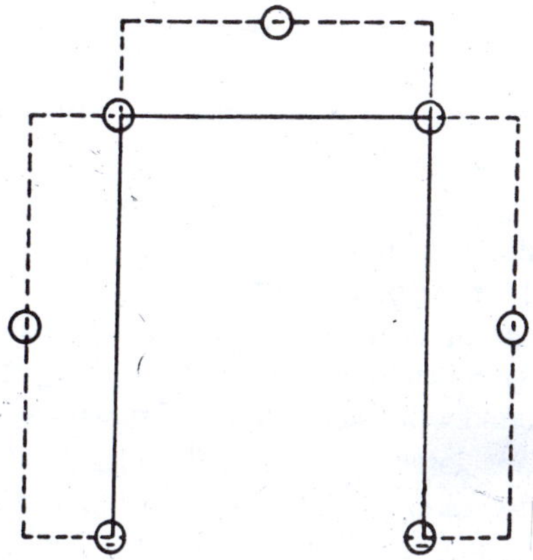

Fig. 17.5.

For large, multi-storeyed buildings, the bore holes should be drilled at all the corners and also at important locations. The spacing between the bore holes is generally kept between 10 to 30 m, depending upon the variation in the subsurface conditions and loading (Fig. 17.5.).

For highways, subsurface explorations are usually carried out along the proposed centre line or along the propose ditch line. The spacing of bore holes usually varies between 150 and 300 m. If the sub-strata is erratic, the spacing may be reduced to even 30m.

In case of concrete dams, the spacing of bore holes generally varies between 40 and 80 m.

17.7. OPEN EXCAVATION METHODS OF EXPLORATION

In this method of exploration, an open excavation is made to inspect the sub-strata. The methods can be divided into two categories: (1) Pits and Trenches, (2) Drifts and Shafts.

(1) Pits and Trenches. Pits and trenches are excavated at the site to inspect the strata. The size of the pit should be sufficient to provide necessary working space. IS : 4453—1967 recommends a clear working space of 1.2 m × 1.2 m at the bottom of the pit. The depth of the pit depends upon the requirement of the investigation as already discussed.

Shallow pits up to a depth of 3 m can be made without providing any lateral support. For deeper pits, especially below the ground water table, the lateral support in the form of sheeting and bracing system (Fig. 17.6) is required. As the depth of the pit increases, its cost increases rapidly. For depths greater than 6 m, bore holes are more economical than open pits.

Deep pits should be properly ventilated to prevent accumulation of dead air. If water is encountered in a pit, it should be suitably dewatered.

Trenches are long shallow pits. As a trench is continuous over a considerable length, it provides exposure along a line. The trenches are more suitable than pits for exploration on slopes.

Test pits and trenches can be excavated manually or mechanically. Adequate precautions should be taken against possible accidents due to caving of the ground.

(2) Drifts and Shafts. Drifts are horizontal tunnels made in the hill-side to determine the nature and structure of the geological formation. IS: 4453—1980 recommends that a drift should have the minimum clear dimensions of 1.5 m width and 2.0 m height in hard rock. In soft rock, an arch roof is more advantageous than a flat roof. If the ground is unable to stand of its own, supports have to be provided to carry the load of the roof and the sides of the drift.

Drifts are useful for establishing the minimum excavation limits to reach sound rock and for locating faults and shear zones and buried channels in the river section. However, drifts are generally expensive. These are used only when other methods do not provide the required information. Drifts are also known as *adits*.

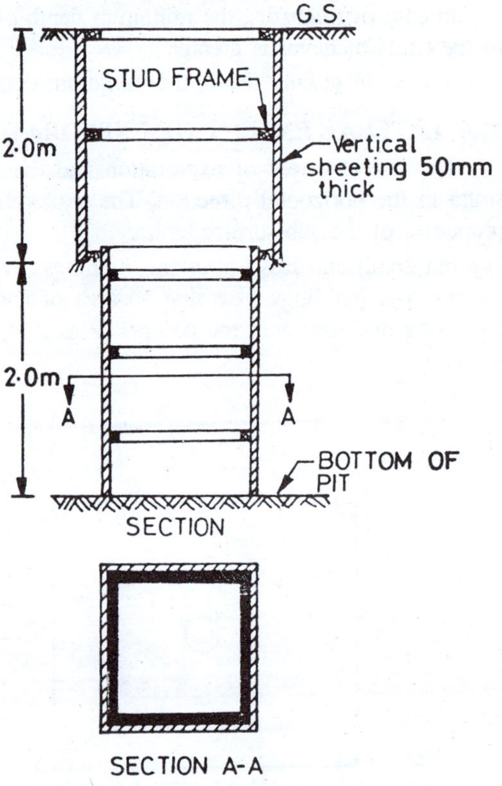

Fig. 17.6.

Shafts are large size vertical holes made in the geological formation. These may be rectangular or circular in section. The minimum width of a rectangular shaft is 2.4 m and for a circular shaft, the minimum diameter is 2.4 m. In weak ground, the sides of the shaft should be properly supported. Deep shafts should be properly ventilated. Shafts are used to reach a particular strata at a depth of 4 m or more. Shafts are also used to extend the exploration below the river bed already done by means of tunnels.

17.8. BORINGS FOR EXPLORATION

When the depth of exploration is large, borings are used for exploration. A vertical bore hole is drilled in the ground to get the information about the sub-soil strata. Samples are taken from the bore hole and tested in a laboratory. The bore hole may be used for conducting in-situ tests and for locating the water table. Extensometers or pressure meter may also be installed in the bore hole for the measurement of deformation in the sub-strata.

Depending upon the type of soil and the purpose of boring, the following methods are used for drilling the holes.

(1) Auger Boring (2) Wash Boring (3) Rotary Drilling
(4) Percussion Drilling (5) Core Boring.

A few holes are drilled during the preliminary investigation. In the detailed investigations, a large number of holes are drilled to thoroughly investigate the sub-soil strata.

The results of boring are presented in the form of boring-log and sub-surface profiles (Sect. 17.22).

17.9. AUGER BORING

An auger is a boring tool similar to one used by a carpentor for boring holes in wood. It consists of a shank with a cross-wise handle for turning and having central tapered feed screw [Fig. 17.7 (*a*)]. The augers can be operated manually or mechanically.

The hand augers used in boring are about 15 to 20 cm in diameter. These are suitable for advancing holes upto a depth of 3 to 6 m in soft soils. The hand auger is attached to the lower end of a pipe of about 18 mm diameter. The pipe is provided with a cross-arm at its top. The hole is advanced by turning the cross-arm manually and at the same time applying thrust in the downward direction. When the auger is filled with soil, it is taken out. If the hole is already driven, another type of auger, known as post-hole auger [Fig. 17.7 (*b*)] is used for taking soil samples.

Mechanical augers are driven by power. These are used for making holes in hard strata to a great depth. However, for depths greater than 12 m, even mechanical augers become inconventient and other methods of boring are used.

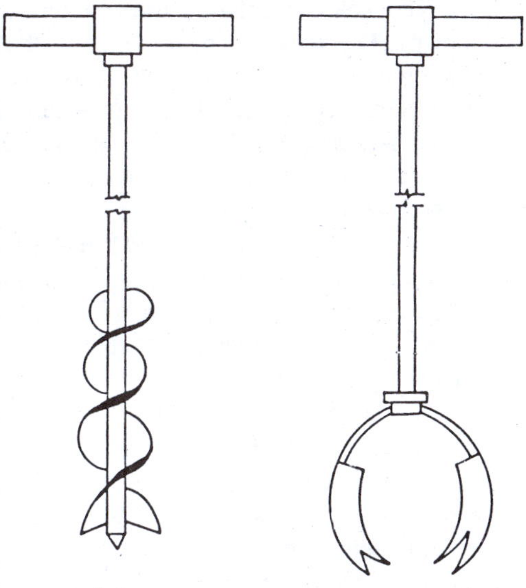

Fig. 17.7. Augers.

Continuous flight augers are special type of mechanical augers which are provided with a central hollow tube. When the hole is advanced, the central tube is kept plugged. As the auger is turned into the ground, the cuttings rise to the surface through the spiral. During sampling, the plug is removed and a sampler is inserted for taking the samples. The main disadvantage of using a continuous flight auger is that it becomes difficult to ascertain the depth from which the cuttings coming on the ground have been removed.

Auger boring is generally used in soils which can stay open without casing or drilling mud. Clays, silts and partially saturated sands can stand unsupported. For soils which cannot stand unsupported, especially for sandy soils below water table, a casing is normally required. For such soils, the method of auger boring becomes slow and expensive. Auger boring cannot be used when there are large cobbles, boulders or other obstructions which prevent drilling of the hole.

Auger borings are particularly useful for subsurface investigations of highways, railways and air fields, where the depth of exploration is small. The investigations are done quite rapidly and economically by auger boring.

The main disadvantage of the auger boring is that the soil samples are highly disturbed. Further, it becomes difficult to locate the exact changes in the soil strata.

17.10. WASH BORING

In wash boring, the hole is drilled by first driving a casing, about 2 to 3 m long, and then inserting into it a hollow drill rod with a chisel-shaped chopping bit at its lower end. Water is pumped down the hollow drill rod, which is known as *wash pipe*. Water emerges as a strong jet through a small opening of the chopping bit. The hole is advanced by a combination of chopping action and the jetting action, as the drilling bit and the accompanying water jet disintegrate the soil. The water and the chopped soil particles rise upward through the annular space between the drill rod and the casing. The return water, also known as wash water, is laden with the soil cuttings. It is collected in a tub through a T-shaped pipe fixed at the top of the casing (Fig. 17.8).

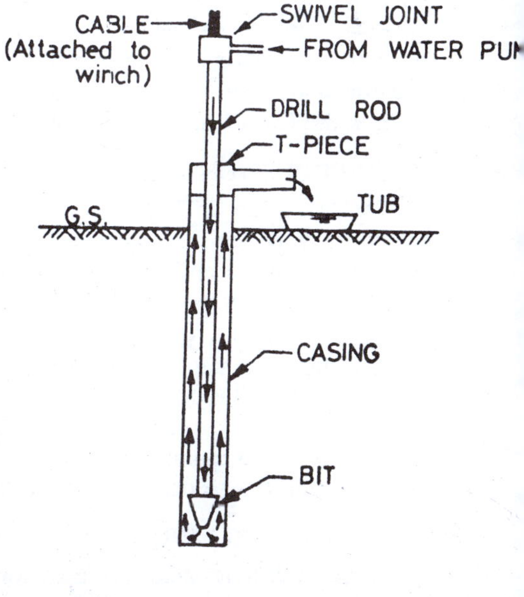

Fig. 17.8. Wash Boring.

The hole is further advanced by alternately raising and dropping the chopping bit by a winch (not shown). The swivel joint provided at the top of the drill rod facilitates the turning and twisting of the rod. The process is continued even below the casing till the hole begins to cave in. At that stage, the bottom of the casing can be extended by providing additional pieces at the top. However, in stable, cohesive soils, the casing is required only in the top portion. Sometimes, instead of a casing, special drilling fluids made of suspension or emulsions of fat clays or bentonite combined with some chemical additives are used for supporting the walls of the hole

The wash samples collected in the tub do not represent the soil in its true condition. There is complete break down of particles by chopping action. There is also mixing of the particles and the loss of fine particles in transportation. The samples are of little practical use. However, some indication about the changes in strata is provided by the reaction of the chopping bit as the hole is advanced. It is also indicated by a change in colour of the wash water.

The wash boring is mainly used for advancing a hole in the ground. Once the hole has been drilled, a sampler is inserted to obtain soil samples for testing in a laboratory.

The equipment used in wash borings is relatively light and inexpensive. The main disadvantage of the method is that it is slow in stiff soils and coarse-grained soils. It cannot be used efficiently in hard soils, rocks and the soils containing boulders. The method is not suitable for taking good quality undisturbed samples above ground water table, as the wash water enters the strata below the bottom of the hole and causes an increase in its water content.

17.11. ROTARY DRILLING

In the rotary drilling method, the bore hole is advanced by rotating a hollow drill rod which has a cutting bit at its lower end. A drill head is provided at the top of the drill rod. It consists of a rotary mechanism and an arrangement for applying downward pressure.

As the drilling rod is rotated, the cutting bit shears off chips of the material penetrated. A drilling fluid under pressure is introduced through the drilling rod to the bottom of the hole. The fluid carries the cuttings of the material penetrated from the bottom of the hole to the ground surface through the annular space between the drilling rod and the walls of the hole. The drilling fluid also cools the drilling bit. In case of an uncased hole, the drilling fluid also supports the walls of the hole.

When the soil sample is required to be taken, the drilling rod is raised and the drilling bit is replaced by a sampler.

Rotary drilling can be used in clay, sand and rocks. Bore holes of diameter 50 mm to 200 mm can be easily made by this method. The method is not well adapted for use in materials containing a large percentage of particles of gravel size and larger. The particles of this size start rotating beneath the drill rod and it becomes difficult to advance the hole.

17.12. PERCUSSION DRILLING

The percussion drilling method is used for making holes in rocks, boulders and other hard strata. In this methods, a heavy chisel is alternately lifted and dropped in a vertical hole. The material gets pulverised. If the point where the chisel strikes is above the water table, water is added to the hole. The water forms a slurry with the pulverised material, which is removed by a sand pump or a bailer at intervals. Percussion drilling may require a casing. Percussion drilling is also used for drilling of tube wells.

The main advantage of percussion drilling method is that it can be used for all types of materials. It is particularly useful for drilling holes in glacial tills containing boulders. One of the major disadvantages is that the material at the bottom of the hole is disturbed by heavy blows of the chisel. It is not possible to get good quality undisturbed samples. Further, the method is generally more expensive than other methods. Moreover, it becomes difficult to detect minor changes in the properties of the strata penetrated.

17.13. CORE DRILLING

The core drilling method is used for drilling holes and for obtaining rock cores. In this method, a core barrel fitted with a drilling bit is fixed to a hollow drilling rod. As the drilling rod is rotated, the bit advances and cuts an annular hole around an intact core The core is then removed from its bottom and is retained by a core lifter and brought to the ground surface. Water is pumped continuously into the drilling rod to keep the drilling bit cool and to carry the disintegrated material to the ground surface.

The core drilling may be done using either a diamond studded bit or a cutting edge consisting of chilled shot. The diamond drilling is superior to the other type of drilling, but is costlier. The core barrel may consist of a single tube or a double tube. A double-tube barrel gives a good quality sample of the rock.

17.14. TYPES OF SOIL SAMPLES

Soil samples are obtained during sub-surface exploration to determine the engineering properties of the soils and rocks. Soil samples are generally classified into two categories :

(1) **Disturbed samples.** These are the samples in which the natural structure of the soil gets disturbed during sampling. However, these samples represent the composition and the mineral content of the soil. Disturbed samples can be used to determine the index properties of the soil, such as grain size, plasticity characteristics, specific gravity.

(2) **Undisturbed samples.** These are the samples in which the natural structure of the soil and the water content are retained. However, it may be mentioned that it is impossible to get truly undisturbed sample. Some disturbance is inevitable during sampling, even when the utmost care is taken. Even the removal of the sample from the ground produces a change in the stresses and causes disturbances.

Undisturbed samples are used for determining the engineering properties of the soil, such as compressibility, shear strength, and permeability. Some index properties such as shrinkage limit can also be determined. The smaller the disturbance, the greater would be the reliability of the results.

17.15. DESIGN FEATURES AFFECTING THE SAMPLE DISTURBANCE

The disturbance of the soil depends mainly upon the following design features :

(1) Area ratio. The area ratio is defined as

$$A_r = \frac{\text{Maximum cross-sectional area of the cutting edge}}{\text{Area of the soil sample}} \times 100$$

Fig. 17.9 shows the lower portion of a sampler. The area ratio can be expressed as

$$A_r = \frac{D_2^2 - D_1^2}{D_1^2} \times 100 \qquad \text{...(17.2)}$$

where D_1 = inner diameter of the cutting edge,

D_2 = outer diameter of the cutting edge.

For obtaining good quality undisturbed samples, the area ratio should be 10 percent or less (Hvorslev, 1949).

(2) Inside clearance. The inside clearance is defined as

$$C_i = \frac{D_3 - D_1}{D_1} \times 100 \qquad \text{...(17.3)}$$

where D_3 = inner diameter of the sampling tube.

The inside clearance allows elastic expansion of the sample when it enters the tube. It helps is reducing the frictional drag on the sample. For an undisturbed sample, the inside clearance should be between 0.5 and 3 percent.

(3) Outside clearance. The outside clearance is defined as

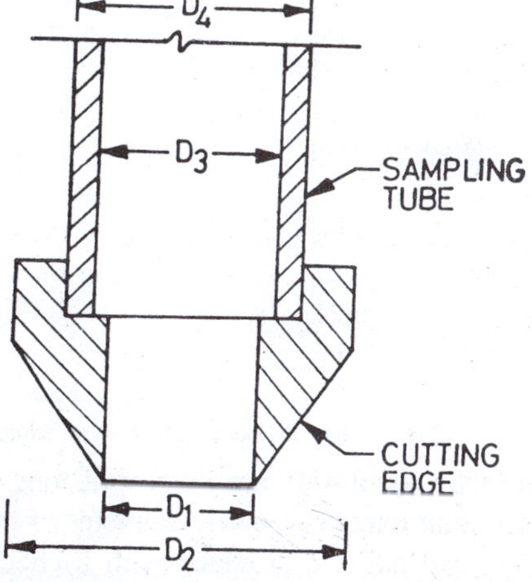

Fig. 17.9. Soil Sampler.

$$C_o = \frac{D_2 - D_4}{D_4} \times 100 \qquad \ldots(17.4)$$

where D_4 = outer diameter of the sampling tube.

For reducing the driving force, the outside clearance should be as small as possible. Normally, it lies between zero and 2 percent.

(4) Inside wall friction. The friction on the inside wall causes disturbance of the sample. The inside surface of the sampler should be smooth. It is usually smeared with oil before use to reduce friction.

(5) Design of non-return valve. The non-return valve provided on the sampler should be of proper design. It should have an orifice of large area to allow air, water or slurry to escape quickly when the sampler is driven. It should immediately close when the sampler is withdrawn.

(6) Method of applying force. The degree of disturbance depends upon the method of applying force during sampling and upon the rate of penetration of the sampler. For obtaining undisturbed samples, the sampler should be pushed and not driven.

17.16. SPLIT-SPOON SAMPLERS

The most commonly used sampler for obtaining a disturbed sample of the soil is the standard split-spoon sampler (Fig. 17.10). If consists mainly of three parts (*i*) Driving shoe, made of tool-steel, about 75 mm long,

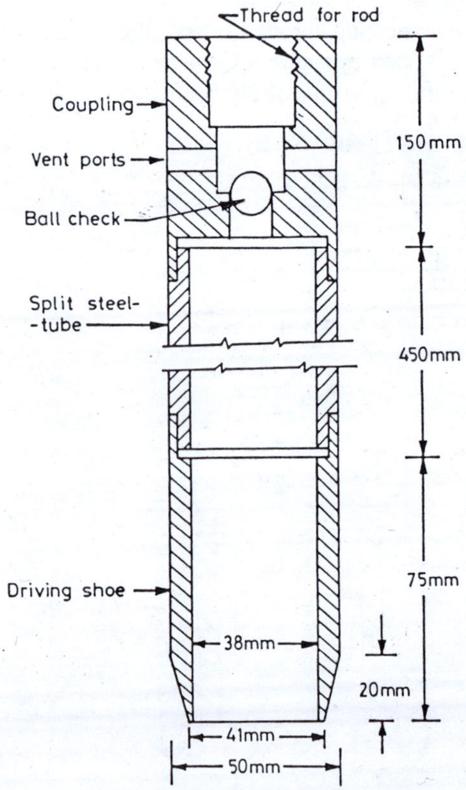

Fig. 17.10. Standard Split Spoon Sampler.

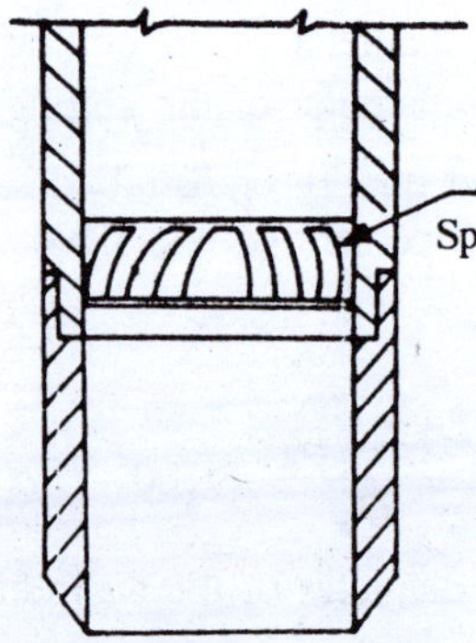

Fig.17.11. Spring; Core Catcher.

(*ii*) steel tube about 450 mm long, split longitudinally in two halves, and (*iii*) coupling at the top of the tube about 150 mm long. The inside diameter of the split tube is 38 mm and the outside diameter is 50.0 mm. The coupling head may be provided with a check valve and 4 venting ports of 10 mm dia to improve sample recovery. This sampler is also used in conducting standard penetration test (Sect. 17.22).

After the bore hole has been made, the sampler is attached to the drilling rod and lowered into the hole.

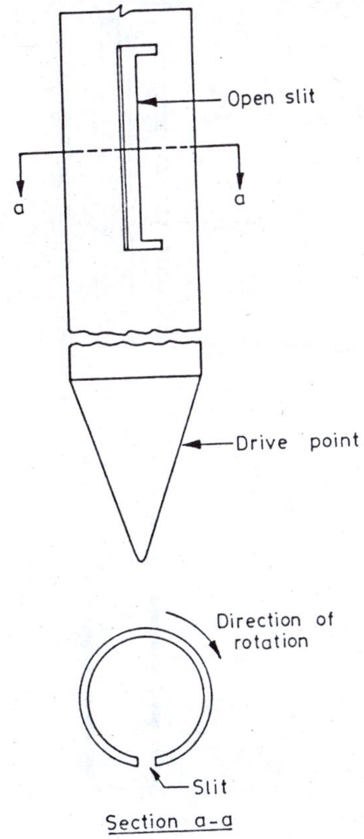

Fig. 17.12. Scraper Bucket Sampler.

The sample is collected by jacking or forcing the sampler into the soil by repeated blows of a drop hammer The sampler is then withdrawn. The split tube is separated after removing the shoe and the coupling and the sample is taken out. It is then placed in a container, sealed, and transported to the laboratory.

If the soil encountered in the bore hole is fine sand and it lies below the water table, the sample recovery becomes difficult. For such soils, a spring-core catcher device is used to aid recovery. As the sampler is lifted, the springs close and form a dome and retain the sample (Fig. 17.11).

While taking samples, care shall be taken to ensure that the water level in the hole is maintained slightly higher than the piezometric level at the bottom of the hole. It is necessary to prevent quick sand conditions.

The split tube may be provided with a thin metal or plastic tube liner to protect the sample and to hold it together. After the sample has been collected, the liner and the sample it contains are removed from the tube and the ends are sealed.

17.17. SCRAPER BUCKET SAMPLER

If a sandy deposit contains pebbles, it is not possible to obtain samples by standard split-spoon sampler or split-spoon sampler fitted with a spring core catcher. The pebbles come in-between the springs and prevent their closure. For such deposits, a scraper bucket sampler can be used.

A scraper bucket sampler consists of a driving point which is attached to its bottom end (Fig. 17.12). There is a vertical slit in the upper portion of the sampler. As the sampler is rotated, the scrapings of the soil enter the sampler through the slit. When the sampler is filled with the scrapings, it is lifted. Although the sample is quite disturbed, it is still representative.

A scraper bucket sampler can also be used for obtaining the samples of cohesionless soils below the water table.

17.18. SHELBY TUBES AND THIN-WALLED SAMPLERS

Shelby tubes are thin wall tube samplers made of seamless steel. The outside diameter of the tube may be between 40 to 125 mm. The commonly used samplers have the outside diameter of either 50.8 mm or 76.2 mm. The bottom of the tube is sharpened and bevelled, which acts as a cutting edge [Fig. 17.13 (*a*)]. The area ratio is less than 15% and the inside clearance is between 0.5 to 3%.

Fig. 17.13 (*b*) shows a thin-walled sampler (IS : 2132-1972). The length of the tube is 5 to 10 times the diameter for sandy soils and 10 to 15 times the diameter for clayey soils. The diameter generally varies between 40 and 125 mm, and the thickness varies from 1.25 to 3.15 mm.

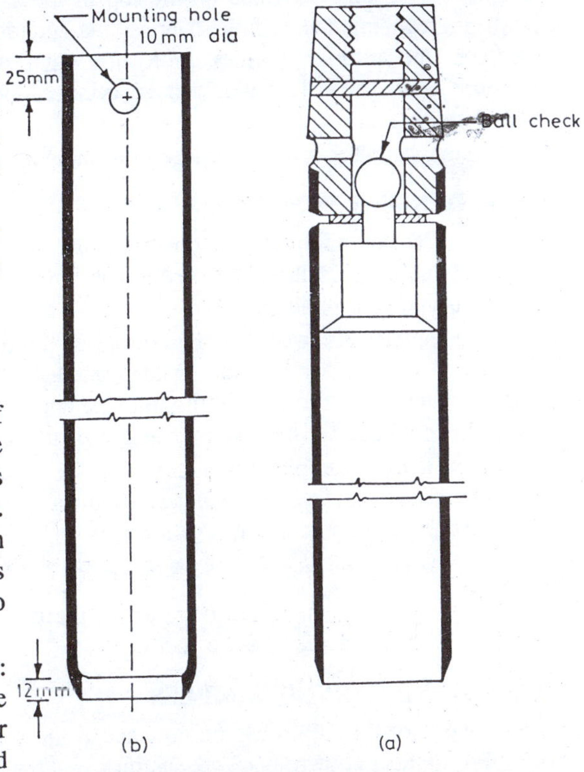

Fig. 17.13. (a) Shelby Tube, (b) Thin-walled Sampler.

The sampler tube is attached to the drilling rod and lowered to the bottom of the bore hole. It is then pushed into the soil. Care shall be taken to push the tube into the soil by a continuous rapid motion without impact or twisting. The tube should be pushed to the length provided for the sample. At least 5 minutes after pushing the tube into its final position, the tube is turned 2 revolutions to shear the sample off at the bottom before it is withdrawn. The tube is taken out and its ends are sealed before transportation.

Shelby tubes are used for obtaining undisturbed samples of clay.

17.19. PISTON SAMPLER

A piston sampler consists of a thin-walled tube with a piston inside. The piston keeps the lower end of the sampling tube closed when the sampler is lowered to the bottom of the hole [Fig. 17.14 (*a*)]. After the sampler has been lowered to the desired depth, the piston is prevented from moving downward by a suitable arrangement, which differs in different types of piston samplers. The thin tube sampler is pushed past the piston to obtain the sample [Fig. 17.14 (*b*)]. The piston remains in close contact with the top of the sample.

The presence of the piston prevents rapid squeezing of the soft soils into the tube and reduces the disturbance of the sample. A vacuum is created on the top of the sample, which helps in retaining the sample. During the withdrawal of the sampler, the piston provides protection against the water pressure which otherwise would have occurred on the top of the sample.

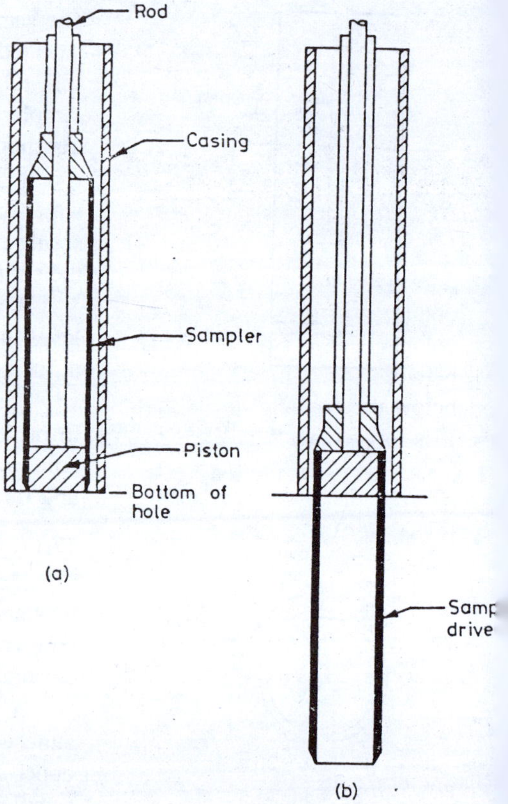

Fig. 17.14. Piston Sampler.

Piston samplers are used for getting undisturbed soil samples from soft and sensitive clays.

17.20. DENISON SAMPLER

The Denison sampler is a double-walled sampler. The outer barrel rotates and cuts into the soil. The sample is obtained in the inner barrel. The inner barrel is provided with a liner. It may also be provided with a basket-type core retainer.

The sampler is lowered to the bottom of the drilled hole. A downward force is applied on the top of the sampler. A fluid under pressure is introduced through the inner barrel to cool the coring bit when the outer barrel rotates The fluid returns through the annular space between the two barrels. The rotation of the outer barrel is continued till the required length of the sample is obtained.

The Denison sampler is mainly used for obtaining samples of stiff to hard cohesive soils and slightly cohesive sands. However, it cannot be used for gravelly soils, loose cohesionless sands and silts below ground water table and very soft cohesive soils.

The Denison sampler gives a sample $5\frac{1}{2}''$ (140 mm) in diameter and 20 inches (508 mm) long. Care is needed in adjusting the speed of rotation, the pressure on drilling bit and the velocity of wash water when drilling in soils and very friable rocks.

17.21. HAND-CARVED SAMPLES

Hand-carved samples can be obtained if the soil is exposed, as in a test pit, shaft or tunnel. Hand-carved samples are also known as *chunk* samples.

The soil should have at least a trace of cohesion so that it can stand unsupported for some time. To

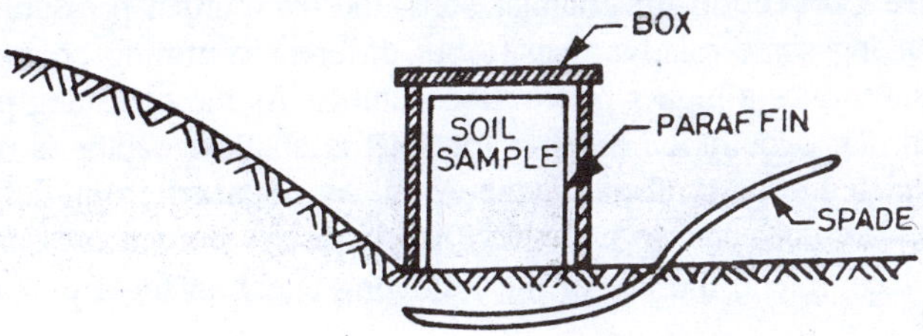

Fig. 17.15. Hand-Carved Sample.

obtain a sample, a column of soil is isolated in the pit. The soil is carefully removed from around the soil column and it is properly trimmed. An open-ended box is then placed over the soil column. The space between the box and the soil column is filled with paraffin. A spade or a plate with sharp edges is inserted below the box and the sample is cut at its base (Fig. 17.15). The box filled with the soil sample is removed. It is turned over and the soil surface in the box is trimmed and any depression is filled with paraffin.

A chunk sample may be obtained without using the box if the soil is cohesive. A column of soil is isolated. The block of soil is carefully removed from the soil column with a sharp knife. The chunk sample is then coated with paraffin wax to prevent loss of moisture.

Samples from open pits can also be obtained by pressing a sampling tube provided with a cutting edge. The soil surrounding the outside of the tube is carefully removed while the tube is being pushed into the soil.

Hand-carved samples are undisturbed.

17.22. STANDARD PENETRATION TEST

The standard penetration test is the most commonly used in-situ test, especially for cohesionless soils which cannot be easily sampled. The test is extremely useful for determining the relative density and the angle of shearing resistance of cohesionless soils. It can also be used to determine the unconfined compressive strength of cohesive soils.

The standard penetration test is conducted in a bore hole using a standard split-spoon sampler, described in Sect. 17.16. When the bore hole has been drilled to the desired depth, the drilling tools are removed and the sampler is lowered to the bottom of the hole. The sampler is driven into the soil by a drop hammer of 63.5 kg mass falling through a height of 750 mm at the rate of 30 blows per minute (IS : 2131—1963). The number of hammer blows required to drive 150 mm of the sample is counted. The sampler is further driven by 150 mm and the number of blows recorded. Likewise, the sampler is once again further driven by 150 mm and the number of blows recorded. The number of blows recorded for the first 150 mm is disregarded. The number of blows recorded for the last two 150 mm intervals are added to give the standard penetration number (N). In other words, the standard penetration number is equal to the number of blows required for 300 mm of penetration beyond a seating drive of 150 mm.

If the number of blows for 150 mm drive exceeds 50, it is taken as refusal and the test is discontinued.

The standard penetration number is corrected for dilatancy correction and overburden correction as explained below.

***(a)* Dilatancy Correction.** Silty fine sands and fine sands below the water table develop pore pressure which is not easily dissipated. The pore pressure increases the resistance of the soil and hence the penetration number (N).

Terzaghi and Peck (1967) recommend the following correction in the case of silty fine sands when the observed value of N exceeds 15.

The corrected penetration number, $N_c = 15 + \frac{1}{2}(N_R - 15)$ ...(17.5)

where N_R is the recorded value, and N_c is the corrected value.

If $N_R \leq 15$, $N_c = N_R$

(*b*) **Overburden Pressure Correction.** In granular soils, the overburden pressure affects the penetration resistance. If the two soils having same relative density but different confining pressures are tested, the one with a higher confining pressure gives a higher penetration number. As the confining pressure in cohesionless soils increases with the depth, the penetration number for soils at shallow depths is underestimated and that at greater depths is overestimated. For uniformity, the *N*-values obtained from field tests under different effective overburden pressures are corrected to a standard effective overburden pressure.

Gibbs and Holtz (1957) recommend the use of the following equation for dry or moist clean sand.

$$N_c = N_R \times \frac{350}{\bar{\sigma}_0 + 70} \qquad \ldots(17.6)$$

where N_R = observed *N*-value, N_c = corrected *N*-value, $\bar{\sigma}_0$ = effective overburden pressure (kN/m^2)

Eq. 17.6 is applicable for $\bar{\sigma}_0 \le 280$ kN/m^2.

The ratio (N_c/N_R) should lie between 0.45 and 2.0. If (N_c/N_R) ratio is greater than 2.0, N_c should be divided by 2.0 to obtain the design value used in finding the bearing capacity of the soil.

The correction may be extended to saturated silty sand and fine sand after modifying the N_R according to Eq. 17.6, *i.e.* N_c obtained from Eq. 17.6 would be taken as N_R in Eq. 17.5.

Thus the overburden correction is applied first and then the dilatancy correction is applied.

Peck, Hansen and Thornburn (1974) give the chart for correction of *N*-values to an effective overburden pressure of 96 kN/m^2, According to them.

$$N = 0.77\, N_R \log\left(\frac{1905}{\bar{\sigma}_0}\right) \quad \text{for } \bar{\sigma}_0 \ge 24 \text{ kN/m}^2 \ldots(17.7)$$

Fig. 17.16 shows the correction diagram. At $\bar{\sigma}_0 = 0.0$, the value of N/N_R is 2.0.

The correction given by Bazaraa (1967), and also by Peck and Bazaraa (1969), is one of the commonly used corrections. According to them,

$$N = \frac{4\,N_R}{1 + 0.0418\,\bar{\sigma}_0} \quad \text{if } \bar{\sigma}_0 < 71.8 \text{ kN/m}^2 \qquad \ldots(17.8)$$

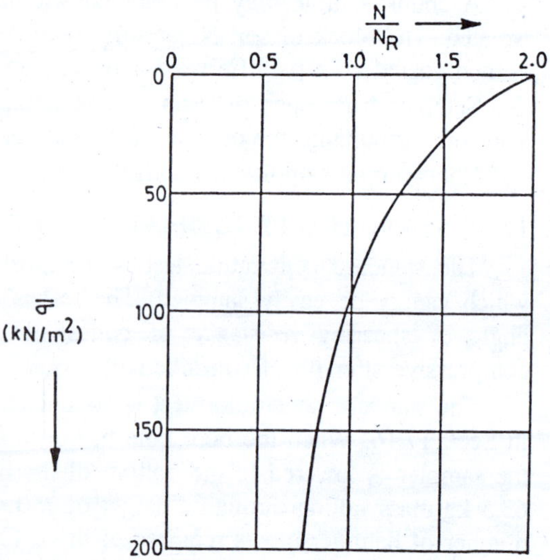

Fig. 17.16. Overburden Correction Diagram.

and
$$N = \frac{4\,N_R}{3.25 + 0.0104\,\bar{\sigma}_0} \quad \text{if } \bar{\sigma}_0 > 71.8 \text{ kN/m}^2 \qquad \ldots(17.9)$$

and
$$N = N_R \quad \text{if } \bar{\sigma}_0 = 71.8 \text{ kN/m}^2 \qquad \ldots(17.10)$$

Correlation of N with Engineering Properties

The value of the standard penetration number *N* depends upon the relative density of the cohesionless soil and the unconfined compressive strength of the cohesive soil. If the soil is compact or stiff, the penetration number is high.

The angle of shearing resistance (ϕ) of the cohesionless soil depends upon the number *N*. In general, the greater the *N*-value, the greater is the angle of shearing resistance (Fig. 17.17). Table 17.1 gives the average values of ϕ for different ranges of *N*.

The consistency and the unconfined shear strength of the cohesive soils can be approximately determined from the SPT number *N*. As the correlation is not dependable, it is advisable to determine the shear strength of the cohesive soils by conducting shear tests on the undisturbed samples or by conducting in-situ vane shear test (Sect. 17.24). Table 17.2 gives the approximate value of the unconfined shear strength for different ranges of *N*. The unconfined compressive strength can also be determined from the following relation.

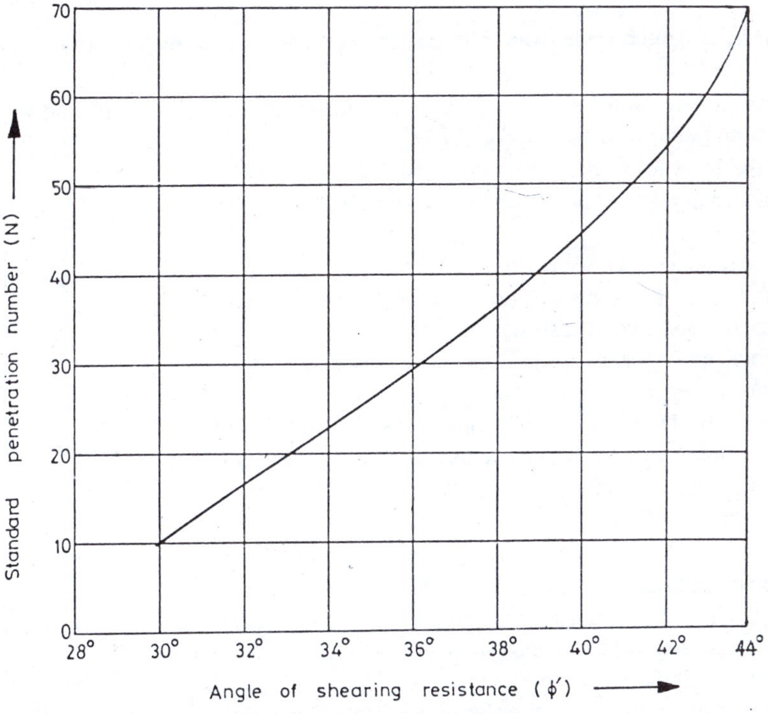

Fig. 17.17. Variation of ϕ′ with *N*.

$$q_u = 12.5\,N \qquad \text{...(17.11)}$$

where q_u is unconfined compressive strength (kN/m^2).

Table 17.1. Correlation between N and ϕ

N	*Denseness*	ϕ
0—4	Very Loose	25°—32°
4—10	Loose	27°—35°
10—30	Medium	30°—40°
30—50	Dense	35°—45°
> 50	Very Dense	> 45°

Table 17.2. Correlation between N and q_u

N	*Consistency*	q_u *(kN/m^2)*
0—2	Very Soft	< 25
2—4	Soft	25—50
4—8	Medium	50—100
8—15	Stiff	100—200
15—30	Very Stiff	200—400
> 30	Hard	> 400

17.23. CONE PENETRATION TESTS

Sounding methods are frequently used to determine the penetration resistance and the engineering properties of the soil. The sounding methods mainly consist of the cone test and the standard penetration test (Sect. 17.22).

The cone test was developed by the Dutch Government, Soil Mechanics Laboratory at Delft and is, therefore, also known as Dutch cone Test. The test is conducted either by the static method or by dynamic method, as discussed below.

(*a*) **Static Cone penetration test.** The Dutch cone has an apex angle of 60° and an overall diameter of 35.7 mm, giving an end area of 10 cm^2 (Fig. 17.18).

For obtaining the cone resistance, the cone is pushed downward at a steady rate of 10 mm/sec through a depth of 35 mm each time. The cone is pushed by applying thrust and not by driving.

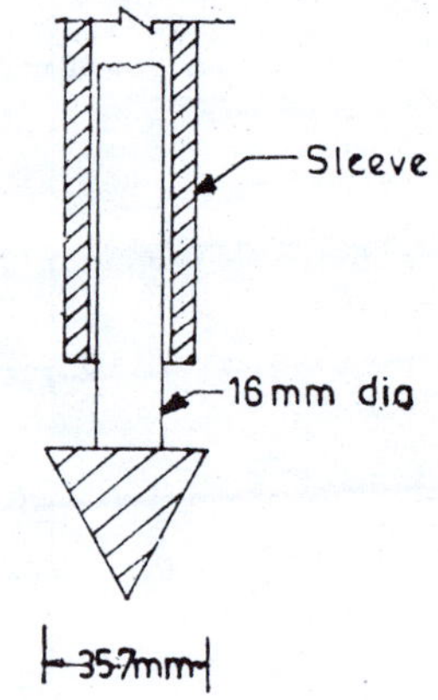

Fig. 17.18. Dutch Cone.

After the cone resistance has been determined, the cone is withdrawn. The sleeve is pushed on to the cone and both are driven together into the soil and the combined resistance is also determined. The resistance of the sleeve alone is obtained by subtracting the cone resistance from the combined resistance.

A modification of the Dutch cone penetrometers is the Refined Dutch cone. It has got a friction sleeve of limited length above the cone point. It is used for obtaining the point resistance of the cone and the frictional resistance of the soil above the cone point.

For effective use of the cone penetration test, some reliable calibration is required. This consists of comparing the results with those obtained from conventional tests conducted on undisturbed samples in a laboratory. It is also convenient to compare the cone test results with the standard penetration test results. As the standard penetration tests have been more commonly conducted in the past, good correlation studies are available between the SPT number (N) and the engineering properties of the soil (Sect. 17.22). If the cone penetration results are related to the SPT number N, indirect correlations are obtained between the cone test results and the engineering properties of soil.

The following relations hold approximately good between the point resistance of the cone (q_c) and the standard penetration number (N).

(*i*) Gravels $q_c = 800\ N$ to $1000\ N$ ...[17.12(a)]

(*ii*) Sands $q_c = 500\ N$ to $600\ N$ ...[17.12(a)]

(*iii*) Silty sands $q_c = 300\ N$ to $400\ N$ ...[17.12(a)]

(*iv*) Silts and clayey silts $q_c = 200\ N$ ...[17.12(a)]

where q_c is in kN/m^2

(*b*) **Dynamic cone Test.** The test is conducted by driving the cone by blows of a hammer. The number of blows for driving the cone through a specified distance is a measure of the dynamic cone resistance.

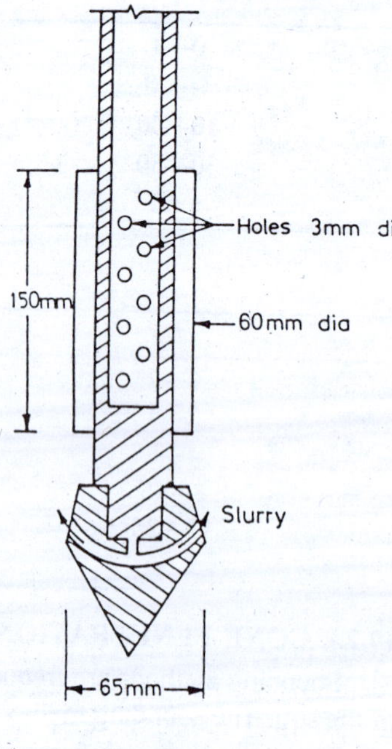

Fig. 17.19. Dynamic Cone Test.

Dynamic cone tests are performed either by using a 50 mm cone without bentonite slurry or by using a 65 mm cone with bentonite slurry (IS : 4968-part I and II-1976). The driving energy is given by a 65 kg-hammer falling through a height of 75 cm. The number of blows for every 10 cm penetration is recorded. The number of blows required for 30 cm of penetration is taken as the dynamic cone resistance (N_{cbr}). If the skin friction is to be eliminated, the test is conducted in a cased bore hole.

When a 65 mm cone with bentonite slurry is used, the set-up should have arrangements for circulating slurry so that the friction on the driving rod is eliminated (Fig. 17.19).

The dynamic cone resistance (N_{cbr}) is correlated with the SPT number N. The following approximate relations may be used when a 50 mm diameter cone is used.

$N_{cbr} = 1.5\,N$ for depths upto 3 m ...[17.13(*a*)]

$N_{cbr} = 1.75\,N$ for depths between 3 to 6 m ...[17.13(*b*)]

$N_{cbr} = 2.0\,N$ for depths greater than 6 m ...[17.13(*c*)]

The Central Building Research Institute, Roorkee, has developed the following correlation between the dynamic cone resistance (N_{bcr}) of 65 mm diameter cone without using bentonite slurry and the SPT number (N)

$N_{cbr} = 1.5\,N$ for depths upto 4 m ...[17.14(*a*)]

$N_{cbr} = 1.75\,N$ for depths between 4 to 9 m ...[17.14(*b*)]

$N_{cbr} = 2\,N$ for depths greater than 9 m ...[17.14(*c*)]

The above relations are applicable for medium to fine sand.

17.24. IN-SITU VANE SHEAR TEST

In-situ vane-shear test is conducted to determine the shear strength of a cohesive soil in its natural condition. The apparatus used is similar to one used in a laboratory (Chapter 13). It consists of four blades, 100 mm (or 150 mm or 200 mm) long, attached at right angles to a steel rod. The steel rod has a torque-measuring device at its top. The height-diameter ratio (H/D) of the apparatus is generally equal to 2 (Fig. 17.20).

For conducting the test, the shear-vane is pushed into the ground at the bottom of the bore hole. When a torque is applied through the handle at the top of the rod, the soil is sheared along a cylindrical surface. The torque required to shear the cylinder of the soil is measured by means of a spring balance. The undrained shear strength s_u of the soil is determined from the equation developed in chapter 13.

$$s_u = \frac{T}{\pi\,(D^2\,H/2 + D^3/6)} \quad ...(17.15)$$

where T = torque applied, H = height of the vane,
D = diameter of the soil cylinder sheared.

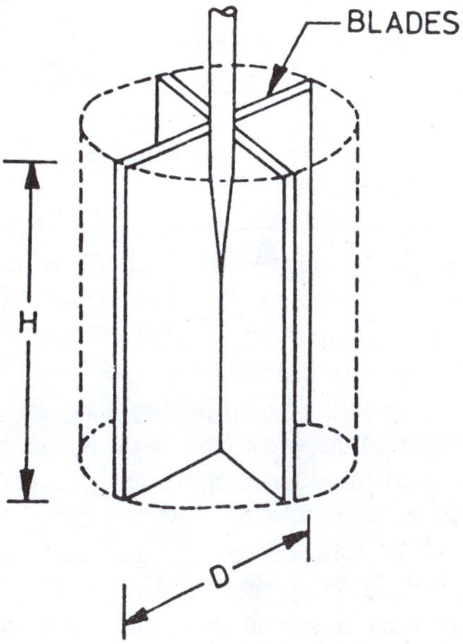

Fig. 17.20. In-situ Vane Shear Test.

The vane-shear test is extremely useful for determining the in-situ shear strength of very soft and sensitive clays, for which it is difficult to obtain undisturbed samples. The test can also be used even for determining the shear strength of stiff, fissured clays. However, the method cannot be used for sandy soils.

17.25. IN-SITU TESTS USING A PRESSURE METER

Menard developed a pressure meter, or sub-soil deformeter, which can be used for determining the stress-deformation characteristics of the soils in the natural conditions. The pressure meter consists of an inflatable cylindrical probe which is connected to a water reservoir (Fig. 17.21). The probe is inserted into the bore hole and it is inflated by applying water pressure. The probe presses against the unlined walls of the bore hole. As the pressure is increased, the soil deforms. The volumetric deformation of the bore hole is obtained by noting the fall in water level in the water reservoir.

Fig. 17.22 shows a typical pressure-volumetric strain curve. The soil is initially in elastic phase but it enters the plastic phase at high pressure. After the plastic stage, there is no change in the volume with further increase in pressure. The pressure-deformation data obtained from the test may be used to determine modulus of deformation, undrained shear strength, angle of shearing resistance and other engineering properties of the soil. IS : 1892-1979 describes the use of pressure meter.

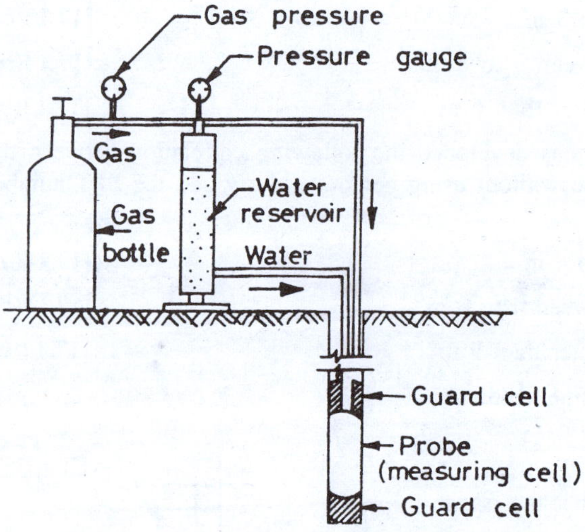

Fig. 17.21. Pressure Meter.

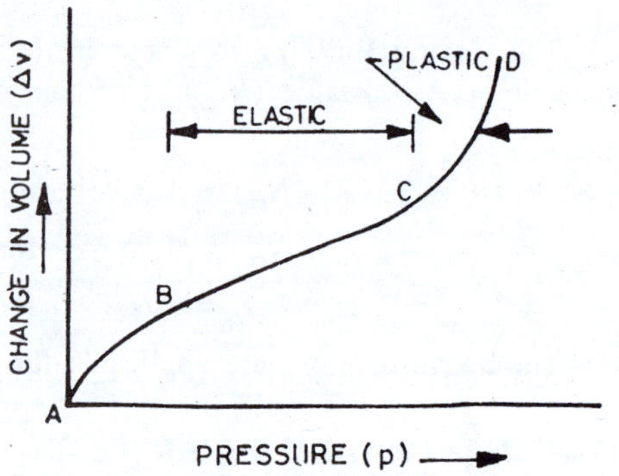

Fig. 17.22.

17.26. OBSERVATION OF GROUND WATER TABLE

The ground water level affects the pore water pressure and hence the shear strength of the soil. As the ground water level changes seasonally, it becomes necessary to establish the highest and the lowest water level for the proper design of structures. The depth of the ground water table is usually determined in an exploratory investigation.

In soils of high permeability, the level of ground water in a bore hole stabilises in about 24 hours. The depth of the water table is measured by lowering a chalk-coated tape in the bore hole. The depth can also be measured by lowering the leads of an electrical circuit. As soon as the open ends of the leads touch the water in the bore hole, the circuit is completed. It is indicated by glow of the indicator lamp.

In soils of very low permeability, the ground water level does not stabilise even after several weeks. Fig. 17.23 shows the Casagrande piezometer commonly used for the determination of the water level in such soils. It consists of a Norton porous tube attached to a plastic tube. The porous tube is carefully placed on a cushion of sand in the bottom of a cased bore hole such that it extends below the casing. The lower end of the porous tube is plugged with a rubber stopper At the top, the porous tube is surrounded by sand. There is an impermeable seal of bentonite clay above the sand. The sand surrounding the porous tube should be kept saturated during and after the installation of the piezometer. The ground water table is determined from the level of the water in the plastic tube. The top of the plastic tube is kept above the ground surface for this purpose.

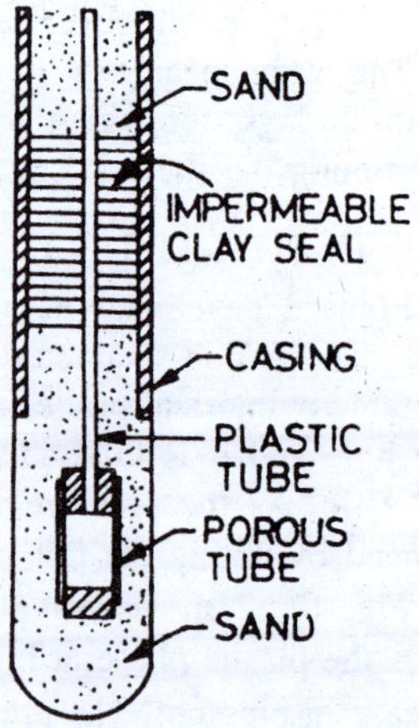

Fig. 17.23 Casagrande Piezometer.

For determination of the water level in silty soils, Hvorslev (1949) gave the following method. The water is bailed out from the bore hole to a level below the expected ground water table. The rise of water level in the bore hole is noted at different time intervals (Fig. 17.24). The figure shows the rise in water level at the time $t = t_1$, $t = t_2$ and $t = t_3$. Let h_1, h_2 and h_3 be the height of the water level above the level 0—0 at time Δt, $2\Delta t$ and $3\Delta t$. The time interval Δt is selected depending upon the type of the soil. The height of the water table above the levels 0—0, 2—2 and 3—3 is calculated from the following equations.

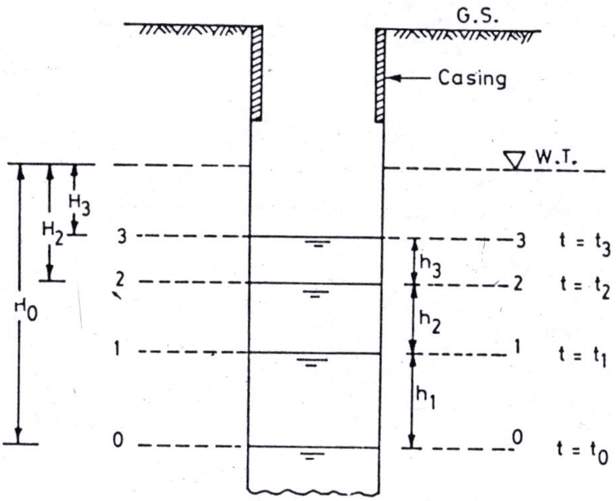

Fig. 17.24.

H_0 = height above level 0 - 0 $= \dfrac{h_1^2}{h_1 - h_2}$

H_2 = height above level 2 - 2 $= \dfrac{h_2^2}{h_1 - h_2}$

and H_3 = height above level 3 - 3 $= \dfrac{h_3^2}{h_2 - h_3}$ ...(17.16)

The depth of the water table is taken as the average of three levels given by heights H_0, H_2 and H_3 *i.e.* level of the water table is given by

$$= \frac{1}{3}\,[(\text{level } 0 - 0 + H_0) + (\text{level } 2 - 2 + H_2) + (\text{level } 3 - 3 + H_3)] \quad ...(17.17)$$

17.27. GEOPHYSICAL METHODS

A number of geophysical methods are used in preliminary investigations of sub-soil strata. The methods can be used for the location of different strata and for a rapid evaluation of the subsoil characteristics. However, the methods are very approximate. For detailed and reliable investigations, the conventional methods of driving a hole, taking a sample and testing it in a laboratory, as already described, must be resorted to. The geophysical methods can be broadly divided into the two categories : Seismic methods and Electrical resistivity methods, described in the following sections.

17.28. SEISMIC METHODS

The seismic methods are based on the principle that the elastic shock waves have different velocities in different materials. At the interface of two different materials, the waves get partly reflected and partly refracted. Seismic methods of subsurface explorations generally utilise the refracted waves.

The shock wave is created by a hammer blow or by a small explosive charge at a point P (Fig. 17.25). The shock wave travels through the top layer of the soil (or rock) with a velocity V_1, depending upon the type of material in layer-I. The observation of the first arrival of the waves is recorded by geophones located at various points, such as A, B, C. The geophones convert the ground vibration into electrical impulses and transmit them to a recording apparatus (not shown).

The basic equations of the refraction survey are derived based on the assumption that the subsurface strata are such that the velocity of the shock waves increases as the depth increases. In other words, it is assumed that $V_3 > V_2 > V_1$ in Fig. 17.25. At geophones located close to the point of impact, such as point A, the direct waves with velocity V_1 reach first.

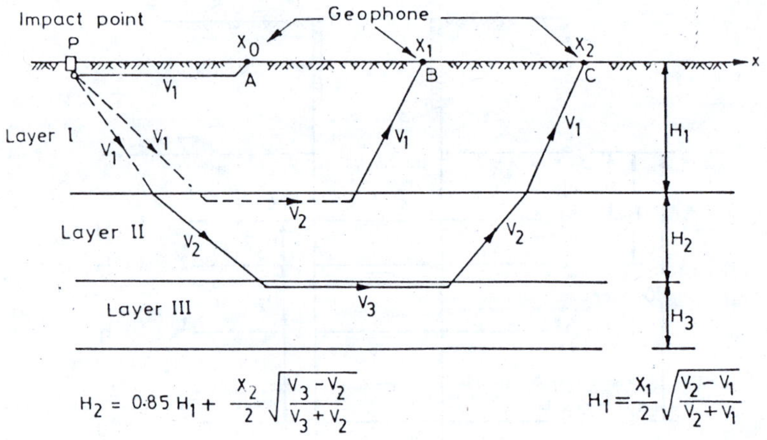

Fig. 17.25. Seismic method.

At points which are located away from the point of impact, such as point B, the refracted waves reach earlier than the direct waves. These waves start from point P, travel with velocity V_1 in the upper layer, get refracted at the interface, move with much higher velocity V_2 in the second layer, emerge again at the interface and travel back to the ground surface at a lower velocity V_1 in the upper layer.

At points further away from the point of impact, such as point C, the waves which are refracted twice, once at the interface of the layers I and II, and once at the interface of the layers II and III, reach earlier.

For the determination of the thickness of different layers, a distance-time graph is plotted (Fig. 17.26). The time (t) of arrival of the first impulse at various geophones is taken as ordinate and the distance (X) of the geophones from the point of impact P is taken as abscissa. Obviously, the velocity in any layer is equal to the reciprocal of the slope of the corresponding line. The slopes of the various lines are determined and the corresponding velocities computed.

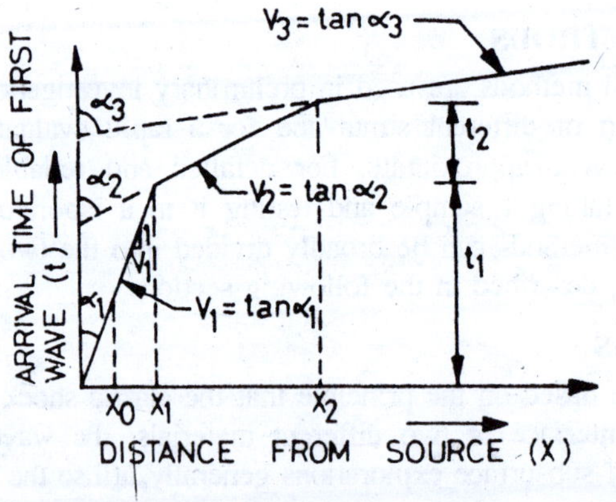

Fig. 17.26.

Upto a certain distance X_1, the direct waves in the layer I reach first. At this point, the first two lines in Fig. 17.26 intersect, which indicates that the direct wave traveling a distance X_1 with a velocity V_1 and the refracted wave traveling with a velocity V_1 in distance $2H_1$ and with a velocity of V_2 in distance X_1 reach simultaneously, where H_1 is the thickness of the layer I. Thus

$$\frac{X_1}{V_1} = \frac{2H_1}{V_1} + \frac{X_1}{V_2}$$

or
$$H_1 = \left(\frac{V_2 - V_1}{V_2}\right) \cdot \frac{X_1}{2} \qquad ...(17.18)$$

Eq. 17.18 gives reliable results when the waves are produced by a sinusoidal force and not by impact. The following empirical equation gives more reliable results for impact shock.

$$H_1 = \frac{X_1}{2} \sqrt{\frac{V_2 - V_1}{V_2 + V_1}} \qquad ...(17.19)$$

Likewise, the thickness of the second layer (H_2) is obtained from the distance X_2 corresponding to the point of intersection of the second and the third line in Fig. 17.26. It is given by the relation

$$H_2 = 0.85\, H_1 + \frac{X_2}{2} \sqrt{\frac{V_3 - V_2}{V_3 + V_2}} \qquad ...(17.20)$$

The procedure is continued if there are more than three layers.

The type of material in various layers can be determined by comparing the velocities obtained with the standard velocities given in Table 17.3.

Table 17.3. Velocities in Different Types of Strata

Type of Rock/Soil	*Granite*	*Sand Stone*	*Shale*	*Hard Clay*	*Loose gravel (Wet)*	*Loose sand (Wet)*	*Loose sand (dry)*
Velocity (m/sec)	4000 to 6000	1500 to 3000	1300 to 3000	600 to 1500	500 to 1000	500 to 1500	250 to 600

Limitation of the seismic methods

(1) The methods cannot be used if a hard layer with a greater seismic velocity overlies a softer layer with a smaller seismic velocity.

(2) The methods cannot be used for the areas covered by concrete, asphalt pavements or any other artificial hard crust, having a high seismic velocity.

(3) If the area contains some underground features, such as buried conduits, irregularly dipping strata, and irregular water table, the interpretation of the results becomes very difficult.

(4) If the surface layer is frozen, the method cannot be successfully used, as it corresponds to a case of harder layer overlying a softer layer.

(5) The methods require sophisticated and costly equipment.

(6) For proper interpretations of the seismic survey results, the services of an expert are required.

Despite above limitations, the method is extremely useful for the determination of the thickness of various strata and their characteristics. These surveys are useful for obtaining preliminary information about the types and depths of various strata at a given site.

17.29. ELECTRICAL RESISTIVITY METHODS

The electrical resistivity (ρ) of a conductor is expressed as

$$\rho = RA/L \qquad ...(17.21)$$

where R = electrical resistance (ohms), A = area of cross-section of the conductor (cm^2),
L = length of conductor (cm), ρ = electrical resistivity (ohm-centimeter).

(It may be noted that the electrical resistivity is the reciprocal of conductance.)

The resistivity of a material depends upon the type of material, its water content and the concentration of dissolved ions and many other factors. Rocks and dry soils have a greater resistivity than saturated clays. Table 17.4 gives approximate values of resistivity of different rocks and soils.

Table 17.4. Resistivity of Different Rocks and Soils

Type of rock/soil	*Sound rock*	*Weathered Rock*	*Gravel*	*Sand*	*Clayey sand*	*Saturated clay and silt*
Resistivity (Ohm-m)	> 5000	1500 ro 2500	1500 to 4500	500 to 1500	200 to 500	2 to 100

The electrical resistivity methods are of the following two types:

(*a*) **Electrical Profiling Method.** The method is also known as the *resistivity mapping method.* Four electrodes are used at a constant spacing *a* (Fig. 17.27). To conduct the test, four electrodes, which are usually in the form of metal spikes, are driven into the ground. The two outer electrodes are known as current electrodes. The two inner electrodes are called potential electrodes. The mean resistivity of the strata is determined by applying a D.C. current to the outer electrodes and by measuring the voltage drop between the inner electrodes. A current of 50 to 100 milliamp is usually supplied.

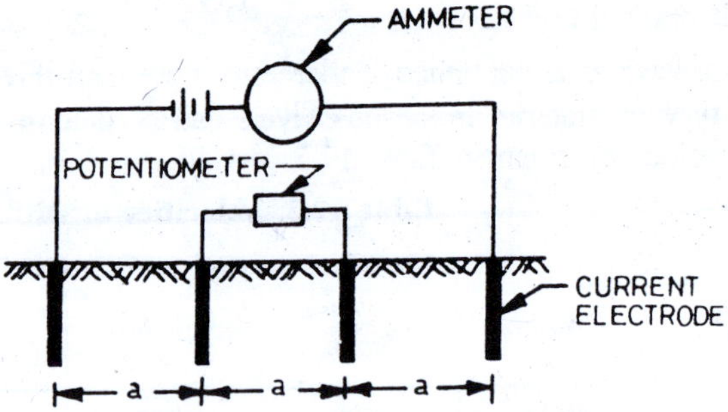

Fig. 17.27. Electrical Profiling Method.

The mean resistivity (ρ) is given by the formula

$$\rho = \frac{2\pi a V}{I} \qquad ...(17.22)$$

where I = current supplied, a = spacing of electrodes, and V = voltage drop.

Eq. 17.22 gives the mean resistivity upto a depth of *a* below the ground surface, as the depth of current penetration below the ground surface is approximately equal to the spacing of electrodes. The electrodes are moved as a group, and different profile lines are run across the area. The test is repeated after changing the spacing and again determining the mean resistivity upto the depth equal to the new spacing. The electrods are moved as a group along different lines, as before.

The method is useful for establishing boundaries between different strata. The method is generally used for locating sand and gravel deposits within a fine-grained soil deposit.

(*b*) **Electrical Sounding Method.** In this method, the electrode system, consisting of four electrodes, is expanded about a fixed location, say *P*, in Fig. 17.28. The spacing in the first setting is a_1, which is increased to a_2 in the second setting and to a_3 in the third setting. The spacing is thus gradually increased to a distance equal to the depth of exploration. As the depth of the current penetration is equal to the electrode spacing, the changes in the mean resistivity is correlated to the changes in strata at that location.

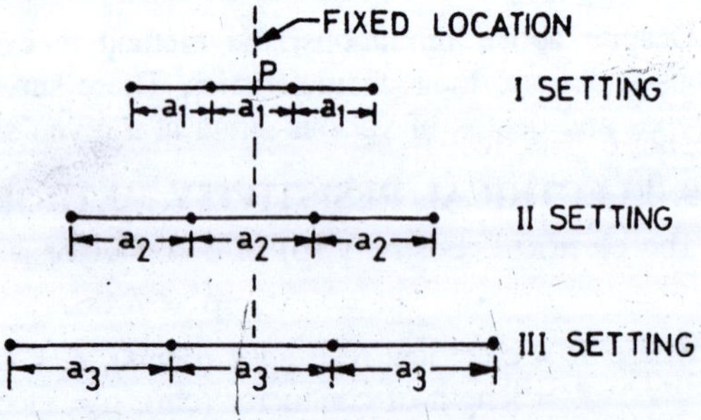

Fig. 17.28. Electrical Sounding method.

The method is useful in studying the changes in the strata with increasing depth at a point. The method can indicate sub-surface variation when a hard layer overlies a soft layer or vice-versa. It can also be used to locate the water table.

Limitation of the electrical resistivity methods.

(1) The methods are capable of detecting only the strata having different electrical resistivity.

(2) The results are considerably influenced by surface irregularities, wetness of the strata and electrolyte concentration of the ground water.

(3) As the resistivity of different strata at the interface changes gradually and not abruptly as assumed, the interpretation becomes difficult.

(4) The services of an expert in the field are needed.

Notwithstanding above limitations, the method is extremely useful for the determination of the average conditions in different strata upto a depth of 30 m or so. The method is very rapid and economical for preliminary investigations.

17.30. SUB-SOIL INVESTIGATION REPORT

A sub-soil investigation report should contain the data obtained from bore holes, site observations and laboratory results. It should also give the recommendations about the suitable type of foundation, allowable soil pressure and expected settlements.

It is essential to give a complete and accurate record of data collected. Each bore hole should be identified by a code number. The location of each bore hole should be fixed by measurement of its distance or angles from some permanent feature. All relevant data for the bore hole is recorded in a boring log (Fig. 17.29). A boring log gives the description or classification of various strata encountered at different depths. Any additional information that is obtained in the field, such as soil consistency, unconfined compression strength, standard penetration test, cone penetration test, is also indicated on the boring log. It should also show the water table. If the laboratory tests have been conducted, the information about index properties, compressibility, shear strength, permeability, etc. should also be provided.

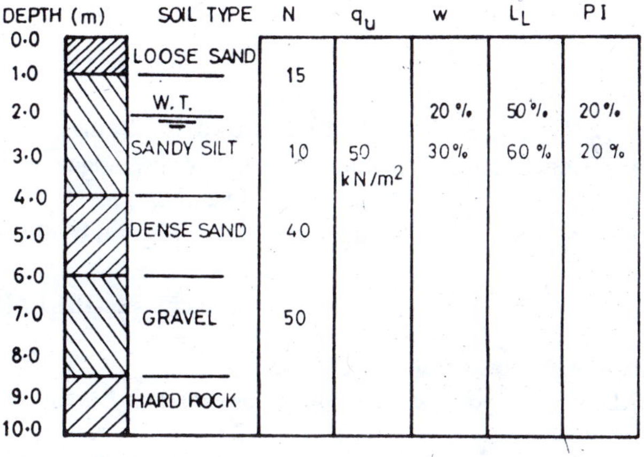

Fig. 17.29. Boring Log. (1 t/m^2 = 10 kN/m^2).

The data obtained from a series of bore holes is presented in the form of a *sub-surface profile* (Fig. 17.30). A subsurface profile is a vertical section through the ground along the line of exploration. It indicates the boundaries of different strata, along with their classification. It is important to remember that conditions between bore holes are estimated by interpolation, which may not be correct. Obviously, the larger the number of holes, the more accurate is the sub-surface profile.

The site investigation report should contain the discussion of the results. The discussion should be clear and concise The recommendations about the type and depth of foundation, allowable soil pressure and expected settlements should be specific. The main findings of the report are given in conclusions.

A soil exploration report generally consists of the following.

(1) Introduction, which gives the scope of the investigation.

(2) Description of the proposed structure, the location and the geological conditions at the site.

(3) Details of the field exploration programme, indicating the number of borings, their location and depths.

(4) Details of the methods of exploration.

(5) General description of the sub-soil conditions as obtained from in-situ tests, such as standard penetration test, cone test.

(6) Details of the laboratory test conducted on the soil samples obtained and the results obtained.

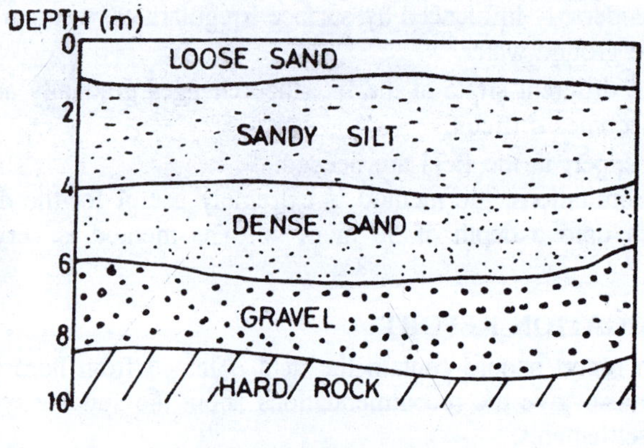

Fig. 17.30. Subsurface Profile.

(7) Depth of the ground water table and the changes in water levels.

(8) Discussion of the results.

(9) Recommendation about the allowable bearing pressure, the type of foundation or structure.

(10) Conclusions. The main findings of investigations should be clearly stated. It should be brief but should mention the salient points.

Limitations of the investigations should also be briefly stated.

PROBLEMS

A. Descriptive and Objective Type

17.1. What do you understand by site investigation ? What are the different purposes for which site investigations are done?

17.2. What is reconnaissance ? What type of information is obtained in reconnaissance? What is its use?

17.3. How would you decided the depth of exploration and the lateral extent of the investigations?

17.4. Describe open excavation methods of exploration. What are their advantages and disadvantages ?

17.5. Describe various methods of drilling holes for subsurface investigations.

17.6. What do you understand about disturbed and undisturbed samples? How would you obtain undisturbed samples?

17.7. What are the factors that affect the sample disturbance? How are these effects minimised?

17.8. Describe the split-spoon sampler. What is its use?

17.9. Discuss various types of soil samplers for obtaining undisturbed samples.

17.10. How would you obtain a hand-carved sample?

17.11. Discuss standard penetration test. What are the various corrections? What is the importance of the test in geotechnical engineering?

17.12. Describe cone penetration tests. How these tests differ from standard penetration test ?

17.13. How would you conduct an in-situ vane shear test? What is its use ?

17.14. Discuss the various methods for determining the level of the ground water table.

17.15. Describe, in brief, various geophysical methods. Discuss their limitations and uses.

17.16. Describe the salient features of a good sub-soil investigation report.

17.17. Write whether the following statements are true or false.

(*a*) A disturbed sample is not a truly representative sample.

(*b*) An undisturbed sample is absolutely undisturbed.

(*c*) The samples obtained from wash water of wash borings are disturbed samples.

(*d*) The sample obtained by a split spoon sampler is an undisturbed sample.

(*e*) The standard penetration test is more useful for cohesionless soil than cohesive soils.

(*f*) The static cone test is more useful for cohesive soils than cohesionless soils.

(*g*) Dilatancy correction of SPT number is done for clayey soils.

(*h*) In-situ vane shear test is extremely useful for obtaining the shear strength of very soft cohesive soils.

(*i*) The Casagrande piezometer is used for determining the water level in cohesionless soils.

(*j*) Seismic methods can be used if the seismic velocity of different strata increases as the depth is increased.

(*k*) Geophysical methods are useful for preliminary investigations.

(*l*) A soil profile gives an accurate profile of different strata.

(*m*) A boring log indicates different strata along the depth of a bore hole.

[Ans. True, (*a*), (*c*), (*e*), (*f*), (*h*), (*j*), (*k*), (*m*)]

B. Multiple Choice Questions

1. The standard penetration test is useful to measure
(*a*) shear strength of soft clays
(*b*) shear strength of sands
(*c*) consistency of clays
(*d*) None of above.

2. For an undisturbed sample, the area ratio of the samples should be
(*a*) zero (*b*) 10% or less
(*c*) 10% to 20% (*d*) more than 20%

3. In-situ vane shear test is used to measure shear strength of
(*a*) very soft and sensitive clays
(*b*) stiff and fissured clays
(*c*) sandy soils
(*d*) All the above

4. The seismic refraction methods cannot be used if the wave velocity in the lower layer is . . . . that in the upper layer.
(*a*) greater than (*b*) less than
(*c*) more than four times (*d*) more than thice

5. Select the incorrect statement:
For a good quality soil sample,
(*a*) The area ratio should be low
(*b*) The cutting edge should be thick
(*c*) The inside clearance should be small
(*d*) The outside clearance should be small

6. If the actual value of the standard penetration number (N) is greater than 15 for fine sands below water table, the corrected value of N is
(*a*) $15 + \left(\frac{N + 15}{2}\right)$ (*b*) $15 - \left(\frac{N + 15}{2}\right)$
(*c*) $15 + \left(\frac{N - 15}{2}\right)$ (*d*) $15 + \left(\frac{15 - N}{2}\right)$

7. The height-diameter ratio for the in-situ vane is
(*a*) 1.0 (*b*) 1.50
(*c*) 2.00 (*d*) 3.0

[Ans. 1. (*b*), 2. (*b*), 3. (*a*), 4. (*b*), 5. (*b*), 6. (*c*), 7. (*c*)]

18

Stability of Slopes

18.1. INTRODUCTION

An earth slope is an unsupported, inclined surface of a soil mass. Earth slopes are formed for railway formations, highway embankments, earth dams, canal banks, levees, and at many other locations. Fig 18.1 shows some of the examples of earth slopes.

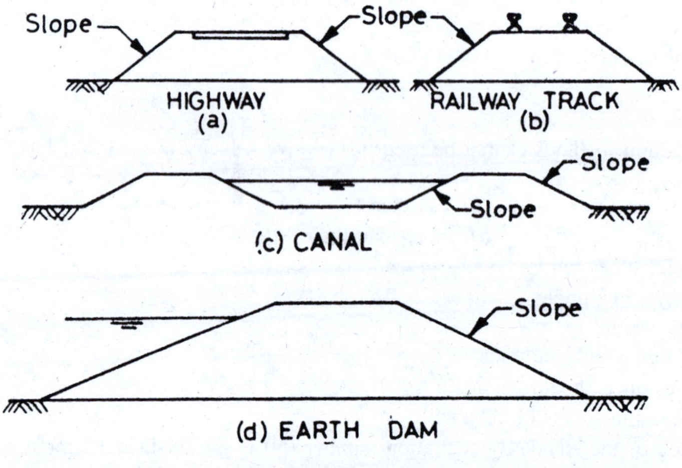

Fig. 18.1. Earth Slopes.

The cost of earth work would be minimum if the slopes are made steepest. However, very steep slopes may not be stable. A compromise has to be made between economy and safety, and the slopes provided are neither too steep nor too flat. In other words, the steepest slopes which are stable and safe should be provided.

The failure of a slope may lead to loss of life and property. It is, therefore, essential to check the stability of proposed slopes. With the development of modern methods of testing of soils and stability analysis, a safe and economical design of a slope is possible. The geotechnical engineer should have a thorough knowledge of the various methods for checking the stability of slopes and their limitations.

The failure of a soil mass occurs along a plane or a curved surface when a large mass of soil slides with respect to the remaining mass. In general, there is a downward and outward movement of the soil mass. A slope failure occurs when the forces causing failure are greater than the shearing resistance (shear strength) developed along a critical surface of failure. The factors leading to the failure of slopes may be classified into two categories.

(1) *The factors which cause an increase in the shear stresses.* The stresses may increase due to weight of water causing saturation of soils, surcharge loads, seepage pressure or any other cause. The stresses are also increased due to steepening of slopes either by excavation or by natural erosion.

(2) *The factors which cause a decrease in the shear strength of the soil.* The loss of shear strength may occur due to an increase in water content, increase in pore water pressure, shock or cyclic loads, weathering or any other cause.

Most of the natural slope failures occur during rainy seasons, as the presence of water causes both increased stresses and the loss of strength.

18.2. BASIS OF ANALYSIS

The soil mass must be safe against slope failure on any conceivable surface across the slope. Although the methods using the theory of elasticity or plasticity are also being increasingly used, the most common methods are based on limiting equilibrium in which it is assumed that the soil is at the verge of failure. The methods of limiting equilibrium are statically indeterminate. As the stress-strain relationships along the assumed surface are not known, it is necessary to make assumptions so that the system becomes statically determinate and it can be analysed easily using the equations of equilibrium. The following assumptions are generally made.

(1) The stress system is assumed to be two-dimensional. The stresses in the third direction (perpendicular to the section of the soil mass) are taken as zero.

(2) It is assumed that the Coulomb equation for shear strength is applicable and the strength parameters c and ϕ are known.

(3) It is further assumed that the seepage conditions and water levels are known, and the corresponding pore water pressure can be estimated.

(4) The conditions of plastic failure are assumed to be satisfied along the critical surface. In other words, the shearing strains at all points of the critical surface are large enough to mobilise all the available shear strength.

(5) Depending upon the method of analysis, some additional assumptions are made regarding the magnitude and distribution of forces along various planes.

In the analysis, the resultant of all the actuating forces trying to cause the failure is determined. An estimate is also made of the available shear strength. The factor of safety of the slope is determined from the available resisting forces and the actuating forces.

18.3. DIFFERENT DEFINITIONS OF FACTORS OF SAFETY

Three different definitions of the factor of safety are used.

(*a*) Factor of safety with respect to shear strength

In common usage, the factor of safety is defined as the ratio of the shear strength to the shear stress along the surface of failure. The factor of safety as defined above is known as the factor of safety with respect to shear strength.

Thus $$F_s = \frac{s}{\tau_m} \qquad \text{...(18.1)}$$

where F_s = factor of safety with respect to shear strength, s = shear strength,

τ_m = mobilised shear strength (equal to applied shear stress).

Eq. 18.1 can be written in terms of the cohesion intercept and the angle of shear resistance as

$$F_s = \frac{c + \bar{\sigma} \tan \phi}{c_m + \bar{\sigma} \tan \phi_m} \qquad \text{...(18.2)}$$

where c_m = mobilised cohesion, ϕ_m = mobilised angle of shear resistance, $\bar{\sigma}$ = effective pressure.

Rearranging Eq. 18.2,

$$\frac{c}{F_s} + \frac{\bar{\sigma} \tan \phi}{F_s} = c_m + \bar{\sigma} \tan \phi_m$$

Therefore, $c_m = c/F_s$ ...(18.3)

and $\tan \phi_m = \tan \phi / F_s$ ...(18.4)

Eqs. 18.3 and 18.4 indicate that the factor of safety with respect to the cohesion intercept and that with respect to the angle of shearing resistance are equal to the factor of safety with respect to the shear strength.

(*b*) Factor of safety with respect to cohesion

The factor of safety with respect to cohesion (F_c) is the ratio of the available cohesion intercept (c) and the mobilised cohesion intercept.

Thus $$F_c = \frac{c}{c_m} \quad ...(18.5)$$

where c = cohesion intercept, c_m = mobilised cohesion intercept,
F_c = factor of safety with respect to cohesion.

(*c*) Factor of safety with respect to friction

The factor of safety with respect to friction is the ratio of the available frictional strength to the mobilised frictional strength. Thus

$$F_\phi = \frac{\bar{\sigma} \tan \phi}{\bar{\sigma} \tan \phi_m}$$

or $$F_\phi = \frac{\tan \phi}{\tan \phi_m} \quad ...(18.6)$$

where F_ϕ = factor of safety with respect to friction, ϕ = angle of shearing resistance,
ϕ_m = angle of mobilised shearing resistance.

For small angles, Eq. 18.6 can be expressed as

$$F_\phi = \frac{\phi}{\phi_m} \quad ...([18.6(a)]$$

In the analysis of stability of slopes, generally the three factors of safety are taken equal, *i.e.* $F_s = F_c = F_\phi$. However, sometimes when greater reliance is placed on the parameter ϕ than the parameter c, the factor of safety with respect to cohesion is taken greater than that with respect to friction. In such a case, the factor of safety with respect to friction is usually taken as unity *i.e.* $\phi_m = \phi$.

18.4. TYPES OF SLOPE FAILURES

A slope may have any one of the following types of failures.

(1) Rotational failure. This type of failure occurs by rotation along a slip surface by downward and outward movement of the soil mass (Fig. 18.2). The slip surface is generally circular for homogeneous soil conditions and non-circular in case of non- homogeneous conditions. Rotational slips are further divided into 3 types.

(*a*) Toe failure, in which the failure occurs along the surface that passes through the toe [Fig. 18.2 (*a*)].

(*b*) Slope failure, in which the failure occurs along a surface that intersects the slope above the toe [Fig. 18.2 (*b*)].

(*c*) Base failure, in which the failure surface passes below the toe [Fig. 18.2 (*c*)].

The slope failure occurs when a weak plane exists above the toe. The base failure occurs when a weak stratum lies beneath the toe. If a strong stratum exists below the toe, the slip surface of the base failure is tangential to that stratum. In all other cases, the failures are generally toe failures. Toe failures are most common.

(2) Translational Failure. A constant slope of unlimited extent and having uniform soil properties at the

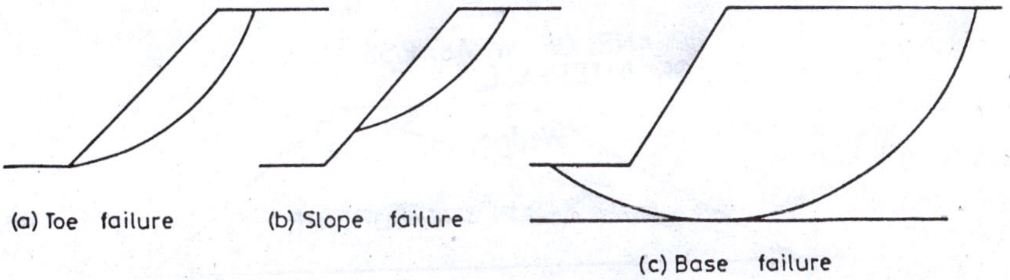

Fig. 18.2. Rotational Failure.

same depth below the free surface is known as an infinite slope. In practice, the slopes which are of considerable extent and in which the conditions on all verticals are adequately represented by average conditions are designated as *infinite slopes.*

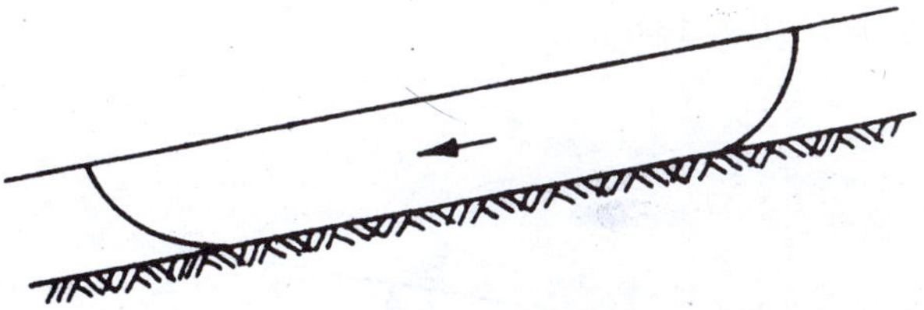

Fig. 18.3. Translational Failure.

Translational failure occurs in an infinite slope along a long failure surface parallel to the slope (Fig. 18.3). The shape of the failure surface is influenced by the presence of any hard stratum at a shallow depth below the slope surface. Translational failures may also occur along slopes of layered materials.

(3) Compound Failure. A compound failure is a combination of the rotational slips and the translational slip (Fig. 18.4). A compound failure surface is curved at the two ends and plane in the middle portion. A compound failure generally occurs when a hard stratum exists at considerable depth below the toe.

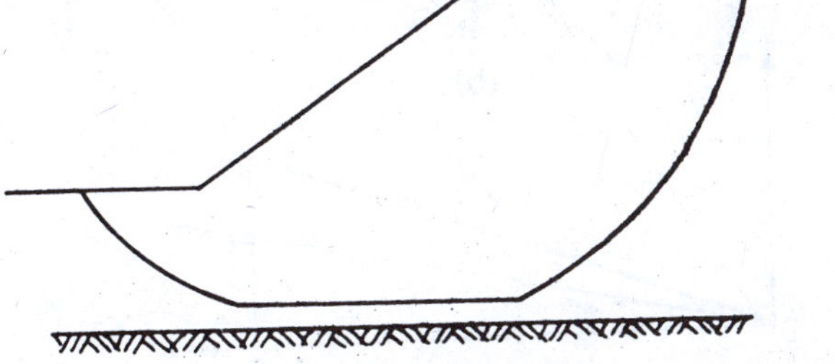

Fig. 18.4. Compound Failure.

(4) Wedge Failure. A failure along an inclined plane is known as plane failure or wedge failure or block failure (Fig. 18.5). It occurs when distinct blocks and wedges of the soil mass become separated.

A plane failure is similar to translational failure is many respects. However, unlike translational failure which occurs in an infinite slope, a plane failure may occur even in a finite slope consisting of two different materials or in a homogeneous slope having cracks, fissures, joints or any other specific plane of weakness.

(5) Miscellaneous Failures. In addition to above four types of failures, some complex types of failures

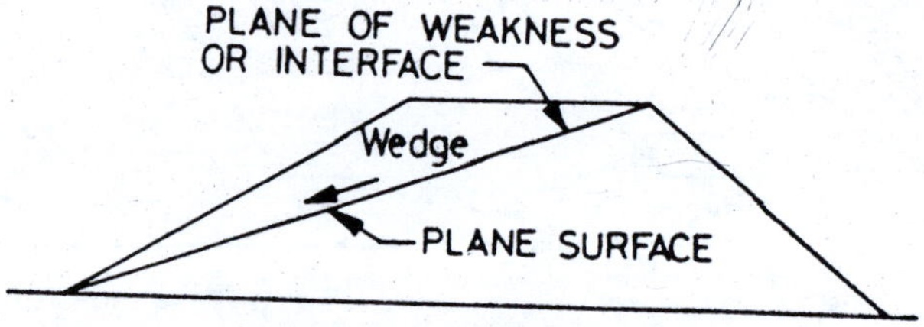

Fig. 18.5. Wedge Failure.

in the form of spreads and flows may also occurs.

18.5. STABILITY OF AN INFINITE SLOPE OF COHESIONLESS SOILS

The stability criteria of an infinite slope of cohesionless soils will depend whether the soil is dry, or submerged or has steady seepage, as explained below.

(1) Dry Soil. Fig. 18.6 (*a*) shows a section of an infinite slope having a slope angle of i. Let us consider the prism $ABCD$ of the soil, with the inclined length AB equal to b. The horizontal length of the prism is $b \cos i$. The height of the prism is H [Fig. 18.6 (*b*)].

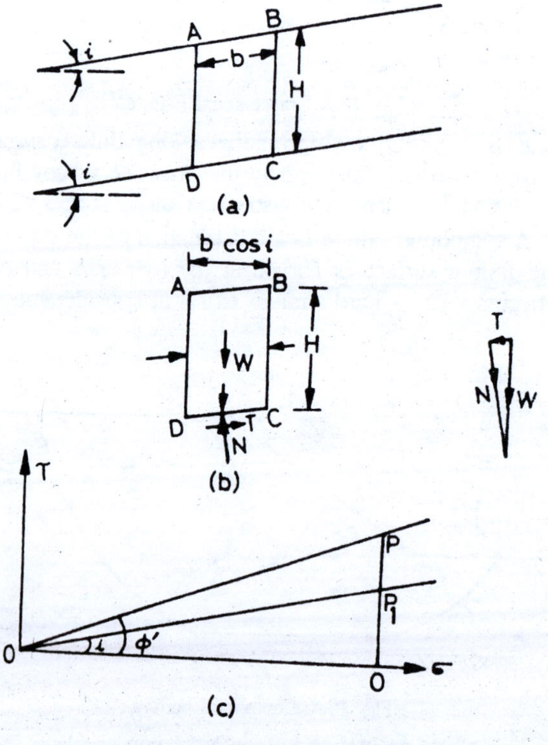

Fig. 18.6. Infinite Slope in Dry Sand.

Volume of prism per unit length $= Hb \cos i$

Weight of prism per unit length, $W = \gamma\,(Hb \cos i)$

The weight of the prism can be resolved into the normal component N and tangential components T to plane CD.

Thus $$N = W\cos i = \gamma\ Hb\ \cos^2 i$$

$$T = W\sin i = \gamma\ Hb\cos i\sin i$$

The normal and shear stresses are given by

$$\sigma = \frac{N}{b} = \frac{\gamma Hb\cos^2 i}{b} = \gamma H\cos^2 i \quad \text{...(18.7)}$$

$$\tau = \frac{T}{b} = \frac{\gamma Hb\cos i\sin i}{b} = \gamma H\cos i\sin i \quad \text{...(18.8)}$$

The shear stresses τ tend to cause the shear failure along *CD*. This tendency is opposed by the shearing resistance developed along the plane *CD*. As the soil is dry, there is no pore water pressure.

Therefore, $$s = \bar{\sigma}\tan\phi' = \sigma\tan\phi'$$

or $$s = (\gamma H\cos^2 i)\tan\phi'$$

The factor of safety against shear failure is given by

$$F_s = \frac{s}{\tau} = \frac{(\gamma H\cos^2 i)\tan\phi'}{\gamma H\cos i\sin i}$$

or $$F_s = \frac{\tan\phi'}{\tan i} \quad \text{...(18.9)}$$

Eq. 18.9 indicates that the slope is just stable when $\phi' = i$. The factor of safety is greater than unity when i is less than ϕ'. For the slope angle i greater than ϕ', the slope is not stable.

It is worth noting that the factor of safety of an infinite slope of a cohesionless soil is independent of the height H of the assumed failure prism.

The angle ϕ' in Eq. 18.9 should correspond to the actual relative density of the soil. As the soil in the surface layers is in a relatively loose state, the angle ϕ' corresponding to the loose state is generally taken.

Eq. 18.9 can be represented graphically [Fig. 18.6 (*c*)]. The ordinate PQ is equal to $\sigma\tan\phi'$ and represents the shear strength. The ordinate P_1Q represents the shear stress τ equal to $\sigma\tan i$. Obviously, the factor of safety is given by

$$F_s = \frac{s}{\tau} = \frac{PQ}{P_1Q} = \frac{\tan\phi'}{\tan i}$$

(2) Submerged Slope. If the slope is submerged under water, the normal effective stress and the shear stress are calculated using the submerged unit weight and not the bulk unit weight as was used for dry soil. Thus, from Eqs. 18.7 and 18.8.

$$\bar{\sigma} = \gamma' H\cos^2 i \quad \text{...(18.10)}$$

and $$\tau = \gamma' H\sin i\cos i \quad \text{...(18.11)}$$

where γ' is the submerged unit weight.

Therefore, the factor of safety is given by

$$F_s = \frac{s}{\tau} = \frac{(\gamma' H\cos^2 i)\tan\phi'}{\gamma' H\sin i\cos i}$$

or $$F_s = \frac{\tan\phi'}{\tan i} \quad \text{...(18.12)}$$

Comparing Eqs. 18.9 and 18.12, it is observed that the factor of safety of a submerged slope is the same as that in dry condition.

(3) Steady Seepage along the slope. Fig. 18.7 (*a*) shows an infinite slope with steady seepage parallel to the surface. Fig. 18.7 (*b*) shows a free-body diagram of the prism *ABCD* of the soil. The forces acting on the vertical sides of the prism due to water and soil are equal and opposite and, therefore, cancel. The weight of the prism W is taken corresponding to the saturated conditions.

Therefore, $W = \gamma_{sat} H b \cos i$

$$N = W \cos i = \gamma_{sat} H b \cos^2 i$$

$$T = W \sin i = \gamma_{sat} H b \sin i \cos i$$

At the base of the prism, there is an upward force due to water pressure (u), given by

$$u = \gamma_w H \cos^2 i$$

Uplift force, $U = (\gamma_w H \cos^2 i)\, b$

Thus, the net normal force $\overline{N}$ is given by

$$\overline{N} = N - U = \gamma_{sat} H b \cos^2 i - (\gamma_w H \cos^2 i)\, b$$

or $\overline{N} = \gamma' bH \cos^2 i$, where γ' is submerged unit weight.

The effective stress is given by

$$\overline{\sigma} = \frac{\overline{N}}{b} = \gamma' H \cos^2 i$$

Shear strength, $s = \overline{\sigma} \tan \phi' = \gamma' H \cos^2 i \tan \phi'$

The shear stress is given by

$$\tau = \frac{T}{b} = \gamma_{sat} H \sin i \cos i$$

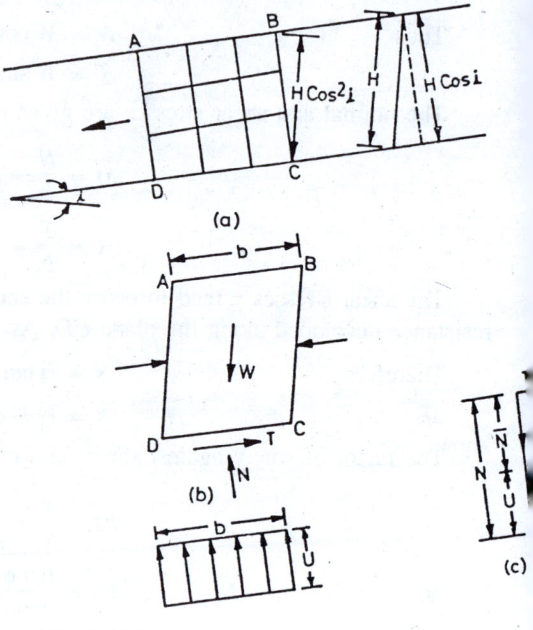

Fig. 18.7. Steady Seepage along the Slope.

Therefore, the factor of safety is given by

$$F_s = \frac{s}{\tau} = \frac{\gamma' H \cos^2 i \tan \phi'}{\gamma_{sat} H \sin i \cos i}$$

or $$F_s = \frac{\gamma' \tan \phi'}{\gamma_{sat} \tan i} \quad \text{...(18.13)}$$

As the submerged unit weight γ' is about one-half of the saturated unit weight, the factor of safety of the slope is reduced to about one-half of that corresponding to the condition when there is no seepage. The angle ϕ' in the wet condition of a cohesionless soil is approximately the same as in dry condition.

18.6. STABILITY ANALYSIS OF AN INFINITE SLOPE OF COHESIVE SOILS

The stability analysis of an infinite slope of cohesive soils is similar to that in the case of cohesionless soils, with one basic difference that the shear strength of a cohesive ($c - \phi$ soil) is given by

$$s = c' + \overline{\sigma} \tan \phi'$$

(a) Dry Soil. Taking values of $\overline{\sigma}$ and τ from Eqs. 18.7 and 18.8, we have

$$s = c' + (\gamma H \cos^2 i) \tan \phi'$$

$$\tau = \gamma H \cos i \sin i$$

Therefore, the factor of safety F_s is given by

$$F_s = \frac{c' + (\gamma H \cos^2 i) \tan \phi'}{\gamma H \cos i \cos i} \quad \text{...(18.14)}$$

Thus, the factor of safety of an infinite slope in cohesive soils depends not only on ϕ' and i but also on γ, H and c'.

Fig. 18.8 represents a graphical method for the determination of the factor of safety. The line RP_2 represents the failure envelope. When the slope angle i is less than ϕ', the slope is always safe as shown by line OP_1. When the slope angle i_2 is greater than ϕ', the slope line cuts the failure envelope. At point P, the slope is just stable. For normal stress greater than that indicated by point P, the shear stress is greater than the shear strength and the slope is not stable.

As the normal stress σ depends upon the height H of the slope, an expression for the height can be found when the slope is just stable. Therefore, equating the shear stress and the shear strength corresponding to point P,

$$\gamma H \cos i \sin i = c' + \gamma H \cos^2 i \tan \varphi'$$

or $$\gamma H \cos^2 i \left(\frac{\sin i}{\cos i} - \tan \varphi'\right) = c'$$

or $$H = \frac{c'}{\gamma (\tan i - \tan \varphi') \cos^2 i}$$

The height at which the slope is just stable is known as the critical height (H_c). Thus

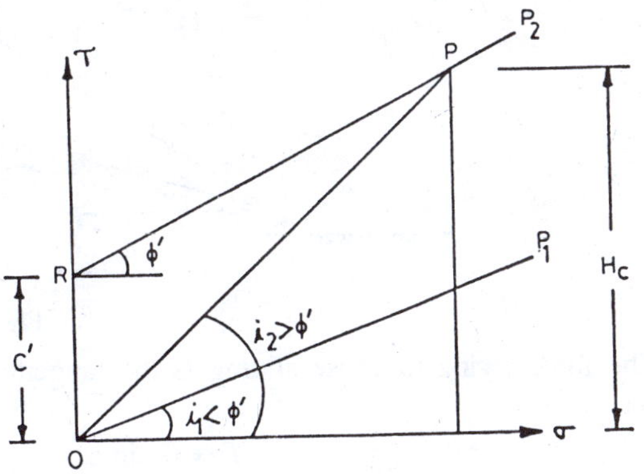

Fig. 18.8. Infinite Slope in Cohesive Soils.

$$H_c = \frac{c'}{\gamma (\tan i - \tan \varphi') \cos^2 i} \quad \text{...(18.15)}$$

For heights less than the critical height, the factor of safety is given by Eq. 18.14.

(b) Submerged Slope

As in the case of cohesionless soils, the normal and tangential components of the weight are taken for submerged unit weights and not for bulk unit weights. Thus, from Eq. 18.14,

$$F_s = \frac{c' + \gamma' H \cos^2 i \tan \varphi'}{\gamma' H \cos i \sin i} \quad \text{...(18.16)}$$

The value of ϕ' in Eq. 18.16 should be taken corresponding to the submerged conditions, which may be quite different from that in dry condition in case of cohesive soils.

(c) Steady Seepage along the Slope

The case is similar to that for a cohesionless soil. In this case, the factor of safety is given by

$$F_s = \frac{c' + \gamma' H \cos^2 i \tan \varphi'}{\gamma_{sat} H \cos i \sin i} \quad \text{...(18.17)}$$

The critical height is obtained corresponding to a factor of safety of unity. Thus

$$c' + \gamma' H_c \cos^2 i \tan \varphi' = \gamma_{sat} H_c \cos i \sin i$$

or $$H_c \cos^2 i (\gamma_{sat} \tan i - \gamma' \tan \varphi') = c'$$

or $$H_c = \frac{c'}{(\gamma_{sat} \tan i - \gamma' \tan \varphi') \cos^2 i} \quad \text{...(18.18)}$$

Sometimes, Eq. 18.18 is written as

$$H_c = \frac{c'}{\gamma_{sat} \left[\tan i - \left(\frac{\gamma'}{\gamma_{sat}}\right) \tan \phi'\right] \cos^2 i} \quad \text{...(18.19)}$$

Eq. 18.19 indicates that the effect of the angle of shearing resistance ϕ' is reduced as compared with Eq. 18.15 of the dry soil.

18.7. WEDGE FAILURE

A wedge failure occurs when a soil deposit has a specific plane of weakness. The stratified deposits generally fail along the interface. Fig. 18.9 shows a soil mass resting on an inclined layer of impermeable soil. There is a tendency of the upper mass to slide downward along the plane of contact AB.

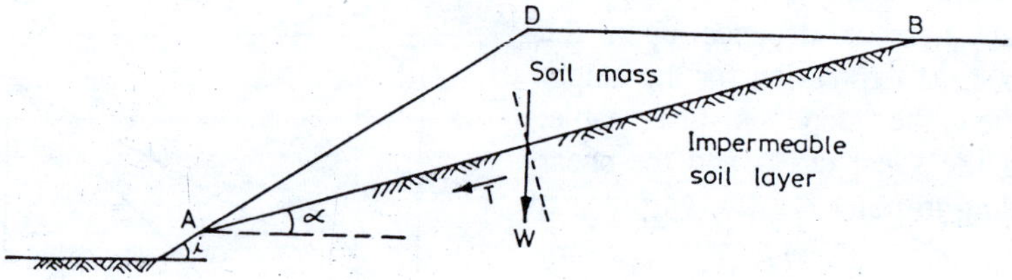

Fig. 18.9.

The force trying to cause sliding is the tangential component T of the weight (W) along the plane of contact.

$$T = W \sin \alpha \quad \text{...(a)}$$

where α is the angle which the plane AB makes with horizontal, W is the weight of wedge per unit length perpendicular to the plane of paper.

The force tending to resist the sliding depends upon the cohesion c and the frictional force and is given by

$$S = cL + (W \cos \alpha) \tan \phi \quad \text{...(b)}$$

where L is the length of the failure surface AB.

The factor of safety against sliding is obtained from (a) and (b) as

$$F_s = \frac{cL + (W \cos \alpha) \tan \varphi}{W \sin \alpha} \quad \text{...(18.20)}$$

18.8. CULMANN'S METHOD

Culmann's method is used for the approximate stability analysis of homogeneous slopes. A plane failure surface passing through the toe is assumed. A plane failure surface is not a correct assumption for a homogeneous soil. However, it is a simple failure mechanism and is described for purpose of illustration and for determination of the approximate value of the factor of safety.

Let us consider the equilibrium of the triangular wedge ABD formed by the assumed failure surface AB (Fig. 18.10) The wedge is in equilibrium under the three forces:

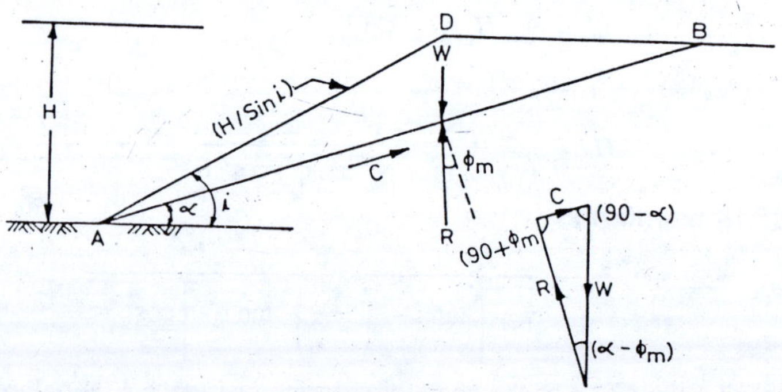

Fig. 18.10.

(1) Weight of the wedge (W).

(2) Cohesive force (C) along the surface AB.

(3) Reaction R. The reaction R is inclined at angle ϕ_m to the normal.

The triangle of forces is also shown in figure. The magnitude and direction of W and C are known. The direction of R is also known. The weight of the wedge is given by

$$W = \frac{1}{2}\gamma L\left(\frac{H}{\sin i}\right)\sin(i - \alpha) \qquad ...(a)$$

and $$C = c_m L$$

where H = height of slope, c_m = mobilised cohesion, ϕ_m = angle of mobilised friction,

and L = length of failure surface AB.

[**Note.** Area of a triangle $ABD = \frac{1}{2} d\, b \sin A$ where d and b are the length of sides BA and AD, respectively.]

From the law of sines, $$\frac{C}{W} = \frac{\sin(\alpha - \phi_m)}{\sin(\phi_m + 90°)} = \frac{\sin(\alpha - \phi_m)}{\cos\phi_m} \qquad ...(c)$$

Substituting the values of W and C from Eqs. (*a*) and (*b*) in Eq. (*c*),

$$\frac{c_m L}{1/2\,\gamma L\,(H/\sin i)\sin(i - \alpha)} = \frac{\sin(\alpha - \phi_m)}{\cos\phi_m} \qquad ...(d)$$

or $$\left(\frac{c_m}{\gamma H}\right) = \frac{1}{2}\operatorname{cosec} i \ \sin(i - \alpha)\ \sin(\alpha - \phi_m)\ \sec\phi_m \qquad ...(18.21)$$

The left-hand side of Eq. 18.21 is known as the *stability number* (S_n). The most dangerous plane is that for which the angle α is such that the stability number becomes a *maximum, i.e.*,

$$\frac{d(S_n)}{d\alpha} = 0$$

or $$\frac{d}{d\alpha}[\sin(i - \alpha)\sin(\alpha - \varphi_m)] = 0$$

or $$\sin(i - \alpha)\cos(\alpha - \phi_m) - \sin(\alpha - \phi_m)\cos(i - \alpha) = 0$$

or $$\tan(i - \alpha) = \tan(\alpha - \phi_m)$$

or $$i - \alpha = \alpha - \phi_m$$

or $$\alpha_c = \frac{(i + \phi_m)}{2} \qquad ...(e)$$

where α_c is the critical slope angle.

From Eqs. 18.21, taking the value of α_c from Eq. (*e*),

$$\left(\frac{c_m}{\gamma H}\right)_{max} = \frac{1}{2}\operatorname{cosec} i \sec\phi_m\left[\sin\left\{i - \left(\frac{i + \phi_m}{2}\right)\right\}\right] \times \left[\sin\left\{\left(\frac{i + \phi_m}{2}\right) - \phi_m\right\}\right]$$

or $$= \frac{1}{2}\operatorname{cosec} i \sec\phi_m \sin\left(\frac{i - \phi_m}{2}\right)\sin\left(\frac{i - \phi_m}{2}\right)$$

or $$= \frac{1}{2}\operatorname{cosec} i \sec\phi_m\left[\frac{i - \cos(i - \phi_m)}{2}\right]$$

or $$\left(\frac{c_m}{\gamma H}\right)_{max} = \frac{1 - \cos(i - \phi_m)}{(4\sin i\cos\phi_m)} \qquad ...(18.22)$$

or $$H = \frac{4 c_m \sin i \cos\phi_m}{\gamma[1 - \cos(i - \phi_m)]} \qquad ...[18.22(a)]$$

where H is the safe height of slope.

The Culmann method gives reasonably accurate results for homogeneous slopes which are vertical or nearly vertical. The critical surface for general slopes is not a plane and, therefore, the critical slope α_c has little practical use for such slopes.

18.9. $\phi_u = 0$ ANALYSIS

In case of fully saturated clay under undrained conditions, the stability of the slope can be checked in terms of total stresses. Such a condition occurs in slopes immediately after construction. In this case, $\phi_u = 0$ and $c = c_u$.

The failure surface is assumed to be a circular arc AB (Fig. 18.11). The figure shows a failure surface with centre O and radius r (*i.e.* $OB = OA = r$).

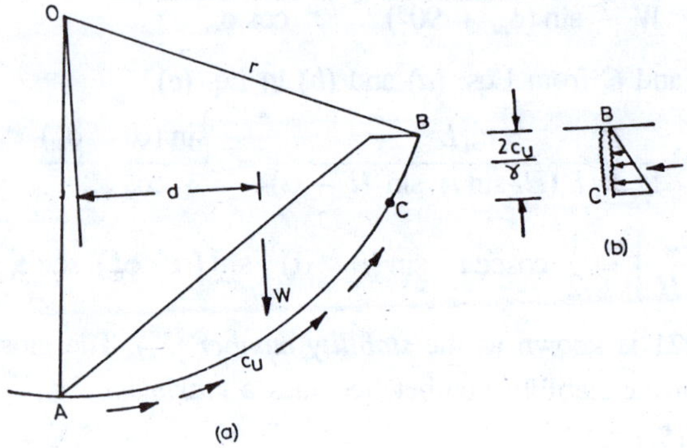

Fig. 18.11.

The total weight W above the failure surface causes instability. For equilibrium, the shear strength to be mobilised along the failure surface can be expressed as

$$\tau_m = \frac{s}{F} = \frac{c_u}{F}$$

where F is the factor of safety.

Taking moments about O,

$$W \times d = (c_u/F)\, L_a\, r \qquad \text{...(18.23)}$$

where L_a is the length of arc AB and d is the lever arm of W about O.

Thus
$$F = \frac{c_u L_a r}{W \times d} \qquad \text{...[18.23(a)]}$$

If a tension crack develops and water enters the crack, the hydrostatic pressure force P_w acts on the portion BC of the arc at a height of $h/3$ from C, where h is the depth of tension crack, equal to $2\, c_u/\gamma$ (see chapter 19). The arc length in that case should be taken equal to AC. Eq. 18.23 may be modified accordingly.

18.10. FRICTION CIRCLE METHOD

The friction circle method is useful for the stability analysis of slopes made of homogeneous soils. In this method, the slip surface is assumed to be an arc of a circle.

(a) Dry Soil

Fig. 18.11 shows a circular failure surface ABC of radius r with its centre at O. The three forces acting on the sliding wedge $AEBD$ are given below:

(1) Weight (W) of the sliding wedge.

(2) Cohesive force C developed along the slip surface AEB.

(3) Reaction R on the slip surface.

The reaction R is inclined at angle ϕ_m to the normal to the slip surface. As the direction of the normal changes, the direction of R also changes. With the centre as O, a small circle, known as *friction circle,* is drawn with a radius $r \sin \phi_m$. All lines which are tangent to the friction circle make an angle ϕ_m with the normal of the slip surface. These lines represent the direction of the combined normal and mobilised frictional forces on the slip surface. The value of ϕ_m is obtained from Eq. 18.6, after choosing a value of F_ϕ. Thus the reaction R is tangential to the friction circle.

[**Note.** Actually, the reaction R is tangential to the friction circle of a slightly larger radius of $K\, r \sin \phi_m$, where K is a factor with a value greater than unity, as it is evident that the two reactions d_R [Fig. 18.12 (*a*)] intersect slightly outside the friction circle of radius $r \sin \varphi_m$. However, this discrepancy is generally disregarded].

The cohesive force C_m is equal to $c_m L_a$ where c_m is the mobilised cohesion and L_a is the length of the circular surface arc. It is convenient to replace this force acting along the arc by an equivalent force C acting along a line. The force along arc AEB is also equal in magnitude to the force $c_m \times L_c$ where L_c is the length of the chord AB. The line of action of this force can be determined by taking moments of the actual force and the equivalent force about O.

$$(L_c \times c_m) \times a = (c_m \times L_a) \times r$$

or

$$a = r \frac{L_a}{L_c} \quad \text{...(18.24)}$$

Obviously, the distance a is greater than r, as $L_a > L_c$.

The intersection of the weight W and the cohesive force C_m establishes a point P through which the

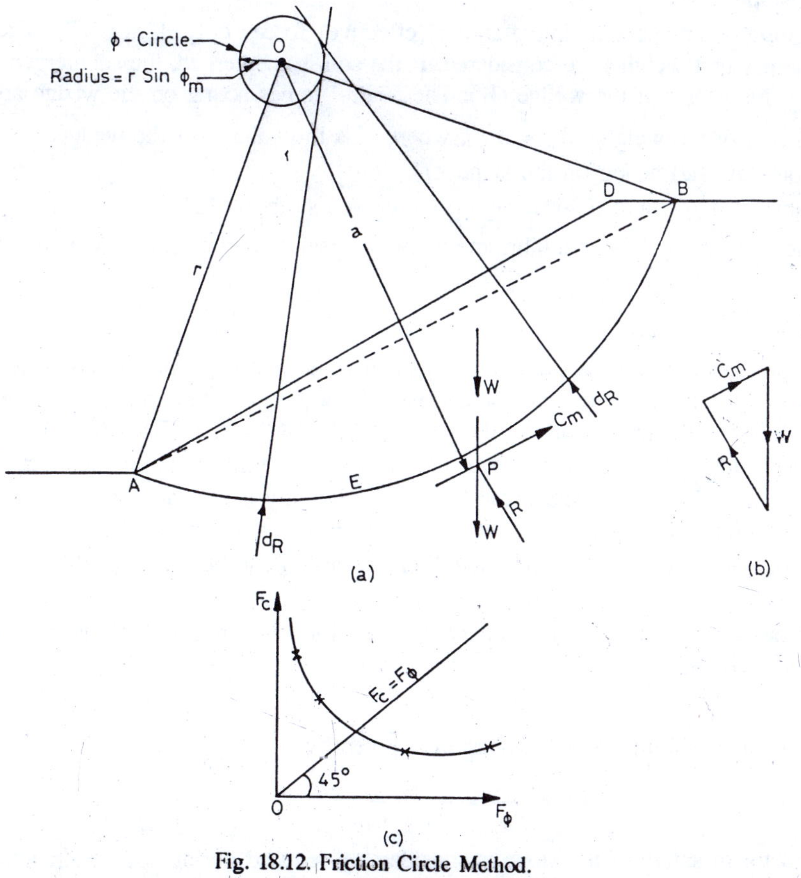

Fig. 18.12. Friction Circle Method.

reaction R must act. The direction of R is obtained by drawing a line tangential to the ϕ-circle. The forces C_m and R can be determined from the force triangle.

Fig. 18.12 (*b*) shows the force triangle. The weight vector W is drawn first. The triangle is completed by drawing the vectors R and C_m along the directions already established. From the force triangle, the value of the cohesive force C_m is determined. The mobilised cohesion is equal to the cohesive force C_m divided by the length of the chord L_c. Thus

$$c_m = \frac{C_m}{L_c} \qquad \text{...(18.25)}$$

The factor of safety with respect to cohesion is given by Eq. 18.5.

If the value of F_c obtained from Eq. 18.5 is not equal to the assumed value of F_ϕ, the analysis is repeated. The procedure is repeated after taking another trial surface. The slip circle which gives the minimum factor of safety (F_s) is the most critical circle. Generally, the analysis is repeated 3-4 times to obtain a curve between the assumed value of F_ϕ and the computed value of F_c, as shown in fig. 18.12 (c). The factor of safety with respect to shear strength F_s is obtained by drawing a line at 45°, which gives $F_c = F_\phi = F_s$.

For a purely cohesive soil, $\phi = 0$ and the friction circle reduces to a point. The factor of safety is determined from the resisting moment due to C and actuating moment due to W (See Sect. 18.9).

Sometimes, the factor of safety with respect to friction (F_ϕ) is assumed to be unity and the factor of safety with respect to only cohesion is obtained.

(*b*) Submerged slope

The above discussion refers to intergranular (effective) forces only. If the soil is dry and there is no submergence. the dry unit weights are considered. If the soil is submerged, the submerged unit weight is used when calculating the weight of the wedge (W). The neutral forces acting on the wedge are given below:

(1) Neutral part of the weight, $W_u = V\gamma_u$, where V is the volume of the wedge.

(2) Water pressure (U) acting on the slope AD.

(3) Resultant water pressure (U_B) acting on the curved surface AEB.

The three neutral forces are in equilibrium among themselves and do not draw upon the shear strength of the soil.

(*c*) Sudden drawdown conditions

When the water standing on the slope AD is suddenly and quickly removed, the water pressure force (U) disappears. However, if there is no time for drainage to occur from the soil in the slope, the soil remains submerged as before and the neutral part of the weight (W_u) is still acting. Thus, the equilibrium of the neutral forces is disturbed, although the equilibrium of the intergranular forces remains unaffected.

The equilibrium of the neutral forces can be maintained only if the soil can mobilise additional cohesion (c_a) such that the additional cohesive force $c_a L_a$, the neutral part of the weight (W_u) and the new resultant water pressure (U_B) are in equilibrium. The additional cohesion can be determined by drawing another force triangle.

The total cohesion mobilised (c_m') in this case is equal to the cohesion mobilised for intergranular forces and that for neutral forces.

$$c_m' = c_m + c_a \qquad \text{...(18.26)}$$

The factor of safety with respect to cohesion is given by

$$F_c = \frac{c}{c_m'} = \frac{c}{c_m + c_a} \qquad \text{...(18.27)}$$

Thus the factor of safety of the slope is considerably reduced during sudden drawdown conditions.

18.11. STABILITY CHARTS

The stability number (S_n), as defined in Sect. 18.8, is given by

$$S_n = \frac{c_m}{\gamma H} = \frac{c}{F_c \gamma H} \qquad \text{...(18.28)}$$

The reciprocal of the stability number is known as *stability factor*. The stability number is a dimensionless quantity.

Taylor determined the values of S_n for finite slopes using the friction circle method. Slopes that are of simple sections and of homogeneous soils may be analysed using the slope stability charts given by Taylor. The charts are prepared indicating the stability number, and slope angle i for various values of φ_m (Fig. 18.13). There are 5 parameter, *viz* c_m, γ, H, i and φ_m. However, if $\phi_m = 0$ (purely cohesive soils), a sixth

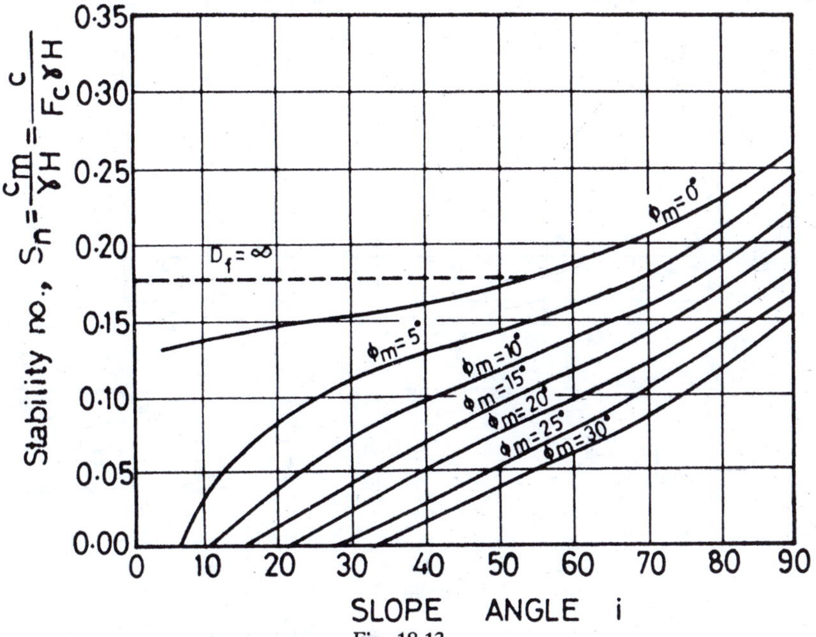

Fig. 18.13.

parameter D_f becomes also important (Fig. 18.14). The parameter D_f depends upon the depth of the hard stratum below the top of the slope, and is given by

$$D_f = \frac{\text{Depth of hard stratum below the top of the slope}}{\text{Height of slope}}$$

When the slope is steep, the failure surface passes through the toe, whereas for the flatter slope, the failure extends below the toe. The chart in Fig. 18.13 is based on the most critical circle passing through the toe of the slope.

For slope angle i greater then 53°, the toe failure occurs. For $i \leq 53°$, and small values of φ_m, a more critical surface may pass below the toe.

The chart in Fig. 18.14 is applicable for $\varphi_m = 0$. In soils with $\varphi_m = 0$ and the slope angle less than 53°, the failure surface extends below the toe as deep as possible. The stability number also depends upon the parameter D_f.

Uses

(1) The stability number can be used to determine the factor of safety of a given slope. For the known values of i and φ_m, the value of stability number (S_n) is determined from the chart in Fig. 18.13 (or Table 18.1) and the factor of safety is determined as

$$F_c = \frac{c}{c_m} = \frac{c}{S_n \gamma H} \quad \text{...(18.29)}$$

If $\varphi_m = 0$, the chart in Fig. 18.14 (or Table 18.2) is used to determine the stability no. (S_n) for the given values of i and D_f. The chart can also be used to determine the distance nH from the toe where the slip circle cuts the horizontal line.

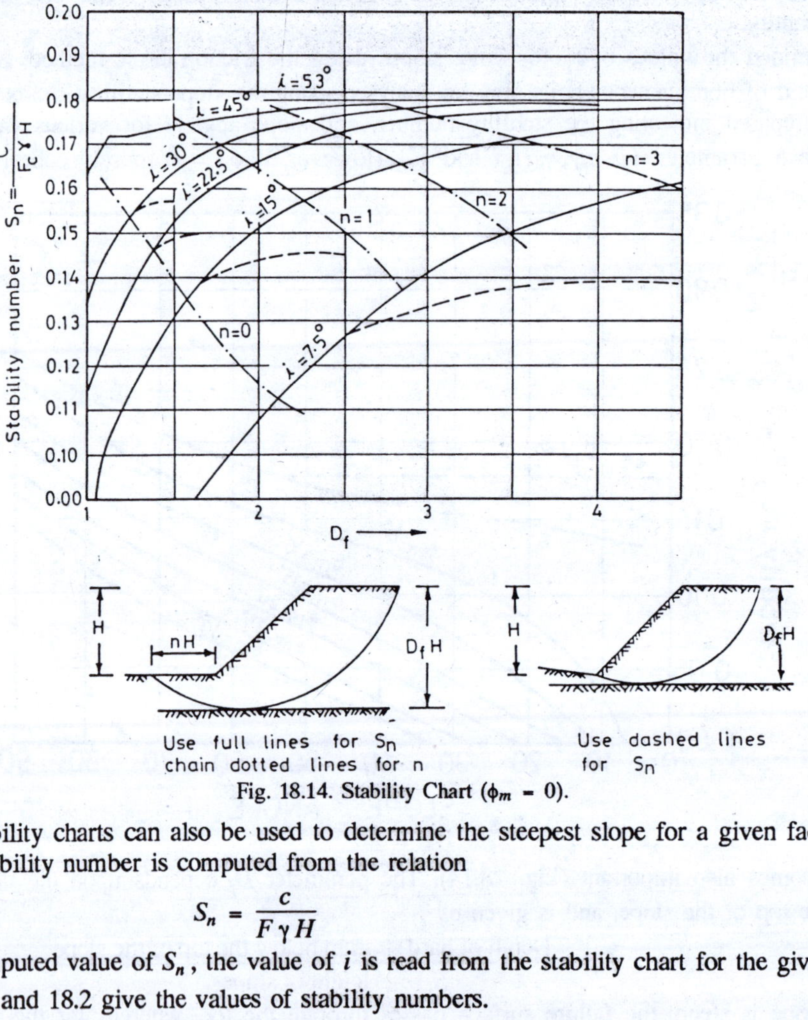

Fig. 18.14. Stability Chart (ϕ_m = 0).

(2) The stability charts can also be used to determine the steepest slope for a given factor of safety. In this case, the stability number is computed from the relation

$$S_n = \frac{c}{F \gamma H}$$

For the computed value of S_n, the value of i is read from the stability chart for the given value of ϕ_m. Tables 18.1 and 18.2 give the values of stability numbers.

Table 18.1. Stability Numbers (see Fig. 18.13)

i \ ϕ_m	0°	5°	10°	15°	20°	25°
90°	0.261	0.239	0.218	0.199	0.182	0.166
75°	0.219	0.195	0.173	0.152	0.134	0.117
60°	0.191	0.162	0.138	0.116	0.097	0.079
45°	(0.170)	0.136	0.108	0.083	0.062	0.044
30°	(0.156)	(0.110)	0.075	0046	0.025	0.009
15°	(0.145)	(0.068)	0.070	(0.023)	—	—

[**Note.** Figures in brackets are for the most dangerous circles through the toe when a more dangerous circle exists below the toe].

Table 18.2. Stability Numbers for Cohesive soils ($\phi_m = 0$) and $i \leq 53°$ (see Fig. 18.14)

i \ D_f	1.0	1.50	2.0	3.0	∞
53°	0.181	0.181	0.181	0.181	0.181
45°	0.164	0.174	0.177	0.180	0.181
30°	0.133	0.164	0.172	0.178	0.181
22.5°	0.113	0.153	0.166	0.175	0.181
15°	0.083	0.128	0.150	0.167	0.181
7.5°	0.054	0.080	0.107	0.140	0.181

The following points should be carefully noted.

(1) If the factor of safety with respect to friction F_φ is unity, $\phi_m = \phi$.

(2) If the factor of safety with respect to shear strength is required, a trial and error procedure is adopted. A value of F_ϕ is assumed, and the value of ϕ_m is used to determine F_c from the stability chart. If F_c is not equal to assumed value of F_ϕ, another value of F_φ is assumed and the procedure is repeated. At least 3-4 trials are required to obtain a curve between F_c and F_ϕ, and to get the correct value of F_c from the curve, as shown in Fig. 18.12 (*c*).

(3) For a submerged slope, the stability number is computed using the submerged unit weight (γ'). The angle of shearing resistance should also be for the submerged conditions.

(4) For a sudden drawdown case, the stability number is computed using the saturated unit weight (γ_{sat}). The weighted angle of internal friction as obtained below is used for finding out the stability number.

$$\tan \phi_w = \frac{\gamma'}{\gamma_{sat}} \left(\frac{1}{F_\phi} \tan \varphi' \right) \quad \ldots(18.30)$$

or

$$\phi_w = \frac{\gamma'}{\gamma_{sat}} \phi_m = \frac{\gamma'}{\gamma_{sat}} \left(\frac{\phi'}{F_\phi} \right) \quad \ldots[18.30(a)]$$

where ϕ' is the effective angle of internal friction.

(5) For purely frictional soils, the cohesion intercept (*c*) is zero. As the stability number reduces to zero, the stability charts cannot be used for such soils.

(6) The values of shear strength parameter (*c* and ϕ) should be obtained from the tests conducted in the laboratory simulating the drainage conditions in the field. For example, for checking the stability of a slope of cohesive soils just after the constructions, the relevant drainage conditions are unconsolidated undrained conditions For long term stability, consolidated drained conditions are relevant.

18.12. SWEDISH CIRCLE METHOD

The actual shape of a slip surface in the case of finite slopes is curvilinear. For convenience, it is approximated as circular. The assumption of a circular slip surface and its application for stability analysis of slopes was developed in Sweden. The method is known as the Swedish circle method or the method of slices.

Fig. 18.15 (*a*) shows a slope. Let *AB* be a circular surface with radius *r* and centre *O*. The trial failure wedge above the slip surface is divided into vertical slices by drawing vertical lines, as shown. The slices are usually of equal width, but not necessarily so. In case of non-homogeneous slopes where the slip surface passes through more than one type of material, a vertical line is always located at the point where the slip surface passes from one material to the other.

Let us consider the equilibrium of one slice (say, No. 4). The slice is in equilibrium under the following forces.

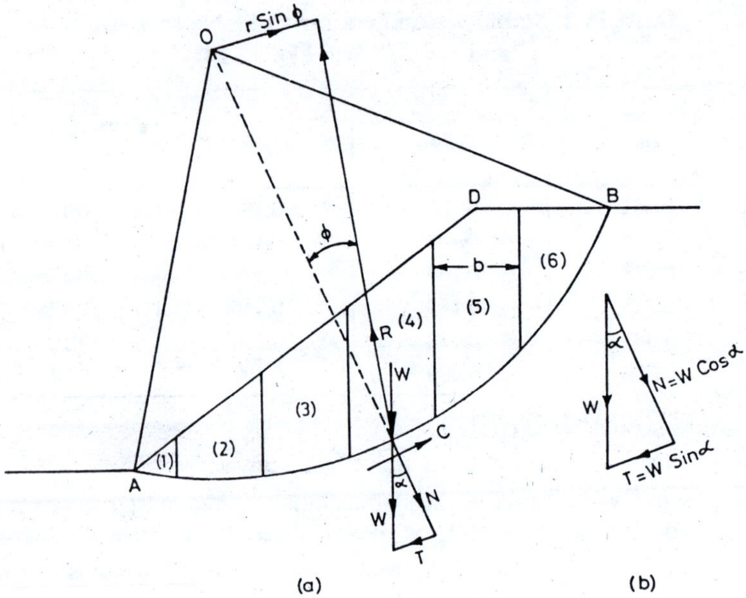

Fig. 18.15. Swedish Circle Methods.

(1) Weight (W) acting vertically through its centre of gravity.

(2) Cohesive force (C) acting along the curved surface in the direction opposite to the direction of probable movement of the wedge.

(3) Reaction (R) at the base inclined at angle ϕ to the normal, assuming the slippage is imminent.

(4) Reactions on the two vertical sides of the slice due to adjacent slices. However, in the Swedish circle method, it is assumed that the reactions on the two sides are equal and opposite and are, therefore, in equilibrium and do not affect the stability of the slice. Accordingly, only the first three forces are considered for the analysis.

The weight W is resolved into its normal component (N) and tangential component (T). Let us take the moments about the centre of rotation O of all the 3 forces.

Actuating or overturning moment, $M_0 = T \times r$ ...(a)

The moment due to N—components is zero, as those components always pass through O.

Resisting moment, $M_R = (C\,\Delta L) \times r + R\,(r \sin \phi)$ ...(b)

where ΔL is the length of the curved surface of the slice.

Resolving the forces is radial direction,

$$N = R \cos \phi \quad \text{or} \quad R = N/\cos \phi \qquad ...(c)$$

or $R \sin \phi = N \tan \phi$ ...(d)

From Eqs. (b) and (d), $M_R = (C\,\Delta L)\,r + N\,r \tan \phi$ ...(e)

The factor of safety for the slice is equal to the ratio of the resisting moment (M_R) and the overturning moment (M_0). Thus

$$F_s = \frac{r\,[c\,\Delta L + N \tan \phi]}{T\,r} = \frac{c\,\Delta L + N \tan \phi}{T}$$

The factor of safety of the entire wedge is given by

$$F_s = \frac{\Sigma c\,\Delta L + \Sigma N \tan \phi}{\Sigma T} \qquad ...(18.31)$$

If c and ϕ are constant, $$F_s = \frac{c\,L_a + \tan\phi\,\Sigma N}{\Sigma T} \quad ...18.32)$$

where L_a = length of the entire slip surface = $\Sigma\,\Delta L$

The components N and T are determined by drawing force triangles as shown in Fig. 18.15 (*b*). If the angle α which the normal makes with the vertical is measured, the components can be computed as under.

$$N = W\cos\alpha, \quad \text{and} \quad T = W\sin\alpha$$

The length ΔL of the arc is given by $b\sec\alpha$, where b is the width of the slice.

Therefore, Eq. 18.31 can be written as

$$F_s = \frac{\Sigma\,c\,b\sec\alpha + \Sigma\,(W\cos\alpha)\,\tan\phi}{\Sigma W\sin\alpha} \quad ...(18.33)$$

It may be noted that the tangential component T may be negative *i.e.* in the direction opposite to that of movement for some of the slices near the toe.

The procedure can be summarised as under:

(1) Take a trial wedge and divide it into 6 to 12 vertical slices.

(2) Determine the weight of each slice and its line of action.

For convenience, the weight is generally taken proportional to the middle ordinate of the slice and it is assumed to have line of its action through the middle of the slice.

(3) The weight is resolved (analytically or graphically) into normal and tangential components.

(4) The curved length ΔL of each slice is measured or computed.

(5) The factor of safety is determined from Eq. 18.33 or Eq. 18.31.

The calculations are generally done in a tabular form. The stability analysis is repeated for a number of trial surfaces. The circle which gives the *minimum* factor of safety is the most critical circle.

Location of Most Critical Circle

In order to reduce the number of trials to locate the most critical circle, the Fellenius line AB can be drawn (Fig. 18.16). Fellenius has shown that the centre of the most critical circle lies on this line. For drawing the Fellenius line AB, the point B is located at a depth H and at a distance 4.5 H from point P at the toe of the slope, where H is the height of the slope. The point A is located by drawing two lines PA and QA, where PA makes angle α with the slope line PQ and QA makes angle β with the horizontal at Q. The angles α and β are obtained from the table given in Fig. 18.16. The angles depend upon the slope.

The centre of the most critical circle may lie anywhere on the line AB or its extension. The centres of

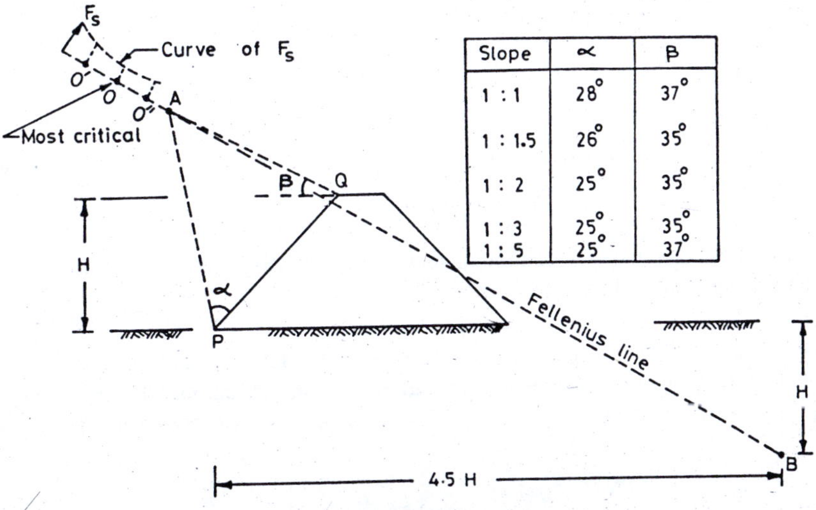

Slope	α	β
1 : 1	28°	37°
1 : 1.5	26°	35°
1 : 2	25°	35°
1 : 3	25°	35°
1 : 5	25°	37°

Fig. 18.16. Fellenius Line.

trial circles are taken on this line shown as O'', O and O'. The factors of safety obtained when these were the trial centres are generally plotted as normals to the line AB to obtain a curve of F_s. The centre corresponding to the minimum factor of safety indicates the most critical circle.

For a purely cohesive soil ($\phi = 0$), the point A itself represents the centre of most critical circle.

The Swedish circle method is a general method of slope stability analysis. It can be used for non-homogeneous soil masses, stratified deposits, fully submerged or partly submerged conditions. The method is also applicable when seepage occurs and pore pressure develops in the soil masses, as explained later.

However, the method is necessarily an approximate one, as it neglects the effect of forces acting on the sides of the vertical strips. Forunately, the method errs on the safe side, *i.e.* the factor of safety obtained is generally less than that obtained from the more accurate methods, such as Bishop's method, which also consider the forces on the sides of the vertical strips.

Use of N-Curve and T-Curve

The values of ΣN and ΣT in Eq. 18.31 can also be determined by drawing N and T curves. The sector ABD (Fig. 18.17) has been divided into 6 slices having equal width. The weight W of each slice is proportional to its central ordinate (*i.e.* height at mid-width), shown dotted. The weight of each slice is

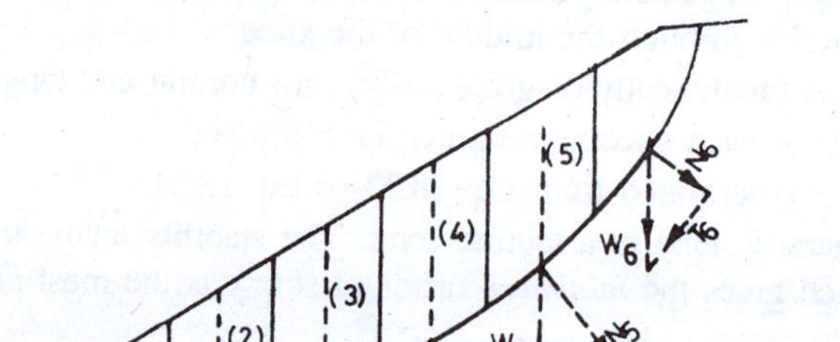

Fig. 18.17. *N* and *T* curves.

resolved graphically into normal and tangential components by drawing a normal to the slip circle through the point where the central ordinate of the slice intersects the circle, and drawing the force triangles (shown dotted).

A horizontal base line is selected and N components of the various slices are plotted directly below the vertical lines of the sector. The extremities are joined by a smooth curve, and thus N-curve is obtained. Likewise, T-curve is obtained. It may be noted that T_1-component in this case is negative and plotted below the base.

Now $$\Sigma N = A_N \times \gamma \qquad ...(18.34)$$

where A_N = area of the N-diagram (m^2), γ = unit weight of soil.

Likewise, $$\Sigma T = A_T \times \gamma \qquad ...(18.35)$$

[**Note.** If A_N and A_T are measured in cm^2, they are multiplied by x^2 where x is the scale (1 cm = x metres)].

The areas A_N and A_T can be measured by means of a planimeter or by using a graph paper.

Use of Rectangular Plot

Prof. Alam Singh (1962) devised a simple method for determination of ΣN and ΣT without using a planimeter. Let $Z_1, Z_2, ... Z_5$ be the *end ordinates* of the 6 slices shown in Fig. 18.17 (*a*). Let b the width of each slice. The total weight ΣW of the sliding wedge may be written as

$$\Sigma W = b\gamma\left[\left(\frac{O + Z_1}{2}\right) + \left(\frac{Z_1 + Z_2}{2}\right) + \left(\frac{Z_2 + Z_3}{2}\right) + \left(\frac{Z_3 + Z_4}{2}\right) + \left(\frac{Z_4 + Z_5}{2}\right) + \left(\frac{Z_5 + O}{2}\right)\right]$$

or $$\Sigma W = b\gamma\,[Z_1 + Z_2 + Z_3 + Z_4 + Z_5] \qquad ...(18.36)$$

If the last slice is of smaller width equal to $m \times b$, the above expression is modified as

$$\Sigma W = b\gamma\left[Z_1 + Z_2 + Z_3 + Z_4 + \left(\frac{1 + m}{2}\right) Z_5\right] \qquad ...(18.37)$$

The value of ΣW can be determined by drawing a rectangular plot of width b and having boundary ordinates Z_1, Z_2 ... etc. as abscissae. In the case of last slice being of width *mb*, the last ordinate is multiplied by (1 + m)/2 before plotting. The area of the diagram is proportional to ΣW.

If the vertical ordinates are resolved along the normal and tangential directions, the *N*-components and T-components are obtained. The rectangular plot of N-components is then drawn, taking the width of the plot equal to the width b of the slice [Fig. 18.18 (*b*)]. The normal components N_1, N_2, etc. are plotted as abscissae.

Likewise, the rectangular plot of *T*-components is drawn. In *T*-plot, as T_1 component is negative, it is plotted in the opposite direction. The net area of *T*- plot is shown hatched [Fig. 18.18 (*c*)]. The values of

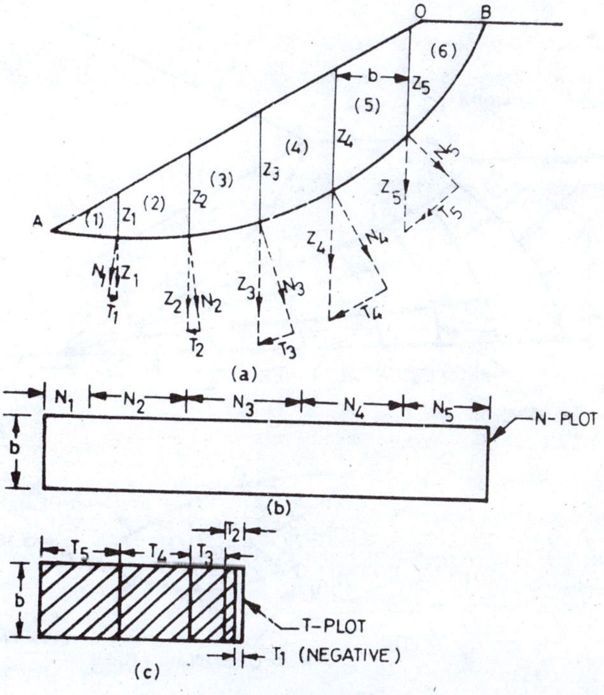

Fig. 18.18. Rectangular Plots.

ΣN and ΣT are obtained using Eqs. 18.34 and 18.35, where γ is the unit weight and A_N and A_T are the areas of the N-plot and T- plot, respectively.

The rectangular plot method greatly simplifies the calculations for determinations of the areas of N and T- diagrams. It may be noted that the weights in the rectangular plot are proportional to end ordinates, and not the mid-ordinates.

18.13. STABILITY OF SLOPE UNDER STEADY SEEPAGE CONDITIONS

The stability of slopes of an earth dam is investigated for the steady seepage conditions, sudden drawdown conditions and during construction conditions. The steady seepage condition is considered in this section. When the reservoir on the upstream of the dam is filled, water starts seeping through the dam. After sometime, the steady seepage conditions are established and a well-defined phreatic line is formed. The soil below the phreatic line is saturated and subjected to pore water pressure.

On the upstream slope, the seepage forces are directed inwards and hence tend to increase the stability. However, on the down-stream slope, the direction of the seepage forces is such that they decrease the stability. The steady seepage condition is, therefore, critical for the downstream slope of an earth dam. Fig. 18.19 (*a*) shows the downstream slope of an earth dam provided with a horizontal filter at its toe. The flow net is drawn using the methods discussed in Chapter 9.

The boundary pore pressures acting on the slip surface are obtained from the flow net. Fig. 18.19 (*b*) shows the enlarged view. The pore pressure u_1 at point-1 where the first equipotential line ($h/9$) cuts the slip surface is equal to the vertical distance between the point-1 and the point P where the equipotential line intersects the phreatic line As the pore pressure acts normal to the surface, a line equal to u_1 is drawn normal to the slip surface at point-1. Likewise, the pore pressure u_2 at point-2 is found. The pore pressure diagram is then drawn joining the extremities of all these lines. The pore pressure diagram is shown hatched in Fig. 18.19 (*b*). Eq. 18.32 can be used to determine the factor of safety. However, in this case the total weight of the slice is due to bulk unit weight above the phreatic line and the saturated unit weight below the phreatic line. The N-component of the weight is reduced due to pore pressure below the phreatic line. Therefore,

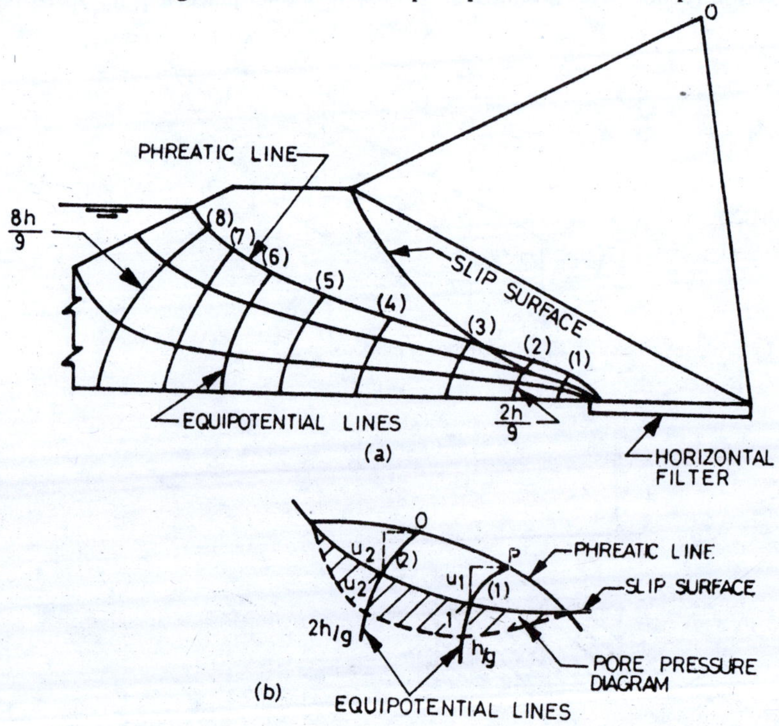

Fig. 18.19. Steady Seepage Conditions.

$$F = \frac{cL_a + \tan\phi\ \Sigma(N - U)}{\Sigma T} \qquad ...(19.38)$$

where ΣU is obtained from the area of the pore pressure diagram using a plaimeter or by rectangular plot method. It represents the total force due to pore pressure.

Alternatively, Eq. 18.38 can be written as

$$F_s = \frac{\Sigma c\ \Delta L + \Sigma(N - ul)\tan\phi}{\Sigma T} \qquad ...(18.39)$$

where u is the average pore pressure on the slice and l is the curved length of the base of the slice. (**Note.** $l = \Delta L = b \sec\alpha$).

18.14. STABILITY OF SLOPE UNDER SUDDEN DRAWDOWN CONDITIONS

The critical condition for the stability of the upstream slope of an earth dam is when there is a sudden drawdown in the reservoir upstream. If the soil is of low permeability, no appreciable change in the saturation level inside the slope takes place when the reservoir level goes down. The weight of water which is still present in the soil tends to cause sliding of the wedge, as the water pressure which was acting on the upstream slope to balance this weight has been suddenly removed. According to another interpretation, the shearing resistance of the soil is considerably reduced due to pore pressure existing in the soil, whereas the disturbing force due to saturated weight of the soil remains the same.

The flow net for the sudden drawdown condition can be drawn using the electrical analogy method (Chap. 9), and the pore pressure acting on the base of the various slices determined. The stability of the slope is investigated using Eq. 18.38 or Eq. 18.39. However, the method is not convenient.

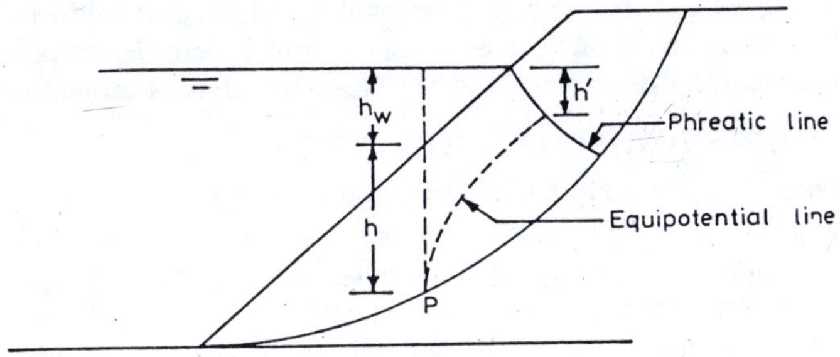

Fig. 18.20.

The pore water pressure (u) can also be estimated using Skempton pore pressure coefficients (Chapter 13). Fig. 18.20 shows the upstream slope of a dam. The pore-water pressure at any point P before drawdown is given by

$$u_0 = \gamma_w(h + h_w - h') \qquad ...(18.40)$$

where h = height of soil above P, h_w = height of water column above P,

and h' = loss of head due to seepage, indicated by the equipotential line passing through P.

It is assumed that the total major principal stress (σ_1) at P is due to the weight above.

When there in a drawdown, the major principal stress decreases by

$$\Delta\sigma_1 = -\gamma_w h_w$$

and the change in pore water pressure is given by

$$\Delta u = \bar{B}\,\Delta\sigma_1 = -\bar{B}\,\gamma_w h_w$$

where $\bar{B}$ is the overall pore pressure coefficient, related to coefficients A and B, as

$$\bar{B} = B\left[1 - (1 - A)\left(1 - \frac{\Delta\sigma_3}{\Delta\sigma_1}\right)\right] \qquad ...(18.41)$$

Therefore, the pore water pressure at P immediately after drawdown is given by

$$u = u_0 + \Delta u = \gamma_w (h + h_w - h') - \bar{B} \gamma_w h_w$$

or

$$u = \gamma_w [h + h_w (1 - \bar{B}) - h'] \quad \text{...(18.42)}$$

The value of $\bar{B}$ is slightly greater than unity. A conservative value of $\bar{B}$ equal to unity is generally taken.

Thus

$$u = \gamma_w [h - h'] \quad \text{...(18.43)}$$

Thus the pore water pressure u at various points on the slip surface is determined.

The factor of safety is then obtained from Eq. 18.39.

Approximate method

An approximate alternative method for investigating the stability of the upstream slope under sudden drawdown conditions is to consider the saturated unit weight of soil for calculating the driving forces (*i.e.* ΣT) and the submerged unit weight for calculating the resisting forces (*i.e.* $\Sigma N'$) in Eq. 18.32.

Thus

$$F_s = \frac{cL_a + \tan \phi \, \Sigma N'}{\Sigma T} \quad \text{...(18.44)}$$

In other words, it is assumed that the full pore pressure acts even after the drawdown and the soil is fully saturated.

Below the drawdown level, as the water pressure on the slope is still acting, the submerged unit weight is used for both the driving forces and the resisting forces. This is similar to the case of a submerged slope as discussed in Sect, 18.5.

If the slope material has high coefficient of permeability, it drains as fast as the reservoir level goes down, then both the driving and resisting forces are calculated using the bulk unit weight.

If the slope material is of medium permeability, the saturation line moves downward at a rate depending on the permeability of the soil. A series of flow nets can be drawn for different positions of saturation line and the corresponding pore pressures determined and the factor of safety is estimated.

18.15. STABILITY OF SLOPES DURING CONSTRUCTION

When an earth dam is built of the soil of low permeability, excess pore pressures develop in the air and water voids due to compaction carried out during construction or due to its own weight. The pore pressure developed depends upon the placement water content, method of compaction, weight of the overlying layers and the rate of dissipation of pore pressure during construction.

When the placement water content is more than the optimum and there is no proper drainage, initial pore pressure at any point may be as high as 100 percent of the weights of the overlying layers. An estimation of initial pore pressure may be made using Bishop's method of predicting pore pressure. According to which, the pore pressure (u) at any point can be written as

$$u = u_0 + \Delta u \quad \text{...}(a)$$

where u_0 = initial pore water pressure, and Δu = change in pore water pressure.

In terms of the change in total major principal stress $\Delta \sigma_1$, Eq. (a) can be written as

$$u = u_0 + \bar{B} \, \Delta \sigma_1$$

where $\bar{B}$ is the overall pore pressure coefficient.

The increase in total major principal stress $\Delta \sigma_1$ is approximately equal to the fill pressure (γh). Thus

$$u = u_0 + \bar{B} (\gamma h) \quad \text{...(18.45)}$$

As the soil is partially saturated when compacted, the initial pore water pressure (u_0) is generally negative. The actual values of u_0 and $\bar{B}$ depend upon the placement water content. For high water content, u_0 may be zero. Thus

$$u = \bar{B} (\gamma h) \quad \text{...(18.46)}$$

The value of $\bar{B}$ in Eq. 18.46 must be that corresponding to the stress condition in the dam. It can be determined from undrained triaxial tests on compacted specimens with pore pressure measurements.

Alternative method

The pore pressure during construction can be determined from Hilf's equation :

$$u = \frac{p_a\, V_a}{h_c\, V_w} \qquad \ldots(18.47)$$

where p_a = air pressure in the voids of a soil mass after initial compaction (absolute pressure).

V_a = volume of air after initial compaction, in percent,

V_w = volume of water after initial compaction, in percent,

h_c = Henry's constant of solubility of air in water by volume (= 0.02 at 20°C)

and u = pore water pressure when the soil mass has been consolidated to complete saturation.

When the soil is not consolidated to complete saturation, the pore pressure can be determined using the following equation:

$$u = \frac{p_a\, \Delta}{V_a + h_c\, V_w - \Delta} \qquad \ldots(18.48)$$

where Δ = embankment compression, in percent, of the original total embankment volume.

A graph is generally plotted between the effective stress $\bar{\sigma}$ and the percent consolidation Δ, from the results of consolidation tests conducted on the samples. The values of u are obtained from Eq. 18.48 for different values of Δ. The total stress is given by

$$\sigma = \bar{\sigma} + u$$

A plot is then made between the total stress σ and the pore pressure u. This plot is used for the determination of pore pressures at various points in the dam during construction from the values of the total stress σ. Obviously, the total stress at any point in the dam is equal to the bulk unit weight multiplied by the depth of the soil above.

18.16. BISHOP'S SIMPLIFIED METHOD

The conventional Swedish circle method (Sect. 18.11) satisfies only the overall moment equation of equilibrium. It neglects the moment equilibrium of the individual slices. It also disregards the effect of the forces acting on the sides of the individual slices and, therefore, only approximates the force equilibrium of each slice. Methods of analysis which satisfy all the equilibrium equations are complicated and not convenient to use.

Bishop (1955) gave a simplified method of analysis which considers the forces on the sides of each slice. The requirement of equilibrium are applied to the slices. The factor of safety is defined as the ratio of the maximum shear strength (s) possessed by soil on the trial surface to the shearing resistance mobilised (τ_m).

Thus $$F_s = s/\tau_m \qquad \ldots(18.49)$$

The forces acting on the slices are given below [Fig. 18.21(a)].

(1) Weight of slice, $W = \gamma\, h\, b$, where h is the average height.

(2) Normal force on the base, $N' = N - ul$, where u is the pore pressure, and l is the length of the base ($= b \sec \alpha$).

(3) Shear force on the base, $T = \tau_m\, l$.

(4) Normal forces on the sides E_1 and E_2.

(5) Shear forces on the sides X_1 and X_2.

(6) Any external force acting on the slice.

The problem is statically indeterminate. To solve the problem, a number of simplifying assumptions are made regarding the interslice forces X_1, X_2, E_1 and E_2.

Fig. 18.21 (b) shows a slope. Taking moments about O.

$$\Sigma\, T\, r = \Sigma\, W\, r \sin \alpha$$

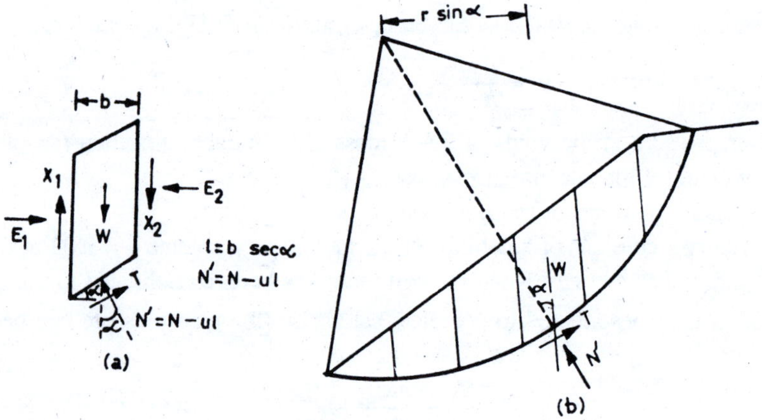

Fig. 18.21.

But $$T = \tau_m \; l = \frac{s}{F_s} \times l$$

Therefore, $$\Sigma \frac{s}{F_s} \times l \times r = \Sigma W \times r \times \sin\alpha$$

or $$F_s = \frac{\Sigma s \times l}{\Sigma W \sin\alpha} = \frac{\Sigma (c' + \bar{\sigma} \tan\phi')\, l}{\Sigma W \sin\alpha}$$

or $$F_s = \frac{\Sigma c' l + \tan\phi' \, \Sigma N'}{\Sigma W \sin\alpha} \quad \text{where } N' = \bar{\sigma}\, l \qquad ...(18.50)$$

Eq. 18.50 is exact. However, approximations are introduced in the determination of the normal component N'. In the conventional Swedish circle method, the resultant of inter-slice forces is taken as zero.

Therefore, $$N' = W \cos\alpha - ul$$

Thus, $$F_s = \frac{c' L_a + \tan\phi' \, \Sigma (W \cos\alpha - ul)}{\Sigma W \sin\alpha}$$

$$F_s = \frac{c' L_a + \tan\phi' \, \Sigma (N - U)}{\Sigma T} \qquad \text{(same as Eq. 18.38)}$$

In Bishop's simplified method, it is assumed that the resultant forces on the sides of the slice are horizontal, *i.e.* $X_1 - X_2 = 0$.

Resolving the forces in the vertical direction,

$$W = N' \cos\alpha + ul \cos\alpha + T \sin\alpha + X_1 - X_2$$

Substituting $T = \frac{s}{F_s} \times l$ and $X_1 - X_2 = 0$, we have

$$W = N' \cos\alpha + ul \cos\alpha + \frac{(c' l + N' \tan\phi')}{F_s} \sin\alpha$$

or $$W = N' \cos\alpha + ul \cos\alpha + \frac{c' l}{F_s} \sin\alpha + \frac{N'}{F_s} \tan\phi' \sin\alpha$$

or $$N' \left(\cos\alpha + \frac{\tan\phi'}{F_s} \sin\alpha \right) = W - ul \cos\alpha - \frac{c' l \sin\alpha}{F_s}$$

or
$$N' = \frac{W - ul\cos\alpha - \dfrac{c' l \sin\alpha}{F_s}}{\cos\alpha + \dfrac{\tan\phi'}{F_s}\sin\alpha} \qquad \ldots(18.51)$$

Substituting the above value of N' in Eq. 18.50,

$$F_s = \frac{\Sigma c' l + \tan\phi' \, \Sigma \left(\dfrac{W - ul\cos\alpha - \dfrac{c' l \sin\alpha}{F_s}}{\cos\alpha + \dfrac{\tan\phi'}{F_s}\sin\alpha} \right)}{\Sigma W \sin\alpha}$$

$$= \frac{1}{\Sigma W \sin\alpha} \Sigma \left[c' b \sec\alpha + \frac{\tan\phi' \left(W\sec\alpha - ul - \dfrac{c' l \tan\alpha}{F_s} \right)}{1 + \dfrac{\tan\phi' \tan\alpha}{F_s}} \right]$$

$$= \frac{1}{\Sigma W \sin\alpha} \Sigma \left[\left\{ c' b \sec\alpha - \frac{c' l \tan\alpha \tan\phi'/F_s}{1 + \dfrac{\tan\phi' \tan\alpha}{F_s}} \right\} + \frac{\tan\phi' (W\sec\alpha - ul)}{1 + \dfrac{\tan\phi' \tan\alpha}{F_s}} \right]$$

$$= \frac{1}{\Sigma W \sin\alpha} \times \Sigma \left[\frac{\left\{ c' b \sec\alpha (1 + \tan\alpha \tan\phi'/F_s) - c' b \sec\alpha \tan\alpha \dfrac{\tan\phi'}{F_s} \right\}}{1 + \dfrac{\tan\phi' \tan\alpha}{F_s}} + \frac{\tan\phi' (W\sec\alpha - u\, b \sec\alpha)}{1 + \dfrac{\tan\phi' \tan\alpha}{F_s}} \right]$$

$$= \frac{1}{\Sigma W \sin\alpha} \Sigma \left[\frac{c' b \sec\alpha}{1 + \dfrac{\tan\phi' \tan\alpha}{F_s}} + \frac{\tan\phi' (W\sec\alpha - u\, b\sec\alpha)}{1 + \dfrac{\tan\phi' \tan\alpha}{F_s}} \right]$$

$$= \frac{1}{\Sigma W \sin\alpha} \Sigma \left[\{ c' b + (W - u\, b) \tan\varphi' \} \frac{\sec\alpha}{1 + \dfrac{\tan\phi' \tan\alpha}{F_s}} \right]$$

or
$$F_s = \frac{\Sigma \dfrac{1}{m_\alpha} [c' b + (W - u\, b) \tan\phi']}{\Sigma W \sin\alpha} \qquad \ldots(18.52)$$

where $m_\alpha = (1 + \tan\phi' \tan\alpha / F_s) \cos\alpha$ ...(18.53)

Sometimes, the pore pressure u is expressed in terms of pore pressure ratio r_u. In that case,

$$u = r_u \gamma h = r_u (W/b)$$

Eq. 18.52 gives the factor of safety of the assumed failure surface. As the factor of safety (F_s) appears on both the sides, a process of successive approximation is required. A value of F_s is assumed and the analysis is done. The value computed from Eq. 18.52 is compared with the assumed value. If the two values differ, the process is repeated till convergence. As the convergence is rapid, only 3-4 trials are required. A computer may also be used.

The effective stress analysis is generally done, but the total stress analysis is also possible. The factor of safety determined by Bishop's simplified method is an underestimate and, therefore, it errs on the safe side. The error is generally less than 2% and not more than 7% even in an extreme case.

18.17. OTHER METHODS OF ANALYSIS

A number of investigators have developed different methods for the analysis of slopes. The methods are similar in nature to the Bishop methods but differ in handling of interslice forces.

(1) Spencer (1967) assumed that the forces on the sides of the slices are parallel [Fig. 18.22 (*a*)]. The method satisfies the requirement of equilibrium of forces and moments.

(2) Morgenstern and Price (1965) developed a method in which all boundary and equilibrium conditions are satisfied. The analysis imposes normal and shear forces on the sides and also includes water pressure effects [Fig. 18.22 (*b*)]. The method satisfies the requirement of equilibrium of forces and moments which act on a slice. The failure surface may be of any shape.

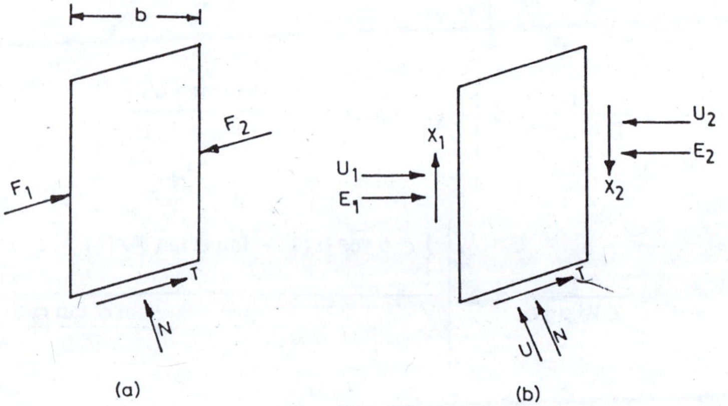

Fig. 18.22

The soil mass above the failure surface is divided into sections by a number of vertical planes. The problem is made statically determinate by assuming a relationship between the forces E and X on the vertical sides, as

$$X = \lambda f(x) E$$

where $f(x)$ is an arbitrary function describing the pattern in which the ratio X/E varies across the soil. The value of the scale factor λ is obtained as a part of the solution along with the factor of safety.

The values of the forces E and X and the point of application of E can be determined at each vertical boundary. For any assumed value of $f(x)$, it is necessary to check that the solution is physically reasonable and no shear failure or tension develop within the soil mass above the failure surface. As the choice of $f(x)$ does not influence the computed values of the factor of safety by more than 5%, for convenience $f(x)$ is taken as unity. The analysis involves a complex process of iteration for the values of λ and F_s. The use of a computer is essential. However, slope stability charts which simplify the solution have been developed for the design office use.

(3) Bell (1968) developed a method in which all the conditions of equilibrium are satisfied. The soil mass is divided into vertical slices. The failure surface may be of any shape. The statical determinacy is obtained by means of an assumed distribution of normal stresses along the failure surface. The soil mass is considered as a free body, as in the case of the friction circle method.

(4) Cousins (1978) developed charts which have application to a wide range of field problems. The charts are quite convenient, as these deal in easily calculated parameters. Cousins charts are based upon the friction

circle method of analysis discussed earlier. However, the charts are for a general case. These can be used for the evaluation of long-term stability analysis in terms of effective stresses. The charts can be used even in case of tension cracks.

(5) Stability charts for the analysis of earth dams under sudden drawdown conditions have been developed by Morgenstern (1963). The charts are based upon an effective stress analysis using the method of slices.

18.18. IMPROVING STABILITY OF SLOPES

The slopes which are susceptible to failure by sliding can be improved and made usable and safe. Various methods are used to stabilise the slopes. The methods generally involve one or more of the following measures, which either reduce the mass which may cause sliding or improve the shear strength of the soil in the failure zone.

(1) Slope flattening reduces the weight of the mass tending to slide. It can be used wherever possible.

(2) Providing a berm below the toe of the slope increases the resistance to movement. It is specially useful when there is a possibility of a base failure.

(3) Drainage helps in reducing the seepage forces and hence increases the stability. The zone of subsurface water is lowered and infiltration of the surface water is prevented.

(4) Densification by use of explosives, vibroflotation, or terra probe helps in increasing the shear strength of cohesionless soils and thus increasing the stability.

(5) Consolidation by surcharging, electro-osmosis or other methods helps in increasing the stability of slopes in cohesive soils.

(6) Grouting and injection of cement or other compounds into specific zones help in increasing the stability of slopes.

(7) Sheet piles and retaining walls can be installed to provide lateral support and to increase the stability. However, the method is quite expensive.

(8) Stabilisation of the soil helps in increasing the stability of slopes.

In the interest of economy, relatively inexpensive methods, such as slope flattening and drainage control, are generally preferred.

ILLUSTRATIVE EXAMPLES

Illustrative Example 18.1. *A long natural slope in an overconsolidated clay ($c' = 10 kN/m^2$, $\phi' = 25°$, $\gamma_{sat} = 20\ kN/m^2$) is inclined at 10° to the horizontal. The water table is at the surface and the seepage is parallel to the slope. If a plane slip had developed at a depth of 5 m below the surface, determine the factor of safety. Take $\gamma_w = 10\ kN/m^3$.*

Solution. From Eq. 18.17,
$$F_s = \frac{c' + \gamma' \ H \cos^2 i \tan \phi'}{\gamma_{sat} H \cos i \sin i}$$

$$= \frac{10 + (20 - 10) \times 5 \times \cos^2 10° \tan 25°}{20 \times 5 \times \cos 10° \sin 10°} = \mathbf{1.90}$$

Illustrative Example 18.2. *A vertical cut is made in a clay deposit ($c = 30\ kN/m^2$, $\phi = 0$, $\gamma = 16\ kN/m^3$). Find the maximum height of the cut which can be temporarily supported.*

Solution. From Eq. 18.28,
$$S_n = \frac{c}{F_c \ \gamma H}$$

Taking $F_c = 1.0$,
$$S_n = \frac{30}{1.0 \times 16 \times H} \quad \ldots(a)$$

From stability chart (Fig. 18.13 or Table 18.1), $S_n = 0.261$.

Substituting the value of S_n in Eq. (a),

$$H = \frac{30}{1.0 \times 16 \times 0.261} = \mathbf{7.18\ m}$$

Illustrative Example 18.3. *A cut of depth 10 m is made in a cohesive soil deposit (c = 30 kN/m², ϕ = 0 and γ = 19 kN/m³). There is a hard stratum under the cohesive soil at a depth of 12 m below the original ground surface. If the required factor of safety is 1.50, determine the safe slope.*

Solution. In this case, $D_f H$ = 12 m and H = 10 m. Therefore, D_f = 1.20.

From Eq. 18.28, $$S_n = \frac{c}{F_c \gamma H} = \frac{30}{1.5 \times 19 \times 10} = 0.105$$

From Fig. 18.14, for S_n = 0.105 and D_f = 1.20, we have i = **15°.**

Illustrative Example 18.4. *Determine the factor of safety with respect to cohesion for a submerged embankment 25 m high and having a slope of 40°. (c = 40 kN/m², ϕ = 10° and γ_{sat} = 18 kN/m³).*

Solution. From Eq. 18.28, using submerged unit weight,

$$S_n = \frac{40}{F_c \times (18 - 9.81) \times 25} = \frac{0.195}{F_c} \quad \text{...(c)}$$

For ϕ = 10° and i = 40°, Table 18.1 gives S_n = 0.097 by interpolation.

Therefore, $$0.097 = \frac{0.195}{F_c} \quad \text{or } F_c = \mathbf{2.01.}$$

Illustrative Example 18.5. *Determine the factor of safety with respect to cohesion if an embankment 25 m high and having a slope of 40° is subjected to sudden drawdown. (c = 40 kN/m², ϕ = 10°, γ_{sat} = 18 kN/m³).*

Solution. From Eq. 18.30, $$\tan \phi_w = \frac{\gamma'}{\gamma_{sat}} \left(\frac{1}{F_\phi} \tan \phi' \right)$$

Taking F_ϕ = 1.0, $$\tan \phi_w = \frac{(18 - 9.81)}{18} (\tan 10°) \quad \text{or } \phi_w = \mathbf{4.59°}$$

From Table 18.1 (or Fig. 18.13), S_n = 0.13 for φ_m = 4.59° and i = 40°.

From Eq. 18.28, using saturated unit weight,

$$0.130 = \frac{40}{F_c \times 18 \times 25}$$

or $$F_c = 0.684 < 1.00$$

The slope is not safe.

Illustrative Example 18.6. *Determine the factor of safety with respect to shear strength of a slope 10 m high and having an inclination of 40° of a soil with c = 30 kN/m², ϕ = 10° and γ = 19 kN/m³.*

Solution. The factor of safety is obtained by trial and error. Let us assume F_φ = 1.30 for the first trial.

From Eq. 18.4, $$\tan \phi_m = \frac{\tan \varphi}{F\phi} = \frac{\tan 10°}{1.30}$$

or $$\phi_m = 7.72°$$

From Table 18.1, for ϕ_m = 7.72° and i = 40°,

$$S_n \approx 0.115$$

From Eq. 18.28, $$0.115 = \frac{c}{F_c \gamma H}$$

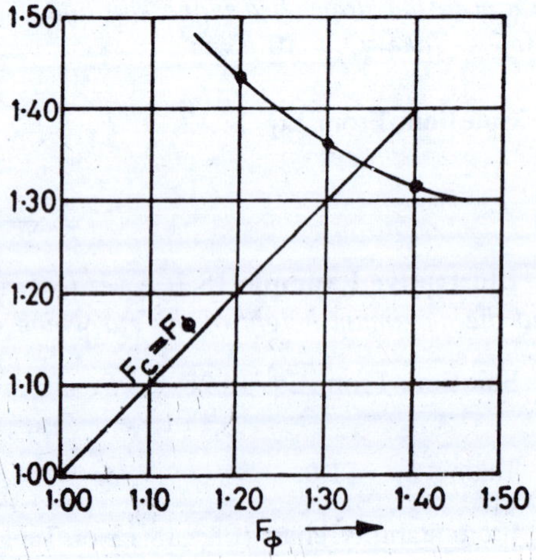

Fig. E-18..

or $$0.115 = \frac{30}{F_c \times 19 \times 10} \quad \text{or } F_c = 1.37$$

Likewise, for $F_\phi = 1.40$, the value of F_c is found to be 1.32 and for $F_\phi = 1.20$, the value of F_c is computed as 1.44. Fig. E-18.6 shows the plot between F_c and F_ϕ. The line drawn at 45° cuts the plot at the required point.

From the plot, $$F_c = F_\varphi = F_s = \mathbf{1.34.}$$

Illustrative Example 18.7. *A soil mass is resting on an inclined impermeable clay layer, as shown in fig. E-18.7. Determine the factor of safety against wedge failure along the interface. The soil has $c = 6$ kN/m^2, $\phi = 20°$ and $\gamma = 17$ kN/m^3.*

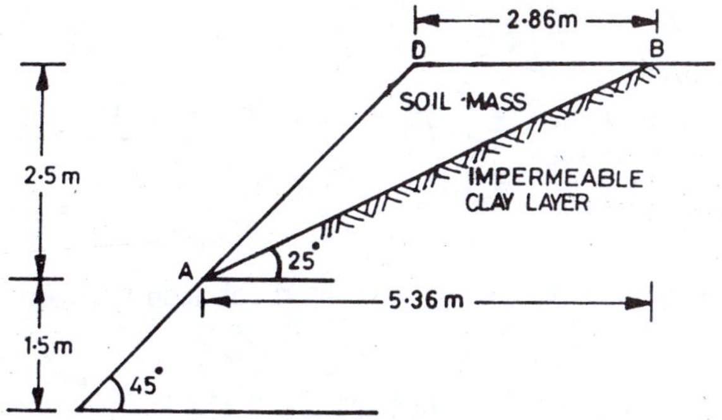

Fig. E-18.7.

Solution. Weight of triangular wedge *ABD*,

$$W = \frac{1}{2} \times 2.86 \times 2.5 \times 17 = 60.8 \text{ kN}$$

Sliding force T = component of W parallel to the plane AB

$$= W \sin 25° = 25.7 \text{ kN}$$

Length of plane *AB*, $$L = 2.5/\sin 25° = 5.91 \text{ m}$$

Resistance to sliding, $$S = cL + (W \cos 25°) \tan \phi$$

$$= 6.0 \times 5.91 + 60.8 \times 0.906 \times 0.364 = 55.51 \text{ kN}$$

Factor of Safety, $$F_s = \frac{S}{T} = \frac{55.51}{25.7} = \mathbf{2.16}$$

Illustrative Example 18.8. *A vertical cut is made through a homogeneous soil mass ($c = 20$ kN/m^2, $\phi = 20°$, $\gamma = 16.5$ kN/m^3). Using Culmann's method, determine the safe depth of the cut, taking a factor of safety of 2.0. (b) Also determine the safe depth using stability charts.*

Solution. From Eq. 18.22 (*a*), $$H = \frac{4c_m \sin i \cos \phi_m}{\gamma [1 - \cos (i - \phi_m)]}$$

Now $$c_m = c/F_c = 20/2 = 10 \text{ kN/m}^2$$

$$\tan \phi_m = \tan \phi/2 = \tan 20°/2 = 0.364/2 = 0.182 \quad \textit{or } \phi_m = 10.31°$$

Therefore, $$H = \frac{4 \times 10 \sin 90° \cos 10.31°}{16.5 [1 - \cos (90 - 10.31°)]} = \mathbf{2.91 \text{ m}}$$

(*b*) For $i = 90°$, $\phi = 10.31°$, we have $S_n = 0.217$ (See Table 18.1)

Therefore, $$H = \frac{c_m}{\gamma S_n} = \frac{20/2}{16.5 \times 0.217} = \mathbf{2.79 \text{ m}}$$

Illustrative Example 18.9. *An unsupported slope is shown in Fig. E-18.9. Determine the factor of safety against sliding for the trial slip surface. Take c = 50 kN/m², and φ = 0. The weight of the wedge ABD is 2518 kN and acts at a horizontal distance of 11 m from the vertical AO.*

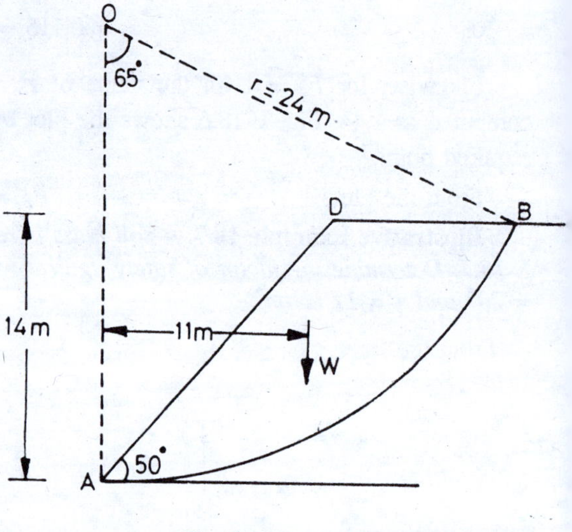

Fig. E-18.9

Solution. Length of arc AB, $L_a = \frac{2\pi \times 24}{360} \times 65$

$= 27.21$ m

Actuating moment, $M_0 = 2518 \times 11$

$= 27698$ kN–m

Resisting moment, $M_r = c L_a r = 50 \times 27.21 \times 24$

$= 32652$ kN–m

Therefore, factor of safety F_c (= F_s)

$$= \frac{32652}{27698} = \mathbf{1.18}$$

Alternative method

The factor of safety can be determined from the mobilised cohesion (c_m). According to this method,

$$c_m L_a r = 27698$$

or

$$c_m = \frac{27698}{27.21 \times 24} = 42.41 \text{ kN}$$

Therefore,

$$F_c = c/c_m = 50/42.41 = \mathbf{1.18}$$

Illustrative Example 18.10. *Fig. E-18.10 shows a slip surface with a radius of 22 m in a slope with a height of 14 m and an angle of inclination of 45°. If ϕ_m = 15°, γ = 18 kN/m³ and c = 40 kN/m², determine the factor of safety with respect to cohesion using the friction circle method.*

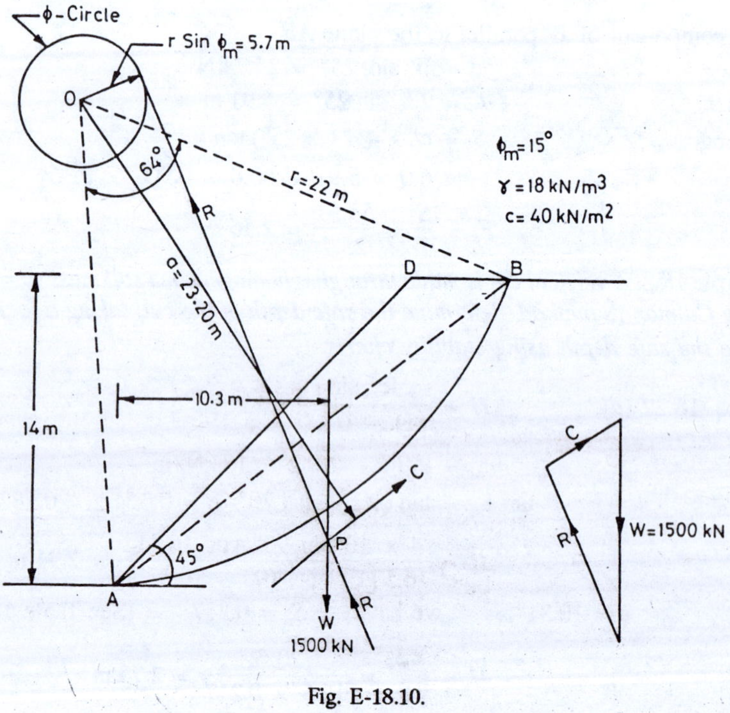

Fig. E-18.10.

The weight of the soil wedge is 1500 kN and it acts at a horizontal distance of 10.3 m from A.

Solution. $r \sin \phi = 22 \times \sin 15° = 5.7$ m.

The friction circle is drawn with a radius of 5.7 m.

From Eq. 18.24,
$$a = r \times \frac{L_a}{L_c} = 22 \times \frac{(2\pi \times 22/360) \times 64}{2 \times 22 \times \sin 32°} = \mathbf{23.20\ m}$$

A line is drawn parallel to the chord at a distance of 23.20 m. The total cohesive force passes through P which is at the intersection of the vertical line through W. The reaction R also passes through P and is tangential to the friction circle.

The figure also shows the force triangle,

From the force triangle, C = 600 kN

$$c_m = \frac{C}{L_c} = \frac{600}{23.32} = 25.7 \text{ kN/m}^2$$

From Eq. 18.5,
$$F_c = \frac{c}{c_m} = \frac{40}{25.7} = \mathbf{1.56}$$

Illustrative Example 18.11. *Fig. E-18.11 shows a trial slip surface through a soil mass (c = 20 kN/m², ϕ = 30°, γ = 20 kN/m³). Determine the factor of safety using Swedish circle method.*

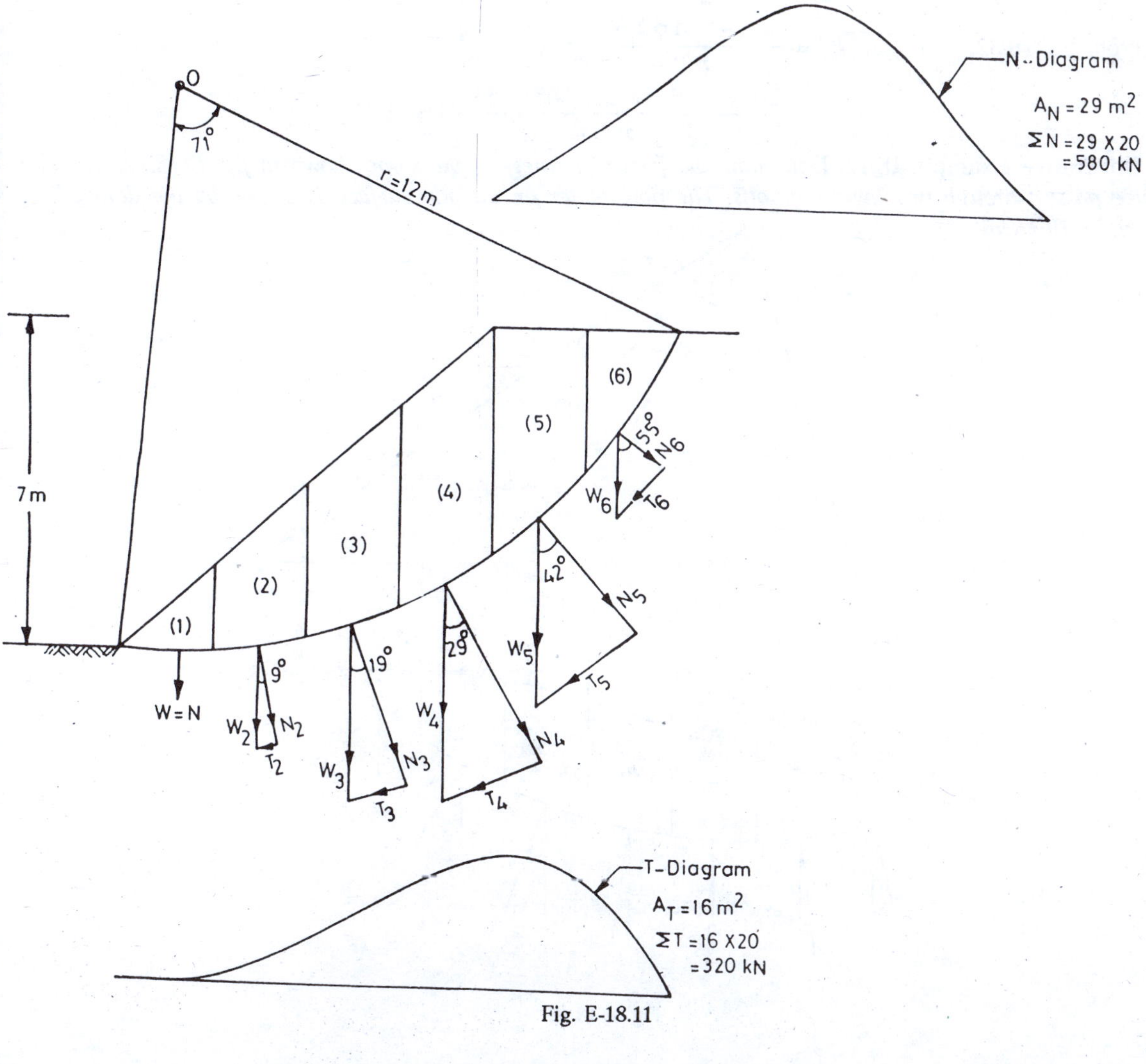

Fig. E-18.11

Solution. The sliding wedge is divided into 6 vertical slices of equal width of 2 m. The calculations are shown in tabular form below.

Table Example 18.11.

Slice No.	*Weight (W)*				α	*T = W sin α (kN)*	*N = W cos α (kN)*
	Average ordinate	*Width*	*Volume*	*Weight W (kN)*			
1.	0.9 m	2 m	1.8 m^3	36	0	0	36
2.	2.4 m	2 m	4.8 m^3	96	9°	15	94.8
3.	3.6 m	2 m	7.2 m^3	144	19°	46.9	136.2
4.	4.5 m	2 m	9.0 m^3	180	29°	87.3	157.4
5.	4.0 m	2 m	8.0 m^3	160	42°	107.1	118.9
6.	1.80 m	2 m	3.6 m^3	72	55°	59	41.3

$\Sigma T = 315.3$ kN $\Sigma N = 584.6$ kN

[Alternatively, from diagrams, $\Sigma T = 320$ kN and $\Sigma N = 580$ kN]

Length of arc, $L_a = (2\pi \times 12/360) \times 71° = 14.86$ m.

From Eq. 18.32,
$$F_s = \frac{cL_a + \tan\phi\, \Sigma N}{\Sigma T}$$
$$= \frac{20 \times 14.86 + \tan 30° \times 584.6}{315.3} = \mathbf{2.01.}$$

Illustrative Example 18.12. *Determine the factor of safety of the slope shown in fig. E-18.12. The slip surface passes through two layers of soils. The pore water on the slip surface is shown by the dotted line. Take $\gamma_w = 10$ kN/m^3.*

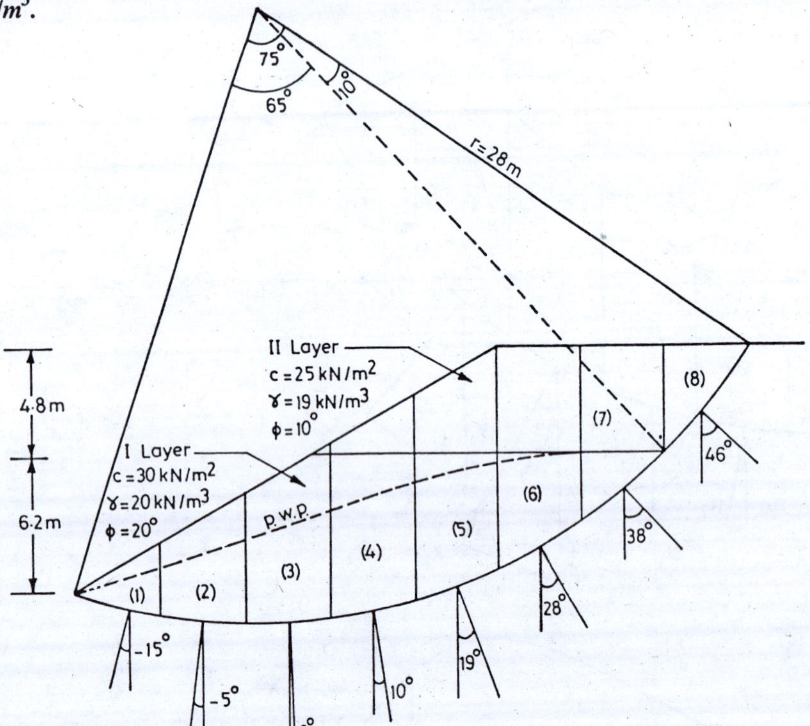

Fig. E-18.12.

Solution. The sliding wedge is divided into 8 slices of equal width 4 m. One vertical line is drawn at the interface of the layers between slices (7) and (8). The weight of each slice is determined from the area of the slice in the layer and the corresponding unit weight of the soil.

The pore water pressure is determined as the vertical ordinate of the dotted line above the curved surface.

The calculations are shown in tabular form.

Total length of arc $(L_a) = (2\pi \times 28/360) \times 75 = 36.63$ m

Length of arc in I-layer $= (36.63/75) \times 65 = 31.75$ m

Length of arc in II-layer $= (36.63/75) \times 10 = 4.88$ m

$\Sigma (N - ul)$ in I layer (Ist to 7th slice) = 2388 kN

$\Sigma (N - ul)$ in II layer (8th slice) = 198 kN

From Eq. 18.39,
$$F_s = \frac{\Sigma c \Delta L + \Sigma (N - ul) \tan \phi}{\Sigma T}$$

or
$$F_s = \frac{(30 \times 31.75 + 25 \times 4.88) + 2388 \times \tan 20° + 198 \tan 10°}{1102.1} = 1.81$$

Illustrative Example 18.13. *Find the factor of safety against sliding for the slope shown in Fig. E-18.13, using Bishop's simplified method.*

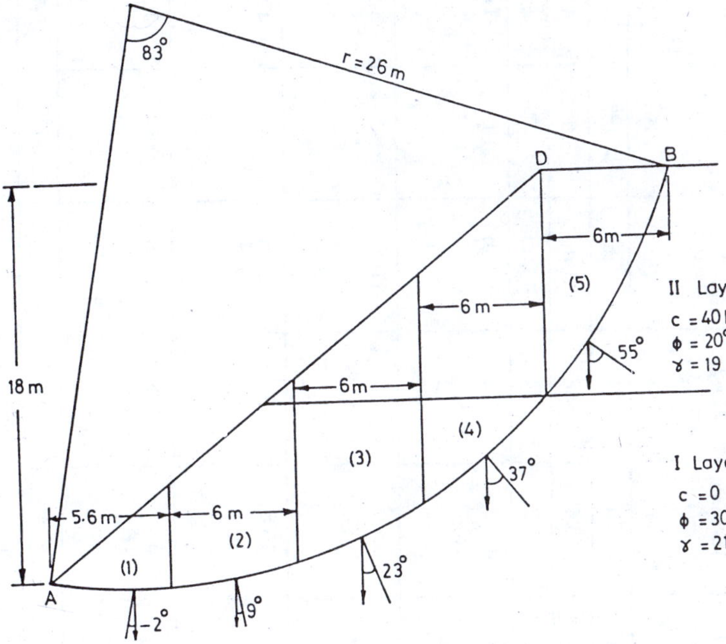

Fig. E-18.13.

Solution. The slip circle has been divided into 5 slices; the first four slices are in Layer I. The width of the first slice is 5.6 m and that of other slices is 6 m.

Calculations are done in tabular form. The value of m_α in column (11) has been computed using Eq. 18.53.

$$m_\alpha = (1 + \tan \phi' \tan \alpha / F_s) \cos \alpha$$

For first trial, F_s is assumed to be 1.20.

From Eq. 18.52,
$$F_s = \frac{\Sigma \frac{1}{m_\alpha} [c' b + (W - ub) \tan \phi']}{W \sin \alpha}$$

Table Example 18.12.

Slice No.	Weight (W)										
	Average ordinate	Width (b)	Volume (m^3)	Weight W(kN)	u metres	α	l = b sec α	ul (kN)	N = W cos α (kN)	N-ul (kN)	T = W sin α (kN)
(1)	(2)	(3)	(4)	(5)	(6)	(7)	(8)	(9)	(10)	(11)	(12)
1.	1.60 m	4 m	6.40	128	1 m	– 15°	4.14	41.4	123.6	82.2	– 33.1
2.	4.6 m	4 m	18.4	368	3 m	– 5°	4.02	120.6	366.6	246.0	– 32.1
3.	6.8 m	4 m	27.2	544	4.2 m	3°	4.01	168.4	543.2	374.4	+ 28.4
4.	7.2 m 1.4 m	4 m 4 m	28.8 6.6	576 106 682	5.2 m	10°	4.06	211.1	671.6	460.5	+ 118.4
5.	6.0 3.6	4 m 4 m	24.0 14.4	480 274 754	5.0 m	19°	4.23	211.5	712.9	501.4	+ 245.5
6.	4.2 4.8	4 m 4 m	16.8 19.2	336 365 701	4.2 m	28°	4.53	190.3	618.8	428.5	+ 329.1
7.	1.6 4.8	4 m 4 m	6.40 19.2	128 365 493	1.8 m	38°	5.08	91.4	386.4	295	+ 303.5
8.	2.6	4 m	10.4	198	0	46°	5.75	0	198	198	+ 142.4

$\Sigma T = 1102.1$

Slice No. (1)	Weight (W)				α	c'	ub	tan φ'	W sin α	mα	(W – ub) tan φ'
	Average ordinate (2)	Width (b) (3)	Volume (m^3) (4)	Weight (kN) (5)	(6)	(7)	(8)	(9)	(10)	(11)	(12)
1.	2.6	5.6	14.56	305.8	– 2°	0	0	0.577	– 10.7	0.984	176.4
2.	6.6	6.0	39.60	831.6	9°	0	0	0.577	+ 130.0	1.063	479.8
3.	6.2	6.0	37.20	781.2	23°	0	0	0.577	+ 465.6	1.107	687.6
	3.6	6.0	21.60	410.4							
				1191.6							
4.	3.0	6.0	18.0	378	37°	0	0	0.577	+776.3	1.088	744.3
	8.0	6.0	48.0	912							
				1290							
5.	6.4	6.0	38.4	729.6	55°	4.0	0.0	0.364	+ 597.6	0.820	265.5
									Σ 1958.8		

$$= \frac{1}{1958.8}\left[\frac{1}{0.984} \times 176.4 + \frac{1}{1.063} \times 479.8 + \frac{1}{1.107} \times 687.6 + \frac{1}{1.088} \times 744.3 + \frac{1}{0.82}(40 \times 6 + 265.5)\right]$$

$$= \mathbf{1.30}$$

The assumed value of F_s is not correct. The process may be repeated after taking F_s = 1.25.

PROBLEMS

A. Numerical

18.1. What inclination is required where a filling 12 m high is to be constructed having a factor of safety of 1.25? The soil has c = 20 kN/m^2, ϕ = 15°, γ = 17.0 kN/m^3. The stability number for ϕ_m = 12° is equal to 0.063 when the slope is 30° and 0.098 when the slope is 45°. **[Ans. 36°].**

18.2. A cutting of depth 10 m is to be made in soil which has c = 30 kN/m^2, γ = 19 kN/m^3 and ϕ = 0. There is a hard stratum below the original soil surface at a depth of 12 m. Find the safe slope of cutting if the factor of safety is 1.50. For D_f = 1.20, S_n = 0.143 for i = 30° and S_n = 0.101 for i = 15°. **[Ans. 17°].**

18.3. A vertical cut is to be made in clayey soil for which tests gave c = 30 kN/m^2, γ = 16 kN/m^3 and ϕ = 0. Find the maximum height for which the cut may be temporarily unsupported. For ϕ = 0, and i = 90°, the value of the stability number is 0.261. **[Ans. 7.18 m]**

18.4. When is the factor of safety for a 45° slope 12 m high in a clay (c = 50 kN/m^2, γ = 18 kN/m^3 and ϕ = 0) having a rock stratum at a depth of 12 m below the toe? For D_f = 2.0 and i = 45°, the value of S_n is equal to 0.177. **[Ans. 1.3]**

18.5. Determine the factor of safety with respect to cohesion only for a submerged embankment 25 m high whose upstream face has an inclination of 45°. The soil has the following properties; c = 40 kN/m^2, ϕ = 10°, γ_{sat} = 18 kN/m^3. The relevant stability number is equal to 0.108. **[Ans. 1.85]**

18.6. What is the factor of safety if the embankment in Prob. 18.5 experiences the effect of sudden drawdown ? For ϕ_m = 4.5° and i = 45°, the value to the stability number is 0.136. **[Ans. 0.65]**

18.7. A cut 10 m deep is to be made in a stratum of cohesive soil (c = 35 kN/m^2, γ = 18.5 kN/m^3 and ϕ = 0). The bed rock is located 15 m below the original ground surface. Determine the factor of safety against failure if the slope is 30°, For D_f = 1.5 and i = 30°, the stability number is equal to 0.164. **[Ans. 1.15]**

18.8. An embankment 10 m high is inclined at 35° to the horizontal. A stability analysis by the method of slices gave the following forces:

$\Sigma N = 900$ kN, $\Sigma T = 420$ kN, $\Sigma U = 200$ kN.

If the length of the failure arc is 23.0 m, find the factor of safety. The soil has $c = 20$ kN/m^2 and $\phi = 15°$.

[Ans. 1.54]

18.9. A dam of homogeneous section is 25 m high with upstream slope of 2.5 to 1.0 and downstream slope of 2 to 1. There is a 12 m long horizontal filter at the downstream end. Taking a free board of 3 m (*i.e.* water depth 22 m), determine the (*a*) factor of safety of downstream slope under steady seepage conditions. (*b*) factor of safety of upstream slope under sudden drawndown conditions.

18.10. For the dam in Prob. 18.9, determine the factor of safety of downstream slope under steady seepage condition using Bishop's simplified method.

18.11. A 40° slope is excavated to a depth of 8 m in a deep layer of saturated clay ($c = 70$ kN/m^2 and $\phi = 0$, $\gamma = 19$ kN/m^3). Determine the factor of safety for the trial failure surface shown in Fig. P. 18.11. **[Ans. 2.87]**

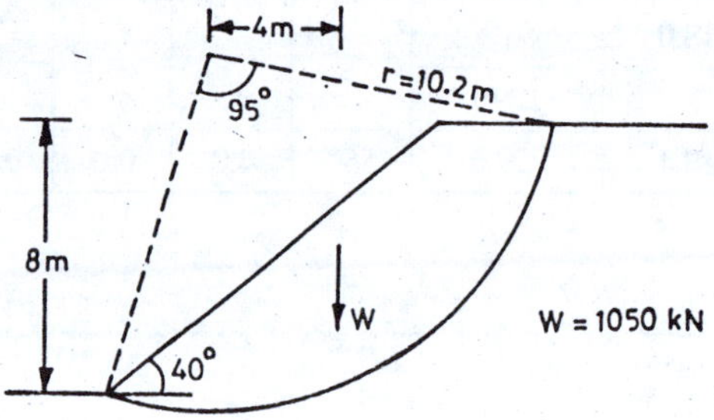

Fig. P. 18.11

18.12. Determine the factor of safety of the slope given in Prob. 18.11. If the soil has $c = 20$ kN/m^2; $\phi = 15°$, $\gamma = 20$ kN/m^3. Use the friction circle method.

B. Descriptive and Objective Type

18.13. What are the assumptions that are generally made in the analysis of the stability of slopes? Discuss briefly their validity.

18.14. What are different factors of safety used in the stability of slopes?

18.15. What are different types of slope failures?

18.16. Derive an expression for the factor of safety of an infinite slope in a cohesionless soil. What is the effect of steady seepage parallel to the slope on the stability?

18.17. Discuss the method for checking the stability of an infinite slope in a cohesive soil. What is critical height?

18.18. Describe Culmann's method for the stability analysis of homogeous slopes. What are its limitations?

18.19. Discuss the friction circle method for the stability analysis of slopes. Can this method be used for purely cohesive soil?

18.20. What is a stability number? What is its utility in the analysis of stability of slopes? Discuss the uses of stability charts.

18.21. How a slope is analysed using Swedish circle method? Derive an expression for the factor of safety.

18.22. Describe Bishop's simplified method. What are its advantages over conventional Swedish circle method? Derive an expression for the factor of safety.

18.23. Discuss the various methods for improving the stability of slopes.

18.24. Write whether the following statements are true or false.

(*a*) The friction circle method can be used for a non- homogeneous soil mass.

(*b*) The stability numbers can be used for the analysis of purely cohesionless soil slopes.

(*c*) The factor of safety of an infinite slope of a cohesive soil depends upon the height H of the slope.

(*d*) Culmann's method assumes that the failure surface is a plane.

(*e*) The upstream slope of an earth dam is critical during sudden drawdown conditions.

(*f*) Bishop's simplified method considers all the forces acting on the sides of the slices.

(*g*) In Bishop's simplified method, a process of successive approximations is required.

(*h*) The total stress analysis can be used for the stability of slopes.

(*i*) The conventional Swedish circle method always errs on the safe side.

[**Ans.** True. (*c*), (*d*), (*e*), (*g*), (*h*), (*i*)].

C. Multiple Choice Questions

1. The method of slices for the stability of slope
(*a*) can be used for stratified soils.
(*b*) can be used when seepage occurs and the pore pressure exists within the soil.
(*c*) gives the factor of safety based on moments and not the forces.
(*d*) All the above.

2. Taylor's stability charts are based on the total stresses using the
(*a*) friction circle method
(*b*) method of slices
(*c*) $\phi_u = 0$ analysis
(*d*) none of the above

3. In stability analysis, the term mobilised shear strength is referred to as
(*a*) shear strength
(*b*) maximum shear stress
(*c*) applied shear stress
(*d*) none of the above.

4. Bishop's simplified method of slices satisfies
(*a*) only the moments equilibrium
(*b*) only the vertical forces equilibrium
(*c*) only the horizontal forces equilibrium
(*d*) all the statics equations, except the horizontal forces equilibrium.

5. The following assumption is not made for the friction circle method of slope stability analysis :
(*a*) Friction is fully mobilised
(*b*) Total stress analysis is applicable
(*c*) The resultant is tangential to the friction circle
(*d*) The resultant passes through the centre of friction circle

6. The factor of safety of an infinite slope in a sand deposit is 1.732. If the angle of shearing resistance is 30°, the safe slope is
(*a*) 19.45° (*b*) 75.4°
(*c*) 18.4° (*d*) 71.6°

7. Identify the incorrect statement
The stability of a slope is decreased by
(*a*) Removal of a part of slope by excavation
(*b*) Shock caused by an earthquake
(*c*) Pore water pressure in the soil
(*d*) Providing a berm at the toe

8. For the computation of N-component for sudden drawdown conditions by approximate method, the weight is
(*a*) Saturated unit weight (*b*) Submerged unit weight
(*c*) Bulk unit weight (*d*) Dry unit weight

[**Ans.** 1. (*d*), 2. (*a*), 3. (*c*), 4. (*d*), 5. (*d*), 6. (*c*), 7. (*d*), 8. (*b*)]

19
Earth Pressure Theories

19.1. INTRODUCTION

As discussed in the preceding chapter, a soil mass is stable when the slope of the surface of the soil mass is flatter than the safe slope. At some locations where the space is limited, it is not possible to provide flat slope and the soil is to be retained at a slope steeper than the safe one. In such cases, a retaining structure is required to provide lateral support to the soil mass. Generally, the soil masses are vertical or nearly vertical behind the retaining structure. Thus, a retaining wall maintains the soil at different elevations on its either side. In the absence of a retaining wall, the soil on the higher side would have a tendency to slide and may not remain stable.

The design of the retaining structure requires the determination of the magnitude and line of action of the lateral earth pressure. The magnitude of the lateral earth pressure depends upon a number of factors, such as the mode of the movement of the wall, the flexibility of the wall, the properties of the soil, the drainage conditions. It is a *soil-structure interaction* problem, as the earth pressure depends upon the flexibility of wall. The earth pressure theories which consider soil-structure interaction are complicated and require a computer. For convenience, the retaining wall is assumed to be rigid and the soil-structure interaction effect is neglected.

The lateral earth pressure is usually computed using the classical theories proposed by Coulomb (1773) and Rankine (1857). The general wedge theory proposed by Terzaghi (1941) is more general and is an improvement over the earlier theories. However, the theory is quite complicated.

The design of rigid retaining walls and flexible retaining walls, such as sheet pile walls and bulk heads, is discussed in the next chapter.

19.2. DIFFERENT TYPES OF LATERAL EARTH PRESSURE

Lateral earth pressures can be grouped into 3 categories, depending upon the movement of the retaining wall with respect to the soil retained. The soil retained is also known as the backfill.

(1) **At-rest pressure.** The lateral earth pressure is called at-rest pressure when the soil mass is not

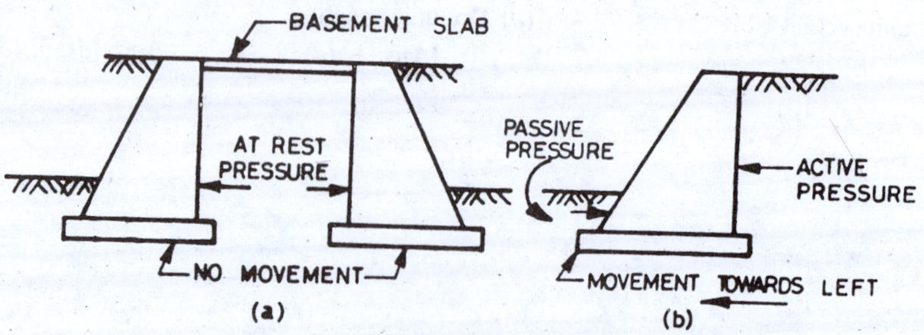

Fig. 19.1.

subjected to any lateral yielding or movement. This case occurs when the retaining wall is firmly fixed at its top and is not allowed to rotate or move laterally. Fig. 19.1. (*a*) shows the basement retaining walls which are restrained against the movement by the basement slab provided at their tops. Another example of the at-rest pressure is that of a bridge abutment wall which is restrained at its top by the bridge slab. The at-rest condition is also known as the *elastic equilibrium*, as no part of soil mass has failed and attained the plastic equilibrium.

(2) Active pressure. A state of active pressure occurs when the soil mass yields in such a way that it tends to stretch horizontally. It is a state of plastic equilibrium as the entire soil mass is on the verge of failure. A retaining wall when moves away from the backfill, there is a stretching of the soil mass and the active state of earth pressure exists. In Fig. 19.1 (*b*), the active pressure develops on the right-hand side when the wall moves towards left.

(3) Passive pressure. A state of passive pressure exists when the movement of the wall is such that the soil tends to compress horizontally. It is another extreme of the limiting equilibrium condition. In Fig. 19.1 (*b*), the passive pressure develops on the left-side of the wall below the ground level, as the soil in this zone is compressed when the movement of the wall is towards left. Another example of the passive earth pressure is the pressure acting on an anchor block.

Variation of Pressure

Fig. 19.2 (*a*) shows the variation of earth pressure with the wall movement. Point *B* represents the case when there is no movement of the wall. It indicates the at-rest pressure.

Point *A* in Fig. 19.2 (*a*) indicates the active pressure. When the wall moves away from the backfill [Fig. 19.2 (*b*)], some portion of the backfill located immediately behind the wall tries to break away from the rest of the soil mass. This wedge-shaped portion, known as the failure wedge or the sliding wedge, moves downward and outwards. The lateral earth pressure exerted on the wall is a minimum in this case. The soil is at the verge of failure due to a decrease in the lateral stress.

The horizontal strain required to reach the active state of plastic equilibrium is very small. Lambe and

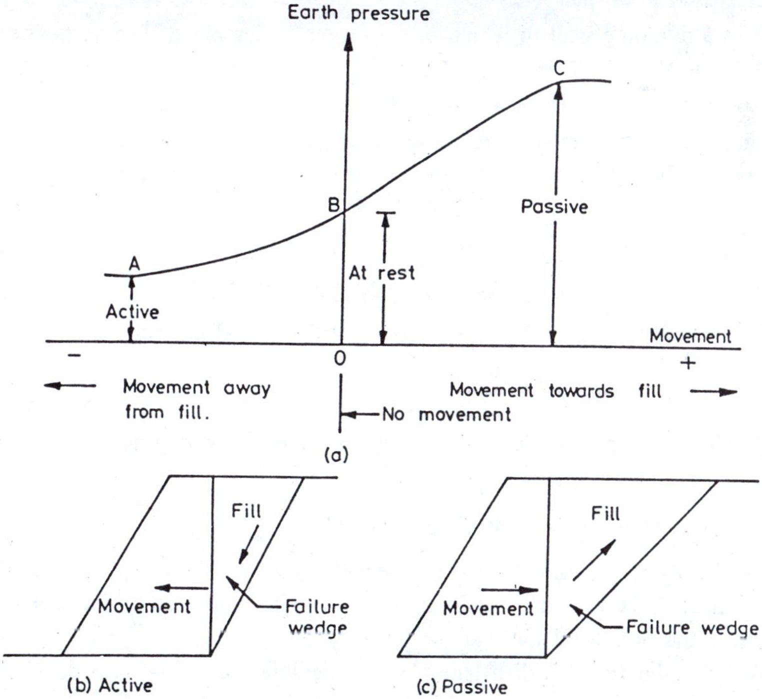

Fig. 19.2

Whitman (1969) have shown that in dense sand, the horizontal strain required is about 0.5%. For example, for a wall of 5 m height, a movement of 0.0144 m would develop active earth pressure.

Point *C* in Fig. 19.2 (*a*) indicates the passive pressure. When the wall moves towards the backfill [Fig. 19.2 (*c*)], the earth pressure increases. The failure wedge moves upward and inwards. The maximum value of the earth pressure is the passive earth pressure. The soil is at the verge of failure due to an increase in the lateral stress.

Lambe and Whitman (1969) found that very little horizontal strain (about 0.5%) is required to reach one-half the maximum passive pressure in dense sand, but much more horizontal strain (about 5% in dense sand, and 15% in loose sand) is required to reach the full passive pressure. However, their data are based on the triaxial shear test results and the magnitude of the horizontal strains required in the field may be somewhat different.

It may be summarised that the state of shear failure corresponding to the minimum earth pressure is the active state and that corresponding to the maximum earth pressure is the passive state. These are the two extreme conditions of plastic equilibrium. For all intermediate states when the soil is not in plastic equilibrium, it is said to be an elastic equilibrium. The at-rest condition is a special case of an elastic equilibrium when the state of stress corresponds to the condition where there is no movement. It indicates the in-situ conditions.

19.3. EARTH PRESSURE AT REST

The earth pressure at rest was discussed briefly in Chapter 11. However, the emphasis there was on the determination of the horizontal stresses in the soil mass. The expressions for earth pressure at-rest would be used in this chapter for the determination of the magnitude and line of action of the total forces due to earth pressure on the retaining structures. The methods for estimation of the coefficient of earth pressure at rest (K_0) have been discussed in chapter 11.

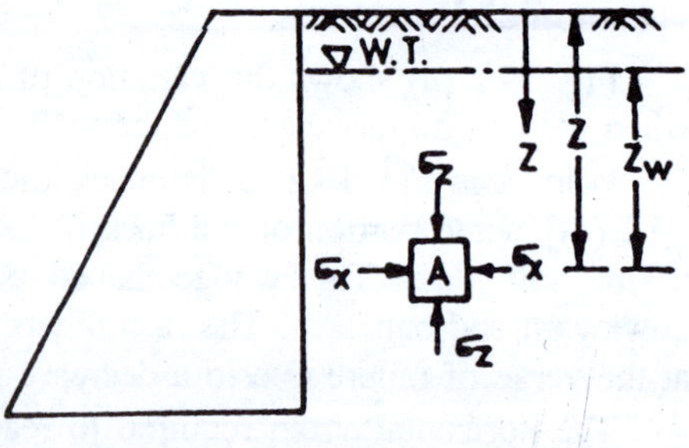

Fig. 19.3.

Fig. 19.3 shows a retaining wall in which no movement takes place. The vertical effective stress at point *A* at a depth *Z* is given by

$$\bar{\sigma}_z = \gamma Z - \gamma_w Z_w \qquad ...(19.1)$$

The horizontal intergranular (effective) stress can be obtained using the coefficient of earth pressure at-rest (K_0), which is equal to the ratio of the horizontal stress to the vertical stress,

Thus $$K_0 = \frac{\bar{\sigma}_x}{\bar{\sigma}_z}$$

or $$\bar{\sigma}_x = K_0\, \bar{\sigma}_z = K_0\, (\gamma Z - \gamma_w Z_w) \qquad ...(19.2)$$

The stress $\bar{\sigma}_x$ is usually represented as p_0, indicating the lateral pressure at rest.

Thus $$p_0 = K_0 \bar{\sigma}_z \qquad ...(19.3)$$

It may be noted that the coefficient of lateral pressure at rest (K_0) relates the effective stresses. The total lateral pressure (p_h) is equal to the sum of the intergranular pressure (p_0) and the pore water pressure (u).

Thus $$p_h = p_0 + u \qquad ...(19.4)$$

In Fig. 19.3, the lateral pressure at depth Z is, therefore,

$$p_h = K_0\,(\gamma Z - \gamma_w Z_w) + \gamma_w Z_w \qquad ...(19.5)$$

As Eq. 19.5 indicates, the pressure distribution is triangular with zero pressure at the top (Z = 0), and the maximum pressure at the bottom of the wall.

Fig. 19.4 (*a*) shows the pressure distribution when the soil is dry. The pressure at the bottom of the wall at depth *H* is given by

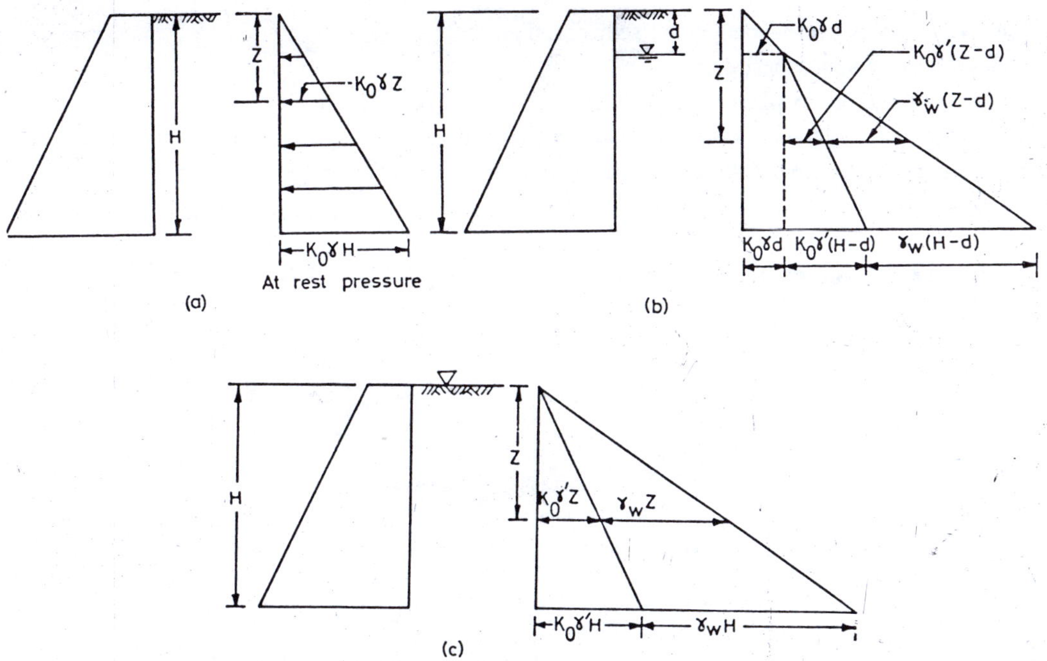

Note : For active pressure, substitute K_a for K_0 and for passive pressure, substitute K_p for K_0.

Fig. 19.4.

$$p_h = K_0 \gamma H$$

The total pressure force per unit length of the wall is given by

$$P = \int_0^H K_0 \gamma z \, dZ$$

or $$P = \frac{1}{2} K_0 \gamma H^2 \quad ...(19.6)$$

In Fig. 19.4 (*b*), the depth of water table is at depth *d* below the surface. The pressure at deptn $Z > d$ is given by

$$p_h = K_0 [\gamma Z - \gamma_w (Z - d)] + \gamma_w (Z - d)$$

or $$p_h = K_0 \gamma d + K_0 \gamma' (Z - d) + \gamma_w (Z - d)$$

The pressure at the bottom ($Z = H$) of the wall is given by

$$p_h = K_0 \gamma d + K_0 \gamma' (H - d) + \gamma_w (H - d) \quad ...(19.7)$$

The total pressure force (P) can be determined from the pressure distribution diagram.

If the water table is at the ground surface [Fig. 19.4 (*c*)], the pressure at the bottom of the wall is given by, taking $d = 0$ in Eq. 19.7,

$$p_h = K_0 \gamma' H + \gamma_w H \quad ...(19.8)$$

The resultant pressure (P) acting on the wall is determined from the pressure distribution diagram.

The point of application of the resultant pressure P is determined from the pressure distribution diagram. For triangular pressure distribution, it acts at height $H/3$ from the base.

19.4. RANKINE'S EARTH PRESSURE THEORY

Rankine (1857) considered the equilibrium of a soil element within a soil mass bounded by a plane surface. The following assumptions were made by Rankine for the derivation of earth pressure.

(1) The soil mass is homogeneous and semi-infinite.

(2) The soil is dry and cohesionless.

(3) The ground surface is plane, which may be horizontal or inclined.

(4) The back of the retaining wall is *smooth and vertical.*

(5) The soil element is in a state of plastic equilibrium, *i.e.*, at the verge of failure.

Expressions for the active earth pressure and the passive earth pressure are developed as explained below:

***(a)* Active Earth Pressure.** Let us consider an element of dry soil at a depth Z below a level soil surface [Fig. 19.5 *(a)*]. Initially, the element is at-rest conditions, and the horizontal pressure is given by

$$\sigma_h = K_0 \sigma_v$$

where σ_v is the vertical stress at C, and σ_h is the horizontal stress at C. Of course, $\sigma_v = \gamma Z$.

The stresses σ_h and σ_v are, respectively, the minor and major principal stresses, and are indicated by points A and B in the Mohr circle [Fig. 19.5 *(b)*].

Let us now consider the case when the vertical stress remains constant while the horizontal stress is decreased. The point A shifts to position A' and the diameter of the Mohr circle increases. In the limiting condition, the point A shifts to position A'' when the Mohr circle [marked (3)] touches the failure envelope. The soil is at the verge of shear failure. It has attained the Rankine active state of plastic equilibrium. The horizontal stress at that state is the active pressure (p_a).

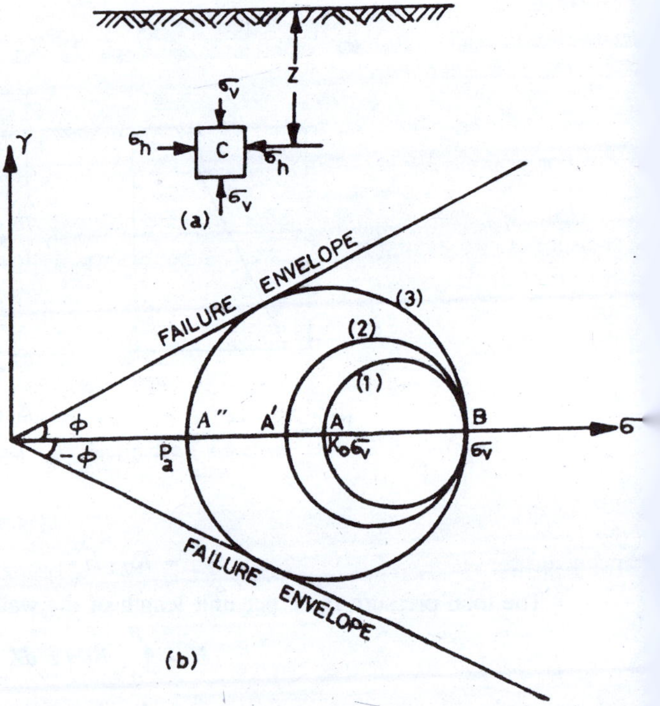

Fig. 19.5.

Fig. 19.6 shows the Mohr circle when active conditions are developed. Point E represents the active condition. From the figure,

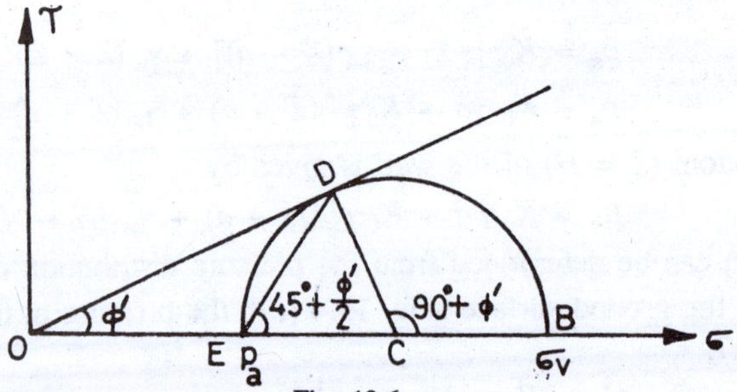

Fig. 19.6.

$$p_a = OE = OC - CE$$

As $$CE = CD = OC \sin \phi',$$

$$p_a = OC - OC \sin \phi' = OC (1 - \sin \phi') \qquad \ldots(a)$$

Also $$\sigma_v = OB = OC + CB = OC + OC \sin \phi'$$

or $$\sigma_v = OC (1 + \sin \phi') \qquad \ldots(b)$$

From Eqs. (*a*) and (*b*),
$$\frac{p_a}{\sigma_v} = \frac{1 - \sin\phi'}{1 + \sin\phi'}$$

or
$$p_a = \left(\frac{1 - \sin\phi'}{1 + \sin\phi'}\right)\sigma_v$$

or
$$p_a = K_a \gamma Z \qquad \text{...(19.9)}$$

where K_a is a coefficient, known as the coefficient of active earth pressure. It is a function of the angle of shearing resistance (ϕ'), and is given by

$$K_a = \frac{1 - \sin\phi'}{1 + \sin\phi'} = \tan^2\left(45° - \frac{\phi'}{2}\right) \qquad \text{...(19.10)}$$

For example, if $\phi' = 30°$, $K_a = \frac{1}{3}$

Eq. 19.9 can be used to determine the active earth pressure on the retaining wall. The pressure distribution is similar to one shown in Fig. 19.4 (*a*) in which K_a is substituted for K_0.

Fig. 19.7 shows the failure planes. These are inclined at (45° + $\phi'/2$) to the major principal plane which is horizontal.

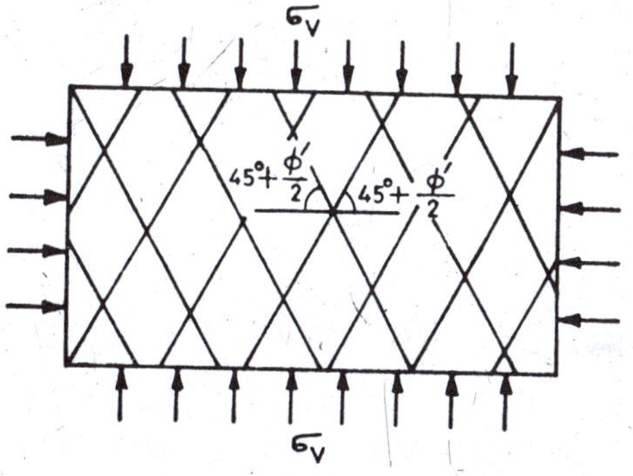

Fig. 19.7

When the wall moves away from the back fill, the failure wedge moves downward and the resisting force due to the shearing strength of the soil is developed in the upward direction along the failure plane (Fig. 19.8). The resisting force causes a decrease in the earth pressure acting on the wall. The decrease in earth pressure continues till the maximum resistance has been mobilised. The earth pressure does not decrease beyond this point and the active state is reached and the soil has attained plastic equilibrium.

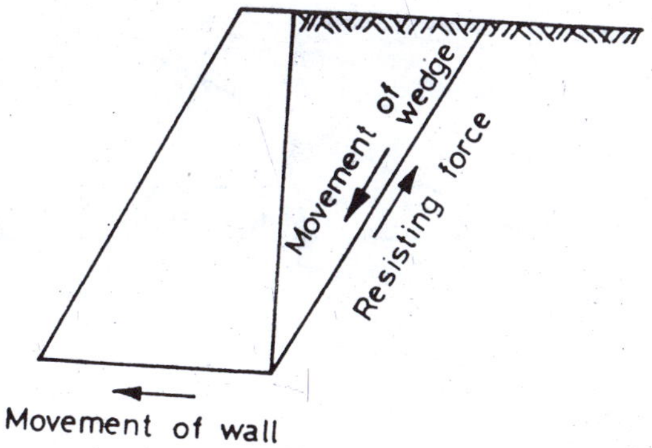

Fig. 19.8.

(*b*) Passive Earth pressure

The passive Rankine state of plastic equilibrium can be explained by considering the element of soil at a point at a depth of Z below the soil surface [Fig. 19.9 (*a*)]. As the soil is compressed laterally, the horizontal stress (σ_h) is increased, whereas the vertical stress (σ_v) remains constant. Fig. 19.9 (*b*) shows the Mohr circles. The circle-1 indicates the insitu condition, in which point *A* indicates the horizontal stress and point *B*, vertical stress. With lateral compressing of the soil, the horizontal stress increases until it reaches a limiting value greater than the vertical stress, indicated by point *A″* and the Mohr circle [marked (3)] touches the failure envelope. The expression for the passive pressure p_p can be obtained as follows. Fig. 19.10, shows the Mohr circle at failure.

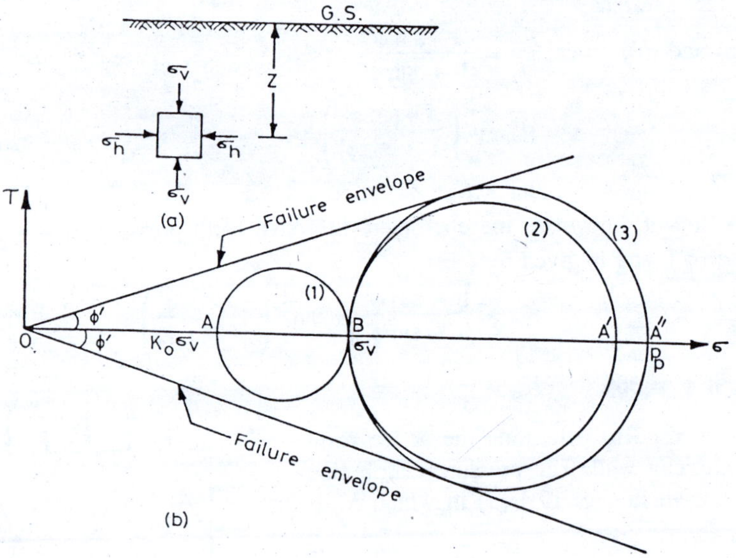

Fig. 19.9.

Now $$p_p = OC + CE = OC + CD = OC + OC \sin \phi'$$

or $$p_p = OC(1 + \sin \phi') \qquad \ldots(a)$$

Also $$OB = OC - BC = OC - CD = OC - OC \sin \phi'$$

or $$\sigma_v = OC(1 - \sin \phi') \qquad \ldots(b)$$

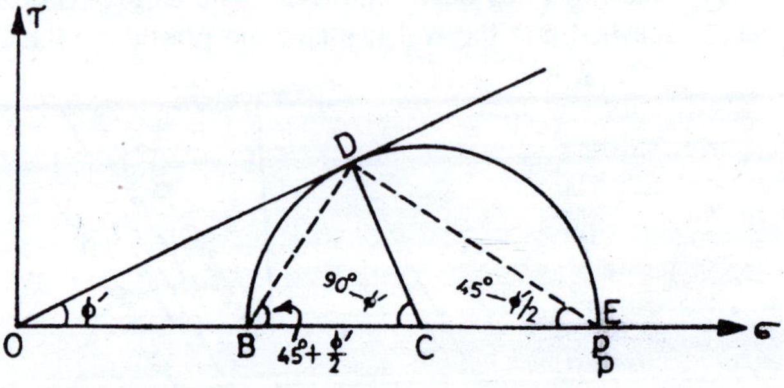

Fig. 19.10.

From Eqs. (*a*) and (*b*), $$\frac{p_p}{\sigma_v} = \frac{1 + \sin \phi'}{1 - \sin \phi'}$$

or $$p_p = \left(\frac{1 + \sin \phi'}{1 - \sin \phi'}\right) \sigma_v$$

or $$p_p = K_p \gamma Z \qquad \ldots(19.11)$$

where K_p is the coefficient of passive earth pressure, given by

$$K_p = \frac{1 + \sin \phi'}{1 - \sin \phi'} = \tan^2 (45° + \phi'/2) \qquad \ldots(19.12)$$

For example, if $\phi' = 30°$, $K_p = 3$

The coefficient of passive pressure (K_p) depends upon ϕ'. The pressure distribution is similar to that shown in Fig. 19.4, in which K_p is substituted for K_0.

The angle which the failure plane makes with the major principal plane is equal to (45° + $\phi'/2$). As the major principal plane is vertical, the failure plane make an angle of (45 – $\phi'/2$) with the horizontal (Fig. 19.11) which is the minor principal plane.

When the wall moves towards the back fill, the lateral earth pressure increases because the resistance builds up in the direction towards the wall. The pressure reaches a maximum value when the full shearing resistance has been mobilised (Fig. 19.12). Further movement of the wall does not increase the pressure, and the passive state is reached and the soil has attained plastic equilibrium.

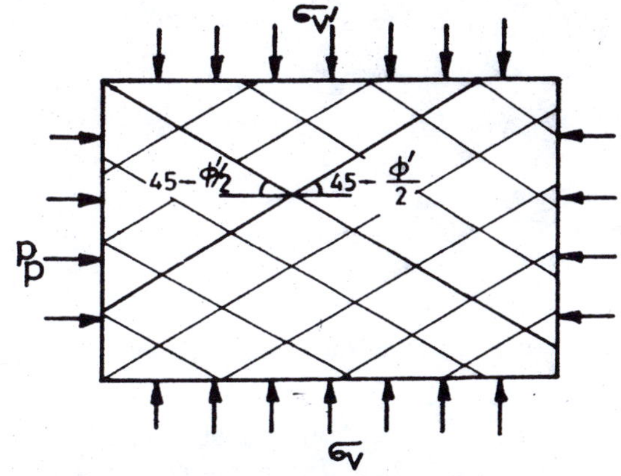

Fig. 19.11.

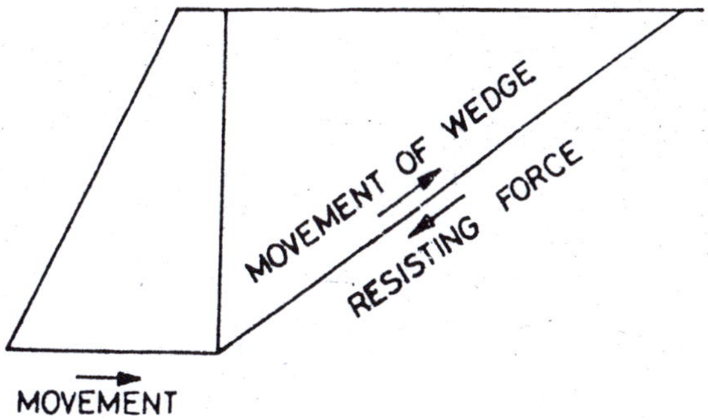

Fig. 19.12.

19.5. RANKINE'S EARTH PRESSURE WHEN THE SURFACE IS INCLINED

Two stresses are called conjugate stresses when the direction of one stress is parallel to the plane on which the other stress acts. Rankine assumed that the vertical stress on an element of the soil within the inclined backfill and the lateral stress on the vertical plane of the element are conjugate stresses. In other words, he assumed that the lateral stress is parallel to the inclined backfill.

Let us consider an element of soil at depth Z below the soil surface inclined at angle i to horizontal (Fig. 19.13). The angle i is known as the angle of surcharge. The intensity of vertical stress (σ_v) on the element is given by

$$\sigma_v = \frac{\gamma (Z\, b \cos i)}{b}$$

or $\sigma_v = \gamma Z \cos i$ ...(19.13)

The other conjugate stress is the lateral stress (σ_x).

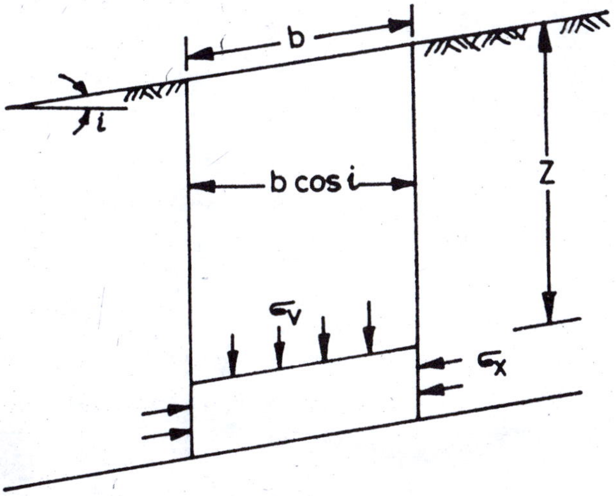

Fig. 19.13.

It may be mentioned that the vertical stress σ_v is not the principal stress, as a shear stress also exists on the inclined plane at the top of the element. Likewise, the lateral stress σ_x is also not a principal stress. A relationship between the lateral pressure and the vertical stress can be obtained for the active and passive cases as given below.

(a) Active Earth Pressure. Fig. 19.14 shows the Mohr circle corresponding to the active limiting

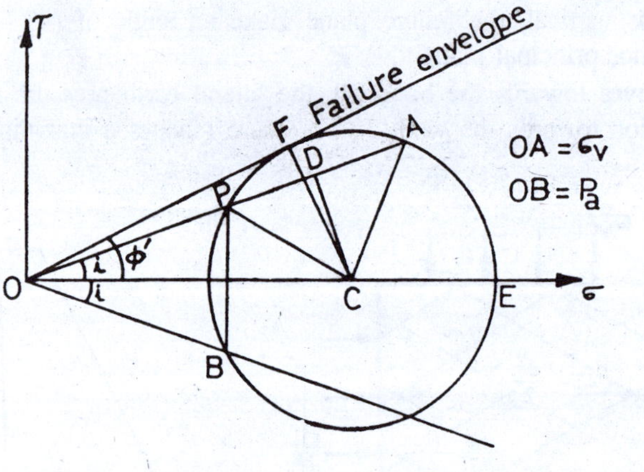

Fig. 19.14.

conditions. The vertical stress σ_v is represented by the line *OA* making an angle *i* with the horizontal. At any depth, the value of σ_v is constant and equal to that given by Eq. 19.13.

If the lateral expansion of the soil is sufficient to induce the state of active plastic equilibrium, the Mohr circle must pass through *A* and it should be tangential to the failure envelope. The origin of planes *P* is obtained as the point of intersection of *OA* with the Mohr circle. As the reader would recollect (chapter 13), the origin of planes is located by drawing from the point representing a stress (vertical stress, in this case) a line parallel to the plane on which it acts (plane inclined at *i*, in this case). A vertical line through *P* cuts the circle at *B* below the σ-axis. The conjugate stress, which is the active pressure (p_a), is represented by *OB*. Numerically, the conjugate stress is also equal to *OP*.

From the figure,
$$\frac{p_a}{\sigma_v} = \frac{OB}{OA} = \frac{OP}{OA} = \frac{OD - DP}{OD + DA} \quad \ldots(a)$$

The ratio p_a/σ_v is known as the *conjugate stress ratio.*

Now
$$OD = OC \cos i \quad \ldots(b)$$

and
$$DA = DP = \sqrt{PC^2 - DC^2}$$

But
$$PC = FC = OC \sin \phi'$$

and
$$DC = OC \sin i$$

Therefore,
$$DA = DP = \sqrt{(OC \sin \phi')^2 - (OC \sin i)^2}$$
$$= OC \sqrt{\sin^2 \phi' - \sin^2 i} \quad \ldots(c)$$

From Eq. (*a*),
$$\frac{p_a}{\sigma_v} = \frac{OC \cos i - OC\sqrt{\sin^2 \phi' - \sin^2 i}}{OC \cos i + OC\sqrt{\sin^2 \phi' - \sin^2 i}}$$

or
$$\frac{p_a}{\sigma_v} = \frac{\cos i - \sqrt{\sin^2 \phi' - \sin^2 i}}{\cos i + \sqrt{\sin^2 \phi' - \sin^2 i}} \quad \ldots(d)$$

$$= \frac{\cos i - \sqrt{(1 - \cos^2 \phi') - (1 - \cos^2 i)}}{\cos i + \sqrt{(1 - \cos^2 \phi') - (1 - \cos^2 i)}}$$

or
$$\frac{p_a}{\sigma_v} = \frac{\cos i - \sqrt{\cos^2 i - \cos^2 \phi'}}{\cos i + \sqrt{\cos^2 i - \cos^2 \phi'}}$$

or
$$p_a = \frac{\cos i - \sqrt{\cos^2 i - \cos^2 \phi'}}{\cos i + \sqrt{\cos^2 i - \cos^2 \phi'}} (\gamma Z \cos i)$$

or
$$p_a = K_a \gamma Z \qquad \text{...(19.14)}$$

where K_a is the coefficient of active pressure, given by

$$K_a = \cos i \times \frac{\cos i - \sqrt{\cos^2 i - \cos^2 \phi'}}{\cos i + \sqrt{\cos^2 i - \cos^2 \phi'}} \qquad \text{...(19.15)}$$

In must be noted that p_a is parallel to the inclined surface.

For the special case, when $i = 0$

$$K_a = \frac{1 - \sin \phi'}{1 + \sin \phi'} \qquad \text{(same as Eq. 19.10)}$$

The conjugate stress in that case become principal stresses, as already discussed.

Direction of failure planes

The inclination of the major principal plane is indicated by the line *PE* in Fig. 19.15. It makes an angle θ_1 with the horizontal. The lines *PF* and *PF'* indicate the failure planes. The failure plane *PF* makes an angle α_1 with the horizontal.

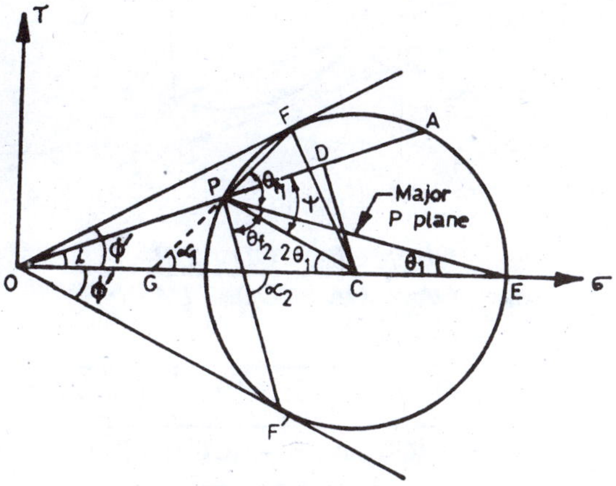

Fig. 19.15.

From the triangle *PEG*, the exterior angle, $\theta f_1 = \alpha_1 + \theta_1$

or
$$\alpha_1 = \theta f_1 - \theta_1$$

where θf_1 is the angle which the failure plane *PF* makes with the major principal plane, and is equal to $(45° + \phi'/2)$.

Thus
$$\alpha_1 = (45° + \phi'/2) - \theta_1$$

In the triangle *OPC*, the exterior angle ψ is given by

$$\psi = 2\theta_1 + i \quad \text{or} \quad \theta_1 = \frac{\psi - i}{2}$$

Therefore,
$$\alpha_1 = (45° + \phi'/2) - \left(\frac{\psi - i}{2}\right) \qquad \text{...(19.16)}$$

The angle ψ is obtained from the triangle *PCD* as

$$\sin \varphi = \frac{CD}{CP} = \frac{OC \sin i}{CF} = \frac{OC \sin i}{OC \sin \phi'} = \frac{\sin i}{\sin \phi'}$$

or
$$\psi = \sin^{-1}\left(\frac{\sin i}{\sin \phi'}\right) \qquad ...(19.17)$$

The other failure plane PF' makes an angle α_2 with the horizontal, which is given by

$$\alpha_2 = \theta_{f2} + \theta_1$$

or
$$\alpha_2 = (45° + \phi'/2) + (\psi - i)/2 \qquad ...(19.18)$$

(*b*) **Passive Earth Pressure.** This case is similar to the one for the active case with one basic difference that the vertical stress is the smaller of the two conjugate stresses. In Fig. 19.16, OA represents the vertical stress (σ_v). The point P shows the origin of planes, and OB represents the passive pressure, From the figure,

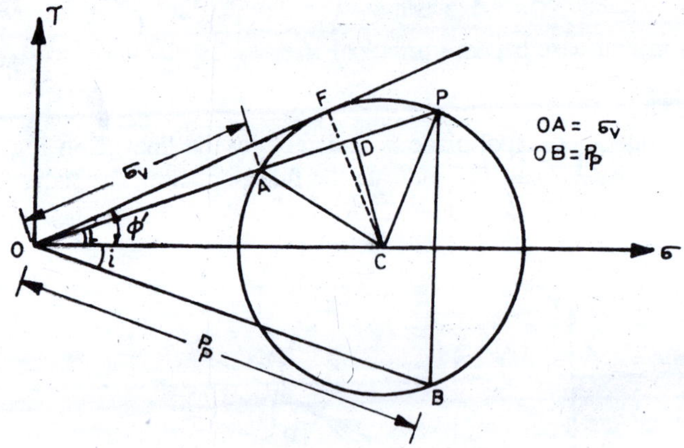

Fig. 19.16.

$$\frac{p_p}{\sigma_v} = \frac{OB}{OA} = \frac{OP}{OA} = \frac{OD + DP}{OD - DA}$$

Now
$$OD = OC \cos i$$

and
$$DP = DA = \sqrt{(AC)^2 - (DC)^2} = \sqrt{(FC)^2 - (DC)^2}$$

$$= \sqrt{(OC \sin \phi')^2 - (OC \sin i)^2}$$

$$= OC \sqrt{\sin^2 \phi' - \sin^2 i}$$

Therefore,
$$\frac{p_p}{\sigma_v} = \frac{OC \cos i + OC\sqrt{\sin^2 \phi' - \sin^2 i}}{OC \cos i - OC\sqrt{\sin^2 \phi' - \sin^2 i}}$$

$$= \frac{\cos i + \sqrt{\sin^2 \varphi' - \sin^2 i}}{\cos i - \sqrt{\sin^2 \varphi' - \sin^2 i}}$$

$$= \frac{\cos i + \sqrt{\cos^2 i - \cos^2 \varphi'}}{\cos i - \sqrt{\cos^2 i - \cos^2 \varphi'}}$$

or
$$p_p = (\gamma Z \cos i)\frac{\cos i + \sqrt{\cos^2 i - \cos^2 \varphi'}}{\cos i - \sqrt{\cos^2 i - \cos^2 \varphi'}}$$

or
$$p_p = K_p \gamma Z \qquad ...(19.19)$$

where K_p is given by

$$K_p = \cos i \frac{\cos i + \sqrt{\cos^2 i - \cos^2 \varphi'}}{\cos i - \sqrt{\cos^2 i - \cos^2 \varphi'}} \quad \text{...(19.20)}$$

For the special case, when $i = 0$

$$K_p = \frac{1 + \sin \phi'}{1 - \sin \phi'} \quad \text{(same as Eq. 19.12)}$$

Inclination of failure planes.

In Fig. 19.17, the line *PB* indicates the major principal plane, and the line *PE* shows the direction of the minor principal plane. The minor principal plane makes an angle θ_2 with the horizontal. From the triangle *PCO*,

$$2\theta_2 = \psi + i$$

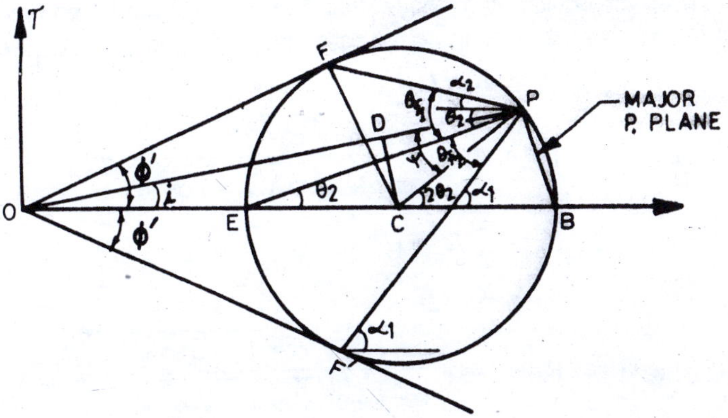

Fig. 19.17.

or

$$\theta_2 = \left(\frac{\psi + i}{2}\right)$$

The inclination of the failure plane *PF* with the horizontal is equal to α_2, obtained from the relation

$$\theta f_1 = \alpha_2 + \theta_2$$

where θf_1 is the angle which the failure plane *PF* makes with the minor principal plane

Thus

$$\alpha_2 = \theta f_1 - \theta_2$$

or

$$\alpha_2 = (45° - \phi'/2) - \left(\frac{\psi + i}{2}\right) \quad \text{...(19.21)}$$

The angle ψ is obtained from the triangle *PCD* as

$$\sin \psi = \frac{CD}{CP} = \frac{OC \sin i}{CF} = \frac{OC \sin i}{OC \sin \phi'} = \frac{\sin i}{\sin \phi'}$$

or

$$\psi = \sin^{-1}\left(\frac{\sin i}{\sin \phi'}\right) \quad \text{...(same as Eq. 19.17)}$$

The other failure plane *PF'* makes an angle α_1 with the horizontal, given by

$$\alpha_1 = \theta f_2 + \theta_2$$

or

$$\alpha_1 = (45° - \phi'/2) + \left(\frac{\psi + i}{2}\right) \quad \text{...(19.22)}$$

(c) Pressure Distribution. Fig. 19.18 shows the pressure distribution when the soil surface is inclined at

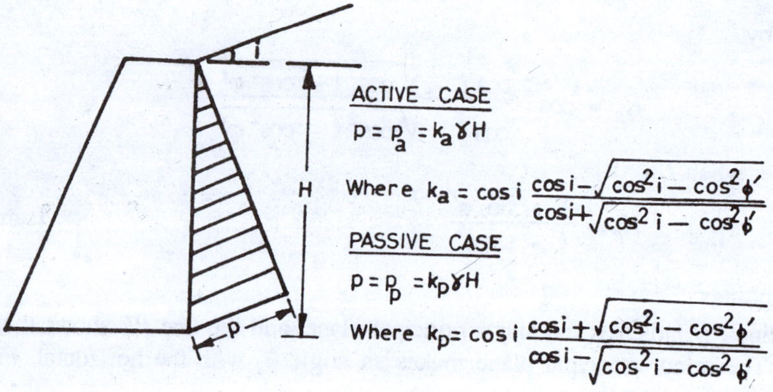

Fig. 19.18.

an angle i. The lateral pressure is parallel to the soil surface. The maximum pressure occurs at the bottom of the wall and is given by p, where $p = p_a$ in active case and $p = p_p$ in passive case.

From Eq. 19.14, $$p_a = K_a \gamma H$$

or $$p_a = \cos i \frac{\cos i - \sqrt{\cos^2 i - \cos^2 \phi'}}{\cos i + \sqrt{\cos^2 i - \cos^2 \phi'}} \gamma H$$

Total active pressure, $$P_a = \frac{1}{2} K_a \gamma H^2 \quad \text{...(19.23)}$$

Likewise, $$p_p = K_p \gamma H$$

or $$p_p = \cos i \frac{\cos i + \sqrt{\cos^2 i - \cos^2 \phi'}}{\cos i - \sqrt{\cos^2 i - \cos^2 \phi'}} \gamma H$$

Total passive pressure, $$P_p = \frac{1}{2} K_p \gamma H^2 \quad \text{...(19.24)}$$

For soils below the water table, the submerged weight γ' should by used instead of bulk unit weight γ.

(*d*) Inclined Back of wall. Sometimes, we come across a retaining wall with an inclined back [Fig. 19.19 (*a*)]. Rankine's theory can be used to determine the earth pressure with some modification. A vertical plane AC is taken through A and the active pressure P_a is determined on this vertical plane from Rankine's theory. The total pressure P_1 on the wall is the resultant of the pressure P_a and the weight W of the soil wedge ABC. Thus

$$P_1 = \sqrt{P_a^2 + W^2}$$

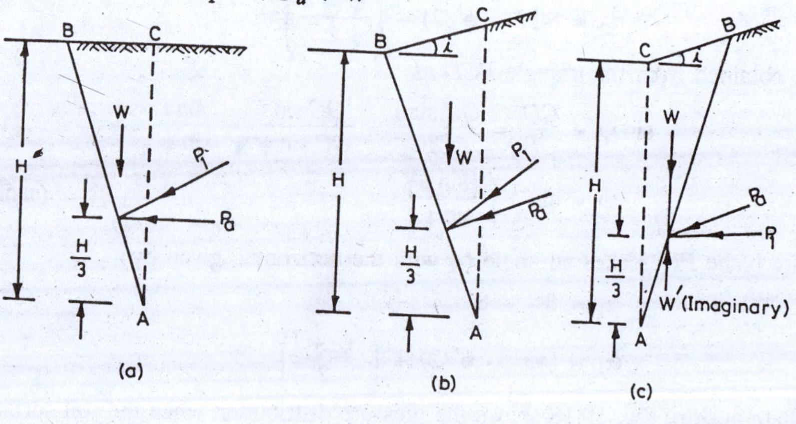

Fig. 19.19.

It acts at a height of $H/3$ from the bottom.

In case of the wall with an inclined back and the backfill with a surcharge angle i [Fig. 19.19 (*b*)], the active pressure P_a is inclined to the vertical plane AC at an angle i, and the total pressure P_1 is the resultant of the pressure P_a and W. Fig. 19.19 (*c*) shows the retaining wall with it back inclined towards the backfill. In this case, the vertical plane AC is on the other side of the back face, and, therefore, W is imaginary.

19.6. RANKINE'S EARTH PRESSURE IN COHESIVE SOILS

Rankine's original theory was for cohesionless soils. It was extended by Resal (1910) and Bell (1915) for cohesive soils. The treatment is similar to that for cohesionless soils with one basic difference that the failure envelope has a cohesion intercept c', whereas that for cohesionless soils is zero.

The following treatment is limited to the case when the backfill is horizontal.

(*a*) **Active Case.** Fig. 19.20 shows the Mohr circle in which point B indicates the vertical stress and point E shows the active pressure. The failure envelope is tangential to the circle. The relationship between p_a and σ_v can be obtained as under.

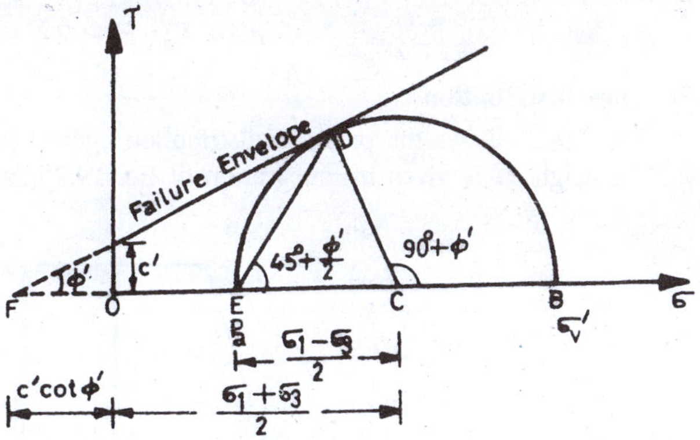

Fig. 19.20.

From triangle FCD, $$\sin\phi' = \frac{CD}{FC} = \frac{CD}{FO + OC}$$

or $$\sin\phi' = \frac{(\sigma_1 - \sigma_3)/2}{c'\cot\phi' + (\sigma_1 + \sigma_3)/2}$$

or $$\frac{(\sigma_1 - \sigma_3)}{2} = \frac{(\sigma_1 + \sigma_3)}{2}\sin\phi' + c'\cos\phi'$$

or $$\frac{\sigma_1}{2}(1 - \sin\phi') = \frac{\sigma_3}{2}(1 + \sin\phi') + c'\cos\phi'$$

or $$\sigma_3 = \frac{1 - \sin\phi'}{1 + \sin\phi'}\sigma_1 - \frac{2c'\cos\phi'}{1 + \sin\phi'} \qquad \ldots(a)$$

or $$\sigma_3 = \sigma_1\tan^2(45° - \phi'/2) - 2c'\tan(45° - \phi'/2)$$

As σ_3 is equal to the active pressure (p_a), and σ_1 is equal to the vertical stress σ_v (= γZ), Eq. (*a*) becomes

$$p_a = \left(\frac{1 - \sin\phi'}{1 + \sin\phi'}\right)\gamma Z - \frac{2c'\cos\phi'}{1 + \sin\phi}$$

or $$p_a = K_a\gamma Z - 2c'\sqrt{K_a} \qquad \ldots(19.25)$$

where $$K_a = \frac{1 - \sin\phi'}{1 + \sin\phi'} = \tan^2(45° - \phi'/2)$$

It can be shown that $$\frac{\cos\phi'}{1 + \sin\phi'} = \tan(45° - \phi'/2) = \sqrt{K_a}$$

Eq. 19.25 indicates that at $Z = 0$, the active pressure is given by

$$p_a = -2c'\sqrt{K_a} \qquad \ldots(19.26)$$

The negative sign shows that the pressure is negative, *i.e.* it tries to cause a pull on the wall. This tensile stress decreases with an increase in depth, and it becomes zero at a depth Z_c, given by

$$0 = K_a \gamma Z_c - 2c' \sqrt{K_a}$$

or
$$Z_c = \frac{2c'}{\gamma \sqrt{K_a}} \qquad \text{...(19.27)}$$

The depth Z_c is known as the depth of tensile crack. The tensile stress eventually causes a crack to from along the soil-wall interface.

The pressure at the depth H is given by,

$$p_a = K_a \gamma H - 2c' \sqrt{K_a}$$

Pressure Distribution

Fig. 19.21 shows the pressure distribution against the retaining wall. The total pressure on the retaining wall of height H is given by integration of Eq. 19.25 as

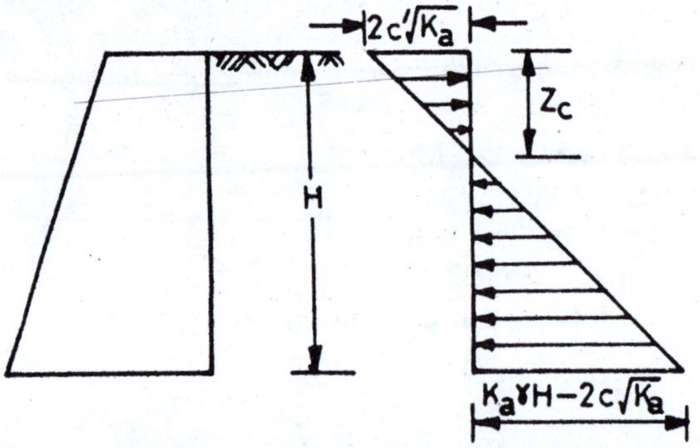

Fig. 19.21.

$$p_a = \int_0^H \left(K_a \gamma Z - 2c' \sqrt{K_a}\right) dZ$$

or
$$P_a = K_a \frac{\gamma H^2}{2} - 2c' \sqrt{K_a}\, H \qquad \text{...(19.28)}$$

Eq. 19.28 is applicable before the formation of crack

After the occurrence of the tensile crack, the force on the wall is caused only by the pressure from $Z = Z_c$ to $Z = H$, *i.e.* tensile stresses are neglected.

Thus
$$P_a = \frac{1}{2}(H - Z_c)(K_a \gamma H - 2c' \sqrt{K_a})$$

Substituting the value of Z_c from Eq. 19.27,

$$P_a = \frac{1}{2}\left(H - \frac{2c'}{\gamma \sqrt{K_a}}\right)\left(\gamma H K_a - 2c' \sqrt{K_a}\right)$$

or
$$P_a = \frac{1}{2}\gamma H^2 K_a - 2c' H \sqrt{K_a} + 2(c')^2/\gamma \qquad \text{...(19.29)}$$

It acts at a height of $(H - Z_c)/3$.

For $\phi = 0$, Eq. 19.29 reduces to

$$P_a = \frac{1}{2}\gamma H^2 - 2c' H + 2(c')^2/\gamma \qquad \text{...(19.30)}$$

For soils below the water table, the submerged unit weight γ' and the corresponding value of ϕ' and c' should be used.

Critical height of unsupported vertical cut.

As shown in Fig. 19.21, the pressure is negative in the top region. It becomes zero at a depth Z_c. If the wall has a height of $2Z_c$, the total earth pressure is zero. This height is known as the *critical height* (H_c).

$$H_c = 2\,Z_c \qquad \text{...(19.31)}$$

If the height of an unsupported vertical cut is smaller than H_c, it should be able to stand. However, the conditions in unsupported vertical cut are different from those near a retaining wall. In the vertical cut, the lateral stress is everywhere zero, whereas in the retaining wall, it varies from $-2c'\sqrt{K_a}$ to $+\;2c\sqrt{K_a}$. Because of this difference in the stress condition, the safe height of the vertical cut is slightly smaller than that given by Eq. 19.31.

Substituting the value of Z_c from Eq. 19.27,

$$H_c = 2 \times 2\,c'/(\gamma\,\sqrt{K_a}) = 4\,c'/(\gamma\sqrt{K_a}) \qquad \text{...(19.32)}$$

For $\phi' = 0$,

$$H_c = 4\,c'/\gamma \qquad \text{...(19.33)}$$

(b) Passive Case

An expression for the passive pressure in a cohesive soil can be determined by referring to Fig. 19.22. The Mohr circle has been drawn such that the point B represents σ_v and point E as p_p. In triangle FCD,

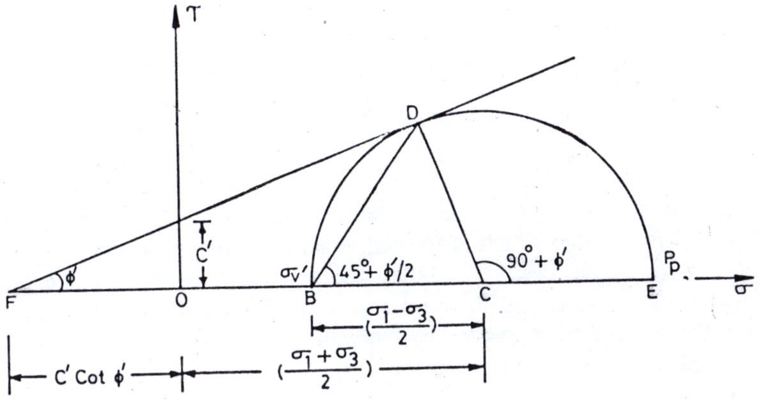

Fig. 19.22.

$$\sin\phi' = \frac{CD}{CF} = \frac{(\sigma_1 - \sigma_3)/2}{c'\cot\phi + (\sigma_1 + \sigma_3)/2}$$

or
$$(\sigma_1 - \sigma_3) = (\sigma_1 + \sigma_3)\sin\phi' + 2\,c'\cos\phi'$$

or
$$\sigma_1\,(1 - \sin\phi') = \sigma_3\,(1 + \sin\phi') + 2\,c'\cos\phi'$$

or
$$\sigma_1 = \sigma_3\,\frac{(1 + \sin\phi')}{(1 - \sin\phi')} + \frac{2\,c'\cos\phi'}{1 - \sin\phi'}$$

or
$$p_p = \sigma_v \tan^2(45° + \phi'/2) + 2c'\,\tan(45° + \phi'/2)$$

or
$$p_p = \gamma\,Z\,K_p + 2\,c'\,\sqrt{K_p} \qquad \text{...(19.34)}$$

where K_p = coefficient of passive pressure, given by

$$K_p = \tan^2(45° + \phi'/2) = \frac{1 + \sin\phi'}{1 - \sin\phi'} \qquad \text{...(19.35)}$$

The failure plane makes an angle of $(45° + \phi'/2)$ with the vertical (major principal plane) and of (45° $\phi'/2$) with horizontal (minor principal plane).

Pressure distribution

Fig. 19.23 shows the pressure distribution obtained from Eq. 19.34.

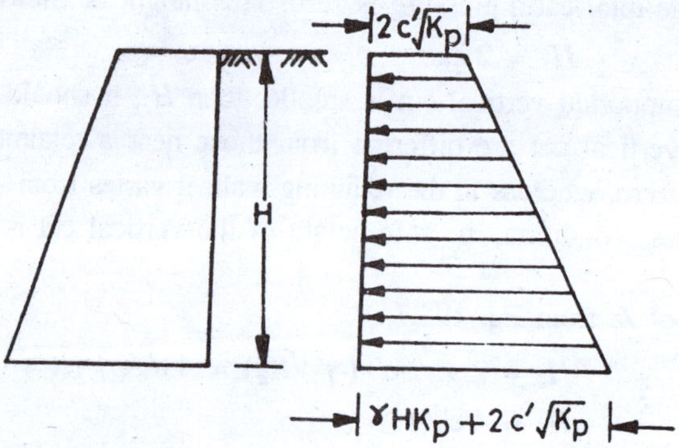

Fig. 19.23.

When $Z = 0$, $p_p = 2\,c'\,\sqrt{K_p}$

When $Z = H$, $p_p = \gamma H K_p + 2\,c'\,\sqrt{K_p}$ ...(19.36)

The pressure, unlike active case, is positive throughout the depth. The total pressure on the retaining wall of height H is given by

$$P_p = \frac{1}{2} H (\gamma H K_p) + 2\,c' \sqrt{K_p} \times H$$

or

$$P_p = K_p\, \gamma H^2/2 + 2\,c'\, H \sqrt{K_p} \quad \text{...(19.37)}$$

19.7. COULOMB'S WEDGE THEORY

Coulomb (1776) developed a method for the determination of the earth pressure in which he considered the equilibrium of the sliding wedge which is formed when the movement of the retaining wall takes place. As discussed before, in the active case, the sliding wedge moves downward and outward relative to the backfill, whereas in the passive case, the sliding wedge moves upwards. The lateral pressure on the wall is equal and opposite to the reactive force exerted by the wall in order to keep the sliding wedge in equilibrium. The analysis is a type of limiting equilibrium method:

The following assumptions are made:

(1) The backfill is dry, cohesionless, homogeneous, isotropic and ideally plastic material.

(2) The slip surface is a plane surface which passes through the heel of the wall.

(3) The wall surface is rough. The resultant earth pressure on the wall is inclined at an angle δ to the normal to the wall, where δ is the angle of the friction between the wall and the backfill.

(4) The sliding wedge itself acts as a rigid body.

The magnitude of earth pressure is obtained by considering the equilibrium of the sliding wedge as a whole.

In Coulomb's theory, a plane failure surface is assumed and the lateral force required to maintain the equilibrium of the wedge is found using the principles of statics. The procedure is repeated for several trial surfaces. The trial surface which gives the *largest* force for the active case, and the *smallest* force for the passive case, is the actual failure surface. The method readily accommodates the friction between the wall and the backfill, irregular backfill, sloping wall, surcharge loads, etc. Although the initial theory was for dry, cohesionless soils, it has now been extended to wet soils and cohesive soils as well. Thus Coulomb's theory is more general than the Rankine theory.

19.8. COULOMB'S ACTIVE PRESSURE IN COHESIONLESS SOILS

Fig. 19.24 (*a*) shows a retaining wall with an inclined back face and a sloping dry granular backfill. In active case, the sliding wedge *ABD* moves downward, and the reaction *R* acts upward and inclined at an angle ϕ' with the normal.

The sliding wedge *ABD* is in equilibrium under three forces:

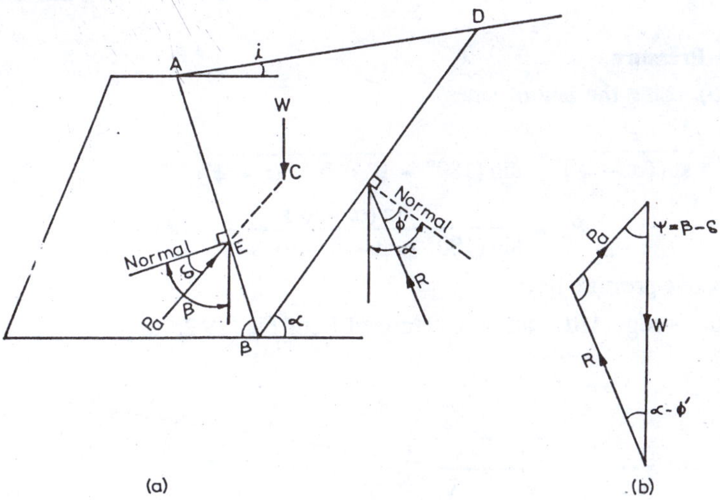

Fig. 19.24.

(1) Weight of the wedge (W).

(2) Reaction R on the slip surface BD.

(3) Reaction P_a from the wall.

It may be noted that, at failure conditions, the shearing resistance on the failure surface is fully mobilised and the reaction R has the maximum obliquity. It is therefore inclined at an angle ϕ' to the normal to the failure plane. Further, P_a acts upward, and it is inclined at an angle δ with the normal as shown in Fig. 19.24 (*a*).

Fig. 19.24 (*b*) shows the force triangle. As the magnitude of one force (*viz.* weight W) and the directions of all three forces are known, the force triangle can be completed. The magnitude of P_a is determined from the force triangle. The pressure acting on the wall is equal and opposite to P_a.

The procedure is repeated after assuming an other failure surface. The surface that gives the maximum value of P_a is the critical failure plane, and the corresponding force is the active force.

Coulomb's method does not give the point of application of the resultant earth pressure (P_a). The point of application is found to be approximately at the point of intersection E of the back of the retaining wall with a line CE drawn from the centroid C of the failure wedge and parallel to the failure surface. As this procedure is cumbersome, for convenience, the pressure distribution is sometimes assumed to be hydrostatic on the back of the wall, and the resultant pressure P_a is assumed to act at one-third the height of the wall from the base.

The following points should be carefully noted while using Coulomb's theory :

(1) For most practical cases, the backfill moves down relative to the wall in the active case, and, therefore, the active force P_a is inclined at angle δ below the normal as shown in Fig. 19.24 (*a*). However, if the wall is supported on a soft, compressible soil, it may settle to such an extent that the movement of the wall will be downward relative to the backfill and the relative movement of the wedge will be upward. In such a case, the force P_a would be inclined at an angle δ above the normal to the wall.

(2) The angle δ is the friction angle between the soil and the wall. It may be determined by means of a direct shear test. For concrete walls, δ is generally taken as $2/3\phi'$. The value of δ cannot exceed ϕ', because in that case the failure will occur in soil.

If the friction angle δ is zero, and the wall is vertical and the ground surface is horizontal, the Coulomb method gives identical results with the Rankine method.

(3) Coulomb's theory assumes the failure surface to be a plane. The actual failure surface is slightly curved. Fortunately, for the active case, the error is small, and therefore the failure surface may be assumed to be plane without any significant error.

Expression for Active Pressure

From Fig. 19.24 (*b*), using the law of sines,

$$\frac{P_a}{\sin(\alpha - \phi')} = \frac{W}{\sin(180° - \beta + \delta - \alpha + \phi')}$$

or $$P_a = \frac{W \sin(\alpha - \phi')}{\sin(180° - \beta + \delta - \alpha + \phi')} \quad ...(a)$$

where P_a = total active pressure force.

The weight W of the wedge ABD can be determined from Fig. 19.25 as

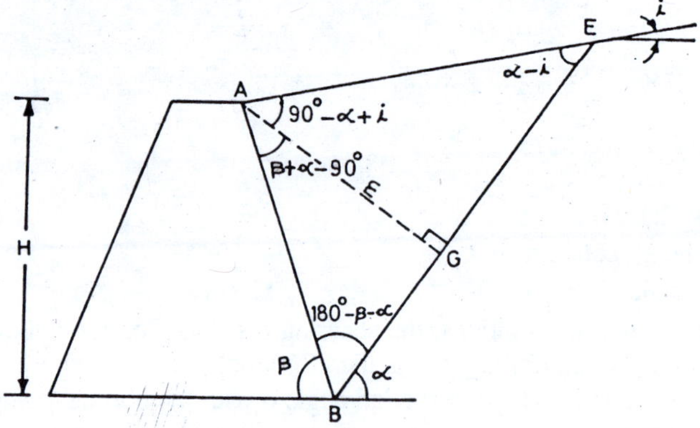

Fig. 19.25.

$$W = \frac{1}{2} \times BE \times AG \times \gamma$$

Taking $AG = m$, and $BE = L$, we have $W = \frac{1}{2} L \times m \times \gamma$

or $$W = \frac{1}{2}(m \times \gamma)(BG + GE) \quad ...(b)$$

Now $AB = H/\sin\beta$.

From triangle ABG, $$\frac{m}{AB} = \sin[180° - (\beta + \alpha)] = \sin(\beta + \alpha)$$

Therefore, $$m = AB\sin(\beta + \alpha) = \frac{H\sin(\beta + \alpha)}{\sin\beta} \quad ...(c)$$

From triangle ABG, $$\frac{BG}{\sin(\beta + \alpha - 90°)} = \frac{m}{\sin(\beta + \alpha)}$$

or $$BG = m\sin(\beta + \alpha - 90°)/\sin(\beta + \alpha) \quad ...(d)$$

From triangle AGE, $$\frac{GE}{\sin(90° - \alpha + i)} = \frac{m}{\sin(\alpha - i)}$$

or $$GE = m\frac{\sin(90° - \alpha + i)}{\sin(\alpha - i)} \quad ...(e)$$

Substituting the values of m, BG and GE from Eqs. (*c*), (*d*) and (*e*) in Eq. (*b*),

$$W = \frac{1}{2} \cdot \frac{H\sin(\beta + \alpha)}{\sin\beta} \cdot \gamma \times \left[m\frac{\sin(\beta + \alpha - 90°)}{\sin(\beta + \alpha)} + m\frac{\sin(90° - \alpha + i)}{\sin(\alpha - i)} \right]$$

or $$W = \frac{1}{2}\,\frac{H^2 \gamma \sin(\beta + \alpha)^2}{\sin^2 \beta}\left[\frac{\sin(\beta + \alpha - 90°)}{\sin(\beta + \alpha)} + \frac{\sin(90° - \alpha + i)}{\sin(\alpha - i)}\right]$$

Substituting the above value of W in Eq. (*a*),

$$P_a = \frac{1}{2}\,\frac{H^2 \gamma \sin(\beta + \alpha)^2}{\sin^2 \beta} \times \frac{\sin(\alpha - \phi')}{\sin(180° - \beta + \delta - \alpha + \phi')}$$

$$\times \left[\frac{\sin(\beta + \alpha - 90°)}{\sin(\beta + \alpha)} + \frac{\sin(90° - \alpha + i)}{\sin(\alpha - i)}\right]$$

The active pressure force P_a will be a maximum when the failure plane makes an angle α with the horizontal such that

$$\frac{\partial P_a}{\partial \alpha} = 0$$

Muller-Breslau gave the solution, which is quite complicated. The maximum value of P_a thus obtained is Coulomb's active force, given by

$$P_a = \frac{1}{2} K_a \gamma H^2 \qquad \text{...(19.38)}$$

where K_a = Coluomb's active earth pressure coefficient, given by

$$K_a = \frac{\sin^2(\beta + \phi')}{\sin^2 \beta \sin(\beta - \delta)\left[1 + \sqrt{\frac{\sin(\phi' + \delta)\sin(\phi' - i)}{\sin(\beta - \delta)\sin(\beta + i)}}\right]^2} \qquad \text{...(19.39)}$$

The line of action of P_a will be at a height of $H/3$ above the base of the wall and it will be inclined at an angle δ to the normal drawn to the back of the wall.

19.9. REHBANN'S CONSTRUCTION FOR ACTIVE PRESSURE

Rehbann (1871) gave a graphical method for the determination of the total active pressure according to Coulomb's theory. It is based on Poncelet's solution (1840), and is, therefore, also known as Poncelet's method.

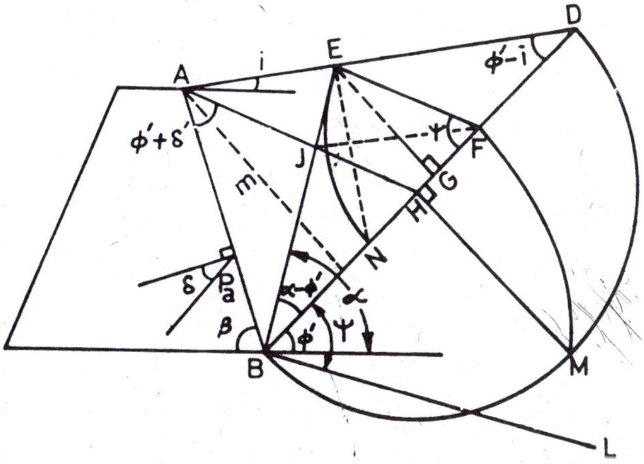

Fig. 19.26.

Fig. 19.26 shows the graphical construction. The line *BD* is drawn at an angle ϕ' to the horizontal. The line *BL*, drawn at an angle ψ with the line *BD*, is known as the *earth pressure line.* The angle ψ is equal to $(\beta - \delta)$.

A semi-circle *BMD* is drawn on *BD* as diameter. The line *AH* is drawn parallel to *BL*, intersecting the line *BD* at *H*. A perpendicular *HM* is drawn at *H*, intersecting the semi-circle at *M*.

With *B* as centre and *BM* as radius, an arc *MF* is drawn, intersecting *BD* at *F*. The line *FE* is drawn parallel to *BL*, intersecting the ground surface at *E*.

With *F* as centre and *FE* as radius, an arc is drawn to intersect *BD* at N. The line *BE* represents the critical failure plane.

The total active pressure P_a is given by

$$P_a = \gamma \text{ (area of triangle } NEF)$$

or

$$P_a = \gamma\,(1/2 \times NF \times x) \qquad \text{...(19.40)}$$

where x is the perpendicular distance *EG* between *E* and *BD*.

Proof. The proof of Eq. 19.40 is as under:

The triangle *BEF* and the force triangle in Fig. 19.24 (*b*) are similar. Therefore,

$$\frac{P_a}{W} = \frac{EF}{BF} \quad \text{or} \quad P_a = W\left(\frac{EF}{BF}\right) \qquad \text{...}(a)$$

From triangle *EFG*,

$$\frac{x}{EF} = \sin\psi, \quad \text{or } EF = x \operatorname{cosec}\psi \qquad \text{...}(b)$$

Now

$$BF = BD - FD = BD - (GD - GF) = BD - DG + GF$$

or

$$BF = BD - x\cot(\phi' - i) + x\cot\psi$$

or

$$BF = L - x\cot(\phi' - i) + x\cot\psi \qquad \text{...}\{c\}$$

where L = Length of *BD*.

Now $W = \gamma$ (volume of triangle *ABE*)

Taking unit length of the wall,

$$W = \gamma \text{ (area of triangle } ABE)$$

$$= \gamma \text{ [area of triangle } ABD - \text{area of triangle } BED]$$

or

$$W = \gamma\left[\frac{1}{2} \times L \times m - \frac{1}{2} \times L \times x\right] = \frac{1}{2}\gamma L (m - x) \qquad \text{...}(d)$$

where m is the perpendicular distance from *A* to *BD*.

From Eqs. (*a*), (*b*), (*c*) and (*d*),

$$P_a = \frac{1}{2}\gamma L(m - x) \times \frac{x \operatorname{cosec}\psi}{[L - x\cot(\phi' - i) + x\cot\psi]} \qquad \text{...}(e)$$

In Eq. (*e*), substituting

$$\operatorname{cosec}\psi = c$$

and

$$\cot(\phi' - i) - \cot\psi = d, \qquad \text{...}(f)$$

$$P_a = \frac{1}{2}\gamma L\,(m - x)\,\frac{cx}{(L - dx)}$$

For maximum value of P_a, $\dfrac{\partial P_a}{\partial x} = 0$

or

$$(L - dx)(mc - 2cx) - (m - x)\,cx\,(-d) = 0$$

$$\text{or } mL - 2xL + dx^2 = 0$$

or

$$mL - xL = xL - dx^2$$

or

$$\frac{L}{2}(m - x) = \frac{x}{2}(L - dx)$$

Substituting the value of d from Eq. (f),

$$\frac{L}{2}(m - x) = \frac{x}{2}[L - x \cot(\phi' - i) + x \cot \psi]$$

Using Eq. (c), $$\frac{mL}{2} - \frac{xL}{2} = \frac{x}{2} \times BF$$

or Area of triangle ABD — Area of triangle BED = Area of triangle BEF

or Area of triangle ABE = Area of triangle BEF

In other words, for the maximum value of P_a, the failure plane BE should be such that the triangles ABE and BEF have equal areas.

Let us examine whether Rehbann's construction satisfies the above criterion. From the properties of the circle,

$$BH \times HD = (HM)^2$$

Adding $(BH)^2$ to both sides,

$$(BH)^2 + BH \times HD = (HM)^2 + (BH)^2$$

$$BH(BH + HD) = (BM)^2$$

or $$BH \times BD = (BF)^2$$

or $$BH/BF = BF/BD \quad \text{...}(i)$$

As AH is parallel to EF, triangles BHJ and BFE are similar.

Therefore, $$BH/BF = BJ/BE \quad \text{...}(ii)$$

From Eqs. (i) and (ii), $$BF/BD = BJ/BE$$

i.e. JF is parallel to AD.

Thus, the figure $AJFE$ becomes a parallelogram.

Therefore, the perpendicular from point A on the diagonal JE and the perpendicular from point F on the same diagonal would be equal in length. Consequently, the areas of triangles ABE and BFE which have the same base BE would be equal. It proves that Rehbann's construction satisfies the required criterion.

Expression for total pressure from Rehbann's construction

From Eq. (a), $$P_a = W \times (EF/BF)$$

$$= \gamma\ (\text{area of triangle } ABE) \times EF/BF$$

or $$P_a = \gamma\ (\text{area of triangle } BFE) \times EF/BF$$

$$= \gamma \times \frac{1}{2} \times BF \times x \times EF/BF$$

or $$P_a = \frac{1}{2} \times \gamma \times EF \times x \quad \text{...(19.41)}$$

or $$P_a = \frac{1}{2} \times \gamma \times NF \times x \quad \text{(same as Eq. 19.40)}$$

Eq. 19.41 can be written in terms of various angles.

From similar triangles EFD and AHD,

$$\frac{EF}{ED} = \frac{AH}{AD}$$

or $$(EF)^2 = \left(\frac{AH \times ED}{AD}\right)^2 \quad \text{...}(iii)$$

From triangle HAB, $$\frac{AH}{\sin(180° - \beta - \phi')} = \frac{AB}{\sin \psi}$$

or
$$AH = AB\,\frac{\sin(\beta + \phi')}{\sin\psi}$$

Also
$$\frac{BH}{AB} = \frac{\sin(\phi' + \delta)}{\sin\psi}$$

or
$$BH = AB\,\frac{\sin(\phi' + \delta)}{\sin\psi}$$

From triangle ABD,
$$\frac{BD}{AB} = \frac{\sin(\beta + i)}{\sin(\phi' - i)}$$

or
$$BD = AB\,\frac{\sin(\beta + i)}{\sin(\phi' - i)}$$

The ratio AD/ED can be written as

$$\frac{AD}{ED} = \frac{AE + ED}{ED} = \frac{AE}{ED} + 1 = \frac{JF}{ED} + 1$$

or
$$\frac{AD}{ED} = \frac{BF}{BD} + 1$$

From Rehbann's construction, $BH \times BD = BF^2$

or
$$BF = \sqrt{BH \times BD}$$

or
$$\frac{BF}{BD} = \sqrt{BH/BD}$$

Therefore,
$$\frac{AD}{ED} = \sqrt{BH/BD} + 1$$

From Eq. (*iii*),
$$EF^2 = (AH)^2 \times (ED/AD)^2$$

or
$$EF^2 = AH^2 \times \frac{1}{(\sqrt{BH/BD} + 1)^2}$$

Substituting the value of AH, BH and BD,

$$EF^2 = AB^2 \left[\frac{\sin(\beta + \phi')}{\sin\psi}\right]^2 \times \frac{1}{\left[\sqrt{\dfrac{\sin(\phi' + \delta)}{\sin\psi} \cdot \dfrac{\sin(\phi' - i)}{\sin(\beta + i)}} + 1\right]^2}$$

From Eq. 19.41, substituting $AB = H \operatorname{cosec}\beta$ and $x = EF \sin\psi$,

$$P_a = \frac{1}{2}\gamma\,(H \operatorname{cosec}\beta)^2 \sin\psi \times \left[\frac{\sin(\beta + \phi')}{\sin\psi}\right]^2 \times \frac{1}{\left[\sqrt{\dfrac{\sin(\phi' + \delta)\sin(\phi' - i)}{\sin\psi \sin(\beta + i)}} + 1\right]^2}$$

or
$$P_a = \frac{1}{2} K_a\,\gamma\,H^2 \qquad \text{...(same as Eq. 19.38)}$$

where
$$K_a = \frac{\operatorname{cosec}^2\beta \sin^2(\beta + \phi')}{\sin\psi \left[\sqrt{\dfrac{\sin(\phi' + \delta)\sin(\phi' - i)}{\sin\psi \sin(\beta + i)}} + 1\right]^2}$$

As $\psi = \beta - \delta$, $$K_a = \frac{\sin^2(\beta + \phi')}{\sin^2\beta \sin(\beta - \delta)\left[1 + \sqrt{\dfrac{\sin(\phi' + \delta)\sin(\phi' - i)}{\sin(\beta - \delta)\sin(\beta + i)}}\right]^2}$$ ...(same as Eq. 19.39)

19.10. CULMANN'S CONSTRUCTION FOR ACTIVE PRESSURE

Rehbann's construction becomes inconvenient when the slope angle i approaches the angle ϕ'. Culmann (1866) developed a method which is more general than Rehbann's method. It can be used to determine Coulomb's earth pressure for ground surface of any configuration, for various types of surcharge loads and for layered back fills. Culmann's construction is, in fact, the method of construction of the force triangle in a rotated orientation. The procedure consists of following steps:

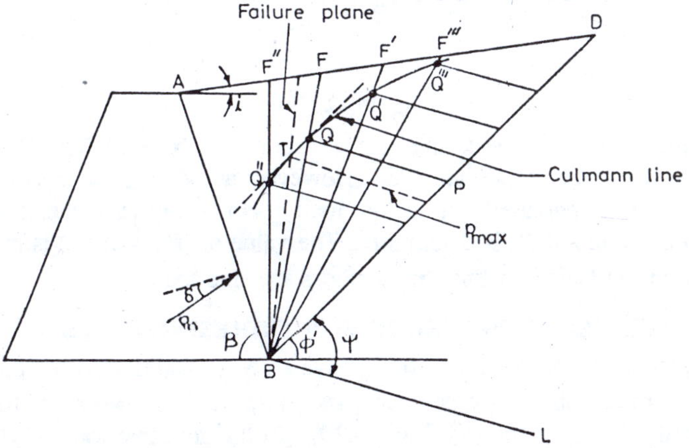

Fig. 19.27

(1) From point B (Fig. 19.27), a line BD is drawn at an angle ϕ' to the horizontal. As the weight of the wedge is plotted along this line, it is also known as the *weight line*.

(2) A line BL is drawn at an angle ψ with the line BD, such that $\psi = \beta - \delta$, where β is the angle which the back face makes with the horizontal and δ is the angle of friction.

(3) A failure surface BF is assumed, and the weight (W) of the failure wedge ABF is computed.

(4) The weight (W) of the wedge is plotted along BD such that $BP = W$.

(5) A line PQ is drawn from point P parallel to BL to intersect the failure surface BF at Q.

(6) The length PQ represents the magnitude of P_a required to maintain equilibrium for the assumed failure plane.

(7) Several other failure planes BF'', BF', BF''', etc. are assumed and the procedure repeated. Thus the points Q'', Q', Q''', etc. are located.

(8) A smooth curve is drawn joining the points Q'', Q, Q', Q''', etc. The curve is known as *Culmann's line.*

(9) A line (shown dotted) is drawn tangential to the Culmann line and parallel to BD. Point T is the point of tangency.

(10) The magnitude of the largest value (P_{max}) of P_a is measured from the tangent point T to the line BD and parallel to BL. It is equal to Coulomb's active pressure (P_a).

(11) The actual failure plane passes through the point T (shown dotted).

Effect of Uniform Surcharge or Line load

Culmann's method can be easily extended to include the effect of uniform surcharge or the line load applied to the backfill. Each such load that falls within the assumed failure wedge is included by adding it to the weight of the failure wedge.

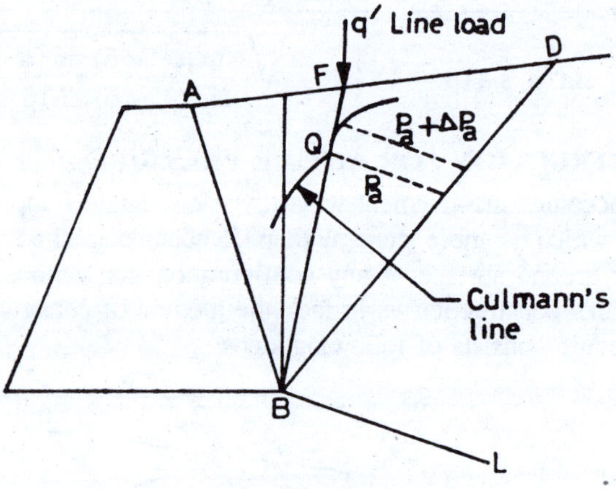

Fig. 19.28.

Fig. 19.28 shows the modification when a line load q' acts on the backfill. The Culmann line upto point Q is similar to the case when there is no line load. However, at point Q, it becomes discontinuous, as suddenly the load of the wedge is increased due to line load q'. In this case, tangents are drawn to the two segments of the curve, and the value of P_a is determined. The value of $(P_a + \Delta P_a)$ is measured from the line BD to the maximum point on the Culmann line that includes the line load.

19.11. COULOMB'S ACTIVE EARTH PRESSURE FOR COHESIVE SOILS

The Coulomb wedge theory can be extended to cohesive soils. In addition to the three forces (W, P_a and R) considered for the cohesionless soils in the preceding sections, there are two more forces, namely, (1) the cohesive force C acting on the failure plane BE (Fig. 19.29), (2) the adhesive force C_a acting on the back of

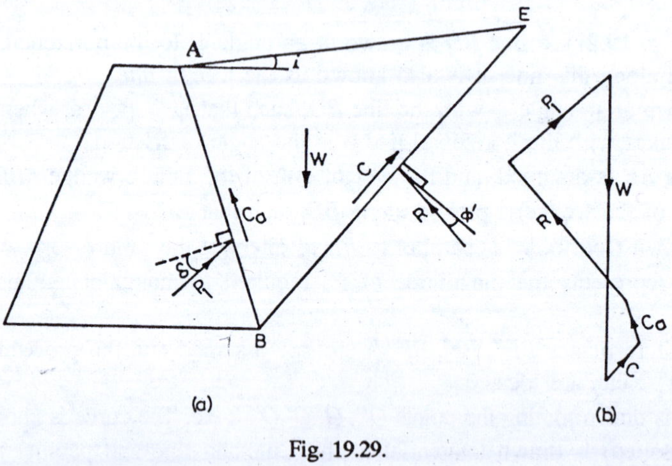

Fig. 19.29.

the wall. In all there are 5 forces which keep the wedge in equilibrium. The resultant pressure P_1 on the back of the wall acts at an angle δ to the normal.

The force polygon for the 5 forces is shown in Fig. 19.29 (*b*). The magnitude of W is determined from the weight of the wedge ABE. The magnitudes of C and C_a are respectively $c \times BE$ and $c_a \times AB$ where c and c_a are unit cohesion and unit adhesion. Knowing the directions of all the five forces and the magnitudes of 3 forces, the force polygon is completed as shown.

The total pressure P_a on the wall is the vector sum of P_1 and C_a. By analysing several trial wedges, the maximum value of P_a can be determined.

The resultant pressure P_a is assumed to act on the back of the wall at the point of intersection of a line drawn parallel to the critical surface and passing through the centre of gravity of the wedge, with the back of the wall.

For convenience, sometimes it is taken at a point at one-third the height of the wall from its base.

Note. (1) Sometimes, the resultant pressure on the wall is taken as maximum value of P_1, without combining it with C_a.

(2) In cohesive soils, tension cracks develop to a depth of Z_c. It is generally assumed that upto depth of tension crack, no cohesion, adhesion or friction acts.

19.12. TRIAL WEDGE METHOD

The trial wedge method assumes that the rupture surface is plane. The trial wedge method is a general method which can be used to compute active pressure and passive pressure for both Rankine's condition and Coulomb's condition. The method can also be applied to the cases when the soils are cohesive. The trial wedge method for computing Coulomb's *active pressure* for cohesionless soils is discussed below.

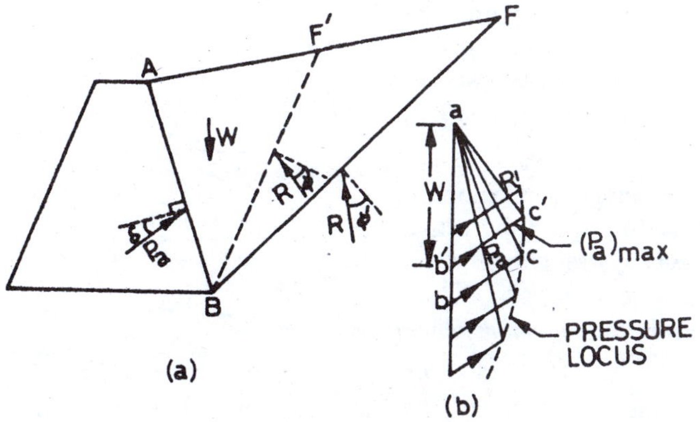

Fig. 19.30.

In Fig. 19.30, the line *BF* shows an assumed failure plane. The wedge *ABF* is held in equilibrium by the reaction *R* acting on the rupture plane *BF* and the force P_a. The force P_a acts at obliquity δ below the normal to oppose the movement of the wedge. The reaction *R* acts with obliquity ϕ' below the normal.

The force triangle *abc* is shown in Fig. 19.30 (*b*). The value of P_a is determined from the force triangle as the distance *bc*.

Another trial surface, such as *BF'* is assumed and the force triangle a *b' c'* is drawn. A common vertical line for the weights is taken for all the force triangles. The value of P_a is again determined.

The procedure is repeated, taking several trial planes, and the corresponding force triangles are drawn and the values of P_a determined in each case.

A curve, called *pressure locus* (shown dotted) is drawn through the points of intersections *c*, *c'*, etc. of P_a and the corresponding *R*. The maximum pressure vector $(P_a)_{max}$ gives the magnitude of the Coulomb active pressure. The failure plane corresponding to this vector is the actual failure plane. However, it is difficult to located the failure plane precisely.

The above procedure is general. Various modifications can be made as under:

(1) If the active pressure corresponding to Rankine's condition is required, the plane *AB* on which the pressure is computed is vertical. The resultant pressure on this plane is parallel to the ground surface. The rest of the procedure remains unchanged.

(2) If the ground surface is irregular, the trial wedge method can still be used. Of course, there would be some difficulty in calculating the weight *W* of the wedge.

(3) The method can also be used to determine the active pressure against the back of the wall if the back fill carries a surcharge load distributed over the ground surface or a line load acting on the fill. The weight of the wedge would include all such loads.

(4) If the soil is cohesive, the force polygon would also include the cohesive force C on the failure plane and the adhesive force C_a on the wall.

It may be noted that the trial wedge method does not give the point of application of P_a. The pressure variation is assumed to be triangular and it is assumed that P_a acts at one-third height. However, in irregular ground surface, the pressure variation is not triangular and it would result in some error.

19.13. COULOMB'S PASSIVE EARTH PRESSURE FOR COHESIONLESS SOILS

The failure surface in Coulomb's passive state is assumed to be a plane. Fig. 19.31 shows the case when the passive conditions develop. In this case, the failure wedge moves upwards. The directions of R and P_p which oppose the movement are also shown. The reaction R acts at ϕ' to the normal in the downward direction and the reaction pressure P_p acts at an angle δ to the normal in the downward direction.

The procedure for computing Coulomb's passive pressure is similar to one for the active case (Sect. 19.8) However, there is one basic difference. In this case, the critical failure surface is that which gives the *minimum* value of P_p.

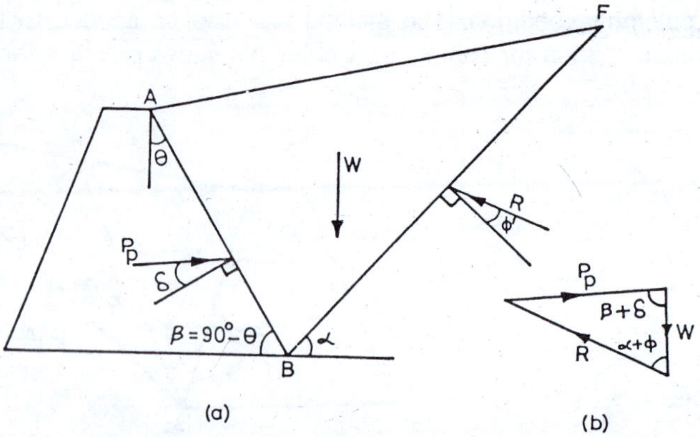

Fig. 19.31.

Fig. 19.31 (*b*) shows the force triangle. The value of P_p is determined from this triangle. The procedure is repeated after assuming a new trial failure surface. The minimum value of P_p is the Coulomb passive pressure. Using the procedure similar to that for the active case, it can be shown that the passive pressure is given by

$$P_p = \frac{1}{2}\gamma H^2 K_p \qquad \text{...(19.42)}$$

where
$$K_p = \frac{\sin^2(\beta - \phi')}{\sin^2\beta \,.\, \sin(\beta + \delta)\left[1 - \sqrt{\dfrac{\sin(\phi' + \delta)\sin(\phi' + i)}{\sin(\beta + \delta)\sin(\beta + i)}}\right]^2} \qquad \text{...[19.42(a)]}$$

The resultant passive pressure P_p acts at a height of $H/3$ measured from the bottom of the wall. It would be inclined at an angle δ to the normal, as shown in Fig. 19.31. However, when the retaining wall moves up relative to the soil, the friction angle δ is measured below the normal and δ is said to be negative. The negative wall friction produces a value of passive pressure lower than that for the usual positive wall friction.

It is worth noting that the wall friction decreases the active pressure, but it increases the passive pressure. Moreover, the wall friction has a greater influence on the passive pressure than on the active pressure. When δ exceeds ⅓ ϕ', Coulomb's assumption of plane failure surface is not justified in the passive case. It gives much greater value of P_p compared to that obtained for the actual curved surface. As the passive pressure is generally required to provide the stability to a retaining wall subjected to the active pressure on the other side, the higher value of P_p obtained from the plane failure surface in unconservative or unsafe. For such cases, the failure surface should be taken as a logarithmic spiral or a circular arc as explained in Sect. 19.14.

Rehbann's construction. Rehbann's construction can be used for the determination of the passive pressure. However, in this case, the ϕ-line is drawn at an angle $-\phi'$ (*i.e.* below horizontal) to intersect the extension of ground surface at point *D*.

Culmann method. The Culmann method and trial wedge method may also be used. From a series of several force triangles corresponding to the various trial surfaces, a pressure locus or a Culmann line for passive pressure is obtained. The Coulomb passive pressure is the minimum value of pressure so obtained.

19.14. PASSIVE PRESSURE BY THE FRICTION CIRCLE METHOD

As discussed above, the rupture surface cannot be assumed to be a plane in the passive case, especially when $\delta > \phi'/3$, as it gives unsafe values of the passive pressure. The actual rupture surface resembles more closely a log-spiral in such a case. Terzaghi's general wedge theory (1943) can be used to determine the passive pressure. However, the method is quite involved and beyond the scope of this text. The friction circle method is somewhat easier and is described below for both cohesionless and cohesive soils.

(*a*) Cohesionless Soil

The lower portion *BF* of the slip surface *BFD* is assumed to be an arc of a circle which joins, without break, the plane slip surface *FD* [Fig. 19.32(*a*)]. The plane slip surface portion *FD* is inclined at $(45° - \phi'/2)$ to the horizontal. It is assumed that the upper portion *CFD* of the sliding wedge *ABD* is in Rankine's passive

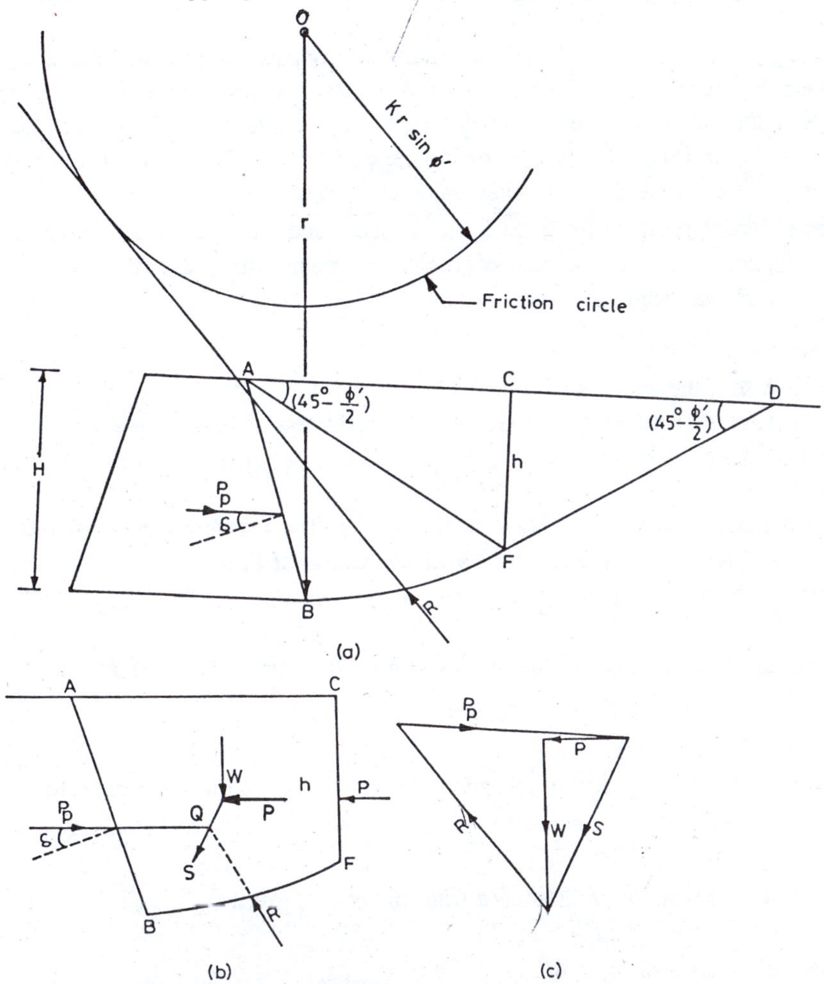

Fig. 19.32 (*a*), (*b*) and (*c*)

state of plastic equilibrium. Planes *AF* and *DF* are the boundary failure planes of the Rankine's passive zone *AFD*. The plane *FD* is tangential to the circular arc *BF* at *F*.

Let us consider the equilibrium of the free body *ABFC* [Fig. 19.32 (*b*)]. The body is in equilibrium under 4 forces.

(1) Weight W of the free body *ABFC* of soil.

(2) Resultant pressure P due to Rankine's passive zone *CFD*.

(3) Reaction R on the curved surface.

(4) Passive force P_p.

The friction circle method uses the concept that the reaction force R acting along the failure arc acts at an angle ϕ' with the normal, and is therefore tangential to a circle of radius $r \sin \phi'$ drawn at the centre of the failure arc. Thus the resultant reaction R is tangential to the friction circle. Actually, R is tangential to a circle of radius $K\, r \sin \phi'$, where K is a factor slightly greater than unity, but for convenience K is taken as unity.

The Rankine passive pressure P acts on plane *CF* at a height of $h/3$, where h is equal to the height *CF* and is given by $P = \frac{1}{2} K_p \gamma h^2$.

The passive pressure P_p acting on *AB* is also assumed to act at a height $H/3$, where H is the height of the wall.

The weight W of the wedge *ABFC* is determined from its area and the unit weight of the soil.

Knowing the magnitudes and directions of W and P, their resultant S can be determined as shown in Fig. 19.32(*c*). Now the system is reduced to 3 forces, namely, P_p, R and S. The three forces must meet at a point Q for the equilibrium [Fig. 19.32(*b*)]. Knowing the directions of P_p and R and the magnitude and direction of S, the force triangle can be drawn and the value of P_p computed.

The procedure is repeated for various trial surfaces and the value of P_p determined. The minimum value of P_p thus obtained is the required passive pressure. Of course, the pressure acting on the wall would be equal and opposite to P_p so found.

(*b*) Cohesive Soil

For a c—ϕ soil, the analysis is done in two parts :—

(1) First P_p is determined as in the case of cohesionless soils, as explained above.

(2) Additional pressure P_p' is determined considering only the cohesive and adhesive forces, as described below.

The total passive force in this case is $(P_p + P_p')$. The minimum value of the force $(P_p + P_p')$ is the required passive pressure. It is determined after evaluation of P_p'.

Fig. 19.33 (*a*) shows a retaining wall with a cohesive fill. The free body *ABFC* is in equilibrium under the additional forces P_p', C, R' and P_c.

The Rankine passive force P_c due to cohesion on the plane *CF* of height h is given by

$$P_c = 2c\sqrt{K_p}\, h$$

It acts at the mid-point of *CF*.

The passive pressure P_p' acts at the mid-point of the face *AB* at an angle δ to the normal.

The force C is equal to the sum of the cohesive forces parallel to the chord *BF* of length L_c.

$$C = c \times L_c$$

The force C is located at a distance a from the centre, given by

$$a = r L_a / L_c$$

where L_a is the length of the arc *BF*.

The adhesive force C_w on the wall is given by

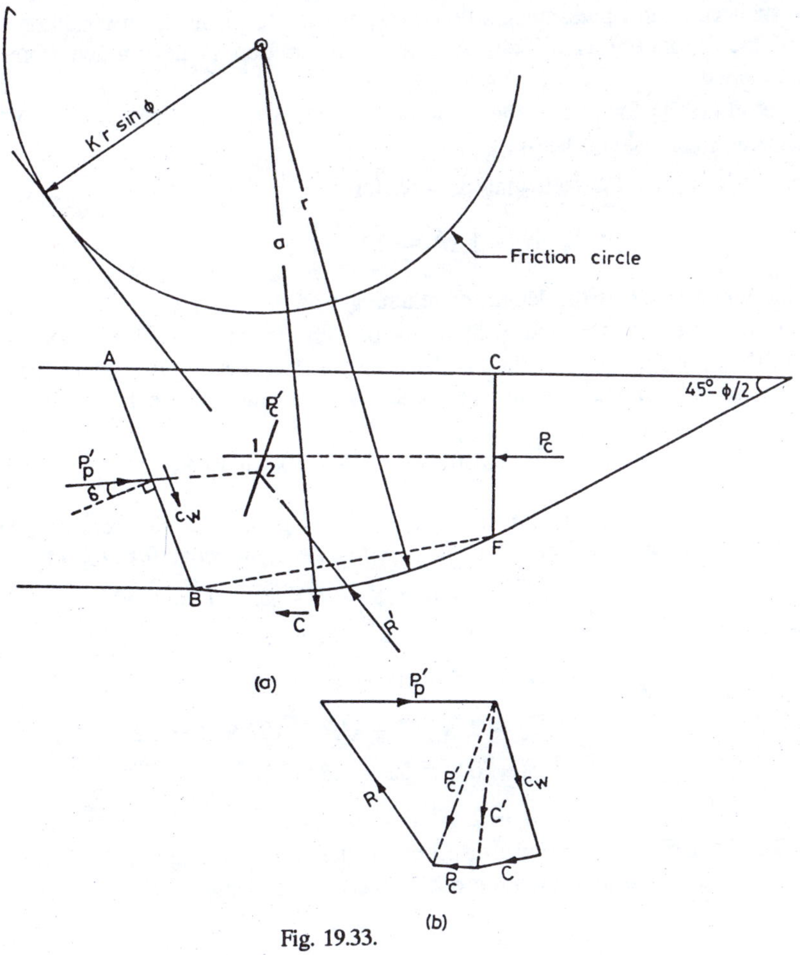

Fig. 19.33.

$$C_w = c_w \times AB$$

where c_w is the adhesion along the wall.

From the force diagram [Fig. 19.33(*b*)], the resultant C' of the force C and the wall cohesion C_w is first determined.

The resultant P_c' of the forces C' and P_c is then determined as shown. Point (1) in Fig. 19.33 (*a*) shows the position through which P_c' passes.

Now the forces P_p', P_c' and R' are considered. Point (2) indicates the position through which the three forces act. It lies at the intersection of P_c' and P_p'. The direction of R' is now determined as it passes through point (2) and is tangential to the friction circle. Once the direction of R' is determined, the force polygon in Fig. 19.33 (*b*) is completed and the value of P_p' is determined.

The total passive pressure is equal to the sum of P_p and P_p', where P_p is the frictional part as calculated for cohesionless soils and P_p' is the cohesion part.

The above procedure is repeated taking different failure surfaces. The failure surface which gives the minimum value of $(P_p + P_p')$ is the required surface, Thus the passive pressure force is found.

19.15. DETERMINATION OF SHEAR STRENGTH PARAMETERS

Retaining wall problems represent plain strain conditions in which intermediate principal stress has a value in-between the major and minor principal stresses. The angle of internal friction (ϕ') used in the computation of lateral earth pressure should correspond to the plane strain condition. As the determination of this angle in the laboratory is difficult, the angle of friction is determined from the triaxial test and a correction is applied.

Bjerrum et al (1961) found that the angle of internal friction in plain strain (ϕ_p) is about 10% more than that in the conventional triaxial test (ϕ_t).

Meyerhof (1963) gave the following relation for ϕ_p.

$$\phi_p = \left(1.1 - 0.1\,\frac{B}{L}\right)\phi_t \quad \text{...(19.43)}$$

where B is the width and L is the length of retaining wall.

Shear strength parametes c and ϕ in terms of either total stresses or effective stresses may be used according to the actual drainage conditions. However, as in practice most back fills are of granular pervious materials, consolidated-drained conditions exist and the cohesion intercept is generally zero.

ILLUSTRATIVE EXAMPLES

Illustrative Example 19.1. *Determine the lateral earth pressure at rest per unit length of the wall shown in Fig. E-19.1(a). Also determine the location of the resultant earth pressure. Take* $K_0 = 1 - \sin\phi'$ *and* $\gamma_w = 10\ kN/m^3$.

Solution. $K_0 = 1 - \sin\phi' = 1 - \sin 30° = 0.50.$

At point B $\bar{\sigma}_z = 2 \times 17 = 34\ \text{kN/m}^2, \qquad u = 0$

$$p_0 = K_0\,\bar{\sigma}_z = 0.5 \times 34 = 17\ \text{kN/m}^2$$

At point C $\bar{\sigma}_z = 2 \times 17 + (19 - 10) \times 2 = 52\ \text{kN/m}^2$

$$p_0 = 0.5 \times 52 = 26\ \text{kN/m}^2$$

$$u = 2 \times 10 = 20\ \text{kN/m}^2$$

Fig. E-19.1 (*b*) shows the pressure distribution diagram. The diagram has been divided into four parts. Let P_1, P_2, P_3 and P_4 be the total pressure due to these parts. Thus

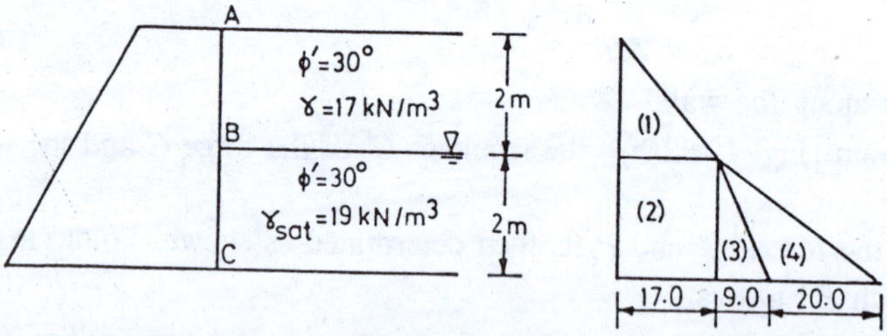

Fig. E-19.1.

$$P_1 = \frac{1}{2} \times 17 \times 2 = 17\ \text{kN}. \qquad P_2 = 2 \times 17 = 34\ \text{kN}$$

$$P_3 = \frac{1}{2} \times 9 \times 2 = 9\ \text{kN}. \qquad P_4 = \frac{1}{2} \times 20 \times 2 = 20\ \text{kN}$$

$$\text{Total } P = 80\ \text{kN}$$

The line of action of P is determined by taking moments about C.

$$P\bar{Z} = 17 \times 2.667 + 34 \times 1.0 + 9 \times 0.667 + 20 \times 0.667$$

or $$\bar{Z} = \frac{45.3 + 34 + 6 + 13.3}{80} = \frac{98.6}{80} = \mathbf{1.23\ m\ from\ base.}$$

Illustrative Example 19.2. *Determine the active pressure on the retaining wall shown in Fig. E-19.2 (a), Take* $\gamma_w = 10\ kN/m^3$.

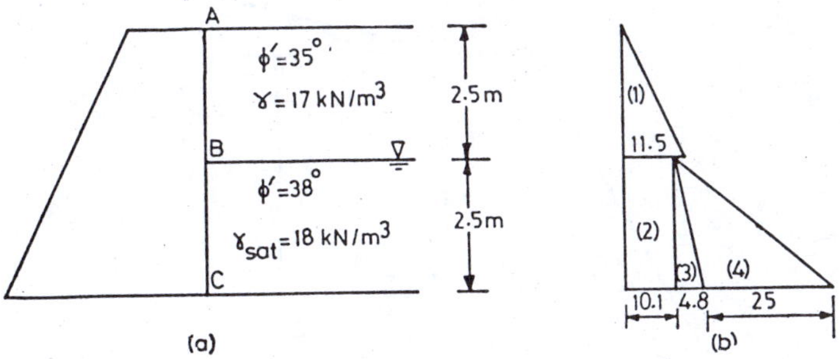

Fig. E-19.2.

Solution. From Eq. 19.10, $K_a = \dfrac{1 - \sin \phi'}{1 + \sin \phi'}$

For the upper layer, $K_a = \dfrac{1 - \sin 35°}{1 + \sin 35°} = 0.271$

For the bottom layer, $K_a = \dfrac{1 - \sin 38°}{1 + \sin 38°} = 0.238$

At point B $\bar{\sigma}_z = 2.5 \times 17 = 42.5\ \text{kN/m}^2, \qquad u = 0$

$p_a = 0.271 \times 42.5 = 11.5\ \text{kN/m}^2$

Below the interface, p_a is given by

$p_a = 0.238 \times 42.5 = 10.1\ \text{kN/m}^2$

At point C $\bar{\sigma}_z = 2.5 \times 17 + 2.5 \times 8.0 = 62.5\ \text{kN/m}^2$

$u = 2.5 \times 10 = 25\ \text{kN/m}^2$

$p_a = 0.238 \times 62.5 = 14.9\ \text{kN/m}^2$

Fig. E-19.2 (*b*) shows the pressure distribution.

The forces P_1, P_2, P_3 and P_4 are determined from the pressure distribution diagram.

$P_1 = \dfrac{1}{2} \times 2.5 \times 11.5 = 14.4\ \text{kN}.$ $\qquad P_2 = 2.5 \times 10.1 = 25.3\ \text{kN}$

$P_3 = \dfrac{1}{2} \times 2.5 \times 4.8 = 6.0\ \text{kN}.$ $\qquad P_4 = \dfrac{1}{2} \times 2.5 \times 25 = 31.3\ \text{kN}$

Total $P = 77.0\ \text{kN}$

Taking moments about C,

$$\bar{Z} = \frac{14.4 \times 3.33 + 25.3 \times 1.25 + 6.0 \times 0.833 + 31.3 \times 0.833}{77.0} = \mathbf{1.44m}$$

Illustrative Example 19.3. *Determine the active pressure on the wall shown in Fig. E-19.3, using Rankine's theory.*

Solution. From Eq. 19.15,

$$K_a = \cos i \times \frac{\cos i - \sqrt{\cos^2 i - \cos^2 \phi'}}{\cos i + \sqrt{\cos^2 i - \cos^2 \phi'}}$$

or

$$K_a = \cos 15° \times \frac{\cos 15° - \sqrt{\cos^2 15° - \cos^2 30°}}{\cos 15° + \sqrt{\cos^2 15° - \cos^2 30°}} = \mathbf{0.373.}$$

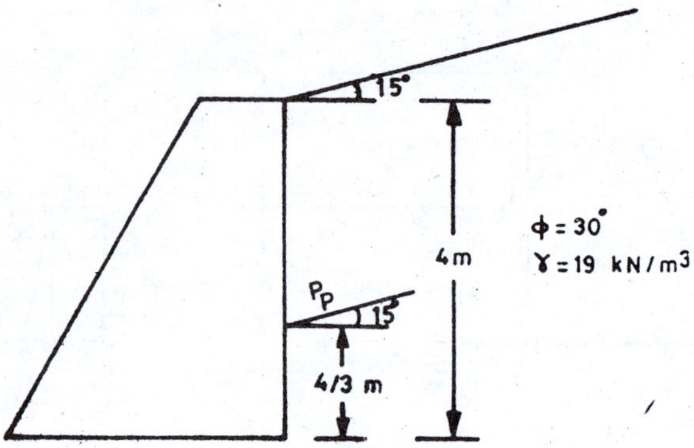

Fig. E-19.3.

From Eq. 19.23, $P_a = \frac{1}{2} K_a \gamma H^2 = \frac{1}{2} \times 0.373 \times 19.0 \times (4)^2 = 56.7$ kN

The active pressure acts at a height of 4/3 m and inclined at 15° to normal.

Illustrative Example 19.4. *Determine the stresses at the top and the bottom of the cut shown in Fig. E-19.4.*

Also determine the maximum depth of potential crack and the maximum depth of unsupported excavation.

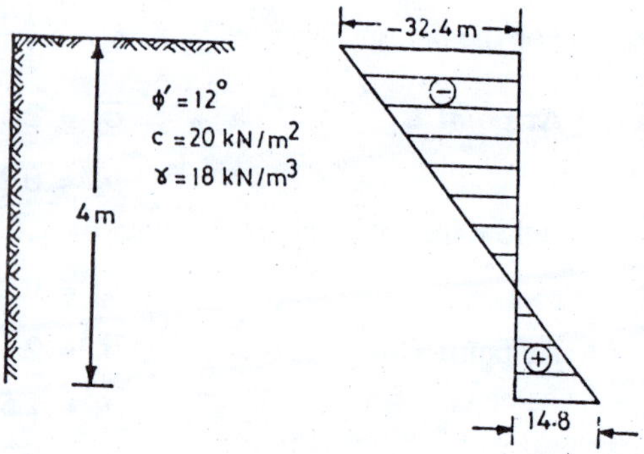

Fig. E-19.4.

Solution. From Eq. 19.25, $p_a = K_a \gamma Z - 2c' \sqrt{K_a}$

where $K_a = \dfrac{1 - \sin 12°}{1 + \sin 12°} = \dfrac{1 - 0.208}{1 + 0.208} = 0.656$

Thus $p_a = 0.656 \times 18Z - 2 \times 20 \times \sqrt{0.656}$

or $p_a = 11.81\,Z - 32.4$

At top $Z = 0$, $p_a = -32.4$ kN/m^2

At bottom $Z = 4$, $p_a = 14.8$ kN/m^2

From Eq. 19.27, depth of crack,

$$Z_c = \frac{2c'}{\gamma \sqrt{K_a}} = \frac{2 \times 20}{18\sqrt{0.656}} = \mathbf{2.745\ m}$$

From Eq. 19.32, maximum depth of unsupported excavation,

$$H_c = \frac{4c'}{\gamma \sqrt{K_a}} = \mathbf{5.490\ m}$$

Illustrative Example 19.5. *A 5 m high retaining wall is shown in Fig. E-19.5. Determine the Rankine active pressure on the wall. (a) Before the formation of the crack. (b) After the formation of the crack.*

Solution. $K_a = \dfrac{1 - \sin \phi'}{1 + \sin \phi'} = \dfrac{1 - \sin 30°}{1 - \sin 30°} = 0.333$

From Eq. 19.25, $p_a = K_a \gamma Z - 2c' \sqrt{K_a}$

or $p_a = 0.333 \times 17.5\,Z - 2 \times 5\sqrt{0.333}$

or $p_a = 5.83\,Z - 5.77$

At top, $Z = 0$, $p_a = -5.77$ kN/ m^2

At point B, where $p_a = 0.0$,

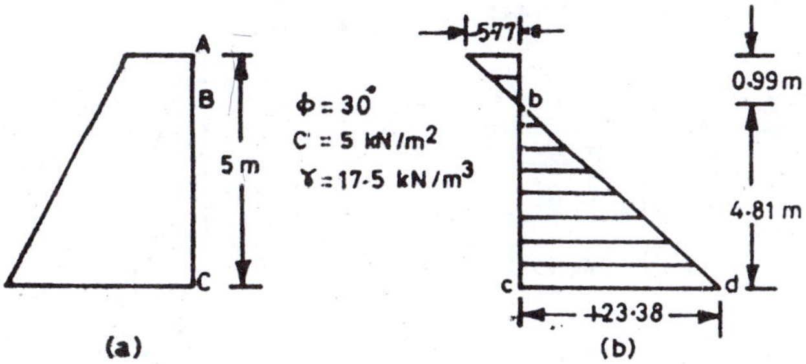

Fig. E-19.5.

$$5.83\,Z - 5.77 = 0 \quad \text{or} \quad Z = 0.99 \text{ m}$$

At bottom $Z = 5$ m, $\quad p_a = 5.83 \times 5 - 5.77 = 23.38 \text{ kN/m}^2$

Before formation of crack

Negative pressure, $\quad P_1 = \frac{1}{2} \times 0.99 \times 5.77 = 2.86 \text{ kN}$

Positive pressure, $\quad P_2 = \frac{1}{2} \times 4.01 \times 23.38 = 46.88 \text{ kN}$

Net $\quad P_a = 46.88 - 2.866 = \mathbf{44.02\ kN}$

Line of action of P_a is determined as under, by taking moments about C.

$$\bar{Z} = \frac{46.88 \times 4.01/3 - 2.86 \times (4.01 + 0.67)}{44.02} = \mathbf{1.12\ m}$$

The same result would be obtained using Eq. 19.28.

After formation of Crack

After the formation of the crack, the negative pressure is eliminated. The pressure distribution is given by the area *bcd.*

$$P_a = \frac{1}{2} \times 23.38 \times 4.01 = \mathbf{46.88\ kN}$$

It will act at a height of 4.01/3 m above base

Alternatively, directly from Eq. 19.29,

$$P_a = \frac{1}{2} \gamma H^2 K_a - 2c' H \sqrt{K_a} + 2(c')^2/\gamma$$

$$= \frac{1}{2} \times 17.5 \times 5^2 \times 0.333 - 2 \times 5 \times 5\sqrt{0.333} + 2 \times (5)^2/17.5$$

or $\quad P_a = 72.84 - 28.85 + 2.86 = \mathbf{46.85\ kN.}$

Illustrative Example 19.6. *Determine the Rankine passive force per unit length of the wall shown in Fig. E-19.6. The water table is at the level of B. Take $\gamma_w = 10$ kN/m³.*

Solution. From Eq. 19.12, $\quad K_p = \frac{1 + \sin \phi'}{1 - \sin \phi'}$

For top layer I, $\quad (K_p)_1 = \frac{1 + \sin 30°}{1 - \sin 30°} = 3.00$

For bottom layer II, $\quad (K_p)_2 = \frac{1 + \sin 24°}{1 - \sin 24°} = 2.37$

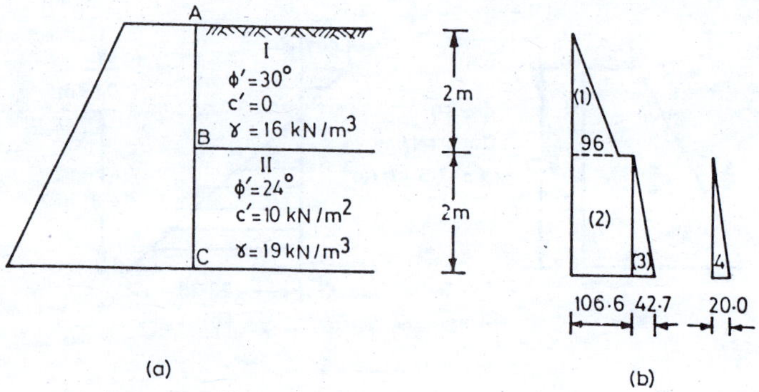

Fig. E-19.6.

From Eq. 19.34, $p_p = K_p \gamma Z + 2c' \sqrt{K_p}$

At point A, $Z = 0, \quad p_p = 0.$

At point B, $Z = 2\,\text{m}, \quad \bar{\sigma}_v = 2 \times 16 = 32\ \text{kN/m}^2$

Top layer, $p_p = 3 \times 32 = 96\ \text{kN/m}^2$

Bottom layer, $p_p = 32 \times 2.37 + 2 \times 10 \times \sqrt{2.37} = 106.6\ \text{kN/m}^2$

At point C, $\bar{\sigma}_v = 2 \times 16 + 2 \times (19 - 10) = 50\ \text{kN/m}^2$

$$p_p = 50 \times 2.37 + 2 \times 10\sqrt{2.37} = 149.3\ \text{kN/m}^2$$

$$u = 2 \times 10 = 20\ \text{kN/m}^2$$

Fig. E-19.6 (b) shows the pressure distribution.

Total pressure $P = P_1 + P_2 + P_3 + P_4$

$$= \frac{1}{2} \times 2 \times 96 + 106.6 \times 2 + \frac{1}{2} \times 42.7 \times 2 + \frac{1}{2} \times 2 \times 20 = \mathbf{371.9\ kN}$$

Illustrative Example 19.7. *Determine the Coulomb active force on the retaining wall shown in Fig. E-19.7.* $\gamma = 17.5\ kN/m^3$.

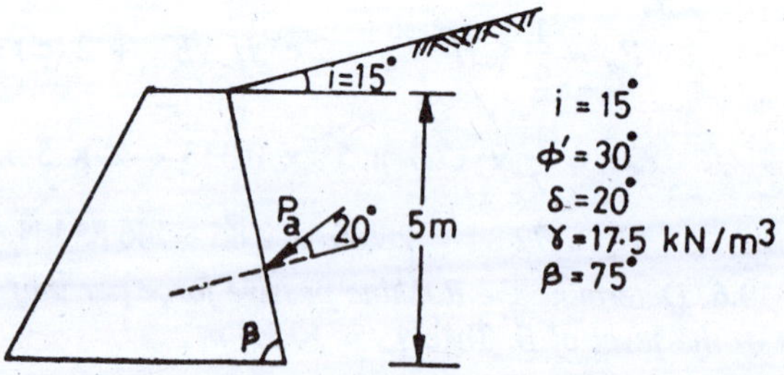

Fig. E-19.7.

Solution. From Eq. 19.39,

$$K_a = \frac{\sin^2(\beta + \phi')}{\sin^2\beta \cdot \sin(\beta - \delta)\left[1 + \sqrt{\dfrac{\sin(\phi' + \delta)\sin(\phi' - i)}{\sin(\beta - \delta)\sin(\beta + i)}}\right]^2}$$

or $$K_a = \frac{\sin^2 (75° + 30°)}{\sin^2 75° \sin (75° - 20°) \left[1 + \sqrt{\dfrac{\sin (30° + 20°) \sin (30° - 15°)}{\sin (75° - 20°) \sin (75° + 15°)}}\right]^2}$$

or $$K_a = \frac{0.933}{0.933 \times 0.819 \left[1 + \sqrt{\dfrac{0.766 \times 0.259}{0.819 \times 1.0}}\right]^2} = 0.548$$

From Eq. 19.38, $$P_a = \frac{1}{2} K_a \gamma H^2$$

or $$P_a = \frac{1}{2} \times 0.548 \times 17.5 \times (5)^2 = \mathbf{119.9\ kN}$$

This will act at a height of 5/3 m and inclined at 20° to normal in the direction shown. The reader should note that the direction of P_a is equal and opposite to that on the wedge, as discussed in Sect. 19.9.

Illustrative Example 19.8. *Determine the active thrust on the retaining wall shown in Fig. E-19.8. The backfill is cohesionless (ϕ' = 30°, γ = 19 kN/m³, δ = 20°).*

Solution. The method of trial wedges is used (Sect. 19.12). Several trial planes such as B—1, B—2, B—3, B—4 are chosen. The sliding wedge is in equilibrium under the three forces W, R and P_a. The weights of the wedges are computed as given below.

Wedge $AB1$, $$W_1 = (6 \times 3/2) \times 19.0 = 171.0 \text{ kN}$$

Wedge $AB2$, $$W_2 = (6 \times 6/2) \times 19.0 = 342.0 \text{ kN}$$

Wedge $AB3$, $$W_3 = \frac{1}{2} \times 2 \times 9 \times 19.0 + 342 = 513 \text{ kN}$$

Wedge $AB4$, $$W_4 = \frac{1}{2} \times 2 \times 9 \times 19.0 + 513 = 684 \text{ kN}$$

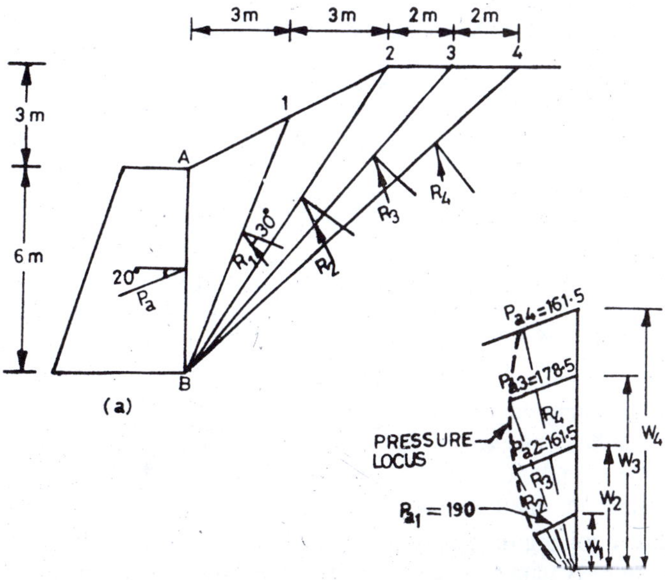

Fig. E-19.8.

Fig. E-198 (*b*) shows the force diagrams drawn along the common weight line. The pressure locus is shown dotted.

The maximum thrust is given by $P_a = 178.5$ kN

Illustrative Example 19.9. *Determine the active thrust on the retaining wall shown in Fig. E-19.9, using Culmann's method.*

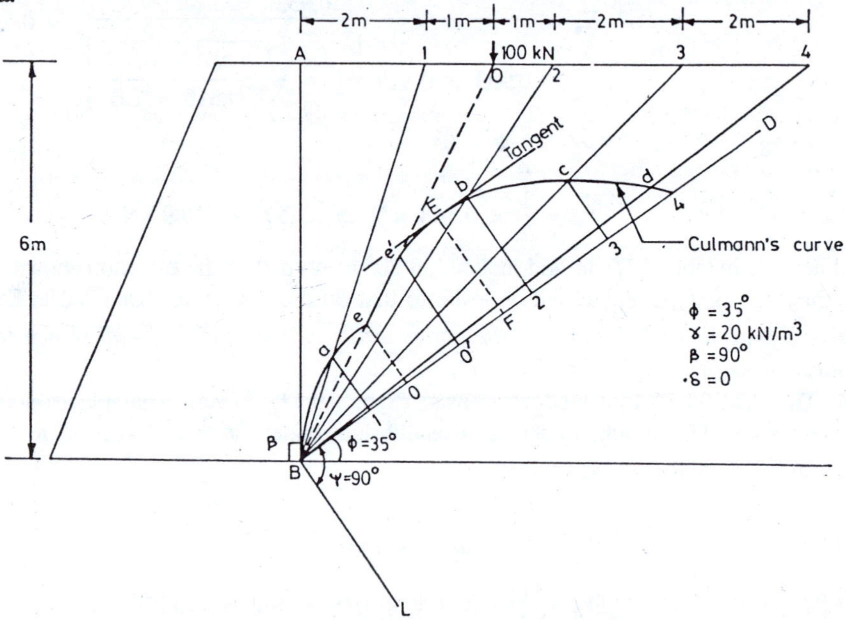

Fig. E-19.9.

Solution. The line *BD* is drawn at an angle of 35° to the horizontal, and the line *BL* at an angle $\psi = 90°$ to the line *BD*.

In this case, $\psi = \beta - \delta = 90° - 0 = 90°$

B—1, *B*—2, *B*—3, and *B*—4 are the four assumed trial failure surfaces. The weights are calculated as under:

$$W_1 = \frac{1}{2} \times 2 \times 6 \times 20 = 120 \text{ kN}$$

$$W_2 = \frac{1}{2} \times 4 \times 6 \times 20 = 240 \text{ kN}$$

$$W_3 = \frac{1}{2} \times 6 \times 6 \times 20 = 360 \text{ kN}$$

$$W_4 = \frac{1}{2} \times 8 \times 6 \times 20 = 480 \text{ kN}$$

The weight (W_0) of wedge *ABO*, upto the line load, is 180 kN.

The weight including line load, $W_0 = 280$ kN

The weights are plotted along the line *BD*.

From each of the points so located on *BD*, lines are drawn parallel to *BL* to cut the corresponding assumed failure surface. On the failure plane *BO*, two points *e* and *e'* are obtained, one for wedge without the line load and one for wedge with the line load.

Culmann's line is drawn joining the points *a*, *e*, *e'*, b, c, d. It is a discontinuous curve because of the line load. A tangent is drawn to the Culmann line such that it is parallel to *BD*. From the point of tangency, a line *EF* is drawn parallel to *BL*. The length of *EF* gives the required P_a.

$$P_a = 156 \text{ kN}$$

PROBLEMS

A. Numerical

19.1. Determine the passive pressure by Rankine's theory per unit run for a retaining wall 4 m high, with i = 15°, ϕ' = 30° and γ = 19 kN/m^3. The backface of the wall is smooth and vertical. [Ans. 380.5 kN]

19.2. For the retaining wall in Prob. 19.1, determine the active pressure per unit run. [Ans. 56.7 kN]

19.3. Determine the active pressure and passive pressure, using Coulomb's theory, on the wall shown in Fig. P. 19.3. [Ans. 73 kN, 1413 kN]

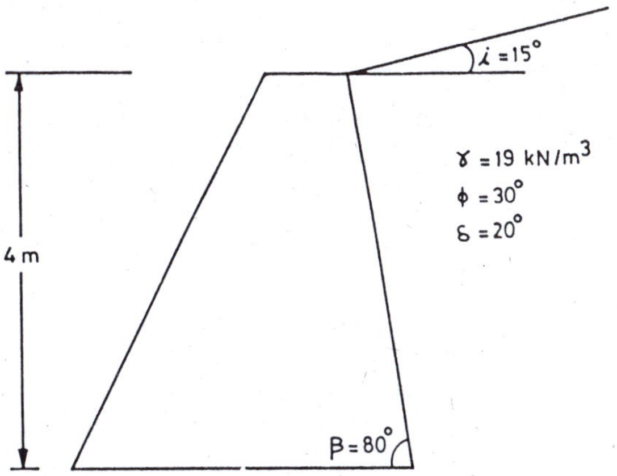

Fig. P-19.3.

19.4. A retaining wall has a vertical back and is 8 m high. The back face of the wall is smooth and the upper surface of the fill is horizontal. Determine the thrust on the wall per unit length. Take c = 10 kN/m^2, γ = 19 kN/m^3 and ϕ = 20°. Neglect tension. [Ans. 181.3 kN]

19.5. A retaining wall with a vertical smooth back is 8 m high. It supports a cohesionless soil (γ = 19 kN/m^3, ϕ = 30°). The surface of the soil is horizontal. Determine the thrust on the wall. [Ans. 202.7 kN]

19.6. Determine the resultant thrust on the wall of Prob. 19.4 if weep holes are blocked and water builds up behind the wall until the water table is 3 m above the base of the wall. [Ans. 204.3 kN]

19.7. A retaining wall is 7m high, with its back face smooth and vertical. It retains sand with its surface horizontal. Using Rankine's theory, determine active earth pressure at the base when the backfill is (*a*) dry, (*b*) saturated and (*c*) submerged, with water table at the surface. Take γ = 18 kN/m^3 and ϕ = 30°, γ_{sat} = 21 kN/m^3.
[Ans. 42 kN/m^2, 49 kN/m^2, 25.7 kN/m^2]

19.8. A vertical retaining wall 10 m high supports a cohesionless soil (γ = 18 kN/m^3). The upper surface of the back fill rises from the crest of the wall at an angle of 15° with the horizontal. Determine the total active pressure by Culmann's method and check the pressure obtained by Rehbann's construction. Take ϕ = 30° and δ = 20°. [Ans. 333.5 kN]

B. Descriptive and Objective Types

19.9. What are different types of earth pressure ? Give examples.

19.10. Define earth pressure at rest. Show the earth pressure distribution on a retaining wall, assuming the soil is dry.

19.11. What are the assumptions of Rankine's theory? Derive the expressions for active pressure and passive pressure.

19.12. Derive an expression for active pressure when the ground surface in inclined.

19.13. What are the assumptions in Coulomb's theory? Compare Rankine's theory and Coulomb's theory.

19.14. Describe Rehbann's construction. What is its use?

19.15. Discuss Culmann's method for the determination of active earth pressure.

19.16. Explain the trial wedge method.

19.17. How would you determine passive pressure when $\delta > 1/3\ \varphi'$?

19.18. Write whether the following statements are true.

(*a*) The active pressure is the minimum pressure which develops when the wall moves away from the fill.

(*b*) The shear resistance developed along the failure surface is a minimum when the active conditions develop.
(*c*) The Rankine pressure is always normal to the wall surface.
(*d*) Culmann's construction is more general than Rehbann's construction.
(*e*) Culmann's construction can be used even for passive case.
(*f*) Coulomb's theory always gives conservative results.
(*g*) For $\phi' = 30°$, the passive pressure is three times the active pressure, according to Rankine's theory.
(*h*) The active pressure increases if a dry soil becomes submerged.
(*i*) The basement walls are generally designed for at-rest pressure.

[**Ans.** True, (*a*), (*c*), (*d*), (*e*), (*i*)]

C. Multiple Choice Questions

1. The inclination of the failure plane behind a vertical wall in the passive pressure case is inclined to the horizontal at
(*a*) $45° - \phi/2$ (*b*) $45° - \phi$
(*c*) $45° + \phi/2$ (*d*) $45° + \phi$

2. The yield of a retaining wall required to reach plastic equilibrium in active case is
(*a*) more than that in the passive case
(*b*) less than that in the passive case.
(*c*) equal to that in the passive case.
(*d*) None of above.

3. The active earth pressure coefficient K_a generally refers to
(*a*) effective stresses (*b*) total stresses
(*c*) neutral stresses (*d*) All the above

4. The active pressure caused by a cohesionless backfill on a smooth vertical retaining wall may be reduced by
(*a*) compacting the backfill.
(*b*) providing a surcharge load on the backfill.
(*c*) saturating the backfill with water.
(*d*) None of the above.

5. The total active pressure after the development of tension cracks is equal to

(*a*) $\frac{1}{2}\gamma H^2 Ka - 2c' H\sqrt{Ka}$

(*b*) $\frac{1}{2}\gamma H^2 Ka + 2c' H\sqrt{Ka}$

(*c*) $\frac{1}{2}\gamma H^2 Ka - 2c' H\sqrt{Ka} - 2\frac{(c')^2}{\gamma}$

(*d*) $\frac{1}{2}\gamma H^2 Ka - 2c'H\sqrt{Ka} + \frac{2(c')^2}{\gamma}$

6. The radius of the friction circle is equal to
(*a*) R sin ϕ (*b*) R cos ϕ
(*c*) R tan ϕ (*d*) R ϕ

7. If a uniform surcharge of 120 kN/m^2 is placed on the backfill with $\phi' = 30°$, the increase in pressure is
(*a*) 12 kN/m^2 (*b*) 30 kN/m^2
(*c*) 40 kN/m^2 (*d*) 120 kN/m^2

[**Ans.** 1. (*a*), 2. (*b*) 3. (*a*), 4. (*a*), 5. (*d*), 6. (*a*), 7. (*c*)]

20

Design of Retaining Walls and Bulkheads

20.1. INTRODUCTION

(*a*) **Design of Retaining Walls.** Retaining walls are relatively rigid walls used for supporting the soil mass laterally so that the soil can be retained at different levels on the two sides. The lateral earth pressures acting on the retaining walls have been discussed in the preceding chapter. The types of retaining walls and their design features are explained in this chapter. However, the design is limited to the determination of the shear forces and bending moments. Actual structural design is outside the scope of this text.

(*b*) **Bulkheads.** Sheet pile walls, or bulkheads, are special type of earth retaining structures in which a continuous wall is constructed by joining sheet piles. Sheet piles are made of timber, steel or reinforced concrete and consist of special shapes which have interlocking arrangements. Sheet pile walls are used for water front structures, canal locks, coffer dams, river protection, etc. Sheet pile walls are embedded in the ground to develop passive resistance in the front to keep the wall in equilibrium. Various types of sheet pile walls and their analysis and design are discussed in this chapter.

20.2. TYPES OF RETAINING WALLS

The most common types of retaining walls are classified as under:

(1) **Gravity Retaining Walls.** These walls depend upon their weight for stability [Fig. 20.1 (*a*)]. The walls are usually constructed of plain concrete or masonry. Such walls are not economical for large heights.

(2) **Semi-Gravity Retaining Walls.** The size of the section of a gravity retaining wall may be reduced if a small amount of reinforcement is provided near the back face [Fig. 10.1 (*b*)]. Such walls are known as semi-gravity walls.

(3) **Cantilever Retaining Walls.** Cantilever retaining walls are made of reinforced cement concrete. The wall consists of a thin stem and a base slab cast monolithically [Fig. 20.1(*c*)]. This type of wall is found to be economical upto a height of 6 to 8 m.

(4) **Counterfort Retaining Walls.** Counterfort retaining walls have thin vertical slabs, known as counterforts, spaced across the vertical stem at regular intervals [Fig. 20.1(*d*)]. The counterforts tie the vertical stem with the base slab. Thus the vertical stem and the base slab span between the counterforts. The purpose of providing the counterforts is to reduce the shear force and bending moments in the vertical stem and the base slab. The counterfort retaining walls are economical for a height more than 6 to 8 m.

[**Note :** Counterforts are on the side of the back fill].

20.3. PRINCIPLES OF THE DESIGN OF RETAINING WALLS

Before the actual design, the soil parameters that influence the earth pressure and the bearing capacity of the soil must be evaluated. These include the unit weight of the soil, the angle of shearing resistance, the

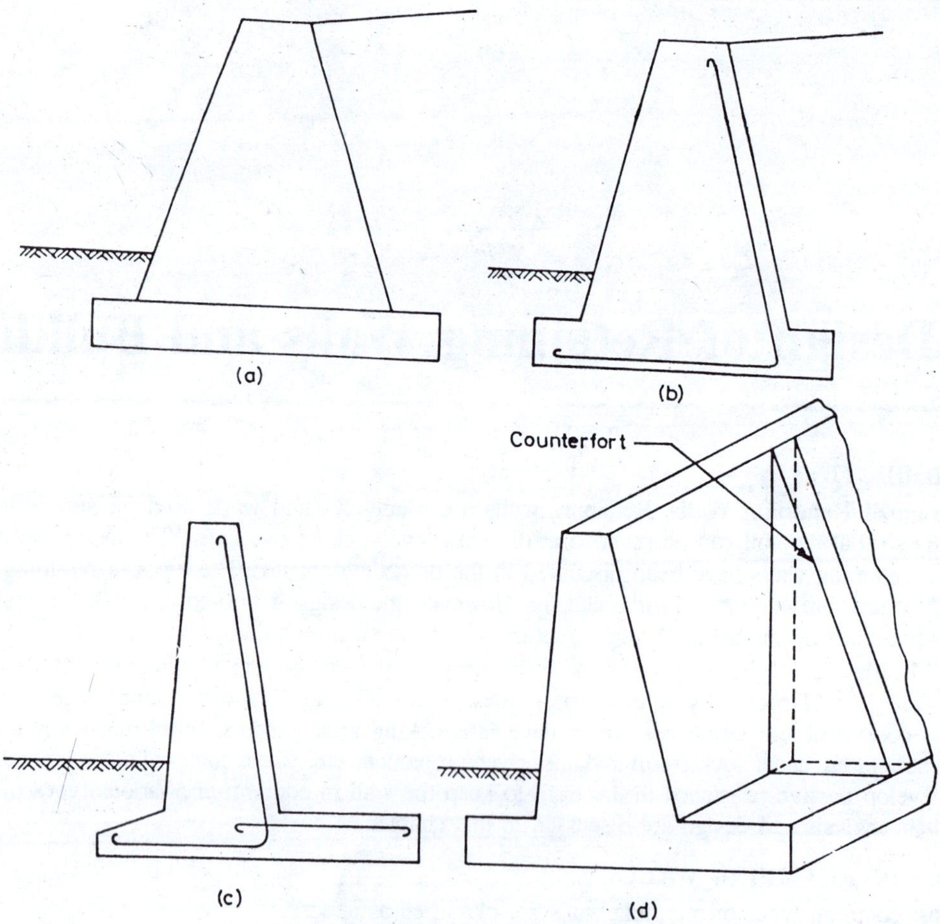

Fig. 20.1. Different Types of Retaining walls.

cohesion intercept and the angle of wall friction. Knowing these parameters, the lateral earth pressure and the bearing capacity of the soil can be determined. Methods for the computation of earth pressure have been discussed in chapter 19. The bearing capacity theories are explained in chapter 23. With the earth pressure known, the retaining wall as a whole is checked for stability.

Fig. 20.2 shows a retaining wall with a smooth back face and no surcharge. The active pressure P_a acts horizontally, as shown. The front face of the wall is subjected to a passive pressure (P_p) below the soil surface. However, it is doubtful whether the full passive resistance would develop. Moreover, often P_p is small and therefore it may be neglected. This gives more conservative design.

The weight W of the wall and the active pressure P_a have their resultant R which strikes the base at point D. There is an equal and opposite reaction R' at the base between the wall and the foundation. For convenience, R' is resolved into the vertical and horizontal components (R_v' and R_H').

From the equilibrium of the system,

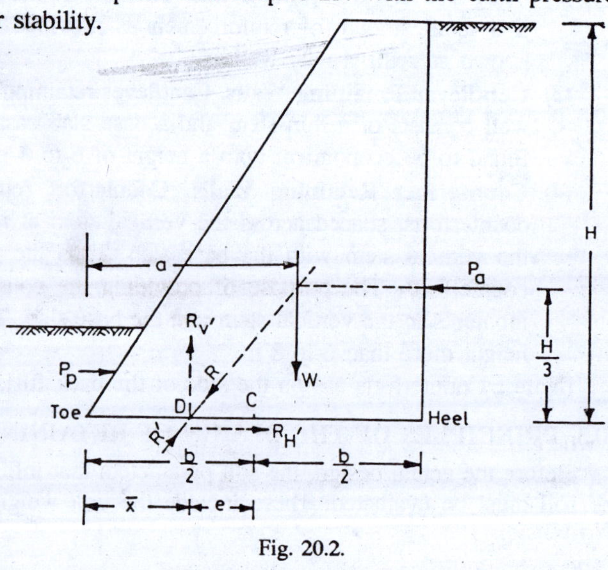

Fig. 20.2.

$$R_V' = W, \quad \text{and} \quad R_H' = P_a$$

The third equation of equilibrium, namely the moment equation, is used to determine the eccentricity e of the force R_v' relative to the centre C of the base of the wall. Obviously, by taking moments about the toe,

$$R_v' \times \bar{x} = W \times a - P_a(H/3)$$

or
$$\bar{x} = \frac{W \times a - P_a \times H/3}{R_V'} \qquad \text{...(20.1)}$$

where $\bar{x}$ is the distance of the point D from the toe.

Thus, eccentricity, $$e = b/2 - \bar{x} \qquad \text{...(20.2)}$$

where b = width of the base.

For a safe design, the following requirements must be satisfied.

(1) No Sliding

The wall must be safe against sliding. In other words,

$$\mu \, R_V > R_H$$

where R_V and R_H are vertical and horizontal components of R, respectively. The factor of safety against sliding is given by

$$F_s = \frac{\mu R_V}{R_H} \qquad \text{...(20.3)}$$

where μ = coefficient of friction between the base of the wall and the soil (= tan δ).

A minimum factor of safety of 1.5 against sliding is generally recommended.

(2) No Overturning

The wall must be safe against overturning about toe. The factor of safety against overturning is given by

$$F_0 = \frac{\Sigma M_R}{\Sigma M_O} \qquad \text{...(20.4)}$$

where ΣM_R = sum of resisting moment about toe,

and ΣM_O = sum of overturning moment about toe.

In Fig. 20.2, $$F_0 = \frac{W \times a}{P_a \times H/3} \qquad \text{...(20.5)}$$

The factor of safety against overturning is usually kept between 1.5 to 2.0.

(3) No bearing capacity failure

The pressure caused by R_V at the toe of the wall must not exceed the allowable bearing capacity of the soil.

The pressure distribution at the base is assumed to be linear. The maximum pressure is given by

$$p_{max} = \frac{R_V}{b}(1 + 6e/b) \qquad \text{...(20.6)}$$

The factor of safety against bearing failure is given by

$$F_b = \frac{q_{na}}{p_{max}} \qquad \text{...(20.7)}$$

where q_{na} = allowable bearing pressure.

A factor of safety of 3 is usually specified, provided the settlement is also within the allowable limit.

(4) No tension

There should be no tension at the base of the wall. When the eccentricity (e) is greater than $b/6$, tension develops at the heel. Tension is not desirable. The tensile strength of the soil is very small and the tensile crack would develop. The effective base area is reduced. In such a case, the maximum stress is given by

$$p_{max} = \frac{4}{3}\left(\frac{R_V}{b - 2e}\right) \quad ...(20.8)$$

20.4. GRAVITY RETAINING WALLS

As in design of all other structures, a trial section is first chosen and analysed. If the stability checks yield unsatisfactory results, the section is changed, and rechecked. Fig. 20.3 shows the general proportion of a gravity retaining wall of overall height H. The top width of the stem should be at least 0.3 m for proper placement of concrete in the stem. The depth (D) of the foundation below the soil surface should be at least 0.6 m. The base width of the wall is generally between 0.5 H to 0.7 H; with an average of $2H/3$.

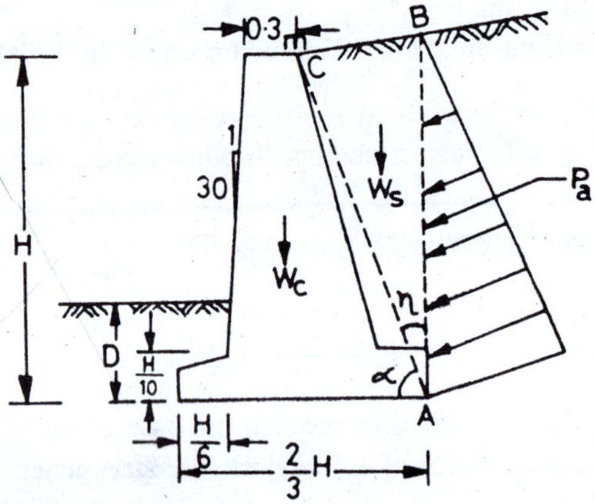

Fig. 20.3. Gravity wall—Rankine Pressure.

The earth pressure can be computed using either Rankine's theory or Coulomb's theory. For using Rankine's theory, a vertical line AB is drawn through the heel point A. It is assumed that the Rankine active conditions exist along the vertical line AB. However, the assumption for the development of Rankine's conditions along AB is theoretically justified only if the shear zone bounded by the line AC is not obstructed by the stem of the wall, where AC makes an angle η with the vertical, given by

$$\eta = (45° + i/2) - \phi'/2 - \sin^{-1}\left(\frac{\sin i}{\sin \phi'}\right) \quad ...(20.9)$$

where i is the angle of surcharge.

The angle α which the line AC makes with the horizontal is given by,

$$\alpha = \left(45° + \frac{\phi'}{2}\right) - \frac{i}{2} + \sin^{-1}\left(\frac{\sin i}{\sin \phi'}\right) \quad ...(20.10)$$

When $i = 0$, the value of α is equal to $(45° + \phi'/2)$ (Fig. 20.4).

While checking the stability, the weight of soil (W_s) above the heel in the zone ABC should also be taken into consideration, in addition to the earth pressure (P_a) on the vertical plane AB and the weight of the wall (W_c).

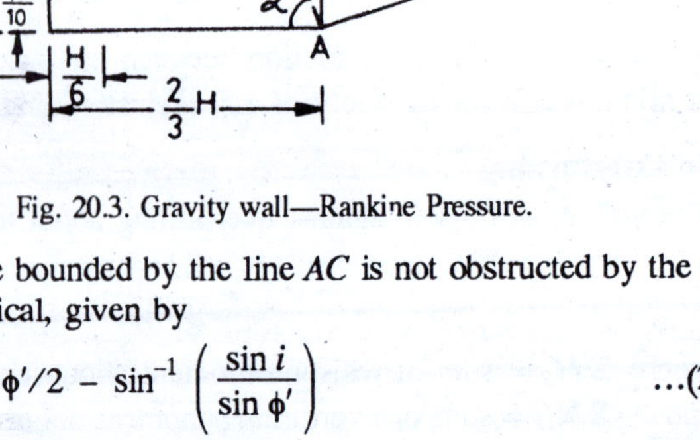

Fig. 20.4.

Coulomb's theory can also be used for the determination of earth pressure (Fig. 20.5). As the Coulomb theory gives directly the lateral pressure on the back face (P_a), the forces to be considered are only P_a (Coulomb) and the weight of the wall (W_c). In this case, the weight of soil (W_s) is not to be considered separately.

Once the forces acting on the wall have been

determined, the stability is checked using the procedure discussed in the preceding section. For convenience, the section of the retaining wall is divided into rectangles and triangles for the computation of weight and the determination of the line of action of the weights.

Semi-gravity Retaining walls. The base width of the semi-gravity retaining walls is slightly smaller than that of a corresponding gravity wall. The rest of the design procedure is the same as that for gravity retaining walls.

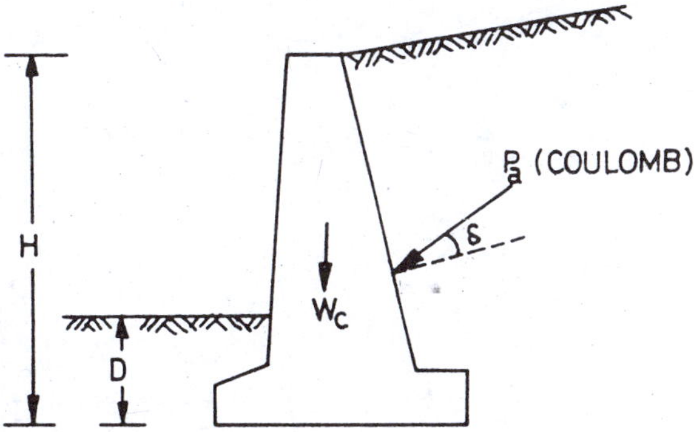

Fig. 20.5. Gravity wall—Coulomb Pressure.

20.5. CANTILEVER RETAINING WALLS

Fig. 20.6 shows a cantilever retaining wall. The general proportions for an overall height of H are also shown. The top width of the stem is at least 0.3 m. The width of the base slab is kept about $2H/3$. The width of the stem at bottom, the thickness of the base slab and the length of the toe projection, each is kept about 0.1 H.

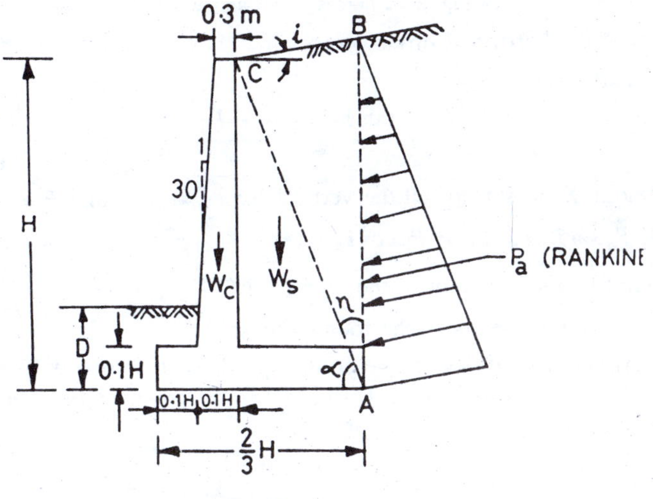

Fig. 20.6. Cantilever Retaining wall.

The earth pressure is computed using Rankine's theory on the vertical plane AB, provided the shear zone bounded by the line AC is not obstructed by the stem of the wall. The line AC makes an angle η with the vertical given by Eq. 20.9.

Fig. 20.7 shows the forces acting on the wall. The Rankine pressure P_a acts at an angle i with the horizontal. It is resolved into the vertical and horizontal components P_v and P_h, as shown. The passive pressure P_p is also shown, but generally it is neglected. For convenience, the weight of soil (W_s) over the slab is divided into two parts (1) and (2). Likewise, the weight of stem is divided into two parts (3) and (4).

(a) Factor of safety against sliding

The factor of safety against sliding may be expressed as

$$F_s = \frac{\Sigma F_R}{\Sigma F_d} \qquad \text{...(20.11)}$$

where ΣF_R = sum of the horizontal resisting forces,

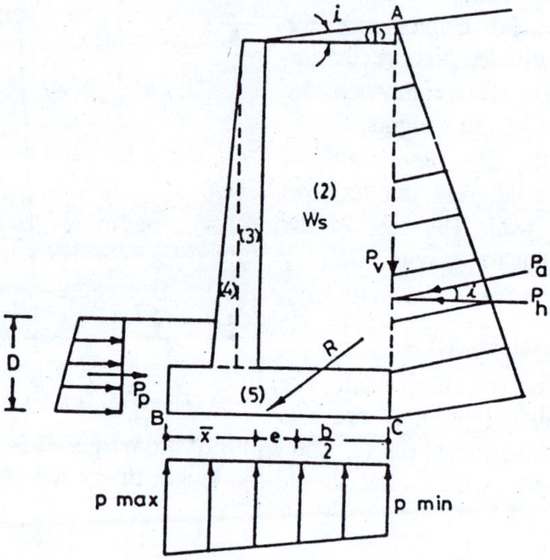

Fig. 20.7. Forces on a Cantilever wall.

and ΣF_d = sum of the horizontal driving forces.

Eq. 20.11 can be written as

$$F_s = \frac{(\Sigma V) \tan \phi_2 + bc_2 + P_p}{P_h} \qquad \ldots(20.12)$$

where b = base width, ΣV = sum of all the vertical forces, W_c, W_s and P_v.

$$P_v = P_a \sin i \text{ and } P_h = P_a \cos i.$$

P_p = passive force in the front of the wall (= $1/2\, K_{p2}\, \gamma_2\, D^2 + 2c_2 \sqrt{K_{p2}}\, D$)

where c_2, γ_2 and ϕ_2 are parameters of the foundation soil.

The factor of safety can also be determined from Eq. 20.3 if μ is given. If the required factor of safety of 1.5 against sliding is not obtained, a base key is generally provided (Fig. 20.8). The key increases the passive resistance to P_p' where

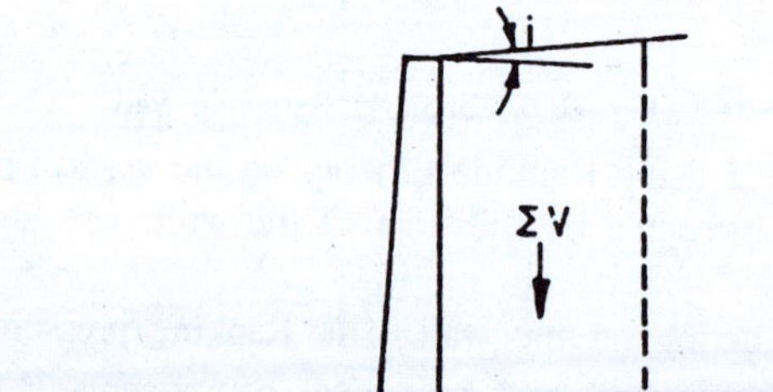

Fig. 20.8. Key in a retaining wall.

$$P_p' = \frac{1}{2}\gamma_2 (D_1)^2 K_{p2} + 2c_2 D_1 \sqrt{K_{p2}} \qquad ...(20.13)$$

where D_1 is the depth of the bottom of the key wall from soil surface.

Generally, the base key is constructed just below the stem and some of the main steel of the stem is extended into the key.

The friction angle ϕ_2 and c_2 are generally reduced to about one-half to two-thirds of the values for extra safety, as the full passive resistance is doubtful.

Factor of safety against Overturning

Eq. 20.4 can be used to obtain the factor of safety against overturning,

$$F_o = \frac{\Sigma M_R}{\Sigma M_o}$$

where ΣM_R = sum of the resisting moments about toe,

ΣM_o = sum of the overturning moments about toe.

The only overturning force is P_h, acting at a height of $H/3$.

$$M_0 = P_h \times H/3 \qquad ...(20.14)$$

The resisting moments (M_R) are due to weights W_1, W_2, W_3, W_4 and W_5 of the soil and the concrete. The vertical component of pressure P_v also helps in resisting moment. Its resisting moment is given by

$$M_v = P_v \times b \qquad ...(20.15)$$

Therefore

$$F_o = \frac{M_1 + M_2 + M_3 + M_4 + M_5 + M_v}{P_h \times H/3} \qquad ...(20.16)$$

where $M_1, M_2, \ldots M_5$ are the moments due to W_1, $W_2 \ldots W_5$ about toe.

Factor of safety against bearing capacity failure

The sum of the vertical forces acting on the base is equal to ΣV. The horizontal force is P_h. The resultant force (R) is given by

$$R = \sqrt{(\Sigma V)^2 + (P_h)^2}$$

The net moment of these forces about toe B is given by

$$\Sigma M = \Sigma M_R - \Sigma M_0$$

The distance $\bar{x}$ of the point E, from the toe, where R strikes the base is given by

$$\bar{x} = \frac{\Sigma M}{\Sigma V} \qquad ...(20.17)$$

Hence, the eccentricity e of R is given by

$$e = b/2 - \bar{x} \qquad ...(20.18)$$

If $e > b/6$, the section should be changed, as it indicates tension. The pressure distribution under the base slab is determined as

$$p_{max} = \frac{\Sigma V}{b}\left(1 + \frac{6e}{b}\right) \qquad ...[20.19(a)]$$

and

$$p_{min} = \frac{\Sigma V}{b}\left(1 - \frac{6e}{b}\right) \qquad ...[20.19(b)]$$

The factor of safety against bearing capacity failure is given by Eq. 20.7.

20.6. COUNTERFORT RETAINING WALLS

For counterfort retaining walls, the general proportions of the stem and the base slab are almost the same as that in the cantilever walls. The counterforts are about 0.3 m thick and have the centre-to-centre spacing of 0.3 H to 0.7 H.

The analysis is also similar to that of a cantilever retaining wall. The pressure p_{max} and p_{min} are determined, as in the case of cantilever walls.

The basic difference between the counterfort retaining wall and the cantilever retaining wall is in the determination of the bending moment and shear forces.

(1) In cantilever retaining walls, the stem acts as a vertical cantilever fixed at base whereas in the counterfort retaining walls, it acts, as a continuous slab supported between the counterforts. The slab has positive moments in the middle and the negative moments at the supports. The reinforcement is provided in the horizontal direction on the front side of the stem in the middle and on the rear side at the supports. In cantilever walls, the main reinforcement is in the vertical direction at the rear face.

(2) In cantilever walls, the toe slab and the heel slab both act as cantilevers subjected to the upward pressure. The reinforcement is provided at the bottom face.

In counterfort retaining walls, although the toe slab acts as a cantilever, the heel slab acts as a continuous slab supported on the counterforts. The main reinforcement is at the top face in the middle portion and at the bottom face near the supports.

(3) In counterfort retaining walls, the counterforts are designed as cantilever of varying section and fixed at the base. The main reinforcement is provided at the back face of the counterfort.

In addition, the vertical and horizontal ties are provided in the counterforts to join the base and the stem to the counterforts.

The structural design of the counterfort and cantilever retaining walls is outside the scope of this text.

20.7. OTHER MODES OF FAILURE OF RETAINING WALLS

In addition to the three types of failures, *viz*, sliding, overturning and bearing failures, a retaining wall may fail in the following two modes if the soil below is weak.

(1) Shallow shear Failure. This type of failure occurs along a cylindrical surface ABC passing through the heel of the retaining wall (Fig. 20.9). The failure takes place because of excessive shear stresses along the cylindrical surface within the soil mass. However, it has generally been found that the factor of safety against horizontal sliding discussed in Sect. 20.3 is lower than that for the shallow shear failure. Consequently, if the factor of safety against sliding (F_s) is greater than about 1.5, shallow shear failure is not likely to occur.

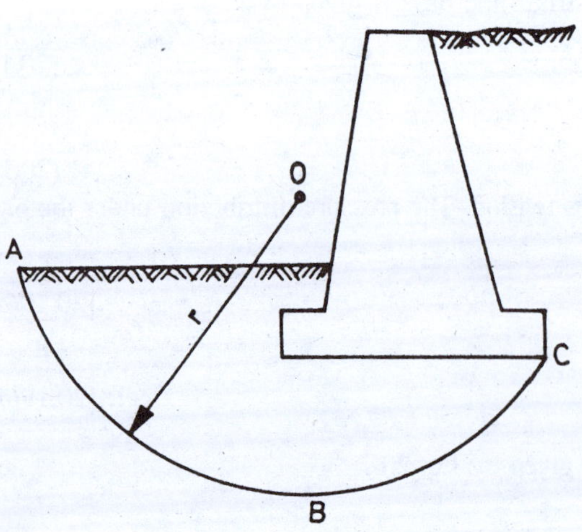

Fig. 20.9. Shallow Shear Failure.

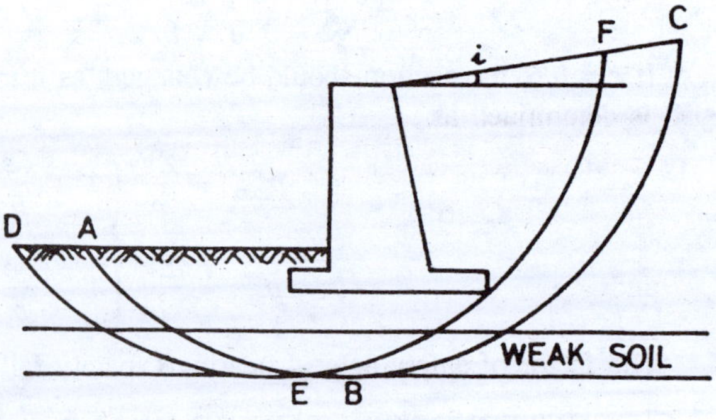

Fig. 20.10. Deep Shear Failure.

(2) Deep shear failure. This type of slope failure occurs along a cylindrical surface *ABC* (Fig. 20.10), when there is a weak layer of soil underneath the wall a depth of about 1.5 times the height of the wall. The critical failure surface is determined by trial and error procedure.

For the backfills having slope *i* less then 10°, it has been found that the critical failure surface *DEF* passes through the edge of the heel slab. The minimum factor of safety is found by trial and error, taking different circles, and determining the resisting forces and the driving forces along the failure surface (See Teng, 1962).

When a weak soil layer is located at a shallow depth below the retaining wall, the possibility of deep shear failure should be investigated. The possibility of excessive settlement should also be looked into. Sometimes, piles are used to transmit the foundation load to a firm layer below the weak layer. However, care shall be taken in the design of piles so that the thrust of the sliding wedge of soil does not cause bending of the piles.

20.8. DRAINAGE OF THE BACKFILL

When the backfill becomes wet due to rainfall or any other reason, its unit weight increases. It increases the pressure on the wall and may create unstable conditions. Further, if the water table also rises, the pore water pressure (u) develops and it causes excessive hydrostatic pressure on the wall. To reduce the development of excessive lateral pressures on the wall, adequate drainage must be provided.

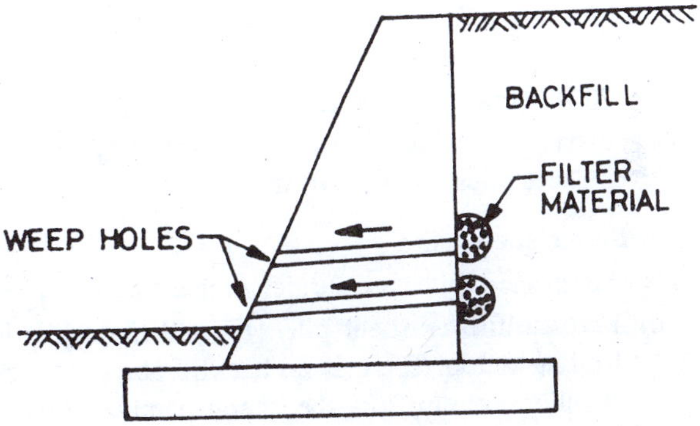

Fig. 20.11. Weep Holes.

Weep holes are generally provided in the walls. The weep holes are of about 0.1 m diameter. The spacings of the holes generally varies between 1.5 m to 3 m in the horizontal direction. As the backfill material may be washed into weep holes and may clog them, filter material is placed around the weep holes (Fig. 20.11).

Perforated pipes are also frequently used for the drainage of the backfill. These pipes are laid near the base (Fig. 20.12). The water is collected from the backfill and discharged at a suitable place at the ends. The filter material is placed around the pipes. These days, a filter cloth or a geotextile fabric is also used to serve the purpose of a filter material. All drain pipes should be provided with clean-outs for cleaning when clogged.

Fine-grained soils cause large earth pressure against retaining walls and are, therefore, rarely used as a backfill material. As far as possible, good draining, granular material should be used, at least in the sliding wedge portion of the wall. In case a fine-grained material cannot be avoided, some form of filter of coarse permeable material is placed behind the retaining walls to prevent the development of excessive pore water pressure. Fig. 20.13 shows two types of drainage filters commonly used. The water percolating into the filter is discharged through the weep holes. The inclined filter is found to be more effective than the vertical filter.

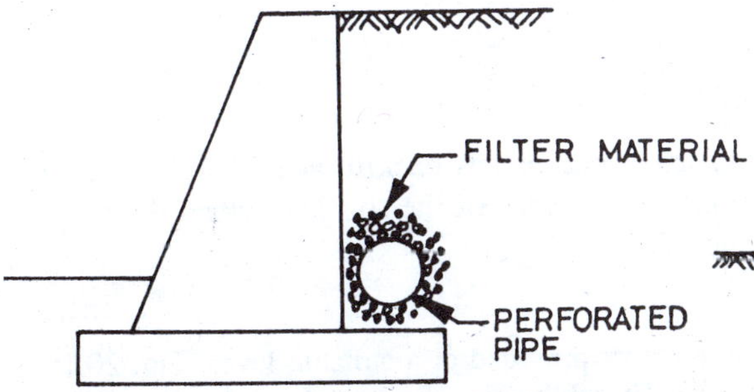

Fig. 20.12. Perforated Pipes.

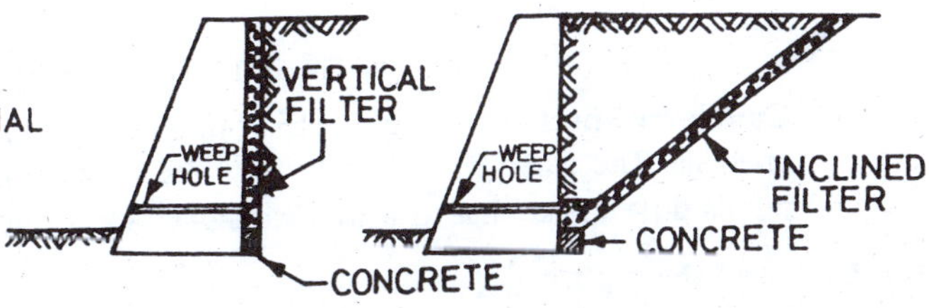

Fig. 20.13. Drainage Filter.

BULKHEADS

20.9. TYPES OF SHEET PILE WALLS

Sheet piles are generally made of steel or timber. However, sometimes reinforced cement concrete sheet piles are also used. The use of timber piles is generally limited to temporary structures in which the depth of driving does not exceed 3m. For permanent structures and for depth of driving greater than 3 m, steel piles are more suitable. Moreover, steel sheet piles are relatively water tight and can be extracted if required and re- used. However, the cost of steel sheet piles is generally more than that of timber piles. Reinforced cement concrete piles are generally used when these are to be jetted into fine sand or driven in very soft soils, such as peat. For tougher soils, the concrete piles generally break off.

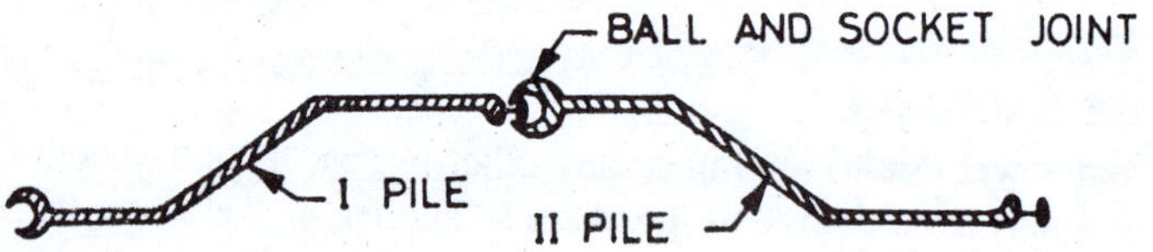

Fig. 20.14. Plan of a Sheet Pile Wall.

Fig. 20.14 shows the plan of a typical steel sheet pile wall, in which 2 sheet piles are shown with joints. Based on its structural form and loading system, sheet pile walls can be classified into 2 types: (1) Cantilever Sheet piles and (2) Anchored Sheet Piles.

(1) Cantilever sheet piles

Cantilever sheet piles are further divided into 2 types :

(*a*) **Free cantilever sheet pile.** [Fig. 20.15 (*a*)]. It is a sheet pile subjected to a concentrated horizontal load at its top. There is no backfill above the dredge level. The free cantilever sheet pile derives its stability entirely from the lateral passive resistance of the soil below the dredge level into which it is driven.

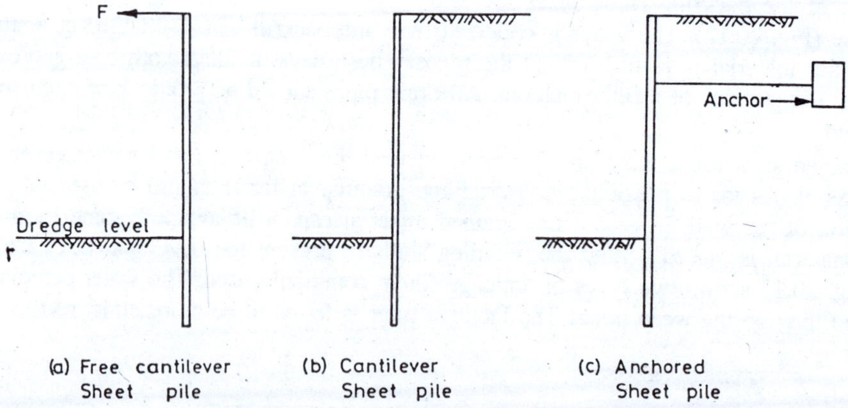

Fig. 20.15. Types of Sheet Pile Walls.

(*b*) **Cantilever sheet pile.** [Fig. 20.15 (*b*)]. A cantilever sheet pile retains backfill at a higher level on one side. The stability is entirely from the lateral passive resistance of the soil into which the sheet pile is driven, like that of a free cantilever sheet pile.

(2) Anchored sheet piles

Anchored sheet piles are held above the driven depth by anchors provided at a suitable level [Fig. 20.15 (*c*)]. The anchors provide forces for the stability of the sheet pile, in addition to the lateral passive resistance of the soil into which the sheet piles are driven. The anchored sheet piles are also of two types:

(*a*) **Free-earth support piles.** An anchored sheet pile is said to have free-earth support when the depth of embedment is small and the pile rotates at its bottom tip. Thus there is no point of contraflexure (or inflexion point) in the pile.

(*b*) **Fixed-earth support piles.** An anchored sheet pile has fixed earth support when the depth of embedment is large. The bottom tip of the pile is fixed against rotations. There is a change in the curvature of the pile, and hence, an inflexion point occurs.

20.10. FREE CANTILEVER SHEET PILE

The free cantilever sheet pile rotates about a point O below the dredge level. The actual pressure distribution is shown in Fig. 20.16 (*a*). Blum (1931) gave a simple solution. The passive resistance of the soil

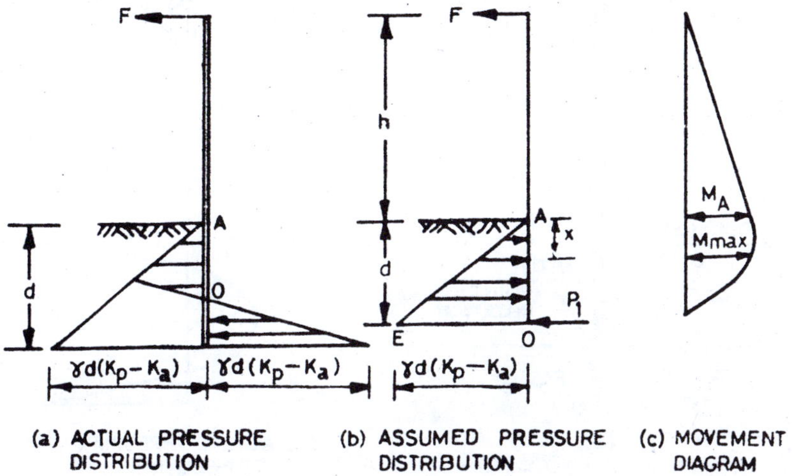

Fig. 20.16. Free Cantilever Sheet Pile.

on the left side is idealized as a right angled triangle AOE [Fig. 20.16 (*b*)]. The distributed pressure acting on the right side below the pivot O is replaced by an equivalent concentrated load P_1 acting at point O. In calculations that follow, however, the magnitude of the force P_1 is not required.

For equilibrium, the moment of all the forces about O must be zero, *i.e.*

$$M_0 = F(h + d) - \left[\frac{1}{2}\gamma d (K_p - K_a) d\right] \times \frac{d}{3} = 0 \qquad ...(20.20)$$

where F is the horizontal force, h is the height of wall above the dredge level,
d is the depth of embedment.

Eq. 20.20 can be solved for d. The actual depth to be provided is generally taken as 1.2 d.

The point of the maximum bending in the sheet pile can be determined as under.

The bending moment at depth x below the dredge level is given by

$$M_x = F(h + x) - \frac{\gamma x^3}{6}(K_p - K_a) \qquad ...(20.21)$$

For maximum bending (M_{max}), $\quad \dfrac{d(M_x)}{dx} = 0$

or $$F - \frac{\gamma (K_p - K_a)}{6}(3x^2) = 0$$

or $$x = \sqrt{\frac{2F}{\gamma (K_p - K_a)}} \qquad ...(20.22)$$

The maximum B.M. (M_{max}) is obtained by substituting the value of x from Eq. 20.22 into Eq. 20.21. The section modulus of the sheet pile can then be determined as

$$S = \frac{M_{max}}{\sigma_a} \qquad \ldots(20.23)$$

where σ_a = allowable bending stress in the pile. Fig. 20.16 (*c*) shows the bending moment diagram.

20.11. CANTILEVER SHEET PILE IN COHESIONLESS SOILS

Fig. 20.17 (*a*) shows a cantilever sheet pile in a cohesionless soil deposit The pile rotates about the point O'. The pressure above O' is passive in the front and active on the back side. However, the pressures below

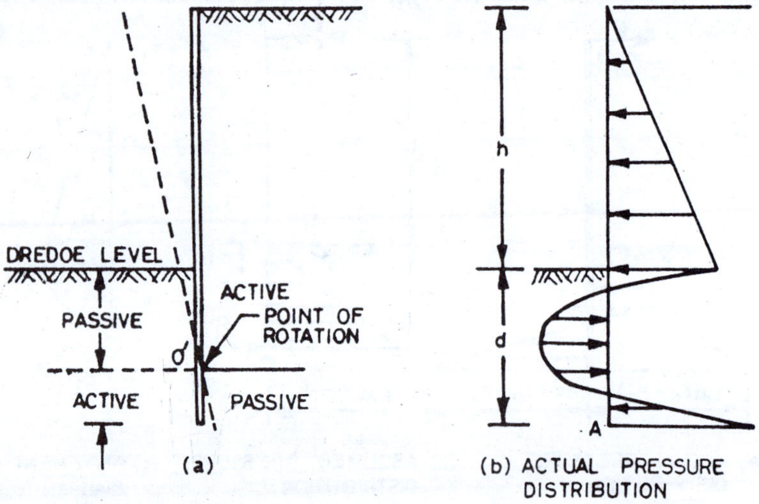

Fig. 20.17. Cantilever Sheet Pile.

the point O' are reversed *i.e.* there is active pressure in the front and passive on the back side. Fig. 20.17 (*b*) shows the actual pressure distribution. As the analysis taking actual pressure distribution is quite complicated, the pressure distribution is generally simplified as shown in Fig. 20.18. In Fig. 20.18, the pressure is zero at point O_1 at a depth a below the dredge level.

The pressure diagram BCO_1 shows the active pressure. The pressure at the dredge level is given by

$$p_1 = \gamma h K_a$$

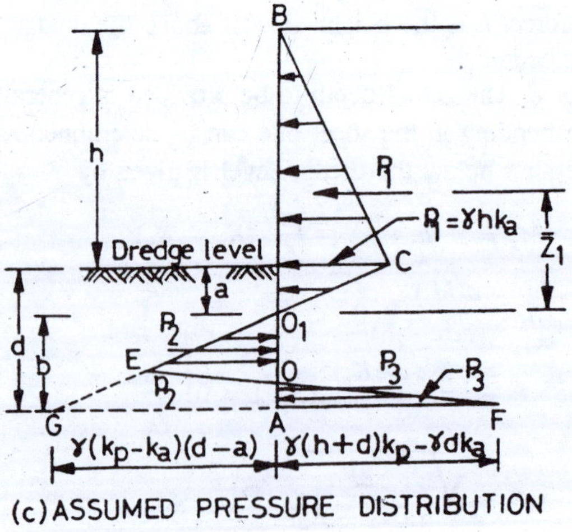

Fig. 20.18. Simplified Pressure Distribution.

The depth a of point O_1 of zero pressure is given by

$$p_1 - \gamma a (K_p - K_a) = 0$$

or
$$a = \frac{p_1}{\gamma (K_p - K_a)} \qquad \text{...(20.24)}$$

Let the total active pressure above point O_1 be P_1 acting at a height of $\overline{Z}_1$ above O_1.

The passive pressure is given by the diagram $O_1 EO$. The passive pressure intensity at the bottom tip A can be expressed as

$$p_2 = \gamma(K_p - K_a)(d - a) = \gamma (K_p - K_a) b$$

where $b = d - a$, in which d is the depth of point A below the dredge level.

The passive pressure is indicated by the diagram OAF on the back side. The intensity of pressure at the tip A is given by

$$p_3 = \gamma (h + d) K_p - \gamma d K_a$$

or
$$p_3 = \gamma (h + b + a) K_p - \gamma (b + a) K_a$$

From the equation of equilibrium in the horizontal direction,

$$P_1 + P_3 - P_2 = 0$$

The total pressure P_3 and P_2 can be expressed in terms of p_3 and p_2 as follows:

$$P_1 + \frac{1}{2} m (p_2 + p_3) - \frac{1}{2} p_2 b = 0 \qquad \text{...(20.25)}$$

In Eq. 20.25, the equivalence of areas has been taken as shown in Fig. 20.19. The height of the point E above the tip A is taken as m.

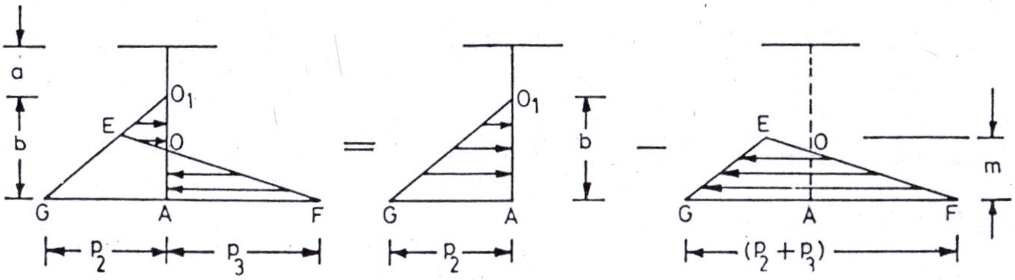

Fig. 20.19.

From Eq. 20.25,
$$m = \frac{1/2\, p_2 b - P_1}{1/2\, (p_2 + p_3)} = \frac{p_2 b - 2P_1}{p_2 + p_3} \qquad \text{...(20.26)}$$

Taking moments of all the forces about A,

$$P_1 (b + \overline{Z}_1) - \frac{1}{2} p_2 b (b/3) + \frac{1}{2} m (p_2 + p_3) \times \frac{m}{3} = 0$$

Substituting the value of m from Eq. 20.26,

$$P_1 (b + \overline{Z}_1) - \frac{p_2 b^2}{6} + \frac{(p_2 + p_3)}{6} \left[\frac{p_2 b - 2P_1}{p_2 + p_3} \right]^2 = 0 \qquad \text{...(20.27)}$$

Eq. 20.27 can be written as
$$b^4 + C_1 b^3 - C_2 b^2 - C_3 b - C_4 = 0 \qquad \text{...(20.28)}$$

where
$$C_1 = \frac{p_4}{\gamma (K_p - K_a)} \; ; \; C_2 = \frac{8P_1}{\gamma(K_p - K_a)}$$

$$C_3 = \frac{6 P_1 [2 \gamma (K_p - K_a) \overline{Z}_1 + p_4]}{[\gamma (K_p - K_a)]^2} \; ; \; C_4 = \frac{P_1 [6 \overline{Z}_1 p_4 + 4 P_1]}{[\gamma (K_p - K_a)]^2}$$

in which $$p_4 = \gamma h K_p + \gamma a (K_p - K_a) \quad ...(20.29)$$

Eq. 20.28 is solved by trial and error to determine b. The value of d is equal to $(b + a)$. The depth d is for a factor of safety of unity. The required depth (D) is usually taken as 1.2 d to 1.4 d. Thus

$$D = 1.2\,d \text{ to } 1.4\,d \quad ...(20.30)$$

This gives a factor of safety of about 1.50 to 2.0.

Alternatively, a factor of safety can be applied to the passive resistance. In that case, the value of K_p is usually taken $\frac{1}{2}$ to $\frac{2}{3}$ of the normal value while computing b from Eq. 20.28, and the required depth D is taken equal to d.

In the above discussions, the depth of water table is not considered. If the water table on the front side is at the same level as on the rear side, the analysis remains unaltered except that the submerged unit weight (γ') should be used for the soil below the water table (see Illustrative Example 20.6). However, if the difference in the two levels is greater than 1 m, the pressure due to water on the sheet pile should be found from the flow net and properly accounted for in the analysis.

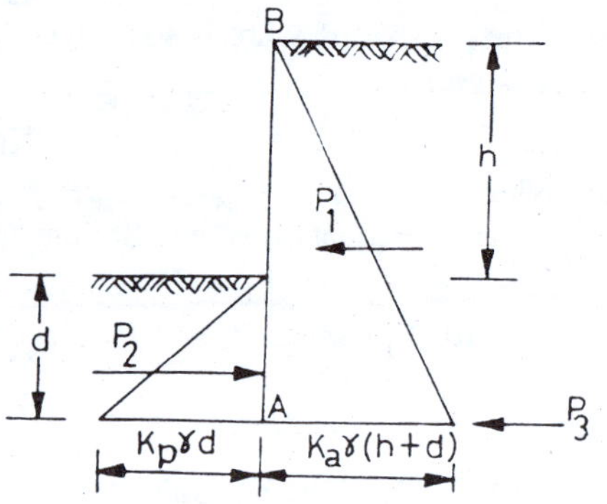

Fig. 20.20. Approximate Analysis.

Approximate Analysis. The exact analysis of the cantilever sheet pile as discussed above is quite involved. An approximate value of d can be obtained using a simplified pressure diagram as shown in Fig. 20.20. In this analysis, the resistance of the pile below the point O is replaced by a concentrated force P_3. (Note that the pressure distribution extends upto tip A).

From the equilibrium in horizontal direction.

$$P_1 - P_2 + P_3 = 0$$

Taking moments about point A, $$P_1\left(\frac{h + d}{3}\right) - P_2 \times \frac{d}{3} = 0$$

Substituting the values of P_1 and P_2,

$$\frac{1}{2} K_a \, \gamma (h + d)^2 \, \frac{(h + d)}{3} - \frac{1}{2} K_p \, \gamma d^2 \times \frac{d}{3} = 0$$

or $$(K_p - K_a)\, d^3 - 3h\, K_a d^2 - 3h^2 K_a d - K_a h^3 = 0 \quad ...(20.31)$$

Eq. 20.31 is is solved by trial and error for d.

The value of d so obtained is usually increased by 20 to 40%. Thus

$$D = 1.2\,d \text{ to } 1.4\,d.$$

20.12. CANTILEVER SHEET PILE PENETRATING CLAY

Fig. 20.21 shows a cantilever sheet pile penetrating clay ($\phi = 0$) below the dredge level. The backfill is of cohesionless soil ($c = 0$). Let the bulk unit weight of the backfill material and clay be, respectively, γ_1 and γ. The cohesion intercept of clay is c.

The pressure p_1 at the dredge line on the back side is given by

$$p_1 = \gamma_1 h K_{a1}$$

Below the dredge level but above the point of rotation O, the passive pressure acts from left to right and the active pressure acts from right to left. Therefore, the pressure at depth Z below the dredge level is given by

$$p_2 = p_p - p_a$$

or $$p_2 = (K_p \gamma Z + 2c\sqrt{K_p}) - [K_a \gamma (Z + h) - 2\,c\sqrt{K_a}\,]$$

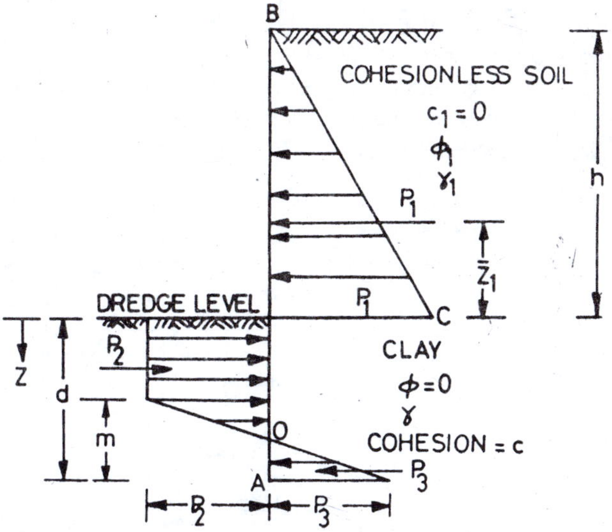

Fig. 20.21. Cantilever Sheet Pile in Clay.

For $\phi = 0$, we have $K_p = K_a = 1.0$.

Therefore, $p_2 = 4c - \gamma h$

Likewise, the pressure p_3 from right to left is given by

or $$p_3 = K_p (h + d)\gamma + 2c\sqrt{K_p} - [K_a \gamma d - 2c\sqrt{K_a}]$$

For $\phi = 0$, we have $p_3 = 4c + \gamma h$

From equilibrium in the horizontal direction, considering equivalent areas as in Fig. 20.19,

$$P_1 - [p_2 \times d] + [p_2 + p_3] \times m/2 = 0$$

or $$P_1 - (4c - \gamma h) d + [8c] \times m/2 = 0$$

or $$m = \frac{(4c - \gamma h) d - P_1}{4c} \quad \text{...(20.32)}$$

Taking moments of all the forces about A,

$$P_1 (\bar{Z}_1 + d) - [4c - \gamma h] \times d \times \frac{d}{2} + \frac{1}{2} \times (8c) \times m \times \frac{m}{3} = 0 \quad \text{...(20.23)}$$

Substituting the value of m from Eq. 20.32,

$$P_1 (\bar{Z}_1 + d) - (4c - \gamma h)\frac{d^2}{2} + \frac{4}{3}c\left[\frac{(4c - \gamma h) d - P_1}{4c}\right]^2 = 0 \quad \text{...(20.34)}$$

The above equation can be written as

$$d^2 (4c - \gamma h) - 2P_1 d - \frac{P_1 (12c\bar{Z}_1 + P_1)}{2c + \gamma h} = 0 \quad \text{...(20.35)}$$

Eq. 20.35 can be solved for d. The actual depth D is kept 40% to 60% more. Thus

$$D = 1.4\,d \text{ to } 1.6\,d$$

Alternatively, the depth d can be computed using a reduced value of $c/2$ or $2c/3$ in Eq. 20.35. In this case, the actual depth D would be equal to the computed value of d, as the factor of safety has already been applied to c.

If the water table exists on both the sides, modification can be done as in the case of cohesionless deposits. The submerged weights are used below the water table (see Illustrative Example 20.8).

20.13. ANCHORED SHEET PILE WITH FREE-EARTH SUPPORT

The stability of anchored sheet pile depends upon the anchor force in addition to that upon the passive earth pressure. The embedment depth is considerably smaller than that in a cantilever sheet pile. Therefore, by this method, the total length of the sheet pile is reduced. Of course, the additional cost of anchors is also to be considered while judging the economy of the two types of construction.

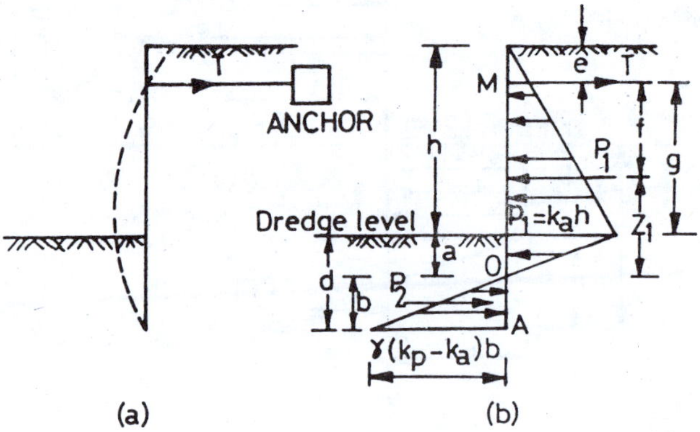

Fig. 20.22. Free Earth Support.

Fig. 20.22 (*a*) shows an anchored sheet pile with free earth support. The deflected shape is also shown. As already mentioned, there is no point of contraflexure below the dredge level. Thus, below the dredge level, no pivot point exists for the statical system. The statical analysis is based on the assumption that the soil into which the pile is driven does not produce effective restraint to induce negative bending moment at its support

The equations for the depth d are derived separately for the cohesionless and cohesive soils.

(*a*) Cohesionless Soils.

Fig. 20.22 (*a*) shows the forces acting on the pile, assuming that the material above and below the dredge level is cohesionless.

From equilibrium, $$T + P_2 - P_1 = 0 \quad \ldots(20.36)$$

where T is the tensile force in anchor.

The depth a to the point of zero pressure can be determined as under.

$$\gamma K_a (h + a) - \gamma K_p a = 0$$

or $$a\gamma (K_p - K_a) = \gamma K_a h$$

or $$a = \frac{h K_a}{(K_p - K_a)} \quad \ldots(20.37)$$

Therefore, $$P_2 = \frac{1}{2} p_2 b = \frac{1}{2} \gamma (K_p - K_a) b^2 \quad \ldots(a)$$

where $$p_2 = \gamma (K_p - K_a) b$$

Taking moments of all the forces about anchor point M,

$$P_1 (a + h - e - \bar{Z}_1) - P_2 (h - e + a + 2b/3) = 0 \quad \ldots[20.38(a)]$$

The distance $\bar{Z}_1$ is determined as in the case of cantilever piles.

Substituting the value of P_2 from Eq. (*a*) in Eq. 20.38 (*a*),

$$P_1 (a + h - e - \bar{Z}_1) - \gamma (K_p - K_a) b \times b/2 (h - e + a + 2b/3) = 0 \quad \ldots[20.38(b)]$$

The above equation can be written as

$$b^3 (K_p - K_a)\gamma/3 + b^2 (K_p - K_a)\gamma/2 (g + a) - P_1 f = 0 \quad ...(20.39)$$

or
$$b^3 + 1.5b^2 (g + a) - \frac{3 P_1 f}{\gamma(K_p - K_a)} = 0 \quad ...(20.40)$$

where $f = a + h - e - \bar{Z}_1$ and $g = h - e$

Eq. 20.40 can be solved for b. Then d is determined as

$$d = b + a$$

The actual depth D is taken equal to 1.2 to 1.4 times d.

The force in anchor rod can be obtained from Eq. 20.36 as,

$$T = P_1 - P_2$$

The values of P_1 and P_2 are obtained from pressure diagrams.

(*b*) Cohesive Soils

Let us now consider the case when the anchored sheet pile is driven in clay ($\phi = 0$), but has the backfill of cohesionless, granular material (Fig. 20.23). The pressure distribution above the dredge line is the same as that in the case of cohesionless soils. However, below the dredge line, the pressure is given by

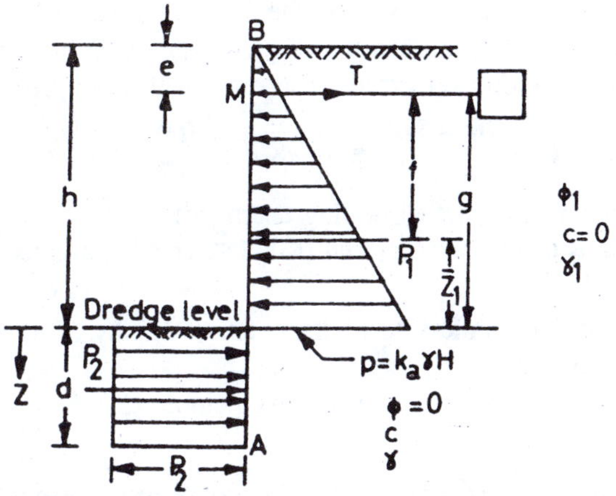

Fig. 20.23. Anchored Sheet Pile driven in Clay.

$$p_2 = (K_p \gamma Z + 2c\sqrt{K_p}) - [K_a(Z + h)\gamma - 2c\sqrt{K_a}\,]$$

For $\phi = 0$, $K_p = K_a = 1.0$. Therefore

$$p_2 = 2c + 2c - \gamma h = 4c - \gamma h$$

From equilibrium of forces, $P_1 - P_2 = T$

or
$$P_1 - p_2 \times d = T \quad ...(20.41)$$

Taking moments of all forces about M,

$$P_1 \times f - p_2 d (g + d/2) = 0$$

Substituting $p_2 = 4c - \gamma h$,

$$P_1 \times f - (4c - \gamma h)\, d\, (g + d/2) = 0$$

or
$$d^2 + 2gd - \frac{2 P_1 f}{4c - \gamma H} \quad ...(20.42)$$

Eq. 20.42 can be solved for d. The actual depth (D) provided is 20 to 40% more than d.

It may be noted that the wall becomes unstable when $p_2 = 0$, *i.e.*, $4c - \gamma H = 0$

or
$$\frac{c}{\gamma H} = \frac{1}{4} = 0.25 \qquad \text{...(20.43)}$$

The left hand side is equal to the stability number (S_n) defined in chapter 18. In other words, the walls becomes unstable when S_n is equal to or less then 0.25. If the adhesion of clay with the sheet pile (c_a) is considered, Eq. 20.43 is modified as

$$S_n = \frac{c}{\gamma H} \sqrt{1 + \frac{c_a}{c}} \qquad \text{...(20.44)}$$

Taking, $\sqrt{1 + c_a/c} = 1.25,$

$$S_n = 0.25 \times 1.25 = 0.31$$

Therefore, the minimum stability number (S_n) required is 0.31. If the factor of safety required is F, the stability number (S_n) should be equal to 0.31 F or more.

20.14. ROWE'S MOMENT REDUCTION CURVES

As sheet piles are relatively flexible, these deflect considerably. Their flexibility causes a redistribution of lateral earth pressure. The net effect is that the maximum bending moment is considerably reduced below the value obtained for the free-earth supports discussed in the preceding section.

Rowe (1952) developed a theoretical relation between the maximum bending moment and the flexibility of the sheet pile and gave moment reduction curves. The relative flexibility (ρ) is defined as

$$\rho = \frac{(h + D)^4}{EI} = 1.1 \times 10^{-6} \frac{(H)^4}{EI} \qquad \text{...(20.45)}$$

where h = retained height (m), D = actual driving depth (m),

E = Young's modulus of the pile material (MN/m^2) and I = moment of inertia of the pile (m^4/m),

H = total length of the pile.

For anchored sheet piles in cohesionless soils, the relative density is important. The relative depth of anchor factor, $\beta = e/H$ is also relevant.

For anchored sheet piles in cohesive soils, the stability number (S_n), as given below, is also required.

$$S_n = 1.25\, c/(\gamma\, h) \qquad \text{...(20.46)}$$

The relative height of piling factor $\alpha = h/H$ is also important for cohesive soils.

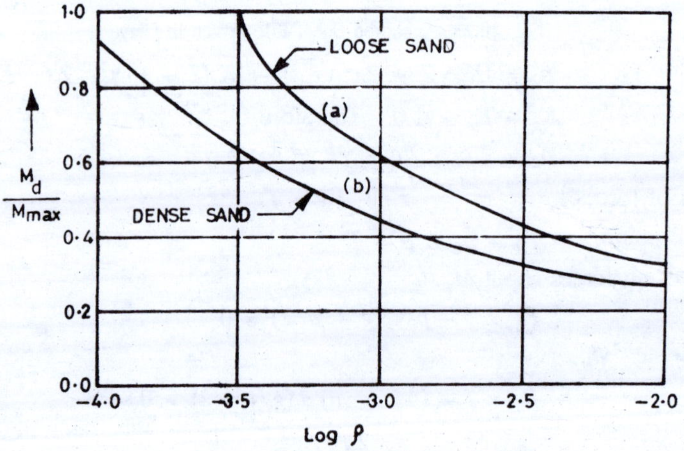

Fig. 20.24.

Fig. 20.24 shows a typical moment reduction curves for cohesionless soils. The ratio M_d/M_{max} is determined directly for the given value of ρ. The curve (*a*) is for loose sand (relative density = 0) and the curve (*b*) for dense sand (relative density = 100%). The value of M_{max} being known from the free-earth support analysis, the design moment M_d can be computed.

(For more details, the original paper may be consulted).

20.15. ANCHORED SHEET PILE WITH FIXED-EARTH SUPPORT

Fig. 20.25 (*a*) shows the deflected shape of an anchored sheet pile with fixed-earth support. The elastic line changes its curvature at the inflexion point *I*. The soil into which the sheet is driven exerts a large

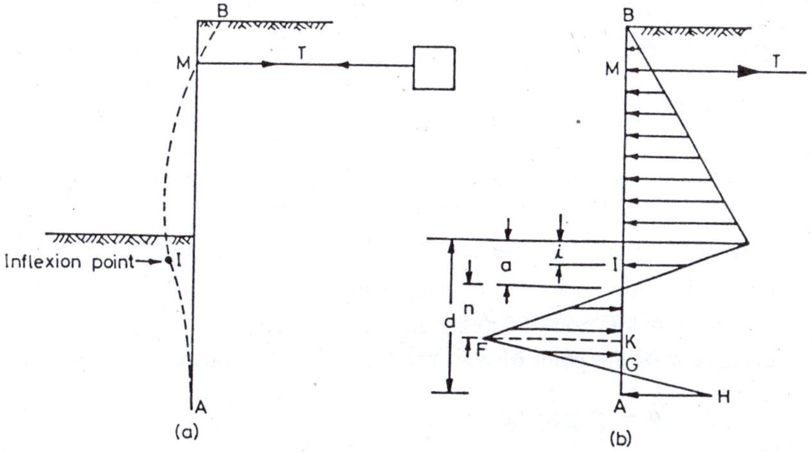

Fig. 20.25.

restraint on the lower part of the pile and causes a change in curvature. Fig. 20.25 (*b*) shows the pressure distribution. Blum (1931) gave a mathematical relationship between (i/h) and ϕ (Fig. 20.26), where i is the depth of the point of inflexion *I* below the dredge level and h is the height of sheet pile above the dredge level. Thus inflection point *I* is located.

For simplicity, the lower portion of the pressure diagram on the right hand side in Fig. 20.25 (*b*) is replaced by a concentrated force R_k at point *K* and the diagram shown in Fig. 20.27 (*a*) is used in the analysis. The magnitude of R_k is initially unknown, but it is automatically excluded from calculations when the moments are taken about *K*. Once the depth has been found, R_k can be determined from the equilibrium equation in the horizontal direction.

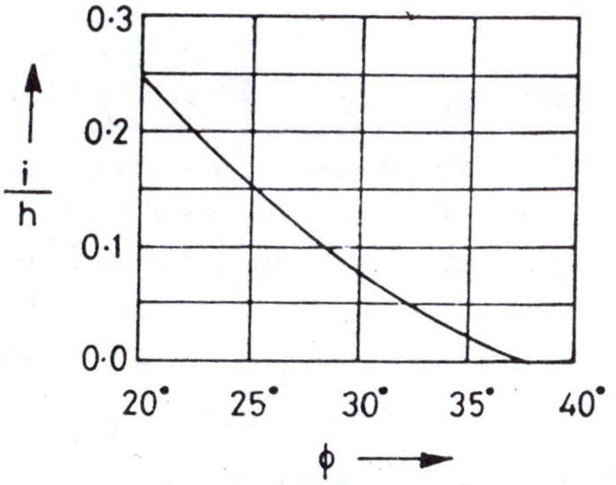

Fig. 20.26.

As the exact analysis of the anchored sheet pile with fixed-earth support is complicated, an approximate method, known as *equivalent-beam method* is generally used. It is assumed that the sheet pile is a beam which is simply supported at the anchor point *M* and fixed at the lower end *K*. Fig. 20.27 (*b*) shows the bending moment diagram. The bending moment is zero at the inflexion point *I*. Theoretically, the lower part *IK* of the pile can be removed and the shear force can be replaced by a reaction R_I. Thus, a simply-supported beam *BI* is obtained [Fig. 20.27 (*c*)].

The following procedure is used for the analysis of the sheet pile with fixed-earth support, using equivalent beam method.

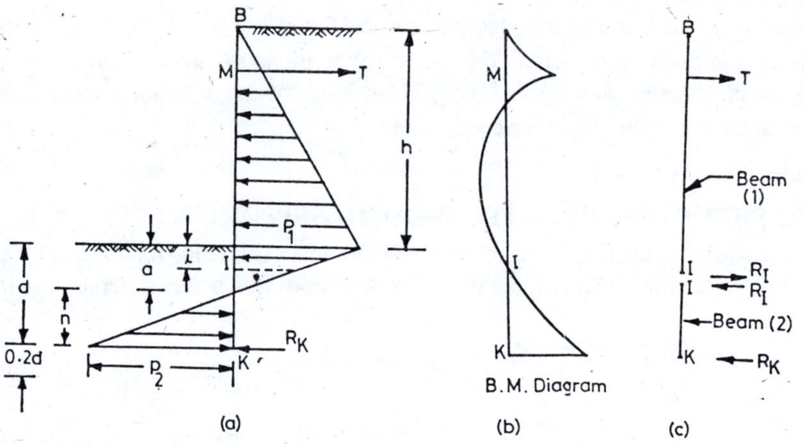

Fig. 20.27. Simplified Pressure Diagram.

(*a*) Upper Beam BI

(1) Determine the pressure p_1 at the dredge level.

(2) Estimate the angle of shearing resistance ϕ' of the soil.

(3) Determine the distance i of the point of inflexion from Fig. 20.26.

(4) Determine the distance a of the point of zero pressure from the equation,

$$a = \frac{p_1}{\gamma (K_p - K_a)} \qquad ...(20.47)$$

(5) Determine the pressure p_o at the point of inflexion from the relation,

$$p_o = \frac{p_1}{a}(a - i) \qquad ...(20.48)$$

(6) Determine the reaction R_I for the beam *IB* by taking moments about the point *M* of anchor of all the forces acting on *IB* [Fig. 20.28 (*a*)].

(*b*) Lower Beam IK

(7) Determine the pressure p_2 from the relation

$$p_2 = \gamma (K_p - K_a)(d - a) \qquad ...(20.49)$$

Alternatively, $$p_2 = \frac{p_o}{(a - i)} \times (d - a)$$

(8) Determine the distance $(d - a)$ by taking moments of the forces on the beam *IK* about *K* [Fig. 20.28 (*b*)] The reaction R_I on the lower beam is equal and opposite to that on the upper beam.

(9) Calculate d from Eq. 20.49 and hence find $D = 1.2\,d$.

(10) Determine the tension T in anchor by considering the equilibrium of beam *IB*. Thus

$$T = P_1 - R_I \qquad ...(20.50)$$

where P_1 = total force due to pressure on *IB*.

20.16. DESIGN OF ANCHORS

The anchors used in sheet pile walls are of the following types:

(1) Anchor plates and Beams (also, known as deadman) (Fig. 20.29).

(2) Tie backs.

(3) Vertical Anchor piles.

(4) Anchor beams supported by batter piles (Fig. 20.30).

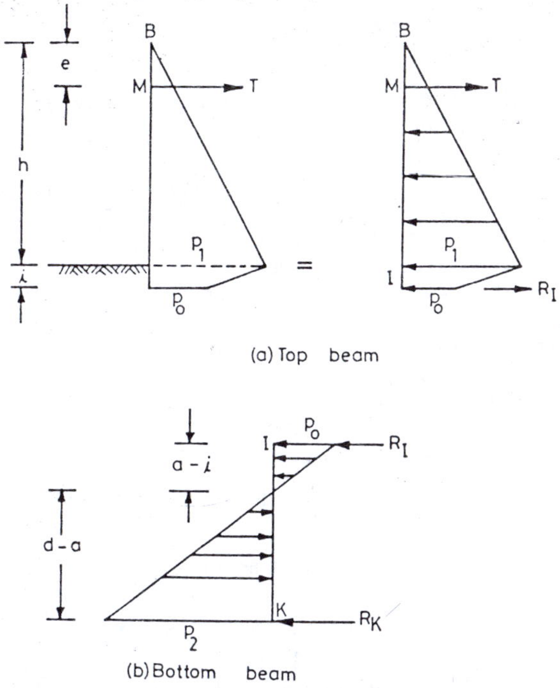

Fig. 20.28.

The design of anchor plates and beams is discussed below.

Anchor plates and beams are made of cast-concrete blocks. A wale (horizontal beams) is placed at the front (or back) face of the sheet pile, and a tie rod is attached to it. The other end of the tie rod is connected to an anchor plate or a beam (Fig. 20.29).

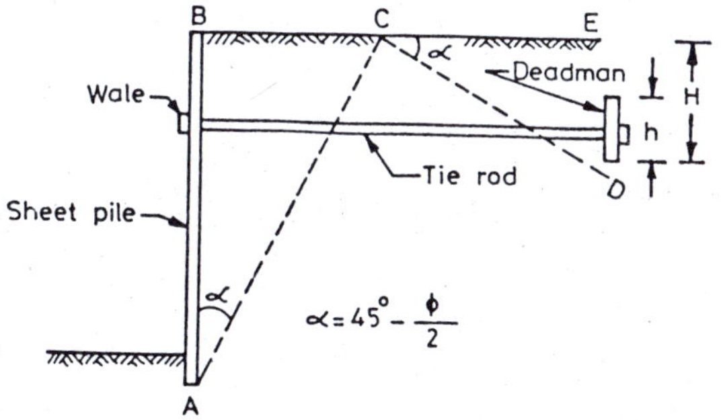

Fig. 20.29. Anchor Plates.

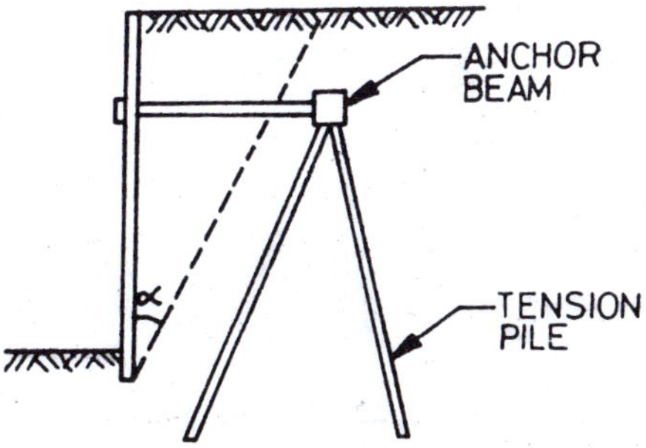

Fig. 20.30. Batter Piles.

The resistance offered by an anchor plate or a beam is derived from the passive resistance of the soil in front of the plate. For full passive resistance to develop, the anchor plate must be located in zone *CDE*. Teng (1962) gave the following equations for the ultimate resistance of anchor plates in granular soils located at or near the ground surface.

Let B be the length of the anchor perpendicular to the cross section and let h be the height of the anchor.

(*a*) For continuous plates or beams with $B/h \geq 5$, the ultimate resistance is given by

$$P_u = B(P_p - P_a)$$

or

$$P_u = B\left(\frac{1}{2}\gamma H^2 K_p - \frac{1}{2}\gamma H^2 K_a\right)$$

or $$P_u = \frac{\gamma H^2 B}{2}(K_p - K_a) \qquad ...(20.51)$$

where H is the depth of the lower face of the anchor beam from the ground surface.

(*b*) For plates or beams with $B/h < 5$, the ultimate resistance is given by

$$P_u = B(P_p - P_a) + \frac{1}{2}K_o \gamma (\sqrt{K_p} + \sqrt{K_o})H^3 \tan\phi$$

where K_o = coefficient of earth at rest (= 0.40).

Thus $$P_u = \frac{\gamma H^2 B}{2}(K_p - K_a) + \frac{1}{3}K_o \gamma (\sqrt{K_p} + \sqrt{K_o})H^3 \tan\phi \qquad ...(20.52)$$

The allowable resistance is taken as

$$P_a = \frac{P_u}{FS} \qquad ...(20.53)$$

where FS = factor of safety (generally taken equal to 2.0).

The centre-to-centre spacing of anchors is obtained from the relation,

$$s = P_a / T \qquad ...(20.54)$$

where T = tension in sheet pile per unit length as obtained from the analysis of anchored sheet pile.

ILLUSTRATIVE EXAMPLES

Illustrative Example 20.1. *Check the stability of the gravity retaining wall shown in Fig. E-20.1. Take allowable soil pressure equal to 600 kN/m². Use Coulomb's theory.*

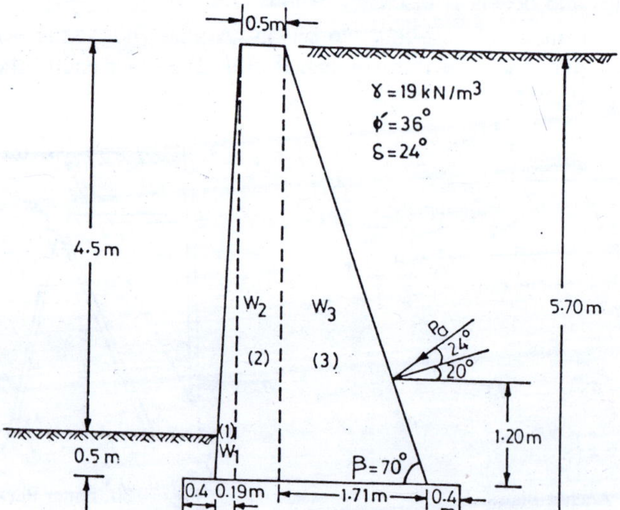

Fig. E-20.1.

Solution. From Eq. 19.38, $P_a = \frac{1}{2} K_a \gamma H^2$

where $$K_a = \frac{\sin^2(\beta + \phi')}{\sin^2\beta \sin(\beta - \delta)\left[1 + \sqrt{\frac{\sin(\phi' + \delta)\sin(\phi' - i)}{\sin(\beta - \delta)\sin(\beta + i)}}\right]^2}$$

or
$$K_a = \frac{\sin^2(70 + 36)}{\sin^2 70° \sin(70° - 24°)\left[1 + \sqrt{\dfrac{\sin(36° + 24°)\sin(36° - 0)}{\sin(70° - 24°)\sin(70° + 0)}}\right]^2}$$

or
$$K_a = \frac{0.924}{0.883 \times 0.719\left[1 + \sqrt{\dfrac{0.866 \times 0.588}{0.719 \times 0.940}}\right]^2} = 0.417$$

Therefore, $P_a = \frac{1}{2} \times 0.417 \times 19 \times (5.7)^2 = 128.7$ kN

The total pressure acts inclined at 24° to the normal.

Horizontal component, $P_h = P_a \cos(20° + 24°) = 92.6$ kN

Vertical component, $P_v = P_a \sin(20° + 24°) = 89.4$ kN

Calculations are shown in the tabular form. The moment are taken about toe. The clockwise moments are taken as positive. The unit weight of concrete is taken as 24 kN/m^3.

S.No.	*Description*	*Forces (kN)*		*Lever arm*	*Moments about toe (kN-m)*	
		Vertical (kN)	*Horizontal (kN)*		*Clockwise*	*Counter Clockwise*
1.	Weight, W_1 $= 1/2 \times 5 \times 0.19 \times 24$	11.4		0.53 m	6.0	
2.	Weight, W_2 $= 5 \times 0.5 \times 24$	60		0.84 m	50.4	
3.	Weight, W_3 $= 1/2 \times 5 \times 1.71 \times 24$	102.6		1.66 m	170.3	
4.	Weight, W_4 $= 3.2 \times 0.7 \times 24$	53.8		1.60 m	86.1	
5.	Vertical component of $P = P_{\alpha v}$	89.4		2.39 m	213.6	
6.	Horizontal component of $P = P_{\alpha h}$		92.6	1.90 m		175.9
		Σ 317.2	Σ 92.6		Σ 526.4	Σ 175.9

$\Sigma M = 350.5$ kN–m

Neglecting passive resistance, the factor of safety against sliding is given by Eq. 20.3 as

$$F_s = \frac{\mu R_v}{R_H} = \frac{\tan 24° \times 317.2}{92.6} = 1.53 \text{ (safe)}$$

The factor of safety against overturning is obtained from Eq. 20.4,

$$F_o = \frac{\Sigma M_R}{\Sigma M_o} = \frac{526.4}{175.9} = 2.99 \text{ (safe)}$$

From Eq. 20.17, $\bar{x} = \frac{\Sigma M}{\Sigma V} = \frac{350.5}{317.2} = 1.10$ m

From Eq. 20.18, $e = b/2 - \bar{x} = 1.60 - 1.10 = 0.50$ m

As $e < b/6$, there is no tension.

The pressures at the base are determined from Eq. 20.19.

$$p_{max} = \frac{\Sigma V}{b}\left(1 + \frac{6e}{6}\right) = \frac{317.2}{3.20}\left(1 + \frac{6 \times 0.5}{3.2}\right) = 192 \text{ kN/m}^2$$

$$p_{min} = \frac{\Sigma V}{b}\left(1 - \frac{6e}{6}\right) = \frac{317.2}{3.20}\left(1 - \frac{6 \times 0.5}{3.2}\right) = 6.2 \text{ kN/m}^2$$

The factor of safety against bearing capacity failure is given by Eq. 20.7 as

$$F_b = \frac{q_{na}}{p_{max}} = \frac{600}{192} = 3.1 \text{ (safe)}$$

Illustrative Example 20.2. *Check the stability of the cantilever retaining wall shown in Fig. E-20.2. The allowable soil pressure is 500 kN/m². $\phi' = 34°$, $\delta = 25°$, $\gamma = 18$ kN/m³, $i = 15°$.*

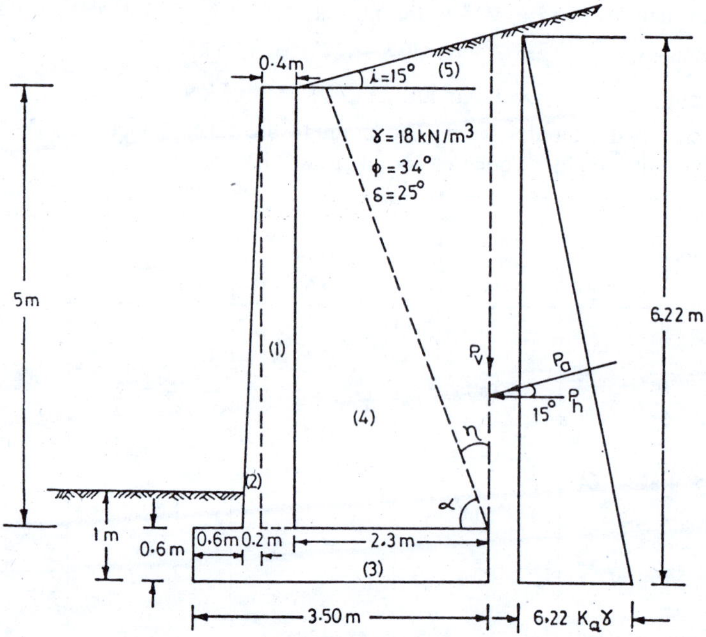

Fig. E-20.2.

Solution. Let us first ascertain whether Rankine's theory is applicable to the cantilever retaining wall. From Eq. 20.9,

$$\eta = (45° + i/2) - \frac{\phi'}{2} - \sin^{-1}\left(\frac{\sin i}{\sin \varphi'}\right)$$

$$= (45° + 7.5) - 17 - \sin^{-1}\left(\frac{\sin 15°}{\sin 34°}\right) = 7.9°$$

The shear zone does not intersect the stem. Therefore the Rankine theory can be applied.

From Eq. 19.23, $$P_a = \frac{1}{2} K_a \gamma H^2$$

From Eq. 19.15, $$K_a = \cos i \times \frac{\cos i - \sqrt{\cos^2 i - \cos^2 \phi'}}{\cos i + \sqrt{\cos^2 i - \cos^2 \phi'}}$$

$$K_a = \cos 15° \times \frac{\cos 15° - \sqrt{\cos^2 15° - \cos^2 34°}}{\cos 15° + \sqrt{\cos^2 15° - \cos^2 34°}} = 0.311$$

Therefore, $P_a = \frac{1}{2} \times 0.311 \times 18 \times (6.22)^2 = 108.3$ kN

$$P_v = P_a \sin 15° = 28.0 \text{ kN}$$

$$P_h = P_a \cos 15° = 104.6 \text{ kN}$$

S.No.	Description	Forces (kN)		Lever arm (m)	Moments about toe (kN-m)	
		Vertical (kN)	Horizontal (kN)		Clockwise	Counter Clockwise
1.	$W_1 = 0.4 \times 5.0 \times 24$	48		1.00	48	
2.	$W_2 = 0.2/2 \times 5.0 \times 24$	12.0		0.73	8.8	
3.	$W_3 = 0.6 \times 3.50 \times 24$	50.4		1.75	88.2	
4.	$W_4 = 2.3 \times 5.0 \times 18$	207.0		2.35	486.5	
5.	$W_5 = 2.30 \times 0.62/2 \times 18$	12.8		2.73	35	
6.	P_v	28		3.50	98	
7.	P_h	—	104.6	2.07	—	216.5
		Σ 358.2	Σ 104.6		Σ 764.5	Σ 216.5

From Eq. 20.3, the factor of safety against sliding,

$$F_s = \frac{\mu R_V}{R_H} = \frac{\tan 25° \times 358.2}{104.6} = 1.60 \text{ (safe)}$$

From Eq. 20.4, the factor of safety against overturning,

$$F_o = \frac{\Sigma M_R}{\Sigma M_O} = \frac{764.5}{216.5} = 3.53 \text{ (safe)}$$

From Eq. 20.17, $$\bar{x} = \frac{\Sigma M}{\Sigma V} = \frac{764.5 - 216.5}{358.2} = 1.53 \text{ m}$$

From Eq. 20.18, $$e = b/2 - \bar{x} = 1.75 - 1.53 = 0.22 \text{ m} < b/6$$ (No tension)

From Eq. 20.19, $$p_{max} = \frac{358.2}{3.50}\left(1 + \frac{6 \times 0.22}{3.50}\right) = 141.9 \text{ kN/m}^2$$

$$p_{min} = \frac{358.2}{3.50}\left(1 - \frac{6 \times 0.22}{3.50}\right) = 63.7 \text{ kN/m}^2$$

From Eq. 20.7, the factor of safety against bearing capacity failure,

$$F_b = \frac{q_{na}}{p_{max}} = \frac{500}{141.9} = 3.52 \text{ (safe)}$$

Illustrative Example 20.3. *Determine the required depth of penetration for the cantilever sheet pile shown in Fig. E-20.3. Take $\gamma = 16$ kN/m³.*

Solution. (Refer to Fig. 20.18 for notations)

$$K_a = \tan^2(45 - 30/2) = 0.333,$$

$$K_p = \tan^2(45 + 30/2) = 3.000$$

$$p_1 = 0.333 \times 16 \times 5.0 = 26.6 \text{ kN/m}^2$$

$$a = \frac{p_1}{\gamma(K_p - K_a)} = \frac{26.6}{16 \times 2.667} = 0.62 \text{ m}$$

$$P_1 = \frac{1}{2} \times 26.6 \times 5 + \frac{1}{2} \times 26.60 \times 0.62$$

$$= 66.5 + 8.2 = 74.7 \text{ kN}$$

Taking moments about O_1 and dividing by P_1.

$$\bar{Z}_1 = \frac{66.5 \times 2.29 + 8.2 \times 0.41}{74.7} = 2.08 \text{ m}$$

$$p_2 = \gamma (K_p - K_a) b$$

$$= 16 \times 2.667 \times b = 42.7 b$$

$$p_3 = \gamma (h + d) K_p - \gamma d K_a$$

$$= 16 (5 + b + 0.62) \times 3.0 - 16 (b + 0.62) \times 0.333$$

or $$p_3 = 266.5 + 42.7 b$$

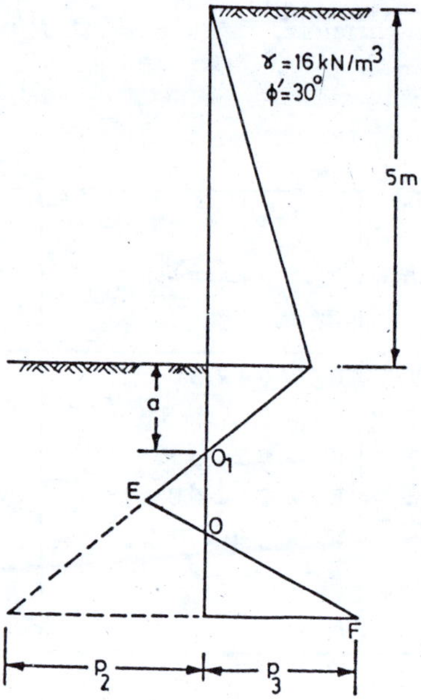

Fig. E-20.3.

From Eq. 20.26, $$m = \frac{p_2 b - 2 P_1}{p_2 + p_3}$$

$$= \frac{42.7 b^2 - 2 \times 74.7}{266.5 + 85.4 b}$$

From Eq. 20.27, $$P_1 (b + \bar{Z}_1) - \frac{p_2 b^2}{6} + \frac{(p_2 + p_3)}{6} m^2 = 0$$

or $$74.7 (b + 2.08) - \frac{42.7 b^3}{6} + \left(\frac{266.5 + 85.4 b}{6}\right) \left(\frac{42.7 b^2 - 149.4}{266.5 + 85.4 b}\right)^2 = 0$$

or $$448.2 (b + 2.08) - 42.7 b^3 + \frac{(42.7 b^2 - 149.4)^2}{266.5 + 85.4 b} = 0$$

Solving by trial and error, $b = 4.4$ m.

Alternative Method for b

Using Eq. 20.28, $$b^4 + C_1 b^3 - C_2 b^2 - C_3 b - C_4 = 0$$

where $$C_1 = \frac{p_4}{\gamma (K_p - K_a)} = \frac{16 \times 5 \times 3.0 + 16 \times 0.62 \times 2.667}{16 \times 2.667}$$

$$= \frac{265.6}{42.67} = 6.24$$

$$C_2 = \frac{8 P_1}{\gamma (K_p - K_a)} = \frac{8 \times 74.7}{16 \times 2.667} = 14.0$$

$$C_3 = \frac{6 P_1 [2 \gamma (K_p - K_a) \bar{Z}_1 + p_4]}{[\gamma (K_p - K_a)]^2}$$

$$= \frac{6 \times 74.7 [2 \times 16 \times 2.667 \times 2.08 + 265.6]}{[16 \times 2.667]^2} = 109.07$$

and $$C_4 = \frac{P_1 [6 \bar{Z}_1 p_4 + 4 P_1]}{[\gamma (K_p - K_a)]^2} = \frac{74.7 [6 \times 2.08 \times 265.6 + 4 \times 74.7]}{(16 \times 2.667)^2}$$

$$= 148.24$$

Therefore, $$b^4 + 6.24 b^3 - 14.0 b^2 - 109.07 b - 148.24 = 0$$

Solving by trial and error, $b = 4.40$ m

Therefore, $d = b + a = 4.40 + 0.62 = \mathbf{5.02\ m}$

$D = 1.30\,d = \mathbf{6.50\ m}$

Illustrative Example 20.4. *Determine the depth of penetration for the cantilever sheet pile wall of Illustrative Example 20.3 by the approximate method.*

Solution. From Eq. 20.31, taking $h = 5$ m,

$$(K_p - K_a)\,d^3 - 3\,h\,K_a\,d^2 - 3\,h^2\,K_a\,d - K_a\,h^3 = 0$$

or $$2.667\,d^3 - 5\,d^2 - 25\,d - 41.67 = 0$$

Solving by trial and error, $d = 4.70$ m.

Therefore, $D = 1.3\,d = \mathbf{6.10\ m}$

Illustrative Example 20.5. *Determine the depth of penetration of the cantilever sheet pile shown in Fig. E-20.5. The water level on both sides is the same.*

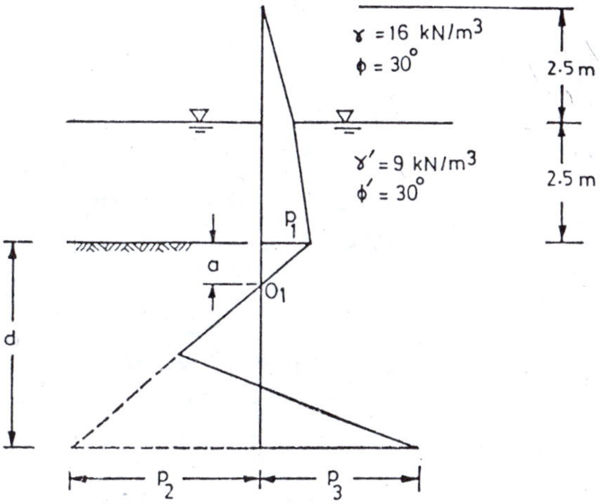

Fig. E-20.5.

Solution. $K_a = 0.333$, $K_p = 3.0$.

$$p_1 = 0.333 \times 16 \times 2.50 + 0.33\,(9.0) \times 2.50$$

$$p_1 = 13.3 + 7.5 = 20.8\ \text{kN/m}^2$$

From Eq. 20.24,
$$a = \frac{p_1}{\gamma'\,(K_p - K_a)} = \frac{20.8}{9.0 \times (3.0 - 0.333)} = 0.87\ \text{m}.$$

$$P_1 = \frac{1}{2} \times 13.3 \times 2.5 + 13.3 \times 2.50 + 7.5 \times 2.5/2 + 20.8 \times 0.87/2$$

$$P_1 = 16.6 + 33.3 + 9.4 + 9.0 = 68.3\ \text{kN}$$

Taking moments about O_1, $$P_1\,\bar{Z}_1 = 16.6 \times 4.20 + 33.3 \times 2.12 + 9.4 \times 1.70 + 9.0 \times 0.58$$

or $$\bar{Z}_1 = \frac{69.7 + 70.6 + 16 + 5.2}{68.3} = 2.36\ m$$

$$p_2 = \gamma'\,(K_p - K_a)\,b = 9.0 \times 2.667 \times b = 24.0b$$

$$p_3 = [16 \times 2.50 + 9.0 \times 2.50] \times K_p + \gamma' d\,(K_p - K_a)$$

$$= (40 + 22.5) \times 3.0 + 9.0 \times (b + 0.87) \times 2.667$$

or $$p_3 = 187.5 + 24b + 21.0 = 208.5 + 24b$$

From Eq. 20.26, $$m = \frac{p_2 b - 2P_1}{p_2 + p_3} = \frac{24b^2 - 2 \times 68.3}{208.5 + 48\,b}$$

From Eq. 20.27, $P_1(b + \bar{Z}_1) - \frac{p_2 b^2}{6} + \frac{(p_2 + p_3)}{6}\left[\frac{24b^2 - 136.6}{(p_2 + p_3)}\right]^2$

or $$68.3\,(b + 2.36) - \frac{24.0b^3}{6} + \frac{(24b^2 - 136.6)^2}{6\,(208.5 + 48.0\,b)} = 0$$

or $$409.8\,(b + 2.36) - 24.0b^3 + \frac{(24b^2 - 136.6)^2}{208.5 + 48\,b} = 0$$

Solving by trial and error, $b = 5.50$ m.

$$d = b + a = 5.50 + 0.87 = 6.37 \text{ m}$$

$$D = 1.3\,d = 8.30 \text{ m}$$

Alternative method

From Eq. 20.28, $$b^4 + C_1 b^3 - C_2 b^2 - C_3 b - C_4 = 0$$

where $$C_1 = \frac{p_4}{\gamma'(K_p - K_a)}$$

where $$p_4 = (16 \times 2.5 + 9.0 \times 2.5)\,K_p + 9.0\,a(K_p - K_a)$$

$$= (40 + 22.5) \times 3.0 + 9 \times 0.87 \times 2.667 = 208.4$$

Therefore, $$C_1 = \frac{208.4}{9.0 \times 2.667} = 8.68$$

$$C_2 = \frac{8P_1}{\gamma'(K_p - K_a)} = \frac{8 \times 68.3}{9.0 \times 2.667} = 22.76$$

$$C_3 = \frac{6P_1\,[2\gamma'(K_p - K_a)\,\bar{Z}_1 + p_4]}{[\gamma'(K_p - K_a)]^2}$$

and $$C_3 = \frac{6 \times 68.3\,[2 \times 9.0 \times 2.667 \times 2.36 + 208.4]}{(9.0 \times 2.667)^2} = 228.81$$

and $$C_4 = \frac{P_1\,[6\,\bar{Z}_1 p_4 + 4\,P_1]}{[\gamma'(K_p - K_a)]^2}$$

$$= \frac{68.3\,[6 \times 2.36 \times 208.4 + 4 \times 68.3]}{[9.0 \times 2.667]^2} = 382.21$$

Therefore, $b^4 + 8.68b^3 - 22.76\,b^2 - 228.81\,b - 382.21 = 0$

Solving by trial and error, $b = 5.50$ m.

$$d = b + a = 5.50 + 0.87 = 6.50 \text{ m.}$$

$$D = 1.3\,d = 8.30 \text{ m.}$$

Illustrative Example 20.6. *Determine the depth of embedment for the cantilever sheet pile shown in Fig. E-20.6.*

Solution. (See Fig. 20.21 for notations) In this case,

$$K_a = 0.333,\; K_p = 3.0$$

$$p_1 = 16.0 \times 5.0 \times 0.333 = 26.7$$

$$P_1 = \frac{1}{2} \times 26.7 \times 5 = 66.8 \text{ kN}$$

$$\bar{Z}_1 = 5/3 = 1.67 \text{ m}$$

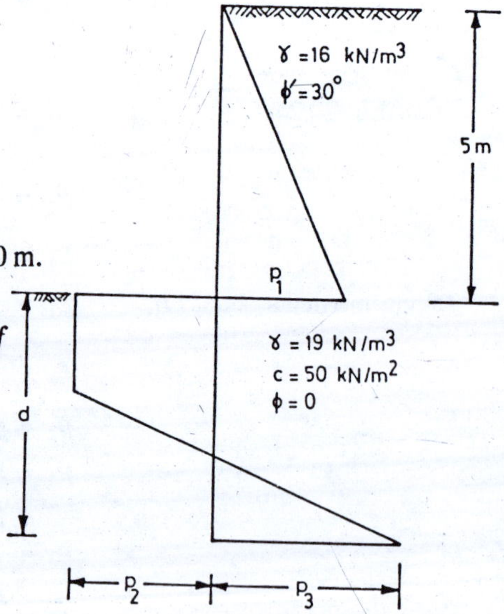

Fig. E-20.6.

$$p_2 = 4c - \gamma h = 4 \times 50 - 19 \times 5 = 105.0$$

$$p_3 = 4c + \gamma h = 4 \times 50 + 19 \times 5 = 295$$

From Eq. 20.32,
$$m = \frac{(4c - \gamma h)\, d - P_1}{4c} = \frac{(4 \times 50 - 19 \times 5)\, d - 66.8}{4 \times 50}$$

$$= 0.53\, d - 0.334$$

From Eq. 20.34,
$$P_1 (\bar{Z}_1 + d) - (4c - \gamma h) \frac{d^2}{2} + \frac{4}{3} \times \frac{1}{16} \frac{[(4c - \gamma h)\, d - P_1]^2}{c} = 0$$

or
$$66.8\,(1.67 + d) - (200 - 19 \times 5) \frac{d^2}{2} + \frac{1}{12} \frac{[(4 \times 50 - 19 \times 5)\, d - 66.8]^2}{50} = 0$$

or
$$111.56 + 66.8\, d - 52.5\, d^2 + \frac{1}{600} (105\, d - 66.8)^2 = 0$$

Simplifying and solving, $d = 2.60$ m

Alternative method

From Eq. 20.35,
$$d^2 (4c - \gamma h) - 2 P_1 d - \frac{P_1 (12\, c\, \bar{Z}_1 + P_1)}{2c + \gamma h} = 0$$

or
$$(4 \times 50 - 19 \times 5)\, d^2 - 2 \times 66.8\, d - \frac{66.8\,(12 \times 50 \times 1.67 + 66.8)}{(2 \times 50 + 19 \times 5)} = 0$$

or
$$d^2 - 1.27\, d - 3.49 = 0.$$

or
$$d = 2.60 \text{ m}$$

$$D = 1.50\, d = 3.90 \text{ m}.$$

Illustrative Example 20.7. *Determine the depth of embedment of the anchored sheet pile shown in Fig. E-20.7. Also determine the force in the anchor per metre of the wall. Assume free-earth support conditions.*

Solution. (See Fig. 20.22 for notations) $K_a = 0.27$, $K_p = 3.69$

$$p_1 = 16 \times 3.0 \times 0.27 + 9 \times 5.0 \times 0.27$$

or
$$p_1 = 12.96 + 12.15 = 25.11$$

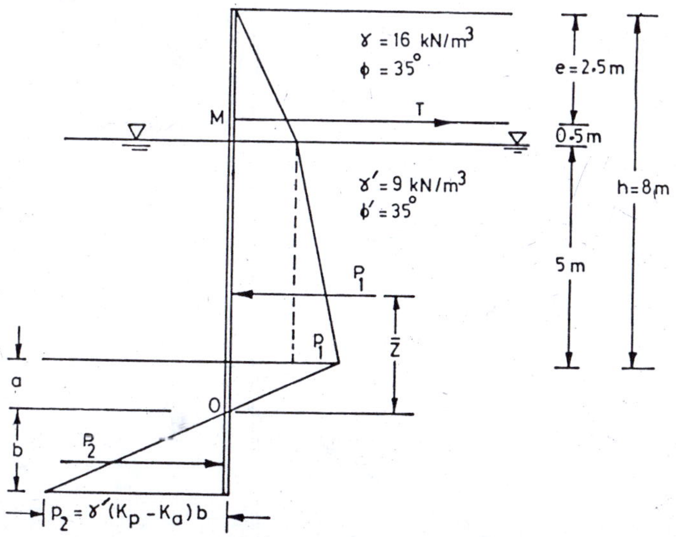

Fig. E-20.7.

$$p_2 = 9(3.69 - 0.27)b = 30.78\,b$$

$$P_2 = 30.78\,b \times b/2 = 15.39\,b^2$$

$$P_1 = 12.96 \times 3/2 + 12.96 \times 5 + 12.15 \times 5/2 + 25.11 \times a/2$$

in which the distance a is determined from the equation, $p_1 - \gamma'(K_p - K_a)\,a = 0$

or $$a = \frac{25.11}{9 \times (3.69 - 0.27)} = 0.82 \text{ m}$$

Thus $$P_1 = 19.44 + 64.8 + 30.38 + 25.11 \times 0.82/2 = 124.92$$

Taking moments about O,

$$P_1 \bar{Z}_1 = 19.44 \times 6.82 + 64.8 \times 3.32 + 30.38 \times 2.49 + 10.3 \times 0.82 \times 2/3$$

or $$\bar{Z}_1 = \frac{132.58 + 215.14 + 75.65 + 5.63}{124.92} = 3.43 \text{ m}$$

Taking moments of all the forces about M, using Eq. 20.38 (a),

$$124.92\,(a + h - e - \bar{Z}_1) - P_2\left(a + h - e + \frac{2b}{3}\right) = 0$$

or $$124.92\ (0.82 + 8.0 - 2.5 - 3.43) - 15.39\,b^2\ (0.82 + 8.0 - 2.5 + 2b/3) = 0$$

or $$361.02 - 97.26b^2 - 10.26b^3 = 0$$

By trial and error, $b = 1.77$ m

Now $d = b + a = 1.77 + 0.82 =$ **2.59 m**

$$D = 1.3 \times 2.59 = 3.40 \text{ m.}$$

Illustrative Example 20.8. *Determine the depth of embedment for the anchored sheet pile shown in Fig. E-20.8. Also determine the force in the anchor per metre run. Assume fixed-end support conditions.*

Solution. (See Figs. 20.28 and 20.27 for notations) $K_a = 0.27$, $K_p = 3.69$

$$p_1 = 16 \times 8.0 \times 0.27 = 34.6 \text{ kN}$$

From Eq. 20.47, $$a = \frac{p_1}{\gamma\,(K_p - K_a)} = \frac{34.6}{16 \times 3.42} = 0.63 \text{ m}$$

From Fig. 20.26, for $\phi' = 35°$, $i/h = 0.025$

Therefore, $i = 0.025 \times 8.0 = 0.2$ m

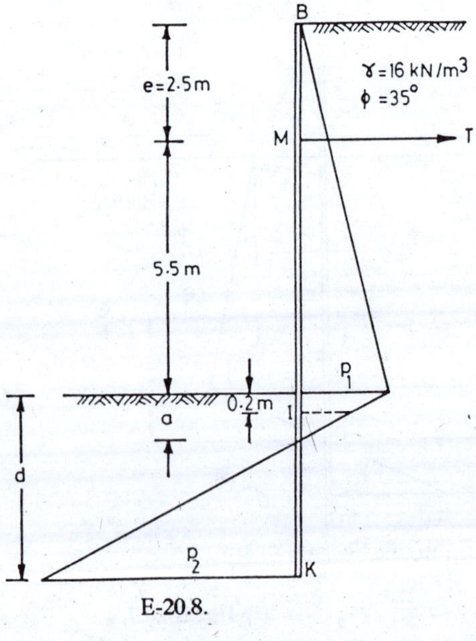

E-20.8.

$$p_0 = \frac{p_1}{a}(a - i) = \frac{34.6}{0.63} \times 0.43 = 23.6$$

Taking moments about M of all the forces acting on beam IB,

$$R_I (5.5 + 0.2) = \frac{1}{2} \times 34.6 \times 8 \times (5.33 - 2.50) + (23.6 \times 0.2) \times 5.60 + (34.6 - 23.6) \times 0.2/2 \times (5.50 + 0.07)$$

or $\quad R_I = 74.4 \text{ kN}$

Now $\quad p_2 = \gamma (K_p - K_a)(d - a) = 16 \times 3.42 \times (d - 0.65)$

or $\quad p_2 = 54.7 (d - 0.63)$

[Alternatively, $\quad p_2 = \dfrac{p_0}{a - i} \times (d - a) = \dfrac{23.6}{0.43}(d - 0.63)$

or $\quad p_2 = 54.7 (d - 0.63)$]

Taking moments about K of all the forces acting on beam IK,

$$R_I (d - a) + \frac{23.6 \times 0.43}{2}(d - a + 0.29) - \frac{54.7 (d - a)^2}{2} \times \frac{(d - a)}{3} = 0$$

or $\quad 74.4 (d - a) + 5.1 (d - a + 0.29) - 9.1 (d - a)^3 = 0$

By trial and error, $\quad d - a = 2.96 \text{ m} \quad$ or $d = 2.96 + 0.63 = 3.59$ m

$$D = 1.2 \times 3.59 = \mathbf{4.31 \text{ m}}$$

From Eq. 20.50, $T = P_1 - R_1 = \dfrac{1}{2} \times 34.6 \times 8 + 23.6 \times 0.2 + 11 \times 0.1 - 74.4$

$$T = \mathbf{69.8 \text{ kN}}$$

PROBLEMS

A. Numerical

20.1. Check the overall stability of the cantilever retaining wall shown in Fig. P. 20.1. Take the allowable soil pressure as 600 kN/m^2.

[**Ans.** $F_s = 1.60$; $F_o = 3.01$; $F_b = 3.48$]

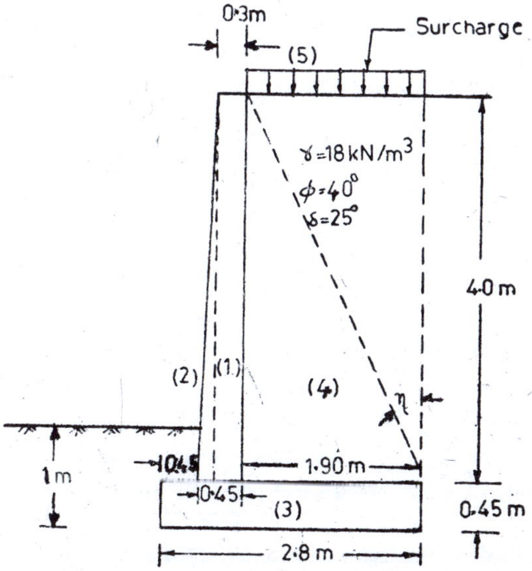

Fig. P-20.1.

20.2. A cantilever sheet pile retains soil to a height of 6 m. Find the depth to which the pile should be driven, assuming two-thirds of the theoretical passive resistance is developed on the embedded length. γ = 19 kN/m^3 and ϕ = 30°. Use approximate method. [**Ans.** d = 7.35 m]

20.3. Determine the depth of embedment for the cantilever pile in clay, shown in Fig. P 20.3. The water table is at a height of 2.5 m above the dredge level on both sides. [**Ans.** d = 2.20 m, D = 3.30 m]

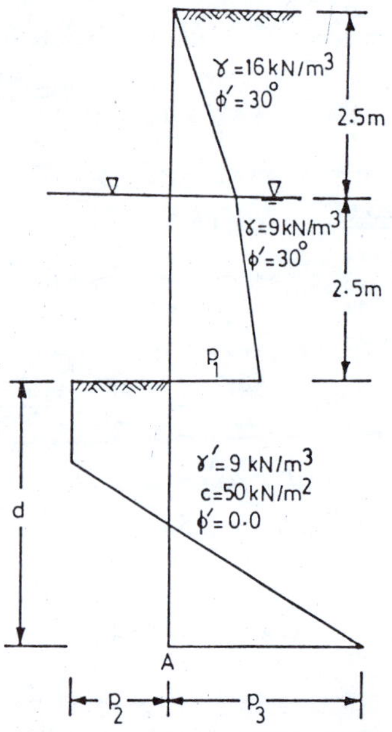

Fig. P-20.3.

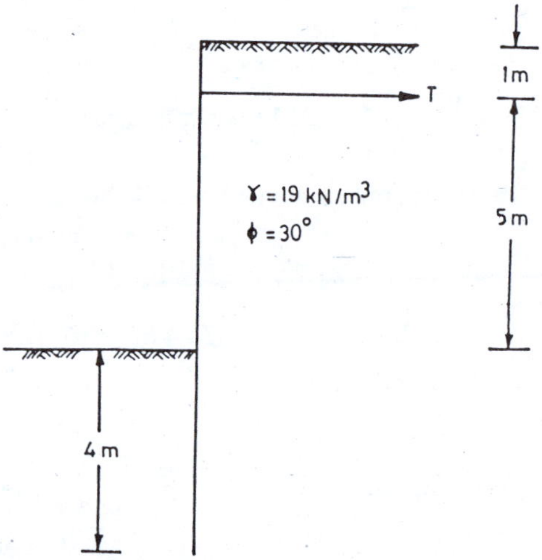

Fig. P-20.4.

20.4. An anchored sheet-pile wall is shown in Fig. P 20.4. Find the fraction of the theoretical maximum pressure on the embedded length which must be mobilised for equilibrium. Use the free- earth support method.

Also determine the force in one of the anchors, assuming that they are spaced at 2.5 centres.

[**Ans.** 0.51; 206.5 kN]

20.5. An excavation 8 m deep is to be made in cohesionless soil (γ = 19 kN/m^3, ϕ = 30°). The sides of the excavation are supported by anchored sheet piles with fixed-end support. Determine the minimum depth of embedment for equilibrium. The anchors are at a depth of 1.5 m below the surface. [**Ans.** 5.0 m]

20.6. For the anchored sheet pile shown in Fig. P-20.6, determine the embedment depth d and the force in the anchor per m run. Assume fixed-end conditions. [**Ans.** 5.00 , 71 kN/m]

20.7. For the anchored sheet pile shown in Fig. P 20.7, determine the theoretical and actual depth of penetration, the anchor force per unit length of the wall and maximum moment. Assume free-earth support.

[**Ans.** 2.40 m; 3.50 m; 39.5 kN, 63.0 kN-m]

B. Descriptive and Objective Type

20.8. What are different types of retaining walls? Discuss the methods for estimation of lateral earth pressure acting on the walls.

20.9. Discuss the principles of the design of retaining walls.

20.10. What are different modes of failure of retaining walls ? Explain with the help of sketches.

20.11. What are different types of sheet pile walls? Draw the sketches showing the pressure distribution.

20.12. Discuss the procedure for checking the stability of a cantilever sheet pile wall.

20.13. How would you check the stability of an anchored sheet pile wall with free-earth support ? What is Rowe's correction ?

20.14. Describe the equivalent beam method for the analysis of an anchored sheet pile.

20.15. Discuss various methods for providing anchors for a sheet pile wall.

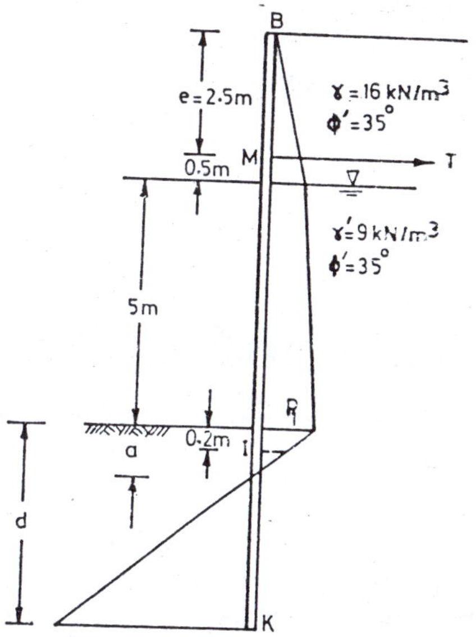

Fig P-20.6.

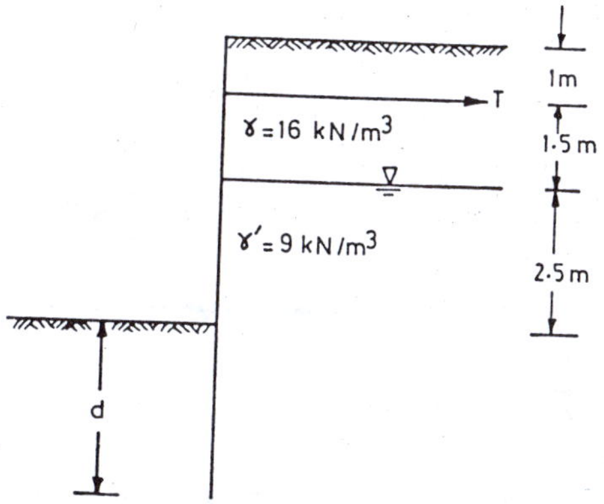

Fig. P-20.7.

20.16. Write whether the following statements are true or false.

(*a*) Gravity walls are more economical than cantilever walls.
(*b*) The lateral pressure on gravity walls is determined using Coulomb's theory.
(*c*) Rankine's theory can always be used for the determination of lateral pressure on cantilever walls.
(*d*) Counterfort retaining walls are more suitable than cantilever retaining walls for greater heights.
(*e*) Tension develops at the base of the retaining wall when the resultant strikes outside the middle-third.
(*f*) Backfill should be preferably of cohesive material.
(*g*) In an anchored sheet pile with free-earth support, there is no point of inflexion.
(*h*) At the point of contraflexure, the shear force is zero.
(*i*) The ultimate resistance of an anchor depends upon the passive resistance.
(*j*) Anchors should be placed as near the wall as possible.

[Ans. True, (*b*), (*d*), (*e*), (*g*), (*i*)]

C. Multiple Choice Questions

1. The minimum allowable factor of safety against sliding in the case of a cantilever retaining wall is
(*a*) 2.0 (*b*) 3.0
(*c*) 1.50 (*d*) 2.50

2. In the case of a counterfort retaining wall, the toe slab acts as a
(*a*) cantilever (*b*) continuous slab
(*c*) simply supported slab (*d*) none of above.

3. In the case of fixed-earth support of a bulkhead, the stability is provided by
(*a*) the passive resistance of the soil
(*b*) the force in the anchor
(*c*) both (*a*) and (*b*)
(*d*) neither (*a*) nor (*b*)

4. Rowe's correction for free-earth support of a bulkhead depends mainly on
(*a*) total height of bulk-head
(*b*) Young's modulus of pile material
(*c*) moment of inertia of the pile
(*d*) all the above.

[Ans. 1. (*c*), 2. (*a*), 3. (*c*), 4. (*d*)]

21

Braced Cuts and Coffer Dams

21.1. INTRODUCTION

(*a*) Braced Cuts

Deep excavations with vertical sides require lateral supports to prevent cave-in of the earth and to protect the adjacent areas against ground subsidence and lateral movement of the subsoil. When excavations are shallow and ample space is available, the sides of the excavation can be sloped at a safe angle to ensure stability. However, in deep excavations, especially in built-up areas, there may not be adequate space for providing safe slopes. Moreover, it becomes uneconomical to provide safe slopes because of large quantities

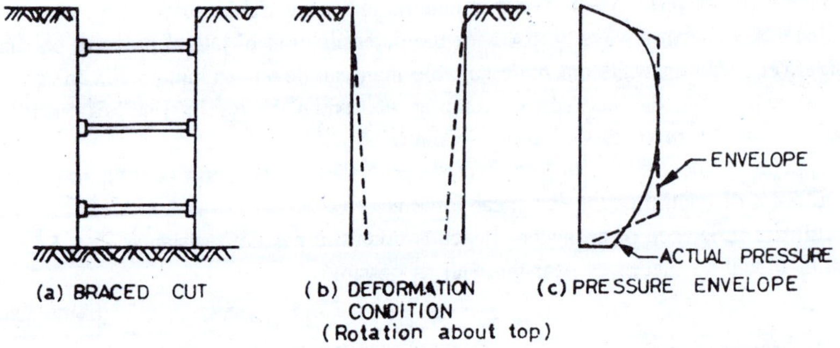

Fig. 21.1. Braced Cut.

of the earth involved. Excavations which are laterally supported are called *braced cuts* [Fig. 21.1. (*a*)]. The vertical sides of excavations are supported by a sheeting and bracing system. It consists of a relatively flexible sheeting placed against the excavation walls. The lateral thrust on the sheetings is resisted by horizontal members in compression (struts), known as bracing system. Bracing is provided as the excavation proceeds and the face of the sheeting becomes exposed. Various types of braced sheeting, the forces acting on them and the design of various components are discussed in this chapter.

(*b*) Coffer Dams

The word 'coffer' means a casket, chest or trunk. A coffer dam is a temporary structure built to enclose an area for excavation of foundation. Coffer dams are used when the size of excavations is very large and sheeting and bracing system becomes difficult, inconvenient or uneconomical. Coffer dams are generally required for foundations of structures, such as bridge piers, docks, locks, and dams, which are built in open water. These are also used for laying foundations on open land where there is a high ground water table. A coffer dam generally consists of a relatively impervious wall built around the periphery of the proposed excavation to prevent the flow of water into the excavation so that the foundation may be laid in dry condition (Fig. 21.2). Various types of coffer dams and their design aspects are considered in this chapter.

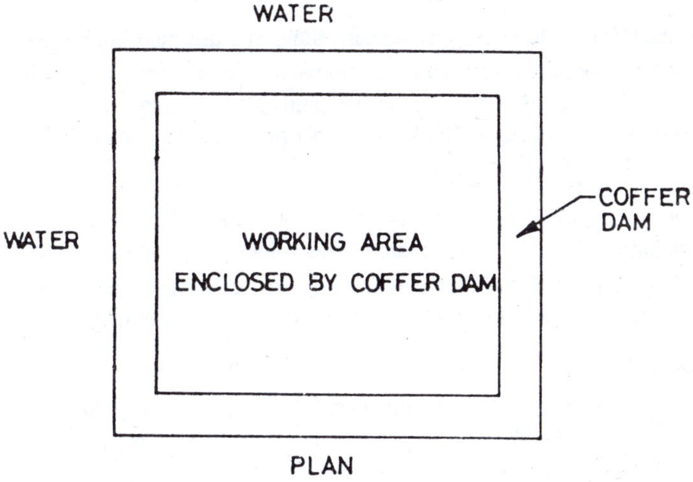

Fig. 21.2. Coffer Dam.

21.2. LATERAL EARTH PRESSURE ON SHEETINGS

Rankine's and Coulomb's theories of earth pressure cannot be used for the computation of lateral earth pressure on sheetings, as those theories are applicable to rigid retaining walls rotating at base. The sheeting and bracing system is somewhat flexible and the rotation takes place at the top of the wall.

Sheetings are placed against the walls of the excavations when these are shallow. The upper strut is placed when the excavation is shallow and little lateral yield of the soil has occurred. As the excavation proceeds downwards, the lower part of the face is free to yield inward before the next strut can be placed. The inward yield of soil increases with an increase in the depth of excavation. Thus, the sheeting tilts about its top [Fig. 21.1 (*b*)].

The method of earth pressure calculations has been developed by Terzaghi based on observations of actual loads in struts in full- scale excavations in sand in Berlin and in soft clay in Chicago. Pressure distributions against the sheeting have been approximated on the assumption that each strut supports a sheeting area. The pressure distribution depends upon the speed at which excavation advances, the care taken in the installation of bracing, the soil type and many other factors. The effect of various factors is not yet fully understood. However, the results of field studies can be used as a basis for developing earth pressure diagram required for the design of bracing system. The pressure distribution diagram recommended for design represents an envelope which encompasses the actual pressure distribution diagrams obtained from

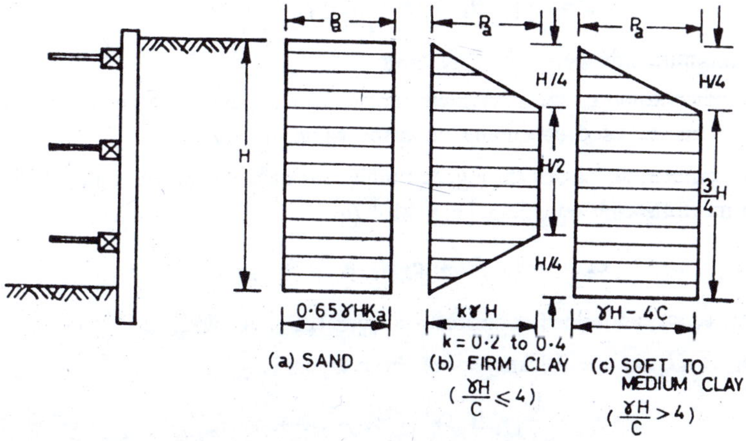

Fig. 21.3. Apparent Pressure Diagrams.

the field tests [Fig. 21.1 (*c*)]. These design pressure diagrams are also known as *apparent pressure diagrams.*

Fig. 21.3. shows the apparent pressure diagrams suggested by Peck (1969). Fig. 21.3 (*a*) gives the pressure distribution for braced cuts in dry or moist sand. The pressure distribution is uniform with a pressure (p_a) equal to 0.65 $\gamma H K_a$, where K_a is Rankine's earth pressure coefficient, given by

$$K_a = \tan^2 (45° - \phi/2) \qquad \text{...(21.1)}$$

Figs. 21.3 (*b*) and 21.3 (*c*) show the pressure diagrams for clay. If ($\gamma H/c$) is less than or equal to 4, the pressure envelope shown in Fig. 21.3 (*b*) is used. The value of p_a varies between 0.2 γH to 0.4 γH, with an average value of 0.3 γH. If ($\gamma H/c$) is greater than 4, the pressure envelope shown in Fig. 21.3 (*c*) is used. The pressure p_a is taken as ($\gamma H - 4c$) or 0.3 γH, whichever is greater.

Non-uniform soils

1. When the braced cut passes through layers of both sand and clay [Fig. 21.4 (*a*)], an equivalent value of cohesion c_e (ϕ = 0 concept) is determined using the following equation suggested by Peck (1943).

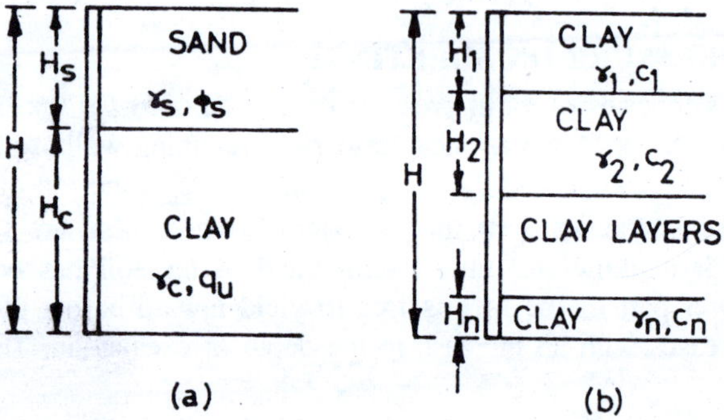

Fig. 21.4.

$$c_e = \frac{1}{2H} [\gamma_s K_s H_s^2 \tan \phi_s + (H - H_s) n' q_u] \qquad \text{...(21.2)}$$

where H = total height of the cut, γ_s = unit weight of sand, H_s = height of the sand layer, K_s = a lateral earth pressure coefficient (~ 1), ϕ_s = angle of friction of sand, q_u = unconfined compressive strength of clay, n' = a coefficient of progressive failure (average value 0.75).

The equivalent unit weight γ_e of the layers is determined from the following equation:

$$\gamma_e = \frac{1}{H} [\gamma_s H_s + (H - H_s) \gamma_c] \qquad \text{...(21.3)}$$

where γ_c = saturated unit weight of clay layer.

Once the equivalent cohesion c_e and equivalent unit weight γ_e have been determined, the diagram in Fig. 21.3 (*b*) or Fig. 21.3 (*c*) can be used depending upon the value of ($\gamma_e H/c_e$).

2. When the braced cut passes through a number of clay layers, the equivalent values of c_e and γ_e are determined from the following equations [Fig. 21.4 (*b*)].

$$c_e = \frac{1}{H} [c_1 H_1 + c_2 H_s + ... + c_n H_n] \qquad \text{...(21.4)}$$

where $c_1, c_2 ... c_n$ are undrained cohesions of layer 1, 2 ... *n* and

$H_1, H_2 ... H_n$ are the thicknesses of these layers [Fig. 21.4 (*b*)].

Likewise, $$\gamma_e = \frac{1}{H} [\gamma_1 H_1 + \gamma_2 H_2 + ... + \gamma_n H_n] \qquad \text{...[21.5]}$$

21.3. DIFFERENT TYPES OF SHEETING AND BRACING SYSTEMS

The following types of sheeting and bracing systems for braced cuts are commonly used.

(1) Vertical Timber Sheeting. In this method, vertical timber sheeting consisting of planks about 8 to 10 cm thick are driven around the boundary of the proposed excavation to some depth below the base of the excavation. The soil between the sheeting is then excavated. The sheeting is held in place by a system of wales and struts. The wales are horizontal beams running parallel to the excavation wall. The wales are

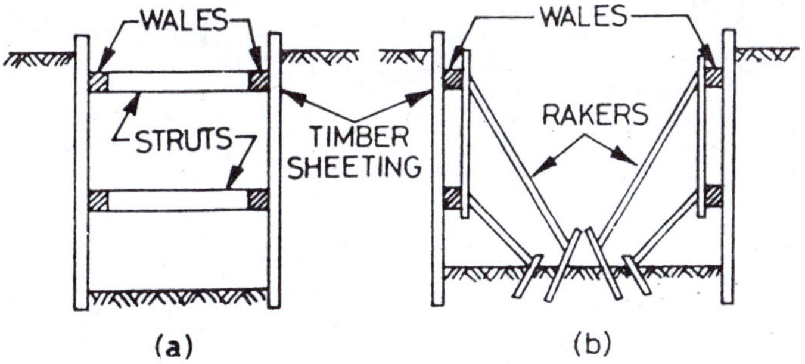

Fig. 21.5. Vertical Timber Sheeting.

supported by horizontal struts which extend from side to side of the excavation [Fig. 21.5 (*a*)]. However, if the excavations are relatively wide, it becomes economical to support the wales by inclined struts, known as *rakers* [Fig. 21.5 (*b*)]. For inclined struts to be successful, it is essential that the soil at the base of the excavation be strong enough to provide adequate reaction.

If the soil can temporarily support itself an excavation of limited depth without an external support, the timber sheeting can be installed in the open or in a partially completed excavation.

Vertical timber sheetings are economical upto a depth of 4 to 6 m.

(2) Steel Sheet Piles. In this method, the steel sheet piles are driven along the sides of the proposed excavation. As the soil is excavated from the enclosure, wales and struts are placed [Fig. 21.6(*a*)]. The wales are made of steel. The struts may be of steel or wood. As the excavation progresses, another set of wales and

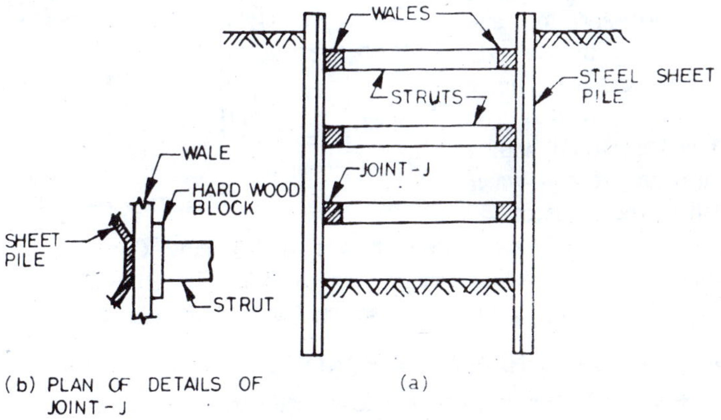

Fig. 21.6. Steel Sheet Piles.

struts is inserted. The process is continued till the excavation is complete. It is recommended that the sheet piles should be driven several metres below the bottom of excavation to prevent local heaves. If the width of a deep excavation is large, inclined bracing may be used. Fig. 21.6 (*b*) shows the details of joint at *J*.

(3) Soldier Beams. Soldier beams are H-piles which are driven at a spacing of 1.5 to 2.5 m around the boundary of the proposed excavation [Fig. 21.7 (*a*)]. As the excavation proceeds, horizontal timber planks called laggings are placed between the soldier beams. When the excavation advances to a suitable depth,

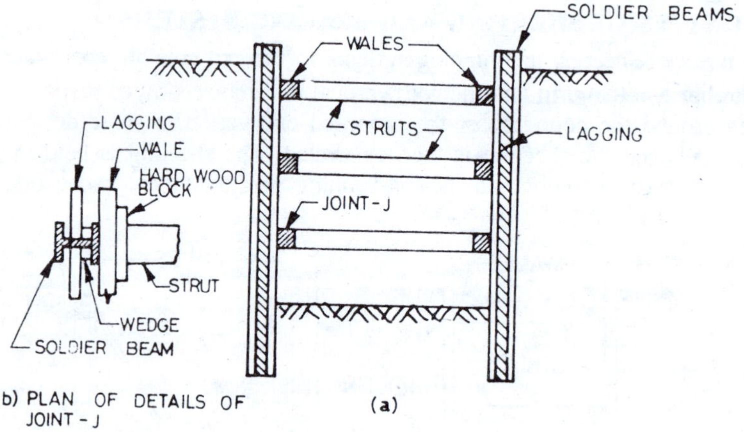

Fig. 21.7. Soldier Beams.

wales and struts are inserted. The lagging is properly wedged between the pile flanges or behind the back flange [Fig. 21.7 (*b*)].

(4) Tie Backs. In this method, no bracing in the form of struts or inclined rakers is provided. Therefore, there is no hindrance to the construction activity to be carried out inside the excavated area. The tie back is a rod or a cable connected to the sheeting or lagging on one side and anchored into soil (or rock) outside the excavation area (Fig. 21.8). Inclined holes are drilled into the soil (or rock), and tensile reinforcement (tendon) is then inserted and the hole is concreted. An enlargement or a bell is usually formed at the end of the hole. Each tie back is generally prestressed before the depth of excavation is increased further to cope with the increased tension.

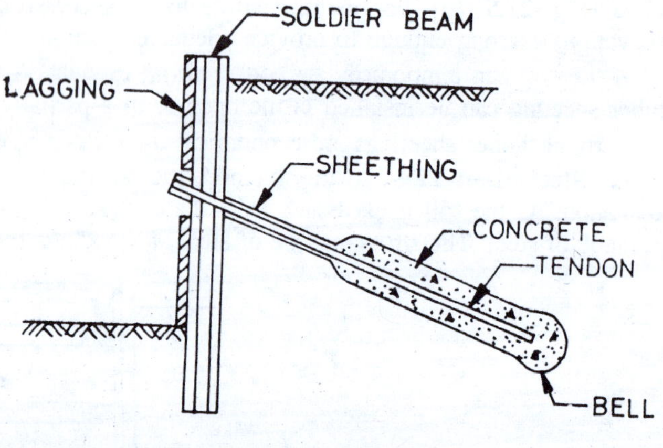

Fig. 21.8. Tie Back.

Use of Slurry Trenches

An alternative to use of sheeting and bracing system, which is being increasingly used these days, is the construction of slurry trenches around the area to be excavated. The trench is excavated and is kept filled with a heavy, viscous slurry of a bentonite clay-water mixture. The slurry stabilises the walls of the trench, and thus the excavation can be done without sheeting and bracing. Concrete is then placed through a tremie. Concrete displaces the slurry. Reinforcement can also be placed before concreting, if required. Generally, the exterior walls of the excavation are constructed in a slurry trench.

21.4. DESIGN OF VARIOUS COMPONENTS OF BRACING

(*a*) Struts. The strut is a compression member whose load-carrying capacity depends upon slenderness ratio, l/r. The effective length l of the member can be reduced by providing vertical and horizontal supports at intermediate points. The load carried by a strut can be determined from the pressure envelope. The struts should have a minimum vertical spacing of about 2.5 m. In the case of braced cuts in clayey soils, the depth of the first strut below the ground surface should be less than the depth of tensile crack (Z_c), which is equal to $2c/\gamma$ (see chapter 19).

While calculating the load carried by various struts, it is generally assumed that the sheet piles (or soldier beams) are hinged at all the strut levels except for the top and bottom struts.

[**Note.** Some designers assume that the sheet piles (or soldier beams) are hinged at all strut levels, except for the top].

Fig. 21.9 (*a*) shows a bracing system with four struts. Figs. 21.9 (*b*) and 21.9 (*c*), respectively, show the pressure envelopes for sand and stiff clays.

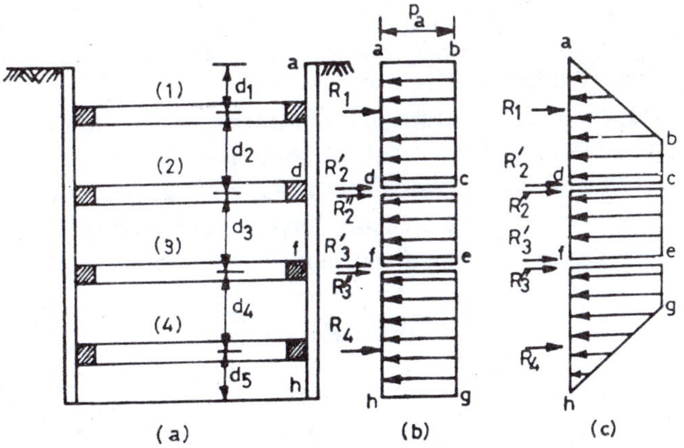

Fig. 21.9.

The reaction R_1 per unit length is determined by taking moments of the forces acting on span *a d* at *d*, and equating them to zero. Once R_1 has been determined, the reaction component R_2' is determined from the equilibrium equation in the horizontal direction for the span *a d*.

The reaction components R_2'' and R_3' are determined considering span *df* as hinged at *d* and *f*.

Thus reaction at *d*, $R_2 = R_2' + R_2''$

The reaction R_4 is determined by taking moments about *f* of the forces acting on *f h*. The reaction component R_3'' is determined from the equilibrium equation for horizontal forces acting on *f h*. The analysis is similar to that of the top strut. Thus reaction at *f*,

$$R_3 = R_3' + R_3''$$

The strut loads are then computed as under.

$$P_1 = R_1 \times s \qquad P_2 = (R_2' + R_2'')\, s \qquad \ldots(21.6)$$

$$P_3 = (R_3' + R_3'')\, s \qquad \text{and } P_4 = R_4 \times s$$

where *s* is the horizontal spacing (perpendicular to plane of paper) of struts.

Proper sections for the struts can be chosen for the respective loads found above.

(*b*) Wales. Wales are considered as horizontal beams pinned at the strut levels. The maximum bending moment will depend upon the span *s* and the loads on the struts. As the strut loads are different at various levels, the maximum bending moments would also be different. For example,

$$M_{max} = \frac{R_1 s^2}{8}, \text{ for the top wale} \qquad \ldots[21.7(a)]$$

and

$$M_{max} = \frac{R_2 s^2}{8}, \text{ for the second wale} \qquad \ldots[21.7(b)]$$

Once the maximum bending moments have been computed, the section modulus (*S*) is computed as

$$S = \frac{M_{max}}{\sigma_{all}} \qquad \ldots(21.7)$$

where σ_{all} = allowable bending stress.

(*c*) **Sheet Piles.** Sheet piles act as vertical plates supported at strut levels. The maximum bending moments in various sections such as *a d*, *d f* and *f h* in Figs. 21.9 (*b*) and 21.9 (*c*), are determined.

Once the maximum bending moments have been computed, the section modulus of the sheet pile can be computed and the section chosen.

Other Criteria for Design of Braced Cuts

(1) Braced cuts in clay may become unstable due to heaving of the bottom of the excavation. This is a type of bearing capacity failure and is discussed in chapter 23.

(2) Braced cuts in sand are generally stable against heaving failure, but these may become unstable due to upward seepage of water into the cut if the water level inside the cut is substantially lower than that outside. Piping failures are discussed later along with the design of coffer dams.

(3) The walls of the braced cut may yield laterally and cause ground settlement in the surrounding area. This effect should be carefully assessed and suitable measures adopted.

COFFER DAMS

21.5. TYPES OF COFFER DAMS

The following are different types of coffer dams commonly used in practice.

(1) Earth Coffer dams. These are the simplest type of coffer dams well-adapted to depths of water upto 3 m. Earth embankments are constructed around the area to be dewatered.

The earth coffer dams are built of local soils, preferably fine sand. These usually have a clay core or a vertically driven sheet piling in the middle. The upstream slope of the bank is covered with a rip rap (Fig. 21.10). A successful coffer dam need not be completely watertight. For reasons of economy, it is not possible

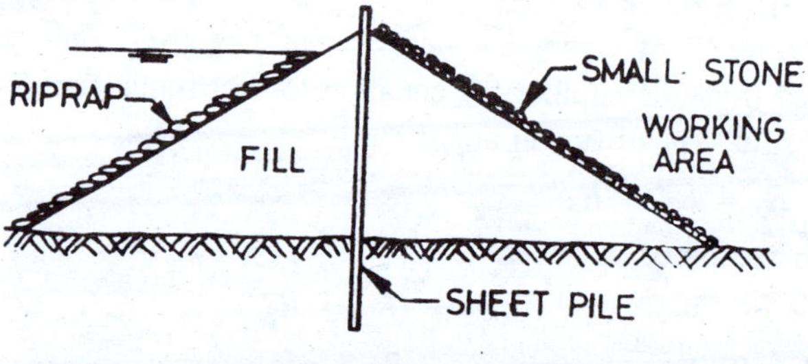

Fig. 21.10. Earth Coffer Dam.

to make it watertight and hence some seepage of water into the excavation is usually tolerated. The water collected is pumped out of the excavation. The embankment should be provided with a minimum free board of 1 m to prevent overtopping by waves.

Sand-bag coffer dams are used in an emergency.

(2) Rockfill Coffer dams. Rockfill coffer dams made of rockfill are sometimes used to enclose the site to be dewatered. These are permeable and are usually provided with an impervious membrane of soil to reduce seepage (Fig. 21.11). The crest and the upper part of the impervious membrane are provided with rip rap to provide protection against wave action. Overtopping does not cause serious damage in case of rockfill coffer dams. The slopes of a rockfill coffer dam can be made as steep as 1 horizontal to 1.5 vertical.

(3) Single-Sheet Pile Coffer dams. Single-sheet piling coffer dams are generally used to enclose small foundation sites in water for bridges at a relatively shallow depth. In this type of coffer dams, there is a single row of cantilever sheet piles. The piles are sometimes heavily braced. Joints in the sheet piles are properly sealed (Fig. 21.12). This type of coffer dams are suitable for moderate-flow velocities of water and for depths upto 4 m. The depth of penetration below ground surface is about 0.25 *h* for coarse sand and gravels, 0.50 *h* for fine sand and 0.85 *h* for silts, where *h* is the depth of water.

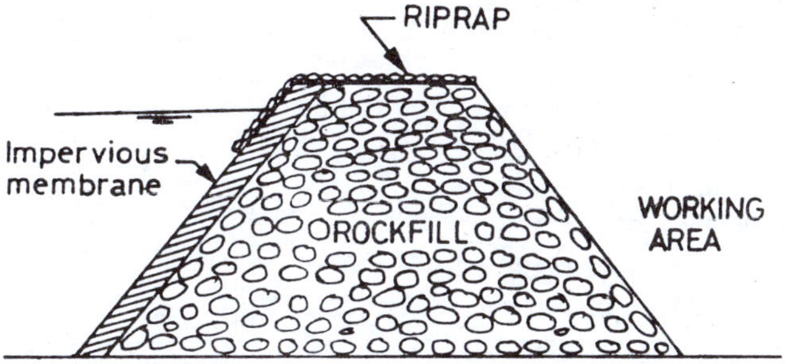

Fig. 21.11. Rockfill Coffer Dam.

Sometimes, single-sheet pile coffer dams are provided with earth fills on one or both sides to increase the lateral stability.

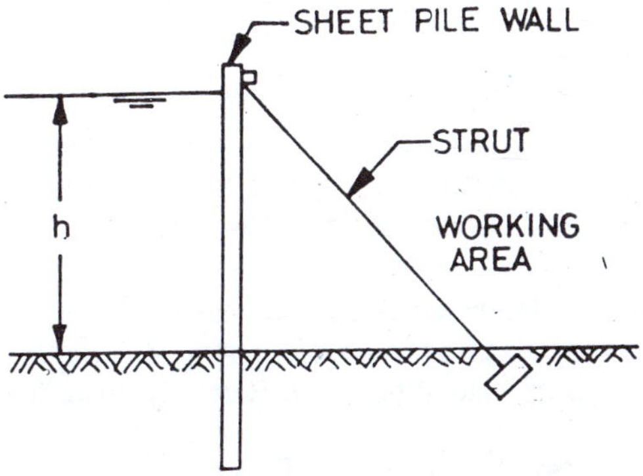

Fig. 21.12. Single-Sheet Pile Coffer Dam.

(4) Double-Wall Sheet Piling Coffer dams. A double-wall sheet piling coffer dam consists of two straight, parallel vertical walls of sheet piling, tied to each other and the space between walls filled with soil. The width between the parallel piles is empirically set as ($h/2$ + 1.50 m), where h is height of water (Fig. 21.13). Double-wall sheet piling coffer dams higher than 2.5 m should be strutted. Sometimes, an inside berm is provided to keep the phreatic line within the berm.

The fill material should have a high coefficient of friction and unit weight so that it performs as a massive body to give the coffer dam stability against sliding and overturning. Suitable measures should be

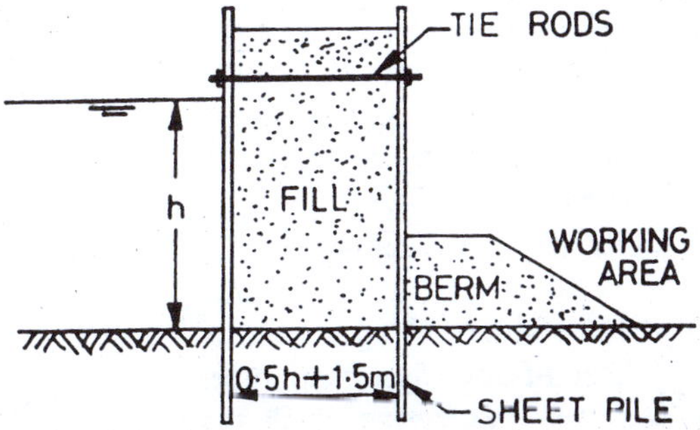

Fig. 21.13. Double-Wall Sheet Piling Coffer Dam.

adopted to reduce the uplift on the coffer dam. This is generally done by driving the sheet piling on the upstream as deep as possible.

The double-wall sheet piling coffer dam has the advantage of having less leakage than that in a single-wall coffer dam. These coffer dams are suitable upto a height of 10 m.

(5) Braced Coffer dams. A braced coffer dam is formed by driving two rows of vertical sheeting and bracing with wales and struts. These are similar to sheeting and bracing system discussed for braced cuts in Sect. 21.3, with one basic difference that braced cuts are required for excavations in dry areas whereas braced coffer dams are used to isolate a working area surrounded by water (Fig. 21.14). The braced coffer dams are susceptible to flood damage.

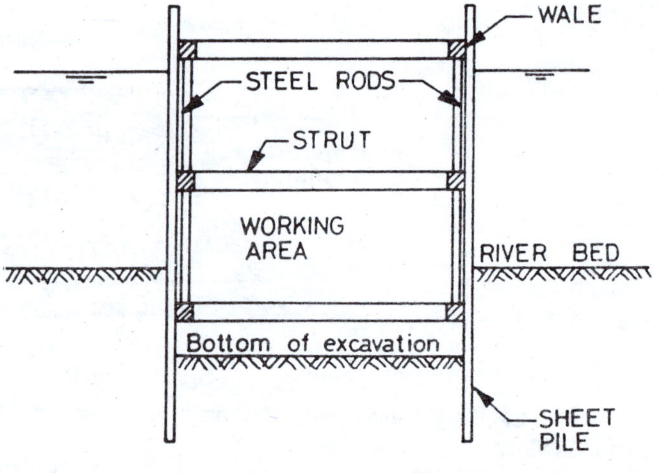

Fig. 21.14. Braced Coffer Dam.

Land Coffer dams. Braced coffer dams are sometimes used as land coffer dams to prevent ground water from entering the foundation pit on land and to support the soil so as to prevent cave in. After the pit is dewatered, the structure is concreted. When concreting has been completed above the water level, the coffer dam is removed.

(6) Cellular Coffer dams. A cellular coffer dam is constructed by driving sheet piles of special shapes to form a series of cells. The cells are interconnected to form a watertight wall. These cells are filled with soil to provide stabilising force against lateral pressure. Basically, there are two types of cellular coffer dams that are commonly used.

***(i)* Diaphragm type** (Fig. 21.15). This type of cellular coffer dam consists of circular arcs on the inner and outer sides which are connected by straight diaphragm walls. The connection between the curved parts and the diaphragms are made by means of a specially fabricated *Y*-element. The coffer dam is thus made from inter-connected steel sheet piles The cells are filled with coarse-grained soils which increase the weight of the coffer dam and its stability. The leakage through the coffer dam is also reduced.

To avoid rupture of diaphragms due to unequal pressure on the two sides, it is essential to fill all the cells at approximately the same rate. One advantage of the diaphragm type is that the effective length of the coffer dam may be increased easily by lengthening the diaphragm.

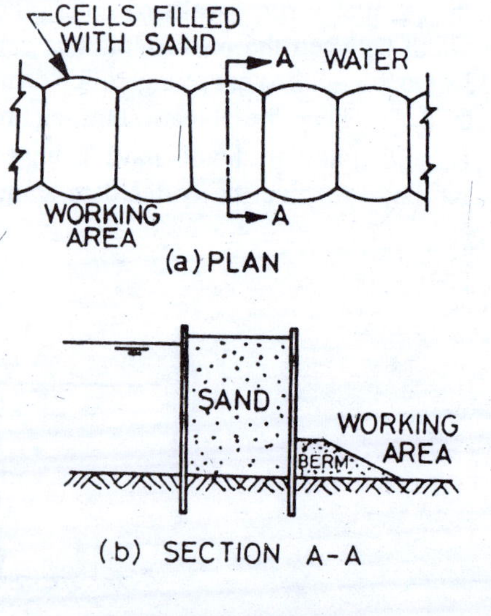

Fig. 21.15. Diaphragm Cellular Coffer Dam.

***(ii)* Circular type** (Fig. 21.16). It consists of a set of large diameter main circular cells interconnected by arcs of smaller cells. The walls of the connecting cells are perpendicular to the walls of the main circular cells of large diameter. The segmental arcs are joined by special T-piles to the main cells.

The circular-type cellular coffer dams are self-sustaining, and therefore independent of the adjacent circular cells. Each cell can be filled independently. The stability of such cells is much greater as compared with that of the diaphragm type. However, the circular cells are more expensive than the diaphragm type, as these require more sheet piles and greater skill in setting and driving the piles. Because the diameter of circular cells is limited by interlock tension, their ability to resist large lateral pressure due to high heads is limited.

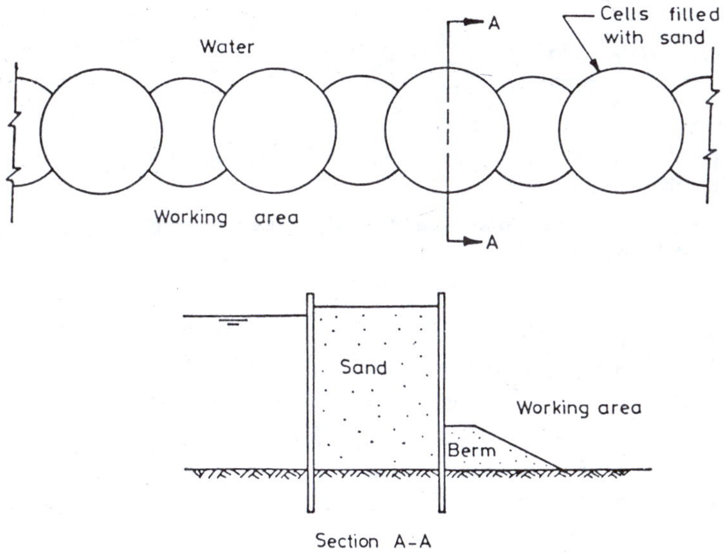

Fig. 21.16. Circular Cellular Coffer Dam.

21.6. DESIGN OF CELLULAR COFFER DAMS ON ROCKS

The design of circular, cellular coffer dams can be done using one of the following methods.

(*i*) TVA method, (*ii*) Cumming's method, (*iii*) Brinch-Hansen method.

The TVA method is more common and is discussed below.

(1) Location of Saturation Line. For determining the weight of the soil in the cell, it is required to locate the saturation line. TVA (Tennessee Valley Authority) engineers gave the approximate location of the saturation line for different types of fill materials, as shown in Fig. 21.17. For a perfectly draining fill, the saturation line is shown in Fig. 21.17 (*a*). The lower half may be assumed saturated for analysis. For other

Fig. 21.17. Location of Saturation Line.

type of fill, the saturation line at a slope of 2 : 1 is assumed [Fig. 21.17 (*b*)]. In case of a berm, the saturation line drops down to the top of the berm [Fig. 21.17 (*c*)]. For stability analysis, two extreme locations (marked 1 and 2) of saturation line should be investigated in this case.

(2) Average Width. The design of a coffer dam is made for a section 1 m long and of uniform, average width. The average width is obtained by equating the section modulus of the equivalent rectangular section to the actual section modulus. An approximate value of the average width may, however, be obtained by equating the equivalent rectangular area to the actual area of the coffer dam between centre to centre. Thus

$$\text{Average width } (b) = \frac{\text{area of one main cell + area of one connecting cell}}{2L} \quad \ldots(21.9)$$

where $2L$ = distance between centre to centre of main cells.

The TVA engineers gave the following relations for computing the average width.

$$b = 0.785\,D \qquad \text{(For 90° T's)} \qquad ...[21.10(a)]$$

and

$$b = 0.875\,D \qquad \text{(For 60° T's)} \qquad ...[21.10(b)]$$

where D is the diameter of the main cell.

The above values of the average width are assumed in the preliminary analysis The actual width to be provided is obtained after the stability analysis.

The diameter (D) of the main cells is chosen depending upon the height (H) of the coffer dam. The diameter (D) of the main cells is given by

$$D = 1.0\,H \text{ to } 1.2\,H \qquad ...[21.11(a)]$$

The diameter of the connecting cells (D_1) is given by

$$D_1 = 0.6\,D \qquad ...[21.11(b)]$$

The circular, cellular coffer dams are economical upto a total height (H) of 15 to 18 m.

(3) Safety against sliding. The coffer dam is subjected to a horizontal sliding force due to water pressure and earth pressure. The sliding is resisted by the frictional resistance at the base of the coffer dam. If berms are provided on the interior side, the passive earth pressure also helps in resisting the sliding (Fig. 21.18). Thus the factor of safety against sliding,

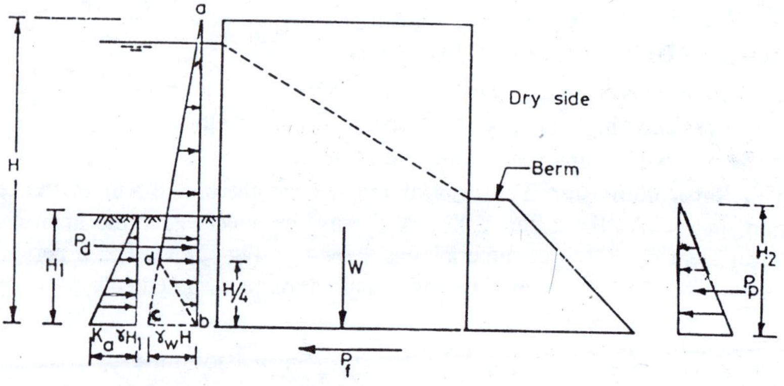

Fig. 21.18. Sliding Analysis.

$$F_s = \frac{\text{Resisting force}}{\text{Sliding force}} = \frac{P_f + P_p}{P_d} \qquad ...(21.12)$$

or

$$F_s = \frac{\mu W + P_p}{P_d} \qquad ...(21.13)$$

where μ = coefficient of friction (= tan ϕ), W = total weight of fill (submerged weight below saturation line),

P_p = passive resistance of the berm on dry side.

P_d = driving force due to water and soil on the water side.

P_f = resisting force at the base.

A factor of safety of at least 1.25 is generally recommended.

(4) Safety against Overturning. The coffer dam should be safe against failure due to overturning at toe. Neglecting the passive resistance of the berm, the factor of safety against overturning is given by

$$F_o = \frac{\text{Resisting moment}}{\text{Overturning moment}} \qquad ...(21.14)$$

The factor of safety (F_o) should be greater than 2.0.

Further, as the soil cannot resist tension, the resultant of the forces must lie within the middle third. The eccentricity (e) is determined after locating the point where the resultant strikes, as the case of retaining walls (see Chapter 20). Thus

$$e \leq b/6 \qquad \text{...[21.15(a)]}$$

or
$$b \geq \sqrt{\frac{6 P_d \bar{Z}}{\gamma H}} \qquad \text{...[21.15(b)]}$$

where $\bar{Z}$ = height of the line of action of P_d above the base.

(5) Safety against slipping. As the cell tends to tip over the toe, the fill material has the tendency to run out (Fig. 21.19). The piles on the water side creep upward as one unit, but the piles on the dry side slip relative to each other. This behaviour occurs because the frictional resistance between the fill and the piles on the water side is smaller than the frictional resistance along the interlock.

On the water side, the force P_d pushes the pile against the fill. The frictional resistance between the pile and the fill is equal to $\mu' P_d$. The factor of safety against slipping is given by

$$F_{sp} = \frac{\text{Frictional resistance against upward movement}}{\text{Upward force}} \qquad \text{...(21.16)}$$

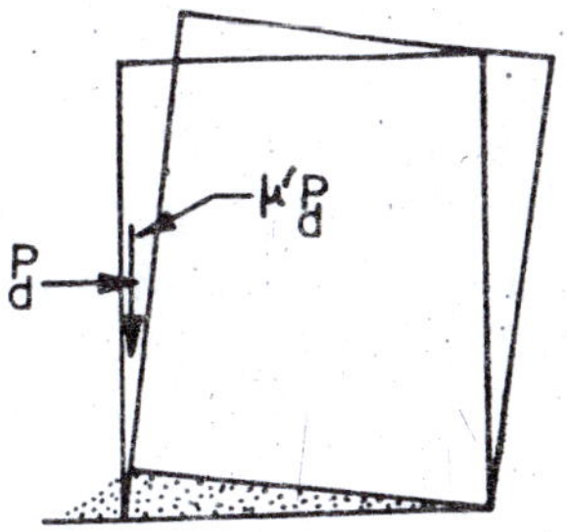

Fig. 21.19. Slipping Analysis.

The value of friction coefficient μ' is equal to $\tan \delta$, where δ is the angle of friction between the fill and the pile.

The minimum width b required can be obtained by taking moments about the toe. Thus

$$P_d \times \bar{Z} \times F_{sp} = P_d \tan \delta \times b$$

or
$$b = \frac{\bar{Z} F_{sp}}{\tan \delta} \qquad \text{...(21.7)}$$

The value of F_{sp} is usually taken as 1.25.

If the sheet pile is embedded in the rock for a substantial depth, the effect of the active pressure and the passive pressure should also be considered when summing up the moments about the toe.

(6) Safety against vertical shear. The cell may fail due to vertical shear developed along a plane through its centre line. The maximum shear (V_{max}) is obtained by computing the maximum bending moment acting on the cell, considering it as a vertical cantilever.

Thus
$$V_{max} = 3M/(2b) \qquad \text{...(21.18)}$$

where M = bending moment due to external forces above the base.

For stability, the shearing resistance developed must be greater than V_{max}.

The shearing resistance is equal to the sum of the shearing resistance of soil (S_1) and the resistance due to interlock (S_2) obtained as follows:

$$S_1 = \frac{1}{2} \gamma_a H^2 \left(\frac{\cos^2 \phi}{2 - \cos^2 \phi} \right) \tan \phi \qquad \text{...[21.19(a)]}$$

or
$$S_1 = \frac{1}{2} \gamma_a H^2 K \tan \phi \qquad \text{...[(21.19(b)]}$$

where K = coefficient of earth pressure having a value greater than that for the active pressure,

ϕ = angle of internal friction of the cell fill,

and γ_a = average unit weight of soil.

The resistance S_2 is equal to the interlock tension T multiplied by the coefficient of friction (f).

$$S_2 = f \cdot T$$

The maximum pressure is developed at a height of (3/4) H_f above the base, where H_f is the height of cell above the point of fixity. Thus

$$T = \frac{1}{2}\gamma_a H (3/4 H_f) K_a$$

Therefore, $$S_2 = \frac{3}{8}\gamma_a H H_f K_a f \qquad \ldots(21.20)$$

The coefficient of interlock friction f is generally taken as 0.30.

The factor of safety against vertical shear is given by

$$F_v = \frac{S_1 + S_2}{V_{max}} \qquad \ldots(21.21)$$

The factor of safety F_v should not be less than 1.25.

(7) Stability against bursting. The cell should be safe against bursting. The stability against bursting depends upon the interlock stresses. The interlock stresses are quite complicated and cannot be determined accurately. TVA engineers gave the following expression for the ring tension (T_{max}) for 90°-Tees.

$$T_{max} = p L \sec\theta \qquad \ldots(21.22)$$

or $$T_{max} \simeq pD/2 \qquad \ldots[21.22(a)]$$

where p = horizontal pressure due to cell fill, given by

$$p = \gamma_a H' K_a + \gamma_w H_w \qquad \ldots(21.23)$$

in which H' = depth of soil upto that level, H_w = water depth,

θ = angle which the T makes with the axis.

L = one-half the distance between centres of main cells.

The maximum ring tension occurs at the lower quarter point (*i.e.* $H_f/4$ above the rock or ground surface).

The computed maximum interlock tension should not exceed the allowable tension (T_{all}). The factor of safety against bursting is given by

$$F_b = \frac{T_{all}}{T_{max}} \qquad \ldots(21.24)$$

The value of T_{all} is usually taken as 1500 kN/m .

The value of F_b should be at least 1.25.

21.7. DESIGN OF CELLULAR COFFER DAM ON SOIL

The procedure for the design of a coffer dam embedded in deep soil is similar to that for a coffer dam resting on rock, as discussed in Sect. 21.6. However, the following additional requirements must be satisfied.

(*i*) The sheet pile in sand must be driven to such a depth that the bearing capacity at that level is greater than the vertical force acting on the pile. A minimum factor of safety of 1.50 is generally recommended.

The maximum vertical force per unit length (Q) developed is equal to the frictional resistance between the fill and the pile and is given by

$$Q = \frac{1}{2}\gamma H^2 K_a \tan\delta \qquad \ldots(21.25)$$

where H = height of cell above top of the stratum, K_a = coefficient of active earth pressure, δ = angle of friction between fill and pile, γ = unit weight of cell fill.

$$\text{Factor of safety} = \frac{Q_{ult}}{Q} \qquad \ldots(21.26)$$

where Q_{ult} = ultimate load capacity against bearing capacity failure (see chapter 23).

(*ii*) If the coffer dam is embedded in clay, the ultimate load capacity is given by

$$Q_{ult} = (5.7\,c)\,b$$

where c = unit cohesion.

The ultimate load capacity should be greater than the fill load. The factor of safety is given by

$$F = \frac{Q_{ult}}{Fill\ load} = \frac{5.7\,c\,b}{\gamma\,b\,H}$$

or $$F = \frac{5.7\,c}{\gamma H} \qquad ...(21.27)$$

or $$H = \frac{5.7\,c}{\gamma F} \qquad ...[21.27(a)]$$

A minimum factor of safety of 1.50 is recommended.

(*iii*) If the coffer dam is embedded in soft to medium clay, it should be safe against tilting caused by unequal settlement. The tilting can be estimated from the compressibility characteristics of the soil.

(*iv*) Cellular coffer dams on a deep sand deposit should have sufficient factor of safety against piping failure.

Fig. 21.20 shows a coffer dam founded on deep sand bed. Water percolates under the base of the coffer dam and rises up in front of the toe. The flow net can be drawn as shown. The flow lines are almost vertical in front of the toe. If the seepage pressure is equal to or greater than the submerged unit weight, quick (boiling) conditions may develop. The factor of safety against boiling is given by (see chapter 10)

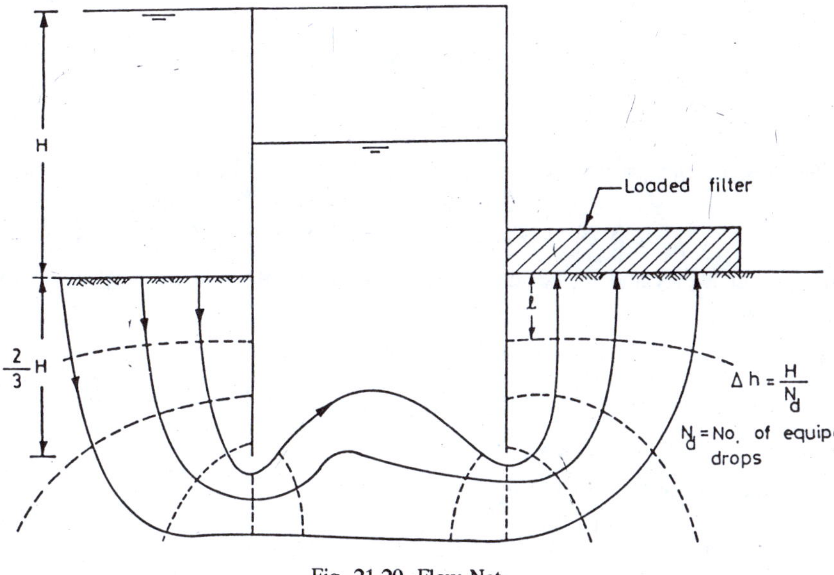

Fig. 21.20. Flow-Net.

$$F = \frac{\gamma'}{i\,\gamma_w} \qquad ...(21.28)$$

where i = hydraulic gradient at exit ($= \Delta h / l$), Δh = drop between last two equipotential lines,

l = length of the last flow field.

If the factor of safety is less than 1.50, a loaded filter is provided as shown to increase the downward force without increasing the seepage pressure.

The factor of safety can also be increased by reducing the gradient i by driving the sheet pile deeper or by reducing the effective head by permitting some water depth on the inner side.

The depth of the sheet pile below the ground surface is generally kept at least equal to two-thirds of the height of the coffer dam.

ILLUSTRATIVE EXAMPLES

Illustrative Example 21.1. *Determine the loads in the three struts shown in Fig. E-21.1 (a). The centre to centre spacing of the struts along the length of the cut is 2.50 m. The soil is stiff, fissured clay (γ = 19 kN/m^3, c = 40 kN/m^2). Also determine the maximum bending moments in wales and sheet piles.*

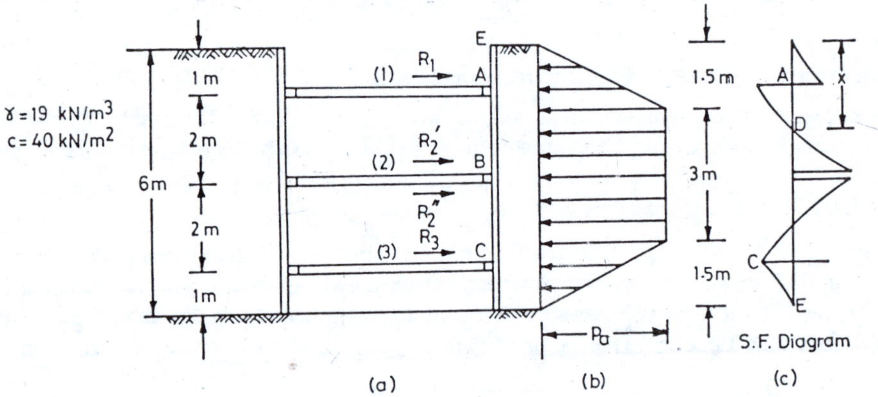

Fig. E-21.1.

Solution.
$$\frac{\gamma H}{c} = \frac{19 \times 6.0}{40} = 2.85 < 4$$

Therefore, the pressure distribution is as given in Fig. 21.2 (*b*). The value of the earth pressure is given by

$$p_a = 0.3\,\gamma H = 0.3 \times 19 \times 6 = 34.2 \text{ kN/m}^2$$

(*a*) **Strut loads.** Let R_1, R_2 and R_3 be the reactions per unit length in three struts at levels *A*, *B* and *C*, respectively.

Taking moments of all the forces acting above point *B*,

$$R_1 \times 2.0 = \frac{34.2 \times 1.50}{2}(1.50 + 0.50) + 34.2 \times 1.5 \times 0.75$$

or
$$R_1 = 44.9 \text{ kN}$$

From the horizontal equilibrium of the portion *AB*,

$$R_1 + R_2' = \frac{34.2 \times 1.50}{2} + 34.2 \times 1.50 = 77.0$$

$$R_2' = 77 - 44.9 = 32.1 \text{ kN}$$

From symmetry, $R_2'' = 32.1\text{kN}$ and $R_3 = 44.9 \text{ kN}$

Thus $R_2 = R_2' + R_2'' = 32.1 + 32.1 = 64.2 \text{ kN}$

From Eq. 21.6, the strut loads are given as

$$P_1 = R_1 \times s = 44.9 \times 2.5 = 112.3 \text{ kN}$$

$$P_2 = (R_2' + R_2'')\,s = (32.1 + 32.1) \times 2.50 = 160.5 \text{ kN}$$

$$P_3 = 44.9 \times 2.50 = 112.3 \text{ kN}$$

(*b*) **Wale.** The maximum bending moment occurs in the wale at level *B*. From Eq. 21.7,

$$M_{max} = \frac{R_2 s^2}{8} = \frac{(32.1 + 32.1)}{8}(2.5)^2 = 50 \text{ kN–m}$$

(*c*) **Sheet pile.** Fig. E-21.1 (*c*) shows the shear force diagram for the vertical sheet pile. The shear force is zero at point *D* at a depth *x* below top point *E*. The depth *x* is determined from the shear force equation.

$$\frac{1}{2} \times 34.2 \times 1.50 + (x - 1.5) \times 34.2 - 44.9 = 0 \quad \text{or } x = 2.06 \text{ m}$$

The maximum bending moment in the sheet pile occurs at D.

$$M_{max} = \frac{1}{2} \times 34.2 \times 1.50 \times 1.06 + (0.56 \times 34.2) \times 0.28 - 44.9 \times 1.06 = 15.0 \text{ kN-m}$$

Illustrative Example 21.2. *Determine the forces in the struts for the bracing system shown in Fig. E-21.2. Assume hinges at levels B, C and D. Take* $\gamma = 18$ *kN/m*^{3}, $c = 30$ *kN/m*^{2} *and* $s = 2.0$ *m.*

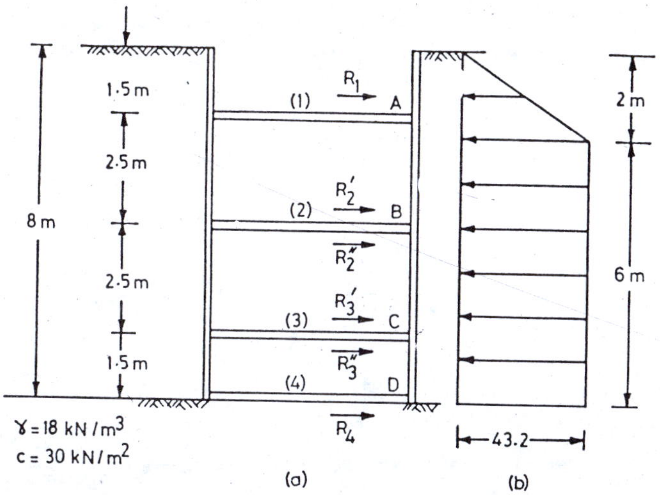

Fig. E-21.2.

Solution.

$$\frac{\gamma H}{c} = \frac{18 \times 8}{30.0} = 4.80 > 4.0$$

The pressure distribution shown in Fig. 21.2 (c) is applicable in this case. Therefore,

$$p_a = \gamma H - 4c = 18 \times 8 - 4 \times 30 = 24 \text{ kN/m}^2$$

Alternatively,

$$p_a = 0.3\,\gamma H = 0.3 \times 18 \times 8 = 43.2 \text{ kN/m}^2$$

Adopt larger of the two values, viz, 43.2 kN/m^2

The reaction in strut (1) is determined by taking moments about B, of the forces acting above that level.

$$R_1 \times 2.5 = \frac{1}{2} \times 2 \times 43.2 \times 2.667 + (2 \times 43.2) \times 1.0$$

or

$$R_1 = 80.6 \text{ kN}$$

Now

$$P_1 = 80.6 \times 2 = \mathbf{161.2\ kN.}$$

From horizontal equilibrium of the portion above B,

$$R_1 + R_2' = \frac{1}{2} \times 2 \times 43.2 + 2 \times 43.2 = 129.6$$

$$R_2' = 129.5 - 80.6 = 49.0 \text{ kN}$$

From horizontal equilibrium of the portion BC.

$$R_2'' + R_3' = 43.2 \times 2.5 = 108.0$$

or

$$R_2'' = R_3' = 54.0$$

Therefore, $R_2 = R_2' + R_2'' = 49 + 54 = 103$ kN

Now $P_2 = 103 \times 2 =$ **206 kN**

Likewise, from equilibrium of the portion *CD*,

$$R_3'' + R_4 = 1.5 \times 43.2 = 64.8$$

$$R_3'' = R_4 = 32.4 \text{ kN}$$

$$R_3 = R_3' + R_3'' = 54.0 + 32.4 = 86.4 \text{ kN}$$

Now $P_3 = 86.4 \times 2 =$ **172.8 kN**

Illustrative Example 21.3. *Design a circular, cellular coffer dam of total height 15 m resting on rock (Fig. E-21.3). Take allowable interlock tension of 1500 kN/m,* $\phi = 30°$, $\delta = 25°$, $K = 0.60$, $f = 0.30$, $\gamma_w = 10$ *kN/m*^{3}.

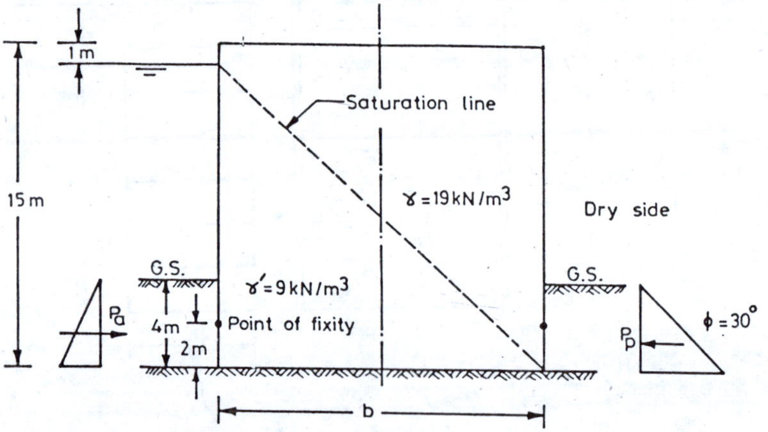

Fig. E-21.3.

Solution. Saturation line is taken as shown in Fig. 21.17 (*a*), assuming a perfectly draining fill.

The average unit weight of soil, taking half of the fill as submerged, is given by

$$\gamma_a = \frac{19.0 + 9.0}{2} = \mathbf{14.0\ kN/m^3}$$

Preliminary section

From Eq. 21.11 (*a*), $D = 1.2\,H = 1.2 \times 15 = 18$ m.

From Eq. 21.10 (*b*), $b = 0.875 \times 18 = 15.75$. Let us take $b = 16$ m.

Safety against sliding

$$\text{Weight of fill} = 14 \times 15 \times 16 = 3360 \text{ kN}$$

$$\text{Weight of steel shell} = 50 \text{ kN (say)}$$

$$\text{Total weight } W = 3360 + 50 = 3410 \text{ kN}$$

Active pressure, $P_a = \frac{1}{2} K_a \gamma' H_1^2 = \frac{1}{2} \times \frac{1}{3} \times 9.0 \times (4)^2 = 24$ kN

Passive pressure, $P_p = \frac{1}{2} K_p \gamma H_1^2 = \frac{1}{2} \times 3 \times 19.0 \times (4)^2 = 456$ kN

Water pressure, $P_w = \frac{1}{2} \times 10 \times 15^2 = 1125$ kN

From Eq. 21.13, factor of safety against sliding,

$$F_s = \frac{W \tan \phi + P_p}{P_d} = \frac{3410 \tan 30° + 456}{1125 + 24.0} = 2.11 \text{ (safe)}$$

Neglecting P_p, $$F_s = \frac{3410 \tan 30°}{1149} = 1.71 \text{ (safe)}$$

Safety against overturning

From Eq. 21.14, $$F_o = \frac{\text{Resisting moment about toe}}{\text{Overturning moment about toe}}$$

or $$F_o = \frac{3410 \times 16/2}{1125 \times 15/3 + 24 \times 4/3} = \frac{27280}{5657} = 4.82 \text{ (safe)}$$

The eccentricity will be within the middle third if

$$b \geq \sqrt{\frac{6 P_d \bar{Z}}{\gamma H}}$$

where $$\bar{Z} = \frac{1125 \times 15/3 + 24 \times 4/3}{1125 + 24} = 4.92 \text{ m}$$

Therefore, $$\sqrt{\frac{6 P_d \bar{Z}}{\gamma H}} = \sqrt{\frac{6 \times 1149 \times 4.92}{14 \times 15}} = 12.70 \text{ m.}$$

As $b > 12.70$ m, the eccentricity is within middle third.

Safety against slipping

From Eq. 21.17, $$F_{sp} = \frac{b \tan \delta}{\bar{Z}} = \frac{16 \tan 25°}{4.92}$$

$$= 1.52 > 1.25 \text{ (safe)}$$

Safety against vertical shear

From Eq. 21.21, $$F_v = \frac{S_1 + S_2}{V_{max}}$$

where $$S_1 = \frac{1}{2} \gamma_a H^2 K \tan \phi = \frac{1}{2} \times 14 \times (15)^2 \times 0.6 \tan 30° = 545.6 \text{ kN}$$

and $$S_2 = \frac{3}{8} \gamma_a H H_f K_a f$$

Taking $H_f = 13$ m, $$S_2 = \frac{3}{8} \times 14 \times 15 \times 13 \times \frac{1}{3} \times 0.3 = 102.4 \text{ kN}$$

From Eq. 21.18, $$V_{max} = \frac{3}{2} \times \frac{M}{b} = \frac{3}{2} \times \frac{5657}{16} = 530.3 \text{ kN}$$

Therefore, $$F_v = \frac{545.6 + 102.4}{530.3} = 1.22 \text{ (slightly unsafe)}$$

Safety against bursting

Pressure at a height of $H_f/4$ (From Eq. 21.23),

$$p = 14 \times (13 \times 3/4) \times 1/3 + (13 - 1) \times 3/4 \times 10 = 135.5 \text{kN/m}^2$$

$$T_{max} = pL \sec \theta \approx p \times D/2 = 135.5 \times 18/2 = 1219.5 \text{ kN}$$

From Eq. 21.24, $$F_b = \frac{1500}{1219.5} = 1.23 \text{ (slightly unsafe)}$$

PROBLEMS

A. Numerical

21.1. For the long braced cut shown in Fig. P 21.1, draw the earth pressure envelope and determine the strut loads, assuming a spacing of 2 m. Take γ = 18 kN/m^3, c = 30 kN/m^2. [**Ans.** 85.1, 60.8, 85.1 kN]

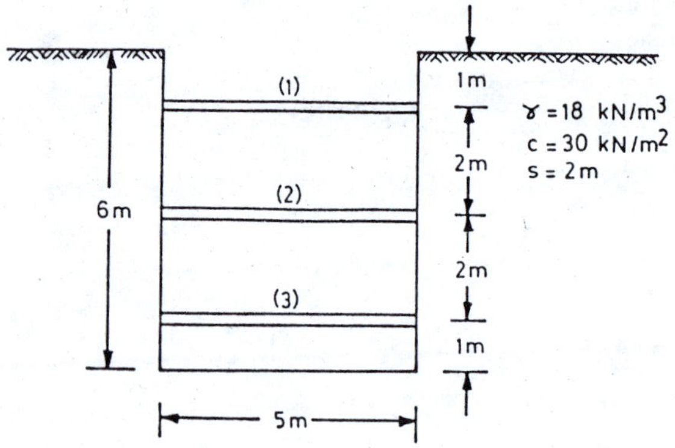

Fig. P-21.1.

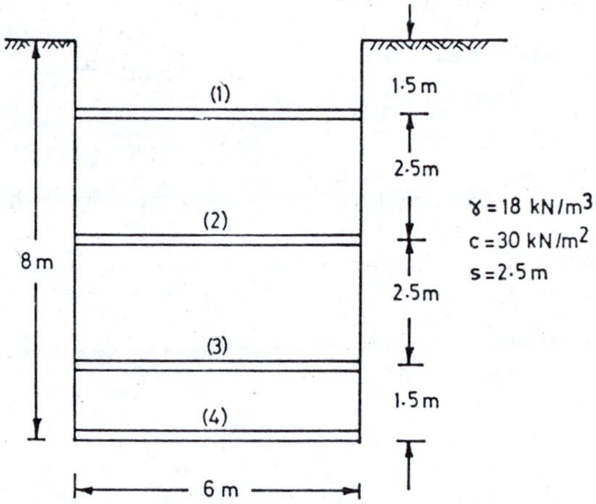

Fig. P-21.2.

21.2. For the braced cut shown in Fig. P 21.2, draw the earth pressure envelope. Determine the loads in struts (1), (2) (3) and (4), assuming that the struts are hinged at levels 2, 3 and 4. The spacing of the struts is 2.5 m. Take γ = 18 kN/m^3 and c = 30 kN/m^2. [**Ans.** 201.8, 257.5, 216.81, 81 kN]

21.3. Check the stability of the circular, cellular coffer dam shown in Fig. P 21.3. Take γ = 18 kN/m^3, ϕ = 30°, δ = 25°, K = 0.6, f = 0.3, T_{all} = 1500 kN/m. The diameter of the main cells is 15 m.

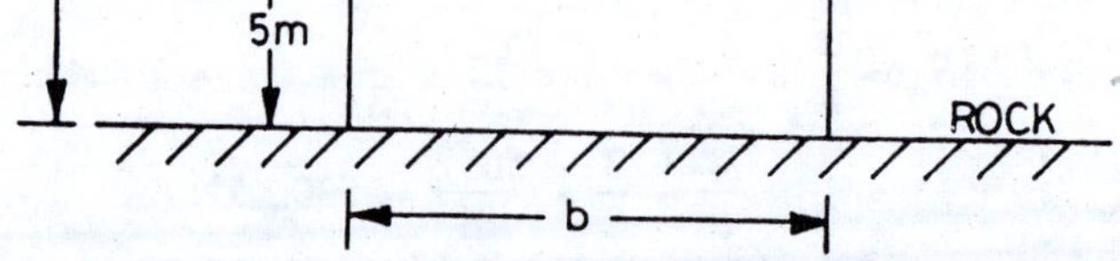

Fig. P-21.3.

21.4. Determine the factor of safety against piping failure of the coffer dam founded on a deep sand bed shown in Fig. P 21.4. Take G = 2.67 and e = 0.67.

B. Descriptive Types.

21.5. What is a coffer dam ? Name the different types of coffer dams and discuss their relative advantages and disadvantages.

21.6. Discuss the method for the design of a circular, cellular, coffer dam on rock.

21.7. Compare diaphragm cellular coffer dams and circular coffer dams.

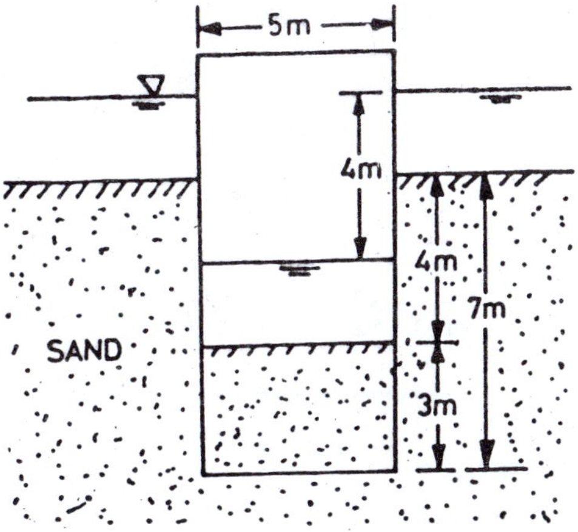

Fig. P-21.4.

21.8. Sketch a typical section of a braced cut and show the various components.

21.9. Draw different types of apparent pressure diagrams used in the design of braced cuts. What are the factors that affect the pressure distribution?

21.10. Describe the methods for the design of various components of a braced cut, stating clearly the assumptions made.

21.11. How the design of a cellular coffer dam on rock differs from that on a soil bed?

C. Multiple Choice Questions.

1. For the design of braced cuts, the earth pressure distribution is based on

(*a*) Coulomb's theory (*b*) Rankine's theory

(*c*) apparent pressure diagram (*d*) none of above.

2. In the design of braced cut, it is generally assumed that the sheet pile is

(*a*) hinged at all the strut levels.

(*b*) fixed at all the strut levels.

(*c*) hinged at all the strut levels except the top and bottom struts.

(*d*) none of above.

3. The diameter of the main cell of a cellular, coffer dam is usually kept

(*a*) 0.5 H (*b*) 0.80 H

(*c*) 1.2 H (*d*) 2.0 H

where H is the height of the coffer dam.

4. Single-sheet pile coffer dams are suitable upto a height of

(*a*) 5 m (*b*) 10 m

(*c*) 15 m (*d*) more than 15 m

5. The factor of safety against piping when $i = 0.75$ and $\gamma' = 9$ kN/m^3 is about

(*a*) 1.20 (*b*) 1.00

(*c*) 0.80 (*d*) 0.75

6. For a cellular coffer dam embedded in clay, the factor of safety against bearing failure when $c = 60\ kN/m^2$, $H = 15$ m and $\gamma = 18\ kN/m^3$ is about

(*a*) 1.10 (*b*) 1.27

(*c*) 1.40 (*d*) 0.80

[**Ans.** 1. (*c*), 2. (*c*), 3. (*c*), 4. (*a*), 5. (*a*), 6. **(*b*)**]

22

Shafts, Tunnels and Underground Conduits

22.1. STRESSES IN SOIL IN THE VICINITY OF A VERTICAL SHAFT

Fig. 22.1 shows the section through a vertical shaft of radius r_o. It is an axis-symmetrical case of stress distribution. The stresses at any point $P(r, z)$ before the excavation of the shaft are given by (see chapter 11).

$$(\sigma_z)_i = \gamma Z \qquad \ldots[22.1(a)]$$

$$(\sigma_r)_i = K_o \gamma Z \qquad \ldots[22.1(b)]$$

and

$$(\sigma_\theta)_i = K_o \gamma Z \qquad \ldots[22.1(c)]$$

where σ_z, σ_r, σ_θ are respectively, the vertical stress, horizontal radial stress and horizontal circumferential stress (hoop stress). The suffix i indicates initial stresses. γ = unit weight of soil, K_o = coefficient of earth pressure at rest.

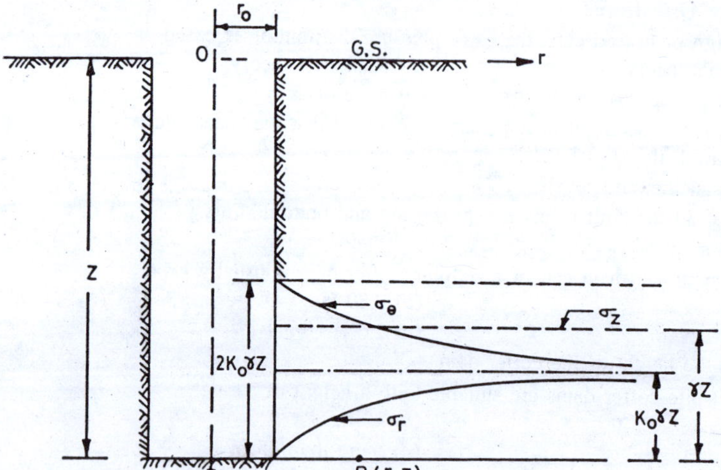

Fig. 22.1. Stresses near vertical shaft.

The shear stresses τ_{rz} are zero, as all the three stresses are principal stresses. Because the shear stresses on all cylindrical surfaces within the soil mass are zero, the material located within the boundaries of the proposed shaft can be replaced by an equivalent liquid of unit weight γ_l equal to $K_o \gamma$, without changing the state of stress in the soil. The horizontal pressure p due to the equivalent liquid is given by

$$p = \gamma_l Z = K_o \gamma Z \qquad \ldots(22.2)$$

As the pressure p is equal to $(\sigma_r)_i$ or $(\sigma_\theta)_i$, there is no effect on the stresses in the soil in the vicinity of the shaft.

The stresses due to equivalent liquid pressure can be found using Lame's formulae for thick cylinders (see any text on strength of materials). According to which,

$$(\sigma_z)_p = 0 \quad ...[22.3(a)]$$

$$(\sigma_r)_p = pr_o^2/r^2 = K_o \gamma Z (r_o^2/r^2) \quad ...[22.3(b)]$$

and

$$(\sigma_\theta)_p = -pr_o^2/r^2 = -K_o \gamma Z (r_o^2/r^2) \quad ...[22.3(c)]$$

where suffix p indicates that the stresses are due to pressure p of the equivalent liquid.

After the shaft has been excavated, the shear stresses and the radial stresses on the interior surface are zero. The effect of excavating the shaft can be considered equivalent to that of pumping the liquid out of a cylindrical hole whose dimensions are identical with those of the shaft (Biot, 1935).

Thus the stresses at any point after the excavation of the shaft can be obtained by superposition of the initial stresses (Eq. 22.1) and those due to pressure (Eq. 22.3). The stresses due to pressure are taken as negative because the liquid is pumped out, which corresponds to a negative value of p.

Therefore,

$$\sigma_z = (\sigma_z)_i - (\sigma_z)_p = \gamma Z \quad ...[22.4(a)]$$

$$\sigma_r = K_o \gamma Z - (\sigma_r)_p = K_o \gamma Z (1 - r_o^2/r^2) \quad ...[22.4(b)]$$

$$\sigma_\theta = (\sigma_\theta)_i - (\sigma_\theta)_p = K_o \gamma Z (1 + r_o^2/r^2) \quad ...[22.4(c)]$$

Fig. 22.1 also shows the variation of stresses on a horizontal plane in the soil mass. The vertical stress (σ_z) is constant and equal to γZ, and is independent of r. The radial stress (σ_r) is zero at $r = r_o$. As r increases, σ_r increases. The curve becomes asymptotic. As r tends to infinity, σ_r approaches a value of $K_o \gamma Z$. The circumferential stress (σ_θ) is equal to $2 K_o \gamma Z$ at $r = r_o$. As r tends to infinity, σ_θ also tends to a value $K_o \gamma Z$. In other words, at infinite radial distance, the stresses reduce to those given by Eq. 22.1 *a* to *c*.

The above stresses have been derived assuming that the soil is elastic, homogeneous and isotropic. These assumptions are seldom justified. Moreover, the circumferential stress (σ_θ) in the vicinity of the shaft is very high ($= 2 K_o \gamma Z$) which may cause plastic flow of the soil. This may lead to a state of plastic equilibrium. Consequently, the actual stresses in the soil in the vicinity of the shaft may be somewhat different from those given by the elastic theory. The elastic theory is used as it is simpler than more advanced theories.

22.2. STRESSES IN SOIL AROUND TUNNELS

Fig. 22.2 (*a*) shows a tunnel of internal radius r_o, with its centre line at a depth of Z_o below the ground surface. Let σ_r and σ_θ be the radial stress and circumferential stress at any point P (r, Z). Before the excavation of the tunnel, the stresses are given by Eq. 22.1 as

$$(\sigma_r)_i = K_o \gamma Z \qquad \text{and } (\sigma_\theta)_i = K_o \gamma Z$$

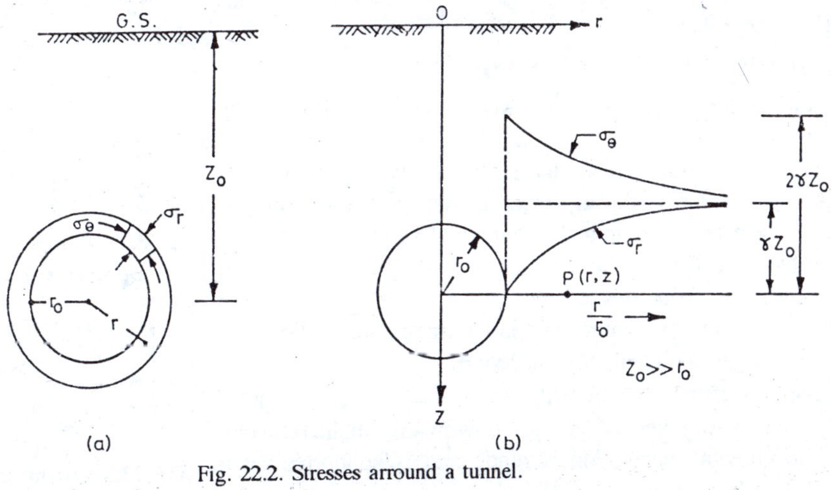

Fig. 22.2. Stresses arround a tunnel.

where γ is unit weight of soil.

Taking $K_o = 1$ and $Z = Z_o$,

$$(\sigma_r)_i = \gamma Z_o \qquad \text{...[22.5(a)]}$$

and

$$(\sigma_\theta)_i = \gamma Z_o \qquad \text{...[22.5(b)]}$$

As in the case of a shaft (Sect. 22.1), let us imagine that proposed tunnel is filled by an equivalent liquid of unit weight γ_l under a pressure given by $p = \gamma_l Z$.

If the depth Z_o is large compared with r_o, the liquid pressure on the tunnel may be taken as constant. Thus

$$p_o = \gamma_l Z_o = \gamma Z$$

The stresses due to the liquid pressure on the tunnel can be determined using Lame's formulae for thick cylinders.

$$(\sigma_r)_{po} = p_o (r_o^2/r^2) = \gamma_l Z_o (r_o^2/r^2) \qquad \text{...[22.6(a)]}$$

and

$$(\sigma_\theta)_{po} = -p_o (r_o^2/r^2) = -\gamma_l Z_o (r_o^2/r^2) \qquad \text{...[22.6(b)]}$$

The stresses after the excavation of the tunnel can be obtained from the superposition of Eqs. 22.5 and 22.6, taking the latter with negative values as it corresponds to pumping out of the liquid. Thus

$$\sigma_r = \gamma_l Z_o - \gamma_l Z_o (r_o^2/r^2) = \gamma_l Z_o (1 - r_o^2/r^2) \qquad \text{...[22.7(a)]}$$

and

$$\sigma_\theta = \gamma_l Z_o + \gamma_l Z_o (r_o^2/r^2) = \gamma_l Z_o (1 + r_o^2/r^2) \qquad \text{...[22.7(b)]}$$

Fig. 22.2(*b*) shows the stress distribution on a horizontal plane in the soil mass. At the interior surface, $r = r_o$.

Therefore, $\sigma_r = 0$ and $\sigma_\theta = 2\gamma_l Z_o = 2\gamma Z_o$

As r tends to infinity, both σ_r and σ_θ tend to approach asymptotically a value of $\gamma_l Z_o = \gamma Z_o$.

The actual stress distribution may be different, as the soil is not a purely elastic material. Further, it has been assumed that the value of K_o is unity which is not justified. The solution is valid when the depth Z_o is very large compared with the radius of the tunnel (r_o). If the tunnel is located close to the ground surface, the result would be erroneous. The solution becomes complicated when all these aspects are considered.

If the support system of the tunnel yields and allows an inward, or radial deflection, of the walls, the radial stress σ_r reduces but the circumferential stress σ_θ increases. The minimum value of σ_r is attained when the circumferential stress becomes equal to the compressive strength of soil (or rock). At this stage, a state of plastic equilibrium is established. The radial deflections may further increase and σ_r is also likely to increase. If the maximum circumferential stress (= $2\gamma Z_o$) is less than the compressive strength of soil (or rock), theoretically no support is required for the tunnel. The deformations must be controlled by providing sufficient restraints to prevent the soil from attaining a plastic equilibrium which would require substantially greater support loadings.

22.3. CONSTRUCTIONS OF EARTH TUNNELS

Earth tunnels are constructed by excavating from below. The procedure is similar to one used in mining. The method of construction used would depend upon the ability of the soil to support itself temporarily during construction and also on the magnitude of the pressure ultimately developed against the support system. The factors affecting these parameters are the tunnel depth, tunnel diameter, the properties of the soil and the ground water table. Most tunnels in soil require support during construction and also permanent support after the construction.

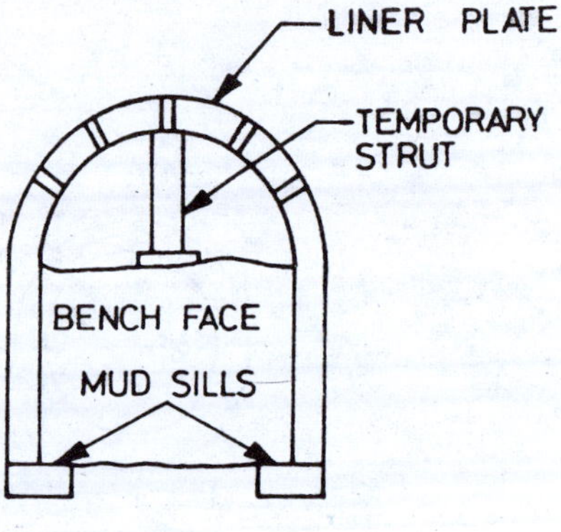

Fig. 22.3. Construction of an earth tunnel.

In a relatively firm soil above the water table, the excavation starts at the tunnel roof and progresses downwards in increments. The initial incremental excavation is made and is supported by a liner plate (Fig. 22.3). The liner plates are flanged steel plates, which when bolted to similar plates form a continuous wall. Sometimes, in place of a liner plate, laggings, which are wooden or steel horizontal planks braced by steel frames, are used. The next increment is then excavated and the process repeated until the complete section has been excavated. If required, the liner can be stiffened with curved arches.

In a very soft soil, a cylindrical temporary support, known as *shield*, is jacked through the soil. The face of the tunnel is supported by a bulkhead fitted with doors through which excavation of a limited portion of the face can be done. The permanent lining is constructed inside the shield after some excavation has been done. The shield is then jacked ahead and the process repeated.

In sand, a tunnel requires support during construction as well as permanent support after construction.

22.4. ARCHING IN SOILS

Arching is a phenomenon that occurs when a yielding part of a soil mass transfers pressure to adjoining soil mass which is less yielding or rigid. The action is similar to one in a structural arch which transfers the load to abutments. When a part of the soil mass yields, it has a tendency to move out of its original position. This tendency is resisted by the shearing resistance at the zone of contact between the yielding and non-yielding parts. Consequently, the pressure on the yielding part is reduced, whereas that on the non-yielding parts is increased. The soil thus arches over the yielding part and transfers the load to the non-yielding parts which act as abutments.

Fig. 22.4 shows a long narrow section of the layer of soil having a yielding part supported on a *deflecting structure*. Although the actual failure surfaces are curved, for simplicity, the analysis is usually based on the assumption of vertical slip surfaces (Terzaghi, 1943). As the structure and the soil above deflect downward, the shear resistance develops between the soil above the yielding part and that in the non-yielding part. It is generally assumed that the shear stress mobilised is equal to the shear strength of the soil. An expression for the vertical stress can be determined as under.

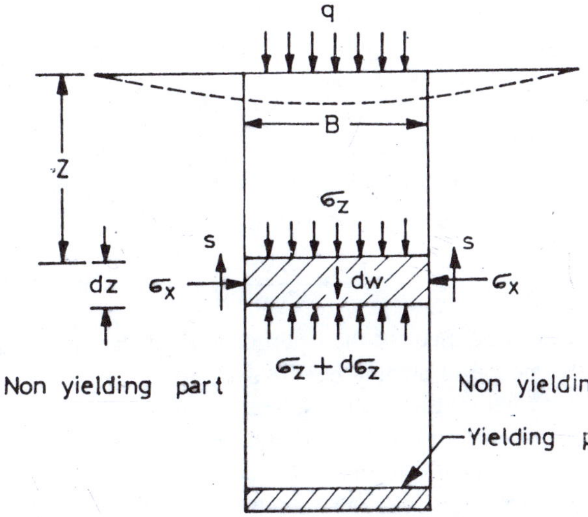

Fig. 22.4. Arching in Soils.

Let us consider the equilibrium of a yielding slice of thickness *dz* at depth *z* from the soil surface. The slice is of width *B* and unit length perpendicular to the plane of the paper. The slice is in equilibrium under its own weight *dW*, the vertical earth pressures on the upper and lower surfaces, and the shear strength (*s*) produced by the lateral earth pressure. Thus

$$dW + B\,\sigma_z = B\,(\sigma_z + d\sigma_z) + 2\,s\,dz$$

Substituting $dW = \gamma\,Bdz$, and $s = c + \sigma_x \tan\phi$,

where $\sigma_x = K\,\sigma_z$,

$$\gamma\,B\,dz + B\,\sigma_z = B\,(\sigma_z + d\,\sigma_z) + 2\,[c + K\,\sigma_z \tan\phi]\,dz$$

or
$$\gamma Bdz + B\,\sigma_z = B\,\sigma_z + Bd\sigma_z + 2c\,dz + 2\,K\,\sigma_z \tan\phi\,dz$$

or
$$B\,d\sigma_z = (\gamma\,B - 2c - 2\,K\,\sigma_z \tan\varphi)\,dz$$

or
$$\frac{d\,\sigma_z}{dz} = \left(\gamma - \frac{2c}{B} - \frac{2K\sigma_z \tan\varphi}{B}\right) \qquad \ldots(a)$$

At the upper surface ($z = 0$), the vertical stress (σ_z) is equal to the surcharge q.

The solution of Eq. (a) with these boundary conditions becomes,

$$\sigma_z = \frac{B(\gamma - 2c/B)}{2K\tan\phi}\left[1 - e^{-2K(z/B)\tan\phi}\right] + qe^{-2K(z/B)\tan\phi} \qquad \ldots(22.8)$$

If $\phi = 0$, Eq. (a) after integration becomes

$$\sigma_z = \left(\gamma - \frac{2c}{B}\right)z + q. \qquad \ldots(22.9)$$

If $c = 0$ and $q = 0$, Eq. 22.8 becomes

$$\sigma_z = \frac{B\gamma}{2K\tan\phi}\left[1 - e^{-2K(z/B)\tan\phi}\right] \qquad \ldots(22.10)$$

At $z = \infty$,

$$\sigma_z = \frac{B\gamma}{2K\tan\phi} \qquad \ldots(22.11)$$

Terzaghi's experimental results indicate that the value of K varies from about unity immediately above the centre of strip to a maximum of about 1.5 at a height of approximately B above the strip.

It has been found by Terzaghi from experiments on dry sand that the shearing resistance of sand is mobilised for a height of about 2.5 B above the yielding strip. Let z_1 be the depth of sand in which there are no shearing stresses and z_2 be the depth in which shearing resistance develops. The vertical stress σ_z on the yielding strip can be obtained by making the following substitution in Eq. 22.8.

$c = 0$, $\quad q = \gamma z_1$, $\quad$ and $\quad z = z_2 = nB$, where $n = 2.5$ for sands.

Thus

$$\sigma_z = \frac{B\gamma}{2K\tan\phi}\left[1 - e^{-2Kn\tan\phi}\right] + \gamma z_1 e^{-2Kn\tan\phi} \qquad \ldots(22.12)$$

Eq. 22.12 can be used for the determination of the vertical stresses.

The phenomenon of arching can be demonstrated by a simple test illustrated by Fig. 22.5. It consists of a platform fitted with a trap door AB. The trap door is mounted on a weighing scale (not shown). The depth H of the dry sand layer over the trap door is very large in comparison with the width of the trap door. The pressure on the trap door and that on the adjoining platform each is equal to γH when the trap door occupies its original position. However, when the trap door is allowed to move slightly in the downward direction, the pressure on the door decreases considerably, as shown by the weighing scale, whereas that on the adjoining parts of the platform increases. As the prism of sand located above the door moves downward, the shear

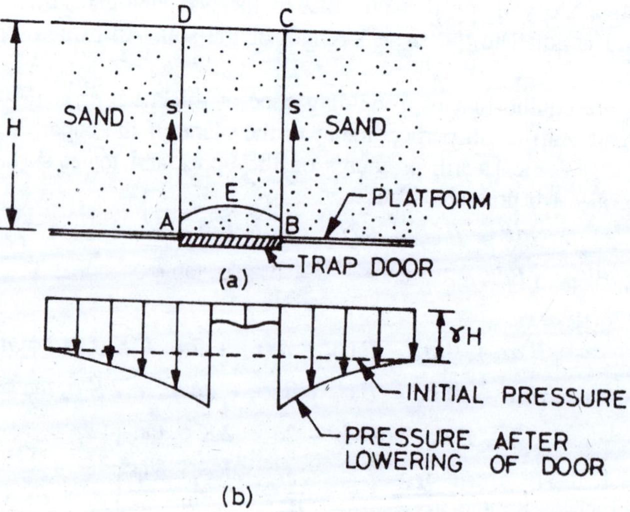

Fig. 22.5. Demonstration of Arching.

stresses along its lateral boundaries *AD* and *BC* resist it and, therefore, the pressure on the trap door is decreased.

The pressure becomes constant after a small movement. It has been established that the ultimate pressure on the trap door is independent of the depth *H* of the layer of sand above it. Only a small portion of the soil prism, shown by the area *ABE,* contributes to the pressure on the trap door. If the sand has even a trace of cohesion, it will not drop out even if the trap door is removed.

22.5. TYPES OF UNDERGROUND CONDUITS

Underground conduits are used for sewers, drains, culverts, water mains, gas lines, electrical cables, telephone lines and many other purposes. The underground conduits should be capable of supporting all external loads. If the conduit is under pressure, it should also be strong enough to withstand internal pressure under extreme conditions. In this text, the forces acting on the conduits due to only external loadings are considered. These forces depend upon a number of factors such as the buried depth, rigidity of the conduit, method of installation, preparation of conduit bedding, live and dead loads acting on the conduit. For purpose of analysis, the conduits are classified into the following types.

(1) Ditch Conduits. (2) Positive Projecting Conduits,
(3) Negative Projecting Conduits, (4) Imperfect Ditch Conduits
(5) Tunneled Conduits.

The loading on different types of conduits are discussed.

22.6. DITCH CONDUITS

A conduit installed in a relatively narrow ditch in undisturbed soil and then covered with earth backfill upto ground surface is called a ditch conduit (Fig. 22.6). This type of conduits are normally installed for sewers, drains, water mains, gas mains, etc. The width B_d of the trench is generally not greater than 2 to 3 times the conduit diameter B_c.

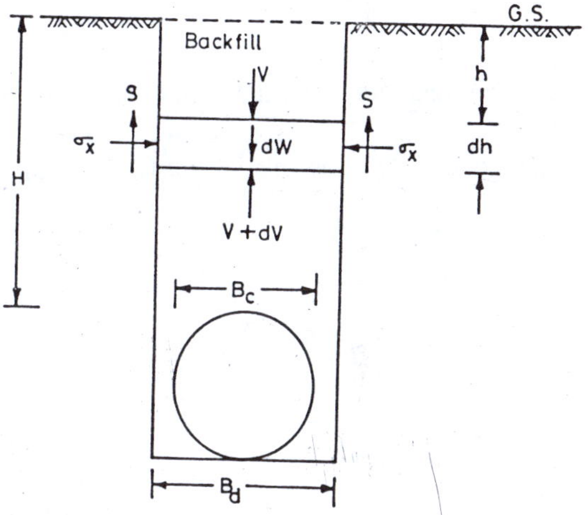

Fig. 22.6. Ditch Conduit.

The loading imposed on to the buried conduit can be obtained by considering the equilibrium of an elemental slice of thickness *dh* at a depth *h* below the ground surface. Let *V* be the vertical load on the top surface of the slice and $(V + dV)$, on the bottoms surface. Let *dW* be the weight of the slice, which is equal to $\gamma B_d\,dh$ per unit length. The horizontal pressure (σ_x) on the vertical sides is equal to *K* times the vertical pressure, where *K* is the coefficient of lateral pressure. Thus

$$\sigma_x = K(V/B_d)\,dh$$

The shearing resistance (*S*) developed along the sides in the vertical direction is equal to μ' times the horizontal pressure, where μ' is the coefficient of sliding friction between the backfill material and the trench wall. Thus

$$S = \mu'\,\sigma_x = \mu' K(V/B_d)\,dh$$

When the elemental slice has a tendency to move downwards, the shear resistance *S* acts upwards. From the equilibrium equation in the vertical direction,

$$V + dV + 2S = V + \gamma B_d\,dh$$

or

$$dV = \gamma B_d\,dh - 2S$$

or

$$dV = \gamma B_d\,dh - 2K\mu'(V/B_d)\,dh$$

The solution of the above differential equations is

$$V = \gamma B_d^2 \left[\frac{1 - e^{-2K\mu'(h/B_d)}}{2K\mu'} \right]$$

At the top of the conduit, $h = H$. Therefore,

$$V = \gamma B_d^2 \left[\frac{1 - e^{-2K\mu'(H/B_d)}}{2K\mu'} \right] \quad ...(22.13)$$

If the ditch has sloping sides, the value of B_d is taken equal to the width of the horizontal tangential plane at the top of the conduit [Fig. 22.7(*a*)]. If the ditch is very wide, the conduit is laid in a sub-ditch to reduce B_d and hence to reduce the load on the conduit [Fig. 22.7(*b*)].

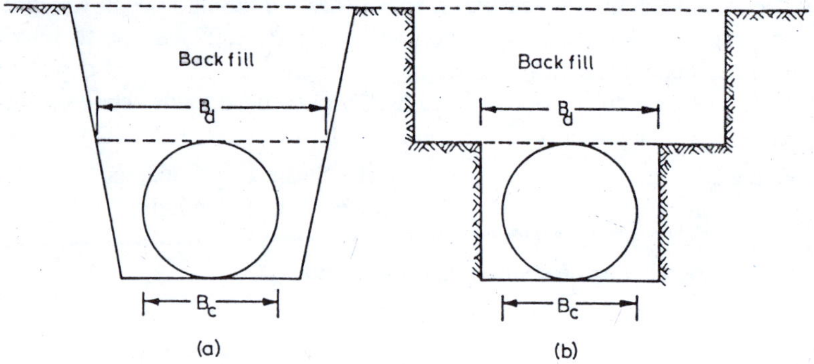

Fig. 22.7.

The load (W_c) carried by the conduit depends upon the rigidity (stiffness) of the conduit in comparison with that of the fill material between the sides of the conduit and the sides of the ditch. In the case of a very rigid conduit, the side fills are relatively compressible and the conduit would carry practically all the load V.

From Eq. 22.13,
$$W_c = \gamma B_d^2 \frac{1 - e^{-2K\mu'(H/B_d)}}{2K\mu'} \quad ...(22.14)$$

or
$$W_c = C_d \gamma B_d^2 \quad ...(22.15)$$

where
$$C_d = \frac{1 - e^{-2K\mu'(H/B_d)}}{2K\mu'} \quad ...(22.16)$$

The coefficient C_d is known as the *load coefficient* for ditch conduits. The values of C_d for different values of H/B_d and $K\mu'$ can be obtained from Fig. 22.8. The ranges for the values of C_d for cohesionless soils and cohesive soils are shown hatched.

If the conduit is relatively flexible and the soil is thoroughly tamped around the conduit, the stiffness of the side fills may approach that of the conduit. The load on the flexible conduit is reduced because some of the load is carried by the side fills. The load carried by the flexible conduit is obtained by multiplying the value given by Eq. 22.15 with the ratio (B_c/B_d), where B_c is the diameter of the conduit. Therefore, for flexible conduits, the load is given by

$$W_c = C_d \gamma B_c B_d \quad ...(22.17)$$

Eq. 22.13 has been derived assuming that the arching effect is achieved without soil cohesion. Actually, some cohesive resistance develops in the vicinity of the trench which would reduce the loading on the conduit. Therefore, Eq. 22.13 is conservative. Field observations indicate that the load at the time when the fill is completed is about 80 to 90% of the final load. The load keeps building up because of strains occurring in the soil in the zones along the vertical boundaries of the trench. This reduces the shearing resistance of the soil and causes transfer of load from the soil to the conduit.

The following points are worth noting :

(*i*) If the trench backfill is compacted to high density, it will not settle relative to the surrounding undisturbed soil, and the arching action would not occur. For such a condition, the load on the conduit is equal to the weight of prism of soil of height *H*.

(*ii*) If the surrounding undisturbed soil settles more than the fill above the conduit, the load on the conduit is computed considering it as a projecting conduit, as explained in the following section.

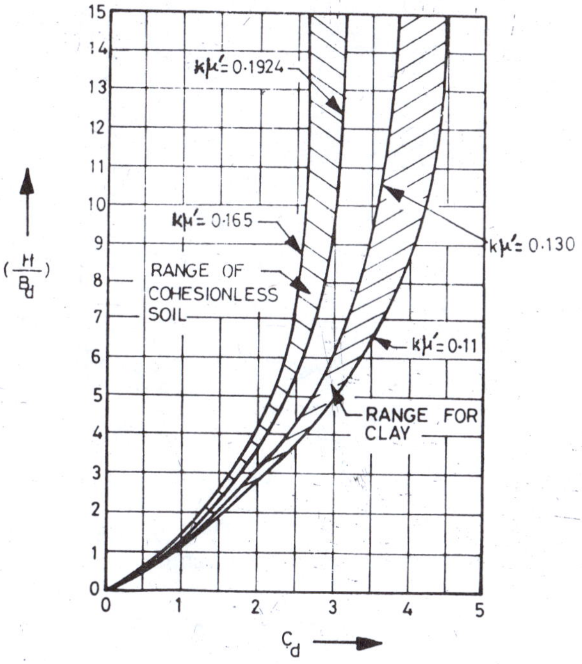

Fig. 22.8. Chart for C_d.

22.7. POSITIVE PROJECTING CONDUITS

A conduit projecting some distance above the natural ground surface (GS) and covered by earth fill is known as a positive projecting conduit [Fig. 22.9(*a*)]. Common examples of positive projecting conduits are culverts for highway, railway and airfield. In positive projective conduits, there is a differential settlement between the central zone directly overlying the conduit and the side zones. Depending upon the relative settlement, the load on the conduit will vary.

For rigid conduits, the total compression of the earth fill in side zones will be greater than the total compression of the central zone of the earth fill directly over the conduit. This is because greater height of fill in side zones. However, if the conduit is flexible and deflects considerably under the weight of the fill, the settlement of the fill in the central zone will be greater than the settlement in the side zones. The differential settlement between the two zones results in the development of shearing stresses on vertical planes passing through the sides of the conduit. Fig. 22.9 (*a*) shows the direction of shearing stresses (upward) when the central zone settles more than side zones, whereas

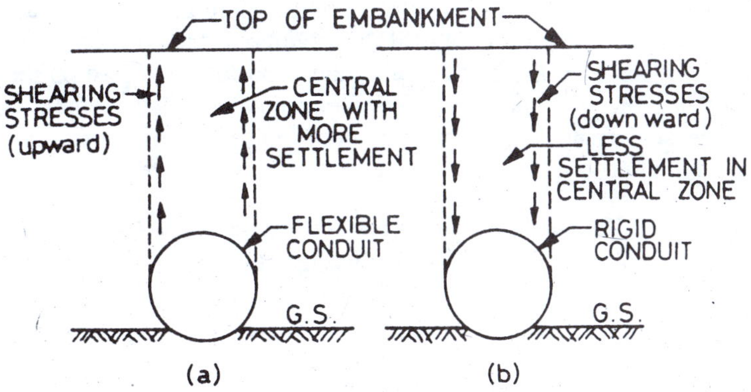

Fig. 22.9. Positive Projecting Conduit.

Fig. 22.9 (*b*) shows the direction (downward) when the central zone settles less than side zones. In the first case, the load on the conduit is less than the weight of fill directly over the conduit, whereas in the second case, the load on the conduit is more than the weight of fill.

The direction of the shearing stresses can be determined from the settlement of a plane, known as a *critical plane.* It is a horizontal plane in the fill material at the level of the top of the conduit at the time when embankment construction is begun (Fig. 22.10). If the critical plane, after the construction, settles less than the top of the conduit, the direction of shearing stresses is upward [Fig. 22.10 (*a*)]. However, if the critical plane settles more than the top of the conduit, the direction is downward [Fig. 22.10 (*b*)]. A ratio, known as

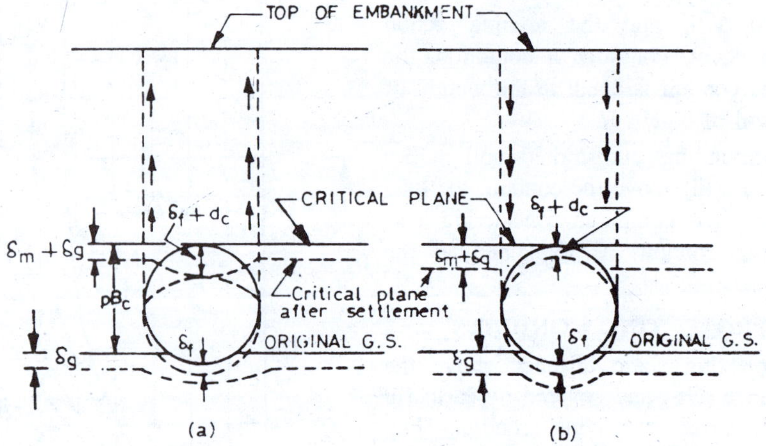

Fig. 22.10. Effect of Settlement of critical plane.

settlement ratio, is used to determine the settlement of critical plane.

The settlement ratio (r_p) is given by

$$r_p = \frac{(\delta_m + \delta_g) - (\delta_f + d_c)}{\delta_m} \quad ...(22.18)$$

where δ_m = compression of the fill on the sides of the conduit in the distance pB_c ,

δ_g = settlement of natural ground surface adjacent to the conduit,

δ_f = settlement of conduit into its foundation,

and d_c = vertical deflection of the conduit,

p = projection ratio. It is the ratio of the distance from the natural ground surface to the top of the conduit to its width B_c ,

B_c = diameter or width of conduit.

The settlement ratio is negative when the conduit settles more than the critical plane. The load on the conduit is less than the weight of the overlying fill. It is similar to the conditions in the ditch conduit and is known as *ditch condition.* However, if the settlement ratio is positive, the load on the conduit is greater than the weight of the overlying fill. It is known as *projection condition.* It is extremely difficult to determine the individual terms in the settlement ratio. However, empirically determined values given by Spangler are generally used for design purposes (Table 22.1).

Table 22.1. Values for Settlement Ratio (r_p)
(After Spangler)

S.No.	*Conduit Condition*	*Settlement Ratio* (r_p)
1.	Rigid conduit of foundation of rock or unyielding soil.	+ 1.00
2.	Rigid conduit on foundation of ordinary soil.	+ 0.5 to + 0.80
3.	Rigid conduit on foundation that yields relative to adjacent ground.	0.0 to + 0.50
4.	Flexible conduit with poorly compacted side fills.	– 0.4 to 0.0
5.	Flexible conduit with well compacted side fills.	– 0.2 to + 0.8 (not well-established)

Plane of Equal Settlement

The shearing stresses acting on the vertical planes do not necessarily extend to the top of the embankment, especially where a high fill exists over the conduit. Differential settlements are greatest at the

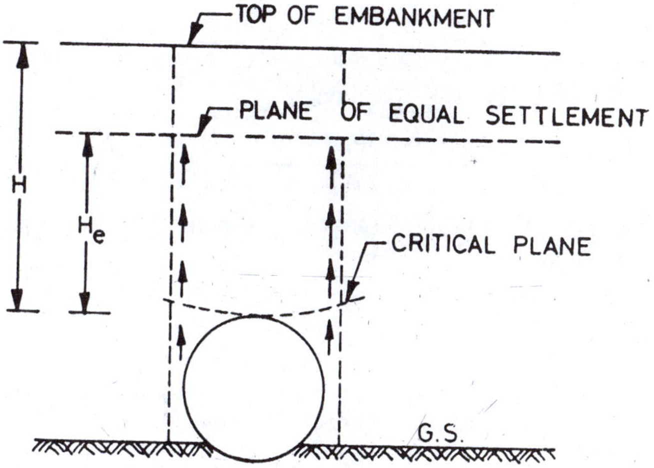

Fig. 22.11. Plane of equal settlement.

top surface of the conduit and progressively decrease towards the top of embankment. At certain height H_e above the conduit, a horizontal plane exist at which the differential settlements are zero. This plane is called the *plane of equal settlement* (Fig. 22.11). Above the plane of equal settlement, the central zone and the side zones settle equally. Thus no differential settlement occurs within the embankment soil above the plane of equal settlement. An expression for the height H_e can be obtained by equating the total vertical settlements in the central zone and the side zones as under.

$$\left[\frac{1}{2K\mu'} \pm \left(\frac{H}{B_c} - \frac{H_e}{B_c} \pm \frac{r_p \times p}{3}\right)\right] \frac{e^{\pm 2K\mu'(H_e/B_c)} - 1}{\pm 2K\mu'} + \frac{1}{2}\left(\frac{H_e}{B_c}\right)^2$$

$$+ \frac{r_p \times p}{3}\left(\frac{H - H_e}{B_c}\right) e^{\pm 2K\mu'(H_e/B_c)} - \frac{1}{2K\mu'}\left(\frac{H_e}{B_c}\right) \mp \left(\frac{H}{B_c}\right)\left(\frac{H_e}{B_c}\right) = \pm r_p p\,(H/B_c) \quad ...(22.19)$$

The upper (plus or minus) signs apply when settlement ratio (r_p) is positive, whereas the lower signs apply when r_p is negative.

Types of Positive Projecting Conduits

The positive projecting conduits can be classified into 4 types, depending upon (1) magnitudes of actual embankment height (H) and the height of plane of equal settlement (H_e) and (2) the sign of the settlement ratio (r_p), as given below.

(*i*) Complete ditch conduits $H < H_e$; r_p = negative

(*ii*) Complete projection conditions $H < H_e$; r_p = positive

(*iii*) Incomplete ditch conditions $H > H_e$; r_p = negative

(*iv*) Incomplete projection conditions $H > H_e$; r_p = positive

It may be noted that for complete ditch or complete projection conditions, the shearing stresses extend upto the top of the embankment; whereas for incomplete ditch or incomplete projection conditions, the shearing stresses extend upto the plane of equal settlement, which is lower than the top of the embankment.

Marston gave the following expression for the load on positive projecting conduit:

$$W_c = C_p \gamma B_c^2 \quad ...(22.20)$$

where the coefficient C_p for the complete projection or ditch condition is given by

$$C_p = \frac{e^{\pm 2K\mu'(H/B_c)} - 1}{\pm 2K\mu'} \qquad \text{...(22.21)}$$

and that for incomplete projection or ditch condition as

$$C_p = \frac{e^{\pm 2K\mu'(H_e/B_c)} - 1}{\pm 2K\mu'} + \left(\frac{H}{B_c} - \frac{H_e}{B_c}\right) e^{\pm K\mu'(H_e/B_c)} \qquad \text{...(22.22)}$$

The positive signs apply when r_p is positive (projection conditions), whereas the negative signs apply when r_p is negative (ditch conditions).

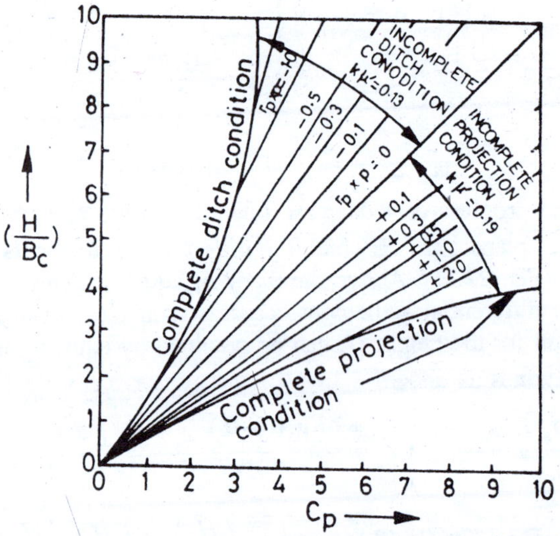

Fig. 22.12. Chart for C_p.

Values of C_p for different (H/B_c) ratios and the products $(r_p \times p)$ can be obtained from Fig. 22.12. The position of the curves is not sensitive to the variation of $K\mu'$ values expected in the field. The curves are for $K\mu' = 0.13$ and $K\mu' = 0.19$.

22.8. NEGATIVE PROJECTING CONDUITS

A conduit installed in a trench such that its top is below the original ground surface and then covered with earth embankment is known as the negative projecting conduit (Fig. 22.13). In case of negative projecting conduits, the critical plane lies below the original ground surface (G.S), and the settlement ratio (r_n) is given by

$$r_n = \frac{\delta_g - (\delta_d + \delta_f + d_c)}{\delta_d} \qquad \text{...(22.23)}$$

where δ_g = settlement of ground surface,
δ_d = settlement of backfill in the trench,
δ_f = settlement of the conduit foundation,
d_c = vertical deflection of the conduit.

The critical plane is taken as the horizontal plane at the ground surface. Generally, the settlement of the critical plane in the trench fill over the conduit is greater than the settlement of the ground surface adjacent

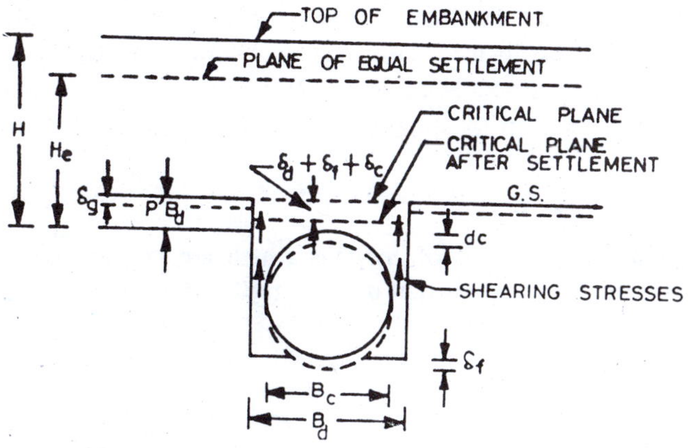

Fig. 22.13. Negative Projecting Conduit.

to the trench. As the central zone moves downward relative to the side zones, the shearing stresses act upward and the load on the conduit is reduced.

The load on the conduit is given by

$$W_c = C_n \gamma B_d^2 \quad \text{...(22.24)}$$

where C_n is the coefficient for negative projecting conduit, given by

$$C_n = \frac{e^{-2K\mu'(H/B_d)} - 1}{-2K\mu'} \quad \text{when } H < H_e \quad \text{...(22.25)}$$

and

$$C_n = \frac{e^{-2K\mu'(H_e/B_d)} - 1}{-2K\mu'} + \left(\frac{H}{B_d} - \frac{H_e}{B_d}\right) e^{-2K\mu'(H_e/B_d)} \quad \text{when } H > H_e \quad \text{...(22.26)}$$

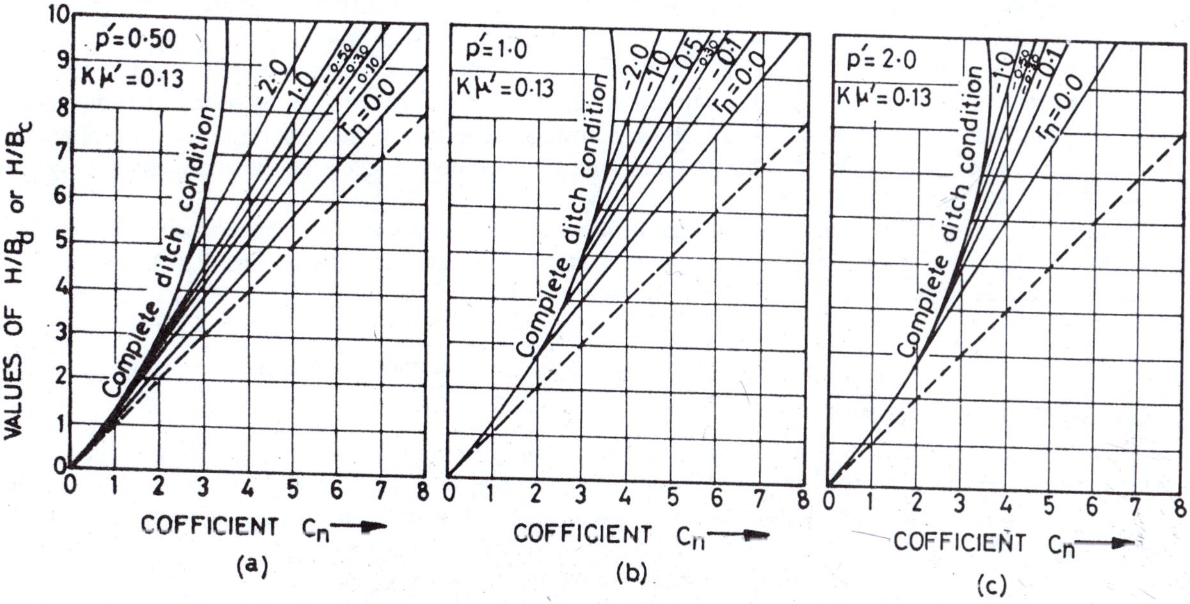

Fig. 22.14. Charts for C_a.

The negative projection ratio p' is the ratio of the distance between the top of the conduit and the natural ground surface to the trench width. Thus the distance between the top of the conduit and the ground surface is $p' B_d$. It is always positive. As the settlement ratio (r_n) is always negative, the product ($r_n \times p'$) is also negative. The settlement ratio (r_n) is generally taken as –0.50 for the purpose of estimating loads on the conduit. The value of C_n may be obtained for different values of r_n from Fig. 22.14 for $K\mu' = 0.13$ and $p' = 0.5$, 1.0 and 2.00.

22.9. IMPERFECT DITCH CONDUIT

Imperfect ditch conduits are special type of conduits which are constructed to reduce the load on a conduit under a high embankment. The construction of imperfect ditch conduits is done in two stages :

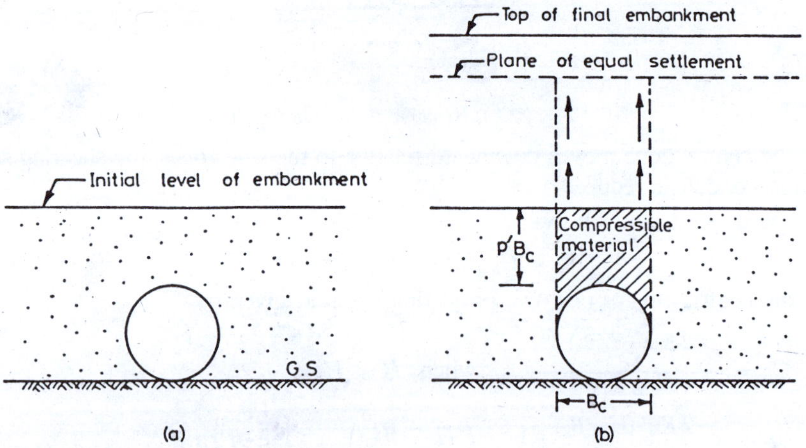

Fig. 22.15. Imperfect Ditch Conduit.

(*i*) The conduit is first installed as a positive projecting conduit above the ground surface and it is covered with earthfill upto a height $p' B_c$ where $p' = 1.0$ to 2.0 [Fig. 22.15 (*a*)]. The fill is well compacted.

(*ii*) In the second stage, a trench is excavated directly above the conduit from the initial level of embankment to the top of the conduit [Fig. 22.15 (*b*)]. The trench is backfilled with some compressible material such as hay, straw, cornstalks. The rest of the embankment is completed as usual upto the final level. Thus this construction procedure creates the situation like a negative projection condition.

As the compressible material settles, arching action develops in the overlying embankment and the shearing stresses act in upward direction. Thus the load on the conduit is reduced. Eqs. 22.23 to 22.26 can be used for the imperfect ditch conduit by substituting B_c for B_d. Fig. 22.14 can be used for determination of C_n.

Imperfect ditch conduits are also known as conduits in *induced trench conditions.*

22.10. TUNNELED CONDUITS

Conduits passing beneath *existing* embankments or hilly areas are known as tunneled conduits. Such conduits are constructed by tunneling or jacking methods instead of the usual open trench methods. The examples of the tunneled conduits are culverts or sewers installed beneath an existing highway or a railway line in service (Fig. 22.16).

The load on a tunneled conduit can be obtained from the following equation:

$$W_c = C_t \gamma B_t^2 - 2 C_t c B_t \qquad \text{...(22.27)}$$

where B_t = width or diameter of the tunnel or jacked opening,

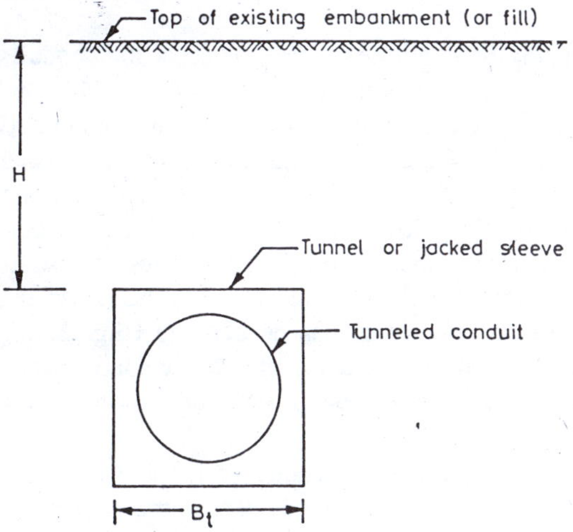

Fig. 22.16. Tunneled Conduit.

c = cohesive strength of the soil overlying conduit,

C_t = coefficient for tunneled conduit. It is equal to C_d, the coefficient of ditch conduit (Eq. 22.16).

The values of C_t can be obtained from Fig. 22.8.

22.11. LOADS ON CONDUITS DUE TO SURFACE LOADS

Loads applied on the surface above a buried conduit are transmitted through soil to the conduit. The stresses on the conduit surface due to concentrated or uniformly distributed loads acting on the soil surface can be determined using the methods discussed in Chapter 11. It is generally assumed that Boussinesq's solution is applicable.

The stresses from loads applied through a highway, runway or railways depend upon the physical characteristics of the pavement (or formation below the track), the magnitude of the load and the conduit area. It is generally assumed that the loads act over some prescribed area whose size increases with depth, somewhat similar to 2 : 1 distribution.

Moving loads also cause impact. Therefore, the effect of moving loads is greater than equivalent dead loads. The effect of impact is more in the case of unpaved formation compared with that of a paved surface. Generally, an impact factor of 1.50 is taken for unpaved formation and of unity for a paved highway.

Martson proposed the following equation for computing the load on the conduit due to loads applied on the surface of the fill.

$$W_t = \frac{1}{L_e} \cdot I_c C_l P \qquad \text{...(22.28)}$$

where W_t = average load on conduit due to wheel load (kN/m),

L_e = length of conduit section,

I_c = impact factor,

C_l = load coefficient,

P = concentrated truck-wheel load on surface (kN).

The value of the load coefficient (C_l) depends upon the length and width of the conduit section and also on the depth of soil cover over the conduit. Its value can be determined from Boussinesq's solution.

The length of conduit section (L_e) is equal to the actual length of a precast segmental section of the conduit which is 1 m or less in length. For continuous conduits or those constructed of segmental sections more than 1 m in length, the effective length (L_e) is the length of the conduit over which the average live load produces the same effect as does the actual load. An effective length of 1 m is generally taken.

22.12. CONSTRUCTION OF CONDUITS

Buried conduits are constructed in different shapes using different materials. The maximum load which a conduit can take depends upon the conduit material, shape of the conduit and the type of support. The safe load which can be imposed on a rigid conduit, such as concrete pipe, is determined from the pipe strength and the quality of foundation bedding. The strength of concrete conduits can be obtained from a three-edge bearing test (Fig 22.17). As the test simulates a very severe loading condition, which rarely develops in the actual conduits, the conduit can support even a load greater than that obtained from the test. Generally, a conduit which has a foundation shaped according to its contour can take a greater load [Fig. 22.18(*b*)] than that which has no bedding [Fig. 22.18(*a*)]. Design charts are supplied by manufacturers which provide load factors for different shaped concrete conduits based upon type of bedding, conduit projection and settlement ratio.

The factors affecting the load-carrying capacity of flexible conduits are not well defined. However, manufacturers of corrugated metal pipes have developed some manuals which can be consulted. Such conduits fail by excessive deflection whereas rigid conduits fail by the rupture of the pipe wall. The deflection of flexible conduits should be limited.

Fig. 22.17. Three-edge bearing test.

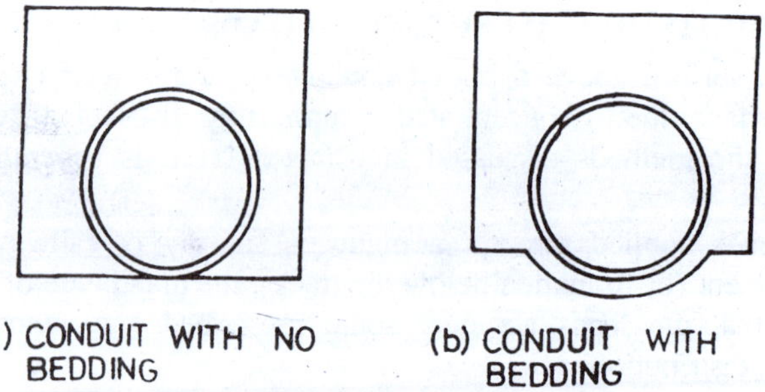

Fig. 22.18.

ILLUSTRATIVE EXAMPLES

Illustrative Example 22.1. *A rigid sewer pipe with an outside diameter of 50 cm is to be laid in a ditch which is 1 m wide at the top of the pipe and is to be covered with 8.0 m of clayey soil back-fill (γ = 19 kN/m^3). Determine the load on the sewer. Take $K\mu' = 0.12$.*

Solution. (See Fig. 22.6.) In this case, $H = 8.0$ m, $B_d = 1.0$ m, $H/B_d = 8.0$.

From Fig. 22.8, for $H/B_d = 8.0$, $C_d = 3.6$

Alternatively, from Eq. 22.16, $$C_d = \frac{1 - e^{-2K\mu'(H/B_d)}}{2K\mu'} = 3.56$$

From Eq. 22.15,
$$W_c = C_d \gamma B_d^2$$
$$= 3.6 \times 19 \times (1.0)^2$$
$$= \mathbf{68.4\ kN/m}$$

Illustrative Example 22.2. *Determine the load on a (positive projecting) square box culvert under a 16 m fill (γ = 19 kN/m^3). Assume that the outside width of the barrel is 2.5 m. Take projection ratio (p) = 0.5 and the settlement ratio (r_p) = + 0.6.*

Solution. $H/B_c = 16/2.5 = 6.4$ and $r_p \times p = 0.6 \times 0.5 = +0.3$

From Fig. 22.12, $C_p = 9.0$.

From Eq. 22.20,

$$W_c = C_p \gamma B_c^2$$
$$= 9.0 \times 19 \times (2.5)^2$$
$$= \mathbf{1068.75\ kN/m.}$$

PROBLEMS

A. Numerical

22.1. A rigid pipe with an outside diameter of 1 m diameter is laid in a ditch which is 1.5 m wide at the top of the pipe. The pipe is covered with 9.0 m of sandy soil backfill ($K\mu' = 0.18$). Determine the load on the pipe. Take $\gamma = 18$ kN/m^3. [**Ans.** 99.6 kN/m]

22.2. A reinforced concrete pipe, 2 m in diameter, is installed as an imperfect ditch conduit, with the negative projection ratio (p') = 1.0, The height of fill above the top of the pipe is 16 m. Determine the load on the pipe. Assume $K\mu' = 0.13$ and $r_n = -0.5$. Take $\gamma = 19$ kN/m^3. [**Ans.** 357.2 kN/m]

B. Descriptive and Objective Type

22.3. Discuss different types of conduits. What are the factors that affect the load on a conduit?

22.4. What do you understand by arching of soil ? Give examples.

22.5. Explain the variation of stresses in soil in the vicinity of a vertical shaft.

22.6. Draw a sketch showing the stress distribution in soil around tunnels.

22.7. Discuss the methods of construction of earth tunnels.

22.8. Define planes of equal settlement. What is their importance ?

22.9. Define settlement ratio. How is it determined ?

22.10. How would you determine the load on a buried conduit due to surface loads ?

22.11. Describe the methods of laying of conduits. What is the effect of foundation bedding on the load carried by the conduit ?

22.12. What do you understand by imperfect ditch condition ? What is its advantage ?

22.13. Write whether the following statements are true.

(*a*) For a vertical shaft constructed in soil, the radial stress becomes infinite when the radius tends to infinity.
(*b*) For a tunnel in soil, as *r* tends to infinity, the radial stress tends to approach a value of zero.
(*c*) The stresses on the yielding portion of a soil mass are increased due to arching of soil.
(*d*) For a ditch conduit, the load carried by a flexible conduit is greater than that by a rigid one.
(*e*) For a ditch conduit, the load load carried by the conduit is reduced if the ditch is made wider.
(*f*) The soil above the plane of equal settlement is not subjected to shear stresses.
(*g*) The plane of equal settlement cannot be above the top of embankment.
(*h*) In imperfect ditch conduits, the excavated trench is backfilled with compact, incompressible material.

[**Ans.** True, (*f*)]

C. Multiple Choice Question

1. The negative projecting conduit is installed

(*a*) above the ground surface in an embankment.
(*b*) in a trench below the ground surface in an embankment.
(*c*) in a ditch below the ground surface in cutting.
(*d*) by tunneling into an embankment.

2. In a vertical shaft driven in a homogeneous soil, the maximum circumferential (hoop) stress is

(*a*) $K_o \sigma_v$ (*b*) $2 K_o \sigma_v$
(*c*) σ_v (*d*) none of above

where K_o is coefficient of earth pressure at rest and σ_v is the vertical stress.

3. In a horizontal tunnel excavated in a homogeneous soil, the maximum circumferential stress is
(a) $2\gamma Z_o$ (b) γZ_o
(c) $0.5\ \gamma Z_o$ (d) zero

where γ is the unit weight of soil and Z_o is the depth of the centre of the tunnel.

4. For complete projection condition in a positive projecting conduit,
(a) $H < H_e$ and r_p should be positive.
(b) $H < H_e$ and r_p should be negative.
(c) $H > H_e$ and r_p should be positive.
(d) $H > H_e$ and r_p should be negative.

where H is embankment height, H_e is height of plane of equal settlement and r_p is settlement ratio.

5. For a ditch conduit of 1.0 m diameter laid in a ditch 2 m wide at its top when $C_d = 3.0, \gamma = 18\ kN/m^3$, the load carried by the conduit is
(a) 54 kN/m (b) 216 kN/m
(c) 108 kN/m (d) 27 kN/m

6. A negative projecting conduit of diameter 1.0 m is laid in a ditch 1.5 m wide. The load carried by the conduit when $C_n = 5.0$ and $20.0\ kN/m^2$ is
(a) 100 kN/m (b) 150 kN/m
(c) 225 kN/m (d) 200 kN/m

7. A positive projecting conduit has a diameter of 2.0 m. The load carried by the conduit when $C_p = 5.0$ and $\gamma = 20\ kN/m^3$ is
(a) 100 kN/m (b) 200 kN/m
(c) 400 kN/m (d) 300 kN/m

8. In a positive projecting conduit, the direction of the shear stresses developed in the fill is
(a) Always upward (b) Always downward
(c) May be upward or downward (d) Always horizontal

[Ans. 1. (b), 2. (b), 3. (a), 4. (a), 5. (b), 6. (c), 7. (c), 8. (c)**]**

23

Bearing Capacity of Shallow Foundations

23.1. INTRODUCTION

A foundation is that part of a structure which transmits the weight of the structure to the ground. All structures constructed on land are supported on foundations. A foundation is, therefore, a connecting link between the structure proper and the ground which supports it. The word 'foundation' is derived from the latin word *fundare,* meaning to set or ground on something solid. In other words, a foundation is an artificially laid base on which a structure is set or built up.

A foundation is required for distributing the loads of the superstructure on a large area. The foundation should be designed such that (1) the soil below does not fail in shear and (2) the settlement is within the safe limits. The pressure which the soil can safely withstand is known as the *allowable bearing pressure.* This chapter gives the methods for the determination of allowable bearing pressure.

Foundations may be broadly classified into two categories: (1) Shallow foundations, (2) Deep foundations. A shallow foundation transmits the loads to the strata at a shallow depth. A deep foundation transmits the load at considerable depth below the ground surface. The distinction between a shallow foundation and a deep foundation is generally made according to Terzaghi's criterion. According to which, a foundation is termed shallow if it is laid at a depth equal to or less than its width. Shallow foundations are discussed in Chapters 23 and 24, and deep foundations, in Chapters 25, 26 and 27.

23.2. BASIC DEFINITIONS

(1) Ultimate Bearing Capacity (q_u). The ultimate bearing capacity is the gross pressure at the base of the foundation at which the soil fails in shear.

(2) Net Ultimate Bearing Capacity (q_{nu}). It is the net increase in pressure at the base of foundation that causes shear failure of the soil. It is equal to the gross pressure minus overburden pressure.

Thus
$$q_{nu} = q_u - \gamma D_f \qquad \text{...(23.1)}$$

where q_u = ultimate bearing capacity (gross),

γ = unit weight of foundation soil, and D_f = depth of foundation.

It may be noted that the overburden pressure equal to γD_f existed even before the construction of foundation.

(3) Net Safe Bearing Capacity (q_{ns}). It is the net soil pressure which can be safely applied to the soil considering only shear failure. It is obtained by dividing the net ultimate bearing capacity by a suitable factor of safety. Thus

$$q_{ns} = \frac{q_{nu}}{F} \qquad \text{...(23.2)}$$

where F = factor of safety, which is usually taken as 3.0.

(4) Gross Safe Bearing Capacity (q_s). It is the maximum gross pressure which the soil can carry safely without shear failure. It is equal to the net safe bearing capacity plus the original overburden pressure. Thus

$$q_s = q_{ns} + \gamma D_f$$

or

$$q_s = \frac{q_{nu}}{F} + \gamma D_f \qquad \text{...(23.3)}$$

Some authors define the gross safe bearing capacity (q_s) as the ultimate bearing capacity divided by a factor of safety (F); that is,

$$q_s = \frac{q_u}{F} = \frac{q_{nu} + \gamma D_f}{F} = \frac{q_{nu}}{F} + \frac{\gamma D_f}{F} \qquad \text{...(23.4)}$$

As the added strength due to γD_f is available in full, it does not seem logical to apply a factor of safety to this term. It is, therefore, more rational to define the gross safe bearing capacity as indicated by Eq. 23.3. This practice will be followed in this text.

(5) Net Safe Settlement Pressure (q_{np}). It is the net pressure which the soil can carry without exceeding the allowable settlement. The maximum allowable settlement generally varies between 25 mm and 40 mm for individual footings.

The net safe settlement pressure is also known as *unit soil pressure or safe bearing pressure.*

(6) Net Allowable Bearing Pressure (q_{na}). The net allowable bearing pressure is the net bearing pressure which can be used for the design of foundations.

As the requirements for the design of foundation are that there should be no shearing failure and moreover the settlements should also be within the limits, the allowable bearing pressure is the smaller of the net safe bearing capacity (q_{ns}) and the net safe settlement pressure (q_{np}). Thus

$$q_{na} = q_{ns} \quad \text{if } q_{np} > q_{ns} \qquad \text{...[23.5(a)]}$$

or

$$q_{na} = q_{np} \quad \text{if } q_{ns} > q_{np} \qquad \text{...[23.5(b)]}$$

The net allowable bearing pressure is also known as the allowable soil pressure or allowable bearing pressure or allowable bearing capacity.

23.3. GROSS AND NET FOOTING PRESSURES

The gross and net footing pressure at the base of a footing can be found as follows.

(1) Foundation Backfilled. Fig. 23.1 shows a footing subjected to a superimposed load Q. If the weight of the footing and the soil above it is W_f, the gross footing pressure is given by

$$q_g = \frac{Q + W_f}{A} \qquad \text{...(23.6)}$$

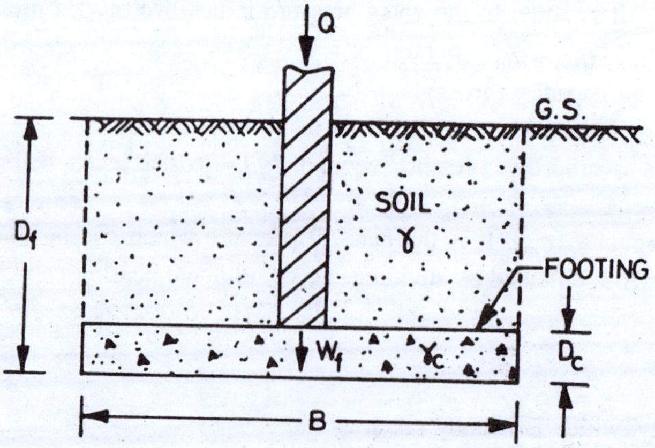

Fig. 23.1.

where A is the base of the footing.

Eq. 23.6 can be written as
$$q_g = \frac{Q}{A} + \frac{(D_c \gamma_c) A}{A} + \frac{(D_f - D_c)\gamma A}{A} \quad \text{...(23.7)}$$

where γ_c = unity weight of concrete, D_c = thickness of footing, γ = unit weight of soil, Q = superimposed load.

The net footing pressure is equal to the gross footing pressure minus the overburden pressure. Thus
$$q_n = q_g - \gamma D_f \quad \text{...(23.8)}$$

Substituting the value of q_g from Eq. 23.7,
$$q_n = \frac{Q}{A} + \frac{(D_c \gamma_c) A}{A} + \frac{(D_f - D_c)\gamma A}{A} - \gamma D_f$$

Simplifying,
$$q_n = Q/A + (\gamma_c - \gamma) D_c$$

If the difference between the unit weight of concrete (about 24 kN/m^3) and the unit weight of soil (about 20 kN/m^3) is neglected, Eq. 23.9 becomes
$$q_n = Q/A \quad \text{...(23.10)}$$

Thus the net footing pressure (q_n) is equal to the superimposed load Q divided by the area A.

For safe design, the net footing pressure (q_n) should be less than or equal to the net allowable bearing pressure (q_{na}), *i.e.*
$$q_n \le q_{na}$$
or
$$Q/A \le q_{na} \quad \text{...(23.11)}$$

(2) Foundation not backfilled. The footing beneath the basement (Fig. 23.2) are not backfilled. For such footings, the gross pressure is given by

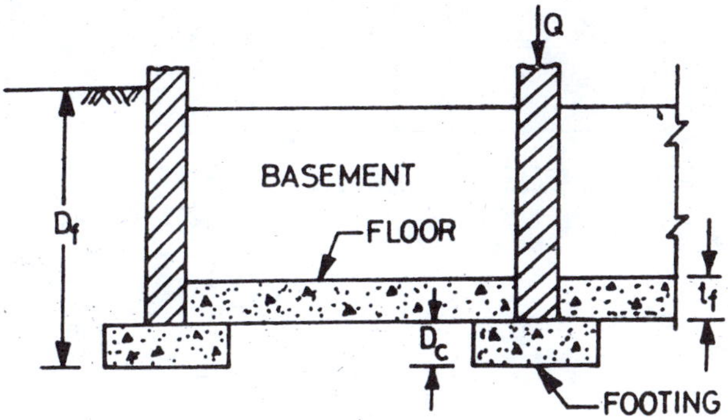

Fig. 23.2.

$$q_g = Q/A + D_c \gamma_c + t_f \gamma_c \quad \text{...(23.12)}$$

where t_f = thickness of basement floor.

The net footing pressure becomes
$$q_n = q_g - \gamma D_f \quad \text{...(23.13)}$$
or
$$q_n = [Q/A + D_c \gamma_c + t_f \gamma_c] - \gamma D_f$$

As the thicknesses D_c and t_f are small in comparison with the depth D_f, the second and third terms in Eq. 23.12 can be neglected.

Thus
$$q_n = Q/A - \gamma D_f \quad \text{...(23.14)}$$

For safe design, the net footing pressure should be equal to or less than the net allowable bearing pressure, *i.e.*,

$$q_n \leq q_{na}$$

or $$(Q/A - \gamma D_f) \leq q_{na} \qquad ...(23.15)$$

or $$Q/A \leq q_{na} + \gamma D_f \qquad ...(23.16)$$

Comparing Eq. 23.11 with 23.16, it is observed that the load- carrying capacity of a foundation is considerably increased if it is not backfilled. The net footing pressure would reduce to zero if in Eq. 23.14,

$$Q/A = \gamma D_f \qquad ...(23.17)$$

This is the principle of compensated raft foundations in which the pressure applied is just balanced by the pressure released (see Chapter 24).

The reader should carefully note the difference between the footing pressure q_n and the allowable bearing pressure q_{na}. The footing pressure depends upon the superimposed load Q acting on the footing, whereas the allowable bearing pressure depends upon the bearing capacity of the foundation and the allowable settlement. The allowable bearing pressure is a function of the type of soil and the footing, as discussed in later sections.

Eq. 23.11 is used for the determination of the area of footing if it is backfilled and Eq. 23.16, if not backfilled.

23.4. RANKINE'S ANALYSIS

Rankine (1885) considered the plastic equilibrium of two adjacent soil elements, one immediately beneath the footing and the other just beyond the edge of the footing (Fig. 23.3). For the element I beneath the footing, the vertical stress is the major principal stress and the lateral stress is the minor principal stress. However, for the element II, the lateral stress becomes the major stress, and the vertical stress becomes the minor principal stress.

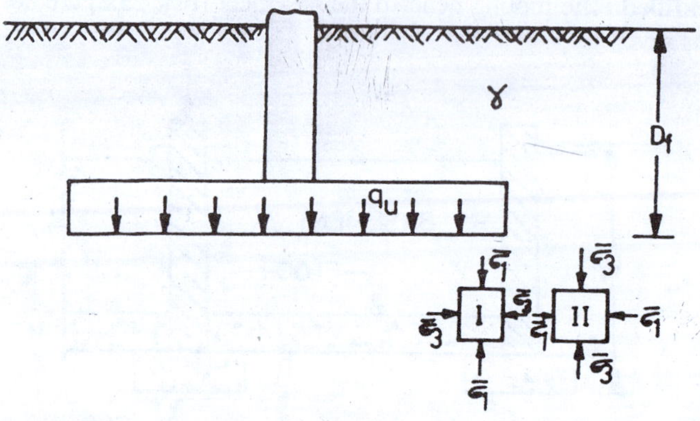

Fig. 23.3. Rankine's Analysis.

When the footing pressure approaches the ultimate bearing capacity (q_u), the element I attains a state of plastic equilibrium. However, the element I can fail only when the adjacent element II also fails. A relationship between the principal stresses acting on the two elements can be obtained using the equations of plastic equilibrium developed in chapter 19.

For the active case, $$\bar{\sigma}_3 = \tan^2(45° - \phi'/2)\,\bar{\sigma}_1$$

For element I, substituting $\bar{\sigma}_1 = q_u$,

$$\bar{\sigma}_3 = \tan^2(45° - \phi'/2)\, q_u \qquad ...(a)$$

For element II, $$\bar{\sigma}_3 = \gamma\, D_f$$

Therefore, $$\bar{\sigma}_1 = \frac{\bar{\sigma}_3}{\tan^2(45° - \phi'/2)} = \frac{\gamma D_f}{\tan^2(45° - \phi'/2)} \qquad ...(b)$$

As $\bar{\sigma}_3$ of element I is equal to $\bar{\sigma}_1$ of element II, from Eqs. (*a*) and (*b*),

$$\tan^2(45° - \phi'/2)\ q_u = \frac{\gamma D_f}{\tan^2(45° - \phi'/2)}$$

or
$$q_u = \gamma D_f \times \frac{1}{\tan^4(45° - \phi'/2)}$$

or
$$q_u = \gamma D_f \tan^4(45° + \phi'/2) \qquad \text{...[23.18(a)]}$$

or
$$q_u = \gamma D_f \left(\frac{1 + \sin\phi'}{1 - \sin\phi'}\right)^2 \qquad \text{...(23.18(b)]}$$

Eq. 23.18 gives an approximate value of the ultimate bearing capacity q_u of the soil. As the equation does not give reliable values, it is rarely used for the determination of the ultimate bearing capacity of the soils. It has been superseded by Terzaghi's theory and other theories which give more dependable values. Rankine did not consider cohesion intercept (c') of the soil. The theory gives the bearing capacity of the soil as zero if $D_f = 0$. This is contrary to experience. These are the limitations of the theory.

Eq. 23.18 is occasionally used to determine the minimum depth of foundation $(D_f)_{min}$. It can be written as

$$(D_f)_{min} = \frac{q}{\gamma}\left(\frac{1 - \sin\phi'}{1 + \sin\phi'}\right)^2 \qquad \text{...(23.19)}$$

where q is the intensity of loading at base.

23.5. HOGENTOGLER AND TERZAGHI'S ANALYSIS

Hogentogler and Terzaghi (1929) approximated the actual curved failure surfaces below the footing with a set of straight lines (Fig. 23.4) for the plastic equilibrium of a long strip footing of width B. At the time of failure, the footing exerts a pressure q_u equal to the ultimate bearing capacity of the soil.

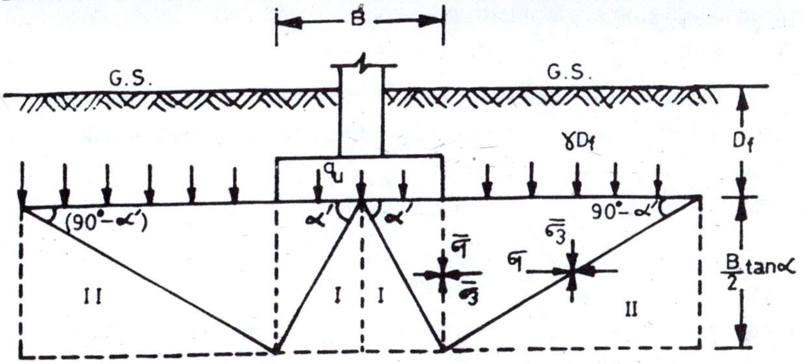

Fig. 23.4. Hogentogler and Terzaghi's Analysis.

The soil in zone I immediately beneath the footing is in compression. The soil in zone I can fail only when the soil in the adjacent zone II also fails. An approximate value of the bearing capacity of the soil can be obtained by considering the stresses at mid-heights of the two failure zones.

The height of the failure zone is $(B/2) \tan\alpha'$, where α' is the angle of the failure surface, equal to $(45° + \phi'/2)$. The overburden pressure at the footing level is equal to γD_f. This pressure is termed as *surcharge.*

Zone II. From the equilibrium of zone II at mid-height

$$\bar{\sigma}_3 = \gamma D_f + (B/2)\tan\alpha' \times \frac{\gamma}{2} \qquad \text{...(a)}$$

where the second term on the right-hand side is the average vertical stress due to self weight.

From the equations developed in Sect. 19.6,

$$\overline{\sigma}_3 = \frac{1 - \sin\phi'}{1 + \sin\phi'}\,\overline{\sigma}_1 - \frac{2c'\cos\phi'}{1 + \sin\phi'}$$

$$\overline{\sigma}_3 = \tan^2\left(45° - \frac{\phi'}{2}\right)\sigma_1 - 2c'\tan\left(45° - \frac{\phi'}{2}\right)$$

or
$$\overline{\sigma}_3 = \overline{\sigma}_1 \cot^2\alpha' - 2\,c'\cot\alpha' \quad \text{...(b)}$$

Substituting the value of $\overline{\sigma}_3$ from Eq. (*a*),

$$\gamma\, D_f + \left(\frac{B}{2}\tan\alpha'\right)\frac{\gamma}{2} = \overline{\sigma}_1 \cot^2\alpha' - 2\,c'\cot\alpha'$$

or
$$\overline{\sigma}_1 = [\gamma\, D_f + (\gamma\, B/4)\tan\alpha']\tan^2\alpha' + 2\,c'\tan\alpha'$$

Zone I. Now $\overline{\sigma}_3$ of zone I is equal to $\overline{\sigma}_1$ of zone II.

Therefore,
$$\overline{\sigma}_3 = [\gamma\, D_f + (\gamma\, B/4)\tan\alpha']\tan^2\alpha' + 2\,c'\tan\alpha' \quad \text{...(c)}$$

and
$$\sigma_1 = q_u + \frac{\gamma\, B}{4}\tan\alpha' \quad \text{...(d)}$$

From Eq. (*b*), for zone I, taking $\overline{\sigma}_1$ and $\overline{\sigma}_3$ values from Eqs. (*c*) and (*d*),

$$\left(\gamma\, D_f + \frac{\gamma\, B}{4}\tan\alpha'\right)\tan^2\alpha' + 2c'\cot\alpha'$$

$$= \cot^2\alpha'\left(q_u + \frac{\gamma\, B}{4}\tan\alpha'\right) - 2\,c'\cot\alpha'$$

or
$$q_u \cot^2\alpha' = \gamma\, D_f \tan^2\alpha' + \frac{\gamma\, B}{4}\tan^3\alpha' - \frac{\gamma B}{4}\cot\alpha' + 2c'\tan\alpha' + 2c' \times \frac{1}{\tan\alpha'}$$

or
$$q_u = \gamma\, D_f \tan^4\alpha' + (\gamma\, B/4)(\tan^5\alpha' - \tan\alpha') + 2c'(\tan^3\alpha' + \tan\alpha') \quad \text{...(23.20)}$$

Eq. 23.20 is a general equation applicable to both cohesive and cohesionless soils.

(*a*) For cohesionless soils, $c' = 0$.

Therefore,
$$q_u = \gamma\, D_f \tan^4\alpha' + (\gamma\, B/4)(\tan^5\alpha' - \tan\alpha') \quad \text{...(23.21)}$$

Eq. 23.21 is known as *Ritter's equation* for bearing capacity of cohesionless soils.

(*b*) For purely cohesive soils, $\phi' = 0$, $c' = c_u$.

Thus
$$q_u = \gamma\, D_f + 4\,c_u$$

or
$$q_{nu} = q_u - \gamma\, D_f = 4\,c_u \quad \text{...(23.22)}$$

where c_u is the undrained cohesion.

As the actual failure surfaces are curved and not plane as assumed in the analysis, the results obtained are approximate. However, the bearing capacity obtained is conservative. The theory has been superseded by other theories as explained later.

23.6. PRANDTL'S ANALYSIS

Prandtl (1921) gave a theory for the penetration of punches into metals. The theory can be used to determine the ultimate bearing capacity of soils The analysis is based on the assumption that a strip footing placed on the ground surface sinks vertically downwards into the soil at failure, like a punch.

Fig. 23.5 shows the failure zones developed below the footing. The soil in the wedge-shaped zone I immediately under the footing is subjected to compressive stresses. As the footing sinks, zone I exerts pressure on side zones II and III. The zones II are assumed to be in plastic equilibrium. The zones II push zones III upward.

Using the theory of plasticity, Prandtl developed expressions for the ultimate bearing capacity for a strip footing, assuming the curved part of the slip surface of the shape of a logarithmic spiral. For purely cohesive

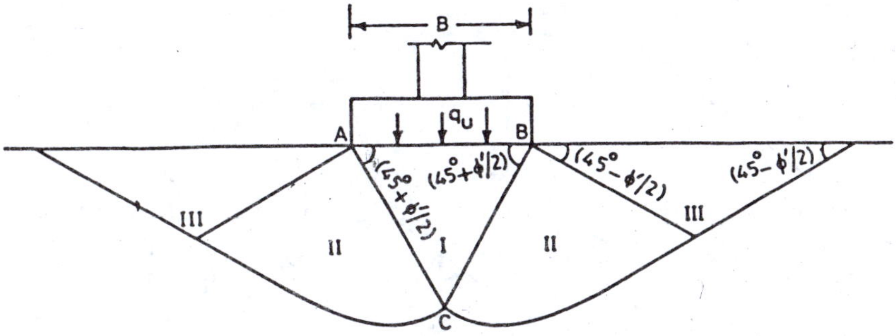

Fig. 23.5. Prandtl's Analysis.

soils ($\phi = 0$), the spiral becomes a circular arc, and Prandtl's analysis gives the following equation for the ultimate bearing capacity,

$$q_u = (\pi + 2)\ c_u = 5.14\, c_u \qquad \ldots(23.23)$$

where c_u is the undrained cohesion of soil.

Eq. 23.23 indicates that the ultimate bearing capacity of a cohesive soil is independent of the width of the footing (B). For cohesionless soils, Prandtl's theory shows that the ultimate bearing capacity increases with the width B.

The theory is applicable for the footings at the surface. For the footing at a depth (D_f) below the surface, an allowance can be made by increasing the bearing capacity by γD_f. Hence for strip footing on cohesive soil,

$$q_u = 5.14\, c_u + \gamma D_f \qquad \ldots(23.24)$$

Prandtl's theory is valid only for footings with perfectly smooth base in contact with the soil. As the actual footings have the rough base, the theory does not give accurate results.

23.7. TERZAGHI'S BEARING CAPACITY THEORY

Terzaghi (1943) gave a general theory for the bearing capacity of soils under a strip footing, making the following assumptions.

(1) The base of footing is rough.

(2) The footing is laid at a shallow depth, *i.e.* $D_f \leq B$.

(3) The shear strength of the soil above the base of the footing is neglected. The soil above the base is replaced by a uniform surcharge γD_f.

(4) The load on the footing is vertical and is uniformly distributed.

(5) The footing is long *i.e.* L/B ratio is infinite, where B is the width and L is the length of the footing.

(6) The shear strength of the soil is governed by the Mohr-Coulomb equation.

Derivation of Equation. As the base of the footing is rough, the soil in the wedge *ABC* immediately beneath the footing is prevented from undergoing any lateral yield (Fig. 23.6). The soil in this wedge (zone I) remains in a state of *elastic equilibrium.* It behaves as if it were a part of the footing itself. It is assumed that the angles *CAB* and *CBA* are equal to the angle of shearing resistance ϕ' of the soil.

The sloping edges *AC* and *BC* of the soil wedge *CBA* bear against the *radial shear zones CBD* and *CAF* (zone II). The curves *CD* and *CF* are arcs of a logarithmic spiral.

Two triangular zones *BDE* and *AFG* are the Rankine passive zones (zones III). An overburden pressure $q = \gamma D_f$ acts as a surcharge on the Rankine *passive zones.*

The failure zones do not extend above the horizontal planes passing through the base *AB* of the footing. In other words, the shearing resistance of the soil located above the base of the footing is neglected, and the effect of soil is taken equivalent to a surcharge of γD_f. Because of this assumption, Terzaghi's theory is valid only for shallow foundations ($D_f \leq B$), in which the term γD_f is relatively small.

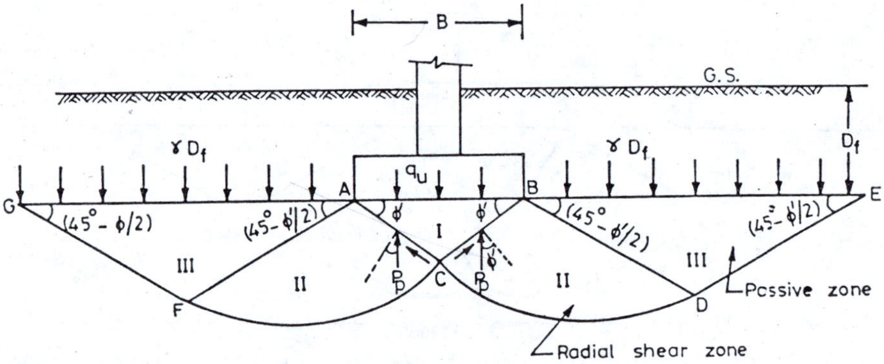

Fig. 23.6. Terzaghi's Analysis.

The loading conditions are similar to that on a retaining wall under passive pressure case. The failure occurs when the downward pressure exerted by loads on the soil adjoining the inclined surfaces *CB* and *CA* of the soil wedge is equal to the upward pressure. The downward forces are due to the load ($q_u \times B$) and the weight of the wedge ($1/4\, \gamma B^2 \tan \phi'$). The upward forces are the vertical components of the resultant passive pressure (P_p) and the cohesion (c') acting along the inclined surfaces. As the resultant passive pressure is inclined at an angle ϕ' to the normal to the surface of the wedge, it is vertical. Therefore, from the equilibrium equitation in the vertical direction,

$$\frac{1}{4}\gamma B^2 \tan \phi' + q_u \times B = 2P_p + 2c' \times L_i \sin \phi'$$

where L_i = length of the inclined surface *CB* [= $(B/2)/\cos \phi'$]

Therefore, $$q_u \times B = 2\,P_p + B\,c' \tan \phi' - \frac{1}{4}\gamma B^2 \tan \phi' \quad \ldots(a)$$

The resultant passive pressure (P_p) on the surface *CB* and *CA* constitutes the following 3 components.

(1) Component $(P_p)_\gamma$ which is produced by the weight of the shear zone *BCDE*, assuming the soil as cohesionless ($c' = 0$) and neglecting the surcharge *q*.

(2) Component $(P_p)_c$ which is produced by the component c' of the soil, assuming the soil as weightless ($\gamma = 0$) and neglecting the surcharge *q*.

(3) Component $(P_p)_q$ which is produced by surcharge (*q*), assuming the soil as cohesionless and weightless ($c' = 0$; $\gamma = 0$).

The three components, $(P_p)_\gamma$, $(P_p)_c$ and $(P_p)_q$ are obtained assuming different surfaces of failures. Although their respective failure surfaces are different from the actual failure surface developed for a footing on a soil possessing weight and cohesion and also having a surcharge, the results can be superimposed without introducing much error. Thus, the resultant passive pressure P_p is taken equal to the sum of the components $(P_p)_\gamma$, $(P_p)_c$ and $(P_p)_q$.

From Eq. (*a*), $$q_u B = 2\,[(P_p)_\gamma + (P_p)_c + (P_p)_q] + B\,c' \tan \phi' - \frac{1}{4}\gamma B^2 \tan\phi'$$

Substituting $$2\,(P_p)_\gamma - \frac{1}{4}\gamma B^2 \tan \phi' = B \times \frac{1}{2}\gamma B N_\gamma$$

and $$2(P_p)_c + Bc' \tan \phi' = B \times c' N_c$$

and $$2\,(P_p)_q = B \times \gamma\, D_f N_q,$$

we get $$q_u\, B = B \times c'\, N_c + B\gamma D_f N_q + B \times \frac{1}{2}\gamma B N_\gamma$$

or $$q_u = c' N_c + \gamma D_f N_q + 0.5\,\gamma B N_\gamma \quad \ldots(23.25)$$

or $$q_u = c' N_c + q_o\, N_q + 0.5\,\gamma\, B N_\gamma \quad \ldots[23.25(a)]$$

where q_o is the overburden pressure = γD_f.

Eq. 23.25 is known as *Terzaghi's bearing capacity equation.* The bearing capacity factors N_c, N_q and N_γ are the dimensionless numbers, depending upon the angle of shearing resistance (ϕ') of the soil. These are defined by the following equations:

$$N_c = \cot\phi' \left[\frac{a^2}{2\cos^2\left(45° + \frac{\phi'}{2}\right)} - 1 \right] \quad ...[23.26(a)]$$

$$N_q = \left(\frac{a^2}{2\cos^2\left(45° + \frac{\phi'}{2}\right)} \right) \quad ...[23.26(b)]$$

where $a = e^{(3\pi/4 - \phi'/2)\tan\phi'}$

and $$N_\gamma = \frac{1}{2}\left(\frac{K_p}{\cos^2\phi'} - 1 \right)\tan\phi' \quad ...[23.26(c)]$$

where K_p = coefficient of passive earth pressure.

Fig. 23.7 gives the values of the bearing capacity factors. The values are also tabulated in Table 23.1. These values are for general shear failure.

If the ground surface on the two sides of the footing is at different levels, D_f is taken as the smaller of the depths on the two sides.

Eq. 23.25 gives the ultimate bearing capacity of a strip footing. The net ultimate bearing capacity and safe bearing capacity can be determined as explained in Sect. 23.2.

Table 23.1. Terzaghi's Bearing Capacity Factors

ϕ'	*General Shear Failure*			*Local Shear Failure*			N_q' (*Vesic*)
	N_c	N_q	N_γ	N_c'	N_q'	N_γ'	
0	5.7	1.0	0.0	5.7	1.0	0.0	1.0
5	7.3	1.6	0.5	6.7	1.4	0.2	1.2
10	9.6	2.7	1.2	8.0	1.9	0.5	1.6
15	12.9	4.4	2.5	9.7	2.7	0.9	2.2
20	17.7	7.4	5.0	11.8	3.9	1.7	3.3
25	25.1	12.7	9.7	14.8	5.6	3.2	5.3
30	37.2	22.5	19.7	19.0	8.3	5.7	9.5
35	57.8	41.4	42.4	25.2	12.6	10.1	18.7
40	95.7	81.3	100.4	34.9	20.5	18.8	42.5
45	172.3	173.3	297.5	51.2	35.1	37.7	115.0
50	347.5	415.1	1153.2	81.3	65.6	87.1	329.10

23.8. TYPES OF SHEAR FAILURE

Vesic (1973) classified the bearing capacity failures into 3 categories:

(1) General Shear Failure. Fig. 23.8 (*a*) shows a strip footing resting on the surface of a dense sand or a stiff clay. The figure also shows the load settlement curve for the footing, where q is the load per unit area and s is the settlement. At a certain load intensity equal to q_u, the settlement increases suddenly. A shear failure occurs in the soil at that load and the failure surfaces extend to the ground surface. This type of failure is known as general shear failure. A heave on the sides is always observed in general shear failure.

(2) Local Shear Failure. Fig. 23.8 (*b*) shows a strip footing resting on a medium dense sand or on a clay of medium consistency. The figure also shows the load-settlement curve. When the load is equal to a certain

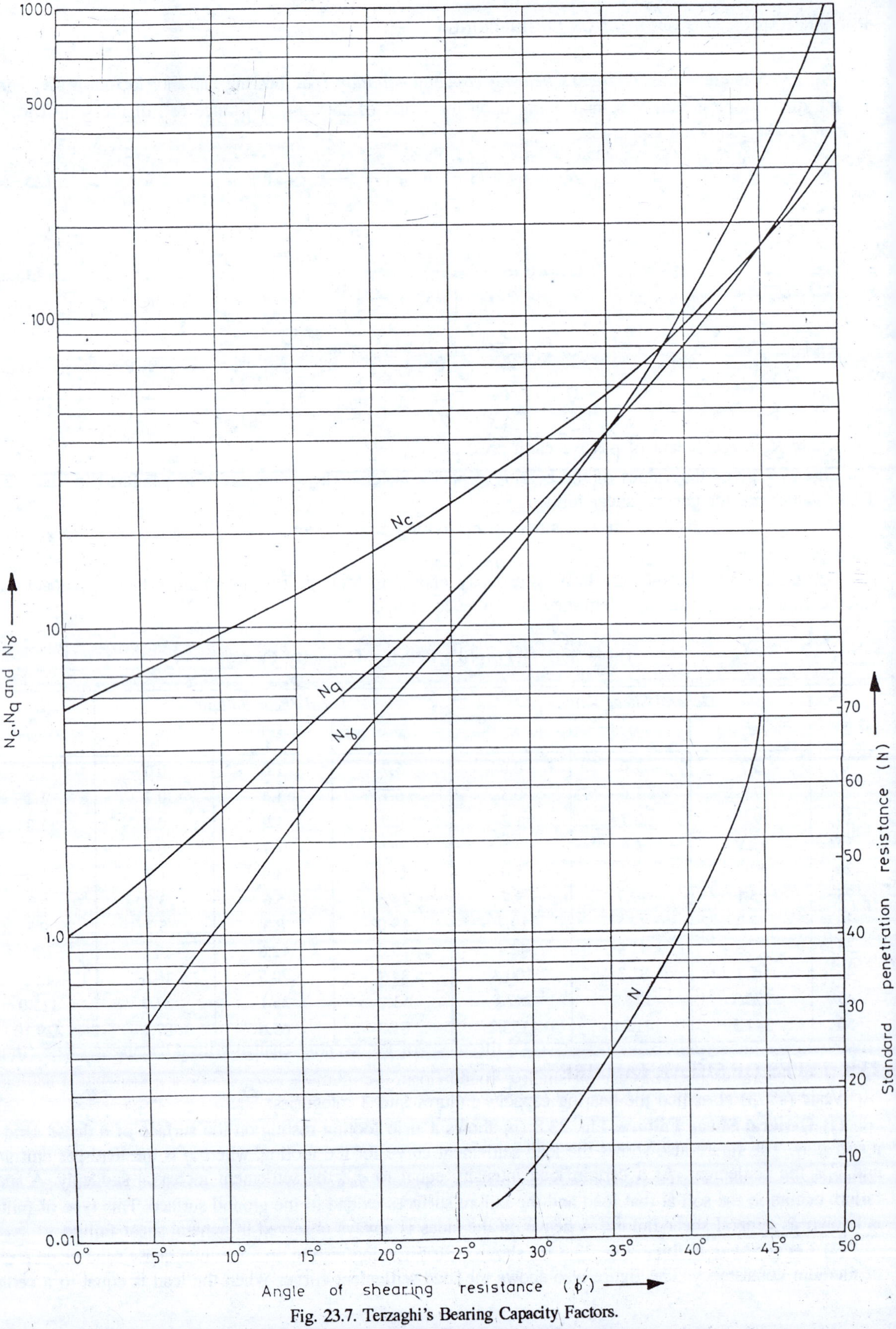

Fig. 23.7. Terzaghi's Bearing Capacity Factors.

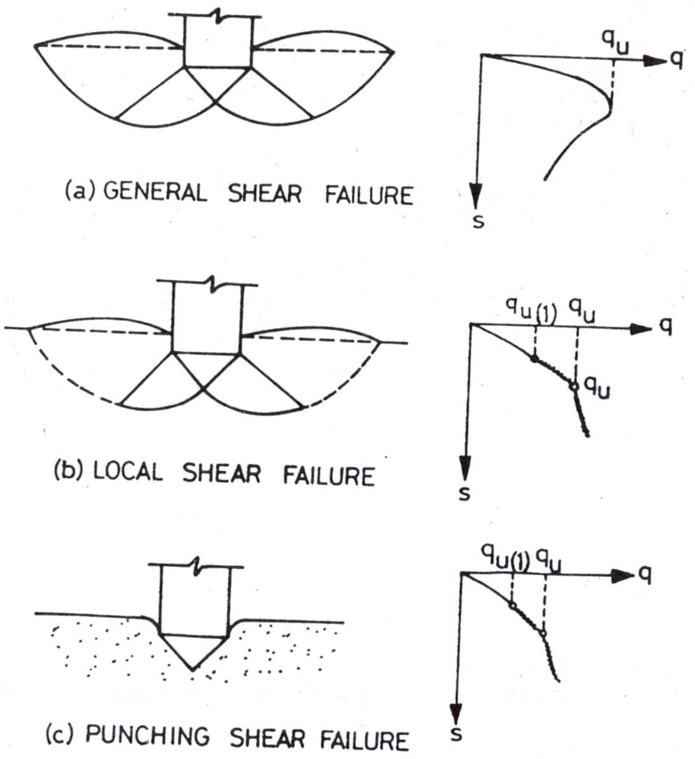

Fig. 23.8. Types of Shear Failure.

value $q_{u(1)}$, the foundation movement is accompanied by sudden jerks. The failure surfaces gradually extend outwards from the foundation, as shown. However, a considerable movement of the foundation is required for the failure surfaces to extend to the ground surface (shown dotted). The load at which this happens is equal to q_u. Beyond this point, an increase of load is accompanied by a large increase in settlement. This type of failure is known as local shear failure. A heave is observed only when there is substantial vertical settlement.

(3) Punching Shear Failure. Fig. 23.8 (*c*) shows a strip footing resting on a loose sand or a soft clay. In this case, the failure surfaces do not extend up to the ground surface. There are jerks in foundation at a load of $q_{u(1)}$. The footing fails at a load of q_u at which stage the load- settlement curve becomes steep and practically linear. This type of failure is called the punching shear failure. No heave is observed. There is only vertical movement of footing.

Vesic proposed a relationship for the mode of failure based on the relative density D_r and D_f/B^*, where $B^* = 2B \times L/(B + L)$, in which B is the width of the footing and L is the length (Fig. 23.9). It is worth noting that even for the same relative density (D_r), the mode of failure may change with a change in D_f/B^* ratio.

It is generally observed that for shallow foundations, the ultimate load occurs at a foundation settlement of 4 to 10% of B in the case of general shear failure, and at a settlement of 15 to 25% of B in local or punching shear failure.

As the footings are seldom constructed on very loose sands, the punching shear failure rarely occurs in practice. It is not of much practical importance. Terzaghi considered only the general shear failure and the local shear failure.

23.9. ULTIMATE BEARING CAPACITY IN CASE OF LOCAL SHEAR FAILURE

Eq. 23.25 gives the ultimate bearing capacity of a strip footing under general shear failure. No theoretical

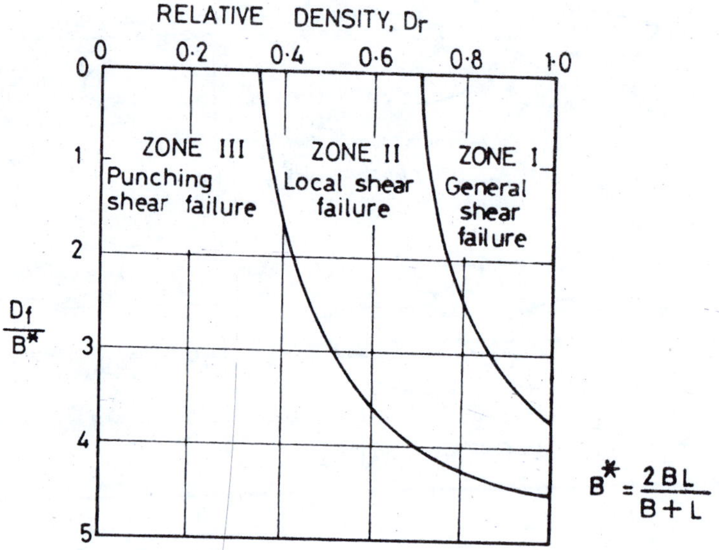

Fig. 23.9. Vesic's Chart.

solution is available for local shear failure and punching shear failure. Terzaghi (1943) has suggested the following empirical reduction to the actual cohesion and the angle of shearing resistance in case of local shear failure.

Mobilised cohesion, $$c'_m = \frac{2}{3}c' \qquad \text{...[23.27(a)]}$$

Mobilised angle of shearing resistance, $\phi'_m = \tan^{-1}\,(2/3 \tan \phi')$ ...[23.27(*b*)]

The reduced values of ϕ' equal to ϕ_m' are used to determine bearing capacity parameters from the values of the general shear failure. The modified bearing capacity factors are indicated as N_c', N_q' and N_γ' for local shear failure. Fig. 23.10 gives the values of these factors for different values of ϕ'. Table 23.1 also gives the values of these factor for values of ϕ'. It has been found from several model tests that the value of N_q' as determined above is underestimated. Sometimes, the following equation for N_q' (Vesic, 1963) is used for local shear failure. This gives more reliable results.

$$N_q' = \left(e^{3.8\,\phi'\,\tan\phi}\right)\tan^2\left(45° + \frac{\phi'}{2}\right) \qquad \text{...(23.28)}$$

Table 23.1 and Fig. 23.10 also give the Vesic's values of N_q'.

The equation for local shear failure for strip footing can be written as

$$q_u' = \frac{2}{3}c'\,N_c' + \gamma\,D_f\,N_q' + 0.5\,\gamma\,B\,N_\gamma' \qquad \text{...(23.29)}$$

where N_c', N_q' and N_γ' are for reduced values of ϕ', equal to ϕ_m'.

It is difficult to ascertain the limiting conditions for which local shear failure should be assumed. Besides, the criteria given by Vesic (Fig. 23.9), the following criteria are also used.

(1) For a cohesionless soil, if ϕ' is greater than 36°, general shear failure is likely to occur. If ϕ' is less than 29°, local shear would be more probable. For intermediate values of ϕ' between 29° to 36°, the values of bearing capacity factors are obtained by interpolation.

For example, if a soil has $\phi' = 35°$, Eq. 23.27 (*b*) gives $\phi_m = 25°$.

The bearing capacity factors are as under.

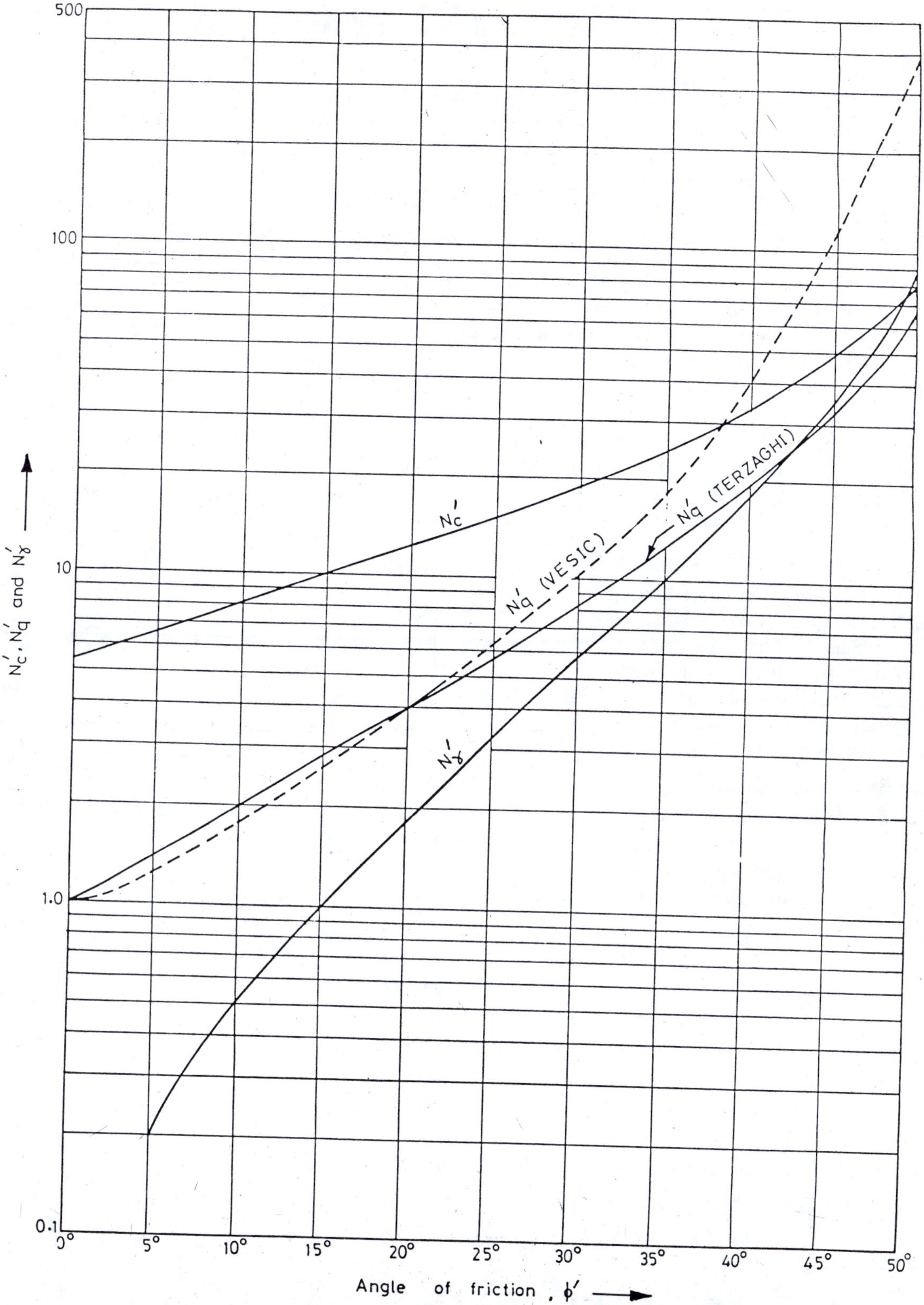

Fig. 23.10. Bearing Capacity factors for Local Shear Failure.

$$N_c = 57.8, \quad N_q = 41.4 \quad \text{and} \quad N_\gamma = 42.4$$

and $$N_c' = 25.2, \quad N_q' = 12.6 \quad \text{and} \quad N_\gamma' = 10.1$$

Difference $(N_c)_d = 32.6$, $(N_q)_d = 28.8$ and $(N_\gamma)_d = 32.3$.

As the actual value of ϕ' is 35° which is 6° more than the value of ϕ' corresponding to local shear failure (*viz.* 29°), the proportional difference to be added to the values of N_c' N_q' and N_γ' is 6/7 times the total difference. Thus, the required values are

$$N_c = 25.2 + 6/7 \times 32.6 = 53.14$$

$$N_q = 12.6 + 6/7 \times 28.8 = 37.29$$

$$N_\gamma = 10.1 + 6/7 \times 32.3 = 37.79$$

(2) If the failure of the specimen of the soil occurs at a relatively small strain, say less than 5%, the failure of the footing would be by general shear failure. If the stress—strain curve does not show a peak and ia a continuously rising curve even upto a strain of 10 to 20%, local shear failure would occur in the footing.

(3) If the relative density (D_r) is greater than about 70%, general shear failure would occur. If it is less than 35%, local shear failure is more likely.

(4) If the standard penetration test (SPT) value is more than 30, the general shear failure would occur. However, if it is less than 5, the local shear failure is more likely.

(5) If e is less than 0.55, the general shear failure occurs. If e is greater than 0.75, the local shear failure occurs.

23.10. EFFECT OF WATER TABLE ON BEARING CAPACITY

Eq. 23.25 for the ultimate bearing capacity has been developed based on the assumption that the water table is located at a great depth. If the water table is located close to the foundation, the bearing capacity equation needs modification, as explained below.

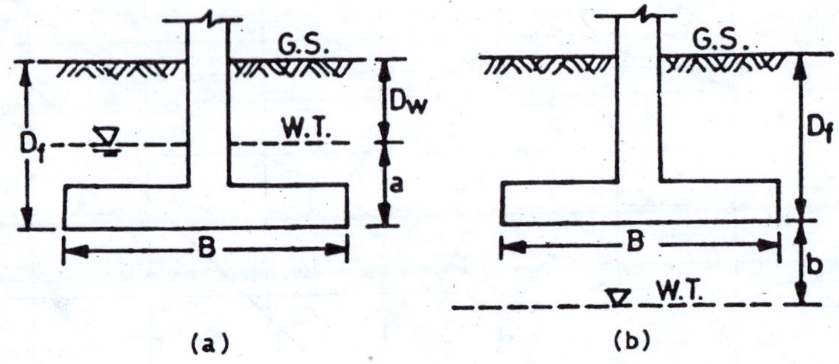

Fig. 23.11.

Case I Water table located above the base of footing [Fig. 23.11 (*a*)]

The effective surcharge is reduced as the effective weight below the water table is equal to the submerged unit. Therefore,

$$q = D_w \gamma + a \gamma' \qquad \text{...(23.30)}$$

where D_w = depthe of water table below the ground surface,

a = height of water table above the base of footing.

Alternatively, Eq. 23.30 can be written as, substituting $a = D_f - D_w$,

$$q = \gamma' D_f + (\gamma - \gamma')D_w \qquad \text{...(23.31)}$$

Moreover, the unit weight in the third term of Eq. 23.25 is equal to the submerged unit weight. Thus Eq. 23.25 becomes

$$q_u = c' N_c + [\gamma' D_f + (\gamma - \gamma') D_w] N_q + 0.5 \gamma' B N_\gamma \qquad \text{...(23.32)}$$

If $D_w = 0$ (*i.e.* $a = D_f$),

$$q_u = c' N_c + \gamma' D_f N_q + 0.5 \gamma' B N_\gamma \qquad \text{...[23.33(a)]}$$

If $a = 0$ (*i.e.* $D_f = D_w$),

$$q_u = c' N_c + \gamma D_f N_q + 0.5 \gamma' B N_\gamma \qquad \text{...[23.33(b)]}$$

Case II Water table located at a depth *b* below base [Fig. 23.11 (*b*)]

If the water table is located at the level of the base of footing or below it, the surcharge term is not affected. However, the unit weight in the third term of Eq. 23.25 is modified as

$$\overline{\gamma} = \gamma' + \frac{b}{B}(\gamma - \gamma') \qquad \text{...(23.34)}$$

where b = depth of water table below the base,
B = base width of the footing.

Therefore, $$q_u = c' N_c + \gamma D_f N_q + 0.5 B\left[\gamma' + \frac{b}{B}(\gamma - \gamma')\right] N_\gamma \qquad \text{...[23.35(a)]}$$

When $b = 0$, *i.e.* W/T at the base,

$$q_u = c' N_c + \gamma D_f N_q + 0.5 B \gamma' N_\gamma \qquad \text{...[23.35(b)]}$$

When $b = B$, *i.e.* W/T at depth B below the base.

$$q_u = c' N_c + \gamma D_f N_q + 0.5 B \gamma N_\gamma \qquad \text{...(same as Eq. 23.25)}$$

Hence, when the ground water table is located at a depth b equal to or greater than B, there is no effect on the ultimate bearing capacity.

General Expression

Taking the submerged unit as roughly one-half of the bulk unit weight, the general equation for the ultimate bearing capacity can be written from Eqs. 23.32 and 23.35 (*a*) as

$$q_u = c' N_c + \gamma D_f N_q W_q + 0.5 \gamma B N_\gamma W_\gamma \qquad \text{...[23.36(a)]}$$

where W_q is water table correction factor for the second term,

$$W_q = 1 - 0.5\, a/D_f \le 1 \qquad \text{...[23.36(b)]}$$

and W_γ is water table correction factor for the third term,

$$W_\gamma = 0.5 + 0.5\, b/B \le 1 \qquad \text{...[23.36(c)]}$$

It may be noted that both the corrections vary linearly.

23.11. BEARING CAPACITY OF SQUARE AND CIRCULAR FOOTING

Eq. 23.25 has been derived for a strip footing. The deformations under a strip footing are two-dimensional. It is known as a plane strain case in theory of elasticity. On the other hand, the deformations of soil under a square or a circular footing are three-dimensional. A rigorous analytical solution for a three-dimensional case is extremely difficult.

Based on experimental results, Terzaghi gave the following equations for the ultimate bearing capacity for square and circular shallow footings.

(*a*) Square Footing

$$q_u = 1.2 c' N_c + \gamma D_f N_q + 0.4 \gamma B N_\gamma \qquad \text{...(23.37)}$$

where B is the dimension of each side of footing.

(*b*) Circular Footing

$$q_u = 1.2 c' N_c + \gamma D_f N_q + 0.3 \gamma B N_\gamma \qquad \text{...(23.38)}$$

where B is the diameter of the footing.

The bearing capacity factors N_c, N_q and N_γ are the same as that for the strip footing.

Note—Terzaghi (1943) took the shape factor as 1.3, which was later changed by Terzaghi and Peck (1967) to 1.2. Both values are used in practice.

For rectangular footing, the following equation is generally used

$$q_u = cN_c(1 + 0.2\, B/L) + \gamma D_f N_q + \frac{1}{2}\gamma B N_\gamma (1 - 0.2\, B/L) \qquad \text{...(23.38 a)}$$

where L = length of the footing.

23.12. MEYERHOF'S BEARING CAPACITY THEORY

Meyerhof (1951) gave a general theory of bearing capacity for a strip footing at any depth. Meyerhof considered the failure mechanism similar to that assumed by Terzaghi, but extended the failure surfaces above foundation level. Thus the shearing strength of the soil above the footing base was also accounted for in the analysis. The curved rupture surfaces in the zone of radial shear were assumed to be logarithmic spirals. Meyerhof's rupture surfaces are more general than those assumed by Terzaghi.

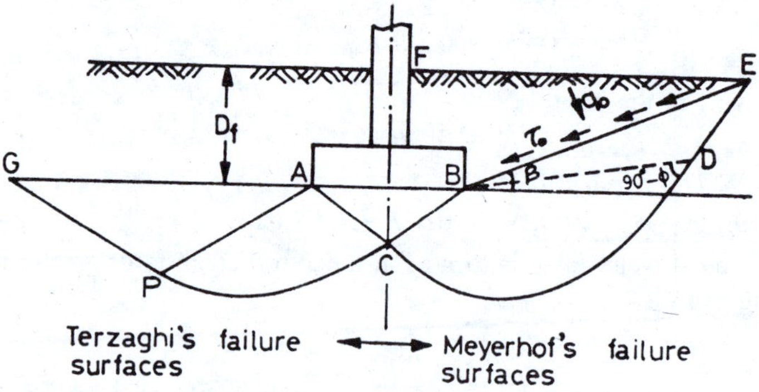

Fig. 23.12. Meyerhof's Analysis.

The right-half of Fig. 23.12 shows the failure surfaces assumed by Meyerhof. The zone *ABC* is the elastic zone but the angle which the inclined surfaces *AC* and *BC* make with horizontal was varied between ϕ' and $(45° + \phi'/2)$. The zone *BCD* is the zone of radial shear. The zone *BDEF* is the zone of mixed shear in which shear varies between radial shear and plane shear. The surface *BE* is known as *equivalent free surface*. It makes an angle β with the horizontal.

The resultant effect of the wedge *BEF* of soil is represented by the normal stress (q_o) and the shear stress (τ_o) on the surface *BE*. The angle β increases with an increase in depth D_f and is equal to 90° for deep foundations. The parameters β, q_o and τ_o are known as foundation depth parameters.

Meyerhof gave the following equation for the ultimate bearing capacity of strip footings.

$$q_u = c'\, N_c + q_o N_q + 0.5\,\gamma\, B\, N_\gamma \qquad \text{...(23.39)}$$

where N_c, N_q and N_γ are the general bearing capacity factors of Meyerhof's theory. These factors depend upon the roughness of base, depth of footing and the shape of footing, in addition to the angle of shearing resistance ϕ'. Meyerhof also gave charts for N_c, N_q and N_γ (Fig. 23.13) for shallow strip footings.

As the equivalent free surface cannot be directly located, the normal stress q_o is determined by a semi-graphical method, which is quite cumbersome. However, for shallow footings, $q_o' = \gamma D_f$.

The ultimate bearing capacity given by Meyerhof's theory is close to the experimental values. For shallow footings, the value lies in-between the general shear value and the local shear value of Terzaghi's analysis. However, for deep footings, Meyerhof's analysis gives values much greater than Terzaghi's analysis. The main advantage of Meyerhof's theory is that it can also be used for deep foundations and for footings on slopes.

(a) Cohesionless soils

For cohesionless soils ($c = 0$), the ultimate bearing capacity is given by a simple equation suggested by Meyerhof as

$$q_u = 0.5\,\gamma\, B\, N_{\gamma q} \qquad \text{...(23.40)}$$

where the parameter $N_{\gamma q}$ is known as the *resultant bearing capacity factor*. It depends upon the coefficient of earth pressure (K) within the failure zone, the D_f/B ratio, the angle of shearing resistance ϕ' and the angle of friction δ between the vertical shaft of the foundation and the surrounding soil. The value of K depends upon

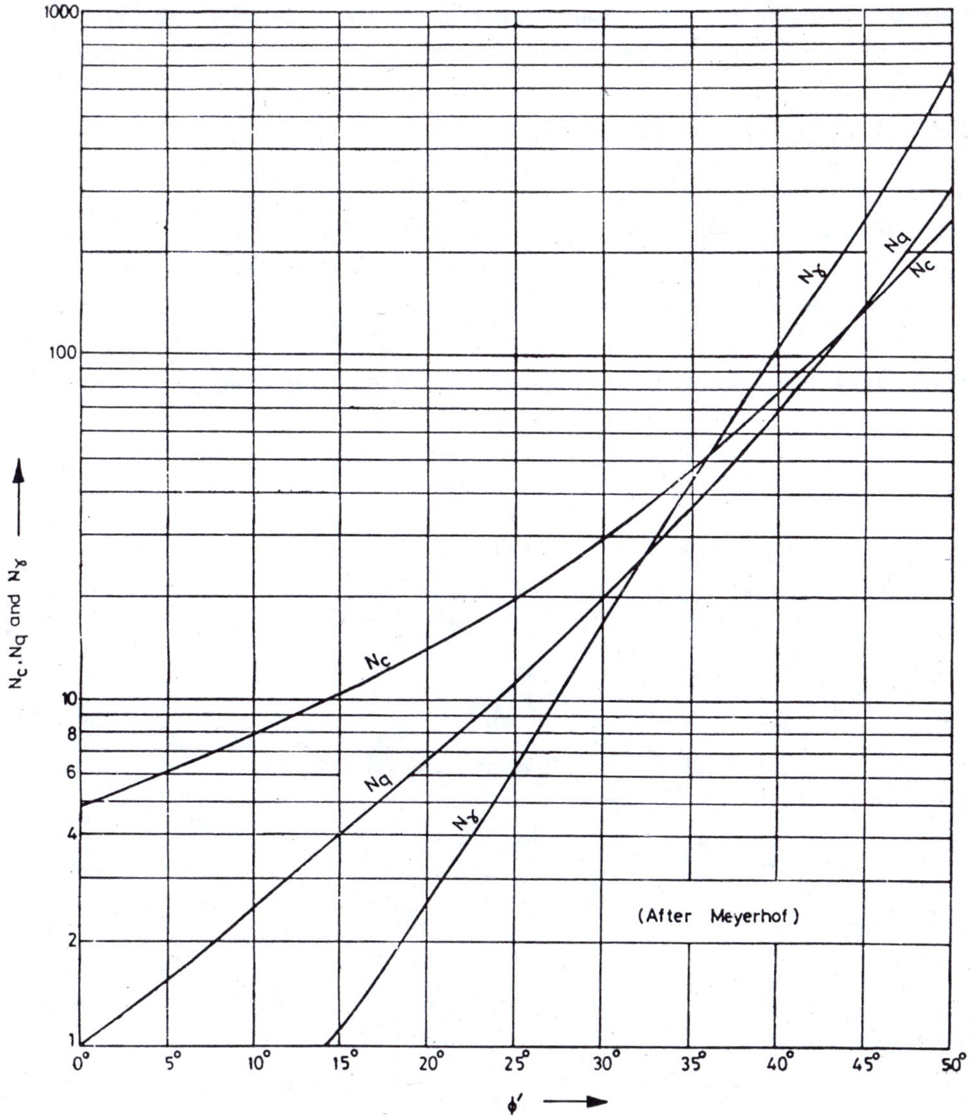

Fig. 23.13. Meyerhof's Chart.

a number of factors, such as the characteristics of the soil, the stress history of the ground, the method of construction and physical characteristics of the foundation. The value of K usually lies between the active and passive earth pressure coefficients. The value of $N_{\gamma q}$ is obtained from the charts for the given values of K, ϕ' and D_f/B ratio, assuming the base and shaft as rough. For smooth surfaces, the values of $N_{\gamma q}$ are reduced to half the values.

(b) Cohesive soils

For cohesive soils ($\phi = 0$) Meyerhof's simplified equations is

$$q_u = c\, N_{cq} + \gamma D_f \qquad \text{...(23.41)}$$

where N_{cq} is the bearing capacity factor depending upon the D_f/B ratio of the footing and on the adhesion on the sides of the footing. For example, for a D_f/B ratio of 2.0, N_{cq} has a maximum value of 8.30 when the adhesion is zero, and a maximum value of 8.8 when the adhesion is equal to the cohesion (c) of the soil.

The above-mentioned equations are for the ultimate bearing capacity of strip footings. The equations can also be used for rectangular, square and circular footings in conjunction with the empirical shape factors given by Meyerhof. The values of $N_{\gamma q}$ and N_{cq} may be obtained from the original paper.

23.13. HANSEN'S BEARING CAPACITY THEORY

It has been shown by Milovic (1965) that Terzaghi's theory gives the values of the ultimate bearing capacity higher than experimental values for cohesive soils and is unsafe. However, for cohesionless soils, the theory gives conservative values of the bearing capacity. For cohesive soils, Hansen (1961) gives the values of ultimate bearing capacity which are in better agreement with experimental values.

According to Hansen, the ultimate bearing capacity is given by

$$q_u = c\, N_c\, s_c\, d_c\, i_c + q N_q\, s_q\, d_q\, i_q + 0.5\, \gamma B\, N_\gamma\, s_\gamma\, d_\gamma\, i_\gamma \quad \ldots(23.42)$$

where N_c, N_q and N_γ are Hansen's bearing capacity factors (Table 23.2), and q is the effective surcharge at the base level ($= \gamma' D_f$), and

s_c, s_q and s_γ are shape factors (Table 23.3),

d_c, d_q and d_γ are depth factors (Table 23.4),

and i_c, i_q and i_γ are inclination factor (Table 23.5).

The bearing capacity factors are given by the following equations.

$$N_q = \tan^2 (45° + \phi')\,(e^{\pi \tan \phi'}) \quad \ldots[23.43(a)]$$

$$N_c = (N_q - 1) \cot \phi' \quad \ldots[23.43(b)]$$

and

$$N_\gamma = 1.80\,(N_q - 1)\, \tan \phi' \quad \ldots[23.43(c)]$$

Table 23.2. Hansen's Bearing Capacity Factors

ϕ'	0°	5°	10°	15°	20°	25°	30°	35°	40°	45°	50°
N_c	5.14	6.48	8.34	10.97	14.83	20.72	30.14	46.13	75.32	133.89	266.89
N_q	1.0	1.57	2.47	3.94	6.40	10.66	18.40	33.29	64.18	134.85	318.96
N_γ	0.0	0.09	0.09	1.42	3.54	8.11	18.08	40.69	95.41	240.85	681.84

Table 23.3. Shape Factors

Shape of Footing	s_c	s_q	s_γ
Continuous Footing (strip)	1.0	1.0	1.0
Rectangular Footing	1 + 0.2 B/L	1 + 0.2 B/L	1 − 0.4 B/L
Square Footing	1.3	1.2	0.8
Circular Footing	1.3	1.2	0.6

where L = length of footing; B = width of footing.

Table 23.4. Depth Factors

d_c	1 + 0.35 (D_f/B)
d_q	1 + 0.35 (D_f/B)
d_γ	1.00

where D_f = depth of footing; B = width of diameter

[**Note.** Take $d_q = d_c$ for $\phi' > 25°$ and $d_q = 1.0$ for $\phi' = 0°$]

Table 23.5. Inclination Factors

i_c	$1 - \frac{H}{2\,c\,BL}$
i_q	$1 - \frac{1.5\,H}{V}$
i_γ	$(i_q)^2$

where H = horizontal component of inclined load.

Limitation : $H \leq V \tan\delta + cBL$

where $\tan\delta$ = coefficient of friction between footing and soil,

c = cohesion of soil between footing and soil, L = length of footing parallel to H.

The revised factors are as follows: (Limitation $H/V < 0.40$)

$$i_c = i_q = \left(1 - \frac{H}{V + Ac\cot\phi}\right)^2 \text{ and } i_\gamma = (i_q)^2$$

23.14. VESIC'S BEARING CAPACITY THEORY

Vesic (1973) confirmed that the basic nature of failure surfaces in soil as suggested by Terzaghi and as given in Fig. 23.6 is correct. However, the angle which the inclined surfaces AC and BC make with the horizontal was found to be closer to $(45° + \phi'/2)$ instead of ϕ'. The values of bearing capacity factors N_c, N_q and N_γ for a given angle of shearing resistance ϕ' change if above modification is incorporated in the analysis, as under

$$N_q = \tan^2(45° + \phi'/2)\, e^{\pi\tan\phi'} \quad \ldots[23.44(a)]$$

$$N_c = (N_q - 1)\cot\phi' \quad \ldots[23.44(b)]$$

and

$$N_\gamma = 2(N_q + 1)\tan\phi' \quad \ldots[23.44(c)]$$

Eq. 23.44(b) was first proposed by Prandtl (1921), and Eq. 23.44 (a) was given by Reissner (1924). Caquot and Kerisel (1953) and Vesic (1973) gave Eq. 23.44 (c). The values of bearing capacity factors are given in Table 23.6.

The bearing capacity equation is similar in form to Hansen's equation,

$$q_u = c'\,N_c\,s_c\,d_c\,i_c + qN_q\,s_q\,d_q\,i_q + 0.5\,\gamma BN_\gamma\,s_\gamma\,d_\gamma\,i_\gamma \quad \ldots(23.45)$$

where s_c, s_q and s_γ are shape factors (Table 23.7),

d_c, d_q and d_γ are depth factors,

and i_c, i_q and i_γ are inclination factors.

As before, q is effective surcharge at the base level ($= \gamma' D_f$).

Table 23.6. Vesic's Bearing Capacity Factors

ϕ'	0°	5°	10°	15°	20°	25°	30°	35°	40°	45°	50°
N_c	5.14	6.49	8.35	10.98	14.83	20.72	30.14	46.12	75.31	133.88	266.89
N_q	1.0	1.57	2.47	3.94	6.40	10.66	18.40	33.30	64.20	134.88	319.07
N_γ	0.0	0.45	1.22	2.65	5.39	10.88	22.40	48.03	109.41	271.76	762.89

Table 23.7. Shape Factors

Shape of Footing	s_c	s_q	s_γ
Strip	1.0	1.0	1.0
Rectangle	$1 + (B/L)(N_q/N_c)$	$1 + (B/L)\tan\phi'$	$1 - 0.4\,(B/L)$
Circle and square	$1 + (N_q/N_c)$	$1 + \tan\phi'$	0.60

The depth factors as proposed by Hansen (1970) are used for $D_f/B \le 1$ as follows.

$$d_c = 1 + 0.4\,(D_f/B) \quad ...[23.46(a)]$$

$$d_q = 1 + 2 \tan \phi' \,(1 - \sin \phi')^2\,(D_f/B) \quad ...[23.46(b)]$$

$$d_\gamma = 1.0 \quad ...[23.46(c)]$$

In Vesic's equation, the following inclination factors proposed by Meyerhof (1963) and Hanna and Meyerhof (1981) are generally used.

$$i_c = i_q = (1 - \alpha°/90°)^2 \quad ...[23.47(a)]$$

and

$$i_\gamma = (1 - \alpha°/\phi')^2 \quad ...[23.47(b)]$$

where $\alpha°$ is the inclination of the load with vertical.

23.15. IS CODE METHOD

IS : 6403—1981 gives the equation for the net ultimate bearing capacity, which is similar to one proposed by Vesic (Eq. 23.45).

$$q_{nu} = c\,N_c\,s_c\,d_c\,i_c + q\,(N_q - 1)\,s_q\,d_q\,i_q + 0.5\,B\,\gamma N_\gamma\,s_\gamma\,d_\gamma\,i_\gamma\,W' \quad ...(23.48)$$

where q = effective pressure at the base.

The second term has been changed, because q_{nu} is given by (See Eq. 23.1)

$$q_{nu} = q_u - \gamma\,D_f = q_u - q$$

The factor W' takes into account the effect of the water table. If the water table is at or below a depth of $(D_f + B)$, measured from the ground surface, $W' = 1.0$. If the water table is likely to rise to the base of the footing or above, the value of W' is taken as 0.50.

If the water table is located at a depth D below the ground surface, such that $D_f < D < (D_f + B)$, the value of W' is obtained by linear interpolation. A little reflection would show that W' is the same as the factor W_γ introduced in Sect. 23.10. The factor W_q is indirectly accounted for by taking q as the effective surcharge in Eq. 23.48.

The bearing capacity factors N_c, N_q and N_γ are the same as those given by Vesic (Table 23.6). The shape factors given in Table 23.3 are used. The depth factors are given below.

$$d_c = 1 + 0.2\,(D_f/B)\,\tan\,(45° + \phi'/2) \quad ...[23.49(a)]$$

$$d_q = d_\gamma = 1.0 \qquad \text{for } \phi' < 10° \quad ...[23.49(d)]$$

and

$$d_q = d_\gamma = 1 + 0.1\,(D_f/B)\,\tan\,(45° + \phi'/2) \quad \text{for } \phi' > 10° \quad ...[23.49(c)]$$

The inclination factors given by Eq. 23.47 are used.

Local shear failure

The net ultimate bearing capacity for local shear failure is given by

$$q_{nu} = \frac{2}{3}\,c\,N_c'\,s_c\,d_c\,i_c + q\,(N_q' - 1)\,s_q d_q\,i_q + 0.5\,\gamma B\,N_\gamma'\,s_\gamma\,d_\gamma\,i_\gamma\,W' \quad ...(23.50)$$

The value of the local bearing capacity factors N_c', N_q', N_γ' are obtained from Table 23.6, for the angle of mobilised friction as

$$\phi_m = \tan^{-1}\left(\frac{2}{3}\,\tan \phi'\right)$$

In case of cohesionless soils, if the relative density is greater than 70% and the void ratio is less than 0.55, the failure is considered as general shear failure. On the other hand, if the relative density is smaller than 20% and the void ratio is greater than 0.75, the failure is local shear failure (or punching failure) and Eq. 23.50 is used. For a relative density between 20% and 70% and a void ratio between 0.55 and 0.75, the bearing capacity factors are obtained by interpolation between the general shear failure and the local shear, as explained below.

For a relative density between 20 to 70% (or $0.55 < e < 0.75$), the value of the net ultimate bearing capacity (q_{nu}) can be interpolated between the general shear failure and the local shear failure conditions depending upon the relative density.

For illustration, let us consider the case when the relative density is 40% and ϕ' is equal to 30°. The bearing capacity factors for general shear from Table 23.6 are as under.

$$N_c = 30.14\ ;\ N_q = 18.40\ ;\ N_\gamma = 22.40$$

Now $\phi_m = \tan^{-1}(2/3 \tan 30°) = 21°$. The corresponding bearing capacity factors for the local shear from Table 23.6 are

$$N_c' = 16.01,\ N_q' = 7.25,\ N_\gamma' = 6.49.$$

Let us find the net ultimate bearing capacity. Let us assume $c' = 0$,

$$B = 2\text{ m}.\ s_q = s_\gamma = d_q = d_\gamma = 1.0, \gamma = 20\text{ kN/m}^2 \text{ and } D_f = 1\text{ m}, W' = 1.0.$$

For general shear failure ($D_r > 70\%$) (Eq. 23.48)

$$q_{nu} = 0 + 20 \times 1\,(18.4 - 1) + 0.5 \times 20 \times 2 \times 22.40 = 796\text{ kN/m}^2$$

For local shear failure ($D_r < 20\%$) (Eq. 23.50)

$$q_{nu}' = 0 + 20 \times (7.25 - 1) + 0.5 \times 20 \times 2 \times 6.49 = 254.8\text{ kN/m}^2$$

For $D_r = 40\%$, by linear interpolation,

$$q_{nu} = 254.8 + \frac{20}{50}(796 - 254.8) = \mathbf{471.28\ kN/m^2}$$

23.16. SKEMPTON'S ANALYSIS FOR COHESIVE SOILS

Skempton (1951) showed that the bearing capacity factor N_c in Terzaghi's equation tends to increase with depth for a cohesive soil ($\phi_u = 0, c = c_u$). Fig. 23.14 shows the variation of N_c with D_f/B ratio for strip and circular (or square) footings. For a strip footing, the value of N_c is equal to 5.14 for the surface footing and has a maximum value of 7.50 for D_f/B ratio ≥ 4.50.

For square and circular footings, the value of N_c is equal to 6.2 for the surface footing. The maximum value of about 9.0 is attained for D_f/B ratio equal to or greater than 4.50. The curve for square and circular footings can also be used for rectangular footings using the following relation.

$$N_c\text{ (rectangle)} = N_c\text{ (square)}\ [0.84 + 0.16\ B/L] \qquad \text{...(23.51)}$$

Alternatively, the curve for the strip can be used, making use of the following relation.

$$N_c\text{ (rectangle)} = N_c\text{ (strip)}\ [1 + 0.2\ (B/L)] \qquad \text{...(23.52)}$$

The following approximate relations can be used for the determination of N_c for different D_f/B ratios.

(*a*) $D_f/B < 2.50$

$$N_c = 5.0\,(1 + 0.2\,D_f/B)(1 + 0.2\,B/L) \qquad \text{...[23.53(a)]}$$

(*b*) $D_f/B \geq 2.50$

$$N_c = 7.50\,(1 + 0.2\,B/L) \qquad \text{...[23.53(b)]}$$

Ultimate Bearing capacity

For $\phi_u = 0$, $N_q = 1.0$ and $N_\gamma = 0.0$

Therefore, Eq. 23.25 gives $q_u = c_u N_c + \gamma D_f$ ...(23.54)

The net ultimate bearing capacity becomes

$$q_{nu} = c_u N_c \qquad \text{...(23.55)}$$

Eq. 23.55 is used for the determination of the net ultimate bearing capacity of footings on cohesive soils, taking N_c values given by Skempton (Fig. 23.14). It may be mentioned that Terzaghi's value of N_c is applicable only for shallow footings ($D_f < B$), whereas Skemptons's value can be used for all values of D_f/B ratio.

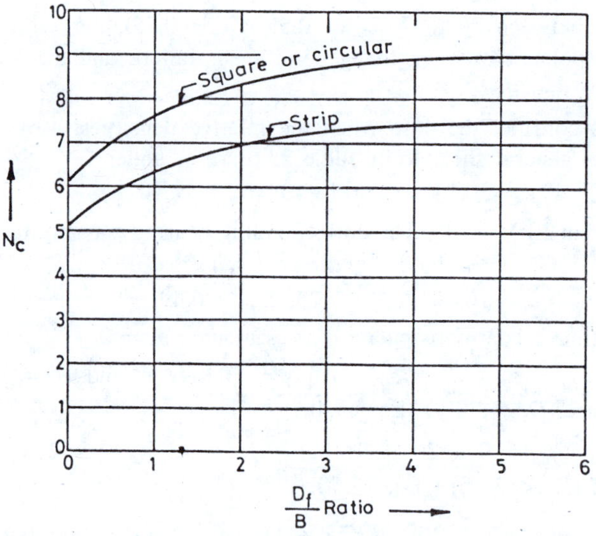

Fig. 23.14. Skempton's Chart.

If the shear strength of the soil for a depth of $2B/3$ beneath the footing does not vary by more than about ± 50% of the average value, the value of c_u in Eq. 23.55 may be taken as the average value.

23.17. IS CODE METHOD FOR COHESIVE SOILS

IS : 6403—1981 gives the following equation for the net ultimate bearing capacity of the footing immediately after construction on a cohesive soil. The equation is obtained by substituting $N_q = 1.0$ and $N_\gamma = 0.0$ in Eq. 23.48. Thus

$$q_{nu} = c_u N_c s_c d_c i_c \qquad \ldots(23.56)$$

where $N_c = 5.14$ and s_c, d_c and i_c are, respectively, the shape, depth and inclination factors (see Sec. 23.15).

The value of c_u is obtained from unconfined compression strength test or it can be derived from the static cone test. The static cone test gives the point resistance (q_c) as explained in chapter 17. For normally consolidated clays, the point resistance q_c is generally less than 2000 kN/m^2 and the value of the undrained cohesion c_u varies between $q_c/18$ to $q_c/15$. For over-consolidated clays, the point resistance is generally greater than 2000 kN/m^2, and the value of c_u varies between $q_c/26$ and $q_c/22$.

23.18. HEAVE OF THE BOTTOM OF THE CUT IN CLAY

The lateral earth pressure acting on a braced cut has been discussed in Chapter 21. A braced cut may also become unstable as a result of the heaving of the bottom of the excavation (Fig. 23.15). Terzaghi (1943) has analysed the braced cut in clay and has given an expression for the factor of safety against bottom heave assuming the failure surfaces as shown in Fig. 23.16.

The vertical load per unit length of the cut at the level BD or AF along the bottom of the cut is given by

$$Q = B_1 \gamma H - cH \qquad \ldots(a)$$

The above load acts on a width B_1 equal to $0.7B$, which can be considered as a continuous foundation. B is the width of the cut.

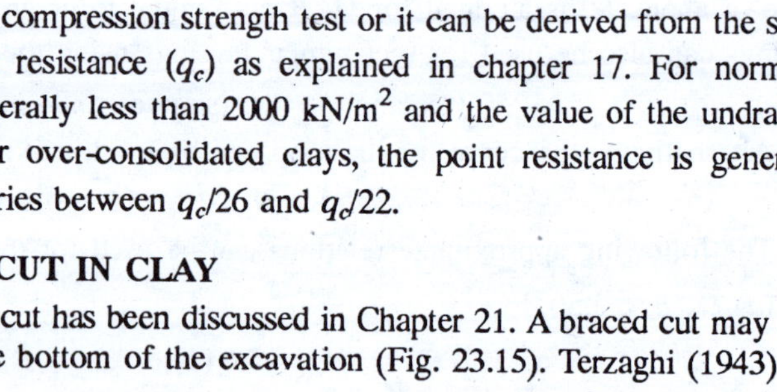

Fig. 23.15. Heave in a Braced Cut.

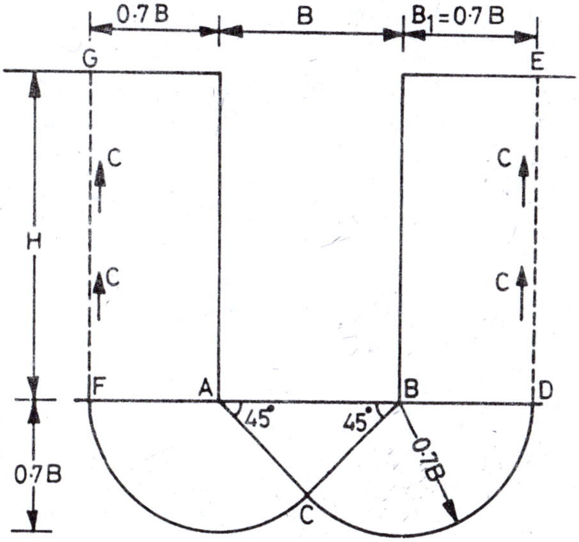

Fig. 23.16.

The ultimate bearing capacity of the continuous foundation can be obtained from Eq. 23.25 as

$$q_u = c\,N_c + N_q\,\gamma\,D_f$$

Taking $N_c = 5.7$, and $N_q = 1.0$,

$$q_u = 5.7\,c + \gamma\,D_f$$

Therefore, $$q_{nu} = 5.7c$$

The net ultimate load is given by $Q_{nu} = 5.7\,cB_1$

The factor of safety is given by

$$\text{F.S.} = \frac{Q_{nu}}{Q} = \frac{5.7\,cB_1}{B_1\,\gamma\,H - cH}$$

or $$\text{F.S.} = \frac{1}{H}\left(\frac{5.7\,c}{\gamma - c/B_1}\right) \qquad \text{...(23.57)}$$

Bjerrum and Eide (1956) proposed the following equation for the factor of safety.

$$\text{F.S.} = \frac{c\,N_c}{\gamma\,H} \qquad \text{...[23.58(a)]}$$

If there is a surcharge q, the factor of safety becomes

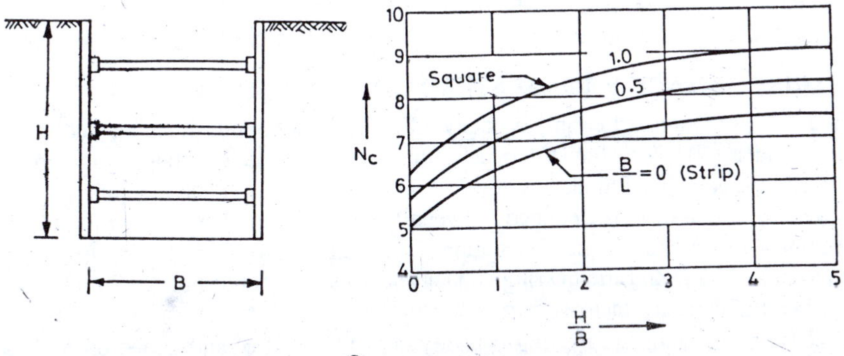

Fig. 23.17. Skempton's Chart for Braced Cut.

$$F.S. = \frac{cN_c}{\gamma H + q} \qquad \ldots[23.58(b)]$$

The bearing capacity factor N_c is obtained from Skempton's chart (Fig. 23.17). The value of c is equal to the undrained cohesion c_u which be determined as explained before.

23.19. FOUNDATION ON LAYERED CLAY

Reddy and Srinivasan (1967) derived the following equation for the bearing capacity of foundations on clayey soils in two layers.

$$q_u = c_1 N_c s_c d_c i_c + q \qquad \ldots(23.59)$$

where c_1 = undrained shear strength of the upper-layer,

s_c = shape factor [$= 1 + (B/L)\, N_q/N_c$]

d_c = depth factor [$= 1 + 0.4\,(D_f/B)$ for $D_f/B \le 1$],

i_c = inclination factor, q = surcharge ($= \gamma\, D_f$).

The bearing capacity factor N_c depends upon the ratio c_2/c_1, where c_2 is the undrained shear strength of the lower layer. It also depends upon the ratio Z/B, where Z is the depth of the interface of the two layers from the bottom of foundation and B is the width of the foundation [Fig. 23.18 (*a*)]. Fig. 23.18 (*b*) shows the

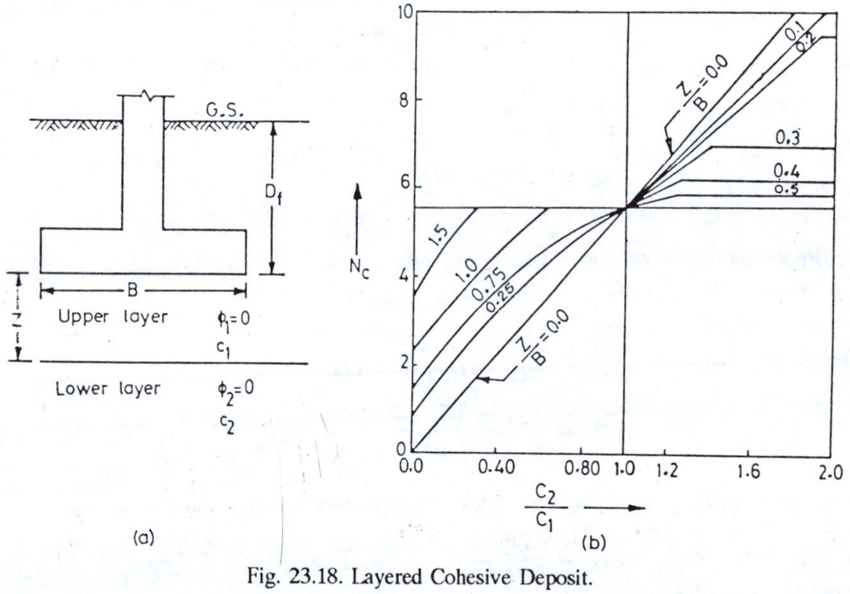

Fig. 23.18. Layered Cohesive Deposit.

variation of N_c with c_2/c_1 ratio and Z/B ratio. It may be noted that for c_2/c_1 ratio less than unity, the value of N_c is smaller than that for a homogeneous soil, (*i.e.*, $c_2/c_1 = 1.0$). It indicates that the ultimate bearing capacity of a clay layer is reduced by the presence of a soft layer beneath the top hard layer.

23.20. BEARING CAPACITY FROM STANDARD PENETRATION TEST

Method I. The ultimate bearing capacity of cohesionless soils may be determined from the standard penetration number (N). The standard penetration test is conducted at a number of selected points in the vertical direction below the foundation level at intervals of 75 cm or at point where there is a change of strata. An average value of N is obtained between the level of the base of the footing and the depth equal to 1.5 to 2.0 times the width of the foundation. The value of ϕ' is obtained from the N value as already discussed (Section 17.22) and the bearing capacity factors are found. Fig. 23.7 may also be used to determine directly the bearing capacity factors from N.

Method II. As the ultimate bearing capacity depends upon ϕ' and hence on N, it can be related directly to N. Teng (1962) gave the following equation for the net ultimate capacity of a strip footing.

$$q_{nu} = \frac{1}{6.0}[3N^2 BW_\gamma + 5(100 + N^2) D_f W_q]$$

or
$$q_{nu} = 0.5 N^2 BW_\gamma + 0.83(100 + N^2) D_f W_q \quad ...(23.60)$$

where q_{nu} = net ultimate bearing capacity (kN/m^2),
B = width of footing, N = average SPT number,
D_f = depth of footing. If $D_f > B$, use $D_f = B$.
and W_γ and W_q are water table correction factor (Sect. 23.10).

For square or circular footings,

$$q_{nu} = \frac{1}{3.0}[N^2 BW_\gamma + 3(100 + N^2) D_f W_q]$$

or
$$q_{nu} = 0.33 N^2 BW_\gamma + 1.0(100 + N^2) D_f W_q \quad ...(23.61)$$

The net allowable bearing capacity can be obtained by applying a factor of safety of 3.0.

For strip footings,

$$q_{ns} = 0.167 N^2 BW_\gamma + 0.277(100 + N^2) D_f W_q \text{ kN/m}^2 \quad ...[23.62(a)]$$

For circular and square footings,

$$q_{ns} = 0.11 N^2 BW_\gamma + 0.33(100 + N^2) D_f W_q \text{ kN/m}^2 \quad ...[23.62(b)]$$

23.21. ECCENTRICALLY LOADED FOUNDATIONS

Foundations are sometimes subjected to moments in addition to the loads (Fig. 23.19). The distribution of footing pressure is not uniform in this case. It is a case of bending combined with thrust, treated in the mechanics of materials. The maximum and minimum pressures are given by

$$q_{max} = \frac{Q}{B \times L} + \frac{M}{I}(B/2)$$

and
$$q_{min} = \frac{Q}{B \times L} - \frac{M}{I}(B/2)$$

where I = moment of inertia (= $LB^3/12$), Q = total vertical load (gross), M = moment on the foundation, B = width of footing, L = length of footing.

Taking the eccentricity e as M/Q, the above equations become

$$q_{max} = \frac{Q}{BL}(1 + 6e/B) \quad ...[23.63(a)]$$

and
$$q_{min} = \frac{Q}{BL}(1 - 6e/B) \quad ...[23.63(b)]$$

The maximum pressure q_{max} should be less than the safe gross bearing capacity.

Meyerhof's Method

The factor of safety of eccentrically loaded foundations against bearing capacity failure can be determined using the method given by Meyerhof (1953), as explained below

(1) Determine the eccentricity of the load, along the width as,
$$e_b = M_b/P$$

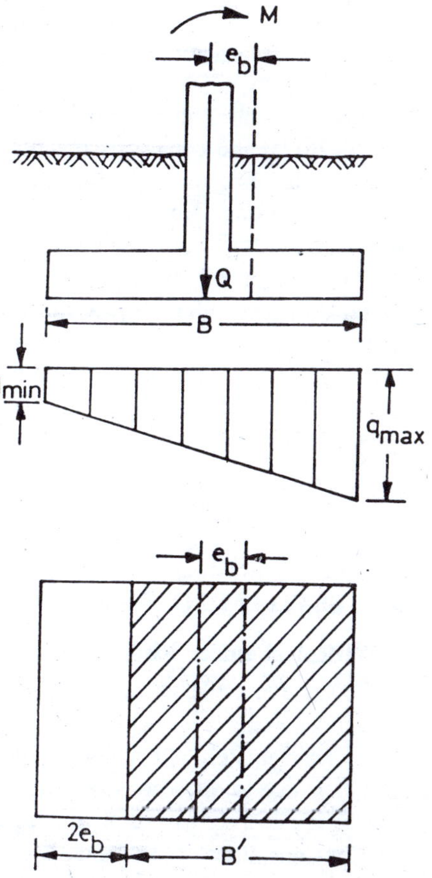

Fig. 23.19. Eccentricaly loaded footing.

It may by noted that the eccentricity is measured from the centre of footing.

(2) Determine the effective width of the footing. $B' = B - 2\,e_b$.

(3) Determine the effective size of the footing as $L \times B'$.

In the above case, the eccentricity has been assumed only along width. However, if the eccentricity is also in the longitudinal direction along length, the eccentricity along the length is given by

$$e_l = M_l/P$$

The effective length of the footing in that case is $L' = L - 2\,e_l$ and the effective size is $L' \times B'$.

The smaller of the two dimensions B' and L' is taken as the effective width for the computation of the ultimate bearing capacity.

(4) The ultimate bearing capacity can be obtained using Eq. 23.42 as

$$q_u = c\,N_c\,s_c\,d_c\,i_c + q\,N_q\,s_q\,d_q\,i_q + 0.5\,\gamma B'\,N_\gamma\,s_\gamma\,d_\gamma\,i_\gamma$$

The shape factors s_c, s_q and s_γ are obtained from Table 23.7. The inclination factors are obtained using Eq. 23.47, in which the effective width B' and effective length L' are used.

For computations of depth factors, Eqs. 23.46 (*a*) to 23.46(*c*) are used, but B' is not substituted for B.

(5) The total ultimate load is computed as

$$Q_u = q_u\,(B' \times L') \qquad \text{...(23.64)}$$

(6) The factor of safety is given by

$$F_s = Q_u/Q \qquad \text{...(23.65)}$$

As the eccentricity causes a reduction in the load-carrying capacity of a foundation, the column is sometimes placed off-centre of the footing to produce a uniform pressure distribution (Fig. 23.20). If the column is not placed off-centre, there is a possibility of tilting of the footing due to higher pressure on one side of the footing. This tendency of tilting can be reduced by allowing a larger factor of safety.

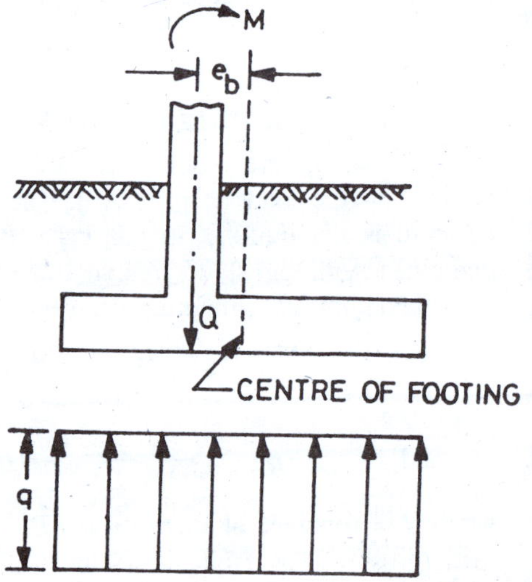

Fig. 23.20.

If the eccentricity is outside the middle third, the minimum soil pressure (Eq. 23.63 *b*), becomes negative, indicating that the tensile stresses develop in soil. As the soil cannot take tension, there is a separation between the footing and the soil. The area of the footing which is in tension is generally neglected. The maximum pressure in such a case is given by

$$q_{max} = \frac{4\,Q}{3L\,(B - 2\,e_b)} \qquad \text{...(23.66)}$$

where e_b is the eccentricity along width of footing.

23.22. SETTLEMENT OF FOUNDATION

(*a*) Settlement under loads

Foundation settlement under loads can be classified into 3 types.

(1) **Immediate or elastic settlement** (s_i).. Immediate or elastic settlement takes place during or immediately after the construction of the structure. It is also known as the distortion settlement as it is due to distortions (and not the volume change) within the foundation soil. Although the settlement is not truly elastic, it is computed using elastic theory, especially for cohesive soils.

(2) **Consolidation settlement** (s_c).. This component of the settlement occurs due to gradual expulsion of water from the voids of the soil. This component is determined using Terzaghi's theory of consolidation (see chapter 12).

(3) **Secondary Consolidation Settlement** (s_s). This component of the settlement is due to secondary consolidation. This settlement occurs after completion of the primary consolidation. It can be determined from the coefficient of secondary consolidation (see chapter 12). The secondary consolidation is not significant for inorganic clays and silty soils.

The total settlement (s) is given by

$$s = s_i + s_c + s_s \quad \text{...(23.67)}$$

(*b*) **Settlement due to other causes**

In addition to settlement under loads, the settlement may also occur due to a number of other causes.

(1) **Underground erosion.** Underground erosion may cause formation of cavities in the subsoil which when collapse cause settlement.

(2) **Structural collapse of soil.** Structural collapse of some soils, such as saline, non-cohesive soils, gypsum, silts and clays and loess, may occur due to dissolution of materials responsible for intergranular bond of grains.

(3) **Thermal changes.** Temperature change cause shrinkage in expansive soils due to which settlement occurs.

(4) **Frost heave.** Frost heave occurs if the structure is not founded below the depth of frost penetration. When thaw occurs, the foundation may settle.

(5) **Vibration and Shocks.** Vibrations and shock cause large settlements, especially in loose, cohesionless soils.

(6) **Mining subsidence.** Subsidence of ground may occur due to removal of minerals and other materials from mines below.

(7) **Land slides.** If land slides occur on unstable slopes, there may be serious settlement problems.

(8) **Creep.** The settlement may also occur due to creep on clay slopes.

(9) **Changes in the vicinity.** If there are changes due to construction of a new building near the existing foundation, the settlement may occur due to increase in the stresses.

Suitable measures are taken to reduce the settlements due to all above causes.

23.23. LOADS FOR SETTLEMENT ANALYSIS

Dead loads include the weight of columns, walls, footings, foundations, the overlying fill, but do not include the weight of the displaced soil.

Live loads depend upon the use of the structure. These loads may be taken from IS : 875.

Wind loads and seismic loads should be considered wherever applicable. However, where wind (or seismic) load is less than 25% of the combined dead and live loads, it may be neglected in design, and only dead load and live loads are considered. When wind (or seismic) load is more than 25% of the combined dead and live load, the foundation is designed such that pressure due to combination of dead, live and wind (seismic) loads does not exceed the allowable bearing capacity by more than 25%.

For foundations resting on coarse-grained soils, the settlements should be estimated corresponding to the full dead load, live load and wind (seismic) load. In such soils, settlements occur in a short time.

For foundations on fine-grained soils, the settlements are estimated corresponding to permanent loads. All dead loads and the loads due to fixed equipment are taken as permanent. Generally, one half of the live load is also taken as permanent. Engineering judgment is required to ascertain the permanent loads.

23.24. IMMEDIATE SETTLEMENT OF COHESIVE SOILS

The linear theory of elasticity is used to determine the elastic settlement of the footings on saturated clay. Schleicher (1926) gave the following formula for the vertical settlement under a uniformly distributed flexible area.

$$s_i = qB\left(\frac{1-\mu^2}{E_s}\right)I \quad \text{...(23.68)}$$

where q = uniformly distributed load, B = characteristic length of the loaded area, E_s = modulus of elasticity of the soil, μ = Poisson's ratio (= 0.50 for saturated clay), I = influence factor.

The value of E_s is determined from the stress-strain curve obtained from a triaxial consolidated-undrained test, with the consolidation pressure equal to the effective pressure at the depth from which the sample was taken, as discussed in chapter 11. It is generally taken as the initial tangent modulus or the secant modulus. For normally consolidated clays, its value varies from 250 c to 500 c, and for over-consolidated clays, from 750 c to 1000 c, where c is undrained cohesion.

The value of the influence factor I for a saturated clay layer of semi-infinite extent can be obtained from Table 23.8.

Table 23.8. Values of Influence Factor I.

Shape	*Flexible footing*			*Rigid footing*
	Centre	*Corner*	*Average*	
Circle	1.0	0.64 (edge)	0.85	0.79
Square	1.12	0.56	0.95	0.82
Rectagle $L/B = 1.5$	1.36	0.68	1.20	1.06
$L/B = 2.0$	1.53	0.77	1.31	1.20
$L/B = 2.0$	1.78	0.89	1.52	1.42
$L/B = 5.0$	2.10	1.05	1.83	1.70
$L/B = 10.0$	2.52	1.26	2.25	2.10
$L/B = 100.0$	3.38	1.69	2.96	3.40

Alternatively, the value of $(1 - \mu^2) I/E_s$ can be determined from the plate load test (Sect. 23.33).

If an incompressible layer exists at a limited depth below the footing, the actual settlement is less than that given by Eq. 23.68. For such a case, Steinbrenner (1936) gave a solution. However, if the depth of the clay layer is more than 2 B, the actual settlement would not change much.

If the foundation is rigid, such as a heavy beam and slab raft, the settlement is about 0.8 times the settlement at the centre of the corresponding flexible foundation. It is approximately equal to the average settlement. Table 23.8 also gives the values of I for rigid footings.

Eq. 23.68 is applicable for the footing located at surface. For the footings embedded in soil, the settlement would be less than the computed values. Fox (1948) gave correction curves. The settlement is obtained by multiplying the computed settlements by a depth factor, which depends upon $(D_f / \sqrt{L \times B})$ ratio.

23.25. IMMEDIATE SETTLEMENT OF COHESIONLESS SOILS

As cohesionless soils do not follow Hooke's law, immediate settlements are computed using a semi-empirical approach proposed by Schmertmann and Hartman (1978).

$$s_i = C_1 C_2 (\bar{q} - q) \sum_{Z=0}^{2B} \frac{I_z}{E_s} \Delta z \qquad \ldots(23.69)$$

where C_1 = correction factor for the depth of foundation embedment = $1 - 0.5 \{q/(\bar{q} - q)\}$.

C_2 = correction factor for creep in soils [= $1 + 0.2 \log_{10}$ (time in years/0.1)].

$\bar{q}$ = pressure at the level of the foundation, q = surcharge (= γD_f),

E_s = modulus of elasticity, I_z = strain influence factor.

The value of the strain-influence factor I_z varies linearly for a square or circular foundation (Fig. 23.21). The value of I_z at depth z = 0, 0.5 B and $2B$ are respectively equal to 0.1, 0.5 and 0.0. For rectangular

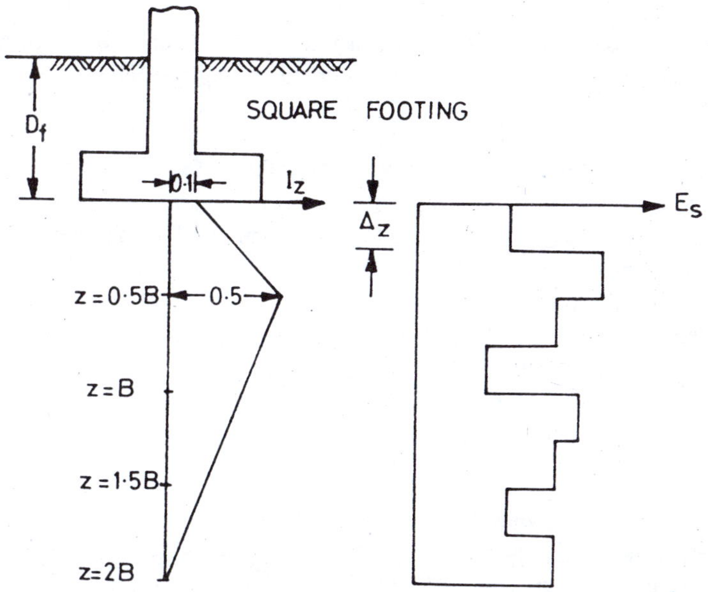

Fig. 23.21.

foundations, with L/B ratio equal to or greater than 10.0, the values at depth Z = 0.0, B, and $4B$ are, respectively, 0.2, 0.5 and 0.0. For intermediate values of L/B ratio, between 1.0 and 10.0, interpolation can be made.

The value of E_s can be determined from the standard penetration number (N) using the following equations given by Schmertmann (1970).

$$E_s = 766\,N \text{ (kN/m}^2\text{)} \qquad \ldots(23.70)$$

Alternatively, it can be estimated from the static cone penetration resistance (q_c) as

$$E_s = 2\,q_c \qquad \ldots(23.71)$$

Procedure. For computation of the immediate settlement, the soil layer is divided into several layers of thickness Δz, upto a depth $z = 2B$, in case of square footings and $z = 4B$, in case of rectangular footings. The immediate settlement of each layer is computed using Eq. 23.69, taking corresponding values of E_s and I_z. The required immediate settlement is equal to the sum of the settlements of all individual small layers.

23.26. CONSOLIDATION SETTLEMENT IN CLAYS

The consolidation settlement (s_c) occurs in saturated, clayey soils when these are subjected to increased loads caused by the foundation pressure. The methods for computation of the consolidation settlement have been discussed in chapter 12.

The settlement (s_t) at any time t after the application of load may be estimated using Terzaghi's consolidation theory as

$$s_t = s_i + U\,s_f \qquad \ldots(23.72)$$

where U = degree of consolidation, expressed as a ratio,

s_f = final consolidation settlement,

s_i = immediate settlement

The degree of consolidation (U) depends upon the time factor T_v, given by

$$T_v = c_v\; t/d^2 \qquad \ldots(23.73)$$

where c_v = average coefficient of consolidation, t = time at which the settlement is required, d = distance of the drainage path.

[**Note.** When considering the drainage path, concrete of foundation is assumed as permeable].

For the final settlement, U = 1.0, and the total settlement is given by

$$s = s_i + s_f \quad \text{...(23.74)}$$

23.27. SETTLEMENT OF FOUNDATION ON COHESIONLESS SOILS

Settlements of foundations on cohesionless soils take place rather quickly after the application of the load. The immediate settlements calculated using Schmertmann and Hartman method (Sect. 23.25) would also be the final settlement in most cases.

Because of the difficulty of sampling cohesionless soils, it is not possible to obtain the stress-strain characteristics of the insitu soils. The settlements are generally determined indirectly using the semiempirical methods.

(1) Static cone penetration method. In this method, the sand layer is divided into small layers such that each small layer has approximately constant value of the cone resistance. The average value of the cone resistance of each small layer is determined.

The settlement of each small layer is estimated using the following equation (De Beer and Martens, 1957).

$$s = \frac{H}{C} \log_e \frac{\overline{\sigma}_0 + \Delta\sigma}{\overline{\sigma}_0} \quad \text{...(23.75)}$$

where $C = 1.5 q_c / \overline{\sigma}_0$

in which q_c = static cone resistance, $\overline{\sigma}_0$ = mean effective overburden pressure, $\Delta\sigma$ = increase in pressure at the centre of the layer due to the net foundation pressure and H = thickness of layer.

The total settlement of the entire layer is equal to the sum of settlements of individual layers.

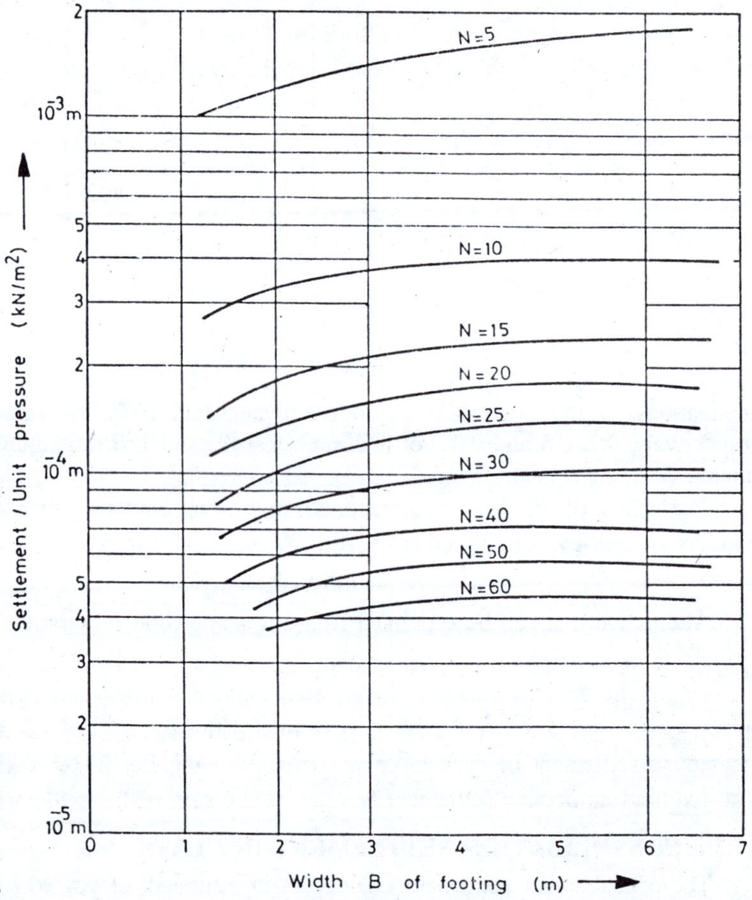

Fig. 23.22.

(2) Standard Penetration Test. Standard penetration test can be used for the determination of the settlement on cohesionless soils. IS : 8009 (Part I)—1976 gives a chart for the calculation of *settlement per unit pressure* as a function of the width of the footing and the standard penetration number (Fig. 23.22). The settlement under any other pressure is computed assuming that the settlement is proportional to the intensity of pressure.

The settlements are in metres per unit pressure in kN/m^2.

If the water table is at a shallow depth, the settlements are divided by the correction factor W_γ [Eq. 23.36 (c)].

(3) Plate Load Test. The settlement of the footing can be estimated from the settlement of the plate in the plate load test (Sec. 23.33).

23.28. ACCURACY OF FOUNDATION SETTLEMENT PREDICTION

The prediction of the foundation settlements as discussed above is rarely accurate. In fact, it is extremely difficult to estimate the probable settlement of a footing because of the following reasons:

(1) The soil deposits are seldom isotropic and linearly elastic. The deposits are generally non-homogeneous.

(2) It is not possible to estimate the increase in stresses caused by loads. The Boussinesq solution gives only approximate results.

(3) For estimation of the settlement due to consolidation, it is not possible to locate exactly the drainage faces.

(4) For computation of immediate settlements, it is not possible to estimate the correct value of the modulus of elasticity

(5) The rigidity of the foundation is usually neglected and the pressure distribution is assumed to be uniform.

(6) It is difficult to obtain undisturbed samples of cohesionless soils. The semi-empirical methods do not give accurate results.

(7) Settlements may occur due to causes other than that due to loads. It is not possible to estimate these settlements accurately.

Despite all the above reasons, the settlements in most cases can be estimated to an accuracy of about 25 to 30%, which is good enough seeing the complexity of the problem.

23.29. ALLOWABLE SETTLEMENT

If there is a large differential settlement between various parts of a structure, damage may occur due to additional moments developed. It is more difficult to predict differential settlement than to predict the maximum settlement. The differential settlements are generally obtained indirectly from the maximum settlements. It has been found from actual observations of various existing buildings that the differential settlements seldom exceed 75% of the maximum settlement. In fact, in most cases, the differential settlement is less than 50% of the maximum settlement. Hence, the differential settlement is automatically controlled if the maximum settlement is controlled.

The allowable maximum settlement depends upon the type of soil, the type of foundation and the structural framing system. The maximum settlement ranging from 20 mm to 300 mm is generally permitted for various structures. Theoretically, no damage is done to the superstructure if the soil settles uniformly. However, settlements exceeding 150 mm may cause trouble in utilities such as water pipe lines, sewers, telephone lines and also in access from streets. If sufficient precautions are taken while designing such utilities, the maximum settlements of even upto 2 m may be permitted in special cases, provided the structure settles uniformly.

IS : 1904 (1966) permits a maximum settlement of 40 mm for isolated foundations on sand and 65 mm for those on clay. The allowable settlement is higher for clays because progressive settlements on clayey soils permit better strain adjustments in the structural members. The maximum permissible settlement for raft foundations on sand are 40 mm to 65 mm and that on clay, 65 to 100 mm. The permissible settlements for rafts are more than those for isolated foundations because the raft bridges over soft patches of the soil, and the differential settlements are reduced.

If the differential settlement between two columns spaced at a distance L is δ, the angular distortion or til T is given by

$$T = \delta/L \qquad \ldots(23.76)$$

Common R.C.C. buildings can sustain angular distortion upto 1/150 without damage. However, to avoid the architectural damage, it should not exceed 1/300. It corresponds to a differential settlement of 20 mm between adjacent columns 6 m centres. According to general practice, the maximum differential settlement is limited to 25 mm in sandy soils and 40 mm in clayey soils.

IS : 1904—1978 gives the safe values of the maximum and differential settlements of different types of

Table 23.9. Maximum and Differential Settlements (IS : 1904—1978)

	Sand and hard clay			*Plastic clay*		
	Max-Settlement	*Diff-Settlement*	*Angular Distortion*	*Max-Settlement*	*Diff-Settlement*	*Angular Distortion*
(*a*) Isolated Foundations						
(i) Steel structure	50 mm	0.0033 L	1/300	50 mm	0.0033 L	1/300
(ii) R.C.C. structures	50 mm	0.0015 L	1/666	75 mm	0.0015 L	1/666
(*b*) Raft Foundations						
(i) Steel Structures	75 mm	0.0033 L	1/300	100 mm	0.0033 L	1/300
(ii) R.C.C. Structures	75 mm	0.002 L	1/500	100 mm	0.002 L	1/500

buildings. The values for isolated footings and rafts are given in Table 23.9. For further details, a reference may be made to the code.

In actual practice, settlements of the structure at salient points such as the centre, the corner, the lightest and the heaviest column locations are determined using the methods discussed. The differential settlements are then determined from the settlements of the various points. Sometimes, the differential settlements are taken as 75% of the maximum settlement without computing the settlements other than that at the point where the maximum settlement is expected to occur. However, for large works, thorough sub-soil investigations are done to determine the characteristics of the soil and the differential settlements are determined taking the properties of the soil into account.

23.30. ALLOWABLE SOIL PRESSURE FOR COHESIONLESS SOILS

The allowable soil pressure (q_{na}) of a shallow foundation is limited either by the net safe bearing capacity (q_{ns}) or the safe settlement pressure ($q_n\,\rho$). The design of shallow foundation on cohesionless soils is generally governed by the safe settlement pressure, as the net safe bearing capacity for a footings of usual size is quite high. However, in the case of narrow footings on water-logged sands, the net safe bearing capacity may be the controlling criterion for the design.

It is the normal practice for the design of footings of usual size to use empirical methods based on N—values for the determination of the allowable soil pressure for cohesionless soils. The plate load tests (Sect. 23.33) are also used in the case of soils having small boulders and stones which obstruct the standard penetration test. The methods using the standard penetration test are preferred to plate load tests for homogeneous soils, as these are more economical.

Footings on granular soils are generally designed using the following empirical relationships for the allowable soil pressure.

(1) Peck Method

Terzaghi and Peck (1967) gave charts for the safe bearing pressures inducing a total settlement of 25 mm and a differential settlement of 19 mm for different sizes of footing. Peck et al (1974) revised the Terzaghi and Peck curves to take into consideration the later research, and gave the following equation for the safe settlement pressure.

$$q_n\,\rho = 0.41\ C_w N s \quad \text{...(23.77)}$$

where $q_n\,\rho$ = safe settlement pressure (kN/m^2),

N = average SPT number, corrected for overburden pressure and dilatancy,
s = settlement (mm), C_w = water table correction factor.

For a settlement of 40 mm,

$$q_n \rho = 16.4 \; C_w N \tag{23.78}$$

Fig. 23.23 gives a chart for s = 40 mm and C_w = 1.0.

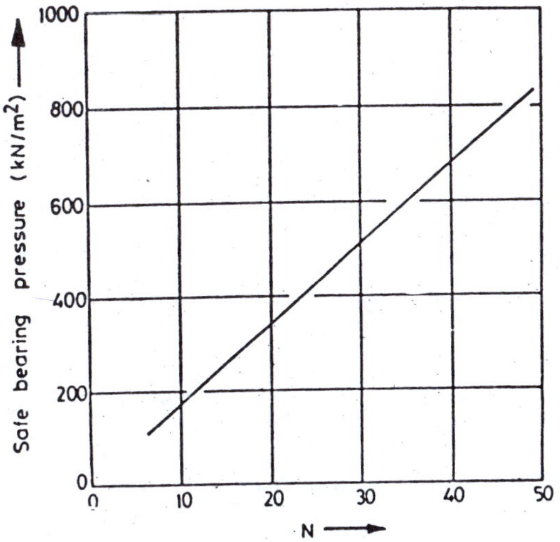

Fig. 23.23.

For a settlement of 25 mm,

$$q_n \rho = 10.25 \; C_w N \quad (\text{kN/m}^2)$$

The water table correction is determined using the relation

$$C_w = 0.5 + 0.5 \; D_w/(D_f + B) \qquad \ldots(23.79)$$

where D_w = depth of water table below the ground surface, D_f = depth of footing, B width of footing.

(2) Teng's Equation

Teng (1962) expressed the charts given by Terzaghi and Peck (1948) in the form of the following formulas. Allowance was made for an increase in pressure with depth by introducing a depth factor.

For a settlement of 25 mm,

$$q_n\rho = 35.0\,(N - 3)\left(\frac{B + 0.3}{2\,B}\right)^2 W_\gamma \; R_d \qquad \ldots[23.80(a)]$$

where $q_n\rho$ = safe settlement pressure (kN/m^2), N = SPT number, B = width of footing (m),

W_γ = water table correction factor (Eq. 23.36 (c)),

R_d = depth correction factor $\left(= 1 + \dfrac{D_f}{B} \leq 2.0 \right)$

The above equation can be written in general form as

$$q_n\rho = 1.40\;(N - 3)\left(\frac{B + 0.3}{2B}\right)^2 W_\gamma \; R_d \; s \qquad ..[23.80(b)]$$

where s = tolerable settlement (mm)

(3) Meyerhof's equation

Meyerhof proposed equations which are slightly different from Teng's equations. According to him, for a settlement of 25 mm,

$$q_n \rho = 12.2\ N\ W_\gamma R_d \quad \text{for } B \le 1.2 \text{ m} \qquad \text{...[23.81(}a\text{)]}$$

and
$$q_n \rho = 8.1\ N \left(\frac{B + 0.3}{B}\right)^2 W_\gamma\ R_d \quad \text{for } B \ge 1.2 \text{ m} \qquad \text{...[23.81(}b\text{)]}$$

where all the terms are the same as in Teng's equation, except R_d, which is given by

$$R_d = 1 + 0.33\ D_f/B \le 1.33.$$

(4) Bowle's equation

Bowles (1977) suggested that the net allowable pressure given by Meyerhof's equation can be safely increased by 50%. Thus, for a settlement of 25 mm,

$$q_n \rho = 18.3 N\ W_\gamma R_d \quad \text{for } B \le 1.2 \text{ m} \qquad \text{...[23.82(}a\text{)]}$$

and
$$q_n \rho = 12.2\ N \left(\frac{B + 0.3}{B}\right)^2 W_\gamma\ R_d \quad \text{for } B \ge 1.20 \qquad \text{...[23.82(}b\text{)]}$$

(5) IS : 6403—1971 equation

IS : 6403—1971 gives the following equation, which is similar to Teng's equation (Eq. 23.80 (*b*)). For a settlement of 40 mm,

$$q_n \rho = 55.4\ (N - 3) \left(\frac{B + 0.3}{2B}\right)^2 W_\gamma \qquad \text{...(23.83)}$$

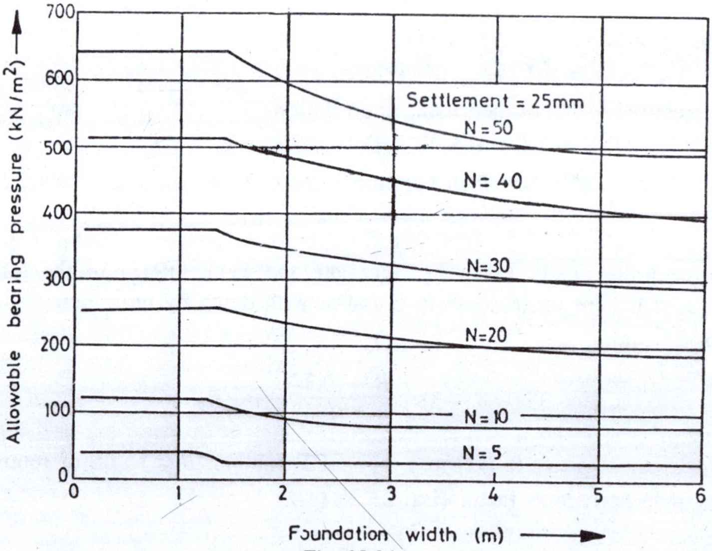

Fig. 23.24.

The depth factor is not considered.

Fig. 23.24 gives the allowable soil pressure for a settlement of 25 mm (Terzaghi and Peck, 1967). Fig. 23.25 gives the allowable soil pressure for a settlement of 40 mm.

Narrow Footings on Sand

The design of narrow footings on sand footings is generally governed by the net safe bearing capacity. The ultimate bearing capacity can be determined from the theories already discussed. The value of the angle of shearing resistance (ϕ') can be determined from shear tests. As it is difficult to duplicate field conditions in the laboratory, ϕ' is usually determined indirectly from standard penetration number (N), as shown in Fig. 23.7.

23.31. ALLOWABLE SOILS PRESSURE FOR COHESIVE SOILS

The allowable soil pressure for cohesive soils is generally controlled by the net safe bearing capacity, although settlement criterion may control in soft clays. In the case of foundations on firm to stiff clays, it is unnecessary to compute settlements, especially when the structure is lightly loaded as the settlements are quite small. A factor of safety of 3 against shear failure would generally ensure that the settlements are within the safe limits. However, it is essential to compute the consolidation settlement in all cases of heavy structures.

The net safe bearing capacity (q_{ns}) depends upon the shear strength of the clay. The minimum value of shear strength should be taken for the computations of the ultimate bearing capacity. On saturated clays, the time of loading is relatively rapid, and the undrained conditions usually apply. The undrained cohesion (c_u) is determined from the vane-shear test, unconfined compression test or unconsolidated-undrained test (Chapter 13). The ultimate bearing capacity is determined using the bearing capacity theories for $\phi = 0$ conditions.

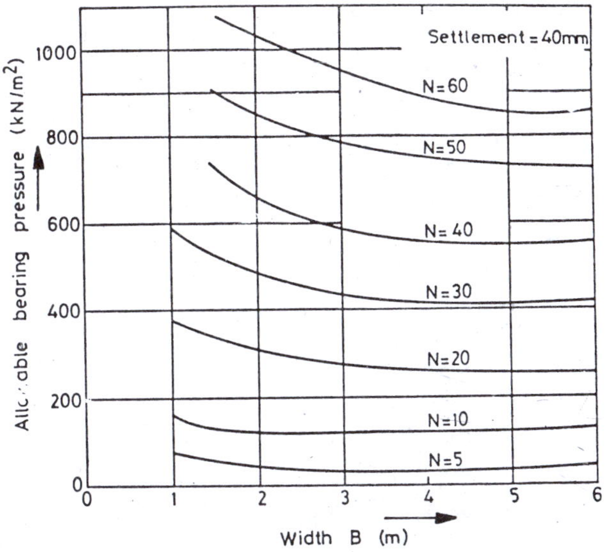

Fig. 23.25. (1 t/m^2 = 10 kN/m^2)

As the settlement calculations are involved, the size of the footing is first selected considering the net safe bearing capacity and then the settlements are checked. It may be noted that the footing width has very little effect on the magnitude of the settlement. The settlements depend mainly on the magnitude of the load. Consequently, if settlements control the design, merely increasing the size of footing may not help solve the problem. An alternative type of foundation such as a mat foundation, a compensating foundation or a pile foundation, would be required.

23.32. PRESUMPTIVE BEARING CAPACITY

Building codes of various organisations in different countries give the allowable bearing capacity that can be used for proportioning footings. These values, known as presumptive bearing capacity, are based on experience with other structures already built. As the presumptive values are based only on a visual classification of the surface soils, they are not reliable. These values do not consider important factors affecting the bearing capacity, such as the shape, width, depth of footing, location of water table, strength and compressibility of the soil. Generally, the values are conservative and can be used for preliminary design or even for the final design of small, unimportant structure.

IS : 1904—1978 recommends that safe bearing capacity should be calculated on the basis of the soil test data. However, in the absence of such data, the values of the safe bearing capacity can be taken equal to the presumptive bearing capacity values given. Table 23.10 gives the values for different types of soils and rocks. The table also gives the values as per New York Building Code for comparison.

It is further recommended that for non-cohesive soils, the values shall be reduced by 50% if the water table is above or near the base of footing.

For further details, a reference may be made to IS: 1904—1978.

23.33. PLATE LOAD TEST

The allowable bearing pressure can be determined by conducting a plate load test at the site. The conduct a plate load test, a pit of the size $5B_p \times 5B_p$, where B_p is the size of the plate, is excavated to a depth equal to the depth of foundation (D_f). The size of the plate is usually 0.3 m square. It is made of steel and is 25 mm thick. Occasionally, circular plates are also used. Sometimes, large size plates of 0.6 m square are used.

Table 23.10. Presumptive Safe Bearing Capacity

Type of Rock/soil	*IS : 1904–1978*		*New York Building Code*	
	t/m^2	kN/m^2	t/m^2	kN/m^2
I. Rocks				
1. Hard Sound Rock	330	3240	199–591	1950–5800
2. Laminated rock	165	1620	77	760
3. Residual Deposits of Shattered and Broken Rocks	90	880	112	1100
4. Soft Rock	45	440	77	760
II. Non-Cohesive Soils**				
1. Compact gravel, sand and gravel	45	440	77–97	760–950
2. Compact and dry coarse sand	45	440	31–77	300–760
3. Compact and dry medium sand	25	245	–	–
4. Fine sand, silt	15	150	–	–
5. Loose gravel or sand	25	245	36	350
6. Loose and dy fine sand	10	100	19	90
III. Cohesive Soils				
1. Hard or Stiff clay, soft shale	45	440	49	480
2. Medium clay	25	245	19	190
3. Moist clay and sand clay mixture	15	150	–	–
4. Soft clay	10	100	10	95
5. Very soft clay	5	50	–	–
6. Black cotton soil, Peat, made up soils	*	*	*	*

* To be determined after soil investigations.

** Compactness or looseness of non-cohesive soils may be determined by driving a cone of 65 mm dia and 60° apex angle by a hammer of 65 kg falling from 75 cm. If corrected number of blows (N) for 30 cm penetration is less than 10, the soil is loose. If more than 30, it is taken as dense. If N is between 10 and 30, it is medium dense.

A central hole of the size $B_p \times B_p$ is excavated in the pit. The depth of the central hole (D_p) is obtained from the following relation :

$$D_p/B_p = D_f/B_f$$

or

$$D_p = (B_p/B_f) \times D_f \qquad \ldots(23.84)$$

where B_f is the width of the pit, and B_p is the size of plate.

For conducting the plate load test, the plate is placed in the central hole and the load is applied by means of a hydraulic jack (Fig. 23.26). The reaction to the jack is provided by means of a reaction beam. Sometimes, trusses are used instead of a reaction beam to take up the reaction. Alternatively, a loaded platform (kentledge) can be used to provide reaction. A seating load of 7 kN/m^2 is first applied, which is released after some time. The load is then applied in increments of about 20% of the estimated safe load or one-tenth of the ultimate load. The settlement is recorded after 1, 5, 10, 20, 40, 60 minutes, and further after an interval of one hour. These hourly observations are continued for clayey soils until the rate of settlement is less than 0.2 mm per hour. The test is conducted until failure or at least until the settlement of about 25 mm has occurred (IS : 1888).

The ultimate load for the plate $q_u(p)$ is indicated by a break on the log-log plot between the load intensity q and the settlements. If the break is not well-defined, the ultimate load is taken as that

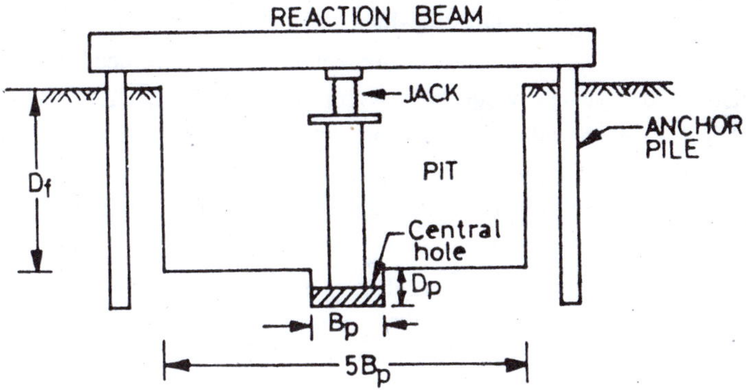

Fig. 23.26. Plate Load Test.

corresponding to a settlement of one-fifth of the plate width (B_p). On the natural plot (Fig. 23.27), the ultimate load is obtained from the intersection of the tangents drawn as shown.

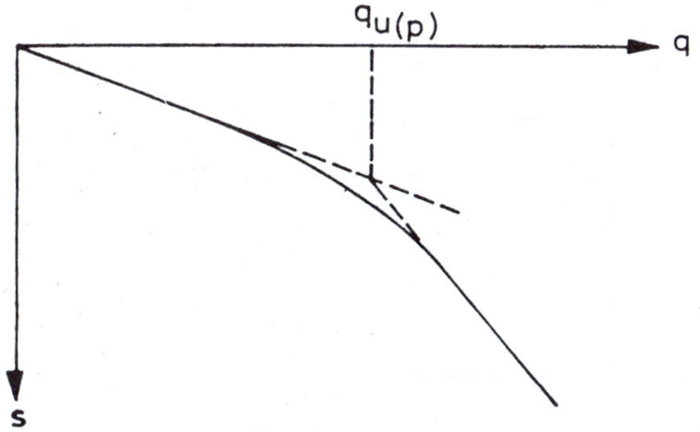

Fig. 23.27.

(1) The ultimate bearing capacity of the proposed foundation q_u (*f*) can be obtained from the following relations:

(*a*) For clayey soils, $$q_u(f) = q_u(p) \quad \text{...(23.85)}$$

(*b*) For sandy soils, $$q_u(f) = q_u(p) \times \frac{B_f}{B_p} \quad \text{...(23.86)}$$

where B_f = foundation width.

(2) The plate load test can also be used to determine the settlement for a given intensity of loading (q_o). The relations between the settlement of the plate (s_p) and that of the foundation (s_f) for the same load intensity are given below.

(*a*) For clayey soils, $$s_f = s_p \times \frac{B_f}{B_p} \quad \text{...(23.87)}$$

where s_p is obtained from the load intensity-settlement curve for q_o.

(*b*) For sandy soils, $$s_f = s_p \left[\frac{B_f(B_p + 0.3)}{B_p(B_f + 0.3)}\right]^2 \quad \text{...(23.88)}$$

In above equations, B_f is the width of foundation in metres and B_p is the width of the plate also in metres.

(3) For designing a shallow foundation for an allowable settlement of s_f, a trial and error procedure is adopted. First of all, a value of B_f, is assumed and the value of q_o is obtained as

$$q_o = Q/A_f \qquad ...(23.89)$$

where A_f is the area of footing, and Q is the load.

For the computed value of q_o, the plate settlement (s_p) is determined from the load-settlement curve obtained from the plate load test. The value of s_f is computed using Eq. 23.87 if the soil is clay and using Eq. 23.88, if sand. The computed value of s_f is compared with the allowable settlement. The procedure is repeated till the computed value is equal to the allowable settlement.

(4) The plate load test can also be used for the determination of the influence factor I (Sect. 23.24). From Eq. 23.68,

$$s = \frac{(1-\mu^2)}{E_s} \times I \times qB \qquad ...(23.90)$$

Fig. 23.28 shows a plot between the settlements and the load qB. The slope of the line is equal to $(1-\mu^2)\, I/E_s$.

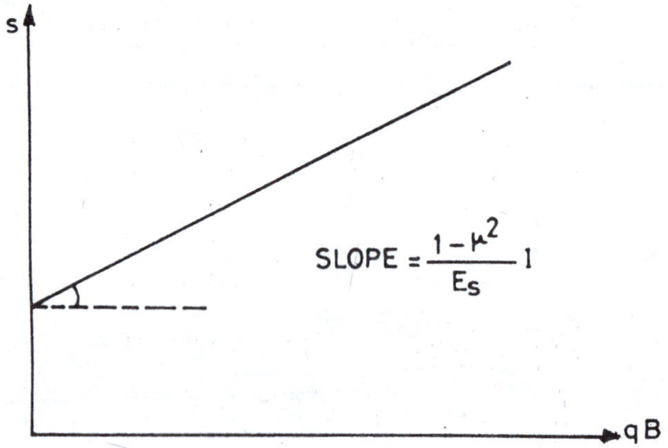

Fig. 23.28.

Limitation of the plate load test

The plate load test has the following limitations:

(1) **Size effect.** The results of the plate load test reflect the strength and the settlement characteristics of the soil within the pressure bulbs (Section 11.6). As the pressure bulb depends upon the size of the loaded area, it is much deeper for the actual foundation as compared to that of the plate (Fig. 23.29). The plate load test does not truly represent the actual conditions if the soil is not homogeneous and isotropic to a large depth.

(2) **Scale effect.** The ultimate bearing capacity of saturated clays is independent of the size of the plate but for cohesionless soils, it increases with the size of the plate (Eq. 23.86). To reduce scale effect, it is desirable to repeat the plate load test with plates of two or three different sizes and extrapolate the bearing capacity for the actual foundation and take the average of the values obtained.

(3) **Time effect.** A plate load test is essentially a test of short duration. For clayey soils, it does not give the ultimate settlement. The load—settlement curve is not truly representative.

(4) **Interpretation of failure load.** The failure load is not well-defined, except in the case of a general shear failure. An error of personal interpretation may be involved in other types of failure.

(5) **Reaction load.** It is not practicable to provide a reaction of more than 250 kN. Hence, the test on a plate of size larger than 0.6 m width is difficult.

(6) **Water table.** The level of the water table affects the bearing capacity of the sandy soils. If the water table is above the level of the footing, it has to be lowered by pumping before placing the plate. The test should be performed at the water table level if it is within about 1 m below the footing.

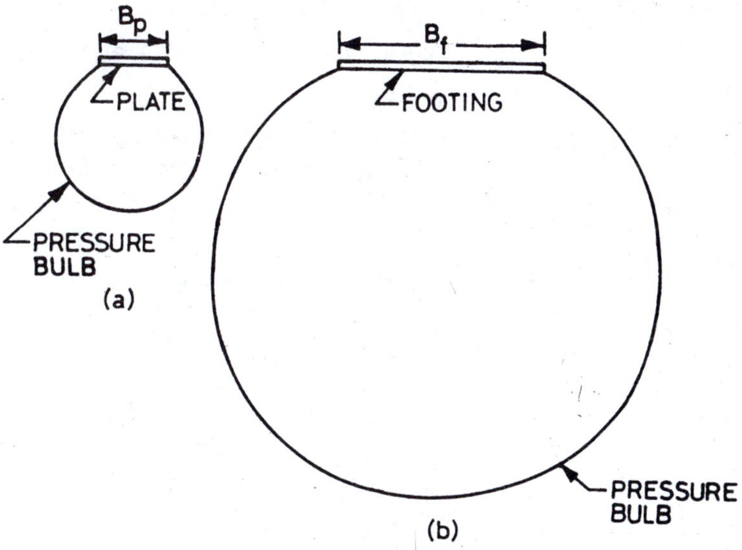

Fig. 23.29.

23.34. HOUSEL'S METHOD FOR DESIGN OF FOUNDATION

Housel's method can be used for the design of a shallow foundation for a given safe settlement. Two plate load tests are conducted, one with a plate of size B_1 and the other, with a plate size B_2. The load-settlement curves are obtained for both the tests. The loads corresponding to safe settlement s_1 are obtained from the load-settlement curves. Let Q_1 and Q_2 be the loads for the plates of size B_1 and B_2 respectively. Housel expressed these loads in the following forms.

$$Q_1 = A_1 m + P_1 n \qquad ...(23.91)$$

and

$$Q_2 = A_2 m + P_2 n \qquad ...(23.92)$$

where A_1, A_2 = areas of plates of size B_1 and B_2, respectively

P_1, P_2 = perimeters of plates of size B_1 and B_2, respectively.

and m, n = constants.

The values of the constants m and n are determined solving equations 23.91 and 23.92.

If A and P are, respectively, the area and perimeter of the given foundation, the safe load is then computed as

$$Q = Am + Pn \qquad ...(23.93)$$

The above load Q is for a safe settlement of s_1. For any other settlement s', the safe load is given

$$Q' = (Q/s_1) \times s' \qquad ...(23.94)$$

Eqs. 23.93 and 23.94 can also be used for the determination of the size of foundation for a given load, as both A and P depend upon the size.

ILLUSTRATIVE EXAMPLES

Illustrative Example 23.1. *Determine the ultimate bearing capacity of a strip footing, 1.20 m wide, and having the depth of foundation of 1.0 m. Use Terzaghi's theory and assume general shear failure. Take* $\phi' = 35°$, $\gamma = 18\ kN/m^3$, *and* $c' = 15\ kN/m^2$.

Solution. From Eq. 23.25, $q_u = c'\ N_c + \gamma D_f N_q + 0.5\, \gamma B N_\gamma$

For $\phi' = 35°$, Table 23.1 gives $N_c = 57.8$, $N_q = 41.4$ and $N_\gamma = 42.4$.

Now $q_u = 15.0 \times 57.8 + 18.0 \times 1.0 \times 41.4 + 0.5 \times 18.0 \times 1.2 \times 42.4$

$= \mathbf{2070\ kN/m^2}$

Illustrative Example 23.2. *Determine the allowable gross load and the net allowable load for a square footing of 2m side and with a depth of foundation of 1.0 m. Use Terzaghi's theory and assume local shear failure. Take a factor of safety of 3.0. The soil at the site has $\gamma = 18\ kN/m^3$, $c' = 15\ kN/m^2$ and $\phi' = 25°$.*

Solution. From Table 23.1, for $\phi' = 25°$

$$N_c' = 14.8,\quad N_q' = 5.6 \quad \text{and} \quad N_\gamma' = 3.2$$

From Eq. 23.37, taking $c_m' = 2/3\ c' = 10\ \text{kN/m}^2$

$$q_u = 1.2 \times 10.0 \times 14.8 + 18 \times 1.0 \times 5.6 + 0.4 \times 18 \times 2 \times 3.2$$

$$= 325\ \text{kN/m}^2$$

From Eq. 23.1, $q_{nu} = 325 - 18 \times 1.0 = 307\ \text{kN/m}^2$

From Eq. 23.2, $q_{ns} = \dfrac{q_{nu}}{F} = \dfrac{307}{3.0} = 102.3\ \text{kN/m}^2$

Net allowable load $= 102.3 \times (2 \times 2) = 409.2\ \text{kN}$

From Eq. 23.3, $q_s = q_{ns} + \gamma D_f = 102.3 + 18 \times 1.0 = 120.3\ \text{kN/m}^2$

Gross allowable load $= 120.3 \times (2 \times 2) = \mathbf{481.2\ kN}$

Illustrative Example 23.3. *A footing 2 m square is laid at a depth of 1.3 m below the ground surface. Determine the net ultimate bearing capacity using IS code method. Take $\gamma = 20\ kN/m^3$, $\phi' = 30°$ and $c' = 0$.*

Solution. For $\phi' = 30°$, Table 23.6 gives

$$N_c = 30.14,\quad N_q = 18.4 \quad \text{and} \quad N_\gamma = 22.4$$

From Table 23.3, $s_c = 1.3$, $s_q = 1.2$ and $s_\gamma = 0.80$

From Eq. 23.49 (*a*), $d_c = 1 + 0.2\,(D_f/B)\tan(45° + \phi'/2)$

$= 1 + 0.2 \times (1.3/2.0)\tan 60° = 1.23$

From Eq. 23.49 (*c*), $d_q = d_\gamma = 1 + 0.1\,(D_f/B)\tan(45° + \phi'/2) = 1.11$

From Eq. 23.48, $q_{nu} = cN_c\,s_c\,d_c\,i_c + q\,(N_q - 1)\,s_q\,d_q\,i_q + 0.5\,\gamma\,B\,N_\gamma\,s_\gamma\,d_\gamma\,i_\gamma W'$

$= 0.0 + 1.3 \times 20 \times (18.4 - 1) \times 1.2 \times 1.11 \times 1.0 + 0.5 \times 20 \times 2.0 \times 22.4 \times 0.8 \times 1.11 \times 1.0$

or $q_{nu} = \mathbf{1000\ kN/m^2}$

Illustrative Example 23.4. *Determine the net ultimate bearing capacity of the footing in Illustrative Example 23.3 if*

(a) the water table rises to the level of the base,

(b) the water table rises to the ground surface, and

(c) the water table is 1 m below the base.

Solution: (*a*) $W' = 0.50$, Therefore, Eq. 23.48 gives

$$q_{nu} = 1.3 \times 20.0 \times (18.4 - 1) \times 1.2 \times 1.11 \times 1.0 + 0.5 \times 20.0 \times 2.0 \times 22.4 \times 0.8 \times 1.11 \times 0.5$$

$$= \mathbf{801\ kN/m^2}$$

(*b*) $W' = 0.50$. The surcharge q is also reduced as the effective stress is reduced, Thus

$$q_{nu} = 1.3 \times (20 - 9.81) \times (18.4 - 1) \times 1.2 \times 1.11 \times 1.0 + 0.5 \times 20 \times 2.0 \times 22.4 \times 0.8 \times 1.11 \times 0.5$$

$$= \mathbf{506\ kN/m^2}$$

(c) W' is obtained by linear interpolation [see Eq. 23.36 (c)].

$$W' = 0.5 + \frac{0.5 \times 1.0}{2.0} = 0.75$$

Therefore,

$$q_{nu} = 1.3 \times 20.0 \times (18.4 - 1) \times 1.2 \times 1.11 \times 1.0 + 0.5 \times 20.0 \times 2.0 \times 22.4 \times 0.8 \times 1.11 \times 0.75$$

$$= \mathbf{901\ kN/m^2}$$

Illustrative Example 23.5. *A square column foundation is to be designed for a gross allowable total load of 250 kN. If the load is inclined at an angle of 15° to the vertical, determine the width of the foundation. Take a factor of safety of 3.0 and use Vesic's equation.* $\gamma = 19\ kN/m^3$, $\phi' = 35°$, *and* $c' = 5\ kN/m^2$. *The depth of foundation is 1.0 m.*

Solution. From Eq. 23.45,

$$q_u = c' N_c s_c d_c i_c + qN_q s_q d_q i_q + 0.5 \gamma B N_\gamma s_\gamma d_\gamma i_\gamma$$

From Table 23.6, $N_c = 46.12$, $N_q = 33.30$ and $N_\gamma = 48.03$

From Table 23.7, $s_c = 1 + 33.30/46.12 = 1.72$

$s_q = 1 + \tan 35° = 1.70$, $s_\gamma = 0.60$

From Eq. 23.46 (a), $d_c = 1 + 0.4 \times 1.0/B$

From Eq. 23.46 (b), $d_q = 1 + 2 \tan 35° (1 - \sin 35°)^2 \times 1.0/B$

or $d_q = 1 + 0.255/B$; $d_\gamma = 1.0$

From Eq. 23.47 (a), $i_c = i_q = (1 - \alpha°/90)^2 = 0.694$

From Eq. 23.47 (b), $i_\gamma = (1 - \alpha°/\phi)^2 = 0.327$

Therefore,

$$q_u = 5.0 \times 46.12 \times 1.72 \times (1 + 0.4/B) \times 0.694 + (19 \times 1.0)\, 33.3 \times 1.7 \times (1 + 0.255/B) \times 0.694 + 0.5 \times 19 \times B \times 48.03 \times 0.6 \times 1.0 \times 0.327$$

$$= 1022.2 + 300.7/B + 89.5\,B$$

From Eq. 23.1,

$$q_{nu} = q_u - \gamma D_f = q_u - 19 \times 1.0$$

$$= 1003.2 + \frac{300.7}{B} + 89.5\,B$$

From Eq. 23.3,

$$q_s = \frac{q_{nu}}{3} + 19 \times 1.0$$

$$= 334.4 + \frac{100.2}{B} + 29.8\,B + 19.0$$

Now gross load $= q_s \times B^2$

or $250.0 = 353.4\,B^2 + 100.2\,B + 29.8\,B^3$

Solving by trial and error, $B = \mathbf{0.7\ m}$

Illustrative Example 23.6. *Determine the ultimate bearing capacity of a square footing 2 m × 2m in a soil with unit weight of 18* kN/m^3, $\phi' = 20°$, $c = 20\ kN/m^2$. *Take the depth of foundation of 1.50 m. Use Hansen's equation.*

Solution. From Eq. 23.42,

$$q_u = cN_c s_c d_c i_c + qN_q s_q d_q i_q + 0.5 \gamma B N_\gamma s_\gamma d_\gamma i_\gamma$$

From Table 23.2, $N_c = 14.83$, $N_q = 6.40$ and $N_\gamma = 3.54$

From Table 23.3, $s_c = 1.2$, $s_q = 1.2$ and $s_\gamma = 0.6$

From Table 23.4, $d_c = 1 + 0.3 \times 1.5/2.0 = 1.225$

$d_q = d_c = 1.225$, $d_\gamma = 1.0$

As $i_c = i_c = i_\gamma = 1.0$,

$$q_u = 20.0 \times 14.83 \times 1.2 \times 1.225 \times 1.0 + (18 \times 1.50) \times 6.40 \times 1.2 \times 1.225 \times 1.0 + 0.5 \times 18 \times 2 \times 3.54 \times 0.6 \times 1.0 \times 1.0$$

$$= \mathbf{728.25\ kN/m^2}$$

Illustrative Example 23.7. *A strip footing of 2 m width is founded at a depth of 4 m below the ground surface. Determine the net ultimate bearing capacity, using (a) Terzaghi's equation, (b) Skempton's equation, and (c) IS Code. The soil is clay ($\phi = 0$, $c = 10$ kN/m^2). The unit weight of the soil is 20 kN/m^3.*

Solution. (*a*) Terzaghi's equation

From Eq. 23.25, $q_u = c_u N_c + \gamma D_f N_q + 0.5 \gamma B N_\gamma$

Taking the values from Table 23.1, $q_u = 10 \times 5.7 + 20 \times 4 \times 1.0 + 0.5 \times 20 \times 2 \times 0.0 = 137$

Therefore, $q_{nu} = q_u - \gamma D_f = 137.0 - 20 \times 4 = \mathbf{57.0\ kN/m^2}$

(*b*) Skempton's equation

From Eq. 23.53 (*a*), for $\frac{D_f}{B} = \frac{4}{2} < 2.5$,

$$N_c = 5.0 (1 + 0.2 D_f/B)(1 + 0.2 B/L)$$

$$= 5.0 (1 + 0.2 \times 4/2)(1 + 0.2 \times 0.0) = 7$$

From Eq. 23.55, $q_{nu} = c_u N_c = 10 \times 7.0 = \mathbf{70.0\ kN/m^2}$

(*c*) IS Code

From Eq. 23.56, $q_{nu} = c_u N_c s_c d_c i_c$

Taking $N_c = 5.14$, $q_{nu} = 10 \times 5.14 \times (1 + 0.2 \times \left(\frac{D_f}{B}\right) \tan 45°) \times 1.0$

$$= 51.4 \times 1.4 = \mathbf{71.96\ kN/m^2}$$

Illustrative Example 23.8. *A square footing (1.5 m × 1.5 m) is located at a depth of 1.0 m in a clay deposit consisting of two layers. The top layer is 1m thick and has $c_1 = 150$ kN/m^2 and $\gamma_1 = 16$ kN/m^3. The bottom layer has $c_2 = 50$ kN/m^2 and $\gamma = 15$ kN/m^2. Determine the net ultimate bearing capacity.*

Solution. From Eq. 23.59, taking $i_c = 1.0$, $q_u = c_1 N_c s_c d_c + q$

From Fig. 23.18, for $c_2/c_1 = 1/3$ and $Z/B = 1.0/1.5 = 0.67$, the value of N_c is equal to 3.50.

$$s_c = 1 + (B/L)(N_q/N_c) = 1 + 1 \times 1/3.5 = 1.29$$

$$d_c = 1 + 0.4 \times 1.0/1.50 = 1.27$$

Therefore, $q_u = 150 \times 3.5 \times 1.29 \times 1.27 + 16.0 \times 1.0 = 876.1$ kN/m^2

$$q_{nu} = 876.1 - 16 = \mathbf{860.1\ kN/m^2}$$

Illustrative Example 23.9. *A square footing (1.5 m × 1.5 m) is located at a depth of 1.0 m. The footing is subjected to an eccentric load of 400 kN, with an eccentricity of 0.2 m along one of the symmetrical axes. Determine the factor of safety against bearing failure. Use Vesic's equation. Take $\gamma = 21$ kN/m^3, $c = 100$ kN/m^2, $\phi = 0$.*

Solution. Effective width $B' = B - 2 e_b = 1.5 - 2 \times 0.2 = 1.1$ m

From Eq. 23.45, taking $N_\gamma = 0.0$, $N_c = 5.14$ and $N_q = 1.0$,

$$q_u = c N_c s_c d_c i_c + q N_q s_q d_q i_q$$

where $s_c = 1 + (B'/L)(N_q/N_c) = 1 + (1.1/1.50) \times 1.0/5.14 = 1.14$

$s_q = 1 + (B'/L) \tan \phi = 1.0 + (1.1/1.50) \tan 0° = 1.00$

$$d_c = 1 + 0.4\,(D_f/B) = 1 + 0.4 \times 1.0/1.5 = 1.27$$

$$d_q = 1 + 2 \tan\phi\,(1 - \sin\phi)^2\,(D_f/B) = 1.0$$

Therefore, $q_u = 100 \times 5.14 \times 1.14 \times 1.27 + 1.0 \times (21.0 \times 1.0) \times 1.0 \times 1.0$

$= 744.2 + 21.0 = 765.2$ **kN/m²**

From Eq. 23.63 (*a*), $$q_{max} = \frac{Q}{BL}\left(1 + \frac{6e}{B}\right) = \frac{400}{1.5 \times 1.5}\left(1 + \frac{6 \times 0.2}{1.5}\right)$$

or $q_{max} = 320 \text{ kN/m}^2$ ...(*a*)

Now $q_{nu} = 765.2 - 21 \times 1 = 744.2.$

Therefore, $$q_s = \left(\frac{744.2}{F} + 21.0\right) \quad ...(b)$$

From Eqs. (*a*) and (*b*), $744.2/F + 21.0 = 320$

or $F = \mathbf{2.49.}$

Illustrative Example 23.10. *A square footing is required to carry a net load of 1200 kN. Determine the size of the footing if the depth of foundation is 2 m and the tolerable settlement is 40 mm. The soil is sandy with N = 12. Take a factor of safety of 3.0. The water table is very deep. Use Teng's equation.*

Solution. From Eq. 23.61,

$$q_{nu} = 0.33\,N^2\,BW_\gamma + 1.0(100 + N^2)\,D_f\,W_q$$

or $q_{nu} = 0.33\,(12)^2 B \times 1.0 + 1.00\,(100 + 12^2) \times 2 \times 1.0$

or $q_{nu} = 47.5\,B + 488.0$

Total net load, $Q_n = (47.5\,B + 488.0)/3 \times B^2$

or $1200 = (47.5B + 488.0)/3 \times B^2$

or $1200 = 15.8\,B^3 + 162.7\,B^2$

Solving, by trial and error, $B = 2.45$ m.

From Eq. 23.80 (*b*), $$q_{np} = 1.40\,(N - 3)\left(\frac{B + 0.3}{2B}\right)^2 W_\gamma R_d\,s$$

or $$q_{np} = 1.40\,(N - 3)\left(\frac{B + 0.3}{2B}\right)^2\left(1 + \frac{2}{B}\right) \times 40$$

or $$q_{np} = 1.40\,(12 - 3)\left(\frac{B + 0.3}{2B}\right)^2 \times 40\,(1 + 0.4/B)$$

or $= 126\,(B + 0.3)^2 \times (1 + 0.4/B)$

Now $Q_n = q_{np} \times B^2$

or $1200 = 126\,(B + 0.3)^2\,(1 + 2/B)$

Solving, $B = 1.90$ m. Adopt $B = \mathbf{2.0\ m}$

Illustrative Example 23.11. *A rectangular footing (3 m × 2 m) exerts a pressure of 100 kN/m² on a cohesive soil ($E_s = 5 \times 10^4$ kN/m² and $\mu = 0.50$). Determine the immediate settlement at the centre, assuming (a) the footing is flexible, (b) the footing is rigid.*

Solution. From Eq. 23.68, $$s_i = qB\left(\frac{1 - \mu^2}{E_s}\right) I$$

As $L/B = 3/2 = 1.5$, from Table 23.8, $I = 1.36$.

Therefore, $$s_i = 100 \times 2\left(\frac{1 - 0.5^2}{5 \times 10^4}\right) \times 1.36 \times 10^{-3} = 4.08 \text{ mm}$$

(*b*) For rigid footing ($I = 1.06$), $$s = (1.06/1.36) \times s_i$$
$$= 1.06/1.36 \times 4.08 = \mathbf{3.18\ mm}$$

Illustrative Example 23.12. *Fig. E-23.12 shows a square footing resting on a sand deposit. The pressure at the level of the foundation ($\bar{q}$) is 200 kN/m². The figure also shows the variation of the elastic modulus with depth. Determine the settlement of the foundation after 6 years of construction.*

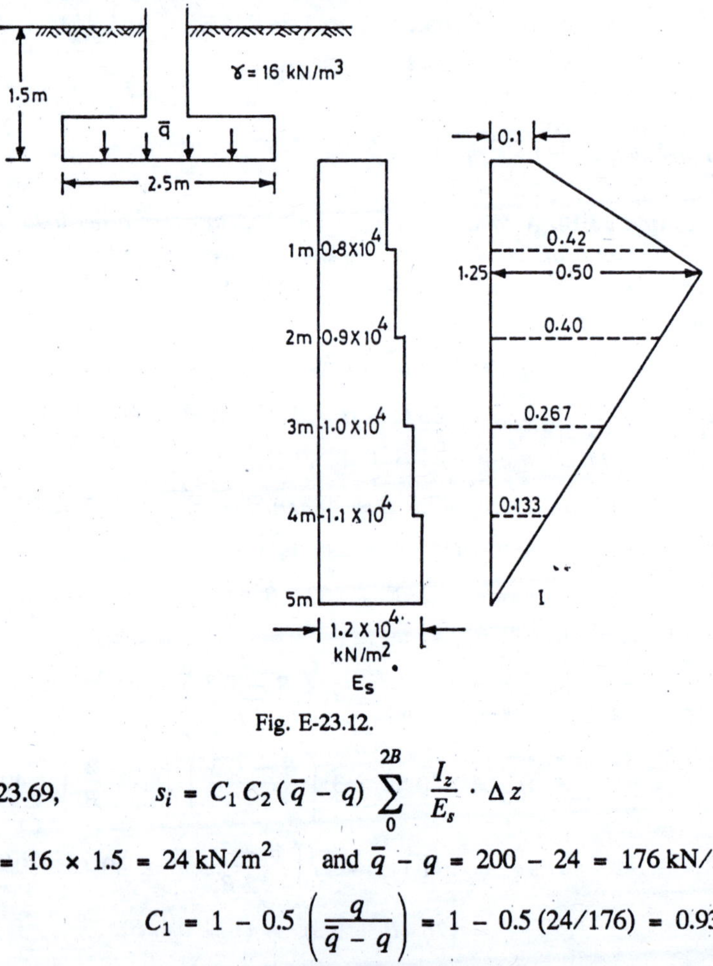

Fig. E-23.12.

Solution. From Eq. 23.69, $$s_i = C_1 C_2 (\bar{q} - q) \sum_0^{2B} \frac{I_z}{E_s} \cdot \Delta z$$

$$q = 16 \times 1.5 = 24 \text{ kN/m}^2 \quad \text{and } \bar{q} - q = 200 - 24 = 176 \text{ kN/m}^2$$

$$C_1 = 1 - 0.5\left(\frac{q}{\bar{q} - q}\right) = 1 - 0.5\,(24/176) = 0.932$$

$$C_2 = 1 + 0.2 \log_{10}(t/0.1) = 1 + 0.2 \log_{10}(6/0.1) = 1.356$$

Therefore, $$s_i = 0.932 \times 1.356 \times 176 \sum_0^{2B} \frac{I_z}{E_s} \cdot \Delta z$$

$$= 222.4 \sum_0^{2B} \frac{I_z}{E_s} \cdot \Delta z$$

The value of $\sum_0^{2B} (I_z/E_s) \cdot \Delta z$ is determined as shown in the table below. It is equal to 13.97×10^{-5}.

z	Δz	E_s (kN/m^2)	I_z	$(I_z/E_s) \cdot \Delta z$
0—1.00	1.0 m	8000	$\frac{0.1 + 0.42}{2} = 0.26$	3.25×10^{-5}
1.0—2.0	"	9000	0.453	5.03×10^{-5}
2.0—3.0	"	10000	0.333	3.33×10^{-5}
3.0—4.0	"	11000	0.200	1.82×10^{-5}
4.0—5.0	"	12300	0.067	0.54×10^{-5}
				$\Sigma\ 13.97 \times 10^{-5}$

Therefore, $s_i = 222.4 \times 13.97 \times 10^{-5}$ m

or $s_i =$ **31.07 mm**

Illustrative Example 23.13. *Fig. E-23.13 shows the load- settlement curve obtained from a plate load test conducted on a sandy soil. The size of the plate used was 0.3 m × 0.3 m. Determine the size of a square column footing to carry a net load of 3000 kN with a maximum settlement of 25 mm.*

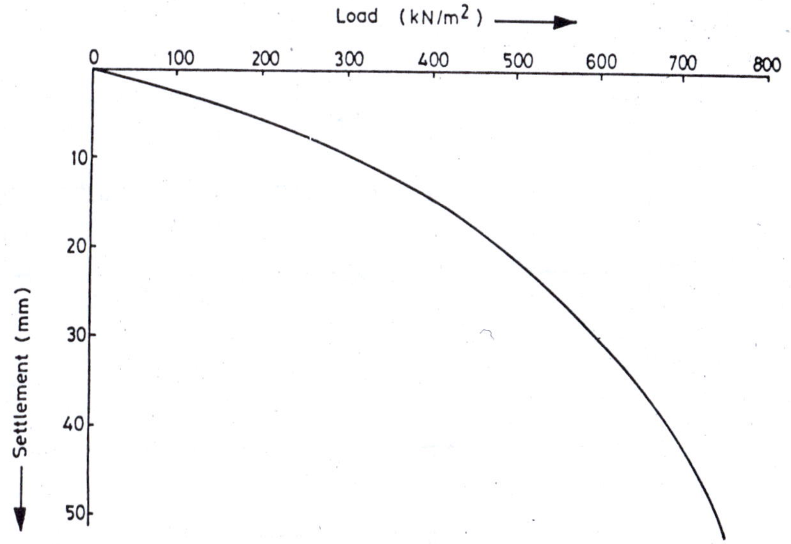

Fig. E-23.13.

Solution. From Eq. 23.88,

$$s_f = s_p \left[\frac{B_f (B_p + 0.3)}{B_p (B_f + 0.3} \right]^2$$

or

$$s_f = s_p \left(\frac{B_f}{B_p} \right)^2 \left(\frac{0.6}{B_f + 0.3} \right)^2$$

The value of B_f is found by trial and error, as shown in the table below.

B_f	$q_0 = \frac{Q}{(B_f)^2}$	s_p from Fig. Ex. 23.13	B_f/B_p	s_f from Eq. (a)
3.80 m	207.7	6 mm	12.67	20.62 mm
3.6 m	231.5	7 mm	12.00	23.85 mm
3.55 m	238.0	7.3 mm	11.83	24.81 mm

Adopt a size of 3.55 m × 3.55 m.

Illustrative Example 23.14. *Two-plate load tests at a site gave the following results.*

Size of plate	*Load*	*Settlement*
0.305 × 0.305 m	*40 kN*	*25 mm*
0.61 × 0.61 m	*40 kN*	*15 mm*

(a) Assuming Poisson's ratio as 0.3, determine the deformation modulus of the soil.

(b) If there are two columns, one of the size 2.5 m × 2.5 m, carrying a load of 2700 kN, and the other of size 3m × 3m, carrying a load of 3900 kN, determine the differential settlement. The columns are 7 m apart.

Solution. (*a*) For the first test, $q_1 = \dfrac{40}{0.305 \times 0.305} = 430 \text{ kN/m}^2$

$$q_1 B_1 = 430 \times 0.305 = 131.1 \text{ kN/m}^2$$

For the second test, $q_2 = \dfrac{40}{0.61 \times 0.61} = 107.5 \text{ kN/m}^2$

$$q_2 B_2 = 107.5 \times 0.61 = 65.6 \text{ kN/m}^2$$

Fig. E-23.14 shows the plot between qB and s.

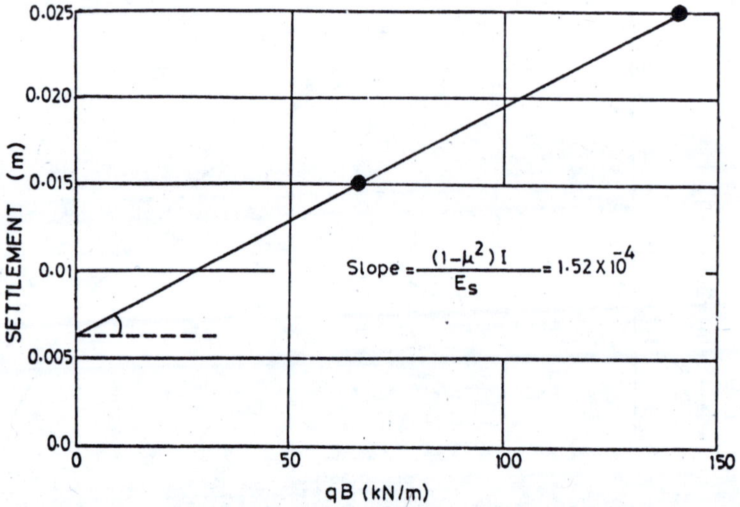

Fig. E-23.14.

From the plot, $\left(\dfrac{1-\mu^2}{E_s}\right) I = 1.52 \times 10^{-4}$

From Table 23.8, $I = 1.12$

As the plate is rigid, $I = 0.8 \times 1.12 = 0.89$

Therefore, $E_s = \dfrac{(1-\mu^2) \times 0.896}{1.52 \times 10^{-4}} = \dfrac{(1-0.3^2)}{1.52 \times 10^{-4}} \times 0.896$

$$E_s = \mathbf{5364 \text{ kN/m}^2}$$

(*b*) For the first column, $q_1 = \dfrac{2700}{2.5 \times 2.5} = 432 \text{ kN/m}^2$

For the second column, $q_2 = \dfrac{3900}{3 \times 3} = 433 \text{ kN/m}^2$

As the settlement of the plate (0.305 m × 0.305 m) at a load intensity of 430 kN/m^2 is 25 mm, it can be used for the determination of the settlement of columns.

Form Eq. 23.88,
$$(s_f)_1 = 25\left[\frac{2.5 \times (0.305 + 0.30)}{0.305 \times (2.5 + 0.3)}\right]^2 = 78.42 \text{ mm}$$
$$(s_f)_2 = 25\left[\frac{3\,(0.305 + 0.30)}{0.305\,(3.0 + 0.3)}\right]^2 = 81.3 \text{ mm}$$

Differential settlement = 81.3 − 78.42 = **2.88 mm**

Illustrative Example 23.15. *The results of two plate load tests for a settlement of 25.4 mm are given.*

Plate diameter	*Load*
0.305 m	31 kN
0.61 m	65 kN

A square column foundation is to be designed to carry a load of 800 kN with an allowable settlement of 25.4 mm. Determine the size using Housel's method.

Solution. From Eq. 23.91 and 23.92,

$$31.0 = (\pi/4) \times (0.305)^2 \times m + \pi\,(0.305) \times n \qquad ...(a)$$
$$65.0 = (\pi/4) \times (0.61)^2 \times m + \pi\,(0.61) \times n \qquad ...(b)$$

Eq.(*a*) can be written as
$$62 = 2 \times \pi/4\,(0.305)^2 \times m + 2\pi\,(0.305) \times n \qquad ...(c)$$

From Eqs.(*b*) and (*c*), by subtraction,
$$3.0 = m\,[\pi/4\,(0.372 - 0.186)] \quad \text{or} \quad m = 20.55$$

From Eq. (*a*)
$$31.0 = 1.5 + 0.9577\,n \quad \text{or} \quad n = 30.80$$

From Eq. 23.93,
$$Q = B^2 \times 20.55 + (30.8 \times 4B)$$

or
$$800 = 20.55\,B^2 + 123.2\,B$$

or
$$B = 3.93 \text{ m say } \mathbf{4\,m \times 4\,m.}$$

PROBLEMS

A. Numericals

23.1. Determine the ultimate bearing capacity of a square footing of size 1.2 m if the depth of foundation is 1 m. Take $\phi' = 25°$, $\gamma = 18$ kN/m^3 and $c = 15$ kN/m^2. Use Vesic's equation. **[Ans. 1050 kN]**

23.2. A circular foundation is of 2.4 m diameter. If the depth of foundation is 1m, determine the net allowable load. Take $\gamma = 19$ kN/m^3, $c' = 30$ kN/m^2, $\phi' = 15°$ and factor of safety as 3.0. Use Terzaghi's equation and assume local shear failure. **[Ans. 418 kN]**

23.3. A square footing is to be designed to carry a load of 500 kN. If the depth of foundation is 1.5 m, determine a suitable size of foundation with a factor of safety of 3.0. The water table is at foundation level. Take $\phi' = 25°$, $\gamma = 16$ kN/m^3, $\gamma_{sat} = 19$ kN/m^3. Use Terzaghi's theory. $c' = 20$ kN/m^2. **[Ans. 2.10 m]** Assume local shear failure.

23.4. A strip footing is required to carry a net load of 1000 kN at a depth of 1m. Taking a factor of safety of 3, determine the width of the footing. Take $\phi = 30°$. $\gamma = 19$ kN/m^3, $c = 20$ kN/m^2. Use Terzaghi's theory. Assume general shear failure **[Ans. 2.4 m]**

23.5. A strip footing is 2 m wide and founded at a depth of 2m in a soil of unit weight 20 kN/m^3 and a cohesion of 10 kN/m^2. Determine the increase in the bearing capacity when ϕ is increased from 20° to 25°. Use Terzaghi's equation. Assume local shear failure. **[Ans. 130 kN/m^2]**

23.6. A purely cohesive soil has a unit weight of 20 kN/m^3 and a cohesion of 150 kN/m^2. Determine the safe bearing capacity for a rectangular footing 8m × 2m founded at a depth of 4 m in clay. (F.S. = 3.0) **[Ans. 448 kN/m^2]**

23.7. A strip footing 2 m wide is to be laid at a depth of 4 m in a purely cohesive soil ($c = 150$ kN/m^2. $\gamma = 19$

kN/m^3). Determine the ultimate bearing capacity from (*a*) Terzaghi's theory (*b*) Skempton's theory.

[**Ans.** 930 kN/m^2; 1125 kN/m^2]

23.8. Estimate the immediate settlement of a concrete footing, 1m × 2m size, founded at a depth of 1 m in a soil with $E = 10^4$ kN/m^2, $\mu = 0.3$. The footing is subjected to a pressure of 200 kN/m^2. Assume the footing to be rigid.

[**Ans.** 22.2 mm]

23.9. A square footing 2.5 m size is founded at a depth of 1.5 m in a sandy soil deposit which has the corrected N value of 30. The water table is at a depth of 2 m from the ground surface. Find the net allowable soil pressure if

(*a*) the desired factor of safety is 3.0.

(*b*) the permissible settlement is 40 mm.

Use Teng's equations. [**Ans.** 605 kN/m^2, 456.5 kN/m^2]

23.10. An excavation 3 m wide is to be made to depth of 5 m in soft clay (c_u = 15 kN/m^2, γ = 19 kN/m^3). The ground surrounding the excavation carries a surcharge of 10 kN/m^2. Determine the factor of safety. Use Terzaghi's equations. [**Ans.** 1.23]

B. Descriptive and Objective Types

23.11. Define the following terms :

(*a*) Net safe bearing capacity,

(*b*) Gross safe bearing capacity,

(*c*) Allowable soil pressure.

23.12. What are the assumptions made in the derivation of Terzaghi's bearing capacity theory ? Write the equation for the ultimate bearing capacity.

23.13. Differentiate between the general shear failure and the local shear failure. How the ultimate bearing capacity in local shear is determined ?

23.14. Discuss the effect of water table on the bearing capacity of the soil.

23.15. Discuss Meyerhof's bearing capacity theory. How does it differ from Terzaghi's theory?

23.16. Write short notes on:

(*a*) Hansen's bearing capacity theory,

(*b*) Vesic's bearing capacity theory.

23.17. Describe Skempton's analysis for bearing capacity of cohesive soils.

23.18. How the bearing capacity of footing on a layered cohesive soil deposit is determined ?

23.19. Discuss the various methods of determination of the allowable soil pressure. What are their limitations ?

23.20. What are different types of settlements which can occur in a foundation ? How are these estimated ?

23.21. Discuss the methods for estimating immediate settlements of foundations on clay.

23.22. How would you estimate the settlements of a foundation on cohesionless soils ?

23.23. Describe plate-load test. What are its limitation and used ?

23.24. Write whether the following statements are true or false.

(*a*) A foundation is considered shallow if its depth is less than 1 m.

(*b*) Terzaghi's theory is applicable when the base of the footing is rough.

(*c*) For footings on cohesionless soils, the bearing capacity, and not the settlement, generally governs the design.

(*d*) The plate-load test is more useful for cohesionless soils than for cohesive soils.

(*e*) The maximum value of N_c is 9.0 by Skempton's chart.

(*f*) The presumptive bearing capacity is the same as the allowable bearing capacity.

(*g*) The immediate settlement of a rigid footing is approximately equal to the average settlement of the flexible footing.

(*h*) The allowable angular distortion in case of steel structures is more than that for R.C.C. structures.

(*i*) The differential settlement is generally three-fourths of the maximum settlement.

[**Ans.** True (*b*), (*d*), (*e*), (*g*), (*h*), (*i*)]

C. Multiple Choice Questions

1. A shallow foundation is usually defined as a foundation which has

(*a*) depth less than 0.6 m
(*b*) depth less than its width
(*c*) depth less than 1.0 m.
(*d*) none of above

2. The ultimate bearing capacity of a shallow foundation on sand is reduced to about ... when the water table rises to the ground surface.
(*a*) 75% (*b*) 50%
(*c*) 25% (*d*) none of above.

3. The allowable soil pressure for foundations in cohesive soils is generally controlled by
(*a*) settlements (*b*) bearing capacity
(*c*) both (*a*) and (*b*) (*d*) neither (*a*) nor (*b*)

4. The immediate settlement of a rigid footing is about ... times the maximum settlement of an equal flexible footing.
(*a*) 0.9 (*b*) 0.8
(*c*) 0.7 (*d*) 0.6

5. The bearing capacity of soil supporting a footing of size 3m × 3m will not be affected by the presence of water table located at a depth below the base of footing of
(*a*) 1.0 (*b*) 1.50 m
(*c*) 3.0 m (*d*) 6.0 m

6. A 2 m wide strip footing rests at a depth of 2 m below the ground surface where water table is at the ground surface. The ultimate load which the strip can carry according to Terzaghi's theory when γ_{sat} = 20 kN/m^3 and c = 30 kN/m^2 is about
(*a*) 171 kN/m (*b*) 342 kN/m
(*c*) 422 kN/m (*d*) 262 kN/m

7. The permissible settlement is the maximum in the case of
(*a*) Isolated footing on clay (*b*) Raft on clay
(*c*) Isolated footing on sand (*d*) Raft on sand

8. If the gross bearing capacity of a strip footing 1.5 m wide located at a depth of 1 m in clay is 400 kN/m^2, its net bearing capacity for γ = 20 kN/m^2 is
(*a*) 370 kN/m^2 (*b*) 380 kN/m^2
(*c*) 390 kN/m^2 (*d*) 360 kN/m^2

[**Ans.** 1. (*b*), 2. (*b*), 3. (*b*), 4. (*b*), 5. (*c*), 6. (*b*), 7. (*b*), 8. (*b*)]

24

Design of Shallow Foundations

24.1. TYPES OF SHALLOW FOUNDATIONS

A shallow foundation, according to Terzaghi, is one whose width is greater than its depth (*i.e.* $D_f/B \leq 1$). Shallow foundations are located just below the lowest part of the wall or a column which they support. Footings are structural members, made of brick work, masonry or concrete, that are used to transmit the load of the wall or column such that the load is distributed over a large area. In fact, a footing is an enlargement of the base of the column or wall it supports. The footings are of the following types:

(1) Strip footing. A strip footing is provided for a load-bearing wall (Fig. 24.1). A strip footing is also provided for a row of columns which are so closely spaced that their spread footings overlap or nearly touch each other. In such a case, it is more economical to provide a strip footing than to provide a number of spread footings in one line. A strip footing is also known as *continuous footing*.

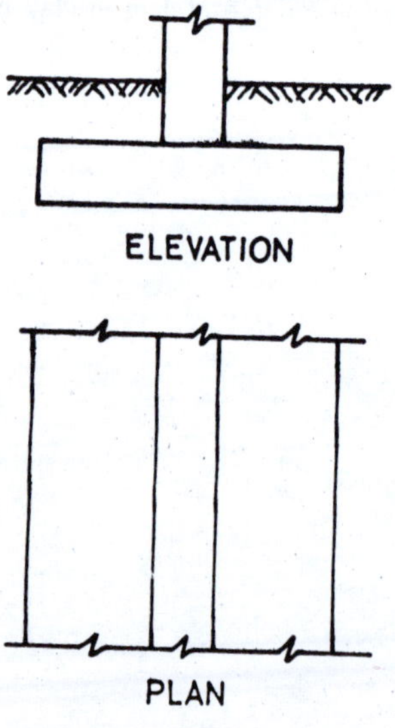

Fig. 24.1. Strip Footing.

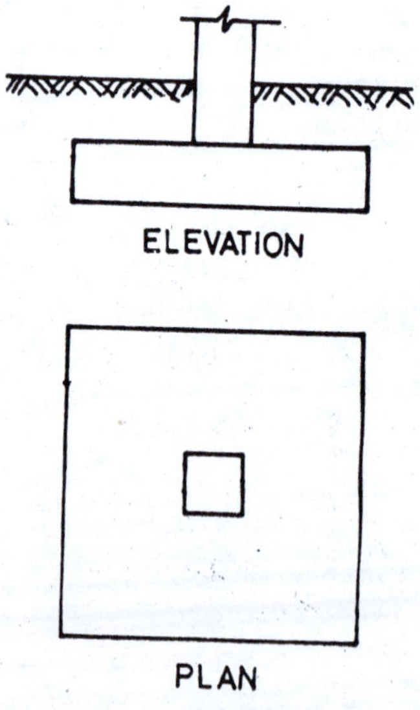

Fig. 24.2. Spread Footing.

(2) Spread or isolated footing. A spread (or isolated or pad) footing is provided to support an individual column (Fig. 24.2). A spread footing is circular, square or rectangular slab of uniform thickness. Sometimes, it is stepped or haunched to spread the load over a large area.

(3) Combined footing. A combined footing supports two columns (Fig. 24.3). It is used when the two columns are so close to each other that their individual footings would overlap. A combined footing is also provided when the property line is so close to one column that a spread footing would be eccentrically loaded when kept entirely within the property line. By combining it with that of an interior column, the load is evenly distributed.

A combined footing may be rectangular or trapezoidal in plan.

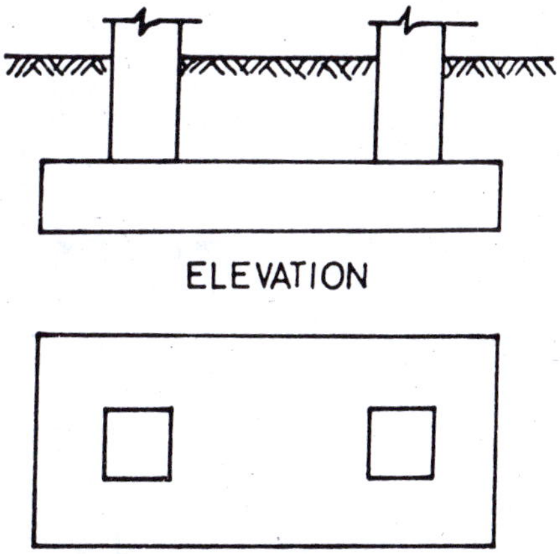

Fig. 24.3. Combined Footing.

(4) Strap or cantilever footing. A strap (or cantilever) footing consists of two isolated footings connected with a structural strap or a lever, as shown in Fig. 24.4. The strap connects the two footings such that they behave as one unit. The strap simply acts as a connecting beam and does not take any soil reaction. The strap is designed as a rigid beam.

The individual footings are so designed that their combined line of action passes through the resultant of the total load. A strap footing is more economical than a combined footing when the allowable soil pressure is relatively high and the distance between the columns is large.

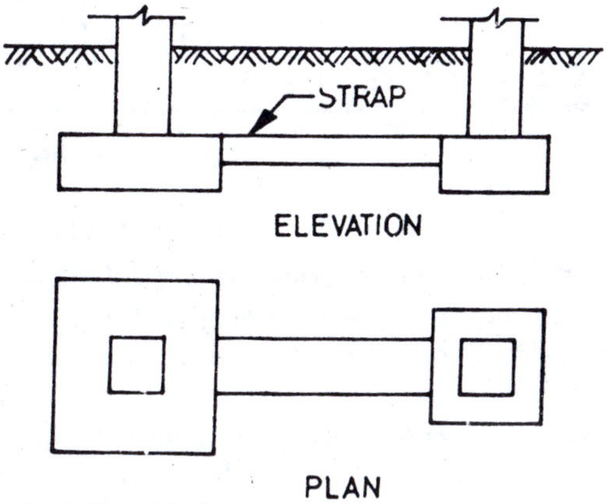

Fig. 24.4. Strap Footing.

(5) Mat or raft foundations. A mat or raft foundation is a large slab supporting a number of columns and walls under the entire structure or a large part of the structure. A mat is required when the allowable soil pressure is low or where the columns and walls are so close that individual footings would overlap or nearly touch each other (Fig. 24.5).

Mat foundations are useful in reducing the differential settlements on non-homogeneous soils or where there is a large variation in the loads on individual columns.

The bearing capacity theories have been discussed in the preceding chapter. The design of shallow foundations is discussed in this chapter. The design is limited to the determination of the depth of footing, area of footing, soil pressure, shear force and bending moments. The structural design of footings is outside the scope of this text.

24.2. DEPTH OF FOOTINGS

To perform its function properly, a footing must be laid at a suitable depth below the ground surface. The vertical distance between the ground surface and the base of footing is known as the depth of footing (D_f). The depth of footing controls the ultimate bearing capacity and the settlement, as discussed in chapter 23. While fixing the depth of footing, the following points should also be considered.

(1) Depth of top soil. The footing should be located below the top soil consisting of organic matters which eventually decompose. The top soil should be removed over an area slightly larger than the footing.

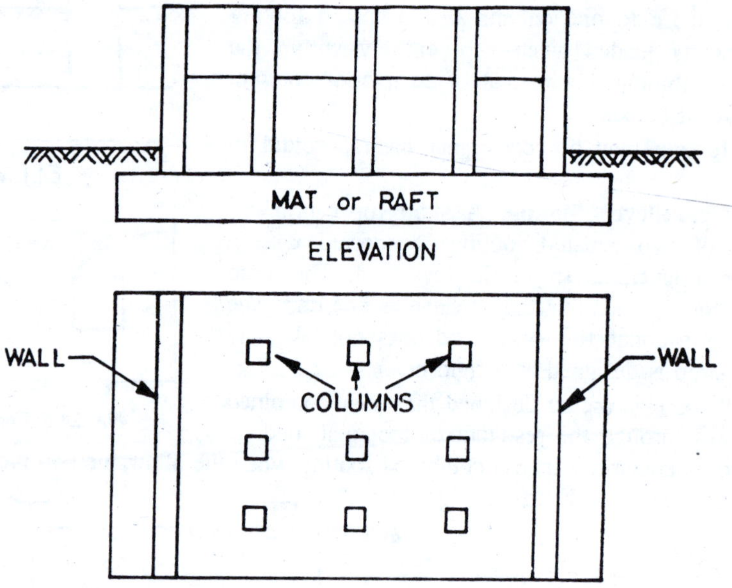

Fig. 24.5. Mat Foundation.

(2) Frost Depth. The footing should be carried below the depth of frost penetration. If the footing is located at insufficient depth, it would be subjected to the frost damage due to formation of ice lenses and consequent frost heave. During summer, thawing occurs from the top downwards and the melted water is entrapped.

As the soil water freezes and melts, the footing is lifted during cold weather and it settles during warm weather. The shear strength of the soil is also decreased during warm weather due to an increase in water content.

To prevent frost damage, the footings should be placed below the frost depth, which may be 1 m or more in cold climates.

(3) Zone of Soil Volume Change. Some soils, especially clays having high plasticity, such as black cotton soil, undergo excessive volume changes. Such soils shrink upon drying and swell upon wetting. The volume changes are generally greatest near the ground surface and decrease with increasing depth.

Large volume changes beneath a footing may cause alternate lifting and dropping. The footing should be placed below all strata that are subjected to large volume changes.

(4) Adjacent footings and property lines. The footing should be so located that no damage is done to the existing structures. The adjacent structures may be damaged by construction of a new footing due to vibrations, undermining or lowering of the water table. The new footing may also impose additional load on the existing footings which may cause settlement. In general, the deeper the new footing and the closer to the existing structure, the greater is the potential damage to the existing structure. This is particularly more severe if the new footing is lower than the existing footing.

As far as possible, the new footing should be placed at the same depth as the old ones, and the sides of excavation adjacent to the existing structure should be suitably supported.

If the footings are placed at different levels, the slope of line joining the two footings should not be steeper than two horizontal to one vertical (Fig. 24.6), as per IS : 1904—1978.

(5) Sloping Ground. If a footing is located adjacent to a sloping ground, the sloping ground surface

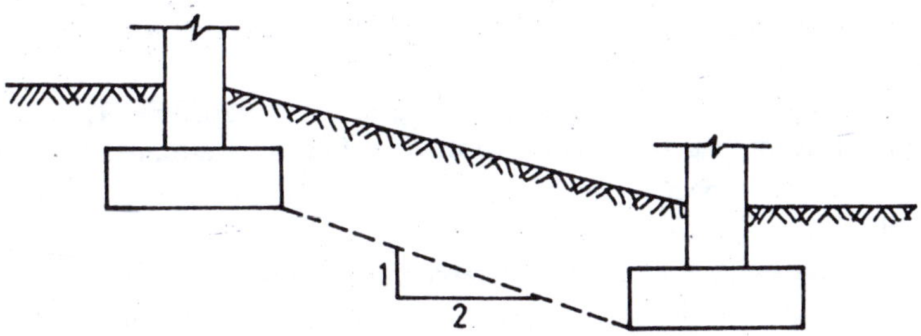

Fig. 24.6.

should not encroach upon a frustum of bearing material under the footing having sides making an angle of 30° with the horizontal. Moreover, the minimum distance from the lower edge of the footing to the sloping ground surface should be 90 cm. (Fig. 24.7).

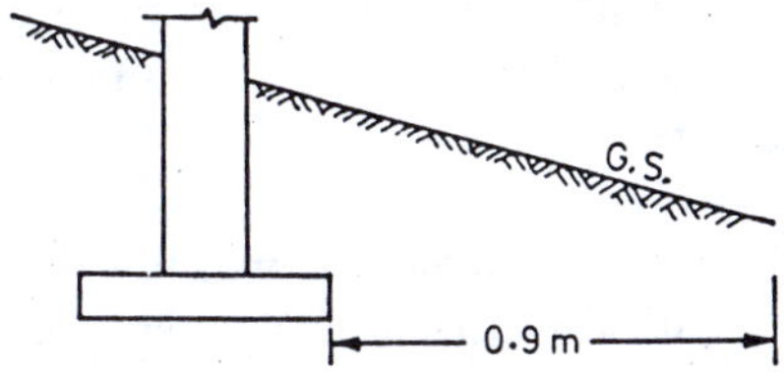

Fig. 24.7.

(6) Water Table. The footing should be placed above the ground water table as far as possible. The presence of ground water within the soil immediately around a footing in undesirable as it reduces the bearing capacity of the soil and there are difficulties during construction. The water proofing problems also arise due to dampness.

(7) Scour Depth. The footings located in streams, on water fronts or other locations where there is a possibility of scouring, should be placed below the potential scour depth.

(8) Underground Defects. The depth of footing is also affected by the presence of underground defects, such as faults, caves and mines. If there are man-made discontinuities, such as sewer lines, water mains, underground cables, these should be shifted or the footing relocated.

(9) Root holes. If there are root holes or cavities caused by burrowing animals or worms, the footing should be placed below such a zone of weakened soil.

(10) Minimum Depth. IS : 1904—1978 specifies that all foundations should extend to a depth of at least 50 cm below the natural ground surface. However, in case of rocks, only the top soil should be removed and the surface should be cleaned and if necessary, stepped.

Sometimes, the minimum depth of foundation is determined from Rankine's formula (Eq. 23.19).

$$(D_f)_{min} = \frac{q}{\gamma}\left(\frac{1 - \sin \varphi'}{1 + \sin \varphi'}\right)^2 \qquad \ldots(24.1)$$

where q is the intensity of loading.

24.3. FOUNDATION LOADING

An accurate estimation of all loads acting on the foundation should be made before it can be properly designed. A foundation may be subjected to one or more of the following loads.

(1) Dead Loads. The dead loads include the weight of materials permanently fixed to the structure, such as beams, floors, walls, columns, and fixed service equipments. The dead load can be calculated if sizes and types of structural materials are known. However, there is a problem in estimating the self-weight of the structure. The usual procedure is to assume the self weight initially and the structure is designed. The weight of the structure is then found from the designed dimensions and compared with the assumed weight. If necessary, the design procedure is repeated using the revised weight.

(2) Live Loads. The live loads are the movable loads that are not permanently attached to the structure. These loads are applied during a part of its useful life. Loads due to people, goods, furniture, equipment, machinery, etc. are the live loads.

It is difficult to estimate the live loads accurately. These are specified by local building codes as uniformly distributed equivalent static loads.

(3) Wind Loads. Wind loads act on all exposed surfaces of structure. These loads depend upon the velocity of wind and the type of structure. Like live loads, wind loads are also specified by building codes.

(4) Snow loads. Snow loads occur due to accumulation of snow on roofs and exterior flat surfaces in cold climates. The unit weight of snow is usually taken as 1 kN/m^3.

(5) Earth pressure. Earth pressures produce lateral force against the structure below the ground surface or fill surface. The earth pressure is determined using the theories discussed in chapter 19. The earth pressure is normally treated as dead load.

(6) Water pressure. Like earth pressure, water pressure also produces a lateral force against the structure below the water level.

Water pressure may also cause an upward force on the bottom of the structure due to uplift pressure. It must be counteracted by the dead load of the structure.

(7) Earthquake loads. The force due to an earthquake may act vertically, laterally or torsionally on a structure in any direction. The worst condition should be anticipated and the relevant code consulted.

The earthquake load is usually assumed as a fraction of the dead load, depending upon the seismicity of the zone.

Computation of Design loads. The dead loads and live loads on columns are usually computed by *tributary area method,* in which it is assumed that a column carries all the load in the floor area enclosed by lines equidistant from its adjacent columns.

As live loads are temporary and transients, only a part of it may act for a duration that may induce the settlement, especially in cohesive soils. Moreover, specified maximum live loads do not occur simultaneously at all the floors. It is usual practice to reduce the live loads in such cases (IS : 875).

It is the common practice to assume that the wind load and earthquake loads do not occur simultaneously.

According to IS 1904—1978, foundations should be proportioned for the following combinations of loads—

(*i*) Dead load + live load, and
(*ii*) Dead load + live load + wind load or seismic load.

The dead load includes the weight of column, wall, footings, foundations, the overlying fill but excludes the weight of the displaced soil. If V is the volume of footing, there is a net increase of load on foundation of $V(\gamma_c - \gamma)$, where γ_c is the unit weight of concrete and γ is the unit weight of soil. If the weight $V\gamma_c$ of the footing is included in the dead weight, the dead load needs a reduction equal to $V\gamma$, equal to the weight of the soil displaced.

If wind load (or seismic load) is less than 25% of that due to dead and live loads, it may be neglected and the foundation should be designed for combination (*i*) given above. However, if wind load (or seismic load) is more than 25% of that due to dead and live loads, the foundation should be designed for combination (*ii*) given above. The foundation pressure should not exceed the safe bearing capacity by more than 25% in the second case.

For foundations resting on coarse-grained soils, the settlements should be computed using the loads given in combination (*ii*), as settlements occur in a very short period. However, for fine-grained soils, the settlements should be computed corresponding to permanent loads. Generally, one half of the design live load is taken as permanent.

24.4. PRINCIPLES OF DESIGN OF FOOTINGS

Before actual design, it is essential to estimate the dead load, live load and other loads. The frequency and duration of various loads should also be available. The bending moment at the base of column (or wall) should also be ascertained if it is subjected to an eccentric load or moment.

As the bearing capacity of the soil depends upon the depth of footing and its length and width, an estimate about these dimensions is required before the actual design. For members carrying axial load combined with bending moments that do not change direction, a rectangular footing is more suitable than a square footing.

The investigation of the site should be first carried out. The samples should be taken to determine the engineering properties of the soil. The safe bearing capacity should be calculated on the basis of soil data obtained from the tests using the methods discussed in chapter 23. For cohesionless soils, as it is difficult to obtain undisturbed samples, the bearing capacity is determined from the standard penetration test number (N) or from the plate load test. The value of N to be used is the average of the N values from the base of footings to a depth equal to width of the footing.

The footing is designed using the following procedure.

(1) The safe bearing capacity is determined using the methods discussed in chapter 23. For small, unimportant structures, these values can be taken as presumptive values (Table 23.10).

(2) The footing is proportioned making use of the safe bearing capacity determined in Step (1).

(3) The maximum settlement of the footing is determined using the methods discussed in chapter 23. An estimate of the differential settlement between various footings is made.

(4) Angular distortion is determined between various parts of the structure.

(5) The maximum settlement, the differential settlement and the angular distortion obtained in the step (3) and (4) are compared with the given allowable values (Table 23.9).

(6) If the values are not within the allowable limits, the safe bearing capacity is revised and the procedure repeated.

(7) The stability of the footing is checked against sliding and overturning.

The factor of safety against sliding should not be less than 1.5 when dead load, live load and earth pressure and wind pressure (or seismic forces) are considered. However, if only dead load, live load and earth pressure are considered, it should not be less than 1.75. The corresponding factors of safety against overturning are 1.50 and 2.00, respectively.

24.5. PROPORTIONING FOOTINGS FOR EQUAL SETTLEMENT

To reduce the differential settlement due to live load variations for footings on fine-grained soils, it is desirable to proportion all the footings in such a way that they have equal pressures under the service loads. Thus all the footings would settle by equal amounts and the differential settlement would be considerably reduced. The following procedure suggested by Peck et al (1974) is usually adopted.

(1) The dead load from each column, including the weight of the footing, is determined.

(2) The maximum live load to which each footing is subjected is determined.

(3) The ratio of the maximum live load to dead load for each footing is computed.

(4) The footing that has the largest live load to dead load ratio is taken as the *governing footing*.

The area (A_g) of the governing footing is determined from the relation.

$$A_g = \frac{\text{Dead load + live load}}{\text{allowable bearing capacity}} \qquad \ldots(24.2)$$

(5) The service loads for all the footings are determined.

(6) The design bearing capacity (q_d) of all the footings, except the governing footing, is determined as under

$$q_d = \frac{\text{Service Load for governing footing}}{A_g} \qquad \ldots(24.3)$$

(7) The area of other footings is determined as

$$A = \frac{\text{Service load for that footing}}{q_d} \qquad \ldots(24.4)$$

24.6. DESIGN OF STRIP FOOTINGS

Plain cement concrete strip footings are provided when the loads are light and the soil is good. If the loads are heavy and the soil conditions are not favourable, plain concrete footings are not economical. Reinforced cement concrete footings are more suitable in such cases.

(a) **Plain Concrete Footings.** The footing is designed so that the contact pressure on the soil does not exceed the allowable bearing pressure. The width (B) of the footing is determined from the relation

$$B = \frac{Q}{q_{na}} \quad \text{...(24.5)}$$

where Q = load per m run,

q_{na} = allowable soil pressure.

If the actual width provided is different from the theoretical width, the actual pressure is given by

$$q_o = \frac{Q}{\text{actual width}} \quad \text{...(24.6)}$$

The thickness at the edge of the footing should be at least 15 cm. On cohesive soils, generally a minimum thickness of 30 cm is specified in order to resist swelling pressure. The thickness of the footing should be adequate to minimise the development of tension on the underside of the projection acting as a cantilever.

The thickness of the projection is generally kept twice the length of the projection from the wall face. A 45° load distribution is also commonly used (Fig. 24.8), which gives a small tension on the underside. For wide and thick footings, sloping of the upper surface is done to effect economy in the quantity of concrete. However, the cost of form work will add to the total cost.

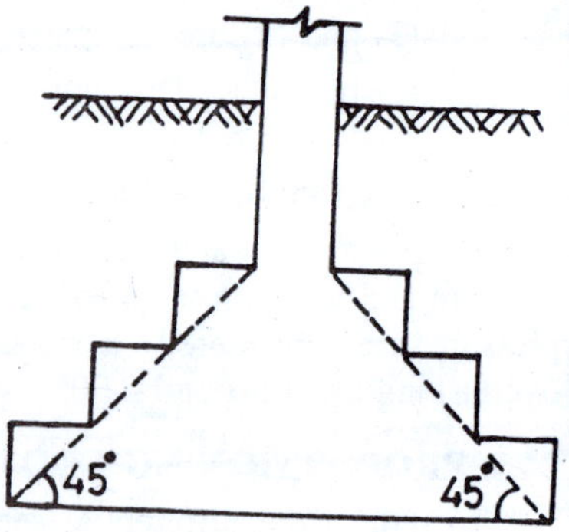

Fig. 24.8.

(b) **Reinforced Concrete Footings.** Footings carrying heavy loads on weak soils are reinforced in the transverse direction. The width of the footing is determined using Eq. 24.5.

For computing the bending moment for which the footing is to be designed, the critical section is taken as follows (IS 456 - 1978).

(i) At the face of the monolithic wall (Fig. 24.9).

(ii) Half-way between the centre line and the edge of the wall for footings under masonry walls.

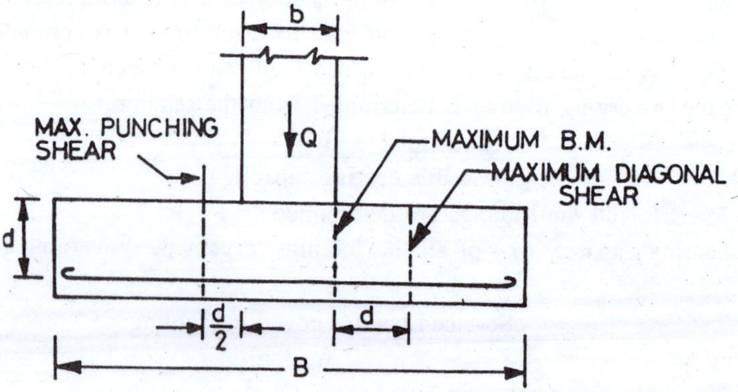

Fig. 24.9. R.C.C. Strip Footing.

For monolithic walls, the maximum bending moment is given by

$$M = q_0 (B - b)^2/8 \quad \text{...(24.7)}$$

where B width of footing, b = width of wall, q_o = actual soil pressure.

For checking the diagonal shear, the critical section is taken at a distance equal to the effective depth (d) of footing from the face of the wall. The diagonal shear is given by

$$F = q_0 [(B - b)/2 - d] \quad \text{...(24.8)}$$

24.7. DESIGN OF SPREAD FOOTINGS

Spread footings are used for distributing concentrated column loads over a large area so that the bearing pressure is less than or equal to the allowable soil pressure. Fig. 24.10 shows general types of spread footings.

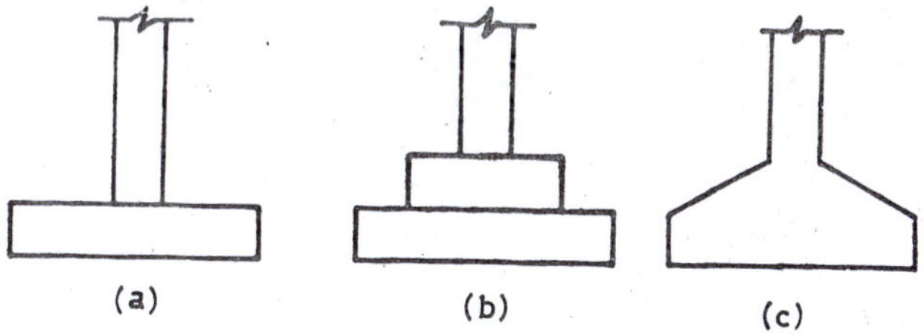

Fig. 24.10.

The area of the footing is given by

$$A = Q/q_{na} \quad \text{...(24.9)}$$

where Q = column load, q_{na} = allowable soil pressure.

If the area actually provided is more, the actual pressure is given by

$$q_0 = \frac{Q}{\text{Actual area}} \quad \text{...(24.10)}$$

(*a*) **Plain Concrete Footings.** The design is similar to that of a plain concrete strip footing. The thickness of the footing is fixed from the consideration of preventing tension on the underside. As in the case of strip footings, the thickness is kept equal to twice the projection or alternatively, the width of the footing is determined by the normal practice of 45° distribution of loading.

(*b*) **Reinforced Concrete Footings.** The area of the footing is obtained using Eq. 24.9. The shape of the footing may be square, circular or rectangular. If the column load is centrally placed, the upward pressure is uniform. As the weight of the footing is directly transferred to the soil below, it does not affect the bending moment and shear force.

The critical section for bending moment is taken as under.

(*i*) At the face of the column or pedestal monolithic with the footing when no metal plate is used.

(*ii*) Half-way between the face of the column or pedestal and the edge of the metal plate on which the column or pedestal rests.

The maximum B.M. for the case (*i*) is given by (Fig. 24.11)

$$M = \frac{q_o B(B - b)^2}{8} \quad \text{...(24.11)}$$

For checking the diagonal shear F, the critical section is taken at a distance equal to the effective depth (d) of the footing from the face of the column.

$$F = q_o B [(B - b)/2 - d] \quad \text{...(24.12)}$$

For punching shear, the critical section is taken at a distance of $d/2$ from the face of the column. Generally the overall depth (d_o) of the footing is determined from the punching shear considerations.

$$d_o = \frac{q_o [B^2 - (b + d)^2]}{4 (b + d) \sigma_{sp}} \quad \text{...(24.13)}$$

where q_o = actual pressure, B = width of footing, b = width of square column, σ_{sp} = safe punching shear.

The depth provided is checked for bending moment, shear and bond.

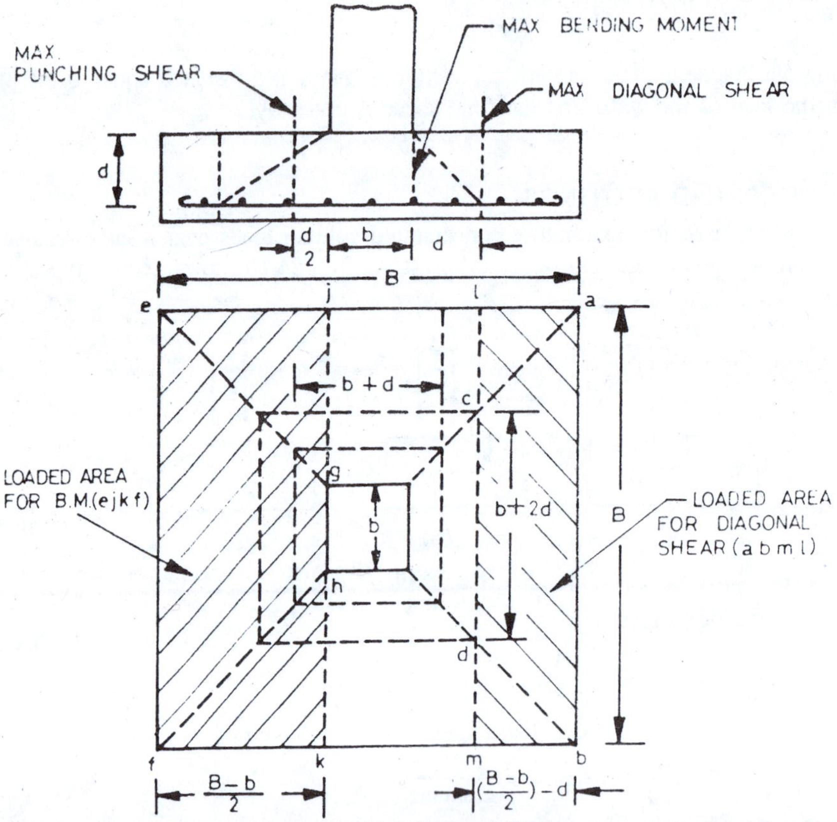

Fig. 24.11. R.C.C. Spread Footing.

The maximum force for bond is given by

$$F_b = q_o B (B - b)/2 \quad ...(24.14)$$

In the case of rectangular footings, the length and the width should be so chosen that the bending moment in each of the adjacent projections is equal to the moment of resistance of the footing. For uniform contact pressure distribution, the centre of gravity of the footing should coincide with the centre of gravity of the column.

The footing of a circular column can be either square or circular. For the design of a square footing for a circular column, the circular column is generally substituted by an equivalent square column of the same area. The design of the footing is then done using the procedure already described for a square footing. The design of the circular footing can also be done using the basic principles. Because of the curved areas, the expressions are more involved than those for a square footing.

The thickness of the footing at the edge shall be not less than 15 cm for footings on soil, nor less than 30 cm above the top of piles for footings on piles.

24.8. DESIGN OF ECCENTRICALLY LOADED SPREAD FOOTINGS

A column is said to be eccentrically loaded when it is subjected to a load which is not centric or when it is subjected to a bending moment in addition to the centric load. If the centre of gravity of the footing is made to coincide with the centre of gravity of the centrically loaded column, the pressure distribution is trapezoidal, with the maximum pressure on one side and the minimum pressure on the other side. However, if the centre of gravity of the footing is placed eccentric with the centre of gravity of the column, the pressure distribution becomes uniform (See Fig. 23.20).

The eccentricity (e) of the footing is given by

$$e = M/Q \quad ...(24.15)$$

when M bending moment, and Q = axial load.

If the footing is provided with the above eccentricity, the resultant of the bearing pressure coincides with the resultant of the loads. The area of the footing is given by

$$A = \frac{Q}{q_{na}} \quad \text{...(24.16)}$$

The longer projection of the footing is designed as a cantilever for the maximum bending moment at the face of the column. Generally, the same reinforcement is provided in the smaller projection as well. The thickness provided is checked for the diagonal shear and the punching shear.

If the bending moment acts temporarily for a short period, a symmetrical footing can also be provided. The maximum pressure is determined using the following equations (see Sect. 23.21)

$$q_{max} = \frac{Q}{L \times B}\left(1 + \frac{6e}{B}\right) \quad \text{...(24.17)}$$

where e is the eccentricity.

If $e > B/6$, the maximum pressure is given by

$$q_{max} = \frac{4Q}{3L\ (B - 2e)} \quad \text{...(24.18)}$$

The dimensions L and B of the footing are chosen such that the maximum pressure (q_{max}) does not exceed that allowable bearing pressure (q_{na}). The bending moment, the diagonal shear and the punching shear are determined considering the trapezoidal pressure distribution.

24.9. COMBINED FOOTINGS

A combined footings supports two columns. When a foundation is built close to an existing building or the property line, there may not be sufficient space for equal projections on the sides of the exterior column. This results in an eccentric loading on the footing. It may lead to the tilting of the foundation. To counteract this tilting tendency, a combined footings is provided which joins the exterior column with an interior column. A combined footing is also required when the two individual footings overlap.

In the conventional design, the combined footing is assumed to be infinitely rigid and the soil pressure distribution to be planar. In the elastic line method, the flexibility of the footing is considered. The elastic line method is more rational but quite involved (see Sect. 24.17). The following discussion is limited to the conventional design.

The footing is proportioned such that the centre of gravity of the footing lies on the line of action of the resultant of the column loads. The pressure distribution thus becomes uniform.

A combined footing is generally rectangular in plan if sufficient space is available beyond each column. If one of the columns is near the property line, the rectangular footing can still be provided if the interior column is relatively heavier. However, if the interior column is lighter, a trapezoidal footing is required to keep the resultant of the column loads through the centroid of the footing. Thus the resultant of the soil reaction is made to coincide with the resultant of the column loads.

24.10. RECTANGULAR COMBINED FOOTINGS

The design of a combined footing consists of selecting length and width of the footing such that the centroid of the footing and the resultant of the column loads coincide. With the dimensions of the footing established, the shear force and bending moment diagrams are drawn. The thickness of the footing is selected from the bending moment and shear stress considerations. The footing is designed as a continuous beam supported by two columns in the longitudinal direction. The reinforcement is provided as in a continuous beam.

The procedure consists of the following steps.

(1) Determine the total column loads.

$$Q = Q_1 + Q_2 \quad \text{...(24.19)}$$

where Q_1 and Q_2 are the exterior and interior columns, respectively.

(2) Find the base area of the footings.

$$A = Q/q_{na} \qquad ...(24.20)$$

where q_{na} is the allowable soil pressure.

(3) Locate the line of action of the resultant of the column loads measured from one of the column, say exterior column (Fig. 24.12).

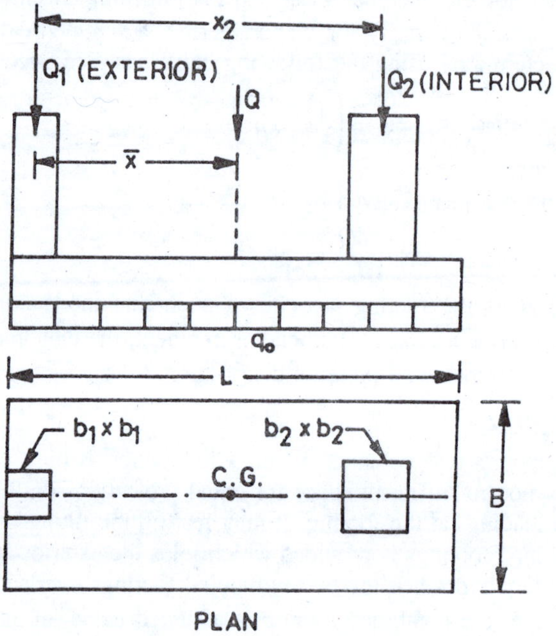

Fig. 24.12. Rectangular Combined Footing.

$$\bar{x} = Q_2 \times x_2/Q \qquad ...(24.21)$$

where x_2 is the distance between columns.

(4) Determine the total length of the footings.

$$L = 2(\bar{x} + b_1/2) \qquad ...(24.22)$$

where b_1 = width of the exterior column.

(5) Find the width of the footing.

$$B = A/L \qquad ...(24.23)$$

(6) As the actual length and width that are provided may be slightly more due to rounding off, the actual pressure is given by

$$q_0 = Q/A_0 \qquad ...(24.24)$$

where A_0 is the actual area.

(7) Draw the shear force and the bending moment diagrams along the length of the footing, considering the pressure q_0.

For convenience, the column loads are taken as concentric loads acting at the centres.

(8) Determine the bending moments at the face of the columns and the maximum bending moment at the point of zero shear.

(9) Find the thickness of footing for the maximum bending moment.

Check for diagonal shear and punching shear, as in the case of an isolated footings.

Check for bond at the point of contraflexure.

(10) Determine the longitudinal reinforcement for the maximum bending moment.

For transverse reinforcement, assume a width of $(b + d)$ to take all the bending moment in the short direction, where b is the column side and d is the effective depth.

24.11. TRAPEZOIDAL FOOTING

Trapezoidal combined footings are provided to avoid eccentricity of loading with respect to the base, as already mentioned. Trapezoidal footings are required when the space outside the exterior column is limited and the exterior column carries the heavier load.

The design procedure consists of the following steps:

(1) Determine the total column loads.

$$Q = Q_1 + Q_2 \quad ...(24.25)$$

(2) Find the base area of the footing.

$$A = Q/q_{na} \quad ...(24.26)$$

(3) Locate the line of action of resultant of the column loads (Fig. 24.13).

$$\bar{x} = Q_2 x_2/Q \quad ...(24.27)$$

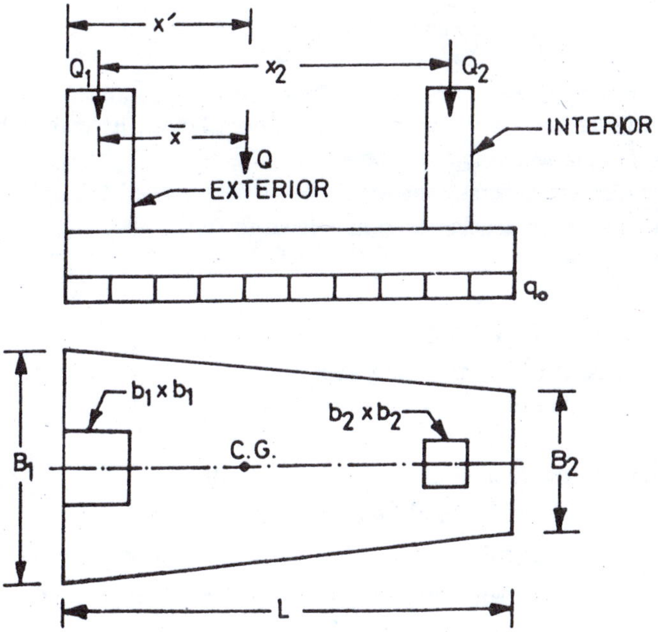

Fig. 24.13. Trapezoidal Combined Footing.

(4) Determine the distance x' of the resultant from the outer face of the exterior column.

$$x' = \bar{x} + b_1/2 \quad ...(24.28)$$

A trapezoidal footing is required if

$$L/3 < x' < L/2$$

where L is the length of the trapezoidal footing, determined from Eq. 24.22.

If $x' = L/2$, a rectangular footing is provided. However, if $x' < L/3$, a combined footing cannot be provided. In such a case, a strap footing is suitable.

(5) Determine the widths B_1 and B_2 from the following relations.

$$\frac{B_1 + B_2}{2} \times L = A \qquad ...(24.29)$$

and
$$\frac{L}{3}\left(\frac{B_1 + 2B_2}{B_1 + B_2}\right) = x \qquad ...(24.30)$$

Solving Eqs. 24.29 and 24.30,

$$B_2 = \frac{2A}{L}\left(\frac{3x'}{L} - 1\right) \qquad ...(24.31)$$

and
$$B_1 = \frac{2A}{L} - B_2 \qquad ...(24.32)$$

(6) Once the dimensions B_1 and B_2 have been found, the rest of the design can be done as in the case of a rectangular combined footing.

24.12. STRAP FOOTINGS

A strap footing is required in the following two cases:

(1) When $x' < L/3$, where x' is the distance of the resultant of column loads from the exterior face of the exterior column, and L is the length of the footing.

(2) When the distance between the two columns is so large that a combined footing becomes excessively long and narrow.

As mentioned before, a strap footing consists of two spread footings joined by a rigid beam known as a *strap*. The strap is not subjected to any direct soil pressure from below. Its main function is to transfer the moment from the exterior footing to the interior footing.

The following assumptions are generally made.

(*a*) The soil pressure is uniform beneath each individual footing.

(*b*) The strap is perfectly rigid.

(*c*) The strap is weightless.

(*d*) The interior footing is centrally loaded.

The design procedure consists of the following steps.

(1) Assume a reasonable value of the eccentricity (e) between the load Q_1 and the reaction R_1 on the exterior column (Fig. 24.14).

(2) Determine the length of the footing of the exterior column.

$$L_1 = 2(e + 0.5\, b_1) \qquad ...(24.33)$$

where b_1 = width of the exterior column.

(3) Compute the reaction R_1, by taking moments about the line of action of R_2.

$$R_1 = Q_1 \times x_2/S \qquad ...(24.34)$$

where x_2 = distance between loads Q_1 and Q_2 and S = distance between reactions R_1 and R_2.

(4) Compute areas A_1 and A_2.

$$A_1 = R_1/q_{na} \qquad ...(24.35)$$

and
$$A_2 = R_2/q_{na} \qquad ...(24.36)$$

The reaction R_2 is obviously equal to $(Q_1 + Q_2) - R_1$.

(5) Find the widths of footings.

$$B_1 = A_1/L_1 \qquad ...(24.37)$$

and
$$B_2 = \sqrt{A_2} \qquad ...(24.38)$$

(6) Design the individual footings as in the case of spread footings.

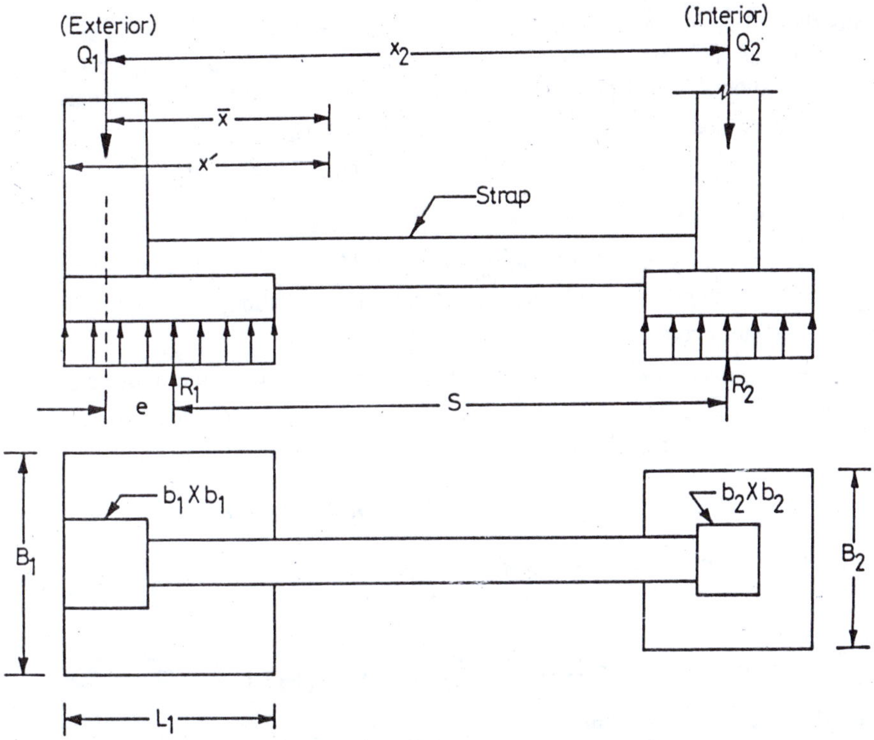

Fig. 24.14. Strap Footing.

(7) Determine the depth of the strap for diagonal shear and bending moment.

[**Note :** It may be mentioned that a number of designs of the strap footing are possible, depending upon the selected value of *e*.]

24.13. PRINCIPLES OF DESIGN OF MAT FOUNDATIONS

A mat (or raft) is a thick reinforced concrete slab which supports all the load-bearing walls and column loads of a structure or a large portion of structure. A mat is required when the loads are heavy and the soil is very weak or highly compressible. A mat is more economical than individual footings when the total base area required for the individual footings exceeds about one-half of the area covered by the structure.

A mat is preferred to individual footings when the soil mass has very erratic properties and contains lenses of compressible soils. In such a case, it would be difficult to control the differential settlements if individual footings are provided. The mat spans over weak patches of the soil and thus the differential settlements are considerably reduced.

Like all other shallow foundations, a mat must be safe against shear failure and the settlements should be within the allowable limits. As the width of a raft is very large, the pressure bulb is quite deep. Thus the loose soil pockets under a raft may be more evenly distributed. This results in a smaller differential settlement than individual footings. It is customary to assume that a differential settlement of 19 mm would occur in a raft when the maximum settlement is twice that in the individual footing. Thus a maximum settlement of 50 mm can be permitted when the differential settlement allowed is 19 mm.

As rafts are generally at some depth below the ground surface, a large volume of soil is excavated and, therefore, the net pressure on the soil is considerably reduced. An advantage of this reduction in the pressure can be taken while designing a raft.

***(a)* Rafts on Cohesionless Soils.** The bearing capacity of a foundation on cohesionless soils depends upon the width, as discussed in chapter 23. As the width of a raft is very large, the bearing capacity is high, and, therefore, the shear failure generally does not occur. Accordingly, the safe settlement pressure, and not the bearing capacity, generally governs the design, except for very loose sands ($N < 5$).

Settlements depend upon the depth of the soil stratum. If a firm stratum exists at a shallow depth below the raft, the settlements are small. However, if the sand deposit extends to a great depth, the settlements would be large. The allowable soil pressure can be found using the following equations.

The safe bearing capacity can be determined as (Teng, 1962),

$$q_{ns} = 0.22 N^2 B W_\gamma + 0.67 (100 + N^2) D_f W_q \qquad \text{...(24.39)}$$

The safe settlement pressure for a settlement of 25 mm is given by

$$q_{np} = 17.5 (N - 3) \; W_\gamma \; \text{kN/m}^2 \qquad \text{...(24.40)}$$

where B = smaller dimension of raft (m), D_f = Depth of foundation,

W_γ, W_q = water table correction factors.

The pressures q_{ns} and q_{np} are in kN/m^2. The smaller of the two values is the allowable soil pressure (q_{na}).

As stated earlier, Teng's equation for the safe settlement (q_{np}) is conservative. Using Bowle's equation [Eq. 23.82 (*b*)], for a settlement of 25 mm.

$$q_{np} = 12.2 N \left(\frac{B + 0.3}{B}\right)^2 R_d W_\gamma \quad \text{kN/m}^2 \qquad \text{...(24.41)}$$

where R_d = depth factor = 1 + 0.33 (D_f/B)

In general,

$$q_{np} = 12.2 N \left(\frac{B + 0.3}{B}\right)^2 R_d W_\gamma \left(\frac{s}{25}\right) \quad \text{kN/m}^2 \qquad \text{...(24.42)}$$

where s = allowable settlement (mm).

As the width of a raft is very large, $\dfrac{B + 0.3}{B} \approx 1.0$

Therefore, $q_{np} = 12.2 N R_d W_\gamma (s/25)$ ...(24.43)

Taking R_d = 1.00 and s = 50 mm,

$$q_{np} = 24.4 N W_\gamma \quad \text{kN/m}^2 \qquad \text{...(24.44)}$$

In case of rafts, as the width B is very large, and the pressure bulb is deep, the water table generally affects the safe settlement pressure.

Taking W_γ = 0.5, $q_{np} = 12.2 N$ kN/m^2 ...(24.45)

Peck et al (1974) express Eqs. 24.44 and 24.45) as

$$q_{np} = 20 N \; W_\gamma \quad \text{kN/m}^2 \qquad \text{...(24.46)}$$

and

$$q_{np} = 10 N \; \text{kN/m}^2 \qquad \text{...(24.47)}$$

The above equations are applicable for $5 \leq N \leq 50$. If the value of (N) after correction is less than 5, the sand is too loose for a raft foundation. The sand should be either compacted or a deep foundation, such as pile foundation, should be provided. For values of (N) greater than 50, the above equations give *unconservative* results.

According to IS : 6403, for a settlement of 65 mm, the safe settlement pressure is given by

$$q_{np} = 25.4 (N - 3) \; W_\gamma \qquad \text{...[24.48(a)]}$$

Taking W_γ = 0.5, $q_{np} = 12.7 (N - 3)$ ...[24.48(*b*)]

The safe bearing capacity (q_{ns}) can also be determined using the equations developed in Chapter 23. Generally, the safe bearing capacity is much greater than the safe settlement pressure and is not of much significance.

As the raft foundations are generally used below basements, the foundations are not backfilled. Eq. 23.16 can be used. Thus

$$Q/A = q_{na} + \gamma D_f \qquad \text{...(24.49)}$$

where Q = superimposed load, A = area of raft, D_f = depth of foundation.

(*b*) **Rafts on Clay.** The net ultimate bearing capacity is generally determined using Skempton's equations (Eq. 23.55).

$$q_{nu} = 5\ (1 + 0.2\ D_f/B)\ [1 + 0.2\,(B/L)]\ c_u \qquad \text{...(24.50)}$$

where c_u = undrained cohesion.

All other notations have been defined above.

The safe net bearing capacity can be obtained as

$$q_{ns} = q_{nu}/F \qquad \text{...(24.51)}$$

where F = factor of safety.

Under normal loading conditions, the factor of safety should not be smaller than 3. IS : 6403 recommends a minimum factor of safety of 2.5.

In case of rafts on clay, as the safe bearing capacity is independent of its size, it generally governs the design. But the safe settlement pressure is also important in some cases. In the case of rafts, the pressure bulb extends to a much greater depth than that for an isolated footing. The settlements of rafts on normally consolidated clays are usually very large. However, in the case of over-consolidated clays, the settlements are small. The settlements are calculated due to the net increase in pressure, given by

$$q_n = Q/A - \gamma D_f \qquad \text{...(24.52)}$$

If the soil stratum extends to a depth greater than about twice the width of the mat, the load on the mat would tend to act as a point load for the soil at large depths, and the settlements would be the same whatever be the type of foundation. In such cases, increasing the width of the mat does not help in reducing the settlements. If the settlements are large, deep foundations, such as piles or drilled caissons, would be more suitable.

The factor of safety against bearing capacity failure can be written as $F = q_{ns}/q_n$.

$$F = \frac{5\,(1 + 0.2\,D_f/B)\,(1 + 0.2\,B/L)\,c_u}{Q/A - \gamma D_f} \qquad \text{...(24.53)}$$

The settlements of a mat foundation can be reduced by decreasing the net increase in pressure, *i.e.*, by increasing D_f. For no increase of the net pressure, Eq. 24.52 gives

$$\gamma D_f = Q/A$$

or

$$D_f = \frac{Q}{\gamma A} \qquad \text{...(24.54)}$$

A foundation that satisfies Eq. 24.54 is known as *fully compensated or a floating foundation.* For such a foundation, the settlement is (theoretically) zero. The factor of safety against bearing capacity failure, as per Eq. 24.53, becomes theoretically infinite.

24.14. COMMON TYPES OF MAT FOUNDATIONS

There are several types of mat foundations.

(1) Flat Plate Type. In this type of mat foundation, a mat of uniform thickness is provided (Fig. 24.15). This type is most suitable when the column loads are relatively light and the spacing of columns is relatively small and uniform.

(2) Flat Plate Thickened Under Columns. When the column loads are heavy, this type is more suitable than the flat plate type. A portion of slab under the

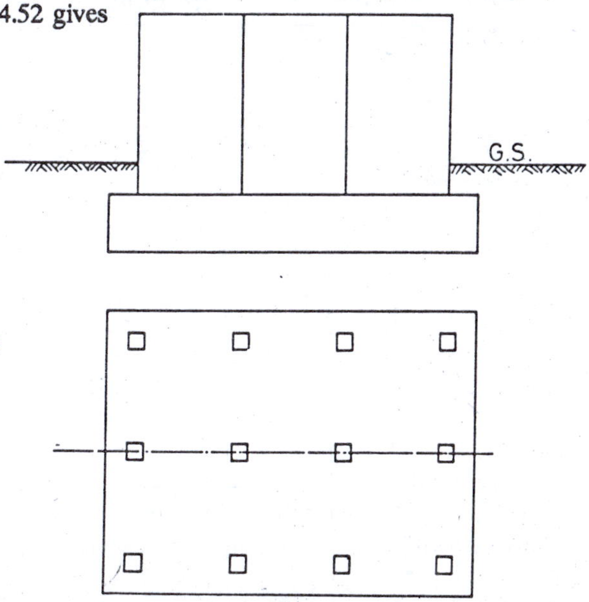

Fig. 24.15. Flat Plate Type Mat.

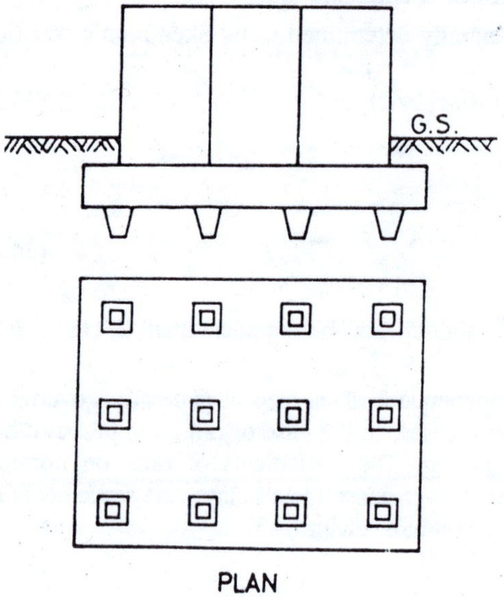

Fig. 24.16. Flat Plate—Thicknened Under Columns.

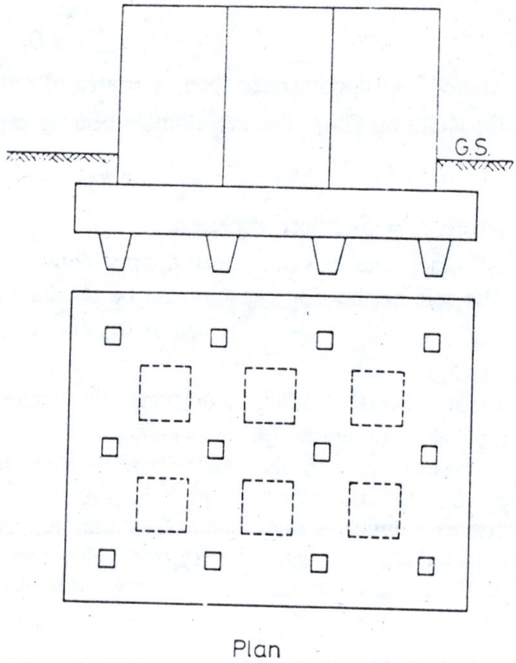

Fig. 24.17. Beam and Slab Construction.

column is thickened to provide enough thickness for negative bending moment and diagonal shear (Fig. 24.16).

Sometimes, instead of thickening the slab, a pedestal is provided under each column above the slab to increase the thickness.

(3) Beam and Slab Construction. In this type of construction, the beams run in two perpendicular directions and a slab is provided between the beams. The columns are located at the intersection of beams (Fig. 24.17). This type is suitable when the bending stresses are high because of large column spacing and unequal column loads.

(4) Box Structures. In this type of mat foundation, a box structure is provided in which the basement

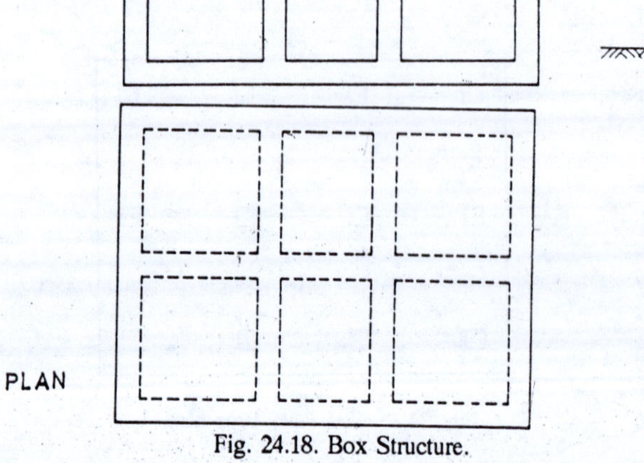

Fig. 24.18. Box Structure.

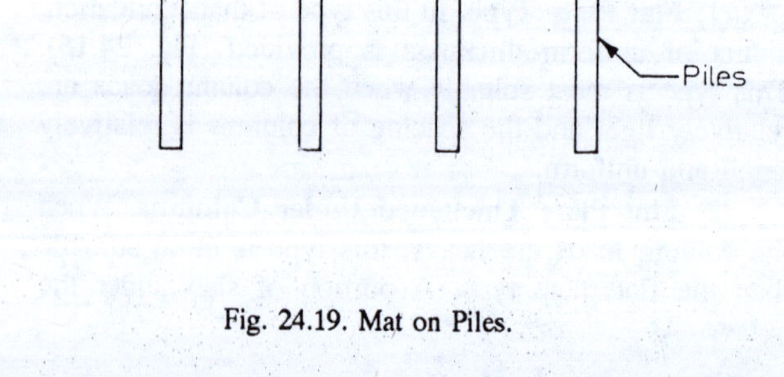

Fig. 24.19. Mat on Piles.

walls act as stiffeners for the mat. Boxes may be made of cellular construction or rigid frames consisting of slabs and basement walls (Fig. 24.18). This type of mat foundation can resist very high bending stresses.

(5) Mats placed on Piles. The mat is supported on piles in this type of construction (Fig. 24.19). This type of mat is used where the soil is highly compressible and the water table is high. This method of construction reduces the settlement and also controls buoyancy.

24.15. DESIGN METHODS FOR MAT FOUNDATIONS

The methods of design of mat foundations can be classified into 4 categories, depending upon the assumptions made.

(1) Rigid Beam Method. In this method of design, the slab is considered to be infinitely rigid as compared with the subsoil. The flexural deflections of the mat in this case do not influence the contact pressure distribution acting on the mat. The pressure distribution is assumed to be planar. The centroid of the soil pressure coincides with the line of action of the resultant of all the loads acting on the mat.

This method is also known as the *conventional method* of design. The design of combined footing discussed in Section 24.10 and 24.11 was also based on this method.

(2) Simplified Elastic Method. This method is based on the assumption that the soil behaves like an infinite number of individual independent elastic springs. The springs are assumed to take tension as well as compression. The assumption was first introduced by Winkler and the foundation model is known as *Winkler's model.* The method takes into account the elasticity of the footing. But as the soil does not behave exactly according to the assumptions made, the method is an approximate one and is a simplification of the actual soil behaviour.

The elastic constant of the spring is related to the coefficient of subgrade reaction, as defined later.

(3) Elastic Method. In this method of design, the soil is considered as a homogeneous, linearly elastic half space. The method uses the solutions provided by the theory of elasticity. As actual soils do not behave as linearly elastic solids, this method also gives approximate solutions. The method is complicated and rarely used in a design office.

(4) Non-linear Elastic Method. The soil is considered to be a non-linearly elastic solid. The method represents the behaviour of actual soil more closely than the elastic method and is more accurate. Numerical techniques, such as finite element method, are required for the design (Arora, 1980). The method has not developed to a stage that this can be used in a design office.

In this text, only the first two methods are discussed.

24.16. CONVENTIONAL DESIGN OF RAFT FOUNDATIONS

In the conventional method of design, the raft is assumed to be infinitely rigid and the pressure distribution is taken as planar (linearly varying). The assumption is valid when the raft rests on a soft clay which is highly compressible, and the eccentricity of the load is small. In case when the soil is stiff or when the eccentricity is large, the method does not give accurate results. The elastic method, which takes into account the stiffness of the soil and raft, is more economical and accurate in the latter case. The simplified elastic method is discussed in Sect. 24.17.

According to the American Concrete Institute Committee 436 (1966), the design of mats should be done using the conventional method if the spacing of the columns in a strip of the raft is less than 1.75/λ metres and using the simplified elastic method when it is greater than 1.75/λ, where λ is given by

$$\lambda = \left(\frac{B_1 k}{4EI}\right)^{1/4} \qquad \text{...(24.55)}$$

where k = coefficient of subgrade reaction (kN/m^3), B_1 = width of the strip (m),

E = modulus of elasticity of the raft material (kN/m^2), I = moment of inertia (m^4).

The coefficient of subgrade reaction of a soil is the pressure required to produce a unit settlement of a plate. It is given by

$$k = q/z \qquad \text{...(24.56)}$$

where q = pressure (kN/m^2), z = settlement (m), k = coefficient of subgrade reaction (kN/m^3).

The coefficient of subgrade reaction is not a constant for a given soil (Terzaghi, 1955). It depends upon a number of factors, such as length, width, depth and shape of foundation (see Sect. 24.20).

Procedure. The procedure for the conventional design consists of the following steps:

(1) Determine the line of action of all the loads acting on the raft. The selfweight of the raft is not considered, as it is taken directly by the soil.

(2) Determine the contact pressure distribution as under.

(*a*) If the resultant passes through the centre of the raft, the contact pressure is given by

$$q = Q/A$$

(*b*) If the resultant has an eccentricity of e_x and e_y in x-and y-directions [Fig. 24.20 (*a*)].

$$q = \frac{Q}{A} \pm \frac{(Q \cdot e_x)}{I_{yy}} x \pm \frac{Q \cdot e_y}{I_{xx}} y \quad \text{...(24.57)}$$

The maximum contact pressure should be less than the allowable soil pressure.

(3) Divide the slab into strips (bands) in x-and y-directions. Each strip is assumed to act as independent beam subjected to the contact pressure and the column loads.

(4) Draw the shear force and bending moment diagrams for each strip.

(5) Determine the modified column loads as explained below.

It is generally found that the strip does not satisfy statics, *i.e.* the resultant of column loads and the resultant of contact pressure are not equal and they do not act in the same line. The reason is that the strips do not act independently as assumed and there is some shear transfer between adjoining strips.

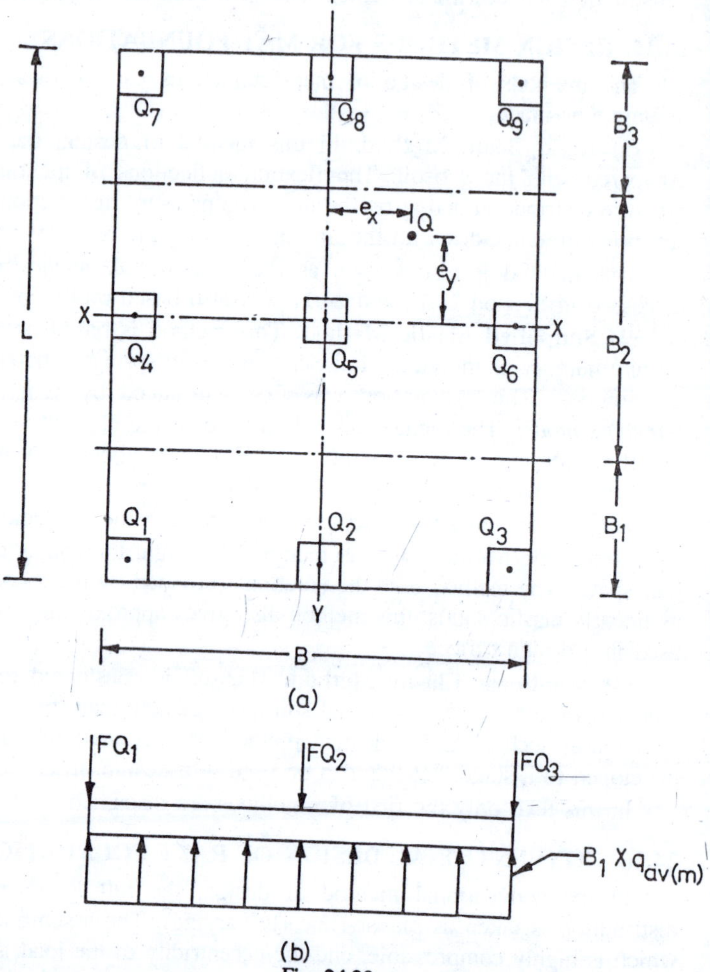

Fig. 24.20.

Let us consider the strip carrying column loads Q_1, Q_2 and Q_3 in Fig. 24.20 (*a*). Let B_1 be the width of the strip. Let the average soil (contact) pressure on the strip be q_{av}. Let B the length of the strip.

Average load on the strip.,

$$Q_{av} = \frac{1}{2} \text{(downward load + upward force)}$$

$$Q_{av} = \frac{1}{2} (Q_1 + Q_2 + Q_3 + q_{av} B_1 B) \quad \text{...(24.58)}$$

The modified average soil pressure is given by

$$\bar{q}_{av} = q_{av} \left(\frac{Q_{av}}{q_{av} B_1 B} \right) \quad \text{...(24.59)}$$

The column load modification factor (F) is given by

$$F = \frac{Q_{av}}{Q_1 + Q_2 + Q_3} \quad \text{...(24.60)}$$

All the column loads are multiplied by F for that strip. For this strip, the column loads are FQ_1, FQ_2 and FQ_3.

(6) The bending moment and shear force diagrams are drawn for the modified column loads and the modified average soil pressure ($\bar{q}_{av}$).

(7) Design the individual strips for the bending moment and shear force found in step 6. The raft is designed as an inverted floor supported at columns.

As the analysis is approximate, the actual reinforcement provided is twice the computed value.

24.17. DESIGN OF COMBINED FOOTING BY ELASTIC LINE METHOD

The design of a combined footing discussed in Sects. 24.9 to 24.11 is the conventional method, in which the footing is assumed to be rigid and the soil distribution is taken as planar. In the elastic line method, the stiffness of the footing is considered.

Whether a footing should be considered as rigid or not depends upon the soil characteristics and the size of the footing. Hetenyi (1946) gave the following criteria whether a beam should be considered as rigid or flexible. As a combined footing acts like a beam on elastic foundation, the same criteria can be used.

Rigid footing $\lambda L < 0.80$

Flexible footing $\lambda L > 3.0$

For the value $0.8 < \lambda L < 3.0$, the footing is of intermediate flexibility.

The value of λL is computed from the following relation

$$\lambda L = \left(\frac{BkL^4}{4EI} \right)^{1/4} \quad \text{...(24.61)}$$

where k = coefficient of subgrade reaction, B = width of footing, L length of footing, E = modulus of footing material, and I = moment of inertia of footing.

The flexible footing is designed as infinite beam on elastic foundation, whereas the footings of intermediate flexibity is designed as a finite beam.

According to the theory of beams on elastic foundations,

$$EI \frac{d^4 y}{dx^4} = -y k B \quad \text{...(24.62)}$$

where y = deflection (settlement), B = width of beam, k = coefficient of subgrade reaction.

The solution of Eq. 24.62 is

$$y = e^{\lambda x} (A \cos \lambda x + B \sin \lambda x) + e^{-\lambda x} (C \cos \lambda x + D \sin \lambda x) \quad \text{...(24.63)}$$

where

$$\lambda = \left(\frac{Bk}{4EI} \right)^{1/4} \quad \text{...(24.64)}$$

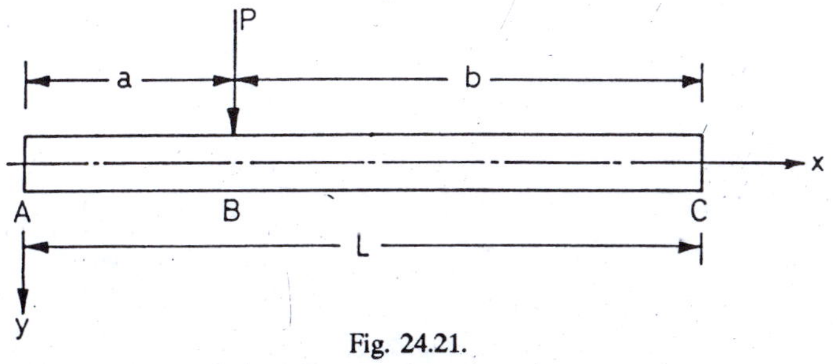

Fig. 24.21.

Applying the boundary conditions (Fig. 24.21), the solution can be expressed as (Winterkorn and Fang, 1975),

$$y = \frac{P\lambda A}{kB} \quad \text{...[24.65 (z)]}$$

$$M = \frac{P}{2\lambda} B \quad \text{...[24.65 (b)]}$$

$$V = PC \quad \text{...[24.65 (c)]}$$

where *A*, *B* and *C* are trigonometric identities, which are quite complicated. A digital computer is required for the design of footing using above method.

24.18. FINITE DIFFERENCE METHOD FOR COMBINED FOOTINGS

As the rigorous solution for a combined footing on elastic foundation is complicated, numerical techniques are used to obtain an approximate solution. A method proposed by Malter (1960) using finite difference is described below. The procedure consists of the following steps (Teng, 1962).

(1) Divide the footing into 4 to 6 equal segments, each of length *h*. For illustration purposes, the beam in Fig. 24.22(*a*) is divided into 3 segments.

(2) Let *a, b, c* and *d* be the deflections (settlements) at nodal points *A, B, C* and *D*.

The soil reactions at these points are, respectively, $a\,k_s$, $b\,k_s$, $c\,k_s$ and $d\,k_s$, where $k_s = k \times B$, in which *B* is the width of the footing.

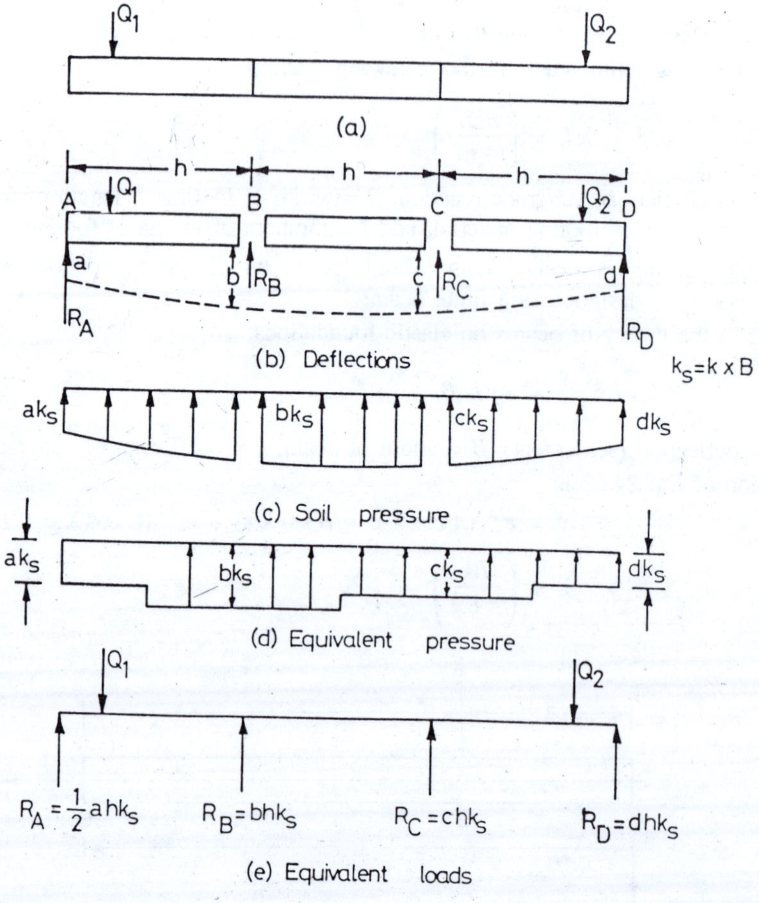

Fig. 24.22.

(3) The soil pressure is continuous [Fig. 24.22 (*c*)]. However, it can be approximated by steps [Fig. 24.22 (*d*)]. For more accurate results, it can be represented by straight lines or parabolas.

Replace the soil reactions by equivalent concentrated reactions R_A, R_B, R_C and R_D, as shown in Fig. 24.22 (*e*).

(4) The footing under the applied load and the equivalent reactions should satisfy the equations of equilibrium, viz.,

$$\Sigma M = 0 \text{ and } \Sigma V = 0$$

Write moment equilibrium equations for each panel point. For example, for the point *B*,

$$M_B = \frac{(a - 2b + c)}{h^2} EI \qquad \text{...(24.66)}$$

In general,
$$M_n = \frac{EI}{h^2} (y_{n-1} - 2y_n + y_{n+1}) \qquad \text{...(24.67)}$$

Also write force equilibrium equation for the whole system.

(5) Solve the simultaneous equations formed in step 4 for values of *a, b, c* and *d*.

(6) Determine the actual pressure distribution from the calculated values of *a, b, c* and *d* in step 5.

(7) Design the beam for the pressure distribution found in step 6.

24.19. ELASTIC PLATE METHOD FOR MATS

The design of a mat (raft) on Winkler's bed can be done using the method proposed by the American Concrete Institute Committee 436 (1956). The method is based on the theory of plates on elastic foundations.

The step-by-step procedure is given below (Das, 1984).

(1) Assume a thickness (t) for the mat either by experience or using the conventional (rigid) method.

(2) Determine the flexural rigidity (D) of the mat as

$$D = \frac{E t^3}{12(1 - \mu^2)} \qquad \text{...(24.68)}$$

where E = modulus of footing material, μ = Poisson's ratio of footing material.

(3) Determine the radius of effective stiffness (L') from the following relation.

$$L' = (D/k)^{1/4} \qquad \text{...(24.69)}$$

The zone of influence of any column load is of the order of $3L'$ to $4L'$.

(4) Find the tangential and radial moments at any point caused by a column load using the following equations.

Tangential moment,
$$M_t = -\frac{Q}{4}\left[A_1 - \frac{(1-\mu)A_2}{r/L'}\right] \qquad \text{...(24.70)}$$

Radial moment,
$$M_r = -\frac{Q}{4}\left[\mu A_1 + \frac{(1-\mu)A_2}{r/L'}\right] \qquad \text{...(24.71)}$$

where r = radial distance from the column load, Q = column load.

A_1 and A_2 are functions of (r/L'), as shown in Fig. 24.23.

In cartesian coordinates, the above equations can be written as

$$M_x = M_t \sin^2\alpha + M_r \cos^2\alpha \qquad \text{...(24.72)}$$

$$M_y = M_t \cos^2\alpha + M_r \sin^2\alpha \qquad \text{...(24.73)}$$

where α is the angle which the radius r makes with x-axis.

(5) Determine the shear force (V) per unit width of the mat caused by a column load as

$$V = -\frac{Q}{4\,L'}\,A_3 \qquad \text{...(24.74)}$$

where A_3 is a function of r/L' (Fig. 24.23).

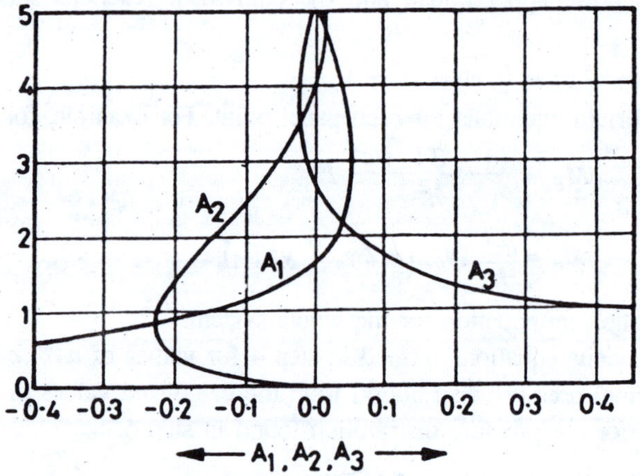

Fig. 24.23.

(6) If the edge of the mat is located in the zone of influence of a column, determine the moment and shear along the edge, assuming that the mat is continuous.

Moment and shear opposite in sign to those determined are applied at the edges to satisfy the known conditions.

(7) If the zones of influence of two or more columns overlap, the method of superposition can be used to obtain the total moment and shear.

24.20. FINITE DIFFERENCE METHOD FOR MATS

The mat is assumed to be a plate supported by a bed of uniformly distributed coil springs with a spring constant k equal to the coefficient of subgrade reaction, *i.e.* Winkler's bed. The differential equation for the deflection of the plate is given below.

$$\frac{\partial^4 w}{\partial x^4} + 2\,\frac{\partial^4 w}{\partial x^2\,\partial y^2} + \frac{\partial^4 w}{\partial y^4} = \frac{q - kw}{D} \qquad \text{...(24.75)}$$

where q = intensity of loading, k = coefficient of subgrade reaction,

w = deflection (settlement), D = flexural rigidity of plate (Eq. 24.68).

Eq. 24.75 can be solved by finite difference method after dividing the mat into a square mesh ($h \times h$). In the case of an interier point a (Fig. 24.24), the deflection w_a can be expressed in the difference form as

$$20w_a - 8(w_t + w_b + w_r + w_l) + 2\,(w_{tl} + w_{tr} + w_{bl} + W_{br}) + (w_{tt} + w_{bb} + w_{ll} + w_{rr})$$

$$= \frac{qh^4}{D} + \frac{Qh^2}{D} \qquad \text{...(24.76)}$$

where w_1, w_b, ... are deflections at adjoining points. The suffixes are shown in Fig. 24.24. The suffixes t, l, b and r are respectively for top, left, bottom and right.

Q = concentrated load at point a. All other notations have been defined earlier.

One difference equation can be written for each grid point.

[**Note** : For points near free edges, the difference equations is modified to account for boundary conditions.]

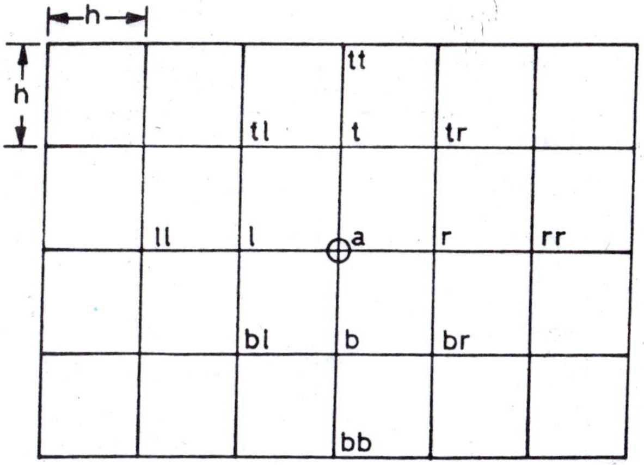

Fig. 24.24.

By solving the simultaneous equations formed, the deflections at all points are determined. A digital computer is normally required.

After the deflections have been determined, the bending moment at any point in each direction can be found using the equations of the theory of elasticity.

$$M_x = M_x' + \mu M_y' \qquad ...(24.77)$$

where M_x = total bending moment in x-direction, M_x' = bending moment in x-direction without including the effect of bending moment in the transverse y-direction, M_y' = bending moment in y-direction without including the effect of bending moment in x-direction.

Using the finite difference operators, the total bending moment on a strip in the l—r direction for an interior point is expressed as.

$$M_{l-r} = \frac{-D}{h^2}\left[(w_l - 2\,w_a + w_r) + \mu\,(w_t - 2\,w_a + w_b)\right] \qquad ...(24.78)$$

Thus the bending moments are computed at all points.

The accuracy of the method depends upon the fineness of the grid. When the squares of the grid are considerably larger than the size of columns, the results are not reliable adjacent to the columns. This difficulty is overcome by introducing subdivisions of the grid adjacent to the columns. However, it increases the computational work considerably.

24.21. COEFFICIENT OF SUBGRADE REACTION

The coefficient of subgrade reaction in the field can be determined by conducting a plate load test, using a square plate of the size 0.3 m × 0.3 m. A load-settlement curve is drawn, as explained in chapter 29 and the coefficient of subgrade reaction is determined using Eq. 24.56.

Table 24.1 gives the approximate values of the coefficient of subgrade reaction (k) for different soils. As the values are for a square plate of size 0.3 m × 0.3 m, these are modified to account for the effect of size, shape and depth as explained below.

1. Effect of the size. (*a*) Cohesionless soils

$$k_{B \times B} = k_{0.3 \times 0.3}\left(\frac{B + 0.3}{2B}\right)^2 \qquad ...(24.79)$$

where B is the size of footing.

Table 24.1. Coefficient of Subgrade Reaction (*k*)

S.No.	*Type of Soil*	*Condition of Soil*	*Value of k*	
			MN/m^3	*kN/m*^{3}
1.	Dry or moist sand	Loose	8 to 25	8×10^3 to 2.5×10^4
		Medium	25 to 120	2.5×10^4 to 1.2×10^5
		Dense	120 to 350	1.2×10^5 to 3.5×10^5
2.	Saturated sand	Loose	10 to 15	1.0×10^5 to 1.5×10^4
		Medium	35 to 40	3.5×10^4 to 4.0×10^4
		Dense	120 to 150	1.2×10^5 to 1.5×10^5
3.	Clay	Stiff	10 to 25	1.0×10^4 to 2.5×10^4
		Very Stiff	25 to 50	2.5×10^4 to 5.0×10^4
		Hard	> 50	$> 5 \times 10^4$

(*b*) Cohesive soils

$$k_{B \times B} = k_{0.3 \times 0.3}\ (0.3/B) \qquad \text{...(24.80)}$$

where $k_{B \times B} = k$ for size $B \times B$ metres, and $k_{0.3 \times 0.3} = k$ for size 0.3 m × 0.3 m.

2. Effect of shape. For a rectangular foundation ($L \times B$), the value of k is given by

$$k_{L \times B} = \frac{2}{3} k_{B \times B} (1 + B/L) \qquad \text{...(24.81)}$$

3. Effect of depth. For cohesionless soils, as the confinement pressure increases with the depth, the modulus of elasticity increases with depth. This causes a reduction in the settlement of the plate. Consequently, the value of k increases with depth.

For cohesive soils, there is no significant change in the value of k with depth.

[**Note:** Refer to chapter 29 for more details about k]

ILLUSTRATIVE EXAMPLES

Illustrative Example 24.1. *Design a reinforced cement concrete footing for a 1 m wide concrete wall carrying a load of 800 kN/m. The allowable soil pressure is 200 kN/m*^{2}.

Solution. From Eq. 24.5, $B = \dfrac{800}{200} = 4$ m

Actual soil pressure $(q_0) = \dfrac{800}{4} = 200$ kN/m^2

Maximum B.M. per m run (Eq. 24.7),

$$M = \frac{q_0 (B - b)^2}{8} = \frac{200\,(4 - 1)^2}{8} = \mathbf{225\ kN\text{-}m/m}$$

From Eq. 24.8, the diagonal shear is given by, taking $d = 0.52$ m,

$$F = q_0 \left[\left(\frac{B - b}{2} \right) - d \right] = 200 \left[\frac{(4 - 1)}{2} - 0.52 \right] = \mathbf{196\ kN}$$

Fig. E-24.1 shows the footing. It also shows the reinforcement.

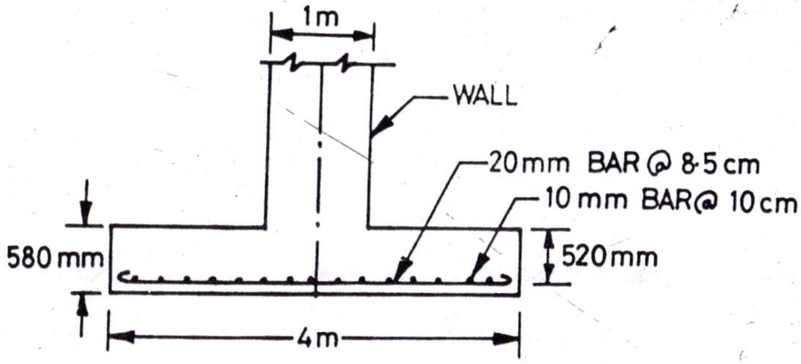

Fig. E24.1.

Illustrative Example 24.2. *Design a square reinforced concrete footing for the following data (Fig. E-24.2).*

Column load	= *800 kN*
Allowable soil pressure	= *200 kN/m²*
Size of column	= *0.4 m × 0.4 m*

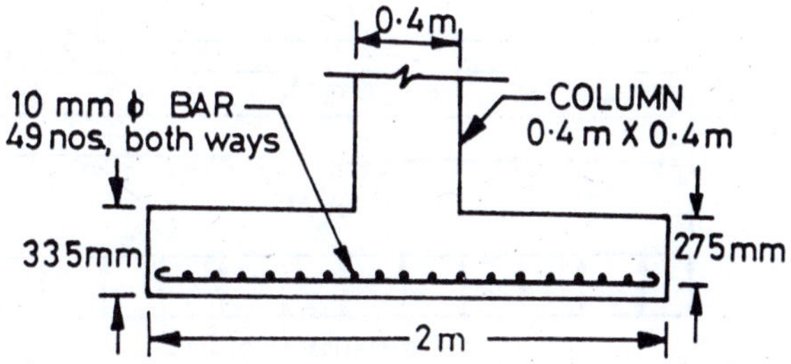

Fig. E-24.2.

Solution. From Eq. 24.9, $A = \frac{Q}{q_{na}} = \frac{800}{200} = 4 \text{ m}^2$

Size of footing, $B = \sqrt{4} = 2\text{m}$

Actual pressure, $q_0 = \frac{800}{2 \times 2} = 200 \text{ kN/m}^2$

From Eq. 24.13, the overall depth (d_0) is given by

$$d_0 = \frac{q_0 [B^2 - (b + d)^2]}{4 (b + d) \sigma_{sp}}$$

Taking σ_{sp} as 1000 kN/m², and cover as 0.06 m,

$$d + 0.06 = \frac{200 [2^2 - (0.4 + d)^2]}{4 (0.4 + d) \times 1000}$$

Solving, $d = 0.23$ m. Adopt 0.275 m.

Therefore, $d_0 = 0.275 + 0.06 = 0.335$ m

From Eq. 24.11, $M = \frac{q_0 B (B - b)^2}{8}$

$$= \frac{200 \times 2 \times (2 - 0.4)^2}{8} = \textbf{128 kN–m}$$

From Eq. 24.12, the diagonal shear is given by

$$F = q_0 B\left[\left(\frac{B-b}{2}\right) - d\right]$$

$$= 200 \times 2\left[\left(\frac{2-0.4}{2}\right) - 0.275\right] = \mathbf{210\ kN}$$

The maximum force for bond is given by Eq. 24.14 as

$$F_b = b_0 B\left(\frac{B-b}{2}\right) = 200 \times 2 \times \frac{1.6}{2} = \mathbf{320\ kN}$$

Illustrative Example 24.3. *Design a rectangular combined footing for two columns shown in Fig. E-24.3 (a). Take allowable soil pressure as 100 kN/m².*

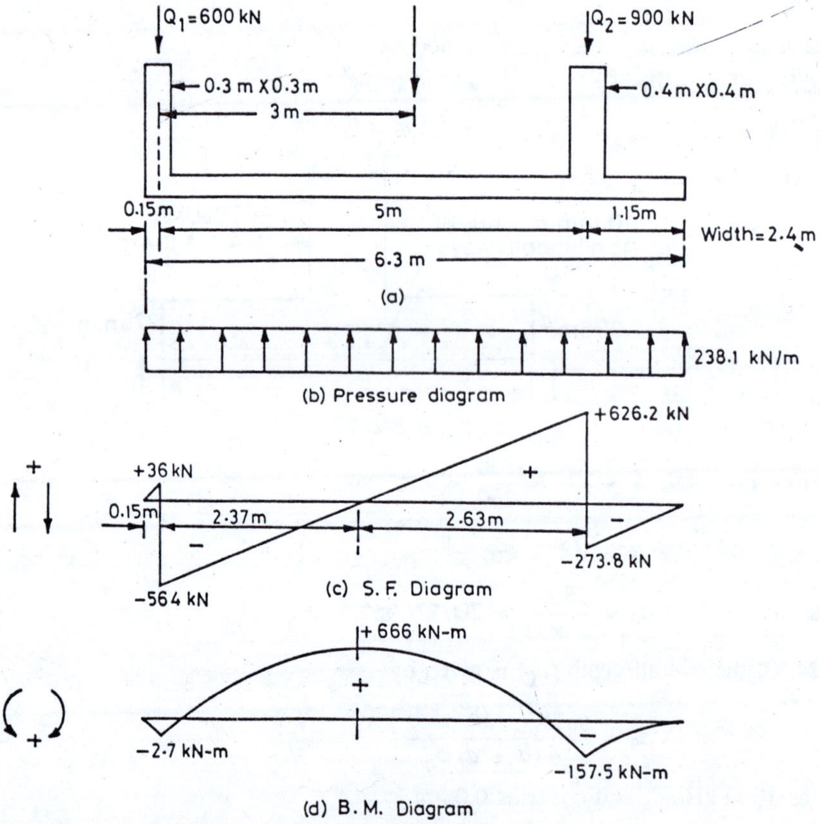

Fig. E-24.3.

Solution. Total load, $Q = 600 + 900 = 1500$ kN

From Eq. 24.21, $\bar{x} = \frac{Q_2 x_2}{Q} = \frac{900 \times 5}{1500} = 3$ m

From Eq. 24.20, $A = \frac{1500}{100} = 15\ m^2$

From Eq. 24.22, $L = 2(\bar{x} + b_1/2)$

or $L = 2(3.0 + 0.3/2) = 6.3$ m

From Eq. 24.23, $B = A/L = 15/6.3 = 2.38$, say 2.4 m.

From Eq. 24.24, $$q_0 = \frac{Q}{A_0} = \frac{1500}{6.3 \times 2.4} = 99.2 \text{ kN/m}^2$$

Actual pressure per metre run $= 99.2 \times 2.4 = 238.1$ kN/m

Fig. E-24.3 (*c*) shows the shear force diagram. The shear force is zero at a distance of 2.37 m from column-1. The maximum shear below column-1 is 564 kN and that below column-1 is 626.2 kN.

Fig. E-24.3 (*d*) shows the B.M. diagram. The maximum B.M. is 666 kN-m.

The footing can be designed for the above values of B.M. and S.F.

Illustrative Example 24.4. *Design a trapezoidal footing for the two columns shown in Fig. E-24.4. Take allowable soil pressure as 200 kN/m².*

Solution. $$Q = 2000 + 1500 = 3500 \text{ kN}$$

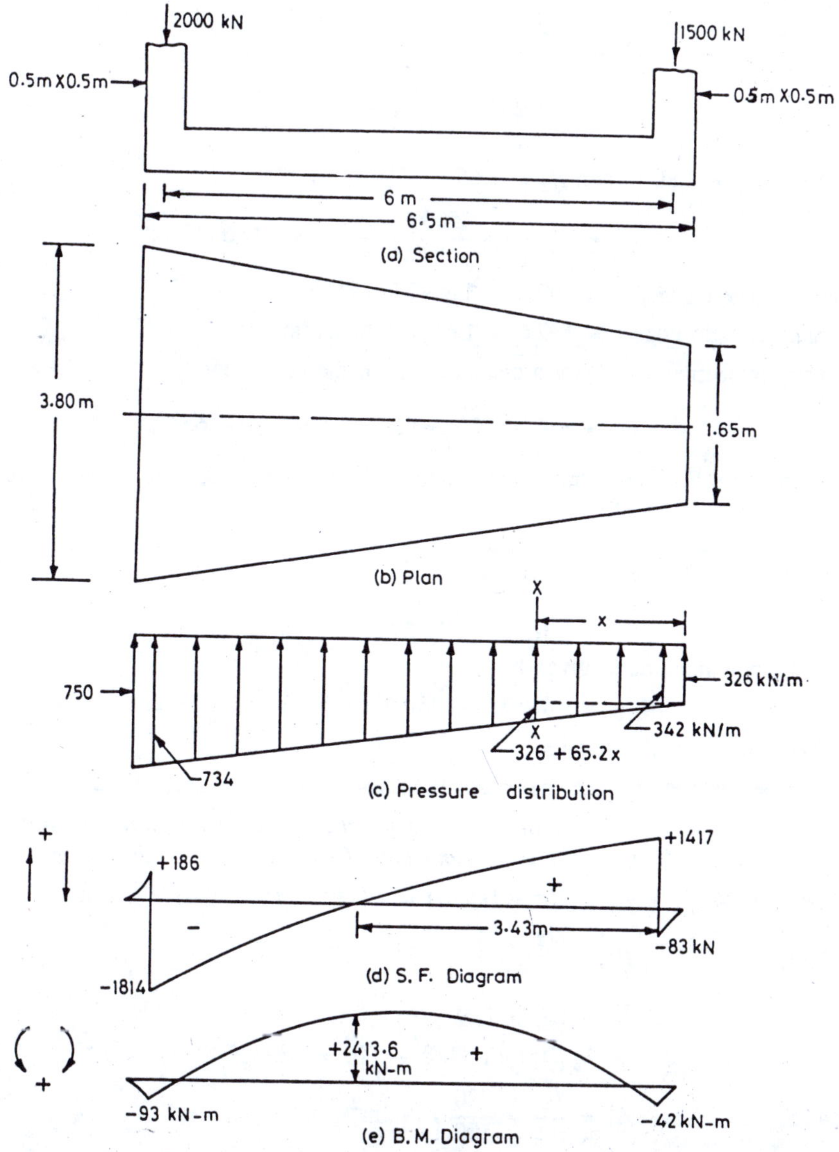

Fig. E-24.4.

$$A = \frac{3500}{200} = 17.5\ \text{m}^2, \qquad L = 6.5\ \text{m}$$

From Eq. 24.27, $$\bar{x} = \frac{Q_2 x_2}{Q} = \frac{1500 \times 6}{3500} = 2.57\ \text{m}$$

From Eq. 24.28, $$x' = \bar{x} + b_1/2 = 2.57 + 0.5/2 = 2.82\ \text{m}$$

$$L/2 = 6.5/2 = 3.25\ \text{m};\ L/3 = 2.17\ \text{m}$$

As $L/3 < x' < L/2$, a trapezoidal footing is required.

From Eq. 24.31, $$B_2 = \frac{2A}{L}\left(\frac{3x'}{L} - 1\right)$$

$$= \frac{2 \times 17.5}{6.5}\left(\frac{3 \times 2.82}{6.5} - 1\right) = 1.62\ \text{m}$$

From Eq. 24.32, $$B_1 = \frac{2A}{L} - B_2$$

$$= \frac{2 \times 17.5}{6.5} - 1.62 = 3.76\ \text{m}$$

Let us provide $B_1 = 3.80$ m and $B_2 = 1.65$

$$q_0 = \frac{3500}{1/2\,(3.8 + 1.65) \times 6.5} = 197.6\ \text{kN/m}^2$$

Pressure intensity at left edge $= 197.6 \times 3.8 = 750$ kN/m

Pressure intensity at right edge $= 197.6 \times 1.65 = 326$ kN/m

Pressure intensity at section X—X at a distance x from the right edge,

$$w_x = 326 + \frac{750 - 326}{6.5} \times x = 326 + 65.2\,x$$

Fig.E-24 (*d*) shows the S.F. diagram. The maximum shear forces under the columns are –1814 kN and 1417 kN.

$$\text{B.M. at } X\text{—}X = -326\frac{x^2}{2} - \frac{65.2 \times x^2}{2} \times \frac{x}{3} + 1500\,(x - 0.25)$$

$$= -163\,x^2 - 10.87\,x^3 + 1500\,x - 375$$

B.M. at $x = 3.43$ m is maximum, given by

$$M_{max} = -1917.7 - 438.64 + 5145 - 375 = 2413.6\ \text{kN–m}$$

Fig. E-24.4. (*e*) shows the BM diagram.

The footing can be designed for the maximum B.M. and checked for shear forces.

Illustrative Example 24.5. *Design a strap footing for the two columns shown in Fig. E-24.5 (a). The allowable soil pressure is 100 kN/m². Take the eccentricity of the footing of column-1 as 1 m.*

Solution. From Eq. 24.33, $$L_1 = 2(e + 0.5\,b_1) = 2\,(1.0 + 0.5 \times 0.4) = 2.4\ \text{m}$$

From Eq. 24.34, $$R_1 = \frac{600 \times 6}{5} = 720\ \text{kN}$$

$$R_2 = (Q_1 + Q_2) - R_1$$

$$= (600 + 1000) - 720 = 880\ \text{kN}$$

From Eq. 24.35, $$A_1 = \frac{R_1}{q_{na}} = \frac{720}{100} = 7.2\text{m}^2$$

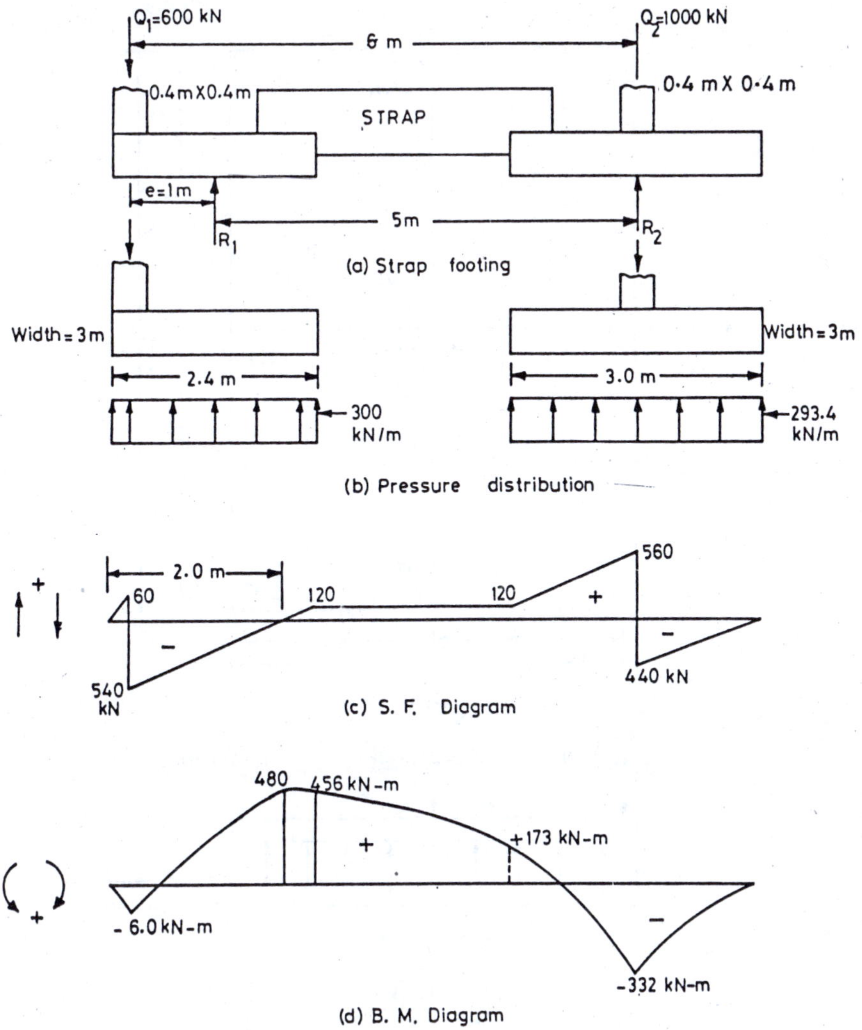

Fig. E-24.5.

From Eq. 24.36, $A_2 = \dfrac{R_2}{q_{na}} = \dfrac{880}{100} = 8.8 \text{ m}^2$

$B_1 = \dfrac{7.2}{2.4} = 3 \text{ m}$

Intensity of pressure, $q_1 = 100 \times 3 = 300 \text{ kN/m}$

From Eq. 24.38, $B_2 = \sqrt{8.8} = 2.96 \text{ m say } 3 \text{ m}$

Intensity of pressure, $q_2 = \dfrac{880}{3 \times 3} = 97.8 \text{ kN/m}^2$

$= 97.8 \times 3 = 293.4 \text{ kN/m}$

Fig. E-24.5 (*c*) and (*d*) show the S.F. and B.M. diagrams. The footings can be designed. The strap should be designed for a maximum B.M. of 480 kN-m and a maximum S.F. of 120 kN.

Illustrative Example 24.6. *The plan of a mat foundation with 9 columns is shown in Fig. E-24.6 (a). Assuming that the mat is rigid, determine the soil pressure distribution. All the columns are of the size 0.6 m × 0.6 m.*

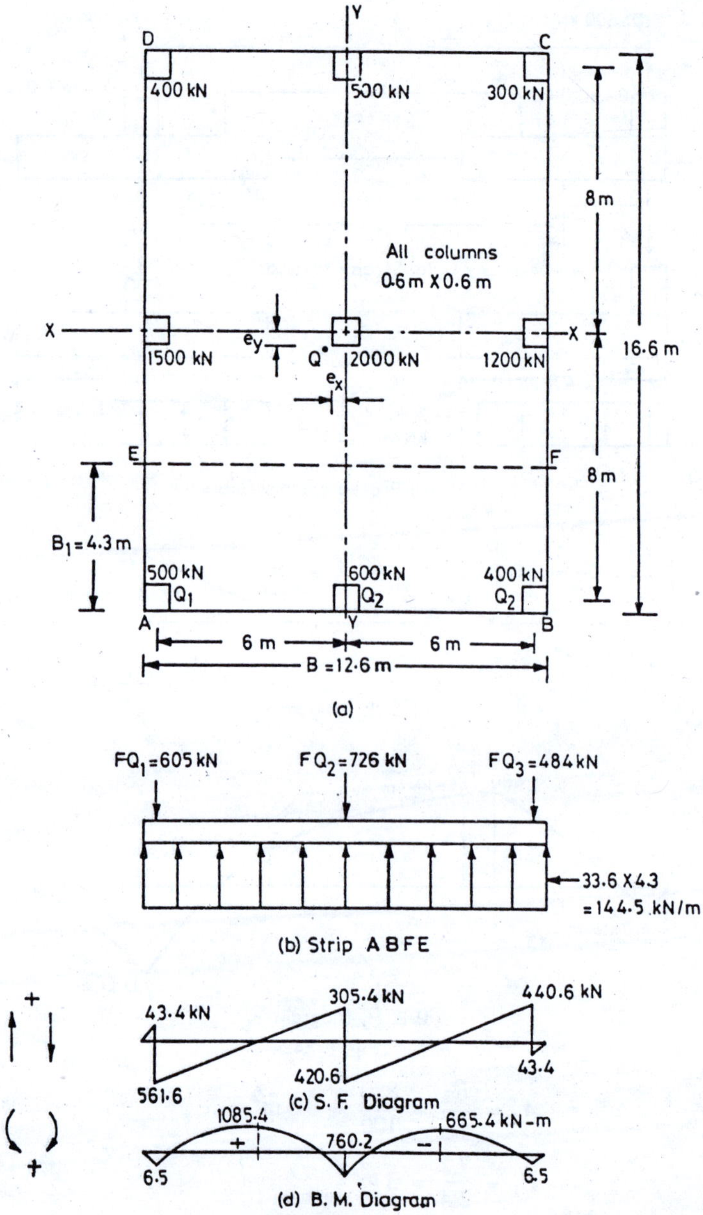

Fig. E-24.6.

Draw the S.F. and B.M. diagrams for the strip ABFE.

Solution. Total loads,

$$Q = 400 + 500 + 300 + 1500 + 2000 + 1200 + 500 + 600 + 400 = 7400$$

Taking moments about the face *AD*,

$$\bar{x} = \frac{(400 + 1500 + 500) \times 0.3 + (500 + 2000 + 6600)\,6.3 + (300 + 1200 + 400)\,12.3}{7400} = 5.89 \text{ m}$$

$$e_x = \frac{12.6}{2} - 5.89 = 0.41 \text{ m } (-ve)$$

Taking moments about the face *AB*,

$$\bar{y} = \frac{(500 + 600 + 400) \times 0.3 + (1500 + 2000 + 1200) \times 8.3 + (400 + 500 + 300) \times 16.3}{7400} = 7.98 \text{ m}$$

$$e_y = \frac{16.6}{2} - 7.98 = 0.32 \,(-ve)$$

From Eq. 24.57,

$$q = \frac{Q}{A} - \frac{Q \cdot e_x}{I_{yy}} \cdot x - \frac{Q \cdot e_y}{I_{xx}} \cdot y$$

or

$$q = \frac{7400}{(12.6 \times 16.6)} - \frac{7400 \times 0.41 \times x}{1/12 \times (16.6)(12.6)^3} - \frac{7400 \times 0.32 \times y}{1/12 \times 12.6 \times (16.6)^3}$$

or

$$q = 35.4 - 1.1\,x - 0.5\,y$$

Pressure at point A ($x = -6.3$ m, $y = -8.3$ m),

$$q_a = 35.4 - 1.1 \times (-6.3) - (0.5)(-8.3) = 46.4 \text{ kN/m}^2$$

Pressure at point B ($x = +6.3$ m, $y = -8.3$ m),

$$q_B = 35.4 - 1.1 \times 6.3 - (0.5)(-8.3) = 32.6 \text{ kN/m}^2$$

Pressure at point C ($x = +6.3$ m, $y = 8.3$ m),

$$q_C = 24.4 \text{ kN/m}^2$$

Pressure at point D ($x = -6.3$ m, $y = +8.3$ m).

$$q_D = 38.2 \text{ kN/m}^2$$

Fro the strip $ABFE$,

$$q_{av} = \frac{46.4 + 32.6}{2} = 39.5 \text{ kN/m}^2$$

From Eq. 24.58,

$$Q_{av} = \frac{1}{2}\,[(500 + 600 + 400) + 39.5 \times 12.6 \times 4.3] = 1820 \text{ kN}$$

From Eq. 24.59,

$$\bar{q}_{av} = q_{av}\left(\frac{Q_{av}}{q_{av} B_1 B}\right)$$

$$= 39.5\left(\frac{1820}{39.5 \times 12.6 \times 4.3}\right) = 33.6 \text{ kN/m}^2$$

From Eq. 24.60,

$$F = \frac{Q_{av}}{Q_1 + Q_2 + Q_3} = \frac{1820}{500 + 600 + 400} = 1.21$$

Fig. E-24.6 (*b*) shows the modified column loads.

$$FQ_1 = 605 \text{ kN}, \; FQ_2 = 726 \text{ kN}, \; FQ_3 = 484 \text{ kN}.$$

Pressure intensity per metre $= 33.6 \times 4.3 = 144.5$ kN/m

Figs. E-24.6 (*c*) and (*d*) show the S.F. and B.M. diagrams.

S.F. at a section X—X at a distance x from the edge AE

$$F_x = 605 - 144.5\,x$$

S.F. is zero at $x = 4.18$ m.

B.M. at section X—X

$$= -144.5 \times \frac{(4.18)^2}{2} + 605 \times 3.88$$

$$= 1085.4 \text{ kN-m}$$

Max. B.M. = **1085 kN-m.** Max. S.F. = **561.6 kN.**

Illustrative Example 24.7. *Determine the pressure distribution below a combined footing shown in Fig. E-24.7 (a). Take $k = 20$ N/cm^3 $EI = 10^{10}$ N-cm^2.*

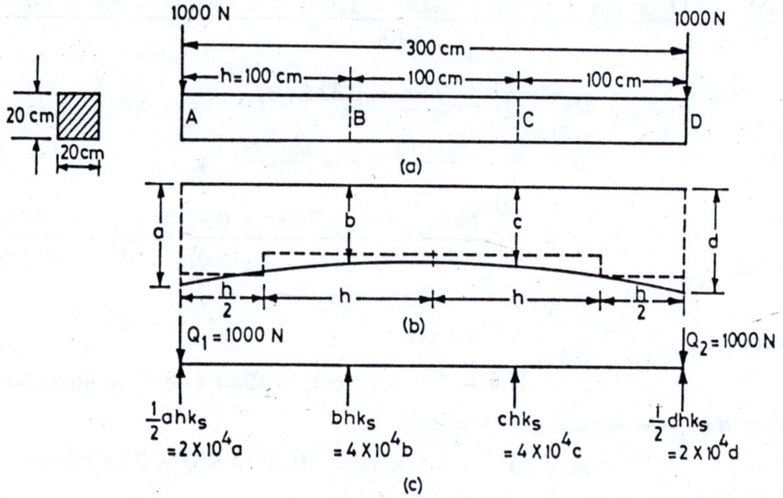

Fig. E-24.7.

Solution. $k_s = B \times k = 20 \times 20 = 400\ \text{N/cm}^2, h = 100\ cm.$

Let the settlements at points *A*, *B*, *C* and *D* be *a*, *b*, *c* and *d* respectively. Fig. E-24.7 shows the settlements and the equivalent loads due to soil pressure.

Eq. 24.67 can be used for writing the moment equations. Taking moments about *B*,

$$\frac{EI}{h^2}(a - 2\,b + c) = Q_1 h - ah^2 (k_s/2)$$

$$\frac{10^{10}}{10^4}(a - 2b + c) = 1000 \times 100 - 20 \times 10^5 a$$

$$a - 2\,b + c = 0.1 - 2\,a \qquad \text{...}(i)$$

Taking moment of forces on left of *C* about *C*,

$$\frac{EI}{h^2}(b - 2c + d) = 2Q_1 h - 2 \times 10^3 \times a \times 200 - 4 \times 10^3 \times b \times 100$$

or $$10^6 (b - 2c + d) = 2 \times 10^5 - 4a \times 10^6 - 4b \times 10^6$$

or $$b - 2\,c + d = 0.2 - 4\,a - 4b \qquad \text{...}(ii)$$

Likewise, taking moment of forces on right of *C* about *C*,

$$\frac{EI}{h^2}(b - 2c + d) = Q_2 h - 2 \times 10^3 d \times 10^2$$

or $$10^6 (b - 2c + d) = 10^5 - 2 \times 10^6 d$$

or $$b - 2\,c + d = 0.1 - 2\,d \qquad \text{...}(iii)$$

From the equilibrium of the vertical forces,

$$2 \times 10^4\, a + 4 \times 10^4\, b + 4 \times 10^4\, c + 2 \times 10^4 d = 1000 + 1000$$

$$a + 2b + 2c + d = 0.1 \qquad \text{...}(iv)$$

Eqs. (*i*) to (*iv*) can be solved for *a*, *b*, *c* and *d*.

In this case, because of symmetry,

$$a = d \qquad \text{and} \qquad b = c$$

Therefore, Eq. (*iv*) can be written as

$$2a + 4b = 0.10 \quad ...(v)$$

and Eq. (*i*) as $3a - b = 0.10$ ...(*vi*)

Solving Eq. (*v*), and (*iv*), $a = 0.0357$ cm and $b = 0.0071$ cm

Therefore, pressure at point *A* and *D*

$$= 0.0357 \times 20 = 0.714 \text{ N/cm}^2$$

and at points *B, C* $= 0.0071 \times 20 = 0.142 \text{ N/cm}^2$

PROBLEMS

A. Numericals

24.1. A raft (20 m × 10m) exerts a gross pressure of 200 kN/m^2 at foundation level. The depth of foundation is 2.5 m. If the soil is clay ($\phi' = 0$, $c_u = 80$ kN/m^2, $\gamma = 19$ kN/m^2), determine the factor of safety. The raft is for a basement. Use Skempton's equation. **[Ans.** 3.03]

24.2. Determine the depth below the ground surface of a raft foundation (30 m × 30 m) to carry a gross load of 150 MN. The soil is clay ($c_u = 15$ kN/m^2, $\gamma = 19$ kN/m^3). Take the factor of safety of 3. Also determine the depth for a fully compensated foundation. **[Ans.** 7.15 m; 8.75m]

24.3. Determine the allowable soil pressure for a raft (10 m × 20 m) if the blow count (*N*) over 20 m depth is 20. Use IS code method. **[Ans.** 216 kN/m^2]

24.4. Determine the allowable soil pressure for a raft (10 m × 10 m) if the depth is 5 m and the undrained cohesion is 40 kN/m^2. Take the factor of safety as 2.5. **[Ans.** 105.6 kN/m^2]

24.5. Design a rectangular combined footing to support two adjacent columns (size 40 cm × 40 cm) at a distance of 5 m and carrying loads of 3 MN and 4 MN. The lighter column is near the property line. The allowable soil pressure is 400 kN/m^2. **[Ans.** 6.20 × 2.9 m]

24.6. Design a trapezoidal combined footing for two columns (30 cm × 30 cm) carrying column loads of 1.2 MN and 0.90 MN, if the spacing between the two columns is 4 m. Take allowable soil pressure as 200 kN/m^2 and the length of footing as 5 m.

24.7. Design the footing in Problem 24.6 as a strap footing if the spacing between the column is 5 m.

24.8. Design a R.C.C. footing to carry a column (50 cm × 50 cm) with 2.5 MN load. The allowable soil pressure is 250 kN/m^2.

24.9. Design a R.C.C. footing for a wall 30 cm wide and having a load of 80 kN/m. The allowable soil pressure is 50 kN/m^2.

B. Descriptive and Objective Type

24.10. What are different types of shallow foundations? Explain with the help of sketches.

24.11. How would you fix the depth of foundation ? Discuss Rankine's formula for the minimum depth.

24.12. Discuss various types of loads that are to be considered in the design of foundations.

24.13. Describe the general procedure for the design of a shallow foundation.

24.14. Discuss the procedure for proportioning of footings for equal settlement.

24.15. Explain the procedure for the design of a (*a*) Strip Footing, (*b*) Spread Footing.

24.16. Where do you provide a combined footing ? Discuss the procedure for the design of the following types of combined footing: (*a*) Rectangular (*b*) Trapezoidal.

24.17. Describe the procedure for the design of a strap footing.

24.18. What are different types of raft foundation ? Discuss the procedure for the design of a raft foundation.

24.19. Describe the procedure for the design of a combined footing by elastic line method.

24.20. Explain the method for the design of a raft on Winkler's bed.

24.21. Define the coefficient of subgrade reaction ? How is it found ? Discuss the various factors affecting its value.

24.22. Write whether the following statements are true or false.

(*a*) A spread footing is provided for an isolated column.

(*b*) The critical section for bending moment in case of spread footing of a monolithic column is at the face of the column.

(*c*) A strip footing can be provided for a row of closely columns spaced.

(*d*) A trapezoidal footing is required when the column near the property line is heavier than the interior columns.

(*e*) The strap footings are useful when the two columns are closely spaced.

(*f*) The elastic line method of a combined footing is used for the design of a flexible footing.

(*g*) The parameter λL is dimensionless.

(*h*) The coefficient of sub-grade reaction of a cohesionless soil is independent of depth.

[**Ans.** True (*a*), (*b*), (*c*), (*d*), (*f*), (*g*)]

C. Multiple Choice Questions

1. Trapezoidal combined footings are required when

(*a*) the space outside the exterior column is limited.

(*b*) the exterior columns is heavier.

(*c*) both (*a*) and (*b*)

(*d*) neither (*a*) nor (*b*)

2. For the design of a strap footing, the following assumption is not made

(*a*) The strap is perfectly rigid.

(*b*) The soil pressure varies linearly.

(*c*) The interior footing is centrally loaded.

(*d*) The strap is not subjected to any direct soil pressure.

3. For conventional design of a rigid combined footing λL should be

(*a*) less than 0.8 (*b*) between 0.8 and 3.0

(*c*) more than 3.0 (*d*) None of above

where λL is characteristic parameter.

4. The coefficient of subgrade reaction depends upon

(*a*) the size of footing

(*b*) the shape of footing

(*c*) the depth of footing

(*d*) all the above.

5. According to Rankine's formula, the minimum depth of foundation when q = 180 kN/m^2, γ = 20 kN/m^3 and ϕ = 30° is

(*a*) 0.50 m (*b*) 0.75 m

(*c*) 1.0 m (*d*) 2.0 m

6. When the wind load is more than 25% of the combined dead and live load, the safe bearing capacity is usually increased by

(*a*) 15% (*b*) 20%

(*c*) 25% (*d*) 30%

7. The value of the factor λL when B = 20 cm, k = 20 N/cm^3, EI = 10^{10} N-cm^2 and L 300 cm is

(*a*) 1.0 (*b*) 2.0

(*c*) 3.0 (*d*) 4.0

[**Ans.** 1. (*c*), 2. (*b*), 3. (*a*), 4. (*d*), 5. (*c*), 6. (*c*), 7. (*c*)]

25
Pile Foundations

25.1. INTRODUCTION

When the soil at or near the ground surface is not capable of supporting a structure, deep foundations are required to transfer the loads to deeper strata. Deep foundations are, therefore, used when surface soil is unsuitable for shallow foundation, and a firm stratum is so deep that it cannot be reached economically by shallow foundations. The most common types of deep foundations are piles, piers and caissons. The mechanism of transfer of the load to the soil is essentially the same in all types of deep foundations.

A deep foundation is generally much more expensive than a shallow foundation. It should be adopted only when a shallow foundation is not feasible. In certain situations, a fully compensated floating raft may be more economical than a deep foundation. In some cases, the soil is improved by various methods to make it suitable for a shallow foundation.

A pile is a slender structural member made of steel, concrete or wood. A pile is either driven into the soil or formed in-situ by excavating a hole and filling it with concrete. A pier is a vertical column of relatively larger cross-section than a pile. A pier is installed in a dry area by excavating a cylindrical hole of large diameter to the desired depth and then backfilling it with concrete. The distinction between a cast in-situ pile and a pier is rather arbitrary. A cast in-situ pile greater than 0.6 m diameter is generally termed as a pier. A caisson is a hollow, watertight box or chamber, which is sunk through the ground for laying foundation under water. The caisson subsequently becomes an integral part of the foundation. A pier and a caisson differ basically only in the method of construction.

Pile foundations are discussed in this chapter. Piers and caissons are dealt with in chapter 26. Well foundations, which are special type of caissons, are discussed in chapter 27.

25.2. NECESSITY OF PILE FOUNDATIONS

Pile foundations are used in the following conditions:

(1) When the strata at or just below the ground surface is highly compressible and very weak to support the load transmitted by the structure.

(2) When the plan of the structure is irregular relative to its outline and load distribution. It would cause non-uniform settlement if a shallow foundation is constructed. A pile foundation is required to reduce differential settlement.

(3) Pile foundations are required for the transmission of structural loads through deep water to a firm stratum.

(4) Pile foundations are used to resist horizontal forces in addition to support the vertical loads in earth-retaining structures and tall structures that are subjected to horizontal forces due to wind and earthquake.

(5) Piles are required when the soil conditions are such that a wash out, erosion or scour of soil may occur from underneath a shallow foundation.

(6) Piles are used for the foundations of some structures, such as transmission towers, off-shore plateforms, which are subjected to uplift.

(7) In case of expansive soils, such as black cotton soil, which swell or shrink as the water content changes, piles are used to transfer the load below the active zone.

(8) Collapsible soils, such as loess, have a breakdown of structure accompanied by a sudden decrease in void ratio when there is an increase in water content. Piles are used to transfer the load beyond the zone of possible moisture changes in such soils.

25.3. CLASSIFICATION OF PILES

Piles can be classified according to (1) the material used (2) the mode of transfer of load, (3) the method of construction, (4) the use, or (5) the displacement of soil, as described below.

(1) Classification according to material used

There are four types of piles according to materials used.

(*i*) **Steel Piles.** Steel piles are generally either in the form of thick pipes or rolled steel H-sections. Pipe steel piles are driven into the ground with their ends open or closed. Piles are provided with a driving point or shoe at the lower end.

Epoxy coatings are applied in the factory during manufacture of pipes to reduce corrosion of the steel piles. Sometimes, concrete encasement at site is done as a protection against corrosion. To take into account the corrosion, an additional thickness of the steel section is usually recommended.

(*ii*) **Concrete Piles.** Cement concrete is used in the construction of concrete piles. Concrete piles are either precast or cast-in situ. Precast concrete piles are prepared in a factory or a casting yard. The reinforcement is provided to resist handling and driving stresses. Precast piles can also be prestressed using high strength steel pretensioned cables.

A cast-in situ pile is constructed by making a hole in the ground and then filling it with concrete. A cast-in situ pile may be cased or uncased. A cased pile is constructed by driving a steel casing into the ground and filling it with concrete. An uncased pile is constructed by driving the casing to the desired depth and gradually withdrawing casing when fresh concrete is filled. An uncased pile may have a pedestal.

(*iii*) **Timber Piles.** Timber piles are made from tree trunks after proper trimming. The timber used should be straight, sound and free from defects.

Steel shoes are provided to prevent damage during driving. To avoid damage to the top of the pile, a metal band or a cap is provided. Splicing of timber piles is done using a pipe sleeve or metal straps and bolts. The length of the pipe sleeve should be at least five times the diameter of the pile.

Timber piles below the water table have generally long life. However, above the water table, these are attacked by insects. The life of the timber piles can be increased by preservatives such as creosote oils. Timber piles should not be used in marine environment where these are attacked by various organisms.

(*iv*) **Composite piles.** A composite pile is made of two materials. A composite pile may consist of the lower portion of steel and the upper portion of cast-in situ concrete. A composite pile may also have the lower portion of timber below the permanent water table and the upper portion of concrete. As it is difficult to provide a proper joint between two dissimilar materials, composite piles are rarely used in practice.

(2) Classification Based on Mode of Transfer of Loads

Based on the mode of transfer of loads, the piles can be classified into 3 categories:

(*i*) **End-bearing piles.** End-bearing piles transmit the loads through their bottom tips. Such piles act as columns and transmit the load through a weak material to a firm stratum below. If bed rock is located within a reasonable depth, piles can be extended to the rock. The ultimate capacity of the pile depends upon the bearing capacity of the rock. If instead of bed rock. a fairly compact and hard stratum of soil exists at a reasonable depth, piles can be extended a few metres into the hard stratum. End-bearing piles are also known as *point- bearing piles.*

The ultimate load carried by the pile (Q_u) is equal to the load carried by the point or bottom end (Q_p).

(*ii*) **Friction piles.** Friction piles do not reach the hard stratum. These piles transfer the load through skin friction between the embedded surface of the pile and the surrounding soil. Friction piles are used when a hard stratum does not exist at a reasonable depth. The ultimate load (Q_u) carried by the pile is equal to the load transferred by skin friction (Q_s).

[**Note:** The term friction pile is actually a misnomer, as in the clayey soils, the load is transferred by adhesion and not friction between the pile surface and the soil].

The friction piles are also known as *floating piles,* as these do not reach the hard stratum.

(*iii*) **Combined end bearing and friction piles.** These piles transfer loads by a combination of end bearing at the bottom of the pile and friction along the surface of the pile shaft. The ultimate load carried by the pile is equal to the sum of the load carried by the pile point (Q_p), and the load carried by the skin friction (Q_s).

(3) Classification based on method of installation

Based on the method of construction, the piles may be classified into the following 5 categories:

(*i*) **Driven piles.** These piles are driven into the soil by applying blows of a heavy hammer on their tops.

(*ii*) **Driven and Cast-in-situ piles.** These piles are formed by driving a casing with a closed bottom end into the soil. The casing is later filled with concrete. The casing may or may not be withdrawn.

(*iii*) **Bored and Cast-in-situ piles.** These piles are formed by excavating a hole into the ground and then filling it with concrete.

(*iv*) **Screw piles.** These piles are screwed into the soil.

(*v*) **Jacked piles.** These piles are jacked into the soil by applying a downward force with the help of a hydraulic jack.

(4) Classification based on use

The piles can be classified into the following 6 categories, depending upon their use.

(*i*) **Load bearing piles.** These piles are used to transfer the load of the structure to a suitable stratum by end bearing, by friction or by both. These are the piles mainly discussed in this chapter.

(*ii*) **Compaction piles.** These piles are driven into loose granular soils to increase the relative density. The bearing capacity of the soil is increased due to densification caused by vibrations.

(*iii*) **Tension piles.** These piles are in tension. These piles are used to anchor down structures subjected to hydrostatic uplift forces or overturning forces.

(*iv*) **Sheet piles.** Sheet piles form a continuous wall or bulkhead which is used for retaining earth or water (see Chapter 20).

(*v*) **Fender piles.** Fender piles are sheet piles which are used to protect water-front structures from impact of ships and vessels.

(*vi*) **Anchor piles.** These piles are used to provide anchorage for anchored sheet piles. These piles provide resistance against horizontal pull for a sheet pile wall (see Chapter 20).

(5) Classification based on displacement of soil

Based on the volume of the soil displaced during installation, the piles can be classified into 2 categories:

(*i*) **Displacement piles.** All driven piles are displacement piles as the soil is displaced laterally when the pile is installed. The soil gets densified. The installation may cause heaving of the surrounding ground. Precast concrete pile and closed-end pipe piles are high displacement piles. Steel H-piles are low displacement piles.

(*ii*) **Non-displacement piles.** Bored piles are non-displacement piles. As the soil is removed when the hole is bored, there is no displacement of the soil during installation. The installation of these piles causes very little change in the stresses in the surrounding soil.

25.4. PILE DRIVING

Piles are driven into the ground by means of hammers or by using a vibratory driver. Such piles are called driven piles. In some special cases, piles are installed by jetting or partial augering.

The following methods are commonly used.

(1) Hammer Driving. Fig. 25.1 shows a pile driving rig. It consists of a hoist mechanism, a guiding frame and a hammer device. The hammers used for pile driving are of the following types:

(*i*) **Drop hammer.** A drop hammer is raised by a winch and allowed to drop on the top of the pile under gravity from a certain height. During the driving operation, a cap is fixed to the top of the pile and a cushion is generally provided between the pile and the cap. Another cushion, known as hammer cushion, is placed on the pile cap on which the hammer causes the impact. The drop hammer is the oldest type of hammer used for pile driving. It is rarely used these days because of very slow rate of hammer blows.

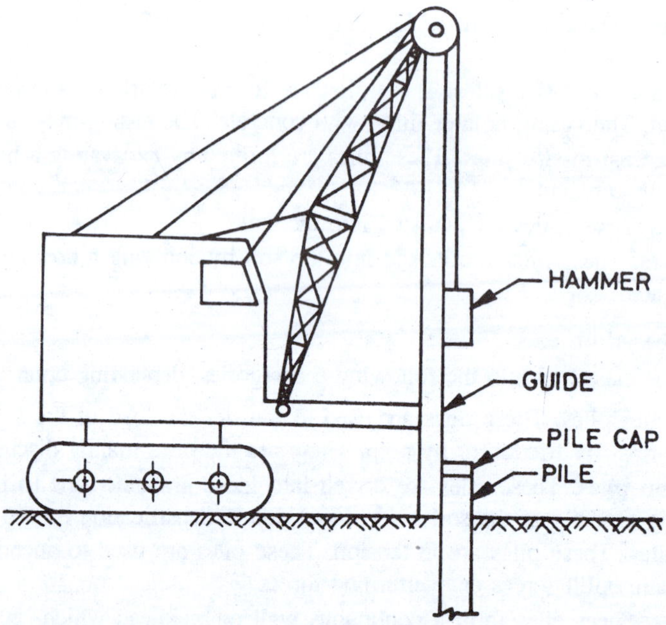

Fig. 25.1. Pile Driving Rig.

(*ii*) **Single-acting hammer.** In a single-acting hammer, the ram is raised by air (or steam) pressure to the required height. It is then allowed to drop under gravity on the pile cap provided with a hammer cushion.

(*iii*) **Double-acting hammer.** In a double-acting hammer, air (or steam) pressure is used to raise the hammer. When the hammer has been raised to the required height, air (or steam) pressure is applied to the other side of the piston and the hammer is pushed downward under pressure. This increases the impact energy of the hammer.

(*iv*) **Diesel hammer.** A diesel hammer consists of a ram and a fuel injection system. It is also provided with an anvil block at its lower end. The ram is first raised manually and the fuel is injected near the anvil. As soon as the hammer is released, it drops on the anvil and compresses the air-fuel mixture and ignition takes place. The pressure so developed pushes the pile downward and raises the ram. The fuel is again injected and the process is repeated.

The ram lifts automatically. It has to be manually raised only once at the beginning.

Diesel hammers are not suitable for driving piles in soft soils. In such soils, the downward movement of the pile is excessive and the upward movement of the ram after impact is small. The height achieved after the upward movement of the hammer may not be sufficient to ignite the air-fuel mixture.

Diesel hammers are self-contained and self-activated.

(2) Vibratory Pile Driver. A vibratory pile driver consists of two weights, called *exciters,* which rotate in opposite directions. The horizontal components of the centrifugal force generated by exciters cancel each other but the vertical components add. Thus a sinosoidal dynamic vertical force is applied to the pile, which forces the pile downward. The frequency of vibration is kept equal to the natural frequency of pile-soil system for better results.

A vibratory pile driver is useful only for sandy and gravelly soils. The speed of penetration is good. The method is used where vibrations and noise of conventional driving methods cannot be permitted.

(3) Jetting Techniques. When the pile is to penetrate a thin hard layer of sand or gravel overlying a softer soil layer, the pile can be driven through the hard layer by jetting techniques. Water under pressure is discharged at the pile bottom point by means of a pipe to wash and loosen the hard layer.

(4) Partial Augering Method. Batter piles (inclined piles) are usually advanced by partial augering. In this method, a power auger is used to drill the hole for a part of the depth. The pile is then inserted in the hole and driven with hammers to the required depth.

25.5. CONSTRUCTION OF BORED PILES

(*a*) Drilling of holes.

Bored piles are constructed after making a hole in the ground and filling it with concrete.

The following methods are used for drilling of the hole.

(1) Hand auger. A hand auger can be used for boring without casing in soils which are self-supporting, such as firm to stiff clays and silts and clayey sands and gravels above the water table. The depth of the hole is generally limited to about 4.5 m. The diameter of the hole is usually not more than 350 mm.

(2) Mechanical auger. For piles of diameter more than 350 mm or depth greater than 4.5 m., a hand auger becomes uneconomical. In such a case, a mechanical auger is used. A mechanical auger can be of rotary type or bucket type. It is power driven. The soil in this case must be self-supporting, with or without bentonite slurry. The soil should be free from tree roots, cobbles and boulders.

A continuous flight auger is also used to drill the bore hole.

(3) Boring rig. A boring rig is used to sink the hole in ground where hand or mechanical augering is not possible, such as water- bearing sand or gravels, very soft clays and silts and the soils having cobbles and boulders.

A specially designed boring rig, known as grab-type bored piling rig, is sometimes used. In this type of rig, the casing is given a continuous semi-rotary motion which causes its sinking as the bore hole is advanced by percussion drilling.

(4) Belling Bucket. Underreamed piles are large diameter bored piles with enlarged bases. Excavation for the underreamed piles is done by a special type belling bucket.

(*b*) Concreting

Before concrete is placed, the bored hole is bailed dry of water. Any loose or softened soil is cleaned out and the bottom of the hole is rammed. A layer of dry concrete is placed and rammed if the bottom of the hole is wet. Then the concrete with a readily workable mix (7.5 to 10 cm slump), not leaner than 300 kg cement/m^3 of concrete, is poured into a hopper placed at the mouth of the hole.

If the hole cannot be bailed or pumped dry before placing the concrete, the hole is lined with a casing throughout its depth. A mass of concrete is then deposited at the base of the hole by a *tremie pipe*. As soon as the concrete has hardened and formed a plug, the hole is pumped free of water. The casing is then gently turned and lifted slightly to break the joint with the plug. The hole is pumped dry. The remainder of the concreting is done by placing it dry upto the ground surface. The casing is then lifted entirely from the bore hole.

If the ground water is under a high pressure, there will be inflow of water between the concrete plug and the inside of the casing. The inflow should be stemmed by caulking. The casing is cut by oxy-acetylene just above the plug. The shaft is then concreted and the casing raised. The cut portion of the casing around the plug is left permanently in place.

Instead of plugging the base of the pile and concreting, an alternative method is to concrete the entire shaft under water using a tremie pipe. Concrete should be easily workable (slump 12.5 to 17.5 cm) and the cement content should be at least 400 kg/m^3. A retarder is added to the concrete if there is a risk of the concrete setting before the casing is lifted out. However, the quality of concreting done under water is not good. This method should be avoided as far as possible.

25.6. DRIVEN CAST-IN-SITU CONCRETE PILES

A driven cast-in-situ concrete pile is formed in the ground by driving a casing with a plug or shoe at its bottom. If the casing is removed after concrete has been placed, it is known as uncased or shell-less pile. On the other hand, if the casing is left in the ground after concreting, it is called a cased pile. In uncased piles, the concrete comes in direct contact with the soil. The concrete may be rammed or vibrated after its deposition. A pedestal may be formed at the lower end of the shell-less pile if required.

Cast-in-situ driven concrete piles can be broadly classified into three types: (*i*) cased pile, (*ii*) uncased pile, and (*iii*) pedestal type. Different types of piles with patent rights are available. The main difference between different patents is in the method of construction, as described below.

(1) The *Franki pile* is a type of driven and cast-in-situ displacement pile. A heavy steel pile is first pitched in a shallow foundation. A plug of lean concrete is then placed in the bottom of the pipe and compacted with a heavy steel rammer. The plug is then rammed and with it the pipe also goes down. This driving operation is continued until the bearing stratum is reached. The concrete is hammered to form a pedestal. A reinforcement cage is then placed in the pipe and the pile shaft is concreted. The pipe is withdrawn as the concrete is rammed.

(2) In *uncased-Western pile*, a heavy steel drive pipe of 35 cm diameter with a steel core is driven. The concrete is deposited in the pipe after removing the core. The concrete is rammed as the pipe is withdrawn. The pedestal is formed after the drive pipe has been lifted to some height.

(3) In *cased-Western pile*, the hole is made using a heavy steel drive pipe as for the uncased-Western type. A shell of 30 cm diameter is lowered inside the drive pipe. After the shell has been filled with concrete, the drive pipe is withdrawn. A pedestal can be formed by placing some concrete before lowering the shell and ramming.

(4) A *Western button-bottom pile* is formed by driving a steel pipe with a 43 cm diameter precast concrete point at its bottom. After reaching the required depth, a shell is lowered into the pipe and locked into the point at its bottom. The shell is then filled with concrete and the drive pipe is withdrawn.

(5) The *Raymond Taper or Step-Taper piles* are steel shell piles driven with a tapered steel mandrel. The mandrel and shells are driven to the required depth. The mandrel is then contracted and withdrawn, and the shell is concreted with or without a reinforcing cage.

(6) A *Simplex pile* is formed by driving a steel tube with a detachable cast iron shoe. After the required depth has been reached, reinforcing cage is lowered. The tube is extracted by wire ropes connected to a winch. At the same time, the concrete is placed and rammed by a falling rammer working inside the cage.

(7) *Alpha piles* are formed by driving a steel tube closed with a detachable cast iron shoe. A concrete-filled mandrel is driven inside the tube. The mandrel is gradually raised and some concrete is allowed to slump down in the tube. The concrete is refilled in the mandrel and it is driven down as the tube is raised. Thus a pedestal is formed. After the formation of the pedestal, the mandrel is raised and refilled with concrete in stages. In each stage, the concrete in the pile shaft is pressed against the soil by the dead weight of the hammer on the mandrel.

25.7. LOAD-CARRYING CAPACITY OF PILES

Like a shallow foundation, a pile foundation should be safe against shear failure and also the settlement should be within the permissible limits. The methods for estimating the load-carrying capacity of a pile foundation can be grouped into the following 4 categories.

(1) Static Methods. The static methods give the ultimate capacity of an individual pile, depending upon the characteristics of the soil. The ultimate load capacity is given by

$$Q_u = Q_p + Q_s \quad ...(25.1)$$

where Q_u = ultimate failure load, Q_p = point (or base or tip) resistance of the pile (Fig. 25.2), Q_s = shaft resistance developed by friction (or adhesion) between the soil and the pile shaft.

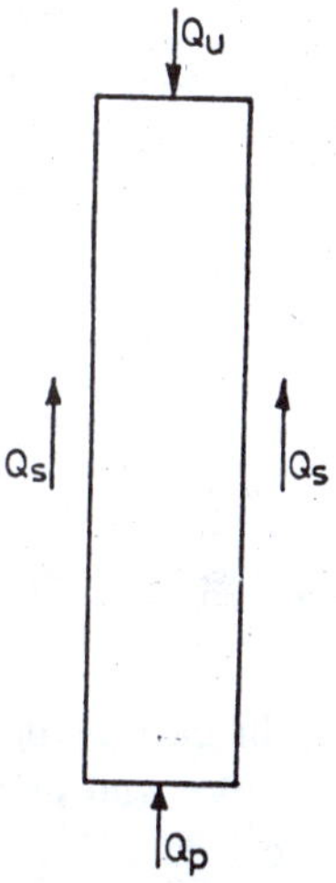

Fig. 25.2.

The methods for the determination of Q_p and Q_s are discussed in Sects. 25.8 and 25.9, respectively, for sand and clay.

The static formulas give a reasonable estimate of the pile capacity if judiciously applied.

(2) Dynamic Formulas. The ultimate capacity of piles driven in certain types of soils is related to the resistance against penetration developed during driving operation. The ultimate load capacity formulas are based on the principle that the resistance of a pile to further penetration by driving depends upon the energy imparted to the pile by the hammer. It is tacitly assumed that the load-carrying capacity of the pile is equal to the dynamic resistance during driving.

The dynamic formulas are not much reliable.

(3) In-situ Penetration Tests. The pile capacity can be determined from the results of in-situ standard penetration test. Empirical formulas are used to determine the point resistance and the shaft resistance from the standard penetration number (*N*). Alternatively, the static formulas can be used after determining the *N*-value, as this value is related to the angle of shearing resistance (ϕ).

Cone penetration tests are also used to estimate the pile capacity.

(4) Pile Load Tests. The most reliable method of estimating the pile capacity is to conduct the pile load test. The test pile is driven and loaded to failure. The pile capacity is related to the ultimate load or the load at which the settlements do not exceed the permissible limits.

All the above methods are discussed in detail in the following sections.

25.8 STATIC METHODS FOR DRIVEN PILES IN SAND

The ultimate capacity of a single pile driven into sand is obtained using Eq. 25.1, $Q_u = Q_p + Q_s$

where $$Q_p = q_p A_p \quad ...(25.2)$$

and $$Q_s = f_s A_s \quad ...(25.3)$$

In above equations, q_p is the ultimate bearing capacity of the soil at the pile tip and and A_p is the area of the pile tip; f_s is the average unit skin friction between the sand and the pile surface, and A_s is the effective surface area of the pile in contact with the soil.

(a) Methods for determination of Q_p. The ultimate bearing capacity (q_p) of the soil at the pile tip can be computed from the bearing capacity equation similar to that for a shallow foundation, as discussed in chapter 23. For sandy soils,

$$q_p = \bar{q} N_q + 0.4 \gamma B N_\gamma \quad \text{...(25.4)}$$

where $\bar{q}$ = effective vertical pressure at the pile tip, B = pile tip width (or diameter),

γ = unit weight of the soil in the zone of the pile tip.

N_q and N_γ = bearing capacity factors for deep foundations.

In driven piles, the second term of Eq. 25.4 is generally small and is, therefore, neglected. Thus

$$q_p = \bar{q} N_q \quad \text{...(25.5)}$$

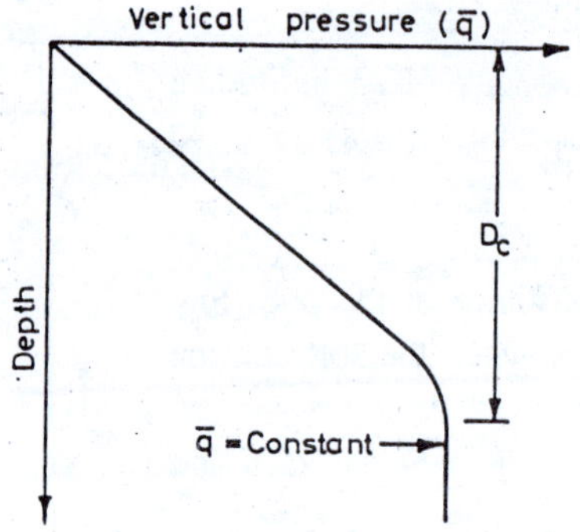

(a)

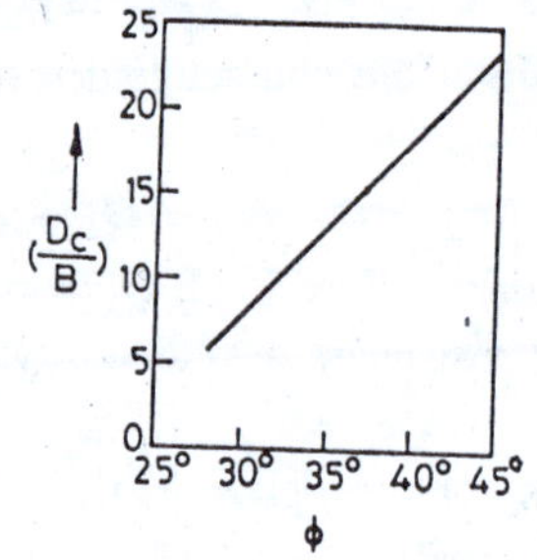

(b)

Fig. 25.3.

In case of driven piles, it has been established that the effective vertical pressure ($\bar{q}$) at the pile tip increases with depth only until a certain depth of penetration, known as the critical depth (D_c). Below the critical depth, the effective vertical pressure remains essentially constant [Fig. 25.3 (a)]. The critical depth

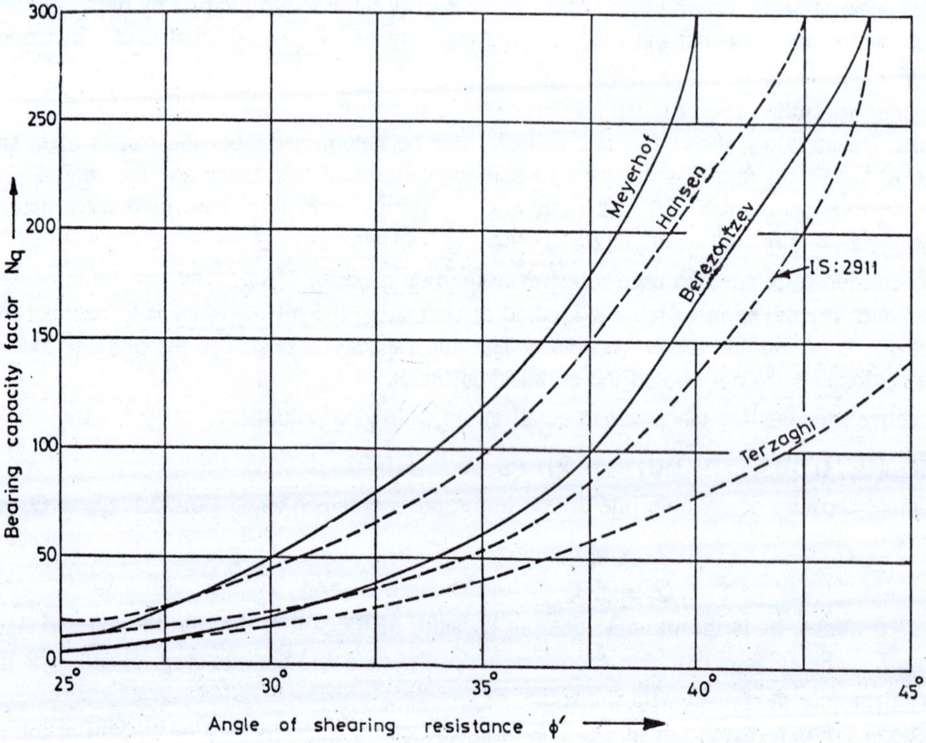

Fig. 25.4.

depends upon the angle of shearing resistance (ϕ') of the soil and the width (or diameter) of the pile [Fig. 25.3 (*b*)]. Its value can be roughly taken as 10 B for loose sands and 20 B for dense sands.

The bearing capacity factor N_q depends upon the angle of shearing resistance (ϕ'). Various investigators gave the expressions for N_q based on theoretical analysis. These values vary over a wide range because of different assumptions made in defining the shear zone near the pile tip. Fig. 25.4 shows the values of N_q given by various investigators and that given by IS : 2911. The values given by Berezontzev are quite dependable,and are generally used.

In the derivation of the value of N_q, it has been assumed that the soil above the pile tip is similar to the soil below the pile tip. If the pile penetrates a compact stratum only slightly and the soil above the tip is loose, it would be more appropriate to use the value of N_q for a shallow foundation given in chapter 23.

If the pile is of relatively large diameter, the second term in Eq. 25.4 becomes significant. The value of N_γ can be conservatively taken as the N_γ value used for shallow foundations, given in chapter 23.

Meyerhof's method for q_p. The point bearing capacity (q_p) of a pile generally increases with the depth of embedment (D_b) in the bearing stratum. It reaches a maximum value at an embedment ratio of $(D_b/B)_{cr}$. For a homogeneous soil, D_b is equal to the actual depth D of the pile, but for a pile which has penetrated into a bearing stratum for a small length, D_b is less than D. Beyond the critical value of $(D_b/B)_{cr}$, the value of q_p remains constant, equal to the limiting q_l. The critical ratio $(D_b/B)_{cr}$ depends upon the soil friction angle (ϕ) (Fig. 25.5).

Once the value of $(D_b/B)_{cr}$ has been determined, the following procedure is used to estimate q_p.

(1) Determine actual (D_b/B) ratio for the pile,

(2) Determine N_q for (D_b/B) ratio from Fig. 25.5.

The value of N_q increases linearly with (D_b/B) ratio and reaches a maximum value at $D_b/B = \frac{1}{2}(D_b/B)_{cr}$.

(3) Determine the point resistance Q_p as

$$Q_p = A_p \bar{q}\, N_q \le A_p q_l \qquad \ldots(25.6)$$

where $q_l = 50\, N_q \tan\phi$, $\bar{q}$ = vertical pressure at the pile tip (kN/m^2), A_p = area of the pile tip.

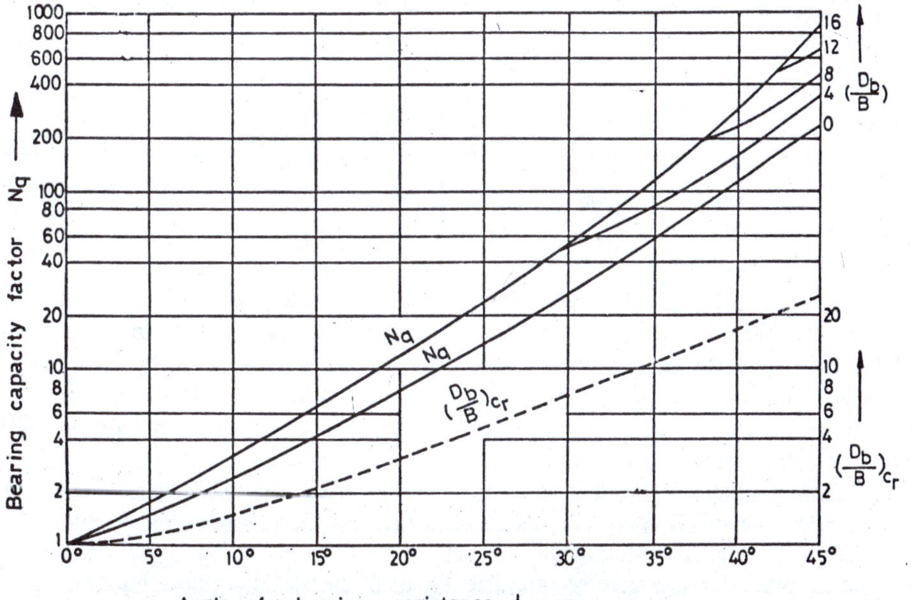

(After Meyerhof ,1976)

Fig. 25.5.

If the pile initially penetrates a loose sand layer and then a dense layer for a depth less than 10 B, the point resistance is given by

$$q_p = q_{l(1)} + \frac{[q_{l(2)} - q_{l(1)}]D_b}{10\,B} \leq q_{l(2)} \qquad \ldots(25.7)$$

where $q_{l(1)}$ = limiting unit point resistance of loose sand (= 50 N_{q1} tan ϕ_1)

$q_{l(2)}$ = limiting unit point resistance of dense sand (= 50 N_{q2} tan ϕ_2)

D_b = depth of penetration in dense sand.

It may be mentioned that the ultimate tip resistance given by Eq. 25.2 is the gross ultimate point resistance. The net tip load is given by

$$Q_p\,(\text{net}) = Q_p - (\bar{q}\,A_p)$$

However, in practice, the deduction of $\bar{q}\,A_p$ is not usually made and Q_p (net) is taken equal to Q_p.

In case of H-piles and open-ended pipe piles, the enclosed soil plug should be considered as the part of the pile for computing the area of the point (A_p).

(b) Methods of determination of Q_s. The frictional resistance Q_s is obtained from Eq. 25.3 after estimating the unit skin friction (f_s). The unit skin friction for a straight-sided pile depends upon the soil pressure acting normal to the pile surface and the coefficient of friction between the soil and the pile material (Fig. 25.6).

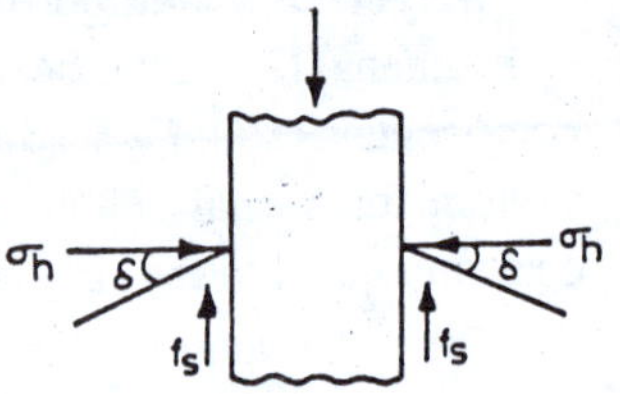

Fig. 25.6.

The soil pressure normal to the vertical pile surface is horizontal pressure (σ_h) and is related to the effective vertical soil pressure as

$$\sigma_h = K\bar{\sigma}_v$$

where K = earth pressure coefficient, $\bar{\sigma}_v$ = effective vertical pressure at that depth.

Thus unit skin friction (f_s) acting at any depth can be written as

$$f_s = \sigma_h \tan\delta \qquad \text{or} \qquad f_s = K\bar{\sigma}_v \tan\delta \qquad \ldots(25.8)$$

where tan δ = coefficient of friction between sand and the pile material.

Selection of suitable values of δ and K requires good engineering judgment. Tomilson (1975) gave the values of δ and K, as given in Table 25.1, based on the studies carried by Broms (1966).

Table 25.1. Values of δ and K.

Pile Material	δ	K *(loose sand)*	K *(dense sand)*
Steel	20°	0.50	1.0
Concrete	0.75 φ	1.0	2.0
Timber	0.67 φ	1.5	4.0

In general, the value of δ generally varies between 0.5 φ and 0.8 φ. In most cases, the value of K varies between 0.6 and 1.25. Meyerhof (1956) recommends that the value of K can be taken as 0.5 for loose sand (φ = 30°) and as 1.0 for dense sand (φ = 45°). According to IS : 2911—1979, the value of δ may be taken equal to φ. For driven piles in loose to medium sands, the recommended value of K is between 1 and 3.

Whether the sand should be considered as loose or dense depends upon not only on the initial relative density, but also on the method of installation. The larger the volume of soil displacement, the higher the value of the resulting friction. For high displacement driven piles, the soil is considered dense. For driven and cast-in place piles, the soil is considered as medium dense if the casing is left in place or if the concrete is compacted as the casing is withdrawn. The sand is considered to be loose, if the concrete is not compacted. Tapered piles develop greater unit friction than the straight piles. Further, the value of K is greater if the pile is driven into undisturbed soil than the one for installed in a predrilled hole.

As stated earlier, the effective vertical pressure ($\bar{\sigma}_v$) increases with depth only upto the critical depth. Below the critical depth, the value of $\bar{\sigma}_v$ remains constant.

The frictional resistance (Q_s) can be expressed as

$$Q_s = \sum_{i=1}^{n} K\ (\bar{\sigma}_v)_i\ \tan\delta\ \ (A_s)_i \qquad \text{...[25.9(a)]}$$

where n = number of layers in which the pile is installed,

$(\bar{\sigma}_v)_i$ = effective normal stress in ith layer,

$(A_s)_i$ = surface area of the pile in ith layer,

Assuming linear variation of σ_v,

$$Q_s = K \tan\delta\ A_p \times (q_0 D) \qquad \text{...[25.9(b)]}$$

where A_p = perimeter of pile; q_0 = average effective pressure = $\left(\frac{\gamma' D}{2}\right) = \frac{\bar{q}}{2}$

and D = depth of pile.

Eq. 25.9 (a) can be written as

$$Q_s = \sum_{i=1}^{n} K \tan\delta \qquad \text{(area of } \bar{\sigma}_v \text{ diagram)} \times \text{pile perimeter} \qquad \text{...[25.9(c)]}$$

Eq. 25.9 (b) is useful when variation of $\bar{\sigma}_v$ with depth has been plotted as $\bar{\sigma}_v$-diagram.

The ultimate load for the pile (Eq. 25.1) can be written as

$$Q_u = Q_p + Q_s$$

or

$$Q_u = \bar{q} N_q A_p + \sum_{i=1}^{n} K(\bar{\sigma}_v)_i \tan\delta\ (A_s)_i \qquad \text{...(25.10)}$$

$$Q_u = \bar{q} N_q A_p + K \tan\delta\ \text{(pile perimeter)} \left(\frac{\bar{q}}{2}\right) D \qquad \text{...[25.10 (a)]}$$

25.9. STATIC METHOD FOR DRIVEN PILES IN SATURATED CLAY

Eq. 25.1 can be used for the determination of the load-carrying capacity of driven piles in saturated clay. The point resistance (Q_p) can be expressed as (Eq. 25.2),

$$Q_p = q_p A_p$$

where q_p is the unit point resistance, equal to the ultimate bearing capacity (q_u) of the soil.

For cohesive soils ($\phi = 0$), the ultimate bearing capacity is found from the following equation, which is similar to that for a shallow foundation.

$$q_u = cN_c + q N_q$$

As $N_q = 1.0$ for $\phi = 0$, the above equation becomes

$$q_u = cN_c + q$$

Therefore, $\quad Q_p \text{ (gross)} = (cN_c + q) A_p$

or $\quad Q_p \text{ (net)} = cN_c A_p \qquad \text{...(25.11)}$

In above equations, c is the cohesion of the clay in the zone surrounding the pile tip, and N_c is the bearing capacity factor for the deep foundation.

The value of N_c depends upon the D/B ratio and it varies from 6 to 9.0. A value of N_c = 9.0 is generally used for the piles. In the case of short piles ($D/B \le 5.0$), the value of N_c is reduced to the values proposed by Skempton (see chapter 23).

The skin resistance (Q_s) of the pile can be expressed as (Eq. 25.3),

$$Q_s = c_a A_s \quad \ldots(25.12)$$

where c_a = unit adhesion (or skin friction) developed between clay and pile shaft.

The unit adhesion (c_a) is related to the unit cohesion by the relation

$$c_a = \alpha \bar{c} \quad \ldots(25.13)$$

where α is the adhesion factor and $\bar{c}$ is the average cohesion along the shaft length.

The value of α depends upon the consistency of the clay. For normally consolidated clays, the value of α is taken as unity. According to IS : 2911—1979, the value of α can be taken as unity for soils having soft to very soft consistency. Fig. 25.7 shows the variation of α with the undrained cohesion c. It may be noted that for normally consolidated clays, with c less than about 50 kN/m^2, the value of α is equal to unity.

As c increases, the value of α decreases. For over-consolidated stiff to hard clays, its value is usually taken as 0.3. For tapered piles, the value of α is generally 20% greater than that for a straight pile.

For very long piles ($D \geq 25$ m), the above method for estimating the skin friction is very conservative. For such soils, the unit skin friction also depends upon the effective overburden pressure. According to Vijayvergiya and Focht (1972), the average unit skin friction can be expressed as

$$f_s = \lambda (\bar{\sigma}_v + 2c) \quad \ldots(25.14)$$

where λ = friction capacity factor, $\bar{\sigma}_v$ = mean effective vertical stress for the embedment length, c = undrained cohesion.

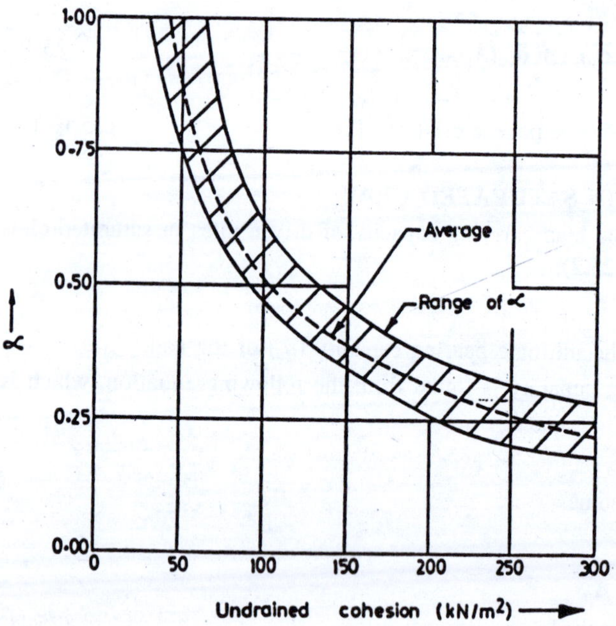

Fig. 25.7.

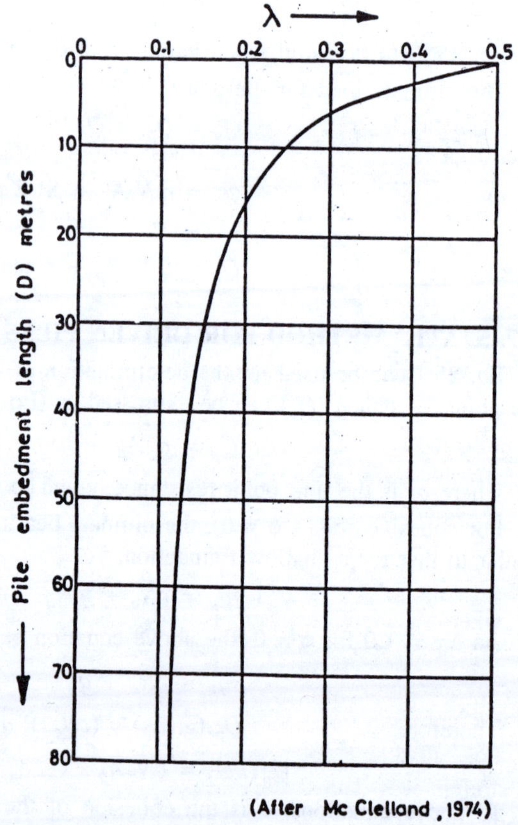

Fig. 25.8.

The value of λ can be obtained from Fig. 25.8 (McClelland, 1974).

Once the unit skin friction has been estimated, the shaft resistance is determined from Eq. 25.3.

For cohesive soils, the ultimate load can be determined by adding the point resistance and the shaft resistance (Eq. 25.1).

Thus $$Q_u = c\,N_c\,A_p + \alpha\,\bar{c}\,A_s \quad \text{...(25.15)}$$

As the clay gets remoulded when the pile is driven, this factor must be taken into account when estimating the load carrying capacity. The remoulded strength is always less than the undisturbed strength, but because of thixotropy, the strength improves with time. The rate of gain of strength depends upon the consolidation characteristics of the soil and the rate of dissipation of excess pore water pressure. When using Eq. 25.15, the value of c and $\bar{c}$ should be judiciously evaluated.

25.10. STATIC METHOD FOR BORED PILES

Bored piles are constructed by drilling a hole into the ground and filing it with concrete. The pile can be straight-sided for its full depth or may be constructed with a bell (or predestal) at its base. The piles with a pedestal are also known as *under-reamed* piles.

The load-carrying capacity of the bored piles can be determined using the procedure similar to that adopted for the driven piles. However, the values of the soil parameters are different, as described below.

***(a)* Bored Piles in Sand.** Eq. 25.10 can be used to determine the ultimate load. The equation can be written as

$$Q_u = (\bar{q}\,N_q)A_p + \sum_{i=1}^{n} (K\,\bar{\sigma}_v\,\tan\delta)\,(A_s)_i \quad \text{...(25.16)}$$

where $\bar{\sigma}_v$ = effective vertical pressure, limited to a maximum value given by the critical depth.

K = lateral earth pressure coefficient for bored foundation.

$\tan\delta$ = coefficient of friction between sand and concrete.

The sand in bored piles is loosened as a result of the boring operation, even though it may initially be in a dense or medium dense state. The value of ϕ to be used to obtain N_q should be for the loose condition.

An approximate value of K can be obtained from the following equation.

$$K = 1 - \sin\phi$$

The value of K generally varies between 0.3 and 0.75. An average value of 0.5 is usually adopted.

The value of $\tan\delta$ can be taken equal to $\tan\phi$ for bored piles excavated in dry soil. If a slurry has been used during excavation, the value of $\tan\delta$ should be reduced.

In general, for a given initial value of ϕ, bored piles have a unit point resistance of $\frac{1}{2}$ to $\frac{2}{3}$ of that of corresponding driven piles. In driven piles, there is densification. Cast-in-place piles with a pedestal show about 50 to 100% greater unit point resistance compared with those without a pedestal. The impact energy of the hammer compacts the soil during the formation of the pedestal.

***(b)* Bored piles in Clay.** Eq 25.15 can be used to estimate the ultimate load. The equation can be written as

$$Q_u = c\,N_c\,A_p + \alpha\,\bar{c}\,A_s \quad \text{...(25.17)}$$

where A_s = area of shaft that is effective in developing skin friction.

The value of α depends upon the pile type and the method of drilling. For straight shafts excavated dry, α is taken equal to 0.5 and that when drilled with slurry is 0.3. For belled shafts, the corresponding values are 0.3 and 0.15.

For calculating the area of shaft that is effective in developing skin friction, the lower 1.5 m (or 2 B) of the straight shaft and the bell section (if provided) are neglected, because of disturbance caused. For the same reason, the top 1.5 m is also neglected.

If a bored pile is installed in stiff, fissured clay, the value of cohesion (c) should be reduced to 75% of the value obtained from the triaxial test.

***(c)* Underreamed Piles in Clay.** The base area of an underreamed pile is increased by underreaming and providing a bulb [Fig. 25.9 (*a*)]. The ultimate load is given by

$$Q_u = c\,N_c\,A_b + \alpha\,\bar{c}\,A_s' \quad \text{...[25.17 (a)]}$$

where A_b is the area of the enlarged base.

The value of N_c is taken as 9.0. The adhesion factor α is taken as 0.40. When the bulb is slightly above the tip, A_b is taken equal to the area of the diameter of the bulb and the projected stem below the bulb is ignored. The average value of c at the bulb is taken. However, if the bulb is quite long, and there is considerable difference in the value of c at the bulb level and the level of the bottom tip of the pile, the ultimate load is given by

$$Q_u = \frac{\pi}{4}(B^2) \times (9\,c) + \frac{\pi}{4}(B_1^2 - B^2) \times 9\,c' + \alpha\,\bar{c}\,A_s \quad \text{...[25.17(b)]}$$

where B = diameter of the pile shaft, B_1 is the diameter of the bulb, c is the unit cohesion at the tip, and c' is the unit cohesion at bulb level. $\bar{c}$ is the average cohesion on the shaft.

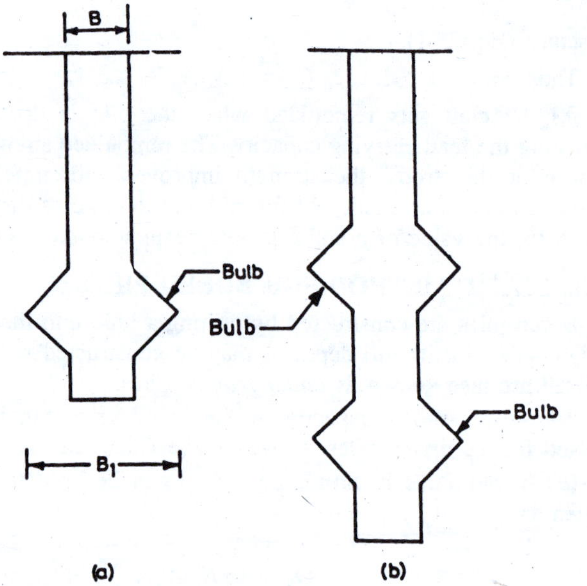

Fig. 25.9.

While calculating the surface area A_s, the length of the shaft equal to 2 B above the bulb is usually neglected. As the pile settles, there is a possibility of formation of a small gap between the top of the bulb and the overlying soil over a length of 2 B, and therefore, this length of the shaft is neglected. The little portion of the shaft projecting below the shaft is also neglected while computing A_s'.

When two or more bulbs are provided, the ultimate load is given by

$$Q_u = \frac{\pi}{4}(B^2) \times (9c) + \frac{\pi}{4}(B_1^2 - B^2) \times (9\,c') + \alpha\,\bar{c}\,A_s + c_a'\,A_{sb} \quad \text{...[25.17(c)]}$$

where A_s = surface area of shaft above the top bulb (ignoring 2 B length), A_{sb} = surface of the cylinder circumscribing the bulbs between top and bottom bulbs, c_a = average cohesion on A_s and c_a' = average cohesions on A_{sb}. For more details, see chapter 34.

25.11. ALLOWABLE LOAD

The allowable load (Q_{all}) is obtained from the ultimate load (Q_u) from the relation

$$Q_{all} = Q_u/FS \quad \text{...(25.18)}$$

where FS is the factor of safety. FS generally varies between 2.5 and 4.0, depending upon the uncertainties involved in the computation of the ultimate load. According to IS : 2911—1979, the minimum factor of safety on static formula shall be 2.5. The final selection of the value of the factor of safety should take into account the load settlement characteristics of the structure as a whole.

25.12. NEGATIVE SKIN FRICTION

When the soil layer surrounding a portion of the pile shaft settles more than the pile, a downward drag occurs on the pile. The drag is known as negative skin friction.

Negative skin friction develops when a soft or loose soil surrounding the pile settles after the pile has been installed. The negative skin friction occurs in the soil zone which moves downward relative to the pile. The negative friction imposes an extra downward load on the pile. The magnitude of the negative skin friction is computed using the same method as discussed in the preceding sections for the (positive) frictional resistance. However, the direction is downwards.

The net ultimate load-carrying capacity of the pile is given by the equation (Fig. 25.10).

$$Q_u' = Q_u - Q_{nsf} \quad \ldots(25.19)$$

where Q_{nsf} = negative skin friction,

Q_u' = net ultimate load.

Where it is anticipated that negative skin friction would impose undesirable, large downward drag on a pile, it can be eliminated by providing a protective sleeve or a coating for the section which is surrounded by the settling soil.

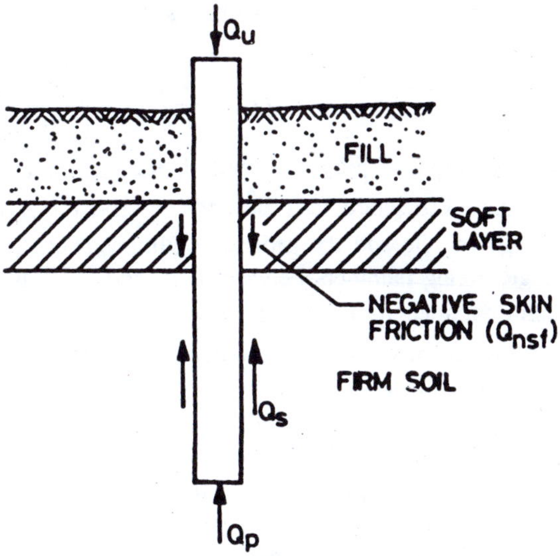

Fig. 25.10.

25.13. DYNAMIC FORMULAE

The load-carrying capacity of a driven pile can be estimated from the resistance against penetration developed during driving operation. The methods give fairly good results only in the case of free-draining sands and hard clays in which high pore water pressures does not develop during the driving of piles. In saturated fine-grained soils, high pore water pressure develops during the driving operation and the strength of the soil is considerably changed and the methods do not give reliable results. The methods cannot be used for submerged, uniform fine sands which may be loose enough to become quick temporarily and show a much less resistance.

The dynamic formulae are based on the assumption that the kinetic energy delivered by the hammer during driving operation is equal to the work done on the pile. Thus

$$W h \eta_h = R \times S \quad \ldots(25.20)$$

where W = weight of hammer (kN), h = height of ram drop (cm), η_h = efficiency of pile hammer, R = pile resistance (kN), taken equal to Q_u, and S = pile penetration per blow (cm).

In Eq. 25.20, no allowance has been made for the loss of energy during driving operation, loss caused by elastic contraction of the pile, soil, pile cap, cushion and due to the inertia of the pile. Some energy is also lost due to generation of heat. Various formulae have been proposed, which basically differ only in the methods for accounting of the energy losses, as described below.

(1) Engineering News Record Formula. According to Engineering News Record (ENR) formula (1888), the ultimate load is given by

$$Q_u = \frac{W h \eta_h}{S + C} \quad \ldots(25.21)$$

where S = penetration of pile per hammer blow. It is generally based on the average penetration obtained from the last few blows (cm), C = constant (For drop hammer, C = 2.54 cm and for steam hammer, C = 0.254 cm)

In Eq. 25.21, the product $W \times h$ can be replaced by the rated energy of hammer (E_n) in kN-cm. Thus

$$Q_u = \frac{E_n \eta_h}{S + C} \quad \ldots(25.22)$$

The efficiency η_h of the drop hammer is generally between 0.7 and 0.9, and that for a single-acting and a double-acting hammer is between 0.75 and 0.85. For diesel hammer, it usually lies between 0.80 and 0.90.

A factor of safety of 6 is usually recommended. However, the pile load tests reveal that the actual factor of safety varies between 2/3 and 30. The formula is, therefore, not dependable.

Modified Formula. The Engineering News Record formula has been modified recently. In the modified formula, the energy losses in the hammer system and that due to impact are considered. According to this formula.

$$Q_u = \frac{W h \eta_h}{S + C} \cdot \left(\frac{W + e^2 P}{W + P}\right) \qquad \text{...[25.21(a)]}$$

where P = weight of pile; e = coefficient of restitution, and η_h = hammer efficiency.

The hammer efficiency (η_h) depends upon various factors, such as pile driving equipment, driving procedure, type of pile and the ground conditions. For drop hammers, it is usually taken between 0.75 and 1.0; for single acting hammers between 0.75 and 0.85; for double-acting or differential hammer, η_h = 0.85 and for diesel hammer, η_h = 0.85 to 1.00.

The representative values of the coefficient of restitution (e) are as under.

Broomed timber pile	= 0.0
Good timber pile	= 0.25
Driving cap with timber dolly on steel pile	= 0.3
Driving cap with plastic dolly on steel pile	= 0.5
Helmet with composite plastic dolly and packing on R.C.C. pile	= 0.4

(2) Hiley Formula. Hiley (1925, 1930) gave a formula which takes into account various losses.

$$Q_u = \frac{W h \eta_b \eta_h}{(S + C/2)} \qquad \text{...(25.23)}$$

where η_h = efficiency of hammer blow, h = height of free fall of the ram or hammer (cm), S = final set or penetration per blow (cm), C = sum of temporary elastic compression of the pile, dolly, packings and ground (= $C_1 + C_2 + C_3$), C_1 temporary compression of dolly and packing (= 1.77 R/A, when the driving is without dolly, = 9.05 R/A, when the driving is with short dolly), C_2 = temporary compression of pile (= 0.657 RD/A), C_3 = temporary compression of ground (= 3.55 R/A), D = length of the pile, A = cross-sectional area of pile, R = pile resistance (tonnes).

The efficiency of hammer blow (η_b) depends upon the weight of hammer (W), weight of pile, anvil and helmet follower (P) and the coefficient of resistution (e).

(a) For W > eP,

$$\eta_b = \frac{W + e^2 P}{W + P} \qquad \text{...(25.24)}$$

(b) For W < eP,

$$\eta_b = \frac{W + e^2 P}{W + P} - \left(\frac{W - eP}{W + P}\right)^2 \qquad \text{...(25.25)}$$

The coefficient of resistution (e) varies from zero for a deteriorated condition of the head of pile to 0.5 for a steel ram of double-acting hammer striking on steel anvil and driving a reinforced concrete pile. For a C.I. ram of a single-acting or drop hammer striking on the head of R.C.C. pile, e = 0.4 and that striking on a well-conditioned driving cap and helmet with hard wood on R.C.C. pile, e = 0.25 (IS : 2911—1979).

(3) Danish Formula. According to Danish formula (1929),

$$Q_u = \frac{W \times h \times \eta_h}{S + 1/2 S_o} \qquad \text{...(25.26)}$$

where

$$S_o = \left[\frac{2 \eta_h (W h D)}{AE}\right]^{1/2} \qquad \text{...(25.27)}$$

in which S_o = elastic compression of pile, D = length of pile, A = cross-sectional area, E = modulus of elasticity of pile material.

The allowable load is found by taking a factor of safety of 3 to 4.

Eq. 25.27 can also be used to determine the final set (S) per blow.

Taking $Q_u = 3\,Q_a$,

$$S = \left(\frac{W h \eta_h}{3Q_a}\right) - \frac{1}{2} S_o \qquad \text{...(25.28)}$$

where Q_a = allowable load.

25.14. WAVE EQUATION ANALYSIS

As the hammer strikes the top of a pile, a stress wave is transmitted through the length of the pile. The wave transmission theory can be used to determine the load carrying capacity of the pile and the maximum stresses that can occur within the pile during driving operation.

In the wave equation analysis (Smith, 1962), the pile is represented by a series of individual spring-connected weights and spring damping resistance (Fig. 25.11). The weight W_1 represents the weight of the ram, and W_2 represents the weight of the pile cap. Weights W_3 to W_{10} correspond to the weights of incremental sections of the pile. The spring constant K_1 represents the elasticity of the cap block; the constants K_2 to K_{11} are for the elasticity of the pile sections. The damping springs R_3 to R_{11} represent the frictional resistance of the soil surrounding the shaft; R_{12} represents the soil resistance at the pile tip.

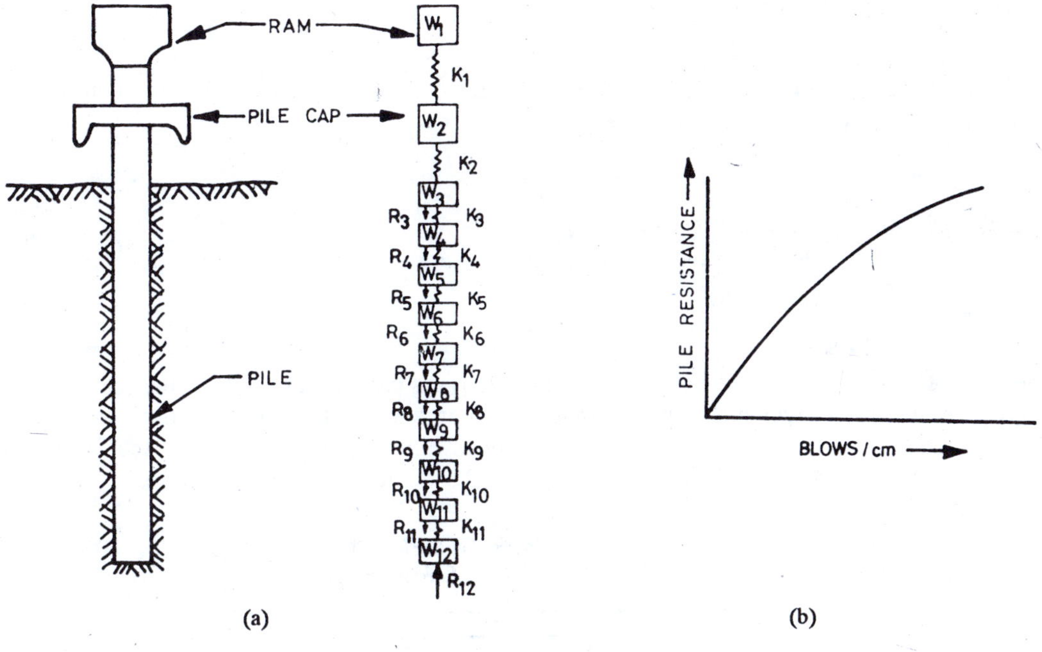

Fig. 25.11.

The propagation of the elastic wave through the pile is analogous to that caused by an impact on a long rod. A partial differential equation is written to describe the pile model shown in Fig. 25.11 (*a*). The equation is solved with the aid of a digital computer, and the pile capacity is determined. The pile capacity is expressed as a function of penetration per blow or blows per cm [Fig. 25.11 (*b*)].

The major drawback of the wave equation analysis for determination of the dynamic resistance is its dependence on a computer. Moreover, the field tests are required to estimate the equivalent spring constant and soil-damping values for the pile under study. Further, the results obtained are valid only for a particular pile driven by a specified pile hammer.

Despite the above shortcomings, the wave equation analysis is a useful tool for determining the pile capacity. The results can also be used for the selection of appropriate pile-driving equipment.

25.15. IN-SITU PENETRATION TESTS FOR PILE CAPACITY

(*a*) **Standard penetration test.** The load-carrying capacity of a pile can be estimated from the standard penetration test value (N).

(*i*) For driven piles in sand, the unit tip resistance (q_p) is related to the *uncorrected* blow count (N) near the pile point (Meyerhof 1976).

$$q_p = 40N(D/B) \leq 400N \qquad \text{...(25.29)}$$

where q_p = point resistance (kN/m^2), D = length of pile, B = width (diameter) of pile.

The value of q_p is usually limited to 400 N.

The average unit frictional resistance (f_s) is related to the average value of the blow count ($\overline{N}$).

For high displacement piles, $f_s = 2.0\overline{N}$ kN/m^2 ...[25.30 (*a*)]

For low displacement piles, $f_s = 1.0\ \overline{N}$ kN/m^2 ...[25.30 (*b*)]

where $\overline{N}$ is average of uncorrected N-values along the length of the pile.

(*ii*) For bored piles in sand, $q_p = 14\ N(D_b/B)$ kN/m^2 ...[25.31)]

where D_b = actual penetration into the granular soil.

For bored piles in sand, the unit frictional resistance (f_s) is given by

$$f_s = 0.67\ \overline{N}\ \text{kN/m}^2 \qquad \text{...(25.32)}$$

(*b*) **Dutch cone test.** Meyerhof (1965) relates the unit point resistance (q_p) and the unit skin traction (f_s) of driven piles to the cone point resistance (q_c).

Point resistance, $q_p = \dfrac{q_c}{10}(D_b/B)$ ...(25.33)

Unit skin friction (a) f_s (dense sand) $= q_c/200$ ...(25.34)

(*b*) f_s (loose sand) $= q_c/400$ ...(25.35)

(*c*) f_c (silt) $= q_c/150$ ...(25.36)

25.16. PILE LOAD TEST

The most reliable method for determining the load carrying capacity of a pile is the pile load test. The set-up generally consists of two anchor piles provided with an anchor girder or a reaction girder at their top (Fig. 25.12). The test pile is installed between the anchor piles in the manner in which the foundation piles are to be installed. The test pile should be at least 3 B or 2.5 m clear from the anchor piles.

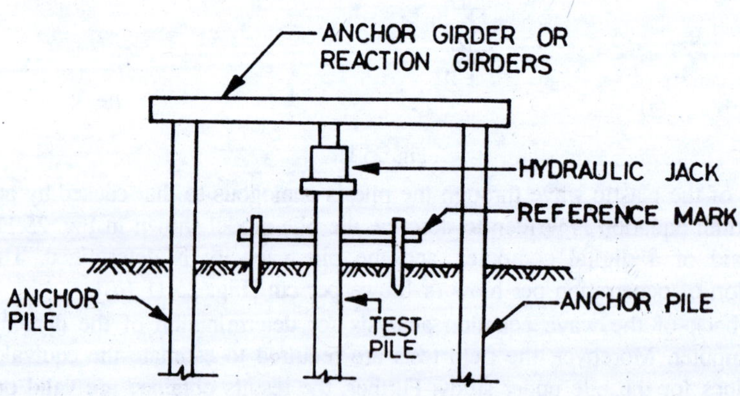

Fig. 25.12. Pile Load Test.

The load is applied through a hydraulic jack resting on the reaction girder. The measurements of pile movement are taken with respect to a fixed reference mark. The test is conducted after a rest period of 3 days after the installation in sandy soils and a period of one month in silts and soft clays. The load is applied in equal increment of about 20% of the allowable load. Settlements should be recorded with three dial gauges. Each stage of the loading is maintained till the rate of movement of the pile top is not more than 0.1 mm per hour in sandy soils and 0.02 mm per hour in case of clayey soils or a maximum of two hours (IS : 2911-1979). Under each load increment, settlements are observed at 0.5, 1, 2, 4, 8, 12, 16, 20, 60 minutes. The loading should be continued upto twice the safe load or the load at which the total settlement reaches a specified value. The load is removed in the same decrements at 1 hour interval and the final rebound is recorded 24 hours after the entire load has been removed.

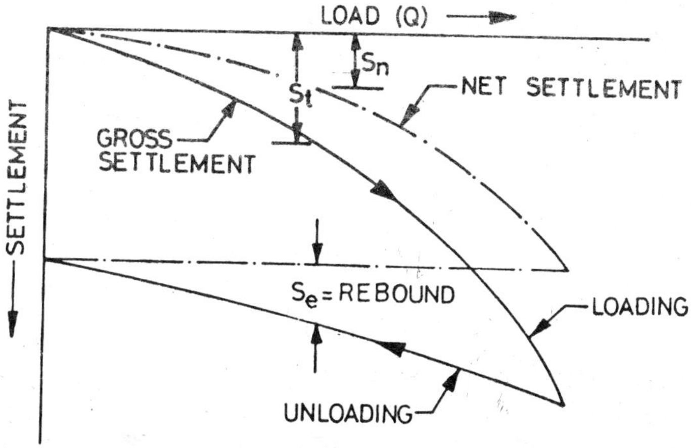

Fig. 25.13. Load Settlement Curve.

Fig. 25.13 shows a typical load-settlement curve (firm line) for loading as well as unloading obtained from a pile load test. For any given load, the net pile settlement (s_n) is given by

$$s_n = s_t - s_e \qquad \ldots(25.37)$$

where s_t = total settlement (gross settlement), s_e = elastic settlement (rebound).

Fig. 25.13 also shows the net settlement (chain dotted line).

Fig. 25.14 shows two load-net settlement curves obtained from a pile load tests on two different soils. At the ultimate load (Q_u), the load-net settlement curve becomes either linear as curve (2) or there is a sharp break as in the curve (1), as shown in the figure. The safe load is usually taken as one-half of the ultimate load.

According to IS : 2911, the safe load is taken as one-half of the load at which the total settlement is equal to 10 per cent of the pile diameter (7.5 per cent in case of under-reamed piles) or two-thirds of the final load at which the total settlement is 12 mm, whichever is less. According to another cirterion, the safe load is taken as one-half to two-thirds of the load which gives a net settlement of 6 mm.

The limiting settlement criteria are also sometimes specified. Under the load twice the safe load, the net settlement should not be more than 20 mm or the gross settlement should not be more than 25 mm.

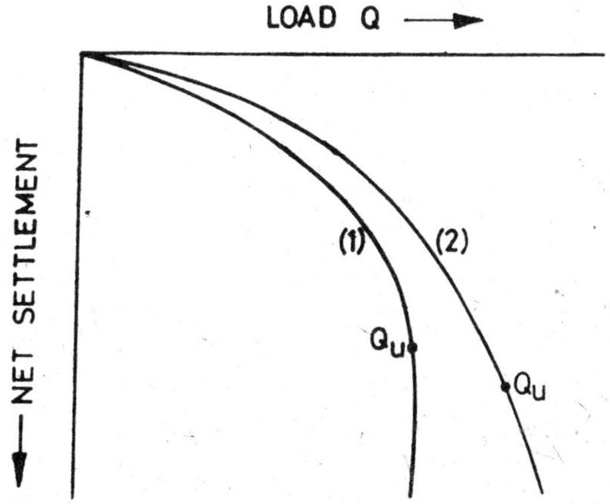

Fig. 25.14.

The test described above is known as *initial test.* It is carried out on a test pile to determine the ultimate load capacity and hence the safe load. The pile load test described in this section is a type of load-controlled test, in which the load is applied in steps. The test is also known as *slow maintained test.*

25.17. OTHER TYPES OF PILE LOAD TESTS

(1) Constant rate of penetration test. In a constant-rate of penetration test, the load on the pile is continuously increased to maintain a constant rate of penetration (from 0.25 to 5 mm per minute). The force required to achieve that rate of penetration is recorded, and a load-settlement curve is drawn. The ultimate load can be determined from the curve.

The test is considerably faster than a load-controlled test.

(2) Routine Load test. This test is carried out on a working pile with a view to determine the settlement corresponding to the allowable load. As the working pile would ultimately form a part of the foundation, the maximum load is limited to one and a half times the safe load or upto the load which gives a total settlement of 12 mm.

(3) Cyclic Load test. The test is carried out for separation of skin friction and point resistance of a pile. In the test, an incremental load is repeatedly applied and removed.

(4) Lateral Load test. The test is conducted to determine the safe lateral load on a pile. A hydraulic jack is generally introduced between two piles to apply a lateral load. The reaction may also be suitably obtained from some other support. The test may also be carried out by applying a lateral pull by a suitable set-up.

(5) Pull out test. The test is carried out to determine the safe tension for a pile. In the set-up, the hydraulic jack rests against a frame attached to the top of the test pile such that the pile gets pulled up.

25.18. GROUP ACTION OF PILES

A pile is not used singularly beneath a column or a wall, because it is extremely difficult to drive the pile absolutely vertical and to place the foundation exactly over its centre line. If eccentric loading results, the connection between the pile and the column may break or the pile may fail structurally because of bending stresses. In actual practice, structural loads are supported by several piles acting as a group. For columns, a minimum of three piles in a triangular pattern are used. For walls, piles are installed in a staggered arrangement on both sides of its centre line. The loads are usually transferred to the pile group through a reinforced concrete slab, structurally tied to the pile tops such that the piles act as one unit. The slab is known as a *pile cap.* The load acts on the pile cap which distributes the load to the piles (Fig. 25.15).

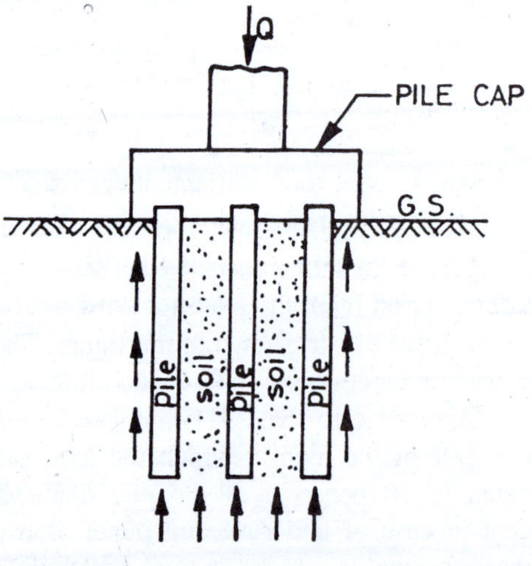

Fig. 25.15.

The load carrying capacity of a pile group is not necessarily equal to the sum of the capacity of the individual piles. Estimation of the load-carrying capacity of a pile group is a complicated problem. When the piles are spaced a sufficient distance apart, the group capacity may approach the sum of the individual capacities. On the other hand, if the piles are closely spaced, the stresses transmitted by the piles to the soil may overlap, and this may reduce the load-carrying capacity of the piles (Fig. 25.16). For such a case, the capacity is limited by the group action.

The efficiency (η_g) of a group of piles is defined as the ratio of the ultimate load of the group to the sum of individual ultimate loads.

Thus $$\eta_g = \frac{Q_{g(u)}}{NQ_u} \times 100 \qquad \ldots(25.38)$$

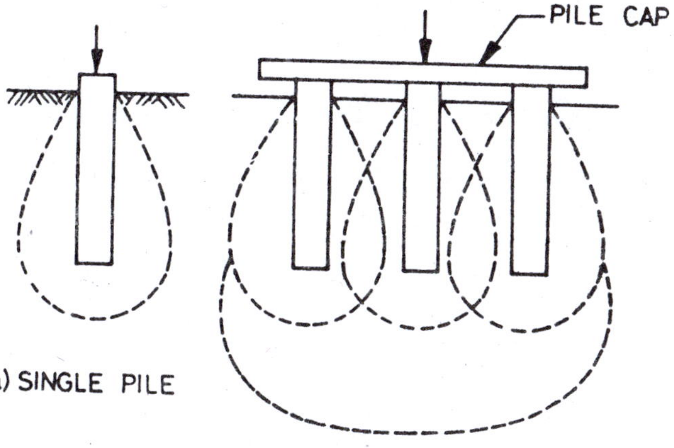

Fig. 25.16.

or $$\eta_g = \frac{Q_{g(u)}/N}{Q_u} \times 100$$

where $Q_{g(u)}$ = ultimate load of the group, Q_u = ultimate load of the individual pile,

N = Number of piles in the group.

Thus the groups efficiency is equal to the ratio of the average load per pile in the group at which the failure occurs to the ultimate load of a comparable single pile,

The group efficiency depends upon the spacing of the piles. Ideally, the spacing should be such that the efficiency is 100%. Generally, the centre to centre spacing is kept between 2.5 B and 3.5 B, where B is the diameter of the pile.

The methods for the determination of the ultimate load of the individual piles have been discussed earlier. The methods for the estimation of the ultimate load of the group are explained in the following sections.

25.19. PILE GROUPS IN SAND AND GRAVEL

For piles driven in loose and medium dense cohesionless soils, the group efficiency is high. The soil around and between the piles is compacted due to vibration caused during the driving operation. For better results, it is essential to start driving the piles at the centre and then work outward.

The piles and the soil between them move together as a unit when subjected to loads. The group acts as a pier foundation having a base equal to the gross plan area contained between the piles.

***(a)* End-bearing piles.** For driven piles bearing on dense, compact sand with a spacing equal to or greater than 3 B, the group capacity is generally taken equal to the sum of individual capacity. Thus

$$Q_g = N\,Q_u \qquad \text{...(25.39)}$$

In this case, the load taken by the group is much greater (η_g > 100%) than the sum of the individual capacities, and the piles fail as individual piles.

For spacing less than 3 B, the group capacity is found for the block of piles group.

***(b)* Friction piles.** The group efficiency of friction piles in sand is obtained from the following expression:

$$\eta_g = \frac{Q_{g(u)}}{NQ_u} \times 100 = \frac{f_s\,(P_g\,D)}{Nf_s\,(pD)} \times 100 \qquad \text{...(25.40)}$$

where P_g = perimeter of the block, p = perimeter of the individual pile, D = length of pile,

f_s = unit friction resistance.

If the centre-to-centre spacing is large, the group efficiency (η_g) may be more than 100%. The piles will behave as individual piles, and the group capacity is obtained from Eq. 25.39.

If η_g is less than 100%,

$$Q_g = \eta_g \frac{(NQ_u)}{100} \qquad ...(25.41)$$

The group efficiency can also be obtained from the Converse- Lebarre equation given below.

$$\eta_g = 1 - \left[\frac{(n-1)\,m + (m-1)\,n}{m\,n}\right] \cdot \frac{\theta}{90} \qquad ...(25.42)$$

where m = number of rows of piles, n = number of piles in a row, $\theta = \tan^{-1}$ (B/s), B = diameter of pile, s = spacing of pile, centre- to-centre, η_g = group efficiency (expressed as a ratio).

Bored piles. For bored piles in sand at conventional spacing of 3 B, the group capacity is taken as 2/3 to 3/4 times the sum of individual capacities for both the end-bearing and the friction piles. Thus

$$Q_g(u) = (2/3 \text{ to } 3/4)\,(N\,Q_u) \qquad ...(25.43)$$

In bored piles, there is limited densification of the sand surrounding the pile group. Consequently, the efficiency is lower.

25.20. PILE GROUPS IN CLAY

As the pile group acts as a block, its ultimate capacity is determined by adding the base resistance and the shaft resistance of the block. The capacity of the block having closely spaced piles ($s \leq 3\,B$) is often limited by the behaviour of the group acting as a block. The group capacity of the block is given by

or $$Q_g(u) = q_p\,(A_g) + \alpha\,c\,(P_g\,D) \qquad ...(25.44)$$

where q_p = unit point resistance (N_c = 9.0), A_g = base area of the block, P_g = perimeter of the block, D = depth of the block, α = adhesion factor (= 1.0 for soft clays), c = undrained cohesion.

As discussed earlier, the individual pile capacity is given by Eq. 25.15,

$$Q_u = q_p A_p + \alpha\,c\,(p \times D) \qquad ...(25.45)$$

The group capacity considering the piles as individual piles is given by

$$Q_g(u) = N\,Q_u \qquad ...(25.46)$$

The lower of the two values, given by Eqs. 25.44 and Eq. 25.46, is the actual capacity.

25.21. SETTLEMENT OF PILE GROUPS

The settlement of a pile group is due to elastic shortening of piles and due to the settlement of the soil supporting the piles. It is assumed that the pile group acts as a *single large deep foundation*, such as a pier

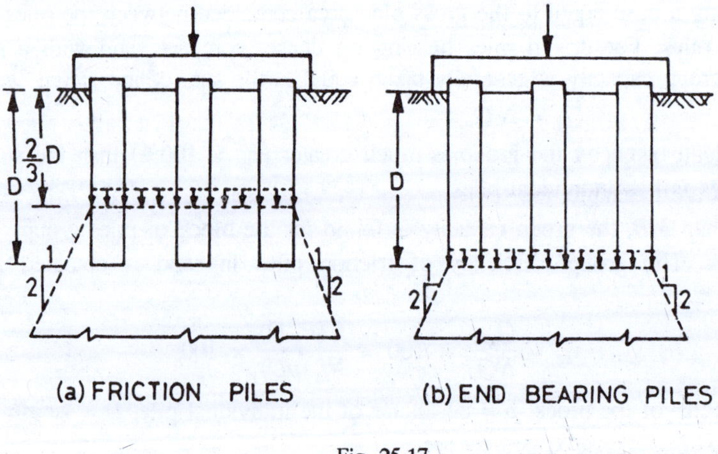

Fig. 25.17.

or a mat. The total load is assumed to act at a depth equal to two-thirds the pile length in the case of frictional piles [Fig. 25.17 (*a*)]. In the case of end-bearing piles, the total load is assumed to act at the pile tips [Fig. 25.17 (*b*)]. In the case of combined action, the frictional component is assumed to act at 2/3 D and the bearing component at the tip.

For determination of the settlements, the compression characteristics of the soil are required. For clayey soils, the characteristics are determined from laboratory tests on undisturbed samples. For cohesionless soils, the characteristics are obtained from empirical correlations developed from in-situ penetration tests.

(*a*) Cohesionless soils

(*i*) Skempton method. The settlement of the pile group is estimated from the settlement of a single pile, as determined in a pile-load test. The settlement of the group is generally very large because the pressure bulb for the group is much deeper than that of a single pile.

Skempton et al (1953) published curves (Fig. 25.18) relating the settlement of the pile group (s_g) of a given total foundation width to that of a single pile (s_o). The curves can be used for both driven and bored piles.

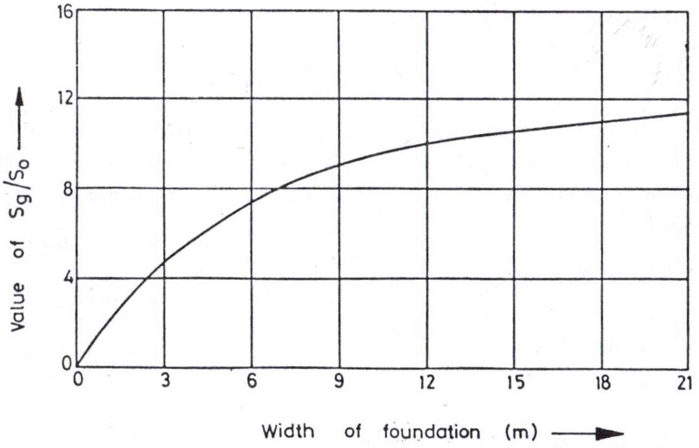

Fig. 25.18.

(*ii*) Meyerhof method. Meyerhof (1976) suggests the following empirical relation for the elastic settlement of a pile group in sands and gravels.

$$s_g = \frac{9.4\,q\,\sqrt{B_g}\;I}{N} \qquad \text{...(25.47)}$$

where s_g = settlement of group (mm), q = load intensity (= Q_g/A_g), B_g = width of the group, I = influence factor [= 1 $-D/(8\;B_g) \geq 0.5$], D = length of pile, N = corrected standard penetration number within the seat of settlement (approximately equal to B_g below the tip).

If static cone results are available, the settlement of the group can be obtained from the relation,

$$s_g = \frac{q\,B_g\,I}{2\,q_c} \qquad \text{...(25.48)}$$

where q_c = average cone penetration resistance within the seat of settlement.

(*b*) Clayey soils

The consolidation settlement of a pile group in clay can be determined using the procedure discussed in chapter 12. Generally, a 2 : 1 load distribution is assumed from the level at which the load acts. Sometimes, the load is assumed to spread outwards from the edge of the block at an angle of 30° to the vertical. For 2 : 1 distribution, the stress increase at the middle of each layer is calculated as (see chapter 11),

$$q_i = \frac{Q_g}{(B_g + Z_i)(L_z + Z_i)} \qquad ...(25.49)$$

where Z_i is the distance from the level of the application of the load to the middle of clay layer i.

The settlement of each layer caused by the increased stress is given by (see chapter 12).

$$\Delta s\,(i) = \frac{\Delta e\,(i)}{1 + e_o\,(i)}\, H_i \qquad ...(25.50)$$

where $\Delta e\,(i)$ = change of void ratio caused by the stress increase, $e_o\,(i)$ = initial void ratio of layer i, H_i = thickness of layer i.

Alternatively,
$$\Delta s\,(i) = C_c\, \frac{H_i}{1 + e_o\,(i)} \log \left(\frac{\bar{\sigma}_o + \Delta \sigma_i}{\bar{\sigma}_o} \right) \qquad ...(25.51)$$

The total consolidation settlement is equal to the sum of the settlement of all layers.

$$s_g = \Sigma\, \Delta s\,(i) \qquad ...(25.52)$$

25.22. SHARING OF LOADS IN A PILE GROUP

All the piles in a group share equal load if the load is central.

$$Q = \frac{Q_g}{N} \qquad ...(25.53)$$

However, if the load is eccentric or if the central load is accompanied by a moment, the sharing of load is computed assuming the pile cap as rigid. As the pressure distribution is planar, the pile reactions also vary linearly with the distance from the centroid of the cap (Fig. 25.19). The axial load in any pile m at a distance x from the centroid is given by

$$Q_m = \frac{Q_g}{N} \pm \frac{(Q_g \cdot e_x)\, x}{\Sigma x^2} \qquad ...(25.54)$$

where e_x = eccentricity of load about Y-Y-axis,

If the load is eccentric about both the axes.

$$Q_m = \frac{Q_g}{N} \pm \frac{(Q_g \cdot e_x)\, x}{\Sigma x^2} \pm \frac{(Q_z\, e_y)\, y}{\Sigma y^2} \qquad ...(25.55)$$

where e_y = eccentricity of load about X-X axis,

In the above equations, the positive sign is taken for the piles on the same side as the eccentricity.

If the load on any pile is negative, it indicates that the pile is in tension. If the pile is not designed for tension, the load in that pile is taken as zero, and the load between other piles is redistributed. This would cause extra compression in other piles.

25.23. TENSION PILES

Piles supporting high structures, such as tall chimneys, transmission towers, water towers, are required to resist uplift forces due to wind. Some of the piles in these structures are required to resist tensile forces and are known as tension piles.

Resistance to uplift forces is provided by the friction between the pile and the surrounding soil. The uplift resistance of a straight-shaft pile can be computed in the same manner as the frictional resistance in frictional piles. However, the unit skin friction (f_s) and adhesion (c_a) for the uplift resistance are considerably less than those for the compressive loads. It is

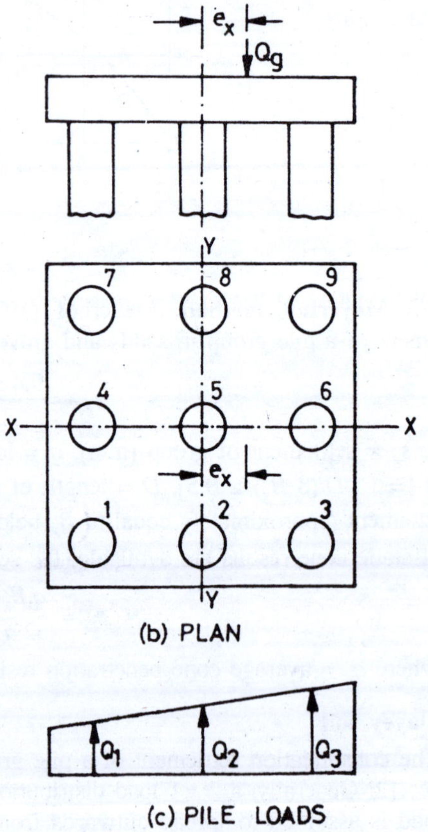

Fig. 25.19.

usual practice to reduce these values to one-half of the normal values if the piles are short. For large structures, it is essential to carry out pull out tests on piles to determine the safe value of the unit skin friction or adhesion for uplift forces.

The uplift resistance of piles can be considerably increased in the case of bored piles by under-reaming or belling out the bottom. A bulb can also be formed in the case of driven and cast- in place piles to increase the uplift resistance.

Mayerhof and Adams (1968) gave the following equations for the pull-out resistance (P_u).

(*a*) **Shallow piles** Fig. [25.20 (*a*)]. Pull-out resistance,

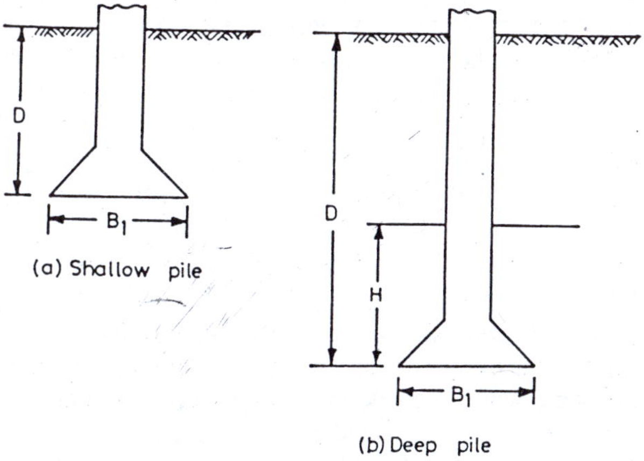

Fig. 25.20.

$$P_u = \text{cohesive resistance} + \text{frictional resistance}$$

or $$P_u = \pi B_1 c_u D + s_f \gamma \left(\frac{\pi}{2} B_1\right) D^2 K_u \tan \phi + W \qquad \text{...(25.56)}$$

where B_1 = diameter of enlarged base, c_u = undrained cohesion, D = length of pile, ϕ = angle of shearing resistance, s_f = shape factor (see Table 25.3), K_u = coefficient of lateral earth pressure (= $K_p \tan \frac{2}{3} \phi$), K_p = coefficient of passive earth pressure $\left(= \frac{1 + \sin \phi}{1 - \sin \phi} \right)$, γ = bulk unit weight, and W = weight of soil and pile in a cylinder of diameter B_1 and height D.

(*b*) **Deep piles** Fig. [25.20 (*b*)], Pull-out resistance,

$$P_u = \text{cohesive resistance} + \text{frictional resistance}$$

or $$P_u = \pi B_1 c_u H + s_f \gamma \left(\frac{\pi}{2} B_1\right) (2D - H) H K_u \tan \phi + W \qquad \text{...(25.57)}$$

where H = maximum height of rupture surface (see Table 25.3) (For deep piles $H \leq D$)

Table 25.3. Values of H/B_1, m and s_f

ϕ	20°	25°	30°	35°	40°	45°	50°
H/B_1	2.5	3.0	4.0	5.0	7.0	9.0	11.0
m	0.05	0.10	0.15	0.25	0.35	0.50	0.60
s_f	1.12	1.30	1.60	2.25	3.45	5.50	7.60

All other notations are the same as before.

For purely cohesive soils, as $\phi = 0$, the second term in Eqs. 25.56 and 25.57 is zero. For cohesionless soils, as $c_u = 0$, the first term is zero. The shape factor (s_f) is equal to $1 + mD/B_1$ for short piles, and equal to $1 + mH/B_1$ for deep piles, where m is a coefficient depending on ϕ.

25.24. LATERALLY LOADED PILES

Piles are sometimes subjected to lateral loads due to wind pressure, water pressure, earth pressure, earthquakes, etc. When the horizontal component of the load is small in comparison with the vertical load (say, less than 20%), it is generally assumed to be carried by vertical piles and no special provision for lateral load is made.

If the horizontal load is large, inclined piles, known as *raking piles or batter piles,* are provided to take the horizontal load. These piles have a high resistance to lateral loads, as a large portion of the horizontal component of the load is carried axially by the pile. Batter piles, along with vertical piles, are provided in situations where the horizontal loads are significant, such as wharves, jetties, bridge piers, trestles, retaining wall and tall chimneys.

Batter piles are driven at a batter ranging from 1 : 12 to 1 : 25. However, driving of batter piles is more expensive than that of vertical piles. The resistance to failure of vertical piles subjected to horizontal loads is provided by the passive resistance of a wedge of soil in front of the piles. In case of batter piles, additional resistance is provided by the skin friction and the end bearing. Therefore, batter piles are more effective than vertical piles in resisting horizontal loads.

It is generally assumed that batter piles can take the axial load equal to that in the corresponding vertical pile. As the axis of the batter pile is inclined, it can resist the horizontal load equal to $Q \cos \theta$, where Q is the axial load capacity and θ is the angle which the pile makes with the horizontal. When piles are oriented in two or three directions, Culmann's method, as described below, is used.

Steps : (1) Group the piles according to their slopes. [In Fig. 25.21 (*a*), the piles are grouped in 3 directions].

(2) Draw the geometry of the pile group to some scale, and mark the directions of the inclined load Q_g and the centre line of each pile group (R_1, R_2 and R_3).

(3) Determine the location of point A which is at the intersection of R_1 and Q_g.

(4) Join A to the point B which is at the intersection of R_2 and R_3.

(5) Draw the force triangle [Fig. 25.21 (*b*)]. Select the line ab parallel to AB. From b draw a line bc parallel to Q_g to some scale. Draw a vertical at c to determine ca which is equal to R_1.

From b draw a line parallel to R_3, and from a, line parallel to R_2, to complete the triangle abd.

(6) Determine forces in piles as follows.

The magnitudes of R_2 and R_3 are, respectively, given by ad and bd. However, R_2 is compressive and R_3 is tensile.

The magnitude of R_1 is given by ca which is compressive.

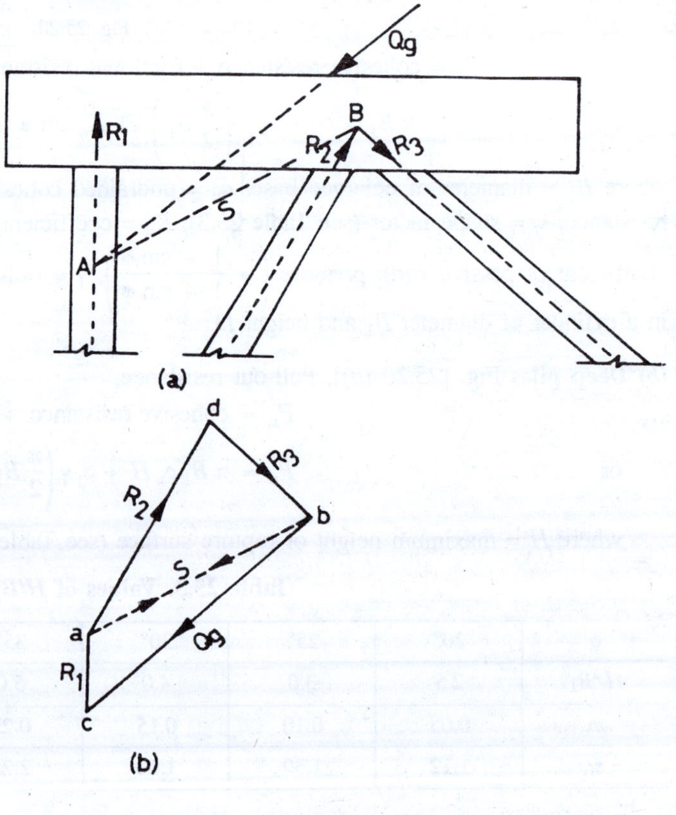

Fig. 25.21.

ILLUSTRATIVE EXAMPLES

Illustrative Example 25.1. *A concrete pile, 30 cm diameter, is driven into a medium dense sand (ϕ = 35°, γ = 21 kN/m^2, K = 1.0, tan δ = 0.70) for a depth of 8 m. Estimate the safe load, taking a factor of safety of 2.50.*

Solution. From Eq. 25.10, $Q_u = \bar{q}N_q A_p + \sum_{i=1}^{n} K(\bar{\sigma}_v)_i \tan\delta\,(A_s)_i$

For ϕ = 35°, $D_c/B \approx 12.0$, from Fig. 25.3.

$$D_c = 12 \times 0.3 = 3.6 \text{ m}$$

$$\text{Maximum value of } \bar{\sigma}_v = 3.6 \times 21 = 75.6 \text{ kN/m}^2$$

The value of N_q is taken from Berezontzev's curve (Fig. 25.4). N_q = 60.

Therefore, $Q_u = 75.6 \times 60 \times \pi/4 \times (0.3)^2 + K \tan\delta$ (area of $\bar{\sigma}_v$ diagram) $\times$ piles perimeter

$$= 320.5 + 1.0 \times 0.70 \times \left(\frac{1}{2} \times 75.6 \times 3.6 + 75.6 \times 4.4\right) \times \pi \times 0.3$$

or $Q_u = 320.5 + 309.2 = 629.7$ kN

$$\text{Safe load, } Q_a = \frac{Q_u}{2.5} = \frac{629.7}{2.5} = \mathbf{251.9\ kN}$$

Illustrative Example 25.2. *Determine the safe load for the pile in Illustrative Example 25.1, if the water table rises to 2 m below the ground surface. Take γ_w = 10 kN/m^3.*

Solution. Vertical pressure at the critical depth,

$$\bar{\sigma}_v = 2 \times 21 + 1.6 \times (21 - 10) = 59.6 \text{ kN/m}^2$$

Therefore, $Q_u = 59.6 \times 60 \times \pi/4 \times (0.3)^2 + \left(\frac{1}{2} \times 59.6 \times 3.6 + 59.6 \times 4.4\right) \times 0.7 \times \pi \times 0.3$

or $Q = 252.6 + 243.7 = 496.3$ kN

$$Q_a = \frac{496.3}{2.5} = 198.5 \text{ kN}$$

Illustrative Example 25.3. *A 30 cm diameter concrete pile is driven into a homogeneous consolidated clay deposit (c_u = 40 kN/m^2, α = 0.7). If the embedded length is 10 m, estimate the safe load (F.S. = 2.5).*

Solution. From Eq. 25.15, $Q_u = c N_c A_p + \alpha \bar{c} A_s$

Taking N_c = 9.0,

$$Q_u = (40 \times 9.0)\,\pi/4 \times (0.3)^2 + 0.7 \times 40\,(\pi \times 0.3) \times 10 = 289.2 \text{ kN}$$

$$Q_a = \frac{Q_u}{2.5} = \frac{289.2}{2.5} = \mathbf{115.7\ kN}$$

Illustrative Example 25.4. *A square concrete pile (30 cm side) 10 m long is driven into coarse sand (γ = 18.5 kN/m^3, N = 20). Determine the allowable load (F.S. = 3.0).*

Solution From Eq. 25.29, $q_p = 40\,N\,(D/B) \le 400\,N$

In this case, $40\,N\,(D/B) = 40 \times 20\ (10/0.3) = 26666.7 \text{ kN/m}^2$

$$400\,N = 400 \times 10 = 8000 \text{ kN/m}^2$$

Adopt the lower value of 8000 kN/m^2

From Eq. 25.30 (*a*), $f_s = 2.0 \times \bar{N} = 2.0 \times 20 = 40 \text{ kN/m}^2$

Therefore, $Q_u = q_p A_p + f_s A_s$

$= 8000\,(0.3 \times 0.3) + 40 \times (4 \times 0.3 \times 10) = 1200$ kN

$$Q_a = \frac{Q_u}{3} = \frac{1200}{3} = \mathbf{400\ kN}$$

Illustrative Example 25.5. *A square concrete pile (35 cm × 35 cm) is driven into a homogeneous sand layer ($\phi = 30°$, $\gamma = 17$ kN/m^3) for a depth of 10 m. Calculate the ultimate load. Use Meyerhof's method. Take $K = 1.3$ and $\delta = 18°$.*

Solution. From Fig. 25.5, $(D_b/B)_{cr} = 7.0$

or $D_c = 7 \times 0.35 = 2.45$

Also $D_b/B = 10/0.35 = 28.57$

$\bar{q} = 2.45 \times 17 = 41.7$ kN/m^2

From Fig. 25.5, $N_q = 55.0$

From Eq. 25.6, $Q_p = A_p \bar{q} N_q \leq A_p q_l$

In this case, $A_p \bar{q} N_q = (0.35 \times 0.35)(41.7 \times 55) = 280.9$ kN

$A_p q_l = (0.35 \times 0.35)(50 \times 55 \tan 30°) = 194.5$ kN

Adopting the lower value, $Q_p = 194.5$ kN

From Eq. 25.8, $f_s = K \bar{\sigma}_v \tan \delta$

Therefore, $Q_s = K \tan \delta$ (area of $\bar{\sigma}_v$ diagram) perimeter

$= 1.3 \tan 18° \times (41.7/2 \times 2.45 + 41.7 \times 7.55) \times 4 \times 0.35 = 216.5$ kN

Thus $Q_u = 194.5 + 216.5 = \mathbf{411\ kN}$

Illustrative Example 25.6. *A concrete pile, 40 cm diameter, is driven 25 m into a soft clay ($c_u = 25.0$ kN/m^2, $\gamma = 19$ kN/m^3). Determine the allowable load using Vijayvergia and Focht method (F.S. = 2.5). The water table is at the ground surface.*

Solution. Taking $N_c = 9.0$,

$Q_p = c_u N_c A_p = 25 \times 9 \times \pi/4 \times (0.4)^2 = 28.3$ kN

From Eq. 25.14, $f_s = \lambda\,(\bar{\sigma}_v + 2c)$

From Fig. 25.8, for D = 25 m, $\lambda = 0.16$

Therefore, $$f_s = 0.16\left[\frac{1}{2} \times 25 \times (19 - 10) + 2 \times 25\right] = 26 \text{ kN/m}^2$$

Thus $Q_s = 26 \times (\pi \times 0.4) \times 25 = 816.4$

$Q_u = 28.3 + 814.4 = 844.7$ kN

$Q_u = 844.7/2.5 = \mathbf{337.9\ kN}$

Illustrative Example 25.7. *A 25 m deep bored pile has a shaft of 1 m diameter and enlarged base of 2.5 m diameter in the lower 1.5 m depth. The undrained cohesion of the soil varies from 100 kN/m^2 at the top to 150 kN/m^2 at the base. Determine the safe load (F.S. = 2.5). Take $\alpha = 0.45$.*

Solution. Total depth of the shaft = 25 – 1.5 = 23.5 m

Assuming no adhesion for a distance 2 B above the bell, the effective depth is 21.5 m.

$$c_u \text{ at that depth} = 100 + \frac{(150 - 100)}{25} \times 21.5 = 143 \text{ kN/m}^2$$

Therefore, $Q_u = (150 \times 9)(\pi/4) \times (2.5)^2 + 0.45 \times (100 + 143)/2 \times \pi \times 1 \times 21.5$

$= 10319.5$ kN

$$Q_a = \frac{10319.5}{2.5} = \mathbf{412.8\ kN}$$

Illustrative Example 25.8. *A precast concrete pile (35 cm × 35 cm) is driven by a single-acting steam hammer. Estimate the allowable load using (a) Engineering News Record Formula (F.S. = 6), (b) Hiley Formula (F.S. = 4) and (c) Danish Formula (F.S. = 4).*

Use the following data.

(*i*)	Maximum rated energy	= 3500 kN-cm
(*ii*)	Weight of hammer	= 35 kN
(*iii*)	Length of pile	= 15 m
(*iv*)	Efficiency of hammer	= 0.8
(*v*)	Coefficient of resistitution	= 0.5
(*vi*)	Weight of pile cap	= 3 kN
(*vii*)	No. of blows for last 25.4 mm	= 6
(*viii*)	Modulus of elasticity of concrete	= 2×10^7 kN/m^2

Assume any other data, if required. Take the weight of pile as 73.5 kN.

Solution. (*a*) From Eq. 25.22,

$$Q_u = \frac{E_n \eta_h}{S + C} = \frac{3500 \times 0.80}{2.54/6 + 0.254} = 4133.9 \text{ kN}$$

Allowable load, $Q_a = 4133.9/6 = \mathbf{689\ kN}$

(*b*) Weight of pile = 73.5 kN

$$P = 73.5 + 3.00 = 76.5 \text{ kN}$$

$$eP = 0.5 \times 76.5 = 38.2 \text{ kN}$$

As $W = 35$ *kN*, $W < eP$

From Eq. 25.25,

$$\eta_b = \frac{W + e^2 P}{W + P} - \left(\frac{W - eP}{W + P}\right)^2$$

$$= \frac{35 + (0.5)^2 \times 76.5}{35 + 76.5} - \left(\frac{35 - 0.5 \times 76.5}{35 + 76.5}\right)^2$$

or $\eta_b = 0.484$

From Eq. 25.23,

$$Q_u = \frac{(Wh)\,\eta_b\,\eta_h}{(S + C/2)}$$

or

$$Q_u = \frac{3500 \times 0.484 \times 0.8}{2.54/6 + C/2} = \frac{1358}{2.54/6 + C/2} \quad \ldots(a)$$

Assuming driving is with dolly,

$$C = (9.05 + 0.657\,D + 3.55)\,R/A$$

$$C = \frac{(9.05 + 0.657 \times 15 + 3.55)\,R}{35 \times 35}$$

or $C = 0.018\ R = 0.018\ Q_u$

where Q_u is in tonnes. For Q_u in kN,

$$C = 0.0018\ Q_u$$

Therefore, Eq. (*a*) gives

$$Q_u = \frac{1358}{0.423 + 0.0009\,Q_u}$$

Solving, $Q_u = 1016$ kN

Allowable load, $Q_a = \frac{1016}{4} = \mathbf{254\ kN}$

(c) From Eq. 25.26, $$Q_u = \frac{(Wh) \times \eta_h}{S + S_o/2}$$

$$= \frac{3500 \times 0.8}{2.54/6 + 0.5\,S_o} = \frac{2800}{0.423 + 0.5\,S_o} \quad ...(b)$$

From Eq. 25.27, $$S_o = \sqrt{\frac{2\,\eta_h\,(Wh)\,D}{AE}}$$

$$= \sqrt{\frac{2 \times 0.8 \times 3500 \times 1500}{35 \times 35 \times 2 \times 10^7 \times 10^{-4}}} = 1.85 \text{ cm}$$

Therefore, Eq. (b) gives $$Q_u = \frac{2800}{0.423 + 0.5 \times 1.85} = 2077.2 \text{ kN}$$

Allowable load, $$Q_a = \frac{2077.2}{4} = \mathbf{519.3\ kN}$$

Illustrative Example 25.9. *A 30 cm diameter pile of length 12 m was subjected to a pile load test and the following results were obtained.*

Load (kN)	0	500	1000	1500	2000	2500
Settlement during loading (cm)	0	0.85	1.65	2.55	3.8	6.0
Settlement during unloading (cm)	4.0	4.6	5.2	5.5	5.8	6.0

Determine the allowable load.

Solution. Fig. E-25.9 shows the load settlement curve, and the rebound curve. The net settlements are calculated below. The figure also shows the net settlements.

Load (kN)	0	500	1000	1500	2000	2500
Gross settlement (cm)	0	0.85	1.65	2.55	3.80	6.00
Rebound (cm)	0	0.60	1.20	1.50	1.80	2.00
Net settlement (cm)	0	0.25	0.45	1.05	2.00	4.00

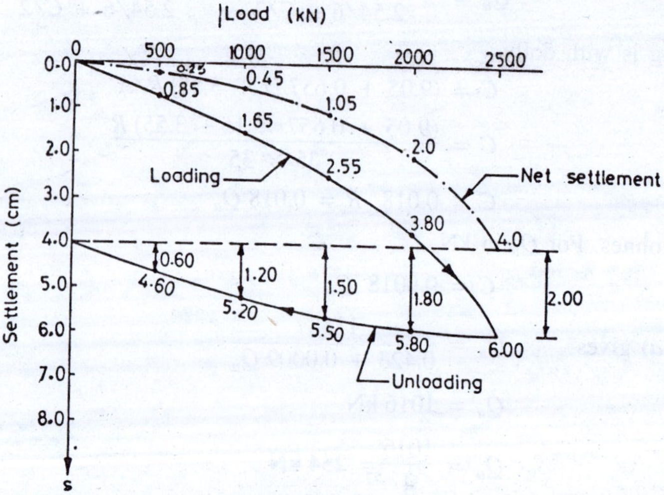

Fig. E-25.9

The allowable load is determined using the following criteria.

(*i*) $\frac{2}{3}$ of load corresponding to a gross settlement of 12 mm,

$$Q_a = \frac{2}{3} \times 750 = 500 \text{ kN}$$

(*ii*) $\frac{2}{3}$ of load corresponding to a net settlement of 6 mm.

$$Q_a = \frac{2}{3} \times 1125 = 750 \text{ kN}$$

(*iii*) $\frac{1}{2}$ of load corresponding to a gross settlement of $B/10$ (= 3 cm),

$$Q_a = \frac{1}{2} \times 1700 = 850 \text{kN}$$

The allowable load is the least of the three values.

$$Q_a = 500 \text{ kN}$$

Illustrative Example 25.10. *A pile group consisting of 12 piles (Fig. E-25.10) is subjected to a total load of 4 MN, with eccentricity e_x = 0.3 m, e_y = 0.4 m. Determine the maximum load in an individual pile.*

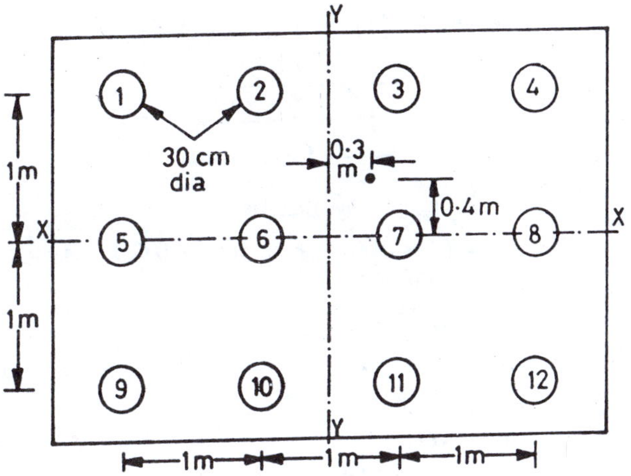

Fig. E-25.10

Solution. From Eq. 25.55, $Q_m = \frac{Q_g}{N} \pm \frac{(Q_g \cdot e_x)\,x}{\Sigma x^2} \pm \frac{(Q_g \cdot e_y)\,y}{\Sigma y^2}$

In this case, $\Sigma x^2 = 6 \times (0.5)^2 + 6 \times (1.5)^2 = 15.0$

$$\Sigma y^2 = 4 \times (1.0)^2 + 4 \times (1.0)^4 = 8.0$$

The maximum load occurs in pile 4.

$$Q_u = \frac{4.0}{12} + \frac{(4.0 \times 0.3)}{15} \times 1.5 + \frac{(4.0 \times 0.4)}{8} \times 1.0$$

$$= 0.6533 \text{ MN} = 653.3 \text{ kN}$$

Illustrative Example 25.11. *A pile group consists of 9 friction piles of 30 cm diameter and 10 m length driven in clay (c_u = 100 kN/m^2, γ = 20 kN/m^3), as shown in Fig. E-25.11. Determine the safe load for the group (FS = 3, α = 0.6).*

Solution. From Eq. 25.44, $Q_g(u) = q_p A_g + \alpha\, c\, (P_g D)$.

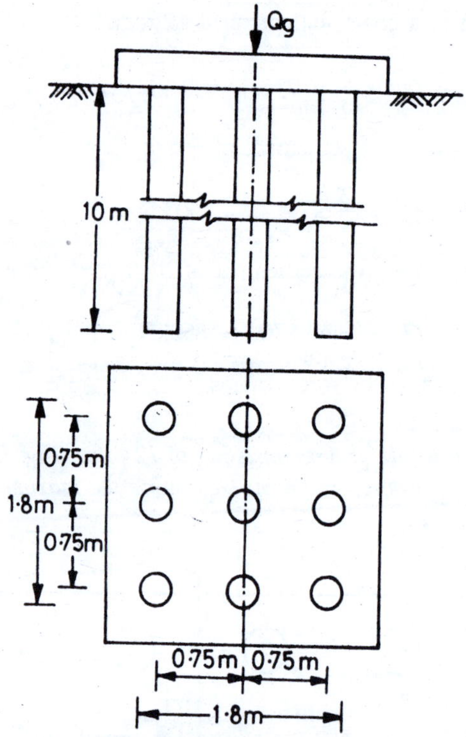

Fig. E-25.11.

$$= (9 \times 100)(1.8 \times 1.8) + 0.6 \times 100 \times (4 \times 1.8 \times 10)$$

or $$Q_g(u) = 7236 \text{ kN}$$

From Eq. 25.45, $$Q_u = q_p A_p + \alpha c (p \times D)$$

$$= (9 \times 100) \times \pi/4 \times (0.3)^2 + 0.6 \times 100 (\pi \times 0.3) \times 10$$

or $$Q_u = 628.8 \text{ kN}$$

From Eq. 25.46, $$Q_g(u) = NQ_u$$

$$= 9 \times 628.8 = 5659.2 \text{ kN}$$

As the ultimate load for individual pile failure is less than the pile group load, the safe load is given by

$$Q_a = \frac{5659.2}{3} = \mathbf{1886.4 \text{ kN}}$$

Illustrative Example 25.12. *A 40 cm diameter pile, 11 m long, has a bell of 2 m diameter and 1 m height. If the soil has* $\phi = 25°$, $c_u = 20$ *kN/m*^{2} *and* $\gamma = 19$ *kN/m*^{3}, *estimate the allowable pull out resistance (FS = 3).*

Solution. From Table 25.3, $H/B_1 = 3.0$,

Therefore, $H = 3 \times 2.0 = 6$ m

As $D > H$, the pile is deep.

From Eq. 25.57,

$$P_u = \pi B_1 c_u H + s_f \gamma (\pi/2 \times B_1)(2D - H) H K_u \tan\phi + W$$

where $$W = \pi/4 \times (2.0)^2 \times 11 \times 19 + \pi/4 \times (0.4)^2 \times 11 \times (23 - 19)$$

or $$W = 662 \text{ kN}$$

From Table 25.3, $$s_f = 1.3.$$

$$K_u = \frac{1 + \sin\phi}{1 - \sin\phi}\left(\tan\frac{2}{3} \times \phi\right) = \frac{1 + \sin 25°}{1 - \sin 25°}\left(\tan\frac{2}{3} \times 25°\right) = 0.737$$

Therefore, $P_u = \pi \times 2.0 \times 20 \times 6 + 1.3 \times 19\,(\pi/2 \times 2)\,(2 \times 11 - 6) \times 6 \times 0.737 \tan 25° + 662$

or $P_u = 3976$ kN

Allowable pull, $P_a = \frac{3976}{3} = \mathbf{1325\ kN}$

Illustrative Example 25.13. *A group of friction piles of 30 cm diameter is subjected to a net load of 2000 kN, as shown in Fig. E-25.13. Estimate the consolidation settlement.*

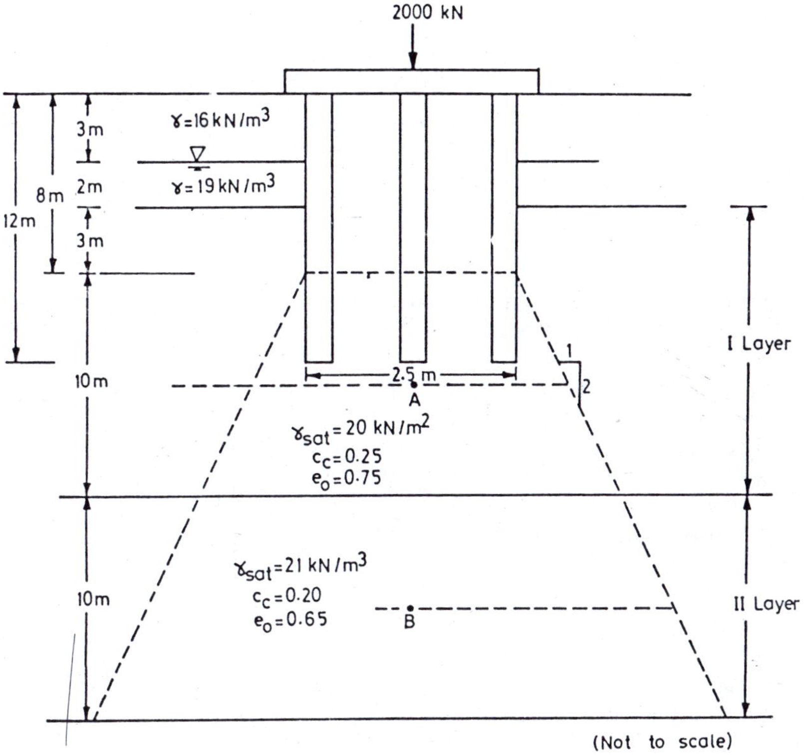

Fig. E-25.12.

Solution. σ_0 at point A, middle of I layer

$$= 3 \times 16 + 2 \times (19 - 10) + 8 \times 10.0 = 146 \text{ kN/m}^2$$

σ_0 at point B, middle of II layer $= 3 \times 16 + 2 \times 9.0 + 13 \times 10.0 + 5 \times 11 = 251 \text{ kN/m}^2$

Cross-sectional area at A $= \left(2.5 + 2 \times 5 \times \frac{1}{2}\right) = 7.5 \text{ m}^2$

$$\Delta\sigma = \frac{2000}{7.5 \times 7.5} = 35.56 \text{ kN/m}^2$$

Cross-sectional area at B $= \left(2.5 + 15 \times 2 \times \frac{1}{2}\right) = 17.5 \text{ m}^2$

$$\Delta\sigma = \frac{2000}{17.5 \times 17.5} = 6.53 \text{ kN/m}^2$$

$$\text{Settlement of I layer} = C_c \left(\frac{H}{1 + e_0} \right) \log \frac{\sigma_0 + \Delta\sigma}{\sigma_0}$$

$$= 0.25 \times \frac{10}{1 + 0.75} \log \frac{146 + 35.56}{146} = 0.135 \text{ m}$$

$$\text{Settlement of II layer} = 0.20 \times \frac{10}{1 + 0.65} \log \frac{251 + 6.53}{251} = 0.014$$

Total settlement $= 0.135 + 0.014 = \mathbf{0.149\ m}$

PROBLEMS

A. Numerical

25.1. A 30 cm diameter concrete pile is driven in a normally consolidated clay deposit 15 m thick. Estimate the safe load. Take $c_u = 70$ kN/m^2, $\alpha = 0.9$ and *F.S.* = 2.5. **[Ans.** 375 kN]

25.2. A 10 m long concrete pile 30 cm diameter is driven into a medium dense sand ($\phi = 30°$, $\gamma = 20$ kN/m^3, K = 1.0, tan $\delta = 0.5$). Take $D_c/B = 10.0$. Estimate the safe load (*F.S.* = 2.5). **[Ans.** 130 kN]

25.3. A 40 cm diameter pile is installed by driving through a 7 m thick layer of normally consolidated silt. The cone tip resistance (q_c) varies from 1500 kN/m^2 at the surface to 2500 kN/m^2 at 7 m depth. The silt layer overlies a dense sand (q_c = 2000 kN/m^2). If the tip is embedded 2 m into the dense sand, estimate the safe load (*F.S.* = 3.0). **[Ans.** 542 kN]

25.4. A concrete pile, 40 cm diameter, 9 m long, is driven through a 6 m thick layer of silty sand ($\phi = 20°$, $\gamma = 17$ kN/m^3) overlying a dense layer of sand ($\phi = 35°$, $\gamma = 19.5$ kN/m^3). If the water table is at the ground surface, estimate the safe load (F.S. = 3). Take K = 1.0, and $\delta = 0.75\ \varphi$. **[Ans.** 168 kN]

25.5. (*a*) The design capacity of a pile is 400 kN with a factor of safety of 4. What should be the average penetration of the pile for the last few blows ?

(*b*) How many blows would be required for the last one metre for the design capacity to be achieved ? Use following data:

Energy rating	= 3500 kN-cm
Efficiency of hammer	= 0.80
Cross-sectional area of pile	= 100 cm^2
Young's modulus of pile material	= 2×10^8 kN/m^2
Length of pile	= 10 m.

Use Danish formula. **[Ans.** 9.1 mm/blow; 110]

25.6. A precast concrete pile is driven with a 50 kN hammer, having a free fall of 1.0 m. If the penetration in the last blow is 0.5 cm, determine the load-carrying capacity of the pile using Engineering News Record formula. (*F.S.* = 6.0). **[Ans.** 274 kN]

25.7. The pile load test on a 40 cm diameter concrete pile in a deposit of sand indicates a settlement of 4 mm under a load of 400 kN. Estimate the settlement of a 4 × 4 pile group. The piles are driven at a spacing of 100 cm. The total load on the group is 6400 kN. **[Ans.** 20 mm]

25.8. A group of nine piles, 8 m long, is used as the foundation for a column. The piles are 30 cm diameter with centre to centre spacing of 90 cm The subsoil consists of clay with unconfined compression strength of 180 kN/m^2. Estimate the safe load. (*F.S.* = 3.0) **[Ans.** 1.8 MN, individual action]

25.9. A 30 cm diameter pile, 12 m long, is driven into a sand deposit. The details of the hammer are as under.

Total weight of hammer	= 20 kN
Length of stroke	= 100 cm
Energy per blow	= 2000 kN-cm
Average penetration blow	= 4 mm

Estimate ultimate resistance of pile using Hiley's formula, assuming that driving is without dolly and cushion is about 2.5 cm thick. **[Ans.** 1150 kN]

25.10. A bored pile in a clayey soil failed at an ultimate load of 400 kN. If the pile is 40 cm diameter and 10 m long, determine the capacity of a group of nine piles, spaced 1 m centre to centre both ways. Take $\alpha = 0.5$.

[Ans. 3.6 MN, individual action; 5.4 MN, group action]

B. Descriptive and Objective Type

25.11. What are the conditions where a pile foundation is more suitable than a shallow foundation ?

25.12. Describe various types of pile foundations.

25.13. Discuss different methods for the installations of piles.

25.14. How would you estimate the load carrying capacity of a pile in (*a*) cohesionless soils, (*b*) cohesive soils ?

25.15. What is negative skin friction ? What is its effect on the pile ?

25.16. Discuss various dynamic formulae. What are their limitations ?

25.17. Discuss the uses of penetration tests for the estimation of load- carrying capacity of piles.

25.18. How would you estimate the group capacity of piles in (*a*) sand (*b*) clay ?

25.19. Discuss the method for the design of (*a*) tension piles. (*b*) inclined piles.

25.20. Write whether the following statements are true or false.

(*a*) Pile foundations are more economical than shallow foundation for moderate loads.

(*b*) The load-carrying capacity of a bored pile is smaller than that of an equivalent driven pile.

(*c*) Negative skin friction occurs when the surrounding soil settles more than the pile.

(*d*) The most reliable method for determining the load- carrying capacity of a pile is the load test.

(*e*) The group capacity of a pile group for closely-spaced piles is generally limited by the behaviour of the group as a block.

(*f*) The group efficiency of the piles can be more than 100%.

(*g*) The load carrying capacity of tension piles depends upon the diameter of the bell.

[Ans. True, (*b*), (*c*), (*d*), (*e*), (*f*), (*g*)]

C. Multiple Choice Questions

1. A pile foundation is used when

(*a*) the loads are heavy.

(*b*) the soil stratum near ground surface is weak.

(*c*) both (*a*) and (*b*)

(*d*) neither (*a*) nor (*b*)

2. The load-carrying capacity of a pile depends upon the

(*a*) skin friction (*b*) point resistance

(*c*) both (*a*) and (*b*) (*d*) neither (*a*) nor (*b*).

3. The negative skin friction on a pile develops when

(*a*) the soil in which it is driven is sandy soil.

(*b*) the soil surrounding it settles more than the pile.

(*c*) the ground water table rises.

(*d*) the soil near the tip is clay.

4. The load-carrying capacity of a bored pile in sand is about ... times that of a driven pile.

(*a*) 1/2 to 2/3 (*b*) 2/3 to 3/4

(*c*) 3/4 to 1.25 (*d*) more than 1.25

5. The group efficiency of driven piles in sand at a close spacing may be

(*a*) equal to 100% (*b*) greater than 100%

(*c*) well below 100% (*d*) None of above

6. A 30 cm diameter pile is driven 10m into a homogeneous consolidated clay deposit. The safe load when the factor of safety is 2.50, unit cohesion is 40 kN/m^2 and adhesion factor is 0.70

(*a*) 150.8 kN (*b*) 105.6 kN

(*c*) 215.4 kN (*d*) 211.2 kN

[Ans. 1. (*c*), 2. (*c*), 3. (*b*), 4. (*a*), 5. (*b*), 6. (*b*)]

26

Drilled Piers and Caissons

26.1. INTRODUCTION

***(a)* Drilled Pier.** A driled pier is a large diameter concrete cylinder built in the ground. For construction of a drilled pier, a large diameter hole is drilled in the ground and subsequently filled with concrete. The difference between a drilled pier and a bored pile is basically of the size. Generally, bored piles are of diameter less than or equal to 0.6 m. The shafts of size larger than 0.6 m are generally designated as drilled piers. A drilled pier is a type of deep foundation constructed to transfer heavy axial or lateral loads to a deep stratum below the ground surface.

***(b)* Caisson.** A caisson is a type of foundation of the shape of a hollow prismatic box, which is built above the ground level and then sunk to the required depth as a single unit. It is a watertight chamber used for laying foundations under water, as in rivers, lakes, harbours, etc. The caissons are of three types : (*i*) Open caissons, (*ii*) Pneumatic caissons, and (*iii*) Floating caissons.

Open caissons are hollow chambers, open both at the top and the bottom. The bottom of the caisson has a cutting edge. The caisson is sunk into place by removing the soil from the inside of the shaft (chamber) until the bearing stratum is reached. Well foundations are special type of open caissons used in India, discussed in chapter 27.

Pneumatic caissons are closed at the top, but open at the bottom. A pneumatic caisson has a working chamber at its bottom in which compressed air is maintained at the required pressure to prevent entry of water into the chamber. Thus the excavation is done in dry.

Floating caissons are open at the top but closed at the bottom. These caissons are constructed on land and then transported to the site, and floated to the place where these are to be finally installed. These are sunk at that place by filling them with sand, ballast, water or concrete to a levelled bearing surface.

This chapter deals with the design and construction of drilled piers and caissons.

26.2. DRILLED PIERS

The transfer of load to the soil from a drilled pier, like a pile, can take place through end bearing, skin friction or a combination of both. Drilled piers in cohesive soils are generally belled or under-reamed to increase the load-carrying capacity. Fig. 26.1 (*a*) shows a straight shaft pier, and Fig. 26.1 (*b*), a belled pier. Belled piers are generally used when the stratum does not have adequate bearing capacity.

The load-carrying capacity of a drilled pier can be estimated using a method similar to that for piles, as explained below.

***(b)* Drilled piers on sand.** The analysis of drilled piers in sand is similar to that for bored piles in sand. As the excavation for a drilled pier is likely to lead to some loosening of the sand deposit, the strength of the sand is considerably reduced. The ultimate load of a drilled pier can be obtained from the following equations.

$$Q_u = q_p A_p + f_s A_s$$

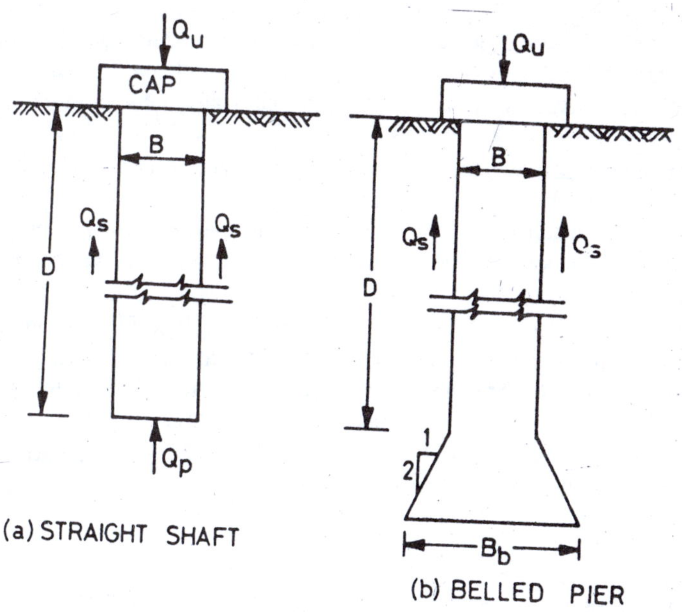

Fig. 26.1. Drilled Piers.

or
$$Q_u = (\bar{q}\, N_q)\, A_p + (K\, \bar{\sigma}_v \tan \delta)\, A_s \quad \text{...(26.1)}$$

where q_p = unit tip resistance, $\bar{q}$ = effective vertical pressure at the base, $\bar{\sigma}_v$ = effective vertical pressure at any depth, tan δ = friction coefficient, K = lateral earth pressure coefficient, A_p = area of base, A_s = surface area of shaft.

As for piles, while calculating the effective vertical pressure at the base ($\bar{q}$), the limits imposed by the concept of the critical depth should be considered, if applicable.

The value of K is approximately equal to the coefficient of earth pressure at rest (K_o). Thus

$$K = K_o = 1 - \sin \phi$$

Generally, the value of K varies between 0.3 and 0.75.

The value of tan δ is taken equal to tan ϕ when the excavation is done dry. However, if slurry is used, some reduction should be applied to the value of tan δ.

The value of N_q for drilled piers is generally lower than that for driven piles. The values given by Vesic (1963) are approximately the lower bound, and are generally recommended for drilled piers in sand (Fig. 26.2). Alternatively, the values of the shallow foundations (chapter 23) can be used conservatively.

A factor of safety of 2.5 to 3.0 is generally applied to the ultimate load to obtain the safe load.

The allowable bearing capacity can also be obtained from N-value obtained from standard penetration test. The bearing capacity of a drilled pier is generally taken one half of the value of an open caisson in identical conditions, obtained from Eq. 26.5.

The settlement of a pier under a given net soil pressure is generally less than that of a comparable

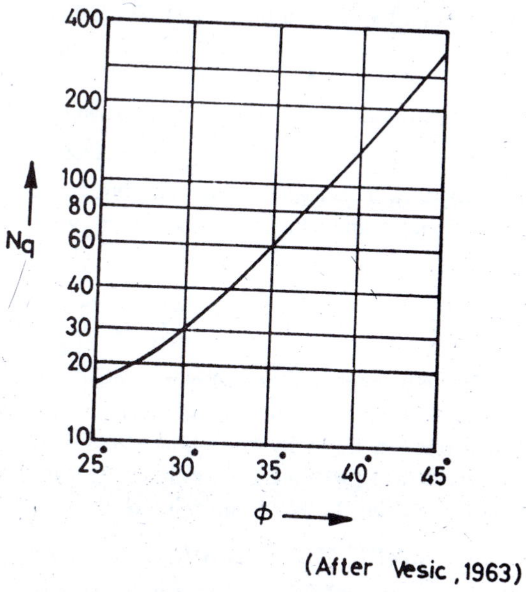

Fig. 26.2.

shallow foundation because of the confining pressure of the surrounding soil. The allowable soil pressure can be obtained from the curves for the shallow foundations given in chapter 23, using the N-values uncorrected for the confining pressure. If the water table is high, the water table correction is made as in shallow foundations.

As the settlement due to self weight occurs before the pier is completed, it is not of much practical significance. The self weight of the pier is usually subtracted from the total load when determining the allowable load for the settlement. However, while computing the factor of safety against bearing failure, the weight of the pier must be considered. The settlement of the pier can be computed using the procedure developed for shallow footings.

According to Terzaghi and Peck (1967), the settlement of a drilled pier in sand at any depth is about one half the settlement of an equally loaded footing covering the same area. Eq. 23.83 can be used to determine the allowable soil pressure. The unit pressure for piers on sand is generally taken twice the value for a footing of the same size under identical conditions, obtained from Eq. 23.83.

(*b*) **Drilled piers on clay.** The analysis of a drilled pier in clay is similar to that of bored piles in clay. The ultimate load is given by the following equations.

$$Q_u = q_p A_p + f_s A_s$$

or

$$Q_u = c N_c A_p + \alpha \bar{c} A_s \quad \text{...(26.2)}$$

where c = undrained cohesion, $\bar{c}$ = average undrained cohesion on the shaft, α = adhesion factor, N_c = bearing capacity factor.

In the case of drilled piers, the value of the N_c depends upon the D/B_1 ratio of the pier, where B_1 is the diameter of the bottom (see Table 26.1).

Table 26.1. Values of N_c (After Teng, 1962)

D/B_1	0	0.5	1.0	1.5	2.0	2.5	3.0	4.0 and above
N_c	6.2	7.1	7.7	8.1	8.4	8.6	8.8	9.0

The value of α generally varies between 0.15 and 0.50, depending upon the drilling method and the type of pier. An average value of 0.4 is usually taken. If the shaft is provided with a bell, only the straight portion is considered for friction (adhesion). For belled shaft drilled dry, the upper limit of unit adhesion is 40 kN/m^2 and that for the belled shafts drilled with slurry is 25 kN/m^2. For straight shafts excavated dry, the upper limit is 100 kN/m^2.

The safe load is determined by applying a suitable factor of safety to the ultimate load.

$$Q_a = Q_u/FS \quad \text{...(26.3)}$$

Generally, a factor of safety (FS) of 3 is taken.

Sometimes, the safe load is obtained by applying a factor of safety only to the tip resistance. Thus

$$Q_a = Q_p/FS + Q_s$$

or

$$Q_a = \frac{(c N_c A_p)}{FS} + \alpha \bar{c} A_s \quad \text{...(26.4)}$$

The settlement of a drilled pier in clay depends upon the load history of the clay. Settlement analysis can be done assuming the bottom of the pier as a footing and applying the consolidation theory (chapter 12). Drilled piers in normally consolidated clays are not economical, as the settlements are excessive. In actual practice, drilled piers are used only in over-consolidated clays. In which case, the settlements are generally small and within the permissible limits.

26.3. CONSTRUCTION OF DRILLED PIERS

Construction of drilled piers can be divided into 3 stages.

(1) Excavation of piers, (2) Providing supports, and (3) Concreting of piers.

(1) Excavation of piers Drilled piers are generally excavated using an auger drill or some other type of drilling equipment. An auger is attached to a shaft and rotated under pressure to dig into the soil. When it is filled with soil, it is raised above the ground and emptied.

For formation of a bell, the auger is replaced by an under- reaming tool. The tool usually consists of a cylinder with the cutting blades that are hinged at the top. The cutting blades are in the folded position when the under-reamer is lowered into the hole. On reaching the bottom of the hole, the blades are spread outward by a mechanism. As the under-reamer is rotated, a bell is formed and the loose soil falls inside the cylinder, which is raised and emptied. The process is repeated till the bell is completely formed. The diameter of the bell is kept two to three times the diameter of the shaft. The angle of the bell is 30° to 45° with the vertical.

The above method of drilling is convenient for hard clays where the hole can be left open for a few hours without a support. In cohesionless soils below the water table, the hole is prevented from collapsing by providing a casing or by drilling in slurry. When rock is encountered during drilling, special machines are required. For boulders and holes socketed in hard rock, special drilling tools are required.

When excavation is in progress, the soil is exposed at the bottom and sides. It is examined carefully to check that the hole is straight and has been drilled to a stratum of adequate capacity. As the hole is of a large diameter, even a man can descend into the shaft for inspection.

2. Providing Supports (*a*) Chicago method. In this method, a circular hole is excavated upto the depth at which the soil will stand unsupported (about 0.5 m for soft clays and 2 m for stiff clay). Vertical boards, known as laggings, are then set in position around the excavated face and are held tightly against the soil by steel rings [Fig. 26.3 (*a*)]. The shaft is then excavated further for 1 to 2 m and another setting of boards and rings is made. The process is repeated until the desired level is reached. The base of the shaft is then belled out.

(*b*) **Gow method.** In this method, excavation of the hole is done manually. Telescopic steel shells are used to support the soil [Fig. 26.3 (*b*)]. The telescopic shells are extended as the hole is deepened. The shells are removed as the concreting progresses. One section of the shell is removed at one time. The minimum diameter of the hole in this method is about 1.25 m.

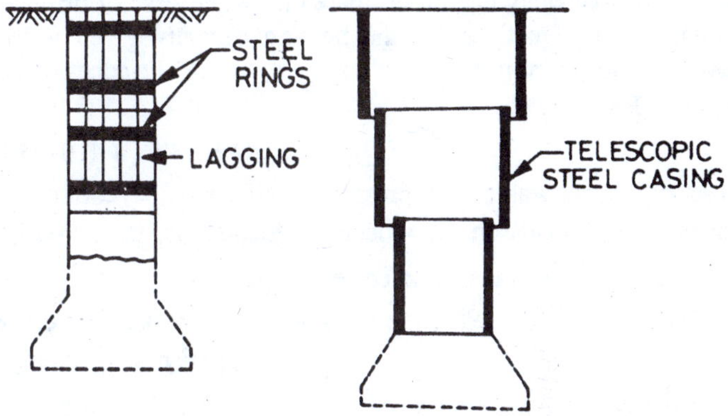

Fig. 26.3.

3. Concreting of piers. After the hole has been drilled to the required depth, the shaft is dewatered and the bottom is cleaned. The casing, if used, is removed before the concreting is done. The reinforcement cage is then inserted and concreting is done. As far as possible, concreting should be done in dry. Concreting can be done in a dry hole by gravity pouring, provided the concrete does not strike the sides. However, if dewatering is not possible or slurry is used to support the shaft sides, concrete is placed using a tremie.

26.4. ADVANTAGES AND DISADVANTAGES OF DRILLED PIERS

Drilled piers have the following advantages and disadvantages, as compared with pile foundations:

Advantages.

(1) As a single drilled pier can take up the load of a group of piles, it is more convenient.

(2) Drilled piers have higher resistance to lateral loads than piles.

(3) Construction of drilled piers generally requires lighter equipment for drilling than that for pile driving.

There is no noise due to hammer blow in the case of drilled piers.

(4) Piles driven by a hammer cause ground vibrations and ground heaving. Such conditions do not exist in the case of drilled piers.

(5) The base and the sides of the drilled pier can be inspected. This is not possible in case of driven piles.

(6) The base of a drilled pier can be enlarged to provide greater bearing capacity and also to provide greater resistance to uplift.

(7) Drilled piers can be used even when the soil contains boulders, etc.

Disadvantages

(1) The concreting operation requires stricter supervision. The quality of concrete obtained is generally inferior to that in precast piles.

(2) Deep excavation of the drilled pier, if not properly supported, can cause substantial subsidence and damage to adjoining structures.

(3) Strict supervision of drilling operation is required in drilled piers.

(4) Subsurface investigations required in the case of drilled piers are more than that in piles.

(5) Load tests in the case of drilled piers are difficult.

26.5 DESIGN OF OPEN CAISSONS

Caissons are carried to a hard stratum, such as compact sand, gravel, hard clay or rock. The load-carrying capacity can be estimated as in the case of drilled piers. As a caisson also acts like a rigid mat foundation, the equations for the bearing capacity given in chapter 24 for mats may also be used. The allowable soil pressure (q_{na}) for an open caisson in cohesionless soils can be obtained from the following equation (F.S. = 3.0).

$$q_{na} = 0.22\,N^2\,BW_\gamma + 0.67\,(100 + N^2)\,D_f\,W_q \qquad \text{...(26.5)}$$

where q_{na} = allowable soil pressure (kN/m^2), N = corrected standard penetration number, B = smaller dimension of the caisson, D_f = depth of foundation, measured below scour level.

W_γ and W_q are water table correction factors.

According to IS : 3955, the allowable bearing pressure can be determined using the following formula,

$$q_{na} = 0.054\,N^2\,B + 0.16\,(100 + N^2)\,D_f \text{ kN/m}^2 \qquad \text{...[26.3(}a\text{)]}$$

If the caisson (well) rests on rock strata, the safe bearing pressure depends upon the crushing strength of rock. The crushing strength can be determined by taking the cores from the field and testing for compression. However, there may be fissures, faults and joints in the rock which would also affect the bearing capacity and which are not detected from the core samples.

Teng (1962) has suggested that the allowable bearing pressure of caissons on bed rock should not exceed that of concrete seal, which is normally taken as 3500 kN/m^2 because the concrete seal is usually placed under water and the quality of concrete is poor.

In case of cohesive soils, undisturbed samples should be tested to determine the unit cohesion (c). The ultimate bearing capacity is determined as

$$q_u = c\,N_c \qquad \text{...(26.6)}$$

where N_c = bearing capacity factor.

The vertical loads acting on the caisson are the loads from the superstructure and the self weight. The buoyant forces should be determined for the lowest water level and deducted from the downward loads. The lateral loads acting on the caisson are due to earth pressure, wind pressure, water pressure and earthquakes. The lateral forces may also act due to tractive forces from traffic, ice pressure, and currents of flow.

The skin friction should be estimated for the most critical condition when the soil has been removed to the maximum depth of scour. The total load is assumed to be carried by the base of the caisson if it penetrates a relatively shallow depth of soil.

Besides the above-mentioned loads, a caisson may also be subjected to large stresses during the sinking operation. When the caisson is hung up near the top by skin friction, the lower portion is subjected to tension. Large stresses also develop if the caisson is dropped suddenly during sinking or when it is pulled to its correct position from the inclined position. If the caisson is supported on one side only or on two opposite corners at some stage during sinking, it is subjected to heavy stresses. The caisson must be safe against all such conditions.

The exterior walls of the caisson are designed to withstand the stresses due to vertical loads and the lateral forces.

Sinking Effort. The caissons are designed to have sufficient self weight in each lift to overcome the skin friction. If the self weight is not sufficient, additional ballast is required during sinking. Occasionally, water jetting is used to reduce the friction.

If it is desired to proportion a circular caisson such that no ballast is required, an expression for the unit skin friction can be obtained by equating the weight of concrete to the frictional force (Fig. 26.4). Therefore,

$$(\pi/4)\,(D_o^2 - D_i^2)\,D\,\gamma_c = (\pi\,D_o)\,D \times f \qquad \text{...[26.7(a)]}$$

where D_o = external diameter of caisson, D_i internal diameter of caisson, γ_c = unit weight of concrete (= 24 kN/m^3 above water level, and 14 kN/m^3 below water level), D = depth of penetration, f = unit skin friction.

Therefore,

$$f = \frac{\gamma_c}{4\,D_o}\,(D_o^2 - D_i^2) \qquad \text{...[26.7(b)]}$$

Similar equations can also be developed for caissons of other shapes.

If the actual unit friction (f) is greater than the value given by Eq. 26.7 (b), additional ballast would be required for sinking. Terzaghi and Peck (1948) give the following value of the unit friction.

Silt and soft clay	= 7.3—29.3 kN/m^2
Very stiff clay	= 49—195 kN/m^2
Loose sand	= 12.2—34.2 kN/m^2
Dense sand	= 34.2—68.4 kN/m^2
Dense gravel	= 49.0—98 kN/m^2

Thickness of concrete seal. Before dewatering the caisson, a concrete seal is placed at the bottom of the excavated shaft to plug it. The concrete seal is also known as the *bottom plug*. It forms the permanent base of the caisson. The thickness of the seal should be sufficient to withstand the upward hydrostatic pressure after dewatering is complete and before the concreting of the shaft has been done.

The seal may be designed as a thick plate subjected to a unit bearing pressure due to the maximum vertical loads. The thickness of concrete seal may be obtained from the following equations.

For circular caissons,

$$t = 0.59\,D_i\sqrt{q/\sigma_c} \qquad \text{...(26.8)}$$

Fro rectangular caissons,

$$t = 0.866\,B_1\sqrt{\frac{q}{\sigma_c\,(1 + 1.61\,\alpha)}} \qquad \text{...(26.9)}$$

where t = thickness of concrete seal, D_i = internal diameter, L_i, B_i = internal length, width, q = unit bearing pressure at the base, $\alpha = B_i/L_i$, σ_c = allowable concrete flexural stress ($\not>$ 3500 kN/m^2).

If H is the depth of water above the base (Fig. 26.4), the value of q can be found from the following equation.

$$q = \gamma_w H - \gamma_c t$$

where q = unit bearing pressure (kN/m^2),
and H and t are in metres.

Taking γ_c = 24 kN/m^3 and γ_w = 10 kN/m^3,

$$q = 10\,\mathrm{H} - 24\,t \qquad \text{...(26.10)}$$

The thickness of the seal should be safe against perimeter shear,

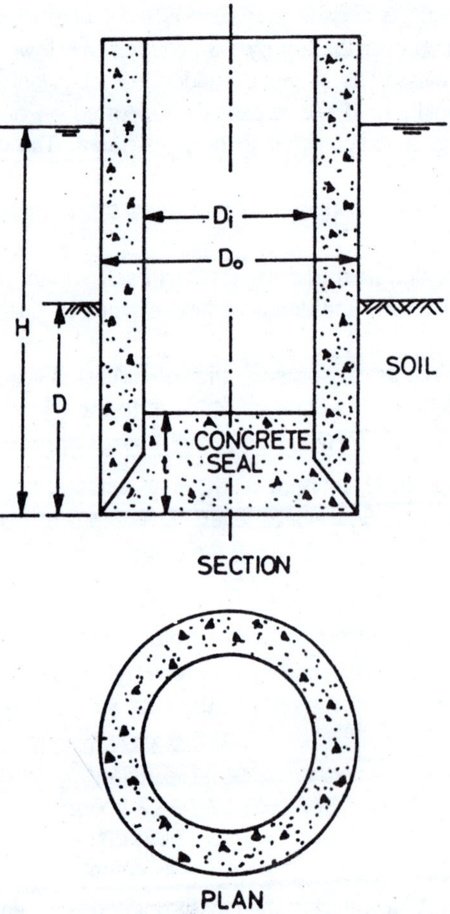

(a) CIRCULAR CAISSON

Fig. 26.4.

$$v = \frac{A_i H \gamma_w - A_i t \gamma_c}{p_i t} \quad \text{...(26.11)}$$

where A_i = inside area (= $\pi/4\, D_i^2$ for circular caissons)

p_i = inside perimeter (= πD_i for circular caissons)

The perimeter shear v should be less than the allowable shear.

There should be overall stability of the caisson against buoyancy. For example, for circular caissons,

Total upward force, $F_u = (\pi/4\, D_o^2)\, H \gamma_w$ ...(26.12)

Total downward force, $F_d = W_c + W_s + Q_s$ ...(26.13)

where W_c = weight of empty caisson, W_s = weight of seal, Q_s = skin friction.

If F_d is greater than F_u, the caisson is safe against buoyancy. However, if F_d is smaller than F_u, the thickness of the seal should be increased to add weight.

Cutting Edge. The cutting edge is provided at the base of the open caisson to facilitate penetration. It also protects the walls of the caisson against impact and obstacles encountered in its path during penetration.

The inside bevel is generally made 2 vertical to one horizontal. The cutting edges are usually made of angles and plates of structural steel or reinforced concrete and steel. As the sharp edges are easily damaged, the blunt edges are commonly used. However, the angle of the edge should not be greater than 35°(Fig. 26.5).

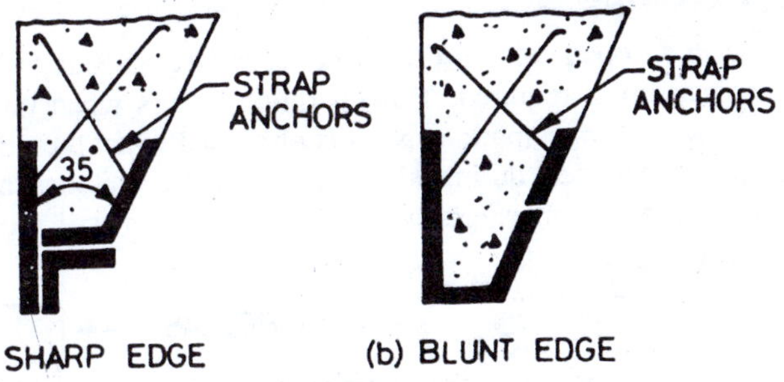

Fig. 26.5.

To avoid tearing off the cutting edge or any lower parts of the caisson, the caisson concrete must be anchored or tied to the cutting edge. The lower portion of the cutting edge is provided with a 12 mm thick steel plate anchored to the concrete by means of steel straps.

26.6. CONSTRUCTION OF OPEN CAISSONS

The sinking of an open caisson is generally done by penetrating it in the dry or from a dewatered construction area or from an artificial island. An artificial island of sand is made for the purpose of raising the ground surface above the water level. Thus a dry area is obtained for sinking the caisson. The size of the sand island should be sufficient to provide working area around the caisson. (Fig. 26.6).

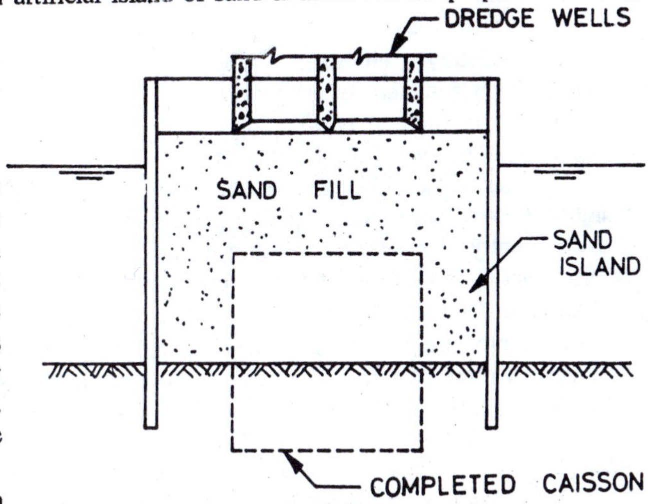

Fig. 26.6.

For the construction of a sand island, a wooven willow mattress is first sunk to the river bottom to provide protection against scour. A timber staging is then constructed around the periphery of the intended island. Sheet piles are driven to enclose the island area. The mattress is cut along the inside face of the shell formed by sheet piles and the inside mattress is removed. The shell is then filled with sand upto the required level.

In case it is not possible to sink the caisson in dry, it is constructed in slipways or barges and towed to its final position by floating. False bottoms are provided for this purpose. Guide piles are generally required for sinking the first few lifts of caisson. Sinking is done through open water and then penetrating it into the soil.

The caisson is sunk by its own weight when the soil is excavated from the dredging well. As sinking progresses, additional lifts of caisson steining are installed. When a hard material is encountered, under-water blasting may be necessary. The excavation is done by dredging with grab buckets. The soil near the cutting edge is removed by hand if it does not flow into the excavation. The sinking operation is, of course, stopped during the period the concrete for the lift is cast and cured. To facilitate sinking, the exterior surface is applied with a film of grease. Alternatively, water jets are used.

When the caisson reaches the final depth, its bottom is plugged by a concrete seal. The concreting for the seal is done by tremie. After the concrete has matured, the water in the caisson is pumped out. The top of the concrete seal is cleaned and more concrete is placed over the seal.

The caisson should be kept in the vertical position during the entire process of sinking. However, it is extremely difficult to sink the caisson perfectly straight and true to its position. Corrective measure are adopted when it becomes inclined (see chapter 27).

26.7. PNEUMATIC CAISSONS

If the soil enclosed in an open caisson cannot be excavated satisfactorily through its shaft during sinking operation, a pneumatic caisson is required. This condition occurs when the soil flows into the open caisson faster than it can be removed. A pneumatic caisson is also used when there is a great influx of water or where difficult obstructions are anticipated during sinking. Pneumatic caissons are suitable in soft, running soils which cannot be excavated in dry or where there is a greater danger of scour and erosion.

A pneumatic caisson is a rigid, inverted box with its bottom open (Fig. 26.7). A working chamber is provided at its bottom to keep the caisson free of water and mud by use of compressed air. The design of a pneumatic caisson is similar to that of an open caisson in many respects. The ultimate load carrying capacity, the design of walls, concrete seal and cutting edge are similar to that of open caissons. However, the following differences should be clearly noted.

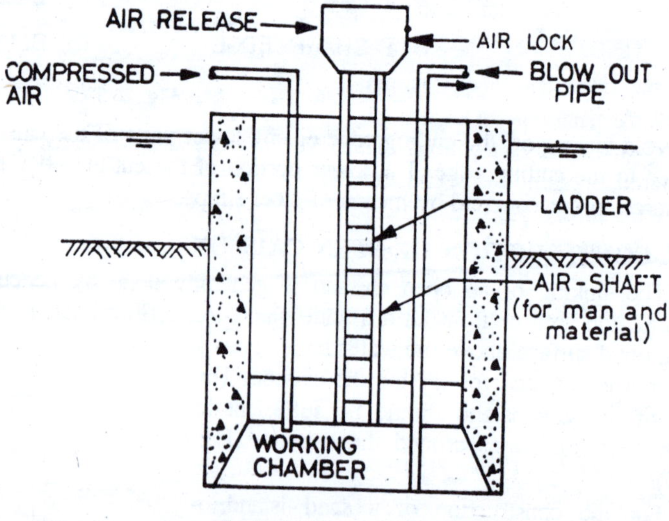

Fig. 26.7. Pneumatic Caisson.

(1) Working chamber. The working chamber is made of mild steel. It is about 3 m high. It consists of a strong roof at its top. The chamber is absolutely air tight. The air in the chamber is kept at a specified pressure to prevent entry of water and soil into it. The walls of the chamber should be thick and leak proof. To keep the frictional resistance low, the outside surfaces of the walls are made smooth. A cutting edge is provided at the bottom to facilitate the penetration of the caisson.

The air pressure must be sufficient to balance the full hydrostatic pressure due to water outside. However, there is a maximum limit to air pressure. Working under a pressure of greater than 400 kN/m^2 is beyond the endurance limit of human beings. Therefore, the maximum depth of water through which a pneumatic caisson can be sunk successfully is about 40 m. Working under a pressure greater than 400 kN/m^2 may cause a special type of sickness, called *caisson sickness.*

(2) Air Shaft. An air shaft is a vertical passage which connects the working chamber with an air lock at the top. It provides an access to the working chamber for workmen. It is also used for the transport of the excavated materials to the ground surface. In larger caissons, two separate air shafts are provided, one for passage of workmen and one for transport of the materials. The shafts are made of steel tubes. The joints of the tubes are provided with rubber gaskets to make them leak proof. Each shaft is provided with its own air lock at its top.

As the caisson sinks, the air shaft is extended to keep the air lock always above water-level. During this period, the working chamber is closed by a gate plate at the lower end of the shaft.

(3) Air Lock. An air lock is a steel chamber provided at the upper end of the air shaft above water level. The purpose of providing an air lock is to permit the workmen and materials to go in or to come out of the caisson without releasing the air pressure in the caisson.

The steel chamber of the ailock is provided with two airtight doors, one of which opens to the shaft and the other opens to the outside atmosphere. When a man enters the airlock through the outside door, the pressure in the chamber is kept equal to the atmospheric pressure. The outside door is closed and the pressure in the chamber is gradually raised till it becomes equal to that in the air shaft and the working chamber. The door to the shaft is then opened and the man descends to the working chamber by a ladder provided inside the air shaft. The procedure is reversed when a man comes out from the caisson. However, the decompression is done much more slowly to prevent caisson disease. A period of about 30 minutes is necessary for decompression from a pressure of 300 kN/m^2 to atmospheric pressure.

To prevent the air in the working chamber from becoming stale, fresh air is circulated into the working chamber by opening a valve in the air lock. The workers should not be kept inside the working chamber for more than two hours at a stretch.

(4) Miscellaneous equipment. Miscellaneous equipment such as motors, compressors, and pressure pumps are usually located on the shore. Pressure to the working chamber is applied through compressed air pipe. In order to cope with an emergency, at least one stand-by unit consisting of all equipment must be provided.

26.8. CONSTRUCTION OF PNEUMATIC CAISSONS

Like open caissons, pneumatic caissons may be constructed at the site or floated and lowered from barges. Artificial sand islands may also be used. The cutting edge of the caisson is carefully positioned. Compressed air is introduced in the working chamber to expel water. After the working chamber has been dewatered, workmen descend through the air lock into the working chamber. The material is excavated by hand tools in dry. As the excavation progresses, the caisson gradually sinks. Concreting of the caisson is then done. The air pressure in the caisson is increased to equalise the increase in the head of water as the caisson goes down. The excavated material is removed by buckets through the air shaft. In granular soils, the excavated material can be removed by the blowout method through the blow-out pipe. When the valve in the blow pipe in opened, the granular material is blown out by high air pressure inside the working chamber.

After the caisson has attained its design depth, the working chamber is filled with concrete. Precautions must be taken to ensure full contact between the concrete fill and the underside of the roof of the working chamber. The fresh concrete is first lowered through the air shaft and a slab about 0.6 m thick is formed on the bottom and well packed under the cutting edge. The air pressure in the chamber is kept constant till the concrete has hardened. A stiff mix of concrete is then packed into the working chamber up to the roof level. Any space left between the roof and the concrete surface is filled with cement grout. There should not be any empty space left in the chamber, as it would lead to settlement when the caisson is subjected to superimposed load. After concreting of the working chamber is completed, the shaft tubes are dismantled. The shaft itself is filled up with a lean concrete.

26.9. ADVANTAGES AND DISADVANTAGES OF PNEUMATIC CAISSONS

Pneumatic caissons have the following advantages and disadvantages, as compared with open caissons.

Advantages

(1) As there is an access to the bottom of the caisson, obstructions can be easily removed.

(2) The soil can be inspected and the soil samples can be taken, if required.

(3) Soil bearing capacity can be determined by conducting in-situ tests in the working chamber.

(4) Excavation and pouring of concrete is done in the dry.

(5) As the position of the ground water table remains unchanged, there is no flow of soil into the excavated area.

(6) There is no settlement of the adjoining structures as the water table is not lowered.

Disadvantages

(1) Pneumatic caissons are highly expensive. Pneumatic caissons should be used only when open caissons are not feasible.

(2) The penetration depth below water table is limited to 30 to 40 m.

(3) There is a lot of inconvenience caused to the workmen while working under high pressure. The workers may develop caisson disease.

(4) In pneumatic caissons, a large amount of manual work is required which increases the cost.

(5) Extreme care is required for the proper working of the system and to maintain the required air pressure. Any slackness may lead to an accident.

26.10. FLOATING CAISSONS

Floating caissons are large, hollow boxes with top open but bottom closed. These are floated to the place where these are to be installed. The caissons are then sunk by filling them with ballast, such as sand, dry concrete, gravel. Unlike the open and pneumatic caissons, a floating caisson does not penetrate the soil. It simply rests on a levelled surface (Fig. 26.8). The load carrying capacity depends solely on the base resistance, as there is no side friction.

After the caissons has been sunk to its final position, it is completely filled with sand or gravel. A concrete cap is cast on its top to receive structural loads. To prevent scour underneath, rip rap is placed around its base.

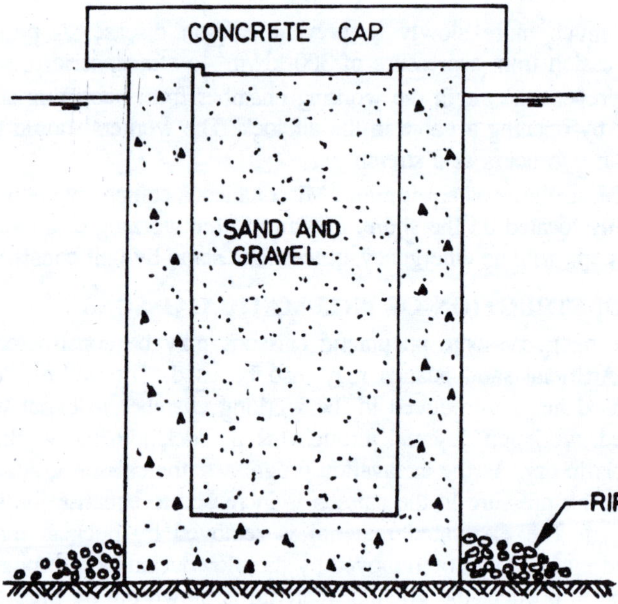

Fig. 26.8. Floating Caissons after sinking.

Floating caissons are usually constructed of reinforced cement concrete or steel. The plan of the caisson may be circular, square, rectangular, or elliptical. It usually contains a number of cells formed by diaphragm walls. If the caisson is to be floated in rough waters, it is designed as a ship and suitable internal strutting is provided.

26.11. STABILITY OF FLOATING CAISSONS

The caisson must be stable during flotation. According to Archimedes' principle, when a body is immersed in water, it is buoyed up by a force equal to weight of the water displaced.

For equilibrium, $$W - U = 0 \qquad \ldots(26.14)$$

where W = weight of the caisson, U = buoyant force.

The weight W acts through the centre of gravity (G) of the body. The buoyant force U acts through the centre of gravity of the displaced water, known as the *centre of buoyancy* (B) (Fig. 26.9). If the caisson is tilted through a small angle θ, the centre of gravity (G) remains at the same location with respect to the caisson itself, but the centre of buoyancy B changes its position as the position of displaced volume is changed. Some portion of the caisson which was not submerged during the vertical position becomes submerged in the inclined position. The point of intersection of the vertical line passing through B and the centre line of the caisson is known as *metacentre* (M).

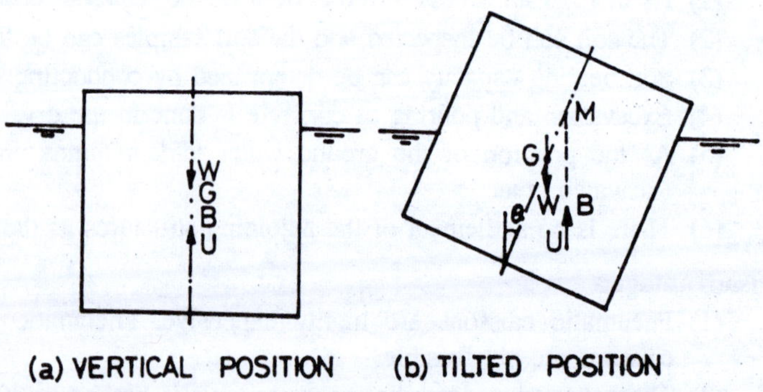

Fig. 26.9.

The caisson would be stable if the metacentre M is above G, *i.e.*, the metacentric height MG is positive.

The metacentric height can be determined analytically (see any text on Fluid Mechanics) as given below.

The distance BM between points B and M is given by

$$BM = I/V \qquad \ldots(26.15)$$

where I = second moment of area of the plan of the caisson at water surface, V = volume of water displaced.

The metacentric height is computed as

$$MG = BM \pm BG \quad ...(26.16)$$

The plus sign in Eq. 26.16 is used when G is below B.

If the caisson is unstable, it should be either redesigned or ballast should be used to make it stable.

The free board when floating should be atleast 1 m.

26.12. ADVANTAGES AND DISADVANTAGES OF FLOATING CAISSONS

Floating caissons have the following advantages and disadvantages as compared with open caissons.

Advantages.

(1) The installation of the floating caisson is quick and convenient.

(2) As the floating caissons are prefabricated or precast, the quality of construction is good.

(3) Floating caissons are less expensive than open caissons.

(4) Floating caissons can be transported by floating at a relatively low cost.

Disadvantages.

(1) The load-carrying capacity of a floating caisson is much lower than that of an equivalent open caisson.

(2) The foundation bed has to be levelled before installation.

(3) The base of the floating caisson is to be protected against scour action.

ILLUSTRATIVE EXAMPLES

Illustrative Example 26.1. *A straight-shaft drilled pier, 1.0 m in diameter, is constructed in a deposit consisting of loose sand overlying dense sand (Fig. E-26.1). Determine the allowable load (F.S. = 3.).*

Solution. Let us take critical depth $D_c = 10\,B = 10$ m

From Eq. 26.1, $Q_u = (\bar{q}\,N_q)\,A_p + (K\,\bar{\sigma}_v \tan\delta)\,A_s$

where $\bar{q}$ = effective vertical pressure at the base

$= 8 \times 17 + 2 \times 21 = 178$ kN/m^2

From Fig. 26.2, $N_q = 140$ for $\phi = 40°$.

Therefore, $$Q_u = 178 \times 140 \times \pi/4 \times (1.0)^2 + 0.5 \times (\pi \times 1.0) \times 0.58\left(\frac{1}{2} \times 136 \times 8.0\right)$$

$$+ 0.4 \times \pi \times 1.0 \times 0.84\left[\frac{1}{2}(136 + 178) \times 2.0 + 178 \times 2.0\right]$$

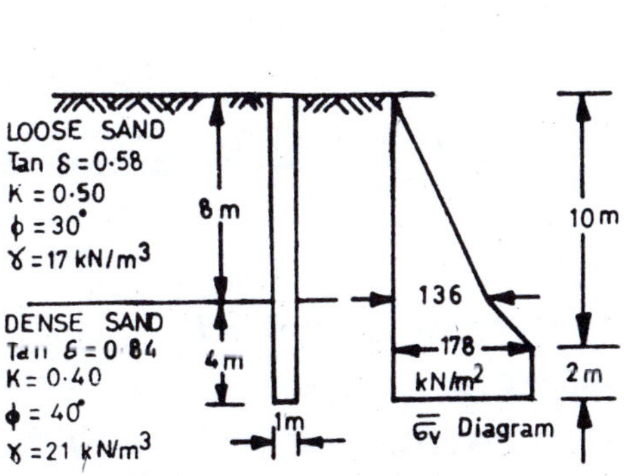

Fig. E-26.1.

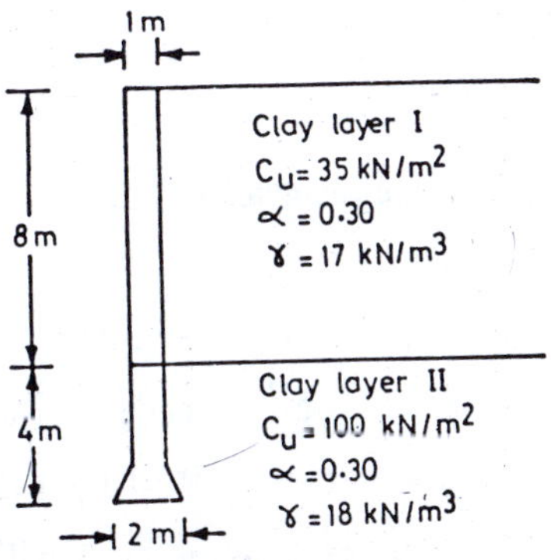

Fig. E-26.2.

or $$Q_u = 20740.00 \text{ kN}$$

$$Q_a = \frac{20740}{3} = 6913 \text{ kN}$$

Illustrative Example 26.2. *Determine the allowable load for the drilled pier constructed in a clayey deposit, shown in Fig. E-26.2. Take F.S. = 3.0.*

Solution. $$D/B_1 = 12/2 = 6$$

From Table 26.1, $N_c = 9.0$.

From Eq. 26.2, $$Q_u = c\,N_c\,A_p + \alpha\,\bar{c}\,A_s$$

$$= 100 \times 9.0 \times (\pi/4) \times (2.0)^2 + (0.3 \times 35)\,(\pi \times 1.0 \times 8.0) + (0.3 \times 100)\,(\pi \times 1.0 \times 3.0) = 3374 \text{ kN}$$

From Eq. 26.3, $$Q_a = \frac{3374}{3} = 1125 \text{ kN}$$

(If Eq. 26.4 is used, the allowable load = 1488.4 kN)

Illustrative Example 26.3. *Determine the outside diameter of an open caisson to be sunk through 40 m of sand and water to bed rock if the allowable bearing capacity is 2000 kN/m². The caisson receives a load of 50 MN from the superstructure. The mantle friction is 30 kN/m².*

Test the feasibility of sinking. Also calculate the thickness of the seal.

Solution. (see Fig. E-2.3)

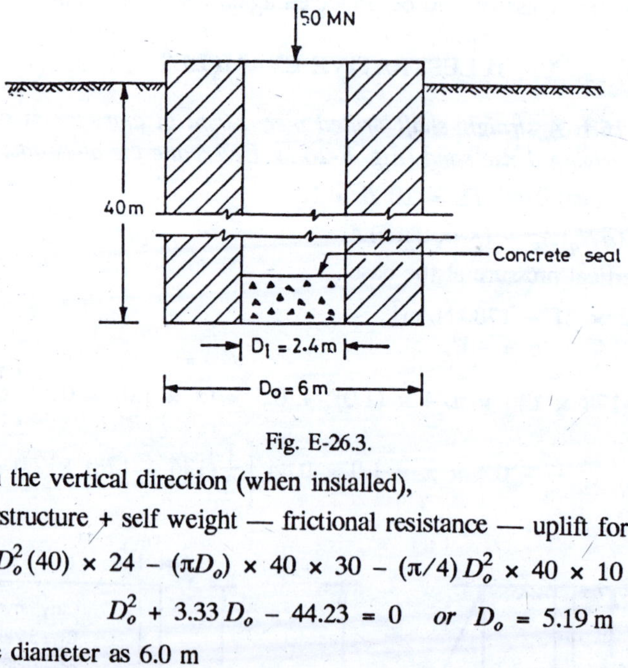

Fig. E-26.3.

From equilibrium in the vertical direction (when installed),

Load from superstructure + self weight — frictional resistance — uplift force — base reaction = 0.0

or $$50000 + (\pi/4)\,D_o^2(40) \times 24 - (\pi D_o) \times 40 \times 30 - (\pi/4)\,D_o^2 \times 40 \times 10 - (\pi/4)D_o^2 \times 2000 = 0$$

or $$D_o^2 + 3.33\,D_o - 44.23 = 0 \quad \text{or} \quad D_o = 5.19 \text{ m}$$

Let us adopt outside diameter as 6.0 m

Feasibility of sinking

From Eq. 26.7 (*a*), $$(\pi/4)\,(D_o^2 - D_i^2) \times 40 \times 24 = (\pi\,D_o) \times 40 \times f$$

Taking $f = 30$ kN/m² and $D_o = 6$ m, we have

$$36 - D_i^2 = 30 \quad \text{or} \quad D_i = 2.4 \text{ m}$$

$$\text{Thickness of wall} = \frac{6.0 - 2.4}{2} = \mathbf{1.8\ m}$$

Thickness of seal

From Eq. 26.8,
$$t = 0.59\, D_i \sqrt{q/\sigma_c}$$
$$= 0.59 \times 2.4\ \sqrt{2000/3500}$$
$$= 1.07 \text{ m} \quad \textbf{say 1.10 m}$$

Illustrative Example 26.4. *Check the stability of a floating caisson 10.5 m high and having a rectangular base 20 m × 9 m. The weight of the caisson is 9 MN and its centre of gravity is 4.0 m above the base. If the caisson is unstable, how would you make it stable ? Take unit weight of water as 10.25 kN/m².*

(b) What is the maximum pressure on the soil when the caisson has been fully installed ? The base is at a depth of 9 m below the water level. The total weight is 50 MN, which acts at an eccentricity of 0.1 m.

Solution. (see Fig. E-26.4)

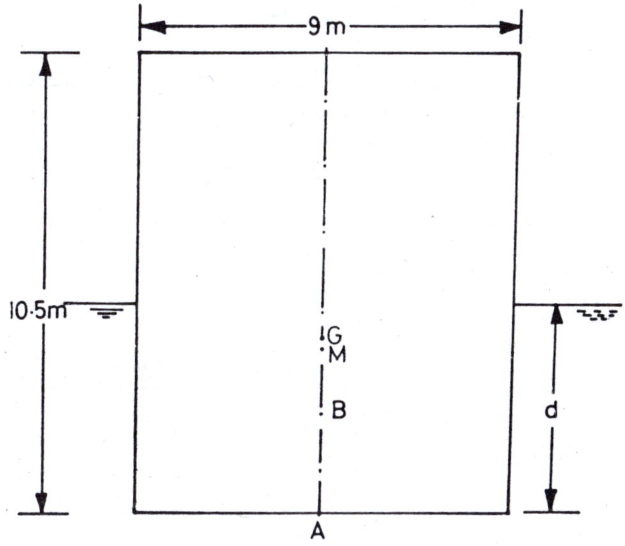

Fig. E-26.4.

Immersed Volume,
$$V = \frac{9000}{10.25} = 878.05 \text{ m}^3$$

Depth of immersion during floating,
$$d = \frac{9000}{10.25 \times 20 \times 9} = 4.88 \text{ m}$$

Height of centre of buoyancy B above base.
$$AB = 4.88/2 = 2.44 \text{ m}$$

As the distance AG is 4 m, the distance BG is given by
$$BG = 4.0 - 2.44 = 1.56 \text{ m}$$

From Eq. 26.15,
$$BM = I/V$$
$$= \frac{(1/12) \times 20 \times (9)^3}{878.05} = 1.38 \text{ m}$$

$$\text{Distance } AM = AB + BM = 2.44 + 1.38 = 3.82 \text{ m}$$

As the metacentre (M) is below the centre of gravity (G), the metacentric height MG is negative and the caisson is unstable. The caisson can be made stable by filling it with ballast, say sand ($\gamma = 22$ kN/m³). Let the thickness of the sand layer be 0.5 m.

The height of the new centre of gravity (G') above the base is given by

$$AG' = \frac{9000 \times 4 + (0.5 \times 9 \times 20) \times 22 \times 0.25}{9000 + 0.5 \times 9 \times 20 \times 22} = 3.32 \text{ m}$$

The new depth of immersion, $$d' = \frac{9000 + (0.5 \times 9 \times 20 \times 22)}{10.25 \times 9 \times 20} = 5.95 \text{ m}$$

AB' = height of new centre of buoyancy = 5.95/2 = 2.975 m

From Eq. 26.15, $$B'M' = \frac{(1/12) \times 20 \times (9)^3}{20 \times 9 \times 5.95} = 1.134 \text{ m}$$

Metacentric height, $$M'G' = (AB' + B'M') - AG'$$
$$= (2.975 + 1.134) - 3.32 = \mathbf{0.789}$$

As the metacentric height is positive (*i.e.* M' is above G'), the caisson is stable.

(*b*) When the caisson has been fully sunk, with its base at a depth of 9 m below the water level, the upward force is given by

$$F_u = (9 \times 20) \times 9 \times 10.25 = 16605 \text{ kN}$$

As the upward force is more than the downward force of 9000 kN, ballast is required. Let h be the thickness of the ballast.

Therefore, $$(9 \times 20 \times h) \times 22 + 9000 = 16605 \quad or \quad h = 1.92 \text{ m}$$

The maximum and minimum pressures when the caisson has been fully installed and subjected to design loads are given by

$$q_{max} = \frac{W}{A}\left(1 + \frac{6e}{b}\right) = \frac{50000}{20 \times 9}\left(1 + \frac{6 \times 0.1}{9.0}\right) = 296.3 \text{ kN/m}^2$$

$$q_{min} = \frac{W}{A}\left(1 - \frac{6e}{b}\right) = \frac{50000}{20 \times 9}\left(1 - \frac{6 \times 0.1}{9}\right) = 259.3 \text{ kN/m}^2$$

Uplift pressure $= 10.25 \times 9 = 92.3 \text{ kN/m}^2$

Net maximum pressure $= 296.3 - 92.3 = 204 \text{ kN/m}^2$

Net minimum pressure $= 259.3 - 92.3 = 167 \text{ kN/m}^2$

PROBLEMS

A. Numerical

26.1. A straight-shaft drilled pier of 1.2 m diameter is constructed in a dense sand deposit ($\phi = 40°$, $K = 0.4$, $\tan \delta = 0.80$ and $\gamma = 21 \text{ kN/m}^3$). The total depth of the pier is 14 m. Estimate the allowable load (F.S. = 3.0). **[Ans. 16.5 MN]**

26.2. A drilled pier of 1 m diameter has a total depth of 15 m. The diameter of the bell is 2 m and its height is 1 m. If $c_u = 80 \text{ kN/m}^2$, $\gamma = 20 \text{ kN/m}^3$ and $\alpha = 0.3$, determine the allowable load (F.S. = 3.0). **[Ans. 850 kN]**

26.3. An open caisson, 19 m deep, has external and internal diameters of 8 m and 6 m, respectively. If the water level is 2 m below the top of the well and the depth of the base below the scour level is 5 m, determine the minimum thickness of the seal that will enable complete dewatering of the caisson. Take $\sigma_c = 2000 \text{ kN/m}^2$, $\gamma_c = 24 \text{ kN/m}^3$ and allowable perimeter shear of 650 kN/m^2. **[Ans. 1.0 m]**

26.4. Determine the allowable soil pressure for a pier, 3 m diameter, 10 m long, on sand with N = 30 for a factor of safety of 3 and for an allowable settlement of 40 mm. Assume full submergence. **[Ans. 450 kN/m^2]**

B. Descriptive Types

26.5. What is the basic difference between a drilled pier and a caisson ? What are the conditions in which a drilled pier is more suitable than a caisson ?

26.6. Describe various methods for the construction of drilled piers.

26.7. How would you estimate the load carrying capacity of drilled pier in (*a*) sand, (*b*) clay ?

26.8. How would you estimate the load carrying capacity of an open caisson ?

26.9. Draw the sketch of an open caisson. How the various components are designed ?

26.10. Describe the various components of a pneumatic caisson with the help of a sketch.

26.11. What are the advantages and disadvantages of pneumatic caissons over open caissons ?

26.12. How would you check the stability of floating caisson during flotation ?

C. Multiple Choice Questions

1. Pneumatic caissons are used where the soil flow into the excavated area of an open caisson is
(*a*) faster than it can be removed
(*b*) slower than it can be removed
(*c*) negligible
(*d*) zero

2. The maximum depth of a pneumatic caisson is usually limited to
(*a*) 10 m (*b*) 20 m
(*c*) 80 m (*d*) 40 m

3. The adhesion factor for drilled piers on clay is usually taken as
(*a*) 1.0 (*b*) 0.80
(*c*) 0.60 (*d*) 0.40

4. The floating caissons generally
(*a*) have greater load-carrying capacity than open caissons
(*b*) have greater depth below the ground surface than open caissons.
(*c*) are less expensive then open caissons
(*d*) have poorer quality of construction than open caissons.

[**Ans.** 1. (*a*), 2. (*d*), 3. (*d*), 4. (*c*)]

D. Objective Type Questions

Write whether the following statements are true or false :

(*a*) The drilled piers can be of diameter less than 0.5 m.
(*b*) A pneumatic caisson is open at the top.
(*c*) The value of Nq for drilled piers is generally smaller than that for piles.
(*d*) The angle that the inclined walls of the belled pier make with the horizontal is usually 45°.
(*e*) The construction of drilled piers generally requires lighter equipment than that for pile driving.
(*f*) The angle of the cutting edge of the caisson with the vertical should not be greater than 35°.
(*g*) The air pressure in the air lock of a pneumatic caisson is usually limited to 600 kN/m^2.
(*h*) At the bottom of a pneumatic caisson, an in-situ test can be conducted.
(*i*) For a floating caisson, the metacentric height must be positive.
(*j*) If the shaft is provided with a bell, only the bell portion is considered for friction.

[**Ans.** True (*c*), (*e*), (*f*), (*h*), (*i*)]

27

Well Foundations

27.1. INTRODUCTION

Well foundation have their origin in India. Well foundations have been used in India for hundreds of years for providing deep foundations below the water level for monuments, bridges and aqueducts. Taj Mahal at Agra (India) has got well foundations.

A well foundation is similar to an open caisson discussed in chapter 26. Well foundations can be constructed on the dry bed or after making a sand island. At locations where the depth of water is greater than 5 m to 6 m and the velocity of water is high, wells can be fabricated on the river bank and then floated to the final position and grounded. Great care is to be exercised while grounding a well to ensure that its position is correct. Once the well has touched the bed, sand bags are deposited around it to prevent scour. The well may sink into the river bed by 50 to 60 cm under its own weight. Further sinking operation is similar to the sinking of wells on dry bed. The well is sunk into the ground to the desired level by excavating through the dredge holes.

Fig. 27.1. shows the section of a well. A strong cutting edge is provided to facilitate sinking. The tapered portion of the well above the cutting edge is known as *well curb.* The walls of the well are known as *steining.* Steining is made of brick masonry, stone masonry, plain or reinforced concrete. As the steining later becomes an integral part of the structure, it should be properly designed for the imposed loads. Further, it should be heavy enough to overcome frictional resistance during sinking.

After the well has been sunk to the final position, the bottom plug is formed by concreting. The bottom plug serves as the base of the well. The well is filled with sand partly or completely. At the top of the well, a top plug is formed by concreting. A R.C.C. well cap is provided at the top to transmit the load of the superstructure (pier or abutment) to the well.

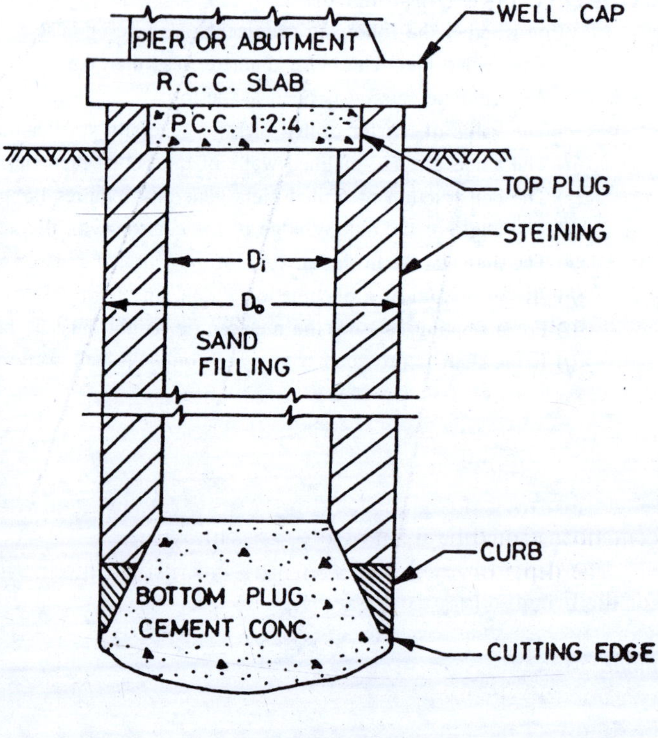

Fig. 27.1. Well Foundation.

27.2. DIFFERENT SHAPES OF WELLS

Different shapes of wells that are commonly used are shown in Fig. 27.2.

(1) The most commonly used shape is *circular* [Fig. 27.2 (*a*)], as it has high structural strength and is convenient in sinking. The chances of tilting are also minimum in this shape. The shape is quite suitable for piers of the single-line railway bridges and the double-lane road bridges. However, when the piers are excessively long, the circular shape becomes uneconomical. The maximum diameter of circular wells is generally limited to 9 m.

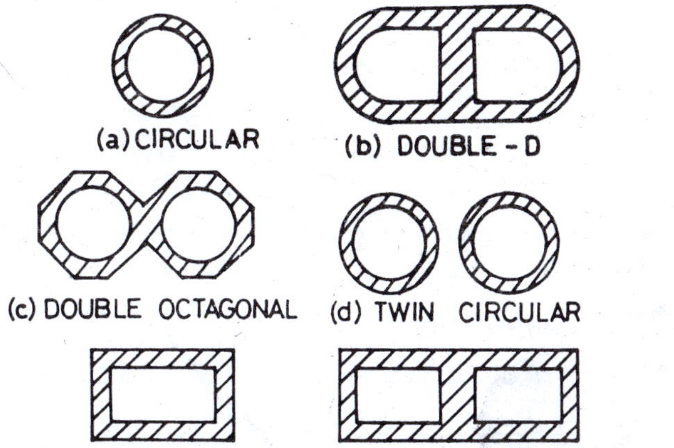

Fig. 27.2. Diferent Shapes of Wells.

(2) *Double-D wells* [Fig. 27.2 (*b*)] are generally used for the piers and abutments of bridges which are too long to be accommodated on a circular well of 9 m diameter. The wells of this shape can also be sunk easily. However, considerable bending moments are caused in the steining due to the difference in pressure between the outside and the inside of the well. Further, the square corners at the partition wall offer greater resistance to sinking.

(3) *Double-octagonal wells* [Fig. 27.2 (*c*)] are better than the double-D wells in many respects. The square corners are eliminated and bending stresses are considerably reduced. However, they offer greater resistance than double-D wells against sinking on account of increased surface area. Moreover, the construction is more difficult.

(4) *Twin-circular wells* [Fig. 27.2 (*d*)] are two independent wells placed very close to each other and having a common well cap. The wells are sunk simultaneously. These wells are suitable where the length of the pier is considerable, which cannot be accommodated on a double-D or double-octagonal well. Twin circular wells are advantageous when the depth of sinking is small and the bearing capacity of the soil is high. The disadvantage of twin circular wells is that there is a possibility of the relative settlement of the two wells even if a heavy R.C.C top cap is provided unless the wells are founded on an incompressible soil.

(5) *Rectangular wells* [Fig. 27.2 (*e*)] are generally used for bridge foundation having depths upto 7-8 m. For large foundations, double-rectangular wells [Fig. 27.2 (*f*)] are used. For piers and abutments of very large size, rectangular wells with multiple dredge holes (not shown in the figure) are used. Bending stresses in the steining are very high in rectangular wells.

27.3. GRIP LENGTH

The well should be sunk below the maximum scour level to sufficient depth such that the resistance from the sides is able to resist the lateral forces acting on the well. The depth of the bottom of the well below the maximum scour level is known as grip length. The depth of foundation should be chosen considering the grip length and the bearing capacity of the soil. The maximum and minimum base pressures under worst loading conditions should be within the permissible values.

The depth of scour in a stream can be ascertained through actual soundings at or near the site proposed for the bridge during or immediately after a flood. The maximum scour would be greater than the measured scour because the design discharge is greater than the flood discharge for which the soundings have been made. Moreover, there would be an increase in the velocity of water due to the obstruction of flow caused by construction of the bridge. An extra allowance should also be made in the measured scour due to the proximity of piers.

In case actual soundings cannot be made, the normal depth of scour in alluvial soils may be calculated by Lacey's formula.

$$d = 0.473\,(Q/f)^{1/3} \qquad ...(27.1)$$

where d = normal scour depth, measured below high flood level (m), Q = design discharge (m^3/sec), f = silt factor.

The silt factor may be calculated from the equation

$$f = 1.76\sqrt{d_m} \qquad ...(27.2)$$

where d_m = mean size of particle (mm)

The regime width can be computed as

$$W = 4.75\sqrt{Q} \qquad ...(27.3)$$

If the actual water way (L) is less than the regime width, the actual scour depth (d') is given by

$$d' = d(W/L)^{0.61} \qquad ...(27.4)$$

The maximum scour depth (d_{max}), as recommended by IRC (1966) and IS : 3955—1967 can be obtained from Table 27.1.

The grip length for wells on railway bridges is generally taken as 50% of the maximum scour depth. For road bridges, a grip length of 30% of the maximum scour depth is generally provided. Thus the base of the well is usually taken at a depth of 2.67 d' below the high flood level.

Table 27.1. Maximum Scour Depth

S. No.	*River Section*	*Maximum Scour* (d_{max})
1.	Straight Reach	1.27 d'
2.	Moderate Bend	1.50 d'
3.	Severe Bend	1.75 d'
4.	Right-angled Bend or at nose of Pier	2.00 d'
5.	Upstream Nose of Guide Banks	2.75 d'
6.	Severe Swirls	2.50 d'

According to IS : 3955 — 1967, the depth of foundation should not be less than 1.33 times the maximum scour depth. The depth of the base of the well below the maximum scour level is kept not less than 2 m for piers and abutments with arches and 1.2 m for piers and abutments supporting other types of structures.

27.4. FORCES ACTING ON THE WELL FOUNDATION

The following forces should be considered in the design of a well foundation.

(1) Dead loads. The dead loads carried by the well include the weight of the superstructure and the self weight.

(2) Live loads. The design live loads for railway bridges are taken according to Indian Railway Bridges Rules. For road bridges, the live loads as specified by the Indian Road Congress Standard Specifications and Code of Practice for Road Bridges—Sect. II should be used.

(3) Impact loads. Impact effect due to live load is considered only in the design of pier cap and the bridge seat on the abutment. For all other members of the well, the effect of impact is ignored.

(4) Wind loads. Wind loads on the live load, superstructure and the part of the substructure located above the water level are calculated according to the provisions of IS : 875. The wind load acts on the exposed area in elevation and thus it acts laterally on the bridge.

(5) Water pressure. Water pressure due to water current acts on the part of substructure which lies between the water level and the maximum scour level. On piers parallel to the direction of the water, the intensity of water pressure is given by

$$p = KV^2 \qquad ...(27.5)$$

where p = intensity of pressure (kN/m^2), K = a constant, depending upon the shape of the well (= max. of

0.788 for square ended pier to a minimum of 0.237 for piers with cut and ease water), V = velocity of current (m/sec).

It is assumed that V^2 is maximum at the free surface of water and zero at the deepest scour level. The variation is assumed to be linear. The surface velocity is taken $\sqrt{2}$ times the average velocity.

Even when the flow is parallel to the pier, a transverse force equal to 20% of that acting parallel to the pier is taken to allow for oblique flow.

If the current makes an angle θ with the axis of the pier, the pressure along the axis of the pier and transverse to it are given by

$$p_a = \text{pressure along axis} = p \cos^2 (20° \pm \theta) \qquad ...(27.6)$$

$$p_t = \text{transverse pressure} = p \sin^2 (20° \pm \theta) \qquad ...(27.7)$$

(6) Longitudinal Forces. Longitudinal forces occur due to tractive and braking forces. These forces depend upon the type of vehicles and bearings. These forces are transmitted to substructure mainly through fixed bearings and through friction in movable bearings. According to IRC code, a longitudinal force of μW is taken on the free bearing and the balance on the fixed bearing, where W is the total reaction and μ is the coefficient of friction.

(7) Centrifugal Force. A centrifugal force is transmitted through bearings if the bridge is curved in plan.

(8) Buoyant Forces. Buoyancy reduces the effective weight of the well. In masonry or concrete steining, 15% buoyancy is considered to account for the porousness.

When the well is founded on coarse sand or shingle, full buoyancy equal to the weight of an equivalent volume of displaced water should be considered. For semi-pervious foundations, it is suitably reduced.

(9) Earth Pressure. The earth pressure is calculated according to Rankine's theory or Coulomb's theory. For the stability of foundations below the scour level, the passive earth pressure of the soil is considered.

To account for the effect of live load placed behind the abutment, an equivalent height of surcharge is considered if no approach slab is provided.

(10) Temperature Stresses. Longitudinal forces are induced due to temperature changes. The movements due to temperature changes are partially restrained in girder bridges because of friction at the moveable end.

(11) Seismic forces. For the wells constructed in the seismic zone, seismic forces should be considered. The forces act on all components of the structure. The force is usually specified as αW, where W is the weight of the component and α is the seismic coefficient. The value of α depends upon the seismic zone (IS : 1893). Its value is generally taken between 0.01 and 0.08. The seismic force acts through the centre of gravity of the component. It may act in any direction, but it is assumed to act in one direction only at a time. The seismic forces are considered separately along the axis of the pier and transverse to it.

Resultant Forces. The magnitude, direction and the point of application of all the above forces are found under the worst possible combinations. The resultant can be replaced by an equivalent vertical force W, and two horizontal forces F and Q, as shown in Fig. 27.3.

The horizontal force Q acting in the transverse direction gives more critical condition for the lateral stability of the well.

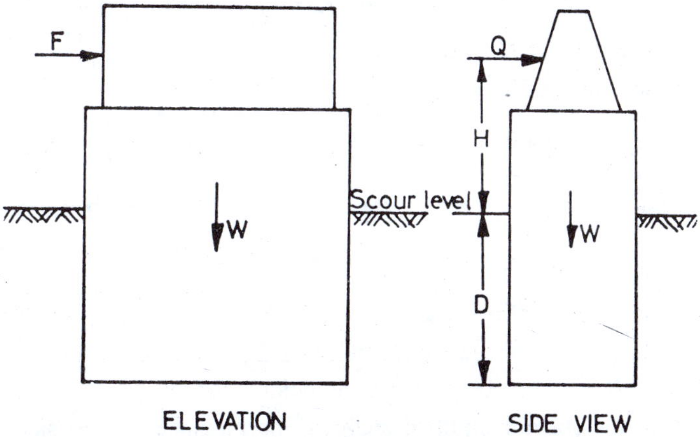

Fig. 27.3. Forces on a pier.

27.5. TERZAGHI'S ANALYSIS

Terzaghi's solution for free rigid bulkheads can be used for an approximate analysis of the well foundation. When a rigid bulkhead embedded in sand moves parallel to its original position, the sand on its front is transformed into passive state whereas that on its rear, into active state. Assuming that both the active and passive resistances are fully mobilised, the net pressure at any depth z below the ground surface is given by

$$p = \gamma z (K_p - K_a) \qquad \text{...(27.8)}$$

A free rigid bulkhead depends for its stability solely on the lateral resistance. Let q_{max}' be the horizontal force per unit length acting on the bulkhead of total height H_1 (Fig. 27.4). The pressure distribution on both sides of the bulkhead at the instant of failure may be represented as shown in the figure. The bulkhead rotates about the point O above the base B. As the soil around the well is generally submerged, the submerged unit weight γ' is used.

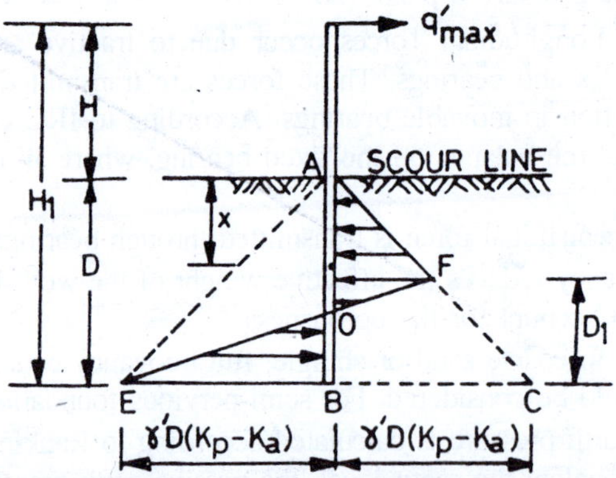

Fig. 27.4. Terzaghi's Analysis.

The applied force per unit length is given by

$$q_{max}' = \text{Area } ABC - \text{Area } FEC$$

$$= \frac{1}{2}\gamma' D^2 (K_p - K_a) - \frac{1}{2}(2\gamma' D)(K_p - K_a) D_1$$

where D_1 is the height of point F above the bottom.

$$q_{max}' = \frac{1}{2}\gamma' D (K_p - K_a)(D - 2D_1) \qquad \text{...(27.9)}$$

Taking moments about the base B,

$$q_{max}' H_1 = \frac{1}{2}\gamma' D^2 (K_p - K_a) D/3 - \frac{1}{2}(2\gamma' D)(K_p - K_a) D_1^2/3$$

where $H_1 = H + D$.

Substituting the value of q_{max}' from Eq. 27.9,

$$\frac{1}{2}\gamma' D (K_p - K_a)(D - 2D_1) H_1 = \frac{1}{2}\gamma' (K_p - K_a) D^3/3 - \frac{1}{2}(2\gamma' D)(K_p - K_a) D_1^2/3$$

or $$(D - 2D_1) H_1 = D^2/3 - 2D_1^2/3$$

or $$D_1^2 - 3D_1 H_1 + (1.5 DH_1 - 0.5 D^2) = 0$$

Solving for D_1, we have $$2D_1 = 3H_1 \pm \sqrt{9H_1^2 - 2D(3H_1 - D)} \qquad \text{...(27.10)}$$

By substituting the above value of D_1 in Eq. 27.9, the value of q_{max}' can be computed. The coefficient K_p and K_a can be obtained from Rankine's theory.

In the above simplified analysis, the moments due to side friction and the base reaction are neglected. The error is on the safe side.

Heavy Wells. A heavy well embedded in cohesionless soil rotates about its base (Fig. 27.5). The force per unit length, q_{max}', can be obtained by taking moments about the base.

$$q_{max}' H_1 = \frac{1}{2}\gamma' (K_p - K_a) D^2 \times \frac{D}{3}$$

or $$q_{max}' = \frac{1}{6}\gamma' (K_p - K_a) \frac{D^3}{H_1} \qquad ...(27.11)$$

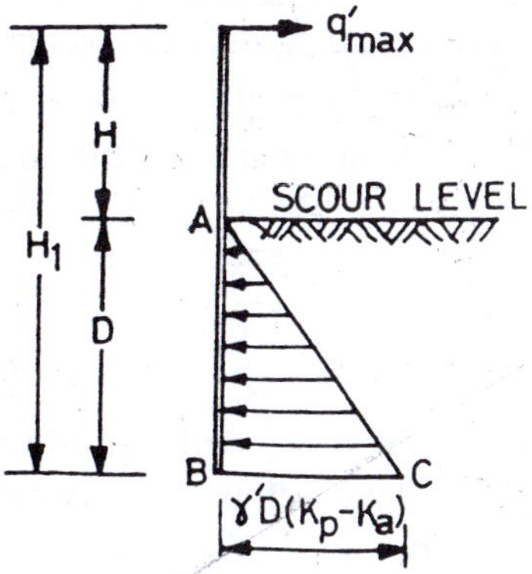

Fig. 27.5. Heavy Well.

Effect of Surcharge. The effect of surcharge due to the weight of soil above the scour line can be taken into account in the analysis. The soil below the maximum scour level is subjected to a surcharge of height Z of the unscoured soil (Fig. 27.6). The height Z may be taken equal to one-half the normal depth of scour, in case it is not possible to ascertain it by actual measurement.

The pressure distribution is shown in the figure. The maximum pressure at the base is equal to $\gamma' (K_p - K_a)(D + Z)$.

In this case, q_{max}' is given by

$$q_{max}' = \frac{1}{6}\gamma' (K_p - K_a) \frac{D^2 (D + Z)}{H_1} \qquad ...(27.12)$$

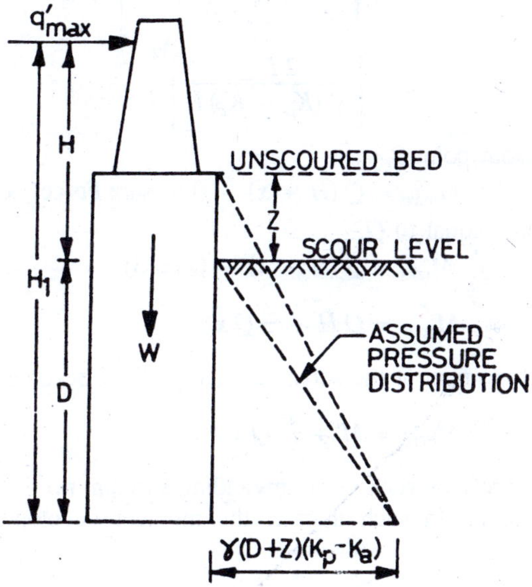

Fig. 27.6. Effect of Surcharge.

Allowable Transverse Load. The total resisting force for the well is equal to the product of q_{max}' and the length L of the well. The allowable load (Q_a) is equal to the resisting force divided by a suitable factor of safety.

Thus $$Q_a = \frac{q_{max}' L}{FS} \qquad ...(27.13)$$

The factor of safety (FS) should not be less than 2.

The length L of the well is generally taken equal to the diameter of the well. Balwant Rao and Muthuswamy (1963) recommend that the allowable load as obtained by Eq. 27.13 should be multiplied by a suitable shape factor. The shape factor is unity in the case of rectangular wells and equal to $\pi/4$ for circular wells. However, for circular wells of the diameter larger than 4.5 m, the shape factor is taken as unity.

Maximum Pressure at Base. If Q is the actual applied transverse horizontal force and Q_a is the allowable equivalent resisting force, the unbalanced force $(Q - Q_a)$ acting at a height of H above the scour level would produce an overturning moment M_B at the base, given by

$$M_B = (Q - Q_a)(H + D)$$

The maximum and the minimum pressure at the base are

$$q_{max} = \frac{W}{A_b} + \frac{M_B}{Z_b}$$

and

$$q_{min} = \frac{W}{A_b} - \frac{M_B}{Z_b}$$

where W = net vertical load on the well base, after making allowance for buoyancy and skin friction, A_b = area of the well base, Z_b = section modulus of the well base.

The maximum pressure should not be more than the allowable soil pressure. The minimum pressure should not be negative (tensile). It is a general practice not to give any relief due to skin friction while calculating the maximum pressure in clays, but to consider it for calculating the minimum pressure.

Maximum Moment in Steining. The maximum moment M_{max} on the steining occurs at point S at depth x below the scour line (Fig. 27.4) where the shear force is zero, *i.e.* the applied force and the earth pressure just balance each other. Taking a factor of safety F,

$$\frac{1}{F}\left[\frac{1}{2}\gamma'(K_p - K_a)x^2 L\right] = Q$$

or

$$x = \left[\frac{2FQ}{\gamma'(K_p - K_a)L}\right]^{1/2} \qquad \text{...(27.14)}$$

Taking moments about point S,

$$M_{max} = Q(H + x) - (\text{Pressure Force}) \times x/3$$

Taking pressure force equal to Q,

$$M_{max} = Q(H + x) - Q(x/3)$$

or

$$M_{max} = QH + \frac{2}{3}Qx \qquad \text{...(27.15)}$$

If M is the applied moment at the scour-level, Eq. 27.15 becomes

$$M_{max} = M + \frac{2}{3}Qx \qquad \text{...[27.15 (a)]}$$

If the well rests squarely on rock or an unyielding stratum, no tilt about the base or rotation about a point above the base is possible. In such a case, the moment developed upto the base is transferred to the foundation bed.

27.6. BANERJEE AND GANGOPADHYAY'S ANALYSIS

Banerjee and Gangopadhyay (1960) derived equations for the lateral load capacity of wells, making the following assumptions:

(1) The well is founded in a sandy stratum.

(2) The well is acted upon by a uni-directional horizontal force Q in a direction across the pier.

(3) The resultant unit pressure on soil at any depth is in simple proportion to the horizontal displacement.

(4) The ratio between the contact pressure and the corresponding displacement is independent of the magnitude of pressure.

(5) The coefficient of vertical subgrade reaction has the same value for every point of the surface acted by the contact pressure.

The method of analysis can be divided into following 6 steps.

(*i*) **Expression for Horizontal Soil Reaction.** When a rigid well embedded in a sandy stratum moves parallel to its original position, the sand on the front face is transformed into passive state and that on the rear face, into active state (Fig. 27.7). The net pressure at a depths z is given by (shown by the firm line).

$$p_1 = \gamma z (K_p - K_a) \qquad \ldots(27.16)$$

Let p be the load per unit area of the vertical surface and ρ be the corresponding displacement. If ρ_1 is the displacement required to increase the value of the net pressure from zero to p_1, we have

$$p/\rho = p_1/\rho_1$$

Substituting the value of p_1 from Eq. 27.16,

$$\frac{p}{\rho} = \frac{\gamma z (K_p - K_a)}{\rho_1}$$

or
$$p = \frac{\rho \gamma z (K_p - K_a)}{\rho_1}$$

or
$$p = m \rho z \qquad \ldots(27.17)$$

where
$$m = \gamma (K_p - K_a)/\rho_1 \qquad \ldots(27.18)$$

The constant m is known as the *coefficient of horizontal soil reaction.* It is an empirical coefficient which depends upon the nature of soil, the size and shape of the loaded area.

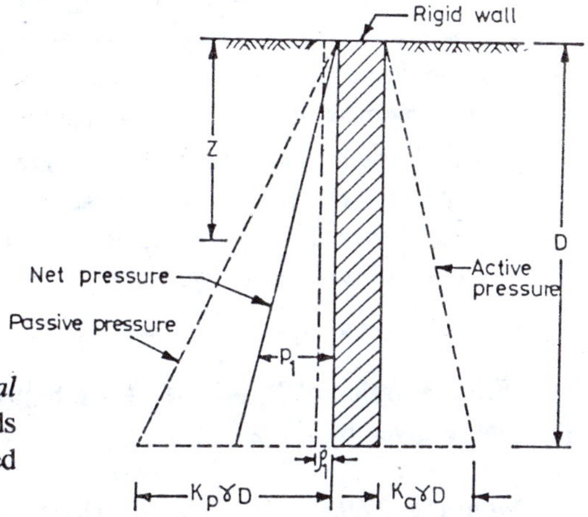

Fig. 27.7.

(*ii*) **Equations of Equilibrium.** The stability of the well can be checked assuming either no plastic flow or partial plastic flow. The following analysis is based on the assumption of no plastic flow. For the other case, a reference may be made to the original paper. Fig. 27.8 shows a well of length L and width B, acted upon by

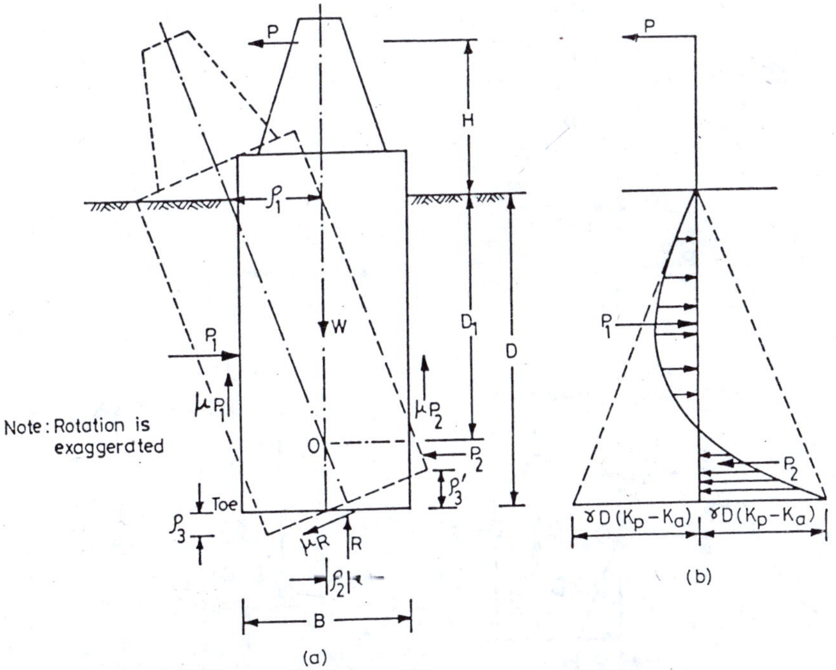

Fig. 27.8.

a horizontal force P per unit length at a height H above the scour line. The well is embedded to a depth D below the scour line.

When the well rotates about a point O located at a depth D_1 below the scour line, the following equations can be written from statics.

$$P = P_1 - P_2 - \mu R \quad \text{...(27.19)}$$

$$PH = M_3 + M_2 - M_1 + \mu RD + \mu (P_1 - P_2) \times B/2 \quad \text{...(27.20)}$$

and

$$W = \mu (P_1 + P_2) + R \quad \text{...(27.21)}$$

where P = horizontal transverse force, P_1 = resultant force on the front face, P_2 = resultant passive force on the rear face, μP_1 = skin friction on the front face, μP_2 = skin friction on the rear face, R = resultant vertical soil reaction at the base, μR = frictional resistance of the soil at the base, M_1 = moment at the scour line due to P_1, M_2 = moment at the scour line due to P_2, M_3 = moment due to the vertical soil reaction at its base. Let ρ_1 = horizontal displacement of the centre line of the well at the scour level, ρ_2 = horizontal displacement of the centre line of the well at the base level, ρ_3 = downward vertical displacement of the well at the toe, ρ'_3 = upward vertical displacement of the well at the heel($\rho_3' \sim \rho_3$), ρ_4 = uniform vertical displacement of the well due to resultant vertical force W.

From Fig. 27.8 (*a*),

$$\frac{\rho_1}{D_1} = \frac{\rho_2}{D - D_1} = \frac{\rho_3}{B/2} \quad \text{...(27.22)}$$

***(iii)* Evaluation of P_1 and M_1.** Let ρ be the horizontal displacement at depth z below the scour line. The pressure as given by Eq. 27.17 is

$$p = m \rho z$$

From Fig. 27.9,

$$\rho = (\rho_1/D_1)(D_1 - z)$$

Therefore,

$$p = mz(\rho_1/D_1)(D_1 - z) \quad \text{...(27.23)}$$

Force per unit length,

$$P_1 = \int_o^{D_1} mz\, (\rho_1/D_1)\, (D_1 - z)\, dz$$

or

$$P_1 = \frac{m \rho_1 D_1^2}{6} \quad \text{...(27.24)}$$

Moment,

$$M_1 = \int_o^{D_1} (pdz)\, z$$

$$= \int_o^{D_1} m\, (\rho_1/D_1)\, (D_1 - z)\, z^2\, dz$$

or

$$M_1 = m \rho_1 (D_1^3/12) \quad \text{...(27.25)}$$

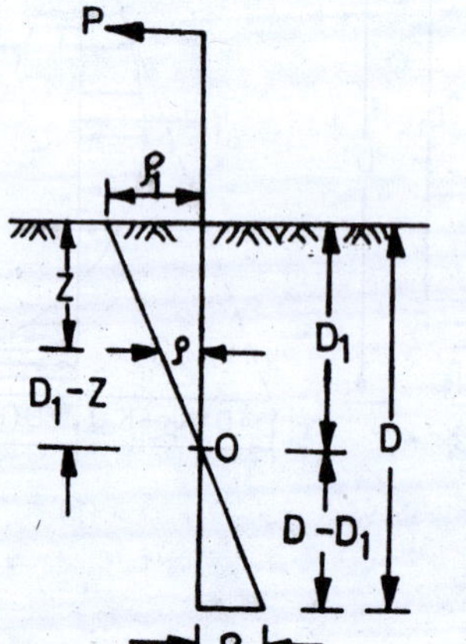

Fig. 27.9.

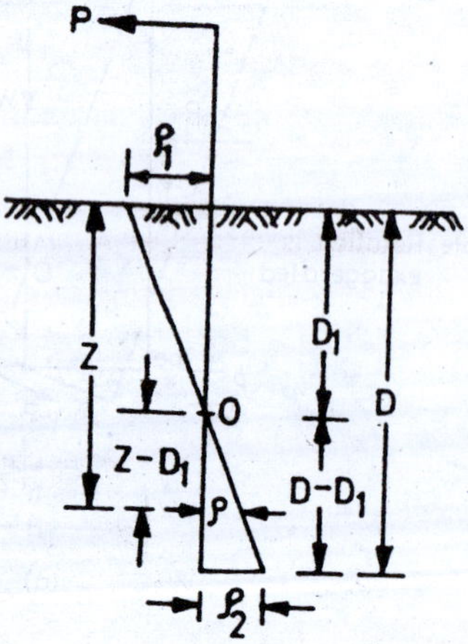

Fig. 27.10.

(iv) Evaluation of P_2 and M_2 From Fig. 27.10,

$$\frac{\rho}{z - D_1} = \frac{\rho_1}{D_1}$$

or
$$\rho = \frac{(z - D_1)}{D_1} \rho_1$$

From Eq. 27.17,
$$p = mz \frac{(z - D_1)}{D_1} \rho_1$$

Force per unit length,
$$P_2 = \int_{D_1}^{D} \frac{m\rho_1}{D_1} z(z - D_1)\, dz$$

$$= \frac{m\rho_1}{6D_1}(2D^3 - 3D_1D^2 + D_1^3) \qquad \text{...(27.26)}$$

Moment,
$$M_2 = \int_{D_1}^{D} \frac{m\rho_1}{D_1} z(z - D_1)\, z\, dz$$

or
$$M_2 = \frac{m\rho_1}{12D_1}(3D^4 - 4D_1D^3 + D_1^4) \qquad \text{...(27.27)}$$

(v) Evaluation of R and M_3 Let K_v be the modulus of vertical subgrade reaction.

$$K_v = p/\rho \qquad \text{...(27.28)}$$

where ρ = vertical deflection (= ρ_4) of soil corresponding to the vertical reaction p.

Therefore, vertical reaction,

$$R = 2\int_0^{B/2} p\, dz = 2\int_0^{B/2} K_v\, \rho_4\, dx$$

or
$$R = K_v B \rho_4 \qquad \text{...(27.29)}$$

The rotation of the well is resisted by a moment M_3 acting at the base due to pressure developed on account of the downward deflection of the toe and the upward deflection of the heel. Fig. 27.11 shows the rotation of the base, with a maximum displacement of ρ_3 at the ends. Let ρ be the deflection at a distance x from the centre of the base. Therefore,

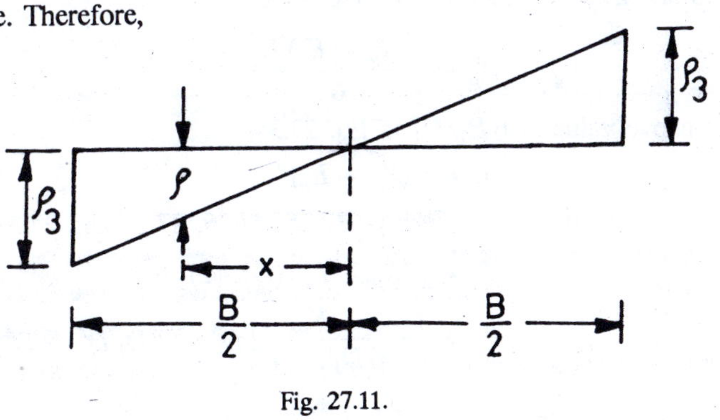

Fig. 27.11.

$$\frac{\rho}{x} = \frac{\rho_3}{B/2} \quad \text{or} \quad \rho = \frac{2\rho_3}{B} x$$

From Fig. 27.28,
$$p = K_v \rho = \frac{2\rho_3}{B} K_v x$$

Moment
$$M_3 = 2\int_0^{B/2} p\, x\, dx = \frac{4\rho_3 K_v}{B}\int_0^{B/2} x^2\, dx$$

or $$M_3 = \frac{\rho_3 K_v B^2}{6} \qquad \text{...(27.30)}$$

Substituting the value of ρ_3 from Eq. 27.22,

$$M_3 = \left(\frac{B/2}{D_1}\right) \rho_1 \frac{K_v B^2}{6} = \frac{K_v B^3 \rho_1}{12 D_1} \qquad \text{...(27.31)}$$

Evaluation of $m\,\rho_1$

The maximum soil pressure at depth z below the scour line is given by

$$(p_z)_{max} = \gamma z (K_p - K_a) \qquad \text{...(27.32)}$$

If no plastic flow is allowed in the soil, the horizontal soil reaction p at any depth z must not exceed $(p_z)_{max}$. Therefore,

$$\frac{d(p_z)}{dz} = \gamma(K_p - K_a) \qquad \text{...(27.33)}$$

The sand starts flowing as soon as the slope of the pressure parabola at scour level becomes equal to the slope of the line whose abscissa represents the value of $(p_z)_{max}$ [see Fig. 27.8 (*b*)].

From Eq. 27.24, $$m\,\rho_1 = \frac{6P_1}{D_1^2} \qquad \text{...(27.34)}$$

From Eq. 27.23, $$p = mz \frac{\rho_1}{D_1}(D_1 - z)$$

Therefore, $$p = \frac{z(D_1 - z)}{D}\left(\frac{6P_1}{D_1^2}\right) = \frac{6P_1 \; (D - z)\, z}{D_1^3}$$

Hence, $$\frac{dp}{dz} = \frac{6P_1(D_1 - 2z)}{D_1^3}$$

At $z = 0$, $$\frac{dp}{dz} = \frac{6P_1 D_1}{D_1^3} = \frac{6P_1}{D_1^2} \qquad \text{...(27.35)}$$

From Eqs. 27.33 and 27.35, $6P_1/D_1^2 = \gamma(K_p - K_a)$

or $$P_1 = \frac{\gamma(K_p - K_a) D_1^2}{6} \qquad \text{...(27.36)}$$

Substituting the above value of $6P_1/D_1^2$ in Eq. 27.34,

$$m\,\rho_1 = \gamma(K_p - K_a) \qquad \text{...(27.37)}$$

Substituting the above value of $m\,\rho_1$ in Eqs. 27.25, 27.26, 27.27 and 27.31,

$$M_1 = \gamma(K_p - K_a)(D_1^3/12) \qquad \text{...(27.38)}$$

$$P_2 = \frac{\gamma(K_p - K_a)}{6D_1}(2D^3 - 3D_1 D^2 + D_1^3) \qquad \text{...(27.39)}$$

$$M_2 = \frac{\gamma(K_p - K_a)}{12 D_1}(3D^4 - 4D_1 D^3 + D_1^4) \qquad \text{...(27.40)}$$

$$M_3 = \frac{K_v \gamma(K_p - K_a)}{m}\left(\frac{B^3}{12 D_1}\right) \qquad \text{...(27.41)}$$

(*vi*) Determination of Base Pressure. The total vertical reaction is given by (see Fig. 27.12)

$$R = \frac{1}{2}(p_t + p_h)\, B \qquad \text{...(27.42)}$$

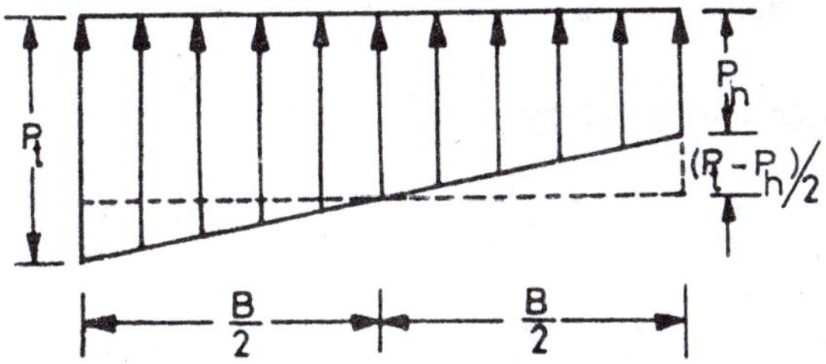

Fig. 27.12.

where p_t = pressure at the toe, p_h = pressure at the heel.

Moment M_3 can be expressed as

$$M_3 = \frac{1}{2}(p_t - p_h) \times \frac{B}{2} \times \frac{1}{2}\left(\frac{2}{3}B\right)$$

$$M_3 = \frac{(p_t - h_h)}{12} B^2 \qquad ...(27.43)$$

The maximum pressure should not be more than the allowable soil pressure. The minimum pressure should not be negative (tensile).

27.7. SIMPLIFIED ANALYSIS FOR HEAVY WELLS

The analysis given in the preceding section is considerably simplified if the well is heavy. For such wells, the reaction R at base is very high, and, therefore, the sliding at the base does not take place (*i.e.* $\rho_2 = 0$). The well however rotates at its base (Fig. 27.13).

As $\rho_2 = 0$, and $D_1 = D$, Eqs. 27.39 and 27.40 give

$$P_2 = 0 \quad \text{and} \quad M_2 = 0$$

The equations of equilibrium (Eqs. 27.19 to 27.21) become

$$P = P_1 - \mu R = P_1 - F \qquad ...(27.44)$$

$$PH = M_3 - M_1 + FD + \mu P_1 (B/2) \qquad ...(27.45)$$

and $$W = \mu P_1 + R \qquad ...(27.46)$$

where F = frictional resistance at the base to prevent sliding.

Also by substituting $D = D_1$, we have

$$P_1 = \gamma (K_p - K_a) D^2/6 \qquad ...(27.47)$$

$$M_1 = \gamma (K_p - K_a) D^3/12 \qquad ...(27.48)$$

and $$M_3 = \frac{\gamma K_v (K_p - K_a)}{m} \times \frac{B^3}{12 D} \qquad ...(27.49)$$

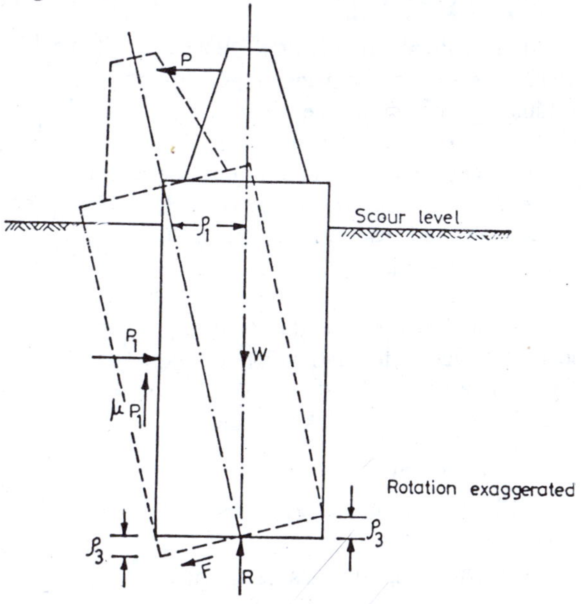

Fig. 27.13.

The above quantities are per unit length of the well and are applicable for a rectangular well. In case of a well with a non- rectangular base, the equations are modified as

$$PL = P_1 L - F_1 \qquad ...(27.50)$$

$$PLH = M_B - M_1 L + F_1 D + \mu P_1 (B/2) L \qquad ...(27.51)$$

and $$W_T = \mu P_1 L + R_1 \qquad ...(27.52)$$

where L = maximum length of the well base, B = width of the well base, M_B = total moment induced in the base due to tilting, F_1 = total horizontal reaction at the base, R_1 = total vertical reaction at the base, W_T = total vertical load of the well.

The pressures at the toe and heel are given by

$$p_t = \frac{R_1}{A_b} + \frac{M_B}{Z_b} \quad ...(27.53)$$

and

$$p_h = \frac{R_1}{A_b} - \frac{M_B}{Z_b} \quad ...(27.54)$$

where Z_b is section modulus.

Combining the above two equations,

$$p_t + p_h = \frac{2 R_1}{A_b} \quad ...(27.55)$$

From Eqs. 27.41 and 27.43,

$$M_3 = \frac{\gamma K_v}{m} (K_p - K_a) \frac{B^3}{12 D} = \frac{(p_t - p_h) B^2}{12} \quad ...(27.56)$$

The values of p_t and p_h can be found from Eqs. 27.55 and 27.56.

Once the value of p_t has been found, the moment M_B can be found from Eq. 27.53, as

$$M_B = (p_t - R_1/A_b) Z_b \quad ...(27.57)$$

27.8. IRC METHOD

Indian Road Congress (IRC : 45—1972) gave recommendations for estimating the resistance of the soil below the maximum scour level. The recommendations are based on the observed behaviour of models of well foundations and research. According to the recommendations, elastic theory can be used to determine the soil pressure at the side and at the base due to design loads. However, for estimation of the factor of safety against shear failure, the ultimate soil resistance is computed.

1. Elastic Theory.

The following assumptions are made in the elastic theory :

(1) The well behaves as a rigid body.

(2) The coefficient of horizontal subgrade reaction K_H increases linearly with depth.

(3) The unit soil reaction increases linearly with the lateral deflection.

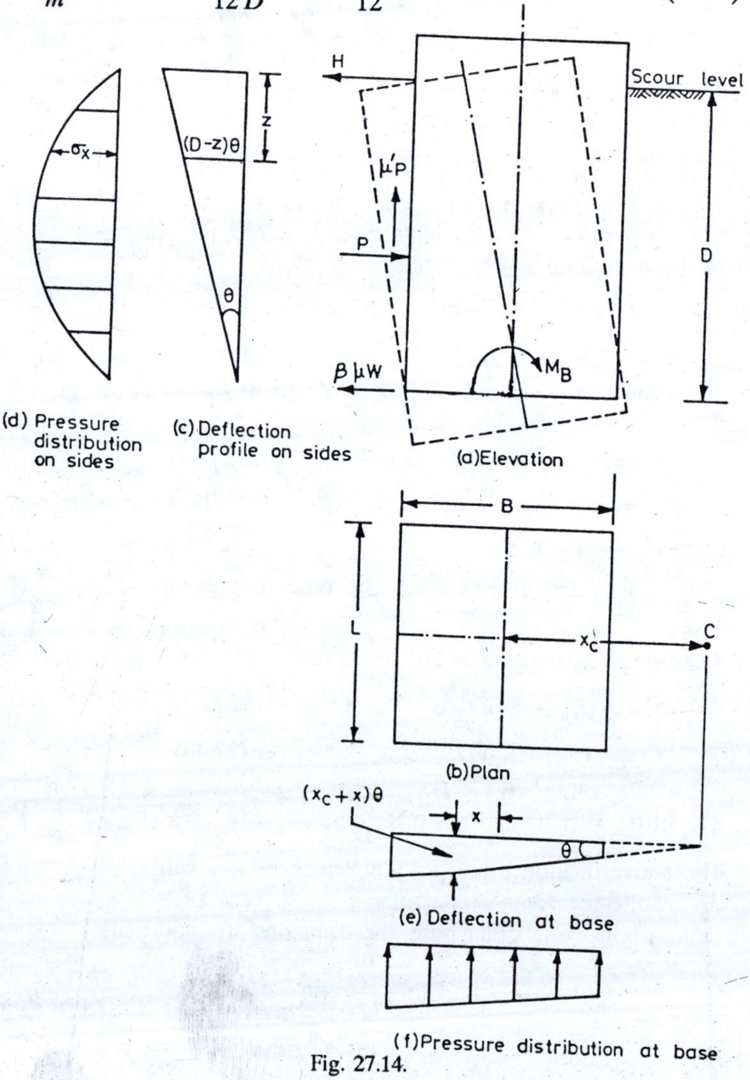

Fig. 27.14.

(4) The well is acted upon by an external horizontal force H and a moment M_0 at the scour level.

Pressure Distribution on Sides. Fig. 27.14 (*a*) shows a rigid well with its base at a depth D below the scour level. The well may rotate at a point above the base, or at the base, or below the base. The base moves toward the centre of rotation if the point of rotation lies above the base and, therefore, the frictional force at the base acts in the direction of the horizontal force H. However, if the point of rotation lies below the base, the direction of the frictional force is in the direction opposite to that of H.

In general, the frictional force is given by

$$F = \beta \mu W \quad \text{...(27.58)}$$

where μ = coefficient of friction, W = total downward load, β = a factor, which lies between –1 and + 1, depending upon the point of rotation.

If the well rotates at point C [Fig. 27.14 (*b*)], the horizontal deflection at any depth z is given by

$$\rho_H = (D - z)\,\theta \quad \text{...(27.59)}$$

The horizontal soil reaction at that level is

$$\sigma_x = K_H (z/D)(D - z)\,\theta \quad \text{...(27.60)}$$

or

$$\sigma_x = m K_v (z/D)(D - z)\,\theta$$

where

$$m = K_H/K_v$$

Total horizontal soil recation acting on the sides.

$$P = \int_0^D L\,\sigma_x\,dz$$

$$= \int_0^D L\,m\,K_v (z/D)(D - z)\,\theta\,dz$$

$$= \frac{m K_v L\,\theta}{D}(D^3/6)$$

or

$$P = \frac{2\,m K_v \theta}{D} \cdot I_v \quad \text{...(27.61)}$$

where $I_v = L D^3/12$ = moment of inertia about the axis passing through the *c.g.* of the vertical projected area.

Moment of P at the base level is given by

$$M_p = \int_0^D L\,\sigma_x (D - z)\,dz$$

$$= \int_0^D L\,m\,K_v (z/D)(D - z)^2\,\theta\,dz$$

or

$$M_p = m K_v \theta I_v \quad \text{...(27.62)}$$

Pressure Distribution at Base. The vertical deflection at a distance $(x + x_c)$ from the centre of rotation is given by [see Fig. 27.14 (*c*)],

$$\rho = (x + x_c)\,\theta$$

Therefore, vertical soil reaction $\sigma_z = K_v (x + x_c)\,\theta$

Moment at the base, $M_B = \int_{-B/2}^{+B/2} K_v (x + x_c)\,\theta\,x\,dA$

or

$$M_B = K_v \theta \int_{-B/2}^{+B/2} x^2\,dA + K_v\,0 \int_{-B/2}^{+B/2} x_c\,x\,dA$$

As the reference axis is taken through the *c.g.* of the base, the second term is equal to zero. Therefore,

$$M_B = K_v \theta I_B \quad \text{...(27.63)}$$

where I_B = moment of inertia of the base about an axis passing through the *c.g.* and perpendicular to the horizontal force H.

For equilibrium, the sum of all the horizontal forces is zero.

Thus $$H + \beta \mu W - \beta \mu \mu' P = P$$

or $$P(1 + \beta \mu \mu') = H + \beta \mu W$$

or $$P = \frac{H + \beta \mu W}{1 + \beta \mu \mu'} \quad ...(27.64)$$

Taking moments of all the forces about the base,

$$M_0 + HD = M_B + M_p + \mu' P(\alpha D) \quad ...(a)$$

where αD is equal to the distance from the axis passing through the *c.g.* of the base to the point at which the resultant vertical frictional force on side acts normal to the direction of the horizontal force (= $B/2$ for rectangular wells and 0.318 B for circular wells).

Eq. (*a*) can be written as

$$M_0 + HD = K_V \theta I_B + m K_V \theta I_v + \mu' \alpha D \times \left(\frac{2 m K_V \theta}{D} \cdot I_V\right)$$

or $$M_0 + HD = K_V \theta [I_B + m I_v (1 + 2 \mu' \alpha)] \quad ...(27.65)$$

or $$K_V \theta = \frac{M_0 + HD}{I_B + m I_V (1 + 2 \mu' \alpha)}$$

or $$K_V \theta = \frac{M}{I_B + m I_V (1 + 2 \mu' \alpha)} \quad ...(27.67)$$

or $$K_V \theta = \frac{M}{I} \quad ...(27.67)$$

where $$M = M_0 + HD$$

and $$I = I_B + m I_V (1 + 2 \mu' \alpha) \quad ...(27.68)$$

From Eqs. 27.61 and 27.64, $$P = \frac{H + \beta \mu W}{1 + \beta \mu \mu'} = \frac{2 m K_V \theta I_V}{D}$$

Using Eqs. 27.67, $$P = \frac{2m I_V}{D} \left(\frac{M}{I}\right)$$

or $$P = \frac{M}{r} \quad ...[27.69\ (a)]$$

where $$r = \frac{D}{2} \cdot \frac{I}{m I_V} \quad ...[27.69\ (b)]$$

Also $$H + \beta \mu W = \frac{M}{r}(1 + \beta \mu \mu')$$

On simplification, $$H + \beta \mu W = \frac{M}{r} + \frac{M \beta \mu \mu'}{r} = \frac{M}{r}(1 + \beta \mu \mu')$$

or $$\beta \left[\mu W - \frac{M \mu \mu'}{r}\right] = \frac{M}{r} - H$$

or $$\beta = \frac{\frac{M}{r} - H}{\mu W - \frac{M}{r}(\mu \mu')} \quad ...(27.70)$$

As $-1 < \beta < 1$, we have

$$\frac{M}{r}(1 + \mu\mu') - \mu W < H < \frac{M}{r}(1 - \mu\mu') + \mu W \quad ...(27.71)$$

The vertical soil reaction is given by,

$$\sigma_z = K_V \theta (x + x_c)$$

Now $$W - \mu' P = \int \sigma_z dA = K_V \theta \int (x + x_c)\, dA$$

or $$W - \mu' P = K_V \theta x_c A$$

or $$K_V \theta x_c = \frac{W - \mu' P}{A}$$

Therefore, $$\sigma_z = K_V \theta x + \left(\frac{W - \mu' P}{A}\right) \quad ...(27.72)$$

The pressures at the toe and heel are given by

$$p_t = \frac{W - \mu' P}{A} + K_V \theta (B/2) \quad ...[27.73(a)]$$

and $$p_h = \frac{W - \mu' P}{A} - K_V \theta (B/2) \quad ...[27.73(b)]$$

Substituting the value of $K_V \theta$ from Eq. 27.67,

$$p_t = \frac{W - \mu' P}{A} + \frac{MB}{2I} \quad ...(27.74)$$

and $$p_h = \frac{W - \mu' P}{A} - \frac{MB}{2I} \quad ...(27.75)$$

For the soil to remain in elastic state, the maximum soil pressure at any depth should not exceed the maximum passive pressure, *i.e.*, $\sigma_x \not> p_p$

or $$m K_V \left(\frac{z}{D}\right)(D - z)\, \theta \not> \gamma (K_p - K_a) z$$

At $z = 0$, the term $\frac{m K_V \theta (D - z)}{D}$ is a maximum.

Therefore, $$m K_V \theta \not> \gamma (K_p - K_a)$$

or $$m \frac{M}{I} \not> \gamma (K_p - K_a) \quad ...(27.76)$$

Eq. 27.76 should be satisfied. Further, the maximum base pressure p_t should not exceed the allowable soil pressure. The minimum pressure p_h should not be negative to avoid tension.

2. Ultimate Soils Resistance

The frictional force mobilised along the surface of rupture can be determined assuming the surface to be cylindrical (Fig. 27.15). For circular wells, the surface of rupture is assumed to be a part of sphere with its centre at the point of rotation and passing through the periphery of the base.

If W_u is the ultimate load, equal to the total vertical load multiplied by a suitable load factor, the load per unit width is W_u/B. It will also be equal to the upward pressure for a rectangular base.

Let us consider a small arc of length $R d\alpha$ at an angle α from the vertical axis.

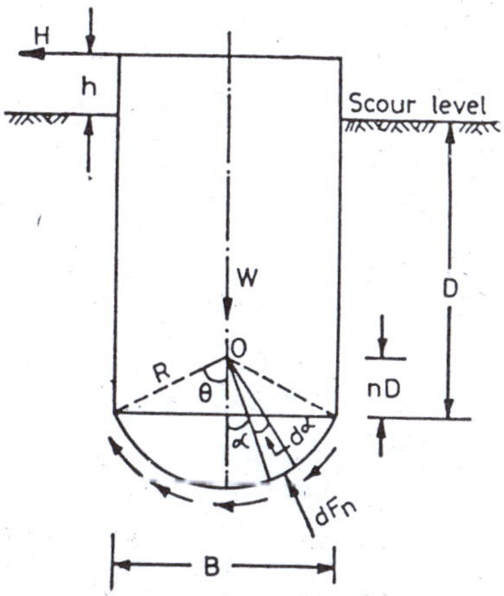

Fig. 27.15. Ultimate Soil Resistance.

Vertical force on the element = $R d\alpha \cos\alpha\,(W_u/B)$ and normal force developed on the element

$$dF_n = (W_u/B)\,R\,d\alpha\,\cos\alpha\,\cos\alpha$$

Total normal force, $F_n = 2\int_0^{\theta} (W_u/B)\,R\cos^2\alpha\,d\alpha$

or $$F_n = \frac{W_u R}{B}(\theta + \sin\theta\cos\theta) \qquad \ldots(a)$$

where $\theta = \tan^{-1}(B/2\,nD)$

and $n\,D$ is the height of point O above the base

and $$R = \sqrt{(B/2)^2 + (nD)^2} = \frac{B}{2}\sqrt{1 + \frac{4n^2D^2}{B^2}}$$

Eq. (a) can be written as $$F_n = \frac{W_u}{2}\sqrt{1+\frac{4n^2D^2}{B^2}}\left[\tan^{-1}\frac{B}{2\,nD} + \frac{B\,(nD)}{2\,R^2}\right]$$

or $$F_n = \frac{W_u}{2}\sqrt{1+\frac{4n^2D^2}{B^2}}\left[\tan^{-1}\frac{B}{2\,nD} + \frac{2\,n\,B\,D}{B^2 + 4n^2D^2}\right] \qquad \ldots(27.77)$$

The moment of resistance of the base about the point of rotation is given by

$$M_b = (F_n \tan\phi)\,R \qquad \ldots[27.78\ (a)]$$

[**Note.** The right-hand side of Eq. 27.78 (a) is multiplied by a shape factor of 0.6 in the case of circular wells.]

Assuming the point of rotation at a height of 0.2 D above the base, the moment of resistance of the base is given by

$$M_b = C\,W\,B\tan\phi \qquad \ldots[27.78(b)]$$

where ϕ = angle of shearing resistance, B = width parallel to the direction of forces or equal to the diameter in circular wells, C = coefficient (see Table 27.2).

Table 27.2. Values of Coefficient C

D/B	0.5	1.0	1.5	2.0	2.5
Rectangular Well	0.41	0.45	0.50	0.56	0.64
Circular Well	0.25	0.27	0.30	0.34	0.38

Side-Resisting Moment

Fig. 27.16 shows the ultimate soil pressure distribution at the front and back faces of the well. As the

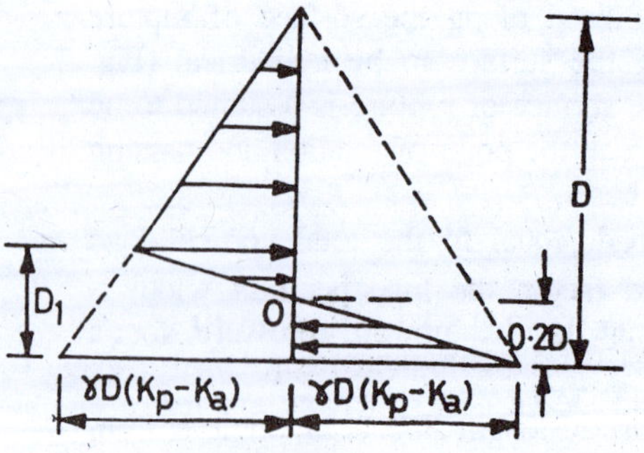

Fig. 27.16.

point of rotation is assumed at a height of 0.2 D above the base, from the similarity of triangles, it can be shown that $D_1 = 1/3\,D$.

Taking moments of the forces about O,

$$M_s' = 0.096\,D^2\,[\gamma\,D\,(K_p - K_a)]$$

$$M_s' \simeq 0.1\,\gamma D^3\,(K_p - K_a) \quad \text{per unit length}$$

Taking submerged unit weight, $M_s' = 0.1\,\gamma'\,D^3\,(K_p - K_a)$ per unit length

For a well of length L, $\quad M_s = 0.1\,\gamma'\,D^3\,(K_p - K_a)\,L$ ...(27.79)

Resisting Moment due to friction on front and back faces

The frictional forces on the faces act in the vertical direction and produce resisting moment M_f.

Force/unit width, $\quad F = \dfrac{1.1}{3}\,\gamma'\;D^2\,(K_p - K_a)\sin\delta$

In the case of a rectangular well of length L,

$$M_f = \left[\frac{1.1}{3}\,\gamma'\;D^2\,(K_p - K_a)\sin\delta\right]\frac{B}{2} \times L$$

or $\quad M_f = 0.183\,\gamma'\,(K_p - K_a)\,LBD^2\sin\delta$

or $\quad M_f \simeq 0.18\,\gamma'\;(K_p - K_a)\,LBD^2\sin\delta$ ...(27.80)

For circular wells, taking shape factor as 0.60,

$$M_f = 0.11\,\gamma'\,(K_p - K_a)\,B^2\,D^2\sin\delta \qquad \text{...(27.81)}$$

Total Resisting Moment

The total resisting moment is given by

$$M_r = M_b + M_s + M_f \qquad \text{...(27.82)}$$

IRC : 45—1972 recommends a reduction factor of 0.7 in the total resisting moment of soil. Thus

$$M_r = 0.7\,(M_b + M_s + M_f) \qquad \text{...(27.83)}$$

The applied moment M should be less than M_r. Thus

$$M \ngtr 0.7\,(M_b + M_s + M_f) \qquad \text{...(27.84)}$$

For the computation of applied moments, the effects of moments due to tilt and shift of wells, if any, about the plane of rotation should also be considered.

If Eq. 27.84 is not satisfied, increase the grip length D and revise the calculations.

Check for maximum average pressure

The maximum average pressure should not exceed one half the ultimate bearing capacity.

$$\frac{W}{A} \ngtr \frac{q_u}{2} \qquad \text{...(27.85)}$$

where q_u = ultimate bearing capacity.

27.9. DESIGN OF INDIVIDUAL COMPONENTS OF THE WELL

The overall design of the well has been discussed in the preceding sections. The individual components are designed as explained below.

(1) Cutting Edge. The cutting edge should have a sharp angle for cutting through the soil. It should be strong enough so that it does not bend when penetrating through a soil containing boulders. A sharp vertical edge having an angle of 30° with the vertical or having a slope of one horizontal to two vertical is generally used [Fig. 27.17 (*a*)]. However, if the sharp edges are likely to be damaged, a cutting edge with a stub nose [Fig. 27.17 (*b*)] is used.

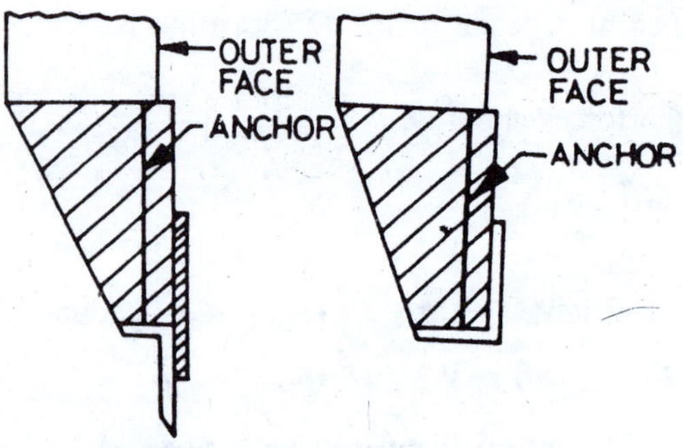

Fig. 27.17. Cutting Edge.

The cutting edge should be properly anchored to the well curb.

(2) Well curb. Fig. 27.18 shows the curb of a well. Curbs are generally made of reinforced concrete. During sinking operation, the curb cuts through the soil. The figure shows the forces acting on the curb when the well has penetrated to a considerable depth below the scour level.

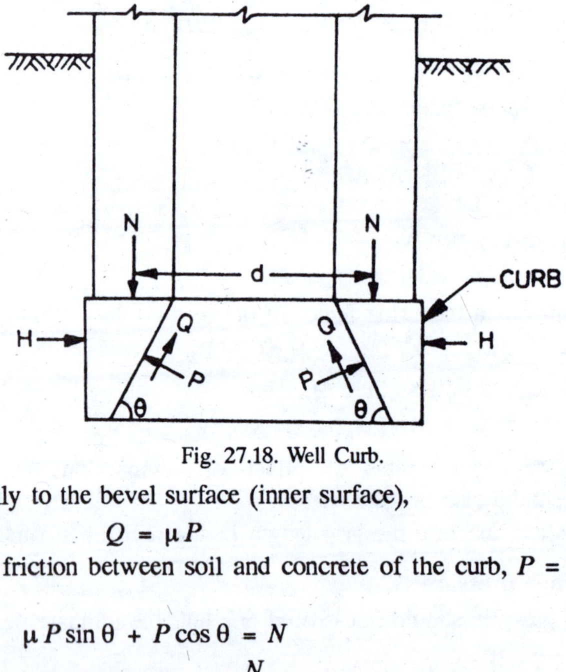

Fig. 27.18. Well Curb.

Forces acting tangentially to the bevel surface (inner surface),

$$Q = \mu P \qquad \text{...}(a)$$

where μ coefficient of friction between soil and concrete of the curb, P = forces acting normal to the bevel surface.

Resolving vertically, $\mu P \sin\theta + P\cos\theta = N$ ...(*b*)

or
$$P = \frac{N}{(\mu \sin\theta + \cos\theta)} \qquad \text{...}(c)$$

where N = vertical force on the curb, θ = angle which the bevel edge makes with the horizontal.

Resolving horizontally, $P\sin\theta - \mu P\cos\theta = H$ ...(*d*)

where H = horizontal force on the curb per m .

From Eqs. (*c*) and (*d*),
$$H = \frac{N(\sin\theta - \mu\cos\theta)}{(\mu\sin\theta + \cos\theta)}$$

Hoop tension, $T = H \times d/2$, where d is the diameter.

Thus
$$T = 0.5N\left(\frac{\sin\theta - \mu\cos\theta}{\mu\sin\theta + \cos\theta}\right) d \qquad \text{...(27.86)}$$

Sometimes, sand-blow may cause sudden descend of the well during sinking and an increase in the hoop tension. To account for such an eventuality, the hoop tension is increased by 50%. Thus

$$T = 0.75N \left(\frac{\sin\theta - \mu\cos\theta}{\mu\sin\theta + \cos\theta} \right) d \qquad \text{...(27.87)}$$

Suitable reinforcement should be provided to resist the hoop tension T developed.

When the cutting edge is not able to move downward due to reaction developed at the curb and the bottom plug, the hoop tension developed is given by

$$T = \left(\frac{q\,d^2}{8\,r} \right) \frac{d}{2} \qquad \text{...(27.88)}$$

where q = pressure at the base = $\dfrac{\text{total weight}}{\text{area of plug}}$

r = vertical height of the imaginary inverted arch (Fig. 27.19).

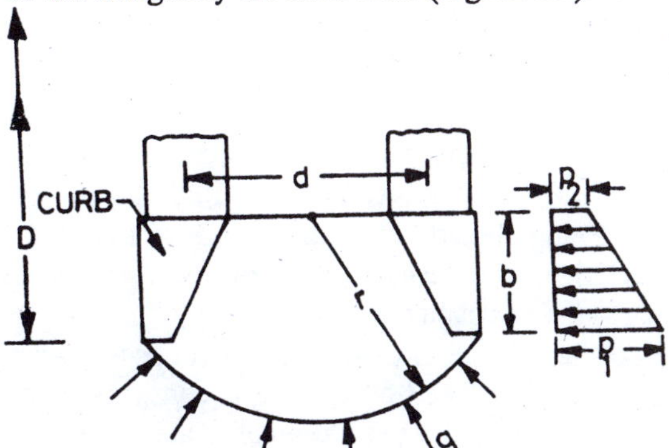

Fig. 27.19.

In case of granular soils, the hoop tension is relieved by active earth pressure around the curb. The net hoop tension is given by

$$T' = \frac{d}{4} \left[\frac{q\,d^2}{4\,r} - (p_1 + p_2)\,b \right] \qquad \text{...(27.89)}$$

where $$p_1 = \frac{1}{2} K_a \gamma' D^2$$

and $$p_2 = \frac{1}{2} K_a \gamma' (D - b)^2$$

in which b = height of the curb, D = depth of the curb below the scour level.

At the junction of the curb and steining, a moment M_o develops due to the horizontal force H caused by bevelled action, given by

$$M_o = H \times b/2 \qquad \text{...(27.90)}$$

Suitable reinforcement is provided at the inner corner to take care of this moment and is anchored into the steining.

IRC : 21—1972 recommends a minimum reinforcement of 72 kg/m^3 in a well curb. The reinforcement should be properly arranged.

The slope of the inner face of the curb should be such as to push forward easily. The angle with the vertical should preferably be not more than 30° in ordinary soil and 45° for sandy soils.

(3) Well Steining. The thickness of the steining should be adequate for the stresses developed during sinking and after installation. The thickness t of the steining may be obtained from the following equation :

$$\pi B(H-h)f = \pi(B-t)\gamma_c t H + P \qquad ...(27.91)$$

where H = depth to which the well has progressed, h = height at which it has got suspended, B = external diameter, P = weight added for sinking.

The design of steining reinforcement depends upon the skin friction and the unit weight of well. It is usual practice to provide reinforcement of about 5 to 6 kg/m^3 of the brick and concrete steining. About 75% of the total reinforcement is in the form of vertical reinforcement and 25% in the shape of laterals or hoop rings. The vertical reinforcement is spread near both (the outer and inner) faces. The laterals should be checked for the moment developed due to eccentric kentledge and half the weight of the well at an eccentricity of one-fourth the width of well in any direction. This condition is generally critical when the well has sunk to about half the designed depth.

The thickness of the steining is usually fixed empirically. For railway bridges in India, it is generally taken as one-fourth of the outside diameter. For road bridges, it is kept as one-eighth of the outside diameter if it is in brick masonry and one-tenth of the outside diameter if in cement concrete. However, the thickness is increased by 12 cm per 3 m of depth after the first 3 m of steining in brick masonry and 15 cm per 6 m of depth after the first 6 m for cement concrete.

A thumb rule commonly used is

$$t = K(B/8 + H/100) \qquad ...(27.92)$$

where B = external diameter of well, H = depth below low water level, K = a constant (= 1.0 for sandy soils; 1.1 for soft clay and 1.25 for hard clay and boulders).

(4) Bottom Plug. The bottom plug should be strong enough to transmit the load to the soil below. It is designed as a thick plate subjected to unit bearing pressure under the maximum vertical load, as already discussed in the case of open caissons in chapter 26.

The bottom plug is given the shape of an electric bulb to produce an arch action, to reduce hoop tension in the curb and to provide larger base area. The bottom plug is constructed in (1 : 2 : 4) cement concrete laid by means of a tremie or a skin box. About 10% extra cement is added because some cement is washed away on account of water. The water in the well must be still and at its normal level. Bottom plugging should always be done in one continuous operation.

While founding the well on rock, it should be properly anchored by taking it 25 cm to 30 cm deep into the rock bed. Adequate dowel bars should be provided.

(5) Well Cap. The bottom of the well cap is generally kept at the low water level. It is designed as a slab resting on the well. The well cap may be extended as cantilevers to accommodate piers of slightly larger size than that of the well.

If the width of the pier is greater than the size of the dredge hole, it is assumed that the weight of a cone of concrete having an apex angle of 60° is carried by the slab and the remaining load is transmitted to the steining.

The well cap should have a minimum reinforcement of about 80 kg/m^3.

(6) Top Plug. The function of the top plug is to transmit the load of the pier to the well steining. If a well cap is provided, there is no need of a top plug. However, it is generally provided as an extra safety precaution. Offsets are provided at the top of the steining to provide bearing to the plug.

Cement concrete (1 : 2 : 4) is used for the construction of the top plug.

(7) Sand filling. The main purpose of sand filling is to provide stability to the well by increasing its weight and to reduce the tensile stresses caused at the base by bending moment. However, sand cannot be depended upon for transmitting the weight of the pier to the bottom plug and, therefore, it does not contribute towards the structural strength of the well.

On the Indian railways, the practice is to do the sand filling upto the top plug. Some of the highway engineers recommend that the sand filling should be done upto the lowest scour level. The actual depth of the filling should be fixed considering the requirement of the dead weight for the stability.

27.22. SINKING OF WELLS

The sinking operation consists of the following steps :

(1) Laying the Well Curb. It the river bed is dry, the cutting edge over which the well curb is to be built is placed at the required position after excavating the river bed to about 15 cm. When the depth of water is upto 5 m, a sand island is made before placing the curb. The size of island should be large enough to accommodate the well with adequate working space all around (Fig. 27.20). In the case of depth of water more than 5 m, it is generally more economical to build the curb on dry ground at the river bank and float it to the site.

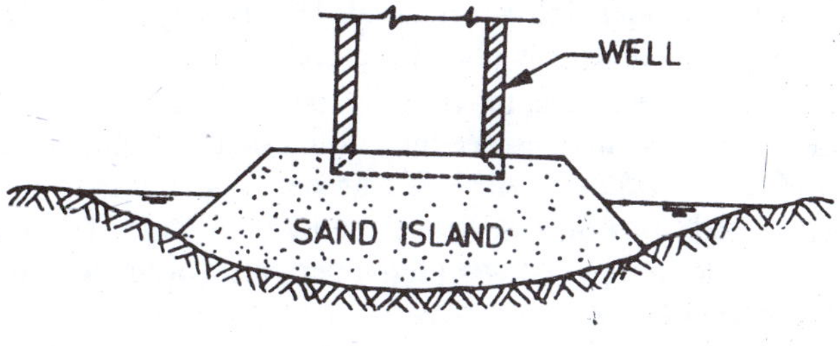

Fig. 27.20. Sand Island.

It is useful to insert wooden sleepers below the cutting edge at regular intervals to distribute the load evenly over the ground. The shuttering of the well curb is then erected. The outer shuttering is generally made of steel (or wood) and the inner shuttering of brick masonry. The reinforcement of the curb is then placed in proper position such that the vertical bars project about 2 m above the top of the curb. Concreting of the curb is done in one continuous operation. The curb concrete is allowed to set at least for one week before the shuttering is stripped off. The sleepers are also then removed.

(2) Well Steining. After sinking the well curb, the steining is raised by about 1.5 m at a time and its sinking done after allowing atleast 24 hours for setting. The steining is built using straight edges, preferably of angle iron. The lower portion of the straight edges is kept butted with the steining of the earlier stages. The steining must not be built in plumb at any intermediate stage when the well has tilted to one side. Once the well has acquired a grip of about 6 m in ground, the steining can be raised by about 3 m at a time. The height of steining built at any stage should be such that the well does not lose stability.

(3) Sinking Process. Sinking process is begun after having cast the curb and the first stage of steining and allowing enough time for curing. The well is sunk by excavating material from inside the curb manually or mechanically. When the depth of water inside the well is upto 1 m, dredging can be done manually. However, beyond this depth of water, excavation is done with the help of 'jhams' (a type of spade). The 'jham' is tied to a rope moving over a pulley. It is pulled by the men. Every time, a diver dives and pushes the jham into the soil and comes up. The jham is then pulled out. In an improved version, the jham has been replaced by an automatic grab operated by diesel (or steam) winches. Straight chisels are used for breaking hard material so that is can be taken out by grab. Under-cutting chisels are used to loosen the material which lies under the steining. Explosives are used for sinking through rock.

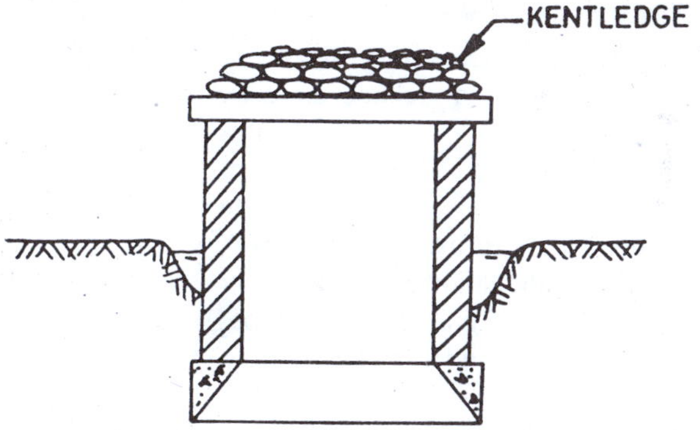

Fig. 27.21. Kentledge on the Well.

As the well sinks, the friction on the sides increases. To accelerate the process of sinking, additional loading, known as kentledge, is applied on the well. Kentledge is generally in the form of sand bags (Fig. 27.21) placed on a suitable platform erected on its top such that it does not interfere with the excavation. Sometimes, even kentledge is not sufficient to sink the well. In such cases, the frictional resistance developed on its outer periphery is reduced by forcing jets of water on the outer face. However, this method is effective only in the case of wells sunk in sandy strata.

In some cases, pumping out the water from inside the well is effective in well sinking. However, this method should be discouraged in early stages when the depth is shallow. It is not desirable to resort to pumping out water unless the well has gone deep enough or has passed through a clayey strata so that the chances of tilt and shift are reduced. Dewatering is not allowed after the well has sunk to about 10 m. After this stage, the sinking is done by usual methods of grabbing, chiselling, applying kentledge or blasting.

Great precaution is necessary if dewatering of the well is done when it is at a shallow depth to avoid blowing of sand from under the cutting edge. If blowing of sand occurs, it results in the loss of time and labour in removing the sand. It also presents danger to the men working inside, as the well may get filled upto a height of a few metres if the blow is large. The well may also tilt suddenly. Scrap gunny bags and grass bundles are placed around the periphery of the well on the outside into the funnel formed. It blocks the passage through which the blow of sand is taking place.

27.23. MEASURES FOR RECTIFICATION OF TILTS AND SHIFTS

The well should be sunk straight and at the correct position. However, it is not an easy task to achieve this objective. Sometimes, the well tilts on one side or it shifts away from the desired position. The following precautions must be taken to avoid tilts and shifts.

(1) The outer surface of the well curb and steining should be regular and smooth.

The diameter of the curb should be kept about 4 to 8 cm lager than the outer diameter of steining, and the well should be symmetrically placed.

(3) The cutting edge should be of uniform thickness and sharpness.

(4) Dredging should be done uniformly on all sides in a circular well and in both pockets of a twin well.

Tilts and shifts must be carefully checked and properly recorded. The correct measurement of the tilt is an important field observation required during well sinking. It is not possible to specify the permissible limits of tilts and shifts. Each case should be examined individually. IS : 3955 — 1967 recommends that tilt should generally be limited to 1 in 60. The shift should be restricted to one percent of the depth sunk. In case they exceed the above limits, the following measures are taken for their rectification:

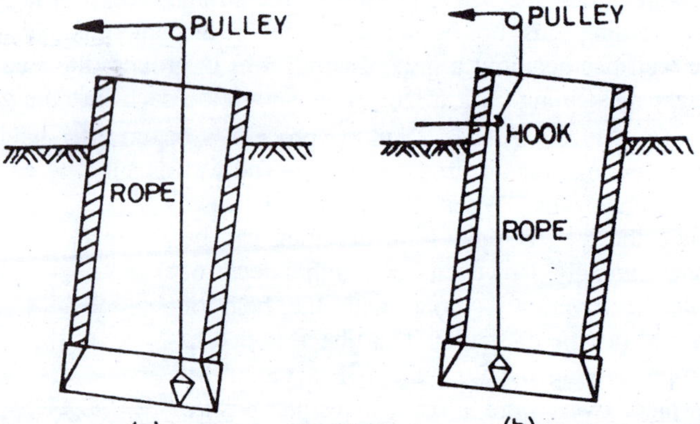

Fig. 27.22.

(1) Regulation of Grabbing. To rectify the tilt, the higher side is grabbed more by regulating the dredging. However, this method is effective in initial stages of sinking. In case of tilted wells, dredging does not work satisfactorily. In such a case, a hole is made in the steining on the higher side and the rope of the grab is pulled through the hook [Fig. 27.22 (*b*)]. Thus dredging with hooking is done.

Alternatively, the well may be dewatered, if possible, and open excavation may be carried out on the higher side.

(2) Eccentric loading. To provide greater sinking effort on the higher side, eccentric loading is applied by adjusting the kentledge. A suitable platform is constructed on the higher side for this purpose (Fig. 27.23). As the sinking progresses, heavier kentledge with greater eccentricity would be required to rectify the tilt.

(3) Water Jetting. If the water jets are applied on the outer face of the well on the higher side, the side friction is reduced and the tilt is rectified.

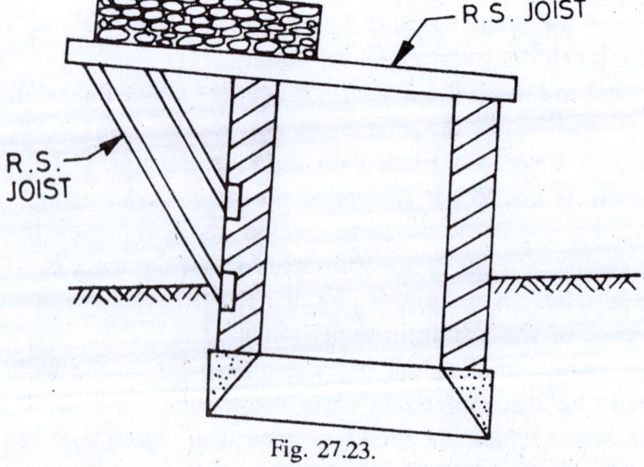

Fig. 27.23.

(4) Excavation under cutting edge. A tilted well in a hard clayey stratum does not straighten due to unbroken hard stratum on the higher side. If dewatering of the well is possible, open excavation is done under the cutting edge. In case dewatering is not possible, divers can be sent to loosen the strata.

(5) Inserting wooden Sleeper under the cutting edge. Sometimes wooden sleepers are inserted temporarily below the cutting edge on the lower side to avoid further tilt [Fig. 27.24 (*a*)]. Alternatively, a hook is inserted below the cutting edge on the lower side and pulled with a wire rope and kept strained [Fig. 27.24 (*b*)].

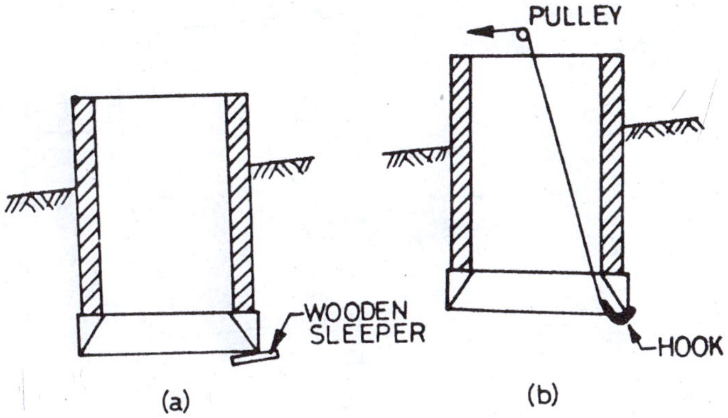

Fig. 27.24.

(6) Pulling the well. In early stages of sinking, pulling the well towards the higher side, by placing one or more steel ropes around the well with vertical sleepers packed in-between to distribute the pressure over larger areas of well steining, is quite effective (Fig. 27.25).

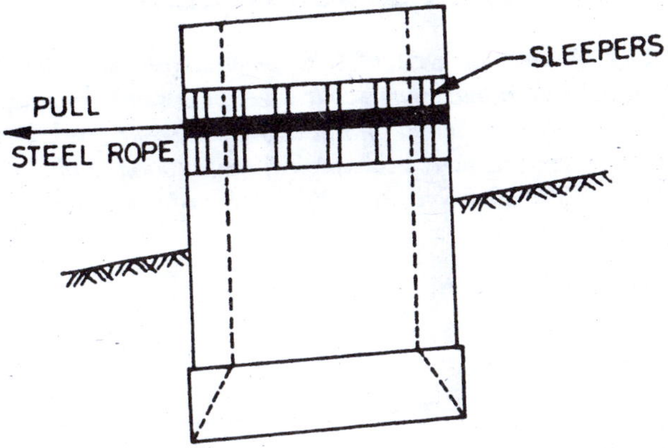

Fig. 27.25.

(7) Strutting the well. The well is strutted on its tilted side with suitable logs of wood to prevent further tilt. The well steining is provided with sleepers to distribute the load from the strut. The other ends of the logs rest against a firm and non-yielding base having driven piles (Fig. 27.26).

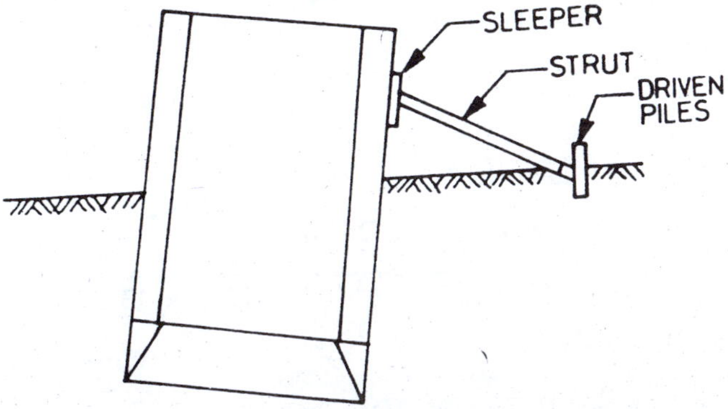

Fig. 27.26.

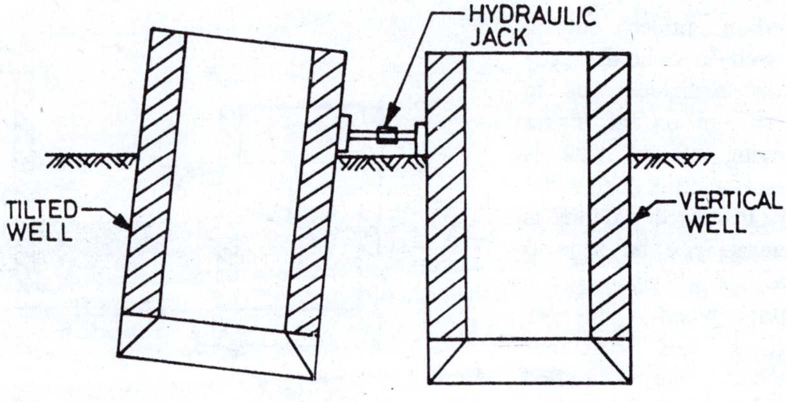

Fig. 27.27.

(8) Pushing the wells by jacks. The tilt can be rectified by pushing the well with a suitable arrangement through mechanical or hydraulic jacks. Fig. 27.27 shows a tilted well being pushed by a jack resting against the vertically-sunk well.

In actual practice, a combination of several methods discussed above is generally used.

ILLUSTRATIVE EXAMPLES

Illustrative Example 27.1. *A circular well of 6 m external diameter and 4 m internal diameter is embedded to a depth of 15 m below the maximum scour level in a sandy soil deposit. The well is subjected to a horizontal force of 800 kN acting at a height of 8 m above the scour level. Determine the allowable total equivalent resisting force due to earth pressure, assuming (a) the rotation is about a point above the base, (b) the rotation is at the base. Take* γ_{sat} *= 20 kN/m*^{3}, ϕ *= 30°; factor of safety for passive resistance = 2.0.*

Use Terzaghi's analysis.

Solution. (See Fig. E-27.1) $K_p = 3.0, K_a = 0.333$

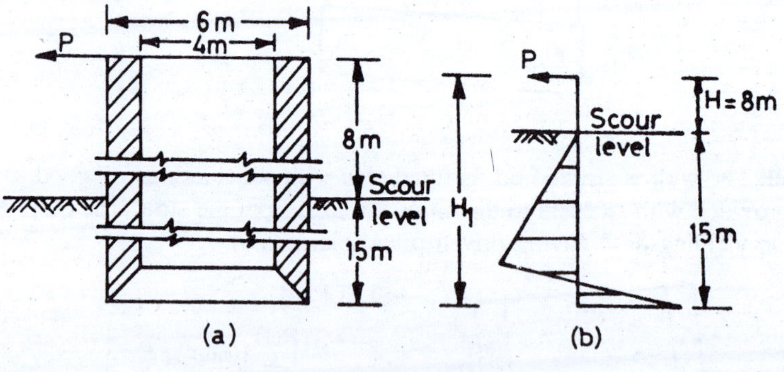

Fig. E-27.1.

(*a*) Rotation above the base. Total height, $H_1 = 8 + 15 = 23$ m

From Eq. 27.10, $$2D_1 = 3H_1 \pm \sqrt{9H_1^2 - 2D(3H_1 - D)}$$

or $$2D_1 = 3 \times 23 \pm \sqrt{9 \times 23^2 - 2 \times 15(3 \times 23 - 15)} = 12.96$$

or $$D_1 = 6.48 \text{ m}$$

From Eq. 27.9, $$q_{max} = \frac{1}{2}\gamma' D(K_p - K_a)(D - 2D_1)$$

$$= \frac{1}{2} \times 10.0 \times 15 \times (3.0 - 0.333)(15 - 2 \times 6.48) = 408.1 \text{ kN/m}$$

From Eq. 27.13, the allowable transverse load,

$$Q_a = \frac{q'_{max} \times L}{FS} = \frac{408.1 \times 6}{2} = \mathbf{1224.2\ kN}$$

(*b*) **Rotation at base.** From Eqs. 27.11, $q_{max}' = \frac{1}{6} \gamma' (K_p - K_a) \frac{D^3}{H_1}$

$$= \frac{1}{6} \times 10 \times (3.0 - 0.333) \times \frac{(15)^3}{23} = 652.3 \text{ kN/m}$$

Therefore, $$Q_a = \frac{652.3 \times 6}{2} = \mathbf{1956.9\ kN}$$

Illustrative Example 27.2. *Fig. E-27.2 shows a well. Determine the base pressure and the lateral load per unit length of the well. The well is subjected to a net downward force of 10 MN. Assume that the horizontal deformation of the well cap at the scour level is 20 mm.*

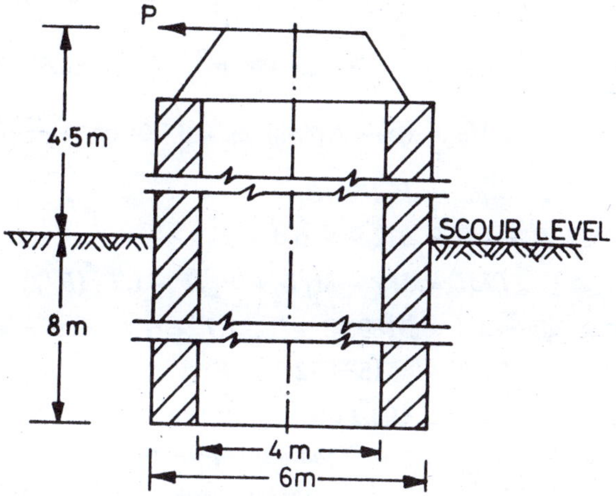

Fig. E-27.2.

Use Banerjee and Gangopadhay's simplified analysis. Take $\mu = 0.50$; $\gamma' = 11$ *kN/m*^{3}; $\phi = 30°$; K_v = *25000 kN/m*^{3}.

Solution. Area of cross-section, $A_b = \pi/4\,(6^2 - 4^2) = 15.70 \text{ m}^2$

Moment of inertia, $I = \pi/64\,(6^4 - 4^4) = 51.03 \text{ m}^4$

Section modulus, $$Z_b = \frac{51.03}{3.0} = 17.01 \text{m}^3$$

From Eq. 27.18, $$m = \frac{\gamma' (K_p - K_a)}{\rho_1}$$

$$= \frac{11 \times (3.0 - 0.333)}{20 \times 10^{-3}} = \mathbf{1466.9}$$

From Eq. 27.47, $$P_1 = \frac{\gamma' (K_p - K_a) D^2}{6}$$

$$= \frac{11 \times (3.0 - 0.333) \times 8^2}{6} = 312.9 \text{ kN}$$

From Eq. 27.48, $$M_1 = \frac{\gamma' (K_p - K_a) D^3}{12}$$

$$= \frac{11.0 \times (3.0 - 0.333) \times 8^4}{12} = 1251.7 \text{ kN}$$

From Eq. 27.49, $$M_3 = \frac{\gamma' K_v (K_p - K_a)}{m} \times \frac{B^3}{12\,D}$$

$$= \frac{11 \times 25000 \times (3.0 - 0.333)}{1466.9} \times \frac{6^3}{12 \times 8} = 1125 \text{ kN-m}$$

From Eq. 27.52, $$W_T = \mu P_1 L + R_1$$

or $$10000 = 0.5 \times 312.9 \times 6 + R_1$$

or $$R_1 = 9061.3 \text{ kN}$$

From Eq. 27.55, $$p_t + p_h = \frac{2R_1}{A_b} = \frac{2 \times 9061.3}{15.7} = 1154.3 \text{ kN/m}^2$$

From Eq. 27.56, $$p_t - p_h = \frac{12\,M_3}{B^2} = \frac{12 \times 1125}{(6)^2} = 375 \text{ kN/m}^2$$

Solving the above two equations, $p_t = 764.65 \text{ kN/m}^2$; $p_h = 389.65 \text{ kN/m}^2$

From Eq. 27.57, $$M_B = (p_t - R_1/A_b)\,Z_b = \left(764.65 - \frac{9061.3}{15.7}\right) \times 17.01 = 3189.38 \text{ kN-m}$$

From Eq. 27.50, $$PL = P_1 L - F_1$$

or $$P \times 6.0 = 312.9 \times 6.0 - F_1 \qquad \ldots(a)$$

From Eq. 27.51, $$PLH = M_B - M_1 L + F_1 D + \mu P_1 (B/2) L$$

or $$P \times 6.0 \times 4.5 = 3189.38 - 1251.7 \times 6 + F_1 \times 8 + 0.5 \times 312.9 \times (6/2) \times 6$$

or $$27\,P = -\,1504.72 + 8\,F_1 \qquad \ldots(b)$$

Solving (*a*) and (*b*), P = **180.2 kN**

Illustrative Example 27.3. *Design a well foundation for the following site conditions using IRC method.*

High flood level = 103.50, Bed level = 100.00, Scour level = 88.20, Base level = 78.00, External diameter = 8 m, Width of pier at top = 1.5 m, Total vertical load = 5000 kN, Total horizontal load at bed level = 750 kN, Moment on well at scour level = 3000 kN-m, Horizontal water pressure on well = 650 kN.

Take $q_{na} = 600 \text{ kN/m}^2$; $\gamma_{sat} = 20 \text{ kN/m}^3$; $\phi = 30°$; $K = 1.25$; $\delta = 20°$; *seismic coefficient* $= 0.10$; $K_p = 6.105$, $K_a = 0.297$; $q_u = 1800 \text{ kN/m}^2$

Solution. (See Fig. E-27.3)

Let us assume that the low water level is at bed level.

From Eq. 27.92, $$t = K(B/8 + H/100)$$

or $$t = 1.25\,(8/8 + 22/100) = 1.53 \text{ say } 1.6 \text{ m}$$

Internal diameter $= 8 - 2 \times 1.6 = 4.8$ m

Design of Well Cap. Let us assume that the thickness of the well cap is 75 cm.

Weight of the well cap $= \pi/4 \times (8.0)^2 \times 0.75 \times 24 = 904.3$ kN

Assuming a buoyancy of 15%, the upward force

$$= 0.15 \times \pi/4 \times (8.0)^2 \times 0.75 \times 10 = 56.5 \text{ kN}$$

Area of cross-section $= \pi/4 \times (8^2 - 4.8^2) = 32.15 \text{ m}^2$

Moment of inertia (I_B) $= \pi/64\,(8^4 - 4.8^4) = 174.92 \text{m}^4$

Section modulus $$= \frac{174.92}{4.0} = 43.73 \text{ m}^3$$

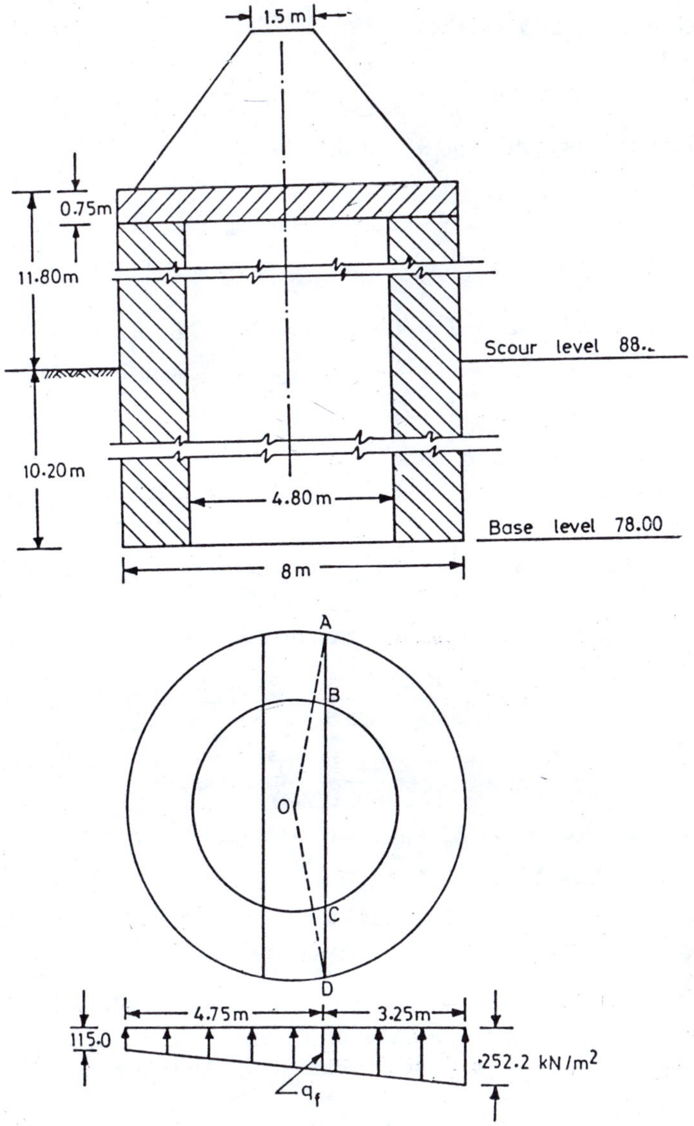

Fig. E-27.3.

Total vertical load, neglecting buoyancy = 5000.0 + 904.3 = 5904.3 kN

The maximum and minimum pressures are given by

$$q_{max} = \frac{5904.3}{32.15} + \frac{3000}{43.73} = 252.2 \text{ kN/m}^2$$

$$q_{min} = \frac{5904.3}{32.15} - \frac{3000}{43.73} = 115 \text{ kN/m}^2$$

Average pressure $$(q_a) = \frac{(252.2 + 115.0)}{8} = 183.6 \text{ kN/m}^2$$

Pressure at the face of the pier [see Fig. E-27.3 (*b*)].

$$q_f = 115.0 + \frac{(252.2 - 115.0)}{8} \times 4.75 = 196.5 \text{ kN/m}^2$$

Intensity of pressure due to self weight,

$$q_c = \frac{904.3}{\pi/4 \times (8)^2} = 18 \text{ kN/m}^2$$

In Fig. E-27.3, let the angle *AOD* be θ_1 and *BOC* be θ_2.

$$\cos \theta_1/2 = \frac{0.75}{4.0} \quad or \ \theta_1 = 158.38°$$

$$\cos \theta_2/2 = \frac{0.75}{2.4} \quad or \ \theta_2 = 143.58°$$

$$\text{Chord } AD = 2 \times 4.0 \times \sin \theta_1/2 = 7.86 \text{ m}$$

$$\text{Chord } BC = 2 \times 2.4 \times \sin \theta_2/2 = 4.56 \text{ m}$$

Area of outer segment,

$$A_1 = \frac{1}{2} R^2 (\theta_1 - \sin \theta_1)$$

$$= \frac{1}{2} \times 4.0^2 \left(\frac{158.38 \times \pi}{180} - 0.368 \right) = 19.16 \text{ m}^2$$

Area of inner segment,

$$A_2 = \frac{1}{2} \times 2.4^2 \left(\frac{143.58 \times \pi}{180} - 0.594 \right) = 5.50 \text{ m}^2$$

The distances of the centroids of these areas are obtained as under.

$$Z_1 = \frac{4 R \sin^3 (\theta_1/2)}{3 (\theta_1 - \sin \theta_1)} = \frac{4 \times 4 \times (0.982)^3}{3 (2.763 - 0.368)} = 2.1 \text{ m}$$

$$Z_2 = \frac{4 \times 2.4 \times (0.95)^3}{3 (2.505 - 0.594)} = 1.44 \text{ m}$$

Moment about face $AD = q_a [A_1(Z_1 - 0.75) - A_2 (Z_2 - 0.75)] - q_c [A_1(Z_1 - 0.75)]$

or

$$M = 183.6 [19.16 (2.1 - 0.75) - 5.5 (1.44 - 0.75)] - 18 \times 19.16 (2.1 - 0.75)$$

$$= 3588.3 \text{ kN-m}$$

Maximum shear force $= q_a (A_1 - A_2) - q_c A_1$

$$= 183.6 (19.16 - 5.5) - 18 \times 19.16 = 2163.1 \text{ kN}$$

It can be checked that the overall thickness of 75 cm of cap is sufficient from the consideration of moment and shear stresses (see any text on R.C.C. design) for M 15 concrete.

Design of Steining

Total vertical load = 5000 + 904.3 = 5904.3 kN

Horizontal load = 750 kN

Moment at scour level = 3000 kN-m

Weight of steining upto scour level

$$= (\pi/4) (8^2 - 4.8^2) \times 24 \times 11.05 = 8527.1 \text{ kN}$$

Seismic force on steining = 0.1 × 8527.1 = 852.7 kN

Water pressure on well = 650 kN

Total horizontal force at scour level

$$Q = 750 + 852.7 + 650.0 = 2252.7 \text{ kN}$$

Moment at scour level,

$$M = 3000 + 750 \times 11.8 + 852.7 \times (11.8/2) + 650 \times (2/3) \times 11.8$$

$$= 21994.2 \text{ kN}$$

The depth of the point of zero shear below the scour level is given by Eq. 27.14.

$$x = \left[\frac{2FQ}{\gamma'(K_p - K_a)L}\right]^{1/2}$$

Taking $F = 2$,

$$x = \left[\frac{2 \times 2 \times 2252.7}{10 \times (6.105 - 0.297) \times 8.0}\right]^{1/2} = 4.4 \text{ m}$$

From Eq. 27.15 *a*, $M_{max} = M + 2/3\, Qx$

$= 21994.2 + 2/3 \times 2252.7 \times 4.4 = 28602.1$ kN–m

Let us assume a tilt of 1 in 60.

Eccentricity due to tilt at the top $= \dfrac{\text{total height}}{60} = \dfrac{22}{60} = 0.367$ m

Eccentricity at scour level $= 0.367 \times \dfrac{10.2}{22} = 0.17$ m

Moment due to tilt $= (5904.3 + 8527.1) \times 0.17 = 2453.3$ kN–m

Let us assume a shift of 1%.

Total shift $= 0.01 \times 22 = 0.22$ m

Moment due to shift $= (5904.3 + 8527.1) \times 0.22 = 3174.9$ kN–m

Total moment $= 28602.1 + 2453.3 + 3174.9 = 34230.3$ kN–m

Buoyancy on the well portion upto the point of zero shear

$= (\pi/4) \times (8.0)^2 (11.80 + 4.4) \times 10 \times 0.15 = 1220.8$

Weight of steining upto the point of zero shear

$= 8527.1 \times \dfrac{(11.05 + 4.40)}{11.05} = 11922.5$ kN

Total vertical weight $= 5904.3 + 11922.5 - 1220.8 = 16606$ kN

The maximum and minimum stresses are given by

$$q_{max} = \frac{16606}{32.15} + \frac{34230.3}{174.92} \times 4 = 1299.3 \text{ kN/m}^2$$

$$q_{min} = \frac{16606}{32.15} - \frac{34230.3 \times 4}{174.92} = -266.3 \text{ kN/m}^2$$

Cement concrete (1 : 3 : 6) can be used for steining.

Lateral Stability of the Well

Total weight of the steining $= \dfrac{8527.1}{11.05} \times 21.25 = 16398.3$ kN

Weight of water in the well $= (\pi/4) \times (4.8)^2 \times 10 \times (11.05 - 1.0) = 1817.7$ kN

Weight of intermediate plug (1 m) thick

$= (\pi/4) \times (4.8)^2 \times 1.0 \times 24 = 434$ kN

Weight of bottom plug (2.5 m thick)

$= \pi/4 \times (4.8)^2 \times 24 \times 2.50 = 1085.2$ kN

Weight of sand fill upto scour level

$= \pi/4 \times (4.8)^2 \times 20 \times 7.7 = 2785.3$ kN

Total weight at the base $= 5904.3 + 16398.3 + 1817.7 + 434 + 2785.3 + 1085.2$

$= 28424.8$ kN

Buoyancy on well $= \pi/4 \times (8.0)^2 \times 22 \times 10 = 11052.8$ kN

Net downward force, $W = 28424.8 - 11052.8 = 17372$ kN

Weight of water and intermediate plug

$= 1817.7 + 434 = 2251.7$ kN

Horizontal seismic force $= 0.1 \times 2251.7 = 225.2$ kN

Moment of seismic forces about scour level

$= 181.77\ (1.75 + 10.05/2) + 43.4 \times 0.50 = 1253.2$ kN

Total horizontal force at scour level,

$H = 2252.7 + 225.2 = 2477.9$ kN

Moment at scour level $= 21994.2 + 1253.2 = 23247.4$ kN–m

Moment at base level, $M = 23247.4 + 2477.9 \times 10.2 + 2453.3 + 3174.9$

$= 54150.2$ kN–m

Elastic Analysis

The design forces and moments acting at the base are as under.

Vertical forces (W) = 17372 kN

Horizontal force (H) = 2477.9 kN

Moment (M) = 54150.2 kN-m

In this case, the grip length, $D = 10.2$ m.

$L = 0.9 \times 8 = 7.2$m

Now $$I_v = \frac{L\,D^3}{12} = \frac{7.2 \times (10.2)^3}{12} = 636.72 \text{ m}^4$$

From Eq. 27.68, $$I = I_B + m\,I_v\,(1 + 2\,\mu'\,\alpha)$$

Taking $m = 1.0$, $$I = 174.92 + 1.0 \times 636.72\,(1 + 2 \times \tan 20° \times \alpha)$$

where $$\alpha = \frac{0.318\,B}{D} = \frac{0.318 \times 8}{10.2} = 0.25$$

Therefore, $$I = 174.92 + 636.72\,(1 + 2 \times \tan 20° \times 0.25) = 927.51 \text{ m}^4$$

From Eq. 27.69 (*b*), $$r = \frac{D}{2} \cdot \frac{I}{m\,I_v}$$

$$= \frac{10.2}{2} \times \frac{927.51}{1.0 \times 636.72} = 7.43 \text{ m}$$

From Eq. 27.71, $$H > \frac{M}{r}(1 + \mu\,\mu') - \mu\,W$$

$$> \frac{54150.2}{7.43}(1 + \tan 30° \tan 20°) - \tan 30° \times 17372$$

or $$H > -1205$$

As H of 2477.9 is greater than –1205, the above condition is statisfied.

Also $$H < \frac{M}{r}(1 - \mu\,\mu') + \mu\,W$$

$$< \frac{54150.2}{7.43}(1 - \tan 30° \tan 20°) + \tan 30° \times 1737.2$$

or $2477.9 < 6760$ (satisfied)

From Eq. 27.76, taking submerged unit weight,

$$m\,\frac{M}{I} \not> \gamma'\,(K_p - K_a)$$

or $$1.0 \times \frac{54150.2}{927.51} \ngtr 10\,(6.105 - 0.297)$$

or $$58.4 \ngtr 58.2 \text{ (almost satisfied)}$$

From Eq. 27.69 (*a*), $$P = \frac{M}{r}$$

$$= \frac{54150.2}{7.43} = 7288 \text{ kN}$$

From Eq. 27.74, $$p_t = \frac{(W - \mu' P)}{A} + \frac{M B}{2 I}$$

$$= \frac{(17372 - 0.364 \times 7288)}{\pi/4 \times (8.0)^2} + \frac{54150.2 \times 8}{2 \times 927.51}$$

or $$p_t = 526.5 \text{ kN/m}^2 < 600 \text{ kN/m}^2 \text{ (safe)}$$

From Eq. 27.75, $$p_h = \frac{W - \mu' P}{A} - \frac{M B}{2I}$$

$$= \frac{(17372 - 0.364 \times 7288)}{\pi/4 \times (8.0)^2} - \frac{54150.2 \times 8}{2 \times 927.51}$$

or $$p_h = 59.5 \text{ kN/m}^2$$

Ultimate Resistance

From Eq. 27.85, $$\frac{W}{A} \ngtr \frac{q_u}{2}$$

Taking $q_u = 1800 \text{ kN/m}^2$,

$$\frac{17372}{\pi/4 \times (8.0)^2} \ngtr \frac{1800}{2}$$

or $$345.8 \ngtr 900 \text{ (satisfied)}$$

From Eq. 27.84, $$M \ngtr 0.7\,(M_b + M_s + M_f)$$

From Table 27.2, $C = 0.29$ for $D/B = 10.2/8 = 1.23$

From Eq. 27.78 (*b*), $$M_b = C W B \tan \phi$$

$$= 0.29 \times 17372 \times 8 \tan 30°$$

$$= 23254.9 \text{ kN–m}$$

From Eq. 27.79, $$M_s = 0.1\ \gamma' D^3 (K_p - K_a) L$$

$$= 0.1 \times 10 \times (10.2)^3 \times (6.105 - 0.297) \times 7.2$$

$$= 44377.3 \text{ kN–m}$$

From Eq. 27.81, $$M_f = 0.1\ \gamma'\ (K_p - K_a) B^2 D^2 \sin \delta$$

$$= 0.1 \times 10\,(6.105 - 0.297) \times 8^2 \times (10.2)^2 \sin 20°$$

$$= 13226.9$$

Now, $$0.7\,(M_b + M_s + M_f) = 0.7\,(23254.9 + 44377.3 + 13226.9)$$

$$= 56601.4 \text{ kN–m}$$

$$M = 54150.2 \text{ kN–m}$$

$$M \ngtr 0.7\,(M_b + M_s + M_f) \text{ (satisfied)}$$

PROBLEMS

A. Numericals

27.1. A well foundation has the following particulars.

Outer diameter = 5.0 m
Inner diameter = 3.0 m
Depth below scour level = 12.0 m
Moment = 5000 kN-m
Horizontal force acting at 8 m above the scour level = 600 kN
Factor of safety = 2.0

Assuming that the well tilts about a certain point above the base, compute the allowable, total equivalent resisting force due to earth pressure. Take γ_{sat} = 20 kN/m^3, ϕ = 30°. **[Ans.** 710.5 kN]

27.2. The following data refer to a well foundation:

(*a*) Net downward load, including self-weight = 12000 kN
(*b*) Height of point of application of horizontal force above the scour level = 4 m
(*c*) Depth of well below scour level = 10 m
(*d*) External diameter = 7.0 m
(*e*) Internal diameter = 5.0 m
(*f*) Vertical subgrade reaction = 3×10^4 kN/m^3
(*g*) Poisson's ratio = 0.5
(*h*) Horizontal deformation of well cap at the scour level (ρ_1) = 20 mm
(*i*) Allowable soil pressure = 600 kN/m^2

Assuming the well to rotate about its base, determine the base pressure and lateral load per unit length of the well. Take γ_{sat} = 20 kN/m^3 and ϕ = 30°.

B. Descriptive Type

27.3. Discuss the situations where a well foundation is more suitable than the other types of foundations.

27.4. What are different shapes of wells ? Discuss the characteristics of each type.

27.5. Discuss the various forces acting on a well foundation.

27.6. What do you understand by grip length ? What is its importance in well foundations ?

27.7. Describe various methods for the design of well foundations. What are their relative merits ?

27.8. What are the various components of a well foundation ? What are their uses ?

27.9. Describe the procedure for construction of wells. Discuss the causes and remedies for tilts and shifts.

27.10. Discuss IRC method for the design of well foundation.

27.11. Explain Terzaghi's analysis for rigid bulkhead applied to well foundation.

27.12. Discuss Banerjee and Gangopadhyay's method for the design of well foundations.

C. Multiple Choice Questions

1. A well foundation is a type of
(*a*) open caisson (*b*) pneumatic caisson
(*c*) floating caisson (*d*) drilled pier

2. The grip length below the maximum scour level for the railway bridges is usually
(*a*) 0.5 R (*b*) 0.25 R (*c*) R (*d*) 2 R
where R is the maximum scour depth.

3. In some well foundation, the following is not provided :
(*a*) R.C.C. well cap (*b*) Top plug
(*c*) Bottom plug (*c*) Curb

4. The most commonly used shape of a well foundation is
(*a*) Double-D well (*b*) Circular well
(*c*) Double octagonal well (*d*) Rectangular well.

5. The thickness of steining for railway bridges is usually kept as .... of outside diameter.
(*a*) one-eight (*b*) one tenth
(*c*) one-sixth (*d*) one-fourth.

[Ans. 1. (*a*), 2. (*a*), 3. (*b*), 4. (*b*), 5. (*d*)]

28

Machine Foundations

28.1. INTRODUCTION

Foundations subjected to static loads have been discussed in the preceding chapters. In some cases, the foundations are subjected to dynamic loads. These loads may result from various causes such as vibratory motion of machines, movement of vehicles, impact of hammers, earthquakes, winds, waves, nuclear blasts, mine explosions, and pile driving. The dynamic loads transmitted to the foundations and their effect on the strata below can be determined using the principles of soil dynamics and theory of vibrations. The analysis is, however, very complex.

Machine foundations are subjected to the dynamic forces caused by the machine. These dynamic forces are transmitted to the foundation supporting the machine. Although the moving parts of the machine are generally balanced, there is always some unbalance in practice which causes an eccentricity of rotating parts. This produces an oscillating force. The machine foundation must satisfy the criteria for dynamic loading, in addition to that for static loading already discussed.

Basically, there are three types of machines:

(*i*) Machines which produce a periodic unbalanced force, such as reciprocating engines and compressors. The speed of such machines is generally less than 600 r.p.m. In these machines, the rotary motion of the crank is converted into the translatory motion. The unbalanced force varies sinusoidally.

(*ii*) Machines which produce impact loads, such as forge hammers and punch presses. In these machines, the dynamic force attains a peak value in a very short time and then dies out gradually. The response is a pulsating curve. It vanishes before the next pulse. The speed is usually between 60 to 150 blows per minute.

(*iii*) High speed machines, such as turbines and rotary compressors. The speed of such machines is very high; sometimes, it is even more than 3000 r.p.m.

This chapter is devoted mainly to the design of foundation for the machines of the first type.

28.2. TYPES OF MACHINE FOUNDATIONS

The following 4 types of machine foundations are commonly used.

(1) Block Type. This type of machine foundation consists of a pedestal resting on a footing [Fig. 28.1 (*a*)]. The foundation has a large mass and a small natural frequency.

(2) Box Type. The foundation consists of a hollow concrete block [Fig. 28.1 (*b*)]. The mass of the foundation is less than that in the block type and the natural frequency is increased.

(3) Wall Type. A wall type of foundation consists of a pair of walls having a top slab. The machine rests on the top slab [Fig. 28.1 (*c*)].

(4) Framed Type. This type of foundation consists of vertical columns having a horizontal frame at their tops. The machine is supported on the frame [Fig. 28.1 (*d*)].

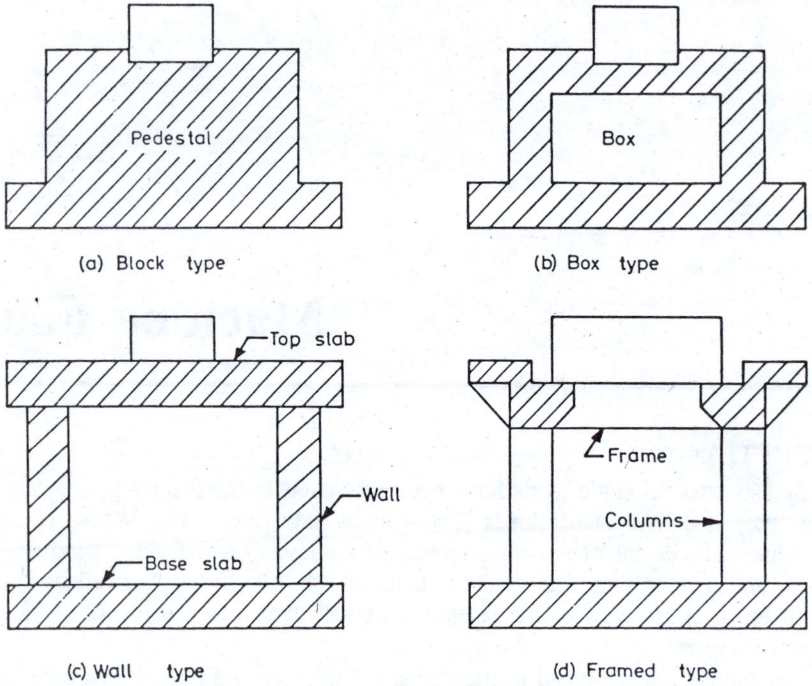

Fig. 28.1. Type of Machine Foundation.

Suitability of various types. Machines which produce periodical and impulsive forces at low speeds are generally provided with a block type foundation. Framed type foundations are generally used for the machines working at high speeds and for those of the rotating types.

Some machines which induce very little dynamic forces, such as lathes, need not be provided with a machine foundation. Such machines may be directly bolted to the floor.

28.3. BASIC DEFINITIONS

The following terms are used in the dynamic analysis of machine foundations.

(1) Vibration (or oscillation). It is the time-dependent, repeated motion of translational or rotational type.

(2) Periodic motion. It is the motion which repeats itself periodically in equal time intervals.

(3) Period (T). The time period in which the motion repeats itself is called the period of motion or simply period.

(4) Cycle. The motion completed in the period is called the cycle of motion.

(5) Frequency (*f*). The number of cycles of motion in a unit of time is known as the frequency of vibration. It is usually expressed in hertz (*i.e.* cycles per second).

The period (T) and the frequency (f) are inter- related as

$$T = 1/f \qquad \text{...(28.1)}$$

Circular frequency (ω) is in radians per second.

(6) Free vibration. Free vibrations occur under the influence of forces inherent in the system itself, without any external force. However, to start free vibrations, some external force or natural disturbance is required. Once started, the vibration continue without an external force.

(7) Forced vibration. Forced vibrations occur under the influence of a continuous external force.

(8) Natural frequency. The system under free vibrations vibrates at the frequency known as natural frequency. The natural frequency is the characteristic of the system. A system may have more than one natural frequency.

(9) Resonance. When the frequency of the exciting force is equal to one of the natural frequencies of the system, the amplitudes of motion become excessively large. This condition is known as resonance.

10. Damping. The resistance to motion which develops due to friction and other causes is known as damping.

Viscous damping is a type of damping in which the damping force is proportional to the velocity. It is expressed as

$$F = c \frac{dz}{dt} \qquad \ldots(28.2)$$

where c = damping coefficient, and $\frac{dz}{dt}$ = velocity

(11) Degree of Freedom. The number of independent co-ordinates required to describe the motion of a system is called the degree of freedom.

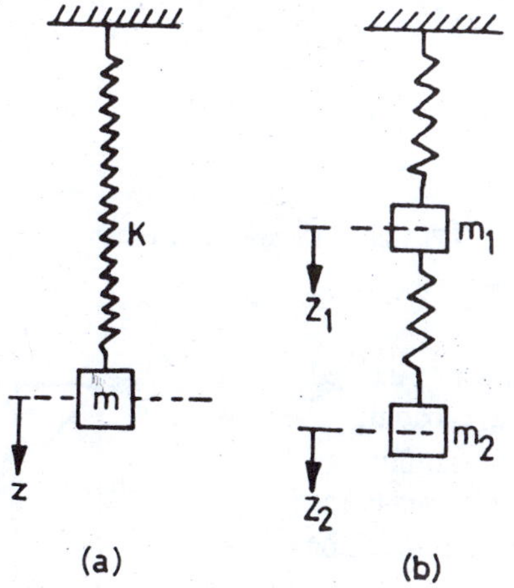

Fig. 28.2.

Fig. 28.2. (*a*) shows a system with one degree of freedom, and Fig. 28.2 (*b*) shows a system with two degrees of freedom. An elastic rod has an infinite degree of freedom. However, for convenience, the rod is divided into segments. The degree of freedom is made finite by considering the masses of these segments.

(12) Principal modes of vibrations. A system with more than one degree of freedom vibrates in complex modes. However, if each point in the system follows a definite pattern of common natural frequency, the mode is systematic and orderly and is known as the principal mode of vibration.

A system with n degrees of freedom has n principal modes and hence n natural frequencies.

28.4. DEGREE OF FREEDOM OF A BLOCK FOUNDATION

A rigid block foundation has 6 degree of freedom (Fig. 28.3). Any displacement can be resolved into 6 independent displacements as under.

(1) Translation along *X*-axis, (2) Translation along *Y*-axis, (3) Translation along *Z*-axis, (4) Rotation about *X*-axis, (5) Rotation about *Y*-axis, (6) Rotation about *Z*-axis.

Translation along *Z*-axes and rotation about *Z*-axis can occur independently of any other motion. However, translations and rotation about *X*-and *Y*-axis are coupled, as these cannot occur independent of one another.

X-, *Y*- and *Z*-axes are called, respectively, pitching, rocking and yawing axes.

The discussions in this elementary text are limited to one-degree of freedom.

29.5. GENERAL CRITERIA FOR DESIGN OF MACHINE FOUNDATIONS

A good machine foundation should satisfy the following criteria.

(1) Like ordinary foundations, it should be safe against shear failure caused by superimposed loads, and

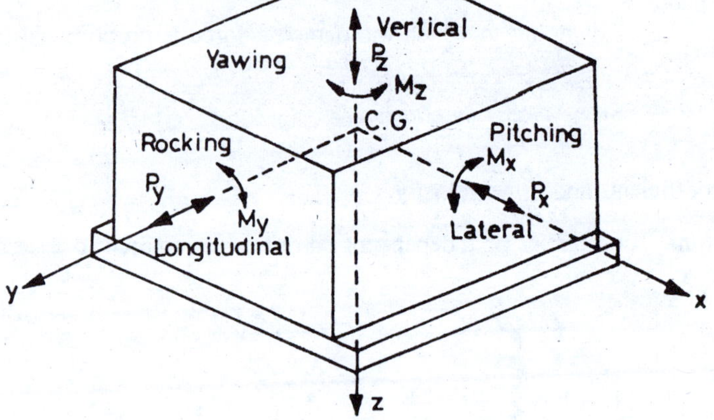

Fig. 28.3. Degree of Freedom.

also the settlements should be within the safe limits.

The soil pressure should normally not exceed 80% of the allowable pressure for static loading.

(2) There should be no possibility of resonance. The natural frequency of the foundation should be either greater than or smaller than the operating frequency of the machine.

(3) The amplitudes under service conditions should be within the permissible limits for the machine.

(4) The combined centre of gravity of the machine and the foundation should be on the vertical line passing through the centre of gravity of the base plane.

(5) Machine foundation should be taken to a level lower than the level of the foundation of the adjacent buildings and should be properly separated.

(6) The vibrations induced should neither be annoying to the persons nor detrimental to other structures.

Richart (1962) developed a plot for vertical vibration, which is generally taken as a guide for various limits of frequency and amplitude (Fig. 28.4).

(7) The depth of the ground-water table should be at least one-fourth of the width of the foundation below the base plane.

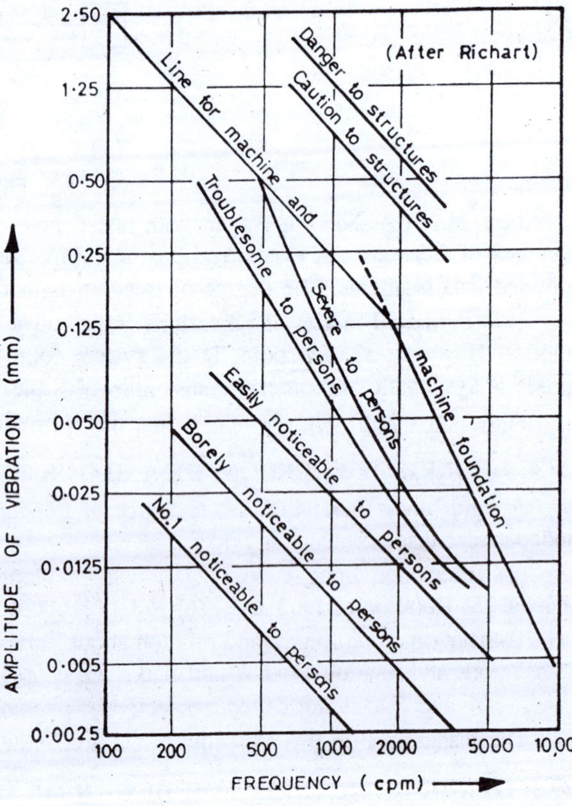

Fig. 28.4. Rchart's Chart.

28.6. FREE VIBRATIONS

The free vibrations may be damped or undamped.

(a) Undamped Vibrations. Fig. 28.5 (*a*) shows a rigid mass (*m*) resting on a spring of stiffness *k*. The system has one degree of freedom. Let us assume that the system has been set in motion and it vibrates in the vertical direction.

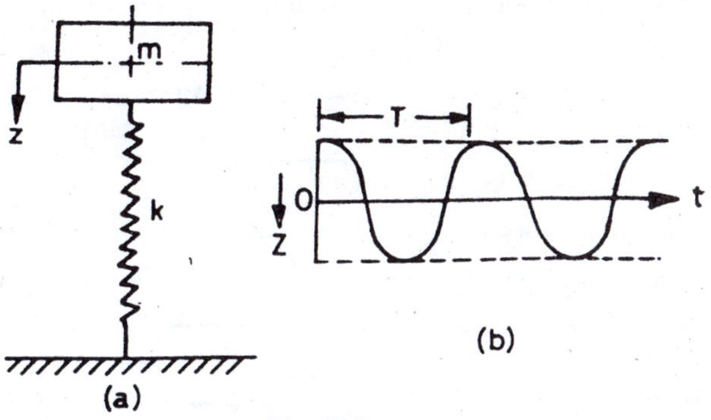

Fig. 28.5.

The equation of motion can be written as under.

$$m \frac{d^2 z}{dt^2} = -kz$$

or $$m \frac{d^2 z}{dt^2} + kz = 0 \quad ...(28.3)$$

The solution of Eq. 28.3 is given by

$$z = A \cos \omega_n t + B \sin \omega_n t$$

where A and B are arbitrary constants, and ω_n is the natural circular frequency (radians/sec).

The solution can also be written as

$$z = A \sin (\omega_n t + \alpha) \quad ...[28.4 (a)]$$

where A and α are constants.

Alternatively, $$z = A \cos(\omega_n t - \alpha) \quad ...[28.4 (b)]$$

From Eq. 28.4 (*a*), $$\frac{dz}{dt} = A \omega_n \cos (\omega_n t + \alpha)$$

and $$\frac{d^2 z}{dt^2} = -A \omega_n^2 \sin (\omega_n t + \alpha) \quad ...(28.5)$$

Substituting these values in Eq. 28.3, $-m A \omega_n^2 \sin (\omega_n t + \alpha) + kA \sin (\omega_n t + \alpha) = 0$

or $$m \omega_n^2 = k$$

or $$\omega_n = \sqrt{k/m} \quad ...(28.6)$$

It may be noted that the greater the mass *m*, the smaller is the frequency.

If f_n is the natural frequency in cycles per second,

$$f_n = \frac{\omega_n}{2\pi} = \frac{1}{2\pi}\sqrt{k/m} \quad ...(28.7)$$

If T is the time period, Eq. 28.1 gives

$$T = 1/f_n = 2\pi/\sqrt{k/m} \quad ...(28.8)$$

Fig. 28.5 (*b*) shows the response curve of the system. As it is evident, the cycle repeats after time *T*.

(*b*) Damped Vibration. Fig. 28.6 (*a*) shows a rigid mass *m* resting on a spring of stiffness *k* and connected to a viscous damper with a damping coefficient *c*. In this case, there is an additional force due to damping. The equation of motion can be written as

$$m\frac{d^2 z}{dt^2} + c\frac{dz}{dt} + kz = 0 \quad \text{...(28.9)}$$

It can be shown that the general solution is of the form

$$z = C_1 e^{s_1 t} + C_2 e^{s_2 t} \quad \text{...(28.10)}$$

where
$$s_1 = \omega_n(-D + \sqrt{D^2 - 1}) \quad \text{...[28.11(a)]}$$
$$s_2 = \omega_n(-D - \sqrt{D^2 - 1}) \quad \text{...[28.11(b)]}$$

and ω_n = natural frequency, D = damping factor (= c/c_c), and c_c = critical damping (= $2\sqrt{mk}$)

If $D > 1$, the system is overdamped and the motion is aperiodic, If $D = 1$, the system is critically damped, which also gives aperiodic motion. If $D < 1$, the system is underdamped and the response is periodic, as shown in Fig. 28.6 (*b*). Only underdamped systems are of practical importance in the design of machine foundations.

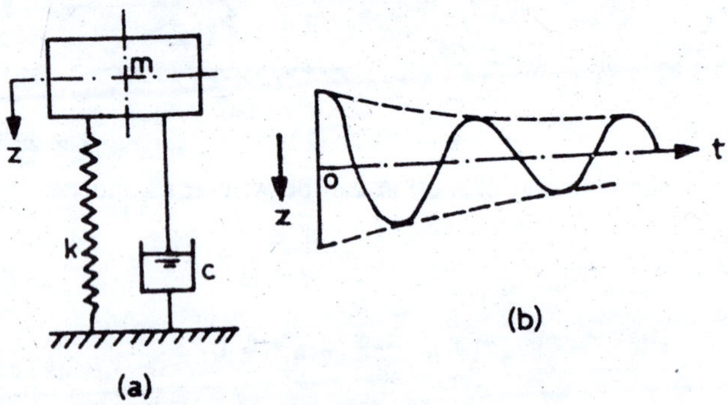

Fig. 28.6.

Eq. 28.11 can be written as

$$s_1 = \omega_n(-D + i\sqrt{1 - D^2}) \quad \text{...[28.12(a)]}$$
$$s_2 = \omega_n(-D - i\sqrt{1 - D^2}) \quad \text{...[28.12(b)]}$$

where $i = \sqrt{-1}$ = iota

Eq. 28.10 can be written as

$$z = e^{-D\omega_n t}\left[C_1 e^{i(1-D^2)^{1/2}\omega_n t} + C_2 e^{-i(1-D^2)^{1/2}\omega_n t}\right] \quad \text{...(28.13)}$$

Let us make the following substitution.

$$\omega_{nd} = \omega_n\sqrt{1 - D^2} \quad \text{...(28.14)}$$

Therefore,
$$z = e^{-D\omega_n t}\left[(C_1 e^{i\omega_{nd} t} + C_2 e^{-i\omega_{nd} t}\right]$$

or
$$z = e^{-D\omega_n t}\left[(C_1 + C_2)\cos\omega_{nd} t + i(C_1 - C_2)\sin\omega_{nd} t\right]$$

or
$$z = e^{-D\omega_n t}\left[A_1\cos\omega_{nd} t + A_2\sin\omega_{nd} t\right] \quad \text{...(28.15)}$$

The term ω_{nd} is known as *damped natural frequency.* Eq. 28.1 can also be written as

$$z = e^{-D\omega_n t}\left[A\cos(\omega_{nd} t - \alpha)\right] \quad \text{...(28.16)}$$

The term $e^{-D\omega_{nd} t}$ gives an aperiodic exponential response; whereas, the term $A\cos(\omega_{nd} t - \alpha)$ indicates a periodic sinusoidal response. The net result is a periodic but gradually decreasing motion [Fig. 28.6 (*b*)].

For the undamped system, as $D = 0$,

$$z = A\cos(\omega_n t - \alpha)$$

This is the same Eq. 28.4 (*b*).

28.7. FORCED VIBRATION

Fig. 28.7 shows a rigid mass with a single degree of freedom. The system is damped and subjected to an exciting force $F(t)$. The equation of motion can be written as

$$m \frac{d^2 z}{dt^2} + c \frac{dz}{dt} + kz = F(t) \qquad \ldots(28.17)$$

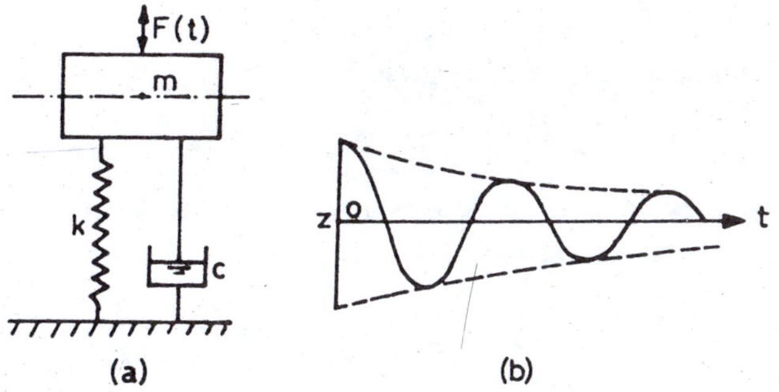

Fig. 28.7.

Let us assume that the exciting force is sinusoidal of the form

$$F(t) = F_o \sin \omega t$$

where F_o = magnitude of the exciting force,

ω = circular frequency of exciting force.

Thus
$$m \frac{d^2 z}{dt^2} + c \frac{dz}{dt} + k z = F_o \sin \omega t \qquad \ldots(28.18)$$

The solution of Eq. 28.18 is

$$z = e^{-D \omega_n t} [A \cos (\omega_{nd} t - \alpha)] + \frac{F_o \sin (\omega t - \beta)}{\sqrt{(k - m \omega^2)^2 + c^2 \omega^2}} \qquad \ldots(28.19)$$

The first part of the solution is transient and dies out after some time. The second part is the steady-state response. Thus

$$z = \frac{F_o \sin (\omega t - \beta)}{\sqrt{(k - m \omega^2)^2 + c^2 \omega^2}} \qquad \ldots(28.20)$$

Substituting $\omega_n = \sqrt{k/m}$, and $D = \frac{c}{2\sqrt{km}}$,

$$z = \frac{F_o \sin (\omega t - \beta)}{\sqrt{k^2 (1 - r^2)^2 + 4 D^2 r^2 k^2}} \qquad \ldots(28.21)$$

where r = frequency ratio = ω/ω_n .

For undamped system, $c = 0$ and $D = 0$. Therefore,

$$z = \frac{F_o \sin (\omega t - \beta)}{k(1 - r^2)} \qquad \ldots(28.22)$$

When $r = 1.0$, *i.e.*, $\omega = \omega_n$, response is inifinite. The condition is known as *resonance*.

As an ideal undamped system is non-existent, damping always exists and the response is finite. However, when operating frequency (ω) is close to the natural frequency (ω_n), the response is very high. To avoid this condition, the operating frequency should not be close to the natural frequency. For a safe design, the frequency ratio is normally kept outside the critical range of 0.4 to 1.50.

The magnitude of the displacement is given by

$$|z| = \frac{F_o}{k(1-r^2)} = \frac{F_o}{m(\omega_n^2 - \omega^2)} \quad \text{...[28.23(a)]}$$

In a general case,

$$|z| = \frac{F_o/k}{\sqrt{(1-r^2)^2 + 4D^2r^2}} \quad \text{...[28.23(b)]}$$

The static displacement under a force F_o is given by

$$z_{st} = \frac{F_o}{k} \quad \text{...(28.24)}$$

The ratio of the magnitude of the steady-state displacement of a forced system to the static displacement is known as magnification factor (M). Thus

$$M = \frac{|z|}{z_{st}} = \frac{F_o/k}{(F_o/k)\sqrt{(1-r^2)^2 + 4D^2r^2}}$$

or

$$M = \frac{1}{\sqrt{(1-r^2)^2 + 4D^2r^2}} \quad \text{(28.25)}$$

Thus $z = M\, z_{st}$

Fig. 28.8 shows the variation of the magnification factor M with r for different values of D. It may be noted that the magnification is high for the value of r between 0.4 and 1.5.

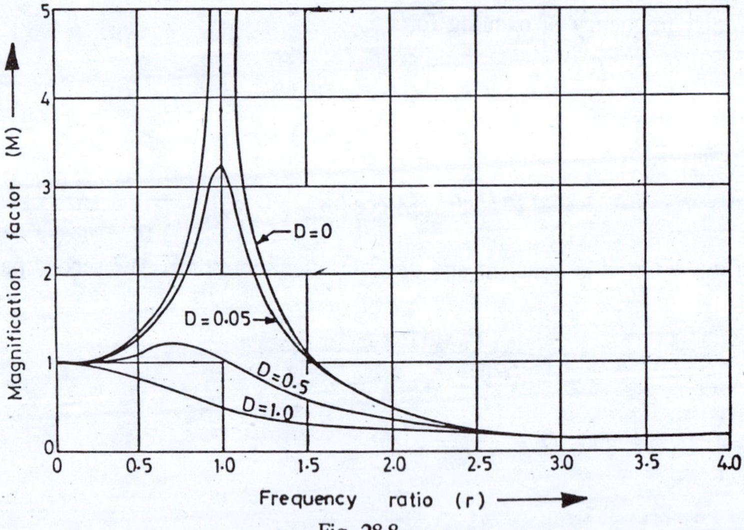

Fig. 28.8.

Force Transmissibility *T*. A damped forced vibratory system derives its support from the foundation. The force transmitted can be expressed as

$$F_T = c\frac{dz}{dt} + kz \quad \text{...(28.26)}$$

Let us express

$$z = \frac{(F_o/k)\sin(\omega t - \beta)}{\sqrt{(1-r^2)^2 + 4D^2r^2}} \quad \text{as } B\sin(\omega t - \beta),$$

where

$$B = \frac{F_o/k}{\sqrt{(1-r^2)^2 + 4D^2r^2}} \quad \text{...(28.27)}$$

Eq. 28.26 can be written as

$$F_T = c\,B\,\omega \cos(\omega t - \beta) + k\,B \sin(\omega t - \beta)$$

or

$$F_T = B\sqrt{k^2 + c^2\omega^2}\;[\cos(\omega t - \beta - \gamma)]$$

where $\gamma = \tan^{-1}(k/c\omega)$.

The magnitude of F_T is given by

$$|F_T| = B\sqrt{k^2 + c^2\omega^2}$$

or

$$|F_T| = \frac{(F_o/k)\sqrt{k^2 + c^2\omega^2}}{\sqrt{(1 - r^2)^2 + 4D^2r^2}}$$

or

$$|F_T| = F_o M\sqrt{1 + (2Dr)^2} \qquad \text{...(28.29)}$$

Force transmissibility (T) is defined as the ratio of the force transmitted to the applied force Thus

$$T = \frac{F_T}{F_o}$$

$$= M\sqrt{1 + (2Dr)^2} \qquad \text{...(28.30)}$$

Like magnification factor (M), the transmissibility is also a function of r and D. The plot is similar in shape to that shown in Fig. 28.8.

28.8. VIBRATION ANALYSIS OF A MACHINE FOUNDATION

Although a machine foundation has 6 degree of freedom, it is assumed to have a single degree of freedom for a simplified analysis. Fig. 28.9 shows a machine foundation supported on a soil mass. In this case, the mass m_f of the system lumps together the mass of the machine and the mass of foundation. The total mass m_f acts at the centre of gravity of the system. The mass is under the supporting action of the soil. The elastic action can be lumped together into a single elastic spring with a stiffness k. Likewise, all the resistance to motion is lumped into the damping coefficient c. Thus the machine foundation reduces to a single mass having one degree of freedom, as shown in Fig. 28.7. The analysis of damped, forced vibration, discussed in Sect. 28.7, is, therefore, applicable to the machine foundation.

Determination of Parameters. For vibration analysis of a machine foundation, the parameters m, c and k are required. These parameters can be determined as under.

(1) Mass (m). When a machine vibrates, some portion of the supporting soil mass also vibrates. The vibrating soil mass is known as the participating mass or in-phase soil mass. Therefore, the total mass of the

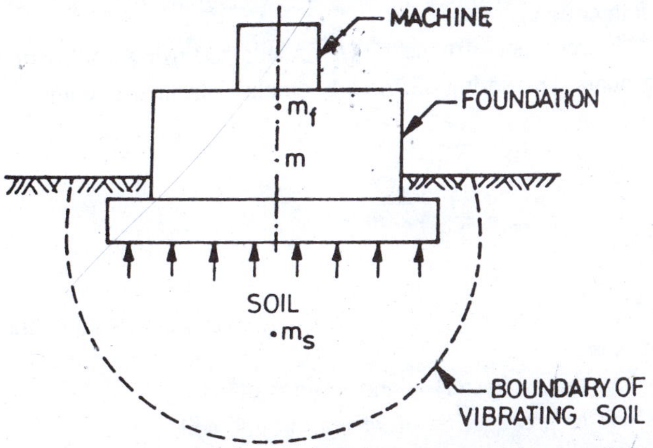

Fig. 28.9.

system is equal to the mass of the foundation block and machine (m_f) and the mass (m_s) of the participating soil. Thus

$$m = m_f + m_s \qquad ...(28.31)$$

Unfortunately, there is no rational method to determine the magnitude of m_s. It is usually related to the mass of the soil in the pressure bulb. The value of m_s generally varies between zero and m_f. In other words, the total mass (m) varies between m_f and $2\,m_f$ in most cases.

(2) Spring stiffness (k). The spring stiffness depends upon the type of soil, embedment of the foundation block, the contact area and the contact pressure distribution. The following methods are commonly used.

***(a)* Laboratory test.** A triaxial test with vertical vibrations is conducted to determine Young's modulus E. Alternatively, the modulus of rigidity (G) is determined by conducting the test under torsional vibration, and E is obtained indirectly from the relation $E = 2\,G\,(1 + \mu)$, where μ is Poisson's ratio.

The stiffness (k) is determined as

$$k = AE/L \qquad ...(28.32)$$

where A = cross-sectional area of the specimen, L = length of the specimen.

***(b)* Barkan's method.** The stiffness can also be obtained from the value of E using the following relation given by Barkan.

$$k = \frac{1.13\,E}{1 - \mu^2}\sqrt{A} \qquad ...(28.33)$$

where A = base area of the machine, *i.e.* area of contact.

***(c)* Plate load test.** A repeated plate load test is conducted and the stiffness of the soil in the test (k_p) is found as the slope of the load-deformation curve. The spring constant k of the foundation is determined as under.

(*i*) For cohesive soils

$$k = k_p\,(B/B_p) \qquad ...(28.34)$$

(*ii*) For cohesionless soils

$$k = k_p\left(\frac{B + 0.3}{B_p + 0.3}\right)^2 \qquad ...(28.35)$$

where B is the width of foundation and B_p is the diameter of the plate.

Alternatively, the spring constant can be obtained from the subgrade modulus (k_s), as

$$k = k_s \cdot A \qquad ...(28.36)$$

where A = area of foundation.

***(d)* Resonance test.** The resonance frequency (f_n) is obtained using a vibrator of mass m set up on a steel plate supported on the ground. The spring stiffness is obtained from the relation

$$f_n = \frac{\omega_n}{2\,\pi}$$

or

$$f_n = \frac{1}{2\,\pi}\sqrt{k/m}$$

or

$$k = 4\,\pi^2 f_n\, m \qquad ...(28.37)$$

(3) Damping constant *c*. Damping is due to dissipation of vibration energy, which occurs mainly because of the following reasons.

(*i*) Internal friction loss due to hysterisis and viscous effects.

(*ii*) Radiational loss due to propagation of waves through soil.

The damping factor D for an underdamped system can be determined in the laboratory. Vibration response is plotted and the logarithmic decrement δ is found from the plot, as

$$\delta = \log (z_2/z_1) \quad ...(28.38)$$

where z_1 and z_2 are amplitudes of two consecutive cycles of an amplitude-response curve.

The damping factor D and the logarithmic decrement δ are related as

$$\delta = \frac{2\pi D}{\sqrt{1 - D^2}} \quad ...(28.39)$$

or

$$D \simeq \frac{\delta}{2\pi} \quad ...(28.40)$$

The damping factor D may also be obtained from the area of the hysterisis loop of the load displacement curve, as

$$D = \frac{\Delta W}{W} \quad ...(28.41)$$

where W = total work done, ΔW = work lost in hysterisis.

The value of D for most soils generally varies between 0.01 and 0.1.

28.9. DETERMINATION OF NATURAL FREQUENCY

The natural frequency of the foundation-soil system can be determined using the theory of vibrations discussed in Sect. 28.7. The equation of motion, neglecting damping, is

$$m \frac{d^2 z}{dt^2} + kz = F_o \sin \omega t \quad ...(28.42)$$

where m = mass of machine, foundation and the participation soil, k = equivalent spring constant of the soil.

The methods for determination of k and m have been discussed in the preceding section.

The natural frequency of the system is given by

$$\omega_n = \sqrt{k/m} \quad ...(28.43)$$

where ω_n is in radians per second.

Also

$$f_n = \frac{1}{2\pi}\sqrt{k/m} \quad ...(28.44)$$

where f_n is in cycles per second.

Thus

$$f_n = \frac{1}{2\pi}\sqrt{\frac{k}{m_f + m_s}} \quad ...(28.45)$$

where m_f = mass of machine and foundation.

and m_s = mass of the participating soil mass.

Barken (1962) gave the following relation for the natural frequency.

$$\omega_n = \sqrt{\frac{C_u A}{m}} \quad ...(28.46)$$

where C_u = coefficient of elastic uniform compression, A = contact area of foundation with soil.

Comparing with Eq. 28.43, $k = C_u \times A$ ...[28.46(*a*)]

The maximum amplitude is given by

$$z_{max} = \frac{F_o}{m\,\omega_n^2 (1 - r^2)} \quad ...(28.47)$$

where F_o = exciting force.

The coefficient of elastic uniform compression (C_u) depends upon the type of soil. It can be obtained from the following relation.

$$C_u = 1.13 \frac{E}{(1-\mu^2)} \cdot \frac{1}{\sqrt{A}} \qquad ...(28.48)$$

As it is evident, the coefficient varies inversely proportional to the square root of the base area of the foundation. Thus

$$\frac{(C_u)_2}{(C_u)_1} = \left(\frac{A_1}{A_2}\right)^{1/2} \qquad ...(28.49)$$

Table 28.1 gives the recommended value of C_u for $A = 10\ m^2$ for different soils (Barkan, 1962).

Table 28.1. Coefficient of Elastic Uniform Compression

Soil Category	*Soil type*	*Allowable soil pressure (q_{na}) kN/m²*	*Coefficient of elastic uniform compression (C_u) kN/m³*
I	Weak soils (clays and silty clays with sand in a plastic state; clayey silts; soils of categories II and III with laminae of organic silts and of peat).	upto 150	upto 3×10^4
II	Soils of medium strength (clays and silty clays with clays and silty clays with sand close to the plastic limit; sand)	150—350	$3 \times 10^4 - 5 \times 10^4$
III	Strong soils (clays and silty clays with sand of hard consistency; gravels and gravelly sands; loess and loessial soils).	350—500	$5 \times 10^4 - 10^5$
IV	Rocks	> 500	$> 10^5$

28.10. DESIGN CRITERIA FOR FOUNDATIONS OF RECIPROCATING MACHINES

The following design criteria for the foundation of reciprocating machines should be satisfied (IS : 2874 (I)—1982).

(1) The machine foundation should be isolated at all levels from the adjoining foundations (See Sect. 28.13).

(2) The natural frequency of the foundation-soil system should be higher than the highest disturbing frequency and the frequency ratio should be less than 0.4, as far as possible.

However, if it is not possible to satisfy above criterion, the natural frequency should be lower than the lowest disturbing frequency and the frequency ratio should not be less than 1.50.

(3) The amplitude of vibration should be within the permissible limits, as given in Fig. 28.4 or any specified permissible limits (For example, chart given in IS : 2874 (I) — 1982).

For most soils, the limiting amplitude for low speed machines is usually taken as 200 micron (= 0.2 mm).

According to another criterion, the amplitude in mm is limited to $4/f$ for frequencies less than 30 hertz and $125/f^2$ for higher frequencies, where f is frequency in hertz (cycles/sec).

(4) Concrete block foundations should be used. However, when the soil is not suitable to support block foundation, cellular foundation may be used.

(5) The size of the block in plan should be larger than the bed plate of the machine. There should be a minimum all-round clearance of 15 mm.

The total width of the foundation measured at right angles to the shaft should be least equal to the distance between the centre of the shaft and the bottom of the foundation.

(6) The eccentricity of the foundation system along X-X and Y-Y axes should not exceed 5% of the length of the corresponding side of the contact area.

(7) The combined centre of gravity of machine and foundation should be as much below the top of foundation as possible. In no case, it should be above the top of foundation.

(8) The depth of foundation should be sufficient to provide the required bearing capacity and to ensure stability against rotation in the vertical plane.

(9) The stresses in the soil below the foundations should not exceed 80% of the allowable stresses under static loads. The base pressure is limited to half the normal allowable pressure (q_{na}) in extreme cases.

(10) Where it is not practicable to design a foundation to give satisfactory dynamic response, the transmitted vibrations may be reduced by providing anti-vibration mountings either between the machine and the foundation or between the foundation and the supporting system.

(11) The machine should be anchored to the foundation block using a base plate and anchor bolts. Bolt holes should be backfilled with concrete and the space below the plate should be filled with 1 : 2 cement mortar.

(12) A number of similar machines can be erected on individual pedestals on a common raft. The analysis for such machines can be made assuming that each foundation acts independently with an area of foundation equal to that obtained by dividing up the raft into sections corresponding to separate machines.

28.11. REINFORCEMENT AND CONSTRUCTION DETAILS

(1) The reinforcement in the concrete block should not be less than 25 kg/m^3.

For machines requiring special design consideration of foundations, such as machine pumping explosive gases, the minimum reinforcement is 40 kg/m^3.

(2) Steel reinforcement around all pits and openings shall be at least equal to 0.5 to 0.75% of the cross-sectional area of the pit or opening.

(3) The reinforcement shall run in all the three directions.

The minimum reinforcement shall usually consist of 12 mm bars at 200 to 250 mm spacing extending both vertically and horizontaly near all faces of the foundation block. The ends of all bars should always be hooked.

(4) If the height of the foundation block exceeds one metre, shrinkage reinforcement shall be placed at suitable spacing in all the three directions.

(5) The cover should be a minimum of 75 mm at the bottom and 50 mm on sides and the top.

(6) The concrete shall be at least M-15 with a characteristic strength of 15 N/mm^2.

(7) The foundation block should be preferably cast in a single, continuous operation.

In case of very thick blocks (exceeding 5 m), construction joints can be provided.

28.12. MASS OF FOUNDATION

Heavy foundations eliminate excessive vibrations. Manufactures of machines sometimes recommend the mass of foundation required for the machines. However, the mass recommended are generally empirical and based largely on experience.

Couzens (1938) gave the ratios of foundation mass to engine mass suitable for various types of machines (See Table 28.2). These ratios can be used for rough estimates.

28.13. VIBRATION ISOLATION AND CONTROL

Vibrations may cause harmful effects on the adjoining structures and machines. Besides, these vibrations cause annoyance to the persons working in the area around the machine. However, if the frequency ratio is kept outside the critical range of 0.4 and 1.50, and the amplitude is within the permissible limits, the harmful effects are considerably reduced, especially if the system is damped.

Transmission of vibrations can be controlled and the deterimental effects considerably reduced by isolating either the source (active isolation) or by protecting the receiver (passive isolation). The following measures are generally adopted.

(1) The machine foundation should be located away from the adjoining structures. This is known as geometric isolation.

Table 28.2. Ratio of Foundation mass to Engine mass

S.No.	*Types of Engine*	*Ratio*
1.	Gas Engines	
	1—Cylinder	3.0
	2—Cylinder	3.0
	4—Cylinder	2.75
	6—Cylinder	2.25
	8—Cylinder	2.0
2.	Diesel Engines	
	2—Cylinder	2.75
	4—Cylinder	2.4
	6—Cylinder	2.1
	8—Cylinder	1.9
3.	Rotary converter	0.5 to 0.75
4.	Vertical compound steam engine coupled to generator	3.8
5.	Vertical triple-expansion steam engine coupled to generator	3.5
6.	Horizontal cross-compound coupled to generator	3.25
7.	Horizontal steam turbine coupled to generator	3.0 to 4.0
8.	Vertical gas engine coupled to generator	3.5
9.	Vertical diesel engine coupled to generator	2.6

The amplitude of surface waves (*R*-waves) reduces with an increase in distance. A considerable reduction in the amplitude is achieved by locating the foundation at a great depth, as the *R*-waves also reduce considerably with an increase in depth.

(2) Additional masses known as dampers are attached to the foundations of high frequency machines to make it a multiple degree freedom system and to change the natural frequency.

In reciprocating machines, the vibrations are considerably reduced by counterbalancing the exciting forces by attaching counterweights to the sides of the crank.

(3) Vibrations are considerably reduced by placing absorbers, such as rubber mountings, felts and corks between the machine and the base.

(4) If an auxiliary mass with a spring is attached to the machine foundation, the system becomes a two-degree-freedom system. The method is especially effective when the system is in resonance.

(5) If the strength of the soil is increased by chemical or cement stabilisation, it increases the natural frequency of the system. The method is useful for machines of low operating frequency.

(6) The natural frequency of the system is modified by making structural changes in foundation, such as connecting the adjoining foundations, changing the base area or mass of foundation or use of attached slabs.

(7) The propagation of waves can be reduced by providing sheet piles, screens or trenches.

ILLUSTRATIVE EXAMPLES

Illustrative Example 28.1. *Determine the natural frequency of a machine foundation having a base area 2 m × 2 m and a mass of 15 Mg, including the mass of the machine. Taking* $C_u = 4 \times 10^4$ *kN/m*^{3}.

Solution. From Eq. 28.46, $\omega_n = \sqrt{\dfrac{C_u A}{m}}$

Expressing C_u in N/m^3 and m in kg,

or $$\omega_n = \sqrt{\frac{4 \times 10^7 \times (2 \times 2)}{15 \times 10^3}} = 103.28 \text{ rad/sec}$$

or $$f = \frac{103.28}{2\pi} = \mathbf{16.43 \text{ cps (hertz)}}$$

Illustrative Example 28.2. *The natural frequency of a machine foundation is 4 hertz. Determine its magnification at the operating frequency of 8 hertz. Take damping factor (D) as 0.30.*

Solution. From Eq. 28.25, $$M = \frac{1}{\sqrt{(1 - r^2)^2 + 4D^2 r^2}}$$

In this case, $r = 8/4 = 2$. Therefore,

$$M = \frac{1}{\sqrt{(1 - 2^2)^2 + 4 \times (0.3)^2 \times (2)^2}} = \mathbf{0.31.}$$

Illustrative Example 28.3. *The exciting force of a machine is 100 kN. Determine the transmitted force if the natural frequency of the machine foundation is 3.0 Hz. Take D = 0.40 and the operating frequency as 5 Hz.*

Solution. From Eq. 28.29, $$|F_T| = F_o M \sqrt{1 + (2Dr)^2}$$

where $$M = \frac{1}{\sqrt{(1 - r^2)^2 + 4D^2 r^2}}$$

$$= \frac{1}{\sqrt{[1 - (5/3)^2]^2 + 4 \times (0.4)^2 \times (5/3)^2}} = 0.45$$

Therefore, $$|F_T| = 100 \times 0.45 \sqrt{1 + (2 \times 0.4 \times 5/3)^2} = 75 \text{ kN}$$

Illustrative Example 28.4. *Determine the coefficient of uniform compression if a vibration test on a block 1 m × 1 m × 1 m gave a resonance frequency of 30 Hz in the vertical direction. The mass of the oscillator used was 60 kg.*

Solution. Mass of foundation block

$$= (1 \times 1 \times 1) \times 2400 = 2400 \text{ kg}$$

$$\text{Total mass} = 2400 + 60 = 2460 \text{ kg}$$

From Eq. 28.46, $$\omega_n = \sqrt{\frac{C_u A}{m}}$$

or $$f = \frac{1}{2\pi}\sqrt{\frac{C_u A}{m}}$$

In this case, $$30 = \frac{1}{2\pi}\sqrt{\frac{C_u \times (1 \times 1)}{2460}}$$

$$C_u = 8.74 \times 10^7 \text{ N/m}^3 = 8.74 \times 10^4 \text{ kN/m}^3$$

Illustrative Example 28.5. *A 2.50 Mg vertical compressor- foundation system is operated at 40 Hz. The soil at the site is medium stiff clay (C_u = 4 × 10^4 kN/m^3). Determine the natural frequency and the magnification factor, assuming $m_s = 0.2\, m_f$. The base area is 2.5 m^2. Take D = 0.*

Solution. Total mass = 2.5 + 0.2 × 2.5 = 3.0 Mg = 3 × 10^3 kg

From Eq. 28.46, $$\omega_n = \sqrt{\frac{C_u A}{m}}$$

or $$f = \frac{1}{2\pi}\sqrt{\frac{C_u A}{m}} = \frac{1}{2\pi}\sqrt{\frac{4 \times 10^7 \times 2.5}{3 \times 10^3}} = \mathbf{29.06\ Hz}$$

From Eq. 28.25, $$M = \frac{1}{\sqrt{(1 - r^2)^2 + 4D^2 r^2}}$$

$$= \frac{1}{\sqrt{\left\{1 - \left(\frac{40}{29.06}\right)^2\right\}^2 + 4 \times 0 \times \left(\frac{40}{29.06}\right)^2}} = \mathbf{1.12}$$

Illustrative Example 28.6. *In a test block of the size 1.5 m × 1.0 m × 0.75 m, resonance occurs at a frequency of 20 cycles per second in the vertical vibration. Determine the coefficient of elastic uniform compression (C_u) if the mass of oscillator is 70 kg and the force produced by it at 15 cycles per second is 1000 N. Also compute the maximum amplitude at 15 cycles per second.*

Solution. $$\omega_n = 2\pi f_n = 2\pi \times 20 = 40\pi$$

Mass of oscillator = 70 kg

Mass of block = 1.5 × 1.0 × 0.75 × 2400 = 2700 kg

Total mass = 70 + 2700 = 2770 kg

Contact area = 1.5 × 1.0 = 1.5 m^2

From Eq. 28.46, $$\omega_n = \sqrt{C_u A/m}$$

or $$40\pi = \sqrt{C_u \times 1.5/2770}$$

or $$C_u = 29.16 \times 10^6 \text{N/m}^2 = 29.16 \times 10^3 \text{ kN/m}^3$$

From Eq. 28.47, maximum amplitude

$$z_{max} = \frac{F_o}{m\,\omega_n^2 (1 - r^2)} = \frac{1000}{2770 \times (40\pi)^2 [1 - (15/20)^2]}$$

$$= 5.23 \times 10^{-5} \text{ m} = \mathbf{0.0523\ mm}$$

Illustrative Examples 28.7. *A foundation block of weight 30 kN rests on a soil for which the stiffness may be assumed as 25000 kN/m. The machine is vibrated vertically by an exciting force of 3.0 sin (30 t) kN. Find the natural frequency, natural period, natural circular frequency and the amplitude of vertical displacement. The damping factor is 0.50.*

Solution. The exciting force $F_o \sin \omega t$ is 3.0 sin (30t).

Therefore, F_o = 3.0 kN and ω = 30 radian/second

Now from Eq. 28.43, $$\omega_n = \sqrt{k/m}$$

or $$\omega_n = \sqrt{\frac{25000 \times 10^3}{(30 \times 10^3/9.81)}} = 90.42 \text{ rad/s}$$

From Eq. 28.44, $$f_n = \frac{1}{2\pi}\sqrt{k/m} = 14.39 \text{ cycles/sec}$$

$$T = \frac{1}{f_n} = 0.069 \text{ second}$$

$$r = \omega/\omega_n = 30/90.42 = 0.33$$

From Eq. 28.25, $$M = \frac{1}{\sqrt{(1 - r^2)^2 + 4D^2 r^2}}$$

$$= \frac{1}{\sqrt{[1 - (0.33)^2]^2 + 4 \times (0.5)^2 \times (0.33)^2}} = 1.05$$

Static deflection, $\delta_{st} = F_o/k = \frac{3.0 \times 100}{25000} = \mathbf{0.012\ cm}$

Amplitude $= M\,\delta_{st} = 1.05 \times 0.012 = \mathbf{0.013\ cm}$

PROBLEMS

A. Numerical

28.1. Determine the natural frequency of a machine foundation having a base area 2 m × 2 m and a mass of 10 Mg, assuming that the soil mass participating in the vibration is (*a*) negligible (*b*) 20% of the mass.
Take $C_u = 3 \times 10^4\ kN/m^2$. **[Ans.** 17.43 cps, 15.92 cps]

28.2. Resonance occurs at a frequency of 20 cps in the vertical vibration of a test block (1m × 1 m × 1 m). Calculate the coefficient of elastic uniform compression (C_u). The mass of the oscillator is 50 kg.
[Ans. 3.87×10^4 kN/m^3]

28.3. In Problem 28.2, if the force produced by the oscillator at 10 cps was 1000 N, compute the maximum amplitude in the vertical direction at 10 cps. **[Ans.** 3.4×10^{-2} mm]

28.4. The foundation for a gas engine with a vertical cylinder and vertically oscillating parts has the following data.

Total mass of engine	= 5 Mg
Speed of rotation	= 300 r.p.m.
Mass of block	= 20 Mg
Mass of participating soil	= 25 Mg
Spring stiffness	= 60×10^4 kN/m

Determine the natural frequency and maximum amplitude. Take $D = 0.1$. The unbalanced vertical force is 12 kN.
[Ans. 17.43 cps; 0.022 mm]

B. Descriptive Type

28.5. Explain the following terms:
Natural frequency; period; resonance; magnification.

28.6. Discuss the use of single-degree-freedom system in the analysis of machine foundations. What are its limitations?

28.7. Describe the methods for the determination of the mass, spring constant, damping factor and the mass of participating soil.

28.8. Briefly explain the Barkan method of machine foundation design.

28.9. What is meant by vibration isolation ? How is it done ?

28.10. Discuss criteria for the design of foundation in the following cases.
(*a*) Free vibration without damping.
(*b*) Free vibration with damping.
(*c*) Forced vibration without damping.
(*d*) Forced vibration with damping.

C. Multiple Choice Questions

1. If the circular frequency (ω) is 30 π radians per second, the time period is
(*a*) 1 second (*b*) 2 seconds
(*c*) 4 seconds (*d*) none of above.

2. The frequency of a system increases with
(*a*) an increase in stiffness of system

(*b*) a decrease in the mass of system
(*c*) both (*a*) and (*b*)
(*d*) neither (*a*) nor (*b*)

3. Magnification factor (*M*) is very high if the value of frequency ratio (*r*) is
(*a*) less than 0.4
(*b*) between 0.4 and 1.50
(*c*) greater than 1.50
(*d*) unity.

4. For most soils, the limiting amplitude for low speed machines is usually
(*a*) 0.1 mm (*b*) 0.2 mm
(*c*) 0.5 mm (*d*) 1.0 mm

5. For diesel engines, the ratio of foundation mass to engine mass is
(*a*) less than 2.0 (*b*) between 2.0 and 3.0
(*c*) between 3.0 and 4.0 (*d*) greater than 4.0

[**Ans.** 1. (*c*), 2. (*c*), 3. (*b*), 4. (*b*), 5. (*b*)]

Objective type

Write whether the following statements are true or false

(*a*) The speed of turbines and rotary compressors is usually less than 600 r.p.m.
(*b*) For lathes, generally block-type foundation is provided.
(*c*) In general, a rigid block foundation has 6 degrees of freedom.
(*d*) The stresses in the soil below the machine foundation should normally be less than 50% of the allowable stress under static loads.
(*e*) The minimum quantity of reinforcement in machine foundation is 15 kg/m^3.
(*f*) The minimum cover for reinforcement at the bottom of the machine foundation is 75 mm.
(*g*) The depth of ground water table below the machine foundation should not be less than the width of foundation.
(*h*) The machine foundations are taken lower than the level of the foundation of adjacent buildings.

[**Ans.** True (*c*), (*f*), (*h*)]

29

Pavement Design

29.1. TYPES OF PAVEMENTS

A pavement is a hard crust constructed over the natural soil for the purpose of providing a stable and even surface for the vehicles. The pavement supports and distributes the wheel loads and provides an adequate wearing surface. Pavement are basically of 3 types : (1) Flexible pavements, (2) Rigid pavements and (3) Semi-flexible pavements.

(1) Flexible Pavement. The flexible pavement is built up in several layers, as shown is Fig. 29.1 (*a*). The natural soil beneath the pavement is known as subgrade. Sub-base is built over the sub-grade and the base is constructed over the sub-base. The top layer is known as surfacing, which is usually bitumen.

The flexible pavement can resist only very small tensile stresses because of limited rigidity. Any derformation of the sub-grade results in a corresponding change in the surface of the pavement.

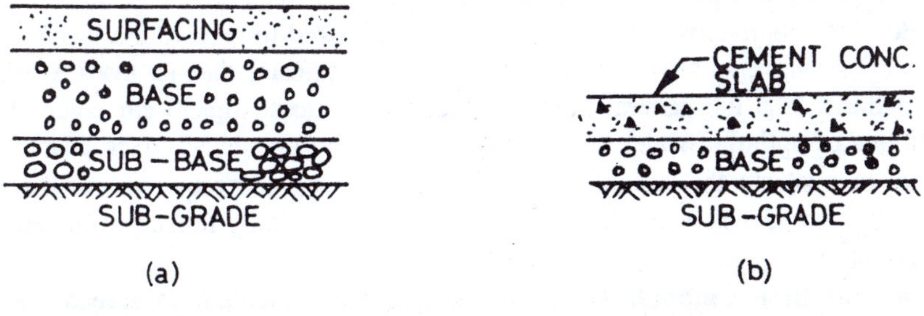

Fig. 29.1 (*a*) Flexible Pavement. (*b*) Rigid Pavement.

(2) Rigid Pavement. The rigid pavement is made of cement concrete. As the concrete layer is quite strong, the sub-base may not be required [Fig. 29.1 (*b*)]. The rigid pavements have high flextural strength and can resist very high tensile stresses. The pavements are capable of bridging small depressions in the subgrade.

(3) Semi-flexible pavement. A semi-flexible pavement has flexural rigidity in-between that of a rigid pavement and a flexible pavement. Such pavements are usually made of pozzolanic concrete, lean concrete, or soil-cement in the base course or sub-base. As the flexural strength of such layers is limited, the pavement can resist only moderate tensile stresses.

The chapter deals with the methods of construction and the design of various types of pavements. As the space is limited, the treatment is necessarily elementary. For more details, a reference may be made to advanced works on pavement design.

29.2. BASIC REQUIREMENT OF PAVEMENTS

The basic requirement of a good pavement is to provide a stable, non-yielding surface for the movement of heavy vehicles. The pavement surface should be even along the longitudinal profile to have least rolling resistance so that fast, heavy vehicles can move safely and comfortably. Unevenness of the surface causes vertical oscillations and thus increases wear and tear and the fuel consumption of vehicles. It also adds to the discomfort and fatigue to the passengers of fast moving vehicles.

The pavement carries the wheel loads and distributes it over a wide area on the subgrade. Consequently, the stresses transferred to the subgrade are considerably smaller than the contact pressure. The distribution of the load depends upon the thickness and the characteristics of the material used in the pavement. A pavement layer that distributes the wheel load through the largest area per unit thickness of the layer is the most efficient.

The pavement layers should not be over-stressed. Even for a well- designed pavement, there would be a small, temporary deformation when loads pass over it. These deformations must be kept within the permissible limits. If the pavement is not properly designed, repeated applications of loads may cause excessive deformations, compaction and consolidation of the subgrade and even failure of the whole pavement.

The pavement may be constructed over an embankment or in a cutting. The pavement should be constructed above the maximum level of the ground water table to keep it dry. Moisture variation and frost action are the main causes of deterioration of the subgrade. On the other hand, when the water content is decreased, shrinkage cracks develop, which cause differential movement in rigid pavements and cracks in flexible pavements. The pavements should be provided with a suitable drainge system.

In case of rigid pavements, the temperature and shrinkage stresses should be properly controlled.

In case of flexible pavements, there should be a good bond between the individual particles of the materials used and also between the surfacing and the base to check stripping or breaking up of the pavement.

29.3. FUNCTIONS OF DIFFERENT COMPONENTS OF A PAVEMENT

Different components of a pavement have the following characteristics and functions.

(1) Subgrade. The subgrade is a layer of natural soil prepared to receive the layers of the pavement. The subgrade should be strong enough to take up the stresses imposed due to loads without shear failure or excessive deformations. It is the general practice to compact atleast top 50 cm layer of the subgrade under controlled conditions of optimum water content at the maximum dry density (see chapter 14).

It is essential to evaluate the strength properties of the subgrade required. As the loads are ultimately received by the subgrade, if it is weak, it will fail. The soil is generally treated to increase its strength and to improve its properties.

(2) Sub-base and Base Courses. These courses provide a medium to spread the wheel loads to the subgrade. The courses usually consist of broken stones, bricks or aggregates. Boulder stones, bricks on edges and stabilised soils are also used for sub-bases. However it is preferable to use small size graded aggregates because large stones and bricks have a tendency to penetrate the wet soil and cause undulation and unevenness in the pavement. As the stresses in a sub-base are much lower than those in the base, the material used is inferior to that in the base. The sub-base is also known as *soling.*

The base and sub-base in a flexible pavement improve the load- supporting capacity of the subgrade by distributing the load over a large area. In a rigid pavement, the base course helps in preventing pumping out of the soil from below. These also protect the subgrade against frost action.

(3) Surface Course. The purpose of the surface course, also known as a wearing course, is to give a smooth riding surface and to resist pressure exerted by wheels. The surface course also provides a water-tight barrier against the infiltration of surface water.

In flexible pavements, a surface course usually consists of a bituminous surfacing. In rigid pavements, the cement concrete may act as a base course as well as a surface course.

There are various types of surface treatments, depending upon the availability of materials, and plants and the magnitude of the load.

29.4. FACTORS AFFECTING PAVEMENT DESIGN

The factors affecting the design of pavements are discussed below.

(1) Wheel Load. The thickness of the pavement depends upon the design wheel load. The design wheel load is selected after considering the actual wheel loads of the various vehicles, contact pressure, load repetition, the dynamic effects and many other factors. The design wheel load will depend upon the multiple-wheel load assembly, as in the dual or multiple wheel loads.

(2) Subgrade. The thickness of the pavement depends upon the properties of the subgrade. A thicker pavement is required over a weaker soil. The thickness also depends upon the stress-strain characteristics of the soil under static and repeated loads.

As the strength and the volume change of the soil depend upon the moisture changes, the worst conditions should be considered in the design.

(3) Climate. The climatic factors, such as rainfall and temperature changes, affect the properties of the soil. The rainfall affects the moisture changes in the soil which affect the strength. The temperature changes affect the pavement. In flexible pavements, the choice of the bituminous binder depends upon the temperature. In rigid pavements, the warping stresses are caused by temperature changes.

If the temperature is likely to fall to the freezing point, the possibility of the frost action is to be considered.

(4) Pavement materials. The stress distribution in the pavement layers depends upon the behaviour of the materials used. The fatigue behaviour of the pavement materials and their durability under adverse conditions should be considered.

Aggregates bear the stresses occurring in the pavement and have to resist wear due to abrasive action of traffic. The aggregate should be hard, strong and of the required size and gradation to bear the stresses.

(5) Location. The design of the pavement should be done considering its location with respect to the ground surface. The height of embankment, the depth of cutting and the level of water table should also be considered. All these factors affect the performance of the pavement.

(6) Miscellaneous factors. There are some special factors which affect the design of pavements. For example, in case of semi-rigid pavements, the formation of shrinkage cracks, the crack pattern and the mode of their propagation should be considered in its design.

29.5. CALIFORNIA BEARING RATIO TEST

California bearing Ratio (CBR) test is a type of test developed by the California Division of Highways in 1929. The test is used for evaluating the suitability of subgrade and the materials used in sub-base and base courses. The test results have been correlated with the thickness of the various materials required for flexible pavements.

The test may be conducted on a prepared specimen in a mould or on the soil in-situ condition. The laboratory CBR apparatus consists of a mould 150 mm diameter and 175 mm high, having a separate base plate and a collar (Fig. 29.2). The load is applied by a loading frame through a plunger of 50 mm diameter. Dial gauges are used for measurement of the expansion of the specimen on soaking and for measurement of penetration.

It may be noted that with the displacer disc inside the mould, the effective height of the mould is only 125mm.

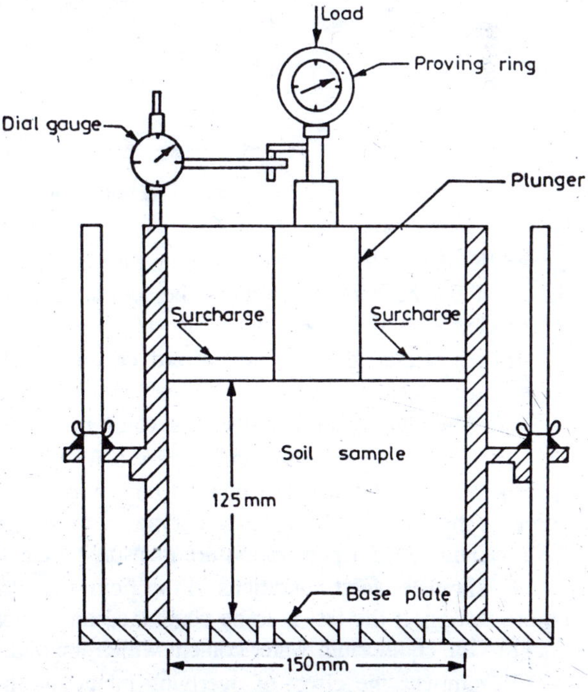

Fig. 29.2. C.B.R. Test Appratus.

The test consists of causing the plunger to penetrate the specimen at the rate of 1.25 mm per minute. The loads required for a penetration of 2.5 mm and 5.0 mm are recorded by a proving ring attached to the plunger. The load is expressed as a percentage of the standard load at the respective deformation level, and is known as the CBR value. Standard load values are for crushed stone and are given in Table 29.1. The CBR value is determined corresponding to both 2.5 mm and 5.0 mm penetration, and the greater value is used for the design of flexible pavement.

$$\text{CBR value} = \frac{\text{Test load}}{\text{Standard load}} \times 100 \qquad \text{...(29.1)}$$

Table 29.1. Standard Load Values

S. No.	*Penetration (mm)*	*Standard Load (kN)*
1.	2.5	13.44
2.	5.0	20.16
3.	7.5	25.80
4.	10.0	31.20
5.	12.5	35.32

The load-penetration curve is drawn as shown in Fig. 29.3. The load corresponding to 2.5 mm and 5.0 mm penetration values are taken from the plot curve (1), and the value of CBR determined using Eq. 29.1.

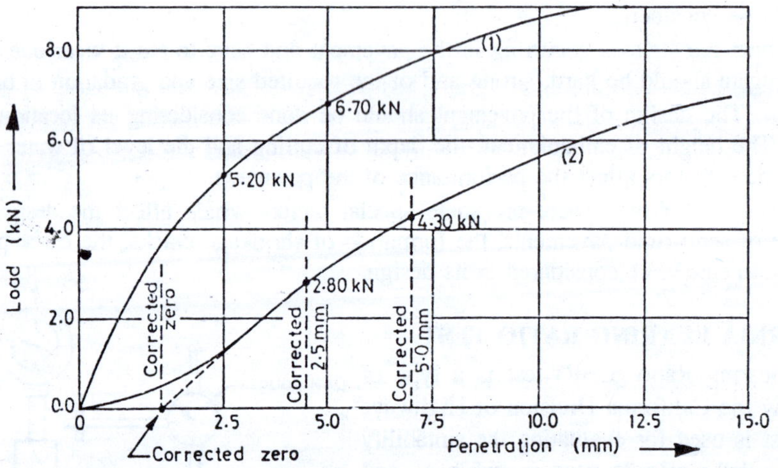

Fig. 29.3.

Generally, the CBR for 2.5 mm penetration is high. However, if the CBR for 5.0 mm penetration is greater than that for 2.5 mm penetration, the test is repeated. If the results are unchanged, the value for 5.0 mm penetration is used for defining CBR value.

In some tests, there is an upward concavity of the load- penetration curve. This may be due to the following reasons.

(*i*) The plunger surface does not come in full contact with the top of specimen.

(*ii*) The top layer of the soil is very soft.

In the case of the initial upward concavity, the corrected zero is obtained by drawing a tangent to the curve at the point of the greatest curvature, as shown for curve-2 in Fig. 29.3. The points corresponding to 2.5 mm and 5.0 mm penetration are measured from the corrected zero.

To simulate worst conditions in the field, the soil specimen is kept submerged in water for about 4 days before testing. If the test is to be conducted on an unsoaked specimen, the moulding water content should be equal to the equilibrium water content which the soil is likely to attain after the construction of the pavement.

To simulate the effect of overlying materials, the specimen is covered with surcharge mass in the test.

The minimum surcharge mass for the tests in the mould is specified as 4.5 kg. The annular surcharge mass of 147 mm diameter are used for this purpose.

[See Chapter 30 (Sect. 30.19) for laboratory experiment].

29.6. DESIGN OF FLEXIBLE PAVEMENTS

A flexible pavement consists of a number of layers in which the stress is transmitted by point-to-point contact. The maximum intensity of stress occurs in the top layer of the pavement. Therefore, the superior materials are used in the top layer.

There is no perfect rational method for the design of flexible pavements. The design methods can be classified broadly as empirical or semi-empirical.

1. Empirical methods

(*a*) Group index method. (*b*) California Bearing Ratio Method, (*c*) California R-value Method, (*d*) McLeod Method.

2. Semi empirical methods

(*a*) Triaxial test Method, (*b*) Burmister Method.

All these methods are discussed briefly in the following sections.

29.7. GROUP INDEX METHOD

The group index (GI) is a parameter used in the classification of soils by AASHTO system (Chapter 5). The group index is used in the grading of soils. The higher the value of the group index, the poorer is the subgrade.

In the group index method (Steel, 1945), the thickness of the base and surfacing is related to the volume of the traffic. Depending upon the number of vehicles, the traffic volume is divided into 3 categories.

Light volume : Less than 50 vehicles/day

Medium volume : 50 to 300 vehicles/day

Heavy volume : over 300 vehicles/day

To determine the thickness of the pavement by this method, the group index and the anticipated traffic volumes are found. The appropriate design curve (*B, C* or *D*) is used according to the traffic volume (Fig. 29.4) and the total thickness of the pavement (surfacing, base and sub base) is determined. Curve *A* gives the thickness of the sub-base required. Curve *E* gives the additional base thickness which may be substituted for sub-base thickness of curve *A*.

The method is essentially empirical, which assumes that the soil with identical group indexes possess

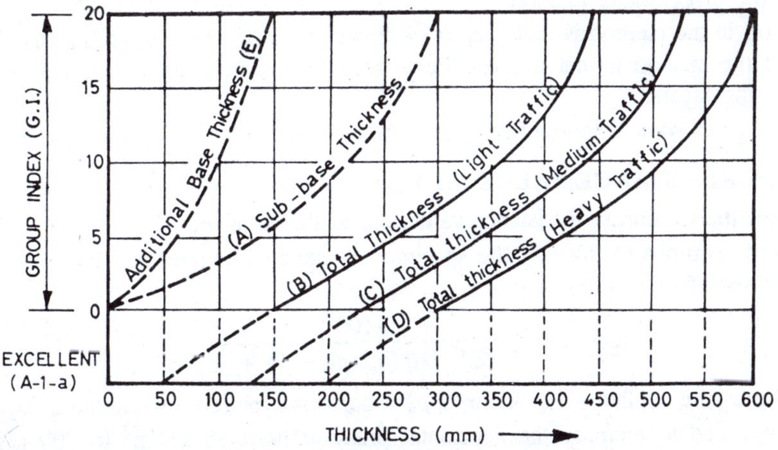

Fig. 29.4.

identical strength after compaction in the field. The method assumes that the subgrade would be compacted to a density not less than 95% of the maximum Proctor density. For base and sub-base materials, the density should not be less than 100% of the maximum Proctor density. The water table should be at least 1 m below the surface.

29.8. CBR METHOD

In this method, the CBR values are used to determine the total thickness of the flexible pavement and the thickness of various layers. Fig. 29.5 give the design curves for different wheel loads and traffic conditions. The design curves are based on the data collected on a large number of pavements which performed satisfactorily. The curves give the required thickness of construction above a material of a certain CBR value. As it is evident, the required thickness of construction above a material decreases as the CBR value increases.

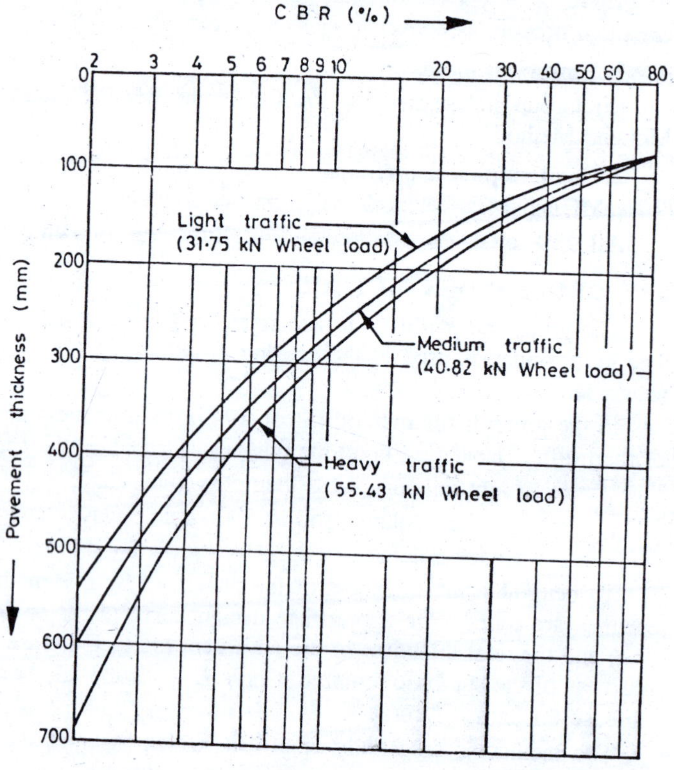

Fig. 29.5.

The Indian Road Congress (IRC) has recommended the design chart given in Fig. 29.6. The chart is similar to one used in U.K. The soaked CBR value of the subgrade is evaluated and the volume of the traffic is estimated. The total thickness of the pavement is determined using the appropriate curve. Likewise, the CBR value of the sub-base material is used to determine the thickness of construction over that material. Obviously, the thickness of the sub-base is equal to the total thickness above the subgrade minus the thickness of construction above the sub-base Likewise, the thickness of the base is determined.

The CBR method is based on strength parameter of the material and is, therefore, more rational than the group index method. The basic assumption in the method is that a layer of pavement is of superior quality than the layer below it. The shortcoming of the method is that it gives the same total thickness above a material irrespective of the quality of the overlying layers.

IRC : 37—1970 gives some important recommendations for the determination of CBR value and its use.

29.9. CALIFORNIA RESISTANCE VALUE METHOD

The method uses the California Resistance value, called the *R*-value.

The R-value is determined by placing the specimen in the stabilometer and by applying the lateral and vertical pressures as specified. The *R*-value is given by

$$R = 100 - \frac{100}{(2.5/D_2)(p_v/p_h - 1) + 1} \qquad \text{...(29.2)}$$

where p_v = vertical pressure applied (1120 kN/m^2), p_h = horizontal pressure transmitted, D_2 = displacement of stabilometer fluid required to increase the horizontal pressure from 35 kN/m^2 to 700 kN/m^2, measured in number of revolutions of the calibrated pump handle.

Hveem and Carmany (1948) gave the following expression for the total thickness of the pavement

$$t = \frac{KT(90 - R)}{C^{1/5}} \quad \ldots(29.3)$$

where t = thickness (cm), K = numerical constant (= 0.166), T = traffic index, R = stabilometer resistance value, C = cohesiometer value, determined from the cohesiometer test.

The traffic index T has been empirically provided to estimate the traffic volume, as

$$T = 1.35\,(EWL)^{0.11} \quad \ldots(29.4)$$

where EWL = equivalent wheel load.

29.10. MCLEOD METHOD

McLeod gave the following formula for the thickness.

$$t = K \log_{10}(P/S) \quad \ldots(29.5)$$

where t = thickness of gravel base (cm), P = gross wheel load (kN), S = total subgrade support (kN), K = base course constant (varying non-linearly from 75 for bearing plate of diameter of 12.5 cm to 210 for diameter of 125 cm)

The repetitive plate bearing tests are conducted using plates of different sizes.

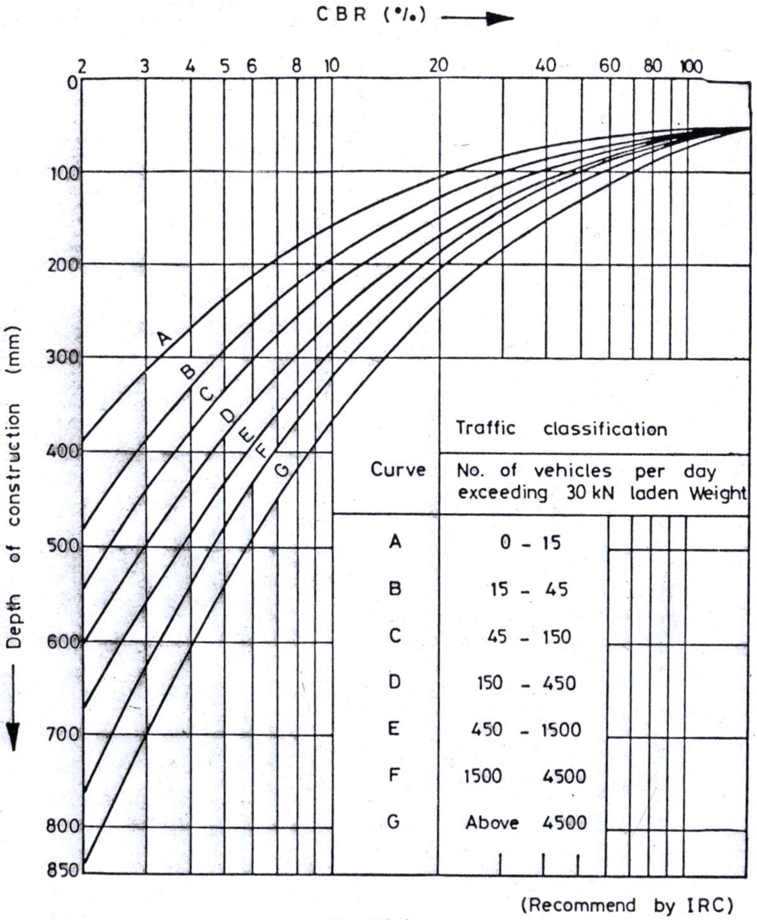

Fig. 29.6.

The subgrade support (S) is determined from the unit support measured or computed for 30 cm diameter plate at 0.5 cm deflection and ten repetitions. Fig. 29.7 can be used for determining the ratio (r) of the unit sub-grade support (S_2) for the

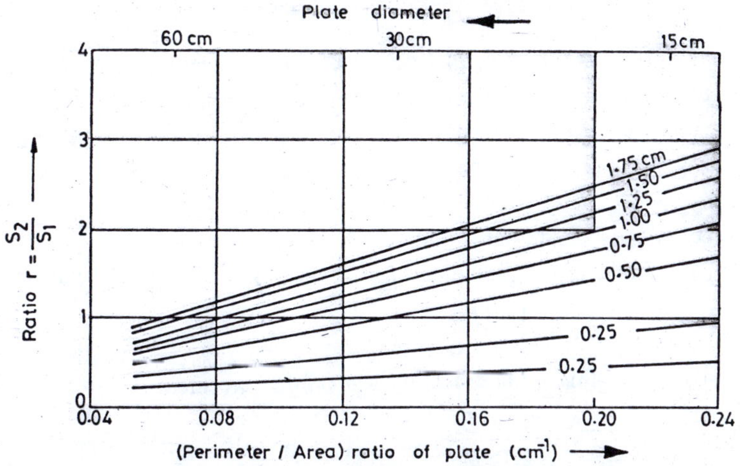

Fig. 29.7.

design wheel diameter to S_1 that on 30 cm diameter plate. The curves for different deflections between 0.25 cm and 1.75 cm are shown. Thus

$$S_2 = r\,S_1 \quad \ldots(29.6)$$

The design subgrade support S is obtained by multiplying the value of S_2 by the contact area of the wheel. Thus

$$S = S_2 \times a \quad \ldots(29.7)$$

where a = contact area of the wheel (cm^2).

29.11. TRIAXIAL TEST METHOD

The triaxial compression test is conducted on a soil specimen under a lateral pressure of 140 kN/m^2 and the value of the modulus of elasticity is determined from the stress-strain curve.

The thickness of the pavement in cm can be determined using the formula given by the Kansas Highway Deptt.

$$t = \sqrt{\left(\frac{3\,P\,x\,y}{2\,\pi\,E\,\Delta}\right)^2 - a^2}$$

where P = wheel load (kN), E = modulus of elasticity (kN/cm^2), x = traffic coefficient (0.5 to 2.0), y = saturation coefficient (= 0.5 to 1.0), a = area of contact (cm^2), Δ = design deflection (= 0.25 cm).

29.12. BURMISTER'S METHOD

Burmister developed a method considering the pavement as a layered system. The Burmister theory is based on the following assumptions.

(1) The material in each layer is homogeneous, isotropic and elastic.

(2) The surface layer is infinite in the horizontal direction and finite in the vertical direction. The underlying layer is infinite in both directions.

(3) The layers are in continuous contact.

(4) The top layer is free of shearing stresses and normal stresses outside the loaded area.

The displacement equation given by Burmister can be written as, assuming $\mu_s = \mu_p = 0.50$,

$$\Delta = 1.5\,\frac{p\,a\,F_2}{E_s} \text{ for flexible plates} \quad \ldots(29.9)$$

and

$$\Delta = 1.18\,\frac{p\,a\,F_2}{E_s} \text{ for rigid plates} \quad \ldots(29.10)$$

where p = uniform pressure, a = radius of plate, F_2 = deflection factor, and E_s = modulus of the soil.

Fig. 29.8 gives the values of the deflection factor F_2. It depends upon the ratio of the modulus of the subgrade (E_s) and the modulus of pavement layer E_p. The ratio (h/a) is equal to the thickness of the base layer divided by the radius of the load. It may be noted that for a single layer, $h = 0$ and $E_s/E_p = 1.0$, and hence $F_2 = 1.0$ and the solution reduces to Boussinesq's settlement equation.

The design procedure can be summarised as under.

(1) Determine E_s by conducting a plate bearing test on a 30 cm diameter plate over the subgrade.

$$E_s = \frac{1.18\,pa}{\Delta} \quad \ldots(29.11)$$

(2) Calculate F_2 by conducting a plate bearing test on the pavement,

$$F_2 = \frac{\Delta \times E_s}{1.18\,pa} \quad \ldots(29.12)$$

(3) For the computed value of F_2 and the given value of (h/a) ratio, determine E_s/E_p from Fig. 29.8.

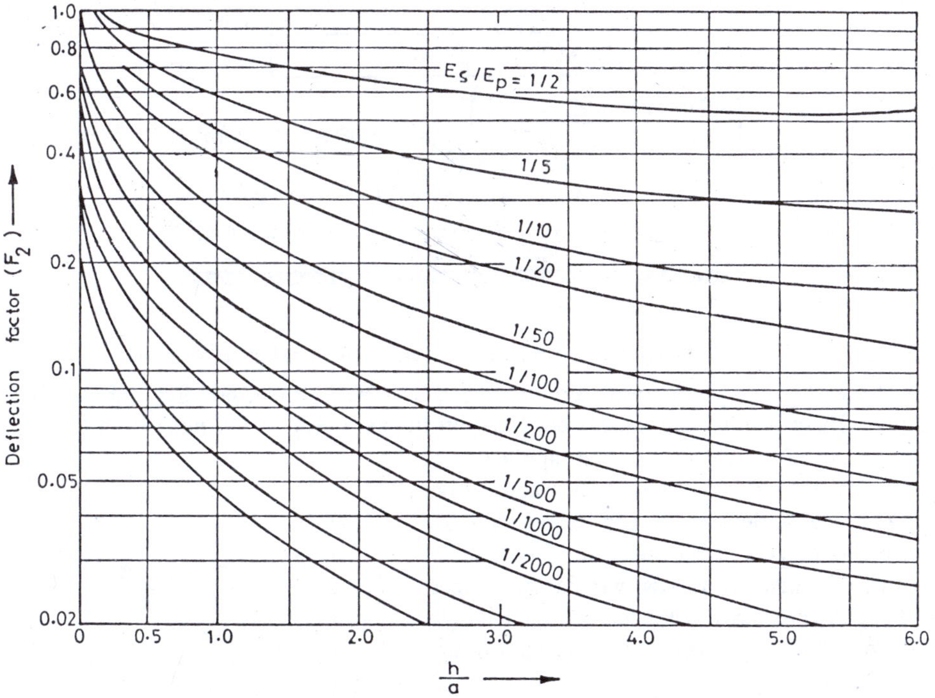

Fig. 29.8.

(4) Determine the contact radius of the design load, as

$$a = \sqrt{\frac{P}{\pi p}} \qquad \ldots(29.13)$$

where P = design load and p = tyre pressure.

(5) Compute the new value of F_2 for the design deflection Δ (= 0.5 cm or 0.25 cm).

$$F_2 = \frac{\Delta \times E_s}{1.5\,pa} \qquad \ldots(29.14)$$

(6) For the computed values of F_2 and E_s/E_p ratio, determine the h/a ratio from Fig. 29.8.

(7) Determine h from the h/a ratio.

29.13. COEFFICIENT OF SUBGRADE REACTION

For the design of rigid pavements, the coefficient of subgrade reaction is required. As discussed in chapter 24, the coefficient of subgrade reaction (k) is defined as the pressure per unit deformation of the subgrade at a specified deformation or pressure level. Thus

$$k = \frac{p}{\Delta} \qquad \ldots(29.15)$$

where k = modulus of subgrade reaction (kN/cm^3), p = pressure (kN/cm^2), and Δ = deformation (cm)

[**Note.** In this chapter, k_s is also used for k].

The test set-up for the determination of the coefficient subgrade reaction is similar to that in a plate load test (chapter 23). It consists of a loading frame which has a hydraulic jack, a proving ring and a reaction beam (Fig. 29.9). The dial gauges are used for the measurement of the settlement of the plate. These gauges rest on a separate datum frame. The standard size of the plate is 75 cm diameter.

The test site is levelled and the plate is properly seated on the prepared surface. The stiffening plates of

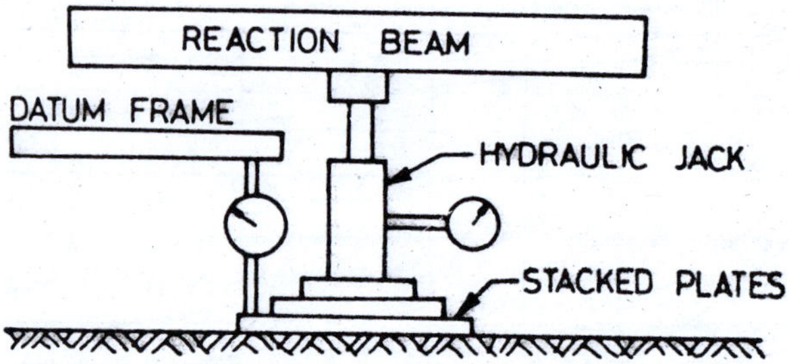

Fig. 29.9. Set-up for Determination of Coefficient of Subgrade Reaction.

progressively decreasing diameters are placed over the plate. The jack and proving ring assembly are fitted to provide reaction. Three or four dial gauges are fixed on the periphery of the plate from the datum frame.

A seating load of 7 kN/m^2 is first applied and released after a few seconds. The load is then applied on the plate to cause a settlement of approximately 0.25 mm. When there is no perceptible increase in settlement or when the rate of settlement is less than 0.025 mm per minute, the dial gauge readings are taken and the average settlement is found. The procedure is repeaed till the settlement is about 0.175 cm.

A graph is plotted between the mean bearing pressure (p) and the settlement (Fig. 29.10). The pressure corresponding to a settlement of 0.125 cm is read from the plot, and the value of k is determined as

$$k = \frac{p}{0.125} \text{ kN/cm}^3 \quad \ldots(29.16)$$

or

$$k = \frac{p \times 10^6}{0.125} \text{ kN/m}^3 \quad \ldots[29.16(a)]$$

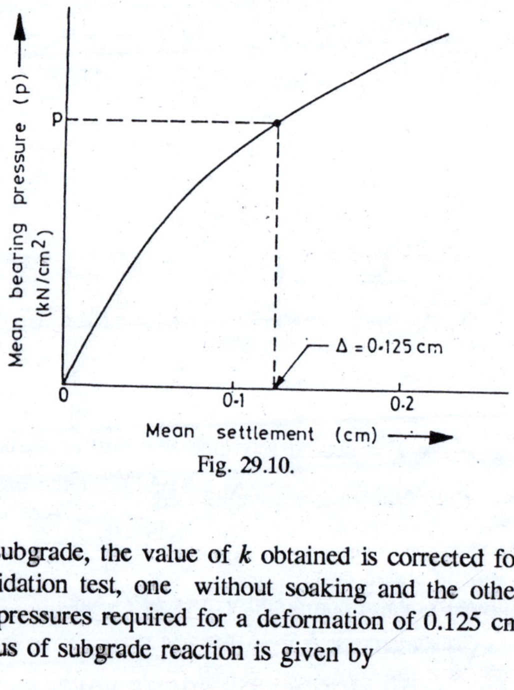

Fig. 29.10.

To allow for the effect of possible future softening of subgrade, the value of k obtained is corrected for full saturation. Two specimens are prepared for the consolidation test, one without soaking and the other fully soaked. The consolidation tests are conducted and the pressures required for a deformation of 0.125 cm are determined for the two specimens. The corrected modulus of subgrade reaction is given by

$$k = k\,(\text{test}) \times \frac{p_s}{p} \quad \ldots(29.17)$$

where p = pressure for the unsoaked specimen,

and p_s = pressure for the soaked specimen.

29.14. WESTERGAARD'S ANALYSIS

The rigid pavements are constructed of cement concrete, which may or may not have reinforcement. These pavements develop slab action due to their high tensile strength. The stresses in the subgrade below a rigid pavement are greatly reduced due to the load-spreading capacity. The failure of rigid pavements usually occurs by overstressing of the concrete and not by overstressing of the subgrade. The thickness determination is generally based on the calculation of stresses in the concrete.

The stresses in the concrete slab due to wheel loads are determined using Westergaard's theory. Westergaard considered the rigid pavement as a thin elastic plate resting on soil subgrade. The upward reaction at any point is assumed to be proportional to the deflection at that point. The slab deflection depends upon the stiffness of the subgrade and the flexural strength of the slab. Thus the pressure-deformation characteristics of a rigid pavement depend upon the relative stiffness of the slab and the subgrade. Westergaard defined the radius of relative stiffness as

$$l = \left[\frac{Eh^3}{12(1-\mu^2)k}\right]^{1/4} \quad ...(29.18)$$

where l = radius of relative stiffness (cm), E = modulus of elasticity of cement concrete (kN/cm^2), μ = Poisson's ratio of cement concrete, h = slab thickness (cm), k = modulus of subgrade reaction (kN/cm^3).

As the pavement slab has a finite length and width, the intensity of maximum stress induced due to a wheel load depends upon the location of the load. The following three locations are critical.

(1) Interior loading. When the load is applied in the interior of the slab at any point away from all edges.

(2) Edge loading. When the load is applied on an edge of the slab at any point away from a corner.

(3) Corner loading. When the centre of the load is located on the bisector of the corner angle formed by two intersecting edges of the slab, and the loaded area is at the corner touching the edges.

Westergaard gave the following equations for the critical stresses based on the assumption that the cement concrete slab is homogeneous and has uniform elastic properties. It is further assumed that the vertical subgrade reaction is proportional to the deflection.

Interior loading $$\sigma_i = \frac{0.316P}{h^2}[4\log_{10}(l/b) + 1.069] \quad ...(29.19)$$

Edge loading $$\sigma_e = \frac{0.572P}{h^2}[4\log_{10}(l/b) + 0.359] \quad ...(29.20)$$

Corner loading $$\sigma_c = \frac{3P}{h^2}\left[1 - \left(\frac{a\sqrt{2}}{l}\right)^{0.6}\right] \quad ...(29.21)$$

where P wheel load (kN), h = slab thickness (cm), l = radius of effective stiffness (cm), a = radius of wheel load distribution (cm), b = equivalent radius of resisting action (cm).

The value of b is given by $b = \sqrt{1.6a^2 + h^2} - 0.675h$ for $a < 1.724h$ ...[29.22(*a*)]

and $b = a$ for $a \geq 1.724h$ ...(29.22(*b*)]

The above equations require trial and error solution if the slab thickness h is to be found for the given allowable values of stresses. Bradbury gave a simplified solution in the form of charts.

Westergaard's equations as given above have been modified by various investigators. The stresses at the edge and the corner are generally found to be critical for the design of rigid pavements. IRC recommends the following equation for the determination of stresses at the edges and the corners.

For edge loading (Teller and Sutherland, 1943)

$$\sigma_e = 0.529\frac{P}{h^2}(1 + 0.54\mu)[4\log_{10}(l/b) + \log_{10}b - 0.4048] \quad ...(29.23)$$

For corner loading (Kelley, 1939)

$$\sigma_c = \frac{3P}{h^2}\left[1 - \left(\frac{a\sqrt{2}}{l}\right)^{1.2}\right] \quad ...(29.24)$$

The above equations are presented in the form of stress charts (Figs. 29.11 and 29.12). The charts given are applicable for a particular set of design parameters, viz; P = 41 kN; a = 15 cm; $E = 3 \times 10^3$ kN/cm^2; and μ = 0.15. However, different curves are given for different value of the coefficient of subgrade reaction (k_s) between 60 N/cm^3 and 300 N/cm^3. The design curves are for slab thickness from h = 15 to 25 cm.

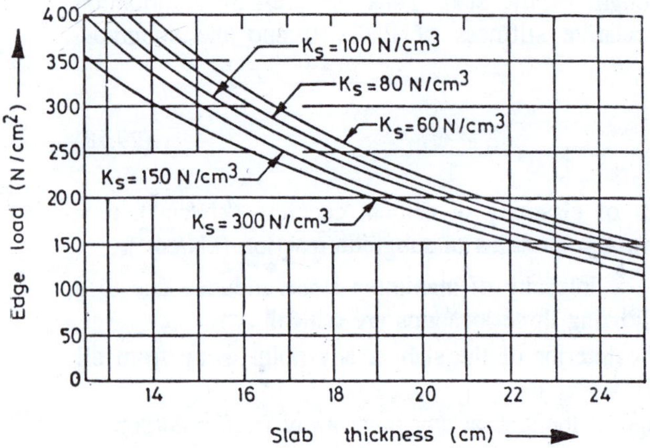

Fig. 29.11.

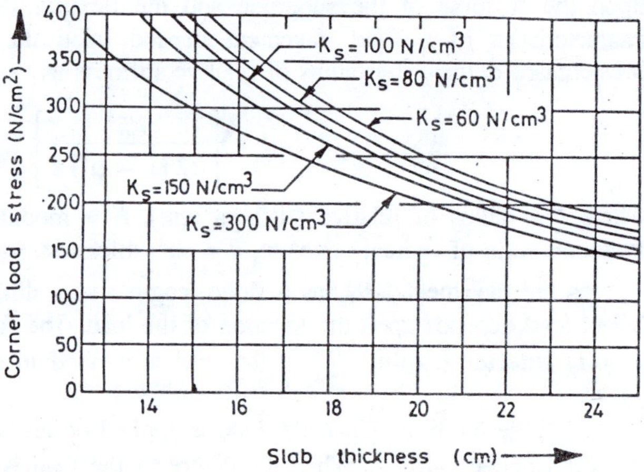

Fig. 29.12.

29.15. TEMPERATURE STRESSES IN RIGID PAVEMENTS

A cement concrete slab expands when the temperature increases and contracts when the temperature decreases. If the slab is free to change its length, there would be no stresses due to temperature changes. In rigid pavements, the slab is restrained against free deformation either by its own weight or by action of frictional forces. Consequently, the stresses are induced due to temperature changes. The stresses due to temperature changes can be divided into two categories.

(1) Warping stresses, (2) Frictional stresses.

1. Warping Stresses. These stresses develop due to differential changes in the top and bottom surfaces of the slab. The temperature change that occurs on the top surface is not transmitted immediately throughout the entire thickness. The temperature difference causes different changes in the top and bottom surfaces and a warping or bending in the slab occurs during this period.

Bradburry gave the following equations for warping stresses.

At interior, $$\sigma_t(i) = \frac{E\,\alpha\,t}{2}\left(\frac{C_x + \mu C_y}{1-\mu^2}\right) \qquad ...(29.25)$$

At edges, $$\sigma_t(e) = \frac{C_x\,E\,\alpha\,t}{2} \quad \text{or} \quad \sigma_t(e) = \frac{C_y E\,\alpha\,t}{2}, \text{ whichever is higher.} \qquad ...(29.26)$$

At corners, $$\sigma_t(c) = \frac{E\,\alpha\,t}{3(1-\mu)}\sqrt{a/l} \qquad ...(29.27)$$

where α coefficient of linear expansion (10×10^{-6} per °C) and C_x and C_y are warping stress coefficients (Fig. 29.13).

2. Frictional stresses. If the temperature of the concrete remains constant for a long time, the temperatures of the top surface and the bottom surface become equal. At that stage, there is a uniform lengthening or shortening of the slab and there is no warping. As the slab is in contact with the subgrade, the slab movements are restrained due to the friction between the bottom surface of the pavement and the subgrade.

The frictional stresses are given by

$$\sigma_f = \frac{\gamma_c L f}{2 \times 10^4} \qquad ...(29.28)$$

where σ_f = frictional stress (kN/cm^2), L = length of the slab (m). γ_c = unit weight of concrete (kN/cm^3), f = coefficient of subgrade restraint (~ 1.5).

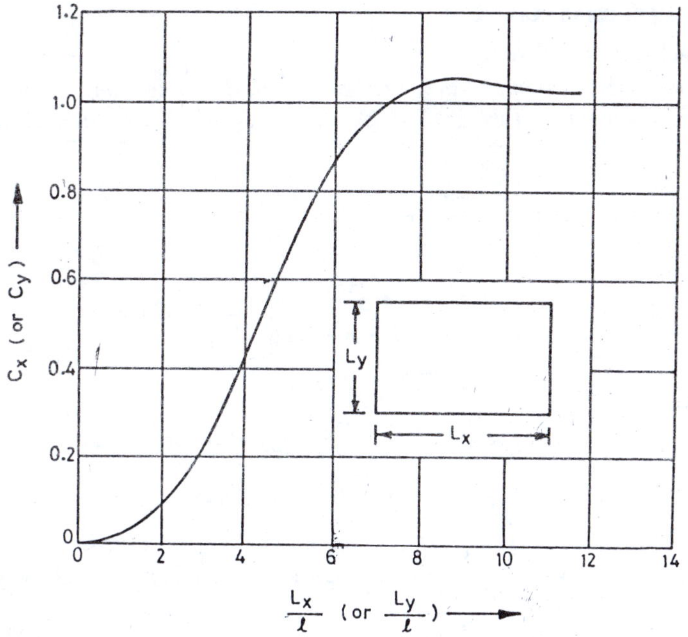

Fig. 29.13.

To reduce temperature stresses, expansion joints, contraction joints and warping joins are provided in the pavement.

29.16. COMBINED STRESSES IN RIGID PAVEMENTS

For proper design of a rigid pavement, it is necessary to consider the most critical combinations of the stresses due to loads and temperature stresses.

(1) During summer, the critical combination for interior and edge regions occurs during mid-day when the slab tends to warp downward.

Maximum stress at the bottom fibre = load stress + warping stress — frictional stress ...(29.29)

(2) However, during winter, the critical combination for the above regions occurs when the slab contracts and the slab warps downward during mid-day.

Maximum stress at the bottom fibre = load stress + warping stress + frictional stress ...(29.30)

Generally, Eq. 29.29 governs, as the differential temperatures are more in summer than in winter.

(3) For corner regions, the most critical combination occurs during the midnight when the slab warps upward.

Maximum stress at top fibre = load stress + warping stress ...(29.31)

ILLUSTRATIVE EXAMPLES

Illustrative Example 29.1. *A sample of subgrade has a group index of 8. Design the pavement for the anticipated traffic volume of over 300 commercial vehicles per day.*

Solution. As the traffic is heavy, curve D of Fig. 29.4 is applicable.

Total thickness = 500 mm

From Curve A, the thickness of sub-base = **200 mm**

Therefore, thickness of base and surfacing = 500 – 200 = **300 mm.**

Illustrative Example 29.2. *The CBR value of subgrade is 10%. Determine the total thickness of the pavement using IRC method. There are more than 4500 vehicles per day.*

Solution. From Fig. 29.6, using curve *G*,

total thickness = **360 mm**

Illustrative Example 29.3. *Determine the stresses at interior, edge and corner regions of a rigid pavement using Westergaard's method. Take* $P = 41$ *kN;* $E = 3 \times 10^3$ *kN/cm²,* $h = 20$ *cm,* $\mu = 0.15$, $k = 40$ *N/cm³ and* $a = 15$ *cm.*

Solution. From Eqs. 29.18, $l = \left[\dfrac{Eh^3}{12(1-\mu^2)k}\right]^{1/4}$

or $$l = \left[\frac{3 \times 10^3 \times (20)^3}{12(1-0.15^2) \times 0.04}\right]^{1/4} = 84.57 \text{ cm}$$

From Eq. 29.22 (*a*), $$b = \sqrt{1.6a^2 + h^2} - 0.675h$$

$$= \sqrt{1.6 \times 15^2 + (20)^2} - 0.675 \times 20 = 14.07 \text{ cm}$$

From Eq. 29.19, $$\sigma_i = \frac{0.316P}{h^2}[4 \log_{10}(l/b) + 1.069]$$

$$\sigma_i = \frac{0.316 \times 41.0}{(20)^2}[4 \log_{10}(84.57/14.07) + 1.069] = \mathbf{0.1355 \ kN/cm^2}$$

From Eq. 29.20, $$\sigma_e = \frac{0.572P}{h^2}[4 \log_{10}(l/b) + 0.359]$$

$$= \frac{0.572 \times 41.00}{(20)^2}[4 \log_{10}(84.57/14.07) + 0.359] = \mathbf{0.2037 \ kN/cm^2}$$

From Eq. 29.21, $$\sigma_c = \frac{3P}{h^2}\left[1 - \left(\frac{a\sqrt{2}}{l}\right)^{0.6}\right]$$

$$= \frac{3 \times 41.0}{(20)^2}\left[1 - \left(\frac{15\sqrt{2}}{84.57}\right)^{0.6}\right] = \mathbf{0.1734 \ kN/cm^2}$$

PROBLEMS

A. Numerical

29.1. A sample of subgrade gave the following results.

(*a*) Soil portion passing 0.074 mm sieve = 55%.

(*b*) Liquid Limit = 45

(*c*) Plastic Limit = 25

Design the pavement by group index method for an anticipated traffic volume of over 300 commercial vehicles per day. [**Ans.** Total thickness = 50 cm, Sub-base = 10 cm]

29.2. The CBR value of subgrade is 8 percent. Determine the total thickness of flexible pavement using design charts recommended by IRC. There are more than 4500 vehicles per day. [**Ans.** 40 cm]

29.3. Determine the stresses at interior, edge and corner of a cement concrete pavement by Westergaard's equations.

(*a*) Modulus of elasticity of concrete = 3×10^3 kN/cm²

(*b*) Poisson's ratio = 0.15

(*c*) Thickness of concrete pavement = 22 cm

(*d*) Modulus of subgrade reaction = 0.03 kN/cm³

(*e*) Wheel load = 41.0 kN

(*f*) Radius of loaded area = 12 cm

[**Ans.** $\sigma_i = 0.1265$, $\sigma_c = 0.1946$, $\sigma_c = 0.1652$ kN/cm²]

29.4. Determine the warping stresses for a concrete pavement of thickness 20 cm with transverse joints at 15 m spacing and longitudinal joints at 3.5 m spacing. For concrete, $E = 3 \times 10^3$ kN/cm^2, $\mu = 0.15$. Take the modulus of subgrade reaction as 30 N/cm^3. Assume temperature differential for day conditions to be 1°C per cm of slab thickness and the coefficient of linear expansion of concrete as 10×10^{-6} per °C. The radius of load area is 15cm.

B. Descriptive Type and Objective Type

29.5. What are different types of pavements ? Explain their salient characteristics.

29.6. What are the various factors that affect the design of pavement ?

29.7. Describe the following methods of the design of flexible pavement.
(*a*) Group index method
(*b*) CBR method
(*c*) Burmister's Method.

29.8. Discuss Westergaard's analysis for the rigid pavements.

29.9. Explain the following terms.
(*a*) Modulus of subgrade reaction.
(*b*) Radius of relative stiffness.

29.10. Discuss the effect of temperature stresses on the design of rigid pavements.

29.11. Write whether the following statements are true or false
(*a*) The flexible pavements are capable of bridging small depressions in the subgrade.
(*b*) Generally, the CBR for 5.0 mm is greater than that for 2.5 mm.
(*c*) The higher the value of the group index, the poorer is the subgrade.
(*d*) The units for the coefficient of subgrade reaction are kN/cm^2.
(*e*) For corner regions of the rigid pavements, the critical stresses usually occur during the midnight.

[**Ans.** True (*c*), (*e*)]

C. Multiple Choice Questions

1. For rigid pavements, generally the following component is omitted
(*a*) Surfacing (*b*) Base
(*c*) Sub base (*d*) Subgrade

2. The minimum surcharge load for the C.B.R. test is
(*a*) 2.5 kg (*b*) 3.5 kg
(*c*) 4.5 kg (*d*) 5.5 kg

3. The standard size of the plate for the coefficient of subgrade reaction test is
(*a*) 50 cm (*b*) 75 cm
(*c*) 100 (*d*) 125 cm

[**Ans.** 1. (*c*), 2. (*c*), 3. (*b*)]

30

Laboratory Experiments

Some of the more commonly conducted laboratory experiments are described below.

30.1. Experiment No. 1. To determine the water-content of a soil sample by oven-drying method.

Theory. The water content (w) of a soil sample is equal to the mass of water divided by the mass of solids.

$$w = \frac{M_2 - M_3}{M_3 - M_1} \times 100$$

where M_1 = mass of empty container, with lid,

M_2 = mass of the container with wet soil and lid,

M_3 = mass of the container with dry soil and lid.

Equipment. (1) Thermostatically controlled oven, maintained at a temperature of 110° ± 5°C; (2) Weighing balance, with accuracy of 0.04% of the mass of the soil taken; (3) Desiccator, with any suitable desiccating agent; (4) Airtight container made of non-corrodible material, with lid; (5) Tongs.

Soil Specimen. The soil specimen should be representative of the soil mass. The quantity of the specimen taken would depend upon the gradation and the maximum size of particles. For more than 90% of the particles passing 425 μ IS sieve, the minimum quantity is 25 g.

Procedure: (1) Clean the container, dry it and weight it with lid (M_1).

(2) Take the required quantity of the wet specimen in the container and close it with lid. Take the mass (M_2).

(3) Place the container, with its lid removed, in the oven till its mass becomes constant (normally for 24 hours).

(4) When the soil has dried, remove the container from the oven, using tongs.

Replace the lid on the container. Cool it in a desiccator.

Data Sheet for Water content by Oven-drying Method

S.No.	*Observations and calculations*	Determination No. 1	2	3
	Observations			
(1)	Container No.	401	402	403
(2)	Mass of empty container (M_1)	20.12 g		
(3)	Mass of container + soil (M_2)	44.32 g		
(4)	Mass of container + dry soil (M_3)	41.18 g		
	Calculations			
(5)	Mass of water, $M_w = M_2 - M_3$	3.14g		
(6)	Mass of solids, $M_s = M_3 - M_1$	21.06 g		
(7)	Water content, w = (5)/(6) × 100	14.91		

Result. The water content of the sample = 14.91%.

(5) Find the mass (M_3) of the container with lid and dry soil sample.

Observations and calculations. Observations and calculations are shown in the data sheet.

30.2 Experiment No. 2. To determine the water content of a soil sample by pycnometer method.

Theory. A pycnometer is a glass jar of about 1 litre capacity, fitted with a brass conical cap by means of a screw-type cover. The cap has a small hole of about 6 mm diameter at its apex.

The water content (w) of the sample is obtained as

$$w = \left[\frac{(M_2 - M_1)}{(M_3 - M_4)}\left(\frac{G-1}{G}\right) - 1\right] \times 100$$

where M_1 = mass of empty pycnometer;

M_2 = mass of pycnometer and wet soil;

M_3 = mass of pycnometer and soil, filled with water;

M_4 = mass of pycnometer filled with water only;

G = specific gravity of solids.

Equipment. (1) Pycnometer; (2) Weighing balance with an accuracy of 1.0 g; (3) Glass rod.

Procedure. (1) Wash and clean the pycnometer and dry it.

(2) Determine the mass of the pycnometer, with brass cap and washer (M_1), accurate to 1 g.

(3) Place about 200 to 400 g of wet soil specimen in the pycnometer and weigh it with its cap and washer (M_2).

(4) Fill water in the pycnometer containing the wet soil specimen to about its half height.

(5) Mix the contents thorougly with a glass rod. Add more water and stir it. Fill the pycnometer with water, flush with the hole in the conical cap.

(6) Dry the pycnometer from outside and take its mass (M_3).

(7) Empty the pycnometer. Clean it thoroughly. Fill it with water, flush with the hole in the conical cap and weigh (M_4).

Observations and calculations

See the data sheet.

Data sheet for water-content by pycnometer method

Specific gravity of solids = 2.67

S.No.	*Observations and calculations*	*Determination No.*		
		1	*2*	*3*
	Observations			
1.	Mass of empty pycnometer (M_1)	580 g		
2.	Mass of pycnometer and wet soil (M_2)	844 g		
3.	Mass of pycnometer soil, filled with water (M_3)	1606 g		
4.	Mass of pycnometer filled with water only (M_4)	1470 g		
	Calculations			
5.	$M_2 - M_1$	264 g		
6.	$M_3 - M_4$	136 g		
7.	$(G - 1)/G$	0.625		
8.	$w = \left[\frac{(5)}{(6)} \times (7) - 1\right] \times 100$	21.32%		

Result. Water content of the sample = 21.32%.

30.3. Experiment No. 3. To determine the specific gravity of solids by the density bottle method.

Theory. The specific gravity of solid particles is the ratio of the mass density of solids to that of water. It is determined in the laboratory using the relation

$$G = \frac{M_2 - M_1}{(M_2 - M_1) - (M_3 - M_4)}$$

where M_1 = mass of empty bottle;
M_2 = mass of the bottle and dry soil;
M_3 = mass of bottle, soil and water;
M_4 = mass of bottle filled with water only.

Equipment. (1) 50 ml density bottle with stopper; (2) Oven (105° to 110°C); (3) Constant temperature water bath (27°C); (4) Vacuum desiccator; (5) Vacuum pump; (6) Weighing balance, accuracy 0.001 g; (7) Spatula.

Procedure.

(1) Wash the density bottle and dry it in an oven at 105°C to 110°C. Cool it in the desiccator.

(2) Weigh the bottle, with stopper, to the nearest 0.001 g (M_1).

(3) Take 5 to 10 g of the oven-dried soil sample and transfer it to the density bottle.

Weigh the bottle with the stopper and the dry sample (M_2).

(4) Add de-aired distilled water to the density bottle just enough to cover the soil. Shake gently to mix soil and water.

(5) Place the bottle containing the soil and water, after removing the stopper, in the vacuum desiccator.

(6) Evacuate the desiccator gradually by operating the vacuum pump. Reduce the pressure to about 20 mm of mercury. Keep the bottle in the desiccator for at least 1 hour or until no further movement of air is noticed.

(7) Release the vacuum and remove the lid of the desiccator.

Stir the soil in the bottle carefully with a spatula. Before removing the spatula from the bottle, the particles of soil adhering to it should be washed off with a few drops of air-free water.

Replace the lid of the desiccator and again apply vacuum.

Repeat the procedure until no more air is evolved from the specimen.

[**Note.** In case a vacuum desiccator is not available, the entrapped air can be removed by heating the density bottle on a water bath or a sand bath.]

(8) Remove the bottle from the desiccator. Add air-free water until the bottle is full. Insert the stopper.

(9) Immerse the bottle upto the neck in a constant-temperature bath for approximately 1 hour or until it has attained the constant temperature.

If there is an apparent decrease in the volume of the liquid in the bottle, remove the stopper and add more water to the bottle and replace the stopper. Again place the bottle in the water bath. Allow sufficient time to ensure that the bottle and its content attain the constant temperature.

(10) Take out the bottle from the water bath. Wipe it clean and dry it from outside.

Fill the capillary in the stopper with drops of distilled water, if necessary.

(11) Determine the mass of the bottle and its contents (M_3).

(12) Empty the bottle and clean it thoroughly. Fill it with distilled water. Insert the stopper.

(13) Immerse the bottle in the constant-temperature bath for 1 hour or until it has attained the constant temperature of the bath.

If there is an apparent decrease in the volume of the liquid, remove the stopper and add more water. Again keep it in the water bath.

(14) Take out the bottle from the water bath. Wipe it dry and take the mass (M_4).

Observations and calculations

See the data sheet.

Data Sheet for Specific Gravity by Density Bottle

S.No.	*Observations and calculations*	*Determination No.*		
		1	*2*	*3*
	Observations			
1.	Density Bottle No.	301	302	303
2.	Mass of emply density bottle (M_1)	41.302 g		
3.	Mass of bottle dry soil (M_2)	54.103 g		
4.	Mass of bottle, soil and water (M_3)	99.082 g		
5.	Mass of bottle filled with water (M_4)	91.112 g		
	Calculations			
6.	$M_2 - M_1$	12.801 g		
7.	$M_3 - M_4$	7.970 g		
8.	$G = \frac{(6)}{(6) - (7)}$	2.65		

Result. Specific gravity of solids = 2.65.

30.4. Experiment No. 4. To determine the specific gravity of solids by pycnometer method.

Theory. The pycnometer method can be used for the determination of the specific gravity of solid particles of both fine-grained and coarse-grained soils. The specific gravity of solids is determined using the relation

$$G = \frac{(M_2 - M_1)}{(M_2 - M_1) - (M_3 - M_4)}$$

where M_1 = mass of empty pycnometer;

M_2 = mass of pycnometer and dry soil;

M_3 = mass of pycnometer, soil, and water;

M_4 = mass of pycnometer filled with water only.

Equipment. (1) Pycnometer of about 1 litre capacity; (2) Weighing balance, with an accuracy of 1 g; (3) Glass rod; (4) Vacuum pump.

Procedure. (1) Clean and dry the pycnometer. Tightly screw its cap. Take its mass (M_1) to the nearest 0.1 g.

(2) Mark the cap and pycnometer with a vertical line parallel to the axis of the pycnometer to ensure that the cap is screwed to the same mark each time.

(3) Unscrew the cap and place about 200 g of oven-dried soil in the pycnometer.

Screw the cap. Determine the mass (M_2).

(4) Unscrew the cap and add sufficient amount of de-aired water to the pycnometer so as to cover the soil. Screw on the cap.

(5) Shake well the contents. Connect the pycnometer to a vacuum pump, to remove the entrapped air, for about 20 minutes for fine- grained soils and for about 10 minutes for coarse-grained soils.

(6) Disconnect the vacuum pump. Fill the pycnometer with water, about three fourths full.

Reapply the vacuum for about 5 minutes, till air bubbles stop appearing on the surface of the water.

(7) Fill the pycnometer with water completely, upto the mark. Dry it from outside. Take its mass (M_3).

(8) Record the temperature of contents.

(9) Empty the pycnometer. Clean it and wipe it dry.

(10) Fill the pycnometer with water only. Screw on the cap upto the mark. Wipe it dry. Take its mass (M_4).

Data Sheet for Specific Gravity by Pycnometer Method

S.No.	Observations and calculations	Determination No. 1	2	3
	Observations			
1.	Pycnometer No.	401	402	403
2.	Room temperature	26°C		
3.	Mass of empty phcnometer (M_1)	580 g		
4.	Mass of pycnometer and dry soil (M_2)	800 g		
5.	Mass of pycnometer, soil and water (M_3)	1707 g		
6.	Mass of pycnometer and water (M_4)	1570 g		
	Calculations			
7.	$M_2 - M_1$	220 g		
8.	$M_3 - M_4$	137 g		
9.	$G = \frac{(7)}{(7) - (8)}$	2.65		

Result. Specific gravity of solids at 26°C = 2.65.

30.5. Experiment No. 5. To determine the dry density of the soil by core cutter methods.

Theory. A cylindrical core cutter is a seamless steel tube. For determination of the dry density of the soil, the cutter is pressed into the soil mass so that it is filled with the soil. The cutter filled with the soil is lifted up. The mass of the soil in the cutter is determined. The dry density is obtained as

$$\rho = \frac{\gamma}{1 + w} = \frac{(M/V)}{1 + w}$$

where M = mass of the wet soil in the cutter;
V = internal volume of the cutter; w = water content.

Equipment (1) Cylindrical core cutter, 100 mm internal diameter and 130 mm long; (2) Steel rammer, mass 9 kg. overall length, with the foot and staff about 900 mm; (3) Steel dolley, 25 mm high and 100 mm internal diameter, (4) Weighing balance, accuracy 1 g; (5) Palette knife; (6) Straight edge, steel rule, etc.

Procedure. (1) Determine the internal diameter and height of the core cutter to the nearest 0.25 mm.
(2) Determine the mass (M_1) of the cutter to the nearest gram.

Data Sheet for Dry Density by Core Cutter Method

S.No.	Observations and calculations	Determination No. 1	2	3
	Observations			
1.	Core cutter No.	501	502	503
2.	Internal diameter	100 mm		
3.	Internal height	129.75 mm		
4.	Mass of empty core cutter (M_1)	1130 g		
5.	Mass of core cutter with soils (M_2)	3120 g		
	Calculations			
6.	Mass of wet soil, $M = M_2 - M_1$	1990 g		
7.	Volume of cutter, V	1019.05 ml		
8.	Water content (determined as in Experiment 30–1), say,	17.75 %		
9.	Dry density = $\frac{(6)/(7)}{1 + (8)}$	1.66 gm/ml		

Result. Dry density = 1.666 g/ml.

(3) Expose a small area of the soil mass to be tested. Level the surface, about 300 mm square in area.

(4) Place the dolley over the top of the core cutter and press the core cutter into the soil mass using the rammer.

Stop the process of pressing when about 15 mm of the dolley protrudes above the soil surface.

(5) Remove the soil surrounding the core cutter, and take out the core cutter. Some soil would project from the lower end of the cutter.

(6) Remove the dolley. Trim the top and bottom surface of the core cutter carefully using a straight edge.

(7) Weigh the core cutter filled with the soil to the nearest gram (M_2).

(8) Remove the core of the soil from the cutter. Take a representative sample for the water content determination.

Determine the water content, as described in Experiment 30.1.

30.6. Experiment No. 6. To determine in-situ dry density by the sand replacement method.

Theory. A hole of specified dimensions is excavated in the ground. The mass of the excavated soil is determined.

The volume of the hole is determined by filling it with clean, uniform sand whose dry density (ρ_s) is determined separately by calibration. The volume of the hole is equal to the mass of the sand filled in the hole divided by its dry density.

The dry density of the excavated soil is determined as

$$\rho_d = \frac{(M/V)}{1 + w}$$

where M = mass of the excavated soil; V = volume of the hole; and w = water content.

Equipment. (1) Sand—pouring cylinder; (2) Calibrating container, 100 mm diameter and 150 mm height; (3) Soil cutting and excavating tools, such as a scraper tool, bent spoon; (4) Glass plate, 450 mm square, 9 mm thick; (5) Metal container to collect excavated soil; (6) Metal tray, 300 mm square and 40 mm deep with a hole of 100 mm in diameter at the centre; (7) Weighing balance; (8) Moisture content cans; (9) Oven; (10) Desiccator.

Clean, uniform sand passing 1 mm IS sieve and retained on 600 micron IS sieve in sufficient quantity.

Part—I Calibration

Procedure (1) Determine the internal volume of the calibrating container by filling it with water and determining the mass of water required. The mass of water in grams is approximately equal to the volume in millilitres. The volume may also be determined from the measured dimensions of the container.

(2) Fill the sand-pouring cylinder with sand, within about 10 mm of its top. Determine the mass of the cylinder (M_1) to the nearest gram.

(3) Place the sand-pouring cylinder vertically on the calibrating container.

Open the shutter to allow the sand run out from the cylinder into the calibrating container till it fills the cone of the cylinder and the calibrating container. When there is no further movement of the sand in the cylinder, close the shutter.

(4) Lift the pouring cylinder from the calibrating container and weigh it to the nearest gram (M_3).

(5) Again fill the pouring cylinder with sand, within 10 mm of its top.

(6) Open the shutter and allow the sand to run out of the cylinder. When the volume of the sand let out is equal to the volume of the calibrating container, close the shutter.

(7) Place the cylinder over a plane surface, such as a glass plate. Open the shutter. The sand fills the cone of the cylinder. Close the shutter when no further movement of sand takes place.

(8) Remove the cylinder. Collect the sand left on the glass plate.

Determine the mass of sand (M_2) that had filled the cone by weighing the collected sand.

(9) Determine the dry density of sand, as-shown in the data sheet, part-I.

Part—II. Dry Density

Procedure. (1) Expose an area of about 450 mm square on the surface of the soil mass. Trim the surface down to a level surface, using a scraper tool.

(2) Place the metal tray on the levelled surface.

(3) Excavate the soil through the central hole of the tray, using the hole in the tray as a pattern. The depth of the excavated hole should be about 150 mm.

(4) Collect all the excavated soil in a metal container, and determine the mass of the soil (M).

(5) Remove the metal tray from the excavated hole.

(6) Fill the sand-pouring cylinder within 10 mm of its top. Determine its mass (M_1).

(7) Place the cylinder directly over the excavated hole. Allow the sand to run out of the cylinder by opening the shutter.

Close the shutter when the hole is completely filled and no further movement of sand is observed.

(8) Remove the cylinder from the filled hole. Determine the mass of the cylinder (M_4).

(9) Take a representative sample of the excavated soil. Determine its water content, as explained in the experiment 30—1.

Determine the dry density of soil as shown in the data sheet, Part—II.

Data Sheet for Sand Replacement Method

Part—I. Calibration for Dry Density of Sand

S.No.	*Observations and calculations*	*Determination No.*		
		1	*2*	*3*
	Observations			
1.	Volume of calibrating cone (V_c)	980 ml		
2.	Mass of pouring cylinder (M_1), filled with sand	11040 g		
3.	Mass of puring cylinder after pouring sand into the calibrating container and cone (M_3)	9120 g		
4.	Mass of sand in the cone (M_2)	450 g		
	Calculations			
5.	Mass of sand in the calibrating container M_c = (2) − (3) − (4)	1470 g		
6.	Dry density of sand $\rho_s = M_c/V_c$	1.5 g/ml		

Part—II. Dry Density of soil.

S.No.	*Observations and calculations*	*Determination No.*		
		1	*2*	*3*
	Observations			
1.	Mass of excavated soil (M)	2310 g		
2.	Mass of pouring cylinder (M_1), filled with sand	11040 g		
3.	Mass of pouring cylinder after pouring into the hole and cone (M_4)	8840 g		
	Calculations			
4.	Mass of sand in the hole $M_s = M_1 - M_4 - M_2$	1750 g		
5.	Volume of sand in the hole $V = M_s/\rho_s$	1166.67 ml		
6.	Bulk density, $\rho = M/V$	1.98 g/ml		
7.	Water content, determined as in experiment 30-1 (w), say,	15%		
8.	Dry density = $\dfrac{\rho}{1+w} = \dfrac{M/V}{1+w}$	1.72 g/ml		

Result. Dry density = 1.72 g/ml.

30.7. Experiment No. 7. To determine the dry density of a soil sample by water- displacement method.

Theory. A soil specimen of regular shape is coated with paraffin wax to make it impervious to water. The total volume (V_t) of the waxed specimen is found by determining the volume of water displaced by the specimen. The volume of the specimen (V) is given by

$$V = V_t - \frac{(M_t - M)}{\rho_P}$$

where M_t = mass of waxed specimen; M = mass of the specimen without wax; ρ_P = density of paraffin.

$$\text{Dry density of specimen} = \frac{M/V}{1 + w}$$

Equipment. (1) Water-displacement apparatus; (2) Weighting balance, accuracy 1 g; (3) Paraffin wax; (4) Cutting knife; (5) Heater; (6) Oven; (7) Measuring jar; (8) Brush; (9) Water content container.

Procedure. (1) Take the soil specimen. Trim it to a regular shape. Avoid re-entrant corners. Weigh the specimen (M).

(2) Take some paraffin wax and melt it on a heater. Apply a coat of melted paraffin wax to the specimen with a brush. When it has hardened, apply another coat.

Take the mass of the waxed specimen (M_t).

(3) Fill the water-displacement apparatus with water. When the overflow occurs, close the valve.

(4) Place a measuring jar below the overflow tube of the apparatus. Open the valve.

(5) Immerse the waxed specimen slowly into the water in the apparatus. Water overflows. Collect the overflowed water in the jar.

Determine the volume of the water collected (V_t).

(6) Take out the waxed specimen from the apparatus. Dry it from outside.

(7) Remove the paraffin wax by peeling it off.

(8) Cut the specimen into two pieces. Take a representative sample for the water content determination. Determine the water content, as in Experiment 30.1.

Data sheet for Water-displacement Method

Density of paraffin (ρ_P) = 0.91 g/ml.

S.No.	*Observations and calculations*	*Determination No.*		
		1	*2*	*3*
	Observations			
1.	Mass of specimen (M)	650 g		
2.	Mass of waxed specimen (M_t)	681 g		
3.	Volume of waxed specimen by water-displacement (V_t)	362 g		
	Calculations			
4.	Mass of wax = $M_t - M$	31 g		
5.	Volume of wax (V_P) = $(M_t - M)/\rho_P$	34.06 ml		
6.	Volume of specimen (V) = $V_t - V_P$	327.06 ml		
7.	Water-content, as in experiment 30-1, (w)	13%		
8.	Dry density (ρ_d) = $\frac{M/V}{1 + w}$	1.75 g/ml		

Result. Dry density of soil = 1.75 g/ml.

30.8. Experiment No. 8. To determine the particle size distribution of a soil by sieving.

Theory. The soil is sieved through a set of sieves. The material retained on different sieves is determined. The percentage of material retained on any sieve is given by

$$p_n = \frac{M_n}{M} \times 100$$

where M_n = mass of soil retained on sieve '*n*',

and M = total mass of the sample.

The cumulative percentage of the material retained,

$$C_n = p_1 + p_2 + \dots + p_a$$

where p_1, p_2, etc, are the percentages retained on sieve '1', '2', etc. which are coarser than sieve '*n*'. The percentage finer than the sieve 'n',

$$N_n = 100 - C_n$$

Equipment. (1) Set of fine sieves, 2 mm, 1 mm, 600 μ, 425 μ, 212 μ, 150 μ, and 75 μ; (2) Set of coarse sieves, 100 mm, 80 mm, 40 mm, 20 mm, 10 mm and 4.75 mm; (3) Weighing balance, with accuracy of 0.1% of the mass of sample; (4) Oven, (5) Mechanical shaker; (6) Trays; (7) Mortar, with a rubber covered pestle; (8) Brushes; (9) Riffler.

Part—I. Coarse sieve Analysis

Procedure. (1) Take the required quantity of the sample. Sieve it through a 4.75 mm IS sieve. Take the soil fraction retained on 4.75 mm IS sieve for the coarse sieve analysis (part I) and that passing through the sieve for the fine sieve analysis (part II).

(2) Sieve the sample through the set of coarse sieves, by hand.

While sieving through each sieve, the sieve should be agitated such that the sample rolls in irregular motion over the sieve. The material retained on the sieves may be rubbed with the rubber pestle in the mortar, if necessary. Care shall be taken so as not to break the individual particles. The quantity of the material taken for sieving on each sieve shall be such that the maximum mass of material retained on each sieve does not exceed the specified value.

(3) Determine the mass of the material retained on each sieve.

(4) Calculate the percentage of soil retained on each sieve on the basis of the total mass of the sample, taken in step (1).

(5) Determine the percentage passing through each sieve.

Part—II. Fine sieve analysis

(6) Take the portion of the soil passing 4.75 mm IS sieve. Oven dry it at 105° to 100°C. Weigh it to 0.1% of the total mass.

(7) Sieve the soil through the nest of fine sieves. The sieves should be agitated so that the sample rolls in irregular motion over the sieves. However, no particles should be pushed through the sieve.

(8) Take the material retained on various sieves in a mortar. Rub it with rubber pestle, but do not try to break individual particles.

(9) Resieve the material through the nest of sieves.

A minimum of 10 minutes of shaking is required if a mechanical shaker is used.

(10) Collect the soil fraction retained on each sieve in a separate container. Take the mass.

(11) Determine the percentage retained, cumulative percentage retained, and the percentage finer, based on the total mass taken in step (1).

Data Sheet for Sieve Analysis

Total mass of dry soil = 400 g
Mass of soil retained on 4.75 mm sieve = 200 g
Mass of soil passing 4.75 mm sieve = 200 g.

S. No.	*Observations*			*Calculations*		
	IS Sieve	*Size of Opening*	*Mass of soil retained*	*Percentage retained*	*Cumulative % retained*	*% finer*
	Coarse Fraction (Part I)					
1.	100 mm	100 mm	—	—		
2.	80 mm	80 mm	—	—		
3.	40 mm	40 mm	—	—		
4.	20 mm	20 mm	30.0 g	7.50	7.50	92.50
5.	10 mm	10 mm	62.0 g	15.50	23.00	77.00
6.	4.75 mm	4.75 mm	108.0 g	27.00	50.00	50.00
	Fine Fraction (Part II)					
7.	2 mm	2 mm	30.5 g	7.62	57.62	42.38
8.	1 mm	1 mm	24.0 g	6.00	63.62	36.38
9.	600 μ	0.600 mm	17.5 g	4.38	68.00	32.00
10.	425 μ	0.425 mm	16.0 g	4.00	72.00	28.00
11.	300 μ	0.300 mm	14.0 g	3.50	75.50	24.50
12.	212 μ	0.212 mm	18.0 g	4.50	80.00	20.00
13.	150 μ	0.150 mm	22.0 g	5.50	85.50	14.50
14.	75 μ	0.075 mm	36.0 g	9.00	94.50	5.50
15.	*Pan*	—	22.0 g	5.50	100.00	

Result. Percentage finer given in the last column can be used to plot the particle size distribution curve with particle size as abscissa on log scale and the percentage finer as ordinate.

[**Note.** If the fine fraction contains an appreciable amount of clay particles, the wet sieve analysis is required. Alternatively, the following method may be used.

Before conducting step (7), add the water containing sodium hexa- metaphosphate at the rate of 2 g per litre of water to the soil fraction. Stir the mix thoroughly and leave for soaking. Wash the soaked specimen on a 75μ IS sieve until the water passing the sieve is clear. Take the fraction retained on the sieve and dry it in an oven. Sieve the oven dried soil through the nest of sieves as discussed in step (7). Perform further steps, as before.

Obviously, the mass of material which would have been retained on pan is equal to the original mass of the soil before washing minus the dry mass of the soil retained on 75μ IS sieve after washing.]

30.9. Experiment No. 9. To determine the particle size distribution by the hydrometer method.

Theory. Hydrometer method is used to determine the particle size distribution of fine-grained soils passing 75μ sieve. The hydrometer measures the specific gravity of the soil suspension at the centre of its bulb. The specific gravity depends upon the mass of solids present, which, in turn, depends on the particle size. The particle size (D) is given by

$$D = M\sqrt{H_e/t}$$

where $M = \left[\dfrac{0.3\,\eta}{g\,(G-1)\,\rho_w}\right]^{1/2}$, in which, η = viscosity of water (poise), G = specific gravity of solids; ρ_w = density of water (gm/ml); g = 981 cm/sec^2, H_e = effective depth (cm); t = time in minutes at which observation is taken, reckoned with respect to the beginning of sedimentation.

The percentage finer than the size D is given by

$$N = \left(\frac{G}{G-1}\right) \times \frac{R}{M_s} \times 100$$

where R = corrected hydrometer reading; M_s = mass of dry soil in 1000 ml suspension.

Equipment. (1) Hydrometer; (2) Glass measuring cylinder (jar), 1000 ml; (3) Rubber bung for the cylinder (jar); (4) Mechanical stirrer; (5) Weighing balance, accuracy 0.01 g (6) Oven (7) Desiccator; (8) Evaporating dish; (9) Conical flask or beaker, 1000 ml; (10) Stop watch; (11) Wash bottle; (12) Thermometer; (13) Glass rod; (14) Water bath; (15) 75 μ sieve; (16) Scale; (17) Deflocculating agent.

Procedure.

Part—I Calibration of hydrometer

(1) Take about 800 ml of water in one measuring cylinder. Place the cylinder on a table and observe the initial reading.

(2) Immerse the hydrometer in the cylinder. Take the reading after the immersion.

(3) Determine the volume of the hydrometer (V_H), which is equal to the difference between the final and initial readings.

Alternatively, weigh the hydrometer to the nearest 0.1 g. The volume of the hydrometer in ml is approximately equal to its mass in grams.

(4) Determine the area of cross-section (A) of the cylinder. It is equal to the volume indicated between any two graduations divided by the distance between them. The distance is measured with an accurate scale.

(5) Measure the distance between the hydrometer neck and the bottom of the bulb. Record it as the height of the bulb (h).

(6) Measure the distance (H) between the neck to each of the marks on the hydrometer (R_h).

(7) Determine the effective depth (H_e), corresponding to each of the mark (R_h), as

$$H_e = H + \frac{1}{2}\left(h - \frac{V_H}{A}\right)$$

[**Note.** The factor V_H/A should not be considered when the hydrometer is not taken out when taking readings after start of the sedimentation at 1/2, 1, 2 and 4 minutes].

(8) Draw a calibration curve between H_e and R_h. Alternatively, prepare a table between H_e and R_h.

The curve may be used for finding the effective depth H_e corresponding to reading R_h.

Part—II Meniscus correction

(1) Insert the hydrometer in the measuring cylinder containing about 700 ml of water.

(2) Take the readings of the hydrometer at the top and at the bottom of the meniscus.

(3) Determine the meniscus correction, which is equal to the difference between the two readings.

(4) The meniscus correction (C_m) is positive and is a constant for the hydrometer.

(5) The observed hydrometer reading (R_h') is corrected to obtain the corrected hydrometer reading (R_h) as

$$R_h = R_h' + C_m$$

Part—III Pretreatment and Dispersion

(1) Weigh accurately, to nearest 0.01 g, about 50 g air-dried soil sample passing 2 mm IS sieve, obtained by riffling from the air-dried sample passing 4.75 mm IS sieve.

Place the sample in a wide mouthed conical flask.

(2) Add about 150 ml of hydrogen peroxide to the soil sample in the flask.

Stir it gently with a glass rod for a few minutes.

(3) Cover the flask with a glass plate, and leave it to stand overnight.

(4) Heat the mixture in the conical flask gently after keeping it in an evaporating dish.

Stir the contents periodically. When vigorous frothing subsides, the reaction is complete. Reduce the volume to 50 ml by boiling. Stop heating and cool the contents.

(5) If the soil contains insoluble calcium compounds, add about 50 ml of hydrochloric acid to the cooled mixture.

Stir the solution with a glass rod for a few minutes. Allow it to stand for one hour or so. The solution would have an acid reaction to litmus when the treatment is complete.

(6) Filter the mixture and wash it with warm water until the filtrate shows no acid reaction.

(7) Transfer the damp soil on the filter and funnel to an evaporating dish, using a jet of distilled water. Use the minimum quantity of distilled water.

(8) Place the evaporating dish and its contents in an oven, and dry it at 105° to 110°C.

Transfer the dish to a desiccator, and allow it to cool.

(9) Take the mass of the oven dried soil after pretreatment, and find the loss of mass due to pretreatment.

(10) Add 100 ml of sodium hexa-metaphosphate solution to the oven-dried soil in the evaporating dish after pretreatment.

(11) Warm the mixture gently for about 10 minutes.

(12) Transfer the mixture to the cup of a mechanical mixer. Use a jet of distilled water to wash all traces of the soil out of the evaporating dish. Use about 150 ml of water. Stir the mixture for about 15 minutes.

(13) Transfer the soil suspension to a 75 μ IS sieve placed on a receiver (pan).

Wash the soil on this sieve using a jet of distilled water. Use about 500 ml of water.

(14) Transfer the soil suspension passing 75 μ sieve to a 1000 ml measuring cylinder.

Add more water to make the volume exactly equal to 1000 ml.

(15) Collect the material retained on 75 μ sieve. Dry it in an oven. Determine its mass. If required, do the sieve analysis of this fraction.

Part—IV. Sedimentation Test

(1) Place the rubber bung on the open end of the measuring cylinder containing the soil suspension.

Shake it vigorously end-over-end to mix the suspension thoroughly.

(2) Remove the bung after the shaking is complete. Place the measuring cylinder on the table and start the stop watch.

(3) Immerse the hydrometer gently to a depth slightly below the floating depth, and then allow it to float freely.

(4) Take hydrometer reading (R_h') after 1/2, 1, 2 and 4 minutes, without removing the hydrometer from the cylinder.

(5) Take out the hydrometer from the cylinder, rinse it with distilled water.

(6) Float the hydrometer in another cylinder containing only distilled water at the same temperature as that of the test cylinder.

(7) Take out the hydrometer from the distilled water cylinder and clean its stem.

Insert it in the cylinder containing suspension to take the reading at the total elapsed time interval of 8 minutes. About 10 seconds should be taken while taking the reading. Remove the hydrometer, rinse it and place it in the distilled water cylinder after reading.

(8) Repeat the Step (7) to take readings at 15, 30, 60, 120 and 240 minutes elapsed time interval.

(9) After 240 minutes (4 hours) reading, take readings twice within 24 hours. Exact time of each reading should be noted.

(10) Record the temperature of the suspension once during the first 15 minutes, and thereafter at the time of every subsequent reading.

(11) After the final reading, pour the suspension in an evaporating dish. Dry it in an oven and find its dry mass.

(12) Determine the composite correction before the start of the test, and also at 30 minutes, 1, 2 and 4 hours. Thereafter, just after each reading, composite correction is determined.

(13) For the determination of the composite correction (C), insert the hydrometer in the comparison cylinder containing 100 ml of dispersing agent solution in 1000 ml of distilled water at the same temperature. Take the reading corresponding to the top of meniscus. The negative of the reading is the composite correction.

Data Sheet for Hydrometer Test

Mass of dry soil (M_s) = 50 g; Meniscus correction (C_m) = +0.4

Specific gravity of solids (G) = 2.67

S. No	*Observations*				*Calculations*					
	Elapsed time (t)	Hydrometer Reading (R_h')	*Tempera-ture*	*Composite correction (C)*	*Corrected Reading* $R_h = R_h' + C_m$	*Height* H_e *(cm)*	*Reading* $R = R_h' + C$	*Factor M*	*Particle size* $D = M\sqrt{H_e/t}$	*Per-centage finer (N)*
1.	1/2 minu.	21.80	22.0	—0.50	22.2	11.8	21.3	1.33×10^{-3}	0.065 mm	68.10
2.	1 "									
3.	2 "									
4.	4 "									
5.	8 "									
6.	15 "									
7.	30 "									
8.	1 *hr.*									
9.	2 "									
10.	4 "									
11.	8 "									
12.	12 "									
13.	24 "									

Result. Particle size distribution curve can be plotted using the last two columns.

30.10. Experiment No. 10. To determine the liquid limit of a soil specimen.

Theory. The liquid limit of a soil is the water content at which the soil behaves practically like a liquid, but has a small shear strength. It flows to close the groove in just 25 blows in Casagrande's liquid limit device.

As it is difficult to get exactly 25 blows in a test, 3 to 4 tests are conducted, and the number of blows (N) required in each test is determined. A semi-log plot is drawn between log N and the water content (w). The liquid limit is the water content corresponding to N = 25, as obtained from the plot.

Equipment. (1) Casagrande's liquid limit device; (2) Grooving tools of both Standard and ASTM types; (3) Oven; (4) Evaporating dish or glass sheet; (5) Spatula; (6) 425 μ IS sieve; (7) Weighing balance, accuracy 0.01 g; (8) Wash bottle.

Procedure. (1) Adjust the drop of the cup of the liquid limit device by releasing the two screws at the top and by using the handle of the grooving tool or a gauge.

The drop should be exactly 1 cm at the point of contact on the base. Tighten the screw after adjustment.

(2) Take about 120 g of the air-dried soil sample passing 425 μ IS sieve.

(3) Mix the sample thoroughly with distilled water in an evaporating dish or a glass plate to form a uniform paste. Mixing should be continued for about 15 to 30 minutes, till a uniform mix is obtained.

(4) Keep the mix under humid conditions for obtaining uniform moisture distribution for sufficient period. For some fat clays, this maturing time may be upto 24 hours.

(5) Take a portion of the matured paste and remix it thoroughly. Place it in the cup of the device by a spatula and level it by a spatula or a straight edge to have a maximum depth of the soil as 1 cm at the point of the maximum thickness.

The excess soil, if any, should be transferred to the evaporating dish.

(6) Cut a groove in the sample in the cup by using the appropriate tool. Draw the grooving tool through the paste in the cup along the symmetrical axis, along the diameter through the centre line of the cam. Hold the tool perpendicular to the cup.

(7) Turn the handle of the device at a rate of 2 revolutions per second.

Count the number of blows until the two halves of the soil specimen come in contact at the bottom of the groove along a distance of 12 mm due to flow and not by sliding.

(8) Collect a representative specimen of the soil by moving spatula width-wise from one edge to the other edge of the soil cake, at right-angles to the groove. This should include the portion of the groove in which the soil flowed to close the groove.

Place the specimen in an air-tight container for the water content determination. Determine the water content.

(9) Remove the remaining soil from the cup. Mix it with the soil left in the evaporating dish.

(10) Change the water content of the mix in the evaporating dish, either by adding more water if the water content is to be increased, or by kneading the soil, if the water content is to be decreased.

In no case, the dry soil should be added to reduce the water content.

(11) Repeat steps 4 to 10, and determine the number of blows (N) and the water content in each case.

(12) Draw the flow curve between log N and w, and determine the liquid limit corresponding to N = 25.

Data Sheet for Liquid Limit Test

S.No.	*Observations and calculations*	*Determination No.*		
		1	*2*	*3*
	Observations			
1.	No. of blows (N)	15		
2.	Water content can No.	101		
3.	Mass of empty can (M_1)	25.15 g		
4.	Mass of can + wet soil (M_2)	36.93 g		
5.	Mass of can + dry soil (M_3)	33.81 g		
	Calculations			
6.	Mass of water = $M_2 - M_3$	3.12 g		
7.	Mass of dry soil = $M_3 - M_1$	8.66 g		
8.	Water content, $w = \frac{(6)}{(7)} \times 100$	36 %		

Result. Draw a flow curve between log N and w. Liquid limit (for $N = 25$) = ...

30.11. Experiment No. 11. To determine the plastic limit of a soil specimen.

Theory. The plastic limit of a soil is the water content of the soil below which it ceases to be plastic. It begins to crumble when rolled into threads of 3 mm diameter.

Equipment. (1) Porcelain evaporating dish, about 120 mm diameter or a flat glass plate, 450 mm square and 10 mm thick; (2) Ground glass plate, about 200 mm × 150 mm; (3) Metallic rod, 3 mm dia. and 100 mm long; (4) Oven; (5) Spatula or palette knife; (6) Moisture content can.

Procedure. (1) Take about 30 g of air-dried soil from a thoroughly mixed sample of the soil passing 425μ sieve.

(2) Mix the soil with distilled water in an evaporating dish or on a glass plate to make it plastic enough to shape into a small ball.

(3) Leave the plastic soil mass for some time for maturing. For some fat clays, this period may be even upto 24 hours.

(4) Take about 8 g of the plastic soil, and roll it with fingers on a glass plate. The rate of the rolling

should be about 80 to 90 strokes per minute to form a thread of 3 mm diameter, counting one stroke when the hand moves forward and backward to the staring point.

(5) If the diameter of the thread becomes less than 3 mm without cracks, it shows that the water content is more than the plastic limit. Knead the soil to reduce the water content, and roll it again into thread.

Repeat the process of alternate rolling and kneading until the thread crumbles, and the soil can no longer be rolled into thread.

[**Note.** If the crumbling occurs when the thread has a diameter slightly greater than 3 mm, it may be taken as the plastic limit, provided the soil had been rolled into a thread of 3 mm diameter immediately before kneading. Do not attempt to produce failure exactly at 3 mm diameter.]

(6) Collect the pieces of the crumbled soil thread in a moisture content container.

(7) Repeat the procedure at least twice more with fresh samples of plastic soil each time.

Data Sheet for Plastic Limit Test

S.No.	*Observations and calculations*	*Determination No.*		
		1	2	3
	Observations			
1.	Moisture content container No.	101	102	103
2.	Mass of empty container (M_1)	24.12 g		
3.	Mass of container + wet soil (M_2)	30.28 g		
4.	Mass of container + dry soil (M_3)	29.12 g		
	Calculations			
5.	Mass of water = $M_2 - M_3$	1.16 g		
6.	Mass of dry soil = $M_3 - M_1$	5.00 g		
7.	Water Content, $w = \frac{(5)}{(6)} \times 100$	23.2 %		

Result. Plastic limit of soil = 23.2%.

30.12. Experiment No. 12. To determine the shrinkage limit of a specimen of the remoulded soil.

Theory. The shrinkage limit is the water content of the soil when the water is just sufficient to fill all the pores of the soil, and the soil is just saturated. The volume of the soil does not decrease when the water content is reduced below the shrinkage limit. It can be determined from the relation

$$w_s = \frac{(M_1 - M_s) - (V_1 - V_2)\,\rho_w}{M_s} \times 100$$

where M_1 = initial wet mass; V_1 = initial volume; M_s = dry mass; V_2 = volume after drying.

Equipment. (1) Shrinkage dish, having a flat bottom, 45 mm diameter and 15 mm height, (2) Two large evaporating dishes about 120 mm diameter, with a pour out and flat bottom; (3) One small mercury dish, 60 mm diameter; (4) Two glass plates, one plain and one with prongs, 75 mm × 75 mm × 3 mm size, (5) Glass cup, 50 mm diameter and 25 mm high; (6) IS sieve 425 μ; (7) Oven; (8) Desiccator; (9) Weighing balance, accuracy 0.01 g; (10) Spatula; (11) Straight edge; (12) Mercury.

Procedure. (1) Take a sample of mass about 100 g from a thoroughly mixed soil passing 425 μ sieve.

(2) Take about 30 g of the soil sample in a large evaporating dish. Mix it with distilled water to make a creamy paste which can be readily worked without entrapping the air bubbles.

(3) Take the shrinkage dish. Clean it and determine its mass.

(4) Fill mercury in the shrinkage dish. Remove the excess mercury by pressing the plain glass plate over the top of the shrinkage dish. The plate should be flush with the top of the dish, and no air should be entrapped.

(5) Transfer the mercury of the shrinkage dish to a mercury weighing dish and determine the mass of the

mercury to an accuracy of 0.1 g. The volume of the shrinkage dish is equal to the mass of mercury in grams divided by the specific gravity of mercury (*viz.* 13.6).

(6) Coat the inside of the shrinkage dish with a thin layer of silicon grease or vaseline.

Place the soil specimen in the centre of the shrinkage dish, equal to about one-third the volume of the shrinkage dish.

Tap the shrinkage dish on a firm, cushioned surface and allow the paste to flow to the edges.

(7) Add more soil paste, approximately equal to the first portion and tap the shrinkage dish as before, until the soil is thoroughly compacted.

Add more soil and continue the tapping till the shrinkage dish is completely filled, and excess soil paste projects out about its edge.

Strike out the top surface of the paste with a straight edge. Wipe off all soil adhering to the outside of the shrinkage dish. Determine the mass of the wet soil (M_1).

(8) Dry the soil in the shrinkage dish in air until the colour of the pat turns from dark to light. Then dry the pat in the oven at 105° to 110°C to constant mass.

(9) Cool the dry pat in a desiccator. Remove the dry pat from the desiccator after cooling, and weigh the shrinkage dish with the dry pat to determine the dry mass of the soil (M_s).

(10) Place a glass cup in a large evaporating dish and fill it with mercury. Remove the excess mercury by pressing the glass plate with prongs firmly over the top of the cup. Wipe off any mercury adhering to the outside of the cup.

Remove the glass cup full of mercury and place it in another evaporating dish, taking care not to spill any mercury from the glass cup.

(11) Take out the dry pat of the soil from the shrinkage dish and immerse it in the glass cup full of mercury. Take care not to entrap air under the pat. Press the plate with prongs on the top of the cup firmly.

(12) Collect the mercury displaced by the dry pat in the evaporating dish, and transfer it to the mercury weighing dish. Determine the mass of the mercury to an accuracy of 0.1 g. The volume of the dry pat (V_2) is equal to the mass of the mercury divided by the specific gravity of mercury.

(13) Repeat the test atleast 3 times.

Data Sheet for Shrinkage Limit Test

S.No.	*Observations and calculations*	*Determination No.*		
		1	*2*	*3*
	Observations			
1.	Mass of empty mercury dish	74.2 g		
2.	Mass of mercry dish, with mercury equal to volume of the shrinkage dish	361.1 g		
3.	Mass of mercury = (2) – (1)	286.9 g		
4.	Volume of shrinkage dish, $V_1 = \frac{(3)}{13.6}$	21.1 ml		
5.	Mass of empty shrinkage dish	23.5 g		
6.	Mass of shrinkage dish + wet soil	68.4 g		
7.	Mass of wet soil, M_1 = (6) – (5)	44.9 g		
8.	Mass of shrinkage dish + dry soil	57. 3 g		
9.	Mass of dry soil, M_s = (8) – (5)	33.8 g		
10.	Mass of mercury dish + mercury equal in volume of dry pat	304.3 g		
11.	Mass of mercury displaced by dry pat, = (10) – (1)	230.1 g		
12.	Volume of dry pat, $V_2 = \frac{(11)}{13.6}$	16.92 ml		

S.No.	Observations and calculations	Determination No.		
		1	2	3
	Calculations			
13.	Shrinkage limit, $w_s = \left(\frac{(M_1 - M_s) - (V_1 - V_2)\,\rho_w}{M_s}\right)$	20.5 %		
14.	Shrinkage ratio, $SR = \frac{M_s}{V_2\,\rho_w}$	2.0		
15.	Volumetric shrinkage, VS = $\left(\frac{V_1 - V_2}{V_2}\right) \times 100$	24.70		

Result. Shrinkage Limit = 20.5%.

30.13. Experiment No. 13. To determine the permeability of a soil specimen by the constant-head permeameter.

Theory. The coefficient of permeability is equal to the rate of flow of water through a unit cross-sectional area under a unit hydraulic gradient. In the constant head permeameter, the head causing flow through the specimen remains constant throughout the test. The coefficient of permeability (k) is obtained from the relation

$$k = \frac{qL}{Ah} = \frac{QL}{Aht}$$

where q = discharge; Q = total volume of water; t = time period; h = head causing flow; L = length of specimen; A = cross-sectional area.

Equipment. (1) Permeameter mould, internal diameter = 100 mm, effective height = 127.3 mm, capacity = 1000 ml; (2) Detachable collar, 100 mm diameter, 60 mm high; (3) Dummy plate, 108 mm diameter, 12 mm thick; (4) Drainage base, having a porous disc; (5) Drainage cap, having a porous disc with a spring attached to the top; (6) Compaction equipment, such as Proctor's rammer or a static compaction equipment; (7) Constant- head water-supply reservoir; (8) Vacuum pump; (9) Constant-head collecting chamber; (10) Stop watch; (11) Large funnel; (12) Thermometer; (13) Weighing balance, accuracy 0.1 g; (14) Filter paper.

Procedure. (1) Remove the collar of the mould. Measure the internal dimensions of the mould. Weigh the mould, with dummy plate, to the nearest gram.

(2) Apply a little grease on the inside to the mould.

Clamp the mould between the base plate and the extension collar, and place the assembly on a solid base.

(3) Take about 2.5 kg of the soil sample, from a thoroughly mixed wet soil, in the mould. Compact the soil at the required dry density, using a suitable compacting device.

(4) Remove the collar and base plate. Trim the excess soil level with the top of the mould.

(5) Clean the outside of the mould and the dummy plate. Find the mass of the soil in the mould.

(6) Take a small specimen of the soil in a container for the water content determination.

(7) Saturate the porous discs (stones).

(8) Place a porous disc on the drainage base, and keep a filter paper on the porous disc.

(9) Remove the dummy plate, and place the mould with soil on the drainage base, after inserting a washer in between.

(10) Clean the edges of the mould. Apply grease in the grooves around them.

(11) Place a filter paper, a porous disc and fix the drainage cap using washers.

(12) Connect the water reservoir to the outlet at the base, and allow the water to flow upwards till it has saturated the sample. Let the free water collect for a depth of about 100 mm on the top of the sample.

[Alternatively, the soil of low permeability can be saturated by subjecting the specimen to a gradually increasing vacuum with bottom outlet closed so as to remove air from the voids. Increase the vacuum gradually to 700 mm of mercury and maintain it for 15 minutes or more, depending upon the type of soil. Follow the evacuation by a process of slow saturation of the sample from the bottom upward under full vacuum. When the sample is saturated, close both the top and bottom outlets].

(13) Fill the empty portion of the mould with deaired water, without disturbing the soil.

(14 [illegible] connect the reservoir from the outlet at the bottom.

(1[illegible] [illegible]onnect the constant-head reservoir to the drainage cap inlet.

(16) Open the stop cock, and allow the water to flow downward so that all the air is removed.

(17) Close the stop cock, and allow the water to flow through the soil till a steady state is attained.

(18) Start the stop watch, and collect the water flowing out of the base in a measuring flask for some convenient time interval.

(19) Repeat this thrice, keeping the interval the same. Check that the quantity of water collected is approximately the same each time.

(20) Measure the difference of head (h) in levels between the constant head reservoir and the outlet in the base.

(21) Repeat the test for at least 2 more different intervals.

Data Sheet for Constant Head Permeameter

Diameter = 100 mm; Length = 120 mm; Volume = 942.48 ml, G = 2.67; Area = 7854 mm^2.

S.No.	*Observations and calculations*	*Determination No.*		
		1	*2*	*3*
	Observations			
1.	Mass of empty mould with base plate	5101 g		
2.	Mass of mould, soil and base plate	6918 g		
3.	Hydraulic head (h)	150 mm		
4.	Time interval (t)	600 s		
5.	Quantity of flow (Q)	1210 ml		
	(a) First time in period t	1205 ml		
	(b) Second time in period t	1215 ml		
	(c) Third time in period t	1210 ml		
	Average Q	1210×10^3 mm^3		
	Calculations			
6.	Mass of soil = (2) – (1)	1817 g		
7.	Bulk density, $\rho = \dfrac{\text{Mass}}{\text{Volume}}$	1.93 g/ml		
8.	Water-content, w, determined as in Test 30 – 1	14 %		
9.	Dry density, $\rho_d = \dfrac{\rho}{1 + w}$	1.69 g/ml		
10.	Void Ratio, $e = \dfrac{\rho_w G}{\rho_d} - 1$	0.58		
11.	$k = \dfrac{QL}{Aht}$	0.205 mm/sec.		

Result. Coefficient of permeability = 0.205 mm/sec.

30.14. Experiment No. 14. To determine the permeability of a soil specimen by the variable-head permeameter.

Theory. The variable-held permeameter is used to measure the permeability of relatively less pervious soils. The coefficient of permeability is given by

$$k = \frac{2.30\, a L}{A t} \log_{10} (h_1/h_2)$$

where h_1 = initial head; h_2 = final head; t = time interval; a = cross sectional area of the stand pipe, A = cross-sectional area of the specimen, L = length of specimen.

Equipment. All the equipment required for the constant-head permeability test (Experiment 30—13), and the following.

(1) Graduated glass stand pipe, 5 to 20 mm diameter.

(2) Supporting frame for the stand pipe, and the clamp.

Procedure. Steps 1 to 14, same as in Experiment 30-13.

(15) Connect the stand pipe of suitable diameter to the inlet at the top. Fill the stand pipe with water.

(16) Open the stop cock at the top, and allow the water to flow out till all the air in the mould is removed.

(17) Close the stop cock, and allow the water from the stand pipe to flow through the soil specimen.

(18) Select the heights h_1 and h_2 measured above the centre of the outlet such that their difference is about 300 to 400 mm.

Mark the level corresponding to a height $\sqrt{h_1 h_2}$.

(19) Open the valve and start the stop watch. Record the time interval for the head to fall from h_1 to $\sqrt{h_1 h_2}$, and also from $\sqrt{h_1 h_2}$ to h_2. These two time intervals will be equal if the steady conditions have established.

(20) Repeat the step (19) atleast twice, after changing the heights h_1 and h_2.

(21) Stop the flow of water. Disconnect all the parts.

(22) Take a small quantity of the soil specimen for the water content determination.

Data Sheet for Variable-head Permeameter Test

Length of specimen	= 120 mm. Diameter = 100 mm
Area of specimen	$= \pi/4 \times (100)^2 = 7853.98 \text{ mm}^2$.
Volume of specimen	$= \pi/4 \times (100)^2 \times 120 = 942.48 \times 10^3 \text{ mm}^3$
Water content	= 18%
Diameter of stand pipe	= 10 mm
Area of stand pipe,	$a = \pi/4 \times (10)^2 = 78.54 \text{ mm}^2$
Specific gravity of solids	= 2.67.

S. No.	*Observations and Calculations*	*Determination No.*		
		1	*2*	*3*
	Observations			
1.	Mass of mould + base plate	5090 g		
2.	Mass of mould + base plate + soil	7120 g		
3.	Initial head, h_1	500 mm		
4.	Final head, h_2	200 mm		
5.	Head $\sqrt{h_1 h_2}$	316 mm		
6.	Time Interval			
	h_1 to $\sqrt{h_1 h_2}$	25 s		
	$\sqrt{h_1 h_2}$ to h_2	25 s		
	h_1 to h_2	50 s		
	Calculations			
7.	Mass of soil = (2) – (1)	2030 g		

S. No.	Observations and Calculations	Determination No.		
		1	2	3
8.	Bulk density, $\rho = \dfrac{\text{Mass}}{\text{Volume}}$	2.15 g/ml		
9.	Dry density, $\rho_d = \dfrac{\rho}{1 + w}$	1.83 g/ml		
10.	Void ratio, $e = \dfrac{G \rho_w}{\rho_d} - 1$	0.46		
11.	$k = \dfrac{2.30\, a\, L}{A\, t} \log_{10} (h_1/h_2)$	0.022 mm/s		

Result. Coefficient of permeability = 0.022 mm/sec.

30.15. Experiment No. 15. To determine the consolidation characteristics of a soil sample.

Theory. Consolidation of a saturated soil occurs due to expulsion of water under a static, sustained load. The consolidation characteristics of soils are required to predict the magnitude and the rate of settlement. The following characteristics are obtained from the consolidation test. As per usual notations (see chapter 12),

Coefficient of compressiblty, $a_v = -\Delta e/\Delta \bar{\sigma}$

Coefficient of volume change, $m_v = \dfrac{-\Delta e}{1 + e}\left(\dfrac{1}{\Delta \bar{\sigma}}\right)$

Compression index, $C_c = \dfrac{-\Delta e}{\log_{10} \dfrac{(\bar{\sigma}_o + \Delta \bar{\sigma})}{\bar{\sigma}_o}}$

Coefficient of consolidation, $C_v = T_v\, d^2/t$

Equipment. (1) Consolidometer, with a loading device; (2) Specimen ring, made of a non-corroding material; (3) Water reservoir to saturate the sample; (4) Porous stones; (5) Soil trimming tools, like fine wire saw, knife, spatula, etc.; (6) Weighing balance, accuracy 0.01 g; (7) Oven; (8) Desiccator; (9) Pressure pad; (10) Steel ball; (11) Dial gauge, accuracy 0.002 mm; (12) Water content cans; (13) Large container.

Procedure. (1) Clean and dry the metal ring. Measure its diameter and height. Take the mass of the empty ring.

(2) Press the ring into the soil sample contained in a large container at the desired density and the water content. The ring is to be pressed with hands.

(3) Remove the soil around the ring. The soil specimen should project about 10 mm on either side of the ring.

Any voids in the specimen due to the removal of large size particles should be filled back by pressing the soil lightly.

(4) Trim the specimen flush with the top and bottom of the ring.

(5) Remove any soil particles sticking to the outside of the ring. Weigh the ring with the specimen.

(6) Take a small quantity of the soil removed during trimming for the water content determination.

(7) Saturate the porous stones by boiling them in distilled water for about 15 minutes.

(8) Assemble the consolidometer. Place the bottom porous stone, bottom filter paper, specimen, top filter paper and the top porous stone, one by one.

(9) Position the loading block centrally on the top porous stone.

Mount the mould assembly on the loading frame. Centre it such that the load applied is axial. In the case of the lever-loading system, counterbalance the system.

(10) Set the dial gauge in position. Allow sufficient margin for the swelling of the soil.

(11) Connect the mould assembly to the water reservoir having the water-level at about the same level as the soil specimen.

Allow the water to flow into the specimen till it is fully saturated.

(12) Take the initial reading of the dial gauge.

(13) Apply an initial setting load to give a pressure of 5 kN/m^2 (2.5 kN/m^2 for very soft soils) to the assembly so that there is no swelling.

Allow the setting load to stand till there is no change in the dial gauge reading or for 24 hours.

(14) Take the final gauge reading under the initial setting load.

(15) Apply the first load increment to apply a pressure of 10 kN/m^2, and start the stop watch.

Record the dial gauge readings at 0, 0.25, 1.0, 2.25, 4.0, 6.25, 9.00, 12.25, 16.00, 20.25, 25.00, 36, 49, 64, 81, 100, 121, 144, 169, 196, 225, 256, 289, 324, 361, 400, 500, 600 and 1440 minutes.

(16) Increase the load to apply a pressure of 20 kN/m^2, and repeat the step (15). Likewise, increase the load to apply a pressure of 40, 80, 160, 320 and 640 kN/m^2 or upto the desired pressure.

(17) After the last load increment had been applied and the readings taken, decrease the load to 1/4 of the last load, and allow it to stand for 24 hours. Take the dial gauge reading after 24 hours.

Further reduce the load to 1/4 of the previous load and repeat the above procedure. Likewise, further reduce the load to 1/4 of the previous load and repeat the procedure. Finally, reduce the load to the initial setting load, and keep it for 24 hours, and take the final dial gauge reading.

(18) Dismantle the assembly. Take out the ring with the specimen. Wipe out the excess surface water using a blotting paper.

(19) Take the mass of the ring with the specimen.

(20) Dry the specimen in the oven for 24 hours, and determine the dry mass of specimen.

Data Sheet for consolidation Test

Specific gravity of solids, G =

Diameter of ring =

Area of ring (A) =

Volume of ring =

Mass of ring + wet soil =

Mass of ring + dry soil =

Mass of water =

Initial height, H_0 =

Water content before test =

Mass of dry soil (M_s) =

Height of solids, $H_s = \dfrac{M_s}{G A \rho_w}$

Initial void ratio, $e_0 = \dfrac{H_0}{H_s} - 1$

Height of ring =

Mass of ring =

Degree of saturation $S = \dfrac{wG}{e}$

Water content after test =

***(a)* Coefficient of Compressibility**

$\bar{\sigma}$ (kN/m^2)	*Initial dial Reading*	*Final dial Reading*	*Change in hight* (ΔH)	*Height* $H = H_0 \pm \Delta H$	*Height of voids* $(H - H_s)$	*Final void ratio* $e = (H - H_s)/H_s$
10						
20						
40						
80						
160						
320						
640						

Plot a curve between $\bar{\sigma}$ as abscissa and final void ratio (e) as ordinate for determination of a_v and m_v.

Plot a graph between log $\bar{\sigma}$ as abscissa and final void ratio as ordinate for determination of C_c.

(b) **Coefficient of consolidation**

Dial gauge readings

$\bar{\sigma}$ / t	10 (kN/m^2)	20	40	80	160	320	640
	(R)						
0.0 min.	—						
0.25	—						
1.0	—						
.	.						
.	.						
.	.						
1440	—						

For each load increment, plot $\sqrt{t}$ as abscissa and the dial gauge reading (R) as ordinate. Determine the value of t_{90} from the plot.

Now
$$C_v = 0.848d^2/t_{90}$$

30.16. Experiment No. 16. To determine the shear parameters of a sandy soil specimen by direct shear test.

Theory. Shear strength of a soil is its maximum resistance to shearing stresses. The shear strength is expressed as

$$s = c' + \bar{\sigma} \tan \phi'$$

where c' = effective cohesion; $\bar{\sigma}$ = effective stress; and ϕ' = effective angle of shearing resistance.

The shear tests can be conducted under three different drainage conditions. The direct shear test is generally conducted on sandy soils as a consolidated-drained test.

Equipment. (1) Shear box, divided into two halves by a horizontal plane, and fitted with locking and spacing screws; (2) Box container to hold the shear box; (3) Base plate having cross grooves on its top surface; (4) Grid plates, perforated, 2 nos; (5) Porous stones, 6 mm thick, 2 nos; (6) Loading pad, (7) Loading frame; (8) Loading yoke; (9) Proving ring, capacity 2 kN; (10) Dial gauges, accuracy 0.01 mm, 2 nos.; (11) Static or dynamic compaction device; (12) Spatula.

Procedure. (1) Measure the internal dimensions of the shear box. Also determine the average thickness of the grid plates.

(2) Fix the upper part of the box to the lower part using the locking screws. Attach the base plate to the lower part.

(3) Place the grid plate in the shear box keeping the serrations of the grid at right angles to the direction of shear. Place a porous stone over the grid plate.

(4) Weigh the shear box with base plate, grid plate and porous stone.

(5) Place the soil specimen in the box. Tamp it directly in the shear box at the required density. When the soil in the top half of the shear box is filled upto 10 to 15 mm depth, level the soils surface.

(6) Weigh the box with the soil specimen.

(7) Place the box inside the box container, and fix the loading pad on the box. Mount the box container on the loading frame.

(8) Bring the upper half of the box in contact with the proving ring. Check the contact by giving a slight movement.

(9) Fill the container with water if the soil is to be saturated; otherwise omit this step.

(10) Mount the loading yoke on the ball placed on the loading pad.

(11) Mount one dial gauge on the loading yoke to record the vertical displacement and another dial gauge on the container to record the horizontal displacement.

(12) Place the weights on the loading yoke to apply a normal stress of 25 kN/m^2.

Allow the sample to consolidate under the applied normal stress. Note the reading of the vertical displacement dial gauge.

(13) Remove locking screws. Using the spacing screws, raise the upper part slightly above the lower part such that the gap is slightly larger than the maximum particle size.

Remove the spacing screws.

(14) Adjust all the dial gauges to read zero. The proving ring should also read zero.

(15) Apply the horizontal shear load at a constant rate of strain of 0.2 mm/minute.

(16) Record readings of the proving ring, the vertical displacement dial gauge and the horizontal displacement dial gauge at regular time intervals. Take the first few readings at closer intervals.

(17) Continue the test till the specimen fails or till a strain of 20% is reached.

(18) At the end of the test, remove the specimens from the box, and take a representative sample for the water content determination.

(19) Repeat the test on identical specimens under the normal stresses of 50, 100, 200, 400 kN/m^2, etc. (The range of stresses selected should correspond to the actual field conditions).

Data Sheet for Direct Shear Test

Size of box = Area of box =
Thickness of specimen = Volume of specimen =
Mass of soil specimen = Bulk density =
Water content = Dry density =
Void ratio = Tare mass of hanger =
Mass on hanger = Total mass =
Normal stress =
Mass of box + base plate + porous stones + grid plate =
Mass of box + base plate + porous stone + grid plate + soil specimen =

S. No.	*Observations*				*Calculations*			
	Elapsed time	*Horizontal dial gauge*	*Vertical dial gange*	*Proving ring*	*Shear displacement*	*Vertical displacement*	*Shear force*	*Shear Stress*

Use separate data sheet for tests under different normal stresses. Determine the shear stress at failure in each case. Summarise the results as follows.

Test No.	*Normal Stress*	*Shear Stress at failure*	*Shear displacement at failure*	*Initial water content*	*Final water content*
1.	25 kN/m^2				
2.	50				
3.	100				
4.	200				
5.	400				

Plot the Coulomb envelope between the normal stress as abscissa and shear stress at failure as ordinate.

Result. From the plot, c' = ; ϕ' =

30.17. Experiment No. 17. To determine the unconfined compressive strength of a cohesive soil.

Theory. The unconfined compressive strength (q_u) is the load per unit area at which the cylindrical specimen of a cohesive soil fails in compression.

$$q_u = \frac{P}{A}$$

where P = axial load at failure; A = corrected area = $\frac{A_0}{1 - \varepsilon}$, where A_0 initial area of the specimen; ε = axial strain = change in length/original length.

The undrained shear strength (s) of the soil is equal to one half of the unconfined compressive strength, $s = q_u/2$.

Equipment. (1) Unconfined compression apparatus, proving ring type; (2) Proving ring, capacity 1 kN, accuracy 1 N; (3) Dial gauge, accuracy 0.01 mm; (4) Weighing balance; (5) Oven; (6) Stop watch; (7) Sampling tube; (8) Split mould, 38 mm diameter, 76 mm long; (9)Sample extractor; (10) Knife; (11) Vernier callipers; (12) Large mould.

Procedure. (1) Prepare the soil specimen at the desired water content and density in the large mould.

(2) Push the sampling tube into the large mould, and remove the sampling tube filled with the soil. For undisturbed samples, push the sampling tube into the clay sample.

(3) Saturate the soil sample in the sampling tube by a suitable method.

(4) Coat the split mould lightly with a thin layer of grease. Weigh the mould.

(5) Extrude the sample out of the sampling tube into the split mould, using the sample extractor and the knife.

(6) Trim the two ends of the specimen in the split mould.

Weigh the mould with the specimen.

(7) Remove the specimen from the split mould by splitting the mould into two parts.

(8) Measure the length and diameter of the specimen with a vernier callipers.

(9) Place the specimen on the bottom plate of the compression machine.

Adjust the upper plate to make contact with the specimen.

(10) Adjust the dial gauge and the proving-ring gauge to zero.

(11) Apply the compression load to cause an axial strain at the rate of 1/2 to 2% per minute.

(12) Record the dial gauge reading, and the proving ring reading every thirty seconds upto a strain of 6%. The reading may be taken after every 60 seconds for a strain between 6% to 12%, and every 2 minutes or so beyond 12%.

(13) Continue the test until failure surfaces have clearly developed or until an axial strain of 20% is reached.

(14) Measure the angle between the failure surface and the horizontal, if possible.

(15) Take the sample from the failure zone of the specimen for the water content determination.

Data Sheet for Unconfined Compression Test

Initial length of the specimen, L_0 =

Initial diameter of the specimen, D_0 =

Initial area of the specimen, A_0 =

Initial volume of the specimen, V_0 =

Mass of empty split mould =

Mass of split mould + specimen =

Mass of the specimen, M =

Bulk density, $\rho = M/V_0$

Water content, w =

Dry density, ρ_d =

Specific gravity of solids, G =

Void ratio, $e = \frac{G\rho_w}{\rho_d} - 1$

Degree of saturation, $S = \frac{wG}{e} \times 100$

S. No.	Observations					Calculations		
	Elapsed time	Dial gauge		Proving ring		Strain	Corrected	Compressive
		Reading	Deformation (ΔL)	Reading	Load (P)	$\varepsilon = \frac{\Delta L}{L_0}$	area $A = A_0/(1 - \varepsilon)$	stress $(\sigma) = P/A$

Plot a curve between the compressive stress as ordinate, and axial strain, as abscissa.

Results. From the plot, unconfined compressive strength, q_u =

$$\text{Shear strength, } s = \frac{q_u}{2} =$$

20.18. Experiment No. 18. To determine the compaction characteristics of a soil specimen by Proctor's test.

Theory. Compaction is the process of densification of soil by reducing air voids. The degree of compaction of a given soil is measured in terms of its dry density. The dry density is maximum at the optimum water content. A curve is drawn between the water content and the dry density to obtain the maximum dry density and the optimum water content.

$$\text{Dry density} = \frac{M/V}{1 + w}$$

where M = total mass of soil; V = volume of soil; and w = water content.

Equipment. (1) Compaction mould, capacity 1000 ml; (2) Rammer, mass 2.6 kg; (3) Detachable base plate; (4) Collar, 60 mm high; (5) IS sieve 4.75 mm; (6) Oven; (7) Desiccator; (8) Weighing balance, accuracy 1 g; (9) Large mixing pan; (10) Straight edge; (11) Spatula; (12) Graduated jar; (13) Mixing tools, spoons, trowels, etc.

Procedure. (1) Take about 20 kg of air-dried soil. Sieve it through 20 mm and 4.75 mm IS sieves.

(2) Calculate the percentage retained on 20 mm sieve, and 4.75 mm sieve, and the percentage passing 4.75 mm sieve. Do not use the soil retained on 20 mm sieve. Determine the ratio of fraction retained and that passing 4.75 mm sieve.

(3) If percentage retained on 4.75 mm sieve is greater than 20, use the larger mould of 150 mm diameter. If it is less than 20%, the standard mould of 100 mm diameter can be used. The following procedure is for the standard mould.

(4) Mix the soil retained on 4.75 mm sieve and that passing 4.75 mm sieve in the proportions determined in Step (2) to obtain about 16 to 18 kg of soil specimen.

(5) Clean and dry the mould and the base plate. Grease them lightly.

(6) Weigh the mould with the base plate to the nearest 1 gram.

(7) Take about 16—18 kg of soil specimen. Add water to it to bring the water content to about 4% if the soil is sandy and to about 8% if the soil is clayey.

(8) Keep the soil in an air-tight container for about 18 to 20 hours for maturing. Mix the matured soil thoroughly. Divide the processed soil into 6 to 8 parts.

(9) Attach the collar to the mould. Place the mould on a solid base.

(10) Take about $2\frac{1}{2}$ kg of the processed soil, and place it in the mould in 3 equal layers.

Take about one-third the quantity first, and compact it by giving 25 blows of the rammer. The blows should be uniformly distributed over the surface of each layer.

The top surface of the first layer should be scratched with a spatula before placing the second layer. The second layer should also be compacted by 25 blows of rammer. Likewise, place the third layer and compact it.

The amount of the soil used should be just sufficient to fill the mould and leaving about 5 mm above the top of the mould to be struck off when the collar is removed.

(11) Remove the collar, and trim off the excess soil projecting above the mould using a straight edge.

(12) Clean the base plate and the mould from outside. Weigh it to the nearest gram.

(13) Remove the soil from the mould. The soil may also be ejected out.

(14) Take the soil samples for the water content determination from the top, middle and bottom portions. Determine the water content, as in Experiment 30—1.

(15) Add about 3% of the water to a fresh portion of the processed soil, and repeat the steps 10 to 14.

Data Sheet for Compaction Test

Diameter of mould = 100 mm Height of mould = 127.3 mm

Volume of mould, $V = \pi/4 \times (10.0)^2 \times 12.73 = 1000$ ml

Specific gravity of solids, $G = 2.67$

S. No.	*Observations and Calculations*	*Determination No.*					
		1	2	3	4	5	6
	Observations						
1.	Mass of empty mould + base plate	5105 g					
2.	Mass of mould + base plate + compacted soil	6710 g					
	Calculations						
3.	Mass of compacted soil, $M = (2) - (1)$	1605 g					
4.	Bulk density, $\rho = \frac{M}{V}$	1.61 g/ml					
5.	Water content, w (Determined as in Sect. 30-1)	9 %					
6.	Dry density, $\rho_d = \frac{\rho}{1 + w}$	1.48 g/ml					
7.	Void ratio, $e = \frac{G \rho_w}{\rho_d} - 1$	0.80					
8.	Dry density at 100% saturation (ρ_d) theomax $= \frac{G \rho_w}{1 + wG}$	2.15 g/ml					
9.	Degree of saturation, $S = \frac{wG}{e} \times 100$	30 %					

Plot a curve between w, as abscissa, and ρ_d as ordinate.

Result. Max. dry density (from plot) =

Optimum water content (from plot) =

30.19. Experiment No. 19. To determine the California Bearing Ratio (CBR) of a soil specimen.

Theory. The California Bearing Ratio test is conducted for evaluating the suitability of the subgrade and the materials used in sub-base and base of a flexible pavement.

The plunger in the CBR test penetrates the specimen in the mould at the rate of 1.25 mm per minute. The loads required for a penetration of 2.5 mm and 5.0 mm are determined. The penetration load is expressed as a percentage of the standard loads at the respective penetration level of 2.5 mm or 5.0 mm.

$$\text{CBR value} = \frac{\text{Penetration load}}{\text{Standard load}} \times 100$$

The CBR value is determined corresponding to both penetration levels. The greater of these values is used for the design of the pavement.

Equipment. (1) CBR mould, inside diameter = 150 mm, total height = 175 mm, with detachable extension collar, 50 mm high, and detachable base plate, 10 mm thick.

(2) Spacer disc, 148 mm diameter, 47.7 mm high.

(3) Rammers, light compaction, 2.6 kg, drop 310 mm; heavy compaction, 4.89 kg, drop 450 mm.

(4) Slotted masses, annular, 2.5 kg each, 147 mm diameter, with a hole of 53 mm diameter in the centre.

(5) Cutting collar, steel, which can fit flush with the mould both outside and inside.

(6) Expansion measuring apparatus, consisting of a perforated plate, 148 mm diameter, with a thread screw in the centre and an adjustable contact head to be screwed over the stem, and a metal tripod.

(7) Penetration piston, 50 mm diameter, 100 mm long.

(8) Loading device, capacity 50 kN, equipped with a movable head (or base) at a uniform rate of 1.25 mm minute.

(9) Two dial gauges, accuracy 0.01 mm.

(10) IS sieves, 4.75 mm and 20 mm size.

Procedure. (1) Sieve the sample through 20 mm IS sieve. Take the material passing 20 mm IS sieve for the test. However, make allowance for large size material by replacing plus 20 mm size material by an equal amount of material which passes 20 mm IS sieve, but is retained on 4.75 mm IS sieve.

(2) Take about 4.5 to 5.5 kg of the material, as obtained in step (1). Mix it thoroughly with the required quantity of water.

If the sample is to be compacted at optimum water content and the corresponding dry density, as found by compaction test (light compaction or heavy compaction), take exact quantity of water and the soil to make sure that the water content is equal to the optimum water content.

(3) Fix the extension collar to the top of the mould. Also fix the base plate to the bottom.

(4) Insert the spacer disc over the base, with the central hole of the disc at the lower face. Place coarse filter paper disc on the top of the displacer disc.

(5) Take the soil sample in the mould. Compact it using either the light compaction rammer or the heavy compaction rammer, as desired. For light compaction, the soil is to be compacted in 3 equal layers, each layer is given 56 blows by 2.6 kg rammer with drop of 310 mm. For heavy compaction, the soil is compacted in 5 equal layers, each layer is given 56 blows by 4.89 kg rammer with drop of 450 mm.

(6) Remove the extension collar. Trim even the excess compacted soil carefully with a straight edge with the top of the mould. Any hole that may form on the surface of the compacted soil by the removal of the coarse particles should be patched with small size particles and levelled.

(7) Loosen the base plate. Remove the base plate and the spacer disc.

(8) Weigh the mould with the compacted soil.

(9) Place a filter paper disc on the base plate. Invert the mould with the compacted soil. Clamp the base plate. Place a perforated disc fitted with an extension stem on the specimen top after placing a filter disc.

(10) Place annular masses to produce a surcharge equal to the mass of the base material and wearing cost of the pavement expected. Each 2.5 kg annular mass is equivalent to 70 mm of construction material. However, a minimum of two annular masses should be placed.

(11) Immerse the mould assembly in a tank full of water. Allow free access of water to the top and bottom of the specimen.

(12) Mount the tripod of the expansion measuring device on the edge of the mould, and take the initial reading of the dial gauge.

(13) Keep the mould in the tank undisturbed for 96 hours. Take the readings of the dial gauge every 24

hours, and note the time of reading.

Maintain water level constant in the tank. Take the final reading at the end of 96 hours.

(14) Remove the tripod. Take out the mould from the tank. Allow the specimen to drain off for 15 minutes. Remove all the free water on the mould, without disturbing the surface of the specimen.

(15) Weigh the mould with the soaked specimen.

(16) Place the mould containing the specimen, with the base plate in position but the top face exposed, on the lower plate of the loading machine. Place the required surcharge masses on the top of the soaked specimen.

To prevent upheaval of soil into the hole of the surcharge mass, one 2.5 kg annular mass shall be placed on the soil surface prior to seating the penetration plunger. After that the remaining masses are placed.

(17) Seat the penetration plunger at the centre of the specimen to establish full contact between the plunger and the specimen. The seating load should be about 40 N.

(18) Set the load dial gauge and the displacement dial gauge to zero. The initial load already applied to the plunger should be considered as zero.

(19) Apply the load on the plunger. Keep the rate of penetration as 1.25 mm/minute.

Record the load corresponding to penetrations of 0.0, 0.5, 1.0, 1.5, 2.0, 2.5, 3.0, 4.0, 5.0, 7.5, 10.0 and 12.5 mm. However, record the maximum load and the corresponding penetration if it occurs at a penetration of less than 12.5 mm.

(20) At the end of the test, raise the plunger, and remove the mould from the loading machine.

(21) Take about 20 to 50 g of soil specimen from the top 30 mm layer for the water content determination. If the water content of the whole specimen is required, take soil specimens from the entire depth.

Data Sheet for California Bearing Ratio Test

Optimum water content =
Mass of empty mould =
Mass of mould + compacted soil =
Mass of compacted soil =
Bulk density =
Dry density =

Soaking and Swelling

Dry density before soaking =
Bulk density before soaking =
Bulk density after soaking =
Surcharge mass during soaking =

Date and time					
Dial gauge reading					
Total expansion					

$$\text{Final expansion ratio} = \frac{\text{Final reading} - \text{initial reading}}{\text{Initial height}}$$

Penetration Test

Surcharge mass used =

Water content after penetration test =

S. NO.	*Penetration dial gauge*		*Load dial gauge*		*Corrected load*
	Dial gauge reading	*Penetration (mm)*	*Dial gauge reading*	*Load (kN)*	
1.		0.0 mm			
2.		0.5			
3.		1.0			
4.		1.5			
5.		2.0			
6.		2.5			
7.		3.0			
8.		4.0			
9.		5.0			
10.		7.5			
11.		10.0			
12.		12.5			

Plot the load-penetration curve. Find the corrected loads, after zero correction, corresponding to penetrations of 2.5 mm and 5.0 mm.

Result.

$$\text{CBR (2.5 mm)} = \frac{\text{Corrected load at 2.5 mm}}{13.44} \times 100$$

$$\text{CBR (5.0 mm)} = \frac{\text{Corrected load at 5.0 mm}}{20.16} \times 100$$

31
Introduction to Rock Mechanics

31.1. INTRODUCTION

Rock mechanics is the science which deals with the properties of rocks and the special methods related with the design and construction of engineering works on the rock, through the rock and below the rock. The basic knowledge of rock mechanics is useful for civil engineers. Some of the common civil engineering applications are bridges, dams and buildings constructed on the rock foundations, various underground installations and tunneling, deep cuts for spillways, stone quarries, etc.

Rock mechanics is an inter-disciplinary subject covering various disciplines such as geology, mining, petroleum industry and civil engineering. As defined in chapter 1, rock is the consolidated, coherent and relatively hard portion of the earth's crust. It is a naturally formed, solidly bonded mass of mineral matter which cannot be readily broken by hand, and which does not disintegrate on its first wetting and drying cycle.

In general, rocks are strong materials and can take up much more loads as compared to that by soils. However, the strength of the rocks is also limited. The rock may fail when the loads are excessively high. The actual behaviour of a rock mass subjected to a change in stress is governed by the mechanical properties of the rock material and the geological discontinuities such as faults, joints, fissures, etc.

Large bridges, high dams, tall buildings, long tunnels and deep mines induce large stress changes in the rock mass. For exact analysis and proper prediction of the rock behaviour, a complete understanding of the engineering properties and the behaviour of rock masses under different loading conditions is essential. This chapter gives a brief introduction to the engineering properties and behaviour of rocks.

31.2. GEOLOGICAL CLASSIFICATION OF ROCKS

According to the geological classification, the rocks can be broadly classified into 3 groups.

1. Igneous rocks
2. Sedimentary rocks
3. Metamorphic rocks.

1. Igneous rocks. Igneous rocks are formed by solidification of molten or liquid material called *magma*. The igneous rocks have the minerals augite, feldspar, horne blende, mica, quartz, etc. Before solidification into rocks, all these mineral are in the molten state. The igneous rocks may be crystalline or glossy (vitreous) or a combination of both. The igneous rocks may be further subdivided according to the grain size and colour.

Example of igneous rocks are granite, basalt, diorite, etc.

2. Sedimentary rocks. Sedimentary rocks are formed from mineral particles which had been transported by water, wind, etc., or which had been precipitated in water and then solidified. Thus the sedimentary rocks are formed by the deposition of fragments of materials like sand, clay, disintegrated rocks, etc. on the pre-existing rocks. The process of deposition continues in regular layers. In due course of time, the deposited mass becomes a sedimentary rock. Because the sediments get consolidated in horizontal or nearly horizontal layers, the sedimentary rocks show characteristic different layers distinctly. The sedimentary rocks can be easily split up along the bedding planes. Sedimentary rocks come from many sources and include distinctly different families.

Examples of sedimentary rocks are sand stone, shale, lime stone, etc.

3. Metamorphic rocks. Metamorphic rocks are formed from igneous and sedimentary rocks by very large heat and pressure. Sometimes, metamorphic rocks are also formed from already existing metamorphic rocks. The process of change to metamorphic rock due to heat and pressure is called *metamorphism*. Due to metamorphism, the original rocks change their character and the resulting mass of rock change into a hard and durable foliated material.

Examples of metamorphic rocks are quartzite, marble, gneiss, slate, schist, etc.

31.3. BASIC TERMINOLOGY

The following terms are commonly used.

1. Rock material. The consolidated aggregate of mineral particles which form the solid material of the rock is called rock material. In other words, rock material is the matrix of the solid phase of the rock mass.

2. Intact rock. The intact rock consists of rock material which is free from major discontinuities. An intact rock is generally quite strong. Moreover, it can be readily sampled and tested in the laboratory.

3. Major discontinuity. The major discontinuity is a relatively large discontinuity in the rock which is fairly well-developed and continuous. However, the shear failure along a major discontinuity does not involve any shearing of intact rock material.

4. Structural discontinuity. The structural discontinuity of the rock is a fracture which makes the rock mass a discontinuous mass and separates solid blocks of a rock mass.

The structural discontinuities are generally is the form of joints, bedding planes, fissures, faults, shear zones, cleavage planes, and solution cavities. These are planes of weakness due to which the strength of rock mass is considerably reduced.

5. Rock mass. The rock mass is defined as the aggregate of regular or irregular blocks of rock material. These blocks are separated from one another by structural features such as bedding planes, joints, cavities and fissures. The rock mass has anisotropy and structural discontinuity.

The rock mass is conceived of discrete intact blocks separated by thin joints. The behaviour of the rock mass is governed by a combination of blocks and joint characteristics.

6. Block size. The block size is defined as the average diameter of a typical rock. It is usually expressed as a range and a typical values. For example, the block size (10-40 cm), 25 cm.

The concept of block size in rock mechanics is analogous to that of grain size on a microscopic scale.

7. Joint Set. A joint set consists of individual joints which have similar physical and mechanical characteristics and which occur in a nearly parallel array. The joints of a rock mass are usually subdivided into two or more such sets, which together constitute the jointing system. The characteristics of different joint sets are generally different because of different geological origin and history.

8. Faults and shears. These are joints along which there has been shearing movement in the rock mass.

9. Rock quality designation. Rock quality designation (RQD) is frequently used to indicate the quality of rock mass. The rock quality designation is defined as the sum total of lengths of the cores of the length 10 cm and longer recovered from the drilling, expressed as a percentage of the total length of the hole drilled (Deere *et al*, 1967). Thus

$$\text{RQD} = \frac{\text{Total length of cores in pieces of 10 cm and longer}}{\text{Length of run}} \times 100 \qquad \text{...(31.1)}$$

Eq. 31.1. is for core sticks of NX cores (57.2 mm diameter).

The RQD of a rock mass indicates the quality of the rock. It is a measure of the degree of fracturing and the amount of weathering in the rock mass. The rocks with RQD from 100 to 90 are classified as excellent rocks, and those with RQD from 90 to 75 as good rocks. The rocks with RQD from 75 to 50 are fair rocks, whereas those with 50 to 25 are poor rocks. The rocks with RQD less than 25 are very poor rocks.

10. Joint and Bedding Description. As already mentioned, the strength of a rock mass is considerably reduced due to the presence of discontinuities. In general, the greater the number of discontinuities, the more is the reduction in strength. Moreover, the degree of such reduction depends upon the spacing, orientation and size of the discontinuities.

Table 31.1. gives the general quality of rock mass depending upon the spacing of discontinuities in the form of joints and bedding (Deere, 1968 and Singh, 1990).

TABLE 31.1. Quality of rock based on spacing of discontinuities.

Average spacing of discontinuties	*Joint Description*	*Bedding Description*	*Rock quality*
Greater than 3m	Very wide	Very thick	Solid rock
1 to 3 m	Wide	Thick	Massive rock
0.3 to 1 m	Moderately close	Medium	Blocky (seamy) rock
0.05 to 0.30 m	Close	Thin	Fractured rock
Less than 0.05 m	Very close	Very thin	Crushed rock

31.4. INDEX PROPERTIES OF ROCKS

Because of a vast range in properties of rocks due to different structures, fabrics and materials, it is difficult to determine the properties and to describe rocks quantitatively. Certain properties that are relatively easy to measure are used to classify the rocks. These properties are known as the index properties of rocks.

The following are the main index properties of rocks :

1. Unit weight (or mass density)
2. Porosity
3. Permeability
4. Point-load index
5. Slaking and durability
6. Sonic velocity

All these properties are discussed in the following sections.

31.5. UNIT WEIGHT (OR MASS DENSITY)

The mass density (ρ) of a rock is the mass per unit volume. It is expressed as kg/m^3. In rock mechanics, the term unit weight is commonly used. The unit weight (γ) is the weight per unit volume. It is expressed as kN/m^3. Sometimes, the term density is also used for unit weight.

The specific gravity of solids (*G*) is the ratio of the mass density of the rock to the mass density of water. The mass density of water is usually taken as 1000 kg/m^3. The specific gravity of solids is also equal to the ratio of the unit weight of rock to the unit weight of water. The unit weight of water is 9.81 kN/m^3 or approximately 10 kN/m^3. Thus the rock with a specific gravity of the solids of the rock of 2.7 will have a unit weight of 27 kN/m^3. The specific gravity of the grains of the rock material can be determined by grinding it and using the methods discussed in chapter 3 for soil particles.

Rocks show a wider range of unit weights as compared to soils. The unit weight of the rock can be determined from the volume and the weight of a carefully drilled core specimen. Table 31.2. gives the average values of the dry unit weights of some common rocks.

TABLE 31.2. Dry unit weights of some common rocks

Rock	*Gabbro*	*Diorite*	*Basalt*	*Marble*	*Granite*	*Shale*	*Gypsum*	*Dense lime stone*
Dry unit weight (kN/m^3)	29.4	27.9	27.1	27.0	26.0	22.1 to 25.7	22.5	20.9

The dry unit weight can be obtained from the wet unit weight by the relation

$$\gamma_d = \frac{\gamma}{1 + w} \quad \text{...(31.2)}$$

where γ_d is the dry unit weight, γ is the wet unit weight and w is the water content.

A knowledge of unit weight of rock is important for various engineering applications. For example, cement concrete made from heavy aggregates would require a smaller volume of concrete in a retaining wall than that from light aggregates. In oil shale deposits, the unit weight indicates the amount of the mineral commodity, as the oil yield is related directly to the unit weight. In coal deposits, the unit weight is related to the ash content.

31.6. POROSITY

The porosity (n) of a rock is defined as the ratio of the void space to the total volume in the rock. It is expressed as a percentage. Thus

$$n = \frac{V_a}{V} \times 100 \quad \text{...(31.3)}$$

where V_a is the volume of voids (or pores) and V is the total volume.

The porosity of a rock depends upon a number of factors, such as particle-size distribution, sorting, grain shape, degree of compaction and cementation, fabric, solution effects and mineralogical composition. The porosity is high when all the particles are of the same size. However, in the case of irregularly shaped particles, there is a wide range of porosity.

In igneous rocks, porosity is usually less than 1 or 2%, unless weathering has occurred. As weathering takes place, the porosity gradually increases and may become 20% or even more. Hence, the value of porosity in igneous rocks serves as an index to the quality of rock.

In sedimentary rocks, porosity is usually high. It may vary from almost zero to as much as 90% in some extreme cases. For an average sand stone, its typical value is 15%. Chalk is one of the most porous of all rocks, with a porosity in some cases even greater than 50%. In sedimentary rocks, porosity generally decreases with the age of rock. It also decreases with an increase with the depth below the surface.

In unweathered rocks, the mechanical properties such as the unconfined compressive strength and the modulus of elasticity can be related to porosity, but there is a large scatter, and correlation is not perfect.

The porosity (n) of a rock can be determined from the water content (w) by the relation

$$n = \frac{wG}{1 + wG} \quad \text{...(31.4)}$$

where G is the specific gravity of solids.

If the pores of the rock are filled with mercury, the porosity can be determined from the relation

$$n = \frac{w_m (G/G_m)}{1 + (w_m \cdot G/G_m)} \quad \text{...(31.5)}$$

where w_m is the mercury content, expressed as a proportion of the weight of dry weight of the rock before mercury injection, and G_m is the specific gravity of mercury (usually taken as 13.546).

31.7. PERMEABILITY

As already defined in the case of soils, the permeability is the ease with which a fluid can flow through it. The permeability of rock depends upon the interconnections of its pores and fissures. It has been found that Darcy's law is valid for most of rocks. For flow of water at about 20°C, the discharge (q) can be obtained from the relation

$$q = k\left(\frac{dh}{dx}\right) \cdot A \quad \text{...(31.6)}$$

where k is the coefficient of permeability, dh is the hydraulic head acting on the distance dx and A is the area of cross-section normal to the direction of flow in which dx is measured.

For flow of water at a temperature much different from 20°C or flow of fluids other than water, Eq. 31.6 is modified as

$$q = \left(\frac{K}{\mu}\right)\left(\frac{dp}{dx}\right) A \qquad \text{...(31.7)}$$

where K is the coefficient of absolute permeability, with a dimension of $(L^2]$. As already mentioned in chapter 8, K is independent of the properties of the permeant (fluid). It is usually expressed in terms of darcy, where 1 darcy = 0.987×10^{-8} cm^2.

μ is the coefficient of viscosity of the permeant. For water at 20°C, μ = 1.005 mN- s/m^2.

dp is the change in pressure (Note. $p = \gamma h$).

The coefficient of permeability (k) of a rock specimen can be determined in the laboratory by measuring the volume (Q) of fluid passed through the specimen under a constant air pressure acting over the surface of the fluid for a known period of time t (*i.e.* $q = Q/t$).

Alternatively, the permeability of the rock can be determined by the radial-flow test. In this test, a hollow cylindrical specimen of the rock is prepared by drilling a coaxial central hole in a core. The coefficient of permeability is determined from the relation.

$$k = \frac{q \log_e (R_2/R_1)}{2\pi L \Delta h} \qquad \text{...(31.8)}$$

where L is the length of the specimen, Δh is the difference of head across the test region, R_1 is the inner radius of the specimen and R_2 is the outer radius of the specimen.

In the radial permeability test, a very large flow gradient can be easily generated to measure even very small permeabilities quite accurately. Fissured rocks show higher permeability for outward flow as compared to that for inward flow. It is due to the fact that when the flow is outward, a tensile body force is developed which opens out the fissures. On the other hand, when the flow is inward, a compressive force is developed which closes the fissures.

The measurement of permeability of the rock is useful for estimating flow of water, oil or gas into or out of rock formation. It is also useful for determining the water tightness of the reservoir, dewatering a deep chamber and predicting the flow of water into a tunnel.

Table 31.3. gives some typical values of the coefficient of permeability of different rocks for water at 20°C.

TABLE 31.3. Typical values of the coefficient of permeability

Type of rock	*Basalt*	*Shale*	*Granite*	*Schist*	*Lime stone*	*Sand stone*
Coefficient of permeability (cm/sec.)	1×10^{-12}	1×10^{-9} to 5×10^{-13}	1×10^{-7} to 1×10^{-11}	1×10^{-8}	1×10^{-5} to 1×10^{-13}	3×10^{-3} to 8×10^{-8}

31.8. POINT LOAD STRENGTH

The point load strength is frequently used to estimate the strength of the rock. The point load strength is usually determined by a test given by Broch and Franklin (1972). In this test, a rock specimen is loaded between hardened steel cones till failure occurs by the development of tensile cracks parallel to the axis of loading. The point load strength or point load index (I_s) is given by

$$I_s = \frac{P}{D^2} \qquad \text{...(31.9)}$$

where P is the load at rupture and D is the distance between the point loads (Fig. 31.1).

The length of the pieces of drill core used for the test should be at least 1.4 times the diameter. It has been found that the strength depends upon the size of the specimen. The point load strength reduces to 1/2 to

1/3 as the diameter of the core is increased from 10 mm to 70mm. For comparison purposes, it is the standard practice to report the point load strength for a core of 50 mm diameter. If the diameter of the test specimen is different from 50 mm, a suitable size-correction is applied.

For reliable results, it is essential that the specimen should be carefully prepared. The results are highly sensitive to the method and style of loading. The point load strength test is quick and simple and is fairly reliable. It can even be conducted in the field at the site of drilling. For rocks with strong anisotropy, such as slates, shales, schists and laminated stones, the test should be conducted to determine the strength along as well as perpendicular to the bedding planes.

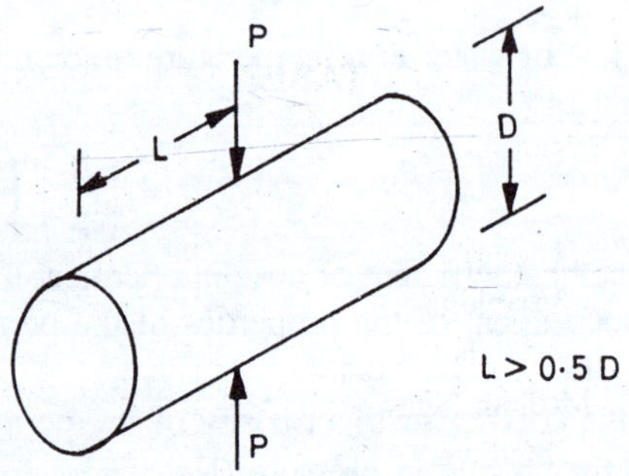

Fig. 31.1.

The point load strength is related to the unconfined compressive strength by the relation

$$q_u = 24\, I_{s\,(50)} \qquad \ldots(31.10)$$

where q_u is the unconfined compressive strength of a cylinder of rock with a length to diameter ratio of 2 to 1, and $I_{s(50)}$ is the point load strength corrected to a diameter of 50 mm. Eq. 31.10 does not give correct results for weak rocks for which special correlation studies are required.

Table 31.4 gives some typical values of the point load index for different rocks.

TABLE 31.4. Point Load Index

Type of rock	*Dolomite*	*Volcanic flow rocks*	*Lime stone*	*Shale*	*Tertiary sand stone and clay stone*
Point load index	6.0-11 MPa	3.0-15 MPa	0.25-8 MPa	0.2-8 MPa	0.05-1 MPa

31.9. SLAKING AND DURABILITY

To indicate the durability of a rock, some sort of index is required. However, because natural processes affecting the durability of the rock are quite large in number and they are also varied, no index can completely describe the durability.

Durability of the rock is essential for various engineering applications. For long life, the rock should be durable. It should not disintegrate; otherwise its life and strength would be reduced. The rock may disintegrate by various process occurring in nature, such as exfoliation, hydration, slaking, solution, oxidation, abrasion, etc.

The following two indexes are commonly used to determine the durability of a rock.

1. Slake durability index 2. Change in liquidity index.

1. Slake durability index. The slake durability index can be determined using the apparatus suggested by Franklin and Chandra (1972). It consists of a drum 140 mm in diameter and 100 mm in length. A sieve mesh forms the cylindrical walls, with a 2 mm size openings in it.

About 500 g of rock is broken into 10 lumps and placed inside the drum. The drum in turned at 20 r.p.m. in a water bath for about 10 minutes. The rock retained inside the drum is then weighed. The slake durability index (I_d) is determined as follows:

$$I_d = \frac{\text{Rock retained in the drum}}{\text{Total dry weight of the rock}} \times 100 \qquad \ldots(31.11)$$

Table 31.5 gives the classification of rocks based on the slake durability index, as proposed by Gamble (1971).

TABLE 31.5. Gamble's Slake Durability Classification

Durability	*Slake durability index after one 10-minute cycle*	*Slake durability index after two 10-minute cycles*
Very high	Greater than 99	Greater than 98
High	98 — 99	95 — 98
Medium high	95 — 98	85 — 95
Medium	85 — 95	60 — 85
Low	60 — 85	30 — 60
Very low	Less than 60	Less than 30

2. Change in Liquidity Index. Morgenstern and Eigenbrod (1974) used a water absorption test to determine the amount and rate of slaking of argillaceous materials. They came to the conclusion that non-cemented claystone or shale absorbed water faster than other materials. Moreover, all materials eventually attained a final water content equal to their liquid limits. It has been established that the materials with high liquid limit are more severely affected by slaking as compared to those with low liquid limit.

Table 31.6 gives the amount of slaking for different values of liquid limit.

TABLE 31.6. Amount of slaking for different liquid limits

Liquid Limit	*Less than 20%*	*20 — 50%*	*50 — 90%*	*90 — 140%*	*Greater than 140%*
Amount of slaking	Very low	Low	Medium	High	Very high

The rate of slaking depends upon the change in the liquidity index (ΔI_L) after immersion in water for 2 hours from the relation

$$\Delta I_L = \frac{\Delta w}{w_L - w_p} \qquad \ldots(31.12)$$

where Δw = change in water content of the rock after soaking for 2 hours on filter paper in a funnel,

w_L = water content at the liquid limit,

w_p = water content at the plastic limit.

All the water contents in Eq. 31.12 must be expressed as a percentage of dry weight. The methods for the determination of Δw, w_L and w_p are the same as those for soil, discussed in chapter 4. Based on the rate of slaking, the rocks are classified slow if ΔI_L is less then 0.75, fast when ΔI_L is between 0.75 and 1.25 and very fast when ΔI_L is greater than 1.25.

31.10. SONIC VELOCITY

The sonic velocity is the velocity with which stress waves travel through an intact rock. The sonic velocity for an intact rock depends upon its elastic properties and density (unit weight). If there are fissures in the rock, the sonic velocity is reduced. Therefore, the extent of fissures and discontinuities in a rock mass can be assessed by comparing the in-situ wave velocity with the sonic velocity of an intact rock core obtained from the same rock mass as determined in the laboratory.

Fourmaintraux (1976) gave a classification system based on longitudinal wave velocity to indicate fissuring in rock specimens. The theoretical longitudinal wave velocity (V_l^*) that the specimen could have if there were no pore or fissures can be estimated from the mineral composition of the rock from the relation

$$\frac{1}{V_l^*} = \sum_i^n \frac{C_i}{V_{l,i}} \qquad \ldots(13.13)$$

where $V_{l,i}$ is the longitudinal wave velocity in mineral constituent i which has a volume proportion C_i in the rock and n is the number of such minerals.

Table 31.7. gives the typical values of V_l^* for a few rocks.

TABLE 31.7. Typical values of longitudinal velocity

Rock	*Gabbro*	*Basalt, dolomite*	*Lime stone*	*Sand stone, quartzite*	*Granite*
Longitudinal Velocity (V_l^*) (m/s)	7000	6500 — 7000	6000 — 6500	6000	5500 — 6000

The actual longitudinal wave velocity (V_l) in the rock specimen as determined in the laboratory is less than the theoretical velocity (V_l^*). The quality index (IQ) is the ratio of the actual velocity to the theoretical velocity, expressed as a percentage. Thus

$$IQ = \left(\frac{V_l}{V_l^*}\right) \times 100 \qquad ...(31.14)$$

Fournmaintraux established that IQ is affected by pores (spherical holes) in the rock, and can be expressed as percentage

$$IQ = 100 - 1.6\, n_p\% \qquad ...(31.15)$$

where n_p is the porosity of the non-fissured rock expressed as a percent.

It may be noted that Eq. 31.15 does not give reliable results even when there is a small fraction of flat cracks (fissures) in the rock. Based on laboratory measurements and microscopic observations of fissures in the rock, Fourmaintraux proposed a plot between IQ and porosity n which can be used to determine the degree of fracturing of a rock specimen (Fig. 31.2).

There are 5 zones in the plot. For example, zone I is for non- fissured to slightly fissured rocks, and zone II for slightly fissured to moderately fissured rocks.

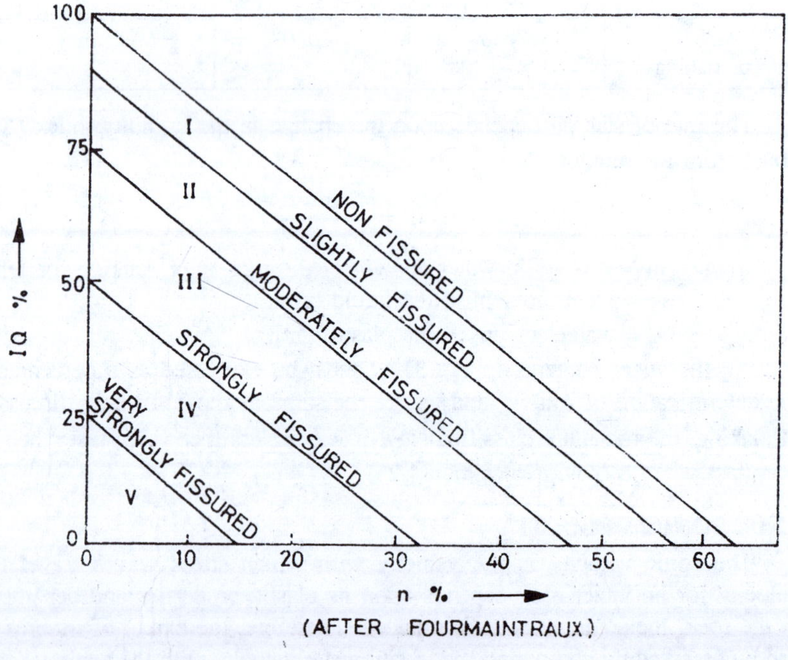

Fig. 31.2

31.11. CLASSIFICATION OF ROCKS FOR ENGINEERING PURPOSES

Bieniawski (1974) gave the geomechanics classification system, which is commonly used for engineering purposes. It provides rock mass rating (RMR) values for different rocks. The RMR value depends upon the rock quality and it may vary from 0 to 100. The RMR is derived from 5 universal parameters given below (Goodman, 1980).

1. Strength of rock
2. Drill core quality
3. Joint spacing
4. Joint characteristics
5. Ground water conditions

Sometimes, orientation of joints is also considered as the sixth parameter, especially for specific applications in tunneling, mining and foundations.

Increments of rock mass rating corresponding to each of the above parameters are determined. These depend upon the characteristics of the rock as explained below. All these increments are added up to determine the RMR value.

1. Strength of rock. The RMR value depends upon the unconfined compressive strength of the rock. The unconfined compressive strength (q_u) of the rock can be determined from a laboratory compression test on a prepared core of the rock. However for the classification of rocks, an approximate value of the compressive strength as determined from the point-load test on intact pieces of drill core can be used. The following equation is commonly used.

$$q_u = 25\, I_s \qquad \ldots(31.15)$$

where I_s is the point load strength.

Table 31.8 gives the rock mass rating increments depending upon the point load index and unconfined compressive strength of the rock. The increments vary from 15 to 0.

TABLE 31.8. Rock mass rating increments depending upon compressive strength

Point load Index I_s (MPa)	*Unconfined Compressive Strength* (MPa)	*Rock mass rating increment*	
Greater than 8	Greater than 200	15	
4 — 8	100 — 200	12	
2 — 4	50 — 100	7	
1 — 2	25 — 50	4	
—	10 — 25	2*	* Do not use this rock.
—	3 — 10	1*	
—	Less than 3	0*	

2. Drill core quality. The drill core quality is related to the rock quality designation (RQD). As already discussed, the RQD of a rock is determined from the percentage recovery of core in lengths greater than 10 cm for NX cores (57.2 mm diameter). For other types of cores, the percentage recovery is determined in lengths greater than twice the corresponding diameter.

The rock mass rating increments for drill core quality depend upon the RQD value. The increments vary from 20 to 3 (see Table 31.9).

TABLE 31.9. Rock mass rating increments for drill core quality

RQD (%)	91 — 100	76 — 90	51 — 75	25 — 50	Less than 25
Rating increment	20	17	13	8	3

3. Joint spacing. The rock mass rating depends upon the spacing of joints. The rating increment should reflect the joint set which is the most critical for a particular application. If the rock mass has fewer sets of joints, the rating is increased.

The rock mass rating increment depends upon the spacing of the most influential joint (Table 31.10). The increments vary from 30 to 5.

TABLE 31.10. Rock mass rating increment for spacing of joints

Joint spacing (m)	Greater than 3	1 — 3	0.3 — 1	0.005 — 0.3	Less than 0.005
Rating increment	30	25	20	10	5

4. Joint characteristics. The rock mass rating also depends upon the characteristics of joints. The condition of joint sets most likely to influence a particular application should be considered for the rating. The increment of rock mass rating for joint characteristics depends upon the roughness of the joint. The description of joint surface rougness and coating material is weighed towards the smoothest and weakest joint set.

Table 31.11 gives the rating increments depending upon the joint characteristics. The increments vary from 25 to 0.

TABLE 31.11. Rock mass rating increment for joint condition

S.No.	*Joint Description*	*Rating increment*
1.	Very rough surfaces of limited extent; hard wall rock	25
2.	Slightly rough surface; aperature less than 1 mm; hard wall rock	20
3.	Slightly rough surface; aperature less than 1 mm; soft wall rock	12
4.	Smooth surface, or gauge filling 1-5 mm thick or aperatu., of 1-5 mm; joints extending more than several meters.	6
5.	Open joints filled with more than 5 mm of gouge, or open more than 5 mm; joint extending more than several metres	0

5. Ground water condition. Ground water conditions can influence the rock mass behaviour to a large extent. The rock mass rating depends upon the general condition of the rock, which, in turn, depends upon the ground water. The rock is assigned the category dry, moist, water under moderate pressure or severe water problems and the rating increments vary from 10 to 0 (Table 31.12).

The rating increment can also be related to the inflow per 10 m length in lit/min if an exploratory adit or pilot tunnel is available at the site for measurement of water inflow. It can also be related to the joint water pressure divided by the major principal stress, as given in Table 31.12. It varies from 10 to 1.

TABLE 31.12. Increments of rock mass rating due to ground water condition

S. No.	*General condition*	*Inflow per 10 m tunel length (lit/min)*	*Joint water pressure divided by major principal stress*	*Rating increment*
1.	Completely dry	None	0	10
2.	Moist	25	0.0 — 0.2	7
3.	Water under Moderate pressure	25 — 125	0.2 — 0.5	4
4.	Severe water problems	125	0.5	1

6. Orientation of joints. The orientation of the joints relative the the work under consideration can have an effect on the behaviour of the rock. Accordingly, the sum of the first 5 rating numbers discussed above is adjusted. If the orientation of joints is very favorable for the work under consideration, no points are subtracted from the sum. However, for unfavorable orientations, there are negative ratings as given in Table 31.13. For tunnels, the rating decrements are from 0 to 12 and for foundation, the rating decrements are from 0 to 25, depending on the orientation of joints. For assessment of influence of orientation of the joints for a particular work, the advice of an engineering geologist is usually required.

TABLE 31.13. Adjustments in rock mass rating for joint orientation

Assessment of Influence of orientation of the work	*Rating increments for tunnels*	*Rating increments for foundations*
Very favourable	0	0
Favourable	– 2	– 2
Fair	– 5	– 7
Unfavourable	– 10	– 15
Very unfavourable	– 12	– 25

Rock mass Rating (RMR). As already mentioned, the rock mass rating is obtained by adding up the rating increments given in Tables 31.8 to 31.12 and adjusting the sum for orientation of joint as per Table 31.13. Table 31.14 gives the geomechanics classification of rock masses based on RMR.

For example, a rock mass with the particulars given below will have the RMR of 84, and the rock will be classified as very good rock as per Table 31.14.

1. Point load index of 6 MPa	= 12 (Table 31.8)
2. RQD of 80%	= 17 (Table 31.9)
3. Joint spacing 2 m	= 25 (Table 31.10)
4. Very rough surface	= 25 (Table 31.11)
5. Moist condition	= 7 (Table 31.12)
Total	= 86
6. Favourable orientation for foundation	= –2 (Table 31.13)
RMR	= 84

TABLE 31.14. Geomechanics classification of Rock masses

Class	*Description of Rock mass*	*RMR*
I	Very good rock	81 — 100
II	Good rock	61 — 80
III	Fair rock	41 — 60
IV	Poor rock	21 — 40
V	Very poor rock	0 — 20

31.12. STRENGTH CLASSIFICATION OF INTACT ROCKS

Intact rocks can be classified according to the uniaxial compressive strength. Deere and Miller (1966) suggested the classification given in Table 31.15. The uniaxial compressive strength is determined by testing a rock specimen with a length to diameter ratio of 2.

TABLE 31.15. Strength classification of Intact rocks

Uniaxial compressive strength (MPa)	*Class*	*Description*
Greater than 224	A	Very high strength
112 — 224	B	High strength
56 — 112	C	Medium strength
28 — 56	D	Low strength
Less than 28	E	Very low strength

Deere and Miller also classified the intact rocks according to the modulus ratio. The modulus ratio is the ratio of the modulus of elasticity to the uniaxial compressive strength. The modulus of elasticity used in the classification is the tangent modulus obtained from the stress-strain curve of the specimen in the uniaxial compression test at a stress level equal to one half the ultimate strength of the rock. Table 31.16 gives the classification based on the modulus ratio.

TABLE 31.16. Rock Classification Based on Modulus Ratio

Modulus ratio	*Class*	*Description*
Over 500	H	High modulus ratio
200 — 500	M	Medium modulus ratio
Less than 200	L	Low modulus ratio

Combined classification based on strength and modulus ratio

The classification based on compressive strength and that on the modulus ratio can be combined to indicate the quality of rock. Two letters are used for classification. The first letter indicates the strength rating and the second letter indicates the modulus ratio rating. For example, a high strength rock with a medium modulus ratio will be designated as BM.

31.13. LABORATORY TESTS FOR DETERMINATION OF STRENGTH OF ROCKS

The following laboratory tests are commonly used for determination of the strength of rocks :

1. Unconfined compression test
2. Triaxial compression test
3. Splitting tension test
4. Beam bending test
5. Ring shear test

All these tests are briefly described below (Goodman, 1980)

1. Unconfined compression test. The unconfined compression test is the most commonly used strength test on rocks. For accurate results, the test should be performed carefully. The specimen should be in the form of a cylinder of length to width ratio varying from 2 to 2.5. The ends of the specimen should be flat, smooth and parallel. The ends should be exactly perpendicular to the axis of the cylinder. Cores obtained during explorations are usually trimmed for this purpose. The specimen is subjected to compression between the cross-head and the platen of a compression testing machine (Fig. 31.3)

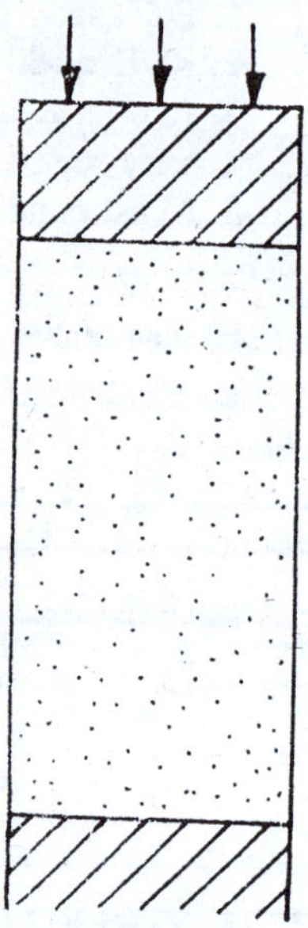

Fig. 31.3.

According to IS : 9143-1979, the specimen should preferably have a diameter of 45 mm. In no case, the diameter should be less than 35 mm. The load should be applied continuously with a stress rate of 0.5 to 1.0 MPa per second.

Table 31.17 gives some typical values of compressive strength of some rocks.

The compressive strength (q_u) is determined from the relation

$$q_u = P/A \qquad ...(31.16)$$

where P is the peak load and A is the initial area of cross-section of the specimen.

2. Triaxial compression Test. The triaxial compression test in principle is similar to that used for soils (chapter 13). The cylindrical specimen is first subjected to the lateral pressure and then deviatoric stress is applied. As the stresses are quite large, a special type of equipment is required for conducting the test (Fig. 31.4).

The usual procedure is to first apply the confining pressure (p) all around the cylinder and then to apply the deviatoric stress when the confining pressure is kept constant.

The all round pressure p increases the strength of the rock. However, the increase in strength is realised only when the specimen is enclosed in an impervious jacket. Normally, hydraulic oil is used as a confining fluid. The jacket is usually made of oil-resistant rubber such as polyurethane.

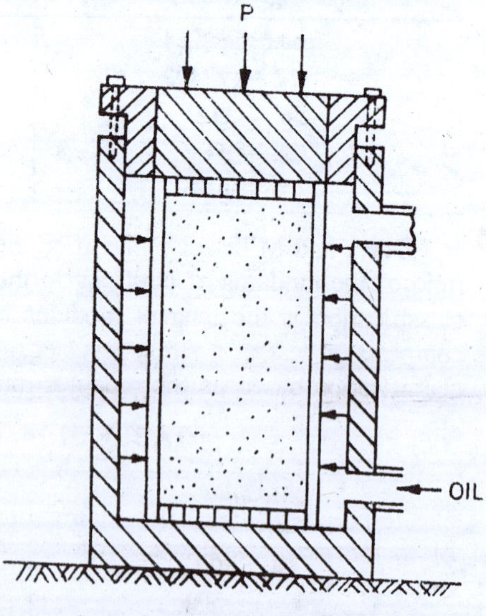

Fig. 31.4

3. Splitting Tension Test. The splitting tension test is also called the Brazilian test. In this test, the rock specimen is split by applying the load along the diametric plane. The rock core having a length-diameter ratio of unity when loaded on its side in a compression testing machine splits along the diameter and parallel to the cylindrical axis (Fig. 31.5). The horizontal stresses perpendicular to the loaded diameter are uniform and tensile. The tensile stress is given by

$$\sigma_t = \frac{2P}{\pi d t} \qquad ...(31.17)$$

where P is the applied load, d is the diameter of the cylinder and t is the thickness of the disc (*i.e.* the length of the cylinder).

IS : 10082–1981 recommends that the diameter of the specimen should be at least 45 mm and the thickness of the disc shall be approximately equal to half the diameter. The load should be applied at the rate of 0.2 kN/sec. The load should be measured at least to an accuracy of 1%.

The splitting tension test generally gives a value of tensile strength higher than that obtained from a direct tension test. This is probably due to the less effect of fissures in the splitting tension test as compared to that in a direct tension test.

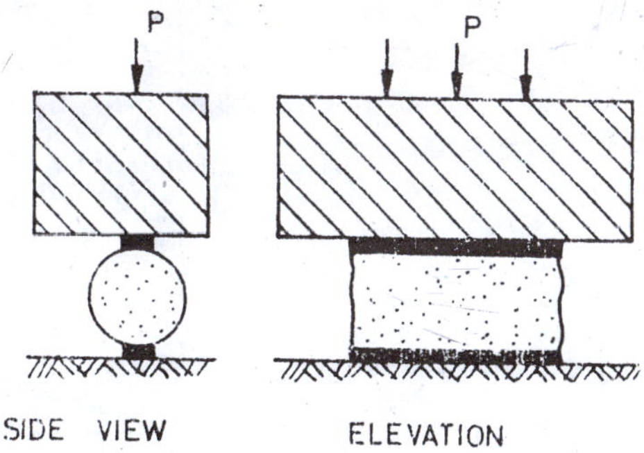

Fig. 31.5.

4. Beam Bending Test. The beam bending test is also called the *flexural test.* In this test, the rock beam is subjected to bending till failure occurs. It is usually conducted on rock cores obtained from the site exploration. Generally, 4-point flexural loading system is used (Fig. 31.6). The bottom surface of the beam is supported at two points, one near each end. The top of the beam is loaded at the third points. This system produces pure bending (without shear) in the middle third of the beam.

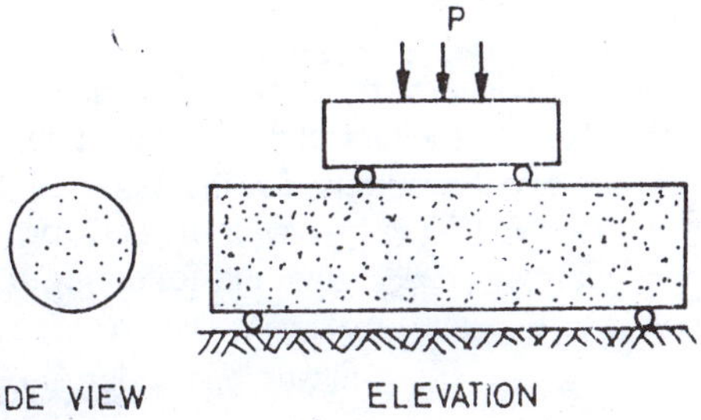

Fig. 31.6

The flexural strength (or modulus of rupture) is the maximum tensile stress at the bottom surface of the core corresponding to the peak load. It is calculated from the simple beam theory assuming that the material remains elastic right upto the failure. The flexural strength (T_{mr}) is given by

$$T_{mr} = \frac{16PL}{3\pi d^3} \quad \ldots(31.18)$$

where P is the maximum load at failure, L = span *i.e.* distance between reactions on the lower surface and d is the diameter of the core

5. Ring shear test. The ring shear test is generally used to test in-situ rocks. It gives the shear strength of the rock as a function of the confining pressure. The rock core specimen in this test does not require perfectly square and smooth ends.

The confining pressure is applied by the load parallel to the axis of core. As the load is applied to the plunger, two sets of complex fracture surfaces form along the two planes of the imposed shear (Fig. 31.7). The shear strength is calculated as

$$\tau_p = \frac{P}{2A} \quad \ldots(31.19)$$

where P is the peak load and A is the area of cross-section of the specimen.

As in the case of the triaxial test, there is a substantial increase of strength due to the confining pressure.

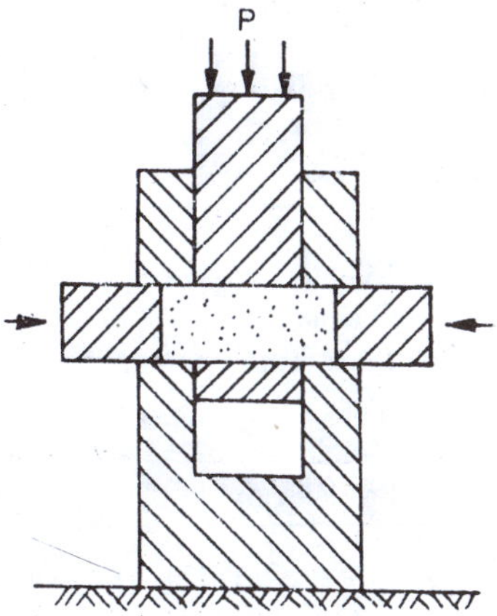

Fig. 31.7

31.14. STRESS-STRAIN CURVES

The modulus of elasticity and the Poisson ratio are two important elastic properties of a rock. The elastic properties depend upon a number of factors such as the size of the rock mass, constituent minerals, the degree of fracture, etc.

1. Modulus of elasticity. The modulus of elasticity of rock is determined from the stress-strain curve obtained from a uniaxial compression test. Most of the rocks have the stress-strain curve of the *S*-shape. At low stress levels, the modulus of elasticity increases with an increase in stress. The initial tangent modulus is

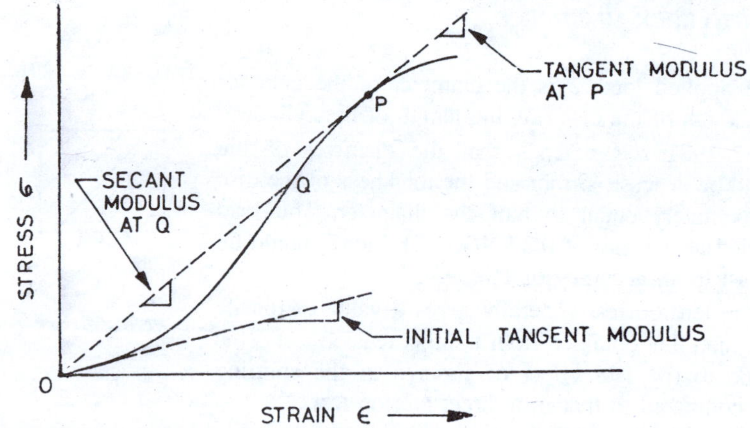

Fig. 31.8.

the slope of the tangent drawn to the curve at the origin. The secant modulus in a particular range is obtained by joining the stress point with the origin and determining the slope of that line. The tangent modulus at any stress is equal to the slope of the tangent drawn at that point. The initial tangent modulus, the secant modulus and the tangent modulus are shown in Fig. 31.8. Generally, the modulus of elasticity is taken equal to the 50% tangent modulus, *i.e.* the slope of the tangent drawn at one-half the failure stress.

IS : 7317-1974 gives a method for the determination of the modulus of elasticity.

Farmer (1968) gave the following three types of elastic behaviour in rocks.

1. Quasi-elastic behaviour   2. Semi-elastic behaviour   3. Non-elastic behaviour

1. The quasi-elastic behaviour is observed in fine-grained compact and massive rocks. The stress-strain curve in this case is almost linear right upto failure. The initial tangent modulus ranges from 6×10^4 to 11×10^4 MPa.
2. The semi-elastic behaviour is observed in coarse grained igneous rocks and also in fine-grained compacted sediments. In this case, the stress-strain curve shows a decrease in slope with increasing stress. The initial modulus ranges from 4×10^4 to 7×10^4 MPa.
3. The non-elastic behaviour is observed in rocks which are less cohesive and which have large pore space. The initial modulus is usually less than 5×10^4 MPa. Table 31.17 gives some typical values of modulus of elasticity of some rocks.

TABLE 31.17. Typical values of the compressive strength and modulus of elasticity of some rocks.

Type of rock	*Compressive strength (q_u)* (MPa)	*Modulus of elasticity (E)* (MPa)
Diabase	490.0	1,00,000
Quartzite	460 — 200	78,000
Dolerite	330	84,000
Granite	230.0	56,000
Lime stone	225.0	53,000
Shale	170.0	68,000
Marble	153 — 77	48,000 — 65,000
Sand stone	37.0	9,700

2. Poisson's ratio. The Poisson ratio (ν) is the ratio of the lateral strain to the axial strain of a rock. Thus

$$\nu = \frac{\varepsilon_d}{\varepsilon_a}$$

where ε_d is the diametric strain (or circumferential strain) of the rock and ε_a is the axial strain. The value of ν generally varies between 0.125 and 0.340 for most rocks.

In rocks subjected to uniaxial compression, ν remains more or less constant with an increase in stress, and it reaches the theoretical maximum value of 0.50 at failure.

The laboratory uniaxial compression test can be conducted on a rock to obtain the circumferential and axial strain. For the measurement of strains various devices, such as electrical resistance strain gauge, compressometer, optical instruments, are used.

IS: 9221-1979 recommends that at least two circumferential strain and two axial strain measurements should be taken for each increment of load. Moreover, the gauge length should be at least 5 times the grain size diameter.

31.15. MODES OF FAILURE OF ROCKS

The following modes of failure may occur in rock (Goodman, 1980):

1. Flexure failure
2. Shear failure
3. Direct tension failure
4. Crushing or compression failure

However, in actual practice, there are a large number of load configuration and no single mode of failure predominates. The actual failure pattern is usually highly complex in which two or more modes of failure may occur simultaneously.

1. Flexure failure. The flexure failure occurs because of the bending action in the rock. Because of bending, tensile cracks develop and propagate and cause failure. For example, flexure failure occurs in the rock layers above the roof of a mine [Fig. 31.9 (*a*)]. A gap is formed in the roof of the mine and a beam of

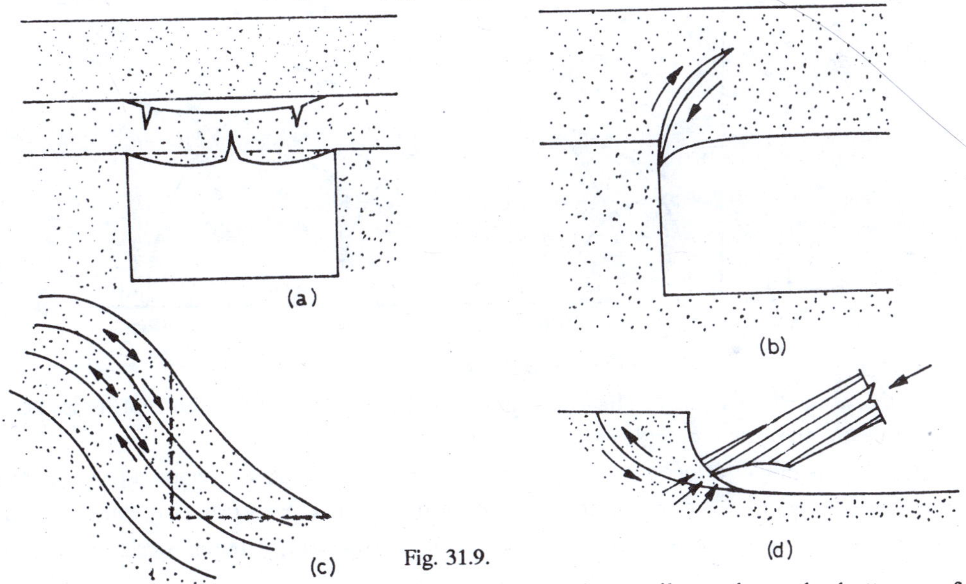

Fig. 31.9.

rock sags downward under its own weight. As the sag increases, the tensile cracks at the bottom surface of the beam propagate upward and the beam ultimately fail. It causes the rock to become loose and fall.

2. Shear failure. This type of failure occurs when a surface of rupture is formed in the rock because of the shear stresses becoming critical. After the formation of the rupture surface, there is release of shear stresses as the rock suffers a displacement along the rupture surface. This type of failure may occur in slopes cut in weak rocks such as weathered clay shale and crushed rock of fault zones. Fig 31.9 (*b*) slows shear failure in a mine which has a stiff ore and a softer (soil-like) roof. The shear stresses developing in the roof or pillar base allow the pillar to punch relatively upward in the roof. If the floor is weak, the pillar may punch relatively downward into the floor.

3. Direct tension failure. Direct tension occasionally occurs in rock. For example, the rock layers resting on convex upward slope surfaces are subjected to direct tension [(Fig. 31.9 (*c*)]. In this case, the layers at the base of the slope are inclined more steeply than what is allowed by friction. To resist the gravity forces, the balance of support is provided by the stable part of the slope above by direct pull. As the pull is increased in the rock layers, the tension failure may occur in which two adjacent parts of the rock are pulled apart.

When the rock breaks in direct tension, a surface of rupture is formed. The rupture surface is quite rough and free from the crushed rock particles and fragments. The surface is quite different from that found in shear failure, which is slick and has more powder formed from the crushing of rock.

4. Crushing or compression failure. Very high direct compression occurs in intensely shortened rock when penetrated by a stiff punch. When the compressive stresses reach the limiting value, the crushing failure may occur.

The crushing failure mode is highly complex. It includes formation of tensile cracks and their growth and interaction through flexure. Fig. 31.9 (*d*) shows an example of crushing and tensile cracking which is followed by shear failure.

31.16. MOHR-COULOMB CRITERION FOR ROCKS

The Mohr-Coulomb criterion is commonly used for determining the shear strength of rocks. As discussed in chapter 13, the Mohr-Coulomb criterion represents a linear envelope which touches all Mohr's circles drawn for critical combination of principal stresses (σ_1 and σ_3) at which failure occurs (Fig. 31.10). The criterion may be written as (Goodman, 1980).

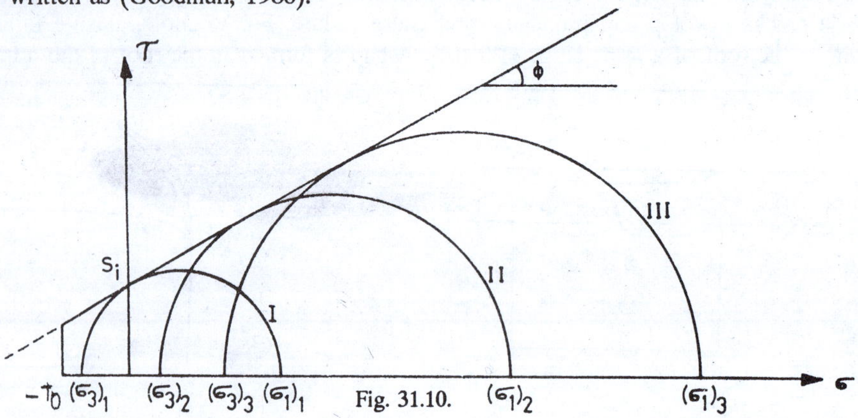

Fig. 31.10.

$$\tau_p = S_i + \sigma \tan \phi \qquad \text{...(31.20)}$$

where τ_p is the peak shear stress or shear strength, S_i is the shear strength intercept, σ is the normal shears on the shear plane, and φ is the angle of internal friction.

The Mohr-Coulomb envelope is extrapolated into the tensile stress region upto the point where σ_3 becomes equal to the uniaxial tensile strength (T_o). The minor principal stress (σ_3) can never be less than T_o.

In terms of the principal stresses at peak load condition, the Mohr-Coulomb criterion can be written as

$$\sigma_{1,p} = q_u + \sigma_3 \tan^2 (45° + \phi/2) \qquad \text{...(31.21)}$$

where $\sigma_{1,p}$ is the major principal stress corresponding to the peak of the stress-strain curve and q_u is the unconfined compressive strength.

The unconfined compressive strength (q_u) is given by

$$q_u = 2 S_i \tan (45° + \phi/2) \qquad \text{...(31.22)}$$

From Eqs. 31.21 and 31.22,

$$\frac{\sigma_{1,p}}{q_u} = 1 + \left(\frac{\sigma_3}{q_u}\right) \tan^2 (45° + \phi/2) \qquad \text{...(31.23)}$$

In general,

$$\frac{\sigma_{1,p}}{q_u} = 1 + N(\sigma_3/q_u)^M$$

or

$$\frac{\sigma_{1,p}}{q_u} - 1 = N(\sigma_3/q_u)^M \qquad \text{...(31.24)}$$

The constants M and N in Eq. 31.24 can be determined by fitting a curve to the family of points drawn between (σ_3/q_u) and $(\sigma_1/q_u - 1)$ obtained from the test.

The maximum tension criterion must be superimposed on the Mohr- Coulomb criterion (Eq. 31.21). It means that the failure will occur because of tensile stress wherever the minor principal stress σ_3 becomes equal to $-T_o$, whatever may be the value of σ_1.

Empirical Failure Envelope

The Mohr-Coulomb criterion is quite simple and easy to use. But it gives unconservative results when one of the principal stresses is negative (*i.e.* tensile) A more accurate failure criterion can be obtained by fitting an envelope to Mohr circles corresponding to stresses at failure as obtained from four types of tests,

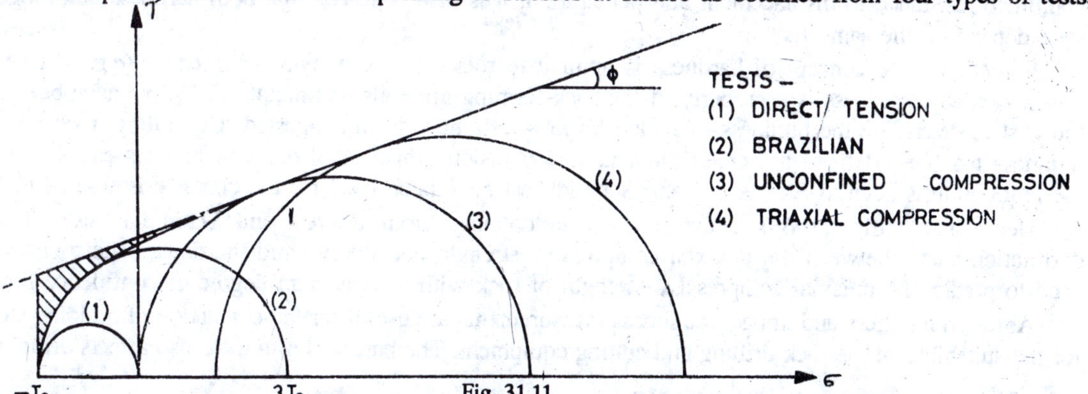

Fig. 31.11

namely, direct tension test, Brazilian test, unconfined compression test and triaxial test. Fig. 31.11 shows the empirical failure envelope. The envelope is generally curved downward and the shape is in-between a straight and a parabola.

It may be noted that the empirical failure envelope lies beneath the Mohr-Coulomb criterion with the superimposed tension cutoff (shown dotted). Inside the hatched portion, the Mohr-Coulomb criterion with tension cutoff overestimates the strength. Therefore, it is necessary to reduce the tensile strength T_o and the shear strength intercept S_i when applying the Mohr-Coulomb criterion to practical problems involving such conditions.

31.17. SHEAR STRENGTH OF ROCKS

The shear strength of rocks is generally found from the triaxial compression test and adopting the Mohr-Coulomb criterion (Eq. 31.20). As already mentioned, the peak stresses are considered for drawing the envelope. Hence the shear strength parameters S_i and ϕ in Eq. 31.20 are for the peak strength.

As in the case of dense sands discussed in chapter 13, the Mohr-Coulomb criterion can also be adopted for the residual stresses, *i.e.* the stresses at the ultimate failure conditions which occur after the peak stresses. The residual shear strength is the minimum strength reached by the rock when subjected to deformations beyond the peak stress. In this case, the subscript r is generally used to identify them as parameters for the residual strength. Thus Eq. 31.20 becomes

$$\tau_r = S_{i,r} + \sigma \tan \phi_r \qquad \ldots(31.25)$$

The residual shear strength intercept $(S_{i,r})$ may approach a value of zero. However, the value of the residual frictional angle ϕ_r usually lies between zero and the peak friction angle (ϕ).

When water is added to some rocks, the strength is considerably reduced because of the chemical deterioration of cement or clay binder of the rocks. Friable sand stone may even lose 15% of its strength by mere saturation.

A substantial reduction in shear strength occurs when pore water pressure (p_w) develops inside the pores and fissures of the rock. For such cases, Terzaghi's effective stress law (as discussed in chapter 13) can be used. Thus Eq. 31.20 becomes

$$\bar{\tau}_p = S_i + \bar{\sigma} \tan \phi \qquad ...(31.26)$$

where $\bar{\sigma}$ is the effective stress ($= \sigma - p_w$).

Eq. 31.26 indicates that the pore water pressure p_w will cause the same reduction in the peak normal stress as that caused by a reduction of confining pressure by an equal amount of p_w.

31.18. HARDNESS OF ROCKS

Hardness of a material is its ability to resist scratching. The relative hardness is used as an index for the identification of minerals. Moh's hardness scale is commonly used for this purpose. The use of hardness for identification of rocks is however quite complex and not well-established. In fact, rock hardness indicates the combined resistance to displacement and penetration. It is also observed that both hardness and toughness of rock depend on the same factors.

Sometimes, the concept of hardness is applied to rocks as a synonym for strength (*e.g.* soft rock, hard rock). Actually, hardness is a property of the rock-forming minerals and metals. Attempts have been made in the past to determine the hardness of rocks. Various tests have been suggested. Depending upon the test, the hardness has been defined as scratch, indentation, abrasion, impact and rebound hardnesses. However, only the rebound tests such as Shores test and Schmidt test have been used for the characterisation of rocks.

Deere and Miller (1966) analysed the data obtained from Shores' and Schmidt's tests. They gave correlation charts between the uniaxial compressive strength and the Schmidt hardness. These charts can be used to predict the uniaxial compressive strength of rock with a fairly high degree of confidence limits.

Abrasion hardness and impact toughness measurements are useful for the evaluation of building stones and for the suitability of the rock drilling and cutting equipment. The hardness of a rock also affects drillability.

31.19. IN-SITU STRESSES IN ROCKS

At a point below the rock surface of the undisturbed rock mass, there are stresses due to weight of the overlying materials, and also due to the confinement and past stress history. These stresses are known as in-situ stresses. The in-situ stresses vary considerably from one point to the other. The in-situ stresses may be almost zero at some points, whereas at some other points, they may be very high, even approaching the failure stress. When the in-situ stresses are almost zero, rock mass may fall from the surface and underground excavations because joints are open and weak. On the other hand, when the in-situ stresses are very high, any small disturbance to the stress field by tunneling or even excavation may trigger violent release of stored energy in rock and cause failure.

The in-situ stresses may be approximated and the order of their magnitude determined by various methods, but the accuracy of estimating is always doubtful. Accurate field measurements are usually required to ascertain the margin of error. Field measurements of in-situ stresses are quite common in the mining practice. However, in the civil engineering practice, the field measurement are rarely done because of high cost of measurements.

A basic knowledge of the in-situ stresses in quite useful in several civil engineering problems such as follows (Goodman, 1980).

1. Orientation for a cavern. While selecting the orientation for a cavern, care is taken to avoid alignment along the longer dimension perpendicular to the major principal stress σ_1.

2. Selection of shape. The shape is selected to minimise the stress concentration if the initial stresses are very high.

3. Layout of complex underground works. Cracks in rocks tend to extend in the plane perpendicular to the minor principal stress σ_3. A knowledge of the direction of the stresses is useful in selecting a layout to reduce the risk.

4. Saving in lining of tunnels. If the in-situ stress in the rock in which a pressure tunnel (or penstock) is constructed, is greater than the internal water pressure, lining need not be provided.

5. Large surface excavation. Substantial economy can be effected when making large surface excavation with pre-splitting technique by orienting the excavation perpendicular to the minor principal stress.

In-situ Stresses

1. *Vertical Stress.* The in-situ vertical stress at any point in the rock can generally be taken equal to the weight of the overlying rock per unit area. Thus

Vertical stress, $\sigma_v = \gamma z$ ...(31.27)

where z is the depth of the point and γ is the unit weight of the rock, which is usually taken as 27 kN/m^3.

When the ground surface is horizontal, the principal stress directions are vertical and horizontal at points near the surface. It is generally assumed that the directions are the same even for points at large depths below the surface, although the actual directions are somewhat different. However, this assumption is not justified in the case of hilly terrains. Beneath a valley side, there is one principal stress normal to the slope, which is equal to zero as the stresses normal to a free surface cannot exist. The other two principal stresses lie in the plane of slope (Fig. 31.12). These stresses approach zero when the rock slope is convex upward, but have high values when the slope is concave upward. In the case of a steep, *V*-shaped valleys, these stresses may even approach the failure stress or the strength of the rock.

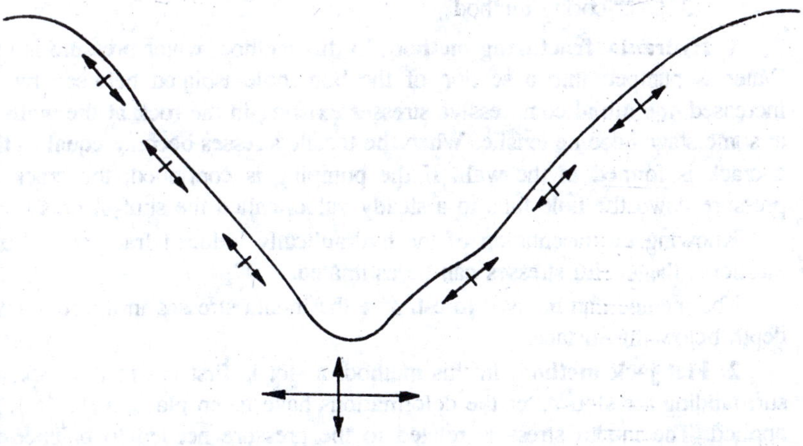

Fig. 31.12

2. *Horizontal stress.* The magnitude of the horizontal stress (σ_h) is usually estimated from the relation

$$\sigma_h = K\sigma_v \qquad ...(31.28)$$

where K is the lateral stress coefficient.

In the case of recent sedimentation, the value of K can be taken equal to $\nu/(1-\nu)$, as obtained from the theory of elasticity (chapter 11). In general, the value of K usually lies between K_a and K_p, where K_a corresponds to conditions for normal faulting and K_p corresponds to conditions for reverse faulting.

In the normal faulting, the vertical stress is the major principal stress and failure occurs due to horizontal extension. (This is similar to the active conditions discussed in chapter 19). Assuming Coulomb's law, the value of K_a is given by

$$K_a = \cot^2(45° + \phi/2) - \left[\left(\frac{q_u}{\gamma}\right)\cot^2(45° + \phi/2)\right] \times \frac{1}{z} \qquad ...(31.29)$$

In the reverse faulting, the vertical stress is the minor principal stress and the failure occurs due to horizontal compression. (This is similar to the passive conditions discussed in chapter 19). The value of K_p is given by

$$K_p = \tan^2(45° + \phi/2) + \frac{q_u}{\gamma} \times \frac{1}{z} \qquad ...(31.30)$$

In case there is no existing fault, the range of possible values of K is quite wide. However, q_u can be assumed as zero near a pre-existing fault and the range of K is considerably reduced.

Brown and Hoek (1978) suggested a hyperbolic relation for the limits of K. According to them,

$$\left(0.3 + \frac{100}{z}\right) < \overline{K} < \left(0.5 + \frac{1500}{z}\right) \qquad ...(31.31)$$

where z is the depth of the point in metres and $\overline{K}$ is the ratio of the average horizontal stress to the vertical stress.

The range of values of $\overline{K}$ is considerately less than the range of K given by Eqs. 31.29 and 31.30 even when q_u is not zero. It may be noted that in Eq. 31.31, the average horizontal stress is considered, whereas in Eqs. 31.29 and 31.30, the maximum and minimum values of the horizontal stresses are considered.

31.20. MEASUREMENT OF IN-SITU STRESSES

The following methods are commonly used for the measurement of in-situ stresses in rocks (Goodman, 1980).

1. Hydraulic fracturing method
2. Flat jack method
3. Over-coring method

1. Hydraulic fracturing method. In this method, water pressure is used to create a crack in a bore hole. Water is pumped into a section of the bore hole isolated between two packers. As the water pressure is increased, the initial compressive stresses existing in the rock at the walls of the bore hole start reducing, and at some stage become tensile. When the tensile stresses become equal to the tensile strength of the rock ($-T_o$), a crack is formed in the wall. If the pumping is continued, the crack extends further, and eventually the pressure down the hole falls to a steady value, called the shut-in pressure.

Knowing the orientation of the hydraulically induced fracture and using the equations of the theory of elasticity, the in-situ stresses can be estimated.

The method can be used to estimate the in-situ stresses in the rock only when the point is at considerable depth below the surface.

2. Flat jack method. In this method, a slot is first cut in the rock. The deformation occurs in the rock surrounding the slot. After the deformations have taken place, a flat jack is inserted into the slot and load is applied. The in-situ stress is related to the pressure needed to balance the deformations which occur as a result of slot cutting.

The first step is to mark the location of the slot on the wall of the rock. Then one or more sets of the measuring points are installed such that the location of the slot is midway between them. The spacing d of the measuring points is kept equal to the gauge length of the extensometer, which is generally 15.24 cm (6′).

A deep slot is then cut perpendicular to the rock face between the measuring points (Fig. 31.13). As a result of the slot cutting, the pins installed at the measuring points move and the spacing between them is reduced if the rock was under initial compression normal to the plane of slot.

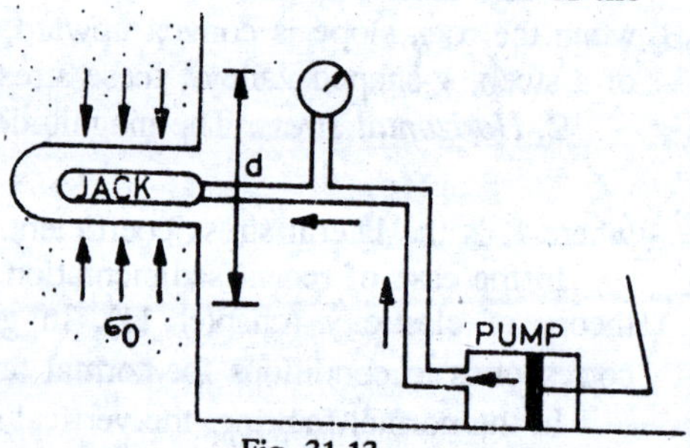

Fig. 31.13.

The flat jack is now inserted into the slot and cemented in place. The pressure in the jack is gradually increased. When the pins have been returned to the initial spacing, the pressure in the jack is equal to the in-situ stress.

The method can be used only when there is an access to a rock face; for example, in an underground gallery in the rock.

3. Over-coring method. In this method, a small diameter bore hole is first drilled into the rock. A suitable deformation gauge is then inserted into the bore hole to measure the change in diameter. A larger diameter hole is then cored concentrically over the small diameter bore hole (Fig. 31.14). Thus a thick cylinder of rock is formed, which is detached from the rock mass and which is free of stress. If the rock was under initial compression, there would be enlargement of diameter, which is measured with the help of the deformation gauge. Analytical methods based on the unloaded thick-walled cylinder model are used to determine the in-situ stresses.

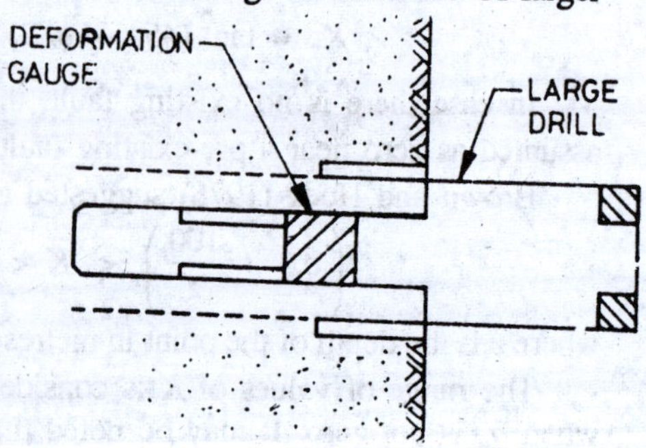

Fig. 31.14.

The over-coring method can be used only to measure the in-situ stresses at some distance away from the rock face. The test is normally discontinued beyond about 5 m from the rock face.

PROBLEMS

A. Descriptive type

31.1. Define rock mechanics. What is its importance for a civil engineer ?

31.2. Discuss geological classification of rocks. Give at least 3 examples of each type of rock.

31.3. Explain the following terms :

(*a*) Rock material (*b*) Intact rock

(*c*) Discontinuity (*d*) rock mass

(*e*) Joint set (*f*) Rock quality designation (RQD)

31.4. What are different index properties of rocks ? How would you determine them ? What is their importance ?

31.5. Explain the geomechanics classification of rocks. How would you determine the rock mass rating (RMR) of a given rock ?

31.6. Discuss the strength classification of rocks.

31.7. Explain various laboratory tests for the determination of the strength of rocks.

31.8. Draw a typical stress-strain curve of a rock. How would you determine the modulus of elasticity and the Poisson ratio ?

31.9. What are different modes of failure of rocks ? Give one example of each.

31.10. Discuss the Mohr-Coulomb criterion for rocks. What is an empirical failure envelope ?

31.11. Write short notes on

(*a*) Shear strength of rocks

(*b*) Hardness of rock.

31.12. What is the importance of in-situ stresses in rock ? How would you measure them ?

B. Objective Type

Write whether the following statements are true or false.

(*a*) Slate is a sedimentary rock.

(*b*) Faults are joints along which there had been shearing movements.

(*c*) The RQD of a rock cannot be greater than 100%.

(*d*) The point load strength of a rock specimen is independent of the diameter.

(*e*) The RMR value is reduced because of unfavourable joint orientation.

(*f*) The shear strength of the specimen as determined by the ring shear test is equal to P/2A.

(*g*) Generally, the modulus of elasticity of the rock is taken equal to the 50% secant modulus.

(*h*) The empirical failure envelope is based on four different types of tests.

(*i*) The in-situ stresses in rocks are not significant for mining operations.

[Ans. True (*b*), (*c*), (*e*), (*f*), (*h*)]

32

Geotechnical Earthquake Engineering

32.1. INTRODUCTION

An earthquake is sudden shaking of the earth surface caused by a source of disturbance inside the earth. Earthquakes occur due to sudden mass shifting in bedrock caused by forces within the earth. Shifting in bedrock occurs along faults. Such earthquakes are called tectonic earthquakes. Besides tectonic earthquakes, there are other types of earthquakes such as volcanic earthquakes, nuclear explosion earthquakes and mine collapse earthquakes, but such earthquakes are small. The discussions herein are limited only to tectonic earthquakes. There is sudden release of strain energy in the zone where the shifting in bedrock occurs. Shock waves are generated due to release of energy. These shock waves are propagated outward from the earthquake source. When these waves reach the earth surface, its shaking occurs.

The movement of the bedrock along the fault occurs at a point called the focus (or hypocenter) (Fig. 32.1). The focus is generally 5 to 50 km below the earth surface but may be deeper in some cases. The *epicenter* is the point on the earth surface vertically above the focus. The distance between the epicenter and the focus is called the *focal depth*. The distance between any point on the earth surface and the epicenter is known as the *epicentral distance*.

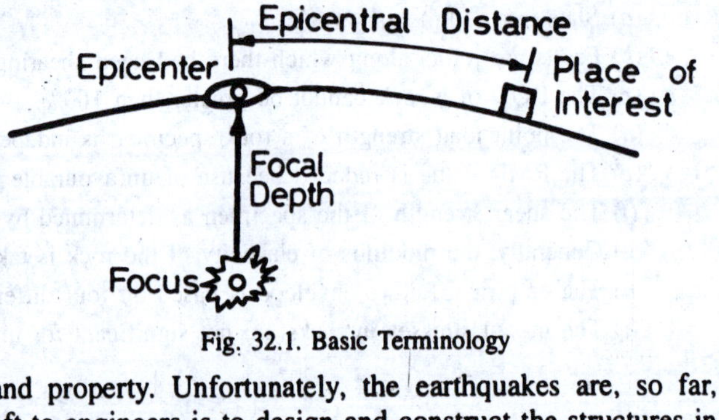

Fig. 32.1. Basic Terminology

Earthquakes cause one of the most destructive natural disasters leading to heavy losses of life and property. Unfortunately, the earthquakes are, so far, unpredictable and unpreventable. The only course left to engineers is to design and construct the structures in such a manner that the damages caused by the earthquakes are minimum.

Earthquake engineering deals with the design and construction of structures that are earthquake resistant. The various structural components are designed to withstand the earthquake forces. *Geotechnical earthquake engineering* is a young branch of earthquake engineering that developed in the last two decades or so. It is concerned with geotechnical aspects of earthquake engineering, such as the type and depth of foundation soil, amplification of earthquake's intensity by soil deposits, liquefaction of soils, etc.

This chapter discusses the basic concepts of the geotechnical earthquake engineering.

32.2. HISTORY OF EARTHQUAKES IN INDIA

India is one of the seismic countries of the world. Some of the severe earthquakes of the world occurred in India. The earthquake of the highest magnitude, represented as M and defined later, occurred in Assam (India) on the 12th of June, 1897. The earthquake was of a magnitude of 8.7 on Richter's scale. That earthquake caused extensive damages in Assam.

In the period from 1905 to 1950, four severe earthquakes of magnitude greater than 8.0 occurred in India.

These earthquakes occurred in Kangra in 1905 (M = 8.6), Bihar in 1934 (M = 8.4), Andaman in 1941 and Assam in 1950 (M = 8.6). Besides these earthquakes, there were some less severe earthquakes. In the period from 1951 to 2000, seven moderate earthquakes of magnitude between 6 and 7 occurred. These earthquakes occurred at Anjar in 1956 (M = 6.1), Koyna in 1967 (M = 6.5), Bihar-Nepal border in 1988 (M = 6.6), Uttarkashi in 1991 (M = 6.4), Killari (Latur) in 1993 (M = 6.2), Jabalpur in 1997 (M = 6.0), and Chamoli in 1999 (M = 6.6). The Killari (Latur) earthquake was the worst, which caused about 10,000 deaths.

In the beginning of this century, on 26th January, 2001, a strong earthquake with a magnitude of 7.9 occurred in Bhuj (Gujarat). It caused heavy damages and more than 20,000 persons died and about 1,67,000 persons were injured. About 3,00,000 houses were destroyed and huge losses of property occurred.

The Bhuj earthquake has demonstrated the vulnerability of various Indian regions to earthquakes. It has very clearly shown that most construction in the seismic regions do not comply with the required safety requirements. The general public has become more concerned about earthquakes and the great damage they can cause. It is a major challenge to ensure that our constructions are made earthquake resistant to limit the damages in future.

32.3. SEISMIC ZONES OF INDIA

The severity of earthquakes in different parts of India is different. The seismicity at a place is assessed by its distance from the active fault in the rock formation. It is also decided from the earthquakes already occurred in that region. In general, the seismic zoning of a country depends upon the seismic history of different regions unless some significant changes in tectonic features occur.

As per IS: 1893 (Part 1): 2002, India has been divided into four zones, designated as Zone II, III, IV and V. The erstwhile zone I has now been merged in zone II. The zone numbers increase as the seismic severity increases. Thus zone II has the lowest seismicity and zone V has the highest seismicity. Zones IV and V are located in the northern and north eastern regions of the country, the Kutch and Maharashtra area and Andaman. The rest of the country has only zones II and III.

About 60% area of India is in zones IV and V.

For details, see IS: 1893-2002.

Various zones indicate the regions in which the seismicity would be approximately same. However, the severity of the earthquake in a zone may be different because of local site conditions.

32.4. MAGNITUDE OF AN EARTHQUAKE

The magnitude (M) of an earthquake is a *quantative* measure of its strength in terms of energy released at focus. It has a unique value for an earthquake.

Charles Richter developed the first scale for magnitude in 1935. Magnitude is defined as the logarithm (to base 10) of the maximum trace amplitude (A) of the ground motion recorded in microns ($=10^{-6}$m) at a distance of 100 km from the epicenter on a Wood-Anderson Type Torsion Seismograph having damping equal to 80% of the critical damping, natural period of 0.8*s* and magnification of 2800. Thus

$$M = \log_{10} A \qquad \ldots(32.1)$$

where,

M is magnitude on Richter's scale,

A is amplitude of ground motion at a distance of 100 km from the epicenter, in microns.

Since the distance of the instrument from the epicenter may not be always 100 km, the following modified equation can be used to estimate the magnitude.

$$M = \log_{10} A - \log_{10} A_0 \qquad \ldots(32.2\ a)$$

or

$$M = \log_{10} A + \text{distance correction} \qquad \ldots(32.2\ b)$$

where *A* and A_0 are usually measured in millimetres ($= 10^{-3}$ m).

The quantity ($-\log_{10} A_0$) is known as the distance correction. The distance correction can be obtained

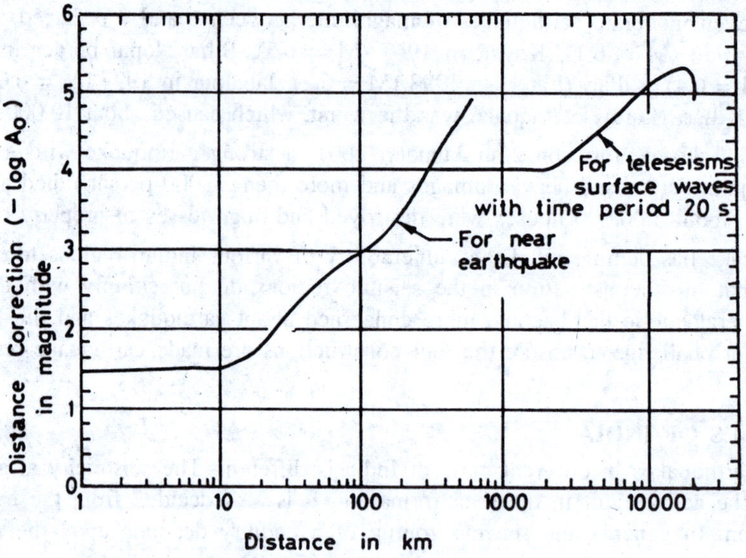

Fig. 32.2. Distance Correction

from Fig. 32.2. For a distance of 100 km, the distance correction is +3. Alternatively, the following equation may be used:

$$M = \log_{10} A - 2.48 + 2.76 \log \Delta \qquad ...(32.3)$$

where Δ is the epicentral distance (km)

For illustration, if the amplitude of the ground motion at a distance of 100 km is 10^4 μm (= 10 mm),

From Eq. 32.1, $M = \log_{10} 10^4 = 4.0$

From Eq. 32.2 (*b*), $M = \log_{10} 10 + 3.0 = 4.0$

From Eq. 32.3, $M = \log_{10} 10 - 2.48 + 2.76 \log_{10} 100 = 4.04$

Table 32.1 gives a typical classification of earthquakes based on their magnitudes. The table also gives the annual average number of earthquakes across the Earth in each of these groups.

Table 32.1. Global occurrence of earthquakes of different magnitudes

Magnitude	*Group*	*Annual average number of global occurrence*
8 and higher	Great	1
7 — 7.9	Major	18
6 — 6.9	Strong	120
5 — 5.9	Moderate	800
4 — 4.9	Light	6200 (estimated)
3 — 3.9	Minor	49,000 (estimated)
Less than 3	Very Minor	About 9000 per day (= 3.285×10^6 per year)

Earthquakes with a magnitude of 3 or less are not felt. Earthquakes with a magnitude less than 5 normally do not cause significant damage. The maximum magnitude of the earthquake occurred so far was 8.7 (Assam, 1897).

The magnitude corresponding to the original formulation proposed by Richter is also called **the local** magnitude (M_L). In addition to the local magnitude, three more magnitude scales have been proposed. The body wave magnitude (M_b) is based on the compressional body waves (or *P*-waves). The surface wave magnitude (M_s) is based on the surface or Rayleigh (R) waves. The seismic moment magnitude or the wave energy magnitude (M_w) is based on the seismic moment.

For small earthquakes, all the four magnitude scales give almost the same value, but for large earthquakes, the values of magnitudes on different scales may be different. For example, for the 1964 Alaska Earthquake, the estimated magnitudes were $M_b = 6.8$, $M_s = 8.6$ and $M_w = 9.2$. Although there are some advantages in these magnitude scales, probably the Richter scale is still quite popular and is commonly used in practice.

Energy released in an earthquake

A relationship between the amount of the strain energy released by an earthquake and its magnitude is given by Richter as

$$\log_{10} E = 11.4 + 1.5\,M \qquad \ldots(32.4)$$

where *E* is the energy released in ergs (1 *erg* = 10^{-7} *J*)
and *M* is the magnitude of the earthquake.

Earthquakes release huge amount of energy. An earthquake of magnitude 6.3 would release energy equal to about 8×10^{20} ergs, which is equal to that of the atomic bomb dropped at Hiroshima in 1945. Fortunately, the most of energy released by an earthquake is dissipated inside the earth into heat and fracturing of rocks. Only a small fraction of the energy goes into seismic waves that cause shaking of the ground surface.

It may be noted that with an increase of one on the magnitude scale would result in an increase in amplitude of ten times and an increase in energy of 31.6 times.

The affected area and the duration of the earthquake increase with an increase in magnitude of the earthquake. The affected area depends upon a number of factors such as the depth of focus, type of strata, etc. The affected areas by the earthquakes of magnitudes 6, 7 and 8 respectively would be of the order of 60,000, 120,000 and 200,000 square kilometres. The corresponding duration of earthquakes would be of the order of 15, 30 and 45 seconds. For example, the Koyna earthquake of 1967 (M = 6.5) affected an area of about 400 km radius, with an area of about 126000 sq. km.

32.5. INTENSITY OF EARTHQUAKES

The intensity of an earthquake is measured *qualitatively* in terms of damages caused by it at a particular location. In general, the larger the earthquake, the greater is the intensity. For the same earthquake, the intensity decreases as the epicentral distance of the location increases. Thus the maximum intensity of an earthquake is near its epicenter.

Several intensity scales have been proposed by different investigators. Generally, modified Mercalli Intensity (MMI) scale is used in practice. IS: 1893-2002 also recommends the use of Comprehensive Intensity Scale (MSK 64). Both MMI and MSK 64 scales are similar. However, MSK 64 scale gives more detailed and specific descriptions of the damages caused. Both the scales categorise the damages in twelve classes or degrees. The intensity is usually expressed in Roman Numerals, I, II, ...XII. Sometimes, the intensity is also expressed in grades 1, 2, ...12. In general, the greater the intensity at a location, the greater would be the damage caused by the earthquake. Table 32.2 gives an abridged version of the MMI scale.

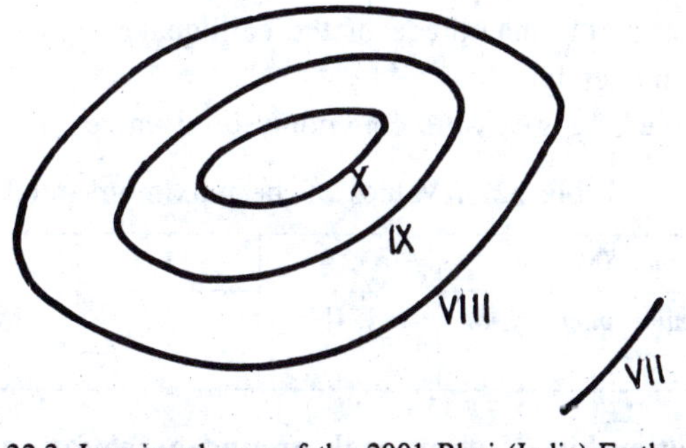

Fig. 32.3. Isoseismal map of the 2001 Bhuj (India) Earthquake (MSK intensity)

An *isoseismal map* represents the affected regions with the same intensity as contours.

Table 32.2. Modified Mercalli Intensity (MMI) Scale

Intensity Class	*Evaluation*	*Description*	*Magnitude on Richter's scale*
I	Insignificant	Not felt by humans, detected by instruments only	1 — 1.9
II	Very light	Felt only by sensitive humans; delicately suspended objects may swing	2 — 2.9
III	Light	Felt noticeably indoors; small vibratory motion	3 — 3.9
IV	Moderate	Felt by many persons; windows and doors make cracking sounds.	4 — 4.9
V	Slightly strong	Felt by almost everyone; some panic; minor damages	5 — 5.9
VI	Strong	Felt by all; many frightened; some damage to ordinary structures	5 — 5.9
VII	Very strong	Everbody runs outdoors; negligible damage to well-built structures; considerable damge to other structures	5 — 5.9
VIII	Destructive	Slight damage to well-built structures; serious damage to ordinary structures	6 — 6.9
IX	Ruinous	Considerable damage to well-built structures	6 — 6.9
X	Disastrous	Serious damage to well-built structures; almost total destruction of non-seismic resistant structures	7 — 7.9
XI	Extremely disastrous	Almost total destruction; broad fissures in ground	7 — 7.9
XII	Catastrophic	Total damage; waves seen on ground; objects thrown into the air.	8 — 8.6

On-the-spot study of the damages caused by an earthquake is made soon after its occurrence. The locations with different intensities are marked on the plan of the area. The isoseismal lines are then drawn by joining the points of the same intensity. The procedure is similar to that of drawing of elevation contours from the elevations of different points. Fig. 32.3 shows the isoseismal map of the Bhuj earthquake of 2001. The maximum intensity observed was X. The contours of intensity IX, VIII, VII are also marked. From the MMI scale, it can be observed that the earthquake was of intensity classified as *disastrous.*

The difference between the magnitude and the intensity should be carefully noted. While the magnitude of an earthquake has a *unique* value, the intensity decreases as the distance from the epicenter increases. In other words, for a certain earthquake, the intensity varies with the distance. Sometimes, a sound source (or an electric bulb) analogy is used to demonstrate the difference between the magnitude and the intensity. While the strength of the sound source (or electric bulb) is constant, the intensity of sound (or light) decreases as the observer moves away from the source.

1. Relationship between magnitude and maximum intensity

The Gutenberg-Richter relation is commonly used in practice; according to which, the magnitude and intensity are approximately related as

$$M = 1.3 + 0.6\, I_0 \qquad \ldots(32.5)$$

where M is the magnitude of the earthquake, and I_0 is the maximum intensity on MMI scale (expressed in Arabic numerals).

Table 32.3 gives the commonly used values of intensities for different magnitudes of earthquakes.

Table 32.3. Values of the maximum intensities for different magnitudes of earthquakes.

Magnitude (M)	2	3	4	5	6	7	8
Maximum intensity on MMI scale	I, II	III	IV, V	VI, VII	VII, VIII	IX, X	XI

2. Relationship between peak ground acceleration and intensity

The earthquake engineers require the value of the maximum ground acceleration so that they can design the structures and assess the liquefaction characteristics of soils. The exact value of the maximum seismic acceleration cannot be theoretically determined with present state of knowledge.

The following expression is sometimes used to estimate the peak ground acceleration from the intensity of the earthquake at any location:

$$\log_{10} a = \frac{1}{3} I - 0.5 \quad \ldots(32.6)$$

where a is the peak ground acceleration in cm/s^2, and I is the intensity on MMI scale (expressed in Arabic numerals).

It is the usual practice to represent ground accelerations in terms of the acceleration due to gravity (g).

Table 32.4 gives the approximate values of the peak ground acceleration for different intensities of earthquake as given by Bolt (1993).

Table 32.4. Peak ground accelerations for different intensities of earthquakes

Intensity on MMI scale	V	VI	VII	VIII	IX	X
Peak ground acceleration (PGA) (*in terms of g*)	0.03–0.04	0.06–0.07	0.10–0.15	0.25–0.30	0.50–0.55	> 0.60

According to IS: 1893: 2002, the basic zone factors (z) given in the code are reasonable estimates of effective peak ground acceleration for the design of various structures. Table 32.5 gives the zone factors for different seismic zones. The intensity as per MSK 64 broadly associated with the various zones is also given.

Table 32.5. Intensity and Zone factors for different seismic Zones

Seismic zone	II	III	IV	V
Intensity as per MSK 64 scale	VI or less	VII	VIII	IX and above
Zone factors (in terms of g)	0.10	0.16	0.24	0.36

32.6. EFFECT OF GROUND MOTION ON STRUCTURES

An earthquake produces seismic waves that cause the earth's crust to vibrate. These waves impart a momentary acceleration to the earth's crust and it starts moving in the direction in which the wave is travelling at that instant. The characteristics of vibration, such as intensity, duration, etc., at any location depend upon a number of factors, including the following:

(*i*) the magnitude of earthquake
(*ii*) the depth of focus
(*iii*) the epicentral distance
(*iv*) the characteristics of the material through which the seismic waves travel.

Inertia Force

When the ground motion occurs, the foundation of the structure must also move with it to avoid its rupture. When the foundation moves, the structure on it tends to stay back because of inertia. Consequently, the structure is subjected to inertia forces. In the earthquake-resistant design of structures, the inertia forces due to earthquake are considered in addition to the normal loads and forces.

To get a clear idea about the inertia forces, let us consider that a person is sitting upright in a railway train not in motion. When the train suddenly starts moving, his lower body on the seat moves with the train but his upper body tends to stay back because of inertia. An inertia force acts on his upper body in the backward direction and his head may strike the back of the seat.

According to Newton's second law of motion, the inertia force is equal to the product of the mass and acceleration, and it acts in a direction opposite to that of the acceleration. Thus the horizontal inertia force is given by the expression

$$F_h = \text{Mass} \times \text{acceleration}$$

or $$F_h = \left(\frac{W}{g}\right)(A_h\,g) = W A_h \qquad \ldots(32.7)$$

where A_h is the seismic coefficient in the horizontal direction, and W is the seismic weight.

When the horizontal shaking of the ground occurs, the horizontal inertia forces are generated at the level of the mass of the structure, which is usually assumed to be concentrated at the floor levels (Fig. 32.4). These inertia forces are transferred from the floor slab, through the walls (or columns), to the foundation, and finally to the strata below the foundation. In the earthquake-resistant design, it is ensured that each of the structural elements, including floor slabs, walls, columns and foundations, can safely transfer the inertia forces through them. Moreover, the connections between the various structural elements are properly designed and constructed.

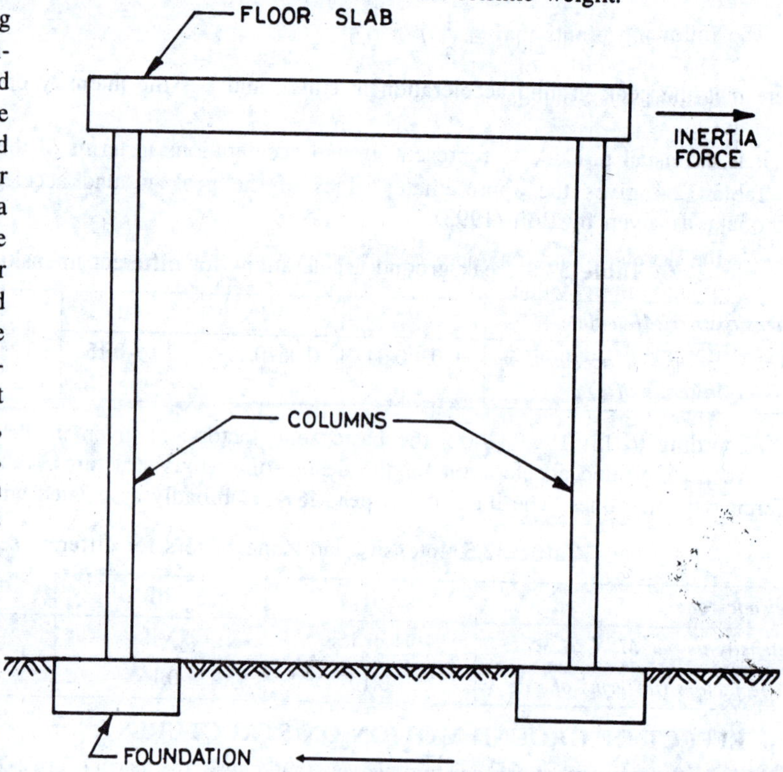

Fig. 32.4. Inertia force on a simple structure

Response spectrum

If the building were rigid, then every point on it would move by the same amount as the ground, and consequently, the inertia force would be equal to the mass multiplied by the ground acceleration. But the buildings are flexible, and different parts move back-and-forth by different amounts during ground shaking.

The natural period of vibration of a structure is the time period of its undamped, free vibration. The *fundamental natural period* of vibration is that for the first (or fundamental) mode of vibration of that structure. Each structure has a unique fundamental period of vibration at which it tends to vibrate when it is allowed to vibrate freely without any external excitation. The fundamental natural period depends upon the form and configuration of structure, the stiffness (or flexibility) of the various structural members, the type and material of construction, etc. Methods for the determination of the fundamental period of a structure are outside the scope of this text. However, IS: 1893-2002 gives the empirical formulae for the estimation of fundamental natural period of vibration for buildings that can be used (see Sec. 32.8).

For the estimation of seismic forces in structures, the response spectra are commonly used in practice. The response spectrum of a structure shows the maximum response induced in the structure by the ground motion. It is usually plotted in terms of the maximum absolute acceleration against natural period, although sometimes the maximum relative velocity or the maximum relative displacements are also used. While plotting the response spectrum, it is assumed that the structure has single degree of freedom and it may have different damping. In other words, the maximum response spectrum represents the maximum acceleration of an idealised single degree freedom systems having a certain natural period of vibration and damping when it is subjected to earthquake ground motion.

Fig. 32.5 shows the maximum response spectrum for a typical structural system with a certain natural period and damping.

The following points may be noted:

1. The response of the system decreases as the damping of the system increases.
   In buildings, usually 5% damping is assumed.
2. As the natural period increases, the acceleration first increases to a maximum value and then it decreases.
   For the structural system shown, the greatest acceleration occurs when natural period is about 0.3 *s*.

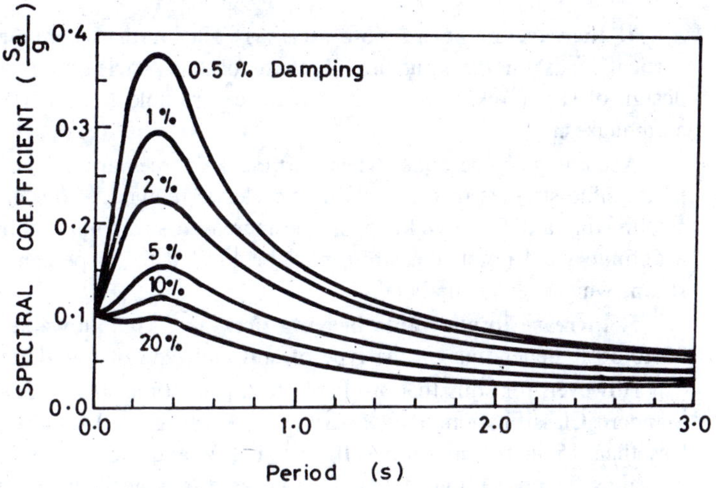

Fig. 32.5. The maximum response spectrum of a typical structural system

3. It is the usual practice in the earthquake-resistant design to represent the structural response by the response factor or spectral coefficient in the normalised form as S_a/g.
4. The response spectrum helps the earthquake engineer to predict how a particular structure with a certain natural period will respond to the earthquake.
5. The response spectra are commonly used in the estimation of seismic forces, as discussed later.

32.7. GENERAL PRINCIPLES OF EARTHQUAKE-RESISTANT DESIGN

Some of the general principles of the design of earthquake-resistant design of structures are briefly discussed below. For more details, a reference may be made to IS: 1893-2002.

1. Design basis earthquake. In the earthquake-resistant design, it is not the objective to make the structure absolutely earthquake proof that will not suffer any damage during the rarest of the earthquakes. A fully earthquake-proof structure will be very huge and highly expensive. Instead, an attempt is made that the structure should be able to withstand the minor earthquakes that occur frequently in that region. Moreover, the structure should be able to resist moderate earthquakes, called design basis earthquakes (DBE), without significant structural damage. Such earthquakes could occur once during the life of the structure. Even a major earthquake, called the maximum considered earthquake (MCE), with its intensity greater than that of the design basis earthquake, would not cause complete collapse of a properly designed and constructed structure and the losses would be limited.

2. Pseudo-static analysis. Earthquakes cause dynamic loading on structures. However, for the design of earthquake-resistant structures, the dynamic analysis is usually not carried out. Instead, a pseudo-static analysis is used in which the earthquake forces are replaced by equivalent static forces. These forces are considered in addition to the normal loads on the structure for its design.

It is assumed that the forces due to earthquake are not likely to occur simultaneously with other occasional forces such as wind loads, maximum flood forces or maximum sea waves forces.

3. Components of acceleration. Earthquakes can cause acceleration in any direction. It is the usual practice to consider the components of acceleration in the vertical direction and in two perpendicular horizontal direction. Moreover, the acceleration components can be either positive or negative in these three directions.

Since the three components of earthquake acceleration may not act at the same time with their maximum magnitude, the code recommends that when the maximum response from one component occurs, the response from the other two components can be 30 percent of their maximum values. All possible combinations, including plus or minus signs, should be considered in the design.

Generally, the horizontal acceleration is the most predominant.

4. Increase in permissible stresses. The vertical component of acceleration can increase the normal vertical loads on the structure. Because of the provision of adequate factors of safety used in the normal design of structures, most of the structures are able to resist the additional momentary vertical loads due to earthquakes.

According to the code, when earthquake forces are considered along with the normal design forces, the permissible stresses in materials in the elastic method of design can be increased by one-third. However, for steel having a definite yield stress, the increased stress may be limited to the yield stress, and for steel without a definite yield point, the stress may be limited to 80 percent of the ultimate strength or 0.2 percent proof strain, whichever is smaller.

5. Increase in allowable bearing pressure. The allowable bearing pressure in soils can be increased by 25 to 50%, depending on the type of soil and type of foundation as per details given in the code.

However, for soils that are liable to liquefaction, this increase is not applicable. Soils falling under Indian Standard Classification category SP (*i.e.*, poorly graded sands), with standard penetration test (SPT) N-values less than 15 in seismic zones III, IV and V and less than 10 in zone II may become liquefied. Suitable measures for prevention of liquefaction, as discussed later, are adopted.

6. Horizontal and Vertical inertia forces. The predominant direction of ground motion is usually horizontal. Therefore, the horizontal seismic forces are most important for the earthquake-resistant design. The methods for the estimation of horizontal seismic forces are discussed in the next section.

According to the code, the vertical inertia forces are to be considered in the design unless checked and proven that they are not significant. When effects due to vertical earthquake loads are to be considered, the design vertical acceleration spectrum is taken as two-thirds of the design horizontal acceleration spectrum.

Vertical inertia forces are important for structures in which stability is a criterion for design, such as retaining walls, cantilevered members and prestressed horizontal members. Reduction in gravity forces due to vertical component of ground motion can be detrimental in some structures.

7. Resonance. According to the code, resonance of the type as visualized under steady-state conditions will not occur because the earthquakes have irregular motion of short duration in which there is not adequate time to build up the required amplitudes.

However, if the structure's fundamental period is close to that of site, resonance may occur. Such conditions have been observed for some tall buildings on deep soft soils.

8. Base shear. Inertia forces generated in the structure due to an earthquake are assumed to be transferred to the base of the structure as the base shear. The base transfers these forces to the foundation, which, in turn, transfers them to the ground.

All the components of the structure and foundations are designed to resist these forces in addition to the normal forces.

The distribution of the seismic forces in different structural elements is done by the methods of structural analysis beyond the scope of this text.

32.8. DESIGN SEISMIC COEFFICIENT

The design horizontal coefficient A_h for a structure can be determined from the following expression:

$$A_h = \left(\frac{Z}{2}\right)\left(\frac{I}{R}\right)\left(\frac{S_a}{g}\right) \quad \ldots(32.8)$$

where,

Z = Zone factor of the seismic zone (For seismic zones II, III, IV and V, the values of zone factor are taken as 0.10, 0.16, 0.24 and 0.36, respectively).

I = Importance factor. It depends upon the functional use of the structure. The value of *I* is 1.5 for important service community buildings, such as hospitals, schools, railway stations, fire stations cinema halls, etc.

For all other buildings, its value is 1.0.

R = Response reduction factor. It depends on the perceived seismic damage performance of the structure. The value of R varies from 1.5 to 5.0 for different types of buildings. For ordinary R.C. moment-resisting frames, its value is 3.0.
The ratio I/R should not be greater than 1.0.

S_a/g = Average response acceleration coefficient. It indicates the average smoothened plot of maximum acceleration. It depends upon the type of strata and natural period.

The code gives the chart for values of (S_a/g) for rock (or hard soil), medium soil and soft soil sites for 5% damping. For other values of dampings, the code gives the multiplying factors. The maximum value of S_a/g is about 2.50 between a time of period about 0.1 s to 0.5 s for rock. As the of time period increases, the value of S_a/g decreases. (For more details, see IS: 1893-2002.)

Natural period

For determining the value of (S_a/g) for a given structure, its natural period of vibration is required. The following approximate formulae for the computation of the fundamental natural period of vibration (T_s) are used.

(*a*) *Moment-resisting frame buildings without brick infil-panels*

(*i*) R.C. frame buildings

$$T_s = 0.075\ h^{0.75} \qquad \ldots(32.9)$$

(*ii*) Steel frame buildings

$$T_s = 0.085\ h^{0.75} \qquad \ldots(32.10)$$

(*b*) *All other buildings, including moment-resisting frames with brick infil-panels*

$$T_s = \frac{0.09\ h}{\sqrt{d}} \qquad \ldots(32.11)$$

where,

T_s = natural period of vibration, in seconds.

h = height of building, as specified in the code, in metres.

d = base dimension of the building at the plinth level along the direction of the lateral force, in metres.

For any structure with $T_s \le 0.1\ s$, the value of A_h is not taken less than (Z/2), whatever be the value of I/R.

32.9. DESIGN SEISMIC FORCES

The design seismic forces are computed as explained below:

(*a*) Design lateral force

The total design lateral force acting on the structure is equal to the product of the design horizontal coefficient (A_h) and the seismic weight of the structure. The seismic weight of a building is equal to the sum of the seismic weights of all the floors of the building. The seismic weight of each floor is equal to its full dead load plus an appropriate fraction of the imposed load, as specified in the code. While computing the seismic weight of any floor, the dead weight of columns and walls in any storey is equally distributed to the floors above and below them.

The design seismic base shear acting on the structure is equal to the total design lateral force and is given by the expression

$$V_B = A_h\ W \qquad \ldots(32.12)$$

where

A_h = design horizontal seismic coefficient

W = seismic weight of the building

V_B = design base shear

The design base shear is distributed to different floor levels, as specified in the code. The horizontal force acting on the foundation is equal to the design base shear.

(b) Design vertical force

Since the design vertical coefficient is equal to the two-thirds of the design horizontal coefficient (A_h), the design vertical force is given by the expression

$$F_v = \frac{2}{3} A_h W \qquad \text{...(32.12 a)}$$

where W is the seismic weight of the building, and F_v = design vertical force.

The design vertical force is transferred from the structure to its foundations.

(c) Design of structure

Buildings and portions thereof should be designed and constructed to resist the effect of design seismic forces. Since the seismic forces occur suddenly and without warning, it is essential to avoid construction practices that lead to brittle failure. The members should be designed to behave in a ductile manner so that complete collapse of the structure is avoided even during severe earthquakes. Actual design is beyond the scope of this text.

(d) Design of foundation

Since the earthquake-resistant designs are generally performed by pseudo-static analysis, the seismic loads on the foundation are considered as static loads: These loads are capable of producing settlements like other loads.

The code permits an increase in allowable bearing pressure depending on the soil-foundation system. The increase in allowable bearing pressure is 25 or 50 percent. In soil-foundation systems, where small settlements are likely to occur during earthquake, the increase in allowable bearing pressure is larger, and *vice-versa*. For example, for all types of foundations on rocks or hard soil, the permissible increase is 50%. For soft soils, the permissible increase for piles resting on hard rock or hard soils and the raft foundations is 50%, whereas that for other types of foundations, it is 25%.

32.10. SITE-SPECIFIC RESPONSE SPECTRA

The response spectra of a structure are site-specific, as they depend upon the type of strata at the site.

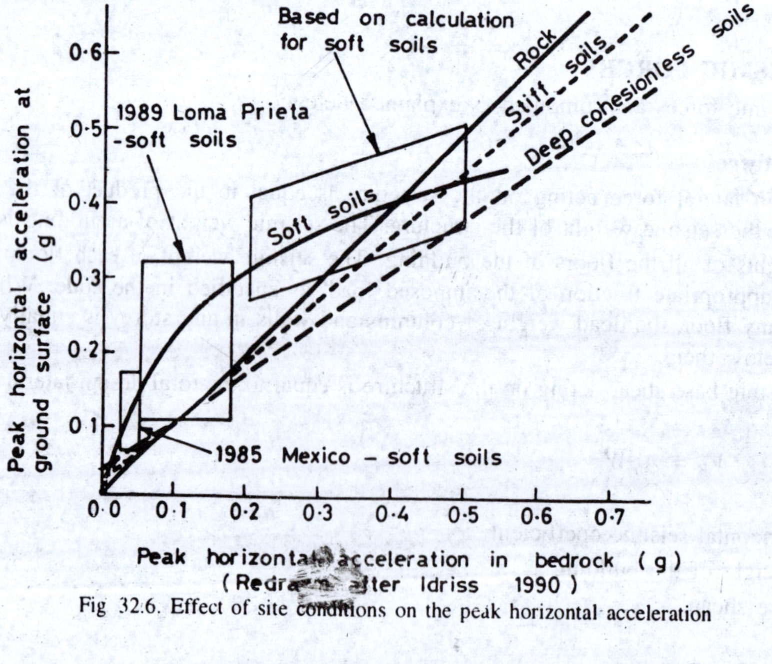

Fig 32.6. Effect of site conditions on the peak horizontal acceleration

Seismic waves travel differently through soils and rocks. Therefore, the ground shaking at sites underlain by soil is different from those underlain by rock.

It was observed during the 1989 Loma Prieta Earthquake that the sites underlain by deep deposits of soft soils experienced peak ground acceleration two to three times greater than that at nearby sites on stiff soils or rock. The structures founded on soft soils collapsed, whereas adjacent buildings founded on stiffer soils withstood the shock without collapsing.

Similarly, in the 1985 Michoacan Earthquake, the damages to buildings in Mexico city, 350 km from the epicenter, were much worse than in other cities that were much closer. The reason for this difference is attributed to the difference in site conditions. The buildings in Mexico city were underlain by deep deposits of soft clay.

Fig. 32.6 gives a simplified, approximate relationship for different soils (Seed, et al 1976, and Idriss, 1990).

Even in Mexico city, the spectral acceleration for different sites was different (Fig. 32.7). The observed response at rock and hard soils was fairly low, and the damage caused was negligible.

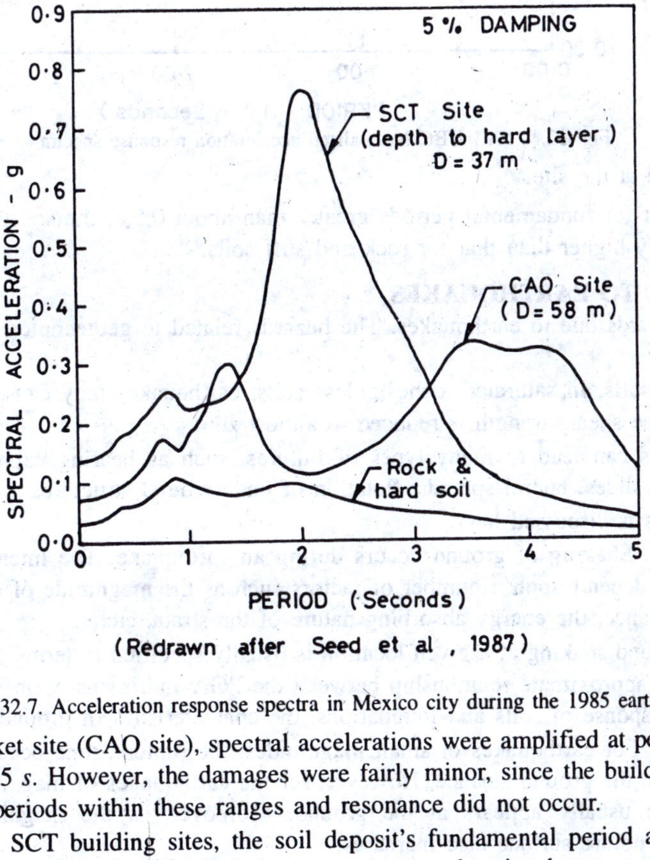

(Redrawn after Seed et al, 1987)

Fig. 32.7. Acceleration response spectra in Mexico city during the 1985 earthquake

At the central market site (CAO site), spectral accelerations were amplified at periods of 1.3 *s* and within the range of 3.5 *s* to 4.5 *s*. However, the damages were fairly minor, since the buildings at the CAO site did not have fundamental periods within these ranges and resonance did not occur.

However, near the SCT building sites, the soil deposit's fundamental period at which amplification of spectral acceleration occurred matched with those of structures and major damages occurred due to resonance.

Uniform Building Code (1991) gives the normalised acceleration response spectra for different site conditions (Fig. 32.8). The spectral acceleration of a structure can be estimated if the following parameters are known:

(*i*) System's fundamental period (T)

(*ii*) Peak ground acceleration of the design earthquake

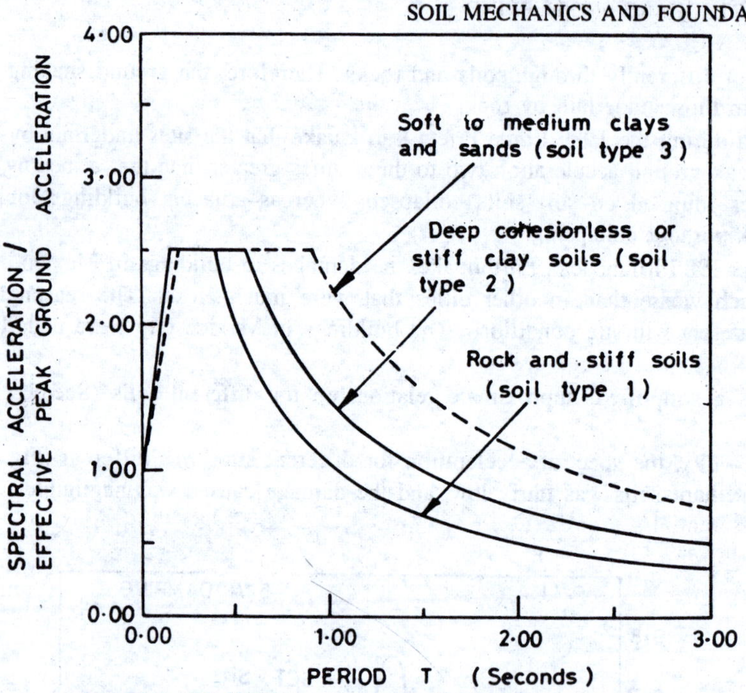

Fig. 32.8. 1991 UBC Normalised acceleration response spectra

(*iii*) Type of the soil at the site.

It may be noted that for fundamental periods greater than about 0.5 *s*, the spectral acceleration for deep soil sites are considerably higher than that for rock and stiff soils.

32.11. HAZARDS DUE TO EARTHQUAKES

There are many hazards due to earthquakes. The hazards related to geotechnical engineering include the following:

1. Liquefaction of soils. In saturated, cohesionless soils, earthquakes may cause liquefaction (or quick-sand) conditions when the shear strength is reduced to almost zero.

Liquefaction of soils can lead to many types of failures, such as bearing capacity failures, sinking or tilting of buildings, land slides, lateral spreads, flotation of underground structures.

Liquefaction of soils is discussed later.

2. Ground shaking. Shaking of ground occurs during an earthquake. The intensity, duration and wave form of ground shaking depend upon a number of factors such as the magnitude of earthquake, the depth of focus, the epicentral distance, the energy absorbing nature of the strata, etc.

The intensity of ground shaking at a given location is usually specified in terms of peak ground acceleration (PGA). There is an approximate relationship between the PGA and intensity on MMI scale, as discussed earlier. To predict the response of soils and foundations, the characteristics of ground shaking are required.

3. Surface rupture. For earthquakes of small magnitudes, the rupture zone occurs deep inside the earth and it does not extend to the ground surface. However, for the earthquakes of magnitudes greater than about 6.0, the rupture surface usually appears at the ground. Moreover, as the magnitude of the earthquake increases, the length of rupture surface also increases.

Surface rupture creates a number of problems for buildings, bridges, railways, highways, water supply lines, etc. that are located directly above the fault. It is the usual practice to locate the buildings and structures susceptible to heavy damage not directly over the active faults.

4. Land slides. Earthquakes can cause land slides and failure of slopes. Even when the slope failures do not occur, tension cracks may appear in the soil. These tension cracks may lead to the formation of water channels, causing softening and saturation of soils.

5. Lateral spreading. Lateral spreading of soil is the massive horizontal movement of soil layers in a direction parallel to the ground slope. Lateral spreading due to earthquakes is usually observed on very gentle slopes.

If the lateral spreading occurs beneath a structure, it can tear it apart, causing heavy damages. In the design of such structures, suitable provisions are made to withstand tensile stresses.

6. Shear failure of foundations. Seismic loads acting on a structure are transmitted to the ground through the foundation. This results in an increase in the stresses acting on the soil. However, the soils can withstand seismic stresses without failure if they are not very large. In some cases, when the stresses exceed the shear strength of the soil, the bearing capacity failure may occur.

However, shear failures of foundation are not common so long as liquefaction of the soil does not occur.

7. Settlements of structures. The earthquakes cause cyclic loading on soils, which may lead to large settlements. Loose sandy soils often compact during the earthquake.

Differential settlements can also be substantial. Adequate provisions should be made in the design of structures so that differential settlements do not cause damage.

8. Failure of retaining walls. If the movements caused by the earthquake are large, the retaining walls may fail. When a retaining wall fails, the support it was providing to the soil mass is removed. It may result in heavy damage to adjacent structures such as ports, etc.

9. Tsunamis and Seiches. A tsunami is a huge wave generated in the ocean by an earthquake. These waves travel very fast and cannot be easily detected in open sea. However, when they approach the beach, they suddenly appear as tall waves. These tall waves may cause heavy damages to onshore structures and can lead to loss of life.

A *seiche* is similar in nature to a tsunami, but it is of much smaller size and it occurs in lakes or rivers. The seiche usually occurs when resonant conditions occur in the lake due to an earthquakes, *i.e.*, when the natural frequency of the lake and that of the earthquake are equal. Sometimes a seiche occurs in a lake when the rupture surface is beneath the lake bed.

Early warning systems should be installed so that suitable actions may be taken by the people at the places where tsunamis and seiches can occur.

32.12. LIQUEFACTION PHENOMENON

Liquefaction is a state of saturated cohesionless soil when its shear strength is reduced to zero due to pore water pressure caused by vibration during an earthquake. The soil starts behaving like a liquid. To understand the liquefaction phenomenon, let us consider a soil element of the soil deposit at a depth of z below the ground surface (G.S.). Let the depth of water table (W.T.) be h below the ground surface (Fig. 32.9).

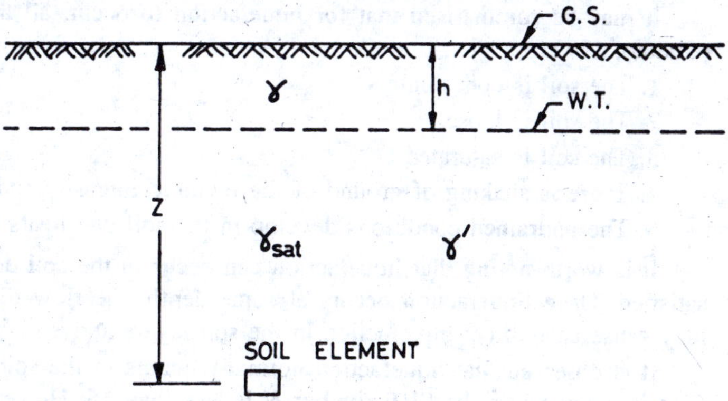

Fig. 32.9. Stresses in a soil element

As discussed in Chapter 10, the effective stress (represented as σ' in this chapter) at a depth z below the G.S. is given by

$$\sigma' = \gamma h + \gamma_{sat}(z-h) - \gamma_w(z-h)$$

$$\sigma' = \gamma h + \gamma'(z-h)$$

where γ' is the submerged density ($= \gamma_{sat} - \gamma_w$)

The shear strength of a cohesionless soil is due to internal friction and is given by

$$s = \sigma' \tan \phi'$$

If the soil deposit is subjected to ground vibrations, it tends to compact and decrease in volume. However, if the drainage of pore water is prevented, this tendency to decrease in volume results in an increase in pore water pressure. Let u_d be the excess dynamic pore water pressure developed due to ground vibration. The dynamic shear strength is expressed as

$$s_d = (\sigma' - u_d) \tan \phi'$$

For sandy soils, the angle of internal friction ϕ' in the dynamic conditions is almost equal to that in static conditions.

The dynamic shear strength (s_d) will become zero when

$$\sigma' = u_d$$

or

$$\frac{u_d}{\sigma'} = 1$$

Expressing u_d in terms of the dynamic hydraulic head (h_d),

$$\frac{\gamma_w h_d}{\sigma'} = 1$$

Thus liquefaction in a cohesionless soil will occur when the dynamic pore water pressure is equal to the effective stress.

It may be noted that because of the dynamic pore water pressure, the shear strength of soil is decreased. It results in transfer of intergranular stress to pore water pressure. If this transfer of stress is incomplete, there is partial loss of strength and partial liquefaction occurs. However, if the transfer of stress is complete, the shear strength becomes zero and complete liquefaction occurs. The sand-water mixture behaves like a viscous liquid after complete liquefaction.

Large settlements occur after liquefaction and the structures resting on such a soil deposit start sinking. This sinking process continues till the sand remains in the liquefied state.

It may be summarised that for liquefaction to occur, all the following five conditions must be satisfied (Coduto, 1999).

1. The soil is cohesionless.
2. The soil is loose.
3. The soil is saturated.
4. There is shaking of ground of the required intensity and duration.
5. The undrained conditions develop in the soil due to its limited permeability.

It is worth noting that liquefaction can occur in the soil deposit at any depth where these conditions are satisfied. Once liquefaction occurs at some depth, the flow of water occurs in the upward direction, and it may cause an indirect liquefaction in the soil layers above.

It is observed that liquefaction normally occurs in the soil classified as SP according to Indian Standard Classification when the SPT number N is less than 15. However, sometimes liquefaction may also occur in the soils classified as SW, SM and ML.

Fine-grained soils do not compress readily under dynamic loadings to cause a high dynamic pore water pressure. Moreover, they also possess the shear strength due to cohesion. Therefore, liquefaction does not occur in the fine-grained soils. However, in quick clays, there is a possibility of liquefaction because their structure is destroyed by shaking.

When liquefaction occurs, sand boils (or mud spouts or sand fountains) appear at the ground surface. A large number of sand fountains were observed during the Dhubri Earthquake in Assam (1934), the Bhuj Earthquake in Gujarat (2001), and many other earthquakes throughout the world.

Liquefaction of soil is responsible for many failures of earth structures, slopes, and foundations. It may also lead to land slides, lateral spreads, sinking of structures and flotation of underground structures.

32.13. FACTORS AFFECTING LIQUEFACTION

The following factors mainly affect the liquefaction of soils in the field.

1. Soil type. As already discussed, liquefaction usually occurs in cohesionless soils, especially soils of type SP.

On the other hand, liquefaction does not occur in fine-grained, cohesive soils. However, highly sensitive, quick clays may liquefy.

2. Particle size and gradation. Fine, uniform sands are more prone to liquefaction than coarse, well-graded sands. Since the permeability of coarse sands is greater than that of fine sands, the pore water pressure is rapidly dissipated in such sands and liquefaction normally does not occur.

Liquefaction potential of sands depends on percentage of fines (size < 0.075 mm) present in it. For a sand of a given initial relative density, as the percentage of fine increases, the liquefaction potential is decreased.

3. Initial relative density. Liquefaction of sands depends to a large extent on the initial relative density (or density index). In dense sands, both pore water pressure and settlements are considerably less than those in loose sands. Hence the proneness of sand to liquefaction is reduced with an increase in relative density indicated by SPT number (or cone penetration resistance).

4. Length of drainage path. If the length of drainage path is large, a sand deposit would behave as undrained when the pore water pressure is suddenly increased due to earthquake and the liquefaction may occur.

The length of the drainage path is sometimes reduced by providing drains of highly pervious materials in sand deposits.

5. Surcharge loads. The initial effective stress in the sand deposit can be increased by the application of a surcharge load on it. With an increase in the effective stress, the transfer of stress from the soil particles to the pore water is delayed. Thus the sand deposit would require higher intensity vibrations for greater duration when the surcharge loads are applied to it.

6. Characteristics of vibration. The main characteristics of vibration are its acceleration, frequency, amplitude and velocity. For liquefaction of soils, the first two characteristics, namely, acceleration and frequency, are more dominant.

Acceleration during vibration is the most important characteristics affecting liquefaction of soils. In general, the greater the acceleration, the greater are the chances of liquefaction. Liquefaction usually occurs only after a certain number of vibration cycles are repeated.

Frequency of vibration is important if it is close to the natural frequency of the soil-foundation system and resonance occurs.

7. Age of soil deposit. If the soil deposit is very old, its proneness to liquefaction is relatively low as compared to that of a recent soil deposit. In old deposits, some form of cementation occurs at the contact points of sand particles and the transfer of interparticle stresses to pore water is delayed.

8. Trapped air. If air is trapped in a sand deposit, a part of the pore water pressure is dissipated due to its compression. Therefore, the possibility of liquefaction is reduced to some extent.

9. Miscellaneous factors. Some miscellaneous factors such as soil structure, method of soil deformations, etc. also affect the liquefaction characteristics of soils but their effect is not significant.

32.14. ASSESSMENT OF SUSCEPTIBILITY OF A SOIL TO LIQUEFACTION

It is important to known whether the soil at the site is susceptible to liquefaction or not, so that suitable measures may be adopted, if required. Recent research has been directed in this direction. Various methods have been proposed by different investigators. The cyclic stress approach is generally used in most of the methods.

The cyclic stress ratio is defined as the ratio of the cyclic shear stress (τ) to the initial effective stress (σ_0'). The cyclic stress ratio produced by the earthquake $(\tau/\sigma_0')_d$ is compared with that required to induce liquefaction in the soil $(\tau/\sigma_0')_l$. The factor of safety (F) against liquefaction is estimated as

$$F = \frac{(\tau/\sigma_0')_l}{(\tau/\sigma_0')_d} \quad ...(32.13)$$

Liquefaction would occur if F is less than unity. However, it has been observed that significant pore water pressures occur even at the values of F slightly greater than 1.0. Generally, a minimum factor of safety of 1.25 to 1.50 is specified.

The proneness of a soil deposit to liquefaction is assessed at various depths in a soil deposit. The factors of safety at all the points should be within the required limit.

(a) Estimation of the cyclic stress ratio produced by the earthquake

The average cyclic shear stress imparted by the earthquake in the top 12 m of a soil deposit can be estimated as (Seed and Idriss, 1982)

$$\left(\frac{\tau}{\sigma_0'}\right)_d = 0.65\left(\frac{\alpha_{max}}{g}\right)\left(\frac{\sigma_0}{\sigma_0'}\right) r_d \quad ...(32.14)$$

where,

α_{max} = maximum horizontal acceleration (MHA) at the ground surface

g = acceleration due to gravity (= 9.81 m/s²)

σ_0 = total vertical stress at the point of interest

σ_0' = effective vertical stress at the same point

$(\tau/\sigma_0')_d$ = cyclic stress ratio produced by the design earthquake.

r_d = stress reduction factor.

The main problem is in the estimation of the maximum horizontal acceleration, (α_{max}). It depends upon a number of factors such as the magnitude of the earth-quake, the focal depth, the distance to the fault trace, the response of the site, etc.

The energy released by an earthquake attenuates as the waves travel away from the zone of fracture. Consequently, the maximum horizontal acceleration decreases as the epicentral distance increases. Fig. 32.10 shows a typical attentuation curve for the rock sites in the central U.S. regions (Nuttli and Herrmann, 1989).

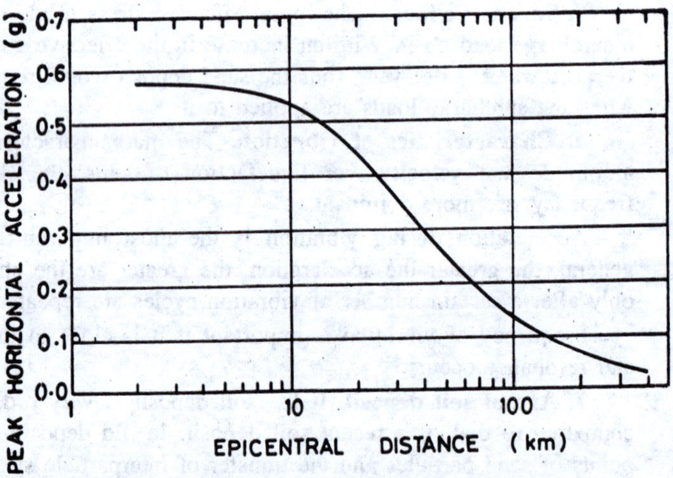

Fig. 32.10. MHA attenuation relationship for $m_B = 7.5$

The values of the maximum horizontal acceleration (MHA) at rock sites get further modified due to local soil conditions. Fig. 32.8 (see Sect. 32.10) shows the variation of MHA at different soil sites with respect to that at rock sites, both the accelerations are expressed in terms of acceleration due to gravity. It may be noted that the MHA at sites on deep clay deposits may be upto 3 times greater than that at rock sites.

The stress reduction factor (r_d) depends upon the depth of the point of interest below the ground surface. The average curve (Fig. 32.11) is generally used in practice (Seed and Idriss, 1971).

The relationship can be approximate as

$$r_d = 1 - 0.008 \times \text{depth in metres} \quad ...(32.15)$$

According to IS: 1893-2002, the basic zone factors (z) are reasonable estimate of effective peak ground acceleration in different seismic zones of India.

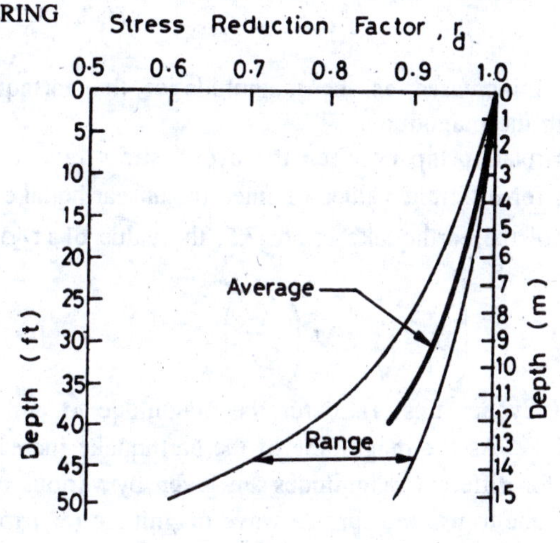

Fig. 32.11. Variation of stress reduction factor with depth

(*b*) Cyclic stress ratio to induce liquefaction in a soil

The cyclic stress ratio $(\tau/\sigma_0')_l$ required to induce liquefaction in a given saturated sand deposit depends upon a number of factors. In the simplified analysis given below, only the following three factors are considered (Seed, et al., 1985).

1. The corrected standard penetration test (SPT) number indicated as $(N_1)_{60}$, obtained from the SPT number $(N)_{60}$ measured in the field.

$$(N_1)_{60} = N_{60}\sqrt{\frac{100}{\sigma_z'}} \qquad \ldots(32.16)$$

where

$(N_1)_{60}$ = standard SPT Number corrected for overburden pressure

$(N)_{60}$ = standard SPT Number measured in the field (duly corrected for field procedures)

σ_z' = vertical effective stress at the test section (kPa).

2. The particle size distribution, expressed as percentage of fines *i.e.*, percent passing the US No. 200 sieve (= 0.075 μ). Alternatively, the mean grain size (D_{50}) can be used.

Soils with less than 5 percent fines are most susceptible to liquefaction. As the percent of fines increases, the liquefaction potential decreases.

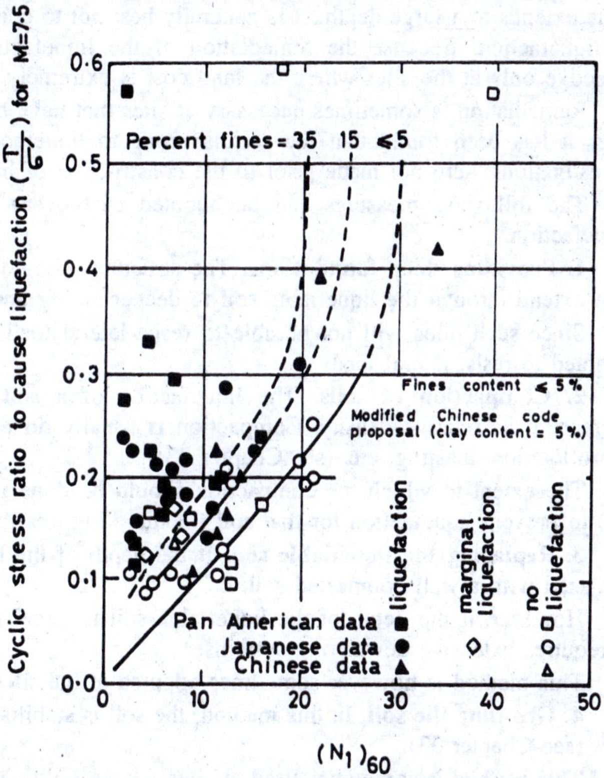

Fig. 32.12. Cyclic stress ratio to cause liquefaction as a function of corrected SPT $(N_1)_{60}$.

3. The duration of shaking or the magnitude of the earthquake. As the duration of the earthquake increases, the susceptibility of a soil to liquefaction also increases. In other words, the possibility of

liquefaction is increased as the magnitude of the earthquake is increased because the duration increases with the magnitude.

Fig. 32.12. gives relationship between the cyclic stress ratio causing liquefaction $(\tau/\sigma_0')_l$ and the SPT Number $(N_1)_{60}$ values for different values of fines for an earthquake of magnitude 7.5.

If the magnitude of the earthquake is not 7.5, the value of $(\tau/\sigma_0')_l$ obtained is to be corrected using the relation

$$\left(\frac{\tau}{\sigma_0'}\right)_M = \psi\left(\frac{\tau}{\sigma_0'}\right)_{7.5} \quad \text{...(32.17)}$$

where $(\tau/\sigma_0')_M$ is the cyclic stress ratio for the magnitude M and $(\tau/\sigma_0')_{7.5}$ is that for the magnitude 7.5 obtained from Fig. 32.12. As the magnitude of the earthquake increases, ψ decreases.

The values of ψ for different magnitudes are given by various investigators.

It is more appropriate to use the surface wave magnitude (M_s) for earthquakes of magnitudes greater than 6.5.

Once the values of $(\tau/\sigma_0')_d$ and $(\tau/\sigma_0')_l$ have been obtained, the factor of safety against liquefaction can be estimated from Eq. 32.13. Thus the zone where the soil is likely to liquefy can be located. Suitable measures can be adopted to prevent liquefaction, if required.

32.15. PREVENTION OF LIQUEFACTION

It is extremely difficult and impractical to prevent liquefaction of soil in the field when the susceptible zone extends to a large depth. It is generally best not to construct buildings and other structures at sites prone to liquefaction. Because the remediation of the liquefaction hazard is quite expensive, it would be cost effective only at the sites where the land cost is extremely high.

Remediation is sometimes necessary at sites that have been developed and the buildings already exist and later it has been found that the soil is prone to liquefaction. This generally happens when adequate site investigations were not made prior to the construction of buildings.

The following measures can be adopted to prevent liquefaction or to limit the damages caused by liquefaction.

1. Providing deep foundations. The structures should be supported on deep foundations, such as piles, that extend through the liquefiable soil to deeper strong and stable strata.

Since such piles will not be able to resist lateral loads in the liquefiable soil, other measures should be adopted to resist lateral loads.

2. Compaction of soils. The liquefaction of a soil can be prevented by compacting the soil and increasing its relative density. Compaction is usually done by means of vibratory rollers, compaction piles, vibroflotation, blasting, etc. (see Chapter 33).

The extent to which the compaction should be done is ascertained by estimating the required SPT No. (N) to prevent liquefaction for that soil for the design earthquake.

3. Replacing the liquefiable soil. If the depth of the liquefiable soil is limited, it can be excavated and replaced with a well compacted soil.

However, if the depth of the liquefiable soil is large, it becomes impractical to replace the soil because it requires extensive dewatering systems.

This method is however sometimes adopted at the sites of critical projects such as earth dams.

4. Grouting the soil. In this method, the soil is stabilised by injecting chemicals or cement grout into the soil (see Chapter 33).

This method is sometimes used at sites where buildings already exist and other methods cannot be used.

5. Ground water pumping. As already discussed, the effective stress at a point increases as the water table is lowered. By restoring to extensive ground water pumping, the liquefaction can be prevented to some extent. However, this method is cost effective only when the water that is pumped can be used for municipal and industrial purposes.

6. Drainage of soils. The liquefaction hazard can be reduced to some extent by providing coarse sand blankets and drains in the soil deposit. The dynamic pore water pressure is thus easily dissipated and the effective stress is increased.

7. Providing stone columns. In this method, a number of holes are bored in the soil deposit and later filled with gravel and stones. Thus stone columns are formed.

The stone columns have high permeability and are quite effective for rapid drainage of pore water. Thus the effective stress is increased. (The behavior of stone columns is somewhat similar to that of the sand drains used for consolidation of soils, discussed in Chapter 12). See Chapter 33 for more details.

8. Application of surcharge. As already discussed, when a surcharge load is applied to a soil deposit, the effective stress is increased. Thus the possibility of liquefaction is reduced.

Monitoring of the pore water pressure will be required to assess the magnitude of the surcharge load and its duration.

32.16. SEISMIC STABILITY OF A SLOPE

Stability of an earth slope under static forces has been discussed in chapter 18. The earth slope is subjected to dynamic forces during an earthquake. In seismic areas, the slope should be designed such that it is also stable under earthquake forces.

The earthquake forces are transient in nature. They act for a fraction of second in one direction and then change the direction. Moreover, the magnitude of the earthquake forces also changes with time. The complete dynamic analysis of a slope is extremely difficult. A simplified method, known as *pseudo- static method,* is generally used for checking the seismic stability of a slope.

In pseudo-static analysis of a slope, the effect of the earthquake is represented by a constant horizontal force acting in one direction. For minimum stability, the force acts horizontally outward through the centre of gravity of the sliding mass. It is assumed that the horizontal force due to an earthquake is of the same nature as the static forces and its effect can be superimposed on that of the static forces in the conventional stability analysis.

The horizontal force due to inertia is given by

$$F_h = \text{mass} \times \text{acceleration due earthquake}$$
$$= k_h W$$

where W = weight of the sliding soil mass.

$$k_h = \text{seismic coefficient} = \frac{\text{acceleration due to earthquake}}{\text{acceleration of due to gravity (g)}}$$

The value of k_h depends upon the ground acceleration. Because of the transient nature of the earthquake, the peak acceleration (α_{max}) acts only for a fraction of a second. During this period, some displacements may occur but complete failure of slope may not occur. The geotechnical engineers, therefore, do not take actual α_{max} in the seismic stability of the slope but they take a smaller value. The value of the seismic coefficient k_h is generally taken between 0.1 and 0.2. A value of 0.15 is quite common.

While checking the seismic stability of a slope, the conventional limiting equilibrium methods for checking the stability of a slope discussed in chapter 18 are used. In the analysis, the fore F_h due to earthquake is considered in addition to the static forces.

The following two types of failures of slopes are considered.

1. Wedge failure
2. Failure along a circular surface

1. Wedge failure

The wedge failure occurs along a plane. Such failures take place when the slope has a specific plane AB of weakness (Fig. 32.13). The factor of safety for static forces is given by Eq. 18.20 as

$$F_s = \frac{cL + (W \cos \alpha) \tan \phi}{W \sin \alpha} \quad ...(a)$$

where c = cohesion, ϕ = angle of shearing resistance

W = weight of wedge ADB

α = angle which the plane AB makes with the horizontal

L = length of the failure surface AB.

When the additional force F_h due to the earthquake is considered, the magnitude of the normal and tangential components are modified.

The component of F_h is the normal direction,

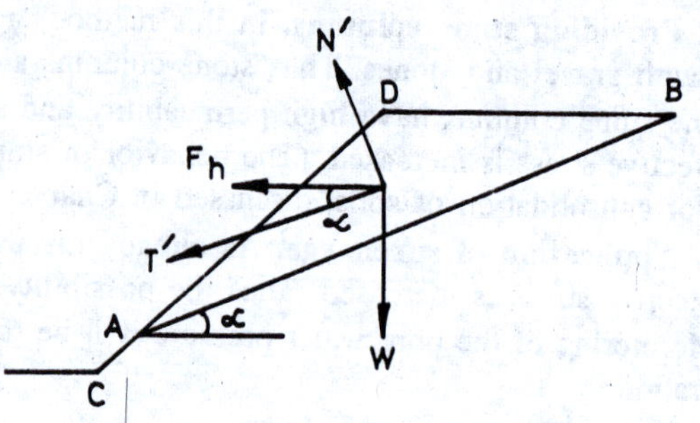

Fig. 32.13.

$$N = -F_h \sin \alpha = -(k_h W) \sin \alpha$$

where k_h = seismic coefficient.

The component of F_h in the tangential direction,

$$T' = F_h \cos \alpha = k_h W \cos \alpha$$

Thus the dynamic factor of safety can be written as

$$DFS = \frac{cL + (W \cos \alpha - F_h \sin \alpha) \tan \phi}{W \sin \alpha + F_h \cos \alpha}$$

Obviously, the value of DFS is less than the factor of safety for static forces. The DFS is generally taken between 1.1 and 1.2. A typical value is 1.15. Because the earthquake is of transient nature, a low value of DFS is permitted.

The values of the strength parameters c and ϕ should correspond to the dynamic material strength. Conservative values of c and ϕ are normally used.

2. Failure along a circular surface

The actual shape of the slip surface in the case of finite slopes in homogeneous soils is curvilinear. In the Swedish circle method, the slip surface is assumed to be an arc of a circle. Fig. 32.14 shows the forces in the case of a circular surface.

W = weight of the sliding soil mass.

F_h = horizontal inertia force (= $k_h W$)

s = shear strengths of the soil along failure surface AB.

The dynamic factor of safety (DFS) is obtained as

$$DFS = \frac{\text{Resisting moment}}{\text{Driving moment}}$$

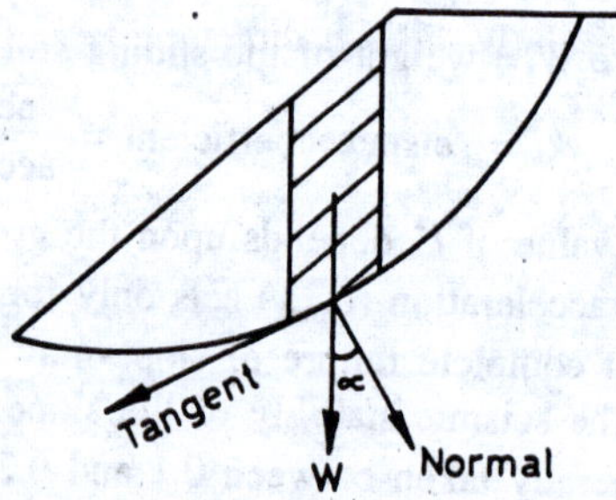

Fig. 32.14

As discussed in Section 18.12, the factor of safety against sliding is given by Eq. 18.33 as

$$FS = \frac{\Sigma cb \sec \alpha + \Sigma (W \cos \alpha) \tan \phi}{\Sigma W \sin \alpha}$$

As in the case of wedge failure discussed above, the normal component is decreased due to the earthquake force F_h but the tangent component is increased (Fig. 32.15). Thus the dynamic factor of safety can be written as

$$DFS = \frac{\Sigma cb \sec \alpha + \Sigma (W \cos \alpha - F_h \sin \alpha) \tan \phi}{\Sigma (W \sin \alpha + F_h \cos \alpha)}$$

where α is the angle which the normal to the curved surface makes with the vertical.

W = weigh of slice.

F_h = the earthquake force in the horizontal direction (= $k_h W$).

c = cohesion intercept.

ϕ = angle of shearing resistance.

b = width of the slice of the slope.

$b \sec \alpha = \Delta l$ = curved length of the slice.

DFS can also be written as

$$DFS = \frac{\Sigma\, cb \sec \alpha + \Sigma\, W(\cos \alpha - k_h \sin \alpha) \tan \phi}{\Sigma\,(W(\sin \alpha + k_h \cos \alpha))}$$

$$= \frac{\Sigma\, cb \sec \alpha + \Sigma\,(N - k_h T) \tan\phi}{\Sigma\,(T + k_h N)}$$

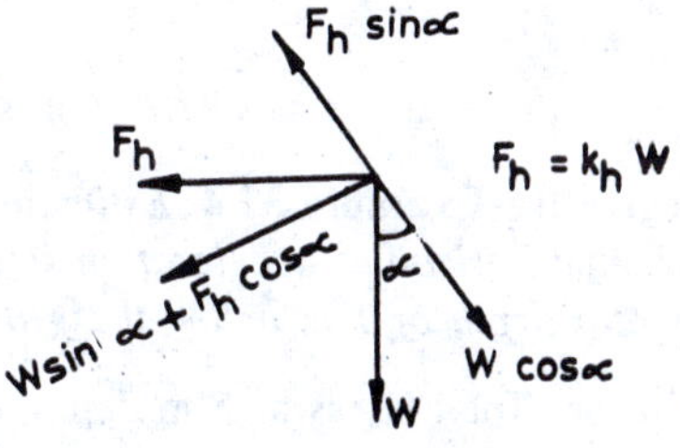

Fig. 32.15

In the above equations, it has been assumed that the pore water pressure does not exist. If the pore water pressure also acts, the equation is modified as

$$DFS = \frac{\Sigma\, cb \sec \alpha + \Sigma\,(N - k_h T - u\, b \sec \alpha) \tan \phi}{\Sigma\,(T + k_h N)}$$

where u is the pore water pressure.

ILLUSTRATIVE EXAMPLES

Illustrative Example 32.1. *The standard torsion seismograph records a trace amplitude of 12.5 mm in E-W direction and a trace amplitude of 11.4 mm in N-S direction at an epicentral distance of 110 km. If the station correction is + 0.1, estimate the magnitude of the earthquake.*

Solution. From Eq. 32.2 (*b*),

$$M = \log_{10} A + \text{distance correction}$$

The distance correction for epicentral distance of 110 km is + 3.1 from Fig. 32.2.

In E-W direction, $M = \log_{10} 12.5 + 3.1 = 4.2$

Applying the station correction, $M = 4.2 + 0.1 = 4.3$

Similarly, in N-S direction, $M = \log_{10} 11.4 + 3.1 + 0.1 = 4.26$

$$\text{Average value of } M = \frac{4.3 + 4.26}{2} = 4.28, \text{ say } \mathbf{4.3}$$

Illustrative Example 32.2. *Estimate the energy released by an earthquake of magnitude 8.0. Compare the energy released to that by the Hiroshima atom bomb.*

Solution. From Eq. 32.4,

$$\log_{10} E = 11.4 + 1.5\,M = 11.4 + 1.5 \times 8.0 = 23.4$$

or $E = 2.51 \times 10^{23}$ *ergs.*

Energy released by the Hiroshima atom bomb = 8×10^{20} *ergs.*

$$\text{Therefore, equivalent number of atom bombs} = \frac{2.51 \times 10^{23}}{8 \times 10^{20}} = \mathbf{314}$$

Illustrative Example 32.3. *Estimate the maximum intensity on MMI scale and the maximum ground acceleration due to an earthquake of magnitude 7.0.*

Solution. From Eq. 32.5, the maximum intensity I_0 on MMI scale is given by

$$M = 1.3 + 0.6\, I_0$$

$$7.0 = 1.3 + 0.6\, I_0$$

or $I_0 = 9.5$

i.e. MMI intensity will be X.

From Eq. 32.6, the maximum ground acceleration a is given by

$$\log_{10} a = \frac{1}{3} I - 0.5 = \frac{1}{3} (10.0) - 0.50 = 2.833$$

$$a = 681.29 \text{ cm/s}^2 = \frac{681.29}{981} g = \mathbf{0.69\ g}$$

Illustrative Example 32.4. *Estimate the cyclic stress ratio produced by an earthquake at a place where the maximum horizontal acceleration is 0.15 g at a depth of 5 m in a sandy soil deposit. Assume the water table is at a depth of 1 m below the ground surface. Take* $\gamma = \gamma_{sat} = 18 \text{ kN/m}^3$.

Solution. Total stress at 5 m depth, $\sigma_0 = 18 \times 5 = 90 \text{ kN/m}^2$

Effective stress at that depth $\sigma_0' = \sigma_0 - u = 90 - (4 \times 9.81) = 50.76 \text{ kN/m}^2$

From Eq. (32.15), $r_d = 1 - 0.008 \times \text{depth} = 1 - 0.008 \times 5 = 0.96$

From Eq. 32.14, $$(\tau/\sigma_0')_d = 0.65 \left(\frac{\alpha_{max}}{g}\right)\left(\frac{\sigma_0}{\sigma_0'}\right) r_d = 0.65 \left(\frac{0.15\ g}{g}\right)\left(\frac{90}{50.76}\right)(0.96)$$

$$\left(\frac{\tau}{\sigma_0'}\right)_d = \mathbf{0.17}$$

Illustrative Example 32.5. (*a*) *A deposit of saturated fine sand has 5% fines. If the corrected SPT number for the sand is 12, estimate the cyclic stress ratio required to cause liquefaction due to an earthquake of magnitude 7.5.*

(*b*) *If the magnitude of the earthquake were 8.0, what would have been the required cyclic stress ratio? Assume a scaling factor (ψ) of 0.65.*

Solution. (*a*) From Fig. 32.12, the cyclic stress ratio to cause liquefaction for $(N_1)_{60} = 12$, percentage of fines = 5% and for the magnitude of 7.50 is **0.12.**

(*b*) For the earthquake of magnitude 8.0, Eq. 32.17 gives

$$\left(\frac{\tau}{\sigma_0'}\right)_M = \psi \left(\frac{\tau}{\sigma_0'}\right)_{7.5} = 0.65\ (0.12) = \mathbf{0.08}$$

Illustrative Example 32.6. *The maximum horizontal acceleration in bed rock is estimated to be 0.20 g. What would be the maximum horizontal acceleration at a project site located over*

(a) a deposit of deep cohesionless soil.

(b) a deposit of deep soft soil deposit.

Solution. (*a*) From Fig. 32.6, for a bed rock acceleration of 0.20 g, the maximum horizontal ground acceleration for a deep cohesionless soil deposit is **0.18 *g*.**

(*b*) For soft soil deposit, the maximum horizontal ground acceleration is **0.30 g.**

Illustrative Example 32.7. (*a*) *A five-storey building, 20 m high, has steel frames without brick infil panels. It is constructed on a deep soil site consisting of loose, saturated sand overlying a thick stiff clay deposit. The building is to be designed to resist an earthquake of magnitude 7.5 with maximum bed rock horizontal acceleration of 0.20 g. Estimate the peak ground acceleration at the site.*

(*b*) *Estimate the spectral response of the building as per 1991 UBC.*

Solution. (*a*) From Fig. 32.6, the bed rock horizontal acceleration of 0.20 g will be amplified to about 0.30 *g*.

(*b*) From Eq. 32.10, the fundamental natural period for steel frame,

$$T_s = 0.085\ h^{0.75} = 0.085\ (20)^{0.75} = 0.80\ s$$

From Fig. 32.8, as per 1991 UBC, for a period of 0.80 s and deep stiff clay, the ratio

$$\frac{\text{spectral acceleration}}{\text{effective peak ground acceleration}} = 1.80$$

Therefore, the spectral acceleration = $1.80 \times 0.30\,g$ = **0.54 g**

Illustrative Example 32.8. *Estimate the design horizontal seismic coefficient (A_h) for a building from the following data, as per IS: 1893:2002.*

1. *Height of building = 30 m*
2. *Base dimension of the building along the considered direction of the lateral force = 10 m.*
3. *The building has moment–resisting R.C. frame with brick infil panels.*
4. *Importance factor = 1.5*
5. *Response factor = 3.0*
6. *The building is situated in seismic zone IV of India.*
7. *The substrata consists of medium soil (type II).*
8. *Spectral acceleration coefficient = 1.50*

Solution. From Eq. 32.11, the fundamental natural period (T_s) is given by

$$T_s = \frac{0.09\,h}{\sqrt{d}} = \frac{0.09 \times 30}{\sqrt{10}} = 0.85\ s$$

Zone factor (z) for seismic zone IV from Table 32.5 = 0.24

From Eq. 32.8,
$$A_h = \left(\frac{Z}{2}\right)\left(\frac{I}{R}\right)\left(\frac{S_a}{g}\right)$$

Taking
$$\frac{Sa}{g} = 1.50,$$

Therefore,
$$A_h = \left(\frac{0.24}{2}\right)\left(\frac{1.5}{3.0}\right)(1.50) = \mathbf{0.09\ g}$$

Illustrative Example 32.9. *The subsoil strata at a site consists of 4 m of soft to stiff clay underlain by 16 m deep fine to medium sand with fines content of 15%. The SPT values at different depths are given in the table. Determine the zone of liquefaction due to an earthquake with M = 7.5 causing a maximum horizontal acceleration of 0.18 g at the ground surface.*

Assume γ_{sat} = 18 kN/m³. The water table is at the ground surface. Take γ_w = 10 kN/m³.

Solution. Calculations are given in a tabular form below:

Depth (m)	*SPT* (N)	σ_0	σ_0'	r_d (Eq. 32.15)	$(\tau/\sigma_0')_d$ (Eq. 32.14)	$(\tau/\sigma_0')_l$ (Fig. 32.12)	Factor of safety (Eq. 32.13)	*Remark*
6	10	108	48	0.95	0.25	0.15	0.60	Liquefiable
8	12	144	64	0.94	0.25	0.17	0.68	Liquefiable
10	9	180	80	0.92	0.24	0.14	0.58	Liquefiable
12	11	216	96	0.90	0.24	0.16	0.67	Liquefiable
14	15	252	112	0.89	0.23	0.23	1.0	Just safe
16	18	288	128	0.87	0.23	0.25	1.09	Not liquefiable
18	24	324	144	0.86	0.23	0.38	1.65	Not liquefiable
20	25	360	160	0.84	0.22	0.40	1.82	Not liquefiable

Conclusion

The zone between depth of 4 to 14 m is liquefiable. Below that depth, the soil is not liquefiable.

Illustrative Example 32.10. *The exploratory borings at a project site show that there is a deposit 5 m thick of saturated sand with a SPT value of 15 overlying a 20 m thick deposit of clay on bed rock.*

If the maximum horizontal acceleration in bed rock due to an earthquake (M = 7.5) is estimated as 0.06 g, is there any possibility of liquefaction in the sand deposit?

Assume the water table is at the ground surface, percentage of fines = 10% and γ = 19 kN/m^3.

Solution. Since the sand deposit overlies a thick clay deposit, the maximum horizontal acceleration will be amplified.

From Fig. 32.6, for a peak horizontal acceleration of 0.06 g in bed rock, peak horizontal acceleration at ground surface will be 0.14 g.

Now for N = 15 and Fines = 10%, the cyclic stress ratio $(\tau/\sigma_0')_l$ required to cause liquefaction from Fig. 32.12 for earthquake of magnitude 7.5 is 0.18.

The total stress σ_0 at the depth of 5 m is

$$\sigma_0 = 5 \times 19 = 95 \text{ kN/m}^2$$

The effective stress at that depth is

$$\sigma_0' = 5\,(19 - 9.81) = 45.95 \text{ kN/m}^2$$

Stress reduction factor $r_d = 1 - 0.008 \times 5 = 0.96$

From Eq. 32.14, the cyclic stress ratio due to the earthquake is given by

$$\left(\frac{\tau}{\sigma_0'}\right)_d = 0.65 \times \left(\frac{\alpha_{max}}{g}\right)\left(\frac{\sigma_0}{\sigma_0'}\right) r_d = 0.65 \times (0.14)\left(\frac{95}{45.95}\right)(0.96) = 0.18$$

From Eq. 32.13, the factor of safety against liquefaction,

$$F = \frac{(\tau/\sigma_0')_l}{(\tau/\sigma_0')_d} = \frac{0.18}{0.18} = \mathbf{1.0}$$

The soil is just liquefiable. However, since the factor of safety is less than 1.25, it not safe. To prevent liquefaction, suitable measures should be adopted.

PROBLEMS

A. NUMERICAL

32.1. If the standard torsion seismograph records an average trace amplitude of 17 mm at an epicentral distance of 100 km, what is the probable magnitude of the earthquake? **[Ans.** 4.2]

32.2. What would be the magnitude of an earthquake to release 8×10^{13} J of energy? **[Ans.** 6.3]

32.3. A six-storeyed building has moment-resisting R.C. frames without brick infil panels and is 25 m high. What would be its natural period of vibration? **[Ans.** 0.84 s]

32.4. Estimate the cyclic stress ratio produced by an earthquake at a site from the following data:

Maximum horizontal ground acceleration = 0.1 g

Stress reduction factor = 0.95

Total stress at the given depth = 120 kN/m^2

Effective stress at that depth = 66 kN/m^2 **[Ans.** 0.113]

32.5. Determine the factor of safety against liquefaction at a depth of 2 m in a sand deposit due to an earthquake of magnitude 7.5 for the following site conditions:

Maximum horizontal acceleration = 0.18 g

Saturated unit weight of sand = 18 kN/m^3

Percentage of fines = 5%

Corrected SPT number = 20

Assume the water table is at a depth of 1.0 m below the ground surface. **[Ans.** 1.39]

32.6. Estimate the design horizontal seismic coefficient as per IS: 1893-2002 for a building with the following site conditions:

Height of building = 28 m

The building has moment-resisting R.C. frames without brick infil panels.
Importance factor = 1.0
Response factor = 3.0
The building is situated in seismic zone V on a substrata consisting of soft soil (type III). **[Ans.** 0.102 g]

B. DESCRIPTIVE AND OBJECTIVE TYPE

32.7. Differentiate between the magnitude and intensity of an earthquake. How would you estimate the maximum intensity if the magnitude of the earthquake is known?

32.8. Explain the following terms:
(*a*) Focus
(*b*) Epicentre
(*c*) Focal depth
(*d*) Epicentral distance

32.9. What is an isoseismal map? How would you prepare an isoseismal for a place after the occurrence of the earthquake.

32.10. What do you understand by the earthquake-resistant structure? What are the criteria for its design?

32.11. What is liquefaction of soils? What are the necessary conditions for its occurrence?

32.12. What are the factors that affect liquefaction of soil? Discuss in brief.

32.13. Explain the procedure for the assessment of the susceptibility of a sand deposit to liquefaction.

32.14. Suggest suitable methods for prevention of liquefaction of soils.

32.15. What are the various hazards due to an earthquake? Discuss in brief only those hazards related to geotechnical engineering.

32.16. Explain the following terms:
(*a*) Pseudo-static analysis
(*b*) Attenuation of ground acceleration
(*c*) Base shear
(*d*) Vertical seismic coefficient
(*e*) Fundamental natural period of vibration.

32.17. What is meant by the term site-specific amplification of horizontal acceleration? What is its effect on the spectral acceleration?

32.18. State which of the following statements are false.
(*a*) The inertia forces on a building act in the same direction as the ground acceleration.
(*b*) As the height of the building increases, its natural period of vibration increases.
(*c*) Liquefaction normally occurs in clayey soils.
(*d*) The earthquake of a magnitude 8.0 would release 1000 times the energy that due to an earthquake of magnitude 6.0.
(*e*) According to IS: 1893-2002, India is divided into four seismic zones.
(*f*) The type of soil according to I.S. classification in which liquefaction is most likely to occur is SW.

[Ans. *False* (*a*), (*c*) (*f*)]

32.19. How would check the seismic stability of a slope by pseudo-elastic anglysis?

C. MULTIPLE CHOICE QUESTIONS

1. According to MSK (64) scale of intensity, the total number of intensity classes (or grades) is
(*a*) 4
(*b*) 8
(*c*) 10
(*d*) 12

2. For an earthquake of magnitude 8.0, the maximum intensity on MMI scale would be about
(*a*) IX
(*b*) XI
(*c*) VIII
(*d*) VI

3. In a sand deposit, the effective stress at a point is 100 kN/m^2. The dynamic pore water pressure head required to cause liquefaction is about
(*a*) 4 m
(*b*) 8 m
(*c*) 10 m
(*d*) 100 m

4. At a depth of 10 m below the ground surface, the stress reduction factor is about
(*a*) 0.90
(*b*) 0.92

(*c*) 0.94 (*d*) 0.96

5. According to IS: 1893-2002, the value of the horizontal seismic coefficient for any structure with $T \leq 0.1\ s$, cannot be less than

(*a*) z/2 (*b*) z/3

(*c*) z/4 (*d*) *z*

6. According to IS: 1893-2002, the ratio (I/R) shall not be greater than

(*a*) 0.50 (*b*) 1.0

(*c*) 1.50 (*d*) None of above

7. Tsunamis due to earthquakes occur in

(*a*) Oceans (*b*) Rivers

(*c*) Lakes (*d*) Reservoirs

8. The type of foundation generally suitable for buildings on liquefiable soils deposit is

(*a*) Spread footing (*b*) Raft foundation

(*c*) Well foundation (*d*) Pile foundation

[**Ans.** 1. (*d*), 2. (*b*), 3. (*c*) 4. (*b*), 5. (*a*), 6. (*b*), 7. (*a*), 8. (*d*)]

33
Ground Improvement Techniques

33.1. INTRODUCTION

The geotechnical engineers design foundation and other structures on the ground after investigation of the type of soil, its characteristics and its extent. If the soil is good at a shallow depth below the ground surface, shallow foundations, such as footings and rafts, are generally most economical. However, if the soil just below the ground surface is not good but a strong stratum exists at a great depth, deep foundations, such as piles, wells and caissons, are required. Deep foundations are quite expensive and are cost effective only in the case where the structure to be supported is quite heavy and huge. Sometimes, the soil conditions are very poor even at a great depth, and it is not practical to construct even deep foundations. In such cases, various methods of ground improvement can be used to improve the ground conditions. Although the ground improvement techniques are also quite expensive, yet they may prove to be cost effective in many cases.

The main objective of the ground improvement is to improve the characteristics of the soil at the site. It usually consists of increasing the shear strength and decreasing the compressibility of the soil so that the bearing capacity of the soil is increased and the settlements of the structures built on it are reduced. Sometimes, the aim is to decrease the permeability of the soil.

The methods of compaction and soil stabilisation discussed in Chapters 14 and 15, respectively are used for soil improvement. They are mainly used for the improvement of soil for the construction of highways, runways, embankments, earth dams, etc. Sometimes, they are also used for ground improvement for shallow foundations. This chapter discusses various techniques of ground improvement mainly for foundations.

Ground improvement is a rapidly developing field because good sites for construction are becoming limited day by day. The geotechnical engineer has the challenge of construction of foundations at the sites which were previously considered unsuitable and unacceptable. Most of these techniques were developed in the last 3-4 decades.

33.2. IMPROVEMENT OF COHESIVE SOILS

Cohesive soils such as soft clay have large void ratio and high water content. Ground improvement is required to reduce the void ratio and the water content to increase its strength so that the bearing capacity is increased and the compressibility is decreased. The following methods are commonly used in practice.

1. Precompression
2. Sand drains
3. Wick drains
4. Stone columns

The above methods are briefly discussed in the following sections.

33.3. PRECOMPRESSION

Precompression is also known as preloading or surcharging. The method of precompression of soil is generally used for soft clays and silty soils.

In precompression, the ground is covered with a temporary surcharge before construction. Generally, a surcharge fill of suitable height is raised on the ground to cause precompression of the soil below. The surcharge fill is kept on the ground for a long time to cause consolidation of the soil in the ground. The surcharge fill is removed before the construction of the structure on the ground. The height of the surcharge fill varies from 3 to 10 m in most cases, depending upon the surcharge required. The surcharge fill should extend beyond the perimeter of the site of construction by at least 10 m on all sides so that uniform compression of the soil at the site of construction occurs.

The principle of precompression can be explained with the help of consolidation theory discussed in Chapter 12. The effect of loading, unloading and reloading of the soil is explained with the help of Fig. 33.1. The virgin curve AB indicates the decrease in the void ratio of the soil during initial loading. The curve BEC shows the increase in the void ratio during unloading. The curve CFD indicates the decrease in the void ratio when the soil is reloaded. It is observed that the increase in the void ratio during reloading is much less than the decrease in the void ratio during initial loading from A to B.

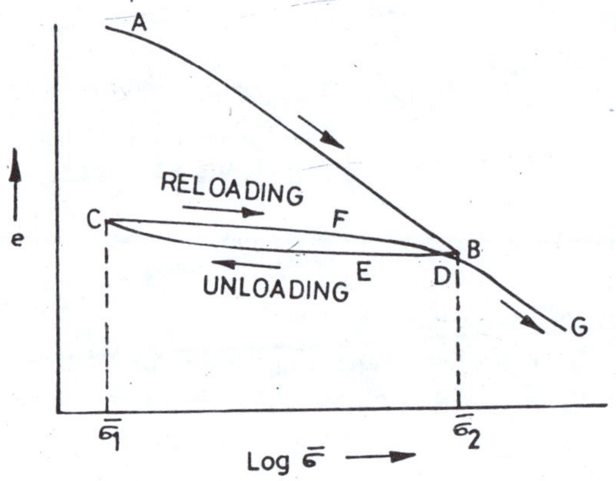

Fig. 33.1

During precompression, when the surcharge fill is applied to the virgin soil in the field, the soil follows the initial loading curve AB. When the surcharge fill is removed, unloading occurs along BEC. When the foundation is constructed on the precompressed soil, the reloading occurs along the curve CFD. The decrease in the void ratio during reloading is considerably smaller than that in the virgin soil in initial loading. The soil properties are also improved. The compressibility of the soil is decreased. Thus settlement of the structure will be decreased. Moreover, the shear strength of the soil is increased and hence the bearing capacity of the soil is improved. In fact, the precompressed soil behaves as an overconsolidated soil.

Advantages of precompression method

1. The method is quite simple and convenient. The conventional earthmoving equipment can be used for raising the surcharge fill.
2. The method is inexpensive if the soil required for surcharge fill is available in the vicinity of the site.
3. Precompression ensures uniformity of improvement of the ground because it eliminates local inhomogeneities and weak spots.
4. The progress of consolidation of the soil can be easily monitored by installing simple settlement plates and piezometers.

Disadvantages

1. The main disadvantage of the precompression method is that the time of consolidation is quite long, sometimes running into several years. If the construction of the structure cannot be delayed that long, other methods of construction are adopted.
2. The method requires a large space around the site of construction to raise surcharge fill. If the site is in the congested area, the method cannot be used.
3. The method is expensive if suitable material for surcharge fill is not easily available at a low cost.

33.4. SAND DRAINS

As already mentioned, the main, disadvantage of the precompression method discussed in the last section is that it takes very long time. Various attempts have been made to accelerate precompression process. Sand drains are commonly used to reduce the time of consolidation.

The theory of sand drains has already been discussed in Sec. 12.20. Because of the provision of sand drains in a soil deposit, the length of the drainage path is decreased. Water flows horizontally to the sand drains. Consequently, the time required to achieve a certain degree of consolidation is considerably reduced.

A sand drain is constructed by driving a casing (or a hollow mandrel) into the ground and then filling the casing with a suitably graded sand. The casing is then gradually withdrawn leaving the sand in the ground. A number of such drains are installed. The typical design parameters are as under:

(*a*) Radius of sand drain = 0.2 to 0.3 m.

(*b*) Spacing of sand drains = 2 to 5 m.

(*c*) Depth of sand drains in embankment = 3 to 35 m.

Fig. 33.2 shows the sand drains installed in position. A sand blanket with a thickness of 0.6 to 1 m is placed over the top of all the sand drains to connect them. On the sand blanket, a surcharge load is placed. The surcharge load is generally in the form of the dumped soil. Due to surcharge load, the consolidation of the clay deposit occurs. The drainage occurs in the horizontal and vertical directions. The sand drains accelerate the process of dissipation of excess pore water created by the surcharge. Thus the rate of consolidation is considerably increased.

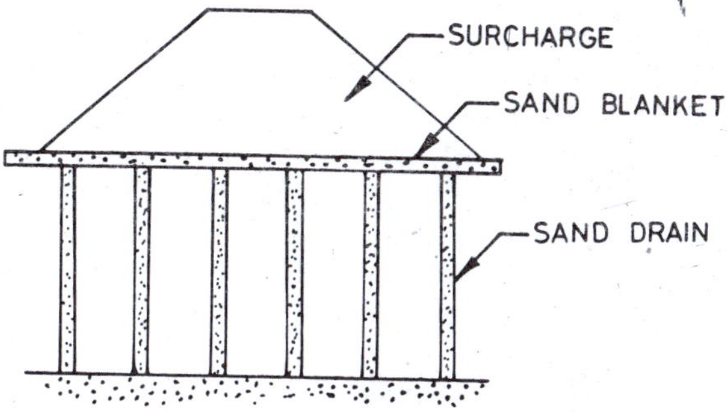

Fig. 33.2

Sometimes the sand drains are provided in the permanent embankments constructed for highways, canals, etc. directly without the surcharge fill. Thus the expense of the surcharge fill is saved.

Sand drains have been used in practice for a long time. The main disadvantage of sand drains is that they are quite expensive. Recently new methods have been developed to reduce cost of construction.

33.5. WICK DRAINS

Because sand drains are quite expensive to construct, the geotechnical engineers have recently developed wick drains (also called band drains).

Wick drains are prefabricated vertical drains consisting of corrugated or fluted plastic ribbons covered by geosynthetic membranes.

At the top of the wick drains, there is a header drains to discharge the pore water (Fig. 33.3). The wick drains are usually of the shape of a strip, about 100 mm wide and 5 mm thick. Because wick drains are less expensive than the sand drains, they have almost replaced the sand drains in modern construction.

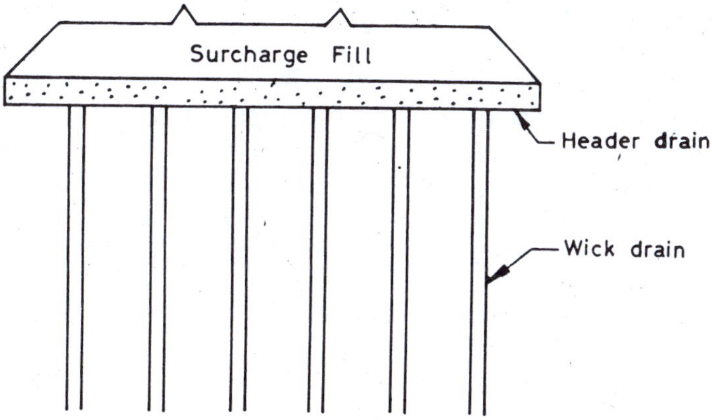

Fig. 33.3

For installation of the wick drains in a clay deposit, a tube is commonly used. The wick drains is held in a tube and the tube is pushed into the soft clay deposit. The tube is then gradually withdrawn leaving behind the wick drains in the clay deposit. Recently special equipment and machines have been developed for the installation of the wick drains to expedite work.

Advantages of wick drains

The wick drains have the following advantages over sand drains:

1. Installation of wick drains is quite easy and convenient. The equipment used is small. Drilling is not required to make holes in the clay deposit.
2. The drainage is good. Once the pore water enters the wick drains, there is little resistance to flow.
3. The wick drains also increase the tensile strength of the clay deposit to some extent.
4. The wick drains are less expensive than sand drains.

33.6. STONE COLUMNS

Ground improvement in case of soft clays can be achieved by installing stone columns in it. Stone columns are constructed by making holes in the deposit and then filing these holes with gravel (or small stones) of size 6 to 40 mm.

For making a hole in the soil deposit, a vibroflot is used (see Sec. 14.14). A vibroflot consists of a cylindrical tube, about 2 to 3 m long and 300 to 500 mm in diameter, fitted with jets at the bottom and at the top. A rotating eccentric mass inside the vibroflot causes a horizontal vibratory motion. The vibroflot is sunk into the clay deposit using the lower water jets. The hole should extend up to a firm stratum below.

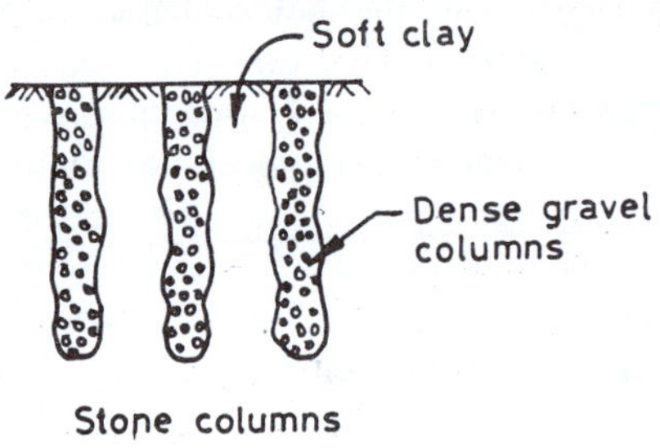

Fig. 33.4

The gravel is then filled into the hole and densified by vibroflot as it is taken out. Thus a stone column is formed (Fig. 33.4). The required number of stone columns are constructed. The spacing of the stone columns is generally between 1.5 to 3 m centre to centre. The diameter of the stone columns is usually 0.5 to 0.75 m.

Stone columns act as vertical drains like sand drains. Consequently, the rate of consolidation is increased.

Besides working as drains, the stone columns reinforce the soft clay deposit because the soft clay is replaced by strong stone columns at discrete points. (The action is somewhat similar to that of vertical steel bars in a R.C.C. column). Because the stiffness of the stone columns is very large as compared to that of soft clay, a large portion of the superimposed load is carried by stone columns. Thus the bearing capacity of the clay deposit is increased and the settlements are reduced.

Stone columns are not effective for thick deposits of highly organic clays and silts.

33.7. IMPROVEMENT OF COHESIONLESS SOILS

Cohesionless soils when in loose condition with N-value less than 10, have a low shear strength, and hence the bearing capacity is quite low. If a deposit of loose sand exists at the site of construction, the ground improvement can be achieved by inducing strong vibrations in the ground to densify the soil. Thus the relative density (or density index) of the sand is increased. Consequently, the N-value and the bearing capacity is increased.

A number of methods have been developed to densify the cohesionless soils. The following methods are commonly used in practice.

1. Vibroflotation
2. Terra Probe
3. Dynamic Impaction
4. Compaction by blasts
5. Compaction piles

These methods are briefly discussed in the following sections:

33.8. VIBROFLOTATION

As already discussed in Sec. 14.14, a vibroflot is a cylindrical tube consisting of a vibrator and jets at the top and bottom. The vibroflot is sunk into the loose soil upto the desired depth using the lower water jet (Fig. 33.5 *a*). As the water comes out of the lower jet, it creates a momentary quick condition ahead of the vibroflot. Because of the quick condition, the shear strength of the cohesionless soil is considerably reduced and the vibroflot settles under its own weight. When the desired depth has been reached, the vibrator is activated. The vibroflot then vibrates laterally and causes the compaction of the soil in the horizontal direction up to a radius of about 1.5 m.

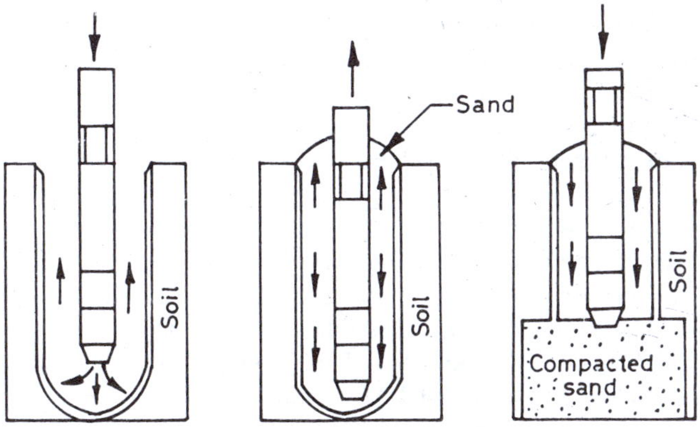

(*a*) (*b*) (*c*)

Fig. 33.5

After the compaction of the soil in the lateral condition, the water from the lower jet is transferred to the top jet and the pressure is reduced so that it has just enough force to carry the sand poured at the top of the hole to the bottom of the hole (Fig. 33.5 *b*). Additional sand is continually dropped into the space (crator) around the vibroflot.

By raising the vibroflot in stages and simultaneously backfilling the hole with sand, the entire depth of the hole is filled with compacted sand (Fig. 33.5 *c*). Repeating the process at a number of points on the ground, the entire soil deposit is compacted. The spacing of the holes is usually kept between 2 to 3 m on a grid pattern.

Vibroflotation method has been effectively used for compaction of loose sand deposits upto 20 m depth. The relative density achieved in most cases is 70% or more.

Vibroflotation is a type of vibrocompaction because compaction is done with the help of vibration. Another method of vibrocompaction is the terra probe method discussed in the following section. Vibro compaction methods are, however, effective only when the silt content in the cohesionless soil is less than 12-15 percent and the clay content is less than 3 percent.

33.9. TERRA PROBE METHOD

Terra probe method, like vibroflotation, is a technique of vibrocompaction of cohesionless soils (see Sec. 14.15). The terra probe consists of an open ended pipe about 75 cm in diameter. It is provided with a vibratory pile driver at its bottom. The vibratory pile driver when activated gives vertical vibrations to the terra probe and it goes down into the sand. After it has reached the desired depth, the terra probe is gradually raised upward while the vibratory driver continues to operate. Thus, the soil within and around the probe is densified. The operation is repeated at a number of locations at the site. The spacing of the holes is usually kept about 1.5 m.

The method can be used upto a depth of 20 m. The method is ideal for the compaction of saturated sand deposits. At the sites where the water table is deep, water jets are fitted to the terra probe to assist the penetration and compaction of soil. The terra probe method is considerably faster than the vibroflotation method but is less effective. As the terra probe method does not require backfilling of sand, it can even be used at offshore locations. In the terra probe method, the zone of influence is considerably smaller than that in the vibroflotation method and the relative density achieved is also lower.

33.10. DYNAMIC COMPACTION

Dynamic compaction is also known as dynamic consolidation, heavy tamping or pounding. In this method, a heavy weight about 2 to 50 tonnes (with mass 2 Mg to 50 Mg) is dropped on the ground from a height of 7 m to 35 m. The heavy weight used is also called a pounder. It consists of a concrete or steel block. A crane is used to lift the pounder to the required height and then it is allowed to fall freely under gravity on the ground surface (Fig. 33.6). A pit (or a crator) is formed at the point where the pounder hits the ground. At each point, 5 to 10 poundings are given. The operation is repeated at other locations. The depth of crator at each location is 1 to 3 m. The ground surface soil is disturbed because of poundings. The ground surface is levelled after the pounding operation is over. The upper layer of the ground surface is compacted and graded properly afterwards.

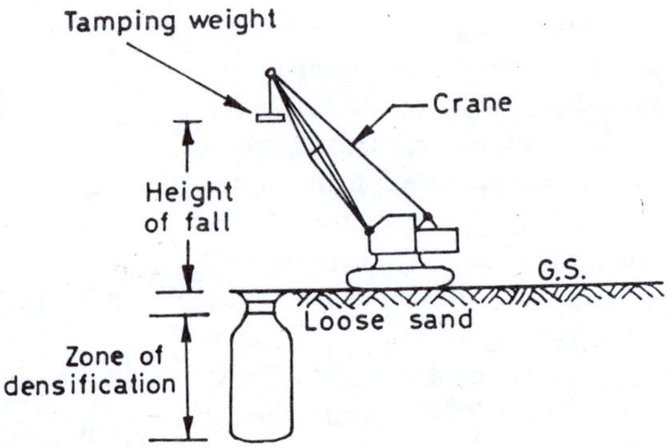

Fig. 33.6

Dynamic compaction method is quite effective for compaction of loose sands and silty soils. The method has also been used for compacting soils containing boulders and sanitary land fills. Good densification usually occurs upto a depth of 5 to 10 m. At greater depths, densification is less. The method is also used for sands prone to liquefaction and for collapsible soils and other soils prone to excessive settlement.

The depth D upto which good densification occurs can be approximately determined from the following relation:

$$D = \frac{1}{2} (MH)^{1/2}$$

where M = mass of pounder (Mg)

H = height of drop (m)

D = depth of densification (m)

For illustration, if M = 25 Mg and H = 20 m, the depth D = 11 m.

In the dynamic compaction method, large shock waves are generated in the ground. Extreme care shall be taken to ensure that harmful vibrations are not transferred to the adjacent structures. Because of this problem, the method cannot be used near existing structure in the urban areas.

The radius of influence (R) beyond which no harmful vibrations are transmitted is approximately given by the relation.

$$R = 130 \sqrt{MH}$$

where R = radius of influence of vibrations (m)

M = mass of pounder (Mg) (1 Mg = 1000 kg)

H = height of drop (m)

For illustration, if M = 25 Mg and H = 20 m, the radius R = 2.9 km.

33.11. COMPACTION BY BLASTS

In compaction by blast, buried explosives are used to densify loose sandy soils. In this method, a series of holes is bored and explosives are placed in them. The holes are filled back with soil. These explosives when detonated cause shock waves in the ground causing densification of loose surrounding soil.

Explosive charges usually consist of about 60% dynamite and 30% special gelatin dynamite and ammonite. The charges are generally placed at two-thirds the thickness of the stratum to be densified. The spacing of the holes is kept between 3 to 8 m. Three to five blasts are generally required at each point.

The radius of influence of the compaction from the blast point is given by the relation.

$$R = (M/C)^{1/3}$$

where R = radius of influence (m)

M = mass of explosive charge (kg)

C = constant (= 0.04 for 60% dynamite)

The surface soil upto a depth of about 1 m gets displaced in a random manner. This surface zone is compacted by the conventional methods, using rollers.

Densification by blasts is effective upto a depth of about 25 m. The method is quite effective when the sand is clean (with no fines) and is fully saturated. The shock waves cause liquefaction of sand. It is followed by densification of sand. The method is not very effective in the case of partially saturated sands because compressive stresses develop due to capillary action, which prevent the sand particles from taking closer position.

Because of harmful vibrations in the adjacent structures and safety precautions required, the method is commonly used for remote sites. Moreover, a large depression is formed at the ground surface after blasts. Surface settlements upto 80 cm have been recorded at some sites.

The method is less common than the vibrocompaction and dynamic compaction methods already discussed.

33.12. COMPACTION PILES

Cohesionless soils can be densified by installing compaction piles. For construction of a compaction pile at the site, a close-ended, hollow tubular pipe is first driven into the ground. The loose soil surrounding the pipe is compacted due to vibration during its driving. The pipe is then extracted and the hole is backfilled with sand to form a compaction pile.

Compaction piles are constructed at several locations on a grid pattern (Fig. 33.7). Thus a uniform deposit much denser than the original soil is formed.

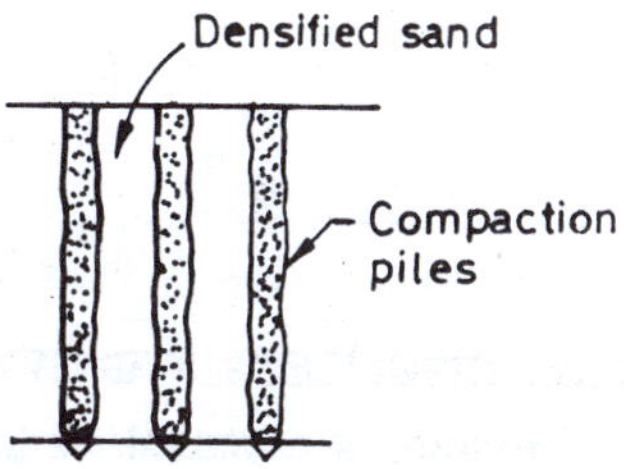

Fig. 33.7

The radius of influence around the compaction pile in which the sand is densified depends upon the diameter of the pile. The radius of influence is about 3 to 4 times the pile diameter. For good results, the silt content should not be greater than 15 percent in the soil. Moreover, the clay content should not be greater than 3%.

33.13. GENERAL METHODS FOR GROUND IMPROVEMENT

The following methods can be used for all types of soils.

1. Removal and replacement of soil
2. Deep mixing of soil
3. Grouting

These methods are discussed below:

33.14. REMOVAL AND REPLACEMENT OF SOIL

When the thickness of the stratum of the poor soil is small and the total quantity of soil to be removed is not very large, it can be excavated and removed. The excavated soil is replaced with a good quality soil in the stratum. The method is cost effective only when good quality soil is available in the vicinity for replacement. In case, the large quantity of soil is involved and the good quality soil is not available near the site, this method becomes expensive.

The method has been successfully used for the replacement of collapsible soils. In some cases, the excavated soil itself is used for replacement after properly treating it and compacting it.

The method is generally practical only when the water table is low so that the earthwork operations can be done in dry condition without pumping of water.

The method is quite simple and convenient. Manual labour or the conventional earthmoving machines can be used for excavation and backfilling.

33.15. GROUND IMPROVEMENT BY DEEP MIXING

Ground improvement can be done by mixing either cement or lime with the soil deep in the ground. Different types of machines have been used for deep mixing. A special type of auger with a hollow drill is commonly used. The auger is inserted by rotating it deep into the ground to the required depth. A rig is normally required for auger driving. The auger is provided with a mixing tool at its bottom tip (Fig. 33.8 *a*).

The auger is later withdrawn and the cement (or lime) slurry is passed through the drill rod to the mixing tool. The slurry is mixed with the soil in the vicinity of the mixing tool. The slurry is mixed mechanically with the soil. Thus a vertical column of the stabilised soil is formed. The diameter of the column is equal to the diameter of the auger. A number of such columns are formed in the ground. The columns are close enough to each other so that there is some overlapping (Fig. 33.8 *b*). Thus the whole of the soil at a large depth is stabilised.

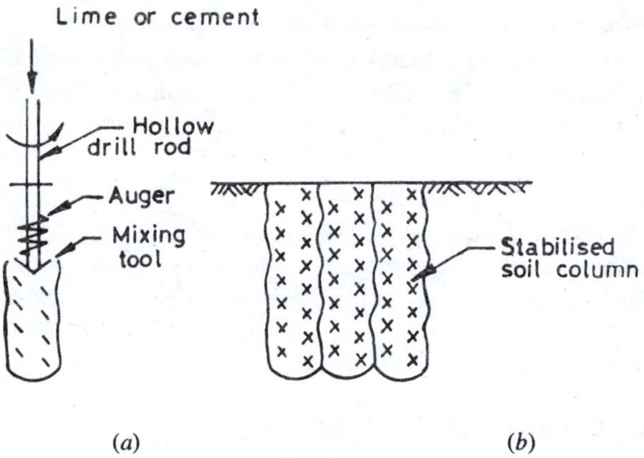

Fig. 33.8

As discussed in Chapter 15, the stabilised soil is more strong, less permeable and less compressible than the natural soil.

The deep mixing method can be used for stabilisation of deep deposits of weak soils such as soft clays.

33.16. GROUTING FOR GROUND IMPROVEMENT

Grouting is a general method of ground improvement which is effective not only for sands and silts but also for rocks. Grouting consists of injection of a grout into the ground. There are mainly two types of grouts which are commonly used, viz., cement grout and chemical grout. Cement grout is formed by adding water to cement and making a slurry that hydrates after it has been injected into the ground. Thus a solid mass is formed around the point of injection. Chemical grout contains chemicals such as silicates and resins. Chemical grouts are used for injecting into fine sand or silt deposits because cement grout is not effective for such soils. However, chemical grouts are quite expensive and are used only in special cases. Moreover, chemical grouts can be toxic and corrosive.

Methods of grouting

There are basically 4 types of grouting used for ground improvement.

1. Intrusion grouting
2. Compaction grouting
3. Permeation grouting
4. Jet grouting

1. Intrusion grouting. In intrusion grouting, joints and fractures in rocks or soils are filled by injecting grout into them (Fig. 33.9 *a*). The pipes are first inserted into the ground from the surface. Sometimes, the pipes are inserted from tunnels or other openings in the ground. The grout is then injected by pumping it into the pipes.

The main advantage of intrusion grouting is that the permeability of the rock or the soil deposit is decreased. Generally, cement grout is used. The method is commonly used for the foundation of water-retaining structures such as dams, to reduce the seepage.

2. Compaction grouting. Compaction grouting is also called displacement grouting. In this method, a thin grout is injected into the ground under a high pressure through a pipe to form a series of intrusions in

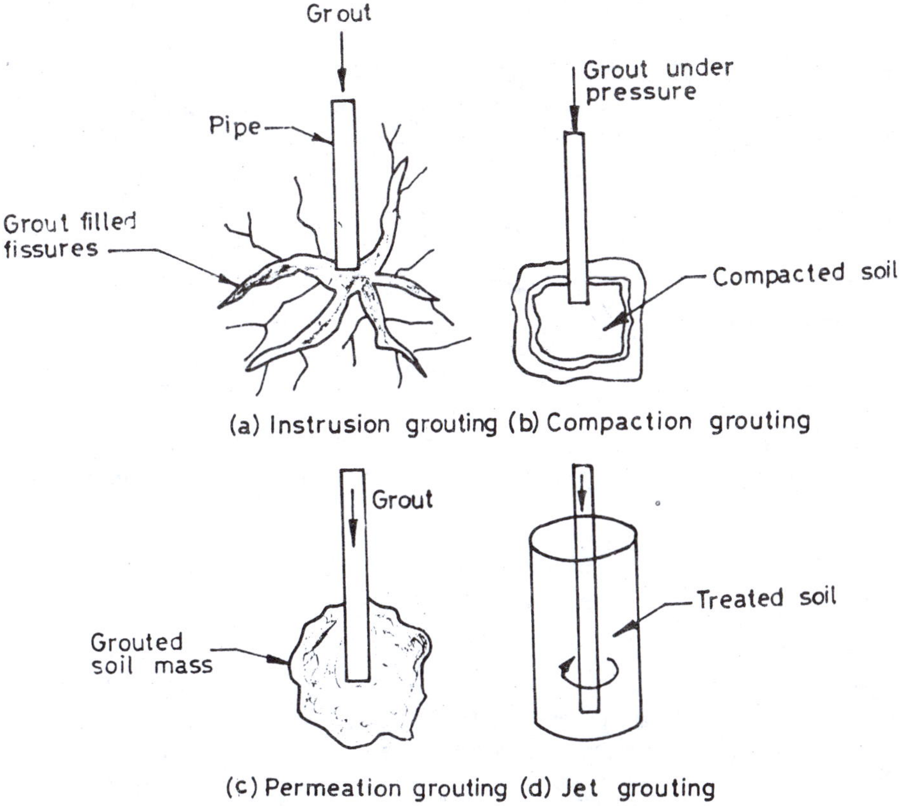

Fig. 33.9

the deposit (Fig. 33.9 *b*). The grout should be stiff, with a slump of about 25 mm. Because the grout is quite thick, it does not penerate deep into the soil. The main effect of the grout in the deposit is to compact the adjacent soil. Consequently, the strength of the soil is increased.

Compaction grouting is quite effective for loose cohesionless soils. It is also used for unsaturated, soft fine-grained soils. The method cannot be used for soft, saturated fine-grained soils because pore water pressure developed in these soils causes uneven, large settlements.

Compaction grouting is commonly used to restore the structures which have suffered excessive settlements. Compaction grouting improves the strength of the underlying soil and even reduces the settlements by raising the existing structure.

3. Permeation grouting. In permeation grouting, a thin grout is injected into the soil deposit so that it permeates into the voids of the soil. Because the voids in the soil are quite small, cement grout is not effective and, therefore, chemical grout is used. After the grout cures, the soil deposit is converted into an almost solid mass (Fig. 33.9 *c*).

After permeation grouting, the soil becomes impermeable, strong and less compressible. This method is commonly used to stabilise soils before starting excavation or tunneling operations.

4. Jet grouting. In jet grouting, a special pipe fitted with horizontal jets is used. The jets inject the grout into the soil under high pressure. The pipe is first inserted into the ground to the required depth and then injection of grout is started. The pipe is then gradually raised and rotated while the grout injection continues. Thus a column of treated soil is formed around the pipe (Fig. 33.9 *d*) The soil in the column becomes quite strong and impermeable.

The method is commonly used for ground water control projects. The method is also suitable for stabilising soils before starting excavation or tunneling operation.

PROBLEMS

A. DESCRIPTIVE AND OBJECTIVE TYPE QUESTIONS

1. What are the site conditions when ground improvement is required for foundations?
2. Describe various methods of ground improvement suitable for cohesive soils. Discuss in brief.
3. What is precompression of soils? How it is done? What are its advantages and disadvantages?
4. Write short notes on:
   (*a*) Sand drains (*b*) Wick drains
   (*c*) Stone columns
5. What are different methods for ground improvement used for cohesionless soils? Describe vibroflotation method.
6. Write short notes on:
   (*a*) Terra probe
   (*b*) Dynamic compaction
   (*c*) Compaction by blasts
   (*d*) Compaction piles
7. Describe the removal and replacement method for ground improvement. What are the conditions when this method can be used?
8. How the ground can be improved by deep mixing? Discuss in brief.
9. Describe various methods of grouting used for ground improvement.
10. State whether the following statements are true or false:
    (*a*) Precompression of soil is generally effective only in clays and silts.
    (*b*) Sand drains may be used for ground improvement even without surcharge loads.
    (*c*) Compaction piles can be used for cohesive soils.
    (*d*) Densification by vibroflotation is effective for cohesionless soils.
    (*e*) For injection in soil, chemical grout is more effective than cement grout.
    (*f*) Deep mixing can be used for rocks. [**Ans.** True, (*a*), (*b*), (*d*), (*e*)]

B. MULTIPLE CHOICE QUESTIONS

1. The following method is not suitable for cohesive soils:
   (*a*) Sand drains (*b*) Stone columns
   (*c*) Wick drains (*d*) Compaction piles
2. The following method is not suitable for cohesionless soils:
   (*a*) Precompression (*b*) Vibroflotation
   (*c*) Terra probe (*d*) Compaction by blasts
3. The radius of the sand drains well is usually from
   (*a*) 0.2 to 0.3 m (*b*) 0.4 to 0.6 m
   (*c*) 0.6 to 0.8 m (*d*) 0.8 to 1.0 m
4. The spacing of the stone columns is generally between
   (*a*) 0.5 m and 1.5 m (*b*) 1.5 m and 2.0 m
   (*c*) 2.5 m and 2.5 m (*d*) 2.5 m and 3.0 m
5. Vibroflotation is best suited for compacting
   (*a*) Coarse sand and gravels (*b*) Silts
   (*c*) Organic soils (*d*) All the above
6. Compaction grouting is suitable for
   (*a*) Soft clays (*b*) Loose cohesionless soils
   (*c*) Organic soils (*d*) Rocks

[**Ans.** 1. (*d*), 2. (*a*), 3. (*a*), 4. (*b*). 5. (*a*), 6. (*b*)]

34

Foundations on Expansive and Collapsible Soils

34.1. EXPANSIVE SOILS

Expansive soils are the soils which expand when the moisture content of the soils is increased. The clay mineral montmorillonite is mainly responsible for expansive characteristics of the soil. The expansive soils are also called swelling soils or black cotton soils.

Expansive soils are generally residual soils left at the place of their formation after chemical decomposition of the rocks such as basalt and trap. The soils are generally dry because the water table there is quite deep. During rainy season, they become wet. The soils expand as the water content is increased. Severe movements of the soil mass may occur. Structures built on such soils may experience cracking and damage due to differential heave.

A large part of the central India and a part of the south India is covered with expansive soils. Although these soils are good for growing cotton, they are treacherous for foundations of structures. Heavy damages may occur to buildings, roads, runways, pipe lines and other structures built on such soils if proper preventive measures are not adopted. The damages can be prevented to a large extent if the characteristics of the expansive soil are properly assessed and suitable measures are taken in the design, construction and maintenance of structures built on expansive soils.

When the water content of the expansive soil is increased, it expands. When the water content is reduced, it shrinks and cracks develop. These cracks may travel deep into the ground. The maximum width of the shrinkage cracks is usually limited to 20mm.

The depth of the expansive soil in which periodic changes of moisture content occur to cause swelling and shrinkage is known as the *active zone.* The soil below the active zone is not affected. The depth of the active zone at various locations is different. In most cases, the depth of the active zone is limited to 3 to 4 m.

The depth of the active zone of a place can be obtained by plotting the liquidity index of the soil against depth for several seasons. In the active zone, there is a large variation in the liquidity index of the soil during different seasons. Below the active zone, the variation in the liquidity index is negligible (Fig. 34.1).

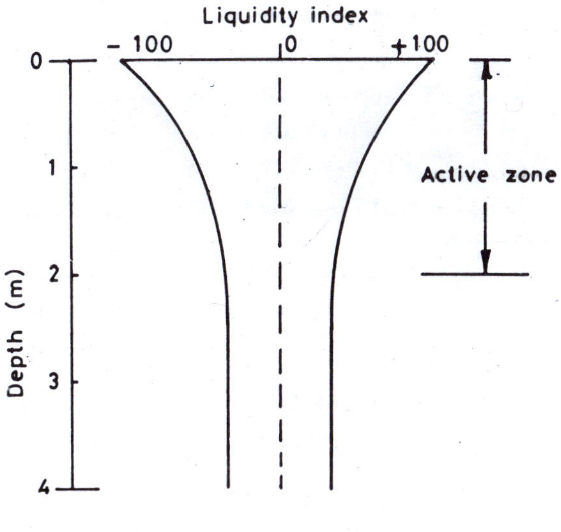

Fig. 34.1.

34.2. PARAMETERS OF EXPANSIVE SOILS

Various parameters are used for identification of expansive soils. Some of the commonly used parameters are discussed below.

1. Free Swell. Free swell (S_f) is defined by the relation

$$S_f = \frac{V_f - V_i}{V_i} \times 100 \quad \ldots(a)$$

where S_f = free swell (percent)

V_f = final volume of soil

V_i = initial volume of soil

For the determination of free swell (S_f), the following method given by Holtz and Gibbs (1956) is normally used.

Ten cubic centimeters (V_i) of dry soil passing through US Sieve No. 40 (425 μ size) is poured into a 100 cm^3 graduated cylinder containing distilled water. The cylinder is kept undisturbed for 24 hours, and then the final volume (V_f) of the settled soil is measured. Free swell is calculated from Eq. (*a*).

The free-swell test is not adequate to predict accurately the swelling characteristics of the soil and is now rarely used in practice.

2. Unrestrained Swell Test. The unrestrained swell test is commonly used to determine the free swell.

In this test, a soil specimen is placed in an oedometer (consolidometer) and a small surcharge of 6.9 kN/m^2 (1 psi) is applied to it. Water is then added to the specimen so that it is fully submerged in water. Because the cross-sectional area of the specimen remains constant, the expansion is also equal to percentage increase in thickness. The increase in thickness is measured after the equilibrium has reached.

The percent free swell (S_{wf}) is calculated from the relation

$$S_{wf} = \left(\frac{\Delta H}{H}\right) \times 100$$

where S_{wf} = free swell (percent)

ΔH = increase in height of specimen

H = initial height of specimen

The free surface swell of an expansive soil deposit in the field can be obtained from the relation given by O'Neill and Poormoayed (1980) as

$$\Delta S = 0.0033\, S_{wf}\, z$$

where ΔS = surface swell of the deposit

z = depth of active zone

S_{wf} = free swell (%)

3. Differential Free Swell. Differential free swell (DFS) is a parameter used for the identification of an expansive soil.

For the determination of the differential free swell of a soil, 20 g of dry soil passing through a 425 μ size sieve is taken. The soil is divided into two samples of 10 g each. One sample of 10 g is poured into a 100 cm^3 capacity graduated cylinder containing water, and the other sample of 10 g is poured into a 100 cm^3 capacity graduated cylinder containing kerosene oil. Both the cylinders are kept undisturbed in a laboratory. After 24 hours, the settled volumes of both the samples are measured.

Differential free swell (DFS) is determined from the relation

$$DFS = \frac{\text{Settled soil volume in water} - \text{settled soil volume in kerosene}}{\text{Settled soil volume in kerosene}} \times 100$$

Because kerosene is a non-polar liquid, it does not cause any swell of the soil.

IS: 2720 (Part III-1980) gives degree of expansion of a soil depending upon its differential free swell as under:

S. No.	*Degree of expansion*	*DFS*
1.	Low	< 20%
2.	Moderate	20 – 35%
3.	High	35 – 50%
4.	Very high	> 50%

If the DFS of a soil is greater than 35%, its expansiveness should be considered in the design of foundation as discussed later. The conventional shallow foundation cannot be used for such soils.

4. Swelling Potential. Swelling potential (S) is defined as the percentage of swell under a pressure of 6.9 kN/m^2 (= 1 psi) surcharge of a laterally unconfined specimen compacted at the optimum moisture content to the maximum dry density in the standard AASHTO compaction test (Modified Proctor test).

Swelling potential of a soil can be determined in a laboratory by conducting a test in an oedometer and measuring the swell. Alternatively, the following empirical relations can be used to determine the swell potential.

(*i*) Swelling potential: $SP = (3.6 \times 10^{-5})\, A^{2.44}\, C^{3.44}$

where SP = swelling potential (percentage)

C = percent of clay (– 2 μ) fraction by weight

A = activity of clay (= plasticity index / C)

(*ii*) The empirical relation based on the plasticity index of soil is

$$SP = (2.16 \times 10^{-3})\, (PI)^{2.44}$$

where PI = plasticity index

According to the USBR, the degree of expansion of soils of different swell potentials is as under:

S. No.	*Degree of expansion*	*SP*
1.	Low	0 – 1.5
2.	Medium	1.5 – 5
3.	High	5 – 25
4.	Very high	> 25

(*iii*) Ranganatham and Satyanarayan (1965) method

For black cotton soils of India, the following relation given by Ranganatham and Satyanarayan (1965) gives good results:

$$SP = 41.13 \times 10^{-5}\, (SI)^{2.67}$$

where SI = shrinkage index = LL – Shrinkage limit.

The degree of expansion based on the shrinkage index is as under:

S. No.	*Degree of expansion*	*SI*
1.	Low	0 - 20%
2.	Medium	20 – 30%
3.	High	30 – 60%
4.	Very high	> 60%

5. Swelling Pressure. The swelling pressure (p_s) is defined as the pressure required to be applied over a swelling soil specimen to prevent its expansion when it is in contact with water. The swelling pressure test is conducted in an oedometer. The test requires continuous adjustment of the pressure on the soil specimen.

Thus the volume of the specimen remains equal to its initial volume throughout the test. Details of the test are given in IS: 2720 (Part XLI)-1977.

The swelling pressure of a soil does not have a unique value because it depends upon a number of factors such as initial moisture content, initially dry density, method of compaction, the height of the specimen, the surcharge pressure, and others. Moreover, the actual swelling pressure of a soil in the field is usually much less than that measured in the laboratory.

If the swelling pressure of a soil is less than 30 kN/m^2, it indicates the expansiveness is low and the conventional design of shallow foundations can be adopted. For higher values of the swelling pressure, special design for expansive soil, as discussed later, is to be used. Some types of clays, such as bentonite, may have the swelling pressure of even 2000 kN/m^2.

34.3. CLASSIFICATION OF EXPANSIVE SOILS

The geotechnical engineer dealing with expansive soils should adopt a systematic method for identification of such soils. An experienced geotechnical engineer may be able to identify an expansive soil from its visual examination. An expansive soil has a large clay content. When dry, it has distinct shrinkage cracks. The soil becomes sticky when wet. Generally, the soils classified as CL, CI or CH are expansive soils. The soils classified as ML, MI or MH may also be expansive. Even the soils classified as SC may be expansive in some cases.

Various investigators have given the classification of expansive soils based on the index properties such as plasticity index, shrinkage limit, linear shrinkage, clay content, potential swell, etc. Unfortunately, there is no fool-proof method of classification. The values of the index properties given by various investigators differ over a large range .

A few selected classifications are given below for general guidance.

1. **Altmeyer (1955)** gave the following classification of expansive soils based on shrinkage limit, linear shrinkage and probable swell:

Shrinkage limit (%)	*Linear shrinkage* (%)	*Probable swell* (%)	*Degree of expansion*
> 12	< 5	< 0.5	Non-critical
10 – 12	5 – 8	0.5 – 1.5	Marginal
< 10	> 8	> 1.5	Critical

2. **O'Neill and Poormoayed (1980)** summarised the criteria given by the U.S. Army Waterways Experiment Station based on plasticity index, liquid limit, and potential swell as under:

Liquid limit (%)	*Plasticity Index* (%)	*Potential swell* (%)	*Swelling potential*
< 50	< 25	< 0.5	Low
50 – 60	25 – 35	0.5 – 1.5	Marginal
> 60	> 35	> 1.5	High

* Potential swell is the vertical swell under a pressure equal to the overburden pressure.

3. **Holtze (1969)** classified the expansive soils based on clay content, plasticity index and shrinkage limit as under:

Colloids (%)	*Plasticity Index* (%)	*Shrinkage limit* (%)	*Swelling potential*
≤ 15	≤ 18	≥ 15	Low
13 – 23	15 – 28	10 – 16	Medium
20 – 31	25 – 41	7 – 12	High
> 28	> 35	< 11	Very high

4. Chen (1988) classified the expansive soils based on the plasticity index as under:

S. No.	*Plasticity Index (%)*	*Degree of Expansion*
1.	≤ 15	Low
2.	10 – 35	Medium
3.	20 – 55	High
4.	≥ 35	Very high

34.4. CAUSES OF MOISTURE CHANGES IN SOILS

Even a soil having a high swelling potential may not cause trouble if there are no moisture changes. The shrinkage or swelling in the soil occurs only if there are moisture changes. Moisture changes can be due to natural processes or due to human activities. Some of the common causes for the moisture changes are as follows:

1. Moisture changes may occur due to change in the water table.
2. Moisture may increase due to land scaping and irrigation of lawns and kitchen gardens around the building.
3. If the surface water drainage becomes defective and the runoff is obstructed, the water will infiltrate the ground and cause moisture changes.
4. If the vegetation around the building is removed, the transpiration will stop and the moisture will increase.
5. If a slab or a pavement is placed directly on the ground, the direct evaporation of water is stopped. Moreover, direct infiltration of rain water is prevented.
6. Moisture changes may occur due to seasonal changes if there is a large variation in temperature in different seasons.

34.5. EFFECTS OF SWELLING ON BUILDINGS

In arid and semi-arid regions, the water content of the soil is considerably reduced due to heat in the summer. The water table also goes down. The soil becomes stiff and shrinkage cracks develop. During the rainy season, the water content of the soil is increased, resulting in swelling. Thus there are large moisture variations in the soil due to seasonal changes.

When a building is constructed on a swelling soil, the soil below the building is protected from excess heat even during summer. This soil swells because its evaporation is obstructed. However, the soil adjacent to the building which is open to atmosphere will experience normal swelling and shrinkage. Thus, the differential movements occur during the hot weather.

The depth of the zone of appreciable ground movement due to seasonal variation in moisture changes depend on the nature of the soil and climate factors. This zone called *active zone* is different at various places. The Central Building Research Institute, Roorkee, has established that the depth of the active zone in India is limited to about 3.5 m.

The movement of the soil with swelling and shrinkage causes damages to the floor slabs and the external walls of the building. Because of swelling, the floor slab is pushed up. It takes a dome shape and cracks

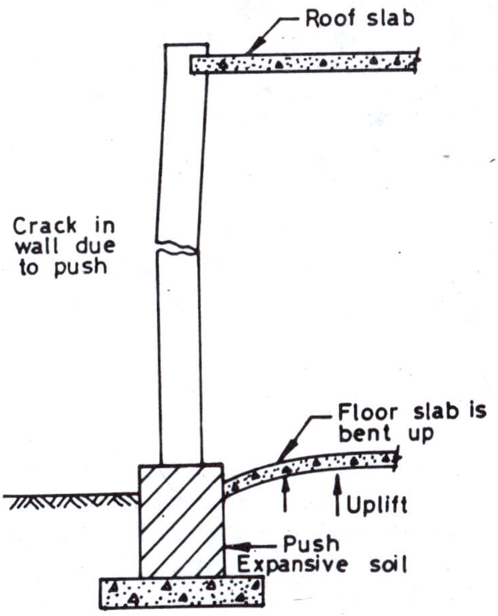

Fig. 34.2

develop in the floor (Fig. 34.2). The footing wall is pushed outward due to swelling. It causes the cracking of the end wall of the building.

Cracks occur at the junctions between the wall and the floor slab and also between the wall and the roof slab because movements are restricted at these points. Cracking also occur at the corners of the window and door openings because of diagonal cracking of walls.

Due to cracking, utilities buried in the soil such as the water pipes, sewage lines, gas lines, telephone lines, etc. may be damaged. If the leakage occurs in the lines carrying water, wetting of the soil may aggravate swelling problems.

All the above damages occur very slowly and are usually visible after several years of the construction. Much of the economic loss and inconvenience can be avoided by taking appropriate preventive measures in the design, construction and maintenance of the structures and utilities built on expansive soils.

34.6. PREVENTIVE MEASURES FOR EXPANSIVE SOILS

If a soil has a high swell potential, preventive measures are required. These measures can be broadly classified into the following categories:

1. Replacement of the expansive soil
2. Modification of the expansive soil
3. Design of foundations to withstand swelling

Replacement of the expansive soil is considered in this section. The other two methods are discussed in the following sections.

Replacement of the expansive soil

If the soil is only moderately expansive and it has a shallow depth, it is economical to remove that soil and replace it with a good quality granular soil. The granular soil should be properly compacted.

Fig. 34.3 shows a strip footing under an outside wall constructed on an expansive soil. The excavation of the expansive soil is carried out for a width greater than the width of the footing proper. Generally, an extra width of 15 cm is excavated on either side.

The trench width is thus equal to (B + 0.30 m) where B is the width of the footing proper. A freely-draining granular soil, such as gravel, sand or a mixture of gravel and sand, is placed in the trench up to the base level of the footing and compacted.

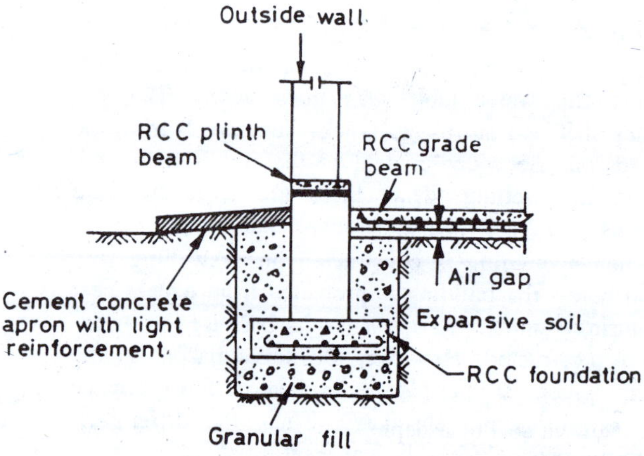

Fig. 34.3

The footing, usually of reinforced concrete, is then constructed at this level and the wall is raised. The freely-draining soil is filled on the sides and the top of the footing up to the ground level.

The foundation on an expansive soil is usually constructed during the hot season when the soil has shrink to its minimum level. A cushion of freely draining soil below the foundation acts as a barrier and the effect of swelling on the foundation is reduced.

A reinforced concrete apron about 2 m wide is provide around the outer walls of the building to prevent the entry of water into the foundation. Suitable arrangements should be made to drain the water away from the granular base during the rainy season.

The RCC grade beam should have an air gap over the ground surface to prevent swelling effect of the soil.

Instead of a RCC apron, a flexible water proof apron made of bituminous concrete about 2 m wide and 75 mm thick can be constructed around the building. The flexible apron will not crack when it is pushed up by the swelling soil. If it cracks or if its connection with the main structure is damaged, it should be properly sealed before the next rainy season. The flexible apron is covered with the soil and given an outward slope of 1 in 30.

34.7. MODIFICATION OF EXPANSIVE SOILS

The characteristics of an expansive soil can be modified to reduce its swelling potential. Some of the commonly used methods are given below:

1. Stabilisation of soil. The expansive soil can be stabilised with lime (or cement). Lime stabilisation is quite effective in reducing the liquid limit and the plasticity index of the soil. Consequently, the swelling potential is also reduced. The quantity of lime required is about 3 to 8 percent. Generally, 5% lime is sufficient in most cases. After mixing the lime thoroughly with the soil, the mixture is compacted at the optimum moisture content. Most suitable percentage of lime for a particular soil can be determined after making a number of trial mixes.

Stabilisation can easily be done up to a depth of 1 to 1.5 m. Generally, high-calcium lime is used in practice.

2. Compaction. As discussed in Chapter 14, the swelling potential of a soil is reduced if it is compacted wet of the optimum water content. Generally, the soil is compacted at a water content 3 to 4% above the optimum water content. The compaction is done to a dry unit weight of about 70 to 90% of the standard Proctor density, depending upon the climate. For extremely unfavorable climates, the compaction is about 70%. However, the strength of the soil is correspondingly reduced.

Compaction is effective only in the cases when the probable expected heave is less than about 40 mm.

3. Installation of moisture barriers. The moisture variation in the soil below the foundation can be controlled by installation of moisture barriers. A trench about 1.5 m deep is excavated around the perimeter of the building and is filled with gravel, or lean pervious concrete. Alternatively, impervious membranes can be placed in the trench to stop flow of water. Thus the effect of the differential heave on the foundation is considerably reduced.

4. Prewetting. If an expansive soil is prewetted by ponding at a water content equal to the equilibrium moisture content before the construction of foundation, most of the potential heave would occur. However, the method is quite time-consuming because the permeability of the soil is extremely low.

Moreover, the effective penetration of water into the soil does not take place and the resultant moisture content is variable. The soil after prewetting becomes unworkable. If 4 to 5% of hydrated lime is added to the top layer of the soil, it becomes more workable because the plasticity of the soil is decreased.

5. Pressure Injection. The soil can be stabilised by injection of lime slurry (or lime-fly ash slurry) into the expansive soil to a depth of 4 to 5 m or up to the active layer zone. Depending on the soil conditions at the site of construction, either single injection or double injection can be used. In double injection, the spacing of the points of injection is reduced by injection at the intermediate points between main injection points.

34.8. DESIGN OF FOUNDATIONS ON SWELLING SOILS

The following types of structures are commonly constructed to withstand heave due to swelling soil.

1. **Strong and rigid structures.** The structures are made strong and rigid to withstand heave. Because the structures are rigid, uniform settlements would occur.
2. **Flexible structures.** The structures are made flexible so that they change shape depending upon the heave. Thus the effect of differential heave is not felt.
3. **Isolating foundations.** Deep foundations such as drilled piers and under-reamed piles are constructed to isolate the foundation from the swelling effect of the soil.

One particular type of structure may not be suitable for all site conditions. The methods may be used singly or a combination of 2 or 3 types.

The method of isolating the foundation is commonly used in practice. The method of replacement of the expansive soil with a granular soil around and below the foundation is also a method of isolating the foundation (Sect. 34.6). The most common method of isolating the foundation is to construct drilled piers. Sometimes, the drilled pier is enlarged at the base to form a bell. The drilled piers have suspended floor slabs at their tops to make the construction independent of the heave effect.

In India, under-reamed piles have proved to be quite successful in expansive soils, especially black cotton soils. A grade beam is provided at the tops of the under-reamed piles. An air gap is provided between the grade beam and the ground surface to permit swelling of the soil.

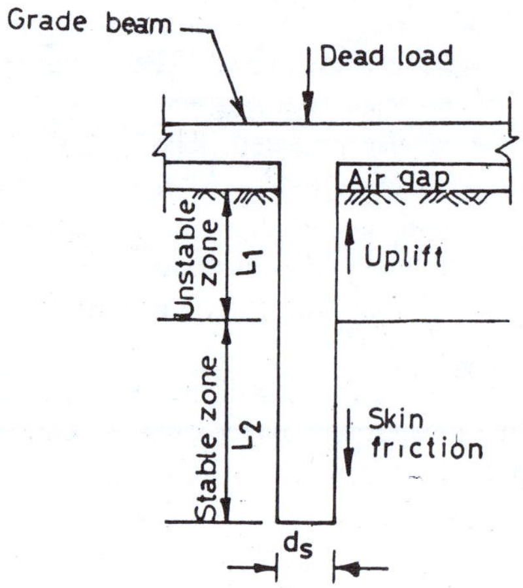

Fig. 34.4. Straight shaft-pier foundation

34.9. DRILLED PIER

Design of a drilled pier for a non-expansive soil has been discussed in Chapter 26. The drilled pier is designed to prevent bearing capacity failure and to keep the settlements within safe limits.

For design of a drilled pier in an expansive soil, in addition to the normal criteria of bearing capacity and settlements, the safety against the uplift force must be checked. The expansive soil tends to stick to the shaft of the drilled pier and it tightly grips the shaft. When the soil expands, it tries to lift the pier along it. Thus an uplift force acts on the pier (Fig. 34.4). If adequate resistance against uplift is not available, the drilled pier may be uplifted.

The uplift force on the drilled pier can be calculated from the relation given by Chen (1988) as

$$U = \pi\, d_s\, L_1\, (\alpha_u\, p_s) \qquad \ldots(a)$$

where U = uplift force

d_s = diameter of the shaft

L_1 = depth of the active zone

α_u = coefficient of uplift between concrete and soil (usually about 0.15)

p_s = swelling pressure for zero swell.

Swelling pressure is taken equal to 480 kN/m^2 for soils with high degree of expansion and equal to 240 kN/m^2 for soils with medium degree of expansion.

Because of difficulty of the measurement of α_u and p_s, the following alternate relation is sometimes used.

$$U = \pi\, d_s\, L_1\, (\alpha_s\, c_u) \qquad \ldots(b)$$

where α_s = adhesion factor between the concrete and soil when the soil is swelled

c_u = unit undrained cohesion of the soil

The value of α_s may be equal to 1.0 or more depending upon the pile roughness, type of soil and climatic conditions. If α_s is taken equal to 1.0, Eq. (*b*) reduces to

$$U = \pi\, d_s\, L_1\, c_u \qquad \ldots(c)$$

The frictional force on the perimeter of the pier embedded in the stable soil below the active zone provides the resisting force. The frictional force acts in the downward direction when the uplift occurs.

The resisting force due to frictional resistance on the shaft is calculated from the relation

$$Q_R = \pi\, d_s\, L_2\, (\alpha\, c_u) \qquad \ldots(d)$$

where L_2 = length of the pier in the stable zone

α = adhesion factor

The value of the adhesion factor can be obtained from Fig. 25.7.

The factor of safety (F_1) against uplift when dead load on the top of the pier is not considered, should be at least 1.20, and the factor of safety (F_2) against uplift when the dead load on the pier is taken into account should be at least 2.0.

(*a*) For no dead load case. The factor of safety F_1

$$F_1 = \frac{Q_R}{U} \geq 1.2 \qquad \ldots(e)$$

In the limiting case, $Q_R = 1.20\, U$

(*b*) For the case when dead load is acting on the pier, the factor of safety F_2,

$$F_2 = \frac{Q_R}{U - Q_D} \geq 2.0 \qquad \ldots(f)$$

where Q_D = dead load.

In the limiting case, $Q_R = 2\,(U - Q_D)$

The length of the pier required in the stable zone (L_2) is the larger of the values given by Eqs. (*e*) and (*f*).

34.10. BELLED DRILLED PIER

A belled drilled pier is used when the uplift acting on the straight shaft drilled pier is large. The uplift force U acting on the belled pier is the same as that in the case of a straight shaft pier discussed in Sec. 34.9.

Thus $U = \pi\, d_s\, L_1\, (\alpha_s\, c_u) \qquad \ldots(a)$

The resisting force on the straight portion of the belled pier is also the same.

Thus Q_{R1} = resisting frictional resistance of the shaft

$= \pi\, d_s\, L_2\, (\alpha\, c_u)$

The resisting force (Q_{R2}) on the belled portion depends upon the bearing capacity of the soil because the bearing pressure acts on the projections of the bell.

The ultimate bearing capacity (q_u) is given by Eq. 23.25 as

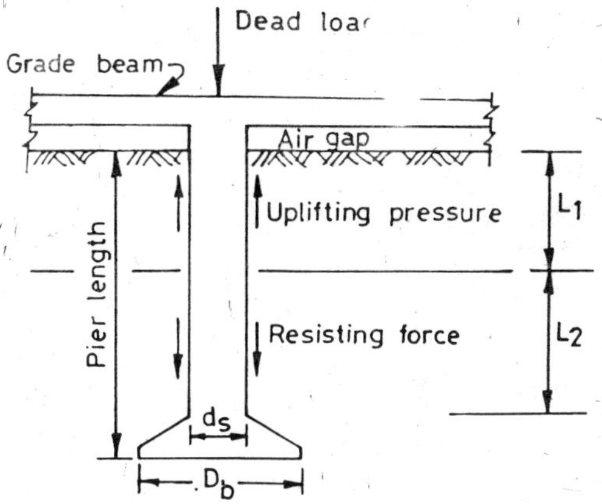

Fig. 34.5. Bell-pier foundation

$$q_u = c\, N_c + \gamma\, L_2$$

The resisting force Q_{R2} can be expressed as

$$Q_{R2} = q_u \times \text{area}$$

$$= (c\, N_c + \gamma\, L_2) \times \frac{\pi}{4}\, (d_b^2 - d_s^2)$$

where N_c = bearing capacity coefficient for $c = 0$

d_b = diameter of the bell

The value of N_c depends upon the ratio L_2/d_b.

O'Neill (1988) gave the following values.

L_2/d_b	1.7	2.5	≥ 5.0
N_c	4	6	9

Sometimes the value of N_c is taken equal to 5.14 as for shallow foundations.

The factor of safety F_1 for no dead load acting on the pier should be atleast equal to 1.2.

Thus $$F_1 = \frac{Q_{R_1} + Q_{R_2}}{U} > 1.20$$

The factor of safety F_2 when the dead load Q_D acting on the top of the pier is considered, should be atleast 2.0.

Thus $$F_2 = \frac{(Q_{R_1} + Q_{R_2})}{U - Q_D} > 2.0$$

where Q_D = dead load on the top of the pier.

Bearing Capacity

The belled pier, like all other foundations, should be safe against bearing capacity failure.

Net ultimate bearing capacity,

$$q_u = c_u N_c$$

Net downward load, neglecting uplift,

$$Q = Q_L + Q_D$$

where Q_L = live load and

Q_D = dead load.

The factor of safety against bearing capacity failure should be at least 3.0. Thus

$$\frac{c_u N_c \left(\frac{\pi}{4} d_b^2\right)}{Q_L + Q_D} \geq 3.0$$

34.11. UNDER-REAMED PILES

The under-reamed piles developed by the CBRI, Roorkee, are commonly used in India for foundations in expansive soils. The under-reamed pile foundations are quite suitable for expansive soils. The under-reamed piles anchor the structure to a stable soil mass well below the unstable soil which is subjected to ground movement because of seasonal changes. The unstable zone is the same as the active zone defined earlier. The depth of the active zone is normally 3 to 4 m at most places in India. The piles are thus seated up in good bearing strata.

As discussed in Chapter 25, the under-reamed piles are bored, cast-in situ concrete piles with a bulb or underream at its bottom. In a double

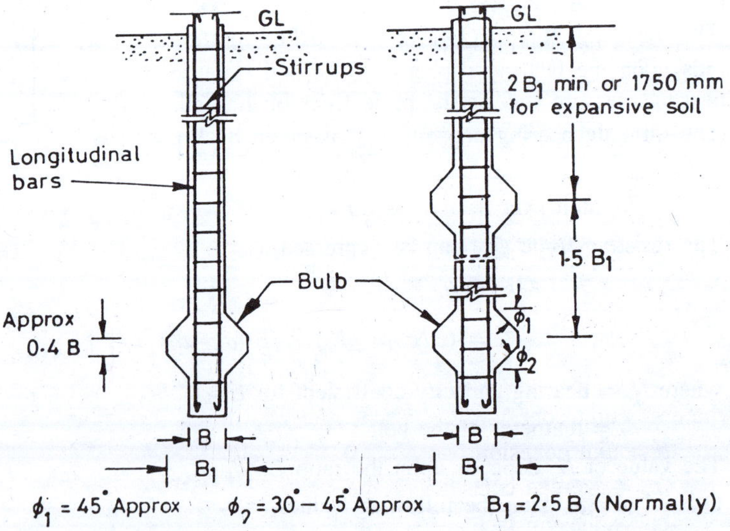

(a) Single under-reamed pile (b) Double under-reamed pile

Fig. 34.6

under-reamed pile, there are two bulbs. IS: 2911-Part 3 (1980) recommends a maximum of two bulbs. Fig. 34.6 shows the details of a single under-reamed and a double under-reamed pile. The bulbs provide adequate bearing and anchorage to the pile.

The under-reamed pile should satisfy the bearing capacity and settlement criteria like all other piles. In addition to the conventional design requirements, it should satisfy the following criteria.

1. Minimum length of the pile in the deposit of expansive soil should be 3 to 5 m below the ground surface. The top 1.2 m (or 2 times shaft diameter) is usually neglected while calculating the skin friction. In shallow deposits, the length of the pile may be reduced but the pile should be at least 60 cm in the stable zone.
2. The diameter of the bulb is taken 2 to 3 times the shaft diameter (normally 2.5 times the shaft diameter). The diameter of the shaft is 20 to 30 cm.

   For heavier loads, double bulbs may be used.
3. The vertical spacing between the two bulbs should not be less than 1.5 times the bulb diameter for the piles upto 30 cm diameter. For larger diameter piles, the minimum spacing is 1.25 times the bulb diameter. The spacing is usually kept 1.5 m to 3 m.
4. Minimum horizontal spacing of piles should be 2 times the bulb diameter under normal loading. The maximum spacing should not exceed 2.5 m to avoid pile cap problems.
5. The depth of the top most bulb from the ground surface should be at least 2 times the bulb diameter. This depth should not be less than 1.75 m.
6. The piles are connected at their top by a RCC plinth beam (grade beam) which supports the superstructure.

There should be an air gap of about 8 cm between the lower surface of the plinth beam and the ground surface. A granular material like sand or gravel is filled in the gap to permit swelling of the soil.

Load Carrying Capacity

The ultimate load and the safe load for an under-reamed pile can be determined by the conventional static analysis discussed in Chapter 25.

The ultimate load for a single bulb, under-reamed pile is given by the relation

$$Q_u = O_p + Q_s$$

or

$$Q_u = [q_{sh}\, A_{sh} + q_b\, (A_b - A_{sh})] + f\, A_{sh} \quad \ldots(a)$$

where q_b = bearing capacity of the soil at the under- ream section

q_{sh} = bearing capacity of the soil at the bulb

A_b = cross-sectional area of the bulb ($= \pi/4\ B_1^2$)

A_{sh} = cross-sectional area of the shaft ($= \pi/4\ B^2$)

f = unit skin friction (adhesion) on the shaft above and below the bulb

A_s = surface area of the shaft of the pile above and below the bulb.

Taking $q_b = 9c$, *i.e.* $N_c = 9.0$,

$$Q_u = \left[\frac{\pi}{4} B^2\,(9\,c) + \frac{\pi}{4}(B_1^2 - B^2)\,(9\,c')\right] + \alpha\, c_a\, A_s \quad \ldots(b) \text{ (same as Eq. 25.17 } b)$$

where c = unit cohesion at the base

c' = unit cohesion at the bulb

c_a = average cohesion on the shaft

α = adhesion factor

IS: 2911-Part 1, 1972 specifies values of α = 1.0, 0.7, 0.4 and 0.3 respectively for soft, medium, stiff and very stiff clays.

B = diameter of the shaft

B_1 = diameter of the bulb

[**Note.** The uplift force on the pile is not considered separately, but the specifications given above are such that the under-reamed pile has an adequate resistance against uplift.]

Double-Under reamed pile

For a double under-reamed pile, the ultimate load is given by

$$Q_u = q_{sh} A_{sh} + q_b (A_b - A_{sh}) + f A_s + \bar{f} \bar{A}_{sb} \qquad \ldots(c)$$

where $\bar{A}_{sb}$ = surface area of the cylinder circumscribing the bulbs. The length of the cylinder is equal to the distance between the centres of the bulbs.

$\bar{f}$ = unit skin friction between soil to soil along the cylindrical surface A_{sb}.

Other notations are the same as explained earlier. Eq. (*b*) can be written as

$$Q_u = \frac{\pi}{4} B^2 (9c) + \frac{\pi}{4} (B_1^2 - B^2)(9\,c') + \alpha\, c_a A_s + c_a' \bar{A}_{sb} \qquad \ldots(d)$$

where c'_a = average unit cohesion on area $\bar{A}_{sb}$

Eq. (*d*) is the same as Eq. 25.17 (*c*).

34.12. CONSTRUCTION OF UNDER-REAMED PILES

Construction of under-reamed piles is quite easy and convenient in most cases. The equipment can be easily handled manually.

In soft soils for the construction of an under-reamed pile upto 3.5 m depth, a bore hole is made using a spiral auger (Sec. 17.9). The auger works through a boring guide frame fixed on the ground. When the desired depth of the bulb is reached, an underreaming tool (underreamer) is inserted into the bore hole. When the underreamer is pressed down and rotated, it gradually widens out and cuts the soil from the sides of the bore hole. The excavated soil enters the underreamer. When the underreamer is full of soil, it is taken out and emptied at the ground. The process is repeated till the complete bulb is formed.

For bore holes deeper than 3.5 m and also for bore holes larger than 30 cm diameters, a portable hoist with a manually operated winch is used to handle the equipment.

Longitudinal reinforcement and shear stirrups are provided in an under-reamed pile as in the case of R.C.C. columns of a building. The design of reinforcement is out side the scope of this text.

After the completion of the bore hole, the reinforcement cage is lowered into it. The cement concrete is poured into the bore hole. The slump of concrete is usually kept between 70 to 150 mm (commonly 120 mm). Compaction of concrete is done by rodding. Curing of concrete is done as usual.

34.13. COLLAPSIBLE SOILS

A collapsible soil suddenly decreases in volume when it becomes saturated. Collapsible soils are generally aeolian (wind-deposited) soils which have low water content and high void ratio in natural state. Such soils usually have a honey-comb structure in which porous structure is maintained by a water-soluble interparticle bond. When the water content of the soil is increased, the interparticle bond is broken and the soil mass suddenly decreases in volume causing its collapse. Buildings and other structures constructed on a collapsible soil have large settlements causing damage. Roads, highways, pipelines and other utilities constructed on such soils have maintenance problems.

Most of the collapsible soils are basically cohesionless but they have a very small coating of clay or other bonding material. Loess deposits are the main collapsible soils. Loess has silt-size particles with clay coatings around the particles to form bond. The bond may also be due to the chemical precipitates leached by rain water. When the soil becomes saturated, the clay bond is weakened and the soil collapses.

Collapsible soils may also be residual soils formed by weathering of parent rocks such as decomposed granite or acid igneous rocks. When the soluble colloidal material forming the interparticle bond is leached out, collapsible soil is formed.

34.14. IDENTIFICATION OF COLLAPSIBLE SOILS

As already mentioned, a collapsible soil has a low water content and high void ratio in its natural state. When it is saturated, its void ratio suddenly decreases.

The behaviour of a collapsible soil under loading is usually explained with the help of a *e*-log *p* plot. If a soil specimen at its natural water content is placed in a consolidometer (oedometer) and the load is applied, the curve *ab* is obtained (Fig. 34.7). At the pressure of p_1, the void ratio is e_1.

After saturation, the equilibrium void ratio at the same pressure suddenly decreases to e_2. The vertical line *bc* represents the collapse of the soil. The curve *cd* is the usual consolidation curve when the load is applied to the collapsed soil.

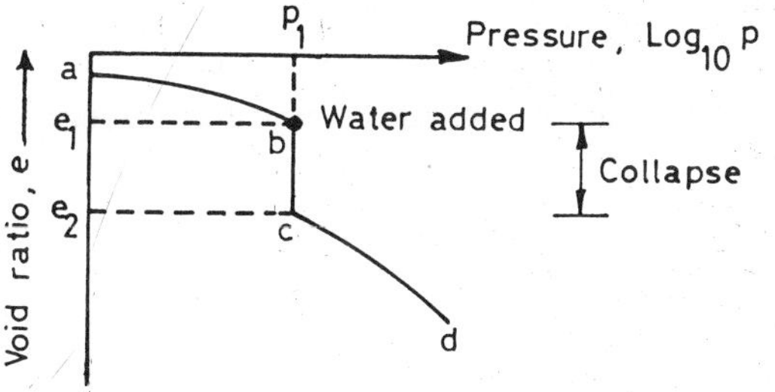

Fig. 34.7

The following criteria are commonly used for the identification of a collapsible soil.

1. Collapse potential. The collapse potential (C_p) is defined as (Jenning and Knight, 1975)

$$C_p = \frac{e_1 - e_2}{1 + e_0} \qquad \text{...(1 a)}$$

where e_0 = natural void ratio of the undisturbed soil

e_1 = void ratio before flooding the specimen

e_2 = void ratio after flooding the specimen

For determination of e_1 and e_2, the undisturbed soil sample at natural moisture content and void ratio e_0 is taken in a consolidometer and steps loads are applied upto a pressure p_1 of 200 kN/m^2. The void ratio e_1 of the specimen is determined.

The specimen is then flooded with water, without changing the load, and left for 24 hours. The void ratio e_2 of the sample is then obtained and the collapse potential is obtained for the soil from Eq. (*a*).

The greater the collapse potential, the greater is the collapse and greater the damage to structures built on such soils. The soils with C_p less than 1.0 generally cause no problems to foundations. For C_p = 1 to 5, there would be moderate trouble to the foundation. For $C_p \geq 5$, there would be excessive trouble if precautions are not taken in the design and construction.

The collapse potential may also be defined in terms of height instead of void ratio. In terms of height,

$$C_p = \frac{\Delta h}{h_i} \qquad \text{...(1b)}$$

where Δh is the change in height upon saturation and h_i is the initial height of the specimen.

2. Liquid limit. Holtz and Hilf (1961) established that if a soil has a high void ratio and upon saturation, its moisture content exceeds the liquid limit, the soil is susceptible to collapse. Thus for a collapsible soil,

$$w_s \geq LL \qquad \text{...(2)}$$

where w_s = water content of the soil after saturation

LL = liquid limit of the soil.

3. Dry unit weight. A collapsible soil has a low dry unit weight. The limit of the dry unit weight below which the soil is likely to be collapsible can be derived from Eq. (2)

For a saturated soil, $e_0 = w_s G$

or $$w_s = \frac{e_0}{G} \qquad \text{...(3)}$$

where G is the specific gravity of the solids and e_0 is the void ratio at saturation.

From Eqs. (2) and (3), for a collapsible soil

$$\frac{e_0}{G} \geq LL$$

or $$e_0 \geq (LL) \times G \qquad \ldots(4)$$

The void ratio e_0 can be expressed in terms of dry unit weight (γ_d) as

$$\gamma_d = \frac{G\,\gamma_w}{1+e_0}$$

or $$e_0 = \frac{G\,\gamma_w}{\gamma_d} - 1 \qquad \ldots(5)$$

From Eqs. (4) and (5),

$$\frac{G\,\gamma_w}{\gamma_d} - 1 \geq (LL)\,G$$

or $$\frac{G\,\gamma_w}{\gamma_d} \geq (LL)\,G + 1$$

or $$\gamma_d \leq \frac{G\,\gamma_w}{1+(LL)\,G} \qquad \ldots(6)$$

A soil will be collapsible if Eq. (6) is satisfied.

34.15. DESIGN OF FOUNDATIONS ON COLLAPSIBLE SOILS NOT SUSCEPTIBLE TO WETTING

If the soil is not susceptible to wetting, the methods discussed below may be used. Before the actual design of foundations on a collapsible soil not susceptible to wetting, it is desirable to conduct the plate load test. Fig. 34.8 shows a typical load-settlement curve for a collapsible soil with low natural water content. The curve is essentially linear upto a certain critical pressure p_{cr} and thereafter suddenly large settlement occurs because of the breakdown of the soil structure. This sudden breakdown of the soil structure is more common for collapsible soils with low natural moisture content. The soils with high moisture content have gradual breakdown of the structure and there is no sudden large increase in settlement.

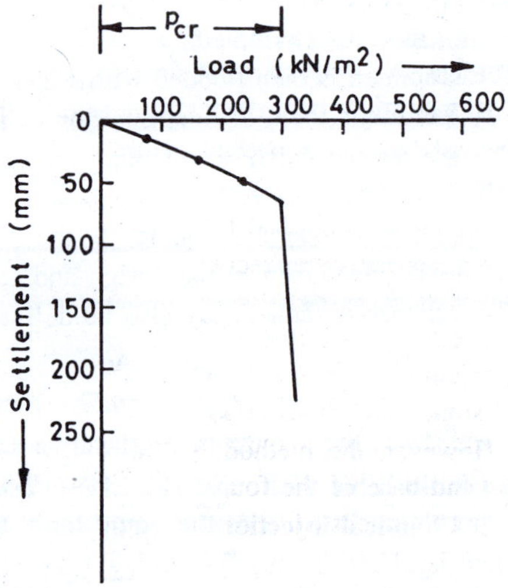

Fig. 34.8

Design Criteria

1. Spread foundations and raft foundations may be constructed on a collapsible soil not susceptible to wetting. Enough precautions must be taken to prevent moisture increase under the foundation. The foundation should be designed for an allowable pressure less than the critical pressure. A factor of safety of 2.5 to 3 is normally provided. Thus

$$p_{all} = \frac{p_{cr}}{FS}$$

where p_{all} = allowable pressure

p_{cr} = critical pressure

FS = factor of safety (2.5 to 3)

Moreover, the differential and total settlements of the foundations should not exceed the limits specified for foundations designed for sandy soils.

2. Continuous footings are better than spread foundations over collapsible soils and should be provided as for as possible. Fig. 34.9 shows a typical foundation design of continuous footings. It consists of the footing beams and loading beams. The footing beams are provided on the tops of the continuous strip footings to connect the footings and to act as reaction beams. The loading beams are at right angles to the footing beams to carry the loads of the structure. The beams are properly reinforced to form a rigid systems.

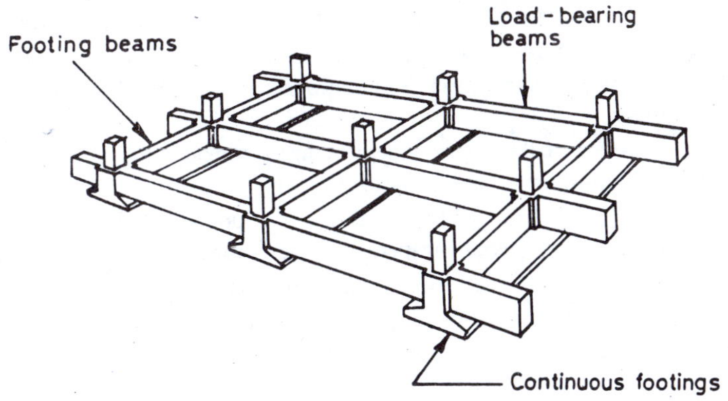

Fig. 34.9

3. For heavy structures, the spread foundations cannot be provided because the settlements are large.

To reduce settlements, drilled piers and pile foundations are used to transfer the load to a strong load-bearing stratum below a collapsible soil.

34.16. DESIGN OF FOUNDATIONS ON COLLAPSIBLE SOILS SUSCEPTIBLE TO WETTING

If the compressible soil is susceptible to wetting, it will collapse after construction some time later. To avoid foundation failure, preventive measures are adopted before the construction. Some of the commonly used methods are given below:

1. Recompaction. If the expected depth of wetting is upto about 1.5 m from the ground surface, the collapsible soil is moistened before the construction of the foundation. The moist soil is compacted by heavy rollers.

Spread footings and rafts may be constructed on the compacted soil.

An alternative to recompaction by rollers is to use heavy tamping (or pounding) discussed in Sec. 33.9. About 8 to 30 m drop of the pounder is generally effective in the densification of the soil.

2. Chemical solution. In this method, the foundation trenches are filled with a solution of sodium silicate and calcium chloride to stabilise the soil chemically. After stabilisation, the collapsible soil behaves like a soft sand stone and resists collapse after saturation.

However, the method is effective only when the solution can penetrate the collapsible soil at the side walls and base of the foundation trench to the desired depth. The method is quite effective for fine sands.

3. Chemical injection. A sodium silicate solution is injected into a collapsible soil deposit. However, this method is effective only for soils that are likely to compress under the weight of the structure to be built on these soils.

This method has been effectively used in the former Soviet Union.

For the compressible soils susceptible to wetting to a depth of greater than 1.5 m, the following methods are quite effective.

1. Ponding. In this method, low dikes are constructed around the construction site and the whole site is flooded with water to cause collapse of the soil before the construction is started.

However, this method is effective only when there is an impervious stratum beneath the collapsible soil deposit to prevent seepage.

2. Vibroflotation. Vibroflotation is used to compact the collapsible soil before construction. Vibroflotation has been discussed in Sec. 33.8.

The method is effective only for free-draining soils.

3. Rock-columns. Rock columns are like stone columns discussed in Sec. 33.6 but instead of gravel, large boulders are used in the construction. The large boulders penetrate the collapsible soil deposit and act like piles. They transfer the load of the structure to stable soil layers beneath the collapsible soil.

4. Drilled piers and piles. For the compressible soils susceptible to wetting to large depths, precollapsing becomes difficult. In such cases, foundations are extended beyond the zone of possible wetting using drilled piers and piles.

The design of drilled piers and piles must consider the negative friction caused due to collapsible soils. The design should also consider the settlements subsequent to wetting.

5. Control of drainage. Potential water sources which cause wetting may be controlled by providing suitable drainage. If necessary impervious membranes can be used to stop seepage.

Sometimes, infiltration wells are used for drainage.

Construction of foundations on a collapsible soil is a challenging task. The geotechnical engineer may use any of the above methods depending upon the experience and the site conditions. Sometimes, combination of two or three methods may be required.

ILLUSTRATIVE EXAMPLES

Illustrative Example 34.1. *A single under-reamed pile is installed in a soft clay deposit. The centre of the underream is located at a depth of 15 m from the ground surface. The diameters of the pile shaft and bulb are respectively 1.0 m and 2.5 m. Determine the allowable load with a factor of safety of 2.5.*

The undrained shear strength of the soil obtained from the vane shear test is given by the relation

$$c_u = 65 + 7D$$

where c_u is in kN/m² and D is the depth in metres. Assume $\alpha = 1.0$.

Solution. The ultimate load Q_u is given by

$$Q_u = \frac{\pi}{4} B^2 \times (9c) + \frac{\pi}{4}(B_1^2 - B^2) \times 9c' + \alpha c_a A_s$$

In this case taking, $c = c' = 65 + 7 \times 15 = 170 \text{ kN/m}^2$

and $c_a = \dfrac{65 + 170}{2} = 117.5 \text{ kN/m}^2$

$$Q_u = \frac{\pi}{4}(1)^2(9 \times 170) + \frac{\pi}{4}(2.5^2 - 1) \times (9 \times 170) + 1.0 \times 117.5 \times (\pi \times 1 \times 15)$$

$$= 1201.66 + 6308.71 + 5537.05$$

$$= 13047.43 \text{ kN}$$

$$Q_a = \frac{Q}{2.5} = \frac{13047.43}{2.50}$$

$$= \mathbf{5218.97 \text{ kN}}$$

Illustrative Example 34.2. *In Illustrative Example 34.1, determine the allowable load if there is another bulb of the same diameter below the first bulb. The spacing between the centres of the bulbs is 4 m. The other data remain the same.*

Solution. The ultimate load is given by

$$Q_u = \frac{\pi}{4} B^2 (9c) + \frac{\pi}{4}(B_1^2 - B^2) \times 9c' + \alpha c_a A_s + c_a' Asb$$

In this case, taking $c = 65 + 7 \times 19 = 198 \text{ kN/m}^2$

$$c_a = \frac{65 + 170}{2} = 117.5 \text{ kN/m}^2$$

$$c_a' = 65 + 7 \times 17 = 184 \text{ kN/m}^2$$

$$A_{sb} = \frac{\pi}{4} \times (2.5)^2 \times 4 = 19.63 \text{ m}^2$$

Therefore,

$$Q_u = \frac{\pi}{4} \times (1)^2 \times (9 \times 198) + \frac{\pi}{4}(9 \times 198)(2.5^2 - 1)$$

$$+ 1.0 \times 117.5 \times (\pi \times 1.0 \times 15) + 184 \times \left(\frac{\pi}{4} \times 2.5^2 \times 4.0\right)$$

$$= 1399.58 + 7347.79 + 5537.05 + 3612.83$$

$$= 17897.25$$

$$Q_a = \frac{17897.25}{2.5} = \mathbf{7158.9 \text{ kN}}$$

Illustrative Example 34.3. *A straight shaft pier is constructed in an expansive soil. If the length of the pier is 6 m and the shaft diameter is 0.75 m, determine the factor of safety for no dead load and that for a dead load of 200 kN. The depth of the active zone is 2.5 m. Assume swell pressure (p_s) = 400 kN/m², coefficient of uplift between concrete and soil (α_u) = 0.15. Coefficient of adhesion (α) = 0.55, undrained cohesion = 100 kN/m².*

Solution. Uplift force, $U = \pi \, ds \, L_1 (\alpha_u \, p_s)$

$$= \pi \times 0.75 \times 2.5 \; (0.15 \times 400)$$

or $U = 353.42$ kN

Resisting force due to friction on the shaft,

$$Q_R = \pi \, ds \, L_2 (\alpha \, c_u)$$

$$= \pi \times 0.75 \times 3.5 \times (0.55 \times 100)$$

$$= 453.57 \text{ kN}$$

Factor of safety for no dead load,

$$F_1 = \frac{453.57}{353.42} = 1.28 > 1.2 \text{ (safe)}$$

Factor of safety with a dead load of 200 kN,

$$F_2 = \frac{453.57}{353.42 - 200}$$

$$= 2.96 > 2 \text{ (safe)}$$

Illustrative Example 34.4. *A drilled pier of length 6 m and shaft diameter of 1 m is constructed in an expansive soil. If the depth of the active zone is 3 m and the bulb diameter is 1.50 m, determine (a) the total uplift force on the pier, (b) total resisting force due to surface friction, (c) the factor of safety against uplift, considering no dead load, (d) the factor of safety against uplift, considering a dead load of 400 kN acting on the top of the pier.*

Assume the undrained shear strength of soil as 75 kN/m² and adhesion factor for side friction (α) = 0.5, γ = 18 kN/m³, N_c = 6.0.

Solution. The uplift force, $U = \pi \, d_s \, L_1 \, c_u$

$$= \pi \times 1.0 \times 3 \times 75$$

$$= \mathbf{706.86 \text{ kN}}$$

Resisting force, $Q_R = Q_{R_1} + Q_{R_2}$

where, $Q_{R_1} = \pi\, ds\; L_2\; \alpha\, c_u$

$= \pi \times 1.0 \times 3.0 \times 0.5 \times 75$

$= 353.43$ kN

$$Q_{R_2} = \frac{\pi}{4}(D_b^2 - D_s^2)\,(cN_c + \gamma L_2)$$

$$= \frac{\pi}{4}(1.5^2 - 1.0^2)\;(75 \times 6 + 18 \times 3)$$

$$= \frac{\pi}{4}(1.25)\,(504) = 494.80 \text{ kN}$$

Therefore, $Q_R = 353.43 + 494.80$

$= \mathbf{848.23\ kN}$

Factor of safety for no dead load,

$$F_1 = \frac{Q_R}{U} = \frac{848.23}{706.86}$$

$= \mathbf{1.2 \geq 1.2\ (safe)}$

Factor of safety considering dead load of 600 kN,

$$F_2 = \frac{Q_R}{U - Q_D}$$

$$= \frac{848.23}{706.86 - 400}$$

$= \mathbf{2.76 > 2\ (safe)}$

PROBLEMS

A. NUMERICALS

34.1. Find the ultimate load and safe load for an underreamed pile installed in an expansive soil using the following data:

Diameter of pile shaft	= 50 cm
Diameter of bulb	= 125 cm
Length of shaft above bulb	= 6 m
Average undrained strength on shaft	= 100 kN/m^2
Undrained strength at the bulb	= 130 kN/m^2
Undrained strength at the base	= 150 kN/m^2
Adhesion factor	= 0.40
Factor of safety	= 2.5

Neglect 1 m of shaft for adhesion. **[Ans.** 1785 kN, 714 kN]

34.2. A drilled pier of length 5 m is constructed in an expansive soil having the depth of active zone as 3 m. If the shaft diameter is 1 m and the bulb diameter is 1.25 m, calculate the factor of safety (*a*) without considering the dead load. (*b*) with a dead load of 300 kN on the pier.

Assume c_u = 100 kN/m^2, swelling pressure = 500 kN/m^2, α_u = 0.15, α = 0.5, γ = 18 kN/m^3.

[Ans. 1.32, 2.31]

B. Descriptive and Objective type

34.3. What is an expansive soil? Where is it found in India? What are its generally characteristics?

34.4. Define the following parameters:

(*a*) Free swell (*b*) Unrestrained free swell

(*c*) Differential swell

34.5. Define swelling potential. How would you classify expansive soils based an swelling potential?

34.6. What is swell pressure? What is its significance?

34.7. How would you identify an expansive soil from its index properties?

34.8. What are the common causes of moisture changes in soils?

34.9. What are the effects of swelling of soils on buildings?

34.10. Describe the method of replacement of the expansive soil with a good quality soil. How would you construct a footing using this method?

34.11. What are the various methods of modification of an expansive soil to improve its characteristics?

34.12. Explain the design of a drilled pier in an expansive soil. How would you determine the uplift force acting on the pier?

34.13. What is an underreamed pile? What are its design criteria?

34.14. How would you estimate the load carrying capacity of an under-reamed pile?
What is the effect of the underream (bulb) ?

34.15. What are collapsible soils? What is the effect of collapsible soils on the structures?

34.16. Define the collapse potential. How would you determine it?

34.17. How would you identify a collapsible soil from its liquid limit and dry unit weight?

34.18. How would you design a foundation on an expansive soil not susceptible to wetting?

34.19. If an expansive soil is susceptible to wetting, what preventive measures would you take?

34.20. Write whether the following statements are true or false.

(*a*) If the shrinkage limit of a soil is less than 20%, it has low expansion.
(*b*) If the differential free swell of a soil is 35%, its expansion is low.
(*c*) According to CBRI, Roorkee, the depth of active zone in India is usually limited to 2.5 m
(*d*) The damages to a building founded on an expansive soil occur very slowly.
(*e*) The lime required for stabilisation of an expansive soil is 3 to 8 percent
(*f*) The expansive soils should be compacted dry of the optimum water content
(*g*) The uplift force on a drilled pier occurs in the active zone.
(*h*) I.S. code permits up to 3 bulbs in an under-reamed pile
(*i*) Loess is an expansive soil **[Ans.** True (*a*), (*d*), (*e*), (*g*)**]**

Miscellaneous Objective-Type Questions

1. Lists I and II contain respectively terms and expressions related to soil classification. Match the two lists and select the correct answer using the codes given below the lists.

List I	List II
(A) Activity number	(1) $\dfrac{\text{Liquid limit} - \text{Water content}}{\text{Plasticity index}}$
(B) Liquidity index	(2) $\dfrac{\text{Plasticity index}}{\text{Percent finer than } 2\ \mu}$
(C) Sensitivity index	(3) $\dfrac{\text{Natural water content} - \text{Plastic limit}}{\text{Plasticity index}}$
(D) Consistency index	(4) $\dfrac{\text{Unconfined compressive strength of undisturbed sample}}{\text{Unconfined compressive strength of remoulded soil sample}}$

Codes

	A	B	C	D
(*a*)	1	3	4	2
(*b*)	1	2	3	4
(*c*)	3	2	1	4
(*d*)	2	3	4	1

2. Match List I with List II and select the correct answer using the codes given below the lists.

List I	List II
(A) Sheep-foot roller	(1) Core of earthen dams
(B) Smooth heavy roller	(2) Dry sand
(C) Pneumatic roller	(3) All types of soils
(D) Vibrating roller	(4) Gravel in W.B.M. road

Codes

	A	B	C	D
(*a*)	3	4	2	1
(*b*)	1	4	3	2
(*c*)	1	3	4	2
(*d*)	2	1	3	4

3. Match List I (different types of soils) with List II (group symbols of I.S. classification) and select the correct answer using the codes given below the lists.

List I	List II
(A) Well-graded gravel sand mixture with little or no fines	1. MI
(B) Silty sand	
(C) Inorganic silts and very fine sands or clayey silts with low plasticity	.. GW
(D) Inorganic clays of high plasticity	4. SM

Codes

	A	3	C	D
(*a*)	3	1	4	2
(*b*)	3	4	1	2
(*c*)	2	4	1	3
(*d*)	2	1	4	3

4. Consider following statements about the properties of the flow nets:
   1 'low s are perpendicular to equipotential lines.
   No tw flow lines or equipotential lines start from the same point.
   3. No tw flow lines cross each other
   4. Phreatic line is a flow line

   Of these statements :

   (*a*) 1, 2, 3, and 4 are correct
   (*b*) 2, 3 and 4 are correct
   (*c*) 1 and 2 are correct
   (*d*) 1 and 3 are correct

5. Match List I (structure) with List II (deformation) and select the correct answer using the codes given below the lists.

List I	List II
(A) Retaining wall	1. The wall moves as a rigid structure
(B) Bridge abutment	2. The wall moves in the form of elastic line with a point of contraflexure
(C) Cantilever sheet pile	3. The top does not move away from the soil
(D) Anchored bulkhead	4. The top of the wall moves away from the soil

Codes

	A	B	C	D
(*a*)	3	4	1	2
(*b*)	3	4	2	1
(*c*)	4	3	1	2
(*d*)	4	3	2	1

6. Match List I with List II and select the correct answer using the codes given below the lists.

List I (*Type of pile*)	List II (*Situation*)
A. Friction pile	1. Stiff clay
B. Batter pile	2. Loose granular soil
C. Tension pile	3. Lateral load
D. Compaction pile	4. Uplift load

Codes

	A	B	C	D
(*a*)	3	1	2	4
(*b*)	1	3	4	2
(*c*)	3	1	4	2
(*d*)	1	3	2	4

(I.E.S.)

7. Match List I with List II and select the correct answer using the codes given below the lists

List I (*Allowable max. settlement* IS: 1904)	**List II** (*Type of foundation and soil type*)
(A) 65 to 100 mm	1. Isolated foundation on sand
(B) 40 mm	2. Isolated foundation on clay
(C) 65 mm	3. Rafts on sand
(D) 40 to 65 mm	4. Rafts on clay

Codes

	A	B	C	D
(*a*)	1	4	3	2
(*b*)	4	1	2	3
(*c*)	1	4	2	3
(*d*)	4	1	3	2

(I.E.S.)

8. Consider the following statements regarding confined aquifer :
   1. The aquifer is bound at top and below by impervious strata.
   2. The pressure of water is greater than atmospheric pressure.
   3. A tube well sunk in such an aquifer starts flowing always by itself.
   4. The aquifer is fully saturated

   Of the above statements :

   (*a*) 1, 2 and 3 are correct (*b*) 1, 2 and 4 are correct
   (*c*) 2, 3 and 4 are correct (*d*) 1, 3 and 4 are correct (I.E.S.)

9. Match List I with List II and select the correct answer using the codes given below the lists

List I (*Type of soil*)	**List II** (*Mode of transportation and deposition*)
(A) Lacustrine soils	1. Transportation by wind
(B) Alluvial soils	2. Transportation by running water
(C) Aeolian soils	3. Deposited at the bottom of lakes
(D) Marine soils	4. Deposited in sea water

Codes

	A	B	C	D
(*a*)	1	2	3	4
(*b*)	3	2	1	4
(*c*)	3	2	4	1
(*d*)	1	3	2	4

(I.E.S.)

10. Consider the following assumptions for slope stability analysis:
    1. Friction is fully mobilised.
    2. Effective stress analysis is adopted
    3. Total stress analysis is used
    4. Resultant R passes through the centre of the circle
    5. Resultant R is tangential to the friction circle

The assumptions necessary for friction circle method of analysis would include

(*a*) 1, 3 and 4 (*b*) 2 and 4
(*c*) 1, 3 and 5 (*d*) 2 and 5 (I.E.S.)

11. Consider the following statements regarding negative skin friction in piles :
 1. It is developed when the pile is driven through a recently deposited clay layer.
 2. It is developed when the pile is driven through a layer of dense sand.
 3. It is developed due to a sudden drawdown of the water table

 Of these statements :

 (*a*) 1 alone is correct (*b*) 2 alone is correct
 (*c*) 2 and 3 are correct (*d*) 1 and 3 are correct (I.E.S.)

12. Consider the following properties for a soil sampler :
 1. Area ratio should be low
 2. Cutting edge should be thin
 3. Inside clearance should be high
 4. Outside clearance should be low

The properties necessary for a good quality soil sampler would include

(*a*) 1, 2 and 4 (*b*) 1, 2, and 3
(*c*) 2, 3 and 4 (*d*) 1, 3 and 4

13. Match List I with List II and select the correct answer using the codes given below the lists

List I (*Soil property measured*)	**List II** (*In-situ test*)
(A) Modulus of subgrade reaction	1. Cyclic pile load test
(B) Relative density and strength	2. Flat jack test
(C) Skin friction and point bearing	3. Plate load test
(D) Insitu stresses	4. Standard penetration test

Codes

	A	B	C	D
(*a*)	1	3	2	4
(*b*)	1	2	4	3
(*c*)	2	4	1	3
(*d*)	3	4	1	2

14. **Assertion (A)** Quick sand is not a type of sand but it is condition arising in a sand mass.
 Reason (R) When the upward pressure becomes equal to the pressure due to submerged weight of a soil, the effective pressure becomes zero.

Codes

(*a*) Both A and R are true and R is the correct ecplanation of A.
(*b*) Both A and R are true but R is not a correct explanation of A.
(*c*) A is true but R is false
(*d*) A is false but R is true (I.E.S.)

15. **Assertion (A)** The safe height (2 Zo) to which an unsupported vertical cut in clay can be made is $4c/\gamma$.
 Reason (R) Active earth pressure of cohesive backfill shows that the negative pressure (tension) is developed at depth Zo and total net pressure upto a depth 2 Zo is zero.

Codes

(*a*) Both A and R are true and R is the correct explanation of A
(*b*) Both A and R are true but R is not a correct explanation of A

(*c*) A is true but R is false

(*d*) A is false but R is true (I.E.S.)

16. **Assertion (A)** Negative skin friction will act on piles in filled up soils, which should be considered in design of pile foundations.

Reason (R) The filled up soils start consoliding and develop a drag force on the pile

Codes

(*a*) Both A and R are true and R is the correct explanation of A

(*b*) Both A and R are true but R is not a correct explanation of A.

(*c*) A is true but R is false

(*d*) A is false but R is true (I.E.S.)

17. Match List I with List II and select the correct answer using the codes given below the lists.

List I (*Flow type*)	**List II** (*Flow characteristics*)
(A) Unsteady flow	1. Seepage flow is a function of time
(B) Turbulent flow	2. Hydraulic gradient varies with square of velocity
(C) Steady state flow	3. Flow at low velocity
(D) Laminar flow	4. Governing equation in 2D is $k_x \dfrac{\partial^2 h}{\partial x^2} + k_y \dfrac{\partial^2 h}{\partial y^2} = 0$

Codes

	A	B	C	D
(*a*)	1	2	3	4
(*b*)	3	2	1	4
(*c*)	1	2	4	3
(*d*)	2	1	4	3

18. Consider the following statements

In subsoil exploration programme, the term significant depth of exploration is upto

1. The width of foundation
2. Twice the width of foundation
3. The depth where the additional stress intensity is less than 20% of overburden pressure
4. The depth where the additional stress intensity is less than 10% of the overburden pressure.
5. Hard rock level

Of these statements :

(*a*) 1, 3 and 5 are correct (*b*) 2, 3 and 5 are correct

(*c*) 1 and 4 are correct (*d*) 2 and 4 are correct (I.E.S.)

19. Match List I with List II and select the correct answer using the codes given below the lists.

List I (*Effect*)	**List II** (*Reason*)
(A) Excessive settlement	1. Fall of water table
(B) High expansivity	2. High compressibility
(C) Increase in bearing capacity	3. Montomorrillonite
(D) Acceleration of consolidation	4. Sand drains

Codes

	A	B	C	D
(*a*)	4	1	2	3
(*b*)	2	3	4	1
(*c*)	4	1	3	2
(*d*)	2	3	1	4

20. Match List I with List II and select the correct answer using the codes given below the lists.

List I	List II
(A) Coefficient of compressibility	1. Stress-deformation
(B) Compression index	2. Stress-void ratio
(C) Coefficient of subgrade reaction	3. Liquid limit
(D) Flow Index	4. Log stress-void ratio

Codes

	A	B	C	D
(*a*)	4	2	1	3
(*b*)	4	3	2	1
(*c*)	2	4	1	4
(*d*)	3	4	1	2

21. Consider the following statements regarding underreamed piles:
   1. They are used in expansive soils.
   2. They are of uniform diameter
   3. The ratio of bulb to shaft diameter is usually 2 to 3.
   4. Minimum spacing between the piles should not be less than 1.5 times the diameter

   Of these statements :

   (*a*) 1, 2 and 3 are correct (*b*) 1, 3 and 4 are correct
   (*c*) 2, 3 and 4 are correct (*d*) 1, 2 and 4 are correct

22. Match List I with List II and select the correct answer using the codes given below the lists.

List I (*Field test*)	List II (*Parameters measured*)
(A) Plate load test	1. Total and frictional resistance
(B) Standard penetration test	2. Load intensity and settlement values
(C) Static Dutch cone penetration test	3. Particle size
(D) Hydrometer test	4. SPT values

Codes

	A	B	C	D
(*a*)	2	4	3	1
(*b*)	4	2	3	1
(*c*)	2	4	1	3
(*d*)	4	2	1	1

23. Consider the following statements:
   1. Constant-head permeameter is best suited for determination of coefficient of permeability of highly impermeable soils.
   2. Coefficient of permeability of a soil mass decreases with increase in viscosity of the pore fluid.
   3. Coefficient of permeability of a soil mass increases with increase in temperature of the pore fluid.

   Of these statements :

   (*a*) 1 and 2 are correct (*b*) 1 and 3 are correct
   (*c*) 2 and 3 are correct (*d*) 1, 2 and 3 are correct (I.E.S.)

24. Consider the following statements :
   1. Relative compaction is not the same as relative density.
   2. Vibrofloatation is not effective in the case of highly cohesive soils.
   3. 'Zero air void line' and 100% saturation line are not identical.

Of these statements :

(*a*) 1 and 2 are correct (*b*) 1 and 3 are correct
(*c*) 2 and 3 are correct (*d*) 3 alone is correct. (I.E.S.)

25. Consider the following statements :
Phreatic line in an earth dam is
1. elliptic in shape
2. an equipotential line
3. the top most flow line with zero water pressure
4. approximately a parabola

Of these statements :

(*a*) 1, 2 and 3 are correct (*b*) 2, 3 and 4 are correct
(*c*) 3 and 4 are correct (*d*) 1 alone is correct (I.E.S.)

26. Consider the following statements :
Rankine's theory and Coulomb's theory give same values of coefficients of active and passive earth pressure when
1. the retaining wall has a vertical back
2. the backfill is cohesionless
3. angle of slope of backfill is equal to the angle of internal friction
4. the angle of slope of backfill is 0°
5. the angle of wall friction δ is 0°
6. the angle of wall friction δ is equal to ϕ

Of these statements :

(*a*) 1, 2, 3 and 5 are correct (*b*) 1, 2, 4 and 5 are correct
(*c*) 2, 3 and 6 are correct (*d*) 1, 4 and 6 are correct (I.E.S.)

27. Consider the following statements:
1. Coulomb's earth pressure theory does not take the roughness of wall into consideration.
2. In case of non-cohesive soils, the coefficient of active earth pressure and earth pressure at rest are equal.
3. Any movement of retaining wall away from the fill corresponds to active earth pressure conditions.

Of these statements :

(*a*) 1 alone is correct (*b*) 1 and 2 are correct
(*c*) 2 alone is correct (*d*) 3 alone is correct (I.E.S.)

28. Consider the following statements :
1. The degree of saturation of a saturated soil mass subjected to pressure remains unchanged during the process of consolidation.
2. Secondary consolidation is due to the plastic deformation of the soil when the pore fluid is not subjected to any excess pressure
3. Primary consolidation is independent of the coefficient of permeability of the soil but depends on the decrease in void volume due to air escape.

Of these statements :

(*a*) 1 and 2 are correct (*b*) 1 and 3 are correct
(*c*) 2 and 3 are correct (*d*) 1, 2 and 3 are correct (I.E.S.)

29. Consider the following statements associated with local shear failure of soils :
1. Failure is sudden with well-defined ultimate load
2. This failure occurs in highly compressive soils.
3. Failure is preceded by large settlements.

Of these statements :

(*a*) 1, 2 and 3 are correct (*b*) 1 and 2 are correct
(*c*) 2 and 3 are correct (*d*) 1 and 3 are correct (I.E.S.)

30. Consider the following statements :

1. Dynamic cone penetration test for site investigations is based on the principle that elastic shock waves travel in different materials at different velocities.
2. Electrical resistivity method of sub-surface investigation is capable of detecting only the strata having different electrical resistivity.
3. In-situ vane shear test is useful for determining the shear strength of very soft soil and sensitive clays and is unsuitable for sandy soils

Of these statements :

(*a*) 1 and 2 are correct (*b*) 1 and 3 are correct
(*c*) 2 and 3 are correct (*d*) 2 alone is correct (I.E.S.)

[**Ans.** 1. (*d*), 2. (*c*), 3. (*b*), 4. (*a*), 5. (*c*), 6. (*b*), 7. (*b*), 8. (*b*), 9. (*b*), 10. (*c*), 11. (*d*), 12. (*a*), 13. (*d*), 14. (*a*), 15. (*a*), 16. (*a*), 17. (*c*), 18. (*d*), 19. (*d*), 20. (*c*), 21. (*b*), 22. (*c*), 23. (*c*), 24. (*a*), 25. (*c*), 26. (*b*), 27. (*d*), 28. (*a*), 29. (*c*), 30. (*c*)]

31. Talus is transported by
(*a*) water (*b*) wind
(*c*) gravity (*d*) glacier

32. Stokes' law cannot be applied to particles of size greater than
(*a*) 0.02 mm (*b*) 0.2 mm
(*c*) 0.002 mm (*d*) 0.05 mm

33. The ASTM tool for liquid limit is more suitable for a clay when it is
(*a*) more sandy (*b*) normal clayey
(*c*) more clayey (*d*) more silty

34. A poorly graded soil has
(*a*) particles of different sizes in good proportion
(*b*) an excess or a deficiency of certain particle sizes
(*c*) particles of uniform size
(*d*) particles of only coarse grains

35. If the plasticity index of a soil lies above A line, its permeability is
(*a*) high (*b*) medium
(*c*) low (*d*) very low

36. The shape of the phreatic line in an earth dam is
(*a*) circular (*b*) hyperbolic
(*c*) elliptical (*d*) parabolic

37. If the liquid limit of a clayey soil increases, its compression index
(*a*) decreases (*b*) increases
(*c*) remains unchanged (*d*) either (*a*) or (*b*)

38. The inclination of the failure plane behind a vertical retaining wall in active state is inclined to the horizontal at an angle of
(*a*) $45° + \phi'$ (*b*) $45° + \phi'/2$
(*c*) $45 - \phi'/2$ (*d*) $45° - \phi'$
where ϕ' = angle of shearing resistance.

[**Ans.** 31 (*c*), 32 (*b*), 33. (*a*), 34. (*b*), 35 (*d*), 36. (*d*), 37 (*b*), 38 (*b*)]

39. Write whether the following statements are true or false:
1. Failure plane in a triaxial test carries the maximum shear stress.

2. Initial compression of an unsaturated soil is rapid.
3. The active earth pressure increases if the magnitude of the angle of friction ϕ decreases.
4. Generally the soils compacted on the dry side of the optimum are less compressible.
5. Halloysites are minerals which have the structure similar to that of a kaolinite.
6. Shrinkage index is the difference between the plastic limit and the shrinkage limit.
7. Soils with high values of the liquid limit have low swelling potential. [**Ans.** True 2, 3, 5]

40. State whether the following statements are true or false:
1. The magnitude of the final settlement of a stratum depends upon the time factor.
2. The value of the angle of shearing resistance decreases with an increase in the plasticity index of clay.
3. The absolute maximum shear stress under a strip load of intensity q is equal to q.
4. The coefficient of lateral earth pressure at rest (K_0) is greater for a loose sand than that for a dense sand.
5. If the failure surface is assumed to be a plane surface, the passive earth pressure is overestimated by Coulomb's theory.
6. Size effects in a plate load tests are more important in the case of cohesive soils as compared to cohesionless soils.
7. The group efficiency of a pile group can be more than 100%.
8. The depth of the well bottom below the bed level is called the grip length.
9. The time period for an undamped vibration increases with the increase in mass.
10. Bentonite is a collapsible soil. [**Ans.** True 2, 4, 5, 7, 9]

41. Match List I with List II and select the correct answer using the codes given below the lists:

List I	**List II**
1. Wind	A. Lacustrine
2. Lake	B. Loess
3. Gravity	C. Till
4. Glacier	D. Talus

Codes:

	1	2	3	4
(*a*)	A	C	B	D
(*b*)	B	A	D	C
(*c*)	C	A	C	D
(*d*)	D	B	A	C

42. Match List I with List II and select the correct answer using the codes given below the lists:

List I (*Type of soil*)	**List II** (*Permeability*)
1. GP	A. Impervious
2. SW	B. Highly pervious
3. ML	C. Semi-pervious
4. CH	D. Pervious

Codes:

	1	2	3	4
(*a*)	A	C	D	B
(*b*)	B	D	C	A
(*c*)	B	D	C	A
(*d*)	D	C	A	B

43. Match List I with List II and select the correct answer using the codes given below the lists:

List I (*Type of soil*)	**List II** (*Dilotancy*)
A. ML	1. Slow
B. OL	2. Quick
C. CI	3. Slow to none
D. MH	4. None

Codes:

	A	B	C	D
(*a*)	2	4	1	3
(*b*)	4	3	2	1
(*c*)	1	2	4	3
(*d*)	2	1	4	3

44. A clay in floculated structure has a
(*a*) Low permeability, Low strength and High compressibility
(*b*) High permeability, High strength and High compressibility
(*c*) High permeability, High strength and Low compressibility
(*d*) Low permeability, High strength and Low compressibility

45. **Assertion A:** The process of compaction is accompanied by the expulsion of water
Reason B: The degree of compaction is characterised by its by dry density.

Codes:
(*a*) Both A and R are true and R is correct explanation of A.
(*b*) Both A and R are true and R is not correct explanation of A.
(*c*) A is true but R is false.
(*d*) A is false but R is true.

46. Range of optimum water content for silty sand is about
(*a*) 6 to 10% (*b*) 14 to 20%
(*c*) 12 to 16% (*d*) 8 to 12%

47. The amount of compaction on site depends on
1. the type of soil
2. the field moisture content
3. method of compaction
4. the amount of compaction

Of the above statements

(*a*) Only 1 and 2 are correct (*b*) Only 1, 2 and 4 are correct
(*c*) All 1, 2, 3 and 4 are correct (*d*) Only 1, 2 and 3 are correct

48. Select the *incorrect* statement:
(*a*) Effective stress is the actual stress transmitted at the points of contact of the soil particles
(*b*) Effective stress is that part of the load that is transmitted at the points of the contact divided by the gross area
(*c*) Effective stress cannot be measured directly in the laboratory
(*d*) Effective stress governs the shear strength of a soil.

49. The contact pressure under a rigid footing on surface of a cohesionless soil is
(*a*) uniform with concavity upward
(*b*) uniform with concavity downward
(*c*) maximum at the centre and zero at the edges

(*d*) minimum at the centre and maximum at the edges

50. Identify the *incorrect* statement.
    (*a*) Flow lines and equipotential lines are orthogonal to each other.
    (*b*) Potential drops between two successive equipotential lines are equal.
    (*c*) Flow through each flow channel is the same.
    (*d*) All flow channels should make complete square.
    (*e*) The flow net will remain the same for all isotropic and homogeneous soils.
51. The quantity of seepage is not directly proportional to
    (*a*) the coefficient of permeability of soil
    (*b*) the head causing flow
    (*c*) the number of equipotential drops
    (*d*) the number of flow channels
52. Identify the *incorrect* statements.
    (*a*) A stress isobar is a curve which connects all points of equal stress.
    (*b*) The pressure bulb under a circular footing is deeper than that under a strip footing.
    (*c*) It is generally assumed that an isobar of 0.1 Q forms a pressure bulbs.
    (*d*) According to Boussinesq's theory, the vertical stress does not depend upon the modulus of elasticity and the poisson ratio of the material.
53. Match List I with List II and select the correct answer using the codes given below the lists:

List I (*Type of soil*)	**List II** (*Type of formation*)
A. Clay	1. aquifuge
B. Impervious rock	2. aquiclude
C. Sand	3. poor aquifer
D. Silt	4. good aquifer

Codes:

	A	B	C	D
(*a*)	1	2	3	4
(*b*)	2	1	4	3
(*c*)	2	1	3	4
(*d*)	1	2	4	3

54. The time factor T_v for consolidation is
    1. directly proportional to the coefficient of consolidation
    2. directly proportional to the elapsed time
    3. directly proportional to the square of the length of the drainage path
    4. inversely proportional to the square of the length of the drainage path

    Of these statements:

    (*a*) 1, 2 and 3 are correct (*b*) 1, 2 and 4 are correct
    (*c*) 1 and 2 are correct (*d*) 1 and 4 are correct
55. The cohesion and angle of shear resistance are not necessarily fundamental properties of the soil because they depend upon
    1. water content
    2. drainage conditions
    3. testing conditions

    Of these statements:

    (*a*) 1 and 3 are correct (*b*) 1 and 2 are correct
    (*c*) 2 and 3 are correct (*d*) 1, 2 and 3 are correct

56. Match List I with List II and select the correct answer using the codes given below the lists:

List I	**List II**
(*Type of test*)	(*Nomenclature*)
A. Consolidated drained test	1. S test
B. Consolidated undrained test	2. Q test
C. Unconsolidated undrained test	3. R test

Codes:

	A	B	C
(*a*)	1	2	3
(*b*)	2	1	3
(*c*)	1	3	2
(*d*)	2	3	1

57. The unconfined compression test is a type of
1. unconsolidated undrained test
2. unconsolidated drained test
3. consolidated drained test

Of these statements

(*a*) only 1 is correct (*b*) only 2 is correct
(*c*) only 3 is correct

58. The maximum dry density in a compaction test is lowest for
(*a*) Low-plasticity silt (*b*) Low-plasticity clay
(*c*) High-plasticity clay (*d*) Well-graded sand

59. Earth-retaining structures that are restrained against lateral movements should be designed for
(*a*) active pressure (*b*) passive pressure
(*c*) at rest pressure (*d*) none of above

60. Select the *incorrect* answer:
(*a*) Most of the natural slope failures occur during rainy season
(*b*) The stability of a slope can be increased by slope flattening
(*c*) The stability of a slope can be increased by drainage control
(*d*) The most critical condition for the upstream slope of an earth dam is during steady seepage.

[**Ans.** 41 (*b*), 42. (*c*), 43. (*d*), 44. (*c*), 45. (*d*), 46. (*d*), 47. (*c*), 48. (*a*), 49. (*c*), 50. (*d*), 51. (*c*), 52. (*b*), 53. (*b*), 54. (*b*), 55. (*d*), 56. (*c*), 57. (*a*), 58. (*c*), 59. (*c*), 60. (*d*)]

61. **Assertion (A):** The aim in machine foundation design is to ensure that the frequency ratio of machine and that of foundation soil system is either less than 0.6 or more than 1.5.
Reason (R): High magnitude of amplitude occurs when the operating frequency equals the natural frequency.

Codes:
(*a*) Both A and R are true and R is the correct explanation of A.
(*b*) Both A and R are true and R is not the correct explanation of A.
(*c*) A is true but R is false.
(*d*) A is false but R is true.

62. If due to temperature change, the unit weight and viscosity of percolating fluid are reduced to 90% and 60% respectively, the change in coefficient of permeability will be
(*a*) 60% (*b*) 50%
(*c*) 40% (*d*) 70%

63. Match List I (Type of structure) with List II (its use) and select the correct answer using the codes given below the lists.

List I	List II
A. Point bearing piles	1. To retain soil filling
B. Bulk heads	2. To transfer heavy loads to a strong stratum beneath a weak stratum
C. Underreamed piles	3. To resist lateral loads
D. Batter piles	4. To carry the loads in expansive soils

Codes:

	A	B	C	D
(*a*)	3	1	4	2
(*b*)	3	4	1	2
(*c*)	2	1	4	3
(*d*)	2	4	1	3

64. Match List I (in-situ test) with List II (measurement) and select the correct answer using the codes given below the lists:

List I	List II
A. SPT test	1. Stiffness coefficient of the soil
B. CPT test	2. Penetration resistance
C. Plate load test	3. Point resistance and skin friction
D. Consolidation test	4. Collapse potential

Codes:

	A	B	C	D
(*a*)	2	3	1	4
(*b*)	3	2	4	1
(*c*)	4	2	3	1
(*d*)	3	4	1	2

65. Which of the following soils is most likely to be subjected to liquefaction under seismic forces?
(*a*) Soft saturated clays
(*b*) Stiff clays
(*c*) Loose saturated sands
(*d*) Dry sands

66. Match List I (test) with List II (property) and select the correct answer using the codes given below the lists:

List I	List II
A. Hydrometer test	1. Flow index
B. Vane test	2. Pavement design
C. CBR test	3. Shear strength
D. Liquid limit test	4. Grain size analysis

Codes:

	A	B	C	D
(*a*)	2	1	3	4
(*b*)	3	4	1	2
(*c*)	4	3	2	1
(*d*)	1	2	4	3

67. If the vertical stress under a concentrated load at a depth of 1 m is 4 kN/m^2, the vertical stress at a depth of 2 m will be
(*a*) 2 kN/m^2
(*b*) 1 kN/m^2
(*c*) 0.5 kN/m^2
(*d*) 8 kN/m^2

68. The method which cannot be used for improvement of a cohesive soil in the field is
    (*a*) Precompression (*b*) Sand drains
    (*c*) Stone columns (*d*) Compaction piles
69. A CD triaxial test was conducted on a granular soil. At failure, the principal stress ratio was 3.0. If the minor principle stress was 50 kPa, the deviator stress was
    (*a*) 100 (*b*) 150
    (*c*) 200 (*d*) 300
70. A machine foundation weighs 50 kN and has a spring constant of 10^4 kN/m^2. The system will be critically damped if the damping coefficient is about
    (*a*) 600 kN-s/m (*b*) 900 kN-s/m
    (*c*) 450 kN-s/m (*d*) 225 kN-s/m

[**Ans.** 61. (*a*), 62. (*b*), 63. (*c*) 64. (*a*), 65. (*c*), 66. (*c*), 67. (*b*), 68. (*d*), 69. (*a*), 70 (*c*)]

71. Write whether the following statements are true of false.
    1. The degree of saturation of a soil can be greater than 100%.
    2. The mass gravity is always less than the specific gravity of solids.
    3. A soil has a porosity of 40% and the specific gravity of solids of 2.65. Its submerged unit weight will be about 12.1 kN/m^3.
    4. A soil has a water content of 20% and the specific gravity of solids of 2.67. Its saturated dry density will be about 16.2 kN/m^3.
    5. If a soil has water content of 20% based on the dry mass, its water content based on the wet mass will be about 16.7%. [**Ans.** True, 1, 2, and 5]
72. Write whether the following statements are true or false.
    1. The radiation method of the determination of the water content of a soil is generally used for in-situ soils.
    2. A saturated soil has a void ratio of 0.40. If the specific gravity of solids is 2.60, its saturated unit will be about 19.0 kN/m^3 (g = 9.81 m/sec^2).
    3. A soil particle of diameter 0.075 mm will travel in still water a distance of 100 mm in about 4 minutes.
    4. The void ratio of a fat clay is generally greater than that of a lean clay.
    5. The mineral illite is more active than the mineral kaolinite. [**Ans.** True, 1, 4 and 5]
73. Write whether the following statements are true or false.
    1. A clay is considered sensitive when its sensitivity is more than 4.0.
    2. As the particle size of the clay increases, its liquid limit decreases.
    3. The toughness index of a soil is a measure of the shearing strength of the soil at the plastic limit.
    4. The value of the toughness index of most soils is zero.
    5. When a clay is in solid state, it has no water. [**Ans.** True 1 and 3]
74. Write whether the following statements are true or false:
    1. The soil with the liquid limit less than 20% is generally sand.
    2. The naked eye can see the particle size of about 0.002 mm (2μ).
    3. If the group index of a soil as per AASHTO system is negative, it is reported as zero.
    4. The symbol M in the soil classification represents organic matter.
    5. Capillary water exists in soil so long as there is an air-water surface. [**Ans.** True 1, 3 and 5]
75. Write whether the following statements are true or false:
    1. The capillary rise in a tube of uniform diameter of 1 cm is about 0.3 cm.
    2. A soil suction P_F of zero indicates that the suction is zero.
    3. Frost heave occurs when a frozen soil thaws.

4. For a soil of effective size 1 mm, the coefficient of permeability according to Hazen's formula is about 10 mm/s.
5. The punching shear failure occurs below a footing when the soil is highly dense.
6. The stability coefficient gives the failure surface along which the factor of safety is a minimum.
7. At the liquid limit, the shear strength of a soil is zero.
8. The darcy is the coefficient of absolute permeability (intrinsic permeability).

[**Ans.** True 1, 4, 6 and 8]

76. Match List I (Type of soil) with List II (Mode of transportation and deposition) and select the correct answer using the codes given below the lists:

List I	**List II**
A. Residual soil	1. Soil transported by wind
B. Loess	2. Organic soil
C. Peat	3. Deposition in lakes during periods of alternately high and low waters
D. Varved clays	4. Soil left in place as a result of weathering of parent rock

Codes:

	A	B	C	D
(*a*)	4	2	1	3
(*b*)	4	2	3	1
(*c*)	4	1	2	3
(*d*)	3	1	2	4

[**Ans.** (*c*)]

77. State whether the following statements are true or false:
1. Coarse grained soils are formed by chemical weathering of rocks.
2. Slopes of cuts made in loess are able to stand nearly vertical without any support.
3. Values of void ratio in coarse-grained soils are generally greater than that in fine-grained soils.
4. Consistency limits can be determined for soils in undisturbed conditions.
5. As the void ratio of a soil decreases, its strength increases.
6. The liquid limit of a soil cannot exceed 100%.
7. The dry density of a soil with zero air content is equal to the saturated dry density.
8. Darcy's law is valid for flow through gravels. [**Ans.** True 2, 5, 6 and 7]

78. **Assertion (A):** Black cotton soils swell when wetted and exert swelling pressure on the foundation.
Reason (R): Black cotton soils have the mineral montmorillonite which exhibits high swelling characteristics when wetted.

Codes:

(*a*) Both A and R are true and R is the correct explanation of A.
(*b*) Both A and R are true and R is not the correct explanation of A.
(*c*) A is true but R is false.
(*d*) A is false but R is true.

79. In the soil specimen of a consolidometer test, pore water is maximum at
(*a*) top of the specimen (*b*) bottom of the specimen
(*c*) centre of the specimen (*d*) one quarter of height

80. According to IS code, a settlement of 65 to 100 mm is permitted in case of
(*a*) isolated foundation on sand (*b*) isolated foundation on clay
(*c*) raft foundation on sand (*d*) raft foundation on clay

81. In a consolidation test, the void ratio of a specimen of height of 25 mm decreases from 1.0 to 0.90. The decrease in height will be
(*a*) 2.5 mm (*b*) 1.25 mm

(*c*) 0.25 mm (*d*) 1.00 mm

82. A sample of clay has a liquid limit of 60%, plastic limit of 30% and natural water content of 40%. Its liquidity index is
(*a*) 33% (*b*) 67%
(*c*) 50% (*d*) 100%

83. In an undrained triaxial test, a saturated clay sample failed at a deviator stress of 100 kN/m^2 when the cell pressure was 50 kN/m^2. The cohesion of the clay is
(*a*) 100 kN/m^2 (*b*) 50 kN/m^2
(*c*) 25 kN/m^2 (*d*) 75 kN/m^2

84. Match List I (*Parameter*) with List II (*Units*) and select the correct answer using the codes given below the lists:

List I	**List II**
A. Coefficient of consolidation	1. m/s
B. Coefficient of subgrade reaction	2. unitless
C. Coefficient of permeability	3. cm^2/sec.
D. Coefficient of curvature	4. kN/m^2

Codes:

	A	B	C	D
(*a*)	1	2	4	3
(*b*)	3	4	2	1
(*c*)	4	3	1	2
(*d*)	3	4	1	2

85. Match List I (*Densities*) with List II (*Expressions, with usual notation*) and select the correct answer using the codes given below the lists.

List I	**List II**
A. Theoretical maximum dry density	1. $\dfrac{(1+w)\, G\, \rho_w}{1+(wG)/S}$
B. Submerged density	2. $\left(\dfrac{G+e}{1+e}\right)\rho_w$
C. Saturated density	3. $\left(\dfrac{G-1}{1+e}\right)\rho_w$
D. Bulk density	4. $\dfrac{G\,\rho_w}{1+w\,G}$

Codes:

	A	B	C	D
(*a*)	3	4	2	1
(*b*)	2	3	4	1
(*c*)	2	3	1	4
(*d*)	4	3	2	1

86. Two soils A and B have their void ratios of 0.50 and 0.60 respectively. The ratio of their porosities will be
(*a*) 5/6 (*b*) 6/7
(*c*) 7/8 (*d*) 8/9

87. Match List I (Equipment) with List II (use) and select the correct answer using the codes given below the lists.

List I	**List II**
A. Sheep foot roller	1. To determine stress-deformation characteristics
B. Geophone	2. To compact sandy soils
C. Terra probe	3. To compact clayey soils
D. Pressure meter	4. Seismic investigation

Codes:

	A	B	C	D
(*a*)	4	3	2	1
(*b*)	1	3	4	2
(*c*)	4	3	1	2
(*d*)	3	4	2	1

88. Consider the following statements:
    1. The permeability of a clay is zero.
    2. The permeability of a soil is independent of temperature.
    3. If the two sandy soils have the same void ratios, their permeabilities will be almost same.
    4. The permeability of a soil decreases with an increase in effective stress.

Which of the above statements are correct?

(*a*) 1 and 2 (*b*) 3 and 4
(*c*) 1 and 4 (*d*) 2 and 3

89. Consider the following statements:
    1. Pile foundations do not have any settlement.
    2. Underreamed piles cannot be used for non-expansive soils.
    3. Driven precast piles are displacements piles.
    4. Pile driving formulae are in general less accurate than the static formulae for load-carrying capacity.

Which of the above statements are correct ?

(*a*) 1 and 3 (*b*) 1 and 2
(*c*) 3 and 4 (*d*) 2 and 4

90. Consider the following statements:
    1. The shear strength of a soil is unique function of the effective stress at failure.
    2. An unconsolidated drained triaxial test is not practically possible.
    3. The value of the pore pressure parameter A_f of a soil can be negative.

Which of the above statements are correct?

(*a*) 2, 3 (*b*) 1, 2
(*c*) All 1, 2 and 3

91. Consider the following statements:
    1. For a loose sand, there is not much difference between the peak strength and the ultimate strength.
    2. A dense sand can show an increase in volume when sheared.
    3. For a normally consolidated clay, the value of the angle of shearing stress ϕ' increases with an increase in its plasticity.

Which of the above statements are correct?

(*a*) 1, 2 (*b*) 2, 3
(*c*) All 1, 2 and 3

92. Consider the following statements:
    1. The shear strength parameters c' and ϕ' do not have unique values for a given soil.
    2. In case the pore water pressure cannot be accurately estimated or measured, the total stress analysis is more convenient than the effective stress analysis.

3. For a heavily overconsolidated clay, the undrained shear strength is less than the drained strength.
4. The angle of shearing resistance ϕ' of a cohesionless decreases with an increase in its relative density.

Which of the above statements are correct?

(*a*) 1, 3 (*b*) 2, 4
(*c*) 1, 4 (*d*) 1, 2

[**Ans.** 78. (*a*), 79. (*c*), 80. (*d*), 81. (*b*), 82. (*a*), 83. (*b*), 84. (*d*), 85. (*c*). 86. (*d*), 87. (*d*), 88. (*b*), 89. (*c*), 90. (*c*), 91. (*a*) 92. (*d*)]

93. Consider the following statements:
1. Secondary consolidation is small for highly plastic clays and organic soils.
2. In Terzaghi's consolidation theory, it is assumed that the consolidation is three-dimensional.
3. A half-closed layer has only one drainage face.
4. The coefficient of volume change is also called the coefficient of volume compressibility.

Of the above statements, which are correct?

(*a*) 1, 3 (*b*) 3, 4
(*c*) 2, 3 (*d*) 1, 4

94. Consider the following statements:
1. Terzaghi's theory of consolidation neglects initial consolidation and secondary consolidation.
2. The coefficient of compressibility (a_v) decreases with an increase in effective stress.
3. The time required for a given percentage of consolidation increases with an increase of the coefficient of consolidation (c_v).
4. When the consolidation pressure is complete, the hydrostatic pressure becomes zero.

Which of the above statements are correct?

(*a*) 1, 4 (*b*) 2, 3
(*c*) 1, 2 (*d*) 2, 4

95. Consider the following statements:
1. According to Boussinesq's theory, the vertical stress at a point is independent of the modulus of elasticity.
2. Westergaard's theory always gives the values of vertical stresses which are smaller than that by Boussinesq's theory.
3. Newmark's chart cannot be used if the point at which the vertical stress is required lies outside the loaded area.
4. If the dimensions of the loaded area are very large, the vertical stress at a point at a large depth will be the same as the applied load intensity.

Which of the above statements are correct?

(*a*) 2, 3 (*b*) 1, 3
(*c*) 2, 4 (*d*) 1 and 4

96. Consider the following statements:
1. At-rest conditions, the soil is in a state of plastic equilibrium.
2. Wall friction is not considered in the Rankine theory.
3. The angle of internal friction in plane strain is less than that in the conventional triaxial test.
4. In an active case, the critical surface has the largest resisting force.

Which of the above statements are correct?

(*a*) 1 and 2 (*b*) 2 and 3
(*c*) 2 and 4 (*d*) 3 and 4

97. Consider the following statements:
1. The resistance offered by an anchor plate depends upon its passive resistance.

2. The active pressure coefficient of a cohesionless soil decreases when it becomes submerged.
3. Basements are designed for active earth pressure.
4. Tension develops at the base of a retaining wall if the resultant strikes outside the middle third of the base.

Which of the above statements are correct?

(*a*) 1 and 4 (*b*) 2 and 4
(*c*) 1 and 2 (*d*) 2 and 3

98. Consider the following statements:
1. In free-earth support, when a pile rotates at its bottom tip, there is a point of contraflexure.
2. The bending moment in a sheet pile decreases with an increase in the flexibility of the pile.
3. Anchor plates and beams are also called deadman.
4. The earth pressure behind a braced cut is the same as that behind a retaining wall.

Which of the above statements are correct?

(*a*) 1 and 2 (*b*) 2 and 3
(*c*) 1 and 4 (*d*) 3 and 4

99. Match List I (Soil type) with List II (Coefficient of permeability) and select the correct answer using the codes given below the lists.

List I	List II
A. Clay	1. 10^{-5} to 10^{-4} mm/s
B. Fine sand	2. 10^{-4} to 10^{-2} mm/s
C. Clean gravel	3. 10^{-8} to 10^{-5} mm/s
D. Dense silt	4. 10^{+1} to 10^{+2} mm/s

Codes:

	A	B	C	D
(*a*)	2	3	4	1
(*b*)	3	2	4	1
(*c*)	2	3	1	4
(*d*)	3	4	2	1

100. In a compaction test if the compactive effort is increased, it
(*a*) increases both maximum dry density and OMC
(*b*) increases maximum dry density but OMC is decreased
(*c*) increases OMC but maximum dry density remains the same
(*d*) increases maximum dry density but OMC remains the same

101. Consider the following statements:
1. Liquidity index can have a negative value
2. The loess is an expansive soil
3. Plasticity index of the soil decreases as the particle size of clay increases
4. The consistency of a soil denotes the degree of firmness.

Which of the above statements are correct?

(*a*) 1 and 2 (*b*) 2 and 3
(*c*) 3 and 4 (*d*) 1 and 4

102. The negative pore pressure can develop in
(*a*) Dense sand and normally consolidated clays
(*b*) Loose sand and normally consolidated clays
(*c*) Dense sand and overconsolidated clays
(*d*) Loose sand and overconsolidated clays

103. Consider the following statements:
1. The soil with the liquid limit less than 20% is generally sand

2. The void ratio of a fat clay is generally smaller than that of a lean clay
3. No two flow lines cross each other in a flow net
4. In an anchored sheet pile with a free earth support, there is a point of contraflexure.

Which of the above statements are correct?

(*a*) 1 and 3 (*b*) 2 and 3
(*c*) 1 and 4 (*d*) 3 and 4

104. Consider the following statements regarding shallow foundations.
1. A heave on the sides is always observed in general shear failure.
2. The foundation settlement at the ultimate load in the case of general shear failure is larger than that in the local shear.
3. Terzaghi did not consider the punching shear failure.

Of the above statements,

(*a*) All 1, 2 and 3 are correct (*b*) 1 and 2 are correct
(*c*) 2 and 3 are correct (*d*) 1 and 3 are correct

105. Consider the following statements regarding shallow foundations
1. A footing on sand with a relative density of greater than 70% will generally fail in general shear.
2. For a fully compensated foundation, the theoretical settlement is zero.
3. The bearing capacity of a footing on clay is independent of its size.

Of the above statements:

(*a*) 1 and 2 are correct (*b*) 2 and 3 are correct
(*c*) 1 and 3 are correct (*d*) All 1, 2 and 3 are correct

106. Consider the following statements regarding pile foundations.
1. Bored cast in-situ piles are displacement piles.
2. If the horizontal load is large, batter piles are generally used.
3. Negative skin friction increases the factor of safety of piles.

Of the above statements:

(*a*) 1 and 2 are correct (*b*) 2 and 3 are correct
(*c*) 1 and 3 are correct (*d*) All 1, 2 and 3 are correct

107. The IS Code recommends that for a well foundation the tilt should not exceed

(*a*) 1/20 (*b*) 1/40
(*c*) 1/50 (*d*) 1/60

[**Ans.** 93. (*b*), 94. (*c*), 95. (*d*), 96. (*c*), 97. (*a*), 98. (*b*), 99. (*b*), 100. (*b*), 101. (*d*), 102. (*c*), 103. (*a*), 104. (*d*), 105. (*d*), 106. (*a*), 107. (*d*)]

Miscellaneous Illustrative Examples

Miscellaneous Illustrative Example 1. *The mass gravity of a fully saturated specimen of clay having a water content of 40% is 1.80. On oven drying, the mass specific gravity drops to 1.70. Determine the specific gravity of clay particles and the shrinkage limit.*

Solution. For a fully saturated clay,

$$e = wG = 0.4\,G \qquad \text{...(a)}$$

From Eq. 2.21,

$$\rho = \frac{(G + Se)}{1 + e}\,\rho_w$$

or

$$\frac{\rho}{\rho_w} = \frac{G + Se}{1 + e}$$

or

$$1.80 = \frac{G + 1 \times e}{1 + e}$$

Since $e = 0.4\,G$, we have

$$1.80 = \frac{G + 0.4\,G}{1 + 0.4\,G}$$

or

$$1.80 + 0.72\,G = 1.4\,G$$

or

$$G = \mathbf{2.65}$$

On oven drying, the mass specific gravity = 1.70.

Assuming it corresponds to the dry density,

$$\frac{\rho_d}{\rho_w} = 1.70$$

From Eq. 4.9, shrinkage limit,

$$w_s = \left(\frac{\rho_w}{\rho_d} - \frac{1}{G}\right)$$

$$= \frac{1}{1.70} - \frac{1}{2.65}$$

$$= 0.21 = \mathbf{21\%}$$

Miscellaneous Illustrative Example 2. *A sampler with a volume of 45 cm^3 is filled with a soil sample. When the soil of the sampler is poured into a graduated cylinder containing water, it displaces 30 cm^3 of water. What are the porosity and the void ratio of the soil?*

Solution. Total volume of soil = 45 cm^2

Volume of solids = volume of water displaced

$= 30$ cm^3

Volume of voids $= 45 - 30$

$= 15 \text{ cm}^3$

Porosity, $n = \dfrac{\text{Volume of voids}}{\text{Total volume}}$

$= \dfrac{15}{45} = 0.33 = \mathbf{33\%}$

Void ratio, $e = \dfrac{\text{Volume of voids}}{\text{Volume of solids}}$

$= \dfrac{15}{30} = \mathbf{0.5}$

Miscellaneous Illustrative Example 3. *A building has to be supported on a R.C.C. raft foundation of dimensions 15 m × 20 m. The subsoil is clay which has an average unconfined compressive strength of 18 kN/m². The gross pressure on the soil due to the weight of the building and the live loads is 150 kN/m² at the base of the raft. Determine the depth of foundation for a factor of safety of 3.0 against shear.* $\gamma = 18$ *kN/m³.*

Solution. $c_u = \dfrac{18}{2} = 9 \text{ kN/m}^2$

From Eq. 23.53 (*a*), assuming $D_f/B < 2.50$,

$$N_c = 5.0\left(1 + \frac{0.2\,D_f}{B}\right)(1 + 0.2\,B/L)$$

$$= 5.0\left(1 + \frac{0.2 \times D_f}{15}\right)(1 + 0.2 \times 15/20)$$

$$= 5.75\,(1 + 0.0133\,D_f)$$

$$= 5.75 + 0.0767\,D_f$$

$$q_{nu} = c_u\,N_c$$

$$= 9\,[5.75 + 0.0767\,D_f]$$

$$q_{ns} = \frac{q_{nu}}{3}$$

$$= \frac{9}{3}\,[5.75 + 0.0767\,D_f]$$

$$= 17.25 + 0.230\,D_f$$

Gross safe bearing capacity,

$$q_s = q_{ns} + \gamma\,D_f$$

$$= (17.25 + 0.23\,D_f) + 18 \times D_f = 17.25 + 18.23\,D_f$$

But $q_s = 150 \text{ kN/m}^2$

Therefore, $150 = 17.25 + 18.23\,D_f$

or $D_f = \mathbf{7.28\ m}$

Miscellaneous Illustrative Example 4. *For a compacted soil, the theoretical maximum unit weight at a water content of 16% is 18.5 kN/m³. Determine the dry unit weight of the compacted soil at the degree of saturation of 80% and water content of 16%.*

Solution. From Eq. 14.4,

$$(\gamma_d)_{theomax} = \frac{G\,\gamma_w}{1 + w\,G}$$

In this case,

$$18.5 = \frac{G \times 9.81}{1 + 0.16 \times G}$$

or $$18.5 + 2.96\,G = 9.81\,G$$

or $$G = 2.70$$

Now dry unit weight, $$\gamma_d = \frac{G\gamma_w}{1+e}$$

$$\gamma_d = \frac{2.70 \times 9.81}{1 + \dfrac{wG}{S}}$$

$$= \frac{2.70 \times 9.81}{1 + 0.16 \times 2.70/0.80}$$

$$= \mathbf{17.2\ kN/m^3}$$

Miscellaneous Illustrative Example 5. *Piles 200 mm diameter and 10 m long are used as foundation for a column in a uniform deposit of medium clay (unconfined compressive strength = 100 kN/m² and adhesion factor = 0.60). There are nine piles in a square pattern of 3 × 3. For a group efficiency = 1.0, find the spacing between the piles (neglect bearing).*

Solution. For a group efficiency of 1.0, the pile group capacity is equal to the sum of the individual piles capacity.

Let the centre to centre spacing be s metres.

Width of the pile block $= 2s + d$

$= 2s + 0.20$ metres

Load carried by group (Eq. 25.44),

$$Q_g = q_p A_g + \alpha c\,(P_g D)$$

Neglecting bearing and remembering $c_u = \dfrac{q_u}{2}$,

$$Q_g = \alpha c\,(P_g D)$$

$$= 0.60 \times \frac{100}{2}(2s + 0.20) \times 4 \times 10$$

$$= 1200\,(2s + 0.2) \quad \ldots(a)$$

Individual pile capacity (Eq. 25.45), neglecting bearing,

$$Q_u = \alpha c\,(p \times D)$$

$$= 0.60 \times \frac{100}{2}(\pi \times 0.2) \times 10 = 188.50 \text{ kN}$$

Since there are 9 piles, total capacity $= 9 \times 188.50$

$$= 1696.46 \text{ kN} \quad \ldots(b)$$

From Eqs. (*a*) and (*b*),

$$1200\,(2s + 0.2) = 1696.46$$

or $$s = \mathbf{0.61\ m}$$

Miscellaneous Illustrative Example 6. *A square footing located at a depth of 1.5 m below the ground surface carries a column load of 500 kN. The soil is submerged having an effective unit weight of 10 kN/m³ and an angle of shearing resistance of 30°. Find the size of footing using Terzaghi's theory if* $F_s = 3$, $N_q = 22.5$, $N_\gamma = 19.7$.

Solution. Net ultimate bearing capacity,

$$q_{nu} = q_u - \gamma' D_f$$
$$= (N_q \gamma' D_f + 0.4\,\gamma' B N_\gamma') - \gamma' D_f$$
$$= (N_q - 1)\,\gamma' D_f + 0.4\,\gamma' B N_\gamma$$
$$= (22.5 - 1) \times 10 \times 1.5 + 0.4 \times 10 \times B \times 19.7$$

or
$$q_{nu} = 322.5 + 78.8\,B$$

Safe bearing capacity $= \dfrac{q_{nu}}{F} + \gamma' D_f$

$$q_s = \frac{322.5 + 78.8\,B}{3} + 10 \times 1.5$$
$$= \frac{337.5 + 78.8\,B}{3}$$
$$= 112.5 + 26.3\,B$$

Now,
$$\frac{\text{Total load}}{q_s} = B^2$$

or
$$\frac{500}{112.5 + 26.3\,B} = B^2$$

or
$$500 = 112.5\,B^2 + 26.3\,B^3$$

Solving by trial and error,

$$B = \mathbf{1.78\ m}$$

Miscellaneous Illustrative Example 7. *A 3.0 m square footing is located in a dense sand deposit at a depth of 2.0 m. Determine the ultimate bearing capacity for the following water table positions: (a) at ground surface, (b) at footing level, and (c) at 1 m below the footing (d) at 1 m above footing. The moist unit weight above the water table is 17 kN/m³ and the saturated unit weight is 19 kN/m³.*

Take $\phi = 35°$, $N_q = 41.4$, $N_\gamma = 42.4$, $c = 0$.

Also determine the net safe bearing capacity (F = 3.0).

Solution. The ultimate bearing capacity,

$$q_u = \gamma D_f N_q + 0.4\,\gamma B N_\gamma$$

(*a*) Water table at ground surface,

$$q_u = (19 - 9.81) \times 2 \times 41.4 + 0.4 \times (19 - 9.81) \times 3 \times 42.4$$
$$= 760.93 + 467.59$$
$$= \mathbf{1228.52\ kN/m^2}$$

Net ultimate bearing capacity,

$$q_{nu} = q_u - \gamma D_f$$
$$= 1228.52 - (19 - 9.81) \times 2$$
$$= 1210.14\ \text{kN/m}^2$$

Net safe bearing capacity, $q_{na} = \dfrac{q_{nu}}{F}$

$$= \frac{1210.14}{3}$$
$$= \mathbf{403.38\ kN/m^2}$$

(*b*) Water table at footing level

From Eq. 23.34, $\bar{\gamma} = \gamma' + \frac{b}{B}(\gamma - \gamma')$

$= (19 - 9.81) + 0 = 9.19 \text{ kN/m}^2$

$q_u = \gamma D_f N_q + 0.4 \bar{\gamma} B N_\gamma$

$= 17 \times 2 \times 41.4 + 0.4 \times 9.19 \times 3 \times 42.4$

$= 1407.6 + 467.59$

$= \mathbf{1875.19 \text{ kN/m}^2}$

$q_{nu} = 1875.19 - 17 \times 2 = 1841.19 \text{ kN/m}^2$

$q_{na} = \frac{1841.19}{3}$

$= \mathbf{613.73 \text{ kN/m}^2}$

(*c*) Water table at 1 m below the footing

From Eq. 23.34, $\bar{\gamma} = \gamma' + \frac{b}{B}(\gamma - \gamma')$

$= 9.19 + \frac{1}{3}(19 - 9.19)$

$= 12.46 \text{ kN/m}^2$

$q_u = 17 \times 2 \times 41.4 + 0.4 \times 12.46 \times 3 \times 42.4$

$= 1407.6 + 633.96 = 2041.56 \text{ kN/m}^2$

$q_{nu} = 2041.56 - 17 \times 2 = 2007.56 \text{ kN/m}^2$

$q_{na} = \frac{2007.56}{3}$

$= \mathbf{669.19 \text{ kN/m}^2}$

(*d*) Water table 1 m above the base of footing

From Eq. 23.31, $q = \gamma' D_f + (\gamma - \gamma') Dw$

$= 9.19 \times 2 + (19 - 9.19) \times 1$

$= 18.38 + 9.81 = 28.19 \text{ kN/m}^2$

$q_u = 28.19 \times 41.4 + 0.4 \times 9.19 \times 3 \times 42.4$

$= 1167.07 + 467.59 = 1634.66 \text{ kN/m}^2$

$q_{nu} = 1634.66 - 28.19 = 1606.47 \text{ kN/m}^2$

$q_{na} = \frac{1606.47}{3}$

$= \mathbf{535.49 \text{ kN/m}^2}$

Miscellaneous Illustrative Example 8. *Determine the magnitude and the line of action of the active thrust on the retaining wall shown in Fig. Neglect tension in soil.*

Solution. From Eq. 19.27, depth of tension crack,

$$Z_c = \frac{2c'}{\gamma \sqrt{K_a}}$$

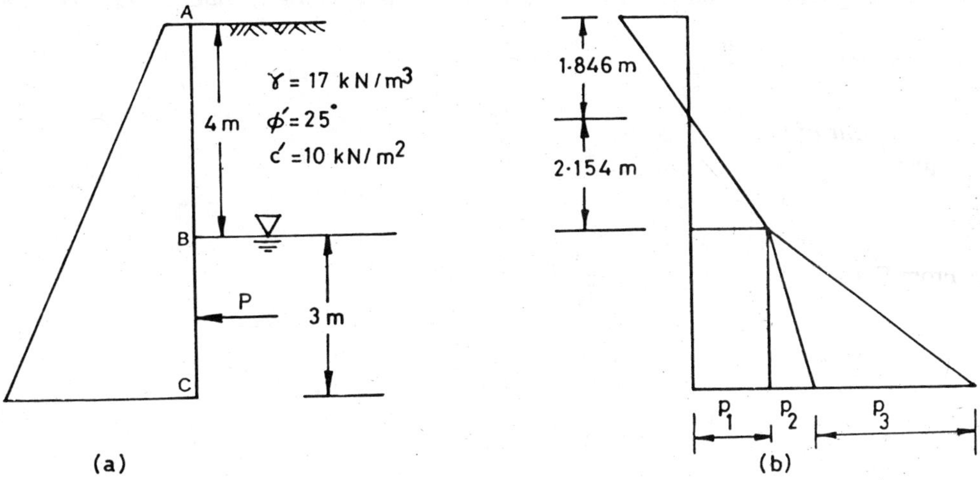

Fig. Miscellaneous Example 8.

$$K_a = \frac{1-\sin\varphi'}{1+\sin\varphi'} = \frac{1-\sin 25°}{1+\sin 25°}$$

$$= \frac{1-0.423}{1+0.423} = 0.406$$

$$Z_c = \frac{2 \times 10}{1.7 \times \sqrt{0.406}} = 1.846 \text{ m}$$

Pressure at $B = K_a \gamma Z - 2c'\sqrt{K_a}$

$$p_1 = 0.406 \times 17 \times 4 - 2 \times 10\sqrt{0.406}$$

$$= 27.61 - 12.74 = 14.87 \text{ kN/m}^2$$

Pressure at C due to soil:

$$p_2 = \gamma' H_2 K_a$$

$$= (17-9.81) \times 3 \times 0.406 = 8.76 \text{ kN/m}^2$$

Pressure at C due to water,

$$p_3 = 9.81 \times 3 = 29.43 \text{ kN/m}^2$$

Total pressure, $P = P_1 + P_2 + P_3$

$$= \frac{1}{2} \times 14.87 \times 2.154 + 14.87 \times 3 + \frac{1}{2} \times 8.76 \times 3 + \frac{1}{2} \times 29.43 \times 3$$

$$= 16.02 + 44.61 + 13.14 + 44.15$$

$$= 117.92 \text{ kN}$$

Height of line of action of P above C is obtained by taking moments about C as

$$\bar{z} = \left[16.02 \times \left(3 + \frac{2.154}{3}\right) + 44.61 \times 1.5 + 13.14 \times 1 + 44.15 \times 1\right] \times \frac{1}{117.92}$$

$$= (59.56 + 66.92 + 13.14 + 44.15) \times \frac{1}{117.92}$$

$$= \mathbf{1.558 \text{ m}}$$

Miscellaneous Illustrative Example 9. *Find the settlement of a footing of width 1.2 m from the following plate load test results for*

(*a*) *Sandy soil, and*

(*b*) *Clayey soil*

Settlement of the plate = *12 mm*

Width of the plate = *400 mm.*

Solution. (*a*) *Sandy soil*

From Eq. 23.88,
$$s_f = s_p \left[\frac{B_f (B_p + 0.3)}{B_p (B_f + 0.3)} \right]^2$$
$$= 12 \left[\frac{1.2\,(0.4 + 0.3)}{0.4\,(1.2 + 0.3)} \right]^2$$
$$= \mathbf{23.52\ mm}$$

(*b*) *Clayey soil*

From Eq. 23.87,
$$s_f = s_p \times \frac{B_f}{B_p}$$
$$= 12 \times \frac{1.2}{0.4} = \mathbf{36\ mm}$$

Miscellaneous Illustrative Example 10. (*a*) *A laboratory constant head permeability test was conducted on a silty sand specimen of void ratio 0.45. The cylindrical specimen had a diameter of 7.5 cm and a height of 16.5 cm. The head during the test was 75 cm. After 1 minute of testing at the room temperature of 20°C, a total 750 g of water was collected. Compute the coefficient of permeability in metres per second.* (*b*) *If the void ratio changes to 0.38, what would be the change in permeability?*

Solution. (*a*) Volume of water $= \frac{750}{1} = 750$ ml

$$q = \text{Discharge} = \frac{750}{60} = 12.50 \text{ ml/second}$$
$$= 12.50 \times 10^{-6}\ \text{m}^3/\text{s}$$

From Eq. 8.5,
$$k = \frac{q\,L}{A\,h}$$
$$= \frac{12.5 \times 10^{-6} \times 0.165}{\pi/4 \times (0.075)^2 \times 0.75} = \mathbf{6.22 \times 10^{-4}\ m/s}$$

(*b*) From Eq. 8.15,
$$\frac{k_1}{k_2} = \frac{e_1^2/(1+e_1)}{e_2^3/(1+e_2)}$$
$$= \frac{(0.45)^3/(1+0.45)}{(0.38)^3/(1+0.38)}$$
$$\frac{6.22 \times 10^{-4}}{k_2} = \frac{0.0628}{0.0398}$$

or
$$k_2 = \mathbf{3.94 \times 10^{-4}\ m/s}$$

Alternative. If k is assumed to vary with e^2,
$$\frac{k_1}{k_2} = \frac{e_1^2}{e_2^2}$$

or
$$\frac{6.22 \times 10^{-4}}{k_2} = \frac{(0.45)^2}{(0.38)^2}$$

or
$$k_2 = \mathbf{4.44 \times 10^{-4}\ m/s}$$

Miscellaneous Illustrative Example 11. *In a Proctor compaction test, the specimen has a bulk density of 18 kN/m³ and a moisture content of 13%. Determine*

(i) degree of saturation of the specimen.

(ii) additional moisture content required for saturating the specimen.

Take G = 2.68.

Solution. (*i*)
$$\gamma_d = \frac{\gamma}{1 + w}$$
$$= \frac{18}{1 + 0.13} = 15.93 \text{ kN/m}^3$$

But
$$\gamma_d = \frac{G\gamma_w}{1 + e}$$

Therefore,
$$15.93 = \frac{2.68 \times 9.81}{1 + e}$$

or
$$e = 0.65$$

But
$$e = \frac{wG}{S}$$

Therefore,
$$0.65 = \frac{0.13 \times 2.68}{S}$$

or
$$S = 0.536 = \mathbf{53.6\%}$$

(*ii*) For $S = 100\%$
$$w = \frac{eS}{G}$$
$$= \frac{0.65 \times 1.0}{2.68} = 0.243$$

Additional moisture content = 0.243 – 0.13 = 0.113 = **11.30%**

Miscellaneous Illustrative Example 12. *A water tower is constructed on a circular raft 4 m in diameter at a depth of 2 m below the ground surface in sand having bulk unit weight of 18 kN/m³. The gross pressure at the base of the raft is 150 kN/m³. What will be the increase in vertical stress at a point 3 m beneath the centre of the raft?*

Solution. Insitu vertical stress at 5 m depth before the construction of tower
$$= 18 \times 5 = 90 \text{ kN/m}^2$$

Vertical stress at 3 m below the centre of the raft, due to load, from Eq. 11.29,
$$\sigma_z = q\left[1 - \left\{\frac{1}{1 + (R/z)^2}\right\}^{3/2}\right]$$
$$= 150\left[1 - \left\{\frac{1}{1 + (2/3)^2}\right\}^{3/2}\right]$$
$$= 150\,[1 - 0.576] = 63.6 \text{ kN/m}^2$$

Increase in stress = 63.6 kN/m²

Miscellaneous Illustrative Example 13. *A deposit of sand has a porosity of 35% and G = 2.68. The soil is dry in top 1.5 m depth, but it has a moisture content of 15% in the next 1.5 m and it is submerged below it. Determine (i) the effective pressure at a depth of 10 m below the ground. (ii) change in effective pressure if the water table suddenly drops to a level of 6 m below the ground surface.*

Solution. (*i*) **Top 1.5 m depth**

$$e = \frac{n}{1-n} = \frac{0.35}{1-0.35}$$

$$= 0.538$$

$$\gamma = \gamma_d = \frac{G\gamma_w}{1+e}$$

$$= \frac{2.68 \times 9.81}{1 + 0.538} = 17.09 \text{ kN/m}^3$$

From 1.5 to 3 m depth

$$\gamma = \frac{(G + S\,e)}{1 + e}\gamma_w$$

$$= \frac{(G + w\,G)}{1 + e}\gamma_w$$

$$= \frac{(2.68 + 0.15 \times 2.68)}{1 + 0.538} \times 9.81$$

$$= 19.66 \text{ kN/m}^3$$

Below 3 m depth

$$\gamma_{sat} = \frac{(G + e)}{1 + e} \times \gamma_w$$

$$= \frac{2.68 + 0.538}{1 + 0.538} \times 9.81$$

$$= 20.53 \text{ kN/m}^3$$

$$\gamma_{sub} = 20.53 - 9.81$$

$$= 10.72 \text{ kN/m}^3$$

Effective stress at a depth of 10 m,

$$\bar{\sigma} = 17.09 \times 1.5 + 19.66 \times 1.5 + 10.72 \times 7$$

$$= 25.64 + 29.49 + 75.04$$

$$= 130.17 \text{ kN/m}^2$$

(*ii*) Effective stress at a depth of 10 m when the water table drops 6 m below the ground surface, assuming that the soil is dry above the water table,

$$\bar{\sigma} = 17.09 \times 6 + 10.72 \times 4$$

$$= 102.54 + 42.88$$

$$= \mathbf{145.42 \text{ kN/m}^2}$$

Change in effective stress = 145.42 − 130.17

$$= \mathbf{15.25 \text{ kN/m}^2}$$

Miscellaneous Illustrative Example 14. *A specimen of fine dry sand when tested in a triaxial compression test failed at a deviator stress of 300 kN/m². Compute the lateral pressure on the specimen in the test if the failure plane makes an angle of 70° with the horizontal.*

Solution. From Eq. 13.39,

$$(\overline{\sigma}_1 - \overline{\sigma}_3) = 2\,c'\cos\varphi' + (\overline{\sigma}_1 + \overline{\sigma}_3)\sin\varphi' \qquad \ldots(a)$$

From Eq. 13.43,

$$\theta_f = 45° + \frac{\phi'}{2}$$

or

$$70 = 45° + \frac{\phi'}{2}$$

or

$$\phi' = 50°$$

Substituting the values in Eq. (*a*), taking $c' = 0$,

$$(\overline{\sigma}_1 - \overline{\sigma}_3) = (\overline{\sigma}_1 + \overline{\sigma}_3)\sin 50°$$

or

$$300 = (\overline{\sigma}_1 + \overline{\sigma}_3) \times 0.766$$

or

$$\overline{\sigma}_1 + \overline{\sigma}_3 = 391.62 \qquad \ldots(b)$$

Also

$$\overline{\sigma}_1 - \overline{\sigma}_3 = 300 \qquad \ldots(c)$$

From Eqs. (*b*) and (*c*) by substration

$$2\,\overline{\sigma}_3 = 391.62 - 300 = 91.62$$

or

$$\overline{\sigma}_3 = \text{lateral pressure}$$

$$= \mathbf{45.81\ kN/m^2}$$

Miscellaneous Illustrative Example 15. *(a) A dry sand specimen failed in a triaxial test when the major and minor principal stresses were respectively 900 kN/m² and 250 kN/m². Determine the angle of shearing resistance of sand.*

(b) If the same specimen would have been tested in a direct shear test at a normal stress of 200 kN/m², what would have been the shear stress at failure?

Solution. (*a*) From Eq. 13.39, taking $c' = 0$,

$$\overline{\sigma}_1 - \overline{\sigma}_3 = (\overline{\sigma}_1 + \overline{\sigma}_3)\sin\varphi'$$

or

$$(900 - 250) = (900 + 250)\sin\varphi'$$

or

$$\phi' = 34.42°$$

In the direct shear test,

$$s = c' + \overline{\sigma}_n \tan\varphi'$$

or

$$s = 0 + 200\tan 34.42° = \mathbf{137.03\ kN/m^2}$$

Miscellaneous Illustrative Example 16. *In an unconfined compression test, a specimen of clay failed at an axial stress of 250 kN/m². If the failure plane was inclined at 50° to the horizontal, determine the parameters c and* ϕ.

Solution. From Eq. 13.40,

$$\overline{\sigma}_1 = \overline{\sigma}_3\tan^2\left(45 + \frac{\phi}{2}\right) + 2\,c\tan\left(45 + \frac{\phi}{2}\right) \qquad \ldots(a)$$

Now

$$\theta_f = 45° + \frac{\phi}{2}$$

or

$$50 = 45° + \frac{\phi}{2}$$

$$\phi = 10°$$

Substituting the values in Eq. (*a*), taking $\sigma_3 = 0$,

$$250 = 0 + 2\,c\tan(45° + 5°)$$

or $$c = \mathbf{104.89\ kN/m^2}$$

Miscellaneous Illustrative Example 17. *(a) A soil stratum has a natural void ratio of 0.80. Determine its relative density if the maximum and minimum void ratios are respectively 1.20 and 0.50.*

(b) If 2.0 m thickness of this stratum is densified to a relative density of 70%, what will be its reduced thickness?

Solution. (*a*) From Eq. 3.20, $D_r = \dfrac{e_{max} - e}{e_{max} - e_{min}} \times 100$

or $$\text{Relative Density} = \frac{1.20 - 0.80}{1.20 - 0.50} \times 100$$

$$= \mathbf{57.14\%}$$

(*b*) For a relative density of 70%, the void ratio e is obtained as

$$70 = \frac{1.20 - e}{1.20 - 0.50} \times 100$$

or $$e = 0.71$$

We know $$V = V_s + V_v$$

or total volume $$= V_s + eV_s = V_s(1 + e)$$

Let us consider a plan area of 1.0 m^2.

Total volume, $$V = 1.0 \times t$$

Therefore, $$t = V_s(1 + e)$$

As the void ratio changes from 0.8 to 0.71, the thickness t_2 is obtained as

$$\frac{t_1}{t_2} = \frac{V_s(1 + e_1)}{V_s(1 + e_2)}$$

$$\frac{2}{t_2} = \frac{1 + 0.80}{1 + 0.71}$$

$$t_2 = \mathbf{1.90\ m}$$

Miscellaneous Illustrative Example 18. *How many cubic metres of fill can be constructed at a void ratio of 0.60 from 1000 m^3 of borrow pit soil that has a void ratio of 1.00?*

Solution. $$e = \frac{\text{Volume of voids}}{\text{Volume of solids}} = \frac{V_v}{V_s}$$

or $$V_v = e\,V_s$$

Total volume, $$V = V_s + e\,V_s$$

or $$V = V_s(1 + e)$$

Using suffix '*f*' for fill and '*b*' for borrow pit,

$$\frac{V_f}{V_b} = \frac{V_s(1 + e_f)}{V_s(1 + e_b)}$$

or $$V_f = \frac{V_b(1 + e_f)}{(1 + e_b)}$$

In this case, $$V_f = \frac{1000\,(1 + 0.60)}{1 + 1.0}$$

$$= \mathbf{800\ m^3}$$

Miscellaneous Illustrative Example 19. *A normally consolidated deposit of clay extends to a depth of 20 m from the ground surface. The ground water table is 8 m below the ground surface. If the plasticity index of the clay is 70%, determine the unconfined compressive strength of the clay sample taken from a depth of 15 m from the ground surface.* $\gamma_{sat} = 19.5 \text{ kN/m}^3$, $\gamma_d = 14.0 \text{ kN/m}^3$.

Solution. The effective stress at a depth of 15 m,

$$\bar{\sigma} = 8 \times 14.0 + 7 \times 19.5 - 7 \times 9.81$$

$$= 112.0 + 136.5 - 68.67$$

$$= 179.83 \text{ kN/m}^2$$

From Eq. 13.24,

$$\frac{c_u}{\bar{\sigma}} = 0.11 + 0.0037\, PI$$

or
$$\frac{c_u}{179.83} = 0.11 + 0.0037 \times 70$$

$$c_u = 66.36 \text{ kN/m}^2$$

Unconfined compressive strength,

$$q_u = 2\, c_u$$

or
$$q_u = 2 \times 66.36 = \mathbf{132.72 \text{ kN/m}^2}$$

Miscellaneous Illustrative Example 20. *(a) A soil sample has a water content of 28.5% when fully saturated. Determine its void ratio, dry unit weight and bulk unit weight. (b) Determine the total volume of water required to saturate a volume of 100 m*^{3} *of the same soil.*

Take $G = 2.65$, $\gamma_w = 9.81 \text{ kN/m}^3$.

Solution. (*a*)
$$e = w\,G$$

$$= 0.285 \times 2.65 = 0.755$$

$$\gamma_d = \frac{G\,\gamma_w}{1 + e} = \frac{2.65 \times 9.81}{1 + 0.755}$$

or
$$\gamma_d = \mathbf{14.81 \text{ kN/m}^3}$$

$$\gamma = \frac{G + e}{1 + e}\gamma_w = \frac{(2.65 + 0.755)}{1 + 0.755} \times 9.81$$

$$= \frac{3.405 \times 9.81}{1.755}$$

or
$$\gamma = \mathbf{19.03 \text{ kN/m}^3}$$

(*b*) Total weight of soil $= \gamma \times V$

$= 19.03 \times 100 = 1903$ kN

Dry weight of soil $= 14.81 \times 100 = 1481$ kN

Weight of water required $= 1903 - 1481$

$=$ **422 kN**

Volume of water required $= \dfrac{422}{9.81}$

$=$ **43.02 m^3**

Miscellaneous Illustrative Example 21. *A rectangular footing 2 m × 3 m is founded at a depth of 1.5 below the ground surface in a cohesive soil* ($c = 50 \text{ kN/m}^2$, $\phi = 0.0$, $\gamma = 18.5 \text{ kN/m}^3$). *The water table is close* *the ground surface. Determine the gross safe bearing capacity using (i) Terzaghi's theory ($N_c = 5.7$, $N_q = 1$), (ii) Skempton's method.*

Take a factor of safety of 3.0.

Solution. (*i*) *Terzaghi's theory*

From Eq. 23.38 (*b*), taking $N_\gamma = 0.0$,

$$q_u = c\,N_c\left(1 + 0.2\,\frac{B}{L}\right) + \gamma D_f N_q$$

$$= 50 \times 5.7\left(1 + 0.2 \times \frac{2}{3}\right) + (18.5 - 9.81) \times 1.5 \times 1.0$$

$$= 323 + 13.04 = 336.04 \text{ kN/m}^2$$

$$q_{nu} = q_u - \gamma D_f$$

$$= 336.04 - (18.5 - 9.81) \times 1.5 = 323 \text{ kN/m}^2$$

Net safe bearing capacity, $q_{ns} = \dfrac{q_{nu}}{F}$

$$= \frac{323}{3.0} = 107.7 \text{ kN/m}^2$$

Gross safe bearing capacity, $q_s = q_{ns} + \gamma D_f$

$$= 107.7 + (18.5 - 9.81) \times 1.5 = \mathbf{120.74\ kN/m^2}$$

(*ii*) *Skempton's method*

From Eq. 23.53 (*a*), $N_c = 5.0\left(1 + 0.2\,\dfrac{D_f}{B}\right)\left(1 + 0.2\,\dfrac{B}{L}\right)$

$$= 5.0\left(1 + \frac{0.2 \times 1.5}{2}\right)\left(1 + 0.2 \times \frac{2}{3}\right)$$

$$= 6.51$$

From Eq. 23.55, $q_{nu} = c_u N_c$

$$= 50 \times 6.51 = 325.74 \text{ kN/m}^2$$

$$q_s = \frac{q_{nu}}{F} + \gamma D_f$$

$$= \frac{325.74}{3} + (18.5 - 9.81) \times 1.5$$

$$= 108.58 + 13.04 = \mathbf{121.62\ kN/m^2}$$

Miscellaneous Illustrative Example 22. *A square footing of width 1.5 m is founded at a depth of 1 m in a clay deposit. The water table is at a depth of 1.0 m below the ground surface. Determine (i) the net load which the footing can safely carry with a factor of safety of 3. (ii) The net load if the water table rises to the ground surface. Take $c_u = 25$ kN/m², $\phi = 0$, $\gamma = 20$ kN/m³.*

Solution. From Eq. 23.53 *a*, $N_c = 5.0\left(1 + 0.2\dfrac{D_f}{B}\right)\left(1 + 0.2\,\dfrac{B}{L}\right)$

$$N_c = 5.0\left(1 + \frac{0.2 \times 1}{1.5}\right)\left(1 + 0.2 \times \frac{1.5}{1.5}\right)$$

$$= 6.80$$

Net ultimate bearing capacity,

$$q_{nu} = c_u N_u$$

$$= 25 \times 6.80 = 170 \text{ kN/m}^2$$

Net safe bearing capacity,

$$q_{ns} = \frac{170}{3}$$

$$= 56.67 \text{ kN/m}^2$$

Net safe load $= 56.67 \times (1.5)^2 = \mathbf{127.5\ kN}$

(*ii*) If there is no change in the value of c_u, the net safe load will remain unchanged.

Miscellaneous Illustrative Example 23. *A square footing on sand failed by general shear under an ultimate load of 6000 kN. The depth of footing was 2.5 m below the ground surface and the water table was at a large depth. Determine the size of the footing.* $\phi = 35°$, $\gamma = 18.5$ *kN/m*^{3}. *For 35°,* $N_q = 41.4$, $N_\gamma = 42.4$.

Solution. From Eq. 23.37,

$$q_u = 1.2\,c' + \gamma D_f N_q + 0.4\,\gamma B N_\gamma$$

or

$$q_u = 0 + 18.5 \times 2.5 \times 41.4 + 0.4 \times 18.5 \times B \times 42.4$$

$$= 1914.75 + 313.76\,B$$

But

$$q_u = \frac{6000}{B^2}$$

where B is the size of footing.

Therefore,

$$\frac{6000}{B^2} = 1914.75 + 313.76\,B$$

or

$$313.76\,B^3 + 1914.75\,B^2 - 6000 = 0$$

$$B^3 + 6.103\,B^2 - 19.12 = 0$$

Solving by trial and error,

$$B = \mathbf{1.58\ m}$$

Miscellaneous Illustrative Example 24. *A circular pier of 2.5 m diameter carries a gross load of 2000 kN. The supporting soil is clayey sand (*$c = 4$ *kN/m*^{2}, $\phi = 30°$, $\gamma = 19$ *kN/m*^{3}*). Determine the depth at which the pier should be located to provide a factor of safety of 3. Use Terzaghi's theory.* $N_c = 37.2$, $N_q = 22.5$ *and* $N_\gamma = 19.7$ *for* $\varphi = 30°$.

Solution. The gross safe bearing capacity (Eq. 23.3),

$$q_s = \frac{q_{nu}}{F} + \gamma D_f$$

From Eq. 23.38,

$$q_u = 1.2\,c\,N_c + \gamma D_f N_q + 0.3\,\gamma B N_\gamma$$

$$= 1.2 \times 4 \times 37.2 + 19 \times D_f \times 22.5 + 0.3 \times 19 \times 2.5 \times 19.7$$

$$= 178.56 + 427.5\,D_f + 280.73$$

$$= 459.29 + 427.5\,D_f$$

$$q_{nu} = q_u - \gamma D_f$$

$$= 459.29 + 427.5\,D_f - 19\,D_f$$

$$= 459.29 + 408.5\,D_f$$

Therefore, $$q_s = \frac{459.29 + 408.5\,D_f}{3} + 19 \times D_f$$

or $$q_s = \frac{459.29 + 465.5\,D_f}{3} \qquad ...(a)$$

But $$q_s = \frac{2000}{\pi/4 \times (2.5)^2} = 407.44 \qquad ...(b)$$

From Eqs. (*a*) and (*b*),

$$\frac{459.29 + 465.5\,D_f}{3} = 407.44$$

or $$459.29 + 465.5\,D_f = 1222.32$$

or $$465.5\,D_f = 763.03$$

or $$D_f = \mathbf{1.64\ m}$$

Miscellaneous Illustrative Example 25. *A rectangular footing is founded at a depth of 1.5 m below the ground surface. If the width of the footing is 2.5 m, determine the length of the footing required to carry a gross allowable pressure of 470 kN/m² with a factor of safety of 3. Use Terzaghi's theory. Take* $c = 20$ *kN/m²,* $\phi = 30°$. $\gamma_{sat} = 20$ *kN/m². The water table is close to the ground surface.* $N_c = 37.2$, $N_q = 22.5$, $N_\gamma = 19.7$.

Solution. From Eq. 23.38 (*a*),

$$q_u = c\,N_c\left(1 + 0.2\frac{B}{L}\right) + \gamma D_f N_q + \frac{1}{2}\gamma B N_\gamma\left(1 - \frac{0.2\,B}{L}\right)$$

$$= 20 \times 37.2\left(1 + \frac{0.2 \times 2.5}{L}\right) + (20 - 9.81) \times 1.5 \times 22.5$$

$$+ \frac{1}{2} \times (20 - 9.81) \times 2.5 \times 19.7 \times \left(1 - \frac{0.2 \times 2.5}{L}\right)$$

$$q_u = 744\left(1 + \frac{0.5}{L}\right) + 343.91 + 250.93\left(1 - \frac{0.5}{L}\right)$$

$$= 1338.84 + 246.53/L \qquad ...(a)$$

But $$q_u = 470 \times 3 = 1410 \qquad ...(b)$$

Therefore, from Eqs. (*a*) and (*b*),

$$1338.84 + \frac{246.53}{L} = 1410$$

or $$L = \mathbf{3.46\ m}$$

Miscellaneous Illustrative Example 26. *The SPT no. (N) for a sandy soil at an average depth of 2 m was 20. The water table was at a depth of 4.0 m below the ground surface. Determine the allowable gross bearing pressure for a 2 m square footing at a depth of 1.5 m. Take a factor of safety of 3.0 against shear and the limiting total settlement of 25 mm. Take* $c = 0$, $\phi = 30°$, $\gamma = 19$ *kN/m²,* $N_q = 22.5$, $N_\gamma = 19.7$.

Solution. Net safe bearing pressure,

$$q_{ns} = \frac{1}{F}[1.2\,c\,N_c + \gamma D_f (N_q - 1) + 0.4\,\gamma B N_\gamma]$$

$$= \frac{1}{3}\,[0 + 19 \times 1.5\,(22.5 - 1) + 0.4 \times 19 \times 2 \times 19.7]$$

$$= \frac{1}{3}\,[612.75 + 299.44] = 304.06\ \text{kN/m}^2$$

Net safe settlement pressure (Eq. 23.77),

$$q_{np} = 0.41\, c_w N s$$

where

$$c_w = 0.5 + \frac{0.5\, D_w}{D_f + B}$$

$$= 0.5 + \frac{0.5 \times 4}{1.5 + 2} = 1.071$$

Taking $s = 25$ mm, $N = 20$,

$$q_{np} = 0.41 \times 1.071 \times 20 \times 25$$

$$= 219.56 \text{ kN/m}^2$$

Since $q_{np} < q_{ns}$, net allowable pressure = 219.56 kN/m².

Gross allowable pressure = 219.56 + 19 × 1.5

$$= \mathbf{248.06\ kN/m^2}$$

Miscellaneous Illustrative Example 27. *A precast concrete pile of size 40 cm × 40 cm is driven into stiff clay (c_u = 80 kN/m²). Determine the length of pile required to carry a safe load of 300 kN with a factor of safety of 2.5. Take adhesion factor as 0.50.*

Solution. Let the length of pile be D.

Now $$Q_u = c\, N_c A_p + \alpha\, c D \times \text{perimeter of pile}$$

$$= 80 \times 9 \times (0.4)^2 + 0.5 \times 80 \times D \times (4 \times 0.40)$$

or $$Q_u = 115.2 + 64\, D \qquad \ldots(a)$$

But $$Q_u = 300 \times 2.5 = 750 \text{ kN} \qquad \ldots(b)$$

From Eqs. (a) and (b),

$$115.2 + 64\, D = 750$$

or $$D = \mathbf{9.92\ m}$$

Miscellaneous Illustrative Example 28. *A square footing is required to carry a net load of 1200 kN. Determine the size of the footing if the depth of foundation is 2 m and tolerable settlement is 40 mm. The soil is sandy with N = 12, FS = 3. Water table is very deep. Use Teng's equation.*

Solution. From Eq. 23.61, net ultimate bearing capacity,

$$q_{nu} = 0.33\, N^2 B\, W_\gamma + 1.0\,(100 + N^2)\, D_f\, W_q$$

$$= 0.33 \times (12)^2\, B \times 1.0 + 1.0\,[100 + (12)^2] \times 2 \times 1.0$$

$$= 47.52\, B + 488$$

Total net load, $$Q_{nu} = 1200 \text{ kN}$$

Therefore, $$1200 = \frac{(47.52\, B + 488)}{3} \times B^2$$

$$1200 = 15.84\, B^3 + 162.67\, B^2$$

Solving by trial and error, B = **2.45 m**

Settlement consideration

From Eq. 23.80 (b),

safe settlement pressure,

$$q_{np} = 1.40\,(N - 3)\left(\frac{B + 0.3}{2B}\right)^2 W_\gamma R_d\, s$$

where $W_\gamma = 1.0$, and $R_d = 1 + \dfrac{D_f}{B} \leq 2.0$

Therefore, $$q_{np} = 1.40\,(12-3)\left(\frac{B+0.3}{2B}\right)^2 \times 1.0\left(1+\frac{2}{B}\right) \times 40$$

$$= 12.6\left(\frac{B+0.3}{2\,B}\right)^2 \times \left(1+\frac{2}{B}\right) \times 40$$

Total net load, $$Q_n = q_{np} \times B^2$$

or $$1200 = (12.6)\left(\frac{B+0.3}{2\,B}\right)\left(1+\frac{2}{B}\right)B^2 \times 40 \qquad \ldots(a)$$

$$2.381 = \left(\frac{B+0.3}{2\,B}\right)^2 \times \left(1+\frac{2}{B}\right) \times B^2$$

$$9.524 = (B+0.3)^2 \times \left(1+\frac{2}{B}\right)$$

Solving by trial and error,

$$B = 1.84 \text{ m.}$$

Therefore, $$R_d = \left(1+\frac{2}{1.84}\right) = 2.087 > 2$$

But maximum value of $R_d = 2.0$

Therefore, Eq. (*a*) is modified as

$$1200 = 12.6 \times \left(\frac{B+0.3}{2\,B}\right)^2 \times 2.0 \times B^2 \times 40$$

$$4.762 = (B+0.3)^2$$

$$B = 1.88 \text{ m} < 2.45 \text{ m}$$

Adopt larger of the widths of 1.88 m and 2.45 m.

Therefore, B = **2.45 m**

Miscellaneous Illustrative Example 29. *A reinforced concrete pile 40 cm × 40 cm size and 12 m length is driven into coarse sand (γ = 18 kN/m³). If the average N value is 12, determine the allowable load on the pile (F = 2.5). Use (a) static formula, (b) formula based on N-value, K = 1.2, F = 4.0, $\delta = 0.75\,\phi$. The ground water table is close to the ground.*

Solution. (*a*) From Fig. 17.7 for N = 12, the value of $\phi = 30.5°$.

From Fig. 25.4, for $\phi' = 30.5°$, the value of N_q is 25.

From Eq. 25.10 (*a*), the ultimate load by static formula,

$$Q_u = \bar{q}\,N_q A_p + K \tan\delta \times p \times D\,(\bar{q}/2)$$

$$= (18-9.81) \times 12 \times (0.4 \times 0.4) \times 25$$

$$+\ 1.2 \times \tan(0.75 \times 30.5°)\ (4 \times 0.4) \times 12 \times \frac{(18-9.81) \times 12}{2}$$

$$= 393.12 + 477.67$$

$$= 870.79$$

$$Q_a = \frac{870.79}{2.5} = \textbf{348.32 kN}$$

(*b*) From Eq. 25.29,

$$q_p = 40\,N\,(D/B) \le 400\,N$$
$$= 40 \times 12 \times (12/0.4)$$
$$= 14400$$
$$400\,N = 400 \times 12 = 4800 < 14400$$

Take $q_p = 4800$ kN/m^2.

From Eq. 25.30 (*a*), $f_s = 2.0\,\overline{N}$

$$= 2.0 \times 12 = 24 \text{ kN/m}^2$$

Therefore,
$$Q_u = q_p\,A_p + f_s \times D \times \text{pile perimeter}$$
$$= 4800 \times (0.4 \times 0.4) + 24 \times 12 \times (4 \times 0.4)$$
$$= 768 + 460.8 = 1228.8$$
$$Q_a = \frac{Q_u}{4.0} = \frac{1228.8}{4}$$
$$= \mathbf{307.2\ kN}$$

Miscellaneous Illustrative Example 30. *The following observations were obtained for determining the bulk density and dry density of an in-situ soil by sand replacement method.*

Initial mass of the cylinder and sand = 3.000 kg

Final mass of cylinder after pouring sand into the hole and cone = 1.254 kg

Mass of sand required to fill only the cone = 0.310 kg

Mass of soil taken out from the hole = 1.665 kg

Mass density of sand = 1539 kg/m^3

Water content of soil = 14.5%.

Solution. Mass of sand in the hole and cone = 3.000 − 1.254

= 1.746 kg

Mass of sand in the hole = 1.746 − 0.310

= 1.436 kg

Volume of sand in the hole $= \dfrac{1.436}{1539}$

or Volume of hole $= 9.331 \times 10^{-4}$ m^3

Volume of excavated soil $= 9.331 \times 10^{-4}$ m^3

Bulk density of excavated soil $= \dfrac{1.665}{9.331 \times 10^{-4}}$

or $\rho = \mathbf{1784.43\ kg/m^3}$

Dry density $= \dfrac{1784.43}{1 + 0.145}$

$= \mathbf{1558.45\ kg/m^3}$

Miscellaneous Illustrative Example 31. *A raft foundation is 20 m × 30 m. The raft is constructed over a soft clay stratum having c_u = 10 kN/m^2 and γ = 19 kN/m^3. If the live load and dead load on the raft are 100 MN, find the depth of foundation. The foundation is fully compensated.*

(*b*) *Determine the depth of foundation for a factor of safety of 3.*

Solution. (*a*) From Eq. 24.54, for a fully compensated foundation,

$$D_f = \frac{Q}{\gamma A}$$

$$= \frac{100 \times 10^3}{19 \times (20 \times 30)} = \mathbf{8.77\ m}$$

(*b*) From Eq. 24.53, the factor of safety is given by

$$F = \frac{5\,(1 + 0.2\,D_f/B)\,(1 + 0.2\,B/L)\,c_u}{Q/A - \gamma D_f}$$

$$3.0 = \frac{5\,(1 + 0.2\,D_f/20) + \left(1 + 0.2 \times \dfrac{20}{30}\right) \times 10}{\dfrac{100 \times 10^3}{20 \times 30} - 19 \times D_f}$$

or $$500 - 57\,D_f = 5\,(1 + 0.2\,D_f/20) + 11.33$$

$$= 5 + 0.05\,D_f + 11.33$$

or $$57.05\,D_f = 483.67$$

$$D_f = \mathbf{8.47\ m}$$

Miscellaneous Illustrative Example 32. *Determine the net allowable load on a raft foundation with dimensions of 20 m × 15 m constructed over a sand deposit (N = 12). Assume that the depth of foundation is 2 m and the allowable settlement is 25 mm. The water table is at a depth of 3 m below ground surface.*

Solution. From Eq. 23.82 (*b*),

The allowable pressure, $$q_{np} = 12.2\,N \left(\frac{B + 0.3}{B}\right)^2 W_\gamma\, R_d$$

Where Eq. 25.36 (*c*) $$W_\gamma = 0.5 + 0.5b/B \le 1$$

$$= 0.5 + \frac{0.5 \times 1.0}{15}$$

$$= 0.533$$

$$R_d = 1 + \frac{0.33\,D_f}{B} = 1 + \frac{0.33 \times 2}{15}$$

$$= 1.044$$

Therefore, $$q_{np} = 12.2 \times 12 \times \left(\frac{15 + 0.3}{15}\right)^2 \times 0.533 \times 1.044$$

$$= 84.76\ \text{kN/m}^2$$

Net allowable load $$= q_{np} \times A$$

$$= 84.76 \times (20 \times 15)$$

$$= \mathbf{25428\ kN}$$

Miscellaneous Illustrative Example 33. *A clay stratum of 10 m thickness lies on bed rock. Because of construction of a building on it, the average effective stress increases from 80 kN/m² to 200 kN/m².*

A consolidation test conducted an on undisturbed sample of the clay gave the following results:

$$e_1 = 0.852, \quad p_1 = 80\ \text{kN/m}^2$$

$$e_2 = 0.741, \quad p_2 = 200\ \text{kN/m}^2$$

Compute the settlement after 10 years of construction. The average coefficient of permeability of the clay is 3.0×10^{-8} mm/s.

Solution. Coefficient of compressibility, $a_v = \dfrac{-\Delta e}{\Delta \bar{\sigma}}$

or $$a_v = \frac{0.852 - 0.741}{200 - 80}$$

$$= \frac{0.111}{120} = 9.250 \times 10^{-4}$$

Coefficient of volume change, $m_v = \dfrac{a_v}{1 + e_0}$

or $$m_v = \frac{9.25 \times 10^{-4}}{1 + 0.852}$$

$$= 4.99 \times 10^{-4} \text{ m}^2/\text{kN}$$

Compression index, $C_c = \dfrac{-\Delta e}{\log (200/80)} = \dfrac{0.111}{0.398}$

or $C_c = 0.279$

Coefficient of consolidation, $c_v = \dfrac{k}{\gamma_w m_v}$

or $$c_v = \frac{3.0 \times 10^{-11}}{9.81 \times 4.99 \times 10^{-4}} = 6.128 \times 10^{-9} \text{ m}^2/\text{s}$$

$$= 6.128 \times 10^{-9} \times 10^4 \times 60 \text{ cm}^2/\text{min}$$

$$= 3.677 \times 10^{-3} \text{ cm}^2/\text{min.}$$

Final settlement, $$s_f = \frac{C_c}{1 + e_0} \times H_0 \times \log \left(\frac{\bar{\sigma}_0 + \Delta \bar{\sigma}}{\bar{\sigma}_0} \right)$$

$$s_f = \frac{0.279}{1 + 0.852} \times 10 \times \log_{10} \left(\frac{200}{80} \right)$$

$$= 0.599 \text{ m}$$

Time factor (T_v) after 10 years,

$$T_v = \frac{c_v t}{d^2} = \frac{3.677 \times 10^{-3} \times 10 \times 365 \times 24 \times 60}{(1000)^2}$$

or $T_v = 0.0193$

For $T_v = 0.0193$, from Table 12.3 by interpolation, we have

$$U = 0.156$$

Alternatively, from Eq. 12.45, $U = \sqrt{\dfrac{4\,T_v}{\pi}}$

$$U = \sqrt{\frac{4 \times 0.0193}{\pi}} = 0.156$$

Settlement after 10 years $= 0.156 \times 0.599 =$ **0.0934 m**

Miscellaneous Illustrative Example 34. *For a normally consolidated clay, the results of the drained triaxial test are as under:*

Chamber pressure = *100 kN/m*^{2}

Deviator stress at failure = *200 kN/m*^{2}

(*a*) *Determine the angle of shearing resistance* ϕ'.

(*b*) *Determine the angle which the failure plane makes with the major principal plane.*

(*c*) *Find the normal stress and the shear stress on the failure plane.*

(*d*) *Find the normal stress and the shear stress on the plane of maximum shear stress.*

Solution. (Refer Fig. 13.35, page 337) $\sigma_3 = 100\ \text{kN/m}^2, \sigma_1 = 200 + 100 = 300\ \text{kN/m}^2$

(*a*)

$$\sin \varphi' = \frac{\left(\dfrac{\sigma'_1 - \sigma'_3}{2}\right)}{\left(\dfrac{\sigma'_1 + \sigma'_3}{2}\right)} = \frac{\left(\dfrac{200}{2}\right)}{\left(\dfrac{300 + 100}{2}\right)}$$

$$\varphi' = 30°$$

(*b*)

$$\theta_f = 45° + \frac{\phi}{2}$$

$$= 45° + \frac{30}{2} = 60°$$

(*c*) The failure plane makes an angle of 60° with the major principal plane (horizontal). From Eq. 13.2, the normal stress on the failure plane

$$\sigma = \left(\frac{\sigma_1 + \sigma_3}{2}\right) + \left(\frac{\sigma_1 - \sigma_3}{2}\right) \cos 2\theta$$

$$= \frac{(300 + 100)}{2} + \frac{(300 - 100)}{2} \times \cos 120°$$

$$= 200 + (-50) = \mathbf{150\ kN/m^2}$$

From Eq. 13.1,

$$\tau = \frac{1}{2}(\sigma_1 - \sigma_3) \sin 2\theta$$

$$= \frac{1}{2}(200) \sin 120°$$

$$= \mathbf{86.60\ kN/m^2}$$

(*d*) The maximum shear stress occurs on a plane making an angle of 45° with the horizontal.

$$\sigma = \left(\frac{300 + 100}{2}\right) + \left(\frac{300 - 100}{2}\right) \cos 90°$$

$$= 200\ \text{kN/m}^2$$

$$\tau = \left(\frac{200}{2}\right) \sin 90° = \mathbf{100\ kN/m^2}$$

Miscellaneous Illustrative Example 35. *Two soils A and B are tested in a laboratory for the consistency tests and the following results were obtained:*

	Soil A	*Soil B*
Liquid limit	35%	65%
Plastic limit	15%	20%
Flow Index	8	4
Water content	30%	30%

Which of the two soils is (a) more plastic, (b) better foundation material, (c) of better strength as a function of water, (d) of better strength at the plastic limit, (d) more compressible?

Solution. (*a*) Plasticity index of soil A = 35 – 15 = 20%

Plasticity index of soil B = 65 – 20 = 45%

Since the plasticity index of soil B is greater than that of soil A, it is more plastic.

(*b*) Consistency index of soil $A = \dfrac{w_L - w}{I_p}$

$$= \frac{35 - 30}{20} \times 100 = 25\%$$

Consistency index of soil $B = \dfrac{65 - 30}{45} \times 100 = 78\%$

Since the consistency index of soil B is greater than that of soil A, it is more stiff and strong. Therefore, it is better foundation material.

(*c*) Since flow index of soil B is smaller than that of soil A, it has a better strength as a function of time.

(*d*) Toughness index of soil $A = \dfrac{I_p}{I_f} = \dfrac{20}{8} = 2.5$

Toughness index of soil $B = \dfrac{45}{4} = 11.25$

Soil B has better shear strength at the plastic limit, because its toughness index is greater than that of soil A.

(*e*) For an undisturbed soil, $C_c = 0.009\,(w_L - 10)$

Compression index of soil $A = 0.009\,(20 - 10)$

or $C_c = 0.09$

Compression index of soil $B = 0.009\,(45 - 10)$

or $C_c = 0.315$

Soil B is more compressible because its compression index is greater than that of soil A.

Miscellaneous Illustrative Example 36. (*a*) *Determine the shear strength of the soil at the base of an embankment just after the height of fill is raised from 5 m to 7 m. Assume that the lateral pressure at any point is one-half of the vertical pressure. The dissipation of pore water pressure is negligible. Assume pore pressure, parameters A = 0.4, B = 0.80, c′ = 40 kN/m², φ′ = 20°, γ = 16 kN/m³.*

(*b*) *Also determine the shear strength after a long time.*

Solution. From Eq. 13.35, pore pressure is given by

$$\Delta_u = B\,[\Delta\sigma_3 + A\,(\Delta\sigma_1 - \Delta\sigma_3)]$$

$$\Delta\sigma_1 = \text{increase in vertical stress}$$

$$= \gamma z = 16 \times 2 = 32 \text{ kN/m}^2$$

$$\Delta\sigma_3 = 0.5\,\Delta\sigma_1 = 0.5 \times 32 = 16 \text{ kN/m}^2$$

Therefore, $\Delta u = 0.8\,[16 + 0.40\,(32 - 16)]$

$$= 17.92 \text{ kN/m}^2$$

Now $\bar{\sigma} = (\sigma_1 + \Delta\sigma_1) - \Delta u$

$$= (5 \times 16 + 2 \times 16) - 17.92$$

$$= 112 - 17.92 = 94.08 \text{ kN/m}^2$$

Shear strength, $s = c' + \bar{\sigma} \tan \varphi'$

$$= 40 + (94.08) \tan 20° = \mathbf{74.24\ kN/m^2}$$

(*b*) After a long time, the whole of pore pressure would have been dissipated. Therefore, $u = 0$

$$\bar{\sigma} = \sigma = 7 \times 16 = 112\ kN/m^2$$

$$s = 40 + 112 \tan 20° = \mathbf{80.76\ kN/m^2}$$

Miscellaneous Illustrative Example 37. *(a) A dry sample of clay has a unit weight of 16.5 kN/m³ and the specific gravity of 2.70. What is the shrinkage limit of the soil? (b) If 1.0 m³ of above clay is allowed to get gradually saturated, what will be the quantity of water at shrinkage limit and what will be the total volume of soil? (c) Find the total volumes at plastic limit of 30% and at the liquid limit of 55%.*

Solution. (*a*) From Eq. 4.9, the shrinkage limit,

$$w_s = \left(\frac{\rho_w}{\rho_d} - \frac{1}{G}\right)$$

$$= \left(\frac{\gamma_w}{\gamma_d} - \frac{1}{G}\right)$$

$$= \left(\frac{9.81}{16.5} - \frac{1}{2.70}\right)$$

$$= 0.594 - 0.370 = 0.224$$

$$w_s = \mathbf{22.4\%}$$

(*b*) Dry weight, $W_s = 16.5\ kN/m^3$

Weight of water at shrinkage limit $= W_s \times w_s$

$$= 16.5 \times 0.224 = \mathbf{3.696\ kN}$$

Total volume of soil does not change for the water content upto the shrinkage limit. Therefore, total volume = 1.0 m³.

(*c*) Weight of water at the plastic limit,

$$W_w = W_s \times w_p$$

$$= 16.5 \times 0.30 = 4.95\ kN$$

Increase in weight of water from w_s to w_p

$$= 4.95 - 3.696 = 1.254\ kN$$

Volume at w_p = volume at shrinkage limit + Increase in weight from w_s to w_p

$$= 1.0 + \frac{1.254}{9.81} = \mathbf{1.128\ m^3}$$

Weight of water at w_L $= 16.5 \times 0.55 = 9.075\ kN$

Increase in weight of water from w_s to w_L $= 9.075 - 3.696$

$$= 5.379\ kN$$

Volume at $w_L = 1.0 + \dfrac{5.379}{9.81} = \mathbf{1.548\ m^3}$

Miscellaneous Illustrative Example 38. *Classify the following soils as per IS classification system.*

(*a*) ***Soil A**: Coarse fraction* = 96%

Coarse fraction larger than 4.75 mm IS sieve = 70%

Fine fraction = 4%

Coefficient of curvature = 2.5

Uniformity coefficient = *10.5*

(*b*) **Soil B:** *Fine fraction* (< 75 μ *size*) = *100%*

Liquid limit = *30%*

Plasticity index = *6%*

Liquid limit when dried = *20%*

General characteristics: dark brown colour, organic odour, weak reaction with hydrochloric acid.

Solution. (*a*) **Soil A**. The soil is coarse grained because it has a coarse fraction greater than 50%.

Since the coarse fraction larger than 4.75 mm size is greater than 50%, the soil is gravel.

Because $C_u > 4$ and C_c between 1 and 3, the gravel is well-graded **(GW).**

(*b*) **Soil B:** The soil is fine grained. From plasticity chart, for the given values of w_L and I_p, it is found that the soil is either *ML* or *OL*.

Because its characteristics are those of an organic soil, it is OL *i.e.* organic silt with low plasticity.

Miscellaneous Illustrative Example 39. (*a*) *Find the safe depth of trench that can be excavated in a stiff clay deposit of 9 m thick underlain by sand if the depth of water table is 3.0 m below ground surface. The saturated unit of clay is 19 kN/m*^{3}.

(*b*) *How much should the water table be covered to carry excavation to a depth of 7 m?*

Assume a factor of safety of unity.

Solution. (*a*) Let z be the safe depth of excavation.

Depth of clay below excavation = $(9 - z)$ metres.

Let us consider a point *P* on the surface of the sand just below the trench (Eq. *a*). Equating the downward force to the uplift,

$$(9 - z) \times 19 = 6 \times 9.81$$

or $$z = \mathbf{5.90\ m}$$

(*b*) For a depth of trench of 7 m, let H be the height of water table above P (Fig. *b*). Therefore,

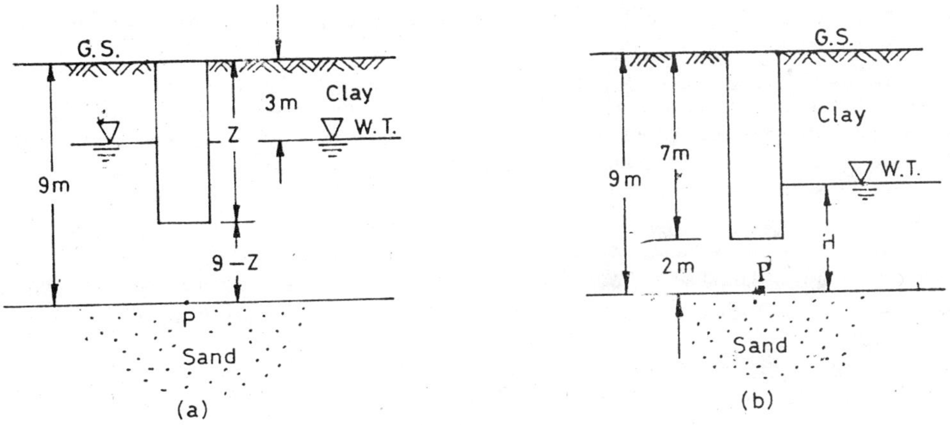

Fig. Miscellaneous Example 39

$$(9 - 7) \times 19 = H \times 9.81$$

or $$H = 3.87 \text{ m}$$

Lowering of the water table required = 6 − 3.87 = **2.13 m**

APPENDIX A
GLOSSARY OF COMMON TERMS

Adsorbed water. It is water bound to clay particles because of the attraction between electrical charges existing on the clay particles and water molecules (dipoles).

Air content. It is the ratio of the volume of air to the volume of voids in soil.

Alluvial soils. These are soils deposited by water. Deposits made in lakes are called lacustrine deposits and those in sea (or ocean) called marine deposits.

Allowable bearing pressure. It is the net allowable bearing pressure which can be used for the design of foundation. It is the smaller of the net safe bearing capacity and the net safe settlement pressure.

Active pressure. It is the pressure developed when the soil mass stretches due to movement of a retaining wall away from the soil.

Aquifer. An aquifer is a pervious stratum which contains water that can be easily drained or pumped out. An aquifer is called an unconfined aquifer when there is an impervious stratum only below it and a confined aquifer when it is sandwiched between two impervious strata.

Arching. It is a phenomenon is which the stresses are transferred from a yielding part of a soil mass to an adjacent non-yielding (or less yielding) part of the soil mass.

At-rest pressure. It is the lateral pressure in a soil mass when there is no movement of the mass.

Atterberg Limits (Consistency limits). The liquid limit, plastic limit and shrinkage limits are known as Atterberg's limits. The water content at which the soil behaviour changes from the liquid to the plastic state is called the liquid limit; from the plastic to the semi-solid state is the plastic limit; and from the semi-solid to the solid state is the shrinkage limit.

Backfill. It is the soil material which is placed into an area that has been excavated, such as against retaining walls and in pipe trenches.

Bearing Capacity. (Ultimate bearing capacity). It is the pressure at the base of the foundation at which the soil below fails in shear. It is called the gross ultimate bearing capacity when the gross pressure is considered and the net ultimate bearing capacity when the net increase in pressure over the existing overburden pressure is considered. The safe bearing capacity is the maximum pressure which the soil in the foundation can carry safely. The safe bearing capacity can be expressed as gross safe bearing capacity or net safe bearing capacity.

Boring. It is the method of investigating subsurface conditions by drilling a hole into the earth. Generally, soil samples are also extracted from the boring for determination of the index and engineering properties.

Borrow. It is soil (or rock) material obtained from another off- site source for use as fill at construction projects.

Braced cut. This is an excavation which is laterally supported. The vertical sides of excavation are supported by sheeting and bracing system.

Bulkheads. These consist of sheet-pile walls constructed to retain earth. These are relatively flexible retaining walls constructed for water front structures, canal locks, coffer dams, etc.

Bulking of sand. The phenomenon of increase in volume of sand (or a cohesionless soil) due to dampness is called the bulking of sand. The effect is predominant when the water content is between 4 and 5%. The increase in volume may be upto 20 to 25%. If the water content is increased, and the sand becomes saturated, the volume of sand mass is decreased.

Caisson. It is a type of foundation in which a large chamber (or box) is built above the ground level and then sunk to the required depth of a foundation as a single unit. The caisson may be an open caisson, a pneumatic caisson or a floating caisson. Open caissons are also known as well foundations.

Capillarity. It is the movement of water due to surface tension and other effects but not the gravity effect. Water moves in very small channels because of the affinity between soil and water.

Chemical weathering. It is the process of weathering in which chemical reactions, such as hydration, oxidation, solution, occur. When chemical weathering or chemical decomposition occurs, original rock minerals are transformed into new minerals by chemical reaction. Clay minerals are formed by chemical weathering.

Clays (clay minerals). These are very small particles (usually smaller than 2μ) which have a crystalline structure developed as the result of the chemical weathering of rocks. The clay particles are flat or plate-like in shape. These are highly surface-active particles.

Cohesion. It is the attraction or bonding force between the particles of fine-grained soils that creates shear strength.

Compaction. It is the process of increasing the density (or unit weight) of a soil by rolling, tamping, vibrating, or other mechanical means.

Consistency. The consistency of a fine-grained soil is the physical state is which it exists. It is indicated by such terms as soft, firm or hard, depending upon the degree of firmness.

Conduit. It is a pipe that is usually buried in a soil mass, or which passes through a soil embankment, and carries water, electrical cables, telephone cables, etc.

Consolidation. The compression of a saturated soil under a steady-state pressure is known as consolidation. It is due to expulsion of water from the voids.

Initially, the stress imparted into the soil is carried by water. The water is gradually forced out and the stress is transferred to the soil skeleton and the compression occurs.

Critical void ratio. The void ratio of the soil at which no change in volume occurs when the soil is subjected to shear strain in a drained test.

Density. The mass per unit volume of soil is called the density of soil. (Sometimes, the weight per unit volume, which is the unit weight, is called density).

Dewatering. The process of removing water from a construction area is known as dewatering. The term dewatering is also used for lowering the water table to obtain a dry area in the vicinity of the excavation.

Deep foundation. It is the type of foundation which transfers the load to deep strata below the ground surface. The common types are piles, caissons, drilled piers, etc. Generally, the foundation is called deep foundation if the depth of foundation is greater than the width of footing.

Dispersive clays. These are types of clays which deflocculate in still water and erode when exposed to a low- velocity flow of water. Dispersivity is due to a high concentration of sodium ions in a clay-pore water system.

Ditch conduits. These are types of conduits which are installed in narrow trenches (or ditches) and subsequently backfilled with soil.

Drawdown. As soon as the pumping is done from a well, the water table is lowered in its vicinity. This drop in water level in the well is called drawdown.

Drilled pier. It is a type of deep foundation in which a large diameter hole is drilled in the ground and subsequently filled with concrete.

Dynamic compaction. It is a method of compacting surface and near-surface zones of soil or fill by dropping a heavy weight from a relatively great height. Multiple poundings are usually done at each location.

Earth pressure. It is the lateral pressure exerted by a soil mass against an earth-retaining structure (or on a fictitious vertical plane located within a soil mass).

Depending upon the movement of the earth-retaining structure, the pressure may be active, passive or at-rest. When the structure moves away from the soil mass, it is active pressure; and when towards the soil mass, it is passive pressure. At-rest pressure acts when there is no movement of structure at all.

The coefficient of earth pressure is the ratio of lateral pressure to vertical pressure existing at a point in the soil mass.

Effective size. It is the size of particle in a soil specimen such that 10 percent of the particles are finer than this size. It is also called the *effective* diameter.

Effective stress. It is the nominal stress transmitted through the particle to particle contact in soil. It is equal to the sum of all the normal components of load divided by the *total* area of cross-section.

The effective stress controls the shear strength and compressibility of the soil.

It is an abstract quantity which is obtained by subtracting the pore water pressure from the total stress.

Electro-osmosis. It is a method of drainage of cohesive soils in which a direct current (DC) is used. Pore water migrates to the cathode, which is usually a well-point. Electro-osmosis also helps in increasing the shear strength of the cohesive soil.

Expansive clays. These are types of clays which show a large volume expansion in the presence of water and a large volume decrease upon drying. The clays contain the mineral montmorillonite.

These are highly difficult soils to work with.

Fill. It is earth placed in an excavation or other area to raise the elevation. It is also called earth fill or soil fill. If fill is required to support structural loading, it is placed in layers and compacted at a suitable water content to achieve a uniform and dense soil mass.

Fines or fine-grained soils. The silt-size and clay-size particles in a soil mass are called fines or fine-grained soils.

Flow net. A flow net consists of two sets of mutually orthogonal lines called the flow lines and equipotential lines. The flow lines indicate the paths of travel followed by the moving water and the equipotential lines indicate the points of equal potential (head).

A flow net is a pictorial representation used to study the flow of water through soil mass.

Footing. It is an enlargement of the base of a column or a wall to spread the load on a large area of the stratum below. It is a type of foundation installed at a shallow depth.

Friction, internal. It is the friction developed in a soil due to soil-to-solid contact. It is responsible for most of the shear strength developed in a soil, especially in a cohesionless soil.

The angle of internal friction (ϕ) is used to represent the internal friction of the soil.

Frost boil. It is a phenomenon which occurs when a frozen soil mass thaws and water is liberated. The strength of soil is reduced due to the softening caused by an increase in water content.

Frost heave. It is the rise in ground surface due to the formation of ice lenses when the temperature falls to the freezing point of water. It may cause lifting of light structure built on the ground.

Geofabrics. These are built of synthetic fibers and are used as filters, drains, or reinforcement in earthwork.

Geostatic stress. The stress at a point due to self weight of the soil above when the ground surface is horizontal and the properties of the soil do not change along a horizontal plane is called the geostatic stress.

Grip length. The depth of the bottom of the well foundation (or caisson) below the maximum scour level is called the grip length. For stability, the well should have adequate grip length.

Graded filter. It consists of layers of sand and gravel which permit flow of water. The size of the particles in different layers increases along the direction of flow. Thus the finest layer is near the soil to be protected against piping.

Ground water table. It is the top surface of the underground water reservoir where the pressure is atmospheric. The water pressure below the ground water table is hydrostatic. It is also called the phreatic surface.

Grouting. It is a process in which the holes are drilled in soil (or rock) and a grout (usually cement and water mixture) is injected into the holes. It improves the bearing capacity and also reduces the permeability and seepage.

Head. It is equal to the difference of water levels on the upstream and downstream. The downstream water levels is usually taken as datum.

Sometimes, the term head is also used for the pressure head indicated by a piezometer inserted at the point.

Hydraulic gradient (i). It is equal to the difference of heads between the points divided by the distance between them.

Heave piping. It is the lifting of a large soil mass downstream of a hydraulic structure due to seepage pressure. The entire soil mass in the affected zone is blown out and heave piping failure occurs.

Hygroscopic water. The amount of water retained in an air- dried soil is called hygroscopic water. It depends upon the type of soil, humidity and temperature. It is removed by oven-drying.

It is the same as adsorbed water.

In-situ. It refers to soil (or rock) in its natural location in the ground when it is in its natural condition.

Isotropic. It pertains to a soil mass having the same properties in all directions.

Isobar. It is a curve joining points of the same stress intensity.

Landslide. It is a relatively rapid lateral and downhill movement of a well-defined earth mass (or land form). It occurs due to gravitational and seepage forces.

Limiting equilibrium. The soil is in the limiting equilibrium when it is at the verge of failure.

The limiting equilibrium methods involve determining the mobilised shear strength of the soil on an assumed failure surface as required to maintain equilibrium (or stability) and comparing this value with the available shear strength.

Liquefaction. It is a phenomenon which may occur in saturated cohesionless loose soil when it is subjected to shocks or vibration. The soil particles momentarily lose contact and the soil behaves as a liquid.

Mechanical Analysis. It is the process of determining the sizes of various particles in soil specimen. It is done by the sieve analysis for coarse particles and the sedimentation analysis for fine particles.

Mechanical weathering. It is the process by which physical forces, such as temperature changes and frost action, breakdown or reduce the rock to smaller fragments and soils. There is no chemical change and the properties of the soil formed are the same as those of the parent rock.

Mineral. It is a naturally formed chemical element (or compound) having a definite chemical composition.

It usually has a characteristic crystal from.

Negative skin friction. It is a down drag on a pile which occurs when the soil in which the pile in driven settles more than the pile. The load-carrying capacity of the, pile is reduced because of negative skin friction.

Natural frequency. A system under free conditions vibrates at a frequency called the natural frequency. It is the characteristic of the system. In general, the natural frequency decreases as the mass increases and the spring constant of the system decreases.

Normally consolidated soil. A soil which had not been subjected to a pressure in the past greater than the present pressure. It is also called a virgin soil. The settlements are large in a normally consolidated soil.

Net allowable bearing pressure. It is the net pressure which can be used for the design of a foundation. It is the smaller of the net safe settlement pressure and the net safe bearing capacity. For cohesionless soils, generally the net safe settlement pressure governs; whereas for cohesive soils, generally the net safe bearing capacity governs.

Optimum moisture content. It is the water content of soil at which the maximum dry density is achieved during compaction.

Over-consolidated soil. These are the soils which had been subjected to a pressure in the past greater than the present pressure. Over-consolidated soils are also called *preconsolidated soils.* The settlements are small for such soils.

Pavement. It is a hard crust constructed on the subgrade (soil) for the purpose of providing a stable and even surface for the vehicles to move on.

The pavement may be a flexible pavement or a rigid pavement. The rigid pavements are made of cement concrete and can take the tensile stresses.

Pile. It is a relatively long, slender column used as a deep foundation. The pile is end-bearing (point bearing) pile when it obtains support from the bottom,and it is a friction pile if it develops resistance due to friction on the sides. In most of the cases, it has resistance from bottom as well as side friction.

Piping. It is a phenomenon which occurs due to erosion by sub-surface water moving through a soil mass. It results in the formation of continuous tunnels or pipe-like formations through which soil is carried by flowing water and piping failure may occur.

Plane strain. It is a state of strain in which all displacements occur in one plane and the displacements perpendicular to that plane are zero. Generally, plane strain conditions occur under a long retaining wall, strip footing, earth dam, etc.

Plasticity. It is a property of fine-grained soils (particularly clays) due to which a soil having adequate water content is able to flow and can be remoulded without breaking apart.

Poisson's ratio. It is the ratio of the lateral strain to the longitudinal strain due to uniaxial stress within the elastic limit.

Pore pressure. It is water pressure developed in the voids of a soil mass. The shear strength of a soil is reduced due to pore pressure as the effective stress is decreased.

Excess pore pressure refers to pressure greater than the normal hydrostatic pore water pressure.

Pressure bulb. It is the zone of the soil mass in which stresses are induced due to superimposed load. Generally, it is assumed that the pressure bulb is confined to the zone in which the stresses are more than 20% (or 10%) of the surface load.

Pressuremeter. It is an instrument used to determine the in-situ strength of a soil (or rock) zone. It is based on the principle of the measurement of the pressure-related lateral expansion of a flexible cylinder inserted in a bore hole.

Projecting conduit. It is a type of conduit over which earth fill or earth embankment is placed.

Quick sand condition. When the head causing upward flow in a cohesionless soil is high, the effective stress is reduced to zero and the shear strength of the soil becomes zero. The condition so developed is known as quick sand condition. The critical gradient at which a cohesionless soil becomes quick is about unity.

Reinforced earth. It is an earth mass strengthened by reinforcement. Earth structures such as embankments, retaining walls and earth dams constructed in layers and reinforced with geofabric, metal strips or fibres to increase the strength of the soil mass are examples of reinforced earth.

Retaining wall. An earth-retaining structure constructed to resist the lateral pressure of soil is called a retaining wall.

Revetment. It is a facing built of stone, concrete blocks, or other durable material to protect an embankment from the wave erosion. It is also called *rip rap.*

Rollers. These are types of construction equipment used for compacting the soil by rolling it. The rollers are of different types.

Sand. It is a type of coarse-grained soil whose particle sizes range between about 0.075 mm and 4.75 mm. Sand is cohesionless and has high internal friction.

Seepage. Seepage is flow through an earth mass under pressure. The term is also used to indicate the quantity of water flowing through a soil deposit or soil structure or foundation.

Seismic exploration. It is a method of subsurface investigation in which a shock wave is created to determine the depth of different strata.

Secondary consolidation. This is consolidation of a cohesive soil which occurs after the primary consolidation is completed. The secondary consolidation is predominant in organic soils.

Settlement. The downward vertical movement experienced by a structure or a soil surface when the soil below compresses. The settlement occurs mainly because of consolidation.

Shear strength. It is the ability of a soil to resist shearing stresses developed within a soil mass as a result of loading imposed onto the soil. It is the maximum resistance which develops just before the failure under shear.

Sheet piling. A type of construction in which piles with flat cross-section are joined to form a thin diaphragm wall or bulkhead to resist the lateral force of retained earth.

Sieve. It is a pan (or tray) having a screen or mesh bottom. It is used to separate particles of a soil sample into various sizes.

Silt. It is a type of fine-grained soil with the particle size smaller than 0.075 mm, but whose mineralogical composition remains the same as that of the parent rock. It does not contain clay minerals.

Soil Stabilisation. It is method of increasing the strength of a soil and improving its properties. It includes mixing of additives and other means of improvement such as compaction and drainage.

Standard Penetration Test. (SPT) It is a type of test in which a sampler is driven into the ground by a hammer. The number of blows required for a penetration of 300 mm is the standard penetration number (N). The standard penetration number is correlated with shear strength. The test is especially useful for cohesionless soils.

Sump. It is a small pit provided to serve as a collecting basin for surface water or near-surface under ground water. The water is pumped out from the sump when full.

Terra-probe. It is an equipment used for compacting the soil. It consists of a vertically-vibrating tubular probe which is vibrated to the desired depth in a soil mass and then slowly withdrawn while continuing to vibrate.

Till. Soil transported by glaciers are called till. These usually consist of a heterogeneous mixture of the fine-grained and coarse-grained soils.

Tube well. It is a type of well in which a pipe (or tube) is driven into the soil to intercept various aquifers. The water flows into the well through the strainer provided. The discharge of a tube well is much greater than that from an open well.

Uniformity coefficient. It is the ratio of D_{60} to D_{10} size of the soil where D_{60} represents the size corresponding to 60% finer and D_{10} to 10% finer material than that size. For a well-graded sand, its value is greater than 6.

Unit weight. It is the weight per unit volume. It is expressed in kN/m^3.

Vibrofloation. It is a method of compacting thick deposit of sand through the use of a horizontally vibrating cylinder called a *vibrofloat.*

Void ratio. It is the ratio of the volume of voids to the volume of solids in a soil. For a fine-grained soil, the void ratio is generally greater than that for a coarse-grained soil.

Water content. It is the ratio of the mass (or weight) of water to the mass (or weight) of solids in a soil. It is expressed as a percentage.

Well point. It is the perforated end section of a well pipe installed in the ground. It permits the ground water to be drawn into the pipe for pumping.

REFERENCES

1. Altmeyer, W.T. (1955), *"Discussion of Engineering properties of expansive clay's"*, Proc. ASCE, Vol. 81, No. SM 2.
2. Arora, K.R., *"Soil-structure Interaction Analysis of the strip and circular Footings on Sand"*. Ph. D. Thesis, Indian Institute of Technology, Delhi, 1980.
3. Arora, K.R., and Varadarajan, A., "Experimental Investigations on Soil-Structure Interaction of circular Footings on Sand", *I.G.J.*, Vol. 14, 1984.
4. Balwant Rao, B. and C. Muthuswamy, "Considerations in the Design and Sinking of Wells for Bridge Piers", Paper No. 238, *Jnl. IRC,* Vol. 27, No. 3, 1963.
5. Banerjee, A. and S. Gangopadhya, "Study on the Stability of well Foundations for major Bridges", *Jnl. IRC*, Vol. 25, No. 22, 1960.
6. Barkan D.D., *"Dynamics of Bases and Foundations",* McGraw-Hill, New York, 1962.
7. Barron, R.A., "Consolidation of Find-grained Soils by Drain wells", *Trans. ASCE,* Vol. 113, 1948.
8. Bieniawski, Z.T., "Engineering classification of Jointed Rock Masses", *Trans. S. Afri. Inst. Civil Engineers,* Vol. 15, 1973.
9. Bieniawski, Z.T., "Geomechanics Classification of Rock Masses and its application in Tunneling". *Proc. 3rd. Int conf. Rock mech,* Denver, Vol. 2, 1974.
10. Bishop, A. W., "The use of the slip circle in the Stability Analysis of Slopes", *Geotechnique,* Vol. 5, No. 1, 1955.
11. Bishop, A.W. and Henkel, D.J., "*The Measurement of Soil Properties in the Triaxial Test*", Edward Arnold, 1962.
12. Borowicka, H., "Influence of Rigidity of a circular Foundation slab on the Distribution of Pressures over the contact Surface", *Proc. Int. Conf. on Soil Mech. & Foundation Engg.*, Vol. 2, 1936.
13. Bowles, J.E., *"Foundation Analysis and Design",* McGraw- Hill, New York, 1968.
14. Brahma, S.P., *"Foundation Engineering",* Tata Mc Graw- Hill Publishing Co. Ltd. New Delhi, 1985.
15. Broch, E. and Franklin, J.A., "The point-load strength test", *Int. J. Rock Mech. & Mini. sci.* Vol. 9, NO. 6, 1972.
16. Broms, B.B., "Methods of Calculating the Ultimate Bearing Capacity of a Pile", Sols-Soils, Vol. 5, No. 18-19, 1966.
17. Brown, E.T. and E. Hook, "Trends in Relationship between measured in-situ stresses and depth", *Int. J. Rock Mech. & Min. Sci,* Vol. 15, 1978.
18. Capper, P.L. and Cassie, W.F., "*The Mechanics of Engineering Soils*", Asia Publishing House, Bombay, 1961.
19. Carman, P.E., "*Flow of Gases through Porous media*", Academic, New York, 1956.
20. Casagrande, A., "*The Hydrometer method for mechanical Analysis of Soils and other Granular materials*", Cambridge, 1931.
21. Casagrande, A., "Research on the Atterberg Limits of Soils", *Public Roads,* Vol. 13, 1932.
22. Casagrande, A., "Classification and Identification of Soils", *Trans. ASCE,* Vol. 113, 1948.
23. Casagrande, A., "Classification of subgrade materials", Proc. Highway Research Board, Washington D.C., 1945.
24. Casagrande, A., "Seepage Through Dams", *Contribution to Soil Mechanics*, Boston Society of Civil Engineers, Boston, 1937.
25. Casagrande, A., "The Determination of the preconsolidation load and its Practical Significance". *Proc. 1st Int. Conf.* Soil mech & Found Engg., 1936.
26. Cedergrem, H.R., "*Seepage, Drainage and Flow nets*", Wiley, New York, 1977.
27. Cernica, J.N., *"Geotechnical Engineering",* Holt- Saunders, International Edition, Japan, 1982.

28. Chellis, R.D., *"Pile Foundation"*, McGraw-Hill Book Co., New York, 1961.
29. Chen, W.E. (Ed.) (1995). *The Civil Engineering Handbook*, CRC Press, Boca Raton, New York.
30. Chen, F.H. (1988), Foundation on Expansive soils, Elsevier Science Publishing company Inc., New York.
31. Coduto, Donald P. (1999), *Geotechnical Engineering, Principles and Practice,* Prentice-Hall of India, Pvt. Ltd., New Delhi.
32. Craig, R.F., *"Soil Mechanics"*, Van Nostrand Reihnold Co. Ltd., London, 1978.
33. Das, B.M., *"Principles of Geotechnical Engineering"*, PWS-KENT Publishing Co., Boston, 1985.
34. Das, B.M., *"Principles of Foundation Engineering"*, Brooks/Cole Engineering Division, Monterey, 1984.
35. Das, B.M., *'Advanced Soil Mechanics"*, Mc Graw-Hill Book Co., New York, 1985.
36. Das, B.M., *"Introduction to Soil Mechanics"*, Galgotia Publication, New Delhi, 1983.
37. Das, B.M., (1994), Advanced Soil Mechanics, McGraw-Hill Book Co.
38. Deere, D.U., "Geological consideration", Chapter in *Rock mechanics in Engineering Practice,* edited by Stagg, M.G. and Zienkiewics, O.C., Wiley, 1968.
39. Deere, D.U., and R.P. Miller, "Engineering Classification and Index Properties for Intact Rock", *Tech. Rep. No. AFWL-TR 65-116, Air Force Weapons Lab., Kirtland* Air Base, New Mexico, 1966.
40. Deere, D.U., Hendron, A.J., Patton, F.D., Cording, E.J., "Design of surface and near-surface construction in Rock", *Proc. 8th Symp. Rock mech. AIME,* Minnesota, 1967.
41. Dunn, I.S., Anderson, L.R., and Kiefer, F.W., *"Fundamental of Geotechnical Engineering"*, John Wiley & sons, New York, 1980.
42. Evans, H.E., "A Note on the Average Coefficient of Permeability for a Stratified Soil mass", *Geotechnique,* Vol. 12, No. 12, London, 1962.
43. Fadum, R.E., "Influence Values for Estimating Stresses in Elastic Foundation", *2nd Int. Conf. on Soil mech & Found. Engg.,* Vol. 3, 1948.
44. Fenske, C.W., "Influence Charts & Vertical Stress Distribution by Westergaard Equation". *Bureau of Engg. Res. Circular No. 21,* University of Texas, Austin, 1951.
45. Fourmaintraux, D., "Characterization of Rocks, Laboratory Tests", Chapter IV in *La mecanique des roches applique and ouvrages du genic civil* by Marc Panet et al, Ecloe Nationale des Points et chaussees, Paris, 1976.
46. Franklin, J.A., and Dusseaut, M.B., *"Rock Engineering"*, Mc Graw-Hill Publishing Co., New York, 1989.
47. Franklin, J.A. and R. Chandra, "The Slake Durability Test", *Int. J. Rock mech. & Min sci.* Vol. 9, 1972.
48. Franklin, J.E. and E. Broch, "The Point-Load Strength Test", *Int. Jnl. Rock mech. and Min. Sci.* Vol. 9, 1972.
49. Goodman, R.F., *"Introductory Rock mechanics"*, Wiley, 1980.
50. Grim, R.E., 'Clay Mineralogy", Mc Graw-Hill Book Co., New York, 1968.
51. Gulati, S.K., *"Engineering Properties of Soils"*, Tata McGraw-Hill Publishing Co. Ltd. New Delhi, 1978.
52. Gulati, S.K. and Datta, M. (2005), *Geatechnical Engineering,* Tata McGraw-Hill Publishing Company Ltd. New Delhi.
53. H.M.S.O, *"Soil Mechanics for Road Engineers"*, London, 1959.
54. Hansen, J.B., "A General Formula for Bearing capacity", *Danish Geotechnical Institute Bulletin,* 11, 1961.
55. Harr, M.E., *"Foundation of Theoretical Soil mechanics"*, McGraw-Hill Book Co., New York, 1966.
56. Harr, M.E., *"Ground Water and Seepage"*, McGraw-Hill Book Co., New York, 1962.
57. Hazen, A., "Some Physical Properties of Sands and Gravels with reference to their use in Filtration", *24 th Annual Report, Massachusetts State Board of Health.*, 1892.
58. Hetenyi, M., *"Beams on Elastic Foundation"*, University of Michigan Press, Ann Arbor, 1946.
59. Hiley, A., "Pile Driving calculations with Notes on Driving Forces and Ground Resistance", *The Structural Engineer*, Vol. 8, London, 1930.
60. Hilf, J.W., "Estimating construction Pore Pressure in Rolled Earth Dams", *Proc. Int. Conf. Soil mech. & Found. Engg.*, Vol. 3, Rotterdam , 1948.
61. Hoek, E. and J.A. Franklin, "Sample Triaxial cell for Field or Laboratory Testing of Rock", *Trans/sect. A, Inst Min Metal,* Vol. 77, 1968.
62. Holtz, W.G. and Gibbs, H.J. (1956), *Engineering Properties of Expansive Clays,* Trans, ASCE, 121.

63. Holtz, W.G., and Hilf, J.W. (1961), "Settlement of Soil Foundations Due to Saturation, Proc Fifth Int. Conf. on Soil Mechanics and Foundaton Engineering, Paris, Vol. 1-1-1961, pp. 673-679.
64. Holtz, W.G. (1969), *Volume changes in Expansive Clay Soils and Control by Lime Treatment,* Proc. of the Second International Research and Engineering Conference on Expansive Clay soils, pp. 157-173, Texas A & M Press.
65. Hough, B.K., "*Basic Soils Engineering*", Ronald Press Co., New York, 1969.
66. Housel, W.S., "A Practical method for the Selection of Foundation Based on Fundamental Research in Soil Mechanics", *Research Bulletin, No. 13, University of Michigan,* Ann Arbor, 1929.
67. Huntigton, W.C., "*Earth Pressure and Retaining Walls*", John Wiley & Sons, New York, 1957.
68. Hvorslev, M.J., "*Subsurface Exploration and Sampling of Soils for Civil Engineering Purposes*". U.S. Waterways Experiment Station, Vicksburg, Miss, 1949.
69. Hvorslev, M.J., "Physical Components of Shear Strength of Saturated clays", *Proc. Resi. Confi. Shear Strength of Cohesive Soils,* ASCE, 1960.
70. Idriss, I.M. (1990), "Response of Soft Soil Sites During Earthquakes", *Proceedings, H. Bolton Seed Memorial Symposium,* J.M. Duncan, Ed; Vol 2, pp 273-289, Bi Tech Vancouver, BC.
71. IRC : 6, "*Standard Specifications and Code of Practice for Road Bridges*", Sect II-Loads and Stresses, IRC, 1966.
72. IRC : 37, "*Guidelines for the Design of Flexible Pavements",* IRC, 1970.
73. IRC : 58, "*Guidelines for the Design of Rigid* Pavements for Highways", IRC, 1974.
74. IRC : 45, "*Rewmmendation for Estimating the Resistance of Soil Below the maximum Scour Level in the Design of Well Foundations",* IRC, 1970.
75. IS: 1893 (Part 1): 2002. *Criteria for Earthquake Resistant Design of structure,* Bureau of Indian Standards, New Delhi
76. IS: 2911-3-1973, *Code of Practice for Design and Construction of Pile Foundations,* Part 3, Under Reamed Pile Foundation, BIS, New Delhi.
77. Jaeger, J.C. and N.G.W. Cook, "*Fundamentals of Rock Mechanics*", Chapman and Hall Ltd. and Science Paperbacks, 1969.
78. Janbu, N., "*An Energy Analysis of Pile Driving with the use of Dimensionless Parameters*", Norwegian Geotechnical Institute, Oslo, Publication No. 3, 1953.
79. Jenning, J.E., and Knight, K. (1975), "A guide to Construction on or with materials Exhibiting Additional Settlements Due to Collapse of Grain Structure," *Proc. Sixth Regional conf. for Afria on S.M. and F. Engineering,* Johannesburg.
80. Jumikis, A.R., "*Soil Mechanics*", D. Van Nostrand Co., 1965.
81. Kasmalkar, B.J., "Foundation Engineering", Pune Vidyarathi Griha Prakashan, Pune, 1983.
82. Khanna, S.K. and C.E.G. Justo, "*Highway Engineering*", Nem Chand & Bros, Roorkee, 1984.
83. Krishna, J., A.R. Chandrasekaran and Chandra, B (1994), *Elements of Earthquake Engineering,* South Asia Publishers, New Delhi.
84. Lambe, T.W. and R.V. Whitman, "*Soil Mechanics*", Wiley Eastern Pvt. Ltd., New Delhi, 1969.
85. Lambe, T.W., "*Soil Testing for Engineers*", Wiley Eastern Ltd., New Delhi, 1977.
86. Lambe, T.W. "Stress Path method". *Jnl. of Soil Mech. and Found. Engg. ASCE,* Vol. 84, 1958.
87. Lambe, T.W., "The Structure of Compacted clay", *Jnl. Soil mech. and Found. Engg., ASCE,* Vol. 84, 1958.
88. Liu, C. and J.B. Evett, "*Soils and Foundations",* Prentice Hall Inc. Englewood cliffs, New Jersey, 1981.
89. Loudon, A.G., "The Computation of Permeability for simple soil tests", *Geotechnique,* Vol. 3, 1952-53.
90. Malter, H., "Numerical Solutions for Beams on Elastic Foundations", *Trans. ASCE, Vol. 125, 1960.*
91. Martson, A., "*The Theory of External Loads on closed conduits in the Light of Latest Experiments*", Bull. 96, Iowa Engg. Exp. Station, Amec, Iowa, 1930.
92. McCarthy, D.F., "*Essentials of Soil mechanics and Foundations*", Reston Publishing Co. Inc., Reston, 1982.
93. McClelland, B., "Design of Deep Penetration Piles for ocean structures", *Jnl. of Geotechnical Engg. Div., ASCE,* Vol. 100, No. GT. 17, 1974.
94. Means, R.E. and J.V. Parcher, "*Physical Properties of Soils*", Prentice Hall of India Pvt. Ltd., New Delhi, 1965.

95. Meyerhof, G.G., "Bearing Capacity and Settlement of Pile Foundation", *Jnl. Geot Engg. Div. ASCE,* GT 3, 1976.

96. Meyerhof, G.G., "Penetration Tests and Bearing Capacity of Cohesionless soils", *Jnl. Soil Mech. Found. Division, ASCE,* SM1, 1965.

97. Meyerhof, G.G. and Adams, J.I.,"The Ultimate Uplift capacity of Foundations", *Canadian Geot,* JS(4), 1968.

98. Mitchell, J.K., "*Fundamentals of Soil Behaviour*", John Wiley and Sons, New York, 1976.

99. Morgenstern, N.R. and V.E. Price, 'The Analysis of the Stability of General Slip Surfaces", *Geotechnique*, Vol. 15, No. 1, 1965.

100. Morgenstern, N.R. and Eigenbrod, K.D. "Classification of Argillaceous Soils and Rocks", J. Geot. Engg. Div, *ASCE,* Vol. 100, No. GT 10, 1974.

101. Newmark, N.M., "Simplified Computation of Vertical Pressure in Elastic Foundation", *Univ. of Illinois, Engg. Exp. Stn., Circular* 24, 1935.

102. Newmark, N.M., 'Influence Charts for computation of stresses in Elastic Foundations", *Univ. of Illinois, Bull. No. 338,* 1942.

103. Murthy, V.N.S. (2007), A Textbook of Soil Mechanics and Foundation Engineering-Geotechnical Engineering Series, CBS Publishers & Distributors, New Delhi.

104. Nuttli, O.W. and Herrmann, R.B. (1984). "Ground motion of Mississippi Valley Earthquakes", *J. Tech. Top. Civ. Eng.,* ASCE 110: (54-69)

105. O'Neill, M.W., (1988) "Special Topices in Foundation", Proc. Geotech, Eng. Div., ASCE National Conf., Nashville.

106. O'Neill, M.W., and Poormoayed, N. (1980), "Methodology for Foundations an Expansive Clays," *Journal of the Geotechnical Engineering Division, American Society of Civil Engineering,* Vol. 106, No. GT 12, p. 1345-1367.

107. Peck, R.B., W.E. Hanson, and T.H. Thornburn, "*Foundation Engineering",* Wiley, New York, 1974.

108. Poulos, H.G. and E. H. Davis, "*Elastic Solutions for Soils and Rock mechanics*", Wiley New York, 1974.

109. Prakash, S., G. Ranjan and S. Saran, "*Analysis and Design of Foundations and Retaining Structures*", Sarita Prakashan, Meerut, 1979.

110. Proctor, E.R., "Design and construction of Rolled Earth Dams"' *Engg. News Record,* Vol. 3, 1933.

111. Purushothama Raj, P., "*Geotechnical Engineering",* Tata McGraw-Hill Publishing Co. Ltd., New Delhi-1995.

112. Ramiah, B.K. and L.S. Chikanagappa, "*Soil Mechanics and Foundation Engineering*", Oxford and IBH Publishing Co., New Delhi, 1981.

113. Rangnatham, B.V. and Satyanarayan, B. (1965), A National method of predicting swelling potential for compacted expansive soils, Proc. 6th Int. Conf. SMFE, Montreal, 1-92-96.

114. Reddy, A.S. and R.J. Srinivasan, "Bearing Capacity of Footings on Layered Clay", *Jnl. of SM & FE, ASCE,* Vol. 98, No. SM 2, 1967.

115. Richart, F.E., "Review of the Theories for Sand Drains", *Trans. ASCE,* Vol. 124, 1959.

116. Richart, F.E., "Foundation Vibrations", *Trans. ASCE, Vol. 127, Part 1, 1962.*

117. Richart, F.E., Jr., J.R. Hall Jr., and R.D. Woodes, "*Vibrations of Soils and Foundations*", Prentice Hall, Englewood Cliff, N.J., 1970.

118. Richter, C.F. (1935). "An instrumental earthquake scale", B. Seism. Soc. Am. 25 (1): 1-32.

119. Rowe, P.W., "Anchored Sheet Pile Walls", *Proc. Institution of Civil Engineers,* Vol. 1, Part 1, 1952.

120. Saran, Swami, (1999) *Soil Dynamics and Machine Foundations,* Galgotia Publication pvt. Ltd., New Delhi.

121. Schmertmann, J.H., "Estimating the True Consolidation Behaviour of clay from Laboratory Test results", *Proc. ASCE,* Vol. 79, 1953.

122. Schmertmann, J.H. and Hartman, J.P., "Improved Strain Influence Factor Diagrams", *Jnl. of Geotechnical Engg. Division,* ASCE, Vol. 104, No. GT 8, 1978.

123. Seed, H.B., and Idriss, I.M. (1971). 'Simplified Procedure for Evaluating Soil Liquefaction Potential', *Journal of the Soil Mechanics and Foundation Division,* Vol. 107, No., SM 9, pp 1249-1274 ASCE.

124. Seed, H. Bolton, Tokimatsu, K., Harder, L.F., and Chung, Riley M (1985), "Influence of SPT Procedures in Soil Liquefaction Resistance Evaluations," ASCE *Journal of Geotechnical Engineering,* Vol. III, No. 12, pp. 1425-1445.

125. Seed, H.B., Romo, M.P., Sum, J., Jaime, A., and Lysmer, J. (1987) "Relationships between soil conditions and earthquake ground motions in Mexico city in the earthquake of Sept., 19, 1985." *Earthquake Engineering Research Centre,* Report No. UCB/EERC-87/15 University of California, Berkeley.

126. Singh, A., "*Soil Engineering in Theory and Practice,* Vol. I and II, CBS Publishers & Distributors Pvt. Ltd. Delhi, 1994.

127. Singh, A., *"Modern Geotechnical Engineering"*, CBS Publishers and Distributors, Pvt. Ltd. Delhi, 1990.

128. Singh, A and Punmia, B.C., "Some Experiments with Jodhpur Mini- Compactor", *Roads and Road Construction,* Vol. 41, No. 484, London, 1963.

129. Singh, B and Prakash, S., "*Soil Mechanics and Foundation Engineering*", Nem Chand & Bros, Roorkee, 1976.

130. Singh, Vijay, "*Wells and Caissons*", Nem Chand and Bros, Roorkee, 1981.

131. Skempton, A.W., The Pore Pressure Coefficients A and B", *Geotechnique,* Vol. 4, 1954.

132. Skempton, A.W. and Bjerrum, L.I., "A Contribution to Settlement Analysis of Foundation on Clay", *Geotechnique,* Vol. 7, No. 4, 1957.

133. Skempton, A.W., "Effective Stress in Soils, Concrete, and Rocks", *Proc of Conf. on Pore Pressure and Suction, in soils,* Butterworth, London, 1961.

134. Skempton, A.W., Yassin, A.A. and Gibson, R.E., "Theorie de la force portante des pioux dans le sable" *Ann. Inst. Tech. Batim,* 6, 1953.

135. Smith, G.N., "*Elements of Soil Mechanics for Civil and Mining Engineers*", Granada, London, 1978.

136. Smith, E.A.L., "Pile Driving Analysis by the Wave Equations", *Trans. ASCE, Vol. 127, 1962.*

137. Sowers, G.B. and Sowers, G.F., "*Introductory Soil Mechanics and Foundations*", Macmillan, New York, 1970.

138. Som, N.N., and Das, S.C. (2003), *Theory and Practice of Foundation Engineering,* Prentice-Hall of India. New Delhi.

139. Srinivasulu, and Vaidyanathan, "*Hand Book of Machine Foundation*" Tata Mc Graw-Hill, New Delhi, 1976.

140. Stagg, K.G. and O.C. Zienkiewics (Ed.) "*Rock Mechanics in Engineering Practice*", John Wiley Sons, New York.

141. Taylor, D.W., "*Fundamentals of Soil Mechanics*", John Wiley and Sons, Inc, New York, 1948.

142. Teller, L.W. and E.C. Sutherland, "The Structural Design of Concrete Pavements", Part 5, *Public Roads,* Vol. 23, No. 8, Washington, 1943.

143. Teng, W.C., "*Foundation Design",* Wiley, New York, 1962.

144. Tennesses Valley Authority, "*Coeffer Dams on Rocks",* TVA Tech. monograph 75, Knoxville, Tenn, 1957.

145. Terzaghi, K., "*Theoretical Soil Mechanics*", Wiley, 1943.

146. Terzaghi, K., "*From Theory to Practice in Soil mechanics*", Wiley & Sons, New York, 1960.

147. Terzaghi, K., "*General Wedge Theory of Earth Pressure", Trans. ASCE,* Vol. 106, 1941.

148. Terzaghi, K., "Evaluation of Coefficient of Subgrade Reaction", *Geotechnique,* Vol. 5, No. 4, London, 1955.

149. Terzaghi, K. and Peck, R.B., "*Soil Mechanics in Engineering Practice*", John Wiley & Sons New York, 1967.

150. Todd, D.K., *"Ground Water and Seepage",* Wiley, New York, 1959.

151. Tomilson, M.J., "*Foundation Design and Construction*", Wiley Interscience, New York, 1969.

152. Tschebotarioff, G.P., "*Foundations, Retaining and Earth Structures*", McGraw Hill Book co. New York, 1973.

153. U.B.C. (1991), Uniform Building Code, U.S.A.

154. U.S. Bureau of Reclamation, "*Earth Manual*", Washington, 1960.

155. U.S. Bureau of Reclamation, "*Design of small Dams",* U.S. Government Printing Office, Washington, 1961.

156. Varadarajan, A and Arora, K.R., "Non-Linear Analysis of strip Footing on Sand", *Geocon, IGS Conf. on Geotechnical Engg.* New Delhi, 1978.

157. Varadarajan, A and Arora, K.R., "Finite Element Analysis using stress-path dependent parameters", *Symp. on Implementation of computer procedures and stress-strain laws in Geot. Engg.* ASCE, Illionois sect., 1981.

158. Varadarajan, A. and Arora, K.R., "An interaction Study of Strip Footing-sand bed system by Finite Element Method", *3rd Int. Conf. on Numerical Methods in Geomechanics,* Aachen, 1979.

159. Varadarajan, A., and Arora, K.R., "Interaction Analysis of circular Footing-Sand bed system", *4th Int. Conf. on Numerical methods in geomechanics,* Edmonton, 1982.

160. Varshney, R.S., Gupta, S.C. and Gupta, R.L., "*Theory and Design of Irrigation Structures*", Vol. 1, Nem Chand and Bros, Roorkee, 1979.

161. Vesic, A.S., "Bearing Capacity of Deep Foundations in Sand", *Highway Research Board,* National Academy of Science, Washington, 1963.

162. Vesic, A.S., 'Ultimate Load and settlement of Deep Foundations in Sand", Proc. Symp. on Bearing Capacity and Settlement of Foundations", Duke Univ., Durham, 1967.

163. Vijayvergia, V.N. and Focht, J.A., "A new way to Predict Capacity of Piles in Clay", *4th Ann. off-share Tech. Conf, Houston, 1972.*

164. Wagner, A.A., "The use of the vnified soil classification system by the Bureau of Reclamation", *Proc. 4th Int. Conf. SMFE,* London, Vol. 1, Butterworth, 1957.

165. Westergaard, H.M., "Stresses in Concrete Pavements Computed by Theoretical Analysis", *Public Roads,* Vol. 7, No. 2, 1926.

166. Westergaard, H.M., "A Problem of Elasticity Suggested by a problem in Soil mechanics", Soft material Reinforced by Numerous strong Horizontal Sheets", *Contribution to the Mechanics of solids, 60th Anniversary vol.,* Macmillan Co, New York, 1938.

167. Wilson, S.D., 'Suggested Method of Test for Moisture Density Relation of Soils using Harward Compaction Apparatus, *Procedures for Testing soils, ASTM,* 1958.

168. Winterkorn, H.F. and Fang, H., "*Foundation Engineering Handbook*", Van Nostrand Reinhold Co., New York, 1975.

169. Wu, T.H., "Soil Mechanics", Allyn & Boston, Inc, 1966.

170. Yong, R.N., and B.P. Winterkorn, "*Soil Properties and Behaviour*", Elsevier Scientific Publishing Co., Amster dam, 1975.

171. Yoder, E.J. and M.V. Witzak, "*Principles of Pavement Design*", John Wiley & sons, 1975.

Publications of Bureau of Indian Standards,

Manak Bhavan, 9, Bahadur Shah Zafar Marg,
New Delhi, 110002.

1. IS:456-1978, "*Code of Practice for Plain and Reinforced concrete's.*
2. IS:1080-1985, "*Design and construction of Shallow Foundations in Soils (other than raft, ring and shell)*
3. IS:875-1964, "*Indian Standard Code of practice for structural safety of Buildings, Loading Standards*".
4. IS:1498-1970, "*Classification and Identifications of soils for General Engineering Purposes.*
5. IS:1888-1982, "*Method of Load Test on soils*".
6. IS: 1892-1979, "*Code of Practice for Subsurface Investigations for Foundations*".
7. IS: 1893-1975, "*Criteria for Earthquake Resistant Design of Structures*".
8. IS: 1904-1986, "*Design and Construction of Foundations in Soils, General Requirements*".
9. IS: 2131-1981, "*Method for Standard Penetration Test for Soils*".
10. IS:2132-1986, '*Code of Practice for Thin-Walled Tube Sampling of Soils*".
11. IS: 2720-Part-1 1983, "*Preparation of Dry Samples for various Tests*".
12. IS: 2720-Part-2 1973, "*Determination of Water content*".
13. IS: 2720-Part-3 Sect. 1-1980, "*Determination of Specific gravity-Fine-grained soils*".
14. IS: 2720-Part-3 Sect.-2-1981, "*Determination of Specific Gravity-Fine, Medium, and coarse-grained soils*".
15. IS: 2720-Part 4-1975, "*Grain size analysis*".
16. IS: 2720-Part 5-1970, "*Determination of Liquid and Plastic Limits*".
17. IS: 2720- Part 6-1972, "*Determination of Shrinkage Factors*".
18. IS: 2720-Part 7-1983, "*Determination of Water content-Dry density Relation using light compaction*".
19. IS: 2720-Part 8-1983, '*Determination of Water Content-Dry Density Relation using Heavy Compaction.*
20. IS: 2720-Part 9-1971, "*Determination of Dry Density- Moisture content Relation by constant weight of soil method*".
21. IS: 2720 Part 10-1973, "*Determination of Unconfined Compressive Strength*".
22. IS: 2720-Part 11-1971, "*Determination of shear strength parameters of soils from consolidated-undrained triaxial compression test with measurement of Pore-water Pressure*".
23. IS: 2720-Part 12-1981, "*Determination of Shear Strength Parameters of Soils from Consolidated-Undrained Triaxial Compression Test with measurement of Pore-Water Pressure*".
24. IS: 2720-Part 13-1972, "Direct Shear Test".
25. IS: 2720-Part 14-1983, "Determination of Density Index *(Relative Density) of cohesionless soils*".
26. IS: 2720-Part 15-1986, "*Determination of Consolidation Properties.*
27. IS: 2720-Part 17-1977, "*Determination of Linear Shrinkage*".
28. IS: 2720-Part 28-1974, "*Determination of Dry Density of Soil in-place-by the sand-replacement method*".
29. IS: 2720-Part 29-1975, "*Determination of Dry Density of Soils in place-by the core cutter method*".
30. IS: 2720-Part 30-1980, "*Laboratory Vane Shear Test*".
31. IS: 2720-Part 33-1971, "*Determination of the density in place, by the Ring and Water Replacement Method*".
32. IS: 2720-Part 34-1972, "*Determination of the Density of Soil in place, by the Rubber-Balloon Mehtod*".
33. IS: 2720-Part 35-1974, "*Measurement of Negative Pore Water Pressure*".
34. IS: 2720-Part 36-1975, "*Laboratory Determination of Permeability of Granular Soils (Constant Head)*".
35. IS: 2720-Part 38-1976, "*Compaction control Test (Hilf Method)*".
36. IS: 2720-Part 39-Sect. 1-1977, "*Direct Shear Test for Soils containing gravel, Laboratory Test*".
37. IS: 2720-Part 40-1977, "*Determination of Free Swell Index of Soils*".

38. IS: 2911-Part 1-Sect. 1-1979, *"Design and Construction of Pile Foundations-Driven Cast in-situ concrete Piles"*.
39. IS: 2911-Part 1-Sect. 3-1979, *~~Design and construction of Pile Foundation-Driven Precast Piles"*.
40. IS: 2911-Part 3-1980, "Code of Practice for Design and Construction of Pile Foundation- Under-reamed Piles".
41. IS: 2911-Part 4-1974, *"Load Test on Piles"*.
42. IS: 2950 A-I (1974), *"Code of Practice for Design and Construction of Raft Foundations"*.
43. IS: 2968-Part 1-1976, "Dynamic method using 50 mm Cone without Bentonite Slurry".
44. IS: 2974-Part 1-1982, *"Foundation for Reciprocating Type Machines"*.
45. IS: 2974-Part 2-1980, *"Foundation for Impact Type machines (Harmmer Foundation)"*.
46. IS: 2974-Part 3-1975, *"Foundation for Rotary Type machines (Medium and Highway Frequency)"*.
47. IS: 2974-Part 4-1979, "Foundations for Rotary Type Machines for Low Frequency".
48. IS: 2974-Part 5-1970, *"Foundations for Impact Type Machines other than Harmmers (Forging and Stamping Press, Pig-breaker, Elevator and Hoist Tower)*.
49. IS: 3764-1970, "Safety Codes for Excavation work".
50. IS: 3955-1967, *"Indian Standard Code of Practice for Design and Construction of well Foundations"*.
51. IS: 4434-1978, *"Code of Practice for In-situ vane Shear Test for soils"*.
52. IS: 4453-1980, *"Code of Practice for Sub-surface Exploration by Pits, Trenches, Drifts and Shafts."*
53. IS: 4968-Part 2-1976, *"Dynamic Method using cone and Bentonite Slurry"*.
54. IS: 4968-Part 3-1976, *"Static cone Penetration Test"*.
55. IS: 5121-1969, *"Safety code for Piling and other Deep Foundations"*.
56. IS: 5529-Part 1-1985, *"Code of Practice for In-situ Permeability Tests-Tests in Overburden"*.
57. IS: 6403-1981, *"Code of Practice for Determination of Bearing Capacity of Shallow Foundations"*.
58. IS: 6955-1973, "Code of Practice for Subsurface Exploration for Earth and Rock fill Dams".
59. IS: 7317-1974, *"Code of Practice for Uniaxial Jacking Test for Deformation Modulus of Rock"*.
60. IS: 8009-Part 1-1976, *"Shallow Foundation Subjected to Symmetrical Static Vertical Loads"*.
61. IS: 8009-Part 2-1980, *"Code of Practice for calculations of settlement of Foundation-Deep Foundation subjected To Symmetrical Static Vertical Loading*.
62. IS: 8763-1978, *"Guide for Undisturbed Sampling of Sands"*.
63. IS: 8764-1978, *"Method for Determination of Point-Load Strength Index of Rocks"*.
64. IS: 9143-1979, *"Method for Determination of Point Load Strength Index of Rocks"*.
65. IS: 9221-1980, *"Method of Determination of Modulus of Elasticity and Poisson's ratio of Rock materials in Uniaxial Compression*.
66. IS: 9259-1979, *"Specifications for Liquid Limit Apparatus for Soils"*.
67. IS: 9640-1981, *"Specifications for Split-Spoon Sampler"*.
68. IS: 10082-1981, *"Method of Test for Determination of Tensile Strength by Indirect Tests on Rock Specimens"*.
69. IS: 10379-1982, *"Code of Practice for Field Control of Moisture and Compaction of Soils for Embankment and Sub- grade"*.
70. IS: 11385-1985, *"Code of Practice for Sub-surface Exploration for Canals and Cross-Drainage Works"*.
71. IS: 11594-1985, *"Specifications for mild Steel Thin-Walled Sampling Tubes and Sampler heads"*.

INDEX